Introduction to Human Factors and Ergonomics

Ergonomics and human factors impact how humans interact with the world around them. Understanding these factors can be difficult. To cut through the tricky aspects of the subject, this bestselling textbook offers a comprehensive and up-to-date introduction to the field. This title places the subject matter into a systems context using a human-machine model to structure the chapters and a knowledge application model to structure the organization of material in each chapter.

Every chapter covers core concepts, basic applications, tools and processes, and system integration issues regardless of topic. This updated fifth edition provides new material on current occupational health issues, such as obesity, menopause, and other modern work-related medical concerns. It has been updated to include coverage of new technological developments such as self-driving cars, exoskeletons, AI, hybrid working, and cell phone ergonomics. Examples where tools are used, including the Strain Index and the Lifting Fatigue Failure Tool, have been fully updated, featuring signposting to additional resources and toolkits. Readers will gain a full and thorough grounding in the need-to-knows of ergonomics and human factors.

Introduction to Human Factors and Ergonomics, Fifth Edition is the premier textbook for any student where ergonomics and human factors play a part in their discipline, including those in aviation, medicine and healthcare, energy, engineering, health and safety, and the sciences.

Also included in this updated new edition are an instructor's manual and a guide to tutorials and seminars. Over 500 PowerPoint slides are available for academic use from the publisher (www.routledge.com/9781032717210).

Introduction to Human Factors and Ergonomics

Fifth Edition

R.S. Bridger

CRC Press
Taylor & Francis Group
Boca Raton London New York

CRC Press is an imprint of the
Taylor & Francis Group, an **informa** business

Designed cover image: Shutterstock

Fifth edition published 2026
by CRC Press
2385 NW Executive Center Drive, Suite 320, Boca Raton FL 33431

and by CRC Press
4 Park Square, Milton Park, Abingdon, Oxon, OX14 4RN

CRC Press is an imprint of Taylor & Francis Group, LLC

First edition published by Willan 2008
Second edition published by Routledge 2009
Third edition published by Routledge 2009
Fourth edition published by Routledge 2009

ISBN: 978-1-032-69450-4 (hbk)
ISBN: 978-1-032-71719-7 (pbk)
ISBN: 978-1-032-71721-0 (ebk)

DOI: 10.1201/9781032717210

Typeset in Times
by codeMantra

It includes instructor resources and power points at URL www.routledge.com/9781032717210

This book owes its existence to the opportunities made available to me while working, at the beginning of my career in Human Factors and Ergonomics, at the University of Cape Town. Professor George Jaros and Mr H.M (Mike) Cooke believed in me and gave me the support and encouragement to produce the first edition. Here we go again!

Contents

Preface

R. S. Bridger

There has never been a better time to embark on a career in *Human Factors and Ergonomics* (HFE). A lot has happened since the fourth edition of this book was published in 2017. We have seen a variety of new, sometimes disruptive, developments such as exoskeletons, self-driving cars, artificial intelligence (AI), hybrid working, and the Internet of Things, to name but a few. There are also new challenges—a growing problem of physical inactivity at work coupled with an epidemic of overweight and obesity in the populations of many countries. This book has been updated to report recent research in these areas and to reflect the current state of scientific knowledge in established fields such as musculoskeletal disorders and occupational biomechanics. HFE continues to be integrated into systems design via systems engineering and into systems management via industrial engineering, safety management, and occupational health (Figure P.1), creating exciting job opportunities for new graduates and senior specialists alike. HFE experts are employed in most high-risk industries, including the nuclear industry, aviation, healthcare, manufacturing, energy, and transport.

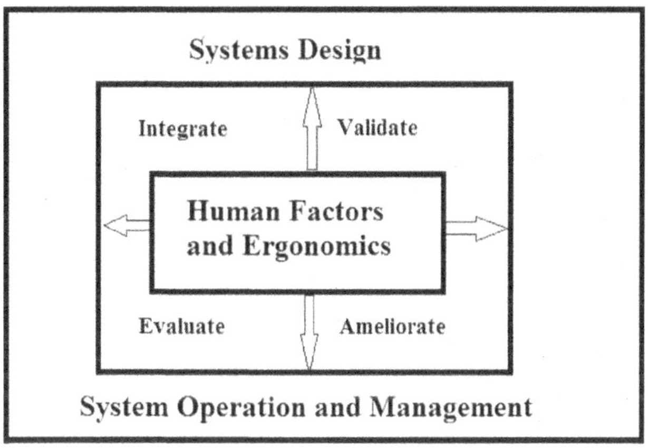

HFE WORKS BEST IN A BIGGER BOX

HFE is never applied on its own but as part of a team effort within a system of management. For this reason, I have included a Prologue, which deals with Human Systems Integration and shows how HFE fits into a "bigger box" at work.

A HUMAN–MACHINE MODEL (ADAPTED FROM PREN 14386: 2002(E))

In accordance with the model, this book has chapters on body mechanics and dimensions, the senses, human information processing, and so on. The interactions between these elements and the corresponding controls, displays, and other elements of the technical subsystem are described. Requirements for optimal design of systems are derived from an understanding of how people interact with technology.

Each chapter begins with a list of general human requirements that must be met when a system is designed. These requirements will help HFE specialists put HFE "on the table" early in the design cycle.

Within each chapter, the knowledge is organized using the standard model below in order to make it easier for readers to navigate HFE across the core disciplines.

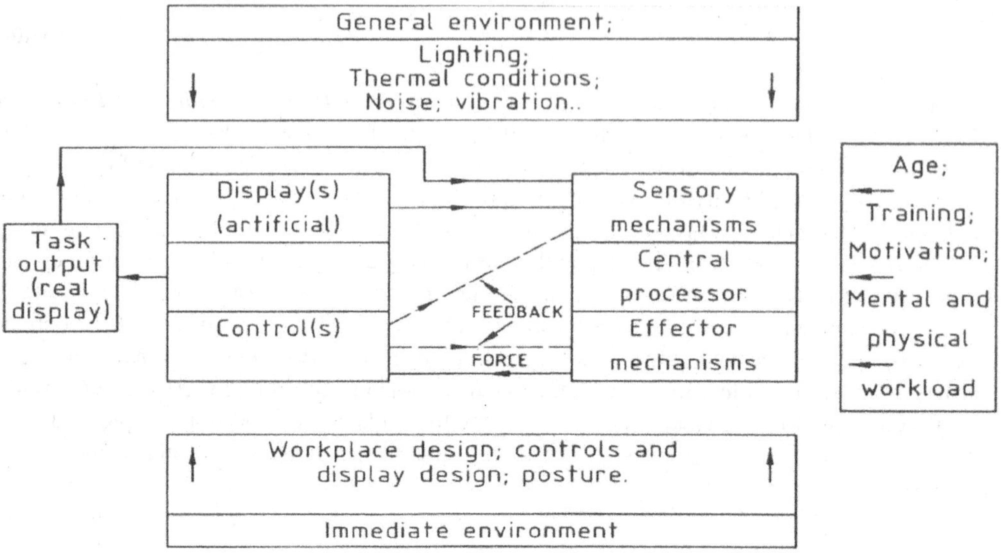

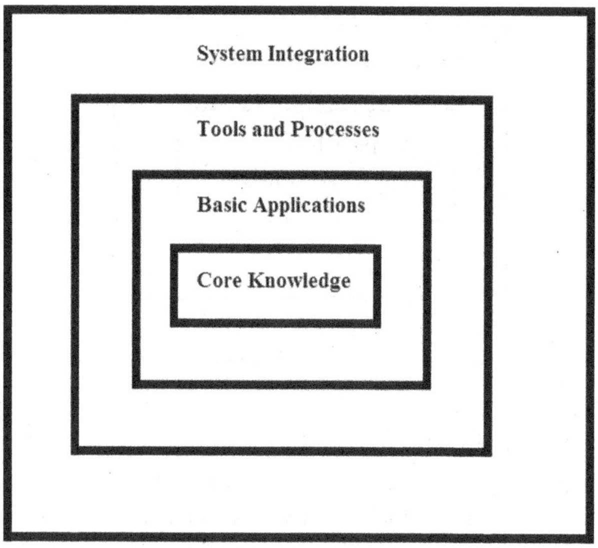

STANDARD MODEL FOR THE ORGANIZATION OF HFE MATERIAL IN EVERY CHAPTER

Each chapter begins with a review of the core scientific knowledge, the terminology, definitions, units and descriptions, and discussions of human physical and psychological subsystems. In updating the core knowledge, I have made extensive use of recently published systematic literature reviews (where experts assess the body of research in a particular area) to draw some general conclusions about each domain. Some basic applications of this knowledge are then described, followed by examples of tools and processes and links to online toolkits. The tools and processes can be used for assurance purposes at *any* point in the system lifecycle, but preferably before the system is accepted for use. Each chapter ends with a discussion of some of the wider issues of systems integration, including cost-effectiveness and considerations of how the basic concepts of HFE can interact with each other when systems are put together. Cost-effectiveness arguments can be used during the design process and afterward to promote proper integration of HFE into the system and to counter the arguments that inevitably arise when requirements are "traded off" against each other.

INSTRUCTOR'S MANUAL, TUTORIAL GUIDE, POWERPOINT PRESENTATIONS, AND VIDEO LINKS[1]

For course providers, a revised and updated manual (with a "Quick-Quiz" at the end of each chapter) and a tutorial guide are available from the publisher, offering advice on the preparation of lectures and demonstrations, as well as the solutions to the problems at the end of the chapters. At the end of each chapter, there is a link to a short video that can be used to enliven lectures and stimulate discussion. A full set of PowerPoint presentations is available from the publisher.

GUIDANCE FOR EDUCATORS AND USERS

The sustainability of educational courses in HFE has recently been discussed.[2] HFE is a "niche" discipline in academia, often in competition for resources with high-volume competitors. The trend has been to shift training to the master's level or to integrate HFE into cognate degrees such as industrial engineering, occupational psychology, and occupational medicine/health. With these developments in mind, this book can be used as a standard HFE text for a dedicated course, or selected "stand-alone" chapters can be integrated into other degrees.

This book can be used in different ways: as a course text, it can be studied in a linear fashion, with students completing some or all of the essays and exercises at the end of each chapter. Alternatively, individual chapters can be studied independently to refresh knowledge of particular topics. In many industries, a "Human Factors Specialist" has a wide brief and is expected to have basic competence in all of the areas described in this book; therefore, this book can be used as a reference text by professionals seeking examples of assessment tools and methods or in support of HFE requirements or interventions.

Following requests from academic users, this book ends with an Epilogue on "The Future of Work."

R. S. Bridger
Consultant in Human Factors and Ergonomics, Hampshire, United Kingdom

NOTES

1 URLs are current at the time of publication.
2 Oakman, J., Hignett, S., Davis, M., Read, G., Aslanides, M., Mebarki, B., and Legg, S. 2020. Tertiary education in ergonomics and human factors: Quo vadis?. *Ergonomics*, *63*(3): 243–252.

Author

Dr R. S. Bridger is an independent consultant and educationalist in Human Factors and Ergonomics. He is a qualified fitness instructor and expert witness. He has published over 200 scientific articles, papers, and official reports. He is the author of the book *A Guide to Active Work in the Modern Office: Homo Sedens in the 21st Century.* Recently, he published a new book, *Introduction to Human Factors in Accident Investigation*, which is available on Amazon worldwide. He is an experienced consultant, teacher, and public speaker and was awarded "Ergonomics Practitioner of the Year" by the Federation for Professional Ergonomists in 2019. He has supported the academic programs of universities in the United Kingdom, Norway, and South Africa and regularly reviews articles for journals and acts as an external examiner of theses. He is a past President of the Chartered Institute of Ergonomics and Human Factors and, at the time of writing, is Honorary General Treasurer of the Institute. He can be contacted via LinkedIn.

Prologue

Human Systems Integration[1]

GENERAL REQUIREMENTS FOR SUCCESSFUL INTEGRATION OF HUMAN FACTORS AND ERGONOMICS IN SYSTEMS DESIGN AND MANAGEMENT

1. Define the capability required of the system and the human component of the capability.
2. Develop a Human Factors and Ergonomics (HFE) Integration Plan.
3. Specify the Requirements for Humans in Systems to:
   a. Make the best use of human capabilities.
   b. Understand and provide for human needs.
   c. Provide mitigation for human limitations.
   d. Use people in ways that maximize system safety.
   e. Use people cost-effectively.
   f. Optimize whole-life costs through the appropriate use of people.
3. Understand How to Test Whether Requirements have been met.

> To think about anything requires a concept or model, not just of the thing itself, but also of the context in which it is being observed.
>
> *Gharajedaghi, 1985, p. 269*

CORE KNOWLEDGE: SYSTEMS THINKING IN HFE

HFE is never applied on its own. HFE specialists work in teams with engineers and others to design new systems or products, or to assess and minimize risk in existing systems.

For this reason, we will begin this book by looking at HFE in a "bigger box"—as an enabling discipline that helps teams build better systems and manage existing systems more effectively. In the subsequent chapters, we will "look inside the box" and learn more. This prologue stands outside the rest of this book, and the rest of this book fits inside it. You can read it any time you like—but make sure you do read it before you finish the rest of this book or get a job in HFE!

What Is a System?

There is a difference between a system and an assembly of "things." Systems are purposeful, and their purpose is to deliver capability of some kind—to achieve a purpose. A system is a set of elements, the relations between these elements, and the boundary around them. Most systems consist of people and devices that perform one or more functions (activities) on their inputs to produce some form of output. Inputs are received in the form of matter, energy, and information. All work systems have a physical or functional boundary around them, which separates them from adjacent systems. *Systems analysis* is the name of the discipline that studies the structure and function of work systems and provides the means by which simple systems may be combined to form more complex systems. Systems analysis is an integral part of all advanced work in HFE.

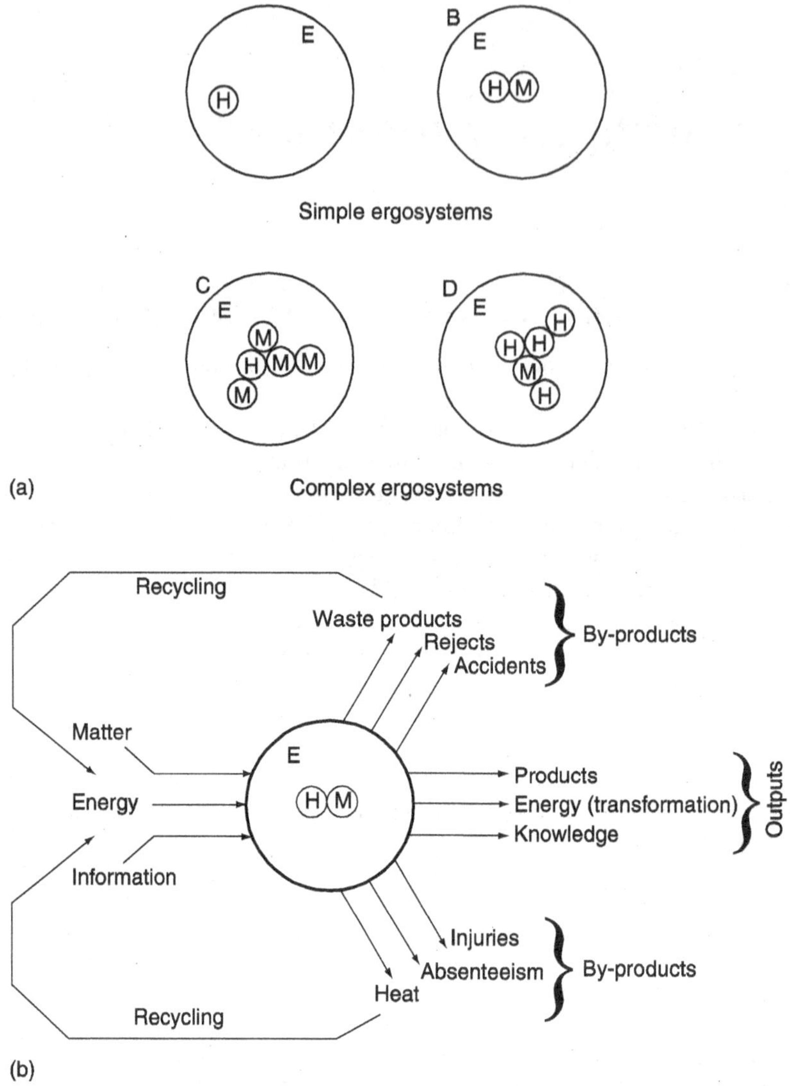

FIGURE P.1 (a) A structural ergonomic view of the work system showing the components. The focus is on how the components fit together and interact and (b) a process ergonomic view of the work system. The focus is on system performance parameters such as productivity, capacity, and effectiveness. Note: E, environment; H, human; and M, machine.

When you design a system, you design the interactions between the elements as well as the elements themselves (Figure P.1). Much of the complexity of systems lies in these interactions because there are many more interactions than elements. In Table P.1, for example, there are six interactions arising from three elements. People are elements in systems (H), and HFE studies the interactions involving people (H<>E, H<>M, H<>H). Complexity arises as we add more elements because the number of interactions increases very quickly. Even in a simple system consisting of one person, one machine, and an environment, six *directional* interactions are possible (H>M, H>E, M>H, M>E, E>H, E>M), and four of these involve the person. Each of the components of a particular work system may interact either directly or indirectly with the others (Table P.1). For example, the machine may change the state of the environment (by emitting noise or heat, for example), and this may affect the user.

TABLE P.1
Basic Interactions in a Work System and Their Evaluation

H > M: The basic control actions performed by the human on the machine. Application of large forces, "fine tuning" of controls, stocking raw materials, maintenance, etc.

Evaluations:

Anatomical: Body and limb posture and movement, size of forces, cycle time, and frequency of movement, muscular fatigue

Physiological: Work rate (oxygen consumption, heart rate), fitness of workforce, physiological fatigue

Psychological: Skill requirements, mental workload, parallel/sequential processing of information, compatibility of action modalities

H > E: Effects of human on the local environment. Humans emit heat, noise, carbon dioxide, etc.

Evaluations:

Physical: Objective measurement of working environment. Implications for compliance with standards

M > H: Feedback and display of information. Machine may exert forces on the human due to vibration, acceleration, etc. Machine surfaces may be excessively hot or cold and a threat to the health of human

Evaluations:

Anatomical: Design of controls and tools

Physical: Objective measurement of vibrations, reaction forces of powered machines, noise, and surface temperatures in the workspace

Physiological: Does sensory feedback exceed physiological thresholds?

Psychological: Application of grouping principles to design faceplates, panels, and graphic displays. Information load. Compatibility with user expectations

M > E: Machine may alter the working environment by emitting noise, heat, and noxious gases

Evaluations: Mainly by industrial/site engineers and industrial hygienists

E > H: The environment, in turn, may influence the human's ability to interact with the machine or to remain part of the work system (due to smoke, noise, heat, etc.)

Evaluations:

Physical/Physiological: Noise, lighting, and temperature surveys of the entire facility

E > M: The environment may affect the functioning of the machine. It may cause overheating or freezing of components, for example. Many machines require oxygen to operate. Oxygen is usually regarded as an unlimited and freely available, rather than part of the fuel

Evaluations: Industrial/site engineers, maintenance personnel, facilities management, etc.

Note: E, environment; H, human; M, machine; >, causal direction.

HFE and Systems

HFE is the study of the interactions between people and technology and the factors that affect these interactions. Its purpose is to assure the performance of the system so that *defined users* with *defined skills and knowledge* can carry out *defined tasks, using defined equipment*, to *defined standards* under *defined conditions*. Thus, HFE is concerned with identifying and defining user characteristics; understanding and documenting existing skills and knowledge; designing and understanding the tasks; defining and operationalizing the required level of task performance; and understanding the environmental and psychosocial conditions in which work takes place (the "context of use").

HFE can be applied by:

1. Designing the user interface to make it more compatible with the task and the user. This makes work easier, can improve efficiency, and may render the system more resistant to the kinds of errors that people are known to make.

2. Changing the work environment so that people can work productively and safely.
3. Changing the task to make it more compatible with user characteristics, education, and skills, reducing the need for training.
4. Changing the way work is organized to accommodate people's psychological and social needs, helping to develop and retain a skilled workforce.

In order to make these improvements, designers must understand the requirements for usability, safety, and so on, and a large part of this book is devoted to describing the knowledge needed to understand these requirements in detail. The rationale behind requirements and the methods of *assurance* (confirming that the requirements have been met) are also explained.

HFE and Risk

HFE is about managing risk by designing the hardware and software interfaces, and the where and how work takes place, to avoid adverse outcomes such as

- Inefficiency—when worker effort produces suboptimal output
- Fatigue—in badly designed jobs people tire unnecessarily
- Accidents, injuries, and errors—due to hazards such as badly designed interfaces, poor organization, and/or excessive mental or physical stress
- User difficulties—due to inappropriate combinations of subtasks, making the dialogue/interaction cumbersome and unnatural
- Low morale and apathy

Emergent Properties and Emergent Problems

The *Gestalt* psychologists of the early twentieth century held that:

The whole is greater than the sum of its parts meaning that a system has a capability and properties over and above that of the aggregate capability of its constituent elements—an *emergent property.* At one level, this is merely a pleasant way of stating the obvious because usually the emergent property is the very reason the system was built in the first place. If the wings fall from an aircraft in flight, it falls to the ground, as do the wings because they cannot fly by themselves either.

In practice, integration of elements during systems design gives rise to *emergent problems.* For example, the solution to a problem in one subsystem may affect the solution to other problems in other subsystems when the subsystems are integrated. Emergent problems are often found at the human interfaces, arising when technical subsystems are integrated with people or teams of people. Designers then have to find a way of solving these emergent problems and can choose from "top-down," "bottom-up," and "same-level" interventions, or a combination of these (Figure P.2). Real work systems are hierarchical. This means that the main task is made up of subtasks (the next level down) and is governed by higher-level constraints, which manifest as production targets, style of supervision, type of work organization, working hours, shift work, etc. If we want to optimize a task in practice, we can focus on the task itself and redesign it entirely (e.g., automate a manual handling task), or we can change or reorganize the elements of the task (at the next level down), or change the higher-level variables (Figure P.2). For example, to optimize a data entry task, we might look at the style of human-computer dialogue in use. We might find there are aspects of the dialogue that cause errors to be transmitted to the hardware and change the dialogue to eliminate these errors (e.g., the operator mistakenly reverses two numbers in a code and it is recognized as a different code rather than being rejected—change the codes). Alternatively, we might find there are insufficient rest periods, that most errors occur during the night shift, or that the whole system needs reorganizing and new tasks designed.

To optimize a manual handling task, we first have to identify the level of the task itself, the next level down (the weight and stability of the load and the container), and the next level up

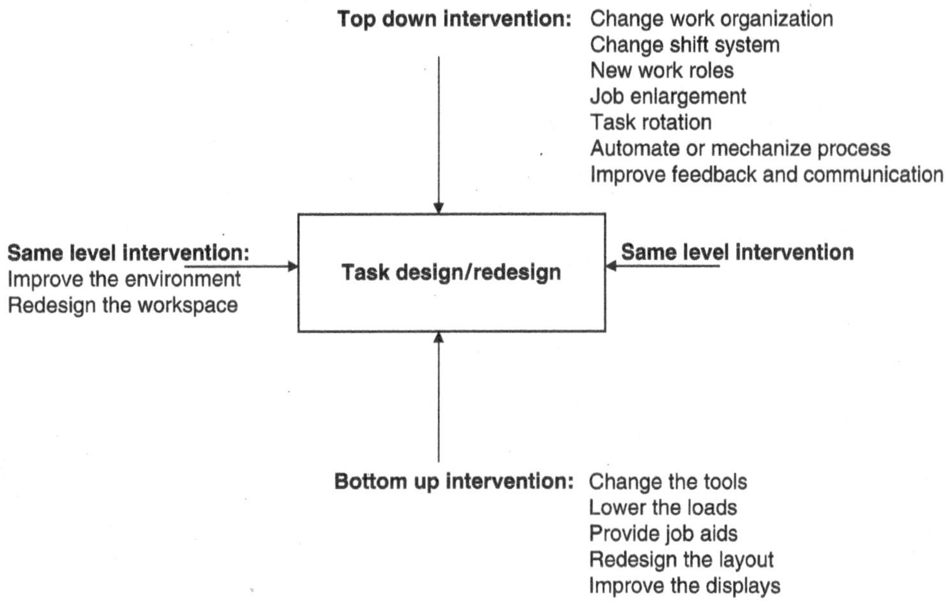

FIGURE P.2 HFE can be applied at one level or at several levels simultaneously.

(the workload and work organization). We can either redesign it from the bottom up (e.g., use lighter containers and stabilize the load) or the top down (introduce job rotation or more rest periods), or both. At the same time, we can look at extraneous or environmental factors at the level of the task but external to it—factors that also degrade performance (e.g., slippery floors in the lifting example, or bad lighting and stuffy air in the data entry example). Having carried out an intervention, we then monitor it over time to detect improvements in system performance.

In summary, *micro ergonomic* task design involves one or more of the following:

- Design of the procedures (rules and steps that govern how the task is carried out)
- Design of the context (workspace and surroundings)
- Design or selection of the task objects (equipment and tools used to perform the task)

In the wider context, *macro ergonomic* task design involves one or more of the following:

- Design of work organization (combining small systems into larger systems)
- Design of jobs (packaging system functions together into jobs)
- Optimal choice of technology (in relation to the needs of the system and the individual)
- Work roles, communication, and feedback (information flow throughout the larger system)

In practice, solutions to emergent problems involve trade-offs, which typically require some consideration of the costs and benefits of meeting HFE requirements. Being effective is not enough, and cost-effectiveness is a key criterion of successful systems integration.

Design of Complex Systems

Complex systems have:

- Long development times
- Many components and subcomponents

TABLE P.2
HFE in the CADMID Cycle

- *Conception:* statement of system/user requirements
- *Assessment:* HFE in design concepts
- *Design:* application of HF guidelines and tools
- *Manufacture:* build to meet the standards and test for acceptance
- *In-use:* identification/solution of operational problems
- *Decommission:* plan for safe decommission/disposal

- "Roles" rather than "people"
- Iterative "top-down" processes as the system concept develops
- No "reverse gear"

And they are underpinned and constrained by complex contractual and legal processes.

Systems engineering provides the overall context for systems design and operation over the life of the system. By "fitting the HFE" to the system's lifecycle, the use of expertise in HFE can be optimized (the "CADMID" cycle in Table P.2).

SYSTEMS ENGINEERING

The International Council on Systems Engineering (INCOSE 2023) defines systems engineering as follows:

> Systems engineering is an interdisciplinary approach and means to enable the realization of successful systems. It focuses on defining customer needs and required functionality early in the development cycle, documenting requirements, then proceeding with design synthesis and system validation while considering the complete problem. Systems engineering integrates all the disciplines and specialty groups into a team effort forming a structured development process that proceeds from concept to production to operation. Systems engineering considers both the business and the technical needs of all customers with the goal of providing a quality product that meets the user needs.

Systems engineering brings together disciplines under a common framework. It takes a "top-down" approach, which, in practice, means that it begins with an analysis of the capability to be delivered by the system, the requirements that must be met to deliver the capability, and the functions (activities) that must take place to meet the requirements. Table P.3 summarizes a systems engineering design process and the corresponding HFE activities that should take place at each stage of the process. Note that the activities of other disciplines, such as industrial engineering, mechanical and aeronautical engineering, and logistics, are also integrated into the systems engineering process in a similar way. Thus, in complex system development, systems engineering provides the overall context for the application of HFE.

SYSTEMS ENGINEERING AND HFE

According to ISO/DIS 6385 (Ergonomic Principles in the Design of Work Systems), the steps needed for an ergonomic approach to design are:

- Formulation of the system's goals
- Analysis and allocation of functions
- Design concept
- Detailed design

TABLE P.3
Stages in a Systems Engineering (SE) Process and Common HFE Activities by Stage

SE Stage	HFE Contribution
Identification of capability need/user	Identify users and task requirements
Concept of operation	HFE in mission analysis
General system requirements	General human requirements
Detailed requirements	Generate specific HFE requirements
Functional design	Allocation of function
Technical design	Human engineering and ergonomics[a]
Component integration	Human engineering and ergonomics[a]
Unit/device testing	User trials and workplace design
Subsystem verification	Usability and workload assessment
System verification	Safety HAZOPS analysis, etc.
System validation	System-level trials
In-use	Post installation assessment/user feedback

[a] Includes task analysis and application of standards.

- Realization, implementation, and validation
- Evaluation

These are described below.

Formulation of Goals and Requirements Specification

Deciding what the system has to do is the first stage in design. This often takes the form of a mission statement or a statement of the required capability and consists of a concise statement of what needs to be done without mention of the means by which it will be done. This is illustrated in the quotation below:

> This nation should commit itself to achieving the goal, before this decade is out, of landing a man on the Moon and returning him safely to the Earth.
>
> *(President John F. Kennedy, May 25, 1961)*

This statement is then decomposed into a set of User Requirements (what must happen for the mission to be accomplished) and System Requirements (what the system must deliver in order to meet the user requirement). In the moon mission example above, some key user requirements might be:

User Requirement

1. Send astronaut to the Moon
2. Land on surface
3. Explore locality and collect samples
4. Return astronaut to Earth

Examples of Systems Requirements: Level 1

1. Provide a habitable environment on the surface of the Moon from which exploration can take place.

2. Provide personal protective clothing (PPC) to protect the astronaut while exploring.
3. Provide a vehicle to extend the range of surface exploration.

As the concept develops, more detailed requirements can be derived:

Examples of Systems Requirements: Level 2

1. PPC must enable the wearer to walk.
2. PPC must enable the wearer to sit and operate the exploration vehicle.
3. PPC must provide protection for the duration of a sortie on the Moon's surface.
4. PPC must be durable.

At the requirements specification stage, the focus is on what is needed, not how it is to be achieved. When existing systems are being upgraded, the requirements may already be known, and only the parameter values may need to be changed (e.g., the new requirement is to deliver 100 tons of product per hour instead of 50). With new systems, information about the performance requirements of the work process may need to be gathered, including information about the role of human operators. In both cases, design should be regarded as an opportunity to solve any existing HFE problems and to avoid generating new ones.

Two central issues at this stage concern the choice of techniques used to specify the requirements and the choice of people to sit on the design team. A common mistake is to leave requirements specification to technical managers, engineers, or business experts and ignore the potential contributions of operators or users. This can be avoided using a participatory approach. The involvement of people closer to the actual work situation provides more information because it takes into account the informal structure and actual working arrangements of the organization (how work is really carried out, as opposed to how it is supposed to be carried out). Critical information regarding health and safety and productivity may be fed into the requirements at this stage.

Gathering Requirements

Some general guidelines for gathering requirements are:

1. Define the capability, as needed by the customer.
2. Take a participatory approach early on.
3. Define and agree on the scope of the project.
4. Ensure that requirements are SMART.
5. Check if there is any doubt.
6. Create a clear, concise, and thorough requirements document to manage the process.
7. Confirm your understanding of the requirements with users.
8. Avoid "solution-eering" until the requirements are fully understood.
9. Get the requirements agreed upon with the stakeholders before the project starts.
10. Create a prototype or model, if necessary, to confirm or refine the customer's requirements.

Requirements Elicitation

Some examples of methods for requirements elicitation are given in Table P.4.

The idea is not to describe how the system is to work so much as its purpose—what it must do. In the design of a new airplane or motor car, the requirements specification is not that the airplane flies or that the car's wheels stay on, but the characteristics of the product with respect to a defined section of the market. For example, the number of passengers the airplane should carry, the routes it would operate, its performance characteristics, fuel consumption, etc. Thus, the requirements for an airplane would be designed through an analysis of the functioning of the parent system—the

TABLE P.4
Data Capture Methods for Requirements Elicitation

1. Analysis of legacy data: safety records, production and performance, maintenance schedules
2. Focus groups
3. Unstructured interviews
4. Structured interviews
5. Unobtrusive observation of the current system
6. Brainstorming sessions
7. Questionnaire surveys
8. Market research
9. Crowdsourcing and serious games

air transport system. Similarly, the requirements for an electrically powered automobile would be arrived at through an understanding of the personal transport system.

Writing Requirements

Well-written requirements are SMART (specific, measurable, achievable, realistic, and timely):

- Explicit, quantified, and testable
- Use single-sentence statements
- Can be prescriptive or proscriptive
- Do not use narrative descriptive paragraphs, other than for contextual and supporting information
- Define for each requirement: who the owner is, what the need is, how much, in what circumstances, and how acceptance is to be tested

In the examples below, some badly written requirements are shown. They are vague, use undefined or poorly defined terms, or are open to interpretation in ways that would make acceptance difficult to test. For example, the first requirement has no definition of usability, who the users are, how usability might be tested, or what the context of use is. The second requirement is badly phrased because 5th and 95th percentile users do not exist and cannot be found to test the system. The "relevance" of any legislation in the third requirement is open to interpretation. SMARTER examples follow.

Bad

- The system shall be usable.
- The system shall fit the 5th to the 95th percentile users.
- The system shall comply with all relevant legislation.

Better

- The system shall be compatible with user training, expectations, and work demands, and shall facilitate the attainment of user goals during the execution of all tasks.
- Where human physical variability places constraints on the design, the system shall be designed to accommodate the 5th to 95th percentile on all relevant user dimensions for males and females in the user population.

- The system shall comply with the Control of Vibration at Work Regulations, 2005 under all conditions of use.

Generation of Detailed Requirements

As the system concept develops, more detailed requirements can be derived from the general requirements above. For example, as the functional design is converted into a technical design, we will gain a better understanding of how human physical variability will place constraints on the solution and the particular anthropometric dimensions that must be accommodated. The specification of detailed requirements requires an understanding of the *context of use* of the system and a description of the *target population.*

Understanding the Context of Use

Context of use is a usability term, used to frame the assessment of the usability of equipment. The context of use is defined within ISO 9241-11 and can be considered in terms of:

- Users
- Tasks
- Equipment
- Environment

The context of use of a device places constraints on its design. For example, if the device is to be operated by users wearing protective clothing or in poor lighting, designers must take this into account. An understanding of the other devices and equipment used in a job may help prevent errors—for example, administering the wrong medication when different drugs have similar names or packaging.

Define the Target Population

The target population is simply the employees who will be operating the system. Their characteristics may place constraints on the design that have to be understood early on. These characteristics include: age range, sex, anthropometry, education, training, first language, and educational level. For example, the level of literacy will influence the wording of instructions and user manuals. Age and sex may influence the physical design of tasks, protective clothing, and equipment.

Degrees of Compliance

The degree of compliance is reflected in the way the requirement is written. The word "must" is used to denote requirements mandated by law, the word "should" is used to reflect requirements mandated by policy, and the word "may" provides discretion for the interpretation of policy.

Analysis of Functions

A system's functions are the activities that enable it to produce its outputs. In the case of new systems, they may first have to be identified before they can be allocated to different operators or machines. Functions are often defined in the simplest terms possible, using only one verb and one noun. For a cash withdrawal from an ATM, for example:

Enable Transactions

- Accept card
- Read card
- Identify user

- Determine user need
- Check status (as required by need)
- Accept command(s)
- Return card
- Dispense cash

Identification and Description of Functions: Information on functions can be obtained in several ways. Panel discussions, interviews with users and experts, and observation can provide function-related information. System design methods often stress that functions initially be identified and described in abstract terms. Some advantages of abstract functional thinking (Singleton, 1974) are:

1. It encourages the search for new methods.
2. It provides a common language across disciplines.
3. The range of solutions is not limited by particular components or methods.

Avoidance of Solution-eering: "Solution-eering" is the practice of deciding upon a technical solution to a problem before the requirements and functions to deliver them have been specified. When designing complex systems, the proper focus should be on requirements rather than solutions, because complex systems take a long time to build, and rapid technological change may render early solutions obsolete. In general, we are not good at predicting the future, as illustrated in the following quotations:

- "Where a calculator like the ENIAC today is equipped with 18,000 vacuum tubes and weighs 30 tons, computers in the future may have only 1000 vacuum tubes and perhaps weigh only 1.5 tons." —Hamilton (1949)
- "I think there is a world market for maybe five computers." —Thomas Watson, CEO of IBM, 1940s
- "640K ought to be enough for anybody." —Bill Gates, 1981
- "There is no reason anyone would want a computer in their home." —Ken Olson, president, chairman, and founder of Digital Equipment Corp., 1977

Figure P.3 shows a product that was developed during the heyday of Victorian inventing. Following the success of a range of products such as steam engines, flush toilets, and bicycles, which had made fortunes for their inventors, new products appeared in the absence of any kind of market research. The Nautilus rocking and wave bath in Figure P.3 is an example. Believing that trips to the seaside were healthy, the product was designed to deliver the benefits by bringing the seaside into the home—an example of "solution-eering," a product in search of a market.

Allocation of Function

Allocation of function between operators and technology determines the level of automation in a system. In practice, functions usually fall into one of several categories:

1. Those that *must* be carried out by machines (because it is impossible or unacceptable for humans to do them).
2. Those that *must* be carried out by humans (because no adequate machines are available or machine execution of the function is not appropriate).
3. Those that might be carried out by either humans or machines, or both.

Allocation of function dates from 1951, when Paul Fitts produced a list of functions best conducted by humans or machines (Fitts, 1951, below). Humans appear to surpass present-day machines in respect to the following:

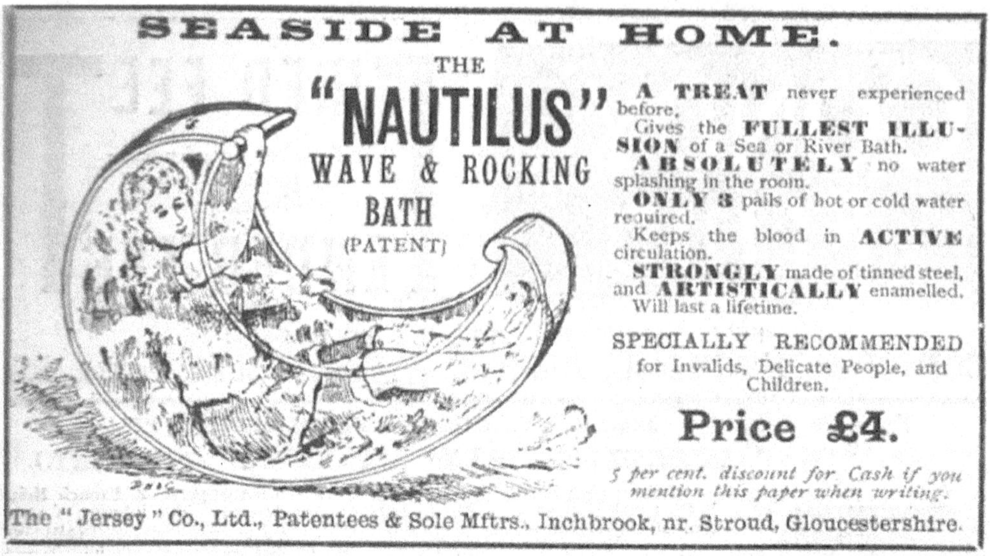

FIGURE P.3 Seaside at Home? "In this bath the water can be set in motion by rocking, producing a sensation very much like the waves of the sea, which will delight and benefit especially invalids, delicate people and children. Only three pails of cold or hot water are required, and there is no splashing in the room to be apprehended. By placing a wedge under the curve of the back the bath can be made to serve the ordinary purposes of the tub." The Nautilus wave and rocking bath—a solution in search of a requirement? (Courtesy of Seaside at Home: The "Nautilus" Wave and Rocking Bath, advert from 'Pick-me-Up' Magazine, 1891 (http://Victorianlondon.org).)

1. Ability to detect a small amount of visual or acoustic energy
2. Ability to perceive patterns of light or sound
3. Ability to improvise and use flexible procedures
4. Ability to store very large amounts of information for long periods and to recall relevant facts at the appropriate time
5. Ability to reason inductively
6. Ability to exercise judgment

Present-day machines appear to surpass humans in respect to the following:

1. Ability to respond quickly to control signals and to apply great force smoothly and precisely
2. Ability to perform repetitive, routine tasks
3. Ability to store information briefly and then erase it completely
4. Ability to reason deductively, including computational ability
5. Ability to handle highly complex operations, i.e., to do many different things at once.

Since then, the list has been criticized on the grounds of being too reductionistic. It has also been expanded. Technology has advanced greatly since 1951—machines' can do a lot more than they used to. The processing power of computers has increased greatly, and with artificial intelligence (AI), natural language processing, and an abundance of sensor technologies, the original list is dated (Fitts knew that it would become dated). While it is often straightforward to identify functions that can only be done by humans or machines, it may be more useful with much modern technology to consider how humans and technologies can work together to maximize system capability—a shift from allocation of function to *complementarity* of action through new joint functions. For example, many modern cars have a lane assist function that nudges the steering wheel if the car drifts out of the lane. One of the most important issues for HFE in the Twenty-first century is the extent to which human tasks should be automated—should operators be left in a supervisory role, or should the task be designed so that the operator remains in the loop? Fitts himself (quoted in De Winter and Dodou, 2014) stated:

> Human tasks should provide activity. The roles of the human operators in the future air navigation and traffic control system should be active rather than passive ones. Activity in any task is conducive to alertness, and helps to ensure that the human will keep abreast of the situation

This applies equally to self-driving cars and any other activity where humans have to control a process.

Affordances

Of the three categories at the beginning of this section, there is increasing potential for category 3, with a change in focus from 'allocation of function' to '*complementarity*'. wherein functionality is shared between humans and technology. Another way to think about this is that the internet and new technologies such as natural language processing, pattern recognition and artificial intelligence surround us, providing '*affordances*' for new ways of performing work functions.

> The affordances of the environment are what it offers the animal, what it provides or furnishes, either for good or ill. The verb 'to afford' is found in the dictionary, the noun 'affordance' is not. I have made it up. I mean by it something that refers to both the environment and the animal in a way that no existing term does. It implies the complementarity of the animal and the environment.
>
> *(J.J. Gibson, 1979)*

Figure P.4 provides a simple example. In the figure, we can see a shop owner talking to a customer. Notice the shop owner's right shin is resting on a stool. The stool provides the affordance for the posture even though that is not the stool's main function.

Elizabeth Churchill, Head of User Experience at Google once commented that, "*We live inside the Internet.*" Complementary allocation of function between people and technology is becoming the norm in many aspects of life and is an area of increasing importance for HFE. In the same way that the stool, and its affordances are part of the shop owner's work environment, many new technologies are now part of the world in which we live. The affordances they make possible will affect the design of new products and devices and, indeed, the future of work itselfa topic that will be returned to in the Epilogue.

Allocation of Function and AI

AI already operates in the background of our lives during the performance of many tasks. Much of the time, it acts as an advanced form of automation, enhancing our performance of everyday tasks such as document creation. Siemens et al. (2022) prefer the term "cognition" to "intelligence," stating that the challenge is to combine two types of cognition that operate in different ways (Figure P.5). They distinguish between intelligent systems that can learn on their own and cognitive systems that are automated cognitive processes. In the case of the latter, one challenge is to represent the output of the automated system in a way that complements human cognition.

In some ways, the limitations of human and artificial cognition determine how functions are allocated. In others, artificial intelligence can replace humans in both low-level tasks (sorting apples into different categories depending on their quality) and high-level tasks such as determining the structure of proteins based on their constituent amino acids. Artificial cognition also exceeds human cognition both in games such as Go and in finding patterns in immense data sets. Siemens et al. consider "human-in-the-loop" interactions, wherein both the artificial system and the human share knowledge about each other's cognition ("hybrid augmented intelligence").

FIGURE P.4 The stool provides the *affordance* for the posture.

In practice, people don't go to work to complete specific tasks; they go to work to perform a job function that includes task completion. Siemens et al. suggest ways in which artificial cognition can be incorporated into job performance:

- Both can identify a problem.
- Humans will identify the emotional component of the problem and understand why a solution is needed.

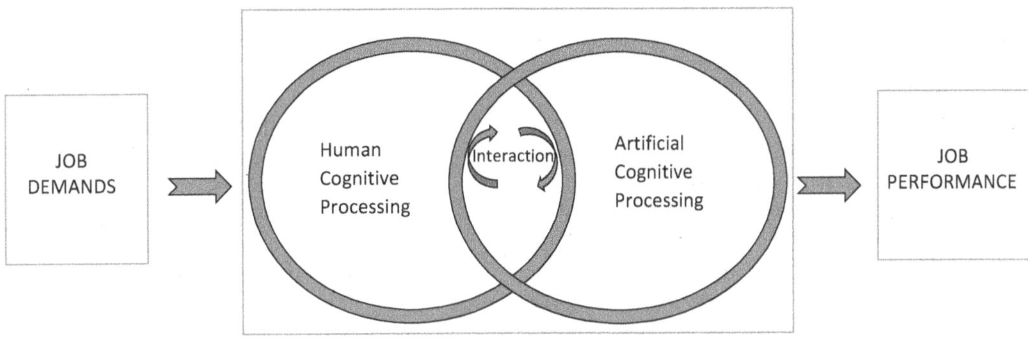

FIGURE P.5 Artificial Cognition in Job Design – Still Room for Humans?

- Artificial cognition can then gather and analyze data relating to the problem more quickly than humans.
- Humans can then review and summarize the findings and possibly brainstorm solutions while considering wider issues such as resource requirements and constraints in relation to a wider strategy.
- Essentially, a key role for humans is to integrate the outputs of artificial cognition within the context of the job role.

Sociotechnical Requirements

Piecemeal allocation of several human functions to one operator can create potentially stressful jobs if there is a conflict between the requirements and responsibilities of the different functions. Role ambiguity is an example of a type of job stress in which the worker is unclear about where their real responsibilities lie and what they are really rewarded for by the organization. *Role conflict* occurs when the successful achievement of one function negates the achievement of another.

When designing complex systems, we are also designing jobs, relationships between people, and feedback loops between and within levels of an organization. The sociotechnical view takes us beyond allocation of function and the traditional H<>M interactions of HFE and provides a perspective on the interactions between people in the system (H<>H).

Most human–machine systems are sociotechnical systems. There is a technical subsystem (which integrates the machines themselves and the interconnections between them, the infrastructure, power supply, etc.) and a social subsystem (which integrates the activities of the people involved in operating and maintaining the machines and the relationships between them).Work organization is the link between machines and the social organization of the individuals who operate them. The optimal utilization of technology depends on an appropriate system of work organization, which itself determines the social organization of the workforce and the relations and interdependencies between individuals. For example, in a classical production line system of manufacture, workers form a homogeneous group in terms of status and skill. The relationship between them is one of linear dependency in one direction (the direction of the production line). Control is exercised by supervisors, foremen, and managers. This can be contrasted with a patient care team in a hospital. Here, the members of the team differ in status, individuals perform different functions, have different expertise, and are mutually interdependent. Individuals can use their own initiative about matters falling within their own area of expertise and can contribute to decision-making at the group level.

Sociotechnical systems theory was developed by members of the Tavistock Institute of Human Relations in the years following World War II. Cherns (1976) has summarized the design principles derived from sociotechnical systems theory (Table P.5).

TABLE P.5
Principles of Sociotechnical Job Design

1. *Compatibility:* The way that sociotechnical design is carried out must be compatible with the changes that are made. For example, if one of the reasons for change is to make the most of the cognitive skills of the workforce, these skills must be utilized in the process of change itself.
2. *Minimal Critical Specification:* In the design of new systems, no more should be specified than is absolutely essential. This leaves individuals with freedom to determine the precise details of how a task is to be carried out. It also leaves open options for change and improvement and enhances adaptability to unanticipated events. Strictly specified rules and procedures can inhibit an organization's ability to adapt.
3. *The Sociotechnical Criterion:* This states that variance in a system (the occurrence of unspecified and unprogrammed events) should be reduced by controlling it as close to its source as possible. Much of the system of supervision, inspection, and maintenance in industry is an attempt to reduce variance from a distance—to correct the consequences of variance rather than control it. If operators carry out their own inspection, the size of the variance control "feedback loop" is reduced.
4. *The Multi-Function Principle:* This is aimed at increasing the adaptability of the organization by allowing workers to fulfil more than one role, while designing jobs so as to reduce the interchangeability of people.
5. *Boundary Location:* In all organizations, boundaries between different departments have to be drawn up. This is usually on the basis of function, technology, territory, or time. For example, engineering products often pass through several different departments, such as milling or grinding shops, in the manufacturing process. The rest of the time is taken up by transport, storage, etc. Deciding where to locate the boundaries is as important as the processes themselves (e.g., whether operators should also carry out routine maintenance).
6. *Information Flow:* This principle states that information should go where it is needed in the first instance rather than via senior management to subordinates. The tendency for information to "filter down" from above can have disadvantages as well as being inefficient. Managers may become preoccupied with matters for which their subordinates should be responsible. An information system designed according to sociotechnical principles should direct information efficiently to those parts of the organization where it is needed and acted upon. It should also support two-way communication.
7. *Support Congruence:* Administrative and management systems should be designed to reinforce those behaviors that the organization wishes to encourage. If an aim is to encourage employees to take a more responsible attitude to their jobs, then supervision and payment systems should be designed to be congruent with this.
8. *Design and Human Values:* This principle emphasizes that systems must be designed to provide high-quality jobs. This is a difficult principle to apply—individuals may respond differently to changes in job design and have different needs.
9. *Incompletion:* Design never ends.

Source: Cherns (1976).

Design Concept

The translation of the requirements specification and the function allocation into a design concept may be facilitated using brainstorming techniques or structured methods of concept generation, such as value engineering. The concept itself should be detailed enough to describe the structure of the work system and the interactions between the components. Those functions allocated to workers are organized into tasks, roles, and jobs as part of the emerging design concept. The concept is reified iteratively using methods such as task description and analysis, the construction of scale models and mock-ups, and discussions with users and consultants.

Detailed Design

The key to modern systems design is that the human and machine components are designed in parallel to save development time. Traditional ergonomics acts as a link between these activities. It is not necessary for the system to be complete before training and job design can take place.

The Apollo space program is an example where all training and development took place offline using simulators and prototypes.

It is during the detailed design phase that jobs are designed, either implicitly or explicitly. Occupational psychologists, however, have proposed alternative ways of designing jobs. Job-centered approaches attempt to optimize the content of jobs so that they will be perceived as satisfying to do. Person-centered approaches are based on the idea that people are motivated to work well if the work satisfies their needs. Thus, it is thought that successful job design requires some consideration of people's needs. The sociotechnical systems approach attempts to optimize the design of a social system of work to support the technical system. These approaches are described in more detail below.

ISO 6385 suggests that detailed design should include all of the following:

- Design of work organization
- Design of work tasks
- Design of jobs
- Design of the work environment
- Design of the work equipment, hardware, and software
- Design of workspaces and workstations

HFE principles for the design of the above are described in this book.

Realization, Verification, and Validation

Realization involves the procurement and installation of the new system on-site. Implementation involves management of the changeover from the old system and the introduction and training of operators to verify that the components deliver the intended functions. If necessary, a temporary backup system may be installed, or there may be a handover period in which both systems run in parallel. Validation requires that the system be shown to function according to the requirements specified earlier—that it delivers the required capability.

Evaluation

Evaluation occurs after the system is up and running. The integration and implementation phase is often characterized by "teething troubles," which are temporary phenomena, not characteristic of the final design. It is for this reason that in activities such as facilities design, a delay is introduced between implementation and final evaluation. This enables teething troubles to be overcome so that the true pros and cons of the facility can be more easily identified. For example, post-occupancy evaluations of new buildings are usually only carried out after one year of occupation.

Evaluation seeks to determine how well the system achieves its goals in practice. Many criteria can be used in evaluation and these are well summarized by simple human–machine models. For example, evaluations of noise, lighting, temperature, the health and safety of the workforce (including the prevalence of any medical conditions), and the occurrence of errors and accidents may all be carried out at this stage.

Safety

Safety is dealt with in a later chapter, and there is much information in this book that can be used to ensure that systems are safe to use and are used safely. For present purposes, it is sufficient to say that *risk* in systems design can never be eliminated completely and that the principle of *ALARP* should apply—risk should be "as low as reasonably practicable." Risk must also be considered in relation to *hazards*—what happens if something goes wrong. The risk of meltdown in a nuclear power station may be very low, but the hazards, if it does melt down, are very high (Table P.6).

TABLE P.6
Both Risk and Hazard Must Be Considered in Safety Cases

		Hazard	
		Minor	**Severe**
Risk	Low	Benign system	Safe system
	High	Unstable system	Dangerous system

Quality Control and Quality Assurance

A system is of good quality if it meets the standards that were set at the requirements specification stage. A high-quality system is simply one that meets high standards. There is a growing body of standards, such as those of the ISO, quoted throughout this book, which can be used to design systems with high-quality HFE. Quality control takes place throughout the design stage to ensure that all relevant standards are being used as the design progresses. Once the system has been built and is functioning, assessments are made to check compliance. If the system does comply with the HFE standards, then we are assured that it is of the appropriate ergonomic quality. Quality control is not possible without reference to some kind of standard.

Acceptance testing is a key quality assurance activity and is part of system verification. In complex systems, it depends on clear *traceability* of requirements. For example, suppose we want to design wearable information and communications technology for mountaineers and explorers of isolated and extreme environments. The overall *capability* requirement is to link the explorer to the World Wide Web for the duration of the mission with minimal physical or psychological intrusion. Clearly, there would be many user requirements over and above access to the web itself. The *context of use* would demand a high level of *resilience* because of the mechanical and environmental stressors to which the system would be exposed. There would be many user requirements, almost all of which would consist of integrating the equipment with the user in the extreme environments—both when being worn while other tasks were carried out and when being used (Table P.7).

There are many kinds of acceptance tests that can be used in systems verification. As HFE matures as a discipline, the findings of research become translated into legislation and standards, which greatly simplifies the assurance process. There is a hierarchy of acceptance in terms of the forcefulness of the criteria that apply:

- *Legal Requirements:* Does the system comply with the law?
- *Standards:* Does the system comply with standards such as ISO standards? If not, has a case been made and a waiver granted?
- *Guidelines:* Does the system comply with published guidelines on, for example, usability, manual handling of loads, and so on?
- *Risk Assessments:* Has risk assessment been conducted using a reliable and valid tool?
- *Best Practice:* Does the system comply with best practice in HFE? If not, what mitigation has been put in place?
- *User Trials:* Are the findings of user trials favorable?
- *Experiments (Rarely):* Experimental research is rarely needed to demonstrate acceptance, except in cases where there is a high level of novelty or exposure to extreme conditions or isolated environments.

Acceptance testing is part of system verification and examples of simple methods of acceptance testing are given throughout this book in the worked examples and workshop sections.

TABLE P.7
Relationship between User Requirements, System Requirements, and Acceptance Tests

User Requirement	System Requirement	Acceptance
1. Minimal effect on agility and joint mobility	1. Restriction of joint mobility less than 5%	User trials in testing room
	2. No reduction in time to climb test wall	Indoor tests on wall
	3. Comfort in use	User feedback
	4. Minimize battery mass	Measurement
2. Ease of use in all conditions	1. Controls are operable in extreme climates	Climate chamber test
	2. Controls can be operated with finger temperatures of 5°C	Climate chamber test
	3. Auditory as well as visual display options	User trials
	4. Interface compatible with past experience	User trials
	5. Easily detachable	
3. Minimal effect on sustained performance	1. No more than 10% increase in aerobic demands	Treadmill testing

Managing Complexity in HFE

HFE is a mature discipline and much of the scientific knowledge is already embedded in existing technology and products. An intelligent facilities management team does not need to know how to design an office chair, for example. It needs to decide *whether there is a requirement for sedentary work and, only if there is such a requirement, how to distinguish between appropriate and inappropriate solutions* (in Figure P.6, much of the science of HFE related to seating is already embedded in the product). It can be seen from this example that functional HFE occurs at the requirements specification stage, whereas much of traditional HFE is used for acceptance testing.

In effect, the top-down approach of systems engineering combined with the iterative, user-centered approach of HFE enables design teams to take a stepwise approach to the management of complexity.

Trade-Offs, Costs, and Benefits

The need for trade-offs is almost inevitable when complex systems are designed due to the length of the development times, changes to requirements, or the introduction of new requirements. Changes to the external environment may also introduce the need for trade-offs. Although there is a growing literature on HFE in complex systems, less is said about the complex contractual relationships that the design of such systems requires. There is normally a contractual relationship between a "customer" (the organization that wants the system) and the "solution provider" (the organization building it). These parties have the same objective (to build a successful system), but their interests may differ, and when this happens, it may be reflected in the ways in which requirements are "traded off" during design. Intangible requirements are likely to be traded off against more tangible requirements (which is why requirements in HFE must be SMART—written as simply and objectively as possible, with minimal scope for interpretation). Costs may be pushed "downstream" as design and manufacturing costs are traded off against increased operational and maintenance costs. Instead of high usability and reliability, the system may require ongoing investment in training and maintenance throughout the rest of its life.

Capability and Competence in HFE

One way of reducing the likelihood of such trade-offs is for the customer organization to employ its own specialists in HFE to ensure that requirements are properly written and met, and also to ensure that the solution provider employs properly qualified people. In the United States, the Human

FIGURE P.6 A modern office chair that meets the HFE requirements for sedentary work. The "Science of Seating" is embedded in the technology and tested at acceptance. NB. The *requirement* for sedentary work is considered much earlier in the design process. (Courtesy of Maria Andreu, Herman Miller Inc, 2007.)

Factors and Ergonomics Society (HFES) has a Board of Certification of Professional Ergonomists (BCPE) and can therefore endorse the competence of its members. In the United Kingdom, the Chartered Institute of Ergonomics and Human Factors (CIEHF) has a Royal Charter and has Chartered Members and Fellows with proven competence. In the European Union, there is a Centre for Registration of European Ergonomists (CREE), and the International Ergonomics Association has a list of member societies worldwide that have acceptable professional accreditation procedures. The employment of competent specialists is key to developing effective working relationships in HFE with a clear separation of responsibilities: the customer specifies and owns the requirements and acceptance process, and the supplier designs and tests the solution. Once the requirements have been successfully identified, gathered, specified, and communicated to a supplier who understands both HFE and the meaning of the acceptance tests, both then share the responsibility to ensure that the roles assigned to people in the solution enable the required system performance to be achieved in the context of use.

Key Performance Indicators

How will we decide whether our new system was worth the investment? Identification of Key Performance Indicators (KPIs) is essential but can sometimes be problematic. When offices were first automated in the 1970s and 1980s, the intent was to improve productivity. However, most of the traditional KPIs relating to industrial productivity showed no benefits from office automation. Why was this?

1. New technologies take time to mature. Computers were first used to automate existing processes and failed to exploit the advantages of automation. The real benefits lay in using new technologies as a platform for new and better ways of working (as in, "when you automate a mess, you get an automated mess").
2. Constraints on IT productivity. Initial productivity gains were consumed by the additional costs of owning, maintaining, and replacing technology, which rapidly became obsolete.
3. Underinvestment or misplaced investment. Decisions about how and where to automate offices were influenced by the costs of doing so. Those aspects of the business that were easiest or cheapest to automate may not have been those that would benefit the most from automation.
4. Time lags—new technologies can take time to mature. The real benefits start to appear with the emergence of networks as everyone, including customers and suppliers, starts using the technology.
5. Growth in administrative activity—activity increases to use up the increased capacity (the new bottle gets filled with old wine).
6. Nonproductive uses of IT include:
   a. Waiting for programs to run or for help to arrive
   b. Excessive drafting and redrafting
   c. Computer games
   d. Overemphasis on quality out of proportion to the value of what is being produced
   e. Employees coping with ever more complex programs. New packages contain ever more features—many of which will only ever be used by a small percentage of users. Generic user interfaces may not map well onto specific jobs, causing unanticipated rigidities and constraints.

The real benefits of office automation were less easy to measure and perhaps related more to better control of quality and work flow and, for many competitive businesses, the ability to respond to customer needs, plan ahead more quickly, and access and process performance data more rapidly.

A CHECKLIST FOR INTEGRATING HFE IN SYSTEMS DESIGN

It was stated at the beginning of this chapter that the application of HFE does not just happen. It only happens if the right processes are in place to link HFE with what everyone else is doing—to integrate it with the rest of the project team. These processes are often described as *Human Systems Integration* (*HSI*) and are distinct from HFE itself. HSI is a management process; it does not generate any new knowledge of its own. It is essentially an exercise in knowledge management that ensures the right knowledge gets used in the right way, at the right time. It can be argued that many examples of badly designed systems, accidents, and failures occurred not because the solution or the knowledge to implement it was absent, but because poor knowledge management and failure to find an exploitation pathway for the knowledge resulted in a suboptimal design. HSI seeks to remedy this state of affairs, and the checklist for HFE is given below:

1. Is there a mission statement or statement of what is required and how HFE can contribute?
2. Is the technical subsystem understood (constraints and dependencies)?
3. Is the human subsystem understood (teams, structure, staffing, and skills)?
   Have users of existing systems been consulted?
4. Has the context of use been defined to enable the definition of the required system?
   Inertia
   Resilience

Scope for succession of elements
Feedback

5. Does the project team include a certified specialist to oversee HFE activities?
6. Has an initial HFE analysis been carried out?
   Review of current or past systems
   HFE issues identified
   HFE process lessons
   Health hazards
7. Have key HFE activities been integrated into project planning?
   Number of specialists required
   Deadlines and milestones
   Link with the systems engineering process
   Detailed HFE integration plan:
      What HFE activities will be conducted?
      When will they be conducted?
      Who will do the work?
      How?
      What assumptions are being made?
      What constraints will apply?
      How will progress be monitored?
      How will findings be reported?
      How will trade-offs be managed and resolved?
   Have human factors activities or inputs been included within project team development activities?
8. Has a forum been defined for discussing and addressing human factors issues within the design team?
   Are there sufficient resources for specialist human factors activities?
   Has an HFI risk register been created?
   Does the forum have formal links with the safety management team?
9. Do project plans, processes, or activities include HFE? Is there a *Human Factors Integration Plan?*
   Specification of the context of use
   Inclusion of user requirements in system or project specifications
   Human factors specialist input into design
10. Is allocation of function considered at the functional design stage?
11. Is there an iterative approach to integrating HFE involving users?
12. Is consideration of HF regulations, standards, and guidelines mandatory?
13. Has an evaluation, testing, and acceptance plan been formulated?
    Who will conduct the testing?
    Will users be involved?
    Have criteria for acceptance been identified?
14. Have human factors activities been included at each stage of the development lifecycle? Are human factors activities compatible with other development activities in the current stage?
15. Are the potential impacts of the design of the new system on the rest of the organization understood?
    Training needs and impact on training budget
    Human resource requirements, recruitment (availability and cost), and retention of trained employees, redundancies, retraining
    Long-term organizational objectives and growth, acquisition, and partnership strategies
    Infrastructure (buildings, IT capacity, bandwidth, etc.)
    Equipment (availability, cost, procurement timescales, and through-life costs)

Information
Organization (structure, interfaces with internal and external stakeholders)
Logistics (supply chain definition and management)
Has mitigation been considered for
Are there implications for recruitment of new skills (e.g., availability and cost)?

16. Have the future requirements for decommissioning and disposal been considered?
17. Is there a plan to archive HFE information?

What does a Human Factors Integration Plan Look Like?

Human Factors Integration plans define the work needed to implement human factors activities in the design process. These plans should be tailored to align with the resources available to execute the plan (i.e., the budget). A typical Human Factors Integration Plan will include:

1. A definition of the HFE goal to be achieved
2. The main milestones and deadlines for their achievement
3. The available resources; and
4. Expertise and responsibilities required and the means of assurance. The Human Factors Integration Plan should include these elements:
   - The HFE issues to be addressed in the project
   - The constraints that limit the design of the new system. There may be legacy equipment from an old system that has to be used in the new one
   - Staffing levels
   - Limitations on operator training capacity
   - The organizational context in which the system is operated
   - Safety constraints falling within the competence of HFE experts
   - The activities that will be conducted to analyze and mitigate HSI issues
   - The planned process of HSI involvement in development. This should include a work breakdown for specialist human factors activities, tied to appropriate activities in the overall project plan (see Table P.3)
   - Constraints or dependencies to and from other development activities: What human factors inputs are needed by developers, and by when
   - What aspects of the design are considered and when?
5. Constraints or dependencies to and from separate development contracts
6. Plans for user involvement, e.g., scheduling of human factors assessments of designs; prototyping activities, user trials; simulations, etc.
7. The method for monitoring and controlling progress against the objective to integrate human factors in the design process (e.g., monthly status reports at team meetings)
8. Processes, mechanisms, and forums for considering human factors trade-offs; and
9. Plans for updating the plan.

A human factors integration plan should be implemented at all organizational levels and divisions, and across management sub systems. Several strategies will increase the probability that human factors integration will be successful, including:

1. Define the purpose and value of implementing human factors;
2. Promote the application of effective human factors principles;
3. Evaluate risk factors;
4. Collect and provide feedback on human factors-related data;
5. Prioritize and choose specific problems to solve;
6. Articulate why these specific problems should be fixed;
7. Implement an organizational-level H program with controls;

8. Ensure that top management openly commits to the plan;
9. Take advantage of locally available improvements;
10. Plan sustained actions—make improvements that will last;
11. Involve and train management and workers; and
12. Maintain involvement and gain feedback on effectiveness.

See, for example, Burgess-Limerick (2020).

CHECKPOINTS FOR INTEGRATING NEW SYSTEMS INTO ORGANIZATIONS

Systems are never designed in a vacuum—they are always part of a larger organization, and project teams must focus not just on how to design a good system but on how to integrate it with its parent. Summary questions include the following:

- How will the new system be maintained—what additional resources will be needed (supplies, procurement of service contracts, etc.)?
- How will the new system be supplied and supported with spare parts and materials?
- Will new skills be needed? Will extra staff have to be recruited to operate the system?
- How will the new system be monitored? What data will be needed? Will new interfaces with existing information systems be required?
- How will the new system be embedded into the organizational structure—what department or group will "own" it?
- What personnel, training, and medical aspects should be taken into account in the development and exploitation of technology, now and in the future? How will the training be designed and embedded into careers?
- What operational guidance will be needed? What will be the policy regarding usage, safety auditing, recordkeeping, and so on?
- How can more effective ways be developed to understand the through-life-cost effectiveness of the system to enable cost-effective delivery of capability?
- How will these considerations change over time with the advent of new technologies and increased human augmentation?
- Are there any ethical, legal, or moral issues around the implementation of the new system?

AUDITING HUMAN FACTORS INTEGRATION

An audit tool for assessing Human Factors Integration has been developed by Greig et al. (2019). This comprehensive tool enables all key organizational functions to be audited to determine the extent to which HFE has been integrated into their operations. The tool enables organizations to determine their level of *maturity* in relation to HFE and to identify areas for improvement. Functions are rated on five possible levels (Infant, Young, Emerging, Front Runner, Ideal) to indicate the development of HF integration and individual HF element maturity. The following functions are assessed in collaboration with one or more subject matter experts in each function, using a detailed list of criteria for each function.

The following functions can be assessed: Health and Safety; Training; Human Resources; Medical Services; Marketing/Communications; Finance; Maintenance; Tooling and Equipment; Construction; Logistics (including warehousing); Product/Service Design; Systems Engineering; Operations Management; Scheduling/Planning; Quality; and Organizational Strategy.

Each function is assessed using a checklist. Table P.8 gives some examples of different levels of HFI maturity for Environmental Health and Safety.

TABLE P.8
Ranking of HFI maturity as applied to selected Health and Safety functions

Level of Human Factors Maturity	Score
No HF applied.	0
HF applied based on legislative requirements.	1
HF applied solely for injury prevention.	2
HF applied for injury prevention and occasionally performance improvement.	3
HF applied as part of culture of organization—maximize employee performance and well-being.	4

Level of Human Factors Maturity	Score	Value
No consideration of HF for performance in the function of EH&S—only related to injury.	0	0
HF rarely considered for performance in the function of EH&S.	1	
HF occasionally considered for performance in the function of EH&S.	2	
HF frequently considered for performance in the function of EH&S.	3	
HF considerations in the function of EH&S always considered for maximizing individual performance and well-being.	4	

Level of Human Factors Maturity	Score	Value
HF justification for change within EH&S is as required by legislation.	0	0
HF justification for change within EH&S occasionally exceeds legislation requirements (injury risk).	1	
HF justification for change within EH&S frequently exceeds legislation requirements and considers human performance.	2	
HF justification for change within EH&S always exceeds legislation requirements and occasionally includes cost-benefit/return on investment.	3	
HF justification for change within EH&S always exceeds legislation requirements and non-injury, cost-benefit/return on investment.	4	

SUMMARY

HFE can be used at any stage in a system's lifecycle:

- *Reactive HFE:* to solve a problem in an existing system using "palliative ergonomics."
- *Proactive HFE:* to specify HFE requirements in the design of new systems.
- *Generative HFE:* to integrate HFE into a larger organizational plan to achieve organizational objectives.

In order to utilize HFE most effectively, it has to be integrated with the rest of the system design activities and incorporated into project planning. In this chapter, we have looked at some of the key concepts in *Human Systems Integration* for complex system design, originally derived from systems engineering. HFE specialists also work in smaller project teams with designers or with health and safety managers. In all cases, though, the activity takes place "in a bigger box," and the main message of this prologue is to show that success depends on integrating HFE activities with the activities of other specialists.

Requirements specification is a good place to start, irrespective of the size of the project or the main driver (system design, product design, health, and safety) because it focuses the team on what needs to be done rather than choosing a "solution" prematurely without considering what need the solution will meet or how it will meet it (the mechanism). Furthermore, HFE requirements are just part of a larger set of requirements from all the other disciplines in the project team. Thus, requirements specification creates a "level playing field" for all disciplines from a management point of view (although this is not to say that requirements might not be "traded off" against one another at some point). The organizational requirements determine the target performance of the system. Other requirements will include health and safety, training times, workstation design and implementation, and running costs.

Technological developments have introduced so many new ways of doing things that "allocation of function" is more than just deciding what the "machine" should do and what the user should do. Rather, the focus is shifting toward complementarity and human-technology symbiosis with shared responsibility for functions. Deciding on the optimal level of user involvement (keeping the person in the loop) is key.

Design solutions are produced iteratively to obtain feedback from multidisciplinary groups. Concrete proposals using mock-ups, storyboards, sample screens, etc., are presented first. Users are allowed to perform simulated tasks at the early stages of design and real tasks once working prototypes have been developed. The design is modified in accordance with feedback received, and the process is managed, often by the HFE specialist. Evaluation should begin as early as possible in the design process. In general, changes are less costly to make at the early, rather than the later stages. Although early evaluations using crude mock-ups may lack realism, large savings can be made. Evaluation is centered on how well the proposed design will enable the goals to be achieved in relation to the performance metrics. Some assessment criteria are that commands function as specified, that the database is secure with modern systems, and that WYSIWYG applies (that "what you see is what you get").

We have also looked at some of the processes and activities that HFE specialists must carry out in order to be effective team members:

- Determine capability need (*user* requirement for HFE)
- Learn from user experiences of legacy systems to strengthen the link between design and operation (consult operators and learn from users)
- Gather and write *system* requirements for HFE (what needs to be done?)
- Evaluate supplier capabilities in HFE (who is qualified to deliver it?)
- Develop and assess system concepts
- Derive detailed requirements from generic requirements as the systems concept develops. At the same time, develop acceptance criteria (what *exactly* needs to be done, and how will we ensure that it has been done?)
- Evaluate prototypes
- Verify system—conduct acceptance tests (traditional ergonomics)
- Validate system (does it actually work when you put it all together?)

In the next chapter, we will review HFE, its history and current scope, and look at some basic HFE tools.

TUTORIAL TOPICS

1. HFE can be studied as a stand-alone discipline but can never be applied this way.
2. What is the difference between a system and a collection of objects?
3. You have just given a lecture on Human Factors Integration. It's question time, and somebody asks, "Can the ergonomics approach be applied to lifeguards?" How are you going to respond?
4. At the link (https://www.youtube.com/watch?v=x-BuADrrmAY) you will find a short video on Human Systems Integration. Summarize the main points of the video. How well do they map onto the contents of this chapter?

NOTE

1 Also known as "Human Factors Integration."

1 Human Factors and Ergonomics from the Earliest Times to the Present

GENERAL REQUIREMENTS FOR HUMANS IN SYSTEMS

1. Good quality of working life
2. Operable and safe equipment
3. Tasks that are compatible with people's expectations, limitations, and training
4. An environment that is comfortable and appropriate for the task
5. Job aids and training that are appropriate to the work
6. A system of work organization that recognizes people's social and economic needs

> In the past, the man has been first; in the future, the system must be first.
>
> *Taylor, 1911*

CORE KNOWLEDGE: HUMAN FACTORS AND ERGONOMICS (HFE)

Human factors and ergonomics (HFE) is growing internationally. Practically, all high-risk industries employ HFE professionals. They are also employed in manufacturing, product and service design, service industries, and government/regulatory agencies. High-risk sectors include nuclear, aviation, energy, shipping, and health care.

Why is this? Twenty-first century technology is becoming increasingly complicated. It is at the boundaries between system components that both problems and opportunities can emerge. The components themselves, be they programs, applications, engines, or vaccines, have become increasingly reliable—the design challenge lies in the interconnections and interfaces between the components. Achieving complementarity between people and technology requires that the operational characteristics of both are understood and are compatible.

Karwowski (2005) put it this way:

> Advanced technologies, with which humans interact today, constitute complex systems that require a high level of integration from both the design and management perspectives. Design integration typically focuses on the interactions between hardware (computer-based technology), organization (organizational structure), information system and people (human skills, training and expertise). Management integration refers to the interactions between various system elements across process and product quality, workplace and work system design, occupational safety and health programmes and corporate environmental protection policies.

A brief review of the history of HFE will explain the development and subsequent growth of the profession over the last 100 years.

DOI: 10.1201/9781032717210-1

Brief History of HFE

HFE came about as a response to the design and operational problems presented by technological advances in the twentieth century. It is a hybrid discipline that emerged when applied scientists came together to solve difficult cross-disciplinary problems, and it owes its development to the same historical processes that gave rise to other disciplines such as industrial engineering and occupational medicine. The core sciences from which HFE is drawn are as follows:

- Psychology
- Anatomy
- Physiology
- Physics (particularly mechanics and environmental physics)
- Engineering

It has also been heavily influenced by other emergent disciplines, notably:

- Industrial engineering
- Industrial design
- Systems theory
- Occupational health and safety

Scientific Management and Work Study

Scientific management, developed by Taylor, and work study, developed by the Gilbreths, are precursors of HFE. Both were developed at the beginning of the twentieth century and were based on the realization that productivity could be improved by redesigning the way work was done and not just by using better machines. Taylor (1911) was a mechanical engineer who is famous for his book, *The Principles of Scientific Management* (although he also wrote a book about concrete, which is not as well acclaimed). Scientific management was a reaction against the prevalent management methods of the Victorians. Factory owners supplied premises, power, and raw materials and hired foremen to organize the work. These foremen acted rather like subcontractors and were left to organize the basic industrial tasks as best they could. Management was concerned only with output and had just a global notion of productivity, regarding the work itself with disdain. Incentives were provided for employees to suggest improvements, and profits depended on getting a "good man" to organize the workers.

Taylor realized that there were many drawbacks to this "incentive and initiative" style of management. Nobody had overall responsibility for productivity, and the system was open to corruption and exploitation of workers. Weekly "kickbacks" to supervisors were common, as was the sexual harassment of women workers (Stagner, 1982). There were few formal ways of generating better designs of systems or work procedures, or of evaluating current practice on a day-to-day basis. Workers, by being too narrowly focused on carrying out daily work activities, might be unaware of the scope for improvement through the implementation of new methods, or even be unaware of the methods themselves. They might be unwilling to suggest changes that might be in the best interests of the company, but not in their own best interests. Management, on the other hand, through its failure to focus on the way basic tasks were carried out, was incapable of maximizing productivity. As Taylor himself put it (1911, p. 7):

> … the remedy for this inefficiency lies in systematic management, rather than in searching for some unusual or extraordinary man

Taylor emphasized that every job, no matter how small, was worthy of study and improvement. Furthermore, he emphasized that it was the management's responsibility to see that this was done in order to maximize returns, for the financial benefit of the company, its shareholders, and its employees. In practice, tasks were to be broken down into their simplest possible form, and by introducing an appropriate bonus scheme, it was assumed that workers would maximize their output. Production targets were set in advance, and production was closely monitored. Workers lost what little autonomy and control they may have had and were reduced to carrying out mundane and highly repetitive tasks throughout the day.

One of the main lessons to be learned from the writings of F.W. Taylor is that it is management's job to manage, and this means having formal systems in place to monitor working practices and their relationship to productivity. However, there are many criticisms of Taylor's approach—the social aspects of work and the intrinsic satisfaction that could be gleaned from carrying out work tasks were ignored. A more fundamental criticism, though, is that throughout Taylor's writings, he assumed that all parties can and do have the same interests. It is assumed that all will cooperate to maximize their economic returns:

> Scientific management, on the contrary, has for its very foundation the firm conviction that they (i.e., the interests of employer and employee) are one and the same.
>
> *Taylor, 1911*

Clearly, the interests of the management and workers are not the same. One of management's main goals is to maximize profits for shareholders, and one of the best ways to do this is to minimize the cost of labor. This is as true today as it was in Taylor's time—companies outsource functions to developing countries or employ gig workers to reduce the cost of labor and the regulatory burden. This conflict of interest between management and workers had implications for the way Taylor's ideas were implemented. It also provided the impetus for the infamous "Hawthorne experiments."

Work study was developed by the Gilbreths around the same time. They developed methods for analyzing and evaluating the way tasks were performed. A task would be broken down into "elements"—the basic movements and procedures required to perform it. Inefficient or redundant movements were eliminated. By redesigning and reconfiguring the remaining elements, productivity was enhanced.

Work study and scientific management were the forerunners of motion and time study and human engineering. Companies employed a new breed of specialists to investigate human—machine interaction and to design tasks. Working practices were no longer at the worker's discretion or determined by tradition or technology, and brought under management control—an essential requirement for the introduction of mass assembly and production line manufacturing.

Time and motion study (methods engineering) can be criticized on many grounds; for example, it only examines the superficial features of task performance, it makes unwarranted assumptions about people, and it is little more than common sense. Many have argued that Taylorism succeeded in "de-skilling" craftsmen and created mundane, repetitive jobs. An alternative view is that it was the reengineering of products for mass production that de-skilled the Victorian craftsmen, giving rise to a whole generation of mass-produced products that were increasingly affordable by large numbers of people, including those who produced them. To the modern consumer, the idea that products such as toothbrushes or keyboards should be assembled using craft skills seems absurd.

Taylorism had many advantages for management:

1. Greater flexibility in allocating operators to easily learnt tasks.
2. Fewer skilled workers were needed. Skill shortages were avoided, and training costs and wages could be more easily contained.
3. Introduction of paced work enabled production schedules to be more rigorously quantified. Better predictions of output could be made.
4. If everyone worked at the same pace, the result would always be a finished product.

A job would be redesigned to make it as simple as possible to learn and perform. A production standard and rate of pay would be set, and a bonus scheme would be introduced as an incentive for workers to exceed the standard. It was assumed that "rational economic men" would maximize their productivity to maximize the bonus. The social context in which work took place was ignored. Despite its advantages, Taylorism also presented management with a dilemma. Continually increasing productivity had to be met with continual increases in pay. To avoid this, new techniques were employed. New and higher production standards were introduced whenever sustained increases in output were achieved; bonuses were only paid when the new standard was exceeded. Not surprisingly, workers reacted by restricting their output to prevent the standard from being raised and placed social pressure on "rate-busters."

One way of summing up Taylorism is that it has always received a lot of criticism, particularly from academics, whereas there has been much less criticism of what it replaced.

Hawthorne Experiments

In the 1920s and early 1930s, a series of experiments was conducted over 12 years by Elton Mayo and his colleagues at the Hawthorne Works of the Western Electric Company in the United States. These were not experiments in the modern sense of the word: they lacked control groups, clear hypotheses, ethical approval, and statistical power.

The most often quoted experiment concerned a relay assembly task. The experimenters began by manipulating the lighting levels to observe the effect on output. Unexpectedly, it was found that output increased even when the illumination was reduced. One interpretation is that the changes in lighting levels reminded workers that they were in an experiment, being observed, and that this motivated them to work harder—this is now known as *participant reactivity.*

In their own words, the experimenters ran into difficulties because they were

> ... trying to maintain a controlled experiment in which they could test for the effects of single variables while holding all other factors constant .... By Period XIII it had become evident that in human situations, not only was it practically impossible to keep all other factors constant but trying to do so in itself introduced the biggest change of all; in other words, the investigators had not been studying an ordinary shop situation but a socially contrived situation of their own making.
>
> *Rothlisberger and Dickson, 1939*

The experiments themselves and their interpretation have always been controversial. Gilson, in 1946, after berating the researchers for failing to "acquaint themselves with the experience of others before they began their experiment," put it this way:

> The elaborate charts indicating production curves of five segregated workers mean little or nothing because the effect of various factors in relation to outside episodes cannot be estimated and because the sample is inadequate .... The question, "Which would you like better, a man or a woman supervisor?" brought an almost unanimous vote for a man, which, considering the fact that it was put to only five girls and a man was supervising the investigation, has doubtful significance .... Someday a study should be made of "Researches into the Obvious Financed by Big Business."

The experiments generated interest in the effect of social rather than physical factors in determining productivity. Elton Mayo had been impressed by the work of Sigmund Freud, in particular Freud's writings on the *Psychopathology of Everyday Life*, and had formed the opinion that uncooperative workers were suffering from a kind of psychopathology amenable to treatment. These ideas were behind his belief that the next wave of productivity increases would be achieved using psychological techniques, rather than economic incentives to prevent industrial discord fueled by "futile hatreds and strife."

One of the consequences of Mayo's experiments was the introduction of a personnel counseling service at the factory. In 1950, Western Electric employed 20,000 workers at Hawthorne and had a counseling department of about 40 people who interviewed over half the workforce over a 10-year period. The method used was nondirective, confidential, and neutral, similar in some ways to the client-centered psychotherapy of Carl Rogers, which dates from the same era. The costs and benefits of the service do not seem to have been assessed, although Wilensky and Wilensky (1951) concluded that counseling appeared to be effective as a supplementary means of exerting management control over workers, by draining off resentment and bitterness that might otherwise have gained expression through militant unionism. Gilson (1946) put it this way:

> From 1933 to 1936 the Western Electric Company paid out $25,825.76 for espionage. The interviewers who engaged in "counselling" service subsequent to the interviewing experiment of 1928–31 surely must have had some echoes of dissatisfaction due to lack of recognition of unions. We know of no instance where spies have been employed without some fear of unionism on the part of management .... We are also surprised that in twenty thousand interviews the workers are reported to have "criticized the company in no instance."

The "Hawthorne Effect": Social Conditioning or Participant Reactivity?

There is no scientific definition of the Hawthorne effect. The term was coined by French (1953) as follows:

> From a methodological point of view, the most interesting finding was what one may call the "Hawthorne effect"—a marked increase in production related only to special social position and social treatment.

It is used as evidence for the importance of social rather than physical factors in determining worker performance. The possibility that productivity might be increased by manipulating the social environment rather than by paying workers more money was obviously attractive to Western Electric management, although there is scant evidence that the Hawthorne experiments provided any support for these claims, because the 'human relations' were never measured in the experiments.

Kompier (2006) launched a stinging attack on this interpretation and the motivations behind it. His is the latest in a long series of articles that have analyzed the original reports and found no evidence of any effect. It is indeed remarkable that the term, derived from 90-year-old experiments lacking both control groups and with insufficient statistical power to demonstrate anything at all, should still be quoted in the twenty-first century.

Kompier offers the following explanations for "what keeps the story going":

- Students are often taught about it at the beginning of their courses and remember it due to the primacy effect (see Chapter 12). Any story in which workers improve their output irrespective of working conditions and pay is too good not to be true.
- The original researchers, particularly Mayo, were selective in the presentation of their data. Subsequent researchers never checked the original reports.
- The Hawthorne researchers wanted to demonstrate a link between social science and everyday working life and were loath to cite evidence that might weaken their arguments.
- The story is in accord with the cognitive world and interests of psychologists. Psychologists tend to be the main peddlers of the myth because they have a vested interest in seeing it perpetuated (management will invest in their services if social and interpersonal factors are perceived to be important).

- The story is in accord with the cognitive world and interests of management. If social factors are so important in determining productivity, then there is no need to improve working conditions or pay.

Participant Reactivity: McCambridge et al. (2014) reviewed 19 studies of the Hawthorne effect. Definitions of the effect varied between studies. Twelve showed some evidence of an effect, but whether or not an effect was observed depended on what the participants were doing. The authors concluded:

> Consequences of research participation for behaviors being investigated do exist, although little can be securely known about the conditions under which they operate, their mechanisms of effects, or their magnitudes. New concepts are needed to guide empirical studies.

Participant reactivity is discussed further in Chapter 16 in relation to research design in HFE. For present purposes, there is no generally accepted scientific definition of the Hawthorne effect, nor is there any unequivocal evidence for its existence. Following McCambridge et al. (2014), the most common use of the term "Hawthorne effect" is as a post hoc interpretation of unexpected study findings, particularly where they are disappointing.

Sociotechnical Systems Theory

Sociotechnical systems theory emerged in the United Kingdom after World War II. Trist and Bamforth (1951) investigated the social and psychological consequences of mechanized coal mining against a background of psychosomatic disorders among miners. They argued that the mechanized method owed its origins and scale to the factory floor rather than to coal mines, where a different form of social organization was necessary due to the intrinsically unpredictable nature of the working environment in a mine. The organization of technology, the social organization, and the local environment had to be seen as interconnected and designed to be mutually compatible if low productivity and psychological stress were to be avoided.

Participation

The idea that workers should participate in decisions about work dates from the same era. Coch and French Jr. (1948) carried out an investigation of participation in a U.S. pajama factory staffed by young, unskilled women. Staff turnover was high, and workers were paid on the basis of individual performance. Management collaborated with two sociologists in an attempt to deal with the problems. Unusually for such studies, it was possible to introduce a degree of experimental control with a control group, where change was introduced by management without the involvement of employees; a "representative group," where change was introduced after discussions with employee representatives; and a "total participation" group in which all employees were involved. Production fell by 10%–20% in the control group immediately after the change, whereas it was maintained in the representative group, rising 10%–15% after a few weeks. In the participation group, production rose immediately, and the gains were maintained for several months. It was argued that the participatory approach had produced better group cohesion, which countered any frustrations brought about during the change. People will accept change more readily if they have some control over its effects. If this explanation is correct, then participation has to become an established management process if its benefits are to be sustained.

Occupational Medicine

Occupational medicine originated in the eighteenth century when Ramazini (1717) wrote his *Treatise on the Diseases of Tradesmen,* but it became more formalized at the beginning of the twentieth century. Around 1914–1918, several government institutions were established in Britain as interest in working conditions spread to scientists and medical doctors. The Health and Munitions Workers Committee studied conditions in munitions factories and the factors influencing productivity, such as the length of the working day. It subsequently became the Industrial Health Research Board, and its area of interest was wide, covering ventilation, the effects of heat, shift work, and training. Recommendations were made at about this time for a variety of aspects of industrial work, including the types of food served in factory canteens, considering the likely nutritional deficiencies of the workforce and the demands of work. This was appropriate in Britain during the early years of the twentieth century and is still appropriate in developing countries today. Attention was also directed at fatigue and the problems of repetitive work. Vernon (1924) investigated postural and workspace factors related to fatigue and concluded that:

> Any form of physical activity will lead to fatigue if it is unvarying and constant.

It is from foundations such as these that occupational health and physical ergonomics have arisen.

Human Performance Psychology

Human performance psychology had its roots in the practical problem of how to reduce the time taken to train a worker to carry out a task. Taylor and the Gilbreths had gone some way toward solving this problem by simplifying tasks. However, the emergence of new, more complex machines and manufacturing processes created new jobs whose skill requirements could not be met from the existing labor pool. Traditional methods of training (such as apprenticeship to a skilled craftsman and "sitting by Nellie") were not always appropriate to the industrial system with its fundamental requirement to standardize all aspects of the production process. Psychologists began investigating the variables influencing the learning of work skills, for example, whether tasks should be broken down into elements to be learnt separately and then brought together, the advantages and disadvantages of "massed" versus "spaced" practice, and the duration of rest periods. After World War II, interest in training continued. The cybernetic approach investigated the use of feedback (knowledge of results) and its effects on learning. The theoretical ideas of BF Skinner and the behaviorist school of psychology were implemented in the form of "programmed learning." Behaviorism saw learning as the chaining together of stimulus–response pairs under the control of reinforcing or rewarding stimuli from the environment. In programmed learning, the material to be learned was presented in a stepwise fashion, and the order of presentation of information to be learned was determined by whether the trainees' previous responses were correct or incorrect.

The pressure for the productive and efficient use of machines was amplified by the demands of World War II and brought psychologists into direct contact with the problems of human–machine interaction. The Cambridge psychologist Sir Frederick Bartlett built a simulator of the Spitfire aircraft and investigated the effects of stress and fatigue on pilot behavior. This led to an increased understanding of individual differences in response to stress and enabled the breakdown of skilled performance to be described in psychological rather than machine-based terms. Perceptual narrowing, which occurs as a result of fatigue or as a maladaptive response to severe stress, is an example. Craik studied a class of tasks known as "tracking tasks" (which involve following a target, as in gunnery or steering a vehicle). The beginnings of user-interface (display and control) design emerged from this in the form of recommendations for gear ratios and lever sizes. This became an important area of HFE, particularly in postwar military applications, as well as in civilian vehicle design and the aerospace industry.

Operations Research

Operations research attempts to build mathematical models of industrial processes. It was also stimulated to grow by the demands for prediction and control brought about to satisfy the requirements of the military during World War II. It had become clear that further advances in system performance would depend on how well technology was used, not just how well it was designed. This shift in attention from the machine to the human–machine system gave birth to the field of HFE.

Fit the Person to the Job (FPJ) versus Fit the Job to the Person (FJP)

In the era of pure engineering, organizations increased their productivity by introducing new methods and machines. This worked because great improvements in machine design and mechanization were happening. Tasks were redesigned to minimize unproductive effort using time and motion study. After World War I, the development of psychometric testing enabled human characteristics such as intelligence or personality to be measured.

Attempts to "fit the person to the job" (FPJ) were based on the idea that productivity and efficiency could be improved by selecting workers with the right aptitudes or characteristics for a particular job. This approach, which forms one of the roots of modern occupational psychology, was based on the assumption that important aptitudes for any particular job did exist and could be identified and objectively measured. Even today, jobs such as firemen, lifesavers, and those in the armed services are restricted to individuals with certain specific aptitudes and physical characteristics. Historically, anthropometric selection criteria were used. Stature is generally a good surrogate measure for strength and size. However, due to changes in labor legislation and the increased emphasis on equal opportunities, static anthropometric criteria have been replaced by functional tests, where selection depends on the applicant's ability to carry out some generic or job-critical task under defined conditions and to meet a preset standard.

An alternative approach, which is the guiding philosophy of HFE, is known as "fitting the job to the person" (FJP). Much of the early human engineering and workspace design attempted to design tasks to suit the characteristics of the worker. The underlying assumptions of the FJP approach are that a suitable set of worker characteristics can be specified around which the job can be designed and that this can be done for any job. A large part of this book is devoted to describing these characteristics at the anatomical, physiological, and psychological levels and explaining their design implications.

FPJ was introduced at a time when the demographics of many countries were very different from those of today. There were many young people, no shortages of the necessary skills, and a great deal of unskilled work. With few options for alternative employment, it was common for workers to remain in a job for life. Management could afford to be selective. In many countries, this is no longer true, due to demographic aging of the workforce and no shortage of vacant jobs. FJP seems to be the most practical of the two approaches, wherever it can be implemented, since many organizations now find themselves with a dwindling applicant pool and are loath to reject otherwise suitable people. There are also political pressures on organizations in many countries to make employment as widely available as possible. Equal opportunities legislation prevents discrimination against women and minorities, and employers are being forced to select based on functional tests of task performance rather than applicant characteristics.

FPJ is still relevant in many sectors of the economy, particularly where specialist skills and aptitudes are required. A study of knowledge workers by Bayona et al. (2020) found that worker knowledge characteristics relating to information processing, problem-solving, and complexity were related to improved performance and job satisfaction through higher levels of engagement with the job. "When there is congruence between the characteristics of the person and the environment, an improvement is expected in the outcome variables, be they job performance or job satisfaction."

FPJ in Extreme Environments

FPJ is still relevant in cases where work is conducted in Isolated and Confined Environments (ICE)—such as serving on research stations at the South Pole or on long-term space missions.

Research is currently underway on "Mindfulness Disposition" as part of the selection process for long-term space missions. Mindfulness Disposition (MD) refers to the awareness that emerges from paying intentional, present-moment, non-judgmental attention to an unfolding experience, moment by moment (Le Roy et al., 2023). It is thought to be a stable personality trait that can be identified using questionnaires. Studies have been carried out on submariners and in simulated sea survival in life rafts. In one study, mindful submariners were more emotionally positive and slept and ate better than non-mindful submariners after a long patrol.

Attempts to Humanize Work

Taylorism was an inevitable consequence of industrialization and mass production in the early twentieth century. Whether it is still appropriate is not obvious with the shift to the service economy and hybrid working.

Despite calls for job reform in Sweden, for the humanization of work in Germany in the 1970s and 1980s, and for a renewed emphasis on the quality of working life in the United Kingdom, many modern jobs in the twenty-first century are intrinsically fragmented, piecemeal, or extremely repetitive.

In the 1960s, 1970s, and 1980s, a number of large-scale programs were initiated in several European countries. These programs were motivated by a variety of factors. For example, successive generations of school leavers in the countries in question had increasingly higher levels of education and higher expectations of work. The programs attempted to provide higher-quality jobs through changes in work organization.

In Sweden, the Volvo motor car company (which was suffering from high absenteeism and labor turnover in the 1960s) tried to find new ways of assembling cars in an attempt to have a more stable, motivated, and productive workforce. Conventional production line methods were replaced by "unit production." Teams of workers manned electric assembly wagons, which moved around the assembly area, stopping at centralized stores to collect the various components.

It is difficult to disentangle the effects of such programs on either productivity or the psychological rewards of work because many different factors are involved. Unit production changes the social relations between people, but it also eliminates pacing (where the rate of work is set by machines) and lengthens cycle times. Both of these latter factors are known to influence job satisfaction. The British Quality of Working Life program (Tynan, 1980) attempted to combine new approaches to job design with technological change under the premise that, since technological change forces job redesign anyway, the opportunities presented by new technologies could best be realized by optimizing the work organization and design of jobs. Management, trade unions, and workers were to be included in a participatory approach toward job design.

HFE from Yesterday to Today

In 1857, Jastrzebowski produced a philosophical treatise on *Ergonomics: The Science of Work*, but it seems to have remained unknown outside Poland until recently (Jastrzebowski, 1857). In Britain, the field of ergonomics was inaugurated after World War II. The name was reinvented by Murrell in 1949, despite objections that people would confuse it with economics.

Human factors, developed in the United States at the same time, and ergonomics have always had much in common, but they developed differently. Human factors focused on the integration of human considerations into the total system design process. It has achieved remarkable success in the design of large systems in the aerospace industry, notably NASA and the U.S. space program. European ergonomics has traditionally been more tied to its basic sciences or a particular topic or application area.

The two terms are interchangeable nowadays.

Sociotechnical Systems Theory (STS) and Quality of Working Life (QWL) in the Twenty-First Century

Taylorism was a "technology-centered" approach to job and system design. Work was designed to get the best out of the technology, with less concern about the quality of working life (QWL) of the people doing it. Sociotechnical systems theory (STS) holds that for any technology, there is an optimal way of arranging the work organization to deliver quality of working life for employees—an optimal *sociotechnical system.* Attempts to implement STS in coal mines and factories had their successes and failures. Management distrust of new ideas, poor industrial relations, and trade union opposition sometimes derailed STS interventions.

QWL is a generic concept relating to all aspects of the employee experience of work, including working conditions, remuneration, the social environment, and the experience of doing the work itself. The digital economy and the Fourth Industrial Revolution (below) have brought great changes to the nature of work, how it is organized and how employees experience it.

Guest et al. (2022) list six principles for the harmonization of QWL in the digital age.

1. QWL is a workplace issue, and the focus should be on the work itself, job-specific well-being and productivity.
2. QWL is an employment issue, and the focus should be on employment issues in view of new developments such as the gig economy.
3. QWL requires institutional support from government and regulatory agencies as new kinds of work and new employment practices emerge and develop.
4. The mutual interests of employers and employees need to be understood and optimized via appropriate channels.
5. Together, all stakeholders in an enterprise (shareholders, managers, employees, government, the local community, and unions) form an ecosystem, and all should have a voice).
6. Renewed focus on "action-research" (rather than academic theorizing or criticism) to implement practical improvements in QWL.

Good job design depends on consultation with users. Find out:

- Who does what?
- What skills are used?
- Under what constraints?
- How will the new system affect this?
- Will new skills be needed? Will de-skilling occur?
- How much training time will be needed?
- What is the line management structure? How will it change? Will anyone lose out?

Many of these themes are returned to in the Epilogue.

BASIC APPLICATIONS

Many HFE specialists work in research organizations or universities and carry out research. This research often leads, directly or indirectly, to the drafting of standards, legislation, and design guidelines. Others work in a consultancy capacity, either privately or in an organization. They work as part of a design team and contribute their knowledge to the design of the human–machine interactions in work systems. This often involves the application of standards, guidelines, and knowledge to specify particular characteristics of the system.

Modern HFE contributes to the design and evaluation of work systems and products. Unlike in earlier times, when an engineer designed a whole machine or product, design is now a team effort. The HFE specialist typically plays an important role in both the conceptual phase and detailed design, as well as in prototyping and evaluating existing products and facilities. Modern HFE contributes in a number of ways to the design of the work system (Table 1.1). These activities should be seen as an integral part of the design and management of systems.

The basic applications of HFE are to meet the general requirements for integrating human operators with technology. Compatibility can best be achieved by focusing design efforts on the interactions between people and the devices they use to achieve work goals, using a participatory approach.

TABLE 1.1
Contribution of Modern HFE to Systems Design and Management

1. A standard format for describing and assessing human–machine systems:
   - System-level model:
     - Organization of subsystems
     - Human–machine-level model (use checklists for assessment):
     - Description of technology
     - Description of interfaces
     - Description of users
       - Physical dimensions
       - Abilities/training/experience
       - Age
       - Fitness
       - Language/literacy
2. Identification, classification, and resolution of design issues involving the human component
   - Workload
   - Work organization
   - Work environment
   - Work psychosocial factors
   - Work scheduling
3. Task and human–machine interaction analysis
   - Description of human–machine interaction
   - Optimum level of description
4. Specification of system design and human behavior. Advice on the implementation of controls. Legislative force
   - Regulations (legal imperative to comply)
   - Standards (contractual/strategic imperative to comply)
   - Guidelines (best practice)
5. Identification of core trends in human and biological science and their implications for system design and management
   - Scientific input
   - Consultative advice on technical matters and trends
   - Strategic planning
   - Advice on employers' liability for accidents and injuries
6. Generation of new concepts for the design and analysis of human–machine systems
   - Human–machine modeling to generate design concepts
   - Physical models (e.g., SAMMIE, JACK)
   - Task models (task analysis and description)
7. Devise and prioritize an implementation plan for ergonomic improvements
   - Implement immediately (current arrangements are hazardous or contravene legislation)
   - Implement soon (no immediate danger, but current arrangements are unsatisfactory)
   - Implement when equipment is shut down (if stoppages are expensive and there is no immediate danger, wait until the system is shut down for regular maintenance)
   - Implement when cost–benefit is acceptable (wait until the financial situation improves or when costs are lower)
   - Implement when new equipment is built or purchased: incorporate ergonomic requirements into the procurement process

TABLE 1.2
Success Factors in Management–Worker Participation

Involvement	Partnerships of stakeholders, especially for the development or use of standards
	Manufacturer involvement
	Trade Union involvement
	Driven by user and company needs
	Multidisciplinary
	Participation through all stages of design
Commitment	Commitment of all stakeholders
	Real support from senior management
	Someone to champion the process
	Clear perceived need for change
	Urgency
Climate	Appropriate knowledge levels among stakeholders
	Track record of success
	Acceptable industrial relations
	Open, communicative organization
Management	Clearly defined actors and roles
	Structured process that matches organization's structure
	Clear, single, well-defined project
Resources	Clear identification of resource availability: time, money, equipment
	"Rich" information from real users

Source: Reproduced from Wilson and Morris (2004), with permission.

Participative practices have been accepted in some organizations and countries (e.g., Germany). The requirements for effective participation, according to Bernoux (1994), are as follows:

- Employees must acknowledge the need for participation.
- Employees must trust that their participation will not have negative effects and that they will have some control over the final decisions.
- Employees must perceive that changes are being introduced in a legitimate way.
- Employees believe that change is being implemented correctly.
- Employees must be given a real role to play in the introduction and testing of new ways of working.

Table 1.2 summarizes the key ingredients for successful participatory design of work equipment, according to Wilson and Morris (2004).

TOOLS AND PROCESSES

HFE is a broad subject, but there are three main classes of tools that span the entire discipline. These are:

1. HFE checklists
2. Task analysis
3. Risk assessment

To use a theatrical analogy, checklists are used to assess the design of the set upon which the performance takes place, and task analysis is used to describe the performance of the actors.

HFE Checklists

The use of checklists has a long history in HFE. The most basic checklists are nothing more than an aide-memoire to ensure that an investigation is thorough and not one that merely reflects the investigator's main area of expertise or interest.

Existing regulations and guidelines are often context- or issue-specific, which is why generic checklists can help to ensure that no possible factors are overlooked. HFE Workshop 1.1 presents a general HFE checklist that is context-independent and conforms to the human–machine model in the Preface. It can be used anywhere to investigate any job and should be seen as complementary to standards and regulations. More detailed checklists for investigating specific issues are available, and some examples appear in later chapters.

HFE WORKSHOP 1.1

GENERAL HFE CHECKLIST

The general HFE checklist is used for the preliminary investigation of working conditions and is derived from a variety of sources: E. Grandjean's original list, that of the International HFE Association, defense standards, and other "mini" checklists that have appeared over the years.

1. Job analysis
   1. What are the main assignments and segments of the job?
   2. Is there a high physical workload?
   3. Is there a high mental workload?
   4. Does the operator have a high level of responsibility?
   5. What are the skill/knowledge requirements of the job?
2. Work organization
   1. Does the operator work alone or with others?
   2. Is the work machine- or self-paced?
   3. What is the system of supervision and accountability?
   4. What shift system is in operation, if any?
   5. What are the hours of work and rest periods?
   6. Is overtime worked?
   7. What are the arrangements for meal breaks and refreshments?
   8. Is there time pressure due to deadlines or meeting production targets?
   9. Is the work carried out on a piece-rate basis?
   10. Does the work organization automatically ensure that periods of work and rest occur naturally, as part of the work itself?
   11. Is the work adapted to the needs of older workers in terms of physical or mental demands?
3. Workspace design
   1. Is there sufficient space for the operator to work?
   2. Can the worker sit while working at all?
   3. Does the position of the body demand static muscular work?4. Does the workspace permit a stable, neutral posture?

(Continued)

HFE WORKSHOP 1.1 (*Continued*)

5. Is the work surface appropriate for the visual and manual requirements?
6. Are foot controls necessary? Do they permit a suitable posture?
7. Are hand controls correctly placed and designed to allow a good upper limb posture?
8. Is the seat height adjustable?
9. Does the chair have a backrest?
10. Can seated workers use footrests, armrests, and lumbar pads, if needed?
11. Are hand tools correctly designed?
12. Are any body parts exposed to constant pressure?
13. Is adequate personal protective clothing provided where needed?
14. Is there any vibration?
15. Are any surfaces hot enough to cause burns or cold enough to freeze skin?
16. Can reasonable adjustments be made for pregnant women or the disabled?

4. Physical demands
   1. Is manual handling required?
   2. Do large forces have to be exerted?
   3. Does the work involve lifting or twisting, bending, stooping, or reaching?
   4. Is muscular work mainly static or dynamic?
   5. Are movements centered around the midpoint of the joint range?
   6. Can static work be eliminated by providing clamps or vices?
   7. Can loads be lifted and carried safely?
   8. Is the workload greater than 40% of maximum aerobic capacity?
   9. Are large or small muscle groups involved?
   10. Can the operator vary the work rate or take rest periods at will?
   11. Are lifting aids or powered tools available?
   12. Are cycle times less than 30 s?
   13. Are job aids provided to reduce high physical demands?
   14. Are the patterns of movement efficient, inefficient, or are any movements unnecessary?
   15. Does the task demand highly accurate movements?
5. Mental demands
   1. Is the mental workload too low, too high, or about right?
   2. Is the task carried out at a predominantly skill-, rule-, or knowledge-based level?
   3. Does the task place high demands on the perceptual or attentional systems or short- or long-term memory?
   4. How must information be processed before a response can be made?
   5. Can mental workload be reduced using external memory aids, predictor displays, decision support systems, navigation aids, etc.?
   6. Does the operator have to carry out more than one task at a time, and are the task modalities compatible?
   7. Are the sequences of mental operations compatible with the physical layout?
   8. Is the representation of the system compatible with the operator's representation?
   9. Does information from different channels/modalities have to be integrated?
   10. Are great demands made on visual search, and can cueing be used to reduce these?
   11. Is proper feedback provided in accordance with task demands?

(*Continued*)

HFE WORKSHOP 1.1 (*Continued*)

12. Can controls, displays, and task demands be recognized easily, and is proper support provided?
13. Does the task require a high level of individual judgment? Can decision support be provided?
14. Are memory aids hardwired into the design of the hardware, software, and work organization?

6. Human–machine interaction
    1. Does the information presented satisfy the operator's requirements?
    2. Is the rate of information flow too high or too low?
    3. How many sources of information does the operator use to work?
    4. Are data readily available, in the right form, and unambiguous?
    5. Are data embedded in noise?
    6. Are there many distractions due to noise, speech, or other disturbances?
    7. Is the layout of controls and displays compatible with prevailing stereotypes?
    8. Are controls close to their corresponding displays?
    9. Are control/display ratios and control dynamics compatible with system order?
    10. Is the "grain size" of feedback appropriate for the control actions and decisions that must be made?
    11. Does the operator have to monitor several channels simultaneously?
    12. Are warnings, instructions, and other displays suitably designed and accessible?
    13. Does the system provide timely feedback or other indications of response adequacy?
    14. Are verbal instructions/displays in the correct language and easily comprehensible?
    15. Are human–computer dialogues user-friendly?
    16. Is the human–computer interaction style appropriate given the expertise of users?
    17. Are colors used in an appropriate way?
    18. Are coding systems compatible with human memory limitations?
    19. Do the task demands form a predictable pattern of stimuli and responses?
7. Work environment
    1. Are temperature, noise, lighting, and vibration within recommended limits?
    2. Is there excessive brightness or glare in the workspace?
    3. Are there sudden loud noises?
    4. Does temperature vary throughout the day and are there hot or cool spots?
    5. Are there reflective surfaces or hot or cold surfaces?
    6. Does the room have an appropriate reverberation time?
    7. Are the colors and reflectance of objects in the environment appropriate for the work?
    8. Are the relative humidity and ventilation satisfactory?
    9. Are protective clothing and devices available for workers in extreme environments?
    10. Can exposure be reduced by taking rest periods in suitable areas or by rotating workers?
    11. Are there toxic or radioactive chemicals or other hazards in the work environment?
    12. Are warning signs or other notices placed in appropriate places?

(*Continued*)

HFE WORKSHOP 1.1 (*Continued*)

8. Workforce characteristics
    1. Is the anthropometry of the workforce known?
    2. Are the workers mainly male or female?
    3. What language(s) are spoken proficiently by the workers?
    4. What is the average age of the workers?
    5. What is the educational level of workers?
    6. Are all workers literate?
    7. What is the level of health and fitness in the workforce?
    8. Are the workers suitably nourished?
    9. Are the workers mainly full-time, part-time, or seasonal?
9. Job stress
    1. Does the user have sufficient resources to do the work?
    2. Are proper systems of appraisal and feedback in place?
    3. Is the employee's work role clear and unambiguous?
    4. Are performance standards clear and unambiguous?
    5. Are standards realistic or constraints too limiting?
    6. Is the system of work likely to cause conflict (work–life balance)?
    7. Is support from peers or supervisors available?
10. Occupational background
    1. Is the work system the result of tradition, custom, or practice?
    2. Has the work been planned from the beginning with the operator in mind?
    3. Is training formal or "on the job"? Is it adequate?
11. Safety
    1. Is there a formal safety management system in place?
    2. Do employees receive adequate safety training?
    3. Are employees aware of the main risks and hazards?
    4. Is there a proper reporting system available to employees?
    5. Are alarms appropriate to the task and the work environment?
    6. Are employees supplied with the necessary safety equipment?
    7. Are sufficient safety constraints in place?
    8. Is there a positive safety culture?
    9. Is the hazard log up to date?
    10. Are items in the hazard log dealt with promptly?

EXAMPLE

Suppose we are called in to advise a company on its cargo handling operations, loading, and unloading specialized equipment onto ships using cranes. The main focus is on the health, safety, and efficiency of the operation, and the avoidance of accidents. The company wishes to consider upgrading the system. The keyword here is *system*. The company has not asked for a health and safety evaluation, although it is interested in health and safety. It does not necessarily wish to buy new cranes, although it would be interested to know of any reasons for doing so. The purpose is to conduct an ergonomic assessment of the system and make suggestions for improvement where possible. The checklist provides a framework for this assessment. Information is gathered by direct and detailed observation of the work, by interviews with workers, and by taking basic measurements of the physical environment, including the forces required to operate heavy controls and the weight of objects

(*Continued*)

HFE WORKSHOP 1.1 (*Continued*)

to be handled. Information about accidents, absenteeism, and job training is also sought. Following the investigation, we might arrive at the following conclusions about the crane operators' jobs.

The job involves routine crane operation with a high mental workload and responsibility. The work is nominally self-paced, but deadlines are rigid. The operator works with others, and rest periods are regular and formalized. Physical demands are low, but awkward static work postures must be adopted due to poor visibility. There are no ergonomic hazards due to the cab design itself. Highly accurate movements are required, which, combined with poor posture and mental demands, increase the risk of musculoskeletal problems. Operators are trained both formally and on the job, have adequate resources and a clear, unambiguous task, and the resources to perform efficiently. Regular feedback is provided, and esprit de corps is good due to well-organized teamwork and a group bonus plan. No serious incidents have been reported in the last 12 months.

The next stage would be to recommend to management that the postural problems be investigated further and resolved. The postural problems could undoubtedly have been identified at the outset; so how has the checklist helped? By providing a solid platform for examining a specific issue, knowing that other, and potentially more serious, problems have now been ruled out.

Task Analysis

Before HFE can be applied to the design of systems, a sound understanding of the task is needed. There are many ways of capturing data on task performance, the simplest being observation in the field. Typically, the interaction is recorded, and any errors are noted. Users may be debriefed after the interaction so that further information can be obtained. Task analysis is widely used and is an essential prerequisite for some of the risk assessment techniques described in later chapters. A task analysis was defined by Snyder (1991) as:

> … an ordered sequence of tasks and subtasks, which identifies the performer or user; the action, activities or operations; the environment; the starting state, the goal state; the requirements to complete the task such as hardware, software or information.

Task analysis drives human-centered design by providing a system-specific context for the application of the fundamental HFE principles. The general procedure for carrying out a task analysis is as follows (Rasmussen, 1983):

1. Identify a prototypical task by collecting detailed descriptions from expert users of what different people do.
2. Identify all the processes that comprise the activity.
3. Analyze the descriptions in terms of the various options for action and the criteria used to select between them.
4. Generate a prototypical task specification by selecting a characteristic set of tasks and specifying the work sequences common to them:
    - Who must be involved?
    - What are the subcomponents of the activity?
    - How are participants involved in the various tasks?

- What information is required at each stage in the task?
- Where does the information come from?
- How is the information exchanged?
- How might any of the above be improved?

The outcome of a task analysis consists of the following:

1. Description of the behaviors required to carry out the task.
2. Description of the system states that occur when the task is carried out.
3. Mapping of the task behaviors onto the system states.

This information can be used for a variety of purposes:

1. Evaluation of the design of the human–machine interface.
2. Identification of the skills needed by an operator of the system.
3. Design of training materials and operating instructions.
4. Identification of critical elements of the task to predict or evaluate the reliability of the system.

Table 1.3 gives an example of the stages involved in the hierarchical decomposition of a job from a job title to specific operations.

TABLE 1.3
Levels of Description in Task Analysis: An Example

		Assignments
A1		Inspection of the environment, equipment, and machinery
A2		Execution of start-up procedures for pumping
A3		Monitoring of the system condition when running
A4		Execution of close-down procedures
A5		General maintenance
		Segments
(A1)	S1	General inspection: safety, lighting, housekeeping
(A2)	S2	Start-up filtration of fuel in the reservoir tank
(A3)	S3	Periodic inspections of pumps
(A4)	S4	Close down pump motors
(A5)	S5	Changing the filter elements
		Tasks
(A1S1)	T1	Replace broken lamps in the station
(A2S2)	T2	Start-up transfer pump
(A3S3)	T3	Check torque settings on the pump retaining clamps
(A4S4)	T4	Close down pump motor 1
(A5S5)	T5	Remove filter element 1
		Operations
(A1S1T1)	O1	Switch off the power to the lighting unit at the wall switch
(A2S2T2)	O2	Press the green button on the transfer pump housing

(Continued)

TABLE 1.3 (*Continued*)
Levels of Description in Task Analysis: An Example

(A3S3T3)	O3	Using a torque wrench, tighten nuts in a clockwise direction until the torque setting is displayed
(A4S4T4)	O4	Press the red button on pump motor 1
(A5S5T5)	O5	With an 8 mm spanner, loosen nuts retaining filter element housing

Note: There are many more segments, tasks, and operations. Those that are shown are included to illustrate the details required at the different hierarchical levels.

TABLE 1.4
Example of Mapping of Task Behavior and System Operation in System Design and Evaluation

Mapping Elements

1. Indications and when to do the task:
   When requested by the captain
   When indicated by the fuel gauge
   With engines switched off
2. Control objects and operations:
   Transfer valves
   Transfer hose
   Hose retaining clamps
   Transfer pump controls
   Slide the hose into the transfer valve orifices on the tanker and storage unit (outlet and inlet)
   Use retaining clamps to secure the hose to the transfer couplings on the tanker and storage unit
   Open transfer valves on the tanker and storage unit
   Press the green button on the transfer pump
3. Precautions: valves in open position and hose secured
   Fuel level in the mobile tanker above the red line
   Tanker on level surface and brakes in "on" position
4. Feedback modality and indication of response adequacy
   Visual/kinesthetic: valves "click" into open position
   Auditory/kinesthetic: pump vibrates/"whines" if flow is impeded
   Visual: transfer pump pressure gauge reads 150–250 kPa
   Absence of leaks
   Fuel level in the storage unit gauge rises
5. Fault diagnosis and maintenance
   If pump "whines" or the flowmeter indicates blockage, press the red button to stop pump
   Check:
   - Transfer valves open
   - Hose not twisted or crushed
   - Fuel level in tanker
   - Pressure gauge on transfer pump

This type of analysis can produce enormous quantities of documentation, as can be imagined. However, in complex systems, it is often essential to describe tasks at this level of detail because even small changes in equipment or procedures can affect the mapping between required human behaviors and system states. Table 1.4 gives examples of the types of questions the systems designer must ask when looking at the mapping between individual task behaviors and system states. A basic question is, "How are operator task behaviors to be invoked, and how will the system respond to their execution?"

HFE Workshop 1.2 gives an example of a basic task analysis using a task flow format.

HFE WORKSHOP 1.2

HIERARCHICAL TASK ANALYSIS

It was Professor J. Annett who developed hierarchical task analysis in the 1970s. It has its origins in time and motion study. F.W. Taylor believed that even the simplest tasks were worthy of study and improvement. This is still true today, especially in light of our increasingly regulated yet highly competitive business environments.

Task analysis must not be allowed to become too complicated, and there are two main ways of ensuring that it does not: define the task and its boundaries, beginning and end, and think hierarchically, but do not go into too much detail in the early stages.

Begin by analyzing the task from the beginning to the end at a high level (i.e., at the simplest level first, before going into detail). Do not go into more detail than you need to make the assessment.

Below is a basic task analysis for making a sandwich (Figure 1.1). The boundaries of the task are clear. Below that, the task is expanded to give details of the main operations required to make the sandwich. The expanded task has two levels of analysis. Running along the top of the page is a high-level description of the entire task (make a cheese sandwich). One of the subtasks is also broken down into its constituent operations. The simplest way to understand the task flow diagram is to begin at the top left of the diagram and read across the page. This is a high-level description of the entire task, which consists of five operations and one decision. Note that, as far as the decision is concerned, the information about whether to make another sandwich comes from outside the task itself. This is often the case, and in our hypothetical example, it might come from a higher-level task of which "make sandwich" is but a subtask (such as "serve customer," for example).

"Prepare work area" is the first subtask of making a sandwich, and it has been analyzed at the next level down (Figure 1.2). As with the main task, the entire subtask has been analyzed in detail from beginning to end. In practice, we would then proceed to analyze the next subtask from beginning to end (prepare ingredients), and so on, until we have analyzed the entire task at

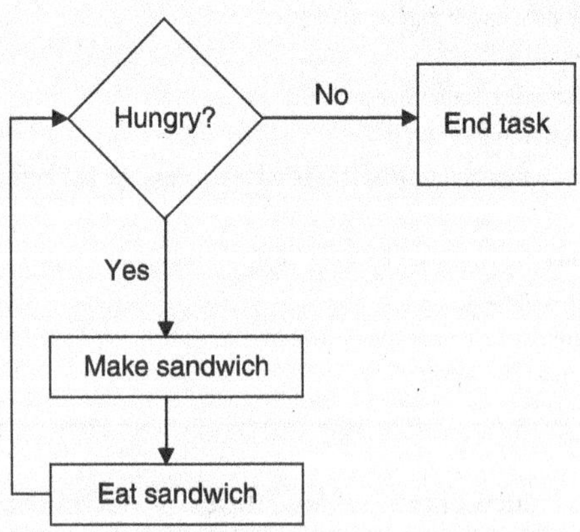

FIGURE 1.1 High-level description of the "make cheese sandwich" task.

(Continued)

HFE WORKSHOP 1.2 (*Continued*)

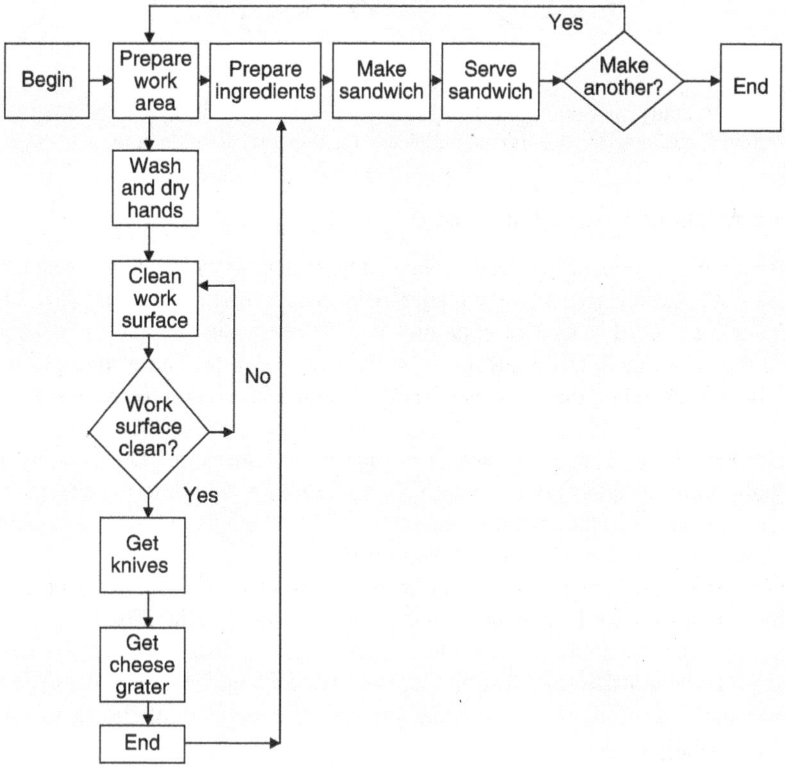

FIGURE 1.2 Detailed analysis of subtask "prepare work area."

that level. Depending on the nature of the project, it might be necessary to analyze one or more of the lower-level operations in more detail; for example, "get knives" might be a good candidate for closer inspection because of potential safety, hygiene, transport, and storage issues.

Risk Assessment

Many tools are available to HFE practitioners for the risk assessment of workplaces. Examples are given in the following chapters with links to online toolkits. Sharan (2023) provided a useful summary.

1. Subjective assessment tools (to assess states such as comfort, sleepiness, situational awareness, anxiety, and so on).
2. Psychophysiological tools (e.g., heart rate, heart rate variability, electroencephalography).
3. Self-report tests of work demand (e.g., physical or mental workload).
4. Physical assessment tools (e.g., posture assessment, electromyography, repetitive work).
5. Physical demand assessment (e.g., NIOSH lifting equation, SNOOK tables, heart rate, and oxygen consumption).
6. Standards (e.g., ISO standards on manual handling, ISO 11228-1, ISO 11228-2, ISO 11228-3, and working postures [ISO 11226]).

These tools are often used in conjunction with checklists and/or task analyses as part of an overall investigation. For further information about HFE methods, see Stanton et al. (2013).

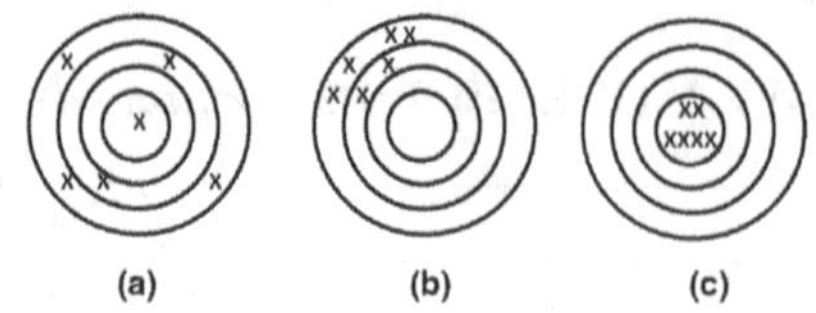

FIGURE 1.3 Validity and reliability (a) neither reliable nor valid, (b) reliable but not valid, and (c) reliable and valid. The tighter the clustering around the bull's-eye, the more accurate the tool.

Status of Risk Assessment and Design Tools

Although there are many tools available, they vary in the extent to which they have been tested. For present purposes, it is sufficient to alert the reader to the criteria that must be satisfied for a tool to be considered the gold standard in its area of application. To be of use, any tool, be it a questionnaire, rating scale, or risk assessment technique, must be *valid* and *reliable.* These concepts are illustrated in Figure 1.3. In later chapters, these issues will be discussed in relation to specific tools.

1. *Validity*: Does the tool measure what it is supposed to measure? Do its contents reflect the domain under investigation (content validity)? Does it adhere to or reflect appropriate theory or knowledge in the domain (construct validity)? Do its predictions or assessments concur with actual outcomes (criterion validity)?
2. *Reliability*: Does the tool give the same result when applied to the same problem under the same circumstances by the same assessors (test–retest reliability)? Does it give the same result under the same circumstances when used by different assessors (inter-tester reliability)? If the tool uses multiple measures of the same underlying problem (or theoretical "construct") at the same time to yield an aggregate measure of risk, is that aggregate measure internally consistent?
3. *Sensitivity*: Is the tool able to resolve differences of the size that occur in the real situation or that are important in determining real outcomes?
4. *Diagnosticity*: Is the tool sufficiently detailed to provide information about the reasons why one situation is different from another?
5. *Intrusiveness*: Can the tool be used without disrupting the process it is designed to assess?
6. *Acceptability*: Is the tool acceptable for general use? Is it harmless or is permission required before it can be used with human subjects (e.g., permission from a research ethics committee or from an occupational medical officer)?
7. *Assessment of Costs*: The real costs, both financial and in terms of human resources and disruption to normal operations, should be reasonable in relation to the value of the information provided.

According to Baber and Young (2022), many HFE methods currently in use lack data on their validity and reliability. The validity and reliability of the methods described in the following chapters are presented in those cases where data are available.

SYSTEMS INTEGRATION

Three key concepts in any discussion of the economic benefits of HFE are

1. *Efficacy*: Whether the applications work under ideal circumstances.
2. *Effectiveness*: Whether the applications work under normal usage conditions.
3. *Efficiency*: Whether the applications save more resources than they consume.

Most of the material presented in this book is of an introductory and explanatory nature and describes the basic concepts used in HFE. At the end of each chapter is a section devoted to effectiveness and efficiency studies, wherever possible in the form of field trials with data on costs and benefits. It is not the purpose of this book to introduce the reader to techniques of cost–benefit analysis, since this is an application of economic theory and is a field of study in its own right. The goal is to illustrate to the reader the kinds of analyses that have been done and, more importantly, to show how the findings demonstrate the efficacy and efficiency of the application of HFE measured by means of "key performance indicators."

In general, there is more information on the cost of problems rather than the economic benefits of solving them.

Oxenburgh Productivity Model

The approach is to identify jobs with a high level of absenteeism, injury, or staff turnover. Ergonomic modifications are made to the workplace, tools, equipment, or work organization. The costs of these modifications can be calculated in terms of the materials and labor charges needed to implement them, as well as any downtime during the redesign. The benefits are expressed in terms of reductions in absenteeism, expressed as a monetary value made up of direct and indirect costs:

$$\begin{aligned}\text{Direct wage costs} &= \text{wage} + \text{obligatory charges to the wage}\\ &\quad + \text{personnel} + \text{administrative costs}\\ \text{Indirect costs} &= \text{costs of losing trained employees}\\ &\quad + \text{costs of hiring and training new employees}\\ &\quad + \text{costs of additional overtime} + \text{over employment costs}\end{aligned}$$

There may be other benefits as well, such as productivity improvements, and these must then be included in the cost–benefit calculations. Oxenburgh (2004) provides a detailed step-by-step procedure for carrying out cost–benefit analyses using the productivity model. In practice, an HFE consultant would need the assistance of the client company's finance department to make detailed calculations of the cost savings brought about by reductions in absenteeism. However, once an estimate has been made, the payback period for the investment in ergonomic improvements is simple to calculate by dividing the costs of the improvements by the monthly savings due to reductions in work loss, plus the value of any productivity gains. Oxenburgh gives some examples of cost-effective modifications to single machines or workplaces (similar to many of the examples described in this book) that paid for themselves in less than 1 year and often in less than 6 months.

The model can be used in a wide variety of industries in different countries. The cost savings will vary between countries due to differences in employment practices—in particular, the social costs of wages depend on whether the country has some kind of worker's compensation and social security scheme, and how it is funded—by private insurance or by general taxation. As a rule of thumb, Oxenburgh suggests that, in businesses in developed countries, 1 day of absenteeism costs up to 3.5 times 1 day's pay when all the direct and indirect costs are included. It is easy to see how ergonomic improvements can be cost-effective when the approach is used to justify change in high-risk jobs. Used correctly, the model should not overestimate the costs of absenteeism because it takes into account the costs of employing temporary workers or paying others overtime. It would seem to be most applicable to blue-collar jobs where output is tightly managed and easily quantified.

The productivity model has a number of limitations. It is only usable in companies or jobs with high absenteeism. It assumes that changes intended to increase the scope for productivity gains and savings will do so in practice, that is, productivity will go up and stay up. Tight control over the life of the exercise is needed to implement the changes, control the costs, and demonstrate the benefits. The "time saved" hypothesis is based on simplistic notions such as "time is money." This may work in rigid production line environments but not in others, because workers may perceive sick leave as part of the package and take time off anyway or in non-paced jobs, the work may "expand to use up the time available." In some occupations, it may be possible to postpone work, or there may be flexibility in the remainder of the workforce (everyone else has to work harder). In service industries, the penalties of failure to deliver (e.g., patients may have to wait longer to be admitted to the hospital) are social rather than financial.

Although time is money, time saved does not always save money. Change needs to be managed to ensure that the extra time made available by ergonomic improvements is used productively. In production line systems, for example, speeding up one part of the line may be ineffective if the rest of the system is unable to keep up with the new pace. Improvements need to be carried out systematically and integrated into the rest of the system.

Costs and Benefits of HFE in Occupational Health and Safety

Goggins et al. (2008) analyzed the findings of 250 case studies of HFE interventions in a variety of industries, including manufacturing, offices, and health care. The interventions typically involved the introduction of one or more elements from a hierarchy of controls (see Figure P.2). The most effective controls involve eliminating hazards 60%–100% effective), followed by controls to reduce the level of exposure to the hazard (40%–60% effective), the time of exposure (such as job rotation, 20%–40% effective), and finally, behavioral controls (such as training, 5%–20% effective). Figure 1.4 summarizes the overall cost-benefit of the interventions as measured by the value of the outcomes of the interventions.

This research shows that HFE interventions are *effective* and *efficient* according to a number of *key performance indicators.*

FIGURE 1.4 Effectiveness of HFE interventions (Goggins et al., 2008).[1] (With permission, Chartered Institute of Ergonomics and Human Factors.)

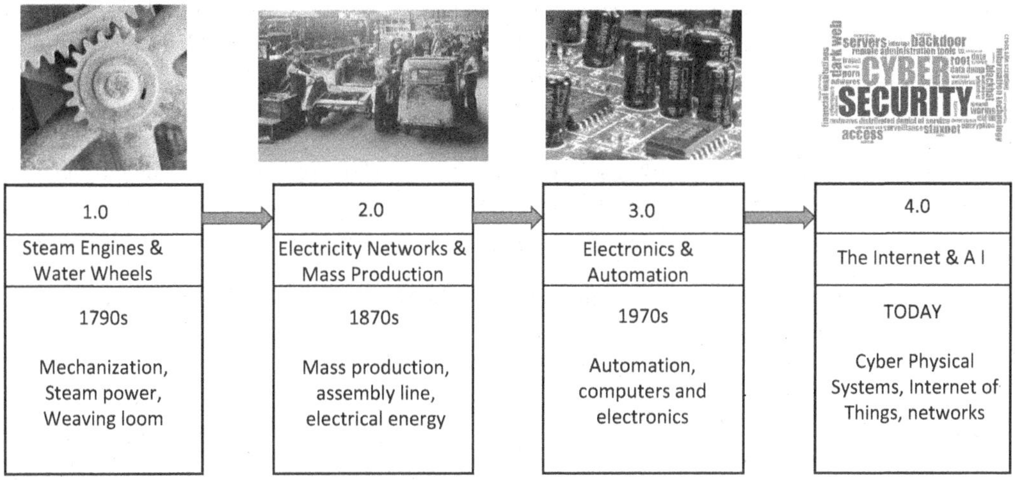

FIGURE 1.5 Stages of industrial evolution.

Drivers of HFE: The Fourth Industrial Revolution

The Fourth Industrial Revolution is the name given to the fusion of different technologies that started at the end of the twentieth century (Figure 1.5): machine learning, speech recognition, natural language processing, and robots.

1. The pace of change has increased due to the confluence of key enabling technologies; principally machine-based learning (essentially neural network technology, which is not new but is a lot more powerful nowadays). Together with near real-time speech recognition, natural language processing (e.g., Siri), and the Internet, these technologies are now far more accessible and usable.
2. Key jobs likely to be automated are those that are routine, rule-based, and process-driven.
3. The link with the Internet, principally the Internet of Things, is that chatbots, cameras with facial recognition technology, voice recognizers, and so on, will be embedded all around us. We will live inside the Internet in a physical sense, not just via our laptops and terminals.
4. Algorithms in search engines already monitor our online behavior and tailor our interactions accordingly. This will increase, and the amount of personal information available to these systems will continue to grow. The same methods can be used in the management of organizations to automate any form of consultative process involving employees.

Developments in the physical, digital, and biological spheres are converging in ways that will have effects that are difficult to anticipate. The Internet has given rise to a network of interconnected "things"—physical goods, people, and information can now be connected. People can communicate with their friends, with databases, and with their domestic appliances at home using their cell phones. Increased automation and integration of transportation may make some jobs redundant (e.g., automated vehicles may make bus drivers redundant). With everyone connected to an Internet of "things," the need to travel anywhere, for any reason, may be reduced, giving people more freedom to work, shop, and socialize at home. At the same time, new forms of relaxation and recreation may emerge. Artificial intelligence may develop to the extent that many middle-class occupations could be automated, with profound implications for the education systems of the future.

Technological change and environmental change are two of the main factors influencing the direction and growth of HFE. Improvements in processing power and speed, coupled with machine learning and artificial intelligence, are bringing major changes to the world of work. While there are many benefits to "living and working inside the Internet," there are also risks. Cybersecurity is an emerging issue for HFE. At the same time, sustainability has become a concern, affecting the way organizations operate and how they use resources, including human resources.

- Technological change makes some jobs extinct (such as 1960s copy typists) while introducing new jobs (such as 1990s call center operators).
- How future jobs are designed has an impact on who can do them and how satisfying they will be to perform.
- We have to consider what people want from work, what they get from it, and what options they believe they have.
- However, the basic principles for designing a good job don't change.

Computationally difficult problems such as supply chain management, manpower planning, and maintenance scheduling are amenable to automation using intelligent systems. These systems might either replace middle managers by taking over the management of the process or aid them through the ability to run large numbers of complex simulations to support planning. Tasks that are of high priority to organizations, amenable to new technologies, and currently challenging are likely to be automated first. With rapid improvements in speech recognition and natural language processing, even more "human-centered" functions, such as human resources, might be automated. Interactive chatbots linked to personnel databases could lead to automated versions of the Hawthorne Counseling program, where employees can voice their opinions and concerns online (computerized industrial espionage or an online continuous attitude survey?). Rapid development of usable systems is a priority in many organizations. Demographic change in industrially developed countries is imposing new constraints. An aging workforce and shortages of skilled people, coupled with the obesity epidemic and unhealthy lifestyles, have resulted in a reduction in the percentage of people capable of normal work. At the same time, equal opportunities legislation demands that employers make work available to all. These trends place pressure on traditional "FPJ" approaches involving personnel selection in favor of "FJP"—redesigning the work so that anyone can do it. System designers are responding by making more use of automation in new "lean-manned" systems. Paradoxically, having fewer people around increases the need for HFE, as the role of those remaining becomes more critical. Physical ergonomic issues are taking on new importance as workforces age and as more women take on jobs previously done by men. In developing countries, there is still a need for basic ergonomic design of factories and offices.

One of the main braking factors in the introduction of AI (and one of the reasons that expert systems were not rolled out in health care 40 years ago) is that somebody has to be legally responsible for the decisions and actions made by the AI system. In other words, it is technologically possible to automate problem-solving, decision-making, car driving, and so on, but it is not possible to automate the legal responsibility for these. Therefore, at the level of the job rather than the task, people are still needed—processes can be automated, but you cannot ascribe responsibility for the conduct of the task to the automation. Another caveat, and a fascinating research problem for HFE, is trust—if drivers do not trust their self-driving cars to drive properly, they will not use the functionality. If the trust is misplaced, they may not take control when they should.

A final braking factor is that many AI systems that use neural networks ("machine learning") are not self-aware and cannot explain their decisions, nor is the process auditable (i.e., there is no traceability except to the database used to train the system).

Drivers of Job Design in the Twenty-First Century

- Develop lean-manned systems (this has been going on for 50 years in car assembly, less so in health care, for example)
- Automate/mechanize functions (this has been happening for 150 years in agriculture, less so in the delivery of social services)

- Changing needs (e.g., reduced size of the armed forces post-1989)
- Shrinking of the recruitment pool as birth rates decline
- A requirement to design jobs that older people can do/will do/want to do
- Minimize the training burden for new systems (encourage positive transfer of learning)
- Replace classrooms with new technologies to train employees (online training, virtual reality, support from artificial intelligence)
- Globalization and the availability of people—pros and cons of using immigration to fill manpower shortages
- Changing job roles requires new skills
- Providing opportunities for hybrid working while maintaining engagement and co-ordination

The historical review tells us that job design doesn't exist in a vacuum, but it is a response to the development of technologies, products, and markets. Hybrid working during the COVID-19 pandemic was made possible in service industries by the Internet and the use of mobile devices. This, in turn, has called into question the rationale for offices—are they where work is done, or are they meeting places for teamwork? Similar questions will arise as AI becomes increasingly infused into those parts of the Internet that we inhabit.

SUMMARY

HFE occupies the no man's land between engineering and medicine, architecture, health and safety, computer science, and consumer product design. It is the only scientific subject that focuses specifically on the interaction between people and technology.

Historically, HFE can be seen to have arisen as a response to the need for rapid design of complex systems. The modern HFE specialist has an important role to play as a member of the design team, providing scientific information about employees (a scarce commodity in many organizations) and ensuring that all aspects of the system are evaluated from the users' or operators' point of view. Many tools are now available for the systematic analysis and specification of system HFE. The participatory approach seems to be the best way to ensure that the implementation of HFE will be effective in the workplace. Further guidance can be found in ISO/TC 159/SC 1 General Ergonomics Principles.

TUTORIAL TOPICS

1. What impact will the Fourth Industrial Revolution have on HFE?
2. Will technological change make HFE obsolete?
3. In the year 2035, a university wishes to introduce a new course on "Human-Robot Symbiosis." Discuss the main lecture topics to be covered in the syllabus, and suggest ideas for lab work and practical (using real robots, of course).
4. Take a look at the video about F.W. Taylor, at the link https://www.youtube.com/watch?v=vNfy_AHG-MU

Do you think Taylorism has a future in the twenty-first century? Discuss why (or why not).

ESSAYS AND EXERCISES

1. Was HFE invented, discovered, or is it just a buzzword for what happens anyway? Discuss.
2. Refer to the cheese sandwich task analysis in Workshop 1.2. Specify the main operations and any decisions required to complete the subtask "Prepare Ingredients." (Hint: Use the description of the subtask "Prepare work area" as your model.)

3. Analyze the task of making a telephone call using a conventional phone (landline). Using task flow format, construct a description of this task.
4. Using a format of your choice, analyze and describe the task of brushing your teeth with a toothbrush.
5. Join an Internet-based HFE discussion group and seek to generate debate on a controversial question of your choice. (Hints, e.g., Does repetitive strain injury (RSI) exist, and if it does, should it be regarded as a compensable workplace injury? Is there such a thing as the Hawthorne effect, and if there is, what is the definition? Do we still need graphic user interfaces?)
6. Search the web for sources of information about HFE. Use the categories in the human–machine model to guide your search.
7. Contact five different manufacturers across a range of industries. Find out whether they use ergonomic principles in the design of their products. For those companies that do use HFE, find out who is responsible and how HFE is applied. (Hint: choose a white goods manufacturer, a car manufacturer, computer hardware and software manufacturers, and one other, such as a manufacturer of medical equipment or garden tools.)
8. Describe the present scope and concerns of HFE. Obtain back issues from the last 5 years of two or more of the journals *Human Factors, HFE*, and *International Journal of Industrial Ergonomics and Applied Ergonomics.* Summarize, using keywords of your choice, the topics and research questions being dealt with in each of the articles in these journals.
9. Use the HFE checklist in Workshop 1.1 to develop a taxonomy of standards for HFE. The standards should provide criteria or limits to support the use of the checklist—try to link each item in the checklist and one or more standards. (Hint: visit the websites of the American National Standards Institute and the International Standards Organization.)
10. To what extent are the following industries aware of the need for HFE in their short- or long-term planning? What mechanisms do they have in place to ensure that HFE principles are applied?
    - Railways
    - Maritime (commercial shipping)
    - Air traffic control
    - NASA
    - Health care
11. The International Ergonomics Association has a professional competencies checklist which can be found at the link (https://m4v211.p3cdn1.secureserver.net/wp-content/uploads/2014/10/Core-Competencies-in-Human-Factors-and-Ergonomics-2021.pdf). Create a matrix to relate these competencies with the contents of this book (hint: for each competence, find the corresponding sections/chapter in this book).
12. Mr. Smith's daughter wants to study for a Master's degree in HFE. Mr. Smith is unconvinced. Construct at least five arguments (50–100 words each) to convince him that HFE is a good career choice.

QUICK QUIZ

1. The word "ergonomics" was first coined by:
    a. F.W. Taylor
    b. W. Karwowski
    c. Jastrzebowski
2. F.W. Taylor believed that productivity could best be improved by getting a "good man" in to supervise workers. True or false?
3. The Hawthorne experiments were designed to prove that better lighting improves productivity. True or false?

4. The Hawthorne experiments were well received when published. True or false?
5. Who developed hierarchical task analysis?
6. Task analyses are the best way to conduct risk assessments. True or false?
7. A reliable tool measures what it is supposed to measure. True or false?
8. The best workspace improvements should be efficacious. True or false?
9. It's not always best to fit the task to the person. True or false?
10. Human factors thrive in times of rapid change. True or false?

NOTE

1 Ontario Ergonomics Intervention Cost-Benefit Calculator available online.

2 Body Mechanics

GENERAL REQUIREMENTS FOR HUMANS IN SYSTEMS

1. Workplaces must be designed to provide postural stability.
2. External forces acting on the human body must not exceed its mechanical tolerance limits.
3. Where the exertion of forces on external objects is required, the magnitude of forces must be within the strength and endurance limits of members of the user population.
4. The task must be designed to minimize postural load and extreme postures (ALARP).
5. Biomechanical risk factors for musculoskeletal injury must be identified and eliminated in the design.

When man assumed the upright position, he immediately made an enemy of gravity and has been fighting this relentless foe ever since.

J.A. Jones (1933)

CORE KNOWLEDGE: BODY MECHANICS

The human body is a mechanical system that obeys physical laws. Many of our postural and balance control mechanisms, essential for even the most basic activities, operate outside of conscious awareness. When these mechanisms break down—as in slipping or losing balance—we are rudely reminded of our physical limitations.

The skeleton plays the major supportive role in the body. It can be likened to the scaffolding to which all other parts are attached. The functions of the skeletal and muscular systems are summarized in Table 2.1.

Like any mechanical system, the body may be stable or unstable and is able to withstand a limited range of physical stresses. Stresses may be imposed internally or externally and may be acute or

TABLE 2.1
Functions of the Skeletal and Muscular Systems

Skeletal System

1. Support
2. Protection (the skull protects the brain, and the rib cage protects the heart and lungs)
3. Movement (muscles are attached to bone, and when they contract, movement is produced by the lever action of bones and joints)
4. Hemopoiesis (certain bones produce red blood cells in their marrow)

Muscular System

1. To produce movement of the body or body parts
2. To maintain posture
3. Heat production (muscle cells produce heat as a by-product and are an important mechanism for maintaining body temperature)

DOI: 10.1201/9781032717210-2

TABLE 2.2
Task Stress and Postural Stress

	Task Stress	
Postural Stress	**High**	**Low**
High	Digging a trench	Painting a ceiling
Low	Competitive weight lifting	Reading a book

chronic. A useful starting point in the discussion on mechanical loading of the body is to distinguish between postural stress and task-induced stress.

- *Posture*: The average orientation of the body at a point in time
- *Postural Stress*: The mechanical load on the body by virtue of its posture
- *Task Stress*: The forces acting on the body while performing the task

There are two different kinds of task-induced stress.

1. *Extrinsic Task Stress*: The mechanical load on the body due to exposure to external forces
2. *Intrinsic Task Stress*: The mechanical load on the body due to exposure to internally generated forces

Task and postural stresses can vary independently of each other (Table 2.2). Some tasks, such as lifting a barbell, are high in task stress but can be performed in non-stressful postures. Painting a ceiling requires little effort to apply the paint but much effort to maintain the posture. Much biomechanical stress is unnecessary because it is postural and can be reduced by redesigning the task to improve posture.

Postural Stability

In order for the body to be stable, the combined centers of gravity (COG) of the various body parts must fall within a base of support (the contact area between the body and the supporting surface). When standing, the weight of the body must be transmitted to the floor through the base of support described by the position of the feet (Figure 2.1). The alignment of the body parts must be maintained to ensure continuing stability, and it is in the maintenance of posture that much stress arises.

Some Basic Body Mechanics

The basic limiting condition for postural stability in standing is that the combined COG of the various body parts falls within the base of support described by the position of the feet (assuming no other external means of support). Ideally, the lines of action of the masses of the body parts should pass through or close to the relatively incompressible bones of the skeleton (Figure 2.2). The jointed skeleton supports the body parts and is itself stabilized by the action of muscles and ligaments, which serve primarily to correct momentary displacements of the mass centers from above their bony supports. Using a rather crude analogy, the skeleton can be likened to an articulated tent pole with guy ropes (postural muscles) on every side. The fabric of the tent corresponds to the soft tissues of the body. Any displacement of the COG of the structure in a given direction leads to tension in the guy ropes on the opposite side. Ligaments can be likened to springs and rubber fittings they stabilize the articulations of the tent pole and tendons to the ends of the guy ropes where they insert into the poles.

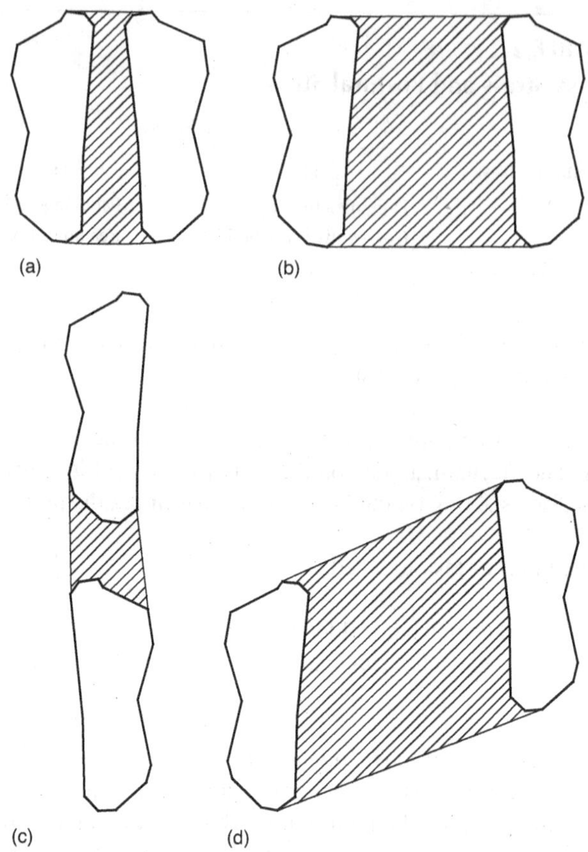

FIGURE 2.1 Stability of the body parts depends on the shape of the base of support described by the position of the feet: (a) unstable, (b) fairly stable in all directions, (c) stable anteroposteriorly, and (d) laterally stable.

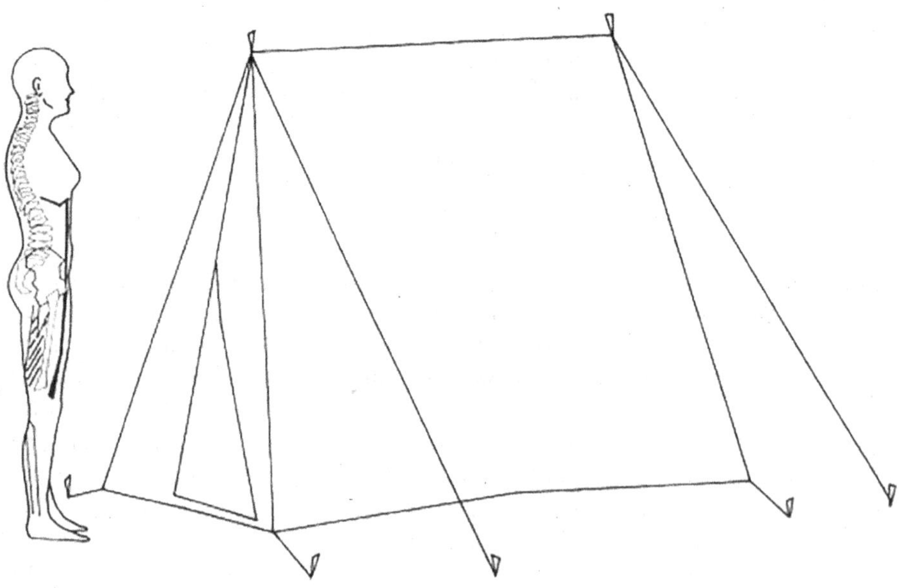

FIGURE 2.2 Tent analogy. The skeleton is the tent pole, the muscles are the guy ropes, and the soft tissues are the canvas.

Postural stress can cause pain. Workers who have to work with the spine flexed forward (by 60° for more than 5% of the day or 30° for more than 10% of the working day), or rotated more than 30°, suffer from back pain (Hoogendoorn et al., 2000).

DEMONSTRATION

To demonstrate the "tent" analogy, stand upright and relaxed with body weight equally distributed between the feet—neither on the heels nor on the balls of the feet. Place one hand on your lower back muscles (you should feel a muscular ridge in the center of your back). Place the other hand on your abdominal muscles. Palpate both sets of muscles; they should feel soft. Next, shift your weight to the balls of your feet and lean forward slightly. As you do this, tension should appear in the lower back muscles as they act to maintain equilibrium. Repeat by swaying backward slightly, placing your weight on your heels. Your abdominal muscles will begin to tense and your back muscles will relax as the weight shifts onto the heels. In the middle, there will be a neutral position, where your upright stance can be maintained with minimal muscular load. This neutral position is associated with low postural load.

Without its associated trunk muscles, the human spine is very weak—it buckle under a compressive load of only 90 N. Cholewicki and Van Vliet (2002) demonstrated that spinal stability depends on the co-activation of all trunk muscles and other loading variables. Although a neutral position of minimal postural stress does exist, no single muscle stabilizes the spine. Muscle activity increases when a person carries a load and is an example of true postural muscle activity. It is not the intrinsic stiffness of muscles alone that stabilizes the spine, but rather the reflex contractions of the muscles that contribute to spinal stability (Moorhouse and Granata, 2007).

Anatomy of the Spine and Pelvis Related to Posture

The spine and pelvis support the weight of the body parts above them and transmit the load to the legs via the hip joints. They are also involved in movement. Almost all movements of the torso and head involve the spine and pelvis to varying degrees. The posture of the trunk may be analyzed in terms of the average orientation and alignment of the spinal segments and pelvis. Figure 2.3 depicts the spine and pelvis viewed sagittally and posteriorly.

Spine

Quadrupedal animals and human babies have a single spinal curve running dorsally from the pelvis to the head. The thorax and abdomen hang from the spine and exert tension, which is resisted by the spinal ligaments, the apophyseal (facet) joints, and the back muscles. In adult humans, the spine is shaped such that it is close to or directly below the COG of the superincumbent body parts, which are supported "axially"—that is, the effect of weight-bearing in the standing posture is to compress the spine, not to bend it (Adams and Hutton, 1980). This compression is resisted by the vertebral bodies and the intervertebral discs. The "cervical" and "lumbar" spines are convex anteriorly—a spinal posture known as lordosis. It is the presence of these lordotic curves that positions the spine close to or directly below the line of gravity of the superincumbent body parts. The effect is to reduce the energy requirements for maintaining the erect posture and to place the lumbar motion segments in an advantageous posture for resisting compression (Klausen and Rasmussen, 1968; Adams and Hutton, 1980, 1983). In a systematic literature review, Arshad et al. (2019) found that lumbar lordosis was greater in young women than in men of the same age, whereas lumbar flexion was greater in men. Age-related reductions in lumbar flexion, extension, axial rotation, and

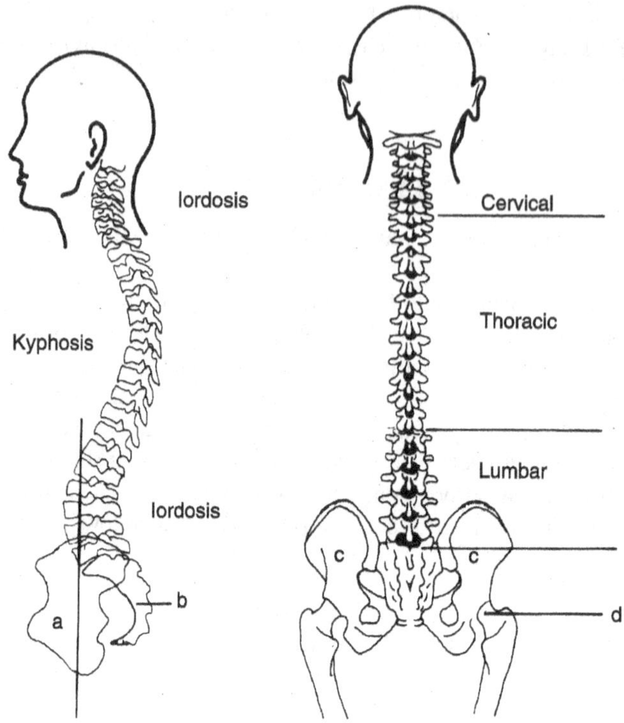

FIGURE 2.3 The lumbar, thoracic, and cervical spines and the pelvis (a) and sacrum (b). The weight of the upper body is transmitted through the lumbar spine, the iliac bones of the pelvis (c) to the hip joints (d) and legs.

lateral bending were found in both men and women. *Thus, older employees have a reduced range of motion in the lumbar spine and less capacity to adopt extreme postures in daily life, which may put them more at risk of injury.*

The thoracic spine is concave anteriorly and is strengthened and supported by the ribs and associated muscles. The term "spinal column," although universally accepted, is something of a misnomer; "spinal spring" might be more appropriate. The "S" shape of the spine in a person standing erect gives the entire structure a "spring-like" quality, enabling it to better absorb sudden impacts, such as mechanical shock when the heel strikes the ground during walking (Schultz, 1969), than if it were a straight column. The loss of the "S" shape during sitting may be one of the reasons why drivers of trucks and farm vehicles who are exposed to vertical-plane vibration, are so prone to back trouble.

The cervical and lumbar spines are not fixed in lordosis. Each vertebral body is joined to its superior and inferior counterparts by muscles, ligaments, and joints. The spine participates in functional movements of the body—part of the postural adaptation required to carry out many activities takes place in the lumbar and cervical spines.

The spine can be considered, simplistically, to consist of three anatomically distinct but functionally interrelated columns (Figure 2.4). The anterior column, consisting of the vertebral bodies, intervertebral discs, and anterior and posterior longitudinal ligaments, is the main support structure of the axial skeleton. It resists the compressive stress of the superincumbent body parts. The two identical posterior columns are positioned astride the neural arch (which forms a bony cavity through which the spinal cord passes), and consist of the zygapophyseal (facet) joints and the associated bony projections, ligaments, and muscles. The posterior elements of the spine act as jointed columns that control spinal movement and provide attachment points for the back muscles.

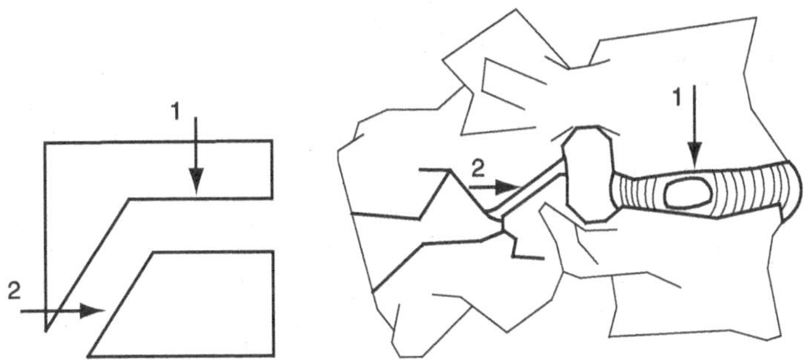

FIGURE 2.4 Function of (1) intervertebral disc and (2) facet joints. The disc resists the compressive load, and the facets resist the intervertebral shear force and protect the discs against torsion. (From Kapandji, 1982, with permission.)

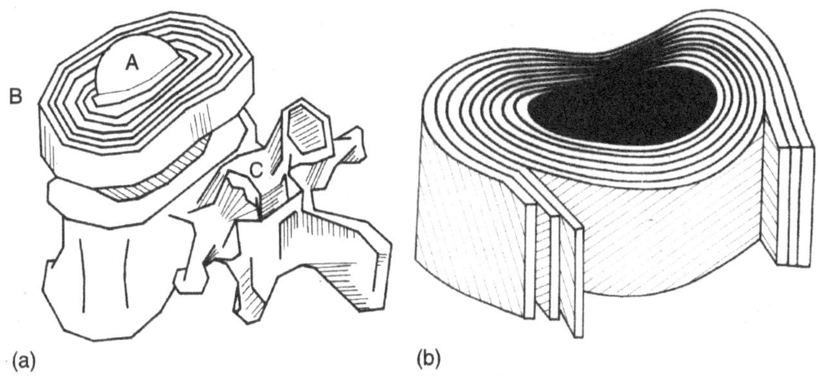

FIGURE 2.5 Intervertebral disc and vertebral body. (a) In this view, the superior vertebral body has been removed to reveal the intervertebral disc below. A is the nucleus pulposus, B is the annulus fibrosus, and C is the inferior facet joints at the rear. (From Kapandji, 1982, with permission.) (b) Details of the structure of the annulus fibrosus. The annulus consists of a number of layers of cartilage. The fibers in the layers run obliquely and in different directions, somewhat like the layers of a cross-ply tire. The outer layers run perpendicular to each other. (From Vernon-Roberts, 1989, with permission.)

The vertebral bodies and their related structures increase in size from the top to the bottom of the spine in accordance with the increased load they must bear. The pars interarticularis is the bony bridge between the inferior and superior facet joints.

The intervertebral discs separate the vertebrae and enable them to coarticulate, whereas the facet joints limit the amount of movement that can occur in any direction. Each disc consists of concentric layers of cartilage whose fibers are arranged obliquely in a manner similar to a cross-ply tire (Figure 2.5).

The layers of cartilage enclose a central cavity, which contains a protein–mineral solution (proteoglycans). Positive osmotic pressure ensures that water is always tending to enter the disc. Thus, the discs are prestressed to withstand loading (in a manner analogous to the reinforced concrete beams used in the construction of modern buildings).

Intervertebral Discs: Form and Function

Intervertebral discs are not shock absorbers (even vertebrate fish have them in their spines). They are spacers that allow movement (co-articulation of the vertebrae) while stiffening the motion segments. The nucleus acts somewhat like a pivot—a fixed point supporting something that turns or balances. According to Yang et al. (2019), compressive stiffness increases with the size of the nucleus (peak stresses are greater with a smaller nucleus). Under axial compression, stresses are transmitted from the nucleus to the annulus; under axial rotation, they are distributed toward the nucleus. In extension, compression is greater in the posterior annulus and the facet joints. In flexion, the nucleus is "wedged" posteriorly. These adaptations show how a healthy spine responds dynamically to posture and movement.

Intervertebral discs exhibit viscoelastic behavior. Forces of rapid onset are resisted in an elastic manner—the disc deforms initially, then rapidly returns to its original shape when the force is removed. Under continuous loading, however, the disc exhibits a type of viscous deformation known as "creep." Creep occurs as a result of loading above or below a threshold level. Under compressive loading, the disc narrows as fluid is expelled and the superior and inferior vertebral bodies move closer together. Under traction (stretching or pulling forces), fluid moves into the disc and the disc space widens (Bridger et al., 1990).

The narrowing and expansion of the disc spaces are natural and occur due to the forces exerted on the spine during daily activities. Since there are 24 vertebral bodies, all with discs between them, the shrinkage and expansion of the disc spaces result in measurable changes in stature—most people are about 1% taller when they wake up in the morning than when they go to bed at night. Healey et al. (2011) report that approximately 30%–40% of the daily loss of stature occurs in the first hour after rising. In subjects with low back pain, discomfort correlated with stature loss, suggesting that the loss of stature was associated with increased loading on symptomatic tissues.

Stature change also occurs with age. After about 30 years of age, intervertebral discs begin to degenerate, developing micro tears and scar tissue in the annular layers. Fluid is lost more readily, and the disc space narrows permanently. At this stage, spinal motion segments lose stability and compressive strength. It is not surprising, then, that most occupationally induced low back pain occurs in middle-aged people. In the elderly, disc degeneration reaches a stage where, together with other degenerative processes, the spine is restabilized—but with a corresponding loss of mobility. There is also evidence that disc degeneration accelerates after menopause, possibly due to estrogen deficiency (Wang and Griffith, 2010).

Prolapsed Intervertebral Discs—"Slipped Discs"

Since the discs have no direct blood supply, the daily ingress and egress of fluid due to variations in loading is the mechanism by which nutritional exchange with the surrounding tissues takes place. Postures that exert static loads on the body interfere with this mechanism and are hypothesized to accelerate the degeneration of the discs. Many other factors may contribute to the degeneration of the discs with age. Damaged cells in the nucleus and annular fibers may be subject to *apoptosis* (a process of programmed cell death that eliminates damaged or non-essential cells without causing local inflammation from cell leakage). Other factors causing cell death include inflammation, matrix degradation, loss of proteoglycan in the nucleus pulposus, disorganization of the concentric lamellae in the annulus, spinal instability, disc height loss, and disc prolapse. Changes to the immune balance of the microenvironment of the disc may also cause immune cells to infiltrate the disc and attack its cells.

Under normal circumstances, the nucleus is hydraulically sealed by the annular fibers. Cell death in the inner layers of the annulus (those farthest from the blood supply) following trauma or exposure to high forces may "break the seal" causing the nuclear material to leak into the annulus. Over time, with repeated trauma, the nucleus is no longer effectively contained, and disc prolapse (a "slipped disc") can occur (Zhang et al., 2016).

Axial overloading of the spine, which compresses the cells in the discs, has been linked to increased rates of cell death (Lotz and Chin, 2000). Dynamic overloading puts the disc at risk of posterior herniation. Forward bending is more hazardous in the early morning than at other times of day. Upon rising, increased fluid in the disc may escape into the outer annulus through radial tears, carrying inflammatory material to nerve endings and resulting in back pain. Snook et al. (1998) reported significant reductions in back pain among a group of individuals trained to avoid early morning flexion of the spine. Examples of avoidance strategies included techniques for rising from bed without flexion, avoiding sitting or squatting for two hours after rising, using reachers to pick things up, etc.

Tasks that place high loads on the spine are obvious candidates for assessment and redesign by human factors and ergonomics (HFE) specialists.

Pelvis

The pelvis is a ring-shaped structure made up of three bones: the sacrum and the two innominate bones. The sacrum extends from the lumbar spine and consists of a number of fused vertebrae. The three bones are held together in a ring shape by ligaments (Figure 2.6). The innominate bones are themselves made from the fusion of three other bones: the ilium, ischium, and pubis. The pubis lies at the anterior part of the pelvis. It joins the other bones together, completing the ring shape and acting like a strut to prevent the pelvis from collapsing under weight-bearing (Tile, 1984). The posterior structures of the pelvis, the sacrum and ilia, carry out the actual weight-bearing function.

The pelvis càn be likened to an arch, which transfers the load of superincumbent body parts to the femoral heads in standing and to the ischial tuberosities (part of the two ischia) in sitting (Figure 2.6).

When viewed from the rear (Figure 2.6), it can be seen that the sacrum resembles the keystone of the arch. The load from above is transmitted through the innominates to the femoral heads. However, when viewed from above, the sacrum has the wrong shape for a keystone—it tends to slide forward, out of the arch (Figure 2.6). Under weight-bearing, the tendency for the sacrum to slide forward anteriorly is resisted by the strong ligaments between the sacrum and the ilia. It is these posterior sacroiliac ligaments that stabilize the joint between the sacrum and the ilia. Many factors can cause sacroiliac joint pain, including leg length discrepancy, gait abnormalities, jogging, scoliosis, and tasks involving axial loading of the spine combined with rotation of the trunk. The pain can also occur in pregnancy due to increased loading of the joint and hyperlordosis of the lumbar spine when the pregnant woman stands (extending the lumbar spine to counterbalance the mass of the uterus) (Cohen et al., 2013). Tasks requiring forward bending from the hip (such as washing dishes at a sink) increase the tendency for the sacrum to be anteriorly displaced, thereby increasing the tension in the sacroiliac ligaments. Small displacements of the sacrum can occur, causing soft tissues to be "pinched" and resulting in pain (Figure 2.7). Sacroiliac joint pain is thought to contribute to up to 38% of low back pain (Falowski et al., 2020).

Lumbo-Pelvic Mechanism

The lumbar spine arises from the sacrum, and the degree of lumbar lordosis depends on the sacral angle, which, in turn, depends on the tilt of the pelvis (Figure 2.8). The pelvis can be represented, as in Figure 2.9, as part of a lever system, with the hip joint regarded as a fulcrum.

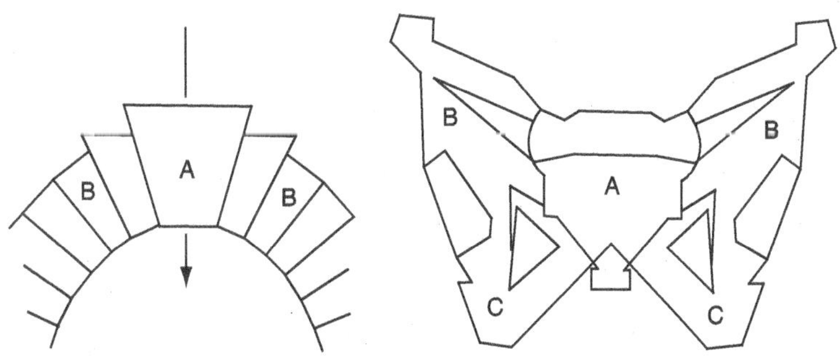

FIGURE 2.6 The pelvis as an arch. The pelvis viewed from the rear. A is the sacrum, B is the ilium, and C is the ischium. The sacrum acts like a true keystone in this plane. (Redrawn from Tile, 1984, with permission.)

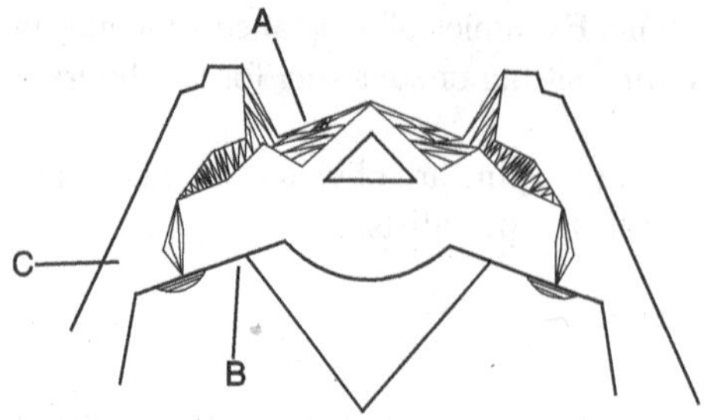

FIGURE 2.7 View of the sacroiliac joint from above. A represents the ligaments, B is the sacrum, and C is the pelvis. The ligaments act like the cables of a suspension bridge, preventing the sacrum from slipping forward. If the joint is deformed by loading, the ligaments can be pinched by bone, causing pain in the very low back, usually on one side. (Redrawn from Tile, 1984, with permission.)

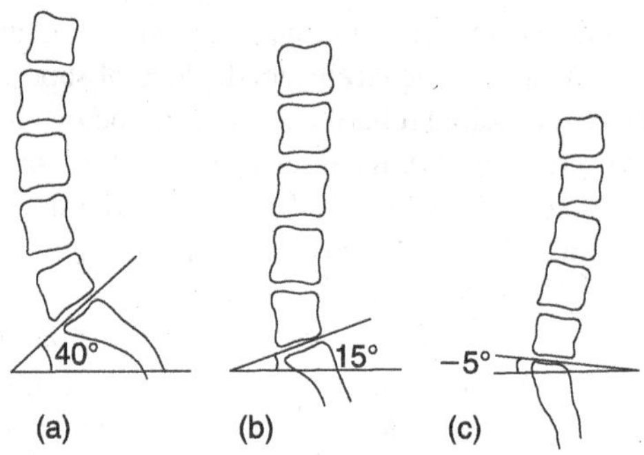

FIGURE 2.8 Relationship between sacral and lumbar angles. (a) Sacral angle and lumbar lordosis, as in standing, (b) Moderate sacral angle and fattened lordosis as in sitting on a chair with a backrest, and (c) Minimal sacral angle and tendency to lumbar kyphosis as in sitting on a low stool.

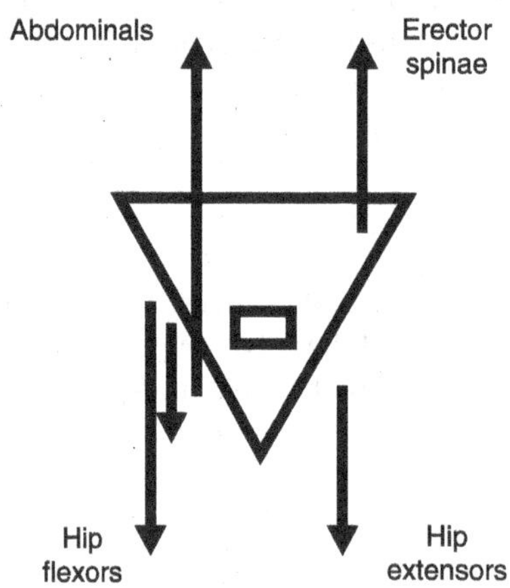

FIGURE 2.9 Schematic representation of the muscular system of the pelvis (sagittal view). When the abdominal or hip extensor muscles shorten, the pelvis tilts backward. The result is a flattening of the lumbar spine to maintain the trunk erect. When the hip flexors or erector spinae muscles shorten, the pelvis tilts forward. This is accompanied by a compensatory increase in the lumbar lordosis.

Many muscles attached to the pelvis can be considered as guy ropes or stays, which fixate the pelvis onto the heads of the femora. These muscles can exert torques on the pelvis, causing changes in pelvic tilt (even though this is not their main function). The hamstring, gluteal, iliopsoas, erector spinae muscles, and other muscles, together with the ligaments of the hip joint, are part of the lumbo-pelvic system. The tilt of the pelvis in the anterior/posterior plane depends on the equilibrium of the torques exerted by the antagonistic muscles in the system.

Imbalances between these muscles cause lumbo-pelvic dysfunction. People with strongly developed rectus abdominis muscles and weak hip flexors may develop a chronic posterior pelvic tilt and corresponding loss of lumbar lordosis (flat-back posture). Weak abdominals, strong iliopsoas, and strong rectus femoris muscles lead to an anteriorly tilted posture and hyperlordosis (hollow-back posture).

Sedentary lifestyles can lead to chronically stretched and weakened gluteal and hamstring muscles and shortened hip flexors, resulting in a hollow-back posture. Ludwig et al. (2024) measured muscle activity in forty-one healthy males who were instructed to tilt their pelvis posteriorly. Four clusters of muscle activation were identified. All were equally effective in the amount of pelvic tilt generated. In group A, the rectus abdominis and the abdominal obliques played a major role. In the remaining groups, gluteus maximus played the main role, either on its own or in conjunction with the hamstrings (biceps femoris). Sustained postures that weaken the gluteal muscles and hamstrings, such as prolonged standing, can lead to an anterior pelvic tilt and backache. The gluteal muscles can be contracted to correct the anterior tilting. Alternatively, placing one foot on a footrest will shorten the hip flexors on one side, particularly the iliopsoas, and help correct the posture.

BASIC APPLICATIONS

Standing

An understanding of biomechanics and basic anatomy enables us to analyze body positions and movements at work and to use the findings from these analyses to assess working arrangements and propose improvements. Standing upright is a position of low postural stress, requiring approximately 6% more energy expenditure than the resting level to maintain postural stability.

A stable posture can only be maintained if the various body parts are supported and kept in an appropriate relationship to the base of support, such as the feet (Figure 2.1). The line of gravity of the body parts must fall within this base of support, and postural reflexes exist to ensure that, as one body part moves, the total body mass remains balanced over the feet.

For example, when a standing person leans forward, as if to touch their toes, the pelvis moves rearward to compensate for the forward displacement of the COG of the upper body (Figure 2.10). It is impossible to perform this movement without falling if there is insufficient space at the rear to allow for the compensatory projection of the buttocks, unless one foot is placed in front of the other. A simple demonstration of this principle can be performed by standing with one's back and heels against a wall and attempting to pick up an object from the floor in front without moving the feet.

It is important to provide "sufficient space" around standing operators and plenty of "room for the feet" to avoid losses of balance.

For the body to be in a condition of static equilibrium, the following conditions must be met:

1. Upward forces (from the floor) = Downward forces (body weight plus any objects held)
2. Forward forces (e.g., bending forward) = Backward forces (extension of back muscles)
3. Clockwise torques (e.g., from an asymmetric load) = Anticlockwise torques (back and hip muscles)

Ideally, the skeleton should play the major role in supporting the various body parts, since this is its primary function. However, muscles, ligaments, and soft tissues can also contribute—though at the cost of increased energy expenditure, discomfort, or risk of soft-tissue injury.

When leaning forward in the manner shown in Figure 2.10, a stable posture can be maintained indefinitely, although the posture itself is uncomfortable. This is because strain is placed on the

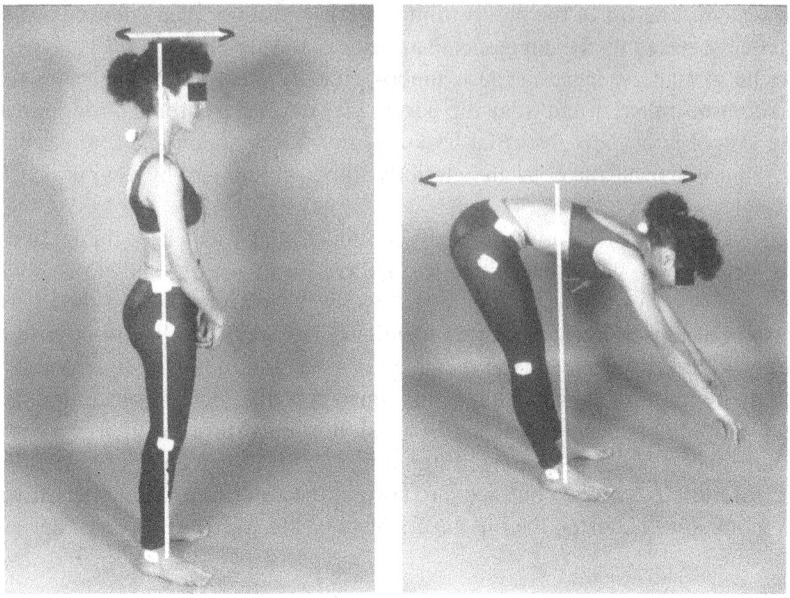

(a) (b)

FIGURE 2.10 When the base of support is constrained, compensatory movements occur automatically to maintain postural stability, demonstrating that the "attitudinal as well as the righting reactions" are indeed involuntary. (a) Balanced erect standing posture and (b) as the hip joints flex and the upper body moves forward, the ankle joints plantar flex to compensate and the lower body moves rearward, maintaining balance.

posterior spinal ligaments and lumbar intervertebral discs—the upper body load is no longer supported by axial compression of spinal structures but rather by tension in ligaments and asymmetric compression (wedging) of the intervertebral discs (Figure 2.11).

Postures can be stable but stressful if support of body mass depends on soft tissues rather than bone. Ligaments are able to resist high tensile forces, particularly if these forces are exerted in the direction of their constituent fibers. They play a major role in protecting joints by limiting the range of joint movement and resisting sudden displacements that might damage the joint. However, injuries can occur if ligaments are exposed to sudden forces when prestressed by extreme joint positions or by complex movements. This is one of the reasons ergonomists stress the importance of the posture of the hands, wrists, elbows, and trunk when tools or controls are operated or when loads are lifted. Poor equipment design that forces the adoption of extreme joint positions while holding an object predisposes the joint to injury. Good design enables equipment to be used with the joints positioned near the middle of their range of movement.

When a person leans forward in the manner described above but "arches" the back in an effort to prevent the spine from losing its shape, the resulting posture is still stable—but a different cost is incurred. In this position, the muscles of the back must carry out static work to maintain the shape of the spine against the pull of gravity, which causes it to flex. Static muscle contractions lead rapidly to fatigue.

Low Back Pain

The spine can be likened to a three-dimensional jigsaw puzzle. As long as all the pieces fit together, the spine will function painlessly. Many degenerative changes can occur over time as a result of wear and tear or from injuries that have healed, but such changes do not necessarily mean that back pain is inevitable. Schmorl's nodes are often seen on spinal radiographs—where the nucleus bulges into the vertebral body above or below. These are often painless, and no treatment is usually needed.

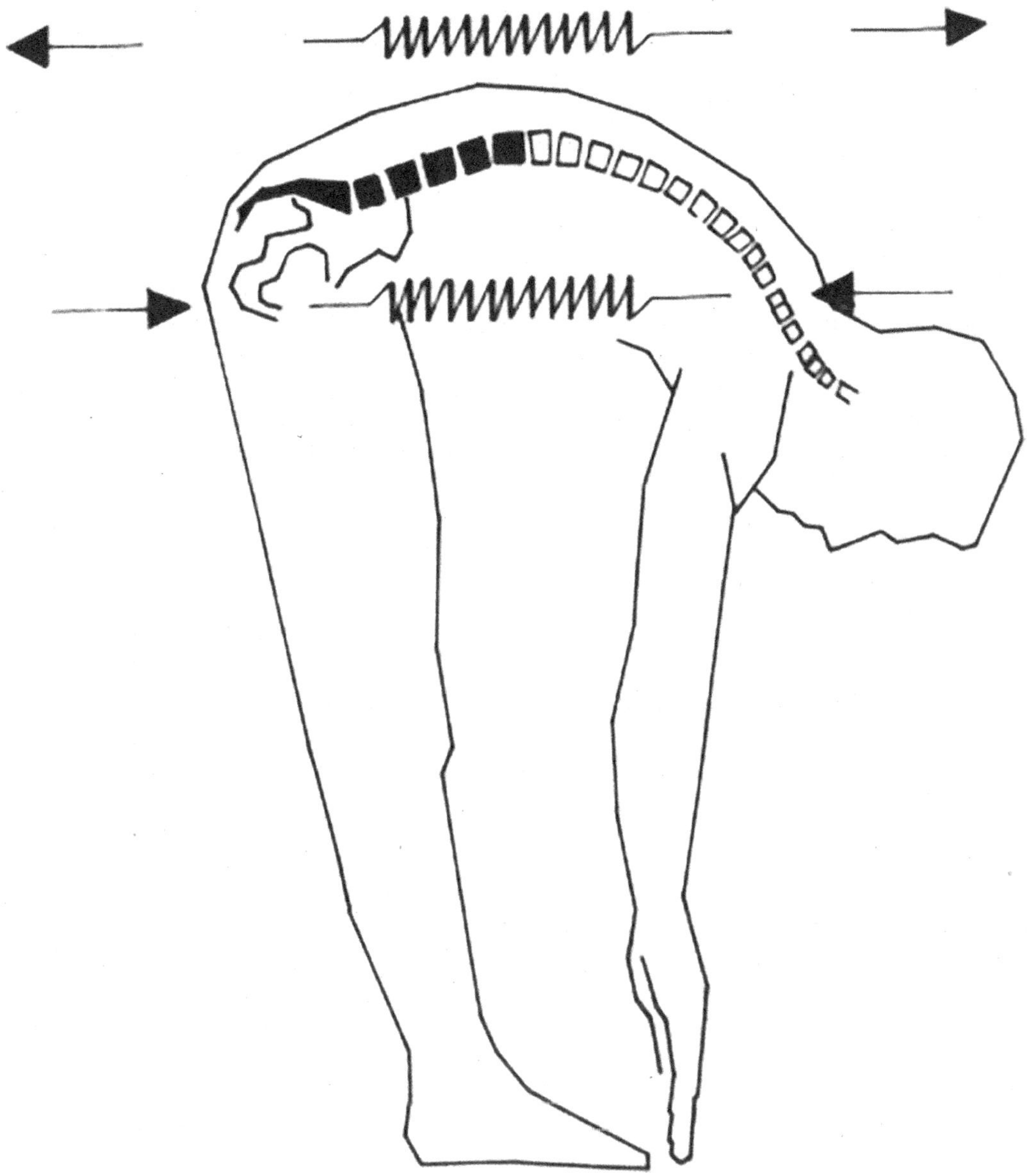

FIGURE 2.11 In this position, postural stress occurs in the form of compression of abdominal contents and intervertebral discs and stretching of the posterior spinal ligaments.

Pain in the low back and/or buttocks can arise from a number of sources: disease or injury to the sacroiliac joint; facet joint arthropathy; coccydynia (strain of the muscles around the coccyx); lumbar spondylolisthesis (displacement of one vertebra over the one below); lumbar spine stenosis (narrowing of the central spinal canal); discogenic pain (damage to the discs leading to invasion of annular tears by nerve endings); prolapsed disc (material from the nucleus pulposus is not contained and irritates adjacent nerve roots); and sciatica (the sciatic nerve can be pinched by a herniated lumbar disc or a bony spur on a vertebra from L4 to the sacrum), which typically develops over time in response to loading. The main symptoms include low back pain; leg pain or tingling in the legs changes in sensation in the leg or foot; low back muscle spasms; and leg or foot pain.

Pain is unlikely to arise from the intervertebral discs themselves, since only the outer parts contain nerve endings in adults. Similar reasoning rules out pain from the capsules of the apophyseal joints. The most likely sources of back pain that HFE specialists are likely to encounter are the posterior ligaments and the back muscles.

Chronic Back Pain

For present purposes, we can distinguish between "back ache" and "back pain." Back ache and its prevention in the workplace are most relevant to HFE professionals. Badly designed workplaces or excessive task loadings may cause fatigue of the back muscles or viscous changes to the soft tissues resulting in pressure "hot spots" on pain sensitive structures in otherwise healthy employees. Following rest, redesign, or time away from work, the pain typically subsides. For employees with existing conditions such as those described above, the symptoms may be more severe following exposure to the same stressors, but they may also be experienced during other activities of daily life. Acute back pain lasts no more than a few weeks and is usually a response to biomechanical loading or its consequences, such as delayed onset muscle soreness. Subacute back pain lasts up to 3 months and may indicate a more complex picture, such as a more severe injury or reinjury of an already affected area.

Chronic pain, defined as pain lasting more than 3 months can be a symptom of a debilitating, degenerative condition in which an initial injury results in tissue damage that changes the mechanical function of one or more motion segments, leading to a cascade of maladaptation, as one problem leads to another. Specialist rehabilitation services may be required in these cases. Psychological and psychosocial factors appear to influence the trajectory from initial injury to chronic illness. Fear avoidance is one such factor in which the patient views the injury as a catastrophe and avoids subsequent physical activity—including beneficial activities that prevent the loss of muscle and are essential for the restoration of normal function (Bunzli et al., 2017). Back problems at work should be dealt with by stressing a return to normal duties as soon as possible, countering negative beliefs with information booklets and advice, and fostering active coping strategies instead of coping strategies based on the avoidance of normal activities. Rest and the avoidance of exertion, together with restricted work duties, may cause muscle wastage and smooth the transition to chronic disability. Roberts et al. (2015) investigated predictors of successful rehabilitation from back injury in military personnel. Patients taking the most medications (for any medical conditions) were less likely to recover, as were those with slow 800-metre run times. High self-efficacy (the belief that one can use the rehabilitation to "get myself better") was a predictor of recovery. By the same token, engaging in physical activity outside work seems to be associated with a lower prevalence of low back pain (Alzahrani et al., 2019).

Clearly, there is not much that HFE alone can do to lessen the burden of chronic back pain or disability in society "as a whole" (the lifetime prevalence rate of back pain in developed countries is around 70%). However, there is strong evidence of associations between workplace risk factors and back pain, and these risk factors can be modified to reduce pain in people exposed to them (Table 2.3). Note that Table 2.3 shows the strength of *associations* between these risk factors and low back pain. There is still no consensus that the effect of exposure is *causal* (Swain et al., 2020).

In a recent systematic literature review of 18 studies involving 19,572 employees, Buruck et al. (2019) concluded that high workload, low job control, and a lack of social support from colleagues and superiors drove the development of chronic low back pain (lasting more than 3 months). HFE can help by designing jobs with optimal work loading that are intrinsically satisfying and rewarding to do in well-designed workplaces, using the guidance provided in subsequent chapters. Back pain sufferers who are motivated to return to work will find ways of coping with their pain and are more likely to get better.

TABLE 2.3
Strength of the Evidence for Major Mechanical Risk Factors for Low Back Disorders

	Source			
	NIOSH	FOM	Adams	McGill
Heavy physical work	xx	xxx	xxx	Xx
Lifting and forceful movements	xxx	xxx	xxx	Xx
Bending and twisting	xx	xxx	xxx	Xx
Whole-body vibration	xxx	xxx	xxx	Xx
Static work postures	I	0	0	Xx
Length of exposure to physical stressors	0	x	x	0
Sedentary work	0	x	0	Xx

Source: From U.S. National Institute of Occupational Safety and Health (NIOSH) (1981), Carter and Birrell (2000), Adams et al. (2000), McGill (2002).

Note: xxx, strong evidence of an association; xx, moderate evidence; and x, weak evidence. I, insufficient evidence; 0, no conclusion reached. Strength of evidence based on author' scrutiny of source documents.

Back Pain at Work and the Role of HFE

The evidence suggests that back pain in the workplace can be prevented by task design. Specific risk factors (Carter and Birrell, 2000) for priority redesign are:

- Repetitive heavy lifting
- Lifting, pulling, and pushing of objects over 11 kg
- Lifting objects from the ground in a twisted position
- Jobs with high physical demands
- Exposure to whole-body vibration
- Prolonged sitting at work (>95% of the workday)
- Nonneutral trunk postures held for >10% of the work cycle in repetitive jobs

Fatigue Failure and Injury

Any factors that reduce the strength of body parts will increase the risk of injury. McGill has argued that fatigue is such a factor and that attempts to specify a maximum "safe" load are limited. Rather, the injury threshold varies throughout the day or over the work shift, depending on the level of fatigue. Figure 2.12 illustrates this concept diagrammatically.

Gallagher and Schall (2020) have proposed that many musculoskeletal injuries occur as a result of fatigue failure. Using concepts from materials engineering, they argue that tissues have an ultimate strength (both tensile and compressive) and that repetitive loading of the tissues will lead to injury if the capacity of the tissues is exceeded. Whereas many risk assessment methods in HFE specify maximum safe loads or maximum repetition rates, the fatigue failure concept indicates that there is a relationship between loading and repetition. This relationship is exponential—in other words, for light loads, the number of cycles (repetitions) to failure is infinite (the activity is not damaging), whereas for very high loads, the tissue will fail after a small number of cycles. There is evidence that tissues such as tendons do fail in this way. The relationship between cycles to failure and applied load depends on the ultimate strength of the tissue – stronger tissues can withstand more loading cycles than weaker tissues.

Tissue loading in the workplace depends not only on the task but also on the posture. Stressful postures will also place a load on the tissues, which, combined with the task load, will reduce the

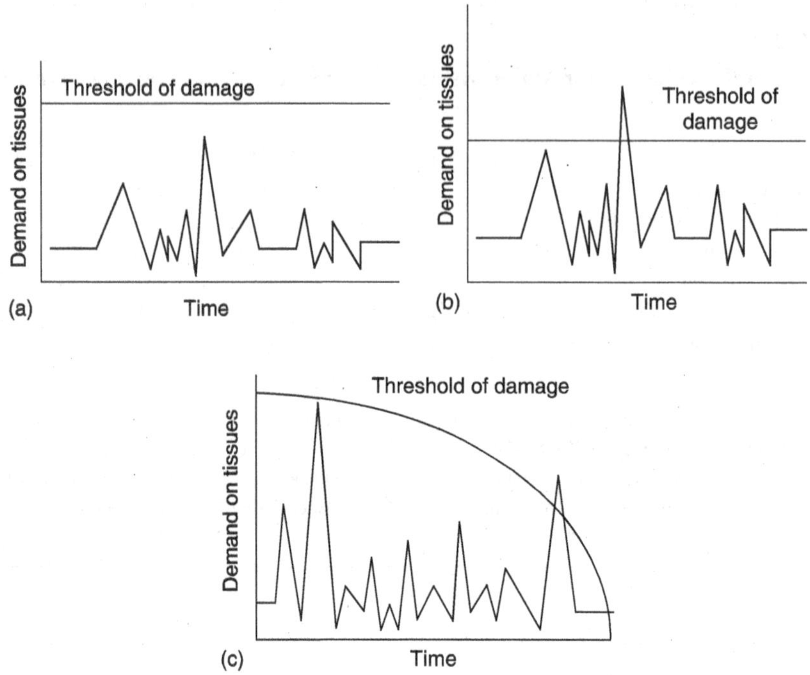

FIGURE 2.12 The specification of force limits for safety (a and b) is complicated by factors such as fatigue that can lower the threshold for injury of the tissues (c). (From Professor S. McGill, University of Waterloo, Canada, with permission.)

number of cycles to failure (effectively reducing the safe workload). Tissues also have the ability to recover and repair damage caused by loading. This is discussed in detail in Chapter 7. Gallagher and Schall point out that many of the injuries seen in the workplace are to tendons, cartilage, and ligaments, all of which have a poor blood supply and heal slowly. Short-term rest periods are unlikely to provide sufficient time for healing to take place. However, short cycles lower the mean loading over time, preventing damage from occurring in the first place.

HFE and the Musculoskeletal System in General

Each joint of the body has freedom for angular motion in one or more directions. A complex linkage such as that between the shoulder, arm, and hand has many degrees of freedom of movement, and power transmission is impossible without accessory stabilization of the joints by muscle action. For example, supination of the wrist may be required to turn a door handle, but this is only possible if the elbow and shoulder joints are stabilized and can counter the reaction at the hand–handle interface. Behaviors such as folding the arms and crossing the legs are postural strategies that turn open chains of body links into approximate closed chains stabilized by friction. Closed chains have fewer degrees of freedom for movement and move in more predictable ways when subjected to destabilizing forces. The net effect of such postural control strategies is to reduce the postural load on the muscles involved. This may be perceived as being "comfortable." Snijders et al. (1995) found that EMG activity of the internal and external oblique abdominals was higher in standing than in sitting and was further reduced when the legs were crossed. Leg crossing stabilizes the sacroiliac joint by compressing it and therefore relieves muscles such as the oblique abdominals. EMG activity in the rectus abdominis (which runs parallel to the joint and is not involved in its stabilization) was not influenced by leg crossing. Leg crossing also increases blood flow in the legs, increases mean arterial blood pressure, and lowers heart rate (Groothuis et al., 2007).

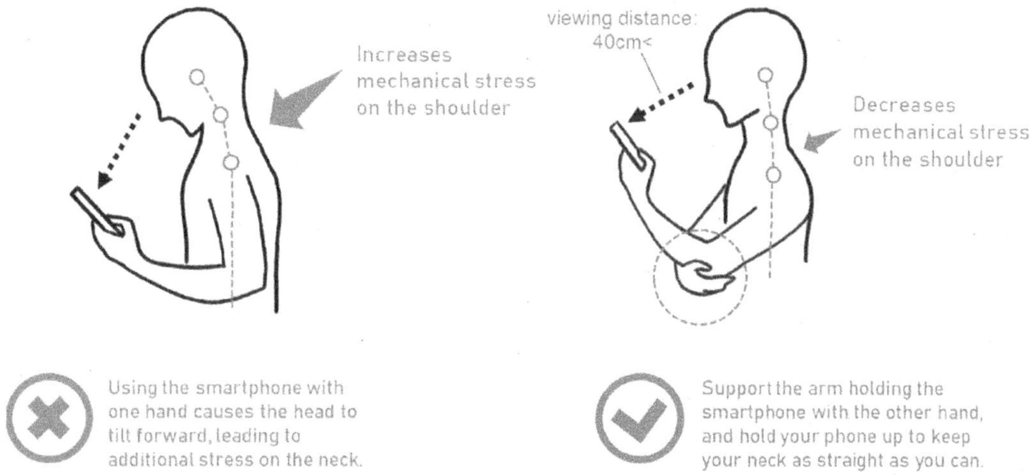

FIGURE 2.13 Closing the postural chain when standing to use a cell phone. (From "*Seven Practical Human Factors and Ergonomics Tips for Teleworking/Home learning using Tablet/Smartphone Devices*" Japan Ergonomics Society (Creative Commons Attribution Licence CC-BY, 2020).)

Figure 2.13 shows the "closed chain" principle applied to the use of cell phones. When holding the phone in one hand, the elbow tends to flex and the user flexes the neck to look at the phone. By cradling the elbow with the other hand, maintenance of the posture is less demanding, and the postural load on the neck and shoulders is reduced.

Individual muscles never work in isolation to produce a movement. Rather, synergistic recruitment of muscles takes place in which contraction of the prime mover is accompanied by contraction of surrounding muscles to position and stabilize the joints. In few activities do the trunk and limb muscles exert direct forces on the environment—rather, they maintain joint postures such that shifts in body weight can be transmitted via the chain of body links to exert a force. The use of body weight when kneading dough is a simple example of the importance of joint stabilization, which enables shifts in body weight to be used to carry out a task.

The same principle also underlies the performance of almost all highly skilled activities in which very large forces are exerted or where body parts or external objects undergo rapid acceleration through pivoting actions. Throwing a discus and swinging a golf club or a tennis racket are excellent examples of this principle. One of the key features of this behavior is the efficient stabilization of joint postures and the use of a backswing involving eccentric contraction of muscles and stretching of tendons to store energy. This backswing can be likened to an archer pulling back the bowstring before releasing the arrow. In the case of club and racket sports and hand tools, it is the hands, rather than the arrow, that are released when sufficient tension has built up. When walking, the stretching and release of the Achilles tendon throughout the gait cycle stores some of the energy when the foot is planted on the ground, which is released during the push-off phase. The deformation of the arch of the foot at these times serves a similar function. The combined effect is to increase the energy efficiency of walking. Excessive tension and compression, or a regime of loading cycles that exceeds the body's repair capability, can occur in industrial jobs as well as in sports if tasks or tools are badly designed. The outcome is pain and diminished capability.

TOOLS AND PROCESSES

There are many tools for the assessment of task demands in the workplace. Kotowski and Gibson (2023)[1] provide an ergonomic assessment toolkit consisting of 27 tools for assessing task stress (risk to the back, shoulders, and hands) as well as overall postural stress. A flowchart for tool selection is also provided. The following sections focus mainly on biomechanical analyses.

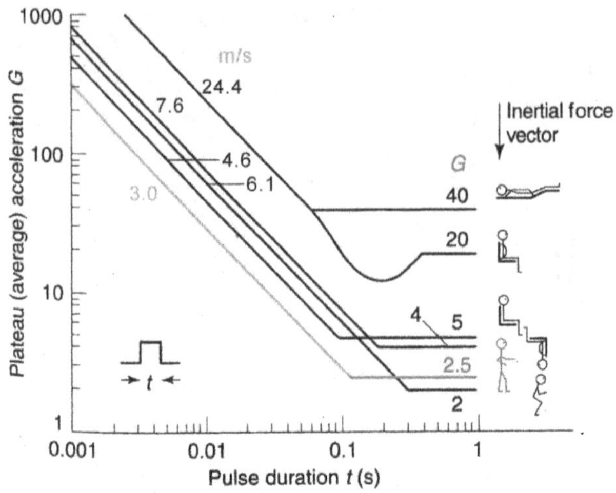

FIGURE 2.14 The tolerance to human whole-body impact for critical velocity change, critical acceleration level, and critical duration. (From Glaister, 1978, with permission.)

Tolerance for Forces of Rapid Onset

Figure 2.14 (from Glaister, 1978) shows human tolerance to whole-body acceleration in different body positions and restraints for pulse durations ranging from one-thousandth of a second to 1 s. In a standing position with no external support, the body can tolerate 2.5 G for pulse durations between 0.1 and 1 s. As can be seen, tolerance is greater when lying down or sitting than when standing.

Falls into Water

LaFave et al. (1995) reported injuries to 297 people attempting to commit suicide by jumping from the Golden Gate Bridge in San Francisco, USA. The bridge is 67 m high on average, and jumpers enter the water at velocities of around 120 kph (~75 mph). Most jumpers died upon entering the water; only 16 (5.1%) survived. All survivors entered the water feet first and were rescued from the generally cold water before they drowned. LaFave et al. quote a maximum survival velocity of 102 kph (63.4 mph) for water entry in the "belly-flop" position and 127 kph (79.1 mph) for feet-first water entry. This explains why so few survive the jump—the velocity at entry is close to the maximum survival velocity, even in the ideal position.

It is not the speed of entry into the water that is the critical risk factor for injury, but the rate of deceleration on entry. When entering the water side-on or in a prone position, the pressure acting on the body is less than when entering water feet or head first. However, the mass of water displaced on entry is much larger, and jumpers decelerate more quickly. Typical injuries include crush fractures of the chest. For those entering the water feet first, the rate of deceleration is less; they come to a standstill over a greater distance and are therefore at greater risk of drowning thereafter. In many of the cases investigated by LeFave et al., the jumpers were severely injured upon entering the water, but the cause of death was drowning. Snyder (1965) reviewed human tolerance limits in water impact. G forces acting on the body are 5–7 times greater in the supine, prone, or lateral positions than in the feet or head-first positions

Velocity on Entry	Deceleration (Feet First vs. Prone or Lateral)
21.9 kph	3.5 versus 21.9 G
32.9 kph	6 versus 40 G
54.9 kph	16 versus 112 G
87.8 kph	43 versus 300 G

FIGURE 2.15 Darren Taylor, also known as Professor Splash, dives from 8 m high into a 30-cm shallow pool of tomato sauce during a promotion at Darling Harbor in Sydney, Australia. Note the "belly-flop" technique. (Courtesy of Darren Taylor.)

Figure 2.15 shows the world champion shallow diver, "Professor Splash" (a.k.a. Darren Taylor), performing a dive from a height of 8 m into a child's plastic swimming pool filled to a depth of 30 cm with tomato sauce. In light of the discussion above, how is he able to survive?

As a first approximation, we can perform some simple calculations to generate data for use with Figure 2.14. Ignoring air resistance:

- Let
- V = velocity on entering the water
- U = initial velocity = 0 m/s
- A = acceleration due to gravity = 9.81 m/s^2
- S = vertical distance = 8 m

$$\begin{aligned} V^2 &= U^2 + 2AS \\ V &= \sqrt{2AS} = 12.5 \text{ m/s} \end{aligned} \tag{2.1}$$

or approximately 45 kph.

Assuming Professor Splash decelerates completely before reaching the bottom of the pool, his deceleration is given by

$$\begin{aligned} -2AS &= U^2 \\ -A &= U^2/2S \\ &= 260.4 \text{ m/s}^2 \end{aligned} \tag{2.2}$$

or approximately 26 G.

And the duration of the deceleration phase is given by

$$\begin{aligned} V &= U + AT \\ T &= V/A \\ &= 12.5/260.4 \\ &= 0.048 \text{ s} \end{aligned} \tag{2.3}$$

or 48 ms.

Figure 2.14d does not provide limits for the prone position, only for its inverse—the supine position—as well as for sitting and standing. However, it does give a broad idea of the kind of pulse durations that can be tolerated and suggests that the forces to which Professor Splash is exposed are within the tolerable range (in the supine position, the critical velocity change is 24 m/s for a pulse duration of 60 ms and 40 G). This conclusion is based on the questionable assumption that deceleration after contact with the tomato sauce is constant to a depth of 30 cm. It may well be that the instantaneous deceleration on contact is much higher due to factors such as the viscosity of the sauce. However, Figure 2.14 shows that much higher values can be tolerated for very short periods.

Success at shallow diving depends on posture at the moment of contact with the liquid. The "belly-flop" posture in Figure 2.15 no doubt maximizes the mass of fluid displaced on contact and leads to maximum deceleration. Failure to maintain this posture would result in impact with the bottom of the pool and could cause very serious injury or even death.

Tolerance for Collisions and Shocks

Figure 2.16 shows road accident data from Richards (2010). An analysis was made of pedestrian casualties in the United Kingdom for 1976 and included 2,333 fatalities, 18,168 serious injuries, and 46,217 slight injuries. These data were weighted to match national statistics, and the result is shown in Figure 2.16. This figure shows that the estimated risk of a pedestrian being killed is approximately 9% at a speed of 30 mph. The risk at an impact speed of 40 mph is much higher, at approximately 50%.

Shock

Figure 2.14 has varied applications. Operators are sometimes exposed to high G-forces when operating vehicles that move over rough terrain or when operating small, fast craft on water. Operators are most at risk when the rate of change of acceleration or deceleration (otherwise known as "shock") is large. The spine is most at risk under G-forces in the vertical axis, and vertebral fracture is a common injury. Figure 2.17 shows an anterior wedge fracture of the 12th thoracic vertebra caused by rapid deceleration when a small, fast craft re-entered the water after cresting a wave. The thorax flexes forward over the lumbar spine, which is held in place by the pelvis—crushing the anterior part of the vertebra like a "nutcracker." The injury is also common in those with osteoporosis, particularly older people, when they fall backward on slippery floors.

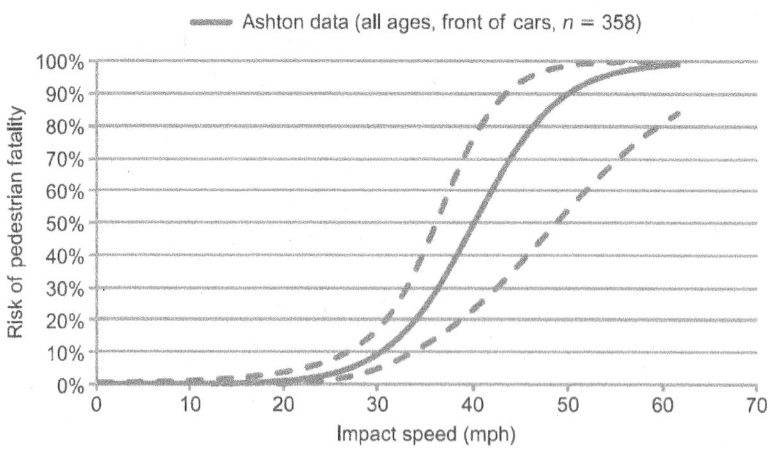

FIGURE 2.16 Risk of pedestrian fatality at different automobile impact speeds.

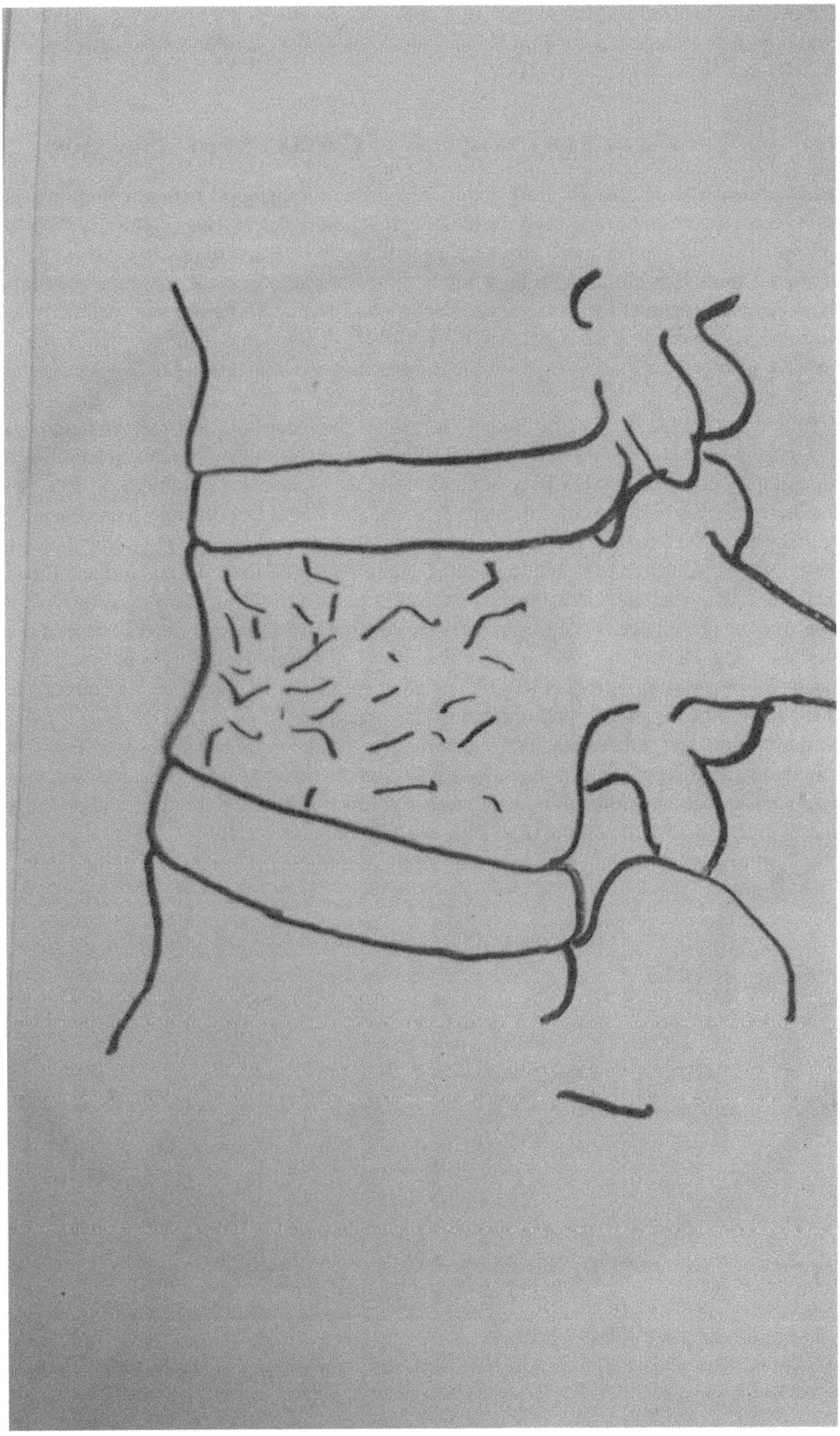

FIGURE 2.17 Wedge fracture at T12–S1.

Such injuries are usually suffered by seated operators. Suspension seats are often fitted to small, fast craft to minimize exposure to vibration and shock. Standing operators often suffer knee injuries because the legs are used to absorb shock.

Occupational Exposure to High Forces: How to Calculate Spinal Compression

The occupational risk factors for back pain, as shown in Table 2.3, have a common feature—exposure increases the compression on the lumbar spinal motion segments (Figure 2.4). The mechanism by which spinal compression causes injury is likely to be fracture of the vertebral end plates, which are the weak link in the chain (and not, as is sometimes supposed, the intervertebral discs). The U.S. National Institute for Occupational Safety and Health (NIOSH) has specified an action limit for spinal compression (NIOSH, 1981) of 3,400 N. This means that work tasks imposing a compressive load greater than 3,400 N on the lumbar motion segments are deemed hazardous and in need of redesign.

Surveys of certified professional ergonomists in the United States by the Liberty Mutual Research Center for Safety and Health (2004) revealed that the most commonly used manual handling assessment techniques are the NIOSH lifting equation (83.1%), biomechanical models (74.3%), psychophysical data (73.1%), body discomfort maps (55.6%), and Rapid Upper Limb Assessment (51.6%). Figure 2.18 shows the output of the Static Strength Prediction Program (SSPP) in the assessment of a potentially hazardous manual handling task. Using data on the worker's posture and the mass of the load lifted, the program calculates the joint forces needed to perform the task, the spinal compression, and the percentage of the U.S. population capable of performing the task. Use of the SSPP obviates the need for tedious mechanical calculations that would otherwise be necessary. HFE Workshop 2.1 provides an introduction to biomechanical modeling for spinal compression estimation that can be used as a first approximation in risk assessment.

The main limitation of these methods is that they are designed for use in the assessment of static tasks. In dynamic situations, the accelerations acting on the body must be measured to obtain accurate estimates of loading, although in most industrial tasks, the accelerations of body parts are low and do not greatly distort the output of the models (Chaffin and Andersson, 1984).

Table 2.4 provides information on the distribution of mass in different parts of the body.

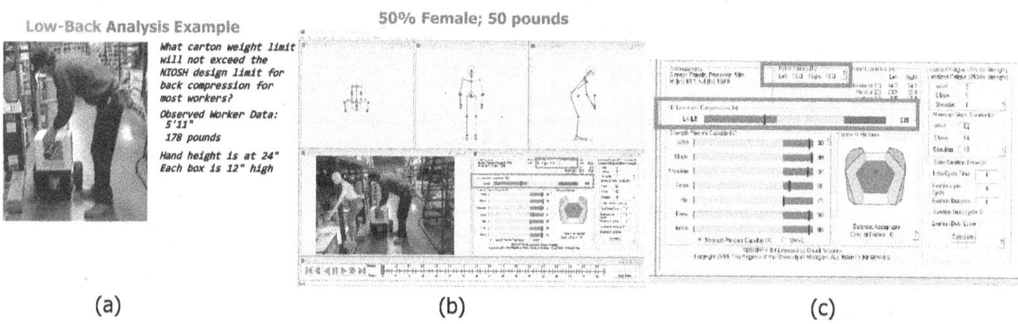

FIGURE 2.18 Sample output of SSPP. Figure developed and provided by the University of Michigan, Center for Ergonomics. (i) Wine picker lifting a carton of wine. (ii) Biomechanical model generated by the system for a 50th percentile female lifting the box. (iii) Detailed output of the system. Findings are based on a female of 1.62 m stature and 70k kg mass, with hand positions specified by the task. Green indicates the percentage of workers for whom the low back compression is acceptable, and red indicates the percentage at risk. The foot diagram shows that, based on the center of pressure at the feet, balance is acceptable.

TABLE 2.4
Percentage Distribution of Body Mass

Body Part	Percentage of Total Body Mass	Percentage of Segment	Average U.S. Male (kg)	Average U.S. Female (kg)
Head and neck	8.4		6.9	5.8
		Head (73.8)	5.1	4.3
		Neck (26.2)	1.8	1.5
Torso	50		41.1	34.7
		Thorax (43.8)	18.0	15.2
		Lumbar (29.4)	12.1	10.2
		Pelvis (26.8)	11.0	9.3
Single arm	5.1		4.2	3.5
		Upper (54.9)	2.3	1.9
		Lower (33.3)	1.4	1.2
		Hand (11.8)	0.5	0.4
Single leg	15.7		12.9	10.9
		Thigh (63.7)	8.2	6.9
		Shank (27.4)	3.5	3.0
		Foot (8.9)	1.2	1.0

Source: Adapted from Chaffin and Anderson (1984), with permission.

ERGONOMICS WORKSHOP 2.1

INTRODUCTION TO BIOMECHANICAL MODELING OF THE SPINE

Suppose we want to estimate the compressive force acting on the intervertebral disc at the L5/S1 level. The simplest model treats L5/S1 as the fulcrum in a first-class lever system. Any loads in front of the fulcrum exert a moment about L5/S1. If this moment is not resisted, the person will either fall forward or the trunk will flex. Resistance is provided by the back extensor muscles, which exert a counter moment to maintain a static posture.

SPINAL COMPRESSION WHEN STANDING ERECT

Let us suppose that a person is standing erect and carrying no load. His upper body, above L5/S1—including the head, trunk, and arms—has a mass of 45 kg. What is the spinal compression at L5/S1?

The compression is found by calculating the force (F) acting on L5/S1, where

$F = m \times a$
m = body mass above L5/S1 = 45 kg
a = acceleration due to gravity = 9.81 m/s^2

Therefore, $F = 45 \times 9.81 = 441.5$ N (N stands for "newton," the SI unit of force)

(Continued)

ERGONOMICS WORKSHOP 2.1 (*Continued*)

SPINAL COMPRESSION WHEN STANDING WITH A 20 KG LOAD ON THE HEAD

Suppose the person places a 20 kg bag of feathers on his head. What is the spinal compression at L5/S1?

The spinal compression at L5/S1 is the sum of the compression due to the upper body weight and the compression due to the load on the head, which is transmitted all the way down the axial skeleton to the floor. Let

m_1 = mass of the upper body
m_2 = mass of the load

$$F = (m_1 + m_2)a$$

$$F = (45 + 20) \times 9.81 = 637.7\text{ N}$$

SPINAL COMPRESSION WHEN STANDING ERECT AND HOLDING A 20 KG LOAD IN FRONT OF THE BODY

Suppose the person removes the bag of feathers from his head and holds it in front of his abdomen. There are three loads that have to be considered: The 20 kg bag in front of the body. The arms, which are now in front of the body rather than hanging by the sides. The head and torso, directly above L5/S1.

In order to calculate the spinal compression, we need to know the value of the moment of flexion acting on the lumbar spine. This moment (L) is the effect of the load acting at a distance from L5/S1 that must be resisted by the back extensor muscles in order for the person to remain standing erect:

$$L = F \times d$$

F = the force in Newtons due to gravity acting on the mass
d = the distance of the COG of the mass from L5/S1
Units are in Newton-meters (N m).

Let us suppose that the COG of the bag of feathers is 40 cm (i.e., 0.4 m) in front of L5/S1. What is the flexion moment about L5/S1 due to the 20-kg bag of feathers?

$$L_{\text{bag}} = F \times d = (20 \times 9.81) \times 0.4 = 78.5\text{ Nm}$$

Similarly, we can estimate the distance of the COG of the arms in front of the body, assuming a symmetrical posture and that the combined mass of the arms is about 12 kg. If the COG of the arms is 20 cm (i.e., 0.2 m) in front of the body, the load moment due to the arms is then

$$L_{\text{arms}} = F \times d = (12 \times 9.81) \times 0.2 = 23.5\text{ Nm}$$

The total flexion moment about L5/S1 is, therefore,

$$L_{\text{total}} = L_{\text{bag}} + L_{\text{arms}} = 78.5 + 23.5 = 102\text{ Nm}$$

In order for the person to maintain a static, erect posture while holding the bag, the back muscles have to exert a counter moment (or moment of extension) of the same size. They act

(*Continued*)

ERGONOMICS WORKSHOP 2.1 (*Continued*)

at an average distance of 7 cm from L5/S1. Therefore, we can calculate the back muscle force needed to exert this counter moment as follows:

$$L_{\text{back muscle}} = F_{\text{back muscles}} \times d$$

Now,

$$L_{\text{back muscle}} = 102 \text{ Nm and}$$
$$d = 0.07 \text{ m}$$

Thus,

$$F_{\text{back muscles}} = 102/0.07 = 1457.1 \text{ N}$$

The total spinal compression, then, is the sum of the compression due to the bag, the upper body, and the compression due to the back muscle force needed to maintain the erect posture:

$$\begin{aligned} C_{\text{bag}} &= 20 \times 9.81 = 196.2 \text{ N} \\ C_{\text{body mass}} &= 45 \times 9.81 = 441.5 \text{ N} \\ C_{\text{total}} &= C_{\text{bag}} + C_{\text{body mass}} + C_{\text{back muscles}} \\ &= 196.2 + 441.5 + 1457.1 \\ &= 2094.8 \text{ N} \end{aligned}$$

INTERPRETATION

By far, the biggest factor contributing to the compression at L5/S1 is the back muscle force needed to maintain posture. When carrying the load on the head, the spinal compression was 637.7 N, whereas carrying it in front of the body results in more than a threefold increase in compression. This example illustrates the importance of carrying loads as close to the body as possible to minimize "load moments."

FORWARD-FLEXED POSTURES

The simple model illustrated above can also be applied to situations where the worker has to lean forward to hold or lift a load. There are two main considerations that must be accounted for when calculating spinal compression in forward-flexed postures. The first is that the mass of the upper body itself is displaced in front of L5/S1, and the second is that L5/S1 experiences both shear and compression forces.

In the example below, we will calculate the compression and shear forces acting at L5/S1 when a person leans forward in a static posture. Let us assume that his upper body, above L5/S1, weighs 45 kg.

First, we have to locate the COG of the upper body, which, viewed sagittally, is approximately at the axilla (the crease at the bottom of the armpit). Next, we have to measure the approximate horizontal distance of the upper body COG from L5/S1. We need the following data:

- Distance (D) between the upper body COG and L5/S1 (measured using a tape measure)
- Angle of forward flexion of the trunk (φ, measured by observation or from photographs)

(*Continued*)

ERGONOMICS WORKSHOP 2.1 (*Continued*)

Let us suppose we get the following results:

$$D = 30 \text{ cm}$$

$$\varphi = 35°$$

The horizontal distance (d) of the upper body COG from L5/S1 is given by

$$\begin{aligned} d &= D \times \cos 55 \\ &= 17.2 \text{ cm} \end{aligned}$$

The load moment is, therefore

$$L_{\text{upper body}} = (45 \times 9.81) \times 0.172 = 75.9 \text{ N m}$$

The back extensor force needed to maintain posture is, therefore

$$F_{\text{back muscles}} = 75.9/0.07 = 1084 \text{ N}$$

The total spinal compression is given by

$$\begin{aligned} C_{\text{total}} &= \left(C_{\text{body mass}} \times \cos 35\right) + C_{\text{back muscles}} \\ &= 361.6 + 1084 = 1445.6\text{N} \end{aligned}$$

The shear is given by

$$C_{\text{shear}} = (45 \times 9.81)\sin 35 = 253.2\text{N}$$

The back muscles run at a very shallow angle relative to the axis of the spine; therefore, they contribute little to the shear component of the loading. As can be seen, large forces are generated when the trunk is in flexed postures.

Postural load can be quantified using biomechanical models and is often found to exceed the task load.

SUMMARY

This simplistic biomechanical model is sometimes called a cantilever model. The lumbar spine is regarded as the fulcrum. In front of the lumbar spine is the load. The total spinal compression is the sum of the component compressive forces: the compression due to body weight, the compression due to any external load, and the compression due to the back muscle force needed to maintain a static posture. In an erect posture, the upper body weight can be regarded as being directly above the spine, imposing a load but no moment. The load moment is the product of the load and its moment arm (the horizontal distance from the lumbar spine). The counter moment produced by the back extensors is the product of the back muscle force and the lever arm (the distance of the back extensors from the lumbar spine). In forward-flexed postures, the COG of the trunk moves in front of the lumbar spine and an additional load moment is created (the product of the weight of the upper body and its horizontal distance from the lumbar spine). The back muscles have to counter two load moments: one due to postural load and one due to task load. At the level of the lumbar spine,

(*Continued*)

ERGONOMICS WORKSHOP 2.1 (*Continued*)

the forward-flexed posture introduces two components of force: one of compression and one of shear.

In order to use biomechanical models in practical situations, we need to measure or estimate the following:

- The magnitude of any loads
- The COG of the load and its distance from L5/S1
- The body mass of the operator
- The distance of the body part COG from L5/S1
- The angle of asymmetry of the trunk (if any)

This can be done using simple equipment, such as spring balances to measure loads and tape measures to measure distances. Body angles can be measured from photographs (e.g., see Figure 4.5).

Spinal Compression Tolerance Limits

Much is now known about the strength of spinal motion segments and their component discs and intervertebral bodies. For a comprehensive review, see Adams and Dolan (1995). These studies typically involve the removal of motion segments from cadavers and testing them for failure in a compression testing machine. Variations on the method include placing the motion segment into a flexed posture before compressing it or combining compression with axial rotation. The load at which failure occurs is recorded, and the specimen is removed from the machine and examined to determine where and how failure has occurred (the vertebral body itself can fail, the bony end plates can fail, or the disc can also fail).

The data obtained from mechanical testing of spinal motion segments enable tolerable loads to be identified, which will not cause failure of lumbar motion segments. The spinal compression tolerance limit (SCTL) is the maximum compressive load that a specified motion segment can be exposed to without failure. In practical situations, manual handling tasks can be evaluated using biomechanical models to estimate the compression load. If the estimated load exceeds the SCTL, then the tasks must be redesigned, either by reducing the load or the load moment.

According to Genaidy et al. (1993), SCTL is greatest in 20–29-year-olds, declining by 22% in the next 10 years, 26% in the next 10, and 42% in the following 10. At 60 years of age or more, the SCTL has declined by 53%. Female SCTLs are approximately 67% of male values. SCTLs are lower when spinal motion segments are loaded in complex ways, such as when compression and bending are combined. A hyper-flexed lumbar spine has a lower SCTL than a flexed lumbar spine. In a flexed position, however, the compressive load on the thicker part of the disc (the anterior annulus) is increased, and the load on the facet joints is reduced. Physical activity seems to strengthen both the vertebral bodies and the intervertebral discs. Women, older workers, and those unaccustomed to lifting should not be expected to carry out forceful exertions at work.

Ayoub and Mital (1997) quote SCTLs of 6,700 N for people under 40 years of age and 3,400 N for people over 60. The regression equation below (Genaidy et al., 1993) can be used to calculate SCTLs for a given lumbar motion segment:

$$CS = -13331.2 - (73.7 \times Age) - (962.6 \times Sex) + (403 \times LMS) + (79.8 \times BW) \tag{2.4}$$

where

CS = compressive strength (N)

Age = age in years

Sex—use 1 for male and 2 for female
LMS = lumbar motion segment (L1 – L2 = 44, L2 – L3 = 45, L3 – L4 = 46, L4 – L5 = 47, L5 – S1 = 48)
BW = body mass (kg)

For risk assessment purposes, a margin of safety is needed. For one-off tasks, the task load should be less than 60% of the SCTL, and for repetitive loads, less than 30% (Sandover, 1988). For example, let us suppose that a 43-year-old female worker, weighing 58 kg, has to lift 25-kg bags of wet laundry from the floor and load them into a trolley. We use the model in Workshop 2.1 to calculate the compression at L5/S1 at the start of the lift and find it to be 4,350 N. Is this safe?

Substituting these data into the equation above, we find

$$CS = -13{,}331.2 - (73.7 \times 43) - (962.6 \times 2) + (403 \times 48) + (79.8 \times 58) = 5546.9\text{N} \qquad (2.5)$$

The estimated compression tolerance limit for this worker's L5/S1 spinal motion segment is 5,546.9 N. Taking a margin of safety of 60%, we obtain a damage load of 60% of 5,546.9, or 3,328.1 N. The applied load is 4,350 N, and the ratio of job demands to biomechanical tolerance is 1.3 (4,350/3,328.1); thus, the applied load is unsafe, and the task needs to be redesigned to bring the ratio below 1 (Genaidy, 1993). There are probably many ways of doing this, but since the load has to be lifted from the floor, a large component of the spinal compression is probably postural, caused by stooping. Thus, we might start by asking why the load has to be lifted from the floor in the first place!

Revised Dortmund Recommendations

Jäger (2018) collated the findings of 25 investigations of lumbar vertebra compressive strength (complete vertebrae and discs plus the endplates obtained from autopsies in specimens over 20 years of age). Large differences due to age and sex were found, and statistical techniques (regression analysis) were used to generate recommendations for the maximum compressive strength of *in vivo* lumbar vertebrae. The recommendations are designed for use in HFE risk assessments of manual work. Table 2.5 summarizes the findings of the analyses.

Measurement of Musculoskeletal Pain in the Workplace

Probably the simplest and most widely used tool available for the assessment of musculoskeletal pain is the pain-rating diagram (Figure 2.19). Employees rate any musculoskeletal pain on a scale of one to ten and write the score on the part of the body diagram corresponding to the body region where the pain is experienced.

The method is valid and reliable, sensitive enough for the purpose, moderately intrusive, acceptable, and inexpensive. It is not diagnostic and tells us nothing about the reasons for any pain experienced.

TABLE 2.5
Lumbar Compressive Strength: Revised Dortmund Recommendations

Age (Years)	Female (kN)	Male (kN)
20	4.1	5.4
30	3.8	5.0
40	3.1	4.0
50	2.4	3.1
60	1.8	2.2

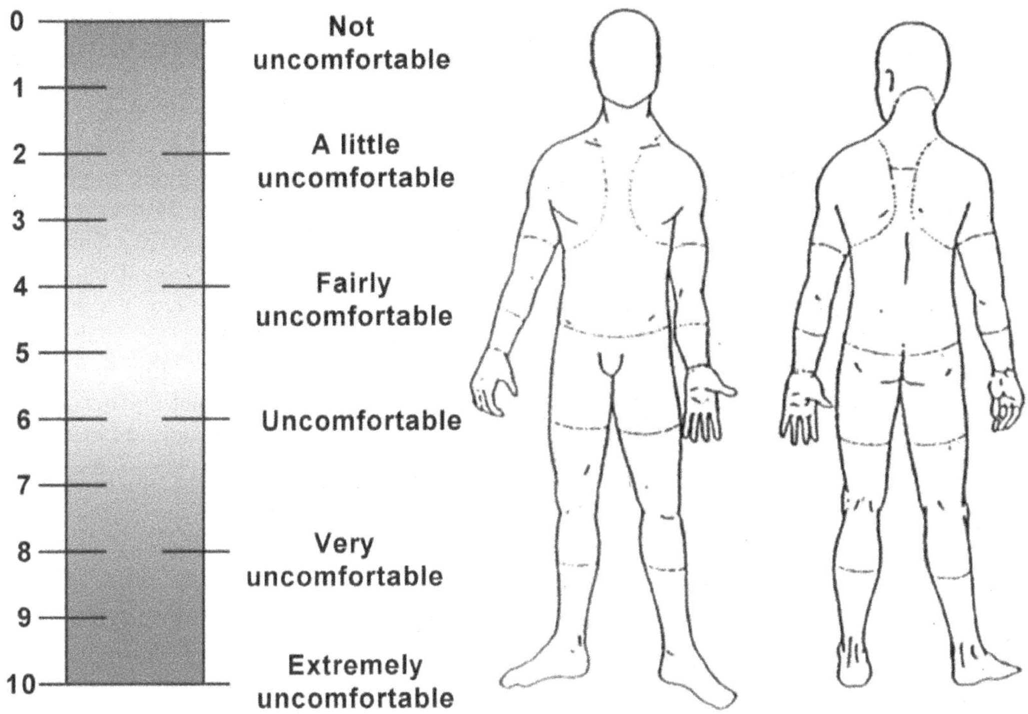

FIGURE 2.19 Body diagram for pain rating. These are often included in questionnaires for use in ergonomic surveys. Pain is often rated on a 10-point scale where 1 = mild discomfort and 10 = the pain could not be worse.

TABLE 2.6
Incident Rates of MSDs and Days Lost in Selected U.S. Industries (2018)

Industry	Number	Incidence Rate	Median
Health Care/Social Assistance	56,360	38.4	8
Retail	41,070	34.8	10
Manufacturing	38,640	30.6	14
Transportation/Warehousing	38,350	77.1	26
Construction	19,380	28.9	10
Wholesale Trade	18,190	32.3	11
Financial Activities	6,330	8.3	21
Information	5,210	20.1	33
Agriculture/Forestry/Fishing/Hunting	2,910	28.5	11

SYSTEM INTEGRATION

Analyze Legacy Data

Designing new systems provides an opportunity to eliminate existing hazards. Legacy data from existing systems can be used to pinpoint areas of risk. The U.S. Bureau of Labor Statistics publishes data on injuries across the United States. Table 2.6 presents an extract for illustrative purposes.

Clearly, the risk of musculoskeletal injury is greatest in industries requiring heavy work and exposure to large objects, such as machines and tools. Repetitive motion is the exception and occurs more in manufacturing and financial services. The data present clear evidence that heavy work is associated with injury. Within any given project, legacy data on occupational injuries should be gathered and risk factors identified using the analytical approach described in this chapter. Harmful exposures can be identified and designed out of new systems.

The tools described in this chapter are valid in the sense that they measure the risk factor of interest, biomechanical stress. The reliability of the methods depends on the reliability of the instruments used to make measurements of the parameters required in the equations. In general, observational methods and simple, manually operated instruments are reliable. In field settings, however, these methods are not very accurate, and so assessments may lack sensitivity. There is a trade-off between intrusiveness, cost-effectiveness, and acceptability on the one hand, and accuracy on the other, because the use of video and electronic devices such as goniometers intrudes into the work situation. Biomechanical models are eminently diagnostic in that they enable task load and postural load to be assessed independently. However, the models are blind to the personal and psychological/psychosocial risk factors that are particularly important in determining whether an injury becomes a chronic health complaint. In conclusion, the models are best used for initial screening and redesign of tasks likely to cause an acute injury due to excessive loading.

Proactive Approach to Prevention

Back injuries often occur at work when people are handling heavy objects manually. They can also occur outside of work as the result of many years of stress, both inside and outside of work. We can use an analogy with noise-induced hearing loss: people exposed to loud noises eventually go deaf. It makes no difference whether the exposure occurs at work, at home, or both. The way to delay or prevent people from going deaf is to lower the total exposure. Similarly, the way to delay or prevent the occurrence of these problems is to lower the total exposure to back stress. At work, this means identifying stressful tasks and redesigning them to reduce the stress.

High Costs of Injury

According to the Occupational Safety and Health Administration (OSHA), back injury claims cost an average of $40,000–$80,000, including medical bills and wage loss benefits, in 2018. The Centers for Disease Control (CDC) reports that about 20% of workers' compensation claims were for back pain, costing employers $1,680 per employee per year (with claims ranging from $40,000 to $80,000). Overall, the loss to U.S. businesses was estimated at $225.8 billion per year.

Ciriello et al. (1993) report that if a manual handling task is acceptable to less than 75% of workers, the probability of compensable back pain is three times greater than if the job is acceptable to more than 75% of the population. Thus, by redesigning jobs to reduce the weight handled, large reductions can be made in back pain compensation claims with immediate and continuing benefits to the company.

Role of Occupational Factors

How much of the high incidence of back pain is actually caused by work is open to debate. However, it is likely that much of the pain, if not directly caused by occupational factors, may well be exacerbated or amplified by work activities. Degeneration of the spine occurs with age in most people, but the process can be accelerated by work stress. Similarly, people with an existing medical condition may find it difficult to cope at work if excessive demands are placed on their musculoskeletal systems. Much of the disabling back pain at work is preventable, can be reduced or delayed, and significant savings can accrue through HFE intervention programs.

Anema et al. (2004) carried out a study on workers sick-listed for 3–4 months due to back pain. Six cohorts, amounting to 1,631 workers, were studied in Denmark, Germany, Israel, the Netherlands, Sweden, and the United States. Time to return to work for workers receiving ergonomic interventions in the workplace was compared with time to return for those not receiving such interventions. Three types of intervention were made: workplace adaptation, adaptation of job tasks, and adaptation of working hours. The outcome variables studied were the date of first return to work and working status 1 and 2 years after date of first injury.

Adaptation of the workplace resulted in earlier return to work in those workers who received it compared to those who did not (median 206 days to return to work compared to 311 days, respectively). Workplace adaptations included changes to chairs and desks, provision of special tools, or lifting aids. Adaptation of working hours resulted in significantly earlier return to work in those who received it than in those who did not (270 days compared to 291 days, respectively). Adaptations included changes in the number and/or pattern of working hours, different shifts, and more variation in hours of work. Adaptation of job tasks did not result in reductions in time taken to return to work. All interventions were applied in the first year after the start of sick leave. The interventions did not differ in their effectiveness in the different countries, suggesting that those interventions that are effective are generally applicable. No details of the costs and benefits of the interventions, or of their cost-effectiveness are reported by Anema et al. However, it is noteworthy that the study was carried out on patients falling into the chronic back pain category, an impressive demonstration of the effectiveness of the interventions. As the Faculty of Occupational Medicine in the United Kingdom has pointed out, this is a particularly difficult group to rehabilitate—the longer the patient is off work, the less likely he or she is to return to work, irrespective of the intervention.

Cost-Effectiveness of Preventing Occupational Back Pain

Lahiri et al. (2005) conducted a large-scale study of the cost-effectiveness of interventions to prevent back pain at work. These included training, engineering and administrative controls, and ergonomics programs. Benefits were assessed in terms of 'healthy life years' gained (the number of years that a person would continue to live in a healthy condition), and costs were expressed as costs (in U.S. dollars) per employee. From these data, cost-effectiveness ratios were calculated to enable the different interventions to be compared.

Training costs consisted of the direct costs of awareness training, hazardous job training, training the trainers, and management costs. Engineering costs included labor and equipment costs. The costs of ergonomics programs consisted of training, engineering costs, management costs, protective equipment, rule reviews, and job analysis.

Analyses were conducted for different regions of the world and for different industrial sectors, where the costs of labor and equipment vary widely, and the reader is referred to the paper for further details. The main conclusions were that training was the most cost-effective intervention (because it was the least expensive) but had the fewest benefits. This was followed by engineering and administrative controls (more expensive but more effective). HFE programs were the most effective. In countries where the cost of labor is inexpensive in relation to the cost of equipment, training is a first option. The authors quote the World Health Organization's view that any intervention that costs less than three times the gross domestic product of a country (per capita) should be considered good value for money.

All of the interventions were found to be good value for money. For example, in the United States and Canada, if \$1.009 billion were spent on training interventions for 10 years, 1.78 million healthy life-years would be gained. Spending \$3.590 billion on engineering controls would gain 4.96 million healthy life-years. Spending \$4.560 billion on training and engineering controls would yield 24.81 million healthy life-years, and \$5.403 billion spent on ergonomics programs would save 30.67 million healthy life-years.

EXOSKELETONS

Exoskeletons are devices attached externally to the body to augment the ability of healthy individuals (unlike braces, which are used to compensate for disability or restore function in injured individuals). They fall into several categories: whole-body, lower-body, and upper-body exoskeletons. Anthropomorphic exoskeletons resemble the body parts they are attached to and mimic their function. Pseudo-anthropomorphic exoskeletons support body parts externally and do not resemble the parts they support. Passive exoskeletons use springs or counterweights to aid performance, whereas active exoskeletons use powered actuators and are normally heavier because of this. There are different types of exoskeletons: holding exoskeletons are used in automobile assembly; trunk stabilizers maintain asymmetric postures of the trunk when working in stressful postures; powered gloves enhance grip force; powered body suits augment human strength; and other types provide "extra" limbs. Exoskeletons have the potential to reduce exposure to the stressors associated with musculoskeletal injuries in industrial jobs, thereby reducing injuries due to fatigue failure. Research on their effectiveness and safety standards for their design is currently underway.

Lower limb exoskeletons have been developed to aid recreational skiers and to assist with the adoption of static semi-squatting postures in industrial settings while carrying out discrete tasks

Another potential area of application is for users of high-speed craft (HSC) who are required to stand and cannot use suspension seats. The legs tend to be used in a manner akin to spring-damper units to absorb shock and vibration, the knee joints are flexed, and there is constant "bracing" of the axial skeleton in readiness for the next exposure to shock or vibration. The physical stress of traveling on HSC is sufficient to degrade physical and cognitive performance at the end of transits lasting several hours (McMorris et al., 2009). Reductions in performance of 15%–25% have been found at sea states 2–4 (smooth wavelets to moderate seas with waves up to 2.5 m). Installation of suspension seats on HSC can prevent these decrements because the shock and vibration are absorbed by the seat mechanism. In those instances where suspension seats cannot be used and the occupants have to stand, alternative methods of reducing exposure to vibration and shock are needed. Exoskeletons that exert an extension moment about the knee joint during flexion are one possibility.

Research on exoskeletons is described in later chapters. Their suitability can be assessed using a variety of performance indicators: Does use of the exoskeleton reduce biomechanical loading? Does it reduce fatigue? Does it reduce the physiological demands of the task? Does it increase human performance capacity? Does it improve productivity? Does it reduce the mental demands of the task? Does it interfere with other tasks? Is it easy to use? Is it acceptable to users; is it compatible with human functional anatomy? And is it comfortable to wear?

Villaroya López et al. (2023) have proposed a process for the evaluation and validation of exoskeletons in industry. Three basic requirements must be met before considering the use of exoskeletons in industrial jobs:

1. Exoskeletons should not replace existing preventive measures (such as eliminating hazards).
2. Exoskeletons should be utilized for specific tasks (match the device to the task to reduce the risk).
3. Use of exoskeletons must be voluntary.

The process for choosing an exoskeleton is as follows (Figure 2.20):

1. Define the need—first conduct a risk assessment of the task. Is the device necessary? Is there a requirement for it?
2. Is the exoskeleton the best preventive measure?
3. Take a participatory approach to determine whether the exoskeleton 'fits' the users and is acceptable to them.
4. Post-implementation—has the risk been reduced and the safety of employees improved? Has productivity improved, and is there a return on the investment?

Several "perverse" outcomes are possible following the introduction of exoskeletons: reduction of loads on the spine may increase loading on the legs, for example; while discomfort due to fatigue may be reduced, discomfort due to chafing, sweating, etc., may increase; while use of the exoskeleton may make difficult tasks easier, it may make easy tasks more difficult. Employees may compensate for the increased safety by taking more risks—this is known as "risk compensation" (see Chapter 15).

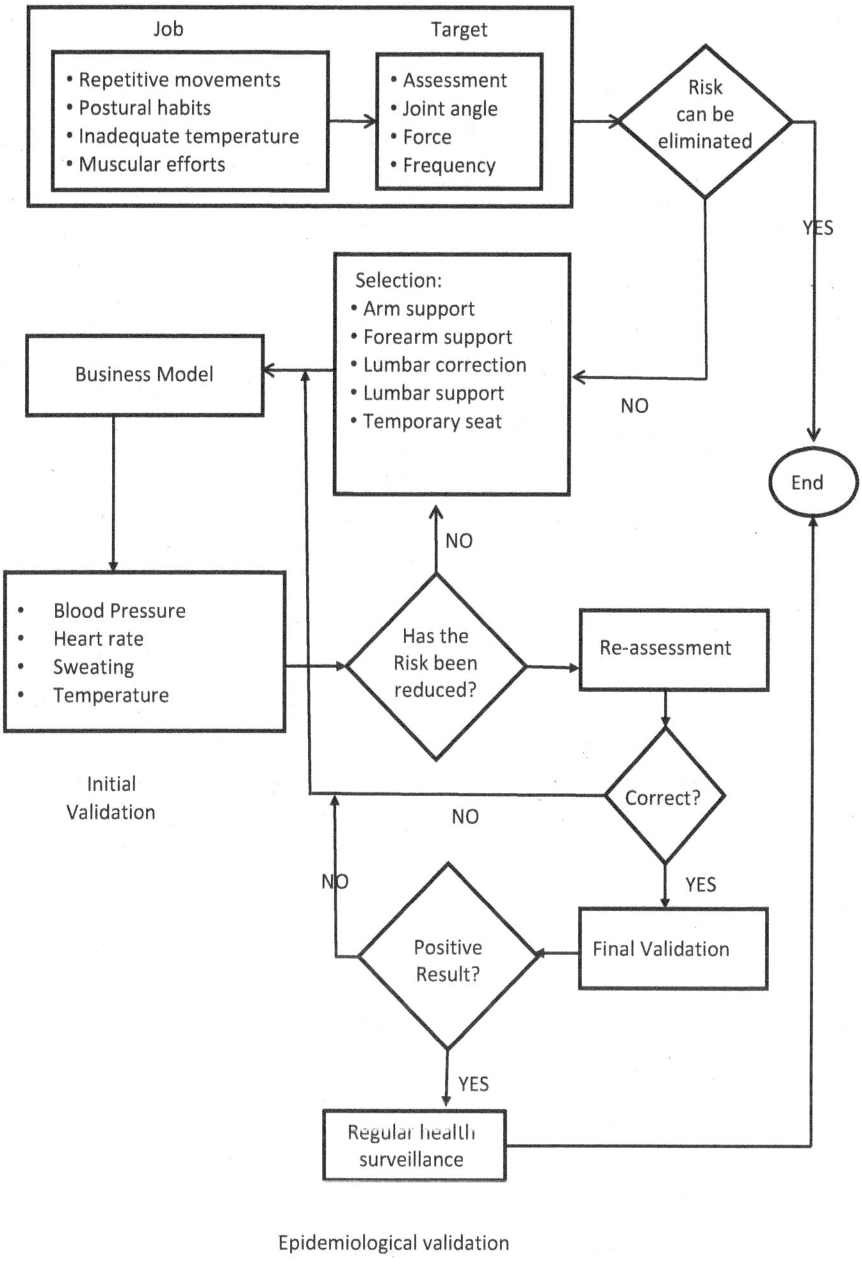

FIGURE 2.20 Exoskeleton implementation process. (Redrawn from Villaroya-Lopez et al., 2023.)

RESEARCH DIRECTIONS

Basic modeling of spine biomechanics at work is now well developed, and many tools are available to assess risk and reduce injury. However, disability due to back injury remains a major cost to the healthcare systems of many countries. The transition from acute injury to chronic disability, when it happens, is not as well understood as the initial injury mechanisms. Further work is needed to understand how biomechanical factors interact with psychosocial and psychological factors that can mediate recovery. Changing demographics, the obesity epidemic and increasing sedentary lifestyles present new challenges for the interpretation of risk assessments.

SUMMARY

Evaluation of the physical workplace requires a basic knowledge of human anatomy and body mechanics. In order to carry out a physical task in a safe and comfortable manner, a number of physical requirements must be met. First, the body must be stable. This depends on the mechanical relationship between the task objects, the body parts, and the base of support provided by the feet, the seat, and any other surfaces in the workplace that can be used to support body weight. The design of a workspace can determine the range of stable postures that can be adopted and evaluated from this point of view. In vehicles, destabilizing external forces due to motion may also have to be accounted for. If a task requires a posture to be held for any length of time, posture analysis is necessary. A starting point for such an analysis is to determine the mechanisms by which the posture is maintained, whether static muscular effort is required, whether ligaments are being strained, and whether parts of the work surface, such as the backrest of a chair or a work surface, are providing support. Knowledge of the anatomy of the spine and pelvis is particularly valuable here, as is an understanding of the mechanisms of physical fatigue. Spinal compression is one of the most important risk factors for back injury.

New and disruptive technologies are emerging all the time. Artificial intelligence (AI) has been applied to the diagnosis of low back pain (Santra et al., 2021), mainly for use in clinical diagnosis and management. However, better and more readily available diagnoses would enable HFE interventions and workplace adjustments to be tailored more precisely to reintegrate and support injured employees.

The literature on occupational exoskeletons is growing rapidly, and recent research is described in the following chapters. To date, the findings appear promising, although the challenge for HFE is to specify the requirement for an occupational exoskeleton within a defined context of use to 'fit the exoskeleton to the person and the task.' Biomechanical models provide important insights into the mechanisms of spinal loading, even if, as is the case with the model presented above, they are nothing more than first approximations.

Key characteristics of a good static posture are symmetry, an erect trunk, minimal static muscle activity, and some kind of external support. Further guidance can be found in ISO/TC 159/SC 3 Anthropometry and Biomechanics.

TUTORIAL TOPICS

1. What is a "good" posture? How would you decide?
2. Are biomechanical models for estimating spinal compression too simplistic?
3. Would it be possible to "re-engineer" the human body when designing life-like robots?
4. At the link (https://www.youtube.com/watch?v=GAI7UAFW7Bw), you will see a video entitled "Builders' Deadlift Challenge." Why are the builders able to deadlift the weights that they do? Comment on:
    - The equipment used
    - The deadlift technique and the main muscle groups involved
    - Is there evidence of a "healthy worker effect"?
    - Any other comments?

ESSAYS AND EXERCISES

1. One of the best ways to develop an understanding of anatomy is to practice drawing anatomical structures. Copy the anatomical drawings in this book and label the various structures using the appropriate terms.
2. Observe people carrying out the following or similar activities:
    - Students sitting for an exam
    - A bus driver behind the wheel

- A person washing dishes
- A computer programmer at work
- An archer aiming at a target
- A gardener digging a hole
- A mother holding a young child
- A person operating a lathe
- Draw stick figures to indicate the position of the main body parts and the base of support at the feet. Indicate where you think the main areas of static and dynamic body load are by considering how the person maintains the posture. Try to visualize the load on underlying structures and indicate whether the task load is greater than the postural load.

3. NIOSH specifies a maximum spinal compression of:
   - 3400 kg
   - 34,000 N
   - 45 kg
   - 85 dB(A)
   - 3400 N
   - None of the above
4. The function of the intervertebral discs is to act as shock absorbers. True or false?
5. Use Equation 2.1 to estimate the SCTL at L3 for a 35-year-old female with a body mass of 61 kg.
6. A 57-year-old male worker weighing 58 kg lifts 25 kg bags of cement onto a conveyor belt. His spinal compression at L5/S1 is 5100 N. Calculate the SCTL and comment on the safety of the task.
7. Define the terms "mass" and "weight" and distinguish between them. Why do domestic weighing scales read in units of mass, such as kilograms?
8. A person is standing erect in an upright, balanced posture with his arms hanging freely by his sides. His upper body, above the L5/S1 intervertebral disc, has a mass of 45 kg. What is the compressive force acting on the disc?
   - 45 kg
   - 45 N
   - 441.45 N
   - 441.45 N m
   - Other
9. What is the difference between the terms "mass" and "weight"?
   - No difference. They mean the same thing.
   - Mass refers to force and the effect of one body on another, whereas weight refers to the amount of matter in a body.
   - Mass is the inertial property of a body, whereas weight is plain English for "static force."
   - For every action there is an equal and opposite reaction, and this applies to mass but not weight.
   - None of the above.
10. An astronaut has a mass of 80 kg. He visits the moon, where gravity is one-fifth of what it is on Earth. What is his mass under these conditions?
    - 16 kg
    - 16 N
    - The same
    - 784.8/5 N = 156.96 N
11. A person, standing upright, holds a mass of 10 kg in front of his body. The moment arm of the load is 50 cm. What force must the back muscles exert to maintain postural stability,

assuming that the back muscles have a lever arm of 7 cm and that the COG of the upper body is located directly above the lumbar spine?
- 701 N
- 700 kg
- Other
- 70 kg

12. What does the term "load moment" mean when used in ergonomics?
13. In which of the following age groups is low back pain most common?
    - Under 20 years
    - 20–40 years
    - 40–60 years
    - Over 60 years
14. A worker has a mass of 85 kg. Use Table 2.4 to estimate the masses of his head and neck, torso, arms, and legs.
15. Following the example in Workshop 2.1, calculate the additional spinal compression due to the load held when a worker holds a 20 kg box of lead in front of his body at a distance of 15 cm from L5/S1.
16. A worker has to carry 20 kg sacks of feathers from a packing machine to a dispatch zone. He is able to walk in an erect posture. The COG of the sack is 70 cm from his lumbosacral (LS) joint. Given that his upper body weighs 40 kg, what is the total spinal compression in Newtons? Comment on the difference between the task load and the postural load and suggest ways of reducing the task load.
17. You have decided to redesign a shovel for shoveling snow. In normal shoveling, people typically flex their trunks by 90°, and the distance from the LS joint to the blade of the shovel is about the same as the distance of the upper body COG from the LS joint. In your redesign, the handle is lengthened so that the person only leans forward by 10°. However, the loaded blade is now 1.2 m from the LS joint. Given the following body dimensions, would you say that your redesign is really an improvement, given that when the blade of the shovel is laden with snow, it weighs 15 kg?
    - Upper body weight = 45 kg.
    - Distance of upper body COG from the LS joint = 45 cm.
18. Mr. Smith is about to drive to the airport before going on holiday. He carries his suitcase to the car and places it on the ground to the right of his body. He opens the trunk of his car and laterally flexes and bends his upper body to reach for the suitcase handle. As he lifts the suitcase in this posture, he suffers an acute herniated intervertebral disc at L3. Review this chapter to determine whether there is any evidence that he placed himself at unnecessary risk of injury. How would you advise Mr. Smith to perform this task in the future?

QUICK QUIZ

1. Define a force.
2. How many vertebrae make up the human spine?
3. Do vertebrate fish have intervertebral discs?
4. The position of the feet when standing provides the:
    a. Appearance of the stance
    b. Base of support
    c. Best posture for the back
5. The main function of intervertebral discs in humans is:
    a. Shock absorbers
    b. Spacers to separate vertebrae
    c. Blood supply

6. Posture is:
    a. A risk factor for injury
    b. The average orientation of the body over time
    c. A way of 'showing off'
7. State a recommended spinal compression tolerance limit.
8. What is the most hazardous industry for days lost due to musculoskeletal injuries?
9. State the key characteristics of good posture.
10. Exoskeletons should be compulsory if a manual handling task imposes spinal compression beyond the tolerance limits of the workers. True or false?

NOTE

1 For a copy of the toolkit, contact infonet@aiha.org.

3 Inclusive Design

GENERAL REQUIREMENTS FOR HUMANS IN SYSTEMS

1. Where human physical variability in size, shape, and strength places constraints on the design, the system shall be designed to accommodate the majority of users on all relevant user dimensions for males and females in the target population.
2. The design solution shall fit *at least* the fifth percentile female to the ninety-fifth percentile male on that dimension.
3. The physical characteristics of the target population of users must be identified and planned for in advance.
4. Subgroups with special anthropometric needs, such as pregnant females, the disabled, or obese employees, must be catered for, either in the design solution or in bespoke arrangements to enable them to work safely and efficiently. Local regulations and guidance must be followed.
5. Prototype designs should be tested with user groups from the target population.

> Routinely forgetting to accommodate the female body in design – whether medical, technological or architectural – has led to a world that is less hospitable and more dangerous for women to navigate. It leads us to injuring ourselves in jobs and cars that weren't designed for our bodies. It leads us to dying from drugs that don't work. It has led to the creation of a world where women just don't *fit* very well.
>
> *Caroline Criado Perez, Invisible Women, page 313.*

CORE KNOWLEDGE: PHYSICAL VARIABILITY IN HUMANS

Anthropometry: Definition

The word "anthropometry" is derived from the Greek words "Anthropos" (man) and "Metron" (measure), meaning measurement of the human body. Anthropometric data are used in human factors and ergonomics (HFE) to specify the physical characteristics of workspaces, equipment, vehicles, and clothing to ensure that these products physically fit the target population. Figure 3.1 provides some examples in which the physical dimensions of workspaces do NOT fit the employees.

Measurements of the Body Used in HFE

Table 3.1 summarizes the main types of anthropometric data used in HFE.

Figure 3.2 provides a basic set of numbered static anthropometric variables along with brief descriptions of their uses. Table 3.2 presents the mean and standard deviation values of the corresponding numbered variables for different populations. The variable numbers in the table correspond to the numbered variable descriptions in the figure (data derived mainly from estimated values from a variety of sources, including Pheasant, 1986; Peebles and Norris, 1998). ISO 7250-1:2017, Basic Human Body Measurements for Technological Design gives detailed definitions of a range of body measurements for use in HFE.

DOI: 10.1201/9781032717210-3

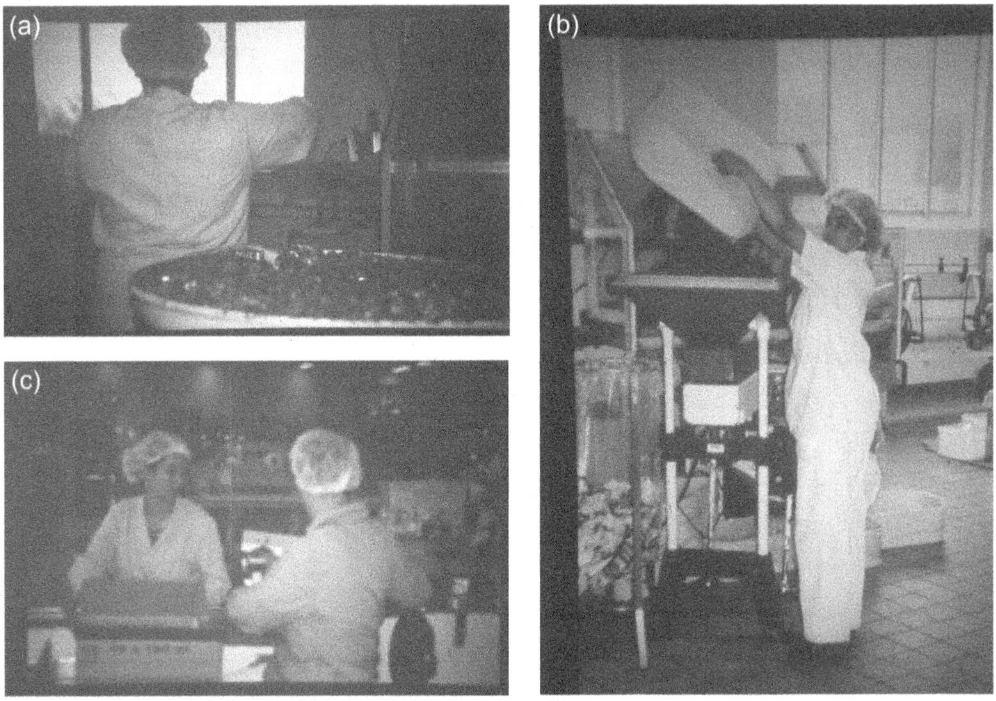

FIGURE 3.1 Avoid anthropometric mismatches in the workplace: (a) conveyor too high, (b) hopper too high, and (c) no clearance space for the knees, causing a twisted sitting posture.

TABLE 3.1
Types of Anthropometric Data Used in Ergonomics

Structural Data: Measurements of bodily dimensions of subjects in static postures. Anatomically rigorous, in that measurements are made from clearly identifiable anatomical sites, usually bony landmarks under the skin. Typically used to optimize furniture, clothing, and vehicle cab dimensions.

Functional Data: Collected from subjects who are allowed to move one or more limbs in one or more planes with respect to a fixed point. The shape of the 3D surface swept by moving the arm with the elbows extended or the amount of forward reach when the subject can bend at the hip. Takes into account the fact that in natural movements, several joints are involved and generates workspace "envelopes" whose size increases with the number of joints allowed to move.

Newtonian Data: This includes both body segment mass data (Chapter 2) and data about the forces that can be exerted in different tasks.

Functional Anthropometry

Figure 3.3 illustrates the basic terminology used to describe the movement of the joints. ISO 11226 (2000-12-15) provides estimates of the limits of motion of different joints of the body (Table 3.3) with reference to a relaxed standing position.

These data are useful for both design and assessment and provide absolute limits for static and dynamic postures. The data also provide the basis for the risk assessment tools described in this chapter and in Chapter 4. Ideally, work postures should vary around the midpoint of the joint ranges given in the table. The exception, of course, is the standing posture, in which the knees are almost fully extended. Functional data are often used to assess clothing assemblies and personal protective equipment (PPE). Well-designed PPE should provide the maximum amount of protection while minimizing restriction of movement.

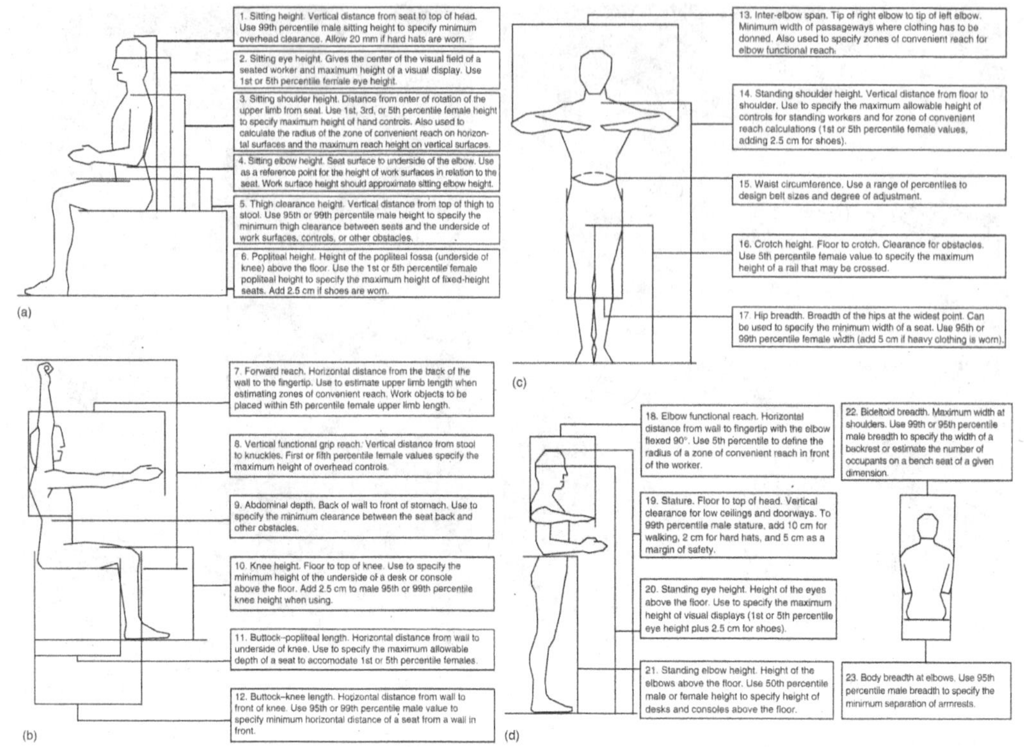

FIGURE 3.2 Static anthropometric measurements commonly used in ergonomics. (Refer to Table 3.2 for means and standard deviations of these in a variety of populations.)

TABLE 3.2
Selected Anthropometric Data: Means and Standard Deviations

	Males				
No.	United States	United Kingdom	China	Brazil	India[a]
1	923(37)	920(36)	906(36)	880(35)	840(25)
2	806(38)	803(37)	785(37)	775(34)	740(26)
3	637(30)	635(30)	624(33)	n/a	555(21)
4	245(28)	244(27)	236(22)	230(28)	205(20)
5	171(24)	169(19)	138(15)	150(16)	135(13)
6	450(28)	449(27)	410(23)	425(24)	415(21)
7	909(40)	811(40)	820(38)	n/a	725(24)
8	1322(54)	1318(53)	n/a	1220(56)	1190(44)
9	287(49)	280(39)	206(24)	245(33)	185(33)
10	546(29)	544(28)	502(28)	530(27)	510(30)
11	524(40)	517(35)	452(29)	480(29)	465(18)
12	620(41)	613(37)	556(31)	595(30)	555(21)
13	949(45)	946(44)	890(51)	925(44)	880(33)
14	1460(64)	1455(63)	1389(53)	1410(60)	1345(49)
15	1011(83)	985(66)	750(67)	n/a	n/a
16	819(49)	816(48)	780(45)	n/a	n/a

(Continued)

TABLE 3.2 (*Continued*)
Selected Anthropometric Data: Means and Standard Deviations

Females

	United States	United Kingdom	China	Sri Lanka[c]	India[a]
17	420(39)	393(31)	336(20)	340(25)	310(16)
18	479(22)	477(21)	450(21)	475(22)	460(20)
19	1755(71)	1755(70)	1691(62)	1700(66)	1620(50)
20	1643(70)	1638(69)	1573(59)	1595(66)	1510(52)
21	1099(51)	1096(50)	1021(42)	1045(49)	1025(40)
22	584(44)	570(35)	442(31)	n/a	410(19)
23	450(40)	449(39)	n/a	n/a	n/a
24[b]	82.1(17)	80(13)	60(8)	66(11)	49(6)

Females

	United States	United Kingdom	China	Sri Lanka[c]
1	861(36)	858(33)	840(35)	775(32)
2	745(36)	742(33)	720(33)	675(30)
3	583(31)	580(29)	573(28)	525(29)
4	233(27)	232(25)	223(19)	185(20)
5	160(39)	154(23)	133(14)	85(9)
6	400(28)	399(26)	378(23)	335(27)
7	811(40)	808(36)	753(35)	670(30)
8	1221(53)	1216(49)	n/a	1090(5)
9	280(80)	270(47)	208(24)	175(28)
10	496(29)	494(27)	456(25)	420(27)
11	503(50)	495(37)	438(28)	445(35)
12	599(56)	589(41)	531(29)	535(29)
13	860(48)	856(44)	775(49)	795(46)
14	1336(67)	1331(61)	1265(49)	1270(60)
15	874(122)	840(71)	692(64)	n/a
16	742(46)	739(43)	717(39)	n/a
17	426(70)	412(41)	351(25)	245(22)
18	433(23)	431(21)	411(20)	410(29)
19	1625(64)	1620(64)	1554(57)	1525(59)
20	1517(64)	1511(62)	1434(55)	1420(59)
21	1000(52)	996(48)	935(39)	940(62)
22	471(75)	455(44)	415(35)	295(14)
23	447(40)	445(37)	n/a	n/a
24[b]	69(23)	67(13)	50(8)	43(7)

[a] Agricultural workers, from Pheasant (1986).
[b] Body mass in kilograms.
[c] Nearest complete data set for females on the Indian subcontinent.
Note: The mean and standard deviation values are in millimeters. See Figure 3.2 for the key to variable numbers.

Sources of Human Variability

Humans vary due to genetic differences (in inherited characteristics), plasticity (the capability of being moulded by the environment when young), acclimatization, and, over the very short term, behavioral adaptation. Only the last two of these forms of adaptation are reversible. Roberts (1995)

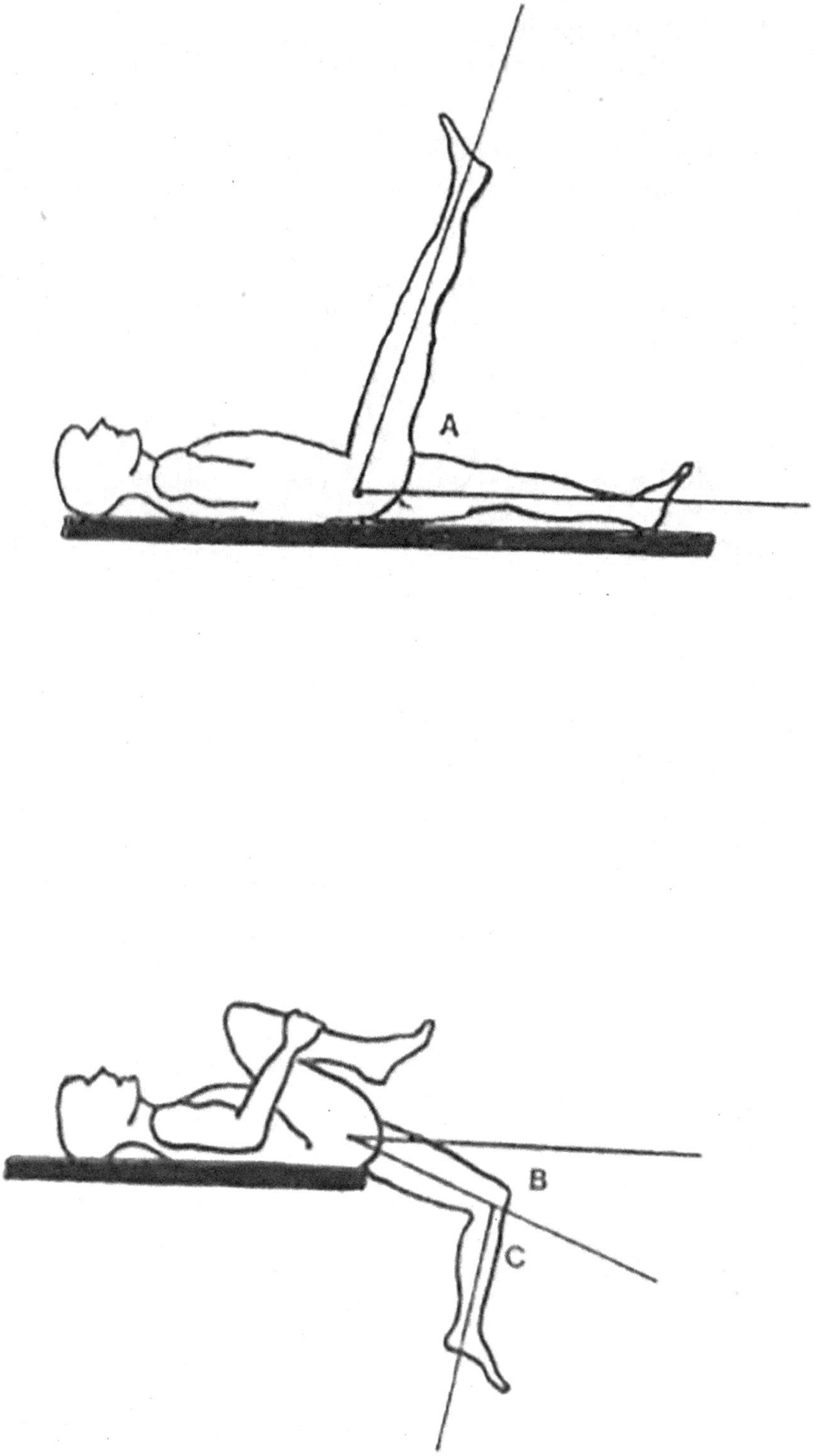

FIGURE 3.3 Terminology used to describe the movement of joints: (a) hip flexion, (b) hip extension, and (c) knee flexion. Adduction and abduction of the shoulder; internal and external rotation of the hip joint.

(Continued)

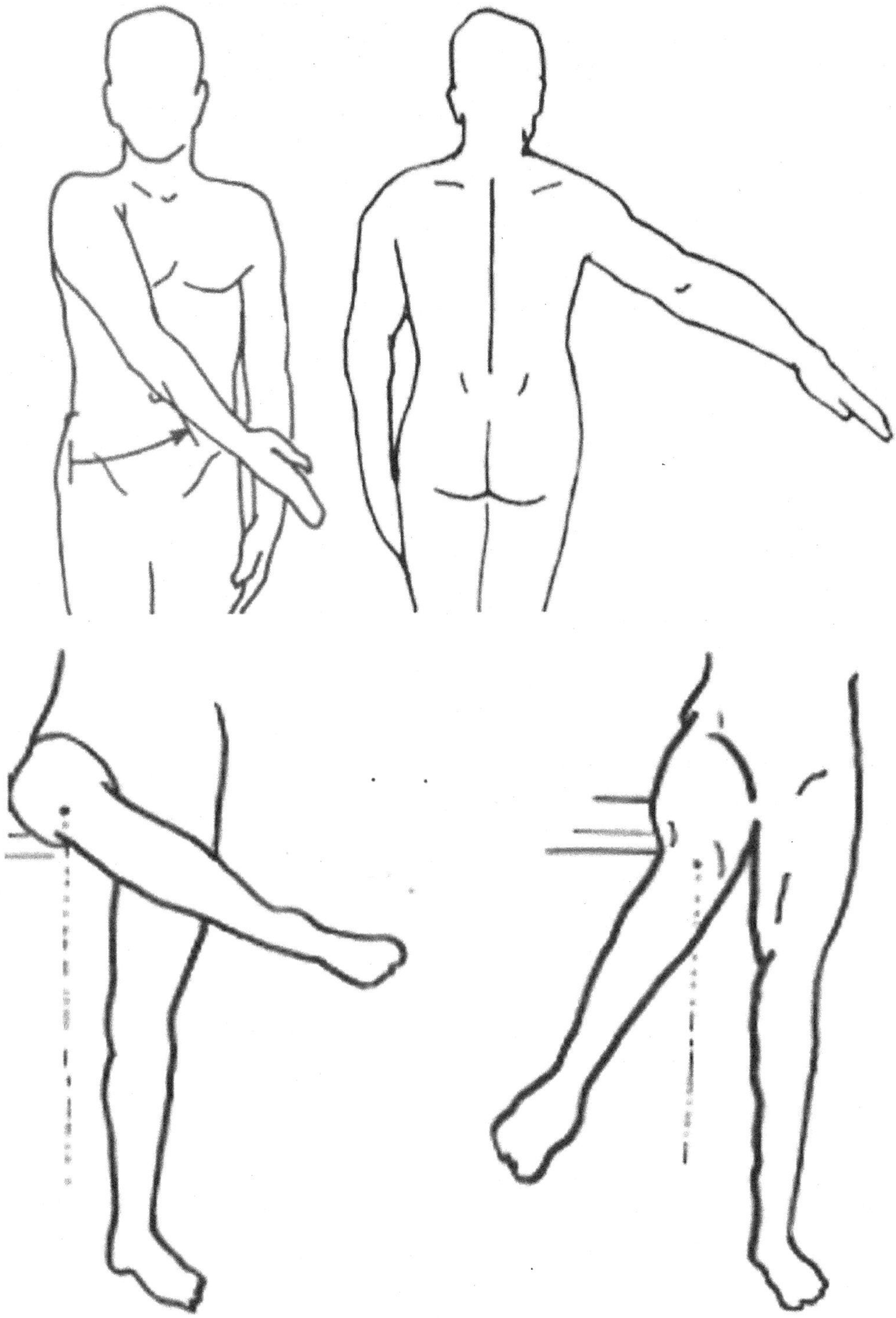

FIGURE 3.3 (*Continued*) Terminology used to describe the movement of joints: (a) hip flexion, (b) hip extension, and (c) knee flexion. Adduction and abduction of the shoulder; internal and external rotation of the hip joint.

TABLE 3.3
Limits of Joint Ranges of Motion

Body Segment/Joint	Range of Motion (Degrees)
Upper arm external rotation	90
Elbow flexion	150
Elbow extension	10
Forearm pronation	90
Forearm supination	60
Wrist radial deviation	20
Wrist ulnar deviation	30
Wrist flexion	90
Wrist extension	90
Knee flexion	40
Ankle dorsiflexion	20
Ankle plantarflexion	50

Source: From ISO 11226 2000-12-15.

cites evidence of plasticity in a population where the heads of people who were habitually placed in a supine position in the first years of life grew to be broader than those who were more often placed on their sides. Boas (1910) found that children born to immigrant parents after arrival in the United States were larger and had differently shaped heads than those born before the parents migrated. This challenged prevailing views about the fixity of racial or ethnic "types."

Factors Influencing the Change in Body Size of Populations

Smallness does not appear to be intrinsic to many groups of people but is instead related to development in biologically stressful environments, a plastic response to deprivation. Many industrialized countries have witnessed an increase in population size over the last 150 years.

Koukli et al. (2023) report data on the secular trend in stature in ancient Greece from 9000 BC to 900 AD. Male and female stature decreased from the Mesolithic (10,000–6500 BC) to the Neolithic period (6500–3000 BC), possibly due to a shift from hunter-gathering to farming, together with genetic changes, greater population density, and a lower quality diet. Stature increased during the Bronze Age (3000–1100 BC) and declined during the Hellenistic period, as did the variability of stature in males and females. This was a period marked by warfare between Greek city-states and external forces. Stature increased again during the Roman and Byzantine periods, which were marked by long periods of peace, improved living conditions (e.g., fresh water and sanitation), and more trade. Men were, on average, 170.6 cm and females 160.4 cm during the early Byzantine period (until 900 AD), compared to mean statures of 179.3 cm for males and 165.8 cm for females today. Overall, Greeks exhibit an approximately 10.0 cm (males) and 6.0 cm (females) increase from the early Byzantine period onwards. Similar patterns have been observed in other European populations. This, in part, is due to better diet, improved sanitation, childhood immunization, refrigerated transportation making available a year-round supply of fresh food, and supplementation of dairy products and cereals with vitamin D. The secular trend in stature seemed to have abated in the USA and Europe by the turn of the century (McCook, 2001), being replaced with an increase in obesity.

Obesity

The World Health Organization defines obesity as "abnormal or excessive fat accumulation that may impair health." Since 1975, in English-speaking countries, the mean body mass index of the population has increased by 1 kg/m^2 per decade in men and slightly more in women. Similar trends in overweight and obesity have been found in many other countries, according to a study of over 19 million participants (NCD, 2016).

Much anthropometric data were gathered many years ago and are often presented today after small corrections to allow for an increase in the stature of the population over time. Measurements of dimensions that depend on the length of the bones are more valid than those that are dependent on body composition (e.g., measures of circumference and body depth). The latter should be used with caution because of the dramatic increase in the number of overweight and obese people in the last 50 years.

Measuring Overweight and Obesity

Body Mass index (BMI) is the term most often used to classify individuals as either "healthy weight," "overweight," or "obese," where:

$$\text{BMI} = \text{Mass(kg)}/\text{Height}^2\text{(m)} \tag{3.1}$$

A BMI >18.49 and <25 is regarded as healthy, 25–29.9 as overweight, and 30+ as obese. One of the main drawbacks of BMI as a measure of obesity is that it ignores an individual's fat-free mass, with the result that very muscular people ("mesomorphs") are misclassified as overweight or obese. Figure 3.4 shows an alternative classification scheme that classifies people according to BMI and waist circumference and therefore accounts better for the proportion of body mass due to abdominal fat. The purpose of the scheme is to assess the risk to health of lifestyle-related diseases such as Type II diabetes and cardiovascular disease. Another measure commonly used in health care is the ratio of the waist circumference to the hip circumference (WHR). According to the World Health Organization, a moderate value of the WHR is 0.9 for men and 0.85 for women. There is a view that WHR is a better indicator of health risk than BMI because it focuses on the abdomen, where most fat accumulates.

Measured manually, WHR may not be the best screening tool for overweight and obesity. Sebo et al. (2008) assessed the reliability and validity of manual measurements of stature, body mass, BMI, waist and hip circumference, and WHR. The inter-observer reliability (similarity of repeated measurements on a group of subjects when made by different observers) of measures of stature, body mass, and derived BMI was good, whereas for waist and hip circumference and derived WHR, reliability was poor, even after observers were given a one-hour training session. For BMI, only 1% of people measured were misclassified, whereas even after training, 9% were misclassified on the basis of WHR.

BMI		Waist circumference (cm)		
		Men < 94	Men 94-101.9	Men ≥ 102
		Women <80	Women 80-87.9	Women ≥ 88
Underweight	< 18.5	Should be informed of risk of underweight		
Healthy weight	18.5-24.9	**None**	**None**	**Increased**
Overweight	25.0-29.9	**None**	**Increased**	**High**
Obese Class I	30.0-34.9	**Increased**	**High**	**Very High**
Obese Class II	35.0-39.9	**High**	**Very High**	**Extreme**
Obese Class III	≥40	**Very High**	**Extreme**	**Extreme**

FIGURE 3.4 Health risk in relation to BMI and waist circumference.

Abdominal fat is associated with an increased risk of adverse health outcomes such as metabolic syndrome. In a study by Bridger et al. (2013a), it was found that 28% of a sample of naval personnel were classified as being "at risk" using the scheme in Figure 3.4, whereas 61% were classified as "overweight or obese" based on BMI alone.

Self-reports of height and weight (from questionnaires, for example) are subject to consistent reporting bias (Spencer et al., 2002). In a study of 4808 adults, it was found that height was over-estimated by a mean of 1.23 cm in men (95% confidence interval 1.11–1.34) and 0.60 cm in women (CI 0.51–0.70); the extent of overestimation was greater in older men and women, shorter men and heavier women. Weight was underestimated by a mean of 1.85 kg in men (CI 1.72–1.99) and 1.40 kg in women (CI 1.31–1.49); the extent of underestimation was greater in heavier men and women, but did not vary with age or height. Caution should be applied when such data are obtained from self-reports.

In a recent systematic review, Sommer et al. (2020) selected studies that compared measures of BMI, waist circumference, waist to hip ratio, and waist to height ratio to detect obesity against reference measures such as magnetic resonance imaging. They found that most of the measures lacked *sensitivity*, detecting as many true positives (people who really were obese) as false negatives (misclassifying obese people as not obese), and were therefore of little use in detecting health risk due to obesity. The measures had better *specificity* (they were better at detecting people not at risk and tended not to misclassify healthy people as obese).

Anthropometry of Obesity

There is little data on the anthropometry of obese workers. In a study of 101 obese volunteers in the UK, Gyi et al. (2019) obtained data on 14 self-assessed body measures. The mean BMI of the volunteers was 40.3 kg/m^2 (sd 8.4) for males and 43.5 kg/m^2 (sd 8.5) for females. Table 3.4 summarizes the findings for comparison with available data for adults in the UK, used as a reference.

The differences are large for circumferential measurements. The authors found that for designs that accommodate up to 95% of British adults, 100% of obese adult males would be excluded for thigh depth; 99% for hip breadth; 98% for hip circumference; 96% for body mass and abdominal

TABLE 3.4

Comparison of Mean Values (mm) of Obese Volunteers with a Reference Population Aged 19–65

	Males			Females		
Measure	**Obese**	**Reference**	**Difference**	**Obese**	**Reference**	**Difference**
Abdominal	537	270	267	498	255	243
Depth of Buttock to the Front of the knee	633	595	38	618	570	48
Fingertip Reach	812	906	-92	735	808	–73
Hip Breadth	590	360	230	609	370	239
Hip Circumference	1296	1047	249	1345	1037	308
Popliteal Height	453	440	13	386	400	–14
Thigh Depth	334	160	174	310	155	155
Shoulder Breadth	570	465	105	537	395	142
Sitting Shoulder height	634	595	39	592	555	37
Stature Waist	1752	1740	12	1604	1610	–6
Circumference	1375	985	390	1308	841	467
Mass (kg)	125	75	50	113	63	50

Source: From Gyi et al. (2019).

depth; 95% for waist circumference; and 64% for shoulder breadth. For females, all would be excluded for body mass, waist circumference, thigh depth, hip breadth, and abdominal depth; 92% would be excluded for hip circumference and shoulder breadth. Obesity also limits forward reach because obese workers have to stand farther away from the worksurface. "Knee splay" is the distance between the knees when sitting and is relevant to the spacing of seats in public areas. Although the sample size was small, the differences are very large.

Sexual Dimorphism: Sex Differences in Form and Function

Homo sapiens is a sexually dimorphic species. In common with other primates, human males are typically larger than human females (about 15% larger overall). Lassek and Gaulin (2022) reviewed multiple studies summarizing the *average* differences between males and females:

1. *Stature*: Females are 7% shorter than males.
2. *Body Mass*: Females are 16% lighter than males.
3. *Lean Body Mass*: It is 36% greater in males.
4. *Muscle Mass*: It is 65% higher in males.
5. *Arm Muscle*: It is 72% greater in males.
6. *Body Fat*: It is 1.6 times greater in females.
7. *Upper Body Strength*: Males have twice the upper body strength of women.
8. *Leg Strength*: It is 60% greater in men.
9. *Bone Mass*: It is greater in males, as is bone density.
10. *Torso Length*: It is greater, in proportion, in females to allow space for pregnancy. Females have shorter legs in proportion to their stature.
11. *Testosterone*: Males have higher levels of testosterone and more "fast twitch" (Type II) muscle fibers for explosive power and strength.

Males tend to distribute body fat in the viscera, whereas females distribute fat distally, typically in the gluteo-femoral region. During menopause, there is an unfavorable increase in central adiposity with more fat stored around the abdomen and less in the periphery (Greendale et al., 2021), leading to a change in body shape post-menopause.

These differences are not due to differences in the sizes of male and female bodies. Controlling for stature and body mass, males still have 9.8 kg more lean mass, 2.3 kg more arm muscle, 2.7 kg more leg muscle, and 4.2 kg more lean trunk mass. The female pelvic body segment is 16% of body mass compared to 13.7% in males. In males, the trunk is 55.1% of body mass compared to 53.2% in females. These are differences in *form and function* that have evolved over millions of years.

The pelvis of human females has evolved to facilitate the birth of neonates with a large cranial capacity. As a result, the female pelvis is wider than the male pelvis, and the hip joints are farther apart than in males—the femurs angle inwards toward the tibia at the knees, placing a strain on the anterior cruciate ligament (ACL). This is known as the *valgus angle*. In effect, women tend to be more "knock-kneed" than men. Typically, women have a 20% larger tibiofemoral angle than men and have between 2 and 8 times more ACL injuries than men when participating in the same sports activities.

Morris et al. (2020) found very large differences in overhead pulling and pushing forces between males and females while controlling for body mass and upper arm fat-free cross-sectional area. They argued that our early hominid ancestors evolved into modern humans as a result of *sexual selection* pressure in males (competition between males for females, favoring males with greater upper body strength) and *natural selection* in females to overcome the *obstetric dilemma*. In order for the birth canal of the female pelvis to be suitable for childbirth and also for bipedal locomotion, hominin females had to be larger (female body size had to increase to allow a larger birth canal). This explains why the difference in body size between human males and females is much smaller than in other primates (male gorillas are twice the size of female gorillas; male chimpanzees are about 35% larger than female chimpanzees). However, sex differences in *form* are much greater in Homo sapiens than in other primates.

Clearly, women are *not* small versions of men with low testosterone and weak muscles. Large *functional* differences in body size, shape, and capacity need to be considered when designing products and workspaces for female and male users.

Anthropometry Surveys

Anthropometric surveys are expensive to conduct, since large numbers of measurements have to be made on sizeable samples of people representative of the population under study. Traditionally, measurements are made using manually operated instruments such as anthropometers and calipers. Such surveys are costly because they require trained technicians, long measurement times, identification of anatomical landmarks, and careful calibration. Whole-body scanners are now available. Modern laser scanners use stereoscopic techniques to create a digital copy of the static human body (from a cloud of points within a coordinate system), from which measurements of defined body dimensions can be made very rapidly. Once calibrated, laser scanning enables accurate and reliable measurements. Rumbo-Rodriguez et al. (2021), in a systematic review of 18 studies, found that laser scanners correlated highly with other methods and were as accurate, or more accurate, and more reliable than manual methods in the measurement of body composition in males and females (intra class correlation coefficients for test-retest reliability were high). Body segment scanners for the head and the feet are also available.

Scanning is used widely in the garment industry for sizing clothing and designing personal protective equipment. The technology is developing rapidly. Dannen and Psikuta (2018) report that smartphones can be used to take frontal and sagittal shots of subjects. Body dimensions are derived from stature and used as a scaling factor. Other main advantages of scanning are that it is non-invasive (the measuring instrument never contacts the body) and that all the data can be captured at the same time. Further information on scanning in relation to human factors and ergonomics can be found in ISO 20685-2:20123 – Ergonomics.

Anthropometric Data

Many anthropometric databases are not publicly available because they are used for commercial purposes. Surveys of military personnel are routinely conducted in many countries, but military personnel are not representative of the general population because of the selection procedures required to join and remain in the service. Much of the publicly available data was derived from surveys conducted many years ago and has been adjusted to take account of secular changes in stature and body mass index. Fryar et al. (2012) present data on children and adults in the USA (2007–2010). Bonin et al. (2023) have published a detailed data set of 39 measures from 2313 subjects in Germany. In Britain, Peebles and Norris (1998) provide a comprehensive data set largely based on data from older surveys that were corrected for secular growth in stature.

Digital human models are described in the following sections; these use a variety of techniques to estimate anthropometric variables in target populations.

Implications for HFE

Extrapolation from an anthropometric database to a target population is always an estimation based on the assumptions made and the data used. When designing for international markets, each target country has to be considered separately. "Self-selection" often occurs in some occupations, and when it does, in no sense are employees representative of the population of the country in which they work. The *Healthy Worker Effect* is an example of self-selection in which, for example, more robust individuals tend to seek more strenuous jobs. Methods of dealing with these challenges are described below.

Statistical Essentials for Using Anthropometric Data in HFE

Anthropometric variables in the healthy population usually follow a normal distribution, as depicted in Figure 3.5.

The Normal Distribution

The two key *parameters* of the normal distribution are the mean and the standard deviation. The mean is the sum of all the individual measurements divided by the number of measurements. It is a measure of central tendency. The standard deviation is a measure of dispersion. It is calculated from the sum of the differences between all the individual measurements and the mean—the larger these differences, the more "spread-out" the distribution is, and the larger the value of the standard deviation.

Thus, the value of the mean determines the position of the normal distribution along the x (horizontal) axis in Figure 3.5. The value of the standard deviation determines the shape of the normal distribution. In a normal distribution with a small standard deviation, most of the measurements are close to the mean value (the distribution has a high peak that tails off rapidly on both sides). A large value of the standard deviation means that the measurements tend to be scattered farther from the mean. The distribution has a flatter shape.

To estimate the parameters of stature in a population (the mean and standard deviation), it is necessary to measure a random sample of people who are representative of that population. The formulae given in Figure 3.6 can then be used to estimate the mean and standard deviation. Estimates of *population parameters* obtained from calculations on data from *samples* are known as *sample statistics.*

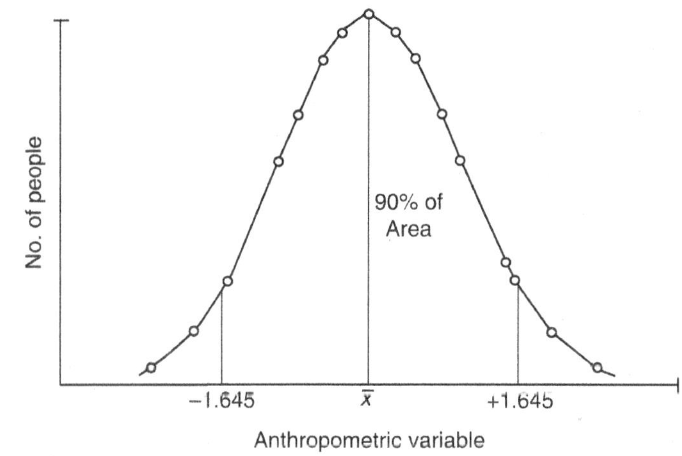

FIGURE 3.5 The normal distribution. Ninety percent of the measurements made on different people will fall within a range whose width is 1.645 standard deviations above and below the average.

$$\bar{x} = \frac{\sum_{i=1}^{n} x_i}{n} \qquad S = \sqrt{\frac{\sum_{i=1}^{n} (x_i - \bar{x})^2}{n-1}}$$

(a) (b)

FIGURE 3.6 Equations used to calculate (a) the mean and (b) the standard deviation from measurements sampled from a population, where x is the individual measurement and n is the number of measurements. The mean is the sum of all the measurements divided by the number of measurements. The standard deviation is the square root of the mean squared difference between each score and the mean.

An important characteristic of the normal distribution is that it is symmetrical—as many observations lie above the mean as below it (or, in terms of the figure, as many observations lie to the right of the mean as to the left). If a distribution is normally distributed, 50% of the scores (and thus the individuals from whom the scores were obtained) lie on either side of the mean.

This simple statistical fact applies to many variables, yet it is often ignored or misunderstood by both specialists and laymen alike. It is common to hear statements such as "your child is below average height for his age" or "you are above average weight for your height," both pronounced and perceived in a negative way. The statistical reality is that half of any normally distributed population is either above or below "average," which is why it is unwise to design for "Mr. Average," a mythical person with the mean stature. Very few individuals will be exactly of "average" height, and therefore it is necessary to design to accommodate a range of people.

Variability and the "Distance" from the Middle

The difference between any particular measurement and the mean of the sample can be thought of as the "distance" of that measurement from the mean (how far away it is along the x-axis). The "distance" of a measurement from the mean can be described in two ways—in terms of the variable being measured (e.g., height in millimeters or weight in kilograms), or as the *number of standard deviations* from the mean, in either direction.

For example, suppose Sarah is 1600 mm tall and the mean height of females in the population is 1650 mm. We can say that Sarah is 50 mm shorter than average. Now, suppose that the standard deviation of the distribution of female height is 100 mm. We can say that Sarah is one-half of a standard deviation below the average height. Using a statistic called the *standard normal deviate*, z, we can calculate how tall Sarah is compared to all other females in the population.

The Standard Normal Deviate: *z*

The table in Appendix A gives the area under the normal curve to the left of a particular z-value. The values of z apply to a normal distribution that has a mean of zero and a standard deviation of one. The total area under the whole of this *standard normal distribution* has a value of one, by definition. As z increases, the distance from the mean increases, and the area under the normal curve to the left of z decreases. We can take any value of z for any normal distribution and draw an imaginary vertical cut-off point through the curve (Figure 3.5). The numbers in the table in Appendix A give us the proportion of the curve to the left of the z cut-off point. For any normal distribution, the value of z can be calculated for any observation (i.e., for any value of x) if the mean and standard deviation are known.

In the example above, we saw that Sarah was one-half of a standard deviation shorter than the average female in her group. For all normal distributions, any measurement (or score or observation) in the distribution can be described in this way. Equation 3.2 defines the "z" score as follows:

$$z = \frac{(x - \bar{x})}{\text{sd}} \tag{3.2}$$

Sarah is 1600 mm tall, and the mean stature was 1650 mm, and the standard deviation was 100 mm. Therefore, for Sarah, her z-score is (1600–1650)/100 or −0.5. The minus sign indicates that Sarah is shorter than average by half a standard deviation.

From the table in Appendix A, we can see that for $z=-0.5$, the area under the normal curve to the left of z is 0.3085. This is the proportion of the normal distribution that lies to the left of $z=-0.5$. In other words, approximately 31% of females are shorter than Sarah—Sarah is of the thirty-first *percentile* stature.

The "z-value" for any observation is just the difference between that observation and the mean, expressed in standard deviations. Thus, if the mean stature in a population was 1700 mm and the standard deviation 100 mm, a person with a stature of 1850 mm would be 1.5 standard deviations above the mean or ninety-third percentile (see Appendix A, where the area to the left of $z=1.5$ is $1-0.0668$, or 0.9332).

Percentiles to Real Measurements and Back Again

To find the measured value of a given percentile, use

$$x = \bar{x} + (z \times \text{sd}) \tag{3.3}$$

If Karen were twenty-fifth percentile stature, and the mean and standard deviation of female stature were 1650 and 100 mm, respectively, how tall would she be?

First, we look up the value of z in Appendix A. The value of z corresponding to the twenty-fifth percentile is $z = -0.68$ (the actual area in the table is 0.2483, rounded up to 0.25). This means that Karen is 0.68 of a standard deviation shorter than the mean. Therefore, Karen's stature, x, is

$$x = 1650 + (-0.68 \times 100) = 1650 - 68 = 1582 \text{ mm} \tag{3.4}$$

Table 3.5 contains z-values and their corresponding areas under the normal curve. Cross-check these values with the table in Appendix A to confirm that you can find z-values for different areas under the normal curve, and vice versa.

Estimating the Range

We can use z-scores to estimate proportions and ranges for all kinds of normal distributions—even those in which the standard deviation is not one and the mean is not zero. For any normal distribution, approximately two-thirds of the observations in the population fall within one standard deviation above and below the mean. Thus, for a population with a mean stature of 1650 mm and a standard deviation of 100 mm, approximately two-thirds of the population would be between 1550 and 1750 mm tall. The remaining one-third would lie beyond these two extremes (one-sixth on either side).

TABLE 3.5
Constants Used to Estimate Population Proportions

Required Percentile	Area to Left of z	z-Value (No. of Standard Deviations to be Subtracted from the Mean)
0.5	0.0049	−2.58
1	0.0102	−2.32
2.5	0.0250	−1.96
3	0.0301	−1.88
5	0.0505	−1.64
10	0.1003	−1.28
15	0.1492	−1.04
20	0.2005	−0.84
25	0.2414	−0.67
75	0.7486	+0.67
80	0.7995	+0.84
85	0.8508	+1.04
90	0.8997	+1.28
95	0.9495	+1.64
97	0.9699	+1.88
97.5	0.9750	+1.96
99	0.9898	+2.32
99.5	0.9951	+2.58

Note: The total area under the normal curve is taken to be 1 (Appendix A).

Table 3.4 gives us a quick method to calculate the percentiles of a body measurement and, from these, how far apart they are (the range). For example, 95% of users will fall within

$$\bar{x} \pm 1.96 \text{ sd} \tag{3.5}$$

Accuracy of the Measurements

The World Health Organization recommends that if anthropometric data are to be used as reference standards, a minimum sample size of 200 individuals is needed (this gives a standard deviation around the fifth percentile of about 1.54 percentiles—a 95% probability that the true fifth percentile falls within ±2.25 percentiles of the estimate, approximately). The more individuals who can be measured in an anthropometric survey, the more accurate the estimates will be. A statistic known as the *standard error of the mean* (se) is calculated to enable accuracy to be estimated, where

$$\text{se} = \text{sd}/\sqrt{n} \tag{3.6}$$

where

sd = the standard deviation of the mean
n = number of people measured in the survey

For example, if the mean stature in a sample was 1650 mm and the standard deviation was 100 mm, and 100 (n = 100) people had been measured, the standard error of the mean would be $100/\sqrt{100} = 100/10 = 10$ mm. We do not know what the real mean stature of the population is (to find the *population parameter*, we would have to measure everyone). However, we do know (from statistical theory) that there is a 95% probability that our statistical estimate of the mean falls somewhere in a range on either side of the sample mean. This range is 1.96 standard errors wide. In other words, the 95% *confidence interval* for the mean is

$$\bar{x} \pm 1.96 \text{ se} \tag{3.7}$$

For our stature example,

$$1650 \text{ mm} \pm 19.6 \text{ mm}$$

Going back to our previous example, we found that Sarah was thirty-first percentile for stature—she was taller than 31% of females in the population. However, if we had only measured 100 females, our estimate of the mean stature, on which our percentile calculation was based, would lack statistical accuracy. What percentile stature would Sarah be if the true mean were not 1650 mm but 1650 + 19.6 mm?

$$\begin{gathered} z = (1600 - 1669.6)/100 \\ z = -0.69 \end{gathered} \tag{3.8}$$

Referring to Appendix A, we find that Sarah is now approximately the twenty-fifth percentile in stature, as opposed to the thirty-first percentile. Similarly, if the true mean had been 1630.4 mm, z would equal −0.34, and Sarah would now be at the thirty-seventh percentile in stature.

Clearly, if we want to use anthropometric data to design a range of clothing or adjustable equipment for large numbers of people, we need large samples and accurate measurements. This is because both large n and low measurement error reduce the size of the standard error; otherwise, our estimates of the true dimensions of different percentiles may be completely wrong. If our statistical accuracy is low, we won't really know how many customers will want to buy different sizes or how many people will need to adjust our equipment by different amounts. Accurate estimates, on the other hand, will provide clear guidance for the design and manufacturing of products.

PATTERNS OF VARIABILITY IN HUMAN BODY SIZE AND SHAPE

Tables of data, such as that in Table 3.2, may appear to represent the values of different (independent) variables. However, these anthropometric variables are abstractions—they are ways of mapping the human body to yield data useful to designers, but the measurements themselves are not independent of each other. People vary according to the length of their long bones and body composition. "Ectomorphs" tend to be slim and gracile, "endomorphs" tend to have a high proportion of body fat, and "mesomorphs" tend to be muscular. The lengths of the long bones covary within a population (people with long legs tend to have long trunks, hands, and feet). People with a high percentage of body fat tend to have a high waist and chest circumference, and muscular people tend to have a narrow waist but large chest and limb circumferences. The covariance of anthropometric measures can be explored statistically in several ways. HFE Workshop 3.1 explains.

HFE WORKSHOP 3.1

COVARIANCE AND CORRELATION IN APPLIED ANTHROPOMETRY

The correlation coefficient is a statistic that describes the amount of covariance between two or more variables that are paired in some way. Correlation coefficients vary between –1.0 and +1.0, where "0" indicates that the variables are independent of each other—there is no covariance between them. A positive value indicates that the variables covary directly in the same direction, and a negative correlation coefficient indicates that they vary in inverse proportion (when one increases, the other decreases). The correlation coefficient is purely descriptive—no causal mechanism is implied or is to be inferred—the covariance might be due to the operation of other, unmeasured or unknown variables.

For the purpose of the present discussion, a positive correlation between two anthropometric variables might indicate that they reflect the operation of the same underlying biological mechanisms or body type. The most commonly used correlation coefficient is the Pearson product–moment correlation coefficient, *Pearson's r.*

The correlation between two sample variables, x and y, is given by

$$r_{xy} = S_{xy} / \sqrt{S_{xx}.S_{yy}}$$

where

$$S_{xy} = \sum (x - \bar{x})(y - \bar{y})$$

$$S_{xx} = \sum (x - \bar{x})^2$$

$$S_{yy} = \sum (y - \bar{y})^2$$

As can be seen from the formula above, if x and y are identical, then the numerator in the expression is the same as the denominator, and the result is +1. If there is perfect negative agreement between x and y (e.g., if $y=-x$), then the expression in the numerator will be negative, and the result will be –1.

Let us suppose we calculate Pearson's r for some anthropometric variables from a large sample of females:

- *Abdominal Depth*: ABDP
- *Head Circumference*: HEDC

(Continued)

HFE WORKSHOP 3.1 (*Continued*)

- *Buttock–Knee Length*: BTKL
- *Chin Height (Sitting)*: CHST
- *Hip Breadth*: HPBR
- *Biacromial Depth*: BIAD
- *Foot Length*: FOOL
- *Chest Circumference at Bust*: CCBU
- *Chest Depth*: CHDP

And generate a matrix of correlation coefficients:

	ABDP	HEDC	BTKL	CHST	HPBR	BIAD	FOOL	CCBU	CHDP
ABDP	1.00	0.18	0.040	0.040	**0.67**	0.06	0.06	0.81	**0.48**
HEDC	0.16	1.00	–0.01	0.00	0.18	–0.3	0.00	–0.01	0.14
BTKL	–0.04	–0.01	1.00	**0.93**	–0.05	0.02	0.02	**0.98**	–0.05
CHST	–0.04	0.00	**0.93**	1.00	–0.04	0.03	**0.94**	–0.04	–0.04
HPBR	**0.67**	0.18	–0.05	–0.04	1.00	0.09	–0.04	**0.63**	0.33
BIAD	0.06	0.00	0.02	0.03	0.09	1.00	0.04	0.11	0.13
FOOL	–0.05	–0.1	**0.98**	**0.94**	–0.04	0.04	1.00	**0.54**	0.40
CCBU	**0.81**	0.14	–0.03	–0.04	**0.63**	0.11	–0.04	1.00	**0.67**
CHDP	**0.48**	0.15	–0.05	–0.04	0.33	0.13	–0.05	**0.67**	1.00

Bold values statistically significant at $p<0.05$

The square of r gives an estimate of the covariance between x and y. For example, the correlation between foot length and chin height (sitting) is 0.94, which means that 88% of the variance between chin height and foot length is shared—probably because both variables are influenced by biological factors that influence the length of the bones.

Further analysis is possible to explore these concepts using a technique known as *principal components analysis* (PCA). PCA is a descriptive technique for reducing a large number of variables into smaller, higher-order components. Variables within a component "belong" together in the sense that they share common variance (they covary). Different components are *independent of each other* (they don't covary).

The table below shows the results of a PCA of the data in this example.

Component	Eigenvalue	% Variance	Cumulative %
1	3.034	33.710	33.710
2	2.767	30.744	64.454
3	1.010	11.219	75.673
4	0.932	10.357	86.030
5	0.693	7.704	93.734

The analysis identified five main components, accounting for nearly 94% of the variance in the dataset. By convention, only components with eigenvalues greater than 1 are selected for further analysis.

(*Continued*)

HFE WORKSHOP 3.1 (*Continued*)

Rotated Component Matrix			
	Component		
	1	2	3
ABDP	−0.019	**0.884**	−0.102
HEDC	0.012	0.257	−0.513
BTKL	**0.984**	−0.023	0.003
CHST	**0.976**	−0.027	0.004
HPBR	−0.022	**0.779**	−0.094
BIAD	0.031	0.199	**0.851**
FOOL	**0.992**	−0.026	0.016
CCBU	−0.011	**0.928**	−0.006
CHDP	−0.031	**0.729**	0.083

Bold values indicate variables with significant component loadings

The relationship between a variable and a component is expressed by its "loading" on the component. Buttock–knee length, chin height, and foot length load strongly (>0.05) onto Component 1. Abdominal depth, hip breadth, chest circumference, and chest depth load onto Component 2, and biacromial depth onto Component 3. These results might be taken to mean that measurements that depend on bone length vary together in a population in support of some of the scaling techniques described later in this chapter (Component 1). Component 2 includes variables dependent on body composition rather than bone length, and component 3 might indicate that muscularity is a separate factor. Interestingly, head circumference does not load onto any of these factors. If we had included more measures of the head, a separate component, "head size," might have emerged, indicating that the size of people's heads does not depend on their height, body composition, or muscularity and cannot be estimated from these variables. The components identified by the PCA may enable us to infer that independent factors influence how human physical variability is expressed in a population. Component 1 may reflect the operation of genetic factors in well-nourished populations; Component 2, the operation of health and lifestyle influences on body composition (particularly obesity in the English speaking world); and Component 3, the expression of different body "types" due to a combination of both. The number of components identified depends, of course, on the set of anthropometric variables measured in the first place.

Principal components analysis can be used to improve our understanding of body shape and size. By varying one component, we can simulate how all the measurements that load onto that component will change, while keeping the other measurements constant. Daanen and Psituka (2018) report that in clothing applications, there are three main components of whole-body size: the first relates to stature, the second to body mass, and the third to arm/leg length. Using PCA, we can vary the body mass of a male or female of a given stature to simulate how body shape will change while keeping stature constant. In effect, this might help designers understand how, for example, trends in obesity impact body shape in the absence of empirical data.

BASIC APPLICATIONS

Design to Fit a Target Population

1. Identify the user population.
2. Define the task and workspace, and identify where human physical variability might place constraints on the design.
3. Identify the body measurements in Figure 3.2 that constrain the design.
4. Specify the range of users to be accommodated.
5. Calculate the range (see "statistical essentials").
6. Use the limits in (5) to specify the dimensions of the workspace.

The word "population" refers to groups of people sharing the following common factors: jobs, ancestors, occupations, geographical locations, age groups, race (ancestors), and ethnicity (culture, dress, customs, language, and so on). The criteria for deciding what constitutes a "target population" are functional and relate directly to the problem at hand—for example, if we want to design a cab for bus drivers in Chile, we require data on the anthropometry of Chilean bus drivers.

Consider product dimensions in human terms, in view of the constraints placed on their design by body size variability; for example, seats should be wide enough for large users and low enough for small users. Body size and proportion vary greatly between different populations. A U.S. manufacturer hoping to export to Central and South America or Southeast Asia would need to consider in what ways product dimensions optimized for a large U.S. population, probably a male user group, would suit Mexican or Vietnamese users.

Information about body size is not, in itself, directly applicable to a design problem. First, the designer has to analyze in what ways (if any) anthropometric mismatches might occur, and then decide which anthropometric data might be appropriate to the problem. In other words, the designer has to develop lear ideas about what constitutes an appropriate match between user and product dimensions. Next, a suitable percentile has to be chosen. In many design applications, mismatches occur only at one extreme (only very tall or very short people are affected, for example), and the solution is to select either a maximum or a minimum dimension. If the design accommodates people at the appropriate extreme of the anthropometric range, less extreme people will be accommodated.

A basic set of strength data for pushing, pulling, and twisting is given in Table 3.6.

TABLE 3.6
Selected Strength Data for Males and Females

Pushing (N)[a]		Mean	Standard Deviation
	Handle Height (m)		
1.7	Males	300	50
	Females	181	75
1.3	Males	337	83
	Females	221	103
0.7	Males	393	134
	Females	185	57
Pulling (N)[a]			
	Handle height (m)		
1.7	Males	263	60
	Females	196	56

(Continued)

TABLE 3.6 (*Continued*)
Selected Strength Data for Males and Females

Pushing (N)[a]		Mean	Standard Deviation
1.3	Males	347	55
	Females	223	80
0.7	Males	541	81
	Females	292	97
Wrist twisting			
Strength (Nm)[b]			
Males only			
Knob diameter (mm)			
9.5		52.75	12.8
12.7		65.21	12.54
19.1		111.65	26.2

[a] Daams (1993).
[b] Swain, Shelton, and Rigby (1970).

TABLE 3.7
Selected Grip and Pinch Strength Data for U.S. Adults 25–75 Years

	Males		Females	
	Mean	Standard Deviation	Mean	Standard Deviation
		Grip Strength		
Right hand	46.2	12.6	27.9	7.6
Left hand	41.4	12.3	24.0	7.0
		Tip Pinch		
Right hand	7.6	1.8	5.0	1.2
Left hand	7.3	1.8	4.8	1.1
		Key Pinch		
Right hand	10.9	2.0	7.2	1.3
Left hand	10.5	2.0	6.8	1.4
		Palmar Pinch		
Right hand	10.4	2.2	7.2	1.7
Left hand	10.2	2.4	7.0	1.6

Source: From Mathiowetz et al. (1985).
Note: The strength is denoted in kilograms (kg).

Table 3.7 provides data on maximum grip and pinch strength for U.S. adults from age 20 to over 75 years. These data may be useful considering the design trend of miniaturization made possible by microelectronics, resulting in smaller products. Figure 3.7 illustrates the three kinds of pinch action: tip pinch, where the thumb contacts the tip of the index finger, as in picking up a small object; key pinch, where the thumb pad touches the lateral aspect of the middle phalanx of the index finger,

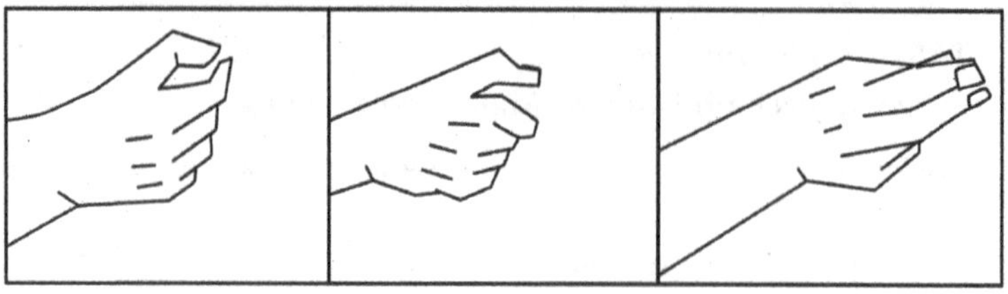

(a) (b) (c)

FIGURE 3.7 (a) Tip pinch, (b) key pinch, and (c) palm pinch.

as when turning a key; and palmar pinch, where the thumb pad touches the pads of the index and middle fingers.

Grip strength drops by about 20% from age 20–60 years and by about 50% by age 75 and above. For pinch strengths, the age-related reductions are not as large—75-year-olds have about 75% of the strength of 20-year-olds.

Anthropometry and Clothing Corrections

Most anthropometric data are captured on nude or semi-nude subjects, whereas most workers wear clothing of some kind. The following corrections are often used:

Shoes: add 25 mm to stature
Hats and helmets: add 90 mm to stature
Protective clothing: add 40 mm to stature

Anthropometric Constraints on Product Dimensions
Finding the Minimum Allowable Dimensions

A high percentile value of an appropriate anthropometric dimension is chosen. When designing a doorway, for example, sufficient headroom for very tall people must be provided, and the ninety-fifth or ninety-ninth percentile (male) stature could be used to specify a minimum height. The doorway should be no lower than this minimum value, and additional allowance must be made for the increase in stature caused by items of clothing, such as shoe heels, protective headgear, and so on.

Seat breadth is also determined using a minimum dimension—the width of a seat must be no narrower than the largest hip width in the target population (Figure 3.8). Minimum dimensions are also used to specify the placement of controls on machines, door handles, and so on. Controls must be sufficiently high off the ground so that tall operators can reach them without stooping—that is, no lower than the ninety-fifth percentile standing knuckle height. In the case of door handles, the maximum vertical reach of a small child might also be considered (to prevent young children from opening doors when unsupervised).

Circulation space must be provided in offices, factories, and storerooms to allow for the ingress and egress of personnel and to prevent collisions. In a female or mixed-sex workforce, the body width of a pregnant woman sould be used to determine the minimum (some anthropometric data for the seated pregnant woman can be found in Culver and Viano, 1990). About 60 cm of clearance space is needed for passages, the separation of machines, and the distance of furniture from walls or other objects in a room.

Anthropometric Constraints on Product Dimensions:
Find the Maximum Allowable Dimensions

A low percentile is chosen, as in determining the maximum height of a door latch, so that the smallest adult in a population will be able to reach it. The latch must be no higher than the maximum vertical grip reach of a small person. The height of non-adjustable seats used in public transport

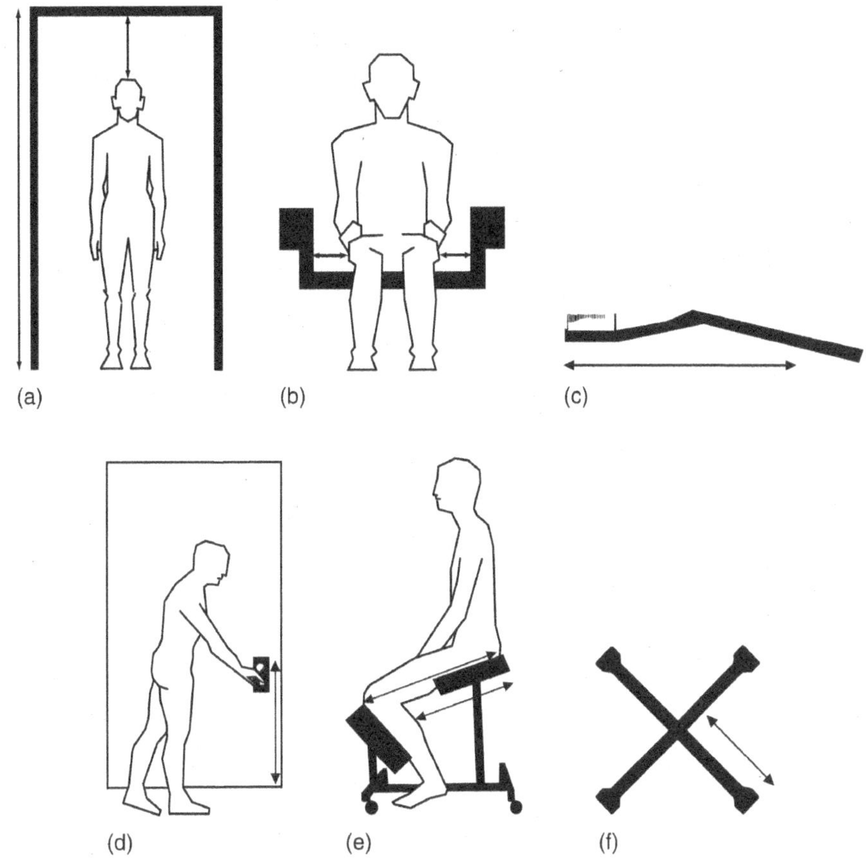

FIGURE 3.8 Some minimum dimensions. (a) The height of a doorway must be no lower than the stature of a tall man (plus an allowance for clothing and shoes), (b) the width of a chair must be no narrower than the hip breadth of a large woman, (c) a toothbrush must be long enough to reach the back molars of someone with a deep mouth, (d) a door handle must not be lower than the highest standing knuckle height in a population so that all users can open the door without stooping, (e) the distance from the kneepad to the back of the seat of a "kneeling chair" must exceed the longest buttock–knee lengths in the population of users, and (f) the length of a wheel brace must provide sufficient leverage for a weak person to generate sufficient torque to loosen the wheel nuts.

systems and auditoria is also determined using this principle—the seat must be low enough so that a short person can rest their feet on the floor when using it. Thus, the seat height must be no higher than the first or fifth percentile popliteal height in the population. Figure 3.9 gives examples of some maximum allowable dimensions.

Anthropometric data must always be used in a cautious manner, with a sound appreciation of the design requirements and practical considerations. In particular, the designer should try to predict the consequences of a mismatch—how serious they would be and who would be affected. The height of the handle on a door leading to a fire escape in an apartment block dramatizes the seriousness of such mismatches—it is essential that a very wide range of users—including children—be able to reach and operate the handle in an emergency. The design of passenger seats for urban transportation systems is also important—although somewhat more mundane. Because the seats are used regularly by a very wide range of users, even small imperfections will affect the comfort of a large number of people every day. When using anthropometric data, the selection of a suitable cut-off point depends on the consequences of an anthropometric mismatch and the cost of designing

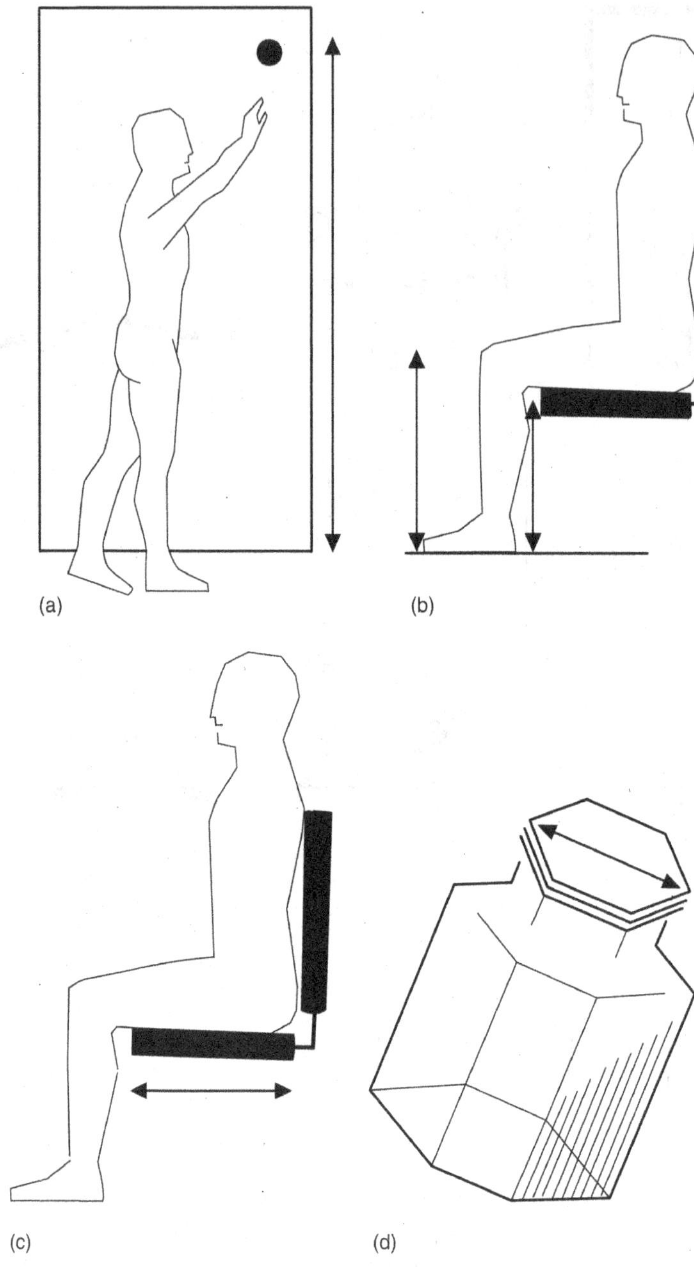

FIGURE 3.9 Some maximum allowable dimensions. (a) A door lock must be no higher than the maximum vertical reach of a small person. (b) Seat heights and (c) depths must not exceed the popliteal height and buttock–knee lengths of small users. (d) Screw-top lids must be wide enough to provide a large contact area with the skin of the hand to provide adequate friction, so that pressure "hot spots" are avoided. They must not exceed the grip diameter of a small person though.

for a wide range of people. One of the most important tasks is to predict and evaluate what any mismatches are likely to be. It is not normally sufficient to specify the required dimensions without also considering other aspects such as usability and misuse.

TOOLS AND PROCESSES

HFE Workshop 3.2 provides more worked examples to illustrate the statistical essentials of anthropometry in HFE. It describes how to use these equations and the z-scores in Appendix A to estimate ranges and body sizes with known accuracy.

HFE WORKSHOP 3.2

USE OF STATISTICS AND STATISTICAL TABLES WITH ANTHROPOMETRIC DATA

Suppose we wish to estimate the fifth percentile popliteal height of Indian males. From Table 3.3, we find that

$$\bar{x} = 415 \text{ mm and sd} = 21 \text{ mm}$$

Table 3.1 indicates that the fifth percentile value will be 1.64 standard deviations below the mean:

$$\text{Popliteal height (5th percentile)} = 415 - (21 \times 1.64) = 415 - 34.4 = 381 \text{ mm}$$

This implies that a fixed-height seat for Indian males should be no higher than about 380 mm. This seat height will fit 95% of males, even if shoes are not worn.

If we want to design to fit down to the first percentile male, then, from Table 3.1

$$\text{Popliteal height (1st percentile)} = 415 - (21 \times 2.32) = 415 - 48.7 = 366 \text{ mm}$$

Let us suppose, in another example, we wish to sell prefabricated fitted kitchen furniture to China for the burgeoning home products market. We want to know whether our kitchen tabletops, cupboards, and work surfaces, designed for sale in the United States and Northern Europe, will fit Chinese housewives. Our kitchen work surface is 1 m above the floor, which is the mean U.S. female standing elbow height. The standard deviation for female standing elbow height is 52 mm, so for 90% of our users, the work surface will be, at most, 1.64×52 mm, or 85 mm, too high or too low. This is unlikely to affect the ability of most of our customers to carry out their kitchen chores comfortably.

What about female Chinese users? From Table 3.3, the standing elbow height is

$$x = 935 \text{ mm and sd} = 39 \text{ mm}$$

This means that for the *average* Chinese female user, the work surface is 1000−935 mm, or 65 mm, too high. For Chinese females, 65 mm above the mean corresponds to

$$65/39 = 1.67 \text{ standard deviations above the mean}$$

From the z-values in Appendix A, a z-score of 1.67 above the mean corresponds to an area of 0.0475 under the normal curve. In other words, only 4.75% of Chinese females have a standing elbow height of 1000 mm or greater—so the worktop is above the elbow height of about 95% of Chinese users. Is this a design fault that needs to be rectified?

For our U.S. users, we were prepared to allow the worktop to be up to 85 mm above or below the mean U.S. female elbow height. We know that for 5% of U.S. females, it will be

(*Continued*)

HFE WORKSHOP 3.2 (*Continued*)

too high and for fifth, too low. For what percentage of Chinese females will the work surface be too high?

$$\text{The criterion for acceptance} = 1000 - 85\ \text{mm} = 915\ \text{mm}$$

This is 935 – 915 = 20 mm below the mean elbow height, or 20/39, or 0.51 standard deviations. From Appendix A, a *z*-score of 0.51 corresponds to an area to the left of 0.305. In other words, for 30.5% of Chinese females, the work surface will be too high.

Should we be concerned? If 30% of customers are potentially inconvenienced, then further investigations should be carried out. We might look into what kind of tasks are carried out. Do task objects raise the functional work surface height even more (e.g., thick chopping boards, deep pots, etc.)? Perhaps we should gather a focus group of users to discuss the issue further (see Chapter 14 for advice on focus groups).

Cost–Benefit Analysis and Trade-Offs

Sometimes it is not necessary to use anthropometric data because there may be no costs incurred in designing to suit everyone (a doorway or entrance is a hole and the bigger it is, the less building materials are used). However, there are often trade-offs between the additional costs of designing to suit a wide range of people and the number of people who will ultimately benefit. As can be seen from the normal distribution, the majority of individuals in a population are clustered around the mean, and attempts to accommodate more extreme individuals soon incur diminishing returns, since fewer people are accommodated into the range.

Example 1: Let us suppose that the minimum height of a car interior has to be specified. Increases in roof height increase wind resistance and construction cost but provide head clearance for tall drivers. Thus, there are both costs and benefits to building cars with plenty of headroom for the occupants. If the mean sitting height for males is 90 cm with a standard deviation of 5 cm, a ceiling height of 91 cm (measured from the seat) will accommodate 50% of drivers, with 1 cm of clearance to allow for clothing or hair (except hats). An increase in ceiling height of 5 cm will accommodate a further 34% of drivers, so that a total of 84% are now catered for. A further increase of 5 cm will accommodate an extra 14% of drivers, bringing the total to 98%. However, a further increase of 5 cm will only accommodate an extra 2% of drivers in the population. Because this 2% probably represents a very small number of potential customers, the additional costs of accommodating their anthropometric requirements in the design of every car built become significant. It may be more cost-effective to exclude these individuals from the generic design, but retrofit the finished product to accommodate one or two extremely tall buyers (e.g., by making it possible to lower the seat).

This example should illustrate why the 5th and 95th percentiles of anthropometric variables are often used to determine the dimensions of products. Ninety percent of potential users are accommodated using this approach, and further sizable alterations to product dimensions will only accommodate a small number of additional users—the point of diminishing returns has been reached.

Example 2: The designers of a new ship wish to specify the minimum deck height to allow sufficient headroom for the crew. The ship will be built in 2 years and service for 30 years. Assume a stature increase due to the secular trend of 1.5 mm/year and recommend an appropriate height, DH.

$$DH = SHx + CA + DA + STA(Yt - Yb) + PA$$

where

DH is the deck height.
SHx is the x percentile stature.
CA, the clothing allowance (shoes, helmets when worn) = 70 mm
DA, the dynamic allowance (accommodates head movements when walking) = 100 mm
STA[1], the secular trend allowance = 1.5 mm/year
Yt, the target year for specifying deck height = 2035
Yb, the base year (in which the height data were captured) = 1986
PA, the psychological allowance (perception of head room) = 50 mm

Use the ninety-fifth percentile male stature for the UK population, measured in 1986, at 1870 mm.

$$\begin{aligned} \text{DH} &= 1870 + 70 + 100 + 1.5(2035 - 1986) + 50 \\ &= 2163.5 \text{ mm} \end{aligned}$$

Using this value, 95% of personnel would not be at risk of head injury during normal operation. However, as can be seen from the shape of the normal distribution, even small increases in deck height will provide a large additional margin of safety. This margin of safety can be expected to last the lifetime of the ship because we have designed to accommodate the ninety-fifth percentile stature at the end of the product life cycle.

Mock-Ups and Models

Traditionally, HFE specialists and designers used physical mock-ups in the early stages of workspace design. These might be scaled-down versions of the proposed workspace made out of wood or cardboard and assessed using manikins to represent users of different sizes. In large projects, full-size prototypes can be built for evaluation with real users at later stages in the design process. Digital human modeling is now advanced, and several tried and trusted systems are available that enable early-stage prototyping and assessment against the requirements.

Digital Human Models

Digital human models facilitate visualization of the physical interaction between users and hardware. In modern systems, the digital models have reasonable functional anatomy and realistic ranges of motion in the joints, enabling basic movements to be simulated within a digital representation of the workspace. SAMMIE (System for Aiding Man–Machine Interaction Evaluation, SAMMIE CAD Ltd.) is an example of such a system (Figure 3.10). The anthropometry of a three-dimensional (3D) "man-model" displayed on the computer screen can be manipulated and the consequences evaluated using a computer-generated representation of the product being designed. Tools such as SAMMIE and JACK have some computer-aided design (CAD) capability. According to Blanchonette (2009), JACK's digital human is composed of 71 individual segments, including a 17-segment torso, made up of 5182 polygons. The segments are joined together by 69 joints that have 135° of freedom and incorporate data on human body contours and strength limits, enabling high-fidelity simulation of human movement.

Several CAD systems are available that incorporate model humans whose body dimensions can be manipulated using anthropometric data.

Figure 3.11 shows digital human modeling being used in a virtual environment to design a recovery vehicle with severe space restrictions.

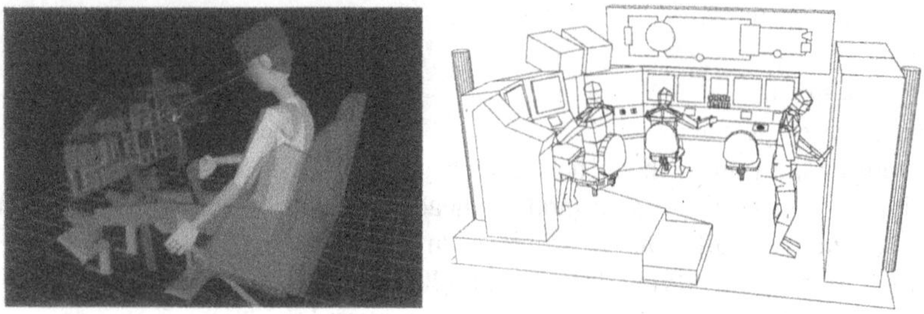

(a) (b)

FIGURE 3.10 Computerized anthropometry: (a) JACK (From JACK, a registered trademark of the University of Pennsylvania, with permission.) and (b) SAMMIE. (From SAMMIE CAD Ltd., Loughborough, United Kingdom, with permission.)

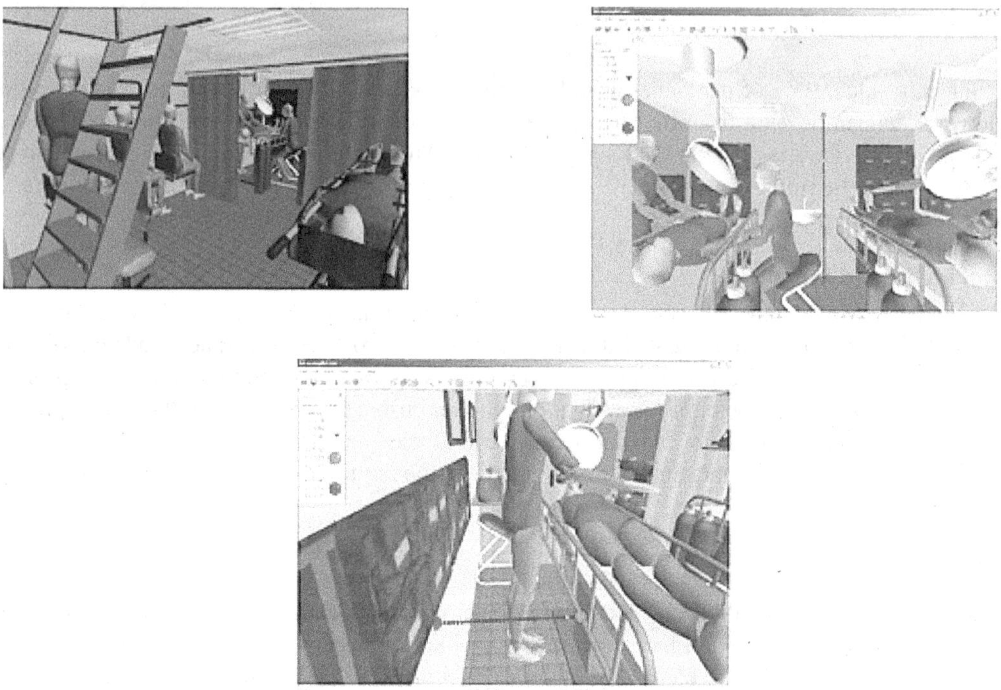

FIGURE 3.11 Digital human modeling in a virtual environment in the design of an emergency recovery vehicle: (a) view of general arrangement to estimate capacity, (b) use of digital tape measure to assess head-room, and (c) use of digital tape measure to assess clearance. (From Bridger et al., 2002. Crown Copyright, contains public sector information licensed under the Open Government Licence v2.0.)

Anthropometric Scaling Techniques

Scaling techniques are used when designing for users whose anthropometry is unknown—for example, a group of workers known to differ from the general population for some reason. HFE Workshop 3.3 gives an example of a scaling technique described by Pheasant (1986), known as "Rapid Anthropometrics Scaled for Height." The method requires that the mean and standard deviation of the target group are known. The mean and standard deviation of linear anthropometric measures (measures that depend on the length of the bones) are estimated using data from a reference population thought to be similar in terms of body proportions.

HFE WORKSHOP 3.3

RASH ANTHROPOMETRY

Pheasant (1986) recommended the use of rapid anthropometrics scaled for height (RASH) technique to estimate the body dimensions of populations whose anthropometry is unknown. The technique requires data on the stature of the target population. These data are normally available from medical records.

RASH estimates are made by calculating scaling factors for the anthropometric variables of interest using data from a known population, and then applying these scaling factors to the height data in the target population. Suppose we want to estimate the sitting height of Chilean bus drivers but have no data on the anthropometry of the Chilean population, where do we begin?

1. Obtain data on the stature of the drivers from medical records and calculate the mean and standard deviation of stature (all data are in centimeters).

   $x = 166$, sd=5.5

2. Obtain data on the mean and standard deviation stature and popliteal height for a similar group of people, assuming that body proportions will be similar. We will use data for Brazilian males.

   Stature: $x = 170$, sd=6.6; sitting height: $x = 88$, sd=3.5

3. Calculate scaling ratios for the mean and standard deviation of stature.

   SR (mean sitting height)=88/170=0.518 SR (sd sitting height)=3.5/6.6=0.53

4. Multiply the mean and standard deviation of stature for the Chilean bus drivers by the scaling ratios.

   SR (mean sitting height)=88/170=0.518
   SR (sd sitting height)=3.5/6.6=0.53

5. Calculate the fifth percentile sitting height by subtracting 1.64 standard deviations from the estimated mean sitting height.

   Estimated fifth percentile sitting height of Chilean bus drivers=86−(1.64×2.92)=81.2

6. Calculate ninety-fifth percentile sitting height by adding 1.64 standard deviations to the estimated mean sitting height.

   Estimated ninety-fifth percentile sitting height of Chilean bus drivers= 86+(1.64×2.92) 90.8

Use of the technique is best limited to body dimensions that depend on the length of bones, sitting eye height, popliteal height, and is not recommended for estimating circumferential dimensions such as girth, or measures, such as hip breadth, which also depend on the distribution of adipose tissue in the body. RASH is easy and quick to use, but it is a crude technique that depends on a number of critical assumptions: namely, that the body proportions of the target population are the same as those of the reference population. Ideally, raw data from the reference population should be available to rebalance the reference population to match

(*Continued*)

HFE WORKSHOP 3.3 (*Continued*)

the target population before using RASH. Suppose, for example, that the target population is much younger than the reference population (e.g., mean age of 35 years compared to 56 years in the reference population). We know that changes in body mass, stature, and body shape occur with age. One way of dealing with this would be to rebalance the reference population by removing data from older subjects so that the mean ages of the reference and target populations are similar, and then recalculate the means and standard deviations in the reference population.

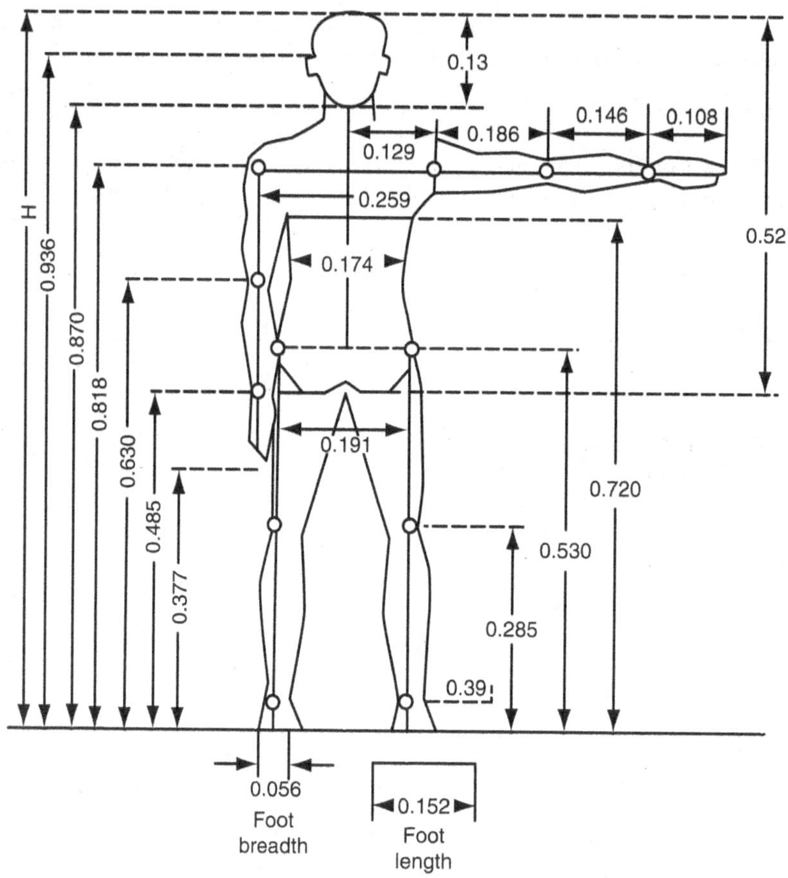

FIGURE 3.12 Linear body dimensions expressed as a percentage of stature. (Redrawn from Keyserling et al., 1988.)

Sometimes we are dealing with a subset of the general population, in which case scaling techniques can be used to make simple proportional estimates of body dimensions. Figure 3.12 (from Keyserling et al., 1988) expresses the lengths of several linear anthropometric dimensions as a proportion of stature for the U.S. population.

If the mean stature of a group of male U.S. workers in a target population were 1800 mm (compared to 1755 mm in the general population), then the mean standing elbow height would be

$$\text{Standing elbow height} = 1800 \times 0.63 = 1134 \text{ mm} \tag{3.9}$$

Similarly, if the mean and standard deviation of stature in the target population were known, we would be able to generate estimates of the smaller and larger percentile statures, as well as scaled estimates of linear dimensions using the proportions in Figure 3.12.

Workstation Design and Reach

One of the most basic applications of anthropometric data in HFE is the specification of reach distances to ensure that controls and other work objects are placed within a "zone of convenient reach" (ZCR). The reach zone is defined such that the worker can reach work objects, either from a standing or sitting position, without having to lean forward. The ability to do this depends on two variables: the forward reach distance (variable number 7 in the table) and the vertical distance (*d*) between the shoulder and the work object. Pheasant (1986) pointed out that the geometry of forward reach follows Pythagoras' theorem, since the arm can be considered the hypotenuse of a right-angled triangle (Figure 3.13).

For a seated worker (Figure 3.13a and b), the radius of the ZCR (ZCR radius) is given by

$$\text{ZCR}^2 = r^2 - d^2 \tag{3.10}$$

And for a standing worker (Figure 3.13c), the maximum horizontal distance (HD) for the placement of controls on a wall in front of the operator is given by

$$\text{HD} = r^2 - d^2 \tag{3.11}$$

In practice, we would usually base our ZCR radius and HD estimates on the fifth percentile female forward reach, since forward reach is normally regarded as a maximum allowable dimension. Figure 3.14 summarizes general recommendations for workstation layout for standing operators.

The problem of designing to suit a range of users can be approached in several other ways.

Make Different Sizes

In clothing and school furniture design, a common solution is to design the same product in *several different sizes*. Anthropometric data can be used to determine the minimum number of different sizes (and the dimensions of each size) that will accommodate all users. Mass production or long production runs often bring economies of scale in product design through reduced retooling and stoppages. This usually has economic benefits and demonstrates why it is important to determine the *minimum number* of sizes in a product range that will accommodate most of the users in the population in question.

Clothing design is the best example of fitting the product to the user. There are standards, such as ISO 3636, for the designation of clothing sizes. It is left to the manufacturer to decide what sizes to make, and this is often done by trial and error or by "knowing the market," rather than by carrying out an anthropometric survey. Head girth, neck girth, chest girth, bust girth, under-bust girth, waist girth, hip girth, stature, outside leg, and inside leg length are dimensions used to designate the size of a piece of clothing. National and international standards only specify which dimensions the manufacturer must use to designate sizes; they neither specify the number of sizes in a range nor what actual size the clothing should be.

Clothing sizing systems divide a variable population into roughly homogeneous subgroups (McCullogh et al., 1998). It is assumed that members of a subgroup are a single size and shape so that a single garment can fit everyone in the group. The simplest approach is to choose one or two dimensions that are thought to be crucial if a good fit is to be achieved, and then divide the range into an equally spaced series of sizes according to these dimensions. The remaining variables

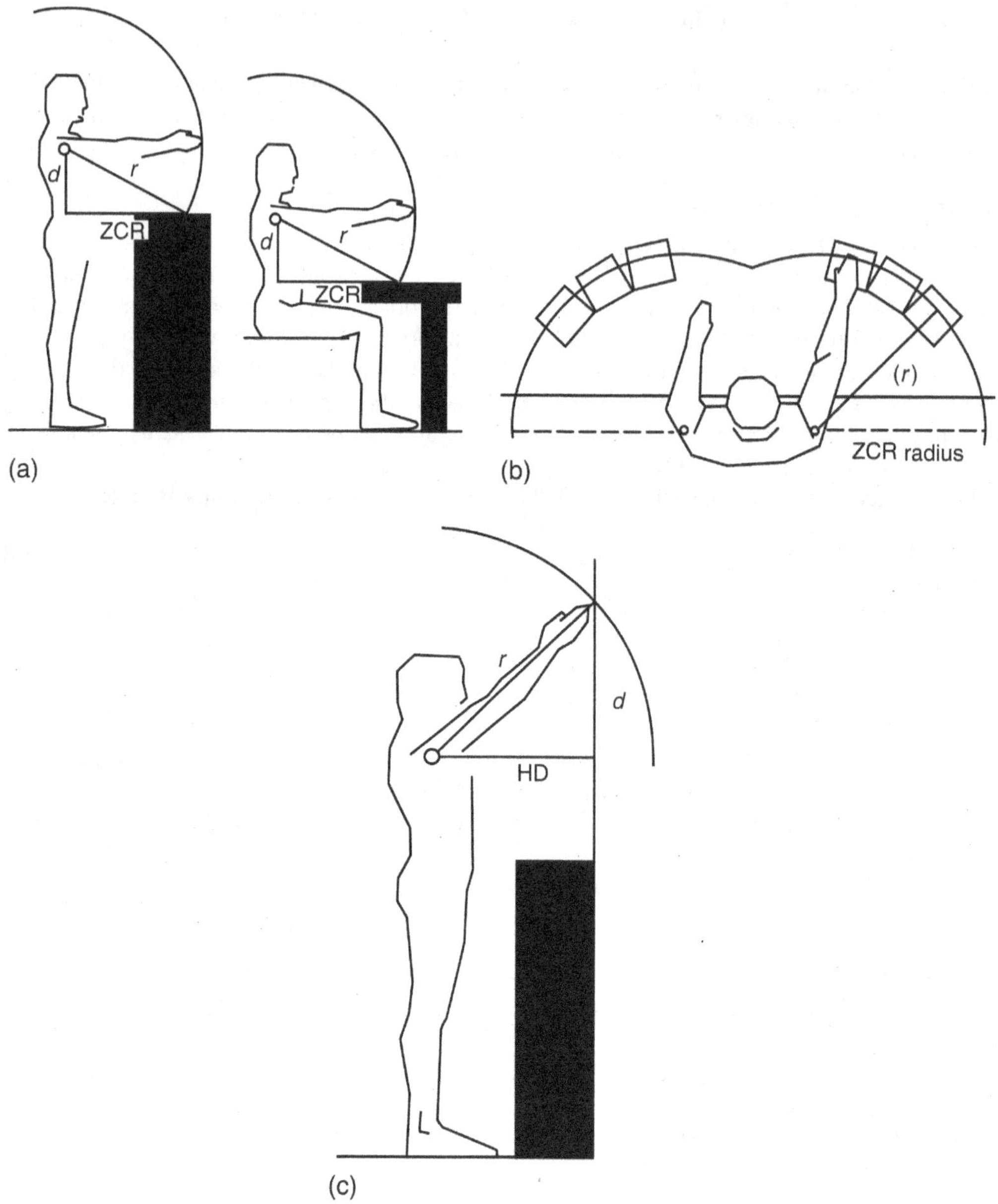

FIGURE 3.13 The radius of the ZCR can be calculated using Pythagoras' theorem. If the reach fingertip or palm reach distance (*r*) and the height of the shoulder above the horizontal work surface (*d*) are known (a and b). For wall-mounted controls, the maximum height of a control above the workers' shoulder can be calculated if reach (*r*), and the HD of the shoulder from the wall are known (c).

are adjusted across the range in proportion to the criterion sizes. Sizing is difficult because many anthropometric dimensions are not highly correlated—long bone measures and girth measures, for example, and users may struggle to find a shirt in which both the collar and the sleeves fit. Because anthropometric variables are usually normally distributed, a far greater number of people will fit clothes in the middle of the range, and only short production runs are needed to accommodate the extreme sizes, which may be difficult to find in shops.

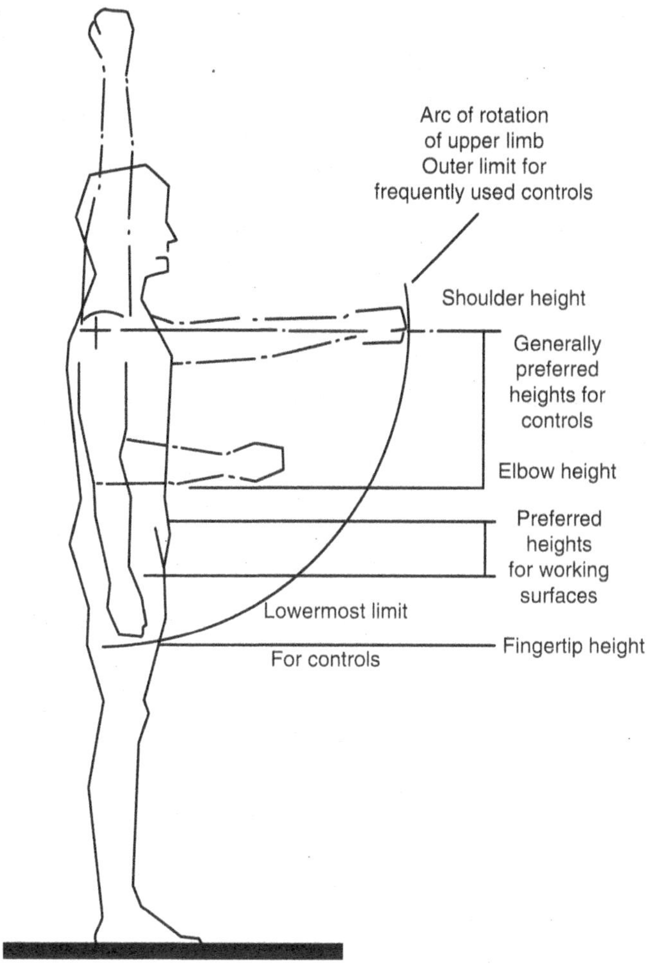

FIGURE 3.14 Recommendations for the height of controls. (Redrawn from DefStan 00–25 [Defence standard 00–25, Ministry of Defence UK].)

Multivariate Techniques for Workspace Design and Sizing Clothing

One of the problems of using anthropometric data in isolation is that fifth and ninety-fifth percentile people do not exist. That is to say, a man of ninety-fifth percentile stature will not have ninety-fifth percentile reach, elbow height, and so on. This is because human anatomy is inherently variable. Blanchonette (2009) describes how, as the number of anthropometric variables increases, the proportion of people fitting a particular percentile on all of them decreases. One variable (such as ninety-fifth percentile sitting height) might enable 95% of users to be accommodated. However, only 47.4% of U.S. Naval aviators could be accommodated when 13 anthropometric variables were considered. In another example, the 5th percentile female stature was estimated by adding seven fifth percentile linear segments, giving a stature of 139 cm. The actual fifth percentile stature was 152 cm. This is why systems such as JACK use principal component analysis to derive components of body sizes. From these components, anthropometric figures (digital mannequins) can be generated according to the strength of the associations between the variables in a component. For example, the analysis might yield nine different mannequins, two of which have similar stature but vary greatly in arm length and foot length. Blanchonette cites evidence that the mannequins generated by JACK are valid. Kozycki et al. (2002) recorded the ranges of motion of the left and right hands of two military helicopter pilots wearing flight clothing. A model of the seat was imported into JACK, and anthropometric data from the pilots were used to create the clothed mannequins. The reach envelopes of the mannequins and the pilots differed by only 1–2 cm.

The goal of achieving as good a fit as possible can be restated as:

- Accommodate as large a percentage of the population as possible with ready-made garments
- For accommodated individuals, provide as good a fit as possible
- Use as few sizes as possible

We saw in the workshop how PCA can be used as a data reduction technique to group anthropometric measures into "components," wherein all the measures in a component "belong together" because they covary. Thus, PCA groups variables measured on people into higher-order components—a data set with a large number of anthropometric dimensions is reduced to a smaller number of dimensions. A statistical technique known as cluster analysis is used to search for groups, or clusters, of anthropometric variables in the target market. In simplistic terms, cluster analysis groups people according to their similarity based on the variables in the data set that describe them. Thus, the output of a PCA is a set of components that group anthropometric variables together, and the output of a cluster analysis is clusters of people grouped according to their similarity on the variables used to describe them. Any number of anthropometric variables can be used in an analysis so that the combination of body dimensions will fit a number of people into a group. Cluster analysis is an example of an "optimization" technique. The optimization criterion is to select a set of values that will minimize the average distance, where distance corresponds to a lack of fit. Because this applies to all the anthropometric variables used to create the clusters, an item of clothing that fits the neck of people in a cluster will tend to fit the chest and sleeves as well. The result is that clothing sizes do not increase in fixed intervals, but according to the clusters identified in the target market. Some jumps between sizes in the range may be bigger than others, and some may change more than others in a single jump. Sizes may sometimes differ in terms of proportion and not in terms of largeness or smallness, and this may mean that potential buyers would have to try on more sizes before finding one that fits. However, from the retailer's point of view, the distribution of people fitted by each size is likely to be more uniform, and this may simplify the ordering of stock.

Kolose et al. (2021) used both PCA and cluster analysis in a study to optimize clothing sizes for the New Zealand military. They analyzed data from 84 measurements on 1003 personnel. The data were captured using scanners and manual measurements of some variables. The data were stratified before analysis (male/female and shirt/trouser), and then a PCA was conducted to find the variables that loaded most strongly onto the components identified. For females, stature and waist girth were chosen for shirt design, and inseam length and hip girth for trousers. For males, stature and waist girth were chosen for shirt design, and inseam length and waist girth for trousers. For males, the cluster analysis yielded 10 clusters for both shirt and trousers, and 6 of each for females. The findings can be illustrated with some examples. For males, cluster 7 consisted of small personnel with low waist girth and stature (<830 mm and <1609–1708 mm). Cluster 10 consisted of large males with waist girth >186 mm and medium stature. From the analysis, digital human models of all the clusters were created. Figure 3.15 shows clusters for the male shirt analysis (upper panel with body height versus waist girth) and the male trouser analysis (inseam length versus waist girth). Ten clusters can be distinguished. The body shape images are derived from each cluster centroid.

The unisex sizing system in use at the time had 12 shirt and trouser sizes, so the study had the potential to reduce the number of sizes needed. For present purposes, the study demonstrates how multivariate data analysis can be used to generate more realistic representations of users by disclosing how large numbers of anthropometric variables covary in a user population. From such analyses, representations of user types can be synthesized digitally.

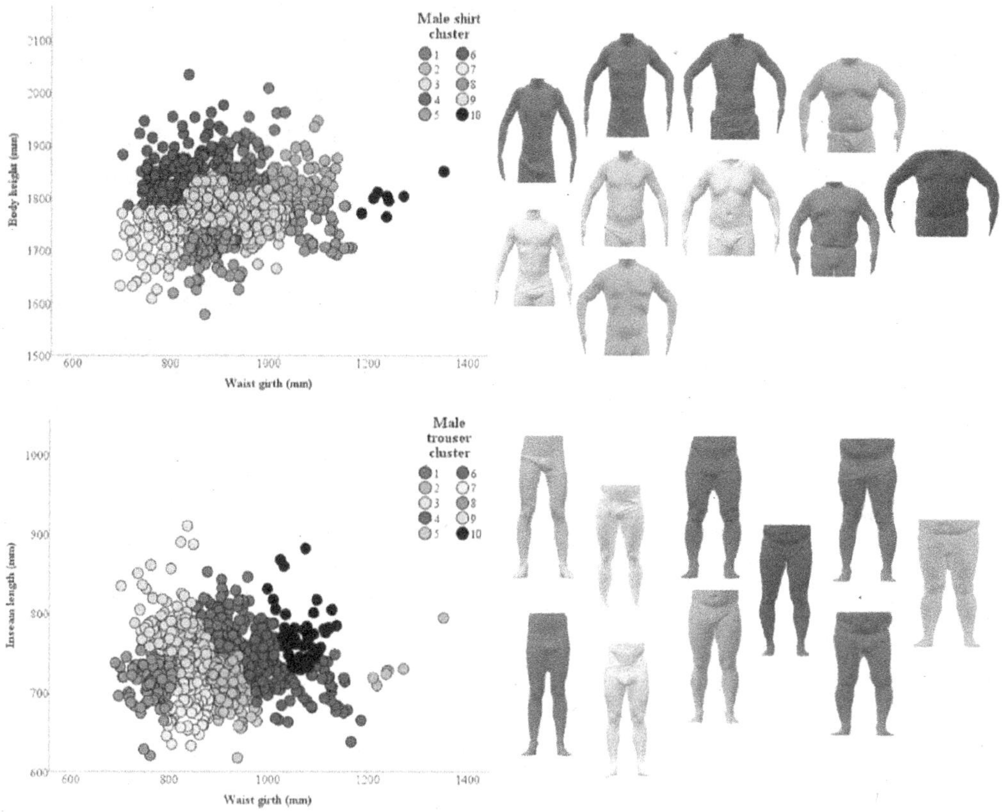

FIGURE 3.15 Clusters for male shirt (upper panel) and male trouser (lower panel) analyses with body shape images derived from the cluster centroids. (With permission, Dr. Kolose.)

SYSTEM INTEGRATION

There are a number of key considerations throughout the design process that can be used to integrate anthropometric considerations with the rest of the design process. Clearly, the specification of the user population comes directly from the description of the target users for whom the system is being designed. Some key design decisions early on are as follows:

Understand the Requirements

The recent COVID pandemic led to an urgent requirement for personal protective equipment for the public and for health care workers. Designing a face mask requires far more than a good fit between the mask and the face. Requirements elicitation is needed to understand the user requirements against an assessment of all the risks (risks due to exposure to the virus and risks due to wearing the mask). This requires stakeholder engagement and an iterative design process. Examples of requirements include a proper seal between the face and the mask for users with and without beards; the ability to carry out critical and generic tasks while wearing the mask; visibility/field of view; effects on communication with patients and colleagues; effects on the performance of tasks requiring fine motor skills; ease of donning and doffing; comfort and thermal comfort; the ability to move easily—especially the head and neck; and sustainability issues relating to reusable and disposable masks. See Boix et al. (2021) for further information.

FIGURE 3.16 Understand the context of use. Escape hatches are used in emergencies when people are wearing protective clothing and apparatus. This adds to the body circumference and must be considered during the design.

Understand the Context of Use

Anthropometric data alone are not sufficient to ensure a good fit. Designers must also consider the context of use, as illustrated in Figure 3.16.

Anthropometry, Workstation Design, and Task Analysis

Anthropometric data help designers to optimize the dimensions of workspaces and furniture but are of limited use in optimizing the final design. Some kind of task analysis is always needed so that the anthropometric recommendations can be considered in relation to the task demands. Figure 3.17 presents the results of a very simple observational task analysis known as "link analysis," used to investigate alternative layouts for domestic kitchens. Two plan views are presented and the black lines show the movements of the user when working in the kitchen. The thicker the lines, the more frequently the user moves between linked objects (e.g., there is much movement between the work surfaces and the sink and less between the refrigerator and the stove). The U-shaped kitchen layout requires less walking but more twisting and turning. In general, objects that are used in succession are positioned close together and vice versa.

Space Planning for Offices

Margaritis and Marmaras (2007) provided guidance for optimizing the use of space in offices for more than one person:

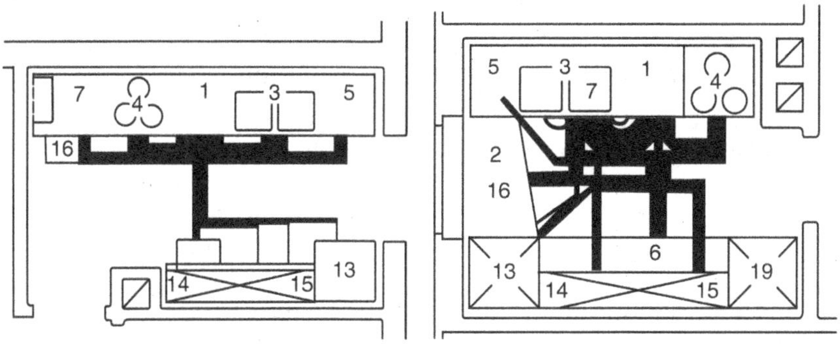

Two-sided kitchen, 7 m^2
1, 5, 7 = Places where things are put down, or work is carried on
4 = Cooker, 3 = Sink
14, 15, 16 = Crockery cupboard and refrigerator

U-shaped kitchen, 6.2 m^2
1, 2, 5, 6 = Places where things are put down, or work is carried on
4 = Cooker, 3 = Sink
13,14,15,19 = Cupboards and refrigerator

FIGURE 3.17 Link analysis of a user's movements when working in two alternative kitchen layouts. (Redrawn from Grandjean, 1973.)

1. Determine the number of people who will work permanently in the office.
2. Determine the organizational structure (see Chapter 16).
3. Describe the main activities, the need for interaction between different employees, the need for privacy, reception of visitors, etc.
4. Determine the equipment needed for each activity (e.g., VDT, printer, storage requirements, etc.).
5. Detailed layout:
   Allow 50 cm of free space in front of windows.
   Allow 3 m to the front and 1 m on either side of the main entrance.
   Allow 1.5 m in front of, and 0.5 m to the side of, all other doors.
   Allow 0.5 m in front of heating units.
   Free space around desks:
   Allow 55 cm along the front of desks for passage.
   Allow 50 cm along the sides of workstations for ingress and egress.
   Allow 75 cm at the back of desks for seating space.
   Allow 100 cm at the back of desks if there are cabinets behind the desk.
6. Determine requirements for proximity: physical separations should be compatible with task separations (the more two workstations collaborate, the closer they should be). Place the units that interact most with other units toward the center of the space.
7. Arrange individual workstations in accordance with local health and safety regulations for office work (see Chapter 4).

"Fit-for-Use" Surveys and Acceptance Testing

There are many ways of assessing how well a work environment fits its users. We can measure the critical features of the equipment and compare them with user anthropometry, we may unobtrusively observe people at work and note any apparent problems. Alternatively, we may carry out some kind of task analysis and consult the users—asking them to report any problems experienced. A minimal set of items for a questionnaire designed to assess "fit" is given in Table 3.8.

TABLE 3.8
Questionnaire Items for a Basic Survey of Anthropometric Fit

Personal Characteristics (Self-Reported or Measured)	Difficulties with Equipment Items
Age	Equipment difficult to operate
Sex	Requires too much force
Height	Parts that are difficult to grip
Weight	Have to reach too far forward
Chest circumference	Have to reach too far up
Department	Have to reach too far down
Job title	Have to stand in an awkward posture
Length of time in job	Have to sit in an awkward posture
Any work-related injuries?	Unable to see parts of the equipment
	Equipment too heavy
	Equipment too bulky
	Straps do not fit/uncomfortable
	Cannot get access
	Problems with ingress

Fit-for-use surveys can be conducted on existing systems to facilitate learning from experience and support the design of better systems, or they can be conducted as part of a "post-implementation evaluation" to determine how well the objectives have been met.

Designing for Female Users: Missing Requirements, Misplaced Requirements, and Misunderstood Requirements

The last 50 years have seen increased participation of women in traditionally male-dominated occupations. Much legacy equipment was designed for men, and the requirements for designing for women have not always been properly specified. Examples include hand tools, personal protective equipment, vehicle interiors, and mobile devices (Criado-Perez, 2019).

Parnell and Plant (2024) reviewed the evidence for a gender data gap. Women are more likely to suffer musculoskeletal injuries in office work and take more long-term sick leave than men. During the COVID-19 pandemic, poorly fitting personal protective equipment reduced the protection afforded to nurses. Stab vests for police officers were modeled on the male body and provided less protection for female officers. A striking example is reported in the design of crash test dummies for vehicle safety testing. These manikins are designed based on the male body form (fiftieth percentile male), which is then scaled up to the ninety-fifth percentile male and down to the fifth percentile female—ignoring the sex differences in form and function mentioned earlier in this chapter and therefore failing to properly account for the protection afforded by seats and seat belts, putting female occupants at increased risk.

As an example, Okholm Kryger et al. (2022) discuss the design issues associated with women's professional football ("soccer"), where female football players, at the time of writing, are forced to use legacy equipment designed for men. They introduce several issues for discussion with sports engineers:

1. First, the design of equipment and uniforms for females is not limited to sport.
2. Some players find that the available shorts are too short and too exposing, and the colors must match those of the men's team. Light shorts may lead to concerns about the visibility of leakage during menstruation.
3. Unisex socks may not fit properly, causing slippage and increased risk of injury.
4. Restrictions on the use of headgear may discriminate against certain religious groups.
5. Many players have to use boots designed for men. Women's feet differ in size and shape from men's feet. Poor fit may lead to blisters and impact performance. Anterior cruciate ligament strain is a major concern.
6. Women play with the same footballs as men. With lighter balls, women kick the ball harder with less exertion, while heart rate and performance are unaffected. There may be concerns about injury while heading the ball, which might be reduced if the ball were lighter.
7. Tracking devices for heart rate and so on often default to male settings.
8. Medical and rescue equipment, such as neck braces, are not optimally designed for small women.

From the above, it can be concluded that:

1. Requirements specified for male users may need to be revised to accommodate females.
2. Much existing design practice is not optimal because of *missing requirements*.
3. Requirements were not elicited, possibly because of inappropriate assumptions, misunderstanding of the context of use, and lack of a participatory approach.
4. Additional requirements need to be specified to accommodate females.

As an example of the first case, we can consider differences in grip strength between males and females. Grip strength is greater in males due to greater forearm muscle mass, larger hands, and thicker fingers, increasing the contact area at the hand-handle interface. From Table 3.7, the mean grip strength for males (right hand) is 46.2 kg (sd 12.6 kg). For females, it is 27.9 kg (sd 7.6). The tenth percentile male grip strength (constants from Table 3.5) is therefore:

$$46.2 - 1.28 \times 12.6 = 30.07 \text{ kg}$$

This is 2.17 kg above the mean female grip strength.

2.17 kg is 2.7/7.6=0.29 standard deviations above the mean female grip strength, which corresponds to the fifty-sixthpercentile female grip strength (refer to Appendix A for the z table). So, a piece of equipment designed to be operable by 90% of males would only be operable by 44% of females. Grip strength places constraints on the design of hand tools, levers, cranks, and the lids on containers such as jars.

Psychosocial Factors: Anthropometry and Personal Space

Personal space can be defined as the area immediately around the body. Two important issues in "psycho-anthropometry" are the volume of space regarded as personal territory and the consequences of an invasion of this space by others. Large individual and cultural differences exist. Invasion of personal space stimulates a defensive reaction in parts of the brain (particularly the amygdala, a part of the brain involved in threat detection and emotional stress reactions). There is experimental evidence that invasion of personal space by both social (people) and non-social (insects or animals) stimuli provokes these reactions (Vieira et al., 2020).

In an experiment that is now infamous, Middlemist et al. (1976) tested the hypothesis that personal space invasion increases arousal. They measured the time to onset and duration of micturition (urination) of 60 men using a public lavatory under one of three conditions—subject alone (control condition), subject standing next to a confederate (adjacent invasion), and subject one urinal removed from the confederate (moderate invasion). The confederate was an accomplice of the experimenters and arrived after the subject had unwittingly taken his place. Onset time of micturition increased from 4.9 s in the control condition to 6.2 (moderate space invasion) and 8.4 s for adjacent invasion. Micturition duration dropped from 24.8 s in the control condition to 23.4 and 17.4 s in the space invasion conditions, respectively. In the context of public lavatory use, the minimum required personal space would appear to be about the distance of one urinal.

The degree of stress depends on the context and the culture of the target population. Invasion of personal space in a library, for example, is much more stressful than in a crowded train or lift. In the workplace, a minimum separation of desks or benches of approximately 1.2 m is thought to be necessary.

Benefits of Protective Clothing That Fits

Ensuring correct fit is one of the most basic requirements for an item to function correctly. When designing protective clothing, there may be a requirement to minimize the restriction of movement of the arms and legs while maximizing protection. Requirements often have to be traded off against one another and it is useful to be able to quantify movement range. Figure 3.18a shows a gravity inclinometer that can be used to measure functional movement range in a variety of applications. The device is essentially a protractor with an embedded spirit level. An angular reading is taken at a starting position and again after a prescribed movement has been made. The difference in the angles gives an indication of the functional movement range and can be used to compare different clothing assemblies. Inclinometers can also be used to quantify movement restriction. In Figure 3.18b, the effectiveness of a neck brace for restricting head movement in rescue stretchers is being assessed.

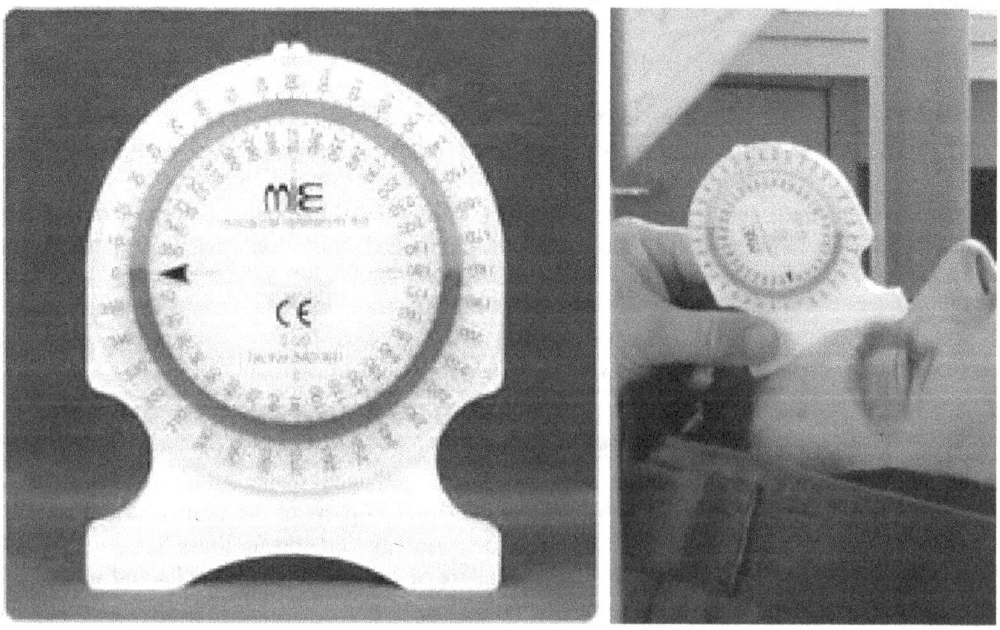

FIGURE 3.18 (a) Gravity inclinometer (™ MIE instruments). (b) Assessing the effectiveness of a neck brace in a rescue stretcher devices.

Industrial Workplace Layout

Lim and Hoffman (1997) investigated the performance of a light assembly task when components were laid out within the "ZCR." ZCR is defined, for each hand, as the area of the table within the arc swept by the hand with the elbow extended (within this zone is the "normal working area," defined by a similar sweep of each hand but with 90° of elbow flexion). An approximately 10% improvement in assembly time was achieved when items were placed within the ZCR and arranged in an ergonomically designed jig. These improvements led to more efficient hand movements (reducing lengthy reaches and contralateral movements and increasing the number of simultaneous hand actions). The fact that the naïve operators participating in the experiment did not spontaneously optimize the working arrangements (even though they were permitted to) highlights the need for this to be done when the task is designed and included in the initial training.

Status of Anthropometry in HFE

Most anthropometric databases are derived from surveys in which the reliability of the instruments was demonstrated in the early stages, at the time when the anthropometric technicians were trained.

HFE Workshop 3.4 illustrates a method used to assess the reliability of measuring instruments.

HFE WORKSHOP 3.4

DETERMINING THE RELIABILITY OF POSTURE MEASURING DEVICES

A study to determine the effect of protective clothing (PPE) on functional movement is carried out. Subjects are measured while wearing light clothing and while wearing the PPE, performing standardized flexion, extension, and abduction of the hip and shoulder using a simple gravity goniometer. The researchers want to know how reliable the procedure is when the measurements are carried out by one researcher only (the "intra-tester" reliability).

They decide to perform a pilot study, taking five repeated sets of measurements from five subjects. The data shown below are from the hip flexion measurement (angles of hip flexion, in degrees, with the goniometer set to "zero" at the starting position—feet flat on the floor, subject looking straight ahead).

	Trial Number					
Subject	1	2	3	4	5	Σ
1	90	91	89	92	94	456
2	85	87	88	88	85	433
3	93	97	96	96	95	477
4	79	81	83	82	84	409
5	87	85	85	89	87	433
Σ	434	441	441	447	445	2208

One way to analyze these data to determine the reliability of the procedure is to calculate the *intraclass correlation coefficient* (ICC).

Theoretically, the ICC can be written as follows:

$$\mathrm{ICC} = \sigma^2(b)/\left(\sigma^2(b)+\sigma^2(w)\right)$$

where

$\sigma^2(b)$ = the variance of the hip flexion measure between the subjects

$\sigma^2(w)$ = the variance of the hip flexion measure within subjects

(Continued)

HFE WORKSHOP 3.4 (*Continued*)

If the measuring procedure (goniometer and the researcher using it) was 100% reliable, the only source of variance in the data would be $\sigma^2(b)$, and so the ICC would equal "1." If the measuring procedure was completely unreliable, differences between subjects would not be detectable and the ICC would equal "0."

Tss is a measure of the total amount of dispersion ("sum of squares") in the data, given by

$$\text{Tss} = \Sigma(x - \bar{x})^2$$

which, for ease of calculation, can be written as

$$\text{Tss} = \Sigma x^2 - (\Sigma x)^2/n$$

To calculate Tss, we would simply sum all the measurements to obtain (Σx=2208) and then square each measurement before summing the squared measurements to obtain Σx^2. These values would then be inserted into the expression above.

We could calculate another sum of squares from the row totals—a measure of the variability between subjects—"B"ss.

$$\text{Bss} = \left(456^2/5 + 433^2/5\ldots\ldots + 433^2/5\right) - (\Sigma x)^2/n$$

and

$$\text{Tss} = \text{Bss} + \text{Wss}$$

WORKED EXAMPLE

We run a one-way analysis of variance (ANOVA) on the data above and find

$$\text{Wss} = 59.20$$

$$\text{Bss} = 534.24$$

$$\text{Tss} = 593.44$$

And the reliability, R, of the method can be estimated as

$$\begin{aligned} R &= 1 - (\text{Wss/Tss}) \\ &= 1 - (59.20/593.44) \\ &= 0.90 \end{aligned}$$

The standard deviation of the method is given by $\surd(\text{Wss}/n - 1) = 3.85°$.

Therefore, our measuring procedure is very reliable. In practice, as long as all measurements are made by the same person, there will be very little measurement variance in the data.

INTERPRETATION

In this example, we had only one rater, so our ICC is a measure of the *intra-rater* reliability of the method—we can only say that our inclinometer is reliable when used by that individual. We cannot say that the inclinometer gives the "right" answer, only that it gives more or less

(*Continued*)

HFE WORKSHOP 3.4 (*Continued*)

the same reading each time a measurement is made. In practice, this means that we can carry out our trials with fewer subjects and fewer repeated measures of the same movements, but only if the same person takes all the measurements).

ADVANTAGES OF THE ICC

Correlation coefficients, such as Pearson's *r*, only tell us about collinearity between paired scores, which may be unrelated in all other respects, whereas the ICC tells us the extent to which repeated measures of the same thing *agree* with each other.

Of course, the ICC does not tell us that our inclinometer is valid—only that it gives a reliable reading when used repeatedly by the same person. If we wanted to compare hip flexion movements of people wearing different types of clothing, the measurement variance due to the inclinometer would not prevent us from detecting a real difference, if it were large enough to be detected. Before carrying out our reliability trial, we would first consider how accurate we needed to be to detect meaningful differences between clothing types in the context of use.

ICCs could also be calculated for different raters, with each rater making one or more measurements, and in practice, we would calculate ICCs for all the different postural measurements we are interested in.

CAVEATS

The example above has been simplified to illustrate the mechanics of the ICC. In practice, there are several different ways of calculating ICCs, depending on the overall measurement model and the assumptions made. Here, we have assumed a "random effects" model and ignored the possibility that trial order might be an additional source of systematic variance (e.g., subjects "loosen up" after repeated testing, causing the movements to increase each time they are measured).

Note: When reliability is expressed in relation to the total variability in the data, a measuring device can appear unreliable if the total variance in the sample is small (as is the case when the subjects are all very similar). The device should, therefore, be tested on a sample of subjects that is representative of the target population.

The software used in whole-body scanners is normally tested for reliability and validity prior to a large survey. The validity of basic data is limited to static situations—structural data may be used for design in situations where people adopt static postures. Caution should be used when applying these data to design problems involving movement, particularly skilled movement. Bilzon et al. (2000) investigated the associations between user anthropometry and the ability to escape through vertical hatches and bulkhead doors on ships. It was hypothesized that subjects with large bi-deltoid breadth and abdominal depth would find it more difficult to move through small spaces. However, even when encumbered by bulky protective clothing, subjects with broad shoulders were found to escape through the hatch more quickly than those with narrow shoulders. There was a strong positive association between escape time and lean body mass. In dynamic tasks such as these, performance appears to depend more on strength and technique than on body size.

Human models in systems such as JACK and SAMMIE are valid for use as visualization aids in engineering design, but their underlying functional anatomy has limited fidelity. They are only as valid as the assumptions used to enter anthropometric data into their databases. However, they can offer good diagnosticity in the evaluation of human–machine fit and are very useful for enabling ergonomists to explore possible design solutions in a cost-effective way.

Anthropometric surveys are expensive to conduct and can be intrusive if subjects have to disrobe and undergo measurements in a specially designed facility. Ethical approval must be obtained before the survey can commence, and personnel must be trained. Digital anthropometry has advanced rapidly with the widespread use of laser scanners and the use of multivariate statistical methods. Much existing data were obtained many years ago and have been "modernized" using assumptions about secular trends in body size, and so may be of limited validity when designing for populations that may have changed in other ways in the intervening years (due to immigration or increasing levels of obesity, for example). The validity of scaling techniques depends to a large degree on whether body proportions are the same in the target and reference populations. Clothing companies often have proprietary anthropometric data, which they make commercially available (Size UK is one such company in the United Kingdom).

RESEARCH DIRECTIONS

In light of the increasingly international nature of business and trade, data are lacking on the anthropometry of many populations. In developing countries, there may be differences in the anthropometry of urban versus rural people. In developed countries, there is a severe obesity crisis, with over half the population overweight or obese. Many existing databases, based on old survey data corrected for the secular increase in stature, will underestimate all measures that depend on body composition.

Parnell and Plant (2024) provide the guidance below to help close the 'gender data gap':

1. Consider how sex differences may impact the research discipline, as well as how the current management of the systems under study may be impacted by biases.
2. Understand the proportions of males and females within your user population by searching databases and statistical resources; do not assume there is an equal split.
3. Conduct representative sampling based on the population breakdown by:
   a. Considering the sociodemographic and other factors that may influence sampling methodology; for example, sampling within universities (or certain departments within them) may not be representative of other populations of people.
   b. Considering the time and accessibility of research trials—"flexibility to dfferent times of day and platforms" (e.g., virtual meetings can enhance sampling efforts).
   c. Allowing a sufficient participant recruitment period to ensure balanced/representative samples can be recruited.
4. Include considerations in system modeling outputs; for example, the composition of organizations, the implications for regulations, and ways to improve equality.
5. Detail the wider systemic considerations for gender in the write-up (intro/discussion).

SUMMARY

Anthropometric data provide the designer with quantitative guidelines for dimensioning workspaces. However, a number of precautions are needed if data are to be used correctly. Always

1. Define the user population and use data obtained from measurements made on that population.
2. Consider factors that might interfere with the assumption of a normal distribution of scores. For example, in some countries, stature may be negatively skewed because many individuals do not attain their potential stature due to disease or malnutrition.
3. Remember that many anthropometric variables are measured using semi-nude subjects. Allowance for clothing is often necessary when designing for real users. Centimeter

accuracy is usually appropriate because the effect of clothing on the estimates of user anthropometry can never be accurately predicted. These considerations are particularly important when using data on stature and leg length—allowances for heel heights of 5 cm or more may be needed depending on the user population and current fashions of dress. The effect of clothing also depends on climate—the colder the climate, the bulkier the clothing, and the greater the importance of allowing for this in design.

In many parts of the world, biological and social changes will cause plastic changes in population parameter values over relatively short time periods. Drift in anthropometric parameter values should be expected. In practical design situations, the data required for a particular body dimension or population may not be available. However, techniques are available for estimating unknown dimensions, and the literature on anthropometry can still be of use to the designer by drawing attention to the various body dimensions that need to be considered in the design and the types of human–machine mismatches that could occur. Anthropometry can provide a very useful perspective on usability issues at the early stages of the design process. Later, the designer might take a more empirical approach and test prototypes using a small sample of users from the extremes of the anthropometric range, taking care to close the gender data gap.

Further guidance can be found in Mrugalska and Karwowski (2024) and in ISO/TC159/SC3 "*Anthropometry and Biomechanics*."

TUTORIAL TOPICS

1. About 30 years ago, only 5% of the U.S. or UK population was obese. Now, more than 20% are obese. Has the 95th percentile waist circumference of 30 years ago now become the 80th percentile?
2. If it has, how might the shape of the distribution have changed?
3. Should HFE specialists advise on how to accommodate obese workers, or should they just tell them to lose weight?
4. At the link (https://www.youtube.com/watch?v=C6vAoD3HA9I), you will find a short presentation by Caroline Criado Perez talking about "Invisible Women." Do you think designers have "unconscious bias" against women? If you do, what will you do about it in your career in HFE?

ESSAYS AND EXERCISES

1. Find the mean and standard deviations of the popliteal height, sitting eye height, and forward reach of U.S. adult females.
2. Refer to Appendix A. Find the value of z that corresponds to the seventeenth percentile value of a normally distributed variable.
3. Calculate the mean and standard deviation of the following data on stature (show all your steps in the calculations to arrive at the answers):

   Stature mm
   1650
   1760
   1665
   1720
   1810
   1690
   1850

4. The ninety-fifth percentile value of a normally distributed variable is found by adding 1.64 standard deviations to the mean. If the mean body mass of U.S. males is 82.1 kg and the standard deviation is 17.1 kg, what is the ninety-fifth percentile body mass?
5. The fifth percentile of a normally distributed variable is found by subtracting 1.64 standard deviations from the mean. If the sitting height of U.S. females is 861 mm and the standard deviation is 36 mm, what is the fifth percentile sitting height?
6. Use the table of z-values in Appendix A to find out the number of standard deviations below the mean that gives the twenty-eighth percentile value of a normally distributed variable.
7. A fifth percentile U.S. male has a forward reach of 843 mm. His shoulder is 450 mm above a horizontal work surface. Calculate the radius of the ZCR on the desktop.
8. Mr. Smith's stature is 0.4 standard deviations below the mean stature for UK males.
   a. What percentile is his stature? (Hint: Refer to the table in Appendix A.)
   b. What percentage of UK males is taller than he is?
   c. Suppose the mean stature is 1700 mm and the standard deviation is 200 mm. How tall would Mr. Smith be?
9. Using the data in this chapter, specify seat and work surface dimensions to fit 95% of VDT users in two different populations. Comment on any design and equipment procurement implications of this exercise.
10. Carry out an anthropometric evaluation of either a commercial kitchen or a number of domestic kitchens. Measure and comment on the following:
    The heights of all major work surfaces, including the sink.
    The heights of all cupboards and shelves.
    The reach requirements of all major work and storage areas.
    The amount of space available for movement and foot positioning.
    Interview users to find out how they spend their time in the kitchen. List the most to least common tasks, and note any problems or difficulties experienced.
    Use all of the above information to suggest improvements.
11. Write a report that explains to a design engineer why anthropometric data should be used in the design of human–machine systems. Give examples from everyday life to support your arguments.
12. You have been asked to help a large financial institution in Cape Town, South Africa, to assist with office chair selection. The workforce is a mixture of different racial and ethnic groups. Because of the cost savings involved in making one large order for 2500 identical chairs, the company wishes to standardize office chairs. One chair must fit all. You decide to get information on the stature of employees from their company medical records (a medical exam is a requirement for membership in the company pension scheme). From this, you decide to use the RASH technique to estimate fifth and ninety-fifth percentile measurements used in chair design.
    a. List the anthropometric variables you would need to estimate to help the company choose a chair.
    b. Given that the mean and standard deviation heights of males and females in the company are (in mm) as follows:

	Mean	sd
Males	1700	80
Females	1580	70

    Estimate fifth and ninety-fifth percentile values of the variables of your choice using scaling factors calculated from the tables in this chapter (RASH technique).

c. You decide that seat depth is a critical factor. The company has a shortlist of 14 chairs, but none of them has adjustable seat depth. You measure the depth of the seats with the following results:

Seat Depth (mm)	Number of Chairs at That Depth
450	3
420	4
400	3
390	1
370	3

What percentage of the company's female workers will be able to sit on the seats without having to adopt a posture in which they are "perched" on the edge of the seat?

13. Specify chair and desk dimensions to fit a ninety-fifth percentile male and a separate set of dimensions to fit a fifth percentile female using anthropometric data from a country of your choice (i.e., all the male's dimensions will be ninety-fifth percentile and all the female's fifth percentile). Comment on the anthropometric mismatches that will occur if the male sits at the female's workstation and vice versa.

    Describe how you would design one set of furniture to suit both of the above users and everyone in between (specify any dimensions and ranges of adjustment).
14. Specify the radius of the ZCR on a conveyor belt for U.S. females of fifth percentile standing shoulder height and reach. The belt is 1000 mm above the floor. Comment on the design of this workstation.
15. A "panic button" is to be placed on a wall above a console in the control room of a nuclear power station (see figure). The console is 500 mm deep from front to back. Calculate the maximum acceptable height above the floor of the panic button that would enable a female with fifth percentile shoulder height and reach to reach it without jumping.
16. You have been asked to advise a bank that wishes to replace all its existing office workstations with new "sit–stand" workstations that enable workers to work in either a standing or a sitting position. Use the RASH technique to estimate appropriate values of elbow and popliteal height using the data below. (Hint: First calculate scale ratios from the data for the reference population.) Using these estimates, specify the maximum allowable height for the desk in its lowest position and the minimum allowable height for the desk in its highest position.

	Reference Population		Company Workers[a]	
	Males	Females	Males	Females
Mean stature	172	168	178	155
sd stature	10	10	7	5
Mean popliteal height	44	40		
sd popliteal height	4	4		
Mean elbow rest height[b]	25	20		
sd elbow rest height	2	2		
Mean elbow height (Standing)	120	95		
sd elbow height	6	5		

[a] All dimensions are in centimeters.
[b] Height of elbows above seat.

17. To answer this question, you will also need to refer to the biomechanics equations in Chapter 2. A nurse has to reach onto a badly designed shelf and pick up, at arm's length,

a defibrillator with a mass of 10 kg. At the moment she picks it up, she pauses, and her arms are horizontal to the ground and fully extended. Assume that the mean and standard deviation of stature in the population of nurses are as follows:

Mean = 1620 mm   Standard Deviation = 64 mm

a. Estimate the reach distance at the moment she picks up the defibrillator, assuming that the nurse is sevety-third percentile in stature and that forward grip reach is 0.42 of stature. (Hint: Use the accompanying tables of z, the standard normal deviate, to calculate her stature.)
b. Use a simple biomechanical model to estimate the L5/S1 spinal compression due to the load. You may assume that the lever arm of her back extensor muscles is 7 cm and that her posture is upright and symmetrical.
c. Calculate the total spinal compression by including the compression due to the nurse's body mass above L5/S1. The nurse is at the fifty-fifth percentile for body mass, and the mean and standard deviation of body mass for the population of nurses are as follows:

   Mean = 66 kg   Standard Deviation = 13 kg

   Assume that the mass of the upper body above L5/S1 is 37% of total body mass. Ignore the contribution caused by the load moment of the extended arms, but include their mass.
d. Calculate the ratio between the spinal compression due to the load and the spinal compression due to body weight, and comment on your findings.

19. Mrs. Smith is on holiday in Sri Lanka with her husband. She persuades him to buy her a new dress in a local village market. She is horrified to discover that all of the dresses on sale are too small for her around the hips. She has always considered herself to be fairly slim. Her hip breadth is 335 mm. Assume that the dresses are designed to fit local females and use Figure 3.5 and the data in Table 3.5 to answer the following questions:
    a. Is Mrs. Smith justified in considering herself to be "fairly slim"?
    b. Why won't any of the dresses fit her?

QUICK QUIZ

1. Anthropometric data follows: (a) a binomial distribution, (b) a Poisson distribution, (c) a normal distribution.
2. Body mass index is a measure of how fat people are. True or false?
3. "Pink it and shrink it." Is this the right way to ensure that products fit female users? Yes or no?
4. What percentage of the population falls one standard deviation either side of the mean?
5. The standard error is a measure of the accuracy of an anthropometer. True or false?
6. You should always design a product to fit the entire user population. True or false?
7. There is no such thing as a fifth percentile user. True or false?
8. The ninety-fifth percentile is 1.64 standard deviations above the mean. True or false?
9. Sample size should always be as large as possible.
10. What is the maximum allowable dimension for the distance from the floor of an emergency stop button: (a) fifth percentile female overhead reach, (b) ninety-fifth percentile male stature, (c) fifth percentile female eye height?

NOTE

1 Repeat calculations depending on whether we wish to assume that the secular trend in stature has now ceased.

4 Standing and Sitting

GENERAL REQUIREMENTS FOR HUMANS IN SYSTEMS

1. Jobs should be designed to allow both sedentary and standing work.
2. Tasks should be designed to minimize the adoption of static standing or seated postures.
3. Where static postures must be held, the limits set out in ISO 11226 must not be exceeded.
4. Computer workstations must be activity-permissive and designed in accordance with national and international standards.
5. Employees should receive training in how to adjust and use their workstations and furniture.
6. Frequently viewed objects must be positioned within the optimum viewing zones such that a neutral neck posture can be maintained.

For most of history, people sat down to rest. Now, they sit down to work. That's the problem.

The Author

CORE KNOWLEDGE: POSTURE

Anatomy of Human Posture and Its Evolutionary Origins

The upright posture and bipedal gait of human beings have a long evolutionary history. Modern humans belong to a family of primates known as "hominin," all of which appear to have anatomical adaptations that enabled them to walk upright on two legs. Humans, like other members of the genus *homo*, differ anatomically from other primates such as chimpanzees. The latter are primarily quadrupedal but are able to walk on two legs for short periods, whereas members of the genus homo, including modern humans, are bipedal. Figure 4.1 depicts the bipedal posture adopted by a modern human and a chimpanzee. The center of gravity (CoG) of the chimp's body lies anterior to its lumbar spine and pelvis, whereas in the human, it lies just above the hip joint. Postural muscle activity is required for the chimp to maintain a standing posture because the masses of superincumbent body parts lie anterior to the joints, causing it to fall forward. In humans, the CoGs of superincumbent body parts lie above or close to the joints, such that the bending moments are minimized and standing can be maintained with little muscle activity. The mechanical advantage of the gluteal muscles makes it easy to stop the trunk from jack-knifing forward over the legs.

To achieve an upright posture, the chimp has to flex its knees because it lacks sufficient hip extension to position the femur perpendicular to the ground. This increases the postural load on the legs when standing and prevents chimps from striding when walking upright (limited hip extension means that the trailing leg has to be lifted early, thereby shortening the stride). Lovejoy (1988) summarizes some of the main anatomical adaptations to bipedalism as follows:

DOI: 10.1201/9781032717210-4

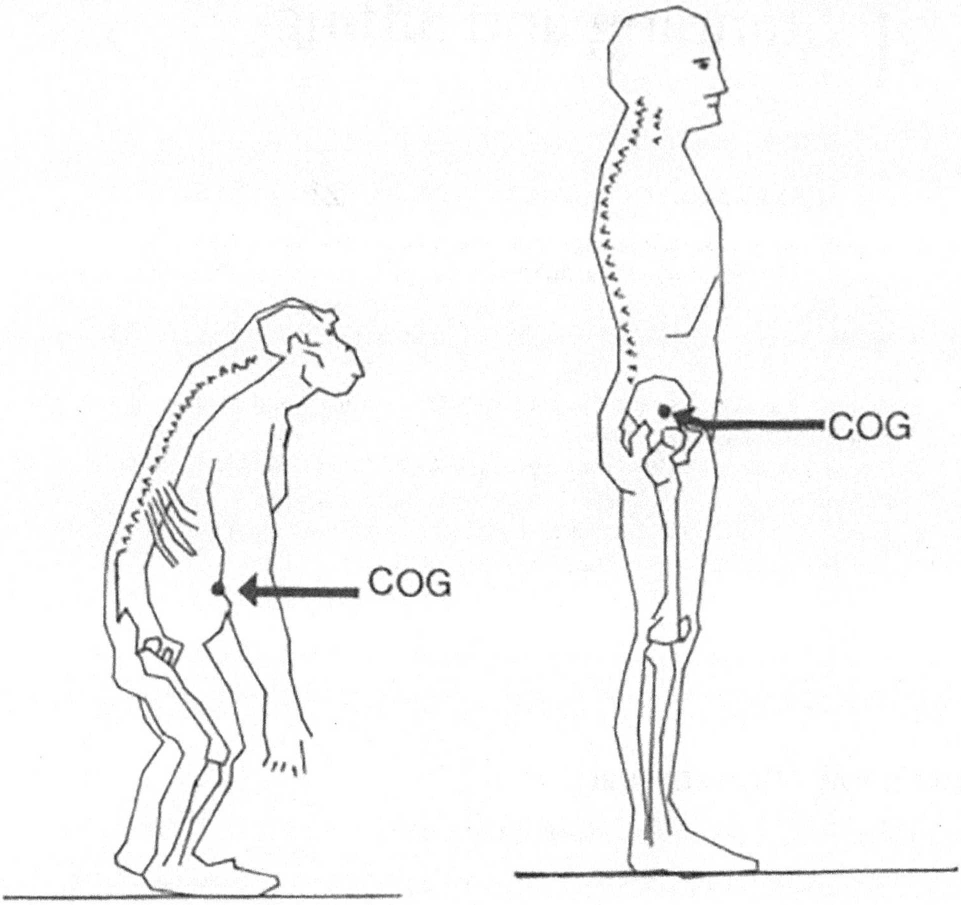

FIGURE 4.1 Upright bipedal posture in a human and a chimpanzee. Note the location of the COG in relation to the hip joint.

1. A lower COG of the body, due to the broadening and flattening of the ilium and sacrum, shortening of the arms, and lengthening of the legs, results in increased postural stability.
2. A change in the shape of the iliac blades and the function of the gluteal muscles. In chimps, the pelvis is flat (Figure 4.2b), and all three gluteal muscles act as hip extensors used in locomotion. In humans, the ilia are curved, and the gluteus medius and minimus muscles are displaced laterally and act as hip abductors.
3. For a major part of the gait cycle, the pelvis is supported on only one leg. To stabilize the upper body, the lateralized medial gluteal muscles on the supporting side prevent the pelvis from tilting to the unsupported side (the hip joint can be considered the fulcrum). At the top of the femur, the bone protrudes away from the hip joint (at the greater trochanter), and the ilium flares outward, giving the anterior gluteal muscles on the supporting side greater mechanical advantage.

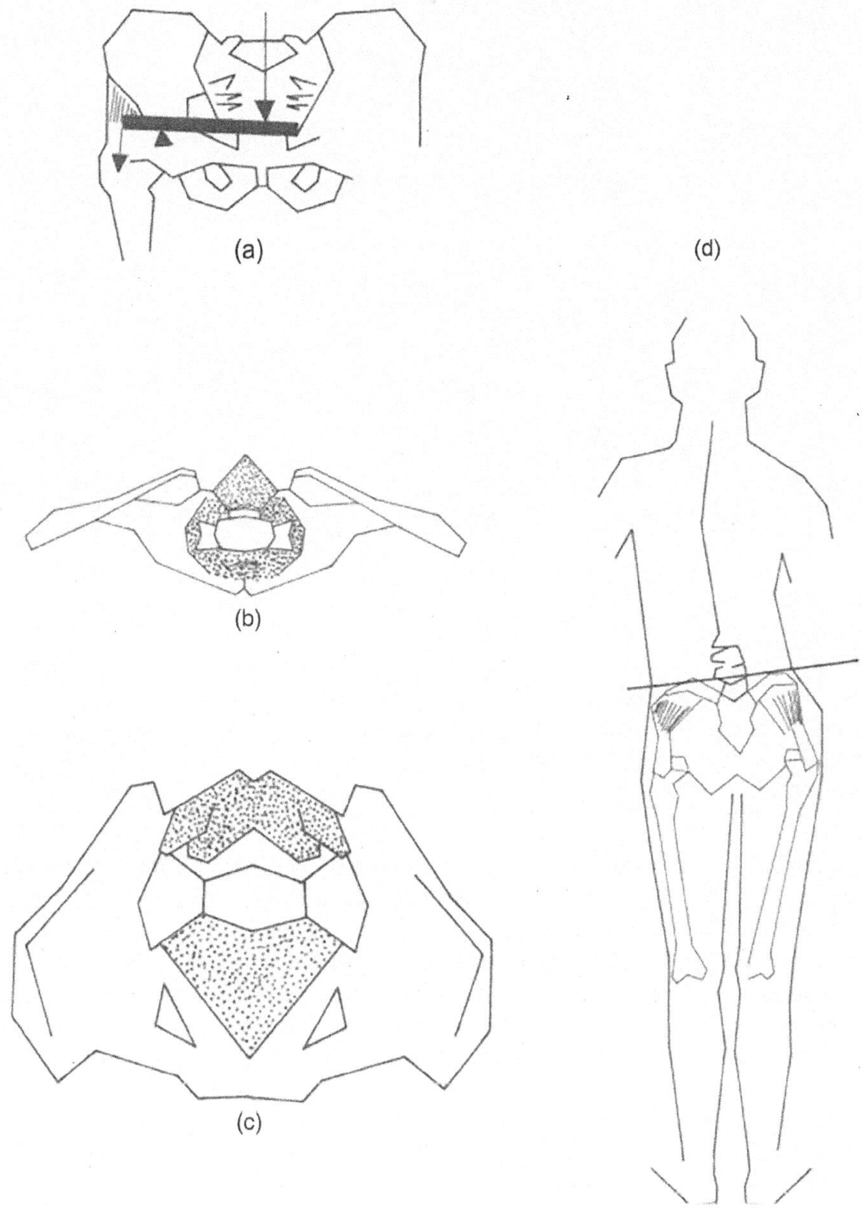

FIGURE 4.2 The hip joint and pelvis related to posture. (a) Frontal view of the human pelvis and hip joint. The joint acts as a fulcrum during walking. The anterior gluteal muscles are hip abductors and counteract the adductor moment exerted by the upper body and swinging leg, preventing the trunk from tilting to the side of the swinging leg. (b) Top view of the pelvis of a chimpanzee. The ilia are flat and lie almost in a single plane. The gluteal muscles are all hip extensors. (c) Top view of a human pelvis. Note the curved ilia that provide lateral attachment points for the anterior gluteals. (d) "Trendelenburg" posture. Weakness in the anterior gluteals causes deviation of the pelvis to the right and tilting to the left. The spine exhibits scoliosis to compensate. Functionally, this can be observed when people stand on an uneven surface to work.

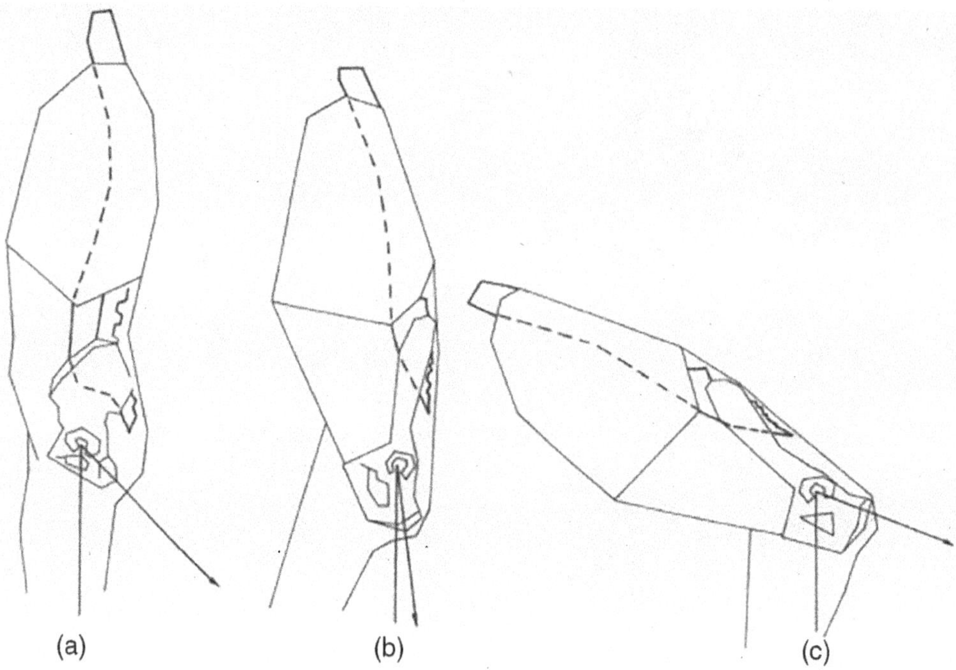

FIGURE 4.3 Orientation of the ischium with respect to the hip joint. (a) Human standing erect. (b) Chimpanzee standing erect. (c) Chimpanzee in a quadrupedal posture. In (b), the ischium prevents extension of the hip joint beyond vertical. Comparison of (a) and (c) reveals the fundamental nature of the lumbar lordosis as an adaptation to bipedalism.

4. When humans stand upright, they do so only partly by extending the hip joint. The upright posture cannot be achieved by extending the hip joint by 90° because this leaves insufficient extension for walking, as movement of the trailing leg beyond vertical is impeded by the ischium (Figure 4.3). The evolutionary solution has been to rotate the sacrum forward on the ischium and increase the extension of the lumbar spine. In effect, humans attain the upright posture by extending the hip joint by about 60° and the lumbar spine by about 30°, with plenty of hip extension left for long strides when walking and running.

These anatomical adaptations account for the origin of lumbar lordosis, which is effectively an adaptation to bipedalism, without which modern human locomotion would be impossible.

The pelvis can be likened to the platform that supports the superincumbent body parts and transmits the weight to the ground via the legs. Several muscle groups, which in other species provide locomotive power, take on new roles. The anterior gluteals stabilize the pelvis in the frontal plane; the iliopsoas muscles initiate "swing-through" of the trailing leg when walking and stabilize the lumbar spine when standing. The hamstrings decelerate the swinging leg prior to heel strike. Only the quadriceps and the plantar flexors provide a ground reaction force for locomotion.

The posture of the spine, particularly the lumbar spine, is of great interest in workspace design. From the above, it can be seen that it is only one of several anatomical adaptations that characterize human bipedalism.

Posture and Spinal Loading

Figure 4.4 depicts the static posture of an astronaut asleep under weightless conditions.

The skeleton is in a "neutral" posture—the internally generated torques acting on the joints as a result of tissue elasticity and muscle tone are in balance. Many readers will recall waking in the morning, lying on one side in bed, in such a posture. The joints are close to the mid-point of their

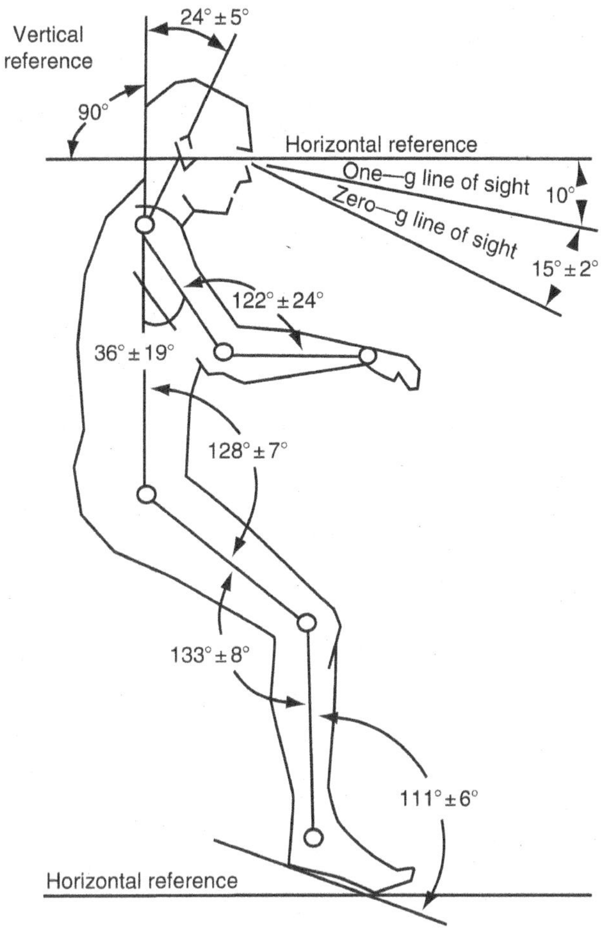

FIGURE 4.4 Posture of an astronaut resting in zero gravity showing the neutral positions of the joints in the sagittal plane. (Redrawn from Niebel and Frievalds, 1993.)

ranges of motion, and there is a slight lumbar lordosis. This restful posture is readily observed in the workplace, when people sit on a stool to work at a bench or recline in a modern office chair when using a computer. When a person flexes the hip and knee joint to sit down, the iliopsoas muscles immediately shorten, and the hip extensors lengthen. The balance of antagonistic muscle forces, which kept the pelvis in its anteriorly tilted position, is changed, and the pelvis tilts posteriorly almost immediately and continues in proportion to the flexion at the hip. To maintain the head erect, the lumbar spine flexes to compensate for the tilting pelvis, and the lumbar lordosis diminishes and eventually disappears (Figure 4.5).

The stiffness of the lumbar spine varies according to its posture, becoming stiffer as it flexes or extends. Buchman-Pearle et al. (2023) investigated the "neutral zone" of lumbar spine stiffness. Prolonged sitting or standing may increase the rate of creep in the stiffened spine, whereas sitting with the spine in the neutral zone may be more comfortable. Beyond the neutral zone, the bending moment required to flex or extend the spine increases. For males, the neutral zone of –22° to 0.2° of flexion was found to occur with a trunk–thigh angle ranging from 124.2° to 159.6°. For females, the neutral zone ranged from –17.8° to –1.3° at trunk–thigh angles of 143.2° to 159.5°. This is similar to the neutral posture in Figure 4.4, supporting the use of perch stools as a default option for seated work.

The lumbar spine is under constant load in all upright body positions, but the load varies depending on the posture. Degeneration of the intervertebral discs is thought to be influenced by exposure

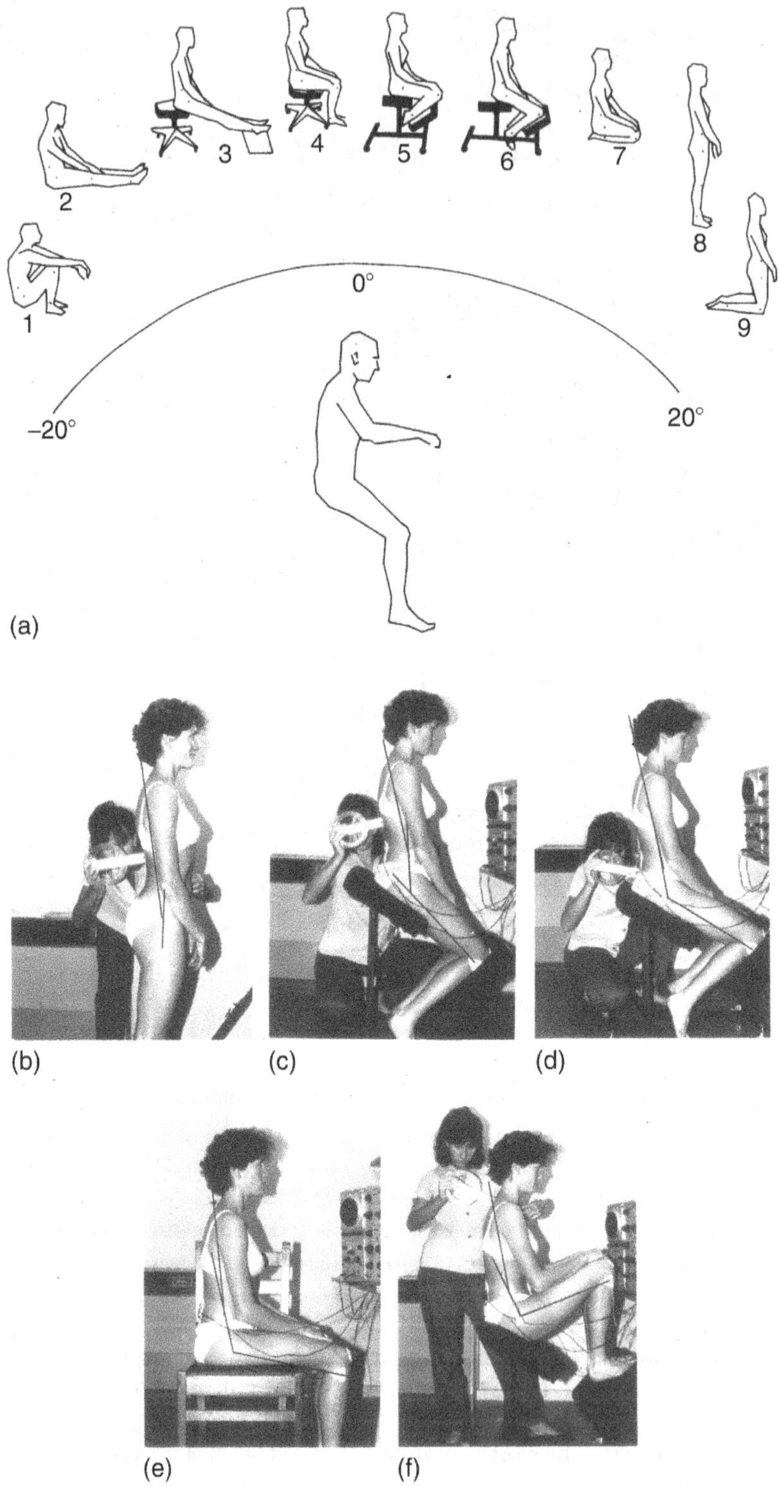

FIGURE 4.5 (a) Angles of pelvic tilt in different body positions. On the left, posterior pelvic tilt with flattened lumbar curve. On the right, anterior pelvic tilt with exaggerated lumbar lordosis. Neutral postures in the center. (b–f) Posterior tilting of the pelvis and flattening of the lumbar curve during the transition from standing to squatting (i.e., increasing hip flexion in the upright position). (From Bridger et al., 1992, with permission. Figure concept by Tom Bendix.)

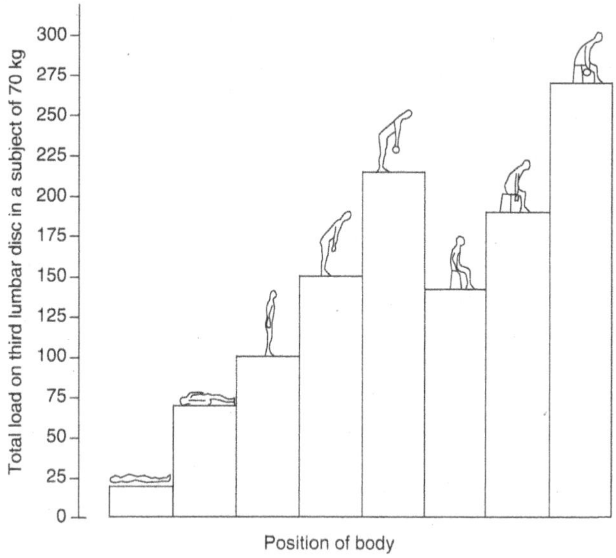

FIGURE 4.6 Lumbar disc pressures in different positions of the body. (Redrawn from Nachemson, 1966.)

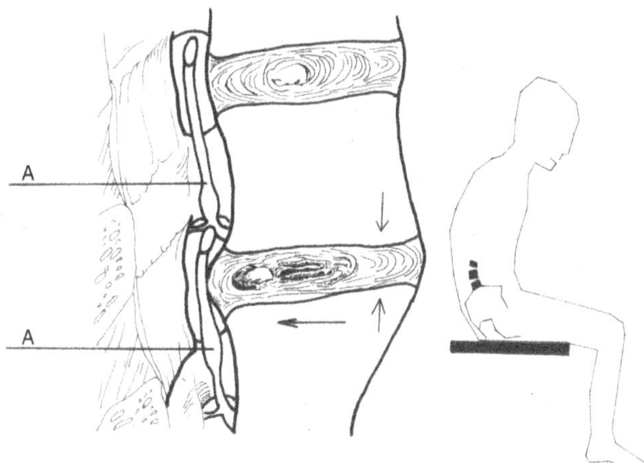

FIGURE 4.7 Anterior wedging of the intervertebral disc occurs in the slumped sitting position (A=posterior ligaments) or when the backrest offers inadequate lumbar support. Soft tissues between the anterior and posterior elements of the spine may be pressurized, resulting in pain. (Adapted from Keegan, 1953, with permission.)

to high loads. McGill concluded that although there are several occupational risk factors for low back pain, increased spinal compression is common to all of them. The spine of a person seated erect exhibits greatly diminished lumbar curvature (mean angle of curvature of 34° compared to 47° in standing, according to Lord et al. 1997) and increased intradiscal pressure. Figure 4.6 shows data from a study by Nachemson, in which intradiscal pressure was measured *in vivo* while subjects adopted a variety of different postures. In a recent systematic review of disc pressure studies, Li et al. (2022) concluded that sitting causes higher pressures in normal discs but not in degenerated discs because the nucleus is no longer fully enclosed by the annulus. When sitting in a flexed posture, due to the increased flexion moment, the posterior spinal ligaments are under tension. The intervertebral discs are "wedged" anteriorly and protrude into the intervertebral foramen (Keegan, 1953; Figure 4.7). This wedging action causes back pain in people with instability in the lumbar motion segments or symptomatic lumbar intervertebral discs.

When sitting, the discs bear more of the load (Adams and Dolan, 1995), whereas when lying, the absolute load on all structures is lowered.

> This implies that seated workers should be able to adopt relaxed posture and recline against a backrest at will.

Forward-flexed postures in which a load is held can more than double the disc pressure, particularly when sitting. *Clearly, handling loads in a seated position is to be avoided.* Further, Beach et al. (2005) demonstrated that lumbar spine stiffness increased after only 1 h of sitting and concluded that this may contribute to low back pain in sitting and increase the risk of injury if people have to handle loads after sitting (such as truck drivers manually offloading a delivery).

In standing, the load is shared between the facet joints and the discs and in many cases the lumbar spine is in extension beyond the neutral zone. Prolonged standing, as when working at a standing desk, may cause fatigue in the hip extensors, which results in an increase in the amount of anterior pelvic tilt. The resulting increase in lumbar lordosis causes compression of the facet joint capsules and leads to pain in those with symptomatic facet joints. These stresses are exacerbated if poor workstation design also leads to hyperlordosis (Figure 4.8). Beaudette et al. (2020) demonstrated that when subjects extend the neck to gaze upwards, extension of the thoracolumbar region is also promoted.

Fundamental Aspects of Sitting and Standing

Anatomy of Standing

The pelvis is held in an anteriorly tilted position by the iliopsoas muscles, and the hip joint is free to extend, as happens during the stance phase of gait. The trunk and head are rotated until they are vertically above the legs. This is achieved by extension of the lumbar and cervical spines, which is why the vertically held spine is S-shaped in humans and C-shaped and held horizontal to the ground in quadrupeds.

Bones and Joints

In the erect posture, the line of gravity of superincumbent body parts passes through the lumbar, sacral, and hip joints, and in front of the knee and ankle joints. This places an extension torque around the knee joint, which is resisted because the joint is already fully extended. The flexion torque around the ankle is resisted by the plantar flexors.

Muscles and Ligaments

The "antigravity muscles" work to maintain posture even during relaxed standing. They consist of the calf muscles (soleus), the knee extensors, the hamstrings and gluteal muscles, the muscles of the back, and the neck extensors (and the muscles of the jaw, of course). The body can be conceived of as a pillar of segments stacked one on top of the other and linked by joints. It is momentarily balanced when the resultant of all forces acting on it is zero. The system is designed to minimize any displacement of the line of action beyond the base of support described by the position of the feet, and compensatory mechanisms come into play to maintain balance immediately after this happens.

Muscles and ligaments play a stabilizing role by means of the active and passive torques they exert around joints to correct small, fleeting displacements of the lines of action away from the joints. A good posture may be defined as one in which the destabilizing moments are minimized and the posture is maintained by the resistance of the relatively incompressible bones (as well as interleaved soft tissues such as the intervertebral discs). When the body is pulled off balance by the requirements of badly designed jobs or workspaces, the antigravity muscles come into play, and a new equilibrium position is established, but with the associated cost of isometric muscle activity.

FIGURE 4.8 Working at a standing desk with the screen too high can cause uncomfortable postural deviations—note the exaggerated lumbar curve (hyperlordosis) and the strained neck posture (chin held high to look at the screen). (Courtesy of Osmond Ergonomics and reproduced from: "*A Guide to Active Working in the Modern Office: Homo sedens in the 21st Century*" by RS Bridger, 2019, with permission CRC Press.)

Erector Spinae Muscles

These are the main extensors of the trunk and are also used to control flexion. During relaxed standing, very little muscle activity occurs since the lumbar lordosis minimizes the trunk flexion moment. When the trunk is flexed even slightly forward, or when a weight is held in front of the body, the erector spinae muscles come into play.

Leg Muscles

The soleus and gastrocnemius muscles are true postural muscles in the sense that they are always switched on when standing. When leaning forward, the activity of the gastrocnemius muscle increases. Prolonged standing causes significant localized leg muscle fatigue and is one of the causes of leg discomfort.

Abdominal Muscles

In standing, the line of gravity falls slightly behind the center of the hip joint, causing the pelvis to "tend" to tilt backward. This relieves the abdominal muscles of a postural role and explains why these muscles are relaxed in standing. (This applies to normal standing; when carrying a load on the back or walking down a steep hill, the abdominal muscles do play a role in helping to maintain a proper relationship between the thorax and pelvis by preventing excessive anterior pelvic tilt and hyperlordosis.)

Hamstring and Gluteal Muscles

The hamstring and gluteal muscles are hip extensors. The gluteal muscles exhibit hypertrophy in humans, and their function is to stop the trunk from jack-knifing forward over the legs—unlike in quadrupeds, where the trunk is already jack-knifed and the gluteals are used for locomotion. The gluteals are, however, used for locomotion in ladder or stair climbing. Activity in the hamstrings is less in the standing position but increases when the stander leans forward, holds a weight, or pulls.

Iliopsoas Muscles

Psoas major and iliacus are hip flexors and are constantly active in normal standing, as they prevent extension of the hip joint (the trunk jack-knifing backward over the legs or the loss of lumbar lordosis if the head position is maintained). The iliopsoas muscles act against the hip extensors.

Adductors and Abductors of the Hip

When standing on two feet, these muscles provide lateral stability, preventing translation of the pelvis in the frontal plane. When standing on one foot (and also during the stance phase of gait), the pelvis tends to tilt in the direction of the unsupported side. The hip abductors on the side of the supporting leg contract to maintain the pelvis level (Figure 4.9).

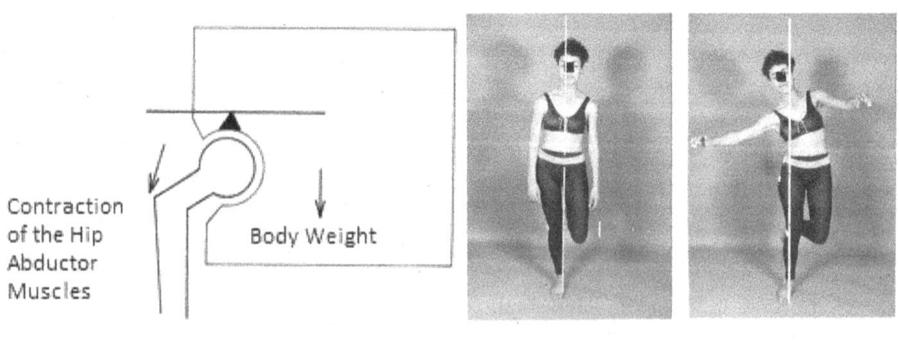

FIGURE 4.9 Humans are well adapted to standing on one leg due to the curved shape of the pelvis and the hip abductor muscles (gluteus medius and gluteus minimus). When we walk, these muscles contract to stop the pelvis from tipping to the unsupported side. If these muscles fail, the upper body has to compensate to maintain balance.

The Extensor Muscles of the Neck

In normal standing and sitting, the neck extensors are active because the COG of the head is anterior to the cervical lordosis, placing a flexion moment around the cervical spine. The greater the angle of flexion of the head, the greater the extensor moment and muscle force required to maintain equilibrium. One of the first signs of sleepiness in sedentary workers is "chin drop," as the extensor muscles relax.

Temporalis

One of the main muscles that controls the jaw keeps it closed.

Static Postures

In everyday life, people rarely adopt static postures for any length of time—if not walking or moving, they adopt a variety of resting positions, which vary depending on the culture (Hewes, 1957). Bridger et al. (1994) list some characteristics of "natural standing behaviour":

People:

- Never stand still.
- Shift body weight onto one leg.
- Lean backwards or sideways against anything.
- Rest the pelvis against counters.
- Maintain contact with fixed objects.
- Use the arms as props to support body mass.
- Place a foot on a raised surface.
- Use thoracic support (e.g., lean on a broom).
- Hang on an elbow.
- Place hands on the hips or in pockets.
- Extend footprint.
- Fold the arms and cross the legs at the ankles.
- Sit down whenever they can.

Short periods of walking and gross body movements are vital to activate the venous muscle pump and assist the return of blood from the lower limbs. Return of blood to the heart from the lower limbs depends on muscle activity. The veins have one-way valves in them, and when the leg muscles contract, blood is squeezed upwards. The calf muscles are sometimes considered to be a "second heart."

Epidemiological research cited in this and other chapters can be summarized in a conceptual diagram (Figure 4.10). Health risks increase when postures are static and when tasks are highly

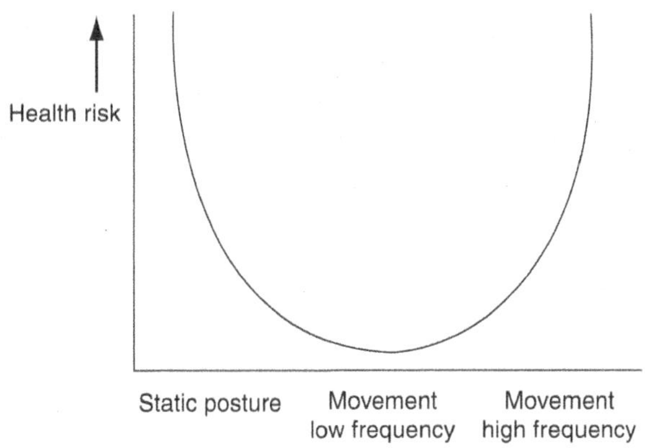

FIGURE 4.10 Health risks increase with highly static and highly repetitive work.

repetitive. Static work should be avoided, and where possible, tasks should be varied to provide relief from static postures (e.g., sitting should be combined with standing and standing should be combined with walking). Where static postures are unavoidable, the risk can be reduced by designing to accommodate the limitations of human anatomy, as described below.

Physiology of Standing and Sitting

Erect standing imposes a hydrostatic handicap that makes humans liable to peripheral circulatory collapse. Peak plantar (foot) pressures of 137 kPa exceed the normal systolic pressure of 17 kPa, resulting in occlusion of blood flow through the foot. Walking and fidgeting temporarily reduce the pressure, allowing fresh blood to pervade the tissues. These movements are important for the absorption of glucose into tissues such as muscles particularly after a meal, helping to lower blood glucose levels. Venous and circulatory insufficiencies in the lower limbs also contribute to the discomfort that results from prolonged standing.

Prolonged standing causes physiological changes including peripheral pooling of blood, a decrease in stroke volume, and an increase in heart rate, diastolic and mean arterial pressure, peripheral resistance, and thoracic impedance. Standing from a supine position is accompanied by an increase in the dimensions of the nasal passages (Whistance, 1996).

Orthostatic hypotension is related to standing still. It is characterized by a chief complaint of marked weakness on standing, blurred vision, dizziness, and fainting. Lowering the head, sitting, or lying down brings complete relief. Warm weather makes things worse because blood is circulated to the skin to cool the body, leaving even less blood available to be pumped to the brain. Dehydration and anemia also increase the risk of fainting.

Constrained standing is particularly troublesome for older workers or those with peripheral vascular disease because the venous muscle pump, which returns blood to the heart, ceases to function. Fidgeting is a preconscious defense against the postural stresses of constrained standing or sitting. Its purpose is to redistribute and relieve loading on bones and soft tissues and to rest muscles.

Musculoskeletal Problems in Standing

Low back pain is common in standing workers, and several authors have suggested various reasons for this. Where does back pain come from? The joint capsules of the facet joints, the peripheral third of the intervertebral discs, and the tendons, muscles, and ligaments around the vertebral column all contain nociceptors (pain receptors). Extended periods of time in postures outside the neutral zone place stress on these tissues, which may result in an inflammatory response triggered by the release of cytokines (part of the body's immune system). This results in pain if not relieved by a change in posture. Sciatic pain (pain in the hip or lower leg) can occur in those with degenerated lumbar discs if the sciatic nerve is placed under stress. Low-level erector spinae activity when sitting can cause compression of capillaries in these muscles and reduced oxygenation, leading to an accumulation of waste products and pain.

In extended postures (e.g., when standing with a pronounced lumbar lordosis), the facet joints may begin to take on some of the compressive load (Figure 4.8). If the lumbar intervertebral discs are degenerated, the space between adjacent vertebrae decreases, and the load on the facet joints increases even more. Excessive loading of the facet joints stresses the soft tissues around the joint and causes low back pain. For these reasons, *excessive lumbar lordosis should be avoided when standing* (Figure 4.8). Extrapolating from this, any workspace or task factors that require workers to arch the back greater than they normally would should be designed out. However, there is evidence to show that hyperextension of the lumbar spine can be beneficial in temporarily relieving the load on the intervertebral discs following the adoption of flexed postures (Magnusson et al., 1996c), and so there are grounds for recommending that short periods of lumbar hyperextension be introduced to relieve disc compression in standing tasks, particularly if forward flexion is also required.

Low back pain can also be caused by *muscular fatigue* if a standing person has to work with the trunk inclined forward (e.g., when washing or ironing). Spinal problems in standing work are often due

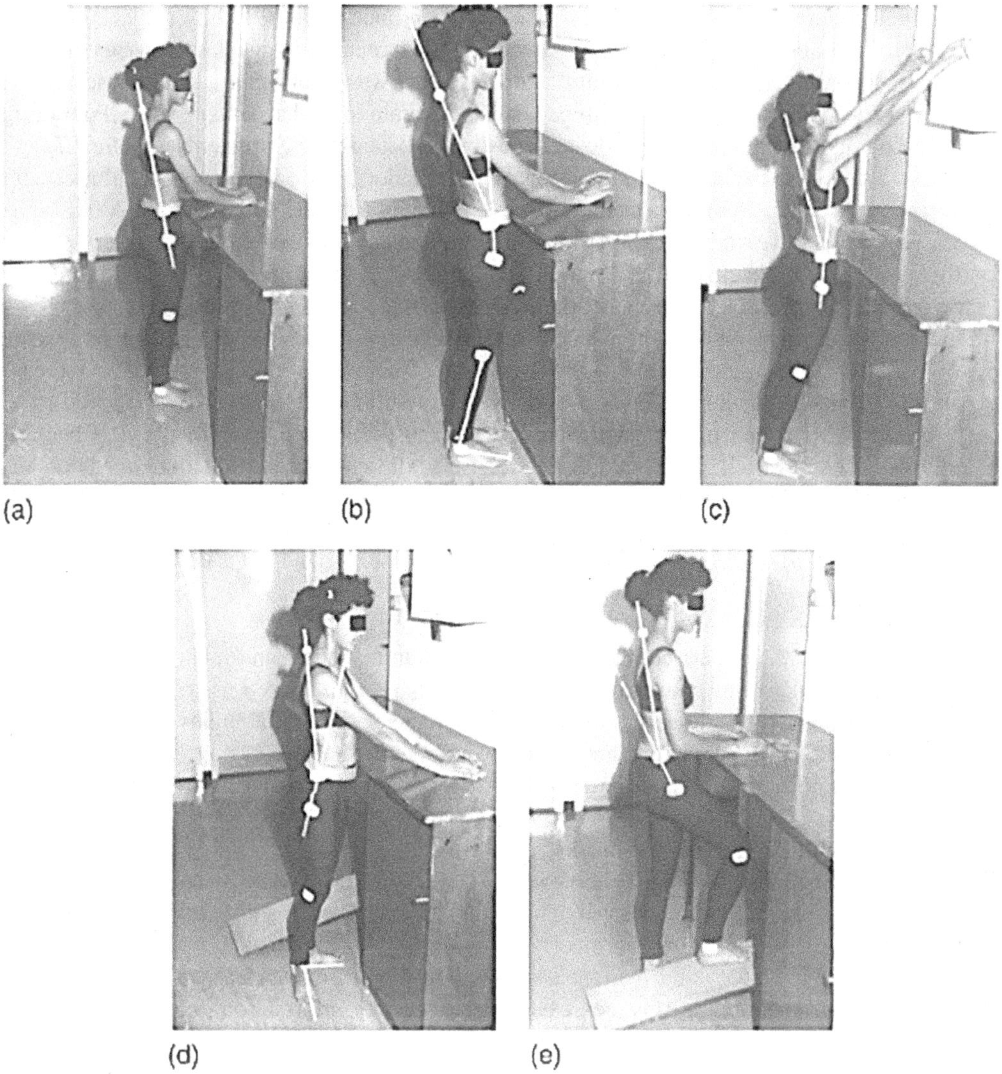

FIGURE 4.11 Spinal posture when standing is affected by workspace design: (a) unconstrained, relaxed standing posture, (b) no toe space, causing compensatory lumbar and hip extension, (c) lumbar extension caused by upward reaching, (d) poor foot position causing extension, and (e) relaxed standing with a footrest to prevent excessive lumbar extension.

to static loading of the back extensor muscles and/or excessive extension, lateral flexion, or twisting of the trunk. Figure 4.11 depicts some stressful standing postures caused by poor workstation design.

Messing et al. (2008) concluded that standing workers should have the freedom to sit, because 9.4% of standing workers reported ankle or foot pain and 6.4% reported lower leg or calf pain in the 12-month period preceding their investigation. Constrained standing (standing in the same place) and older age were associated with an increased risk of pain.

Spinal Problems in Sitting

Despite the popular myth that everyday occupational sitting is a risk factor for low back pain, evidence suggests that sitting at work, in itself, is quite harmless. In a recent systematic literature review of time-controlled trials, De Carvalho et al. (2020a) concluded that prolonged sitting is associated with significant increases in reported back pain after one hour in both healthy subjects and back pain sufferers. Durations longer than 5 h are associated with recurrent episodes. However, although the pain intensity was clinically meaningful, no evidence was found that sitting causes clinical *cases* of low back pain. Many people with bad backs gravitate toward sedentary work. Poor design of workstations causing flexed sitting postures may exacerbate existing problems (Figure 4.7).

Static seated work is the norm for professional drivers, pilots, and operators of machines such as forklift trucks. Back pain in these occupations is endemic. For example, Krause et al. (2004) carried out a 7.5-year prospective study of compensable back pain in San Francisco transit operators. Drivers reporting ergonomic problems in the design of their cabs were approximately 1.5 times more likely to experience a compensable back injury than drivers not reporting problems. The risk increased with the number of problems, which included lack of seat adjustability and back support; reaching across the steering wheel; and reaching for work objects. Total weekly driving hours also increased the risk. For severe back injuries, 30–50 h of driving more than doubled the risk compared to less than 30 h. The risk was over five times greater in drivers working more than 50 h/week. Chen et al. (2005) investigated low back pain in taxi drivers. Drivers who sat with a trunk–thigh angle less than 91° (and outside the neutral zone) were over five times more likely to suffer back pain than those with a trunk–thigh angle greater than 91° (odds ratio 5.11). Drivers who did not use a lumbar support were three times as likely to suffer low back pain compared to those who did (odds ratio 0.33).

What's Best for Back Pain? Sitting or Standing

De Carvalho et al. (2020a) carried out a systematic literature review and meta-analysis of timed studies of standing and sitting (about 1 h in each condition). Overall, there was no difference in back pain between sitting and standing (although there was some evidence that males experienced more pain when standing than when sitting, and females vice versa). The authors concluded that standing to work cannot be considered a replacement for sitting, as both can cause back pain.

Why We Need "Activity-Permissive Workstations"

It is sometimes stated that "Sitting is the new smoking." This is incorrect because, in many jobs, there is an occupational requirement to sit, and the health outcomes of sitting and smoking are different. However, prolonged sitting is harmful. Greer et al. (2015) found that men with middle to high sedentary behavior had a 65% and 76% greater risk of developing metabolic syndrome (inability to control blood sugar levels effectively; see Chapter 8 for further information) than those who were less sedentary. The harmful effects of sedentary lifestyles remained statistically significant when cardiorespiratory fitness and physical activity were controlled for (and vice versa). Van der Ploeg et al. (2012) found no increased risk of mortality in a sample of 222,497 individuals over 45 years of age if they sat for 4–8 h/day or less. The risk increased by 15% for those sitting 8–11 h/day and by 40% for those sitting 11 or more hours per day. Again, sitting and physical activity were independent predictors of mortality. Finally, Wilmot et al. (2012) reviewed 18 studies with a total of 794,577 participants and found that sedentarism was associated with a 112% increased risk of diabetes, a 147% increase in the risk of cardiovascular events, a 90% increase in the risk of cardiovascular mortality, and a 49% increase in the risk of all-cause mortality. Different time periods were used in the different studies to classify participants into sedentary and non-sedentary groups—approximately <3–4 h/day (low sedentary) and >7 h/day (high sedentary).

Diaz et al. (2017) conducted a cohort study in which 7985 participants over 45 years of age wore an accelerometer for 4 or more days, for 10 or more hours per day. Data were captured on total time spent sitting per day and the duration of seated periods. All-cause mortality (deaths for any reason) was the outcome. Participants spent 74% of the time sitting (about 12.5 h/day) on average. Seated periods of less than 1–29 min were associated with lower mortality. A larger number of longer breaks from sitting was also associated with lower mortality. In other words, it is not only the total daily sitting time that is important, but how it is accumulated.

Although the evidence is not yet definitive, there appears to be sufficient evidence to suggest that sitting time at work should be limited—possibly by as much as 4 h/day (Katzmarzyk et al., 2009).

All the evidence to date suggests that it is not sitting per se that is harmful, but the lack of skeletal muscle activity that occurs when sitting. So, it would seem that the biomechanical solutions to the problems of sedentary work have exacerbated the physiological problem of a lack of physical activity. The more comfortable the chair and the more we can recline on it, the more inactive we are. This has led to a change in focus with the design of *activity-permissive workstations*.

BASIC APPLICATIONS

Good workstation design encourages good posture. Figure 4.12 presents a framework for posture. It emphasizes the interaction of three classes of variables (Table 4.1).

Figure 4.13 depicts a forward-flexed sitting posture caused by a lack of fit between visual and manual task demands, workspace layout, and user characteristics.

Some advantages and disadvantages of sitting and standing to work are summarized in Table 4.2.

Walsh et al. (2017) measured blood flow in the popliteal artery with the knees and hips straight or bent at 90° as when sitting. They found that blood flow was impaired due to the angulation of the artery and concluded that time spent sitting with the legs bent should be limited on a daily basis to preserve vascular health. Evidence is mounting from different sources that workstations should be designed to permit a variety of postures and to encourage movement.

Three Steps to Effective Workstation Design for Good Posture

Carry out a task analysis to determine the

1. Visual requirements
2. Postural (effector) requirements
3. Temporal requirements

Visual Requirements

The visual requirements determine the position of the head and, therefore, the posture of the neck. The eye is sensitive to stimuli up to 95° to the left and right, assuming binocular vision—15° either side of the straight-ahead line of sight is the region of binocular overlap (Figure 4.14).

These limits describe a visual field in which objects may be placed such that they can be viewed without moving the head from its comfortable, erect position. Thus, static loading of neck muscles and other soft tissues in the neck can be avoided if the visual component of the task is kept within a cone from the "straight-ahead" line of sight to 30° below and 15° to the left and

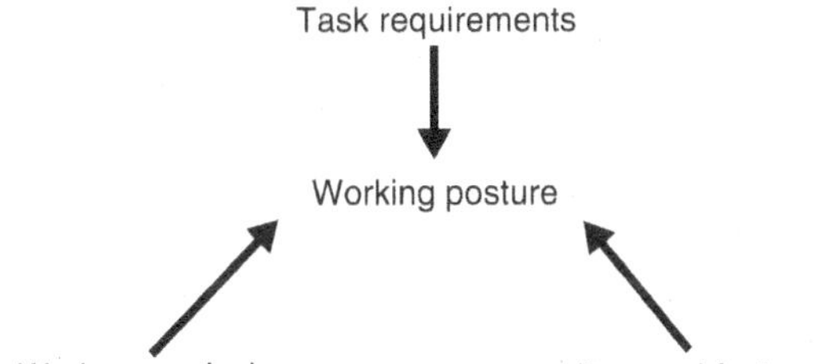

FIGURE 4.12 The postural triangle. A person's working posture is the result of the requirements of the task, the design of the workspace, and personal characteristics such as body size, shape, and eyesight. Consideration of all three components is needed in posture analysis and workspace design.

TABLE 4.1
Examples of Factors that Influence Working Posture

Factor	Example
User characteristics	Age
	Anthropometry
	Body weight
	Fitness
	Joint mobility (range of movement)
	Existing musculoskeletal problem
	Previous injury/surgery
	Eyesight
	Handedness
	Obesity
Task requirements	Visual requirements
	Manual requirements
	Positional
	Forces
	Cycle times
	Rest periods
	Paced/unpaced work
The design of the workspace	Seat dimensions
	Work surface dimensions
	Seat design
	Workspace dimensions
	Headroom
	Legroom
	Footroom
	Privacy
	Illumination levels and quality

right. A gaze angle of 6°–9° below the horizontal appears to be optimal for minimizing postural stress (Delleman and Berndsen, 2002).

When the head is erect, the "straight-ahead" line of sight, parallel to the ground, can be comfortably maintained. However, most people will not maintain the head in this position in order to look at objects on a horizontal work surface if this requires the eyes to be rotated downward by more than 30°. If the main visual area is more than 30° below the "straight-ahead" line of sight, it is viewed by tilting the head forward. The tilted posture places a static load on the neck muscles and displaces the COG of the body anterior to the lumbar spine, causing the characteristic forward "slumped" posture in which the backrest or lumbar support of the chair is no longer used. If objects are placed above the line of sight, the neck is extended to tilt the head backward. Frequently used displays should not be above the standing or sitting eye height of a short worker. When carrying out editing tasks, document holders should be provided and placed orthogonal to the line of sight and adjacent to the Screen. According to Beaudette et al. (2020), the angle of gaze, the posture of the neck, and thoracolumbar spines are linked. Visual targets outside Zone A in the figure may cause postural adaptations outside the neutral zone.

FIGURE 4.13 Application of the postural triangle to workspace evaluation. The illustration depicts St. Gregory the Great transcribing a text (from a ninth century ivory relief). The slumped sitting posture is a result of the low seat, excessive task distance, and the visually demanding task. The right elbow is resting on the thigh to improve stabilization of the writing hand. The left hand is steadying the book and the left elbow is resting on the work surface to close the postural chain and reduce the load on the left-hand side of the body. The left foot is resting on a footrest. How would you redesign this workspace to improve the posture?

TABLE 4.2
Standing versus Sitting to Work?

Advantages of Standing
1. Reach is greater in standing than in sitting.
2. Body weight can be used to exert forces.
3. Standing workers require less leg room than seated workers.
4. Legs are very effective at damping vibration.
5. Lumbar disc pressures are lower.
6. It can be maintained with little muscular activity and requires no attention.
7. Trunk muscle power is twice as large in standing as in semi-standing or sitting.
Disadvantages of Standing
1. Leg and foot swelling
2. Varicose veins
3. Arthritis in the hip joints
Advantages of Sitting
1. Lower energy expenditure
2. Less fatigue
3. Body is stabilized by the chair
4. Can be maintained for long periods
Disadvantages of Sitting
1. Lack of skeletal muscle activity in the lower limbs
2. Weakening of gluteal and hamstring muscles
3. Shortening of hip flexors
4. Varicose veins
5. Foot swelling
6. Impaired endothelial function in the popliteal artery

Sources: Based on Singleton (1972); Nachemson (1966); Hellebrandt (1938); Cartas et al. (1993).

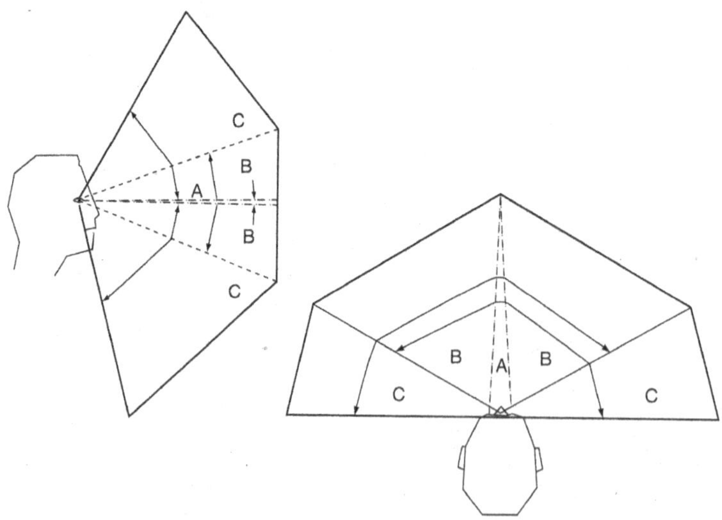

FIGURE 4.14 Regions of high (A), medium (B), and low (C) resolution in the visual field. A and B are preferred regions for the placement of objects. Objects should not, however, be placed above the straight-ahead line of sight, particularly at fixed workstations.

Postural Requirements

Task objects should be placed in the zone of convenient reach—a major determinant of posture and postural load (Chapter 3). In vehicle design, the comfort of the driver's seat depends on its positioning in relation to the foot pedals and manual controls, as well as its design.

Work Surface Design

Tilted desktops (of 15° or even 10°) reduce trunk and neck flexion of seated persons engaged in reading and writing (Bridger, 1988). Burgess and Neal (1989) found that using a document holder when writing on a flat desk significantly reduced the moment of flexion of the head and neck at the C7–T1 level of the spine and was rated by the subjects as more comfortable than not using one.

For both standing and sedentary workers, the main working area should be directly in front of the worker's body, not to the side, to minimize any twisting of the trunk. This is particularly important in seated work. Pearcy (1993) has shown that the twisting mobility of the human back is increased in sitting compared with standing, either in an upright or in a forward-flexed position. The increase occurs because the morphology of the lumbar facet joints permits more axial rotation of the superior vertebral body over the inferior body when the spine is flexed. The increase is considerable (38% more twisting in 90° sitting and 44% more twisting in long sitting over the whole spine, or about 2° more at each intervertebral joint). Because the posterior fibers of the annulus are already stressed when the spine is flexed, the additional stress of twisting may result in very high annular stresses, predisposing the fibers to rupture. Jobs involving the asymmetric handling of loads from a seated position (such as supermarket checkout personnel) are particularly hazardous. Seated workers who must resist sudden, external twisting forces (such as catching an object falling from a supermarket conveyor) have a high risk of injury. Twisting of the trunk while seated is achieved by activating the contralateral external oblique and ipsilateral erectors spinae muscles. Up to 20° of twisting can be achieved with little effort. Beyond this, muscle activity increases progressively, presumably as the forces on the twisted tissues that have to be overcome are increased (Torén, 2001). Workstations should be arranged such that seated workers should not have to twist their trunks more than 30°, or so that chairs should swivel.

Temporal Requirements

The temporal requirements of tasks influence the standard of workstation design. For example, in a multiuser computer workstation in which individuals spend only 15–20 min at a time interrogating the system, a sit–stand workstation using a high bench and stool might be appropriate. At the other extreme, certain data entry jobs require users to carry out a repetitive task over the whole day. These jobs impose a high degree of postural constraint, as the position of the hands and head is fixed by that of the keyboard, documents, and screen. It is particularly important when analyzing physically constraining jobs to characterize task components in terms of frequency and importance of operation and ensure that the workspace is arranged optimally for high-ranking elements.

Prolonged sitting is harmful, and regular short breaks are recommended. Dunstan et al. (2012) compared the effects of taking a break from sitting every 20 min (2-min breaks of light or moderate walking) with uninterrupted sitting on plasma glucose and insulin levels in overweight and obese volunteers. Interestingly, the total time spent walking over the 5 h was negligible in terms of energy expenditure, but was enough to have significant beneficial metabolic effects, providing further evidence that it is the prolonged physical inactivity that is harmful in itself.

Prolonged standing is also harmful. Gregg et al. (2021) recommend that workers should not have to stand in one position for more than 10 min.

Holding Times for Static Postures

Task objects are best placed between hip and shoulder height to minimize postural stress caused by stooping or working with the hands and arms elevated (see Chapter 3). Work surface heights should approximate the standing elbow height of workers, depending on the task. When carrying out fine work, a higher work surface is appropriate to reduce the visual distance and allow the worker to stabilize the forearms by resting them on the work surface. When carrying out heavy work, a lower work surface is needed to permit the worker to apply large vertical forces by transmitting part of the body weight through the arms.

Figure 4.15 presents data on the maximum acceptable durations of 19 static standing postures, without rest, applicable to 95% of workers (Dul et al., 1993). The durations are given in minutes, and the postures are expressed in relation to the position of the hands when holding each static posture. For example, the first posture on the left of the diagram is labeled "75/50." This means that the hands are held at a working height that is 75% of shoulder height and are at a distance from the body of 50% of horizontal forward reach. The maximum acceptable durations were derived by taking 20% of the maximum holding time for each particular posture.

The limits given in Figure 4.15 are used to determine the maximum allowable time for a worker to adopt a static posture (e.g., to hold or steady a work object) without discomfort. As can be seen, for stooped postures or those at the extremes of reach, the maximum durations are less than 1 min. Work surface heights should take into account the height of the worker and the height (vertical depth) of any work objects to enable the hands to be held in a comfortable position in relation to the body. The shoulder angle resulting from the combination of these factors can be assessed using the four-zone rating system presented in Table 4.3 and Figure 4.16.

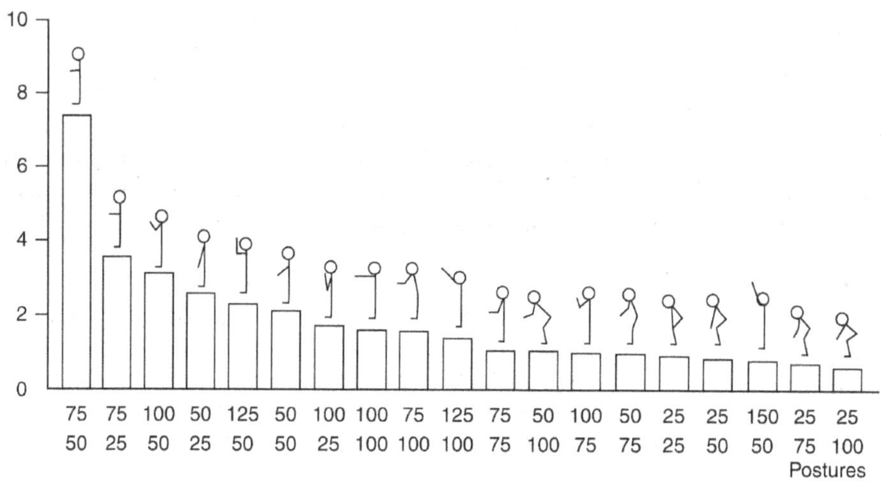

FIGURE 4.15 Maximum recommended holding times for static postures in minutes. (Redrawn from Dul et al., 1993, with permission.)

TABLE 4.3
Evaluation of Upper Arm Posture

	Posture Static	Movement	
Zone	**Low Frequency (<2/min)**	**High Frequency (>2/min)**	
1	Acceptable	Acceptable	Acceptable
2	Conditionally acceptable	Acceptable	Conditionally acceptable
3	Not acceptable	Conditionally acceptable	Not acceptable
4	Not acceptable	Conditionally acceptable	Not acceptable

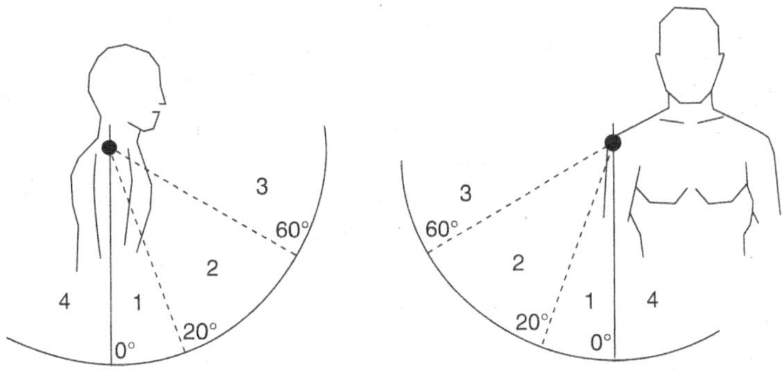

FIGURE 4.16 Static holding times for work objects. (From prEN 614. Safety of machinery. Ergonomic design principles–Part 2: Interactions between the design of machinery and work tasks, with permission.)

Postural constraint in standing workers can be relieved by providing stools to enable workers to rest during quiet periods or to alternate between sitting and standing. Adequate space for the feet should be provided to permit workers to change the position of their feet at will.

Standing Aids

Footrests and Footrails: Bridger and Orkin (1992) determined the effect of a footrest on pelvic angle in standing. The footrest raised the resting foot 250 mm above the level of the floor and resulted in a net posterior rotation of the pelvis of 4°–6° (flexion of the hip on the ipsilateral side releases the iliopsoas muscle and the iliofemoral ligament). Whistance and Bridger (1995) confirmed this finding—use of a footrail reduced anterior pelvic tilt, straightened the supporting leg, and increased the plantar flexion of the supporting foot. The footrest would appear to be a valid way of reducing lumbo-pelvic constraint in standing workers and helping prevent discomfort in the lumbo-pelvic region. Rys and Konz (1994) found that a 100 mm foot platform used by subjects was perceived as more comfortable than normal standing in 9 of 12 body regions, including the neck. Use of either a flat or a 15° tilted platform was perceived to be better than a simple footrail, but all standing aids were preferred to standing on a bare floor. When using a footrest, about 80% of body mass is supported on one leg and the remainder on the rested leg. During a 2-h period of standing, subjects placed one foot on the platform 83% of the time switching their foot from the platform to the floor every 90 s on average. Interestingly, in normal gait, people spend about 80% of the time on one leg as the other swings through, so use of a footrest can be likened to walking on the spot in slow motion. Gregg et al. (2021) following a biomechanical study of footrests while subjects worked at a standing desk, came to similar conclusions finding a weight distribution of 85%–15% irrespective of footrest height (10, 20, 30 cm), greater gluteal muscle activity on the supporting side, and greater lumbar erector spinae activity on the unsupported (footrest) side. They recommended that footrests be made available at standing desks.

Intermittent use of a footrest activates the venous muscle pump because the gastrocnemius muscle relaxes when the leg is resting on the footrest and contracts again as the feet alternate between the floor and the footrest. This may prevent orthostatic hypotension and pooling of blood in the lower limbs.

The freedom to stand with one foot forward and elevated seems to be an important feature of a well-designed standing workplace. A footrest height of 10% of stature seems reasonable.

Anti-fatigue Mats: Yazuli et al. (2022) compared two groups of subjects who carried out a sorting task for two hours with and without anti-fatigue mats to stand on. Electromyographic activity in the tibialis anterior and gastrocnemius muscle activity was lower when the mats were used. There were statistically significant differences in discomfort between the two groups, particularly in the lower limbs but also in the back, with less discomfort when the mats were used. Resilient rubber mats may activate the venous muscle pump. There are many reasons why standing workers should not have to stand on hard and sometimes cold concrete floors to work. Mats, wooden, rubber or plastic platforms, and carpets provide a more yielding surface and better insulation. They may also offer better friction and, therefore, aid postural stability and help prevent accidents. In a comparison of three different mats, Konz et al. (1990) found that a mat with 5.8% compression was perceived as more comfortable than mats of 7.4% and 18.6% compression. All were preferred over concrete.

Compression Stockings and Varicose Veins: Prolonged standing causes insufficient calf muscle pumping and hypertension in the lower legs. Varicose veins are superficial veins, often in the legs, in which the valves function ineffectively, resulting in pooling of blood and painful swelling and inflammation. With deep veins, the problem is more serious and can cause blood to return along abnormal pathways resulting in long-term health problems, including chronic edema and leg ulcers. Risk factors include obesity, cigarette smoking, high blood pressure, and lack of exercise. The disease is one of the 10 leading causes of hospitalization in Denmark. Occupational standing is associated with varicose veins in the lower extremities. Tuchsen et al. (2000) followed a sample of 1.6 million working Danes for 3 years from 1991. Men who worked mostly in a standing position were almost twice as likely to experience a first hospitalization for varicose veins compared to all other men. Women who worked mostly in a standing position were two and a half times at risk than all other women. Tomei et al. (1999) compared the prevalence of chronic venous diseases in office workers, industrial workers, and stoneworkers. The prevalence of the disorders increased with age and the number of hours spent standing at work. Controlling for age, the prevalence was higher for workers who stood for *50% or more* of their shift. Kim et al. (2022) found evidence for a therapeutic effect—the use of compression stockings reduced leg volume and leg circumference after an 8-hour workday but did not reduce discomfort.

Toe space: Panels or obstructions in front of benches cause users to stand further away from the work surface. The postural adaptation is for people to bend forward. Whistance and Bridger (1995) found that this was achieved by a combination of pelvic tilting and lumbar flexion—placing more stress on the spine. Toe space (Figure 4.17) can prevent this from happening (Whistance and Bridger, 1995).

Work Surface Design

Workstations can be improved by considering various aspects of desk and bench design (Table 4.4). Some important considerations are the provision of tilt in the work surface or of document holders, and the provision of free space. Zacharkow (1988) provides many interesting illustrations of Victorian school desks having a 15° slope for writing and an integral book holder for reading, angled at 45°. Mandal (1991) suggested that chairs with forward-tilting seats be used with desktops that tilt toward the user by about 15° to lessen the visual angle and encourage a more upright posture of the trunk.

Several studies have indicated that tilted desktops (of 15° or even 10°) do reduce the trunk and neck flexion of seated persons engaged in reading and writing (de Wall et al., 1991), and thus reduce the load on the corresponding parts of the spine. Significant effects have been found on subjects seated on both conventional and forward-sloping seats. Porter et al. (1992) have reported similar benefits with a sloping computer desk. Burgess and Neal (1989) found that using a document holder

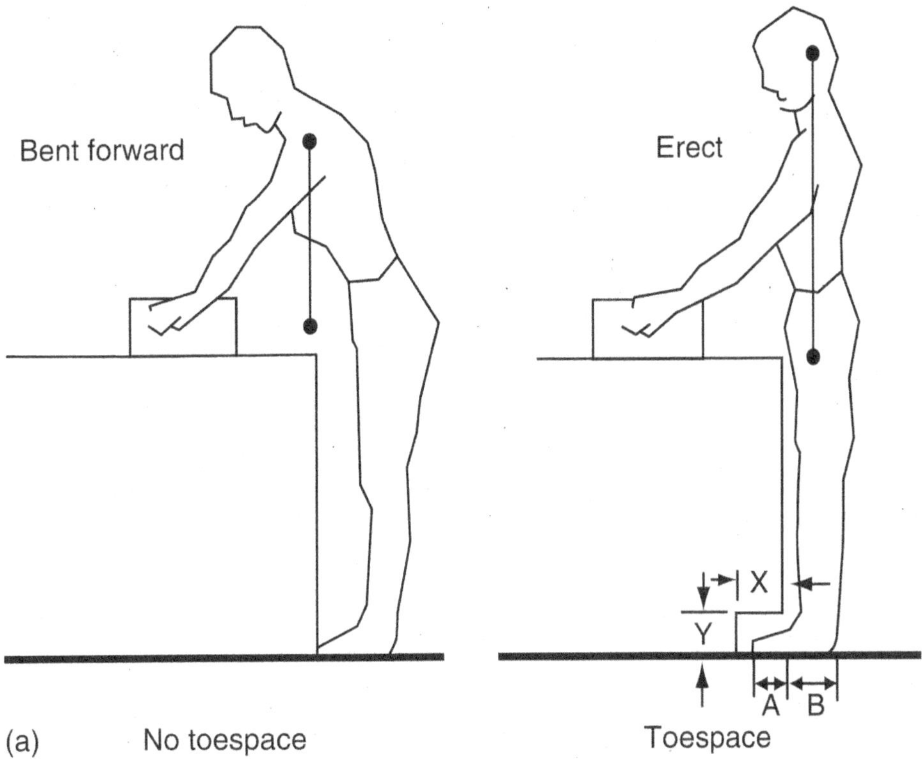

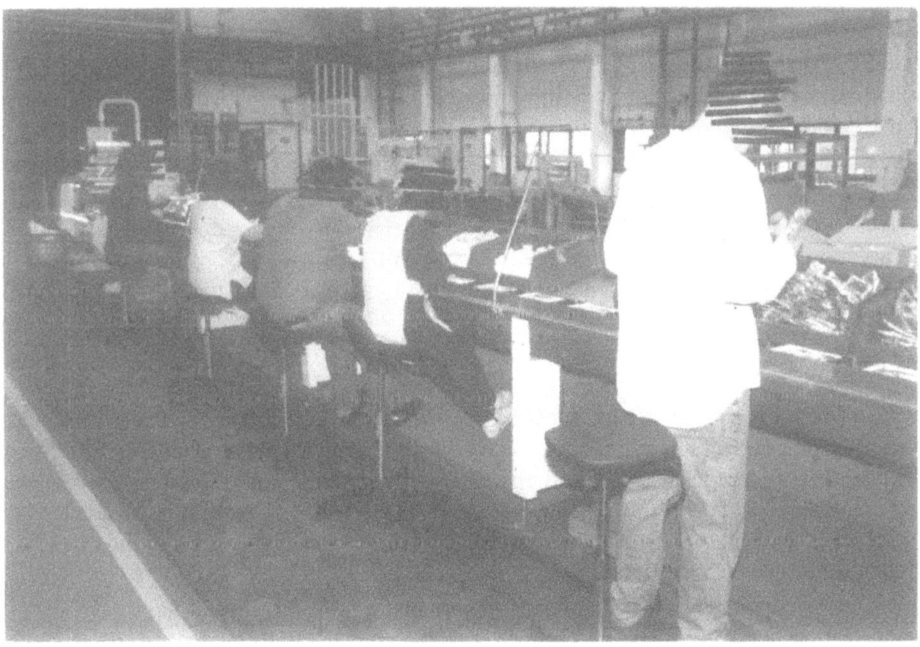

(b)

FIGURE 4.17 (a) Toe space is needed if standing workers are to stand up straight. (b) Sit–stand industrial production line. Note the use of height-adjustable sit–stand stools and a large sloping footrest to accommodate a variety of trunk–thigh angles.

TABLE 4.4
Recommended Work Surface Heights for Standing Workers (cm)

Task Requirements	Male	Female
Precision work	109–119	103–113
Light assembly work	99–109	87–98
Heavy work	85–101	78–94

Source: From Ayoub (1973), with permission.

when writing on a flat desk significantly reduced the moment of flexion of the head and neck at the C7–T1 level of the spine and was rated by the subjects as more comfortable than not using one. The use of the horizontal desktop for reading and writing would appear to be a rather new and inferior development in the history of furniture design.

Shoes

Wearing high-heeled shoes causes people to walk with short steps and with the feet further apart for ankle stability. The plantar flexors are shortened and therefore weakened according to the muscle length–tension relationship, affecting gait, the toe-off phase of gait, and causing muscles further up the legs to contribute more to the gait cycle. The pelvis can be tilted anteriorly for ankle stability, and the lumbar spine extended for balance. Changes to the gait cycle include higher ground reaction forces and greater stress on the knee. In the longer term, the Achilles tendon may become shortened. Other effects of wearing high-heeled shoes include increased trunk extension, increased tibialis anterior muscle activity, increased low back muscle activity, and increased vertical oscillation of the body center of mass while walking. Taken together, these findings suggest that wearing high-heeled shoes increases postural load in standing and reduces walking efficiency. In some occupations, high-heeled shoes are part of the dress code or uniform. Following a systematic literature review (to counter the negative effects of wearing high-heeled shoes), Wang et al. (2021) recommended:

1. Heel-heights should be limited between 3.76 and 4.47 cm.
2. Larger heel bases should be chosen to improve gait stability.
3. Inserts should be used to distribute plantar pressure more evenly.

Ergonomics of Seated Work

The basic postural requirements for sedentary work should be achieved using a combination of modern office chairs with lumbar supports and adjustable backrests, footrests as required, and the appropriate placement of task objects and visual targets within the zone of convenient reach and the optimal visual field. Key features of chair design are summarized in Table 4.5.

Backrests and Lumbar Supports

A lumbar support works by providing space for the buttocks to protrude when a person sits down (Figure 4.18). The convex lumbar support then uses the lumbar spine as a lever to tilt the pelvis forward, preventing a complete loss of the lumbar lordosis. The location of the lumbar support with respect to the level of the lumbar spine (L1/L2 or L4/L5) does not seem to be of importance. This suggests that a height-adjustable lumbar support is not necessary, since the height of a fixed lumbar support can be optimized to contact the lumbar spines of a wide range of users (population differences in stature are largely due to differences in the lengths of the long bones rather than the lumbar vertebrae).

TABLE 4.5
Key Features of Chair Design

Seats should swivel and have heights adjustable between 38 and 54 cm. Footrests should be provided for shorter users.
Free space for the legs must be provided both underneath the seat to allow the user to flex the knees by 90° or more and underneath the work surface to allow knee extension when reclining.
A 5-point base is recommended for stability if the chair has castors.
The function of the backrest is to stabilize the trunk. A backrest height of ~50 cm above the seat is required to provide both lumbar and partial thoracic support.
If the backrest reclines, it should do so independently of the seat to provide trunk–thigh angle variation and consequent variation in the distribution of forces acting on the lumbar–pelvic region.
Lumbar support can be achieved by either using extra cushioning to form a lumbar pad or contouring the backrest. In either case, there must be open space between the lumbar support and the seat pan vertically below it to allow for posterior protrusion of the buttocks.
The seat pan must have a slight hollow in the buttock area to prevent the user's pelvis from sliding forward. This keeps the lower back in contact with the backrest when reclining. The leading edge of the seat should curl downward to reduce the pressure under the thigh.
Armrests should be high enough to support the forearms when the user is sitting erect. These should also end well short of the leading edge of the seat so as not to contact the front edge of the desk. If the armrests support the weight of the arms, less load is placed on the lumbar spine.
Modern chairs tend to have a thin layer of high-density padding. Layers of thick foam tend to destabilize the sitter. The foam can collapse after constant use.
Cloth upholstery provides friction to enhance the stability of the sitter.

Both lumbar supports and forward-tilted seats result in a more neutral posture with the pelvis tilted anteriorly (compared to sitting on a chair with no lumbar support) and less lumbar flexion. A lumbar support also reduces activity in the back muscles (De Carvalho and Callaghan, 2023).

To the extent that lumbar supports extend the lumbar spine, their use may improve breathing. Slumped sitting postures increase intra-abdominal pressure, as the ribs approximate the pelvis and therefore prevent the downward movement of the diaphragm and the expansion of the chest wall. Landers et al. (2003) showed a significant (and over 10%) reduction in tidal volume and minute ventilation of the lungs when subjects sat in a slumped, as opposed to an upright posture. Since breathing frequency did not increase, these findings imply that slumped sitting postures may cause drowsiness, and lumbar support may improve pulmonary efficiency in otherwise healthy people.

Backrest Tilt

Office chairs should have a tilt facility to support the trunk at different angles of repose (Figure 4.18). Ergonomically designed workplaces must be *flexible if postural fixity*, causing static loading of the musculoskeletal system, is to be avoided. Flexibility implies that the worker can carry out the task, at least some of the time, in more than one working posture, with a workspace designed to accommodate both postures. A "designed-in" resting posture is also desirable. Well-designed seats may enable the user to "take the weight off the feet," but will not take the weight off the spine unless the user reclines against a tilted backrest.

Getting the Fit Right

Anthropometric mismatches can have serious consequences for health and efficiency. Short users may have to raise *seat heights* beyond popliteal height in order to gain access to the desk. As a consequence, the feet no longer rest firmly on the floor, and the floor cannot be used by the legs as a fulcrum for stabilizing and shifting the weight of the upper body. The load of stabilization is now

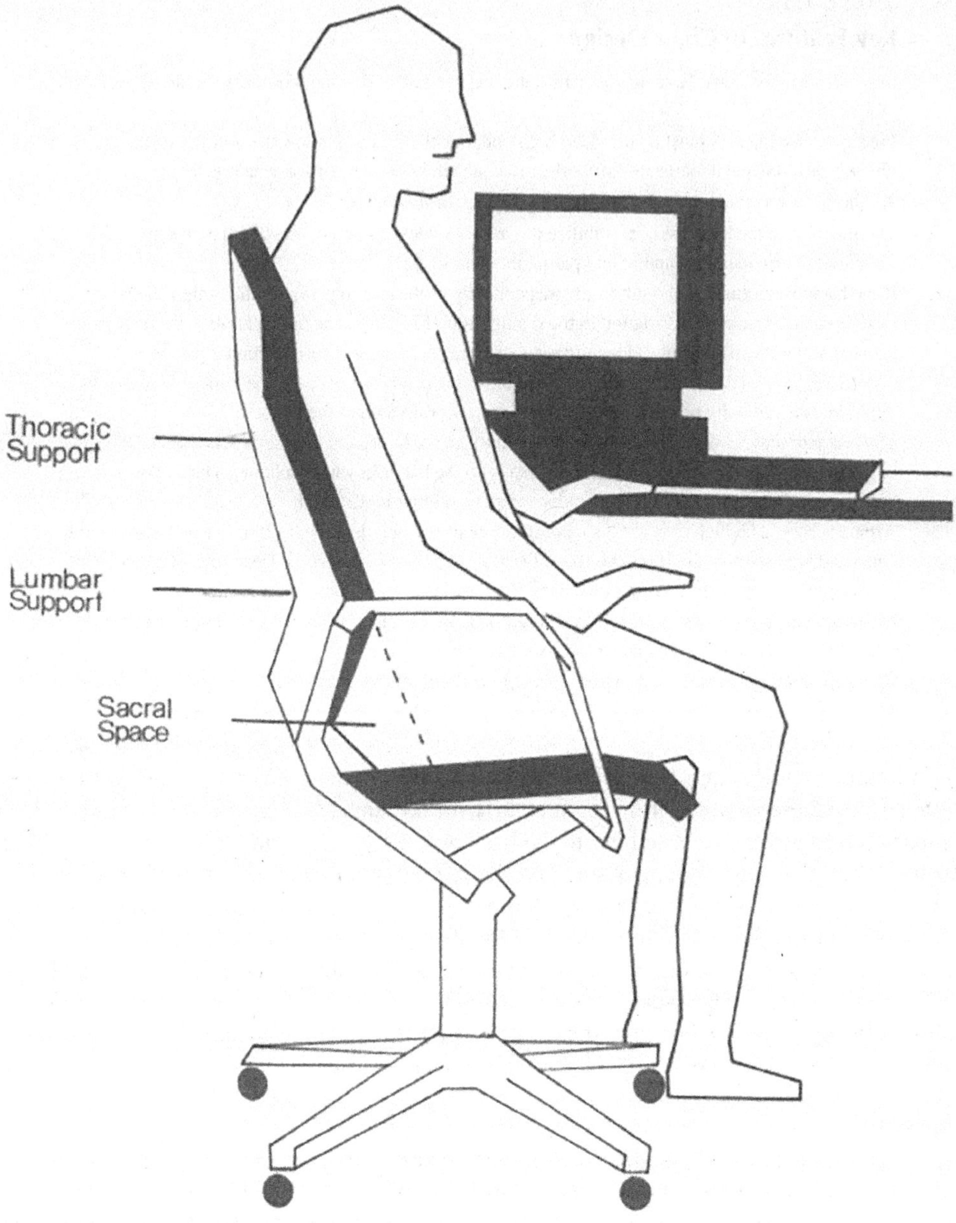

FIGURE 4.18 Modern office chair with lumbar support.

transferred to the muscles of the trunk. The weight of the legs, instead of being borne by the feet, is borne by the underside of the thigh. This can restrict blood flow and is particularly undesirable for those with varicose veins. It can also cause a condition known as "Lipoatrophia Semicircularis" (band-like circular depressions and isolated atrophy of subcutaneous fatty tissue on the thighs). Hermans et al. (1999) found postural differences between LS sufferers and non-sufferers—static sitting postures, less use of lumbar support, chair too high, and higher pressures at the seat surface.

When seats are too high, the feet can no longer be used to extend the base of support beyond the base of the chair. This makes activities such as reaching and picking up heavy objects more hazardous. Continuous compensatory movements of body parts may be necessary to maintain stability. An alternative to raising the chair is for the short user to work with the elbows below desk height, increasing the static loading of the upper body—particularly the shoulder girdle as the elbows are held in an elevated position. Computer and typewriter keyboards increase the effective work surface height; hence, the use of sliding keyboard drawers for programmers. Tall users may find *desk heights* of approximately 70–75 cm too low, since even with the chair at its lowest level, the distance between the eyes and the work surface may be too long for comfortable viewing, causing the user to slump over the desk when writing.

Lack of footspace may also be a problem for tall workers. Short users can be given *footrests as standard equipment* to ensure anthropometric fit. This is similar to the industrial practice of designing high benches to suit taller standing workers and providing raised platforms for those of lesser height. Height-adjustable desks are available and were proposed many years ago (Ostberg et al., 1984) for use with height-adjustable chairs in order to increase the range of users accommodated by a workstation. The correct way to adjust one of these workstations is

1. First adjust the chair height so that the feet are resting firmly on the floor.
2. Adjust the work surface height for comfortable access to the desktop or keyboard.

Dynamic Postures

Van Deursen et al. (2000) designed a dynamic office chair in which rotations about an axis perpendicular to the seat were applied by a motor at an amplitude of 0.6° and a frequency of 0.08 Hz. The application of these gentle twisting motions was found to result in increases in spinal length over a 1-h period of sitting—significantly more so than when subjects sat in a static control chair. These findings support the hypothesis that rotation applied to the vertebrae during sitting reduces pressure in the nucleus pulposus, allows fluid to enter, increases disc thickness, and improves the nutritional status of the disc. When provided with additional workstation flexibility in the form of a tilting seat, subjects do use it and appear to prefer it to similar chairs with fixed seats. Interestingly, the tilt facility was used less frequently, and the muscular load was higher in a typing task compared to the subject's ordinary desk work not involving typing. This indicates that tasks involving typing are constraining and should be combined with other work. Stranden (2000) found that tilting seats produced a significant reduction in leg swelling (note the difference was small—of the order of 1% of calf volume).

Guidance for Office Workstation Design

Table 4.6 summarizes some design solutions for increasing the flexibility of workstations, and Table 4.7 gives dimensions for U.S. workers. A participatory approach is often the most effective way to decide which product features and accessories are likely to have the greatest cost-benefits for different groups of workers in an organization. For example, document holders are often of great benefit to copy typists and data entry clerks, whereas systems designers may regard increased storage space and shelving as a higher priority.

Standards vary between countries, but normally the following requirements apply:

- *Screen*: A stable image, adjustable, legible, and reflection-free.
- *Anthropometry*: Desks should have free space underneath ("leg room"), and the workstation should allow postural changes.
- *Software*: Appropriate to the task, adaptable by the user, providing feedback on system operation and status. No undisclosed monitoring.
- *Adequate Contrast*: No glare or distracting reflections, particularly on the computer screen.
- *Work Chair*: Adjustable and suited to the user, task, and furniture.

TABLE 4.6
Product Features and Accessories for Enhancing Workstation Flexibility

Item	Accessory
Source documents	Document holders
	Bookstands
	Tilted area of work surface
	Shelving
	Notice board
VDT screen	Movable screens
	Tilt and swivel adjustment
	Screen holders
Keyboard	Detached from screen VDT
	Keyboard drawer
Seating	Footrest
	Armrest
	Lumbar pad
	Narrow backrest (permits trunk lateral flexion and rotation)
	Lumbar and thoracic support
	Recline mechanism
Desks	Height adjustable
	Tilt adjustable
	Extensions can be fitted

TABLE 4.7
Workspace Dimensions (cm) for U.S. Users

	Ninety-fifth Percentile Male	Fifth Percentile Female
Seat height	49	41
Elbow rest height	29	18
Work surface height	71	58
Screen height	Adjust screen so that screen top is no higher than the horizontal sight line when reclining by 15°	
Eye height (90° sitting)	130	103

Source: From ANSI/HFS 100 (1988), with permission.

- *Adequate Lighting*: About 300 lux of desktop illumination is needed.
- *Distracting Noise Minimized*: An ambient noise level of about 60 dB(A) is the maximum window covering. Vertical blinds are preferred to provide a view of the outside world.
- *Work Surface*: Sufficient space to allow for flexibility and increased workload. Appropriate job aids such as in-trays and desk lamps. Glare-free and adjustable.
- *Footrest/Footstool*: Normally tilted toward the user at 5°–15°. May have a "treadle" action.

Some important features of office chairs are summarized in Table 4.5.

Foot Pump Devices

There is a 66% reduction in venous blood flow through the lower leg when sitting for long periods, which is why prolonged, inactive sitting is a risk factor for deep vein thrombosis and why lower leg volume can increase by over 5% on an inactive day. According to Stranden, the key to venous pump activation is leg movement. Plantar flexion, in particular, is immediately effective and can be elicited by reclining on a tilting seat or by resting the feet on dynamic footrests that are pivoted to allow a "treadle pumping" kind of action. Wentzel and McKune (2013) evaluated the effects of a foot pump device designed to encourage plantar flexion against resistance, on lower leg swelling of seated workers. The device was found to prevent leg swelling over an 8-h day. In general, leg swelling can be prevented by walking for about 2 min every 15 min during an otherwise inactive day. When employees have to sit for 8 h, foot pump devices may be beneficial.

Office Workstation Requirements

Figure 4.19 summarizes the requirements to achieve an anthropometric match in seated workstations.

Standards for the design of these workspaces include ANSI Standard No. 100-1988, BS 7179, 1990, BS 5940, 1980, BS 3044, 1990, and the Health and Safety Commission (1991). The guidelines proposed in the above documents attempt to specify user space requirements and define appropriate furniture in functional as well as physical terms. Readers are referred to these documents for detailed information.

Standing to Work at Computers and Laptops

Sit–stand workstations for office workers are widely available. To obtain a good posture when standing to work, the screen should be at standing eye height, and the space around the feet should be clear (Figure 4.20). One foot can be placed on a footrest when standing. Foster et al. (1998) measured postural sway when subjects carried out an interactive task at a standing workstation. There was barely any difference in sway compared to when people were asked to stand perfectly still. This implies that venous pooling and discomfort due to calf muscle fatigue and static loading is a major risk factor for computer users who stand. In fact, discomfort in the legs increased steadily after 20 min of standing and rapidly after 45 min, suggesting that standing workers should sit every 30–45 min at least.

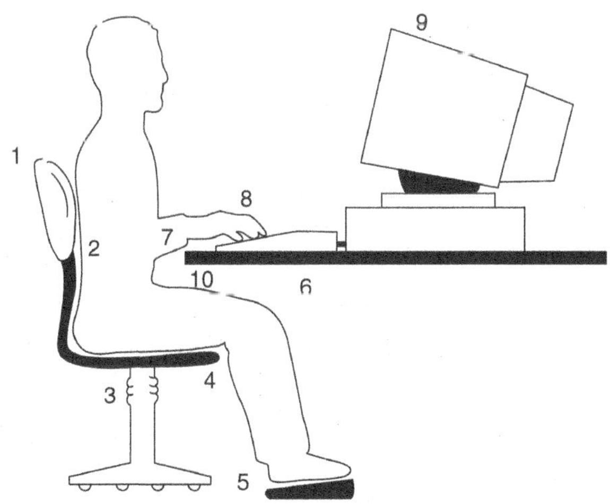

FIGURE 4.19 Essential furniture considerations in fixed computer workstation design. (From Display Screen Equipment—Work Guidance on Regulations, copyright HMSO, 1992, with permission.)

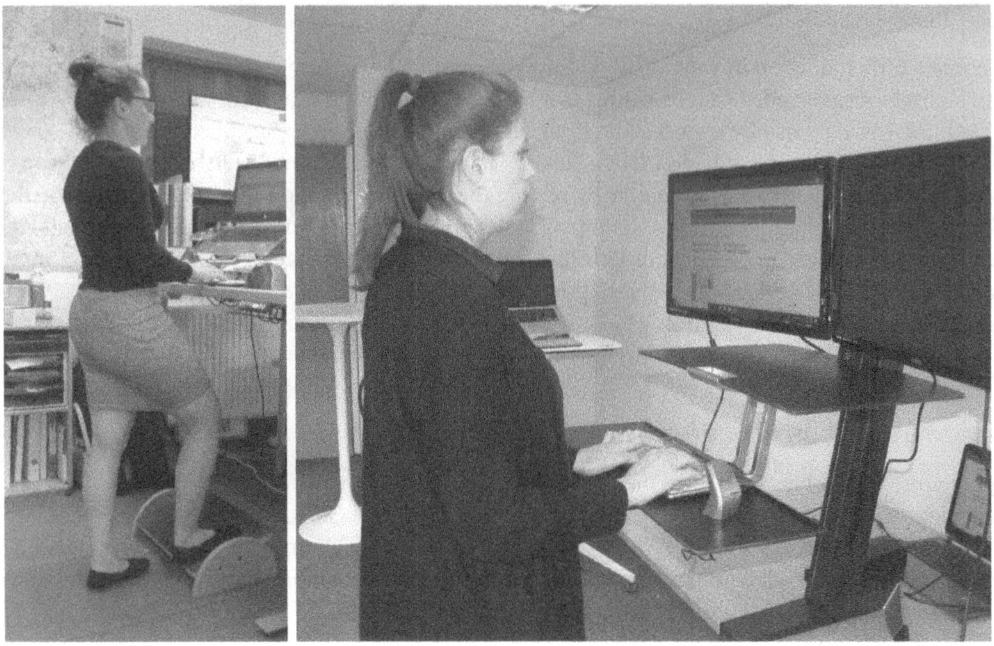

FIGURE 4.20 (a) Standing to work at a computer. (b) A footrest is recommended. (Courtesy Osmond Ergonomics Ltd.). The footrest must be usable by both feet and around 10%–15% of stature. Is she standing on one leg or walking on the spot in slow motion? (b) Height-adjustable platform with inclined pillar; as the screen is raised, it moves away from the front of the desk. (Workfit-S by ERGOTRON®, ergotron.com)

General recommendations for work at height-adjustable standing desks include:

1. Stand in an upright, comfortable position and look straight ahead.
2. Adjust the desk height so that the mid row of keys on the keyboard is at the height of the elbows.
3. For electrically adjustable/automatic desks, standing elbow height is about 63% of stature. So, a female 1.65 m tall would need a desk height of about 1.04 m. A male 1.80 m tall would need a desk height of about 1.13 m.
4. Ensure that there is space for the feet underneath and all around the desk.
5. Check that the footrest (if provided) is approximately 15 cm distant from the front edge of the desk (so that you can stand close to the desk and move one foot forward under the desk and onto the footrest).
6. Check that the footrest can be used with either foot.
7. Check that the footrest is approximately 10% of stature with the footwear you are using.
8. Check that the screen is still a comfortable viewing distance from your eyes when you transition from sitting to standing. When people sit at a desk, they often sit about 15–20 cm from the front of the desk, resting their wrists on the desktop or a wrist rest). When they stand to work, they often move closer to the desk, reducing the distance between the eyes and the screen (Figure 4.21). You may need to move the screen further back when you stand to work.
9. Some height-adjustable platforms for sit–stand computer working have an inclined pillar such that the screen moves away from the user as the height is raised, which may compensate for the user moving closer to the desk when standing (Figure 4.21).
10. Use a perch stool to take the weight off the feet if a conventional office chair is not available or if the desk is not height-adjustable (Figure 4.22).

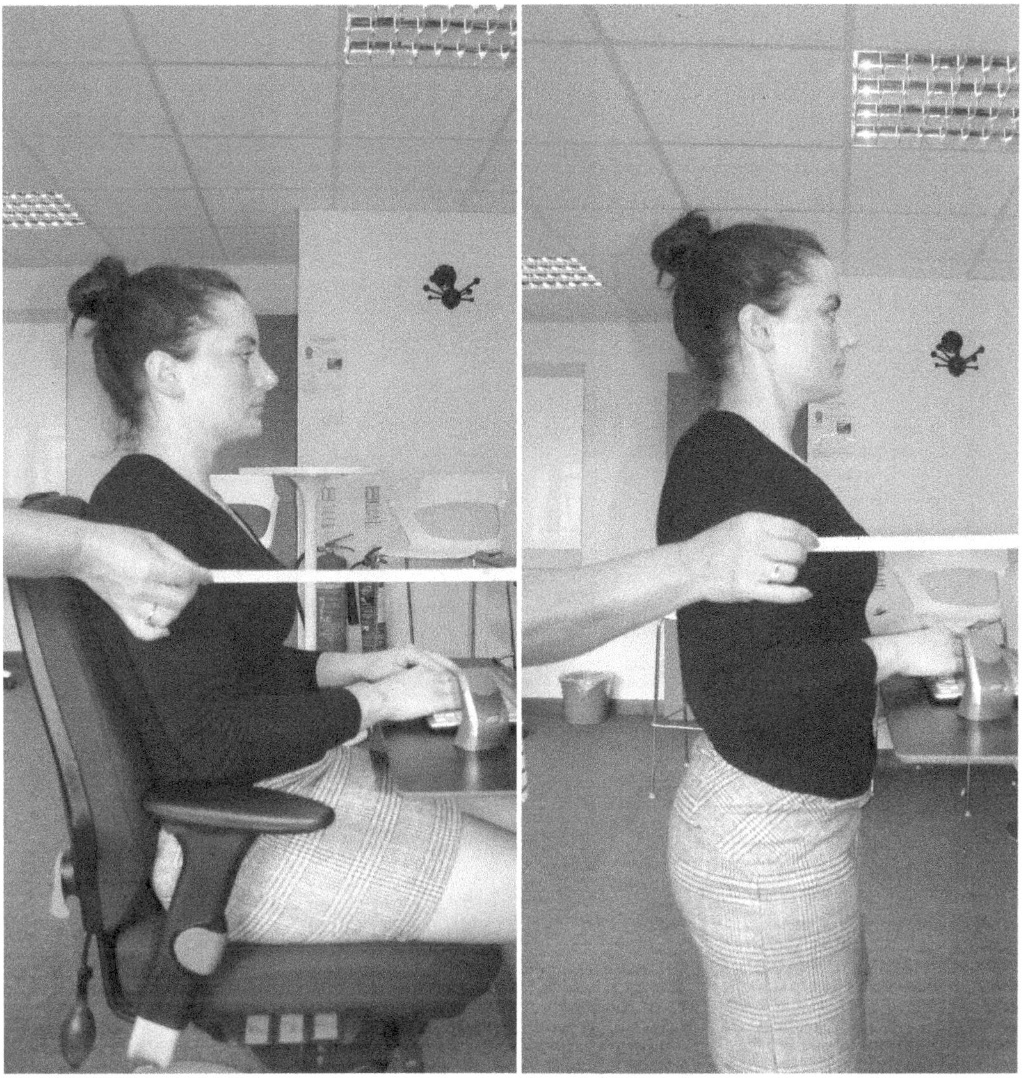

FIGURE 4.21 When sitting, people normally recline. When they stand up, the upper body is more upright and the eyes are closer to the screen.

Sit–Stand Ratios for Office Workers*

Black et al. (2022) investigated sit–stand ratios ranging from 30-min periods of sitting with no standing to 15 min alternating between the two (changing the intervals in steps of 3 min to create six sit–stand combinations). The research was carried out in participants' own workplaces. Although there was no overall preference for a particular ratio, all combinations had lower discomfort due to the postural changes. The least liked ratios were 30:0 (no standing), 27:3, and 24:6. Preference was greater for 15:15, 18:12, and 21:9, although there was no clear winner. These findings suggest that office workers prefer to alternate between sitting and standing if the furniture permits them to do so.

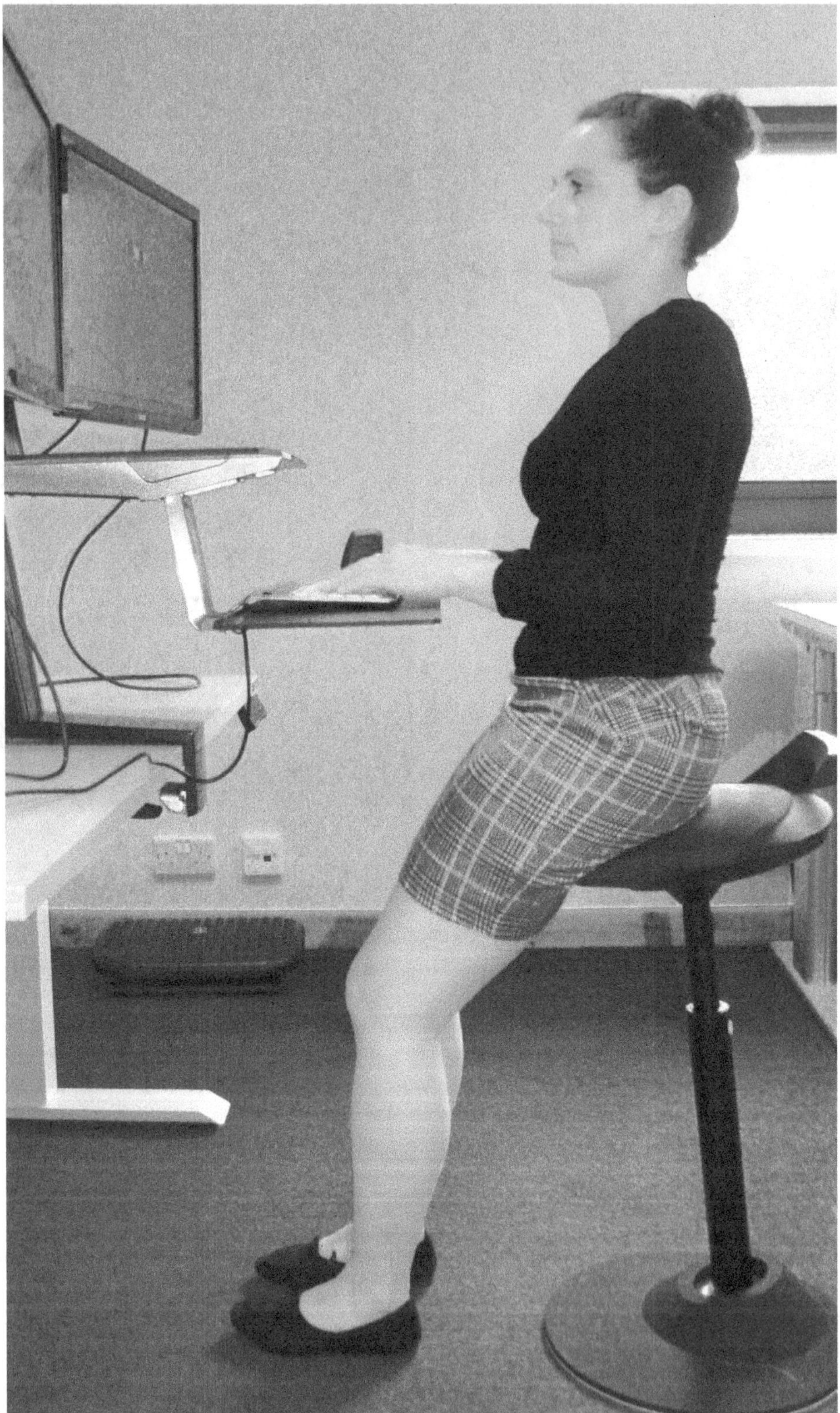

FIGURE 4.22 Use of a perch stool to take a break when standing to work. (Photo courtesy Osmond Ergonomics Ltd.)

Laptops and Tablets

User interaction with laptops and tablets is not the same as with personal computers. Straker et al. (2008) compared the postures of children using a conventional computer screen, a tablet, and paper. Musculoskeletal stresses using the tablet and paper were higher than with the conventional screen with a more asymmetrical trunk posture, greater elevation of the shoulders, and increased muscle activity around the neck. However, this was offset by greater variation in posture and muscle activity. One of the challenges facing employers nowadays is that employees can work at home, in cafes, and on public transport using mobile IT devices. At present, validated guidance is limited, and HFE specialists should adhere to first principles as espoused in this and other chapters. For standing and sitting at desks, laptops should be supplemented with detachable keyboards and laptop stands to enable the keyboard to be at elbow height with the screen elevated to eye height.

TOOLS AND PROCESSES

The American Industrial Hygiene Association has published an extensive toolkit for the assessment of workplace ergonomics. It is freely available online.[1] In this section, some examples will be given, plus some checklists and other techniques not available in the toolkit.

Table 4.8 presents items for a subjective chair evaluation checklist that can be used either as an ergonomist's aide-memoire or as a source of items for questionnaire surveys of sedentary workers. Table 4.9 presents a quick checklist for office workers.

For both standing and sedentary workers, work surfaces should be arranged so that the worker does not have to work continually with objects placed at one side or reach excessively to the side. The main working area should be directly in front of the worker's body to minimize any twisting of the trunk when carrying out task-related movements.

TABLE 4.8
Quick Chair Design Usability Checklist

- Seat design
  - Any discomfort due to
    - The shape of the seat
    - Pressure under the thighs
    - Pressure under the buttocks
    - Pressure from the leading edge of the seat
  - Is the seat
    - Too narrow
    - Too wide
    - Too shallow
    - Too deep
    - Slope too far forward
    - Slope too far backward
    - Too restrictive
    - Not supportive
    - Unstable
    - Too high
    - Too low
  - Can you
    - Get in and out easily
- Chair adjustment
  - Any difficulty in
    - Locating the height controls

(*Continued*)

TABLE 4.8 (*Continued*)
Quick Chair Design Usability Checklist

Adjusting the height controls[a]
Controlling the chair when adjusting the height
Obtaining a comfortable height
Backrest
Does the backrest support your lower back?
When reclining slightly
When reclining fully
Does the backrest support you at shoulder level?
When reclining slightly
When reclining fully
Do you feel cramped when using the backrest?
Does the backrest fit your back?
Lumbar support
Does the lumbar support your lower back?
Does the lumbar support feel comfortable?
When sitting upright
When reclining
Does the lumbar support relieve any backache?
Does the lumbar support help you to sit up straight?
Chair-user fit
Can you recline and keep your feet on the floor[b]?
Do you feel stable when you recline[b]?
Is the seat too deep for you to use the lumbar support?
Backrest articulation
Is the backrest spring tension too high?
Is the backrest spring tension too low?
Can you change the position of the chair easily?
Elbow rests
Are the elbow rests too high?
Are the elbow rests too low?
Do they make you feel locked in?
Are they too close together?
Are they too far apart?
Do they prevent you from moving your chair close to your desk?
Upholstery and material
Does the material provide adequate friction?
Is the padding adequate:
On the seat
On the backrest
Does the chair feel hot in summer?
Others
Does the seat swivel easily?
Does the chair feel stable?
Can you tuck your feet under the seat when you want to?
Can you move easily when seated (chairs with castors)?

[a] Pneumatic controls are generally preferred as manually operated height controls are invariably difficult to adjust and cause problems for less strong users.
[b] Negative response indicates that a footstool is needed.

TABLE 4.9
Quick Office Evaluation Checklist

Office desk
- Have you enough space on your desktop?
- Have you enough storage space?
- Is your desk
  - Too high
  - Too low
  - Too large
  - Too small
- Are the drawers easy to operate?
- Is there enough space for your legs under the desk?
- Is there a modesty board?

Computers
- Can you adjust the screen brightness?
- Can you change the screen colors?
- Is there any glare on the screen?
- Can you adjust the screen position and tilt?
- Are the cables neatly tucked away?
- Are the screen characters the right size?
- Is the screen image stable?
- Does the screen flicker?
- Is there sufficient contrast?
- Is the screen color acceptable?

Keyboard
- Does the keyboard have a QWERTY layout?
- Is the keyboard detachable?
- Can you operate the keyboard with your wrists in a comfortable position?
- Are the characters on the keys legible?

Mouse
- Is the mouse cable long enough?
- Does the mouse move easily?
- Is the mouse comfortable to use?
- Does the cursor stick when using the mouse?
- Is there enough space for the mouse and mouse pad?

Workspace
- Can you rest your feet on something if you need to?
- Can you change posture at will?
- Does the backrest support your back when you are doing your normal work?
- Can you adopt more than one comfortable work posture?
- Does your chair recline?
- Can you look out of a window when you want to?
- Can you reach down and into drawers and trays easily?
- Can you get in and out of your workspace easily?

Office environment
- Are the lighting levels comfortable?
- Are the office lights fitted with diffusers?
- Is there enough natural light in the office?

(*Continued*)

TABLE 4.9 (*Continued*)
Quick Office Evaluation Checklist

Are the lights flicker-free?
Are there pleasant visual objects (e.g., plants)?
Are there enough windows?
Is the temperature comfortable?
Is the air pleasant to breathe?
Are you free of unwanted noise?
Is there enough space in your office?
Job content
Is there sufficient variety in your work?
Can you leave your desk from time to time?
Do you have enough work to keep you busy without being over-loaded?
Do you see other people during the day?
Do you work with others during the day?
Are you happy with the way you are supervised?
Can you get on with your job without distraction?
Do you have enough privacy?
Do you find your job interesting?
Do you have enough responsibility?
Aches and pains
Do you get low back pain at work?
Do you get neck or shoulder pain at work?
Do your eyes feel tired at work?
Do you get pins and needles in any part of your body?
Do you have difficulty focusing your eyes at the end of the day?
Do you get pains in your hands, wrists, or arms at work?
Do you usually feel tired at the end of the day?
Do you get headaches at work?
Do you often suffer colds and flu?
Does your skin ever feel itchy or irritated at work?

Office Environment Standards vary between countries, but normally, the following requirements apply:

- *Screen*: A stable image, adjustable, legible, and reflection-free.
- *Anthropometry*: Desks should have free space underneath (leg room), and the workstation should allow postural changes.
- *Software*: Appropriate to the task, adaptable by the user, providing feedback on system operation and status. No undisclosed monitoring.
- *Adequate Contrast*: No glare or distracting reflections, particularly on the computer screen.
- *Work Chair*: Adjustable and suited to the user, task, and furniture.
- *Adequate Lighting*: About 300 lux of desktop illumination is needed.
- *Distracting Noise Minimized*: An ambient noise level of about 60 dB(A) is the maximum window covering. Vertical blinds are preferred to provide a view of the outside world.
- *Work Surface*: Sufficient space to allow for flexibility and increased workload. Appropriate job aids such as in-trays and desk lamps. Glare-free and adjustable.
- *Footrest/Footstool*: Normally tilted toward the user at 5°–15°. May have a treadle action.

Subjective data obtained by such questionnaires are best interpreted, together with objective measurements made according to applicable standards.

Consulting Users for Furniture Selection: A Structured Approach

Irrespective of whether anthropometric data are available for furniture selection or whether task analyses have been conducted, HFE takes a user-centered approach to gather the opinions and thoughts of users about the choice of furniture. The Display Screen Equipment regulation in the EU specifically requires assessors to look not just for problems but also enquire about the positive aspects of workstations and furniture so that when offices are redesigned, and positive aspects can be carried forward.

HFE Workshop 4.1 describes two methods for analyzing user opinions about office chairs that can be used to make evidence-based purchasing decisions.

HFE WORKSHOP 4.1

USER-CENTERED OFFICE CHAIR SELECTION

A large financial institution wishes to purchase 1000 new office chairs for its headquarters. It wishes to benefit from a bulk purchase of identical chairs, but has concerns about whether a "one size fits all" approach will work. HFE specialists at the company carry out an anthropometric study of available chairs using the methods described in the previous chapter. A shortlist of six chairs (A–F) is drawn up, and these chairs are tested for a day each by three employees—a female of fifth percentile stature and healthy body composition and waist circumference (see Chapter 8), a male of ninety-fifth percentile stature, and an obese female. All employees try each chair for a day in a different order.

TESTING FOR PREFERENCE: DO THE JUDGES AGREE?

The three judges rank the order of the chairs according to their preference (1 = "most" to 6 = "least" preferred). The ranks are summarized below.

	Office Chair					
	A	B	C	D	E	F
Fifth percentile ♀	1	6	2	3	5	4
Ninety-fifth percentile ♂	1	6	2	3	4	5
Obese ♀	2	4	3	1	6	5
R_j	4	16	7	7	15	14

The study has been conducted using the "beauty contest paradigm," where the aim is to determine whether the judges agree on which chair is preferable. If there were complete agreement, then the most preferred chair would have a score of $R_j = 3$, the second-most preferred chair a score of $R_j = 6$, and the least preferred chair a score of $R_j = 18$.

Kendall's Coefficient of Concordance "W" is a statistical test used to determine whether the level of agreement between n judges about k items is statistically significant. Most statistical packages support this test; therefore, details of the calculation of the test statistics are not included here. The test is a nonparametric test that can be used for small samples. The data above yield a value of

$$W = 0.81 (P < 0.05)$$

showing statistically significant agreement between the preferences of three physically different judges.

(Continued)

HFE WORKSHOP 4.1 (*Continued*)

Note, a significant result would not mean that the chair with the lowest ranking was the "best" chair—only that the judges agree on which chairs they prefer. It may well be that the "most" preferred chair was still "unacceptable" for the majority of users (the judges agree that it is the "best of the worst"), so it is always wise to debrief participants after such exercises to find out what lies behind their rankings. If the judges had been found not to agree, then further investigation would be needed to determine the reasons. Close inspection of the rankings might reveal that one or more of the chairs was preferred by the short and tall judges but not by the obese judge.

TESTING FOR ACCEPTABILITY: IS THERE A TREND?

Suppose we carry out further investigations and select the four most preferred chairs as presented in the table above—that is, chairs A, C, D, and F. We want to know whether the chairs differ in acceptability (all or none might be acceptable), and whether there is a trend in acceptability from the most to the least preferred. In other words, given that the judges have their preferences, does it really matter which one we choose? Chairs A and C were preferred, but if they are more expensive than D and F, would it really make a difference if we purchased D or F rather than A or C?

One way of looking at this is to ask users to rate the acceptability of the chairs. Eight users, unfamiliar with the chairs, try them out for a few days, with the order in which they try them varied between users. Each of the eight users tries each of the four chairs for a day and then rates each chair in terms of acceptability on a 20-point scale (1 = "completely unacceptable to me as my office chair" to 20 = "completely acceptable as my office chair," with 10 being "neutral"). The data are shown below, with the chairs ordered in terms of increasing preference.

User	Chair F	Chair D	Chair C	Chair A
1. ♀	5	8	12	15
2. ♀	3	11	11	14
3. ♀	8	8	12	12
4. ♀	3	6	9	13
5. ♂	6	6	6	15
6. ♂	8	10	14	15
7. ♂	5	7	12	13
8. ♂	4	9	15	15

Pages' *L* test is the name of the statistical test used to investigate trends when there is an ordered series of treatments scored using rank (ordinal) data. The procedure for calculating *L* is given below.

PROCEDURE

1. Arrange the chair columns in the order predicted by the preference scores
2. Take each user's acceptability scores in turn and rank them (1–4)
3. Sum the ranks for each chair
4. Multiply each sum by its weighting coefficient
5. Add the weighted sums to calculate "*L*"
6. Refer to tables of critical values of "*L*"[2]

HFE WORKSHOP 4.1 (*Continued*)

User	Chair F	Chair D	Chair C	Chair A
1. ♀	1	2	3	4
2. ♀	1	2.5	2.5	4
3. ♀	1.5	1.5	3.5	3.5
4. ♀	1	2	3	4
5. ♂	2	2	2	4
6. ♂	1	2	3	4
7. ♂	1	2	3	4
8. ♂	1	2	3.5	3.5
Totals	9.5	16	23.5	31
Weighting	1	2	3	4
	9.5	32	70.5	124

$L = 9.5 + 32 + 70.5 + 124 = 236$

With $n = 4$ treatments (chairs) and $n = 8$ subjects (users), $L = 236$ exceeds the critical value of L at the 0.01 level of statistical significance; hence, we can conclude that there is a statistically significant trend in acceptability across the chairs. The preferred chairs were judged to be more acceptable.

Therefore, our small user trial has shown that the users have clear preferences about which chair they prefer and that the most preferred chairs are judged to be more acceptable than the least preferred. A recommendation can be made to the company that Chair A will deliver the most benefit in terms of preference and acceptability. Note also that Chair D was rated close to "neutral" in terms of acceptability, and that none of the chairs was rated as "completely acceptable." This latter fact might be used to make the case for job redesign to increase the amount of time spent standing or away from the desk. We could repeat the trial in other parts of the company and use judges from different departments, with different work roles and seniority levels, or we might test a wider range of chairs.

In this example, we have deliberately used small numbers of judges and raters to demonstrate that even with limited time and resources, HFE specialists can take a structured approach that yields data amenable to statistical testing and external scrutiny. This is the kind of "hard data" on a "soft topic" that might help to persuade a budget manager to buy Chair A, if it turns out to be more expensive than the others.

SYSTEM INTEGRATION

Integrated solutions must take into account the long- and short-term health effects as well as efficiency and effective working relationships.

Activity-Permissive Workstations

Research to date indicates that physical activity should be integrated into the design of sedentary jobs. One way to do this is by introducing activity-permissive workstations. These include treadmill desks, pedal or stepping desks, sit-stand (height-adjustable) desks and "fidget" bars. Neuhaus et al. (2014) carried out a systematic literature review of research on activity-permissive workstations. The main conclusions from a total of 26 studies were as follows: 11 of 14 studies reported reductions in sedentary time (mean of 90 min/8-h shift). Some studies showed lasting effects of workplace interventions (182 min/day from baseline to 3 months, and 88 min from baseline to 9 months). Health outcomes varied across studies, with some reporting improvements in waist circumference and others reporting increases in foot swelling. Overall, work performance was unchanged, and the workstations were accepted.

Nguyen at al. (2020) carried out an "umbrella review" (a review of systematic reviews of interventions to reduce sedentary time at work), concluding that there was strong evidence that activity-permissive workstations were effective in reducing occupational sitting time in offices by at least 30 min/day.

Sit–Stand Workstations

Garrett et al. (2016) compared the productivity and self-reported sitting time of call-center workers using either sedentary workstations or "stand-capable" workstations, where employees could either stand or sit. Productivity (number of successful call outcomes per hour, where a successful call resulted in revenue) was higher in the stand-capable group by 23% in the first month of the study, rising to 53% after 6 months. Employees using stand-capable furniture reported sitting for 72%–73% of the workday, compared to 91% among those using conventional furniture. Those in the stand-capable group reported less discomfort than those in the sedentary group, which may explain the difference in productivity. The study provides tentative support for productivity enhancement in sit–stand working, but the mechanism by which these benefits were achieved is not clear from this study (was it because standing is more productive, or because there was less discomfort?).

Foster et al. (1998) carried out an investigation of standing computer work. One of the interesting findings was that screen height *did* affect the accuracy of a visual checking task. In particular, it was found that accuracy was best when subjects worked with the line of sight approximately 15° below horizontal. Comfort was also greatest at this level, and the best postures were achieved.

Treadmill Desks

Oye-Somefun et al. (2021) carried out a systematic review of treadmill desks and their effects on energy expenditure, sitting time, and cardiometabolic health. Significant increases in energy expenditure and metabolic rate were found with even modest reductions in daily sitting time (1.73 min/h when walking at ≤3mph). No other changes in metabolic outcomes, such as blood pressure. were found. There is evidence that the use of treadmill desks, even when walking at low speeds such as 1.5 kph, reduces typing speed and increases errors without affecting cognitive performance (Podrekar et al., 2020).

McEwan et al. (2015) reviewed research on treadmill desks and identified a number of benefits:

1. Reductions in total and LDL cholesterol in overweight and obese volunteers, and an increase in HDL ("good") cholesterol, were observed over 9 months of using treadmill desks.
2. Reductions in blood pressure and evidence of better glucose control in volunteers with high blood pressure when they started using treadmill desks.
3. Improved blood glucose response in overweight participants using a treadmill desk for 2 min for every 10 min of sitting.
4. Some evidence that using a treadmill desk for 9 months to one year brings about significant weight loss and reductions in waist circumference, particularly in obese volunteers.
5. Walking on a treadmill at 1.6 km/h or at a self-selected speed does not appear to affect typing performance.

6. Although the findings of different studies are mixed, walking at 1.6 km/h seems best if task performance is to be maintained. Different studies do not always agree, and there is evidence for learning effects as performance improves after several months of using a treadmill desk—people get used to "walking while they work."
7. Walking slowly on a treadmill does not appear to affect cognitive performance. This is not surprising, since such walking is likely to be unobtrusive and will not distract the user from the task at hand.
8. In general, using of treadmill desks in the office can bring health benefits, more so than standing desks, because the physical activity of walking is greater (2–3 METS for walking compared to 1–2 for standing) (Figure 4.23).

FIGURE 4.23 A treadmill desk providing the affordance of walking while you work. (Courtesy of TNO, dep. Healthy Living, the Netherlands.)

Pedal and Step Desks

FIGURE 4.24 A pedal desk providing the affordance for cycling at your desk. (Courtesy of TNO, dep. Healthy Living, the Netherlands.)

Pedal desks seem to bring similar benefits in increased energy expenditure to treadmill desks. Schuna et al. (2019) compared the energy expenditure of eight male and eight female volunteers while: sitting at rest; seated in an office chair typing; typing while seated on a pedal desk at their own pace; typing on a treadmill desk while walking at their own pace. Energy expenditure when using the desk was about double that when typing while seated at a conventional desk, with pedaling at about once per second. A study by Rovniak et al. (2022) found that the work performance of physically inactive adults was not affected when pedaling at a light intensity of approximately 60–90 revolutions per minute, compared to sitting (Figure 4.24).

Fidget Bars

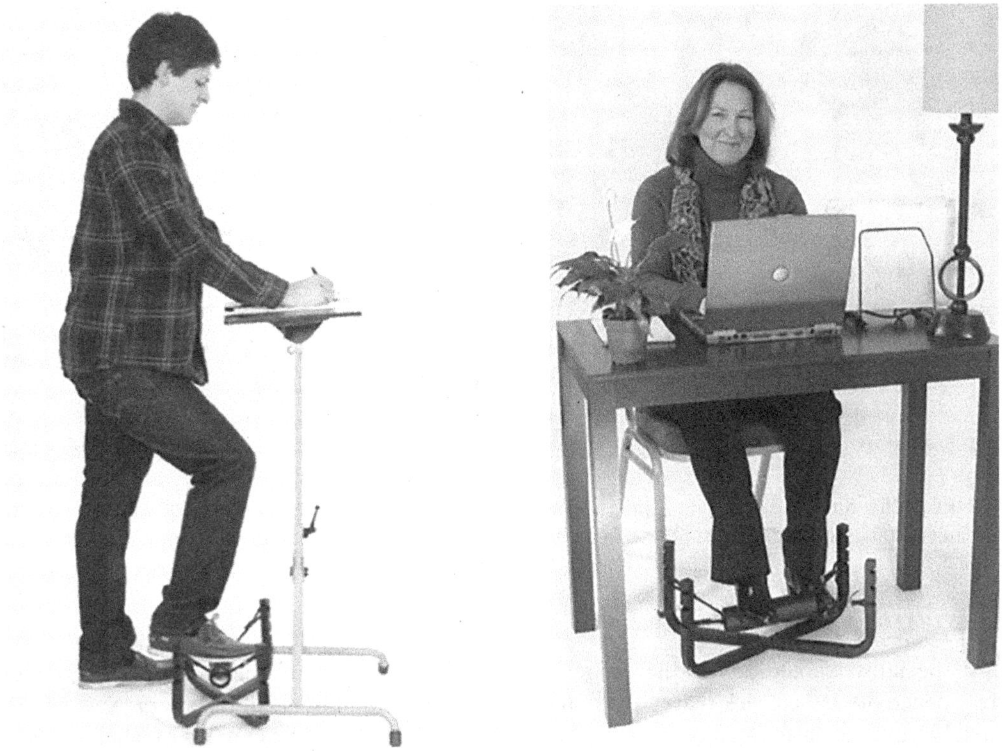

FIGURE 4.25 A "FootFidget" device providing the affordance for fidgeting at your desk. (Courtesy of classroom seating solutions, www.footfidget.com.)

There is some evidence that fidgeting while sitting has physiological and cognitive benefits. Hwang et al. (2023) found that the use of a fidgeting device when seated improved the executive function of subjects in their experiment. The "FootFidget" is an elasticated footrest that fits under the desk of a seated or standing employee. Koepp et al. (2017) compared conventional sitting and sitting with an "under the desk" fidget device. Fidgeting increased energy expenditure by about 20% but did not increase heart rate. The increase in energy expenditure was less than that found when walking slowly on a treadmill. The authors concluded that use of the device might have metabolic benefits but would not improve aerobic fitness.

Split Pad Seats

Noguchi et al. (2023) evaluated a new chair with a split seat pad (Figure 4.26). Subjects worked in each of three conditions during 3-h sessions on separate days: traditional standing, traditional sitting, or sit-standing (supported standing or sitting on the split pad). Lumbar and pelvic angles in the new chair fell between those of traditional standing and sitting, discomfort was lower, and task performance was not affected. Sit–stand changes occurred every 3 min on average when using the new chair.

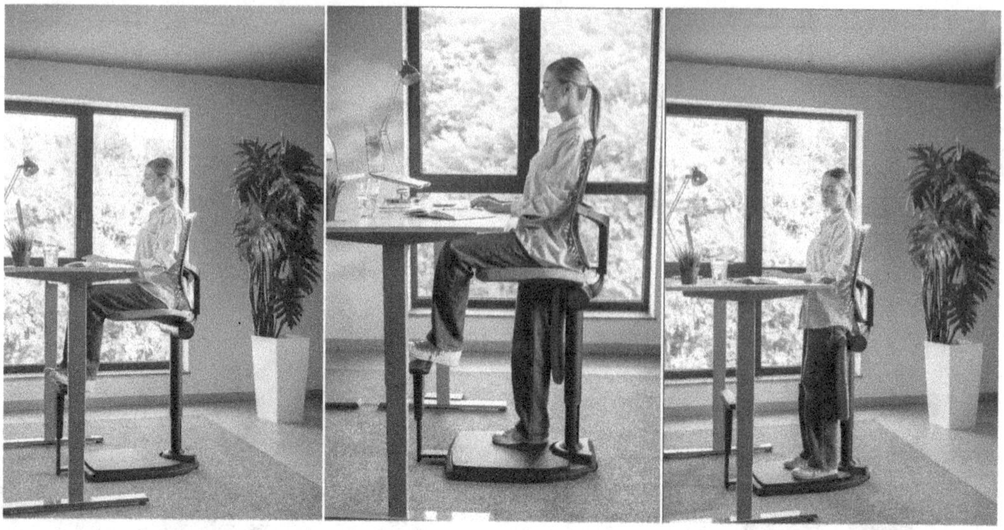

FIGURE 4.26 Sit–stand seat with split seat pad proving affordances for a wide range of working postures. Traditional standing and sitting, supported standing, and semi-standing when the pad is split. (Moveably.com.)

Analyze Legacy Data

HFE specialists should advise the design team on the organization's past performance regarding productivity issues, sick leave, and complaints about musculoskeletal disorders. New systems should be built to ensure that old problems are solved and new problems are not created. Attention should be paid to employee demographics in relation to age, sex, health, and overweight/obesity.

RESEARCH DIRECTIONS

Epidemiological data from over 1 million people link excess sitting to 34 chronic diseases and conditions (Levine, 2014). About 75% of workers in developed countries are sedentary, and the basic requirements for the design of seated workspaces are well understood. Research is needed to demonstrate the effectiveness of redesigning sedentary jobs and to specify how such jobs should be redesigned to make them safe. Activity-permissive workstations are gaining in acceptance, but well-designed prospective studies are needed to demonstrate their long-term acceptance and effectiveness. The psychosocial and interpersonal implications of changing from sedentary to sit–stand and standing office work are less well understood than the health risks of seated work. Knight and Baer (2014) provide suggestive evidence that non-sedentary working promotes increased arousal at the individual level and less territoriality at the group level, leading to better information elaboration and improved team performance. Many office workers now carry out their primary roles using laptops, tablets, and phones, all of which are portable. Once connected to the Internet, they become "Digital Nomads" and can work anywhere. With the introduction of "hybrid working" following the COVID-19 pandemic, many offices are getting smaller or have fewer single-user workstations as employees spend more time working at home. "Third places" are also emerging, small co-working centers close to home, used by employees from different organizations.

At present, there is evidence of diminishing returns in the utilization of activity-permissive workstations, with employees sitting more as time goes on. This may change as a new generation of employees uses such workstations from the beginning and subsequently refuses to sit at a desk all day long.

Even today, for some of us, there may be no need for everyone to have their "own" standardized workstation in an office—opening up the design space to new ways of thinking about office design. The implications of these changes, both for productivity and for health and safety, are not yet clear, and the requirements for future research remain to be specified (Haapakangas et al., 2023).

SUMMARY

The reduction of postural stress has always been fundamental to ergonomic workstation design. Unfortunately, traditional HFE guidelines for seated work with computers have forced users to adopt stereotypical sitting postures in workstations that provide no affordances for physical activity. These postures minimize postural stress in the short term while increasing metabolic risk in the long term. It is now recognized that physical inactivity is harmful and that sedentary work is a risk factor for the development of serious medical problems. Thus, traditional HFE recommendations are harmful, and it is important that we do not change seated workstations for equally inflexible standing workstations (Figure 4.27).

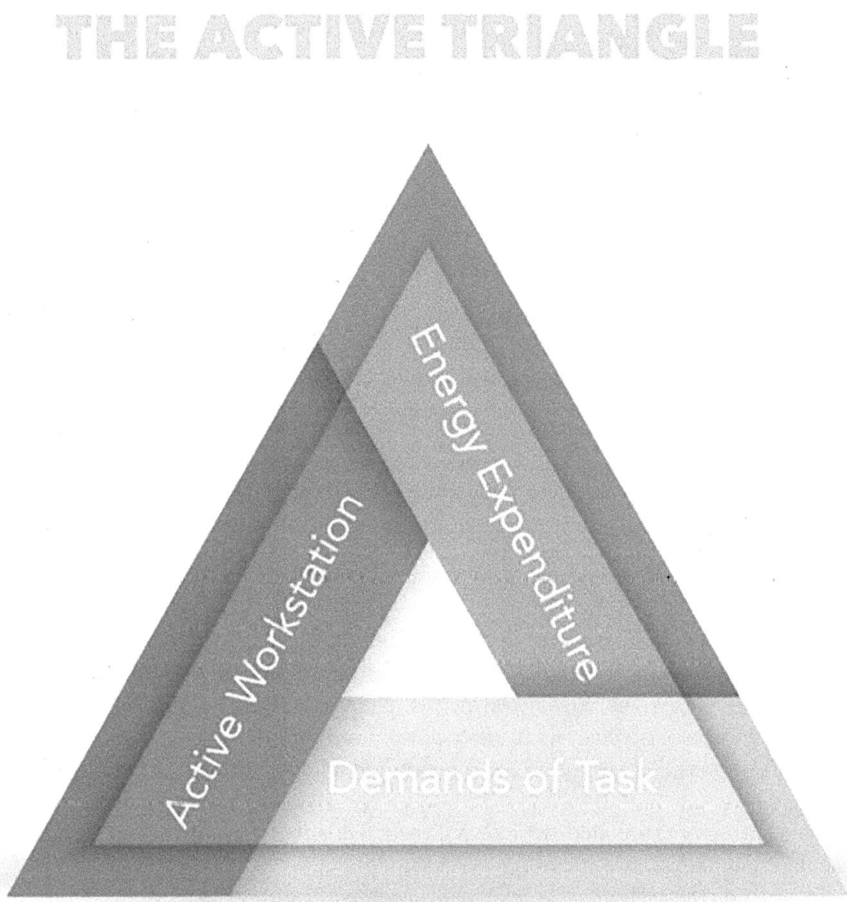

FIGURE 4.27 The active triangle. Choosing an "activity permissive" workstation involves understanding the context of use and finding a balance between a meaningful level of physical activity, the physical and social demands of the task, and the workstation options available. (Graphic courtesy of Mr. A. Almajid.)

TABLE 4.10
General Guidelines for Workstation Design

Provide clearance under desks and benches so that the foot and knee positions are not constrained. Do not use the area around the bench or the desk as storage space.

The feet must never be confined to a small area, as this degrades balance and shifts the load of postural adaptation to vulnerable structures higher up the kinetic chain.

For every task, find an optimal visual and manual distance that will minimize forward flexion of the trunk and flexion of the neck.

Offer pregnant workers seats with a forward slope of 10°.

Provide footrest for standing and seated workers.

Enrich or enlarge the job to increase postural variety. Differentiate new tasks by introducing configurations demanding alternate postures.

Activate the venous muscle pump. Require walking or fidgeting wherever possible. Five minutes per hour may be optimal.

Design activity-permissive workstations. Allow standing workers to sit and work for 50% of the day.

If multiuser work surfaces are of fixed height, provide height-adjustable platforms for standing workers.

For seated workers, choose a height that will fit taller workers and provide footrests for shorter workers.

The height of the floors is always adjustable in the upward direction.

Alternating between asymmetric postures is a strategy for minimizing static loading. Design workstations and jobs to encourage this strategy.

Redesign jobs if the head inclination is
- >85°
- <0° without full head support

Redesign jobs if the neck flexion angle is
- >25°
- <0°

For seated workers, design for at least one working posture and one reclining posture for resting.

For hot desking
- Select a chair that fits the smallest females to the tallest males.
- Provide docking stations for laptops.
- Ensure that the workstation can be adjusted easily to suit all users, including those with disabilities.
- Consider storage requirements of different users and provide mobile storage units where needed.

Source: Based on Lee et al. (1999). See ISO 11226 for further information.

Table 4.10 summarizes the general considerations for designing workstations for standing and sitting.

Criteria for the evaluation of static work can be found in ISO 11226 (International standards on working postures and movements ISO 11226 and EN 1005-4) and ISO 9241: Ergonomic requirements for office work with visual display terminals. Note: ANSI/HFES 100-2007 Human Factors Engineering of Computer Workstations is under development until 2024.

For office workstations, Pearse et al. (2024) conducted a literature review, concluding that:

- Prolonged sitting is associated with several negative adverse health effects, affecting the physical, physiological, and cognitive systems.
- Movement throughout the day helps negate negative health effects associated with prolonged sitting.
- Two types of movement should be introduced into a standard upright seating configuration:
  - Sustained movement over a range of seated postures, occasionally reaching neutral lordosis.
  - High frequency and light contractile activity in the lower limbs.
  - The upright sitting posture provides the greatest biomechanical potential for supported movement.

- The use of active chairs shows significant potential for health benefits at the workplace.
- Combining active chairs with sit–stand workstations may be the ideal coupling to help reduce sedentary behavior in the office.

See Bridger (2019) for further information.[3]

TUTORIAL TOPICS

1. If you were to follow up the Garrett et al. study, how would you find out whether improved productivity was due to standing per se or due to a reduction in discomfort that might be achievable in other ways?
2. "Humans are not designed to stand or sit, they are designed to walk." Discuss.
3. "In 20 years' time, office chairs and desks will be obsolete." Discuss.
4. At the link (https://www.youtube.com/watch?v=H1e4BC41wRU&t=295s), you will find a video from 1944. Ms. Lenore Fenton, the US Navy's champion typist, provides useful advice about posture, seating, and typing technique. Some points for discussion are:
    1. How has the technology for keyboard work changed in 80 years?
    2. How have the requirements for the design of workstations changed?
    3. Is any of Ms. Lenore Fenton's advice still relevant?
    4. What are the differences between operating a mechanical typewriter and a modern computer keyboard?

ESSAYS AND EXERCISES

1. Carry out a survey of the furniture in a modern office. Measure furniture dimensions and relate them to user anthropometry, commenting on any anthropometric mismatches you observe. Find out whether the furniture is adjustable and if users know how to adjust it to optimize their working posture. Examine and classify the furniture and accessories used by different categories of staff and different levels of management. Are the different types of furniture used appropriate for the different users and their task requirements? Is new furniture needed? Are adequate accessories provided?
    a. Use a protractor to measure the trunk–thigh, knee, trunk flexion, and pelvic tilt angles in Figure 4.5. Plot the results on a graph to illustrate how the postural angles change as the trunk–thigh angle changes.
    b. Analyze the posture of the monk in Figure 4.13 using the framework in Table 4.1. Describe the main areas of postural stress in terms of static postural and task workload. Assess the trunk posture using the criteria of ISO 11226.
2. You are the facilities manager of a major financial institution. You need to spend $50 million to upgrade the company's head and regional offices worldwide. Write a report to the top management explaining what constitutes good office furniture and why it is essential in the automated office.
3. Obtain a copy of your national display screen equipment regulations. Carry out an assessment of an office of your choice and determine whether any improvements are necessary
    a. Having identified any improvements, determine the cost of implementing them by obtaining quotations from local suppliers.
    b. Obtain feedback from employees on any problems in the work environment that influence their productivity.
    c. Obtain information from management on absenteeism and any other work losses.
    d. Make a case for the cost-effectiveness of implementing the changes.
4. Use the material in this book to prepare an HFE training program for office workers. Assume that your students have no knowledge of ergonomics or the human sciences and are completely unaware of any of the issues (hint: have several sections beginning with the basic concepts, followed by self-help skills and some practical demonstrations).

5. Specify chair and desk dimensions to fit a ninety-fifth percentile male and a separate set of dimensions to fit a 5th percentile female (i.e., all the male's dimensions will be ninety-fifth percentile and all the female's fifth percentile). Comment on the anthropometric mismatches that will occur if the male sits at the female's workstation and vice versa. Describe how you would design one set of furniture to suit both the above users and everyone in between (specify any dimensions and ranges of adjustment).
6. Below are listed important steps to take when adjusting a computer user's workstation to suit an individual (both chair and desk are height-adjustable). Arrange them in the order you would carry them out in a real office, so as to get the best result most quickly.
   a. Optimize the position of the document holder for easy reading.
   b. Move the desk to minimize glare from luminaries and windows.
   c. Adjust the elbow rest height on the chair.
   d. Adjust backrest tensioner spring.
   e. Adjust the height of the lumbar support.
   f. Adjust the height of the seat to avoid pressure on the popliteal fossa.
   g. Adjust the height of the desk.
   h. Move the screen to the best viewing position for the task.
   i. Position the footrest under the desk.
7. Chair design. What is the difference between a lumbar support and a backrest?
8. For which group of sedentary workers might it be advisable to recommend the use of a kneeling (pelvic tilt) chair at work?
   a. University professors
   b. People over 50 years of age
   c. Obese workers
   d. Pregnant workers
9. Mr. Smith is told (by somebody well-qualified to know) that he needs to spend less time sitting around all day. He is then told that he needs to buy an "activity-permissive workstation" for his home office. Given that he is very inactive, a heavy smoker, and obese, which active workstation would you recommend he use? What other advice would you give him?

QUICK QUIZ

1. The lumbar lordosis is an evolutionary adaptation to bipedalism. True or false?
2. The lumbar lordosis increases when you sit down. True or false?
3. Static neck flexion should be avoided when seated. True or false?
4. Humans are designed to stand on one leg. True or false?
5. A footrest should be about knee height. True or false?
6. Lumbar supports on chairs will prevent all kinds of back pain. True or false?
7. A general-purpose office chair should have a seat depth no greater than the fifth percentile female buttock-popliteal length. True or false?
8. "Sitting is the new smoking." True or false?
9. Which activity-permissive workstation is best for an obese sedentary worker (a) a treadmill desk, (b) a pedal desk, and (c) a standing desk?
10. It's a good idea to get up from your chair every half an hour or so. True or false?

NOTES

1 Ontario Eronomics Intervention Cost-Benefit Calculator. Available online.
2 Page (1963).
3 https://www.routledge.com/A-Guide-to-Active-Working-in-the-Modern-Office-Homo-Sedens-in-the-21st-Century/Bridger/p/book/9780367002282#:~:text=Description,therapists%2C%20practitioners%2C%20and%20managers.

5 Repetitive Tasks

GENERAL REQUIREMENTS FOR HUMANS IN SYSTEMS

1. Repetitive, monotonous work should be minimized.
2. When repetition cannot be minimized, the optimum combination of force, repetition, posture, and duration of task performance must be defined and implemented.
3. The risk of injury to the musculoskeletal system must be ALARP ("as low as reasonably practical").
4. All tasks requiring repetitive movements of the upper limbs should bear a low risk of causing musculoskeletal injury as defined by an SI rating of less than 6 (or an equivalent valid and reliable risk assessment tool).
5. Repetitive tasks involving the operation of cursor control devices (CCDs) should be carried out with neutral wrist postures, low forces, low-repetition rates, and short durations.

Disease is a pathological process that, at least in theory, is amenable to objective, external verification. Illness, on the other hand, is by definition, a subjective state.

Coggan, 2006

CORE KNOWLEDGE: FUNCTIONAL ANATOMY AND EPIDEMIOLOGY OF INJURY CAUSED BY REPETITIVE WORK

Repetitive tasks are of interest because of their potential to cause pain and injury, or to exacerbate pain from existing musculoskeletal conditions. There is a continuum of conditions, differing greatly in the extent to which they are work-related, all of which are characterized by pain. The classification scheme of Schilling and Anderson (1986) is useful in any discussion of the work-relatedness of pain:

Category 1: The work exposure is a necessary cause of the disorder (as in occupational diseases such as silicosis or lead poisoning).
Category 2: The work exposure is a contributory causal factor but not a necessary one.
Category 3: The work exposure provokes a reaction by a latent weakness or aggravates an existing disease.
Category 4: The work exposes the worker to potential dangers that may increase the likelihood of a disease developing (such as alcoholism in liquor industry workers).

At one extreme, certain shoulder, hip, and knee disorders are strongly related to specific work exposures which, if removed by ergonomic redesign, would greatly reduce the risk (Table 5.1 lists common work-related disorders). For example, knee osteoarthritis (OA) in later life is strongly associated with prior participation in knee-straining jobs. Many medical conditions can also cause musculoskeletal pain (Table 5.2) but are work-related only to the extent that they disable otherwise healthy employees (Arthritis Research Campaign, United Kingdom, 2005). The effectiveness of any ergonomic improvements in the workplace may be palliative but will not prevent or cure these conditions.

DOI: 10.1201/9781032717210-5

TABLE 5.1
International Classification of Diseases (ICD) Diagnoses for WMSDs

Nerve Root and Plexus Disorder
Brachial plexus lesions
Unspecified nerve root and plexus disorder
Mononeuritis of the Upper Limb and Mononeuritis Multiplex
CTS (median nerve entrapment)
Lesions of the ulnar nerve (cubital tunnel syndrome)
Lesions of the radial nerve
Mononeuritis of the upper limbs (unspecified)
Other Peripheral Vascular Disease
Raynaud's syndrome
Raynaud's phenomenon (hand–arm vibration syndrome)
Arterial Embolism and Thrombosis
Arteries of the upper extremities (ulnar artery thrombosis)
Other Disorders of the Cervical Region
Cervicobrachial syndrome (diffuse)
Unspecified MSDs and symptoms referable to the neck
Peripheral Enthesopathies and Allied Syndromes
Disorders of the bursae and tendons in the shoulder region (rotator cuff syndrome, supraspinatus syndrome, bicipital tenosynovitis)
Enthesopathy of the elbow region (medial and lateral epicondylitis) (enthesopathies are disorders of the peripheral ligamentous or muscular attachments)
Other Disorders of the Synovium, Tendon, and Bursae
Synovitis and tenosynovitis
Trigger finger (acquired)
Radial styloid tenosynovitis (de Quervain's)
Other tenosynovitis of the hand and wrist
Specific bursitis
Unspecified disorder of the synovium, tendon, and bursa
Disorders of Muscle, Ligament, and Fascia
Unspecified disorder of muscle, ligament, and fascia
Other Disorders of Soft Tissues
Myalgia and myositis (fibromyositis)
Other musculoskeletal symptoms referable to the limbs (cramping, swelling)
Other unspecified disorders of soft tissue

TABLE 5.2
Common MSDs that are not Work Related

Regional sympathetic dystrophy (RSD) is characterized by persistent pain, often with a burning quality, in the hands, wrists, or feet. The cause is unknown, and the pain is often difficult to treat. It sometimes occurs as a complication of a bone fracture, heart attack, or head injury. The affected area may be hypersensitive to touch, swollen, and may have a different color or temperature compared to other areas. RSD may become chronic or resolve of its own accord. Blocking the sympathetic nervous system in the affected area with local anesthetics may help.
Gout is a metabolic disorder. Uric acid builds up and forms crystals in the joints or the skin. It is an inherited disorder requiring treatment by qualified medical practitioners.

(*Continued*)

TABLE 5.2 (*Continued*)
Common MSDs that are not Work Related

Fibromyalgia is a type of rheumatism that affects the muscles and ligaments but not the joints. It can last for months or years, and specific sites of tenderness on the trunk, back, and legs are often found. Pain, tiredness, and sleep disturbance are the main symptoms. Exercise, sound sleep, and coping with physical and mental stress seem to be the key factors in the restoration of normal function.

Rheumatoid arthritis is a chronic disease marked by stiffness and pain in the joints, swelling, and loss of joint mobility. It can lead to total destruction of the joint as it progresses and consists of inflammation of connective tissues such as the synovium.

HFE Workshop 5.1 introduces some basic concepts and tools used in epidemiology. Much of the evidence for the work-relatedness of WMSDs comes from epidemiological studies, and this workshop aids in understanding that evidence.

Specific versus Nonspecific WMSDs

Most of the disorders listed in Table 5.1 are anatomically and occupationally specific. Clinicians can identify some kind of pathology at a particular anatomical site and the likelihood of suffering the disorder increases with exposure to particular physical work conditions. For example, Walker-Bone et al. (2006) found that physical risk factors, such as those found in the workplace, were more strongly related to the presence of specific disorders than to nonspecific pain in the shoulder, elbow, or wrist (i.e., diffuse pain not traceable to a specific site or pathology). Such nonspecific pain is often labeled "repetitive strain injury" (RSI) or "cumulative trauma disorder" (CTD). These terms are problematic—partly because they are essentially process descriptions which, when also used as outcome definitions (as is usually the case), introduce problems of circularity and confusion in the ascription of causation. RSI and CTD are also problematic because they are incomplete descriptions of the pathological process itself. MacFarlane et al. (2000), for example, demonstrated that psychosocial risk factors as well as physical risk factors were important in the causation of forearm pain of new onset. The main physical risk factors were repetitive movements of the hand and wrist, and the main psychosocial risk factor was dissatisfaction with support from colleagues and supervisors.

Variability in Movement Patterns

When there is a lack of variability in exposure to one or more of these factors, the risk of injury increases (Hug and Tucker, 2017). Highly repetitive, fast movements reduce the options for varying muscle recruitment patterns. Workstations that constrain employees into one "correct" posture for carrying out a task, and a lack of task variety, all reduce variability and increase risk. Exposure to more than one factor often causes a large increase in the prevalence of the disorder because the combination reduces movement variability. There are also individual differences in muscle recruitment patterns, which mean that some people are more at risk than others. According to Srinivasan and Mathiassen (2012), variable motor responses in repetitive tasks are protective because they slow the onset of fatigue and prevent the "Cinderella Effect," where small groups of muscle fibers are preferentially recruited during low-level exertions and denied the opportunity to recover. By the same reasoning, more variability in movement patterns may reduce the risk of musculoskeletal disorders.

How Much Risk Is Risky Enough?

Graveling (2018) describes how a disorder can be recognized as being occupationally related and therefore can be compensated for. Pain may be experienced at work but not at home because the sufferer avoids domestic activities that provoke pain. In the UK, an odds ratio of 2 (see Workshop 5.1) is used to assess work-relatedness, in other words, for every 100 cases of the disorder, 50 would have occurred anyway and 50 due to the job exposure. Further:

1. The disease must be serious enough to cause a significant degree of disability.
2. It must be capable of definition and diagnosis with a reasonable degree of confidence.
3. The diagnosis should be possible without the need for extensive or invasive tests.
4. It should be possible to confirm the exposures.

HFE WORKSHOP 5.1

SOME BASIC CONCEPTS USED IN EPIDEMIOLOGY

Exposure: The condition of being subjected to something be it an infectious agent, extremes of weather, or mechanical loading.
Dose: A measure of exposure; the amount of exposure.
Outcome: A change in personal condition, normally as a result of one or more exposures.
Case: An individual who has developed the outcome of interest.
Prevalence: The number of cases of a disorder at any particular time.
Incidence: The number of new cases over a time period.
Odds Ratio: A statistic that describes the likelihood of an outcome given exposure to a risk factor. In the table below, there are some imaginary data on the prevalence of shoulder problems in a group of 125 shopworkers. The occupational exposure is having to lift objects on and off high shelves. The health outcome is the presence of a shoulder problem. As can be seen, there are 25 cases out of a total of 125 workers.

	Shoulder Problems		
	Yes	**No**	**Row Total**
Exposed to loading high shelves?			
Yes	20	20	40
No	5	80	85
Total	25	100	125

The data can be set out, as above, in a 2×2 table. The total number of workers surveyed is 125. Of these, 25 have shoulder problems and 40 carry out work involving high shelves. At first glance, it appears that people who have to stack high shelves are four times more likely to have shoulder problems than those who do not (because, of the 25 with shoulder problems, 20 work with high shelves and 5 do not). However, this ignores the null cases—people in the sample who do not have any shoulder problems, regardless of whether they work with high shelves or not. Clearly, in any assessment of risk, we need to take these individuals into account as well.

The OR calculation does this elegantly, as follows:

Let A = those with shoulder problems and exposed to high shelves = 20
Let B = those without shoulder problems and exposed to high shelves = 20
Let C = those with shoulder problems and not exposed to high shelves = 5
Let D = those without shoulder problems and not exposed to high shelves = 80

(*Continued*)

HFE WORKSHOP 5.1 (*Continued*)

$$\mathrm{OR} = (A\times D)/(B\times C) = (20\times 80)/(20\times 5) = 1600/100 = 16$$

The overall prevalence rate for shoulder disorders is 25/125 or 20%, so there is a high rate of shoulder disorders in the sample. The prevalence for those exposed to the risk factor is 20/5 or 4. We can conclude that there is a greatly increased risk of developing shoulder disorders given exposure to the risk factor—people who stack high shelves at work are 16 times more likely to develop the problem.

Note that the size of the OR in this example is very strongly influenced by the large number of people who do not have shoulder problems and are not exposed to the risk factor (cell D). Thus, we can say that the shoulder problem is very specific to the exposure (it is rare in those not exposed). For many musculoskeletal conditions, such as low back pain, the ORs are much smaller. One of the reasons for this is the lack of specificity—many non-work-related factors can cause back pain (in which case cell C is larger and cell D smaller).

ORs are usually calculated to estimate risk in cross-sectional studies. Such studies are like snapshots of the situation—they tell us who is experiencing a particular outcome and the work conditions they are currently exposed to. ORs greater than 1 indicate a positive association between the health outcome and the work exposure. Ratios less than 1 indicate a negative association (i.e., that the exposure has a protective effect). However, cross-sectional surveys cannot tell us anything about causality—whether the outcome was caused by the exposure. To get evidence of such causality, we need to measure the exposure before the outcome; only then can a causal link be demonstrated.

Suppose we follow a cohort of shop workers for 1 year to find out the risk of shoulder problems in those who stack items onto high shelves.

Of 50 workers who stack high shelves, 30 develop shoulder disorders after 1 year. So, the risk for these workers is 30/50 = 0.6.

However, people develop shoulder problems for all kinds of reasons; thus, what we really need to know is whether workers who load high shelves are more likely to develop shoulder problems than those who do not.

We find that, of 50 workers in the same plant who do not load high shelves, 5 develop shoulder disorders after 1 year of follow-up. So, the risk for these workers is 5/50 = 0.1.

So, workers in the high shelves group are six times more likely to develop shoulder problems than those who do not—evidence of a causal link over time, as the exposure preceded the outcome. Relative Risk (RR) is the risk of an event given an exposure: $\mathrm{RR} = p_{\mathrm{exposed}}/p_{\mathrm{non\text{-}exposed}}$, where p is probability. The analysis is summarized below:

	Shoulder Problems		
	Yes	**No**	**Row Total**
Exposed to loading high shelves?			
Yes	30	20	50
No	5	45	50
Total	35	65	100

Let A = those with shoulder problems and exposed to high shelves = 30
Let B = those without shoulder problems and exposed to high shelves = 20
Let C = those with shoulder problems and not exposed to high shelves = 5
Let D = those without shoulder problems and not exposed to high shelves = 45

(Continued)

HFE WORKSHOP 5.1 (*Continued*)

$$
\begin{aligned}
RR &= [A/(A+B)]/[C/(C+D)] \\
&= A(C+D)/C(A+B) \\
&= (30\times 50)/(5\times 50) \\
&= 6
\end{aligned}
$$

The larger the value of the relative risk (RR) statistic, the greater the risk. As can be seen, the size of RR depends largely on the fact that most of those not exposed to the stressor do not develop the condition and not just on the high prevalence rate in the exposed group.

A common fallacy in much discussion of everyday risk is to assume that a high prevalence of an outcome means high risk. For example, a sudden increase in the observed number of car accidents while drivers are using cell phones might, on emotional grounds, seem to argue for banning cell phone use while driving. However, without knowing the number of mobile phone-using drivers who did not have an accident in the same period, it is impossible to estimate the risk. Furthermore, without data on accidents from some comparable group and the proportion of drivers who did not use mobile phones, it is impossible to tell whether there is any increased risk (i.e., a value of $OR > 1$) due to mobile phone use (see Chapters 12 and 13 for more information on cell phones).

Development of WMSDs

Work-related MSDs develop when there is a mismatch between the capacity of the employee (size, strength, and skill) and the demands of the task (exposure to forces, postures, rest periods, and so on) resulting in short-term responses such as fatigue and longer-term changes such as inflammation when recovery is insufficient. The capacity of the employee also depends on general health (smoking, obesity, infections such as colds and flu, and poor diet), which may already have led to chronic inflammation (an immune response that persists in response to internal factors or external stimuli that can damage healthy tissues).

Fontana et al. (2023) summarize workplace risk factors as an 'Ergo-Index'. Exposure to the job demands in the index is rated according to their importance in the job and the level of exposure to them. The items in the index are summarized in Table 5.3.

Some common musculoskeletal conditions and their work-relatedness are described in the following sections.

Work versus Exercise: Two Sides of the Same Coin

In properly designed weight training programs, the goal is, in fact, to cause muscle damage because the body responds by increasing the size of the contractile elements in the muscle, resulting in improved strength. The goal of resistance training is to stimulate the development of Type II muscle fibers and DOMS is often taken to be a sign of a challenging training session. However, a rest period of at least 48 h is usually an intrinsic part of strength training programs—the time needed to allow the exercised muscles to recover. The pattern of activity in many industrial jobs bears little resemblance to that in proper muscle training programs. One of the main differences is that rest periods are far more frequent and far longer in muscle training regimes than at work, which is why work does not normally have the same beneficial effects as exercise or training. Damage to muscle tissue on a daily basis may exceed the repair capability, leading to a decrease rather than an increase in strength or endurance and chronic pain in the muscle (myalgia). It is thought that the tissue change responsible for chronic pain, in the absence of inflammation, is due to a buildup of tissue in the extracellular matrix (Cutlip and Marras, 2000). The matrix acts as a barrier to nutrients

TABLE 5.3
Exposure Items for an Ergonomics Job Index (Fontana et al., 2023)

Exposure	Domain	Definition
Force Exertion		
Dynamic Strength	Abilities	Ability to exert forces over time: endurance
Static Strength	Abilities	Maximum force to lift, push, pull, etc.
Trunk Strength	Abilities	Abdominal and back strength
Activity Level and Repetitive Movement		
Wrist/Finger Speed	Abilities	Fast, simple hand and finger movements
Finger Dexterity	Abilities	Precise, coordinated movement of both hands
Manual Dexterity	Abilities	Precise gripping and manipulation of objects
Handling and Moving	Abilities	Handling, installing, and positioning objects
Duration of Repetition	Context	Requirement for repetitive movements over time
Hand usage	Context	Handling objects, controlling tools over time
Postures		
Awkward Postures	Context	Working in cramped conditions
Time Standing	Context	Requirement to stand to work
Time Kneeling, Crouching, Crawling or Stooping	Context	Requirement of sustaining these postures
Time Bending or Twisting	Context	Requirement to adopt asymmetric postures
Vibration		
Exposure to Whole Body Vibration	Context	Exposure time—e.g., jackhammer use
Driving Vehicles or Mobile Equipment	Context	Forklifts, trucks, watercraft, aircraft, etc.

and contains pain receptors. There may also be an increase in collagen in the muscles, causing an increase in muscle stiffness. The end result is a loss of strength, endurance, and the ability to absorb shocks or vibration without further damage.

Review of Tissue Patho-Mechanics and WMSDs

The mechanism of WMSDs is thought to be repeated microtrauma at the cellular level. The repair capacity is exceeded due to a lack of rest during the day, and repeated daily exposure prevents recovery after work.

Muscle Pain

In general, human muscle has excellent endurance capacity for loads less than 15% of the muscle's maximum voluntary contraction (Putz-Anderson, 1988). Above this threshold, rest periods are needed if acute or chronic problems are to be avoided.

Cramp: Cramp is a sudden, unpredictable, and painful contraction of skeletal muscle that can occur normally after exercise or repetitive work (Maughan and Shirreffs, 2019). It seems to occur following fatigue in the affected muscles, particularly when working in cramped conditions with the muscles shortened. One explanation is that it occurs following changes in electrolyte balance in the muscles, and another is that it occurs due to abnormal spinal reflex activity following fatigue. As early as 1911, a UK Parliamentary committee, writing about cramp in Morse code telegraphists, wrote:

> Some authorities have regarded it [i.e. telegraphists'] cramp as a muscular disorder; others as a disease of the peripheral nervous system; others as a disease of the central nervous system. After careful consideration of these antagonistic theories of telegraphists' cramp, and examination of a number of telegraphists affected with the disease, the Committee accept the last-named view; namely, that telegraphists' cramp is a disease of the central nervous system, and is the result of a weakening or breakdown of the cerebral controlling mechanism in consequence of strain upon a given set of muscles.

Even today, there is debate about the cause because cramp, being unpredictable, is difficult to study. It may be caused by abnormal reflex activity following fatigue and can be relieved by cessation of the activity and by stretching. Maughan and Shirreffs conclude that in occupational settings, hard work in high temperatures followed by a large intake of fluid may be risk factors, and there is some evidence that the risk is reduced if salt is added to the ingested fluids.

In settings where cramp is common, there may be a requirement to review the work–rest cycles and other workplace arrangements to prevent the onset of muscle fatigue.

Delayed Onset Muscle Soreness: Delayed onset muscle soreness (DOMS) is a natural response to abnormal loading (e.g., unfamiliar tasks or heavier-than-normal loads). Pain, due to inflammation, appears up to 12–24 h after the exposure, peaking 1–3 days after, before gradually decreasing. DOMS is indicative of muscle damage. It can occur after exposure to sudden high forces, particularly during eccentric contractions (as, e.g., when trying to hold a falling object or resist a sudden reaction torque of a powered tool). Range of joint motion and strength in the affected muscle is temporarily reduced until the soreness subsides. At the tissue level, there may be evidence of damage such as muscle fiber z-line rupture, ragged type 1 (red) muscle fibers, decreased intracellular adenosine triphosphate (ATP), and reduced local blood flow. Normally, the muscle will recover and even strengthen, but some researchers believe that chronic exposure to static load prevents proper recovery, leading to permanent damage (see Chapter 7 for explanations of muscle structure and physiology). This primary response may be accompanied by a feeling of soreness in the muscle, which diminishes as the damaged muscle fibers regenerate.

Muscle Tone and "Tightness": In a conscious person, skeletal muscles always have a certain degree of "tightness." There is a baseline level of muscle fiber recruitment even during relaxation. This is known as muscle tone and it is controlled by the central nervous system and by a feedback system involving the spinal cord and the muscle spindles. Muscle tone is essential for the maintenance of posture. People under mental stress may, without realizing it, develop increased tension in their muscles which they cannot control. Psychosocial factors at work, if they cause anxiety or stress, may cause chronic increases in muscle tone, which increase the risk of injury. At work, the prestressed body part may be a source of pain even though the task loading is mild. A chronic, stereotyped pattern of recruitment of motor units may be the dose that leads to damage of the muscle tissues. Roman-Liu et al. (2013) found increased levels of muscle activity (electromyography [EMG]) of approximately 10%–25% when subjects engaged in sustained attention and vigilance tasks compared to control conditions. Increases in EMG activity were greatest in the shoulder muscles (deltoids and trapezius).

Exertional Compartment Syndromes: Exertional compartment syndromes normally occur in the lower limbs with dull aching in a given muscular compartment and increased pressure in the muscle. The pain is triggered by activity. Muscle is known to increase in volume by up to 20% during exercise, and it is thought that the accompanying pressure increase against the relatively noncompliant fascia around the muscle is sufficient to degrade blood flow through the muscle. The pain subsides after cessation of the activity.

Ligament Pain

When highly repetitive movements are required, the increase in blood supply to the muscles may be associated with a decrease in blood supply to the adjacent tendons and ligaments of the associated joints. As Hagberg (1987) puts it, the muscles "steal" blood from the insertions of the ligaments and tendons. Plantar fasciitis is an inflammation of the ligament that supports the arch of the foot. It is sometimes known as "Policeman's Heel," "Jogger's Heel," or "Tennis Heel" and is caused by the repetitive microtrauma of walking or running long distances every day. Occupational risk factors include excessive walking or standing and poorly cushioned footwear. Obesity is also a risk factor, as is having limited dorsiflexion of the ankle (wearing high-heeled shoes may help by reducing dorsiflexion but may cause other problems). Stretching exercises carried out several times a day may also help (Agyekum and Ma, 2015).

Tendon Pain

Tendons have a tensile strength of 50–100MPa. They consist of bundles of collagen microfibrils with many cross-linkages between them. At rest, the fibrils have a "crimped" appearance. Under tension, the tendons elongate nonlinearly as the fibrils straighten. Under further tension, the fibrils elongate by 1%–5% of their length. Beyond this point, individual fibers fail, putting more stress on those remaining. Under constant load, tendons exhibit "creep"—viscous lengthening takes place.

Tendinitis (inflammation of tendons or "tenosynovitis," inflammation of the sheaths around the tendons) can be caused by force, posture, and repetition, and the risk is greater when workers are exposed to these stressors in combination.

Impaired blood supply to the tendons is thought to be the cause of much occupational shoulder pain because it increases the rate of cell death within the tendon. This is thought to provide sites where chalk (calcium carbonate) is deposited. It seems that increased tension in tendons reduces their blood supply, which may explain why static work positions are associated with tendon problems. Armstrong et al. (1993) describe an interesting hypothesis stating that the accumulation of dead cells in tendons can cause an inflammatory response in the tendon by the immune system. Inflammatory responses normally occur when there is an injury such as a cut—the blood supply to the affected region is increased to attack any incoming foreign bodies, and the part of the body in question normally feels hot and swollen. According to Armstrong et al., if the person already has an infection such as influenza, the immune system will have been activated, and the local inflammatory response described above is more likely to be triggered by the accumulation of dead cells in the tendon. This may underlie the popular conception that we are more prone to injury when suffering from colds, flu, or other diseases.

Frequent mechanical loading can cause tendonitis or inflammation of the cartilage surrounding a joint. Extreme positions of the wrist can press the flexor tendons of the fingers against the bones of the wrist, increasing the friction in the tendons. Rapid, repetitive movements of the hand or fingers can cause the sheaths surrounding tendons to produce excess synovial fluid (Figure 5.1). The resultant swelling causes pain and impedes movement of the tendon in the sheath. This is known as "tenosynovitis." Repeated exposure can ultimately leave scar tissue which impedes movement of the tendon in its sheath and thus degrades function. Joint structure may be degraded by the formation of bony spurs around damaged areas. Reduced mobility, pain, and weakness may result. Sudden large forces may cause tendons to separate from bones.

Graveling (2018) concluded that the relationship between occupational demands and tendonitis was not strong—possibly because of a lack of specificity in the measurement of the occupational demands. The proposed mechanisms of injury are biologically plausible, however. More recently, a review of the literature by Kier et al. (2021) only included studies where exposures were quantified (e.g., high versus low repetition as opposed to repetitive versus non-repetitive); outcomes were measured objectively (e.g. medical diagnosis versus self-report); and participant characteristics were well-defined. For hand/wrist tendonitis, the risk was greater for higher force requirements (>40N in one study), with odds ratios of 2.9 for wrist extensor tendonitis at 3-year follow-up. Inconsistent findings were found for repetition and tendonitis. For posture, significant increases in tendonitis were found for jobs with a high prevalence of wrist flexion, but not for wrist extension. There were strong interactions between force and repetition. Low-force/low-repetition (LFLR) jobs (baselines) were compared with those with high force HF/high repetition. The odds ratio for HFLR was 6.1; for LFHR, it was 3.3; and for HFHR, it was 29.4. Regarding tendon-related disorders, the authors concluded:

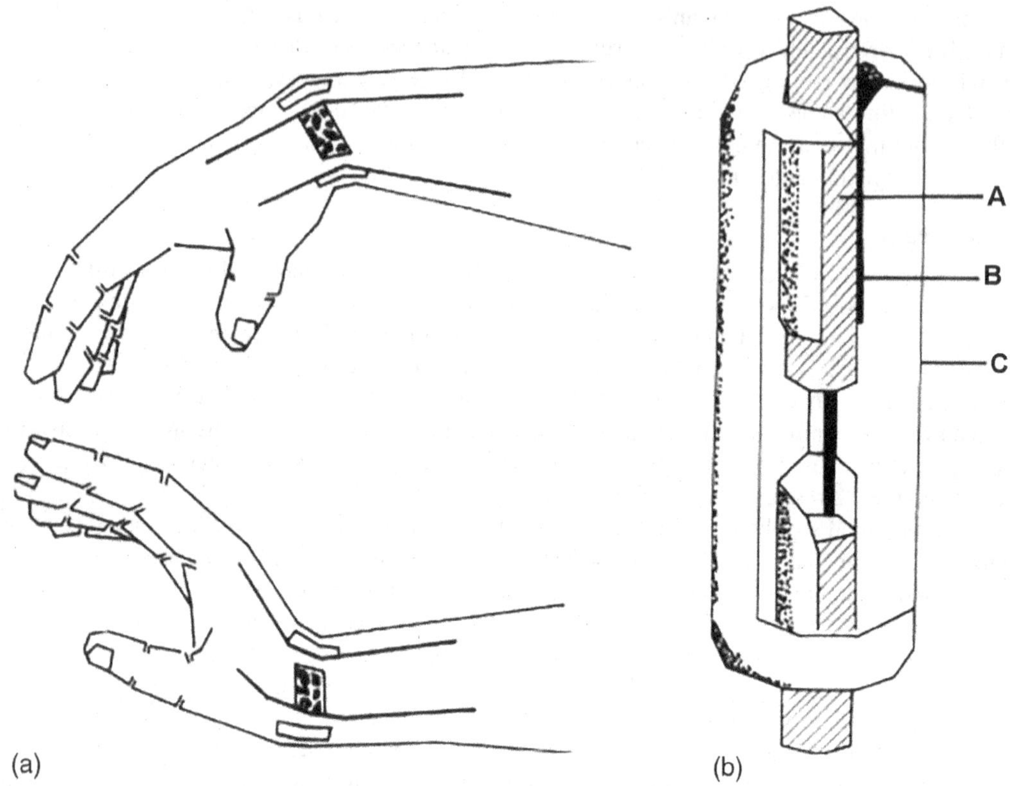

FIGURE 5.1 Wrist posture and tendon function. (a) Extreme postures can preload the finger tendons. (b) Simplified view of a flexor tendon of the fingers. The tendon (A) is surrounded by a synovial sleeve consisting of an outer (C) and an inner layer (B). When the tendon moves, the inner layer glides over the outer layer, lubricated by synovial fluid. If the layers become inflamed or scar tissue builds up, the layers cannot glide smoothly over one another. The tendon then behaves like a rusty brake cable and smooth, pain-free movement is impossible.

> Higher levels of force, repetition, and deviated wrist/finger postures (particularly wrist flexion and pinching grips) increase tendon-tendon sheath frictional work and tensile stresses that can manifest in tendon pathology (animal studies) and is consistent with histological findings among humans. Overall, the biomechanical plausibility is consistent with the controlled lab-based animal experiments and the bulk of epidemiological studies indicative of a dose–response relationship between force, repetition, and posture on hand/wrist tendon–related MSDs.

Trigger Finger (Stenosing Flexor Tendinitis)

Stiffness and "snapping" of the fingers during volitional flexion are thought to be caused by thickening of the fibro-osseous canal through which the finger flexors pass. A higher prevalence has been found in certain meatpacking jobs that require static grasping of powered knives activated by a trigger (Gorsche et al., 1998). However, it is more common in middle-aged women than in men, and nonoccupational factors such as thyroid disease, diabetes, and arthritis are usually the cause (Vasiliadis and Itsiopoulos, 2017).

De Quervain's Tenosynovitis

Characterized by pain on the radial styloid on the thumb side of the wrist due to restricted glide of the tendons, thumb function is impaired. It is more common in the preferred hand of middle-aged women, suggesting that it is activity-related (Moore, 1997). Activities that require heavy use of the thumbs seem to be associated with the disorder (fitting rubber rings on a pipe, sewing, weaving, and cutting). Benites-Zapata et al. (2021) found an increase in de Quervain's symptomatology among young smartphone users. Problematic smartphone use was assessed using the questionnaire, and de Quervain's symptomatology was related to hours/day of smartphone use (exposure time), pain in the last week, and use of "WhatsApp."

The avoidance of ulnar-deviated wrist postures when operating handheld devices would seem to be indicated.

Bursitis

A "bursa" is a sac containing viscous fluid situated at places in tissues where friction would otherwise occur (bursa is the Greek word for wine skin and is related to the English word purse). There are about 150 bursae in the body, and they act like cushions that protect muscles and tendons from rubbing against bones during movements of the body. Overexertion and injury can cause inflammation of bursae, or bursitis. "Housemaid's knee" is a well-known type of occupational bursitis, which is also common in carpet layers. Bunions are also a form of bursitis, caused by wearing ill-fitting shoes—friction of the shoe on the bursa on the joint of the big toe causes it to become inflamed. Bursitis can be distinguished from tendonitis anatomically and because of the dull, aching pain that accompanies it—in contrast to the sharper pain of tendonitis.

Neuritis

Repeated or prolonged exertion can cause damage to the nerves supplying a muscle or passing through it. This can cause sensations of numbness or tingling (pins and needles) in areas of the body supplied by the nerve. In the case of nerves, overexertion can cause increased pressure in a muscle due to edema or scar tissue formation. The increased pressure can itself be a dose that results in impaired nerve function. Impaired nerve function, destruction of fibers, or damage resulting in reduced nerve conduction velocity may cause muscle weakness.

All of these problems are more likely to occur if the joints are held in an awkward posture (at the extremes of the ranges of movement), since this "preloads" tendons and ligaments and stretches muscles and nerves.

There are several common "compression neuropathies"—"bowler's thumb," involving the ulnar digital nerve of the thumb and "handlebar palsy," which involves the ulnar nerve in cyclists. The relative roles of interference with blood supply and direct mechanical trauma to the nerve are not well understood but the pressure threshold limit for nerve viability is approximately 40–50 mmHg. Rydevik et al. (1981) found that blood flow in nerves ceased at a pressure of 60 mmHg. Furthermore, after 2 h of compression at 40 mmHg, blood flow in the compressed segment was severely reduced for up to 7 days afterward. Any activities or conditions that increase the pressure are likely to degrade nerve function.

Peripheral Neuropathies and Nonlocalized Arm Pain

Several researchers have suggested that much of the diffuse and difficult to diagnose arm pain, often labeled RSI, may really be peripheral nerve pain (Quinter and Elvey, 1993). The pain is the result of a disorder of the pain receptors and occurs independently of the person's mental state. Unlike neuritis, there is no inflammation of the nerve itself. Disorders of nerve function are known as neuropathies. Chronic abnormal inputs from peripherally damaged nerves can sensitize nerve cells within the spinal cord, resulting in the hypersensitivity to painful stimuli characteristic of RSI. According to the theory, RSI is really a form of peripheral neuropathy in which "damaged nerves can come to contribute actively to chronic pain by injecting abnormal discharge into the nervous system and by amplifying and distorting naturally generated signals."

Butler (1991) pointed out that nerves are "bloodthirsty" structures—the nervous system constitutes 2% of body weight and consumes 20% of the oxygen in the blood. The nervous system is also the most extended and connected system in the body. What this means is that nerves have to accommodate postural movements. When, for example, a person flexes the elbow, the nerves on the flexion side of the joint shorten and may be pinched by other tissues, while nerves on the opposite side are stretched. It is conceivable that both these accommodations may interfere with the nerve's blood supply. According to Butler, movements at joints such as the ankle can increase the tension in the nerves in distant parts of the body. For example, the angle through which the straight leg of a person lying supine can be flexed depends on whether the person's neck is flexed. Cervical flexion pre-tenses the nervous system and hastens the onset of pain during the straight leg raising maneuver. In these situations, the nerve is susceptible to injury not just at the joint that moves, but at any point along its length where adverse neural tension occurs. Vulnerable areas are where nerves pass through tunnels, where the nervous system branches and is less able to glide over surrounding tissues, where the nerve is relatively fixed (at some points in the spine, for example), and where the nerve passes close to unyielding surfaces. It is interesting to speculate whether the flexed cervical postures of many office workers predispose them to pain in the upper limbs through the mechanism of adverse neural tension.

Pritchard et al. (1999) proposed that nonlocalized arm pain may have vascular origins. They measured the vascular response to muscular work in the radial arteries of patients with diffuse forearm pain compared to control patients who had localized forearm pain. They found that the radial artery was relatively constricted in the patients with diffuse pain and did not vasodilate with exercise, as it did in control patients. The pain, therefore, was the result of restricted of blood supply to the muscles.

Non localized arm pain can also be a sign of other conditions, such as diabetes.

Another puzzling feature of some activity-related upper limb pain is its task specificity. The person may complain of pain only when carrying out small repetitive motions such as typing or playing the piano but be perfectly capable of carrying out similar activities that require a wider range of joint movement. Butler (1991) uses the term "activity-specific mechanosensitivity" to describe this phenomenon. It is hypothesized to occur when, during movement, a small region of scarred nerve tissue moves, in a particular direction, against a damaged or pathological surface such as a bony outgrowth.

Bones and Joints

Repeated heavy loading is essential for the proper formation and maintenance of bone. Wolff's law states that bone grows in proportion to, and in opposition to, the forces imposed on it.

Stress Fractures: Under repeated loading, particularly of the lower extremities, stress fractures can occur. Stress fracturing, like many WMSDs, is a process rather than an event. Repeated loads (below the threshold for acute fracture) damage bone, eventually leading to acute fracture. It is thought that these injuries are more likely to occur if the loading is accompanied by muscle fatigue. O'Leary et al. (2020) report the incidence of stress fractures in women and men in UK infantry training. On entering military training, women had lower aerobic fitness, upper-body strength, lower-body strength, and muscle mass compared with men and were more likely to suffer stress fractures (odds ratio 6.9).

Osteoarthritis: Mechanical trauma seems to be a contributory factor in the development of osteoarthritis (OA) in joints. Intrinsic risk factors include older age, female sex, obesity and previous injury. Obesity not only increases the mechanical loading on the joints but can be accompanied by chronic inflammation. OA is a noninflammatory disease characterized by degeneration of the articular cartilage, hypertrophy of bone, and changes to the synovial membrane, which cause stiffness, pain, and restricted movement in the joints.

In a systematic literature review of studies on osteoarthritis of the hip and knee conducted in the prior 15 years, Canetti et al. (2020) presented odds ratios for the development of hip and knee osteoarthritis. Twenty-eight studies, with 266,227 cases of lower limb OA in construction workers, floor layers, bricklayers, fishermen, farmers, and service personnel (including, but not limited to, salespersons, health care workers, and police officers) were included.

Occupational activities associated with increased risk were kneeling, squatting, standing more than 2 h/day, walking, and lifting heavy loads. For example, women who had lifted more than

1088 tons in their lifetime had an increased risk of knee osteoarthritis (odds ratio 2.13). For men, cumulative exposures to lifting and carrying (from 5120 to 37,000 kg.hrs) yielded an odds ratio of 2.0 for knee osteoarthritis 2.0 and 2.6 for men with a BMI > 25. Lifting contributed to the risk of hip osteoarthritis (odds ratio 1.35). Risk was greater in jobs that combined knee-straining tasks (kneeling, squatting, and heavy lifting).

Cumulative exposure to heavy work increases the risk of both hip and knee OA (Anderson et al., 2012). Farmers, floor and bricklayers, and care assistants all have an increased risk with years of exposure (in farmers, the risk increases after as little as 5 years and is 2–4 times greater after 20 years compared to controls [office workers]). OA has a genetic component and takes time to develop, often after the worker has retired, and has less visibility as an occupational problem. However, the cost to society, in terms of hip replacement surgery and possible disability, should not be forgotten.

Injuries to the Upper Body at Work

When a muscle/joint system is placed in an extreme posture, the muscles on one side of the joint will be lengthened and their antagonists shortened, resulting in a strength imbalance in the antagonistic pair. The ability of the muscles to protect the joint against external forces is degraded, and the joint itself is more easily damaged when the limb is exposed to high forces. An analysis of joint posture is particularly important when evaluating the design of hand tools, particularly in heavy work where the joints may be exposed to high forces. The most clear-cut work-related upper body injuries occur as a result of accidents at work.

Hand Tools: NIOSH reports that there are over 10,000 finger amputations reported annually in the USA among patients over 15 years of age. Many of these involve the use of hand tools either at work or at home. Kim et al. (2019), in a 3-year study of hospital admissions, found that 84% of hand tool injuries were to the fingers; 22% to the forearm; 7% to the wrist; and 1% to the upper arm. Extensor tendon injuries were most common (41.4%), followed by fractures (28%), amputations (11.2%), muscle injuries (8.6%), nerve injuries (4.5%), and vessel injuries (5.6%). Simple lacerations occurred in 57 cases (21.2%). Electric saws were the most commonly used tool that caused the injury, followed by hand-held grinders and motorized saws. Over half of the injuries occurred in non-occupational settings.

The most common injury mechanism seems to be of a catastrophic nature, in which the tool itself suddenly strikes the user, causing a laceration, bruise, or sprain. A second, more pernicious type of injury involves sprains or strains, which appear to result from the handling of the tool itself over longer periods of time. A third type of minor injury occurs to the skin in the form of blisters, due to pressure "hot spots" caused by poor handle design. Attempts to prevent the first type of injury might emphasize training workers in safe tool-handling techniques and thinking ahead—to recognize potentially dangerous situations and prepare the workplace to minimize the likelihood of unforeseen events. Attempts to prevent the second type of injury might concentrate on the redesign of the tool itself and training workers to recognize the onset of fatigue and avoid stressful work postures. Handle redesign can prevent the third type of injury, as can increasing the task variety of the job.

Other hand-powered tools commonly involved in occupational injuries are knives, hammers, wrenches, shovels, and ropes and chains. The powered tools most commonly involved were saws, drills, grinders, hammers, and welding tools.

"Strains and Sprains": A "strain" may be defined as overexercise or overexertion of some part of the musculature, whereas a "sprain" is a joint injury in which some of the fibers of a supporting ligament are ruptured, although the ligament itself remains intact. Many powered tools can cause strains or sprains because of the reaction force they exert on the

user—particularly if these forces are unexpected or occur suddenly due to irregularities at the interface between the tool and the workpiece. Percussive tools, such as paving breakers, exert a reaction force that has to be opposed by the user. In practice, if the tool is well designed and used on a flat surface, the weight of the tool will dampen much of this force. Rotary-powered tools such as drills, sanders, and screwdrivers can exert a reaction torque on the user, which may force the wrist into ulnar or radial deviation, causing a strain or sprain. The design of handles for holding powered tools in place has received the attention of ergonomists, as described below. It should not be forgotten, however, that additional handles may need to be fitted to provide the user with sufficient mechanical advantage to overcome the reaction torque of the tool or to carry it.

Disorders of the Neck

A history of neck injury in a motor vehicle accident has been found to be strongly associated with neck pain. Other factors positively associated with pain include comorbidities (headaches, cardiovascular, and digestive disorders), suggesting that chronic health problems tend to cluster in some individuals and reinforcing the view that personal characteristics also play a role.

Palmer and Smedley (2007) carried out a systematic literature review of physical factors associated with neck pain. Causal factors included repetitive work, static loading, neck flexion, force (five studies), and occupational psychosocial factors. Moderate evidence was found for a causal relationship between repetitive movement at the shoulder and neck flexion allied with repetition. Reinforcing these findings, Korpinen et al. (2013) investigated self-reported neck problems in over 6000 Finns aged 18–65, finding that people suffering neck pain very often were more likely to be blue-collar workers or lower- and higher-level white-collar workers. They were more likely to report pain in the wrist and fingers (21.3% vs. 5.4%), elbows or forearms (14.1% vs. 3.1%), and shoulders (44.8% vs. 8.0%) than people who suffered less neck pain, and reported more use of cell phones and computers. Grooten et al. (2007) followed 803 people with neck pain. Only 36% were free of neck pain 5–6 years later. Those with two of three biomechanical stressors (manual handling, working with the hands above the shoulders, or with vibrating tools) were less likely to be symptom-free (OR = −0.61), whereas sedentary workers were more likely to be symptom-free (OR = 1.32).

Text-Neck Syndrome: Text-neck syndrome is a lay term without an accepted scientific definition. It is thought to be a strain injury resulting from flexing the neck to look at a cell phone or text. Sathya and Tamboli (2020) carried out a survey of students using a neck disability, questionnaire finding a high prevalence of neck symptoms among students. However, many studies such as this one lack control groups or proper measures of exposure or outcome (e.g., a diagnosis of a neck disorder).

Grasser et al. (2023) reviewed 41 studies of text-neck syndrome, finding different definitions in separate studies: flexed posture of the head or neck while using a phone and thought to be harmful; overuse (excessive or repetitive); mechanical tension and stress on the neck; musculoskeletal symptoms; and tissue damage. Although posture appears to be the defining characteristic, no evidence was found of an association between neck posture and neck pain. Overuse was not found to be associated with neck pain in a longitudinal study where in one 5-year study, the number of texts sent did not predict pain. The authors concluded that there is no scientific definition of the term and no evidence for an association with neck pain. This may be because texting, by its very nature, is short term and intermittent, and phone use simply replaces other neck-straining activities in daily life.

Possible Causal Pathways

The cervical spine has several functions—principally to support the weight of the head and to provide a conduit for nerves and attachment points for the muscles that control the position of the head. It consists of seven vertebrae designed to permit complex movements of the head. The first two cervical vertebrae (known as the atlas and the axis) are different from other vertebrae in the spinal column. The remaining vertebrae have the same general structure as vertebrae in other parts of the spine and are surrounded by anterior and posterior ligaments. The cervical spine consists of vertebral bodies and intervertebral discs, facet joints, bony processes for the attachment of ligaments and muscles, and the intervertebral foramen through which the spinal cord passes.

The head can be thought of as being balanced on top of the cervical spine with the fulcrum directly above the first cervical vertebra. The head can be considered to be in balance when a person looks directly forward and a cervical lordosis is present. Since the center of gravity (COG) of the head lies in front of the cervical spine, the head has to be held erect by contraction of the posterior neck muscles. These powerful muscles are true postural (anti-gravity) muscles—they are essential to the maintenance of the erect posture and constantly work to prevent the head from falling forward due to gravity. The role of the posterior neck muscles in the maintenance of posture becomes clear when it is recalled how a sitting person's head droops forward onto the chest when the person is overcome by sleep. It can be appreciated that in ordinary standing and sitting postures, the structures of the cervical spine are "prestressed" by the need to maintain the head in an erect position. They are therefore prone to overexertion by any additional stresses imposed by work tasks that require flexion of the neck. Static flexion of the cervical spine increases the moment arm of the head according to the sine of the angle of flexion. This increases the load on the soft tissues in the cervical region, and the posterior neck muscles are placed under increased static load in order to maintain the forward-flexed head in equilibrium with gravity. The increased static load may cause pressure ischemia and starve the muscle tissues of fuel and oxygen. Pain in the neck and shoulders may result, causing "muscle spasm" (reflex contraction of the muscles). This, in turn, may exacerbate the pain and lead to a vicious circle. The forward-flexed position may subject the cervical intervertebral discs to increased compression and the posterior ligaments to increased tension. Poor workspace design, if it requires workers to adopt flexed cervical postures, may be a cause of reversible pain or may amplify pain due to existing degenerative changes.

When standing upright, looking straight ahead, the compressive force on the cervical spine is about 5 kg. As the neck flexes and the head moves forward, this force increases because the neck extensor muscles have to counteract the flexion moment. With 15° of flexion, compression increases to about 12 kg; at 30°, 18 kg; at 45°, about 22 kg; and at 60° 25 kg. The shear force also increases, placing a load on the facet joints.

Movements of the head are accomplished by the muscles attached to it and to the surrounding parts of the skeleton. The arrangement of the muscles, like the movements of the head, is complex and will only be discussed in greatly simplified form. The deep, short muscles of the neck serve to stabilize the individual vertebrae, whereas the longer, more superficial muscles produce movements of the spine and head as a whole. The posterior muscles, which extend the neck, are stronger than the anterior muscles, which flex the neck, because the latter are assisted by gravity, whereas the former have to work against gravity. The *upper crossed syndrome* is characterized by tightness and weakness of the muscles of the neck, shoulders, and upper back that cross between the dorsal and the ventral sides of the body—the lower part of the cervical spine is flexed and the upper extended (refer to Figure 4.13, Pope Gregory the Great).

The erector muscles of the neck produce extension of the head and neck if they contract bilaterally (i.e., together). If they contract unilaterally (on one side only), lateral flexion and rotation of the head are produced. The trapezius muscle plays a very important role in many work activities. Owing to its oblique orientation, it produces extension, lateral flexion, and rotation of the head toward the side of contraction. It is also involved in elevating the shoulders. Static contractions of the trapezius muscle as low as 10% of maximum voluntary contraction appear sufficient to cause electromyographic changes in the muscle indicative of muscle fatigue. However, the discs can certainly degenerate, as can the intervertebral joints, and this can cause irritation of the nerve roots in the cervical spine (radiculopathy). Pain in the wrist, hand, neck, and shoulder may result. Degeneration of the cervical spine, sometimes known as "cervical spondylosis," can have serious consequences.

Compression of the spinal cord at the level of the cervical spine can take place, resulting in weakness and wasting of the upper limbs. This may then spread to the lower limbs. As is the case with the lumbar spine, some of the degeneration of the cervical spine is part of the natural process of aging. Cervical spine degeneration is a potential cause of neck pain due to the mechanical changes that occur as a result of age-related degenerative processes.

Carpal Tunnel Syndrome

Evidence for Work Relatedness: Carpal tunnel syndrome (CTS) is associated with forceful and repetitive work, alone or in combination with other factors such as non-neutral wrist postures and exposure to hand-arm vibration at work. Combined stressors such as force and extreme posture or repetition are strongly associated with CTS. As with tennis elbow, it seems that CTS is more common in "hand-intensive jobs," such as fish processing and supermarkets (checkout workers). Burt et al. (2011) found a prevalence of 12% in a group of 477 industrial workers. CTS, which was assessed by means of nerve conduction tests, was more common (odds ratio 2.74) in workers exerting forces >70% of maximum voluntary contraction (MVC) compared to those exerting <20% of maximum in their jobs. For fiftieth percentile males, 70% of MVC was 392 N for a power grip and 85 N for pinch force. For fiftieth percentile females, 70% of MVC was 231 N for power grip and 58 N for pinch grip (Figure 5.5). High repetition over 15/min increased the risk of CTS in those with a BMI >30 (odds ratio 3.35). CTS was more common in females than in males, but females tended to have jobs with lower forces and higher hand activity levels. A "Hand Activity Level" questionnaire (HAL) was used to assess exposure—higher HAL scores were associated with increased risk in males only. Hassan et al. (2022) carried out a systematic literature review of occupational factors for CTS involving 9270 patients from a population of 1,051,707 workers. Higher risk was associated with repetition (hazard ratio 1.87); force intensity (HR 1.75); and a Strain Index Score >10 (HR 1.58). No significant associations were found for pinch gripping, hand-arm vibration, or force duration. High computer use was associated with a *lower risk* of CTS.

Apart from age and BMI a number of non occupational factors are associated with CTS, although not strongly. These include pregnancy, oral contraceptive use, use of tobacco, caffeine, or alcohol, wrist dimensions, diabetes, inflammatory arthritis, hypothyroidism, and renal dialysis.

Possible Causal Pathways: The muscles that flex the fingers lie in the forearm and have long tendons that pass through a narrow opening in the wrist before inserting into the fingers. This opening, known as the carpal tunnel, is also traversed by the nerves and blood vessels of the hand (Figure 5.2).

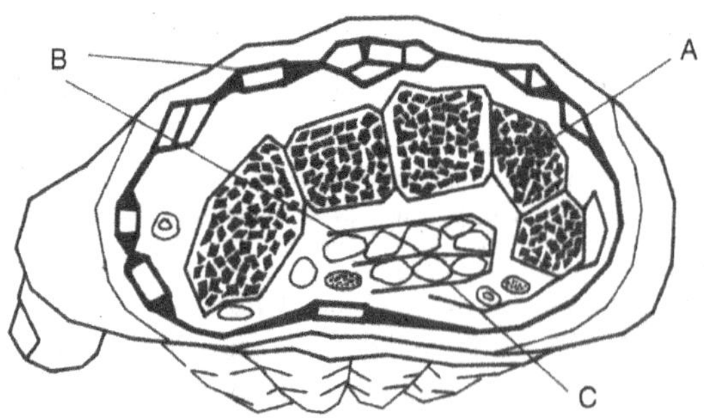

FIGURE 5.2 Carpal tunnel. Section through the wrist showing (a) carpal bones, (b) tendons, and (c) the carpal tunnel containing finger flexor tendons, blood vessels, and nerves. If the wrist is held in extreme postures, the tendons rub against the bone when the fingers move, and friction is increased. (Courtesy of Dr. Douglas A. Bauk of IBM Brazil.)

An increase in the pressure in the carpal tunnel can cause CTS if it entraps the median nerve or reduces the blood supply to the nerve by compressing the capillaries, resulting in nerve damage and reduced conduction velocity of neural signals. The result is a sensation of tingling and numbness in the palm and fingers. (However, degenerative changes in the cervical spine can interfere with the spinal cord and nerve roots, causing referred pain in the arms.) Numbness and tingling in the hands by the same cause might be mistaken for CTS. In severe cases, surgery may be required to relieve the pressure. Concerning sensory conduction of the median nerve, CTS is associated with slower median nerve conduction. It seems that the health of the median nerve is linked to the health of the rest of the body and that median neuropathy is closely related to lifestyle and also to work activities. In some studies, CTS occurs more frequently in the preferred hand, supporting the idea that it is caused by physical activities. As the paper by Burt et al. demonstrates, non-occupational and occupational factors can interact, meaning that some employees are more at risk than others. Obesity seems to be a major non-occupational risk factor. Psychosocial factors may also be important.

You et al. (2004) compared personal and physical risk factors in subjects with work-related CTS, CTS not related to work, and controls with no CTS. CTS not related to work was associated with being hard-driving and competitive, female, and older. Work-related CTS was associated with the use of high pinch grip forces and very high repetitive motions at work.

Tennis Elbow and Golfer's Elbow (Lateral and Medial Epicondylitis)

Evidence for Work Relatedness: Tennis elbow occurs when the extensor tendons of the wrist are irritated at the lateral epicondyle of the elbow (Figure 5.3). Golfer's elbow is an injury at the medial epicondyle. It is thought that overuse or overexertion of the wrist muscles is the cause. In severe cases, the muscle and tendon may separate from the bone. The risk of injury is said to be increased by activities requiring large grasping forces. Walker-Bone et al. (2012) found that 11% of 6038 workers reported pain in the past week, of whom 0.7% were diagnosed with lateral epicondylitis and 0.6% with medial epicondylitis. Tennis elbow was four times more likely in manual workers and was associated with bending or straightening of the elbow at work.

Keir et al. (2021) reviewed studies dating back to 1997 and summarized the following as occupational risk factors for tennis elbow. Hip grip forces and using heavy hand tools

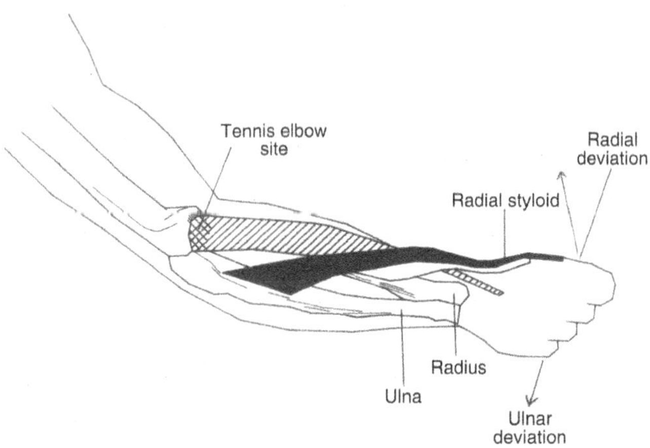

FIGURE 5.3 General view of the forearm, wrist, and hand showing injury sites.

(>1 kg) increased the risk for males (odds ratio 2.2) and females (odds ratio 2.8). An odds ratio of 3.8 was found for workers performing gripping tasks for more than 4 h/day compared to less than 1 h/day. Highly repetitive use of the wrists and fingers (75% or more of the day compared to 25% or less of the day) increased the risk in females (odds ratio 2.8). Repetitive wrist motions in a job for more than 20 years increased the risk of medial epicondylitis (odds ratio 3.6) and lateral epicondylitis (odds ratio 2.8). Increased risk was also found to be associated with repetitive elbow flexion/extension or rotation. Awkward wrist postures for more than 4 h/day have also been associated with increased risk, although the evidence for exposure to vibration was inconclusive.

Possible Causal Pathways: The elbow can be considered a weight-bearing joint. Many of the muscles that control movements of the hand are extrinsic to it and have their origins higher up. Some cross the elbow joint. The act of grasping and holding objects is only possible if the wrist is stabilized by the muscles of the forearm, many of which originate at the elbow. For example, when the finger flexors contract to enable an object to be grasped, the wrist extensors also have to contract to prevent the wrist itself from flexing. These contractile forces are transmitted across the elbow to the distal end of the humerus where the wrist extensors originate. The elbow joint is compressed and the tendons are under tension and may become swollen at the point where they insert into the humerus. Clearly, any activity that requires a strong grip to be maintained for long periods will place a high load on the elbow joint and associated structures. It seems to be the case that elbow problems are not particularly prevalent compared to other conditions and, although they can be caused by occupational exposures, they are not specific to them. From the evidence available, it would seem that a "high-risk" job for epicondylitis is one that requires high grip forces combined with repetitive work in extreme postures or pronation and supination of the wrist—essentially factors that increase the stress on the tendon insertions. Low-friction handles and contaminants such as sweat, oil, and lard that reduce friction (Bobjer et al., 1993) would seem to increase the risk by causing people to grip harder. Perhaps this is why the disorder has been found in occupations such as fish processing where several different factors operate together.

Disorders of the Shoulder

Evidence for Work Relatedness: Punnett et al. (2000) found an increased risk of shoulder disorder when the shoulder was abducted or flexed more than 90°, with the risk increasing in proportion to the percentage of the work cycle that the arm was held in that position. The ORs for these exposures ranged from 1.5 to 6.5, but high risks were found to occur when the flexion or abduction was held for more than 10% of the work cycle. There is evidence for a positive association between highly repetitive work (cycle times less than 30 s or spending more than 50% of the time doing the same task) and shoulder problems. Repeated or static shoulder postures (more than 60° of shoulder flexion or abduction with increased risk when posture is combined with other stressors such as holding a tool) also increase the risk. There is evidence that shoulder problems increase with the duration of employment and with the length of the workday. Dalbøge et al. (2014) found that 24% of all first-time surgeries for impingement syndrome of the shoulder were related to occupational exposures (working with elevated arms, forceful exertions, and repetitive movements). Wærsted et al. (2020) following a systematic literature review, essentially confirmed that working with the arms elevated is a risk factor for shoulder disorders even when non-occupational factors were controlled for (e.g., age, sex, leisure-time sports activity, smoking, obesity, and metabolic syndrome).

Possible Causal Pathways: Most work involving hand tools imposes a combination of repetitive and static loads on the body which usually involve the shoulder, if only indirectly. There is no single shoulder joint—the arms join the scapula at the glenohumeral joint,

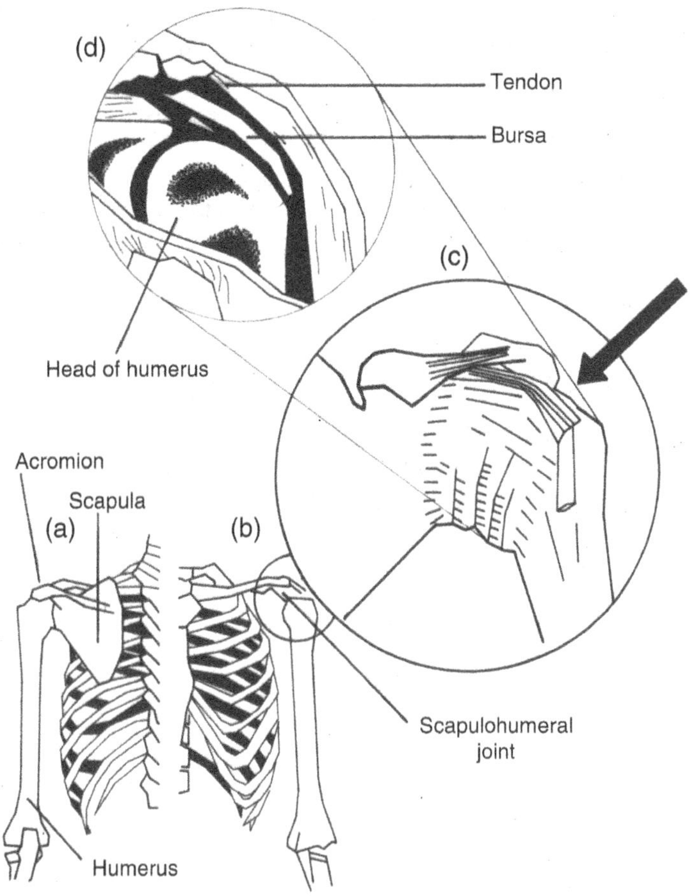

FIGURE 5.4 Shoulder joint: (a) rear view, (b) front view, (c) expanded view, and (d) scapulohumeral joint itself.

> and the scapula joins the body at the scapulothoracic joint. Not surprisingly, the shoulder "joint" is the most mobile in the body and, together with its related soft tissues, is particularly prone to injury in any activities where the arms are held above the horizontal. Working with the hands above shoulder height is stressful and may increase the risk of developing the so-called "impingement syndrome," otherwise known as "swimmer's shoulder," "pitcher's arm," or rotator cuff syndrome. The disorder is known to be more common in sportsmen who use high overhead actions. When the arm is held in front of the body, the muscles around the glenohumeral joint (rotator cuff muscles) contract to stabilize the joint. Figure 5.4 depicts, in broad outline, the anatomy of the shoulder joint and related structures.

The shoulder joint is a kind of ball and socket joint, but the ball part, the head of the humerus, or upper arm bone, represents only a third of the surface of a sphere when it engages the socket. The socket (the glenoid cavity of the scapula, or shoulder blade) is correspondingly shallow. The head of the humerus has to be held in place by tonic muscle activity. This explains why the shoulder joint is so easily dislocated. This can be contrasted with the much more stable hip joint, where over 50% of the femoral head is enclosed by the acetabulum.

That the shoulder joint requires muscle activity to be held in place might alert the ergonomist to its likely susceptibility to rapid fatigue and damage when exposed to static loads or repetitive actions. One of the simplest ways to reduce occupationally induced shoulder stress in many jobs is to provide armrests, slings, or other means of supporting the weight of the arms to enable the shoulder muscles to relax. Whenever the hands or arms are used, muscle activity is necessary to keep the humerus in its socket and to hold the scapula in place on the thorax. The stabilizer muscles of the scapula are at a great mechanical disadvantage when the arms are held forward of the body (or cantilevered), and static muscle contractions are needed to resist the resulting moments. One of these muscles, the serratus anterior, acts to pull the scapula into the thorax. Excessive load or fatigue of this muscle may cause pain in the upper back. Damage to the serratus anterior can result in a condition known as a "winged scapula," where the scapula protrudes because it is no longer held close to the thorax due to muscle weakness. Carrying heavy rucksacks can damage the nerve supply to this muscle, resulting in a condition known as "rucksack palsy."

Above the humeral head lie the acromion and the coracoacromial ligaments, and in between lies a bursa. In this narrow space pass many tendons, nerves, and blood vessels. The space can easily be taken up by the growth of bony spurs, by bleeding, or by soft tissue swelling due to overexertion (the subacromial bursa can become inflamed, for example). If this happens, the range of motion of the shoulder joint is reduced because the impingement of the subacromial structures causes pain when the joint is moved. Sufferers cannot raise their arms above shoulder height. It is likely that jobs which require the hands to be chronically elevated above elbow height can cause short-term changes which, over time, may ultimately lead to disorders of the shoulder joint. Shoulder pain can also be caused by localized muscle fatigue—particularly if the arms have to be held above shoulder height for long periods of time (as in painting a ceiling or pruning a tree)—and by mental stress that increases shoulder muscle tension (Roman-Liu et al., 2013). Some recommendations for preventing shoulder pain are given in Table 5.4.

Lower Limbs

There is extensive literature on lower limb injuries in sports. The main occupational causes of lower limb injury (excluding falls) appear to be walking while carrying heavy loads and jobs that require excessive use of the knees. Reynolds et al. (1999) followed infantrymen on a 160 km march over 5 days. The soldiers carried heavy packs. Of them, 36% suffered one or more injuries and 8% were unable to complete the march. The main injuries were blisters (48%) and foot pain (18%). Other injuries included ankle and knee strain, foot swelling, hip and thigh pain, and metatarsal stress fracture. Risk of injury was higher among smokers and lower among older people.

The 12-month prevalence of knee problems in the Danish population was reported to be 19%. Unsurprisingly, kneeling work was associated with bursitis (housemaid's knee). Table 5.5 summarizes the main risk factors for knee disorders (see e.g., Sandmark et al., 2000).

TABLE 5.4
Methods of Reducing Shoulder Stress

1. If possible, work with the hands near waist level and close to the body.
2. If the hands have to be positioned above shoulder level, their elevation above the shoulders should be no more than 35°. Hand loads should not exceed 0.4 kg, and the posture should be held for no more than 20 s for each minute of work.
3. Avoid shoulder flexion/abduction >90° for >10% of cycle time.
4. Select taller workers for workplaces that cannot be modified.
5. Take regular rest breaks.
6. Minimize handheld weight.
7. Provide external support for the weight of the arms (slings, ledges, etc.).
8. Confine work objects within the zone of convenient reach.
9. Provide wrist rests for keyboard workers.

Sources: Adapted from Punnett et al. (2000); Sommerich et al. (1993).

TABLE 5.5
Summary of Occupational and Personal Risk Factors for Knee Disorders

Occupational Factors Associated with Increased Risk
Occupational knee bending
Kneeling >1 h/day
Squatting >1 h/day
Getting up from kneeling/squatting >30 times/day
Climbing >30 stairs/day
Climbing >10 flights of stairs per day
Jumping from one level to another
Caring for disabled relative at home (females only)
Working in a heavy job >10 years
Occupational Factors Associated with No Increased Risk
Sitting >2 h/day
Standing/walking >2 h/day
Driving >2 h/day
Personal Factors Associated with Increased Risk
Previous knee injury for any reason
Previous meniscectomy
Age
Obesity
Gender
Cigarette smoking (lower prevalence in smokers)
Chondrocalcinosis
Heberden's nodes[a]
High-Risk Occupations
Firefighter
Farmer
Construction worker
Forestry worker
Miner
Carpet and floor layers and tilers
Sports
High risk for soccer and rugby, and knee injury
No evidence for increased risk for running, team sports, or racket sports, in the absence of previous injury

[a] The presence of nodes suggests genetic predisposition to OA.

BASIC APPLICATIONS

Hand Tools

The control of finger movements depends on many small muscles which can easily become fatigued, particularly during prolonged work with inadequate rest periods and poorly designed tools (Figure 5.5). One of the most fundamental problems in hand tool design is optimizing the dimensions of the tool in relation to the hand anthropometry of the population under study.

Ducharme (1977) found that many women working in previously male craft skills in the U.S. Air Force were dissatisfied with the design of the tools they used. Crimping tools, wire strippers, and soldering irons were said to have grips that were too wide or required the use of two hands to operate. Other tools were said to be too heavy and awkward to use. Pheasant and Scriven (1983)

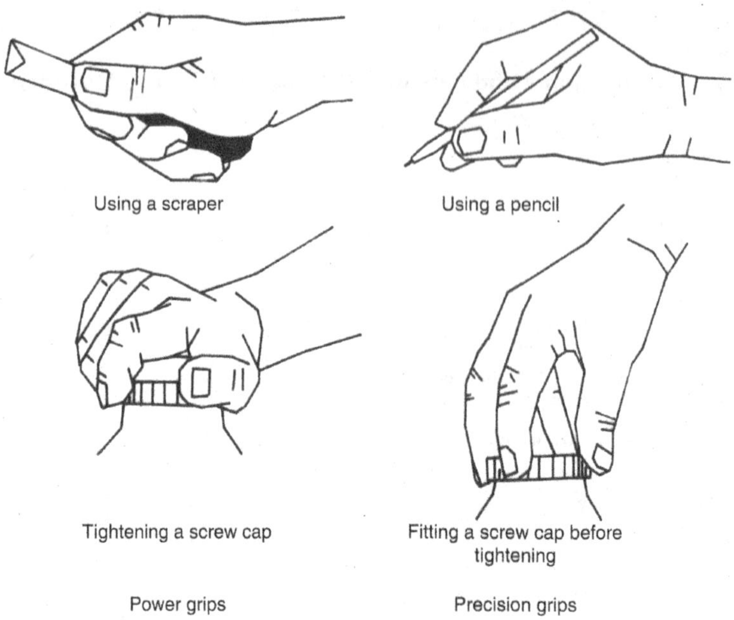

FIGURE 5.5 Precision versus power grips.

reported that motorcar wheel braces provide inadequate mechanical advantage for both male and female users. Although a greater proportion of females than males are affected, both sexes would benefit from improved designs offering greater mechanical advantage.

Handle Design

Pheasant and O'Neill (1975) investigated handle design in a gripping and turning task (such as using a screwdriver). They found that strength deteriorated when handles greater than 5 cm in diameter were used and that, to reduce abrasion of the skin, hand–handle contact should be maximized. Knurled cylinders were found to be superior to smooth cylinders due to the increase in friction at the hand–handle interface. The authors concluded that, for forceful activities, the size of a handle rather than its shape was most important. A useful rule of thumb for evaluating handle diameters is that the handle should be of such a size that it permits slight overlap of the thumb and fingers of a worker with small hands. The larger the handle diameter, the greater the torque that can be applied to it, in principle, but people with small hands must be able to enclose the handle with their fingers. The maximum handle diameter for male users seems to be 40 mm, and smaller still if gloves are to be worn. Cylindrical handles are better than handles with finger grooves, since these cause pressure hot spots and blistering of the skin on hands they do not fit. Handle lengths should be at least 11.5 cm, plus clearance for large (ninety-fifth percentile) hands. An extra 2.5 cm should be added if gloves need to be allowed for (NIOSH, 1981).

Some examples of screwdriver handles designed for one- and two-handed operation are shown in Figure 5.6. Grip strength depends largely on the posture of the wrist. When the wrist is extended, the finger flexors are lengthened and can therefore exert more tension, resulting in a stronger grip. When the wrist is flexed, the opposite occurs, and grip strength is severely weakened. A general requirement of handle design is that the wrist joints should be kept in a neutral position (in the middle of their ranges of movement) when tools are being used. Pheasant (1986) describes how the axis of a handle is at an angle of 100°–110° with respect to the forearm when the wrist is in a neutral position (Figure 5.7).

Tools such as soldering irons can be redesigned using a "pistol grip" handle rather than the more traditional straight handle for the same reason (Figure 5.8). When using straight-handled tools, there is a tendency for the wrist to be bent outward (ulnar deviation). This stretches the tendons

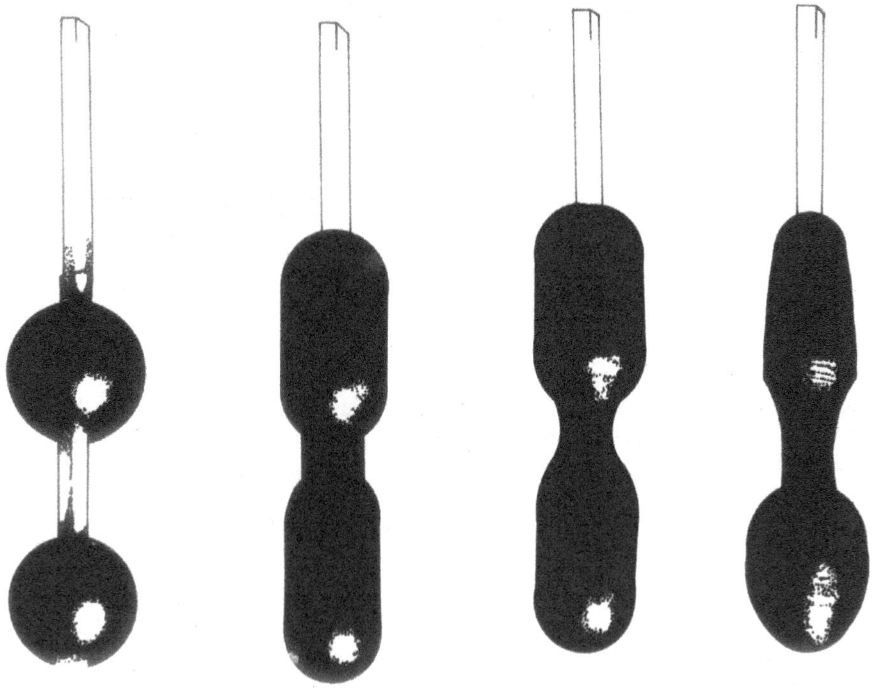

FIGURE 5.6 Some innovative screwdriver handle designs for one- and two-handed operation. (After the Ergonomic Design Group, Stockholm, Sweden.)

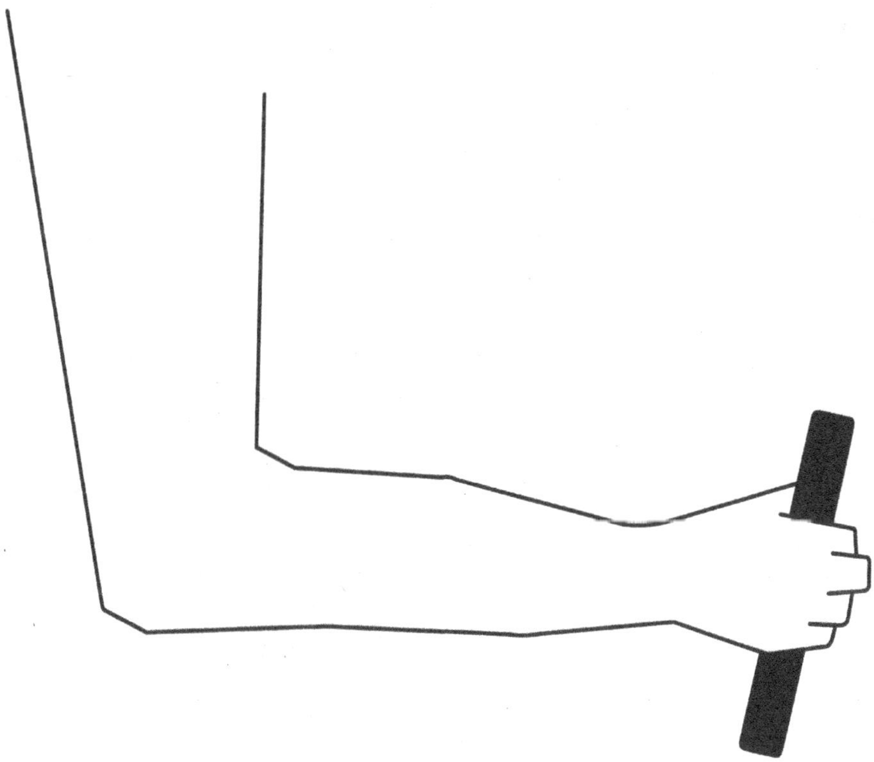

FIGURE 5.7 Angle of grip when the wrist is in the neutral position.

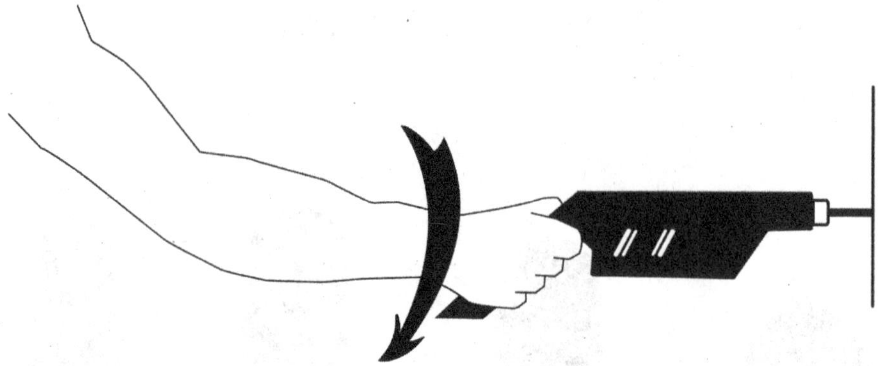

FIGURE 5.8 When working at a vertical surface, an in-line grip reduces wrist strain. Note the large moment of ulnar deviation when the tool is not engaged with the workpiece. This can be minimized by locating handles below the mass center of the tool.

of the forearm muscles on one side, causing them to rub against a bony protrusion on the thumb side of the wrist (known as the radial styloid). Repeated exposure can cause the sheath (synovium) within which the tendon runs to become inflamed. This is known as "de Quervain's syndrome." Inflammation of tendons and tendon sheaths can occur in other parts of the hand and other body structures. In the long term, permanent damage to the tendon and its sheath may result. The buildup of scar tissue in the tendon may ultimately reduce the range of movement of the wrist.

The idea of "bending the handle instead of the wrist" is a valid one and has many potential applications. Konz (1986) sums up the use of bent handles in tool design as follows:

> … When a tool, gripped with a power grip has its working part extend above the hand, then a curve in the handle may be beneficial…. A small bend (5 to 10 deg) seems best.

For powered tools, pistol grips seem most appropriate when the task is oriented vertically with respect to the operator (as when drilling a hole into a wall). When the task is a horizontal one (as in fastening a screw into a horizontal desktop with a powered screwdriver), an in-line tool may be better (Figure 5.9).

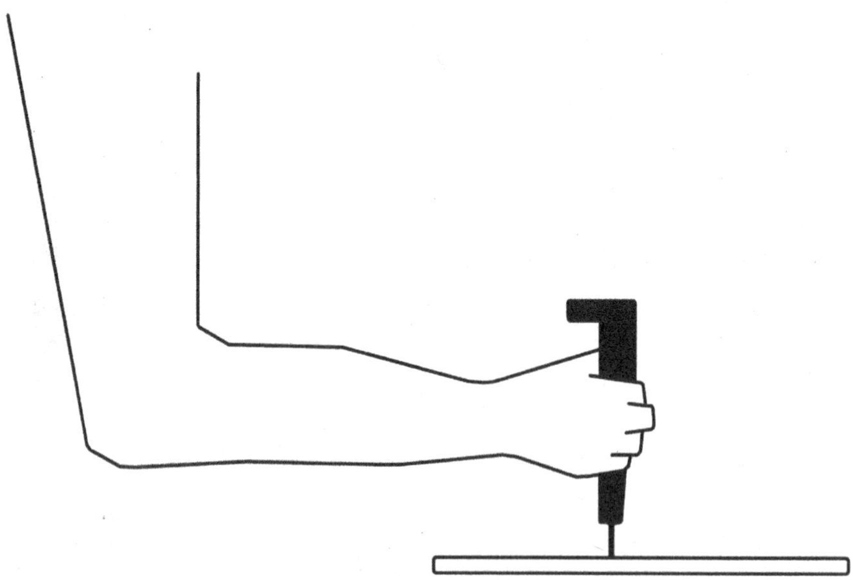

FIGURE 5.9 When working at a horizontal surface, a perpendicular grip may be better.

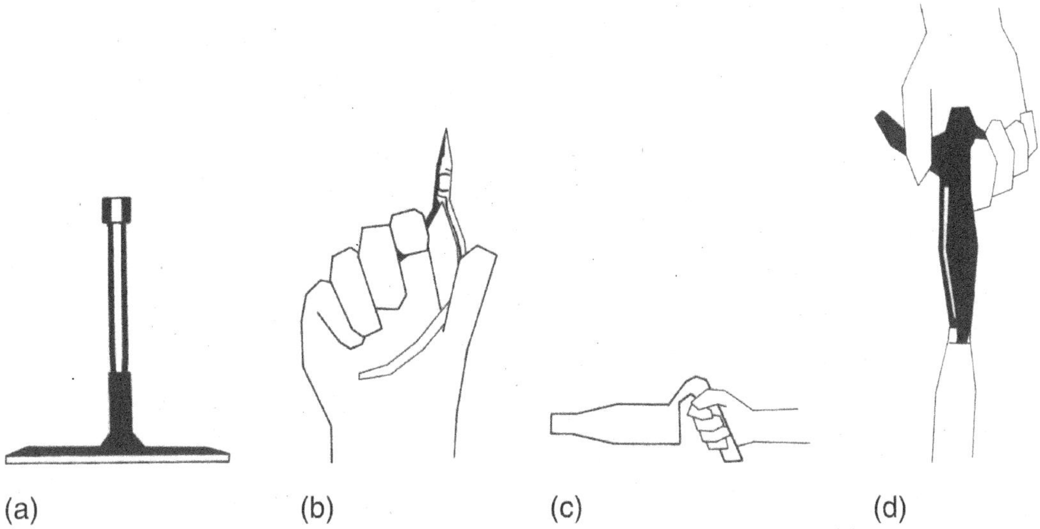

FIGURE 5.10 Redesigned tools with improved hand–handle interfaces: (a) socket wrench with T-bar for greater mechanical advantage, (b) pliers with bent handles to maintain neutral wrist posture, (c) drill with handle at appropriate wrist angle, and (d) paint scraper with thumb stall to relieve pressure on the palm of the hand and prevent blistering of the skin on the palm.

Powered tools tend to be considerably heavier than their non-powered counterparts and a potential source of wrist strain comes from the weight of the tool itself—particularly if the handle is placed at one end of the tool rather than in the middle. Having to hold a heavy drill with a pistol grip while positioning the drill bit exemplifies this problem. Wrist loading can be reduced by fitting the handle at the tool's center of mass so that the tool is counterbalanced.

Finally, tools can be redesigned and fitted with longer handles, or handle extensions can be fitted to increase the worker's vertical reach, obviating the need for the hands to be raised above shoulder height. Some examples are handle extenders for paintbrushes when painting ceilings and long-handled secateurs for pruning the higher branches of trees. Figure 5.10 shows some redesigned tools with improved hand–handle interfaces.

Limits for Hand/Wrist Exertions in Repetitive Work

Ciriello et al. (2002) obtained data on the maximum exertions acceptable to ten females carrying out repetitive hand/wrist exertions for 7 h/day, 5 days/week for 4 weeks. Table 5.6 summarizes the findings for the right hand carrying out the following tasks:

- Screw driving clockwise (handle diameter 31 mm; handle length 155 mm)
- Screw driving clockwise (handle diameter 40 mm; handle length 158 mm)
- Screw driving clockwise with a yoke (T-bar) handle (handle diameter 31 mm; handle length 139 mm)
- Screw driving counterclockwise (handle diameter 31 mm; handle length 155 mm)
- Ulnar deviation with a power grip
- Gripping task with a power grip

TABLE 5.6
Maximum Acceptable Torques (Nm) and Forces (N) for U.S. Females

		Repetition Rate		
	Percentage of Population	**5/min torque**	**20/min torque**	**25/min torque**
	Screw Driving			
Clockwise	75	0.16	0.12	0.07
31 mm handle	50	0.43	0.40	0.33
	25	0.70	0.73	0.59
Clockwise	75	0.14	0.13	0.13
40 mm handle	50	0.49	0.43	0.43
	25	0.84	0.73	0.73
Clockwise	75	0.27	0.20	0.12
Yoke handle	50	0.65	0.53	0.61
	25	1.03	0.86	1.10
Counterclockwise	75	0.19	0.13	0.14
31 mm handle	50	0.47	0.41	0.38
	25	0.75	0.69	0.62
Ulnar deviation	75	0.48	0.43	0.44
	50	1.13	1.10	1.08
	25	1.78	1.77	1.72
Handgrip task	75	2.03	2.25	2.38
	50	4.83	4.80	4.85
	25	7.63	7.35	7.32
		Force	**Force**	**Force**
Handgrip task	75	13.4	14.8	15.7
	50	31.8	31.6	31.9
	25	50.2	48.4	48.2

Source: Adapted from Ciriello et al. (2002), with permission.

Wrist Flexion and Extension

Snook et al. (1995) published tables of maximum acceptable forces for female wrist extension and flexion in the workplace. Table 5.7 summarizes these limits.

Elbow Rests

They appear to be beneficial for both sitting and standing workers—by stabilizing and supporting the arms, they reduce the load on the shoulder musculature (by supporting the weight of the arms, elbow rests would also reduce the load on the lumbar spine).

Keyboard Design

With old-fashioned mechanical typewriters, the typist had to stop to manually return the carriage after each line (see Chapter 4, Tutorial topic no. 4). Errors had to be manually corrected using erasers or correcting fluid, and the paper had to be changed after each page had been typed. All of these secondary tasks provided changes of posture and broke up the continuity of the typing task. Brief periods of rest were intrinsic to the operation of mechanical typewriters. A very large proportion

TABLE 5.7
Maximum Acceptable Forces (N) for U.S. Female Wrist Flexion and Extension

Percentage of the Population	Repetition Rate 2/min	5/min	10/min	15/min	20/min
	Wrist Flexion (Power Grip)				
90	14.9	14.9	13.5	12.0	10.2
75	23.2	23.2	20.9	18.6	15.8
50	32.3	32.3	29.0	26.0	22.1
25	41.5	41.5	37.2	33.5	28.4
10	49.8	49.8	44.6	40.1	34.0
	Wrist Flexion (Pinch Grip)				
90	9.2	8.5	7.4	7.4	6.0
75	14.2	13.2	11.5	11.5	9.3
50	19.8	18.4	16.0	16.0	12.9
25	25.4	23.6	20.6	20.6	16.6
10	30.5	28.2	24.6	24.6	19.8
	Wrist Extension (Power Grip)				
90	8.8	8.8	7.8	6.9	5.4
75	13.6	13.6	12.1	10.9	8.5
50	18.9	18.9	16.8	15.1	11.9
25	24.2	24.2	21.5	19.3	15.2
10	29.0	29.0	25.8	23.2	18.3

Source: Adapted from Snook et al. (1995), with permission.

of today's desktop computer users have never undergone formal training and do not possess these keyboard skills. Gerr et al. (2002) concluded that the risk of musculoskeletal disorders increases with more than 15 h of keyboard use per week.

Zipp et al. (1983) investigated the posture of the hands and wrists, noting marked ulnar variation and fatigue (Figure 5.11).

They concluded that keyboards should be designed with separate banks of keys (one for each hand), each bank being inclined and contoured to be compatible with the functional anatomy of the hand. Keyboards based on this design are commercially available (Figure 5.12), are perceived as comfortable to use, and give fair performance.

Tittiranonda et al. (1999) compared several alternative geometry keyboards and a placebo over a 6-month period. Although there was no improvement in clinical findings, self-reported pain severity was lower and hand function improved with the Microsoft Natural Keyboard compared to the placebo. Typing performance was unchanged. Gerard et al. (2002) provide evidence that conventional computer keyboards can be improved by enhancing the auditory feedback when keys are depressed. Their experimental keyboard was designed to produce a tone at 1.2 dB above the ambient noise whenever a threshold value of either keying force, finger flex, or EMG was exceeded. The tone increased by further increments of 1.2 dB the higher the keying force or EMG was above the threshold. The feedback produced a 10%–20% reduction in 90th percentile keying force in under 3 min. Further exposure to the feedback did not bring about more reductions in keying force. These findings may mean that augmented auditory feedback will help to prevent upper extremity discomfort in keyboard users. A controlled trial would be needed to determine whether these reductions are sufficient to reduce the incidence of disorders.

FIGURE 5.11 Conventional keyboards cause ulnar deviation.

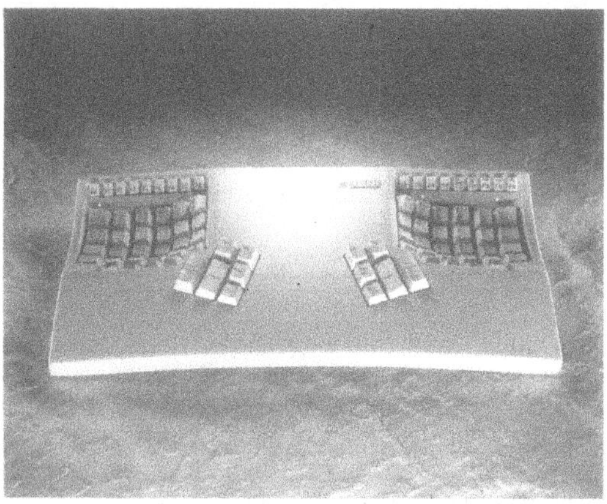

FIGURE 5.12 (a) Separate banks of keys for each hand may relieve wrist strain. (b) Arm support may reduce load on the distal extremities (courtesy of Contour Design). (Courtesy of Kinesis Corporation, Bellevue, Washington, USA.)

Monitor Heights

Laptops need to be supplemented with detachable keyboards and stands ("risers") for daily use in the office or at home (Figure 5.13).

Yadegaripour et al. (2021) compared cervical muscle activity and posture in subjects working at a laptop placed on a desk and a laptop with a separate keyboard and riser to lift the top of the screen to the eye height of their subjects. Keyboard and mouse work was carried out in both conditions. Cervical and thoracic flexion were reduced by 4.53° and 3.14°, respectively, when the riser

FIGURE 5.13 Laptop with stand ("riser") and detachable keyboard for use when working from home or in the office. (Courtesy of Contour Design.)

was used. Cervical erector spinae and multifidus muscle activity was also lower, as was discomfort. The multifidus muscle attaches to the spinous process and has a greater mechanical advantage in maintaining neck posture during cervical flexion.

Cell Phones and E-Games

Between 1995 and 2004, the number of cell phone users increased from 91 million to 1.75 billion worldwide (Xiong and Muraki, 2014). Users of phones with touchscreens often use the thumb to operate the keypad using a combination of flexion/extension and adduction/abduction. The human thumb is capable of *opposition*, where the fleshy part of the thumb pad has contact with that of the opposing fingers. It does this by rotating about its axis as it flexes. Unlike the power grip (Figure 5.5), opposition involves all the muscles of the thumb, which can be likened to guy ropes supporting a tent pole. Clearly, the human thumb did not evolve to press buttons, and the adductor and abductor muscles, in particular, did not evolve to stabilize the thumb in flexion—rather, to position it for grasping large objects in a power grip. Xiong and Muraki (2014) used electromyography to investigate thumb muscle fatigue when using cell phones with 3- and 9-mm buttons. The muscles fatigued more quickly in the directions of flexion–extension and adduction to abduction, and with smaller rather than larger buttons. The first dorsal interosseous muscle (extending from the ventral base of the thumb to the index finger) fatigued most quickly. The findings suggest that buttons should be as large as possible and that pain in the back of the hand between the thumb and the index finger may be a symptom of excessive touchscreen use on cell phones.

Chany et al. (2007) compared a traditional office handset telephone (operated with a power grip) to a clamshell design more reminiscent of a cell phone (and held in a modified pinch grip).

They found that discomfort increased with the clamshell design, as did EMG changes indicative of fatigue in the deltoid and thenar muscles. For long-term office use, the traditional handset design seems best.

For cell phone use, Gustafsson (2011) recommends that pain and injury can be avoided if the forearms are supported, both thumbs are used, sitting with the neck flexed is avoided, and typing at high velocity when texting is minimized. Many users, particularly children and young people, do not appear to be aware of these guidelines.

Straker et al. (2014) provide the following guidance for reducing the risk of musculoskeletal injury to children using e-games:

- Keep the e-game equipment, surrounding furniture, and yourself safe when you play.
- Swap sedentary with active e-games—you will enjoy the feeling of your body moving.
- Minimize the time spent on e-games that require repetitive button clicking or wrist actions.
- Be aware of being in awkward postures (being uncomfortable) when playing—try to be comfortable.
- Break up your e-games every 30 min to give both your body and eyes rest.
- Be aware of how long you play, and try to minimize the time when you are sitting still—make time for active play both inside and outside.
- For any new active e-game, or after a break, start with short turns to allow your muscles to get used to the activity.
- Parents should encourage real-world physical activity and ensure that it is not displaced by e-games.
- Parents should encourage good "technique" and a safe playing space to avoid muscle and bone discomfort and injury.
- Encourage limiting sedentary leisure and enhancing active leisure—virtual and preferably real-world.
- Discourage poor and sustained postures, repetitive actions, high accelerations, and high forces during e-game play.
- Encourage breaking up SEG bouts every 30 min to provide an active break for the body and eyes.

Cursor Control Devices (CCD)

There is *no evidence* that occupational use of computer mice causes CTS (Thomsen et al., 2008). Lee et al. (2007) investigated hand muscle fatigue in several designs of computer mouse, varying in the placement and number of keys. They found that removal of the right-hand key lowered finger extensor muscle activity but there were trade-offs with performance. The short-term benefits of new mouse designs need to be considered in relation to longer-term usability issues, with the caveat that Lee's short-term laboratory study may not have given users sufficient practice with the different designs. Clearly, CCD design has the potential to induce injury in high-demand, long-exposure tasks if the small muscles of the hand are involved in CCD operation. Figure 5.14 shows a CCD with a thumb-controlled trackball—clearly an efficient use of the thumb for short-term, intermittent operations, but also a potential cause of fatigue in intensive work—particularly if the control-display ratio requires significant rotation of the trackball to traverse the cursor across the screen.

Mouse Design

Traditional designs require the forearm to be pronated and the wrist extended and ulnar deviated. Alternative designs are now available that can be operated with 20°–30° less pronation of the forearm (Figure 5.15).

Nodooshan et al. (2023) investigated forearm muscle activity (EMG) and comfort of subjects using a conventional mouse on a flat mouse pad and with the pad tilted at 10°, 20°, and 30° away

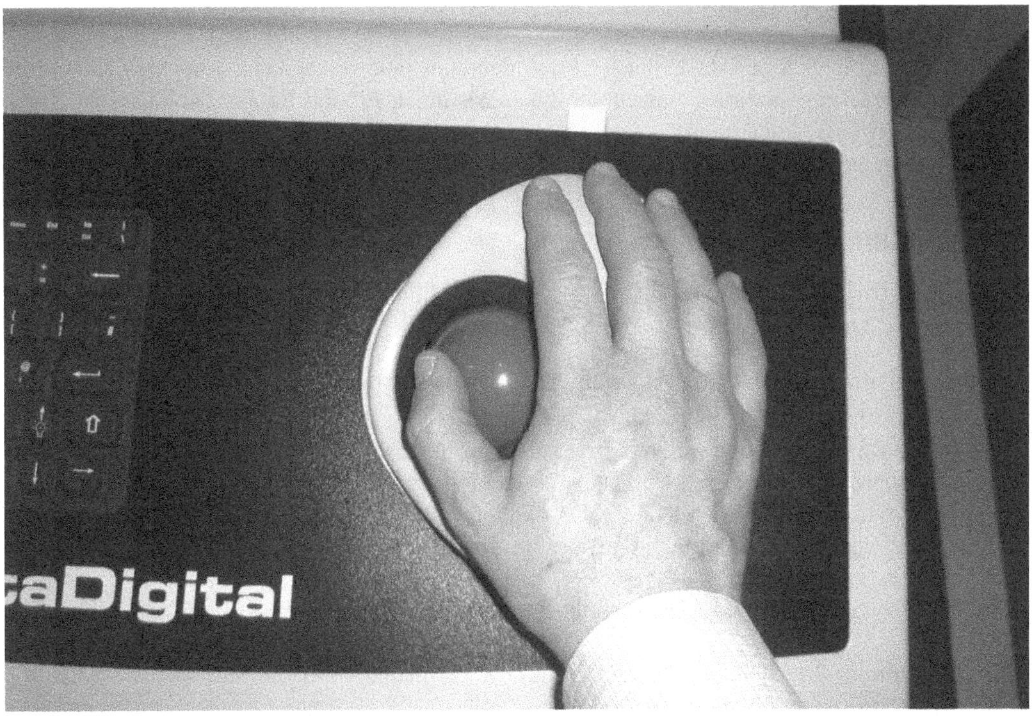

FIGURE 5.14 Thumb-operated trackball.

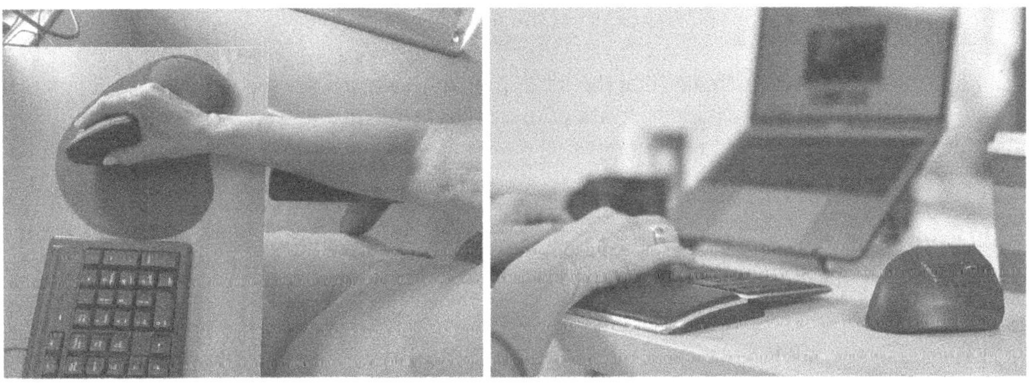

FIGURE 5.15 Alternative mouse designs (a) Pronation of the forearm is reduced. (b) Rollermouse. (Courtesy of Contour Design.)

from the screen (for right-handed users, the pad was tilted to the right). This places the forearm and wrist into a less pronated posture. Pronator quadratus (a forearm pronator) activity was lower when the mouse pad was tilted than when flat. Extensor carpi radialis activity was lowest when the mouse pad was flat. This agrees with the findings of other studies of forearm muscle activity when using alternative mouse designs. Limb discomfort was lower in the 10° and 20° conditions. Shoulder discomfort was greater with 30° of tilt. Ten degrees of tilt was the most preferred angle and 30°, the least preferred. These data suggest that alternative designs that enable the mouse to be used with less pronation may be beneficial for users with existing wrist and/or forearm conditions.

TOOLS AND PROCESSES

There are many tools for the assessment of task demands in the workplace. Kotowski and Gibson (2023) provide an ergonomic assessment toolkit consisting of 27 tools for assessing task stress (risk to the back, shoulders, and hand) as well as overall postural stress. See also Gallagher and Barbe (2022) for details of the fatigue failure mechanism and its use in risk assessment and mitigation.

Identifying Repetitive, Monotonous Work

In Table 5.8, Hedén et al. (1993) define the criteria for defining repetitive, monotonous work, which is to be avoided to reduce the risk of WMSDs.

The Revised Strain Index (RSI): Assessing the Risk of Injury to the Distal Extremities

The RSI is an observational tool for the assessment of jobs for the risk of upper extremity WMSDs.

Thus, the RSI includes the following measures: (a) intensity of exertion (force), (b) efforts per minute (frequency), (c) duration per exertion, (d) hand/wrist posture (wrist flexion or extension), and (e) duration of task per day (Garg et al., 2017). Note that the index measures the risk associated with a job, not the risk to particular individuals performing the job.

- *Intensity of Exertion (IM)*: Measured using the Borg Scale (see Chapter 7), coded from 0 to 10
- *Duration of Exertion per Task Cycle (DM)*: Estimated from the number of exertions divided by the observation time
- *Number of Efforts per Minute (EM)*: Counted by the observer from direct observation or video
- *Wrist Posture (PM)*: Whether the wrist is in neutral, flexion, or extension during the exertion with maximum force
- *Duration of Task/Day (HM)*: Length of the workday (hours)

An RSI score of 3 or less indicates that the job is probably safe, a score greater than 3 but less than 7 indicates that there may be increased risk of musculoskeletal disorders, and a score greater than 7 indicates that the job is probably hazardous.

TABLE 5.8
Criteria for Defining Repetitive, Monotonous Work[a]

	Red	Yellow	Green
Working cycle	Repeated several times per minute	Repeated several times per hour	Repeated some times per hour
Postures and movements	Fixed, uncomfortable postures and movements	Limited positions to alter postures and movements	Good workplace layout and opportunity to vary posture
Freedom of action	Work is completely governed by something or someone else	Work is governed externally to some extent. Opportunities for influencing work are limited	Good opportunities to fit the work to own ability and influence planning and organizing the work
Work content learning	Employee performs one isolated task. Short training time	Several tasks are in process. Job rotation may be present. Training provided	Employee takes several parts of the process, including planning and control. Competence develops continuously

[a] Assumes a full day's work. If work is in the red zone, change is required.

An RSI calculator, courtesy of North Carolina State University, can be found at https://www.ergocenter.ncsu.edu/resources-and-tools-hub/analysis-tools/

The RSI score is calculated from multipliers as follows:

$$RSI = IM * EM * DM * PM * HM$$

An RSI score of 3 or less indicates the job is probably safe, >3<7 indicates that there may be increased risk of musculoskeletal disorders and >7 indicates that the job is probably hazardous.

Example

The best way to understand the application of the index is by means of an example. Figure 5.16 shows a worker in a fruit packing shed performing an apple packing task. Before the task can be analyzed, it must first be understood, and this requires that a quantitative task analysis be carried out to determine not just the main tasks and operations and how they are sequenced, but also the time taken to carry out each operation. Table 5.9 presents the results of such an analysis, with the basic work cycle being all the operations required, from beginning to end, to pack one box of apples.

In practice, the performance of several different packers would have to be measured in order to generate mean times for the different operations. This would probably involve observation of at least several hundred packing cycles. Thus, the times given in Table 5.7 would be averages. In our example, the time taken to fill a box with 96 apples is 3 min. Thus, in an 8 h day, working continuously, a packer would be able to fill 160 boxes—a total of 160×96, or 15,360 apples! In reality, work such as this is normally remunerated on a piece-rate basis, and it is not uncommon to find packers who can pack well over 20,000 apples per day. HFE Workshop 5.2 illustrates how the task analysis data can be incorporated into the SI procedure to assess the risk of this task.

FIGURE 5.16 Apple packing task for SI worked example.

TABLE 5.9
Task Analysis of Apple Packing Task and Times (Seconds) of Main Operations

Operation	Left Hand	Right Hand	Time
1. *Assemble Box*			20 s
Take a flat box from the stack	X	X	
Open and assemble	X	X	
Place in holder	X	X	
2. *Pack Box*			
Layer 1			
Insert cardboard sleeve	X	X	5 s
Pack 24 apples			15 s
Pack 1 apple			
Grasp apple	X		
Take tissue paper		X	
Wrap the apple in paper	X	X	
Place the wrapped apple in the sleeve		X	
Repeat × 23			
Layer 2: Repeat as layer 1			20 s
Layer 3: Repeat as above			20 s
Layer 4: Repeat as above			20 s
3. *Seal Box*			20 s
Close lids	X	X	
Seal with tape	X	X	
4. *Stack Box*			
Lift the box from the holder	X	X	5 s
Carry to the pallet	X	X	25 s
Stack and sign off	X	X	10 s
5. *Return to Carousel (Next Box)*			20 s
			–
			180 s

Note: Times are means of 500 cycles.

Since this is a self-paced task, and the cycle and operation times given in the task analysis are means, we could also calculate the fifth and ninety-fifth percentile packing times (see Chapter 3 for examples of how to calculate fifth and ninety-fifth percentiles from means and standard deviations) and recalculate the SI. We would then be able to estimate the average risk, as well as the risk to slower and faster packers (or the risk during quiet or busy periods).

HFE WORKSHOP 5.2

USING THE SI TO ASSESS THE RISK OF UPPER EXTREMITY DISORDER IN AN APPLE PACKING TASK

Refer to Table 5.9. The task is mainly bimanual, except for the most repetitive part, the packing of apples into the box. The left hand picks up the apple and the right hand picks up the paper, but at the same rate and for the same duration of time. Separate analyses will be carried out for the left and right hands, proceeding as follows.

The intensities of these exertions are light. The weights of the paper and the apple are low in relation to the maximum weight that a female worker might be able to pick up. An apple weighs 150–200 g, less than 10% of the maximum we might expect a female to be able to pick up.

Duration of exertion is given by the total repetitive motion time as a percentage of cycle time. In this case, four layers are packed per box at 15 s per layer. Duration is therefore $4 \times 15/180$ s, or 30% of the cycle time.

Hand/wrist posture is obtained by observation but could be supplemented with high-speed video recording to capture any extremes of wrist posture. The left-hand–wrist posture is very good. It is simple pronation as the apple is grasped, followed by 90° of supination (i.e., back to neutral) when the apple is tossed to the right hand. The right-hand–wrist posture is very poor, as it involves both ulnar deviation, when the paper is flicked from the tray and flexion as the apple is caught. The speed of work is very fast for both hands, as they work at the same pace. This is particularly stressful since the work is remunerated on a piece-rate basis.

Duration per day is limited by the fact that there is no production line. The repetitive task is part of a larger piece of work. Actual packing only takes place for 80/180 s per cycle. So, for an 8 h day, only 2–4 h are spent in high-repetition activities.

	Assessment Summary			
	Left Hand (Apple)		Right Hand (Paper)	
	Exposure	Multiplier[a]	Exposure	Multiplier[a]
Intensity of exertion	1	1.57	3	3.70
Duration of exertion (s)	2	1.07	2	1.07
Efforts/minute	21	5.35	21	5.35
Hand/wrist posture (degrees)	0	1.00	50	1.11
Duration per day	3 h	0.7	3h	0.7

[a] Multipliers are obtained from the RSI calculator at https://www.ergocenter.ncsu.edu/resources-and-tools-hub/analysis-tools/.

$$\text{SI}(\text{left hand}) = 1.0 \times 1.07 \times 5.35 \times 1.0 \times 1.07 = 6.3$$

$$\text{SI}(\text{right hand}) = 3.70 \times 5.35 \times 1.07 \times 1.11 \times 0.70 = 16.5$$

Remarkably, although the apple weighs far more than the tissue paper, it is the hand/forearm that picks up the paper that is at the greatest risk of injury (high as compared to medium risk). This is due to the poor posture. Despite the fact that the task involves very high repetitions, there are natural breaks throughout the cycle that are not repetitive. Modernizing the packing sheds to improve productivity would increase the risk of injury if the natural breaks were removed by simplifying the work cycle (e.g., by using a conveyor to deliver empty boxes and

(*Continued*)

HFE WORKSHOP 5.2 (*Continued*)

remove full boxes). The risk of injury could be reduced by redesigning the packing stand so that the paper tray is lower and placed so as to slope away from the hand at an angle of 45°. Paper might then be lifted by flexing the elbow with the wrist in a neutral position. Further mechanization of this task might improve efficiency, but at the cost of increased unemployment in developing countries where low-level agricultural jobs are still an important part of the economy. One mitigating factor, of course, is that fruit packing is seasonal work, and employees find other jobs out of season.

Preventing Overuse of the Thumb

Sonne and Potvin (2015) provide an equation to enable the maximum acceptable exertion (MAE) of the thumb to be derived for tasks requiring repetitive thumb motions. The MAE is the percentage of a single MVC of the thumb during the operation of a particular piece of equipment, such as the thumb-operated trackball shown in Figure 5.14. The MAE, for an 8-h day of 28,800 s, is given by:

$$\text{MAE} = 1 - [\text{DC} - 1/28{,}800]^{0.24} \tag{5.1}$$

where DC = duty cycle (ratio of the duration of effort to duration of the cycle).

So, if the duty cycle were 0.5, the MAE would be 15.3% of MVC. If the duty cycle were 0.9, the MAE would be 2.5% of MVC and if the duty cycle were 0.10, the MAE would be 42% of MVC. Clearly, whether or not the resistance of a thumb-operated trackball such as that in Figure 5.14 is appropriate depends on the design of the task and how this influences the duty cycle. It will also depend on the design of the trackball itself, such as its size and location in relation to the thumb and the effect on MVC. Generally, the better the design, the greater the MVC and the longer the duty cycle for acceptable operation.

The OCRA Checklist (Occupational Repetitive Actions)

The OCRA checklist (Figure 5.17) was developed for the assessment of the risk of musculoskeletal disorders of the hands, wrists, forearms, elbows, and shoulders (Occhipiniti and Colombini, 2004). It considers forces, postures, repetitions, task duration and other factors such as rest periods. The checklist is used for the initial screening of workstations—there is a more detailed OCRA Index

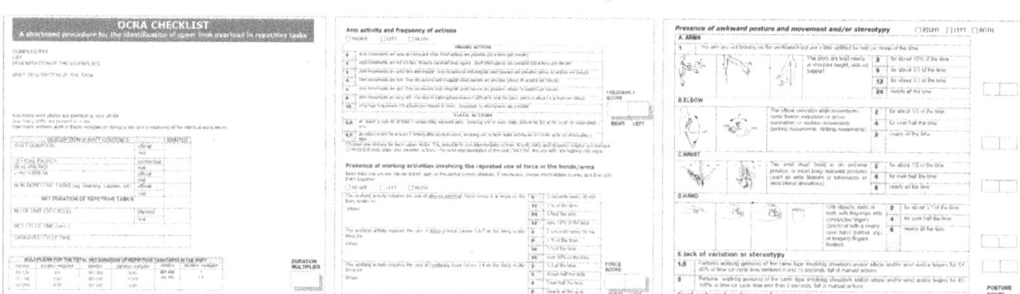

FIGURE 5.17 OCRA checklist.

that can be used for in-depth analysis and task redesign. The checklist provides a summated score, which is essentially an aggregate of the risk factors and can be interpreted as follows:

Checklist Score	Exposure Level	Risk
≤7.5	Green	No Risk
7.6–11.0	Yellow	Borderline Risk
11.1–14.0	Red Light	Light Risk
14.1–22.5	Red Medium	Medium Risk
≥22.6	Red High	High Risk

OCRA is currently being developed in the pr EN 1005-5 and ISO DIS 11128-3. Both the Strain Index and OCRA are reliable and valid methods for assessing the risk of MSDs.

Questionnaires

Questionnaires are available to assess both acute pain during work and the disabling effects of chronic conditions that may be associated with work. Among the former, the Nordic questionnaire is a well-established tool for assessing regional musculoskeletal pain, and there is evidence for its reliability and validity (Descatha et al., 2007).

Neck Disability Index

The Neck Disability Index is a valid and reliable questionnaire for assessing neck pain and its impact on activities of daily living (MacDermid et al., 2009). It is of particular relevance nowadays in view of concerns about mobile working—many of us use mobile devices in unconventional workplaces which place a strain on the neck. The 60-item questionnaire has 10 sections of 6 items each (with mid-range sample items below):

Pain Intensity: The pain is fairly severe at the moment.
Personal Care: It is painful to look after myself; I am slow and careful.
Lifting: Pain prevents me from lifting weights off the floor, but I can manage if weights are conveniently located, i.e., on a table.
Work: I can do most of my usual work, but no more.
Headaches: I have moderate headaches that come infrequently.
Concentration: I have a fair degree of difficulty concentrating.
Sleeping: My sleep is mildly disturbed, up to 1–2 h.
Driving: I can drive as long as I want, with moderate neck pain.
Reading: I can't read as much as I want because of moderate neck pain.
Recreation: I have some neck pain with a few recreational activities.

SYSTEM INTEGRATION

There is now plenty of epidemiological evidence that tasks requiring excessive force, poor posture, high repetition, long duration, and stress carry an increased risk of MSDs. Thus, methods seek to assess risk are valid in that they focus on factors that are theoretically and empirically related to the outcomes of interest (i.e., the methods have both construct and content validity). The RSI itself has predictive (or criterion) validity. In one study of single-task jobs (Rucker and Moore, 2002), its sensitivity, specificity, positive predictive value (PPV), and Negative predictive value (NPV) were 1.00, 0.84, 0.47, and 1.00, respectively (see HFE Workshop 5.3). There is evidence that redesigning high-risk jobs to remove exposure to ergonomic stressors can reduce work loss. However, many jobs expose workers to stressors not currently included in the current ergonomic WMSD assessment tools (such as vibration, cold, and psychosocial stress)—these tools are likely to be less accurate when used to assess more complex jobs involving multiple, non-mechanical exposures (Table 5.10).

TABLE 5.10
Simple Measures for Managing WMSDs of the Upper Limbs

Tool Design and Posture
Bend the handle (5°–10°), not the wrist
Avoid excessive use of "pinch grip" (thumb and fingers)
Maintain neutral wrist posture
Reduce required grip forces:
Use high-friction materials for handles
Longer handles increase mechanical advantage
Maximum allowable grip force (holds for less than 3 s):
Handgrip: 26.2 kg
Pinch grip: 5.8 kg
Maximum allowable grip force (holds for more than 3 s):
Handgrip: 15.6 kg
Pinch grip: 3.6 kg
Add handles for carrying the tool and resisting reaction torque
Damp vibration from powered tools
Encourage the use of large muscle groups
Upper arm posture to be <15° of flexion and <10° of abduction (Aaras, 1994)
Heavy tools should be supported externally (e.g., suspended or supported by props or held by clamps)
Workstation Design
Damp jerk and impact forces
Position work objects so as to eliminate static neck postures
Eliminate static shoulder elevation or provide elbow rests or slings
Provide vices, clamps, or other aids to minimize the need for sustained holding of work objects
Use arm supports in the form of slings or balancers to lower EMG activity in the trapezius and deltoid muscles (Feng et al., 1999)
Task/Job Design
Limit repetitive movements to 2000/h or less
Limit hand loads to <1 kg
Avoid repetitive work with loads >4 kg
Eliminate highly repetitive (<30 s cycle time) jobs
Design-out movements requiring rapid acceleration of body parts
Spend no more than 50% of the time doing the same task
Rotate workers between high-repetition and low-repetition tasks
Avoid mental stress:
Unrealistic deadlines/production targets
Machine pacing with fast cycles
Excessive supervision
Piece-rate systems of remuneration
Design-in "micro breaks" (2–10 s rest every few minutes)
Repetition frequencies greater than 1 Hz may pose a risk to joints or soft tissues
Frequencies less than 1/6th Hz may cause fatigue through the mechanisms of static loading
Increase task variety (e.g., integrate routine maintenance, inspection, and record keeping with basic production tasks)
Limit repetitive and sustained exertions to <10% of maximal effort for all muscle groups

(Continued)

TABLE 5.10 (*Continued*)
Simple Measures for Managing WMSDs of the Upper Limbs

Management of Working Conditions and Worker Exposure
Avoid repetitive work in cold conditions
Use ergonomic criteria when making equipment procurement decisions
Redesign tasks before discomfort becomes a disorder
Eliminate excessive overtime working

Source: Based on U.S. Mil Std-1472, Solomonow and D'Ambrosia (1987).

HFE WORKSHOP 5.3

ASSESSING THE PREDICTIVE POWER OF SCREENING TOOLS

Quite a number of ergonomic assessment tools have been developed recently for the identification of risk factors for WMSDs. These tools are intended to assist in the identification of high-risk jobs and in the identification of the risky components that can then be ergonomically redesigned. Such tools can be treated in exactly the same way as any other kind of screening procedure, be it a blood test, an IQ test, a device for detecting faults in electrical circuits, or a computerized radar system for detecting enemy aircraft. Standard statistics have been developed that describe how well a screening tool works. An ideal tool would identify 100% of all its "targets" as targets. It would also identify all "nontargets" as nontargets—it would never misclassify targets or nontargets.

Suppose we go into a factory "blind," knowing nothing about the injury and absence rates of the factory or any of its jobs. We carry out some ergonomic assessments using a risk assessment tool such as the SI and then classify jobs as high risk or low risk for WMSDs. We then collect injury and sickness data for the next 12 months to identify which jobs pose a risk of WMSDs.

Let us suppose that there are 150 different jobs in the factory and that 40 of these expose workers to a high risk of WMSDs. We have classified all jobs as either "risky" or "safe," but we do not know which jobs really are risky and which really are safe. We want to know how well our assessment tool predicts WMSDs.

The sensitivity of the tool is given by the percentage of high-risk jobs that it correctly identifies as high risk. So, if the tool correctly identifies 35 of the high-risk jobs, its sensitivity is 35/40=87.5%. Note that it has "hit" 35 and "missed" 5 high-risk jobs.

The specificity of the tool is given by the percentage of low-risk jobs that it correctly identifies. If there are 110 low-risk jobs and the tool correctly identifies 100 of these as low risk, then its specificity is 100/110=91%. Since the tool only classifies jobs as risky or safe, the remaining 10 low-risk jobs will have been classified as risky. These are known as "false positives."

The PPV of the tool is the proportion of those jobs identified as being high risk by the tool that really are high risk. Suppose, of the 150 jobs in the factory, the tool rates 45 of them as risky. The PPV is therefore 35/45=78% (only 35 of the 45 jobs predicted to be high risk really were high risk; 10 were safe). To be of any use at all, a tool's PPV must exceed the prevalence rate of the condition it is used to detect. The PPV is the probability that the tool's prediction of risk is true.

The negative predictive value (NPV) of the tool is the proportion of jobs identified as being low risk that really are low risk. If the tool predicts that 105 jobs are low risk, but only 100 really were low-risk jobs, then the NPV is 100/105=95%. The NPV is the probability that the tool's prediction that the job is safe is true.

(*Continued*)

HFE WORKSHOP 5.3 (*Continued*)

Summary

Outcome Assessment of Job	Tool Prediction Risky	Safe	Total
Risky	35 (*A*)	5 (*B*)	40 risky jobs
Safe	10 (*C*)	100 (*D*)	110 safe jobs
Total	45	105	150

A. Hits—correct prediction

B. Misses—false negative

C. False positive

D. Correct prediction

Sensitivity $= A/(A+B)$; Specificity $= D/(C+D)$
PPV $= A/(A+C)$; NPV $= D/(B+D)$

Data on the reliability of different ergonomic techniques for WMSD risk assessment are piecemeal, although there is evidence for good inter-rater and test–retest reliability of the RSI (Moore and Vos, 2005).

The SI appears to be sufficiently sensitive to differences in the degree of exposure to ergonomic stressors that occur in real situations and that are important in determining real outcomes. It is also diagnostic, as it is sufficiently detailed in its content to enable risk assessments to vary depending on the presence or absence of particular ergonomic stressors (e.g., separate risk assessments can be done for each hand performing a single task, as in the workshop example).

Assessment methods can be intrusive if they require the use of EMG or goniometers/videotaping. However, the RSI and other exposure measures rely entirely on observation and can be carried out with minimal disruption to work activities. These methods are also acceptable in a wide variety of situations and are appropriate for general use. The main cost is the assessor's time and, if necessary, training in the particular technique.

Effectiveness and Cost-Effectiveness

Schneider and Wahl (1998) reported the benefits of "early ergonomic intervention" (EEI) on the reduction of lost time injuries (LTIs) among office workers in Colorado. The company filed 2547 "cumulative trauma injuries" in 1995, 25% of which were LTIs (three or more workdays were lost). EEI was carried out when a physician identified an injury as being work-related and then called for EEI. Trained loss control specialists then visited the patient's workstation and carried out ergonomic assessments and employee training, including posture training and the concept of the "neutral" posture. Workstation adjustments were made where needed. Table 5.11 summarizes the main adjustments and recommendations made.

Out of a total of 201 injured employees, 72 received EEI, and the remainder did not. The number of employees who subsequently developed LTIs is given below:

	LTI	NLTI	Total
No EEI	24	105	129
EEI	5	67	72

TABLE 5.11
Main Workstation Adjustments and Recommendations in an EEI Program

Adjustment/Recommendation	Frequency (%)
Explain proper body alignment	100
Recommend breaks, stretching, etc.	100
Adjust chair	40
Adjust keyboard location	33
Adjust monitor location	31
Adjust the location of the phone/calculator	18
Obtain chair	16
Obtain a document holder	16
Obtain a wrist rest	16
Obtain footrest	14
Obtain a telephone headset	12
Adjust the mouse location	12
Obtain keyboard/mousetrap	10
Adjust work surface height	6
Obtain a slanting board	4

The OR for the risk of developing an LTI given no EEI is

$$\begin{aligned} \text{OR} &= \frac{24 \times 67}{105 \times 5} \\ &= 3.1 \end{aligned}$$

Thus, employees not receiving EEI were three times more likely to develop an LTI than those who did receive it.

The cost of the EEI per employee was calculated to consist mainly of the loss prevention specialist's wage at $38 per hour for an average 2.5 h visit, or $95. The average cost of an LTI was $4652 compared to $2959 for an NLTI—a difference of $1693. The estimated benefit/cost ratio is therefore 1693/95, or 17.8.

A more conservative estimate of the return on investment for EEI would also take into account some or all of the average wage of injured employees and the cost of the 2.5 h spent carrying out the EEI, which had to be done by observing the employee at work. The cost of any job aids purchased as a result of the intervention should also be included. Nonetheless, the return on investment would still be favorable.

Productivity and the Use of Bent-Handled Pliers

Duke et al. (2004) investigated the productivity benefits of bent-handled pliers compared to straight-handled pliers in two simulated industrial tasks, one of which required subjects to rotate the pliers and one which did not. Use of the bent-handled pliers resulted in a 5.3% improvement in performance in the latter task (with ulnar deviation significantly reduced). In the former task, performance was 4.9% faster when the straight-handled pliers were used.

Taken together, these findings indicate that the choice of the best tool for a task requires consideration of the task requirements, the workplace design, and worker characteristics, and that bent-handled pliers designed to reduce ulnar deviation may be beneficial in tasks that do not require rotational motions at the wrist.

Managing Musculoskeletal Pain in Aircraft Assembly

Melhorn et al. (2001) evaluated a risk management program for new workers. The program combined traditional occupational medical assessment and treatment of workers with musculoskeletal pain with medical data collection, surveillance to identify problem areas, employee education, ergonomic engineering, and design changes. The risk assessment and medical cost per injured worker in the program was $2468, compared to $3800 in a matched group of controls (the company-wide average cost was $2691). The program yielded a saving of $1332 per patient.

Shorter Workday

Wergeland et al. (2003) evaluated the effects of a reduction in work hours from seven or more per day to six per day (a 30-h week) in care home workers. Large reductions in neck and shoulder pain (but not back pain) were observed compared to a control group that did not receive a shorter workday. Since the reduction in work hours was not accompanied by a reduction in salary and extra personnel had to be employed, the cost-effectiveness of the approach is open to question.

Exoskeletons

Several types of exoskeletons are available for use with the upper limbs. They are designed to support the shoulder muscles, particularly during tasks that require shoulder elevation or the maintenance of shoulder posture when using tools such as drills (Figure 5.18). By supplementing the shoulder moment that elevates the arms, muscle activity around the shoulder is reduced.

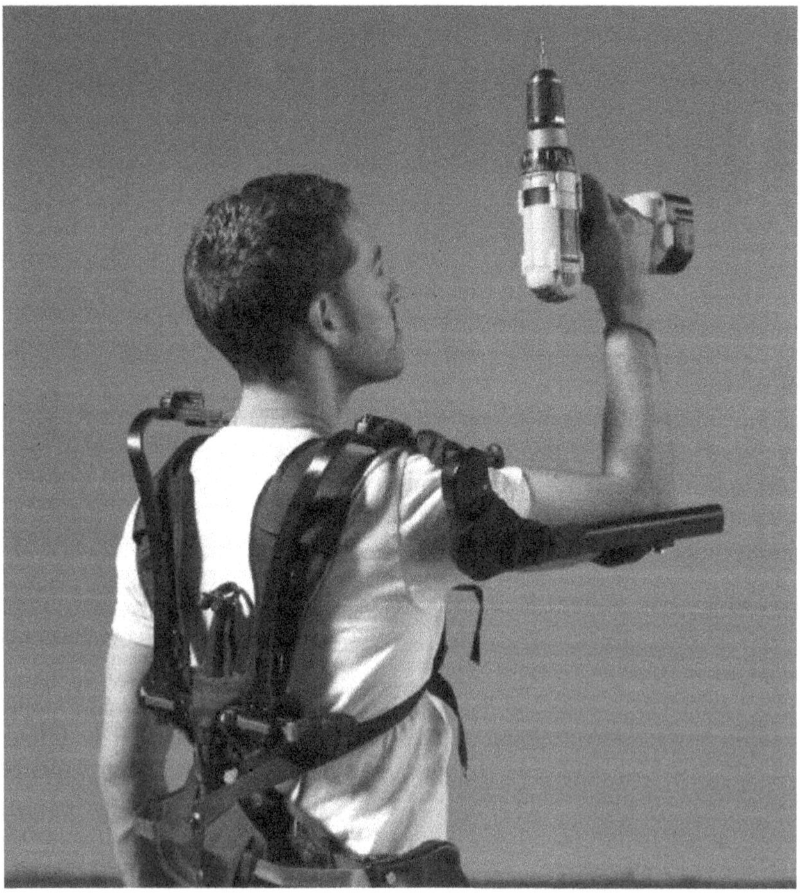

FIGURE 5.18 Arm support exoskeleton to support shoulders when drilling overhead.

Crea et al. (2021) used the following classification scheme:

- *Active*: Actuators supply an assistive torque and sensors and control units synchronize assistance with user movements.
- *Semi-Active*: Have different operational modes—e.g., powered assisted movement or rigid support to maintain a fixed posture.
- *Passive*: Springs or similar materials store and release energy to assist effortful movements.
- *Anthropomorphic*: The exoskeleton fits onto the limbs and/or torso, following the structure of the body.
- *Non-Anthropomorphic*: The exoskeleton is attached to specific limbs or joints by structures that are not attached to the limbs themselves.
- *Soft exo-suits*: These can generate torque by pulling on cables in parallel with muscle and tendon action.

In all cases, kinematic compatibility between the exoskeleton and the user is a requirement—the wearer must be able to move in a natural manner appropriate to the task because the movement of the exoskeleton mirrors the movement of the joint. One possible limitation in some tasks is that the exoskeleton may have fewer degrees of freedom than the joint it supports, causing unnatural movements or interfering with task performance.

De Vries and de Looze (2019) reviewed studies of arm support exoskeletons in realistic situations. Eleven studies were selected for review in which work activities were compared with and without an exoskeleton. Overall, the findings were mixed and depended to some extent on the design of the exoskeleton. Muscle activity in the anterior deltoid (AD, involved in shoulder flexion), the medial deltoid (MD, involved in shoulder abduction), the triceps brachii (TB, an antagonist involved in lowering the arm), and the iliocostalis lumborum pars lumborum (ILL, part of the erector spinae group) was measured.

Reductions in AD and MD activity ranged from 16%–73% in a variety of quasi-static (e.g., drilling) and dynamic tasks (e.g., lifting overhead). Increases in TB activity (up to 107%) were found (greater effort needed to lower the arms against the arm elevation moment due to the exoskeleton). The best results were found for passive, anthropomorphic exoskeletons, where subjective discomfort was rated lower in quasi-static tasks such as drilling.

Passive exoskeletons tend to weigh less than active exoskeletons and therefore place lower loads on other structures when worn. Kong et al. (2023) compared musculoskeletal load when wearing two passive exoskeletons versus a control condition in an overhead drilling task at working heights of 160 and 205 cm (Figure 5.19).

Subjects worked in 60-s cycles (40 s drilling and 20 s rest) using a 1.5 kg drill. EMG activity from the upper limb (upper trapezius, middle deltoid, biceps brachii, triceps brachii), the low back (erector spinae), and the lower limb (rectus femoris, biceps femoris, and tibialis anterior) was recorded. The exoskeletons weighed 2.5 and 3.2 kg, respectively. The upper limb muscles were expected to show lower EMG activity when the exoskeletons were worn. Low back and lower limb activity was expected to indicate postural or other loading changes as a result of wearing the exoskeleton and/or its effects on task performance. Comfort was measured on a 10-point scale. Muscle activity was significantly reduced when wearing the exoskeleton at both working heights. The reductions were greater at 205 cm for the middle deltoid, with the exoskeletons reducing by 38.3%–43.6% compared to no-exoskeleton conditions. At the working height of 160 cm, the reduction in middle deltoid muscle activity was 29.3%–34.5%. The exoskeletons significantly reduced muscle activity of the triceps brachii (54.8%–58.1%) at the 160 cm working height and 51.8%–56.3% at 205 cm. Reductions were greatest in the upper body muscles—from 18.7%–57.5% reductions in EMG in the upper trapezius, middle deltoid, triceps brachii, and biceps brachii). Reductions in EMG activity were also observed in the erector spinae, biceps femoris, and tibialis anterior (from 3.4% to 41.8%), suggesting that there were no adverse postural reactions to wearing the exoskeleton or its effect on task performance.

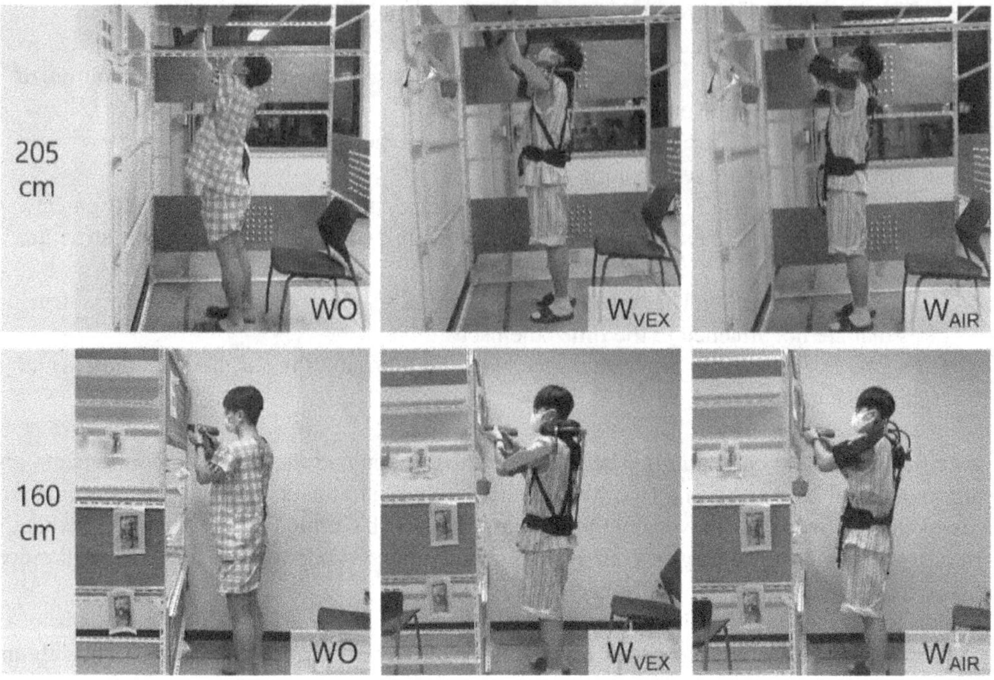

FIGURE 5.19 Evaluation of passive exoskeletons in an overhead drilling task. (With permission.)

Subjective discomfort was greater at 205 cm than at 160 cm and was lower when only one of the exoskeletons was worn. Wearing the exoskeletons had no effect on discomfort in the back or lower limbs. There was evidence that wearing the exoskeletons improved subjects' working postures, resulting in lower EMG activity in the low back and biceps femoris muscles. Although promising, these findings are from short experimental trials outside of an occupational context of use.

In practice, in order to select an exoskeleton for a task, the total movement profile must be considered, and not just the stressful movements or postures for which the exoskeleton is being considered. This includes movements required to reposition a tool or to take microbreaks, since antagonistic muscle activity might increase when these movements are made, canceling out the main benefits. In many jobs, the *context of use* of the exoskeleton needs to be considered, and issues such as ease of donning and doffing, activation/deactivation, and so on will arise.

There is plenty of evidence that upper limb exoskeletons can reduce muscle activity in both laboratory and field studies. However, a systems approach is needed to decide on their use in practice and to select an appropriate exoskeleton. Gull et al. (2020), in a detailed review, suggesting the following as key performance indicators:

- *Duration and Speed*: Maximum time and sustainable speed to perform the set of designed tasks with or without the exoskeleton must be considered and compared.
- *Ergonomics*: Measuring the comfort level while wearing an exoskeleton must be considered.
- *Pose Uncertainty*: It should provide a certain level of positioning accuracy and repeatability.
- *Others*: Cost, range of use, environment, and ease of use.

Finding exoskeletons that fit the task and user may increase productivity and quality while reducing injury and making physically demanding jobs more accessible to older workers and females.

RESEARCH DIRECTIONS

Demographic changes in some countries suggest a change in the capacity of the workforce due to aging, obesity, and a sedentary lifestyle, and these factors will also have to be incorporated into models of WMSDs. Much modern work is physically invariant and undemanding, based on tablets, smartphones, and other devices. Work can take place away from the workplace, and the workplace can be wherever work is done using these devices. There is evidence that, with lower physical demands, WMSDs manifest themselves differently—as a characteristic of people (Bridger and Brasher, 2011) rather than of job demands, either because of an idiopathic difference in the propensity to report problems or due to poor health, indicating that a "person-centered" approach may be needed when WMSDs arise. Emotional stress is thought to cause central sensitization, leading to chronic pain seemingly unrelated to work demands and injury sites. The role of psychological interventions such as relaxation techniques and cognitive behavioral therapy requires further investigation. Perceived muscular tension has been found to predict neck–shoulder and hand–wrist symptoms in office workers (Huysmans et al., 2012) over a period of 2 years (OR 4.4 for neck–shoulder problems in often tensed workers and 2.3 for hand–wrist symptoms).

Exoskeletons have been developed for a variety of tasks. The main research challenge is to "fit the exoskeleton to the worker and task." Reductions in workload are not enough; the key performance indicators of any exoskeleton within its proposed *context of use* have to be specified, and these include task performance, comfort, and safety such that the device is *integrated* into the job.

SUMMARY

Despite automation and computerization of work, WMSDs are still prevalent and repetitive work is common. The occupational risk factors for WMSDs, when work consists of simple physical tasks, are well understood and usually modifiable by better design. In office-based service industries, the role of physical risk factors is less clear. Dick et al. (2011) report that there is little evidence that altered keyboards with different geometry were effective against CTS; limited but good evidence that multidisciplinary interventions for nonspecific arm pain were beneficial; limited evidence that modified keyboards were effective in reducing symptoms of tenosynovitis; and insufficient evidence to inform workplace management of lateral epicondylitis. Valid and reliable tools are available to carry out risk assessments, but they are *tools, not rules*, and their interpretation should be based on a proper understanding of the task and the work environment, including psychosocial factors. The use of exoskeletons may reduce the risk of upper body MSDs in a particular task but must be compatible with all the other requirements of the particular job. User trials will be mandatory. New research indicates that it is the lack of *variability* in movement caused by high-repetition rates and constrained postures that underlies much of MSD development. Contrary to traditional views based on Taylorism, that there is one best way of carrying out a task, modern research indicates that *variety* must be built into task design.

TUTORIAL TOPICS

1. Does "RSI" exist?
2. If a worker experiences pain at work, it must be the work that caused it. Discuss.
3. "People should finish work in the same state of health as when they started." Discuss.
4. Take a look at the video at the link https://www.youtube.com/watch?v=NCzkOqMZDjk. Comment on the exercises shown—can you find any evidence for their effectiveness? Would you advise keyboard users to do these exercises during rest breaks?

ESSAYS AND EXERCISES

1. Obtain a number of commercially available screwdrivers and a large piece of hardwood. Using friends or colleagues as subjects, time how long it takes them to screw in a standard-size screw with each screwdriver. Use five screws per screwdriver per person and take the average time for each person per screwdriver combination (use a different screwdriver after each screw to balance out fatigue effects).

   Can any differences be attributed to handle or blade design? Is previous experience in the use of hand tools important?
2. Find the missing values in the table below (200 office workers are surveyed to determine the prevalence of wrist pain and whether their jobs require them to operate keyboards).

	Wrist Pain	
Keyboard Work?	**Yes**	**No**
Yes		5
No	3	
	30	170
Grand total = 200		

3. Refer to the table in Question 2. Calculate the prevalence rate of wrist pain in this group of workers. Next, calculate the OR for wrist pain when exposed to keyboard work.
4. Refer to the example in Workshop 5.2. Use the data below to recalculate the value of the SI for the fifth and ninety-fifth percentile operation times in the apple packing cycle. Are faster workers at increased risk?

Time to Pack 1 Layer (Seconds)	Mean	SD
	15	4

5. Which of the following tasks might provoke symptoms in a worker suffering from tennis elbow? List them in order from most likely to least likely.
   - Scrubbing debris from a brick wall using a wire brush held in the hand
   - Operating a computer keyboard
   - Painting a ceiling
   - Mopping a floor
   - Playing tennis

   State your reasons very briefly.
6. Which of the following might provoke symptoms in a worker suffering from CTS? List them in order from most likely to least likely.
   - Rapid data entry
   - Plucking a chicken
   - Playing a guitar
   - Kneading dough
   - Painting a ceiling

   State your reasons, very briefly, and justify the order you have chosen. Describe two personal characteristics of a worker that might cause symptoms to develop.

7. Which of the following tasks might provoke symptoms in a worker suffering from rotator cuff syndrome?
    - Trimming lawns with a powered "edge trimmer"
    - Pulling "wheelie bins" from in front of people's houses to a refuse collection vehicle
    - Cleaning "wheelie bins" with a handheld high-pressure water jet
    - Conducting an orchestra
    - Working as a "tree surgeon"
    - Typist

    List in order of risk (highest to lowest) and state your reasons very briefly.
8. Which of the following activities might provoke symptoms in a worker recovering from a neck injury caused by an accident while driving?
    - Word processing
    - Reading a book
    - Mowing the lawn
    - Bird watching
    - Lying in bed

    List in order of risk (highest to lowest) and state your reasons very briefly.
9. Which of the following jobs might be "strongly" associated with anterior knee pain in later life?
    - Sailor
    - Tiler
    - Chef
    - University lecturer
    - Administrator
    - Nurse
    - Farmer
    - Doctor

    State your reasons very briefly.

    State one personal characteristic of the individual that might increase the risk even further in your chosen occupation(s).
10. The occupational medical officer in a poultry processing plant wishes to reduce the high incidence of WMSDs. Which of the following would be best to focus on first of all?
    - Aerobic capacity of the workers
    - Lighting levels
    - Prevalence of repetitive tasks
    - Job content
11. Which of the following alterations to the design of a powered tool would be most likely to reduce the prevalence of WMSDs?
    - Fitting a high-friction sleeve to the handle
    - Changing from a pinch grip to a power grip
    - Switching from foot pedal operation to a hand-operated trigger
    - Designing a control that requires continuous force to operate
    - Increasing the power of the tool
12. List two occupational and two nonoccupational risk factors for CTS.
13. The occupational medical officer in an ice cream factory has asked you to carry out an investigation of shoulder pain in the shop floor workers. After assessing the tasks of all workers (both those who did and did not report to the clinic with shoulder problems), you obtain the following data:

Do You Load High Shelves at Work?	Shoulder Pain Present	Shoulder Pain Absent	
Yes	20	20	40
No	5	80	85
Total	25	100	125

a. What is the prevalence rate of shoulder pain in these workers? Calculate the OR with respect to loading high shelves and shoulder pain, and interpret the findings in a sentence or two.

b. What approach would you take to tackling this problem?

14. Which of the following might be caused by working all day at a computer with the mouse positioned too far to one side of the user and at an elevated position?
    - Pain in the lumbar region of the back
    - Neck pain
    - Pain on the opposite side of the body
    - Rotator cuff tendonitis
    - CTS
    - Wrist tendonitis
    - RSI
    - Nonspecific forearm pain
15. A thousand municipal employees involved in road, domestic refuse collection, and recycling are recruited to a 12-month prospective study of WMSDs. All are screened for existing WMSDs, and any chronic sufferers are excluded from the study. The remaining 863 are followed for a 1-year period, and all new cases of shoulder problems are recorded. Three hundred and thirty-seven of the employees worked, during the year, in the domestic refuse recycling center, where heavy items have to be lifted over a 1.3 m high ledge into a bin prior to being transported. Of these, 31 had at least 1 day off due to shoulder problems, compared to 24 workers in the other group who had not worked in the recycling center. Calculate the overall 12-month incidence rate for shoulder injury.

    Calculate the RR of incurring a shoulder injury in the recycling center.

16 A screening tool for WMSDs, developed in U.S. industry, is tested for its suitability for use on cruise ships by a major tour operator. These ships have crews made up of individuals from developing countries all over the world. The work/rest patterns are also completely different from shore-based U.S. jobs. The company wants to know whether the tool is suitable for risk assessing its operations in order to inform the designers of the next generation of ships.

    The results are as follows:

Actual Risk	Screening Tool Predicted Job Risk: High	Screening Tool Predicted Job Risk: Low
High	43	11
Low	40	85

    Calculate the sensitivity, specificity, PPV, and NPV of the tool and comment on the false positive rate. Would you advise the company to use this tool?

17. Mr. Smith has harvested some apples from his apple tree before going on holiday. He is peeling the apples in order to preserve them for the winter. Is it likely that he will suffer an upper extremity MSD as a result of carrying out this task? What reservations, if any, do you have about the results of your analysis?

QUICK QUIZ

1. Gout is a work-related disorder. True or false?
2. A prevalence of 20 cases means that 20% of the workforce is affected. True or false?
3. Incidence is the number of new cases over a time period. True or false?
4. Which of these is not a risk factor for carpal tunnel syndrome: (a) being female (b) highly repetitive work (c) alcohol consumption (d) obesity
5. The positive predictive value of a test is the proportion of cases identified by the test. True or false?
6. An odds ratio is a measure of risk. True or false?
7. Wrist extension is a risk factor for tendonitis. True or false?
8. Women have a higher risk of stress fractures than men. True or false?
9. Daily use of a computer mouse increases the risk of carpal tunnel syndrome. True or false?
10. Up to how many hours per week can a keyboard be operated without increasing the risk of WMSDs: (a) 10, (b) 15, (c) 25, (d) 40?

6 Lifting and Carrying

GENERAL REQUIREMENTS FOR HUMANS IN SYSTEMS

1. Manual handling of loads must be minimized and, where this is not possible, have a lifting index of less than 1.
2. Manual handling tasks must be designed to
   - Minimize the load moment
   - Allow maintenance of normal posture
   - Allow free and efficient gait
   - Allow minimal interference with breathing
3. In all tasks that require carrying, carrying aids shall be provided.
4. Lifting aids must be provided based on a risk assessment.
5. Push or pull forces shall not exceed the capacity of the target population.
6. Friction at the foot–floor interface must be sufficient to prevent slips both at heel strike and at toe-off.
7. Walkways and exits must be clear of slip, trip, and fall hazards.

Walking is a series of catastrophies, narrowly avoided.

Steindler, 1955

The average annual direct and indirect cost estimates per population for LBP ranged from \$2.3 billion to \$2.6 billion and \$0.24 billion to \$8.15 billion, respectively. The pooled annual rate of hospitalization for LBP was 3.2% (95% confidence interval: 0.6%–5.7%). The pooled direct costs and total costs of LBP per patient were USD 9231 (95% confidence interval: −7126.7125,588.9) and USD 10,143.1 (95% confidence interval: 6083.59–14,202.6), respectively (Fatoye et al., 2023).

CORE KNOWLEDGE: FUNCTIONAL ANATOMY AND BIOMECHANICS OF LIFTING AND CARRYING

Standing and Walking

The erect human body is a tall structure with a narrow base of support—its COG lies at more than half its total height and is at the mercy of environmental perturbations. Adults with fully developed postural reflexes are very stable, given that movement is not restricted.

Walking itself is a form of controlled falling (Figure 6.1). The body teeters on the brink of catastrophe as each leg swings through to save the ever-falling mass. Trips occur when the swinging leg is prevented from reaching its destination, and slips occur when there is insufficient friction between the weight bearing leg and the floor. Both problems result from a violation of a person's expectations or assumptions by some aspect of the workplace—steps, ramps, or ridges may be in unexpected places.

DOI: 10.1201/9781032717210-6

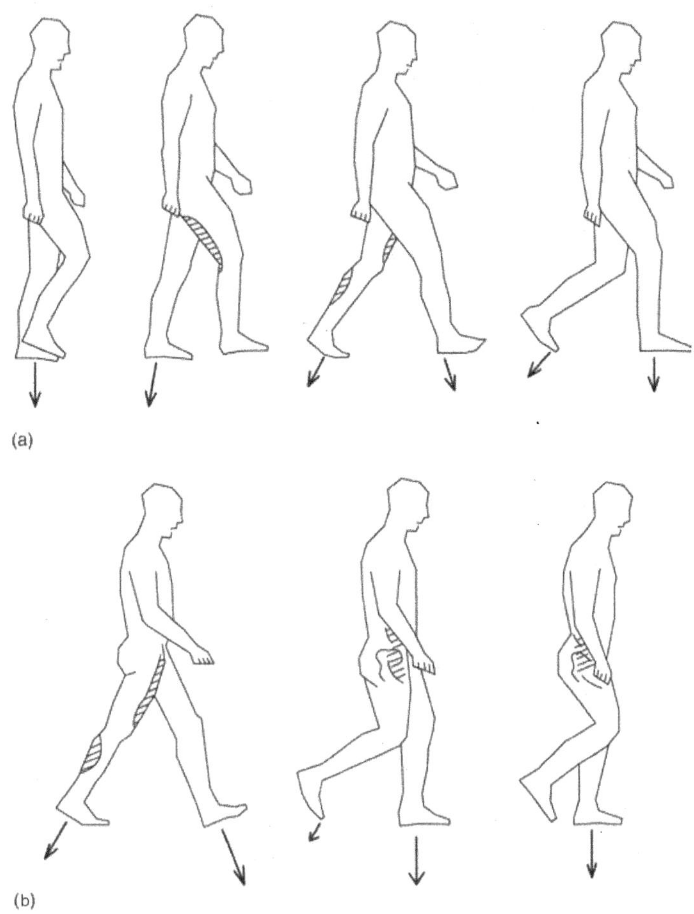

FIGURE 6.1 Mechanism of walking. The calf muscles plantarflex the foot, and the quadriceps straighten the knee at "toe-off" to propel the body forward. The weight shifts to the left foot. The iliopsoas muscle pulls the trailing right leg forward, and its knee bends passively. The hamstring muscles decelerate the swinging right leg so that the heel can be planted on the ground. The body weight passes along the length of the foot from the heel at "heel strike" to the big toe at toe-off—two critical stages of force transmission at the foot–floor interface. It is critical for stability that the swinging leg overtakes the body and is planted ahead of it, and that the heel does not move as it strikes the ground. Throughout the gait cycle, the reaction force must be approximately in line with the supporting leg(s), or the walker will lose balance. (Courtesy of Eadweard Muybridge.)

Biomechanics of Human Walking (Gait)

The phases of the gait cycle in Figure 6.1 are as follows: heel strike, mid-stance, toe-off, pre-swing, mid-swing, and terminal swing. Both feet are on the ground for about 20% of the gait cycle; the rest of the time, the mass of the body is supported by one leg. When the heel strikes the ground, there is a ground reaction force (GRF) according to Newton's third law of motion (the GRF in Figure 6.2). When a person is standing still, the GRF is equal and opposite to the person's weight. During walking, there is a horizontal component to the GRF parallel to the walking surface—the frictional force—and a vertical component, the normal force.

When the ratio between the frictional force and the vertical force is the same as, or exceeds, the coefficient of friction (COF) of the surface, walking is regarded as safe. Note that the ratio of these two forces depends on the angle of the leg with respect to the surface at heel strike. When taking very short steps (as, e.g., when walking with a "mincing" gait), the angle of the leg at heel strike is large; therefore, "F" is small and "N" large, and the ratio between them is lower than when

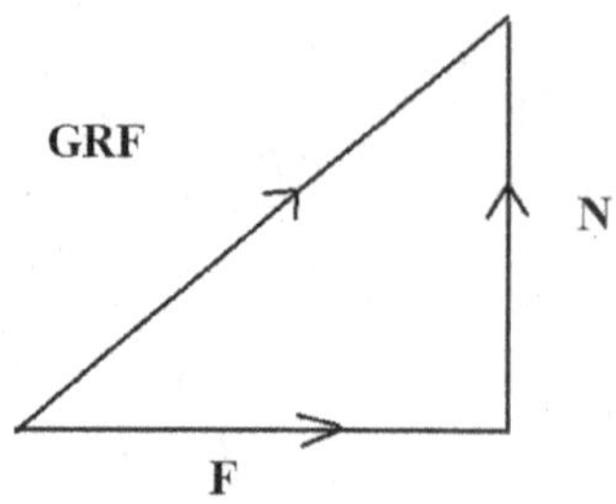

FIGURE 6.2 Resolution of forces at the foot-floor interface during the "toe-off" phase of gait. If "F" is too low, the foot will slip.

walking with a "striding" gait. This explains why we naturally tend to walk with shorter strides when walking on slippery surfaces (as long as we know they are slippery, of course!).

The same principle applies when walking down a slope—in effect, for the same stride length as on a horizontal surface, the angle of incidence of the leg at heel strike is smaller; therefore, "F" is larger and a greater COF is needed for safety.

Postural Control in Dynamic Tasks

The antigravity reflexes are centered in the hindbrain (pons and medulla) and cause automatic bracing of a limb when it is loaded by body weight. Although they are essential for the maintenance of an upright posture, they do not assist in the maintenance of equilibrium. A second set of postural reflexes (located in the basal ganglia of the midbrain) controls the posture of the various body parts in relation to each other (as in the postural fixation of limbs, for example) and of the whole body itself. The cerebellum is also involved in posture through the coordination of body movements.

The reflex control of posture can be distinguished from low-level reflex arcs, such as the "knee-jerk" reflex, and from voluntary movement. Mechanisms exist to protect the body against mechanical instability using feedback from the body itself to maintain balance. Feedback is obtained from both the semicircular canals in the inner ear and from sense organs in the muscles themselves.

In healthy people, the erect posture is characterized by postural sway, which can be monitored by movements of the center of pressure within the base of support determined by foot position. In quiet standing, the plantar flexors counter frontal sway, the tibialis anterior counters rearward sway, and the hip adductors and adductors counter lateral sway. For posterior displacements of the COG of the body (where the center of pressure moves toward the heels), the tibialis anterior and then the quadriceps femoris contract. A "hip strategy" is used to correct large perturbations or when the position of the feet is constrained by, for example, a lack of space. The paraspinal and hamstring muscles contract to thrust the pelvis forward, thereby compensating for the rearward displacement of the upper body and maintaining equilibrium. The hip strategy can be seen when someone is balancing on a narrow surface (such as a gymnast on the beam who momentarily loses balance).

Other Factors Influencing Postural Stability

Many factors influence postural stability and sway in healthy subjects. One cause of falls is the muting of, and delay in initiating, the startle response. The startle response is linked at a neurological level to the emergency reactions involved in recovery from losses of balance (Bisdorff et al., 1999). A delayed response to loss of balance may well make recovery less likely and increase the risk of injury should a fall occur.

Females have better balance than males—possibly due to the fact that their body's COG is lower. Of the various balance tests, standing on one leg while blindfolded is the most difficult, and standing on one leg with a blindfold is more difficult than standing with the feet together with a blindfold. These results are not surprising in view of the reduction in the size of the base of support and the

TABLE 6.1
Fracture Incidence (percent) in Relation to Falls in the Elderly in Relation to Floor Covering

Carpet 7 mm	Carpet 4 mm Loop	Carpet 4 mm Pile	PVC	Concrete
1	1.6	3.2	4	6.2

increased postural load on the supporting leg. Standing on one leg appears to increase the relative importance of visual feedback for postural control—even in young subjects—and has interesting practical implications. For example, *operating foot pedals from a standing position would appear to be contraindicated, particularly when the lighting is poor.*

Effects of Age

Shafizadeh et al. (2021) found that postural sway in children and older adults increased with task difficulty and became less variable. The addition of a cognitive task increased task difficulty, and the changes in sway in older adults were attributed to the increased attentional demands.

Age has been related to sway in some studies but not in others. Trips are common in older people because of age-related changes in gait that result in the feet not being lifted as high as they used to be and due to an increase in postural sway. Visual deprivation has a large effect on elderly subjects, contributing to about 50% of postural stability, possibly due to degraded feedback from stretch receptors in the muscle spindles. The loss of balance in older people seems to be due to increased reliance on visual feedback, which is slower than intrinsic feedback from the muscles. Tanaka and Uetake (2005) found that mediolateral sway increased in older subjects when they stood on soft surfaces with their eyes closed.

Table 6.1 presents data on fracture incidence in the elderly walking on different surfaces. Carpet or PVC floors should be laid in old age homes or in workplaces employing older workers.

Attentional Demands of Maintaining Posture

Maintenance of posture is almost effortless under ideal conditions and places no demands on attention (see Chapter 13). When the base of support is narrowed and when visual feedback is impaired, conscious attention is required to maintain posture. Remaud et al. (2012) found that reaction time to an auditory stimulus was greater when standing than when sitting, and greater when standing with the feet together or on one leg with the eyes closed.

Anatomy and Biomechanics of Manual Handling

During sagittal lifting (Figure 6.3), the trunk can be considered to be cantilevered over the legs at the hip joint. The hip extensors (hamstrings and gluteal muscles) extend the trunk to lift the load held in the hands. There are two sources of load on the body. One set of muscles acts to accelerate the load, and another to maintain the integrity of the spinal column and control the acceleration and deceleration of the trunk itself. The result of all this contraction and co-contraction is increased compression and shear on the vertebral motion segments.

The abdominal and thoracic muscles play a major role in stabilizing the spine when a weight is lifted. On its own, the spine is unstable. In relaxed standing, these muscles exhibit little activity. When a person leans forward to lift a weight, a moment of flexion is placed on the spine. The greater the load, the greater the flexion strain. The back muscles contract to resist the flexion (by exerting a moment of extension about the spine), and this is accompanied by antagonistic contraction of

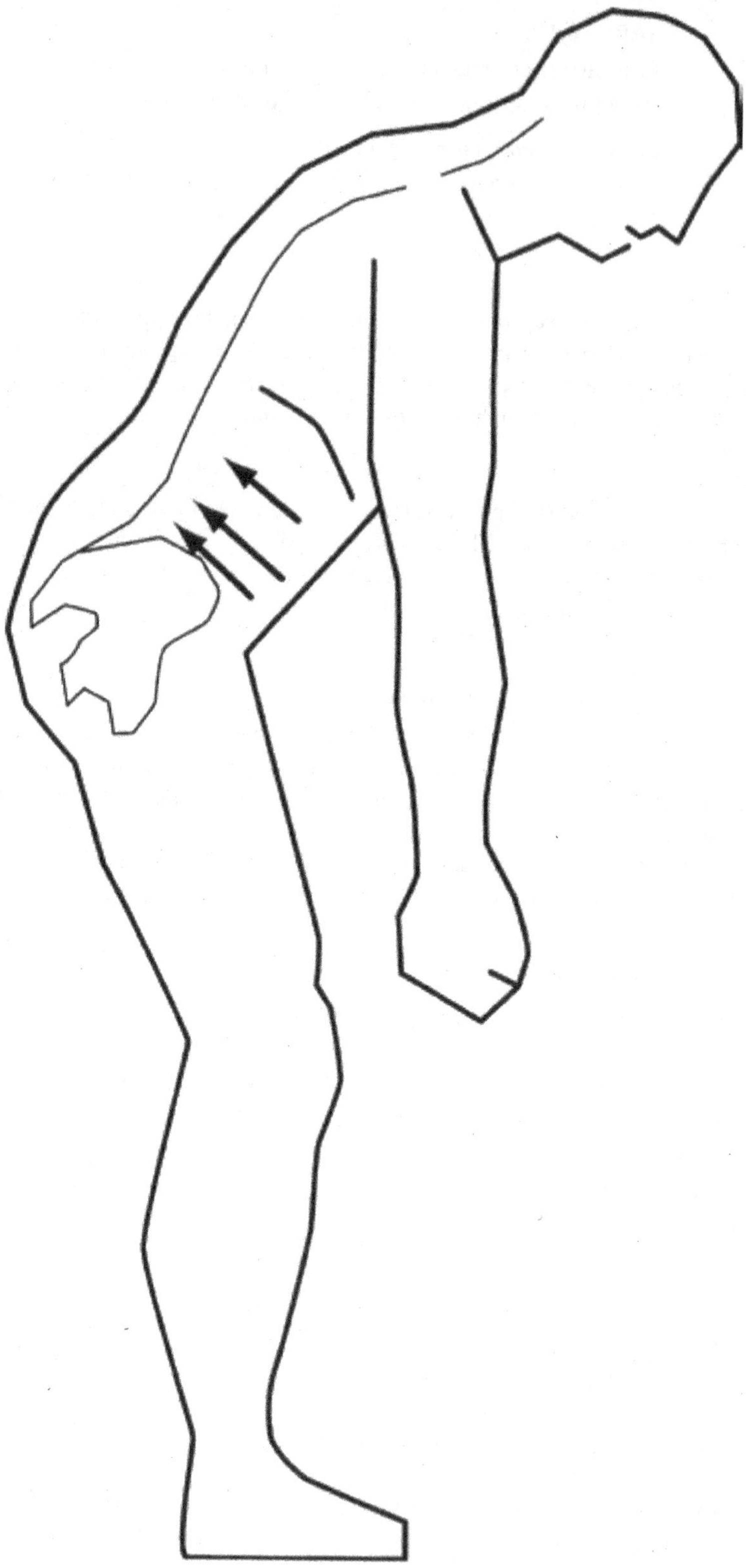

FIGURE 6.3 Abdominal mechanism in lifting.

the abdominal and thoracic muscles. These muscles pressurize the contents of the abdomen and thorax, possibly converting them into approximate hydraulic and pneumatic splints, which stiffen the trunk and stabilize the spine (Figure 6.3). If a deep breath is taken and the glottis is closed, the diaphragm moves downward and is held against the pressurized abdominal contents. The internal oblique and transversus abdominis muscles are spine stabilizers that pass from the anterior trunk to the lumbar spine, where it is thought that they act on angled fibers of the lumbodorsal fascia to extend the trunk (Norris, 1995). Multifidus, the short, deep back muscles, are also spine stabilizers and act to maintain the lumbar lordosis. The precise role of intra-abdominal pressure (IAP) is unclear because, when the abdominal muscles contract, they increase the compression on the spine. Increased IAP may simply be a side effect of trunk muscle co-contraction, wherein it is the contraction of the abdominal muscles that stabilizes the trunk and not the IAP itself. Alternatively, the pressurized abdomen may protect the spine by providing an additional pathway for the transmission of compressive forces to the pelvis. Aspden (1989) suggested that IAP compresses the convex surface of the lumbar lordosis, causing it to stiffen (in the same way that placing a load on a masonry arch increases its stability). Whatever the reason, the reflex increase in IAP when a person lifts a load appears to be a normal response.

It can be concluded that several abdominal and back muscles are spine stabilizers; they do not contribute directly to lifting a load. The gluteals, hamstrings, and erector spinae muscles are the prime movers during lifting (the knees may also be involved if knee flexion is required to access the load). However, neither the knee flexors nor the erector spinae muscles on their own are strong enough to enable loads to be lifted effectively or safely. Hence, the traditional advice to "lift with the knees and not with the back" is highly questionable.

Spinal compression is increased when loads are lifted and is increased even more when they are lifted quickly or when the posture is asymmetric. High load moments and high postural moments increase the forces even more.

Back Injuries and Lifting and Carrying

The trunk can fail in three ways when a weight is lifted:

1. Muscles and ligaments of the back can fail under excessive tension.
2. The intervertebral disc may herniate as the nucleus is extruded under excessive compression.
3. Abdominal contents may be extruded through the abdominal cavity due to excessive IAP.

These injuries are often referred to colloquially as muscle strains or tears, slipped discs, and hernias.

Lumbar Disc Disease (LDD)

Lumbar disc disease (LDD) is diagnosed when degeneration of lumbar discs causes symptoms. The main risk factors are genetics, ageing, smoking, obesity, lack of exercise, and jobs involving manual handling and/or pushing, pulling, and carrying loads or other people (as in health care settings). Schröder and Nienhaus (2020) carried out a systematic literature review of LDD in health care personnel. After reviewing seven studies, they found that health care workers had an increased risk of LDD after 10 years in the job—OR 2.45; 95% confidence interval (CI) 1.41, 4.26.

Discs with LDD lose fluid; the fluid itself may irritate surrounding tissues, and the discs are less able to resist the compressive forces of super incumbent body parts or external loads. The result is increased compression of surrounding structures, some of which have pain receptors. Over time, blood vessels may invade the degenerated disc, which may then become a source of pain. A catastrophic injury such as a disc prolapse is not simply caused by a sudden event such as lifting a heavy weight. It is usually the end product of years of degeneration of the disc and surrounding structures. According to Wolff's law, bone adapts to the mechanical demands placed upon it (being laid down where needed and reabsorbed where not needed) and bony spurs (or osteophytes) may grow around the intervertebral foramen due to LDD. Excessive loading of the facet joints may also cause problems.

A prolapsed intervertebral disc usually causes severe pain for a considerable time. Owing to the viscoelastic nature of the intervertebral disc, the prolapsed material does not return to its original position as soon as the load is released. Sometimes treatments such as lumbar traction can assist in returning the prolapsed material and also relieve pain in the process. Since the intervertebral discs have no direct blood supply in adults, recovery from injury is slow—the damaged tissues never heal completely because previously healthy tissue is replaced by scar tissue.

The term slipped disc is particularly misleading—the correct term is prolapse, as the nucleus pulposus is extruded posterolaterally through ruptured annular fibers. Since the posterior and lateral fibers of the annulus fibrosus are weaker than the anterior fibers (Adams and Dolan, 1995), herniation usually occurs posterolaterally. It would seem that bending to the side to lift objects would place the discs at increased risk, as lateral flexion loads the weaker parts of the disc. Added forward flexion would seem to increase the risk even more. As the load is grasped, the spine is subjected to compression, and when the load is picked up, the spinal extension may "trap" some nuclear material posteriorly, stretching the posterior ligaments and pressurizing nerve roots (causing referred pain in one leg, known as "sciatica"). This may serve as a basic description of the mechanism of disc herniation (Figure 6.4). The injury is one of prolapse of the annular fibers, rather than slippage of the body of the disc itself and to prevent it, from first principles, it would seem best to avoid complex spinal postures and motions in which forward flexion, lateral flexion, and rotation occur simultaneously. There is some evidence that this view is correct. Gagnon et al. (1993) investigated a lifting task in which subjects had to bend forward laterally and twist. They measured trunk postures and load moments when the task was carried out in this way and when a "pivoting" technique was used. The pivoting technique involved minimizing lateral flexion and rotation by making more use of the feet (turning with the lower half of the body). The trunk posture was safer when the pivoting technique was used, and twisting moments were reduced. For the technique to be usable, free space for the feet is needed in the entire lifting zone.

Manual handling injuries seem to be associated with unexpected events such as slips, trips, and falls. Marras et al. (1987) found that spinal loads were significantly greater (up to 70%) when the spine was exposed to sudden, unexpected loading. The lack of opportunity for physical preparation may cause the back muscles to overcompensate or increase the required forces to prevent loss of balance under sudden, unexpected loading. In either case, the implications are clear—workers should

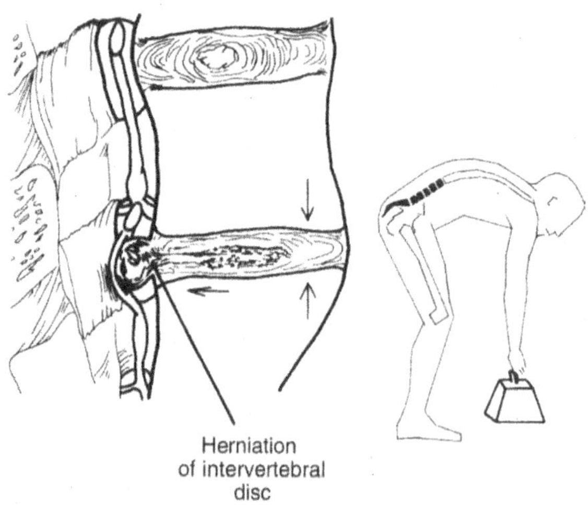

FIGURE 6.4 Basic mechanism of a slipped disc. The annular fibers rupture and the nuclear material is extruded posteriorly. (Adapted from Keegan (1953), with permission.)

be allowed plenty of time to carry out manual handling tasks, should not be distracted or pressurized by conflicting task demands, and should not be exposed to slip, trip, or fall hazards.

There is a strong genetic predisposition to lumbar disc disease (Doraisamy et al., 2021).

Inguinal Hernias

Inguinal (groin) hernias affect men mainly and occur when a portion of the bowel or fatty tissue protrudes through a weak spot in the abdominal wall. Kuijer et al. (2020) carried out a systematic review of studies on occupational risk factors for inguinal hernias. They found evidence for an increased risk when standing or walking for more than 6 h/workday or lifting more than 4000 kg/workday. For groin hernia management, limits of standing/walking for no more than 4 hours/day and lifting no more than 1000 kg/day were recommended.

BASIC APPLICATIONS

Foot–Floor Interface: Coefficients of Friction for Safety

One of the most important factors influencing the incidence of slipping and tripping is the design of the foot–floor interface. Floor materials, such as concrete, steel, earth, tile, and rubber matting, differ in their frictional properties, as do shoe materials such as synthetic rubber, natural leather, etc. The static friction between the shoe and the floor depends on this and on the contact area. High heels are more hazardous than flat shoes for this reason. Rubber and synthetic soles have higher friction on dry floors than leather soles, but not on wet floors. Higher friction is required on slopes, which suggests that, in addition to warning signs, slopes be designed with high-friction materials or lateral ribbing. Concrete floors can be laid with ribbing perpendicular to the intended direction of travel. NIOSH recommends that shoes and work surfaces be matched to result in a minimum coefficient of static friction of 0.5 and that there be smooth transitions between areas with different surface frictional properties. Swenson et al. (1992) estimate that the threshold COF for walking is between 0.2 and 0.4. A COF of 0.56 on a 1:20 slope is equivalent to 0.50 on a horizontal surface, as is 0.63 on a 1:10 slope and 0.86 on a 1:4 slope. Good housekeeping is needed to clear up spills, replace worn carpeting or missing tiles, and repair loose floorboards or cracked or pockmarked concrete surfaces. The U.S. Transportation Barrier Compliance Board recommends a minimum of 0.6 for people with disabilities and 0.8 for ramps. Slips do not always result in falls. People can normally recover and may not even notice "micro slips" of less than 3 cm. In fact, people can walk on ice if they know it is there, and the guidelines given above have a large margin of safety. The relative risk of falls when walking, turning on the left foot, and turning on the right foot is 1 in 20 for coefficients of 0.24, 0.27, and 0.29, respectively, rising to 1 in a million for coefficients of 0.36, 0.40, and 0.36 (see Curry et al., 2004, for further information). It is no surprise that the Liberty Mutual Insurance Company finds that falls are more likely when employees are forced to use an unfamiliar, are engaged in unexpected or non-routine activities, or are distracted or rushing.

Changes in floor materials or slope can be signaled with warning signs. Yellow-on-black stripes are often used to indicate hazards and can be painted on the floor at or close to the transition zone or along the length of ramps. Studies of falls among hospital workers by the Liberty Mutual Research Institute for Safety found that 40% of all falls occurred on contaminated floors (e.g., wet or slippery).

Slips, Trips, and Falls: Catastrophic Failure of the Erect Position

According to the World Health Organization (WHO, 2021), globally, falls result in 684,000 deaths and leave 172 million people with short- or long-term disability. Falls are the second leading cause of unintentional injury deaths and 80% of them occur in low- and middle-income countries. Seventy-five percent of fatal falls are among people over 70 years of age. The main risk factors are personal (e.g., physical characteristics and personal circumstances), environmental (e.g., lack of safety equipment, wet floors, and unsafe steps), and organizational/legislative (e.g., lack of legislation, poor standards, and poor enforcement). Postmenopausal, osteoporotic women are particularly at risk. High-risk occupations include cleaning, maintenance, transport, agriculture, warehousing, and material moving. In the United States, the health care sector has the highest incidence of non-fatal falls.

Lockhart et al. (2005) found that older people were no more likely to slip when walking on slippery surfaces, although they were more likely to fall because their recovery was slower. This may be due to sarcopenia (see Chapter 8), age-related loss of muscle, weaker leg muscles being unable to exert sufficient forces quickly enough to prevent a fall.

Hip fracture is common and very serious in older people when they suffer a fall, with a mortality rate of 17%–25% in the year following the fall. Ilic et al. (2023) report that in 2019, there were 400,000 cases of hip fractures in Central European countries. Females over 70 years of age were nine times more likely to suffer hip fractures after a fall than females aged 50–69 years, while in males, this difference was smaller (three-fold). Some of the worst back injuries are precipitated by sudden impacts transmitted through the musculoskeletal system when normal postural control mechanisms break down.

Approximately 6000 people die annually in the United Kingdom due to injuries caused by slips, trips, and falls. Two-thirds of these are elderly. The importance of visual feedback for postural control in the elderly has clear implications for the design of the visual environment in retirement villages, hospitals, and the workplace, particularly in countries undergoing demographic aging. *Good lighting on stairwells, hallways, and so on is clearly a priority, as is a consistent approach to the placement of light switches inside and outside buildings.*

Several researchers have attempted to establish optimum dimensions for the design of stairs. Irvine et al. (1990) concluded that riser heights (the height of the step) below 152 mm and above 203 mm should be avoided, and run depths (the distance from the front to the back of a step) less than 254 mm and more than 330 mm should also be avoided. Brown et al. (2023), in a field study, found that the addition of high-contrast black vinyl striping to the leading edge of steps (5.5 cm wide) slowed the rate of descent of users and may reduce the risk of falls. Other factors that slowed descent speed included holding something in the hand and interacting with another person. Older people also descended the stairs more slowly than younger people.

Smartphone "SMOMBIES": The use of smartphones by pedestrians is currently under investigation. When problematic smartphone use distracts attention from the external environment, "smombies" (smartphone zombies) are at greater risk of accidents. Texting while walking is thought to be the most distracting smartphone activity. Lai et al. (2024) investigated the effects of smartphone use on gait when walking up and down stairs and escalators. Overall, playing games or texting while walking down escalators resulted in the lowest walking confidence and the largest gait performance decrement: slower step frequency, reduced root mean square, decreased step regularity and step symmetry, and increased step variability.

Preventing Falls from a Height

Falls from a height are particularly common in industries such as construction. Some of the common risk factors that increase the risk of falls include

- Unsuitable equipment, including means of access and exit
- Use of ladders as a temporary means of access when scaffolding is required
- Unsafe foundations for scaffolding and ladders
- Lack of guardrails

- Walkways that are too narrow (lack of planks)
- Insecure fixings
- Equipment in poor condition
- Environmental factors (increased risk of postural instability: ice, wind, rain, etc.)
- Failure to wear safety lines/harnesses when required

Abdominal Belts for Manual Handling Safety: Health or Hoax?

Research has been conducted to investigate the claimed benefits of wearing tight-fitting elasticated or rigid belts during manual handling. The twofold aim of this research is to determine:

- Whether the occupational use of the belts confers some kind of biomechanical advantage that will protect the spine from injury in the workplace.
- Whether the occupational use of the devices has unwanted or unexpected side effects that negatively impact the health of the worker.

Table 6.2 summarizes the possible benefits and possible negative side effects of occupational wearing of abdominal belts in industry.

Do Abdominal Belts Increase IAP When Worn?

McGill et al. (1990) measured back extensor EMG and IAP when subjects lifted weights wearing a competition weight lifter's belt. Although IAP did increase when the belt was worn (from 99 to 120 mmHg), unlike when subjects held their breath, there was no corresponding reduction in back extensor activity.

Does Wearing an Abdominal Belt Reduce Back Muscle Fatigue When Lifting?

Ciriello and Snook (1995) measured fatigue of the back extensors in 13 male industrial workers who lifted average loads of 28.1 kg, 4.3 times per minute for 4 h a day. A belt was worn on two of the days. Maximum isokinetic endurance decreased by 9%–11% after 4 h of lifting. This change was not significantly different when a belt was worn.

TABLE 6.2
Possible Benefits and Potential Negative Side Effects of Abdominal Belts in Industry

Possible Benefits
1. Increased IAP and reduced spinal compression when lifting
2. Stabilization of lumbar motion segments
3. Stiffening of the lumbar spine due to increased IAP
4. Lifter "reminded" to avoid lumbar flexion and lift correctly
5. "Splinting" action of the belt. Dangerous motions, such as excessive sagittal flexion and axial rotation, are restricted because the belt stiffens the trunk
6. Increased sense of security and stability
Possible Hazards
1. Increased IAP but no reduction in spinal compression when lifting
2. Deconditioning of trunk musculature through long-term use
3. Increased blood pressure increases the risk of blackouts, stroke, or heart attack
4. Increased risk of trunk herniation
5. Increased sense of security and stability causes workers to take unnecessary risks
6. Management issues belts to "protect" workers instead of redesigning or mechanizing hazardous operations

Does Wearing an Abdominal Belt Have an Overall Protective Effect?

Reilly and Davies (2013) evaluated a weightlifter's belt by having subjects lift a 30 kg weight for eight sets of 20 repetitions. Spinal shrinkage as a result of exercise-induced loading was reduced by 49% when the belt was worn (from 4.08 to 2.08 mm). Perceived exertion was also lower. Magnusson et al. (1996c) compared lifting with and without a belt when subjects lifted 10 kg from the floor to desk height twice per minute for 5 min. Spinal shrinkage was lower when the belt was used, as was the back muscle EMG (normalized with respect to each subject's maximum voluntary contraction). Miyamoto et al. (1999) demonstrated that abdominal belts raise the intramuscular pressure in the erector spinae muscles and stiffen the trunk.

Does Wearing an Abdominal Belt Give Lifters an Increased Sense of Stability and Security?

McGill et al. (1990) and Magnusson et al. (1996c) report that wearing either competitive weightlifter's belts or abdominal belts for industrial workers increases the sense of security.

Does Occupational Abdominal Belt Wearing Decondition the Trunk Musculature?

The evidence suggests not. McGill et al. (1990) noted that even though belt wearing does reduce abdominal muscle EMG during lifting, the reduction is not large.

Is Abdominal Belt Wearing Hazardous for Workers with Latent Coronary Heart Disease?

It is known that both belt wearing and breath holding while lifting increase IAP and intrathoracic pressure. Hunter et al. (1989) had subjects hold 40% of their maximum weight in the deadlift posture for 2 min. Blood pressure and heart rate were higher when the belt was worn, leading to the conclusion that cardiac-compromised individuals are probably at greater risk when exercising while wearing back supports.

Does Abdominal Belt Wearing Cause Gastric Symptoms?

Over the course of an 8 h working day, belt wearing did not appear to cause gastroesophageal reflux in people with competent lower esophageal sphincters, as measured by objective and subjective means (Forouzandeh and Wright, 1998).

Do Abdominal Belts Protect Industrial Workers in Practice?

Van Duijvenbode (2009) concluded that issuing lumbar supports was not more effective than either training or doing nothing to prevent back pain, and there was no evidence that they were effective either on their own or as supplementary measures to other interventions.

Training People to Lift "Safely"

Figure 6.5 depicts two lifting techniques. Safety propaganda directed at workers usually describes these techniques as "unsafe" and "safe," respectively.

As was explained in the previous section, neither the knee extensors nor the back extensors can exert sufficient force to lift heavy loads. The gluteals and hamstring muscles which make up to 20% of lean body mass are the main extensors of the trunk. Thus, attempts to train workers to lift "safely" by squatting rather than stooping are unlikely to succeed or even be adopted in practice.

Clemes et al. (2010) conducted a systematic review of 53 intervention studies of manual handling training, finding little evidence that it was effective in preventing back pain and injury and considerable evidence that the principles taught were not applied in practice. Strength and flexibility training was found to show some promise. More recently, Hogan et al. (2014) reviewed 13 studies in which the effectiveness of manual handling training had been investigated, concluding that such training was ineffective in changing employees' manual handling behavior or in reducing back injury.

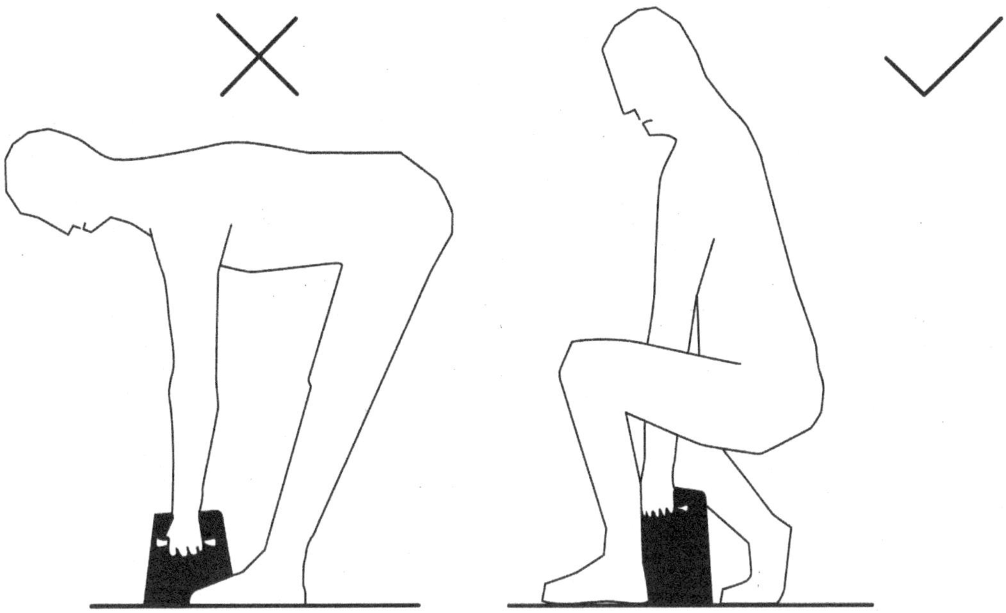

FIGURE 6.5 "Unsafe" and "safe" lifting techniques according to much safety propaganda. However, there is no evidence that they are safer.

Why Is Training in Lifting Technique So Often Ineffective? False Assumptions about Manual Handling Safety

Assumption No. 1

The way that untrained people voluntarily handle loads is unsafe. Allread et al. (2000) compared trunk motions in different jobs, finding that most of the variability was due to the design of the job rather than who was doing it or how frequently. Ferguson et al. (2004) compared trunk motions in 200 low-back-injured workers and 200 asymptomatic workers performing the same job. The risk of back injury was the same in both groups. This suggests that it is the design of the task that determines the risk, not who is doing the task.

Assumption No. 2

The techniques being taught are safer in practice.

Most weights in industry are not designed to be lifted from a squatting position, so it is no surprise that the squat technique is rarely seen. Rabinowitz et al. (1998) compared the squat and stoop lifts in a crate-handling task and found no difference in spinal shrinkage (a measure of spinal strain) between the two techniques.

Assumption No. 3

Safe techniques are usable and have no hidden costs. Squat lifting techniques require greater coordination and control than the alternatives and also place a higher load on the cardiovascular system and the knees. For repetitive lifting (e.g., unloading crates of beer), they soon take their toll. The knees weaken rapidly beyond about 60° of knee flexion, and the knee ligaments are at increasing risk of rupture. Rabinowitz et al. (1998) found that stoop lifting was associated with greater back pain, and squat lifting with greater knee pain. Repetitive squat lifting for 15 min placed an escalating cardiovascular load of an extra 26 heartbeats/min compared with stoop lifting. People rated the task as "somewhat hard," compared with stoop lifting, which was rated as "light."

Assumption No. 4

The training will transfer to the work situation.

The author knows of only one study that demonstrates long-term (6 months) change in lifting technique as a result of manual handling training. The hidden costs of these techniques may negatively reinforce the training, which is quickly unlearned in the workplace.

Assumption No. 5

Any reductions in risk are large enough to protect people from injury.

For most people, the spine is flexed to 50% of its maximum in a squatting position (Adams and Dolan, 1995). Training people to make more use of their legs does not guarantee lower back stress. Paradoxically, lifting with the back, or stoop lifting, is accomplished mainly by bending from the hips—the "hip hinge" method. A lowering of the load will only bring about a reduction in injury rates if the compressive forces are brought below the threshold. A reduction in risk is not the same as an improvement in safety.

Assumption No. 6

There are no perverse outcomes associated with the use of safe handling techniques.

Bridger and Freidberg (1999) interviewed 50 managers in a range of light industries and found that their estimates of maximum acceptable loads for their workers were over 50% higher when the managers were told that the squat lift was to be used. The implication is that manual handling training could act as a barrier to change by creating the impression that "something has been done." Similar arguments might be made about ergonomic improvements intended to reduce back injury—that employees will lift lighter loads with less care, thereby negating the effects of the improvement. Faber et al. (2007) did indeed find that when loads were reduced and pallet heights were increased, subjects in a lifting experiment "consumed" some of the increased safety by lifting from farther away, thereby increasing the load moment. For further discussion of behavioral adaptations to safety measures, refer to "Risk Homeostasis Theory" in Chapter 15.

Content of Safety Training Programs

One possible explanation for the dismal record of training is that the concepts of safe lifting embodied in the programs are incorrect. McGill and Norman (1992) offered some tentative lifting guidelines based on current knowledge of lumbar spine biomechanics (Table 6.3).

TABLE 6.3
Tentative Guidelines for Safe Lifting

1. Maintain normal lumbar lordosis and use the hip joints to flex and extend the trunk
2. Do not lift immediately after prolonged flexion, allow time for the disc nucleus to equilibrate and the posterior ligaments to regain their stiffness
3. Avoid lifting shortly after rising from bed
4. Follow the prestress system. Lightly co-contract the stabilizing muscles of the trunk to remove slack from the system and stiffen the spine
5. Choose a posture that minimizes the load moment on the lumbar spine (i.e., that gets you close to the load), but do not compromise 1
6. Avoid twisting (either to reach the load or to pick it up)
7. Exploit the acceleration profile of the load (experienced lifters only). Grasp the load and extend the trunk rapidly to impart explosive momentum. The "ballistic" effect will keep the load moving through weak postures until it can be "caught" in a strong, upright posture

Source: Adapted from McGill and Norman (1992).

Misconceptions about Lumbar Flexion and Risk of Injury

Saraceni et al. (2020) conducted a systematic literature review to assess evidence of a relationship between lumbar flexion and low back pain, derived from 12 studies. They found no evidence that lifting with a flexed spine is a risk factor for low back pain and cited evidence that spinal compression and shear are not greater when the spine is flexed. In a follow-up laboratory study, Saraceni et al. (2022) compared lumbar flexion of subjects with and without a history of back pain, finding that back pain sufferers lifted with less flexion than non-sufferers, against the view that lifting with flexion causes back pain. In another laboratory study, Mawston et al. (2021) found that subjects were able to exert greater trunk extension moments when adopting a flexed rather than a lordotic lumbar posture. Decreased electromyographic activity was found in the trunk extensor muscles when lifting with a flexed trunk. These findings may indicate the operation of the length–tension relationship, as the trunk extensors are lengthened in flexed postures.

Given the failure of training programs to reduce pain and injury, HFE specialists are advised to focus on task design—to "fit the task to the worker."

Design of Manual Handling Tasks

NIOSH has produced a work practices guide for the design of manual handling tasks and an equation for determining safe loads (NIOSH, 1994). In Europe, a directive for the design of these tasks was issued in 1990. The UK Health and Safety Commission (Health and Safety Commission, 1991) provides proposals for the design of manual handling tasks. Three principles of occupational medicine (first, remove the threat; second, remove the operator; and third, protect the operator) are usefully paraphrased in the context of manual handling (Table 6.4).

Task Requirements

Some factors that increase postural stress in lifting and carrying are given in Tables 6.5 and 6.6 (see Table 6.7 for advice on reducing time pressure in lifting tasks).

Lifting from a seated position deserves special consideration. The flexion moment about the lumbar spine is greater in sitting than in standing, and thus the spine can be considered to be "pre-stressed." Roman Lui et al. (2023), following a review of nine studies, concluded that lumbar

TABLE 6.4
General Principles for the Control of Manual Handling Hazards in the Workplace

1. Avoid hazardous manual handling as much as possible
2. Assess any hazardous operations and redesign the task to obviate the need to move the load or automate or mechanize the process
3. Reduce the risk by providing mechanical assistance, redesigning the load itself, or redesigning the workspace

TABLE 6.5
Some Task Factors Which Exacerbate Postural Stress in Manual Handling

1. Having to grasp or hold the load at a distance from the trunk
2. Having to twist the trunk while supporting or lifting a load
3. Having to lift or lower objects placed below the knee or above shoulder height
4. Having to lift or move the load through large vertical or horizontal distances
5. Having to hold or carry the load for long periods
6. Having to lift or carry frequently
7. Having to lift while seated

TABLE 6.6
Some Work Space Factors Which Exacerbate Postural Stress in Manual Handling

1. *Confined Spaces*: The ability to exert forces decreases when space is restricted. Restricted headroom is a good example. Less use of the legs is possible, which increases the load on the trunk muscles
2. *Height of Object*: Only items placed between the knee and elbow height should be lifted. Conveyors, shelves, and palletizing systems should be designed to allow for this
3. *Flooring*: Space for the feet should be provided both underneath the load and around the worker. Slippery floors should be avoided

TABLE 6.7
How to Increase the Time Available for Lifting

1. Increase the time by relaxing the standard time for the job
2. Reduce the frequency of lifts
3. Introduce job rotation to parcel out lifting between workers
4. Introduce appropriate work–rest cycles

disc compression was higher in upright sitting than in standing (about 30% higher) and increased with forward flexion of the trunk. In sitting, the leg muscles cannot be used to generate a GRF and more is required of the muscles of the trunk, which will increase compression further (the task load is added to the postural load). Li et al. (2022) came to similar conclusions after their systematic literature review but cautioned that the difference only applies to those with healthy intervertebral discs. Stooping from a seated position is particularly hazardous. It is curious that, in light of these facts, furniture designers still design desks with the largest drawer (to contain the heaviest objects) at the bottom, close to the floor.

The characteristics of the load itself should also be considered. Weight is not the only consideration. The size of the object is also important; 20 kg of lead is, in a sense, lighter than 20 kg of feathers since, being smaller, it can be held closer to the body. Containers for one- or two-handed handling should be designed as small as possible so that the load is kept close to the body. The provision of handles is important—well-designed handles can facilitate secure grasping of the load and prevent sudden movements of the combined person/load COG. Conversely, slippery materials or unstable loads (lacking rigidity) can increase the risk of injury.

Personal Characteristics of Workers

When designing lifting tasks, it is essential to consider the characteristics of the worker population. Owing to the requirements for task design described above, straightforward anthropometric considerations are not as useful in the design of lifting tasks as they are in seating, although a tall workforce will be better able to handle objects stacked higher, but not low objects, and vice versa for short workers. The following personal considerations are important.

People whose occupations typically do not require them to lift objects seem to be particularly prone to injury. It is undesirable to expect office workers, drivers, or professionals to lift heavy objects either occasionally or in emergencies. There is still some controversy over whether the level of fitness of workers is related to the probability of them suffering a back injury. Hyper- or hypomobility of the lumbar spine may be a risk factor for back problems, but empirical studies have failed to show a clear relationship between spinal flexibility and future risk of back trouble. Although women are generally less capable of handling heavy loads than men, sex is not always a useful consideration, since there is considerable overlap in the abilities of males and females in the workforce. Age should be considered, since it is known that muscular strength declines from middle age onwards.

If the workforce contains a large proportion of women, some special considerations are relevant. Generally speaking, manual lifting of heavy loads should be avoided because it increases stress on the trunk. Menstrual pain may incapacitate some women and the increased IAP when lifting may increase menstrual flow. It may also increase the risk of miscarriage in the first few months of pregnancy. Menopausal women often suffer from osteoporosis (demineralization of bone leading to a loss of bone strength), particularly in the spine and pelvis. This increases the risk of trunk failure when lifting a load.

Given that lifting is known to be a hazardous activity, and that there are many good reasons why women, in particular, should not have to lift heavy weights, the special case of back pain in the nursing profession deserves some consideration. Nursing is one of the few jobs where women have to lift heavy (and awkward) weights (i.e., patients) on a regular basis. In a survey of 569 nurses, Gilchrist and Pokorná (2021) found the 12-month prevalence of low back pain was 84.7%. The lower back was the most commonly reported body region for pain, followed by the neck, shoulders, and upper back. Lower prevalence was found among nurses who used mechanical lifting aids. Nurses with a higher level of education were more likely to use mechanical patient-lifting equipment (see ISO/TR 12296:2012 for further guidance).

Maximum Loads for Lifting and Lowering

Figure 6.6 from the UK Manual Handling Regulations gives recommended maximum loads for daily sagittal lifting. These can be used for initial screening of manual handling tasks for average workers, and it is permissible to extrapolate between loads if the hands are close to a boundary.

Manual Handling Outdoors: Effects of Wind on Postural Stability

Table 6.8 summarizes the effect of wind (measured at 1.75 m above the surface walked upon) on humans. People are at risk of falling when:

- Wind velocity suddenly increases to 15 m/s.
- They are exposed to a sudden gust of wind at 15 m/s (a condition that should be regarded as dangerous for the elderly and infirm).

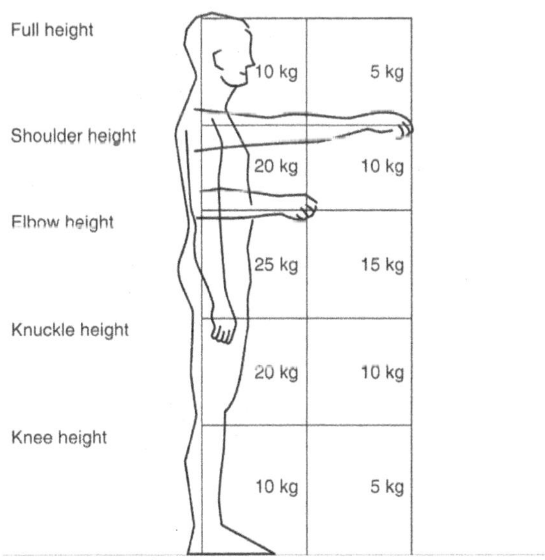

FIGURE 6.6 Recommended loads for lifting. (From the U.K. Manual Handling Regulations, with permission.)

TABLE 6.8
Effects of Wind on Postural Stability

Beaufort Number	Description	Wind Speed at 1.75 m	Effect on People
0	Calm	0.0–0.1	
1	Light air	0.2–1.0	No noticeable wind
2	Light breeze	1.1–2.3	Wind felt on the face
3	Gentle breeze	2.4–3.8	Hair disturbed, clothing flaps, newspaper difficult to read
4	Moderate breeze	3.9–5.5	Raises dust and loose paper, hair disarranged
5	Fresh breeze	5.6–7.5	Force of the wind felt on the body. Danger of stumbling when entering the windy zone
6	Strong breeze	7.6–9.7	Hair blown straight. Umbrellas used with difficulty. Difficult to walk steadily. Lateral wind force similar to the forward walking force. Wind noise in ears unpleasant
7	Near gale	9.8–12.0	Walking with difficulty
8	Gale	12.1–14.5	Progress impeded. Balance problems with gusts
9	Strong gale	14.6–17.1	People blown over

Gusts of 20 m/s are dangerous even for young people, and gusts of 23 m/s will blow most people over. These data can be of assistance when performing risk assessments of manual handling tasks carried out in outdoor workplaces such as construction sites.

Design of Carrying Tasks

Carrying increases the load on the body in two ways. First, the increase in weight increases the physiological cost of walking and the load on the muscles of the legs (the plantar flexors and the quadriceps). Second, the method by which the load is held or attached to the body can be an additional source of postural stress. Fatigue, discomfort, or injury may arise from either or both sources of loading. The second source is of particular interest in ergonomics since, through the design of more efficient carrying methods, the total postural load can, in principle, be reduced. From an energy expenditure viewpoint, an efficient method of load carriage is one that imposes little additional postural load over and above the extra energy cost per kilogram of load (Soule and Goldman, 1969). As a rule of thumb, people can safely carry about one-third of their lean body mass (DefStan 00–25).

Table 6.9 gives maximum acceptable loads for males and females when the load is carried in the hands.

Klausen (1965) investigated several methods of holding a 40 kg load and measured EMG activity in various muscle groups. He also compared spinal and pelvic posture in free standing and while holding a load, and noted any load-induced changes. Holding 20 kg in each hand was found to increase the electrical activity in the thoracic portion of the back muscles. It also placed a static load on the muscles controlling flexion of the fingers used to grip the weights. Holding the weights by means of a yoke over the shoulders caused a slight forward inclination of the spine and stooping. This was accompanied by an increase in back muscle EMG to counter the forward shift in the line of gravity of the superincumbent body parts.

The effects of carrying the load high and low on the back were also investigated using a carrying frame. When the load was carried high on the back, similar results to those using the yoke were observed. The body was inclined forward, and back muscle activity increased to exert an extensor moment on the trunk to counteract the flexor moment introduced by the load. When the load was carried low on the back, the subjects still adopted a forward-inclined trunk posture—presumably to maintain the combined COG of the load plus body weight over the feet. However, decreased EMG

TABLE 6.9
Maximum Acceptable Loads (kg) for Carrying

		4.3 m Carry						
		One Carry Every						
		10 s	16 s	1	2	5	30 min	8 h
Males	90	9	11	15	15	17	19	22
111	75	13	16	21	21	23	26	30
	50	17	20	27	27	30	34	39
	25	20	25	33	33	37	41	48
	10	24	29	38	39	43	48	57
79	90	11	14	18	19	21	23	27
	75	16	19	25	25	28	32	37
	50	20	25	32	33	36	41	48
	25	25	30	40	40	45	50	59
	10	29	35	47	47	52	59	69
Females	90	9	10	13	13	13	13	18
105	75	11	12	15	15	16	16	21
	50	12	13	18	18	18	18	24
	25	14	15	20	20	21	21	28
	10	16	17	22	23	23	23	31
72	90	10	11	14	14	14	14	20
	75	11	13	16	16	17	17	23
	50	13	15	16	16	17	17	23
	25	15	22	22	22	22	22	30
	10	17	19	24	24	25	25	33
		8.5 m Carry						
		One Carry Every						
		10 s	16 s	1	2	5	30 min	8 h
Males	90	10	11	13	13	15	17	20
111	75	13	15	18	18	20	23	27
	50	17	19	23	24	26	29	35
	25	21	24	29	29	32	36	43
	10	24	28	34	34	38	42	50
79	90	13	18	17	18	20	22	26
	75	17	20	24	24	27	30	35
	50	22	26	31	31	35	39	46
	25	27	32	38	38	42	48	56
	10	32	38	44	45	50	56	65
Females	90	10	11	12	12	12	12	16
105	75	12	13	14	14	14	14	19
	50	14	15	16	16	16	16	22
	25	15	17	18	18	19	19	25
	10	17	19	20	20	21	21	28
72	90	12	12	14	14	14	14	19
	75	14	15	16	16	17	17	23
	50	16	17	19	19	20	20	26
	25	18	19	21	22	22	22	30
	10	20	21	24	24	25	25	33

Note: See Snook and Ciriello (1998) for further information on acceptable loads for lifting and carrying.

activity was observed in the back muscles, while increased activity was observed in the iliopsoas muscles. The interpretation of these findings was that carrying the load low on the back introduced an extensor moment on the trunk (i.e., it tended to pull the trunk backward). Although this would reduce the activity of the long back muscles, the forward-inclined position of the trunk would only be maintained if increased activity in the iliopsoas muscles occurred to counteract the backward pull of the load. Thus, carrying a load low on the back would seem to be a relatively efficient method that decreases activity in the back muscles. This, in turn, would be expected to reduce the stress on the lumbar spine.

Cook and Neumann (1987) investigated the effects of load placement on low back muscle EMG when carrying and compared it with EMG activity when walking without an external load. They concluded that carrying with a backpack required the least additional muscular effort of the methods investigated and that carrying a weight in front of the body required the most effort. Carrying a weight in one hand induced asymmetrical loading on the spine—decreasing back muscle activity on the side of the load and increasing it on the unloaded side.

Soule and Goldman (1969) concluded that the energy cost per kilogram of load carried efficiently on the back is identical to the energy cost per kilogram of body weight; that is, no additional postural load is imposed.

Carrying a load in front of the body may also obscure the carrier's view of the floor in front and of potential hazards such as steps or obstructions, thereby increasing the risk of slips, trips, and falls. Thus, for many applications, loads should be carried low on the back and placed symmetrically. Finally, it is worthwhile to consider the practice of carrying loads on the head. In many countries, headload carrying is no longer practiced, but it can be readily observed in many developing countries, mainly by women, to carry heavy objects such as containers full of water, firewood, and cumbersome objects such as large baskets. Datta and Ramanathan (1971) found that carrying 14 kg on the head was approximately half as demanding as carrying 7 kg in each hand in terms of oxygen consumption. Soule and Goldman (1969) reported that headload carrying was only slightly more demanding than carrying a load on the torso at the same speed. The extra cost was due to the increased shoulder and neck muscle activity required to stabilize the head. The limiting factor in headload carrying appears to be the mechanical loading tolerated by the neck musculature rather than the total energy cost. Headload carrying is one of the most efficient of all methods of load carriage.

Rucksack Palsy: Excess pressure on the neck can damage the nerve roots at C5 and C6, causing numbness and pain. Sometimes, the serratus anterior muscle can be weakened, causing the appearance of a "winged" scapula. The serratus anterior muscle forms part of the scapulothoracic joint, which holds the scapula against the thorax. Rucksack palsy can be caused by badly fitting rucksacks and is less likely if the rucksack has a hip strap to enable the load to be transmitted directly to the pelvis rather than via the shoulders.

Metatarsalgia: Nonspecific foot pain near the ball of the foot. Usually caused by ill-fitting footwear, the risk can be increased when walking with heavy loads for long distances, particularly if the load is unbalanced. With heavy loads, people tend to lean forward when walking, which places additional load on the front of the foot. Balancing the load on the front as well as the rear of the trunk can restore a more upright posture and a more even distribution of load throughout the gait cycle.

General Guidelines for Carrying Loads

For increased efficiency and safety:

- Distribute loads evenly over the torso.
- Carry critical items with ease of access.
- When moving over flat terrain, carry the load higher on the back.

- Distribute the load between team members.
- Carry the minimum required for the task.
- Use trolleys, carts, and light vehicles when possible.
- Consider novel methods of delivering to teams to obviate the need to carry nonessential items: remote-guided land vehicles or drones.
- Avoid unnecessary movement.
- Carry loads at cooler times of the day.

Pushing and Pulling

Similar reasoning suggests that when using carrying aids such as trolleys, pushing rather than pulling would also minimize any additional back muscle load and therefore be preferable. When pushing a trolley, the reaction tends to extend the trunk and is resisted by the abdominal muscles and the hip flexors. When pulling a trolley (facing the trolley), the reaction tends to flex the trunk and is resisted by the back muscles. Laboratory investigations confirm the superiority of pushing compared to pulling (Lee et al., 1991). The compressive force on the L5/S1 disc was two or three times larger during pulling than during pushing. Body weight itself contributes more to the compressive load on the disc when pulling a load than when pushing it.

Weston et al. (2018) provide force limits to protect the back during pushing and pulling tasks in Table 6.10 (percentage of the population protected when pushing or pulling straight and when turning).

TABLE 6.10
Risk Limits for Straight and Turning Push/Pull Exertions to Protect against Back Injury (Weston et al., 2018)

			Straight	Turning
Exertion	**Height (cm)**	**% protected**	**Hand force (N)**	**Torque (Nm)**
One hand pull	81.3	90	212	58
		75	252	72
		50	295	87
		25	338	103
		10	378	117
	101.6	90	184	74
		75	217	88
		50	253	104
		25	289	119
		10	322	134
	121.9	90	184	78
		75	198	91
		50	235	108
		25	272	125
		10	306	141
Two-hand pull	81.3	90	203	71
		75	282	86
		50	368	102
		25	454	118
		10	533	133
	101.6	90	260	87

(Continued)

TABLE 6.10 (*Continued*)
Risk Limits for Straight and Turning Push/Pull Exertions to Protect against Back Injury (Weston et al., 2018)

Exertion	Height (cm)	% protected	Straight Hand force (N)	Turning Torque (Nm)
		75	306	104
		50	356	124
		25	406	143
		10	452	161
	121.9	90	301	96
		75	336	113
		50	374	133
		25	412	153
		10	447	170
Two-hand Push	81.3	90	185	72
		75	243	88
		50	295	105
		25	347	122
		10	395	138
	101.6	90	217	94
		75	250	109
		50	286	126
		25	322	142
		10	355	157
	121.9	90	290	96
		75	333	114
		50	380	133
		25	427	152
		10	470	169

TOOLS AND PROCESSES

Low Back Disorder Models, Risk Assessment, and Task Redesign

Marras and colleagues have developed a quantitative approach to back injury prevention using a device known as a lumbar motion monitor or LMM (Figure 6.7).

The LMM captures data on trunk posture, velocity, and acceleration in flexion, lateral flexion, and rotation. Together with data on the lifting rate and the load moment, a quantitative risk analysis can be carried out by feeding all the data into the low back disorder model (Marras et al., 1993). This model was determined empirically using data from 403 jobs differing in back injury rate. A combination of static and dynamic trunk posture variables predicts the risk of back injury:

- Maximum sagittal flexion
- Maximum lateral velocity
- Average twisting velocity
- Load moment
- Lift rate

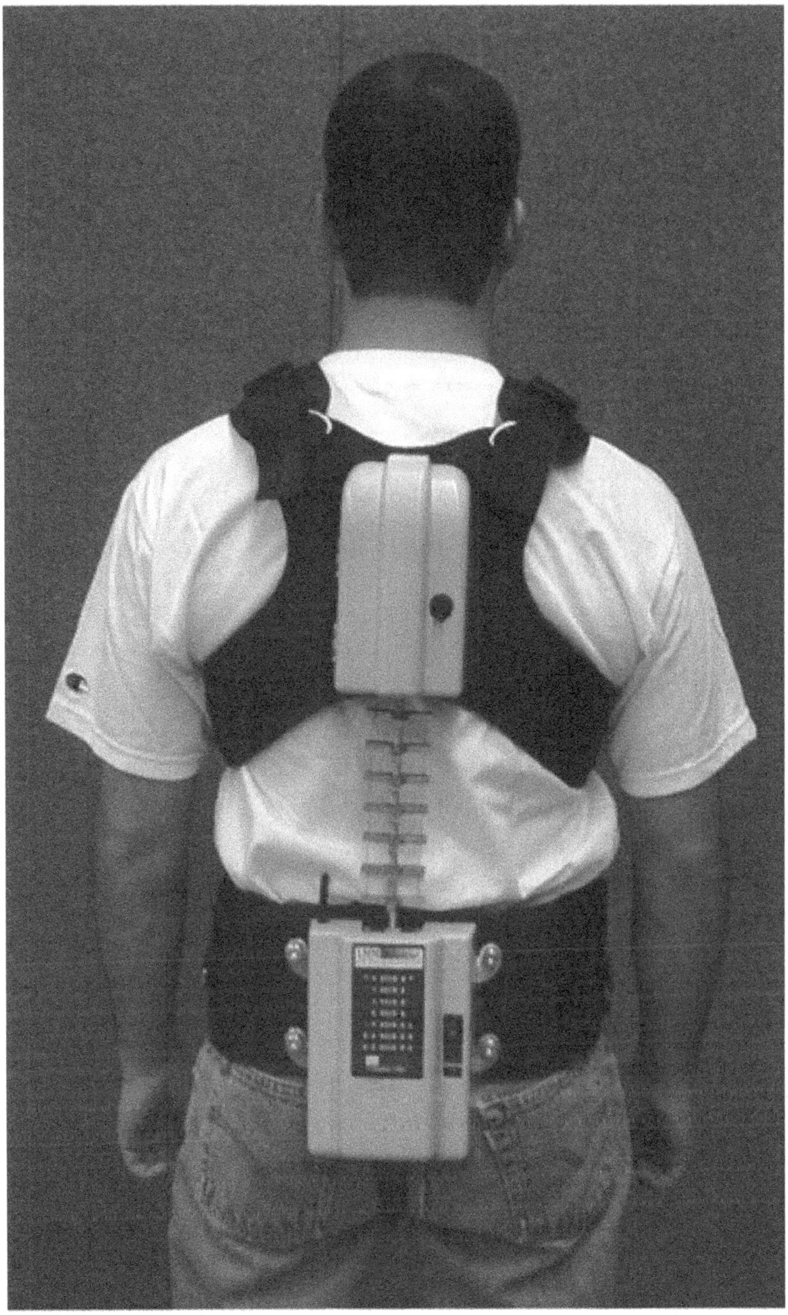

FIGURE 6.7 Lumbar motion monitor. (Courtesy of Professor W.S. Marras, Ohio State University.)

Use of the model in risk assessment is illustrated in a study by Bridger et al. (1998). Two designs of spade were compared—a conventional spade and a spade with a second handle fitted to the neck (Figure 6.8), intended to reduce the amount of stooping needed to dig. The output of the low back disorder model, using data for the two kinds of spades, is given in Table 6.11.

The data indicate that the hazardous components of digging with the conventional spade are the digging rate used in the experiment, maximum sagittal flexion (stooping), and the load moment. Lateral flexion velocity and twisting velocity are not hazardous. Although the redesigned spade does

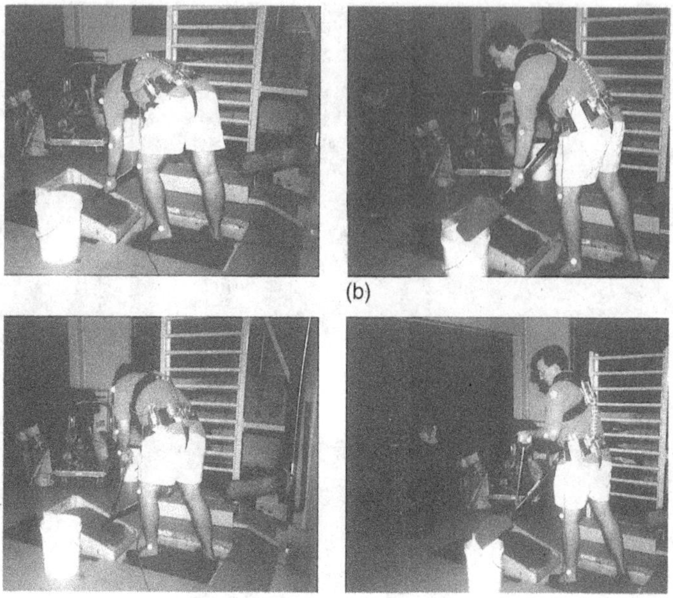

FIGURE 6.8 Using the LMM to evaluate lumbar motions using a conventional spade (a and b) and a spade with a second handle, designed to reduce stooping (c and d).

TABLE 6.11
Biomechanical Risk (Probability of Task Being a Member of a Group of Tasks with High Risk of Low Back Injury) Associated with Five Components of Digging

Components of Digging	Conventional Spade	Two-Handled Spade
Maximum lateral velocity	0.30	0.20
Maximum sagittal flexion	0.96	0.85
Average twisting velocity	0.06	0.13
Load moment	0.55	0.27
Lift rate	0.98	0.98
Overall probability	0.57	0.49

reduce sagittal flexion, the probability of high-risk group membership is still high, and the reduction is not enough to make the task safe. Interestingly, the reduction in stooping is accompanied by an increase in twisting velocity (although not to levels that markedly increase the probability of high-risk group membership), which suggests a compensatory recruitment of the trunk musculature. This example illustrates the use of the LMM and the low back disorder model to break a lifting task down into components of risk. The next step, in practice, would be to redesign the task to reduce the risk, starting with the component having the highest probability. In the digging example, this would be to lower the digging rate, followed by a reduction in sagittal flexion and load moment (Figure 6.9).

Validity of the Low Back Disorder Model

Marras et al. (2000) used the LMM to assess 36 jobs in 16 companies. Thirty-two jobs, deemed hazardous by the model, were redesigned. Significant associations were found between changes

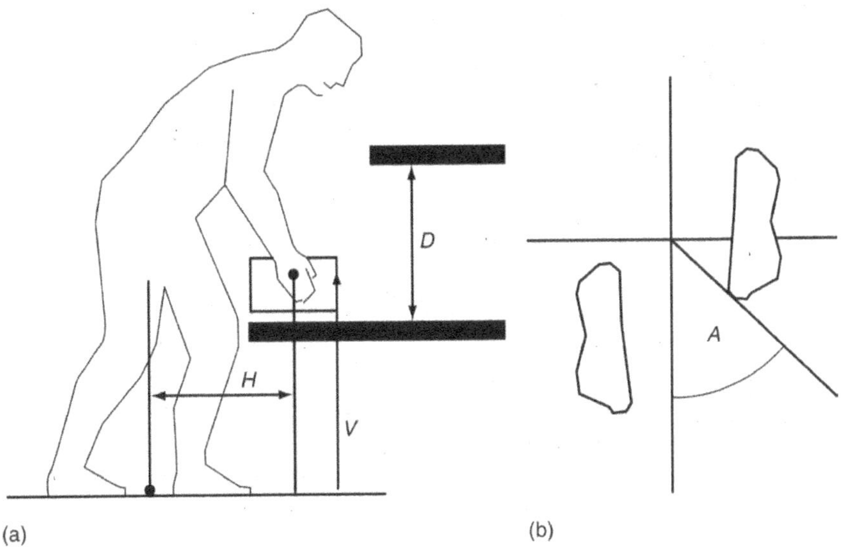

FIGURE 6.9 NIOSH approach to lifting task evaluation. (a) *H* is the distance of the load from the midpoint, *V* is the height of the hands above the floor, and *D* is the distance through which the load is lifted; and (b) angle of asymmetry of lift.

TABLE 6.12
Criteria Used in the NIOSH Lifting Equation

Consideration	Criterion	Cutoff Value
Biomechanical	Maximum disc compression	3.4 kN
Physiological	Maximum energy expenditure	2.2–4.7 kcal/min
Psychophysical	Maximum acceptable weight	
	For 75% of females and 99% of males	

in the low back disorder risk values and changes in low back disorder incidence rates. The most effective interventions seemed to be the introduction of lift tables (which bring the load closer to the worker) and lifting aids (that support the load).

The NIOSH Lifting Equation[1]

NIOSH has developed an equation for calculating the recommended weight limit (RWL) for a specific lifting task, which a worker could perform for a specified period without an increased risk of low back pain (see NIOSH, 1994). The equation has been constructed empirically. Three criteria were used to develop the equation (Table 6.12).

The use of three criteria is essential because different tasks impose different loads on workers. Infrequent handling of heavy, awkward objects is limited by biomechanical rather than physiological factors. Repetitive lifting of light objects is limited by metabolic stress and local muscle fatigue. Workers' perceptions of their capability are also a limiting factor. Sometimes, the different criteria will produce conflicting RWLs, in which case the lowest is used. NIOSH recommends a maximum load of 23 kg under ideal lifting conditions—the RWL is to be lifted in the sagittal plane from a height of 75 cm above the floor and held 25 cm in front of the body. The load is to be lifted no more than 25 cm vertically, there is to be a good coupling between the load and the lifter (which is achievable using handles or occurs due to the shape of the load itself), and the load is only to be lifted occasionally. Thus, the conditions for lifting a maximum load are specified.

TABLE 6.13
1991 NIOSH Lifting Equation

RWL = LC × HM × VM × DM × AM × FM × CM

LC = load constant of 23 kg

HM = horizontal multiplier = (25/*H*)

VM = vertical multiplier = (1 – (0.003|*V* – 75|))

DM = distance multiplier = (0.82 + (4.5/*D*))

AM = asymmetric multiplier = (1 – (0.0032*A*))

FM = frequency multiplier (from Table 6.14)

CM = coupling multiplier (from Table 6.15)

Where

H is the horizontal distance of the hands from the midpoint (0) of the ankles

V is the vertical distance of the hands from the floor

D is the distance through which the load is lifted, and *A* is the angle of asymmetry (Figure 6.9)

F is the frequency of lifting (lifts per minute every 1, 2, or 8 h)

All distances are in centimeters, and the angle of asymmetry is in degrees.

TABLE 6.14
Values of FM for Use in the 1991 NIOSH Equation for Determining RWL

	Work Duration					
	≤1 h		>1 but ≤2 h		>2 but ≤8 h	
Frequency (lifts/min)	V < 75	V ≥ 75	V < 75	V ≥ 75	V < 75	V ≥ 75
0.2	1.00	1.00	0.95	0.95	0.85	0.85
0.5	0.97	0.97	0.92	0.92	0.81	0.81
1	0.94	0.94	0.88	0.88	0.81	0.81
2	0.91	0.91	0.84	0.84	0.65	0.65
3	0.88	0.88	0.79	0.79	0.55	0.55
4	0.84	0.84	0.72	0.72	0.45	0.45
5	0.80	0.80	0.60	0.60	0.35	0.35
6	0.75	0.75	0.50	0.50	0.27	0.27
7	0.70	0.70	0.42	0.42	0.22	0.22
8	0.60	0.60	0.35	0.35	0.18	0.18
9	0.52	0.52	0.30	0.30	0.00	0.15
10	0.45	0.45	0.26	0.26	0.00	0.13
11	0.41	0.41	0.00	0.23	0.00	0.00
12	0.37	0.37	0.00	0.21	0.00	0.00
13	0.00	0.34	0.00	0.00	0.00	0.00
14	0.00	0.31	0.00	0.00	0.00	0.00
15	0.00	0.28	0.00	0.00	0.00	0.00
>15	0.00	0.00	0.00	0.00	0.00	0.00

Six coefficients have been developed that reduce the RWL to account for task factors that cause departures from the ideal situation. The RWL is multiplied by these coefficients to arrive at a new RWL for the specified conditions. The values of the coefficients have been determined using biomechanical models of spinal loading and the findings of epidemiological and psychophysical studies. The 1991 NIOSH equation is presented in Tables 6.13–6.15 together with definitions of the coefficients and the terminology used in the equation (Figure 6.9).

TABLE 6.15
Values of Coupling Multiplier for Use in the 1991 NIOSH Equation for Determining RWL

	Coupling Multipliers	
	V < 75 cm	V ≥ 75 cm
Good	1.00	1.00
Fair	0.95	0.95
Poor	0.90	0.90

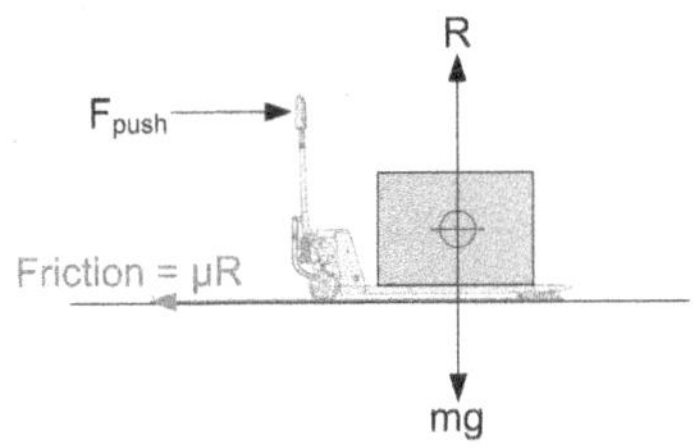

FIGURE 6.10 Relationships between push force, friction force, and weight (mg) when pushing a load (m) along a horizontal surface. (From Pisula (2015), with permission.)

The ratio between the load actually lifted and the RWL is known as the lifting index. An index of less than one is not believed to increase the risk of injury. An index of three or more indicates that many workers will be at increased risk, and the task should be redesigned. NIOSH's goal in developing the equation is to specify controls on industrial lifting that will protect healthy workers. It is not valid to apply the equation to set limits for those with already damaged backs.

Further details of the NIOSH method can be found in the publications of the Institute (e.g., PB91-226274), as well as in Waters et al. (1993). The above should serve to illustrate NIOSH's admirably rigorous approach to the pervasive problem of occupational lifting. The approach has the distinct advantage of lending itself to implementation as an interactive decision aid for the industrial engineer or designer. HFE Workshop 6.1 illustrates the application and interpretation of the NIOSH equation in practice (Figure 6.9, Tables 6.14 and 6.15).

Several Lifting Tasks

The NIOSH equation can be applied to situations where more than one lifting task is carried out, using a more advanced procedure to calculate the composite lifting index (CLI). The steps are as follows:

1. Calculate the single-task RWL and LI for the individual tasks in the usual way (Workshop 6.1) (STRWL).
2. Next, calculate frequency-independent RWLs and LIs for the tasks by setting the value of FM to 1. Call these the FIRWLs and FILIs.
3. Rank the tasks from most to least stressful using the LIs from Step 1.
4. Renumber the tasks according to severity.
5. Calculate the CLI as follows:

$$\text{CLI} = \text{STLI}_1 + \sum \text{delta LI}$$

where $\Sigma\text{delta LI} = \text{FILI}_2 \times \left(1/\text{FM}_{1,2} - 1/\text{FM}_1\right) + \text{FILI}_3 - \left(1/\text{FM}_{1,2,3} - 1/\text{FM}_{1,2}\right) + \ldots$

Ergonomics Workshop 6.2 gives a worked example.

HFE WORKSHOP 6.1

NIOSH EQUATION, A WORKED EXAMPLE

A worker has to unload trays of pork pies as they emerge from an oven. He picks up a tray, turns 45°, and places it on a conveyor. He does this three times per minute for 8 h. You have been asked to investigate the task and take the following measurements with a tape measure:

H = 45 cm, *V* = 60 cm, *D* = 70 cm, *A* = 45° (estimated angle of asymmetry by visual inspection).

Step 1. Calculate multipliers

$$HM = 25/45 = 0.56$$

$$VM = 1-(0.003|V-75|) = 1-(0.003|60-75|) = 1-0.045 = 0.955$$

$$DM = 0.82+(4.5/D) = 0.82+(4.5/70) = 0.82+0.06 = 0.88$$

$$AM = 1-0.0032A = 1-0.0032(45) = 1-0.144 = 0.856$$

FM (from Table 6.14) = 0.55
CM = 1 (from visual inspection, trays are easy to hold)

Step 2. Calculate RWL

$$RWL = 23\times 0.56\times 0.955\times 0.88\times 0.856\times 0.55\times 1 = 5.08 \text{ kg}$$

Step 3. Calculate the lifting index
A full tray weighs 12 kg
Lifting index = 12/5.08 = 2.36

INTERPRETATION

Although less than three, the lifting index is quite high, and the task cannot be described as safe. To redesign the task to make it safer, we should first examine the values of the multipliers, listing them in order from smallest to largest:

$$FM = 0.55$$

$$HM = 0.56$$

$$AM = 0.856$$

$$DM = 0.88$$

$$VM = 0.955$$

$$CM = 1$$

We find that the frequency of lifting and the horizontal distance contribute the most to the reduction in the load constant from 23 kg. Changing the lifting frequency will lower production and might be counterproductive if the weight of the trays has to be increased to compensate. However, redesigning the trays so that a full tray with the same number of pies can be held 25 cm from the body will change the value of HM to 1. The new RWL will then be:

(*Continued*)

HFE WORKSHOP 6.1 (*Continued*)

$$RWL = 5.08 \times 1/0.56 = 9.07$$

and the lifting index now becomes

$$LI = 12/9.07 = 1.32$$

Further reductions could be achieved by removing the asymmetry. As can be seen, the general approach to task redesign should be guided by the numerical values of the multipliers, beginning with the risk factors associated with the smallest multiplier values, as these will have the biggest impact.

HFE WORKSHOP 6.2

CALCULATION OF THE CLI

Suppose a worker has to carry out a composite manual handling task for 2 h—picking up an empty crate, picking up five fragile items, and packing them before loading the filled crate onto a conveyor. These three tasks are interdependent inasmuch as they are contingent upon one another and are subtasks of a main task called "Load Crate." A task analysis provides the following information:

		H	*V*	*D*	*A*	F (Lifts/min)	*C*	Mass kg[a]
1	Start							
2	Get an empty crate 1. Lift the empty crate from the stack 2. Place on the worktop	25	25	75	90	1	1	3
3	Fill crate 1. Pick up the item from the bin 2. Bubble-wrap 3. Place at the bottom of the crate 4. Repeat for the next four items	75	150	50	0	5	0.9	4
4	Load crate 1. Pick up a filled crate 2. Place on conveyor	25	100	25	90	1	1	23
5	Go to 2. Continue until 2 h has passed from Start							

[a] See Figure 6.9 for definitions of terms.

Step 1. Calculate RWLs and LIs for the three tasks

Task "Get crate" RWL = 23 × 1 × 0.85 × 0.88 × 0.712 × 0.88 × 1 = 10.8 kg

LI = 3/10.8 = 0.28

Task "Fill crate" RWL = 23 × 0.33 × 0.78 × 0.91 × 1 × 0.6 × 0.9 = 2.9 kg

LI = 4/2.9 = 1.38

Task "Load crate" RWL = 23 × 1 × 0.93 × 1 × 0.71 × 0.88 × 1 = 13.4 kg

LI = 23/13.4 = 1.72

(*Continued*)

HFE WORKSHOP 6.2 (*Continued*)

Step 2. Renumber tasks according to severity

Load crate = Task 1; Fill crate = Task 2; Get crate = Task 3

Step 3. Calculate frequency-independent RWLs and LIs (FIRWLs and FILIs)

$FIRWL_1 = 10.8 \times (1/0.88) = 12.3$ $FILL_1 = 3/12.3 = 0.24$

$FIRWL_2 = 2.9 \times (1/0.6) = 4.8$ $FILI_2 = 4/4.8 = 0.83$

$FIRWL_3 = 13.4 \times (1/0.88) = 15.3$ $FILL_3 = 23/15.3 = 1.50$

Step 4. Calculate the CLI

$CLI = 1.72 + \Sigma delta\ LI$

$\Sigma delta\ LI = FILI_2 \times (1/FM_{1,2} - 1/FM_1) + FILI_3 \times (1/FM_{1,2,3} - 1/FM_{1,2})$

where

$F_1 = 1$ (from 4 above)

$F_2 = 5$ (from 3b above)

$F_3 = 1$ (from 2 above)

and

$F_{12} = F_1 + F_2 = 1 + 5 = 6$

$F_{123} = F_1 + F_2 + F_3 = 1 + 5 + 1 = 7$

Then,

$FM_{12} = 0.50$

$FM_{123} = 0.42$ (from Table 6.14)

$$\begin{aligned}
\Sigma delta\ LI &= 0.83(1/0.5 - 1/0.88) + 0.24(1/0.42 - 1/0.5) \\
&= 0.83(2 - 1.14) + 0.24(2.38 - 2) \\
&= 0.73 + 0.09 \\
&= 0.82 \\
CLI &= 1.72 + 0.82 = 2.54
\end{aligned}$$

INTERPRETATION

The main component of physical stress in this lifting task is lifting the loaded crate onto the conveyor. This is largely because of the asymmetry involved. Redesign of the task should start with a better way of transferring the loaded crate onto the conveyor—perhaps a gravity-feed mechanism or a feeder conveyor to one side of the worker would suffice. Another large component of risk is the large horizontal distance involved in reaching for the items before packing them. The workspace should be redesigned so that the items are located within the zone of convenient reach (Chapter 3).

Validity of the NIOSH Lifting Equation

Dempsey et al. (2001) reported a prospective evaluation of the equation based on over 600 person-years of manual handling by 449 workers. The outcomes of interest included OSHA 200 reports and workers' compensation claims. The claim rate (claims for 100 full-time equivalent employees) varied according to the LI value (0.82 for LIs between 0 and 1; 1.31 for LIs between 1 and 2; 3.52 for LIs between 2 and 3; and 2.18 for LIs above 3). This is evidence that the equation has criterion validity when worker's compensation claims are the outcome to be predicted, though it may overestimate the added risk of tasks with an LI > 3.

Sensitivity, Diagnosticity, Intrusiveness, and Cost

Both the LMM and the NIOSH equations are sensitive enough to assess risk in simple manufacturing jobs. The LMM is intrusive because it has to be worn, and the actual lifting behavior may change as a result. Both are diagnostic. Compared to other techniques, the NIOSH equation offers greater interpretive capability to support specific ergonomic interventions for task redesign. Both techniques require special training, and the LMM requires special equipment and is therefore more expensive. The NIOSH equation is probably the most acceptable, but it is designed for monotonous, simple tasks, and many modern manufacturing jobs are not suitable for analysis, presumably because they are too varied/complex.

Lifting Fatigue Failure Tool (LiFFT²)

Gallagher et al (2017) use fatigue failure theory as an approach to the assessment of manual handling safety. Low back pain and back injury at work are seen as fatigue failure, failure as a result of repetitive loading over time, and can be assessed by measuring the load weight, the peak horizontal distance of the load from the trunk, and the repetition rate. Studies on cadaveric lumbar motion segments show that they exhibit fatigue failure under repetitive loading in the same way that other materials do (the greater the load, the fewer loading cycles before failure, and vice versa). The relationship between the magnitude of the load and cycles to failure is non-linear—under very high loads, at or beyond the ultimate compressive strength, failure occurs immediately; and at low loads, there is a point at which failure will never occur due to loading. Analysis of data from cadaveric studies at different percentages of the ultimate strength yielded the following relationship:

$$\text{Cycles to failure} = 902{,}416 * e^{-0.1628\%\,\text{US}} \tag{6.1}$$

Where %US is the percentage of the ultimate compressive strength of the motion segment, based on an average for U.S. males and females of 6 kN. To calculate the peak horizontal distance of the load from L5/S1, the greater trochanter is used as a landmark for L5/S1. This enables the peak load moment to be calculated if the mass of the load is known.

LiFFT has been validated by comparing its assessments with those of other models and against an epidemiological database.

Biomechanics of Pushing and Pulling Trucks and Trolleys

When two surfaces are in contact, friction opposes the relative motion between them. A box in contact with the floor can only be moved by pushing or pulling if the push or pull force exceeds the frictional force. There is a coefficient of static friction—which has to be overcome to initiate movement—and a coefficient of dynamic friction, which has to be overcome for movement to continue. When a worker pulls on a stationary box, the GRF at the worker's feet has two components—a vertical component and a shear component parallel to the floor (Figure 6.2). If the shear component exceeds the frictional force, the box moves. The magnitude of the shear force that can be applied depends on the friction at the foot–floor interface. If the COF at the worker's feet is lower than that between the box and the floor, the worker will slip and the box will remain stationary.

There are many good reasons for using trolleys (Figure 6.10) and barrows to transport loads, and one of the reasons is to minimize friction.

$$F = \mu R \tag{6.2}$$

where

F = push, or pull, force of static or kinetic friction (N)
μ = coefficient of static or kinetic friction (dimensionless)
R = normal reaction force between the two surfaces (N)

$$R = mg \tag{6.3}$$

where

m = mass of object being pushed or pulled (kg)
g = acceleration due to gravity (9.81 m/s^2)

If we know the magnitude of the applied force and the weight of the load being pushed or pulled (mass of trolley plus mass of load multiplied by g), we can estimate the coefficients of static and dynamic friction.

$$\mu = \frac{F}{mg} \tag{6.4}$$

In practice, it is usually the case that we do not know either the friction coefficients of trolleys and barrows, so although we will know the load to be carried and the mass of the trolley, we will not know the required push forces. These forces can be measured using load cells (Figure 6.11).

Implications

The HFE literature contains guidance on the maximum push and pull forces for males and females. However, this information does not indicate what kinds of loads can be pushed or pulled on trolleys and other devices. Figure 6.12 (from Pisula, 2015) shows the forces needed to push a loaded trolley (mass of trolley plus load: 845 kg) from rest, along a flat surface, and up a 3.5° ramp. Clearly, the forces are much lower than the load itself, which demonstrates the effectiveness of even the simplest manual handling aids in reducing the effort required.

Equation 6.2 makes it clear that more work can be done using lower forces if friction forces are minimized. In practice, almost anything that lowers the friction at the trolley–floor interface will increase the likelihood of workers slipping and may be self-defeating. More effective options include optimizing the size of the wheels to reduce rolling resistance; proper maintenance of trolleys and lubrication of wheel bearings; optimum handle height or adjustable/pivoting handles so that operators can use their own body mass effectively; and level surfaces and elevators to lift trolleys between different levels.

Identify the Risk Factors in Pushing and Pulling

Task-Trolley-Operator Interaction

- High forces/great effort to initiate movement
- High effort to maintain movement
- Sudden movements—starting and stopping
- Twisting of the body to maneuver the trolley
- One-handed operation
- Hands below waist or above shoulder height
- Movement over long distances

Design of Load

- Adequacy of handholds
- Stability of load—ability to secure the load on the trolley
- Operator's line of sight around the load

Wheels/Castors

- Suitable for the type of load?
- Suitable for the flooring?
- Ease of steering?
- Brakes fitted and effective?

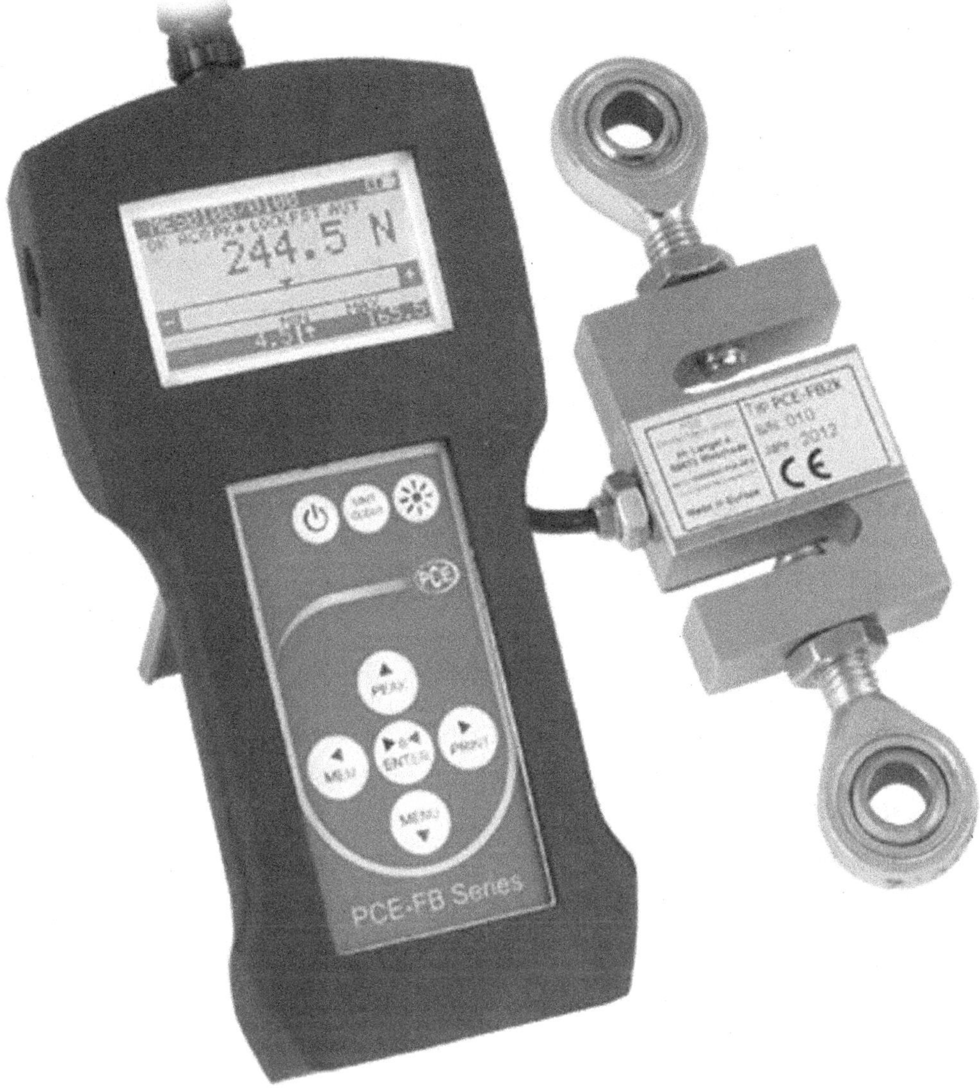

FIGURE 6.11 PCE load cell.

Work Environment

- Constraints on body positioning?
- Passage spacious and clear?
- Rutted, damaged, or slippery floors?
- Ramps, slopes, or uneven ground?
- Trip hazards
- Lighting adequate?
- Wind?

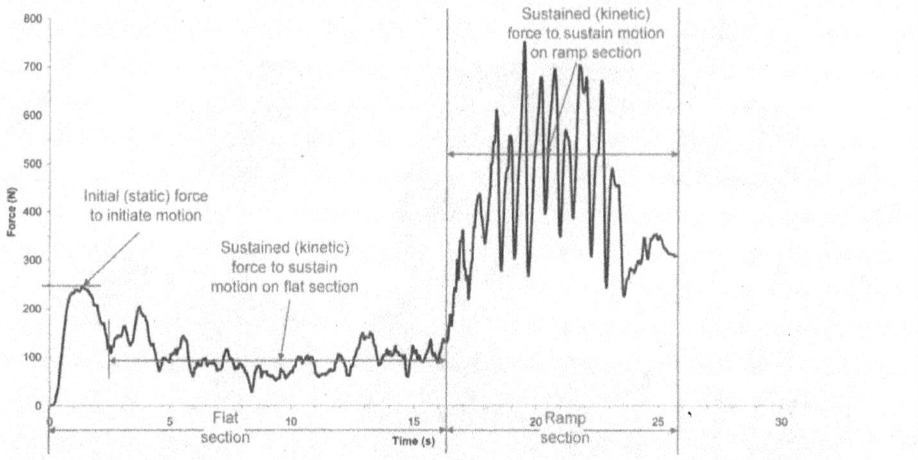

FIGURE 6.12 Plot of parallel- to-surface (horizontal) force as a function of time when pushing a manual pallet truck with a combined load of 845 kg along the flat and ramp section from Pisula, (2015), Crown Copyright (contains public sector information licensed under the Open Government Licence v2.0).

Other

- Compatible with personal protective equipment?
- Operation and operating instructions clear and reasonable?
- Have workers received necessary training to operate the controls?

SYSTEM INTEGRATION

Prevention of Falls

Garrone (2001) provides evidence for the cost-effectiveness of fall prevention programs in retail operations, where 17.1% of workers' compensation claims and 22.4% of costs were due to same-level falls in 1996–1999. Such programs are of general interest and are of particular relevance to the prevention of falls during manual handling operations. The main risk factors for falls in retail operations are

- Floor surfaces (when dry and when wet)
- Footwear (in particular, the interaction of footwear and flooring)
- Water/objects on floor (water, snow, or ice from entrances, drinking fountains, or from leaks in ceilings, clips, bags, or packaging/refuse)
- Obstructions/poor design (transitions from tiles to carpeted areas, from one level to another, display pedestals too low and below the field of view, obstructions that obscure other hazards)
- Distractions (displays designed to attract attention)

The prevention program aimed to identify and eliminate fall hazards. Employees were educated about the causes of falls and held accountable for keeping their workspaces free of hazards. Feedback was provided as improvements were made. Additional safety measures included

- Extending mats at entrances on rainy days to prevent the ingress of water
- Placing mats below drinking fountains

- Enforcing a minimum 75 cm distance between display racks
- Reducing the use of electrical extension cords and running them through covers
- Providing paper towels for employees to clear spillages
- Monthly self-inspections were carried out using a checklist to assist in the identification of hazards

The program was implemented in one retail outlet in 1997 at a time when there were 0.7 falls per $1 million worth of sales, costing the company $17 million in accident losses. In the following 2 years, the fall rate dropped to 0.62 falls per $10 million and then to 0.48 falls per $10 million, yielding an estimated savings of $3 million.

Manual Handling Training (MHT) in Organizations

Manual handling ("how to lift," or HTL) training persists in many countries despite the lack of evidence for its effectiveness and cost-effectiveness. Oakman et al. (2024) conducted a questionnaire survey of organizations in Australia to understand more about the beliefs held by these organizations and how MSD prevention is managed. The legislative background requires employers to consider and manage all the risks associated with manual handling (as described in previous sections of this chapter) and implement controls (Table 6.4). These controls should either eliminate the hazard, mitigate it, or reduce the risk of exposure to it. HTL is not regarded as a control legislatively because it doesn't focus on the working conditions that are hazardous (training workers about the implementation of the controls in place is regarded as a control measure). Eighty percent of employers reported using HTL training in the last 2 years, and 62% of workers had received it. Over half of the respondents believed it to be effective in itself, and over 85% considered it to be part of a wider safety program. Three-quarters believed it to be a legislative requirement, and some believed that the provision of HTL training could be used as a defense in civil court cases. The use of HTL training was more common in large organizations and where it was believed to be a legislative requirement.

Apart from HTL training, 54% of employers used other methods to mitigate exposure to hazards rather than eliminate them. These included mechanical lifting devices; trolleys, pallet jacks, forklift trucks; and job rotation to reduce exposure to repetitive work.

There was also evidence that employees themselves believed the training to be effective. Over 60% of workers responding to the survey had participated in HTL training. Whether they were receiving HTL training or not, workers strongly believed it to be an effective way of reducing injury. For those who had received training approximately 80% believed it had been effective.

Among employer comments, a view was expressed that the legislation requires employers to manage risk, and the provision of training is one way of demonstrating that the risk is being managed. This, of course, is from the perspective of legislation and corporate liability, rather than science. Clearly, this research puts the application of HFE into an organizational perspective. HFE specialists must take a broad view of the *context of use* of their advice, and this includes the organizational and legislative implications of any changes.

Does Manual Handling Add Value?

It can be argued that manual handling of goods does not add value and should be minimized wherever possible. This can be challenged, of course, by arguing that the handling of components or products is part of the production process and adds value all the way through to delivery. If you cannot deliver the finished product to the right person, at the right place, at the right time, how are you going to generate revenue?

The answer lies in the distinction between the process and the way the process is implemented. Manual handling is often the default method of moving products around or delivering a service. HFE cannot prevent everyone who always lifts a load from getting back pain, but it can prevent acute back injury in high-risk tasks. Since manual handling tasks are part of the production process, they are not seen in the same way as other non-value-added costs (such as insurance). Only when such operations are disaggregated from other financial data can we start to make some predictions about the benefits of redesign. There are three types of health- and safety-related costs that need to be considered when looking at manual handling operations (Riel and Imbeau, 1995):

- Insurance-related costs (including legal and medical expenses and direct insurance costs)
- Work-related costs (delays, "dead" time, and productivity loss due to less-than-optimal work conditions)
- Perturbation-related costs (activities and other cost effects that are consequences of poor design, such as accidents or the need to reschedule)

These costs must be disaggregated for the operation in question. For example, we can estimate the total current cost of employing individuals to handle loads manually. From historical data, we can determine any production shortfall in relation to the standard of performance set for that part of the operation involving manual handling. Perturbation costs might have to be estimated in other ways and are an example of "activity-based costing." These costs can be difficult to estimate, but they can be large because accidents, for example, tend to have a ripple effect through an organization and trigger all kinds of non-value-added activities that are not routine and often involve external authorities or the hiring-in of specialized services. Once an estimate has been made of the costs of an operation, as well as of the additional costs of any accidents or stoppages, an economic case can be made for investment in ergonomic improvements.

Exoskeletons

New developments, particularly the use of exoskeletons to assist manual handling of loads (Figure 6.13), may present system designers with new opportunities for improved efficiency and safety. The device shown in Figure 6.13 weighs less than 30 kg and can lift to 30 kg for several hours on one charge.

Kermavnar et al. (2021) reviewed 33 studies of back support exoskeletons (two were field studies). Wearing an exoskeleton (passive or active) resulted in significant reductions in back muscle activity in 22 of 27 studies where EMG was measured. For passive exoskeletons (Figure 6.14), the

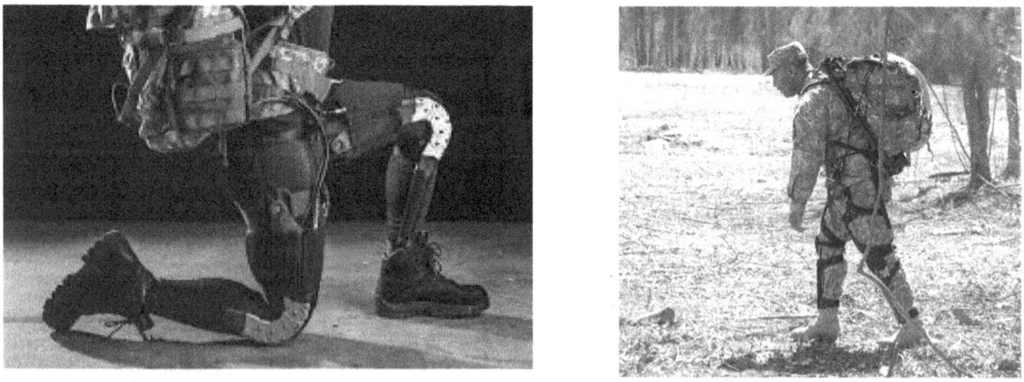

FIGURE 6.13 Powered exoskeleton for assisted manual handling of heavy loads. (US Army Technology: http://armytechnology.armylive.dodlive.mil/index.php/tag/overmatch/page/2/.)

FIGURE 6.14 Passive back support exoskeleton. (Courtesy Hero.)

reductions ranged from −6% to −35%, and for active exoskeletons, −6% to −48% when lifting. During static bending, similar reductions in back muscle EMG were measured. Overall, reductions in load moments and spinal compression were found. In a number of studies, there was evidence that endurance and time on task improved when exoskeletons were worn (the number of lifting cycles until exhaustion increased with an active (+30%) and a passive exoskeleton (+26%), and the number of scoops until exhaustion during snow-shoveling increased by 186% with an active exoskeleton. Snow-shoveling time increased by 149%, and shoveling distance by 269% with an active exoskeleton). Ratings of perceived exertion, fatigue, and task difficulty were lower in 10 studies. Among the authors' conclusions was the requirement for future research on user acceptability.

Farris et al. (2023) reviewed research on exoskeletons to augment performance during load carriage and manual lifting in the military. Evidence for augmentation was considered either as direct evidence (improved task performance) or mechanistic evidence, and 54 studies were reviewed. Passive exoskeletons for use during load carriage aim to reduce the load on the lower limbs by transferring backpack loads to the ground. These exoskeletons brought about reductions of up to 30% of the load. Ratings of perceived exertion (RPE) were also lower, but wearers also experienced lower postural sway and less stability. Powered exoskeletons generate torques at the hip or knee via powered actuators. Studies showed a 15% reduction in metabolic demands and lower EMG in the leg muscles; vastus medialis (one of the quadriceps muscles that inserts at the medial side of the knee joint) and gastrocnemius were lower when carrying light loads. Active exoskeletons that provide assistive torques at the hip or ankle only also led to lower metabolic load reductions (8%–15% lower).

Studies on passive exoskeletons for manual lifting were also reviewed (ten studies). Passive exoskeletons assist the spine and pelvis via stiff elastic elements that are attached to the body, often by straps. These exoskeletons stiffen the trunk and reduce spinal compression during flexion and extension. The energy stored during flexion is stored in the element and released when the trunk extends. Five studies showed reductions in trunk extensor EMG of 15%–30% of EMG during a maximum voluntary contraction. Mixed results on reductions in trunk loading were found, possibly because of differences in the design of the exoskeletons and the assistance they provided. Some studies have also found reductions in metabolic cost.

Powered exoskeletons for manual lifting were also reviewed (11 studies). These exoskeletons used actuators to assist at the pelvis and back to assist trunk extension. Significant reductions in trunk extensor EMG were reported in eight of ten studies that measured EMG. The authors concluded that the exoskeletons would reduce the metabolic cost of load carriage, even though the mass of the exoskeleton increases the cost—in other words, the added cost due to the mass of the exoskeleton is less than the benefits of carrying it. However, the authors also cautioned that an important issue which has not been addressed in these studies is 'proof of concept' against a requirement, which has yet to be specified. In other words, the requirement for a targeted reduction in metabolic cost. Similar arguments apply for manual lifting—the exoskeleton must deliver benefits against a defined loading task (e.g., manual loading of shells into a gun).

In a study of exoskeletons for military use, Slaughter et al. (2023) took a participatory approach. Back overuse injuries are a significant problem in the U.S. Army, responsible for nearly a quarter of musculoskeletal injuries. Low back injuries resulted in 1.2–1.7 million limited or lost duty days for Soldiers each year. Back injuries are also the most common condition resulting in disability-related discharges from the U.S. Army. The researchers conducted a field study with soldiers during training. A passive exosuit (weighing 1.2 kg) consisting of a harness (upper-body interface) and thigh sleeves was trialed. The suit had a clutch-switch system (for turning assistance on and off). Elastic bands arranged along the back of the exosuit acted as artificial back and hip extensor muscles. When carrying a load, the exosuit stiffened the trunk as it flexed, relieving the back muscles of some of the load. All of the soldiers were satisfied, to varying degrees, with the weight, fit, ease of donning, and lifting assistance provided by the exosuit and 91% were satisfied overall. Nine percent reported neutral satisfaction. Two soldiers were only dissatisfied with the range of motion and ease of use, representing <2% of responses. Approximately 90% of soldiers reported that exosuits increased

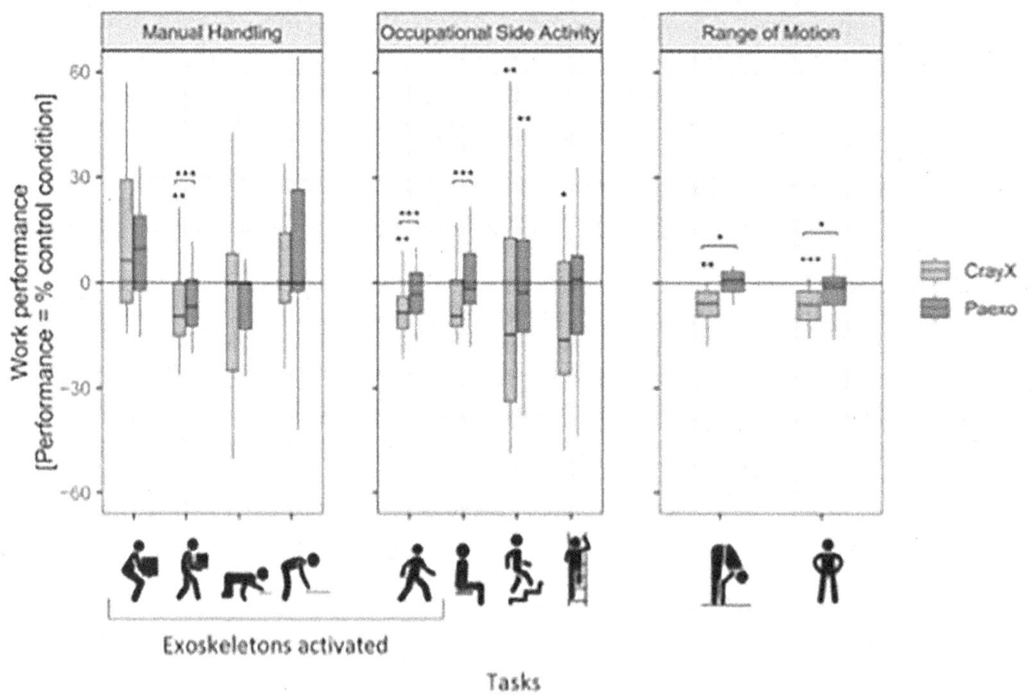

FIGURE 6.15 Changes in performance of selected tasks with passive and active exoskeletons. (From Goeverts et al., 2023, with permission).

their ability to perform their duties, and 100% said that if the exosuit were further developed and made available to them, they would be likely to wear it.

Govaerts et al. (2023) compared an active exoskeleton (7 kg) and a passive exoskeleton (4.5 kg) during a variety of tasks carried out under experimental conditions (Figure 6.15):

- Lifting a 20 kg box as many times as possible in 2 mins
- Walking 10 m with a 20 kg box as fast as possible
- Endurance time for a 3-point kneeling task
- Endurance time for forward bending
- Distance walked in 10 min at a self-selected pace
- Sit-to-stand and stand-to-sit as fast as possible
- Stair climbing as fast as possible
- Ladder climbing as fast as possible
- Trunk rotation
- Fingertip reach to the floor (distance when fully flexed)
- Hip abduction (standing)
- Deep squat × 3

The exoskeleton data was normalized as a change in performance compared to a control condition (the tasks were carried out without an exoskeleton).

For manual handling and bending, both exoskeletons improved performance. The results for the other activities were mixed, and the powered exoskeleton performed worse than the passive exoskeleton (Figure 6.15). Overall, the findings suggest that exoskeletons, while suitable for a specific task, might not be suitable for jobs involving varied work. In other words, the exoskeleton has to be fitted to both the task and the user.

The characteristics of exoskeletons that have been investigated in much of the HFE research summarized above are not the same as those that determine their acceptance and use in organizations. Raghuraman et al. (2023) were interested in learning more about the opinions of industry stakeholders and whether a brief and informal familiarization with a variety of exoskeletons would change their opinions.

User requirements are a key determinant of exoskeleton acceptability. These include but are not limited to fit, comfort, ease of donning and doffing, compatibility with environmental conditions such as temperature and humidity, perceived performance benefits, and safety. Other considerations include perverse outcomes such as pain or injury to other parts of the body when exoskeletons are used, and a false sense of security leading to risk-taking or overexertion.

Twenty-two volunteers from 12 manufacturing industries in the United States briefly trialled multiple exoskeletons. The main themes influencing the adoption of exoskeletons were effectiveness, usability, environmental concerns, unintended impacts, individual factors, and expectations to get support from management. Following brief exposure to exoskeletons, the volunteers' views became more positive. The authors concluded that questions remain about how to introduce exoskeletons to workers, what technology expectations to set, and the best training methods to account for the physical, psychosocial, and other experiences of using them. However, even brief exposure to exoskeletons had a positive influence on stakeholder opinions, although there were lingering concerns about usability and acceptability within their own organizations.

Exoskeletons and Return to Work (RTW) After Injury

Exoskeletons can be used to support return to work (RTW) of injured employees. The Washington State Department of Labor and Industries has drafted a process for doing so.[3] The key criteria are:

1. Employees must be motivated to try an exoskeleton to assist their return to work.
2. The task and the nature of the injury must be compatible with the use of an exoskeleton to assist with RTW.
3. Top management support via occupational health policy is required.
4. A budget for procurement, use, and oversight must be in place.
5. The RTW process must be overseen by a multidisciplinary team.
6. Training must be provided for all involved in RTW.
7. Medical oversight is required.
8. A graduated RTW plan must be in place to progress the employee to the RTW goal.

RESEARCH ISSUES

One of the main challenges is integrating physical approaches to low back injury risk assessment with psychological approaches that take into account employee perceptions of their own work capability (self-efficacy), work organization, and psychosocial factors (Chapters 8 and 16).

Wearable Tech and Sensors

The development of new devices such as the LMM enables researchers and designers to quantify movements of workers and relate them to task requirements and workspace design. These devices will enable more accurate estimates to be made of worker exposure to spinal stress and can be used to identify and redesign high-risk jobs. Donisi et al. (2021) used a sensor known as an inertial measurement unit (IMU), worn on the waist, to capture data on subject motions during a laboratory lifting task. IMUs are based on triaxial accelerometers and gyroscopes, able to measure 3D acceleration and angular velocity of the sensor with respect to gravity. The study aimed to determine whether machine learning algorithms could classify the task as "safe" or "unsafe" as predicted by the NIOSH lifting, using data from the IMUs sampled at 20 Hz. Subjects carried out the lifting task under two conditions with lifting indices of 0.5 ("safe") and 1.3 ("unsafe"). The data were processed by nine different machine learning algorithms. The performance of the algorithms was assessed using the following metrics: accuracy (the ratio of correct predictions to the total number of cases); sensitivity (the percentage of risky cases correctly identified); and specificity (the percentage of safe cases correctly identified). The algorithms varied in accuracy (0.79–0.98); sensitivity (79%–98%); and specificity (79%–98%). These excellent results suggest that the technology could be developed to enable real-time monitoring of manual handling activities for large numbers of workers. The IMUs are non-invasive, and relatively unobtrusive, with data transmitted for processing via Bluetooth.

Several new systems are now available for use by HFE and Health and Safety personnel to conduct risk assessments of manual handling tasks (Figure 6.16). The worker is fitted with wearable sensors (either IMUs or reflective markers) to capture kinematic data on joint angles and positions. These data are fed into a segmented and jointed musculoskeletal model, which can be calibrated to the stature and mass of the wearer. The joints have varying degrees of freedom depending on their anatomy (for example, the elbow joint is modeled with one degree of freedom, whereas the hip joint has three degrees of freedom of translation and three of rotation). Typically, tools such as the NIOSH lifting equation can be embedded in such systems, effectively automating the assessment process.

SUMMARY

Modern thinking about the design of manual handling tasks emphasizes that a detailed analysis of all aspects of the task is necessary. For example, the design of the load, its size and density, is,

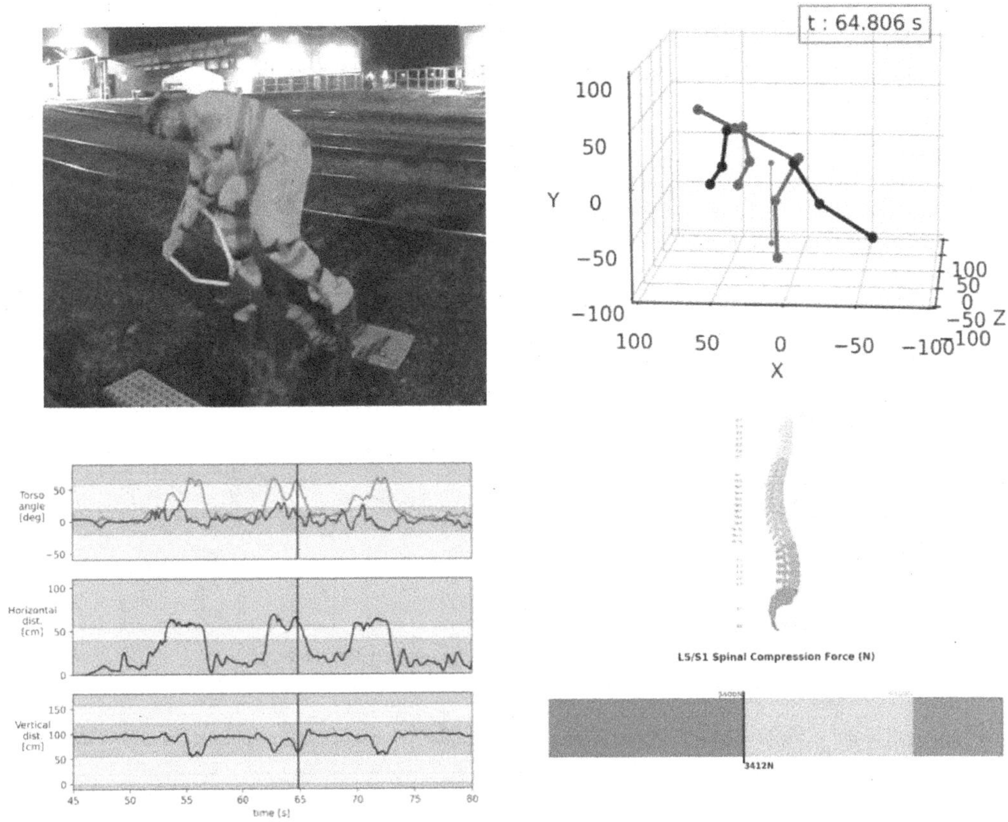

FIGURE 6.16 Wearable moment sensors attached to workwear used to track movement and posture risks. (SpatialCortex MOVA™, spatialcortex.co.uk.)

within limits, as important as its weight. The design of the workspace and the environment in which the work is to be carried out are also important considerations, as are data on the characteristics of the workers themselves. Table 6.16 presents a short hazard checklist for rapid screening of manual handling tasks in industry, and Table 6.17 summarizes guidance for risk reduction.

Further information for the specification and design of manual handling tasks can be found in ISO 11228 parts 1–3, "Ergonomics—Manual handling." A comprehensive set of tables, the "Liberty Mutual Manual Materials Handling Tables," for determining acceptable loads for workers can be found on the web at Liberty Mutual.com.

A systems approach is needed to select exoskeletons for industrial tasks. Fournier et al. (2023) conducted a systematic review, finding 15 high-quality studies using either passive or active exoskeletons. Mixed effects were found for quality improvements in tasks such as drilling and electroplating, with some reductions in errors and fewer visual defects in electroplating. Productivity improvements were observed in shoveling, drilling, and electroplating tasks, with mixed results for manual handling and manual assembly. Exoskeletons are not a solution looking for a problem; a systems approach is needed, based on task analysis, to understand the requirements of the task and the context of use. Comfort and usability should be assessed using a participatory approach.

TABLE 6.16
NIOSH Hazard Evaluation Checklist for Manual Handling Tasks

Risk Factors	Yes	No
1. General		
1. Does the load handled exceed 23 kg?		
2. Is the object difficult to bring close to the body because of its size, shape, or bulk?		
3. Is the load hard to handle because it lacks handles or cutouts or does it have slippery surfaces or hard edges?		
4. Is the footing unsafe? For example, are the floors slippery, inclined, or uneven?		
5. Does the task require fast movement, such as throwing, swinging, or rapid walking?		
6. Does the task require stressful body postures such as stooping to the floor, twisting, reaching overhead, or excessive lateral bending?		
7. Is most of the load handled by only one hand, arm, or shoulder?		
8. Does the task require working in environmental hazards, such as extreme temperatures, noise, vibration, lighting, or airborne contaminants?		
9. Does the task require working in a confined area?		
2. Specific		
1. Does lifting exceed five lifts per minute?		
2. Does the vertical lift distance exceed 1 m?		
3. Do carries last longer than 1 min?		
4. Do tasks that require large, sustained pushing or pulling forces exceed 30 s duration?		
5. Do extended reach static holding tasks exceed 1 min?		

Note: A "yes" response indicates risk of low back pain. The larger the number of "yes" responses, the greater the risk.

TABLE 6.17
Guidance for the Reduction of Risk in Manual Handling Tasks

Minimize the Weight to Be Handled
Assign the job to more than one person
Order goods in smaller packages
If possible, mechanize the process
Use forklift trucks
Machines, rather than employees, should transfer loads between surfaces
Change the job from lifting to lowering, from lowering to carrying, from carrying to pulling, and from pulling to pushing
Use handles, hooks, or similar features to enable workers to get a firm grip on objects to be lifted[a]
Reduce the weight of containers used to transfer objects
Reduce the size of containers to be lifted
Balance and stabilize the contents of containers to avoid sudden shifts in load during a lift
Minimize Reach and Lift Distances
Design containers so that they can be held close to the body
Treat work surfaces to allow for the ease of movement of containers
  Use conveyors
  Air bearings
  Ball caster tables
  Monorails
  Slides or chutes
Provide carrying aids
  Sack barrows
  Dollies
  Four-wheel trucks

(*Continued*)

TABLE 6.17 (*Continued*)

Guidance for the Reduction of Risk in Manual Handling Tasks

Increase the height at which lift is initiated, decrease the height at which it terminates
Stack objects no higher than shoulder height
Store heavy components on shelves between shoulder and knuckle height
Avoid deep shelves
Avoid side-to-side lifting from a seated position
Provide access space around components to cut down on the need for manual repositioning
Storage bins or containers should be fitted with spring-loaded bottoms
Use sloped surfaces to gravity-feed items to the point of lifting
Provide free space around and under the work surface to increase functional reach
Place objects within the zone of convenient reach
Provide sufficient space for the whole body to turn
Provide swivel chairs for seated workers

Increase the Time Available for Lifting

Relax the time standard for the job
Reduce the frequency of lifts
Rotate workers between jobs
Improve work–rest cycles (Chapter 7)

Task Redesign

Reduce the weight of hand tools (e.g., use plastic shovels for moving ice)
Increase object weight and size so that mechanical handling is the only option
Palletize all handling tasks
Increase wheel size on trolleys for hand pushing and pulling
Change the shape of the object for good coupling
Add handles or grips

Maintenance

Check that trolleys, slides, and conveyors run smoothly

Use Assistive Devices

Reduce bending and reaching with
  Lift tables
  Temporary scaffolds
  Footstools for short workers
Reduce carrying with conveyors, chutes, trucks, and trolleys
Reduce pushing and pulling with electric carts, chains, or water conveyors
Ease stacking and storing with forklift trucks, stackers, hoists, and pulleys
Reduce twisting and turning with turntables

[a] Maximum acceptable weights are about 16% lower if no handles are used (Ciriello et al., 1993).

TUTORIAL TOPICS

1. How do competitive weight lifters manage to lift such heavy loads without injury?
2. Should robots do all the lifting in the future?
3. Why do we not carry things on our heads anymore (in twenty-first-century Western countries, at least)?
4. At the link is a video about exoskeletons. How would you decide whether to recommend one to an employer? (Hint: consider the requirements, the context of use, and the cost-benefit) https://www.youtube.com/watch?v=u3yKebhbXkE.

ESSAYS AND EXERCISES

1. Which weighs the most: a 20 kg bag of feathers or a 20 kg bag of lead? Which would place the greatest strain on the back when lifted from the floor?
   a. Bag of feathers
   b. Bag of lead
   c. They are both the same
2. A worker unloads 20 kg sacks of apples from a conveyor and loads them onto a chute from where they are dispatched. He loads for 2 h/day at a rate of 5 sacks per minute. The height of the conveyor is 60 cm and the height of the chute is 100 cm. There is an angle of asymmetry of 45°, and the load is held 30 cm from the body.
   - Use the NIOSH equation to calculate the RWL and the LI
   - Comment on the safety of the task and identify the risk factors
3. Which of the following factors is the most important in reducing the frequency of back injuries?
   - Lifting with the knees instead of the back
   - Lifting with the back kept straight and as vertical as possible
   - Compulsory training in safe lifting techniques
   - Holding the object as close to the body as possible
4. A worker lifts cartons of grapes from a conveyor onto a pallet. The cartons weigh 20 kg each and are 30 cm^2. The worker grasps the middle of each carton, as there are no handles. The center of gravity of the load is 50 cm from his lumbar spine. The height of the pallet adjusts automatically to 80 cm, and the height of the conveyor is 70 cm above the floor. The worker turns by 45° each time a carton is transferred. The worker performs 5 lifts per minute for 1 h/day. Calculate the RWL and the lifting index, where

$$RWL = LC \times HM \times VM \times DM \times AM \times FM \times CM$$

   - $LC = 23$ kg
   - $HM = 25/H$
   - $VM = 1 - (0.003|V - 75|)$
   - $DM = 0.82 + (4.5/D)$
   - $AM = 1 - (0.0032A)$
   - $FM = 0.80$
   - $CM = 1$ for good coupling, 0.95 for fair coupling, and 0.9 for poor coupling

   Comment on the safety of this task and identify the main risk factors. Suggest how the safety of the task can be improved in the short term.
5. "There is no such thing as a safe maximum load in manual handling tasks." Discuss.
6. Discuss the factors that cause the trunk to fail during heavy work.
7. Which of the following tasks might provoke symptoms in a worker suffering from osteoarthritis of the zygapophyseal joints?
   - Making pastry (standing at a table)
   - Carrying a rucksack
   - Sitting at a production line doing a light assembly task (trunk inclined forward slightly, elbows resting on the bench)
   - Kitchen work involving reaching for objects ahead and above
     - Driving a taxi
     - Traffic warden
     - Sitting on a chair with no lumbar support
     - Swimming

8. See Figure 6.17. Assume that the mass of the bucket is 25 kg and it is balanced directly above the lumbar spine. Calculate the additional spinal compression due to the load (see HFE Workshop 2.1 for equations). Comment on the effect on spinal loading, if any, of carrying the baby in the manner shown.

FIGURE 6.17 Efficient load carriage. The bucket is carried on the head, and the child is carried on the lower back. The spine and pelvis are loaded axially.

9. A fisherman has to handle heavy nets and posts on the deck of a small fishing boat at sea. He works in all kinds of weather throughout the year. In addition to the loads lifted, what other risk factors for back injury might he be exposed to in this environment?
10. Mr. Smith has just suffered an acute prolapsed intervertebral disc while tidying his garden before going on holiday. Which of the following recreational pursuits would you advise him to avoid, and which might be beneficial? Briefly state your reasons.
    a. Water skiing
    b. Swimming
    c. Scuba diving
    d. Horseback riding
    e. Pony trekking
    f. Cycling
    g. Hiking
    h. Sunbathing on a lounger
    i. Wine tasting
    j. Off-road motorcycle riding
    k. Driving a car
11. Mr Smith has started a new job in a warehouse. Before starting work, he has to watch a training video on the safe handling of loads (https://www.youtube.com/watch?v=901uQgfiuVk). Given that Mr. Smith is obese, 60 years old, and a smoker, do you think this is the right advice for him? Give your reasons.

QUICK QUIZ

1. "Discs don't slip." True or false?
2. Abdominal belts can protect the spine from injury. True or false?
3. The value of the load constant is: (a) 15 kg, (b) 23 kg, (c) 50 kg
4. What is FM for 10 lifts/min when lifting 1–2 h at $V < 75$ cm?
5. Is "how to lift" training effective?
6. Is a kilogram of feathers lighter than a kilogram of lead?
7. Does manual handling add value to a product? Yes or No
8. Intra-abdominal pressure supports the spine when lifting. Yes or No
9. Walking a lot can cause inguinal hernias. True or False
10. Carrying loads on the head is more efficient than carrying loads in the arms. True or False

NOTES

1 NIOSH has made available an APP that can be downloaded from https://www.cdc.gov/niosh/topics/ergonomics/nlecalc.html

2 Available from Auburn University: https:ergoweb.com.

3 Available from: Washington State Department of Labor and Industries, Process for Considering and Deploying Industrial Exoskeletons for Return to Work.

7 Work and Recovery

GENERAL REQUIREMENTS FOR HUMANS IN SYSTEMS

1. Job demands must match the work capacity of employees.
2. Rest periods should be of sufficient length and quality to allow physiological recovery.
3. Allowances must be made for the effects of environmental conditions and personal characteristics on operator work capacity.
4. Shift systems must minimize the disruption of sleep.
5. At least 6h of uninterrupted sleep every 24h and two additional hours of rest are recommended.

The term (fatigue) should be absolutely banished from precise scientific discussion and consequently that attempts to obtain a fatigue test be abandoned *B*. Muscio (1921). Is a fatigue test possible?

Muscio, 1921

CORE KNOWLEDGE: FUNDAMENTALS OF WORK PHYSIOLOGY

Physical work is achieved by the contraction of muscle, powered by local energy stores dependent on a ready supply of oxygen and glucose.

Muscles, Structure and Function, and Capacity

Muscles exert tension when they contract, making physical work possible. There are three types of muscles in the body:

1. Skeletal muscle
2. Smooth muscle
3. Cardiac muscle

Smooth muscle is found in the intestines and makes possible the movements essential for the digestion of food (peristalsis). It is also found in the walls of blood vessels, where it is involved in the regulation of blood pressure and blood flow. It is not normally under conscious control. Cardiac muscle has a special structure and constitutes the bulk of the heart. The present discussion is limited to work involving skeletal muscle—muscle connected to the bones of the skeleton.

The energy required for muscle contraction is obtained from phosphate compounds in the muscle tissue, formed from the breakdown of food. The contraction of most skeletal muscles is under conscious control, and the rate and strength of contraction can be varied at will. A complete muscle is made up of many bundles of fibers (cells) arranged side by side and covered by connective tissue sheaths. Nerves and blood vessels are located in the connective tissue. Each muscle fiber consists of many smaller myofibrils (Figure 7.1).

DOI: 10.1201/9781032717210-7

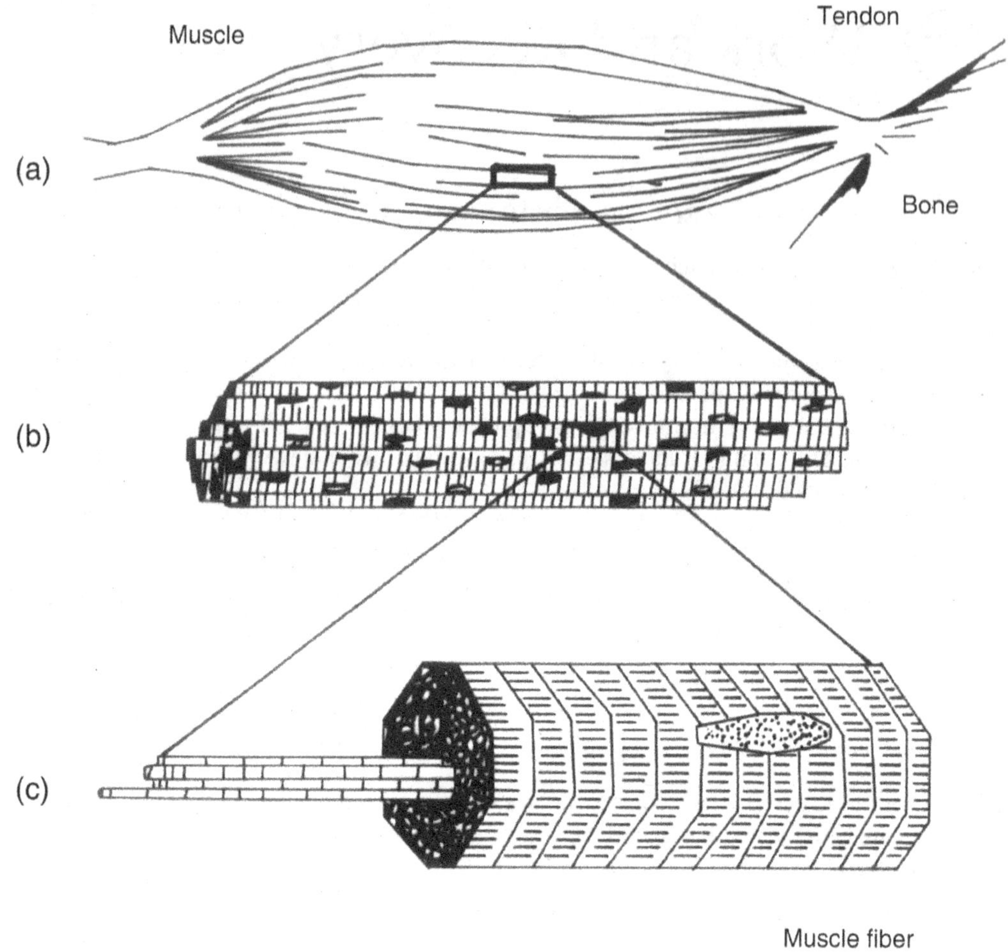

FIGURE 7.1 Basic structure of skeletal muscle. (Adapted from Noakes (1992), with permission.)

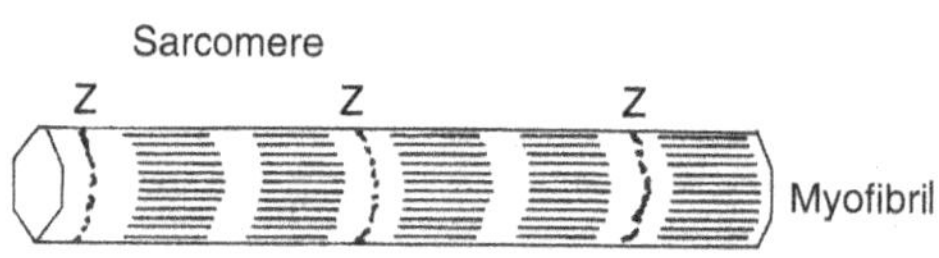

FIGURE 7.2 Myofibril.

A myofibril is split up into a number of sarcomeres arranged in series (Figure 7.2). A sarcomere consists of many filaments layered over each other in alternating bands. There are two types of filaments. Thick filaments consist of about 300 myosin molecules. Thin filaments consist of a globular protein called actin. The filaments are the true contractile elements of a muscle. Muscles can be likened to bundles of strings all joined together. Each individual string (muscle cell) is made up of fibers (myofibrils), each of which is constructed from many alternating filaments of actin and myosin. The whole structure is bathed in intracellular and extracellular fluid and is permeated by blood vessels and nerves.

The mechanism of muscle contraction consists of actin filaments sliding over myosin filaments. Since these filaments are arranged in overlapping, alternating bands, like a multilayered sandwich,

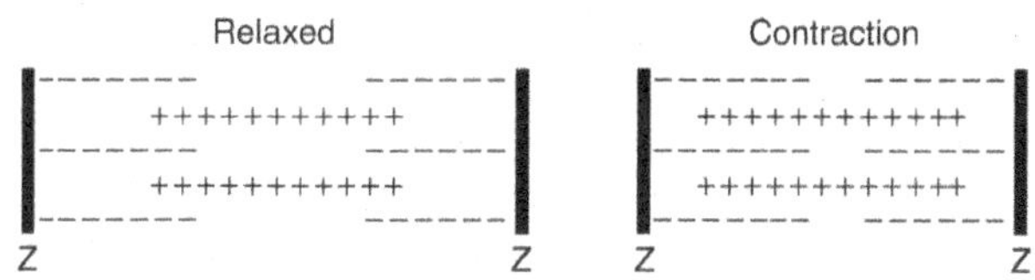

FIGURE 7.3 Sliding filament theory of muscle contraction.

TABLE 7.1
Steps Leading to the Contraction of Skeletal Muscle

1. A nerve impulse travels along a motor nerve to the motor nerve end plates in the muscle fibers
2. A small amount of neurotransmitter is secreted and travels to the membrane of the muscle fiber, initiating an action potential in the muscle fiber
3. The action potential travels along and permeates the muscle fiber until it reaches a structure known as the sarcoplasmic reticulum
4. The sarcoplasmic reticulum, which acts like a store of calcium ions, releases calcium ions into the myofibrils. This increases the attraction between the actin and myosin filaments, causing the sliding action of one filament over another
5. The calcium ions are then actively pumped back into the sarcoplasmic reticulum. ATP is needed for this. Muscle contraction ceases
6. Further nerve impulses cause the process to repeat

sliding of the former over the latter causes the sarcomeres to shorten (the filaments themselves do not shorten, Figure 7.3). The primary stimulus for contraction is the release of calcium ions stored in the sarcoplasm. The calcium ions bind to the actin, which increases the affinity of myosin for actin. The fundamental physiology of muscle contraction is still not completely understood; however, the main steps in the process are summarized in Table 7.1.

The duration of the calcium pulse (from release to reabsorption of calcium ions) is about one-twentieth of a second. For continuous muscle contraction to occur, a new impulse must reach the muscle after the calcium pump has removed the calcium ions. Smooth movements of the body are, in reality, the result of many individual contractions and relaxations of contractile elements throughout the muscle.

The amount of contraction that can take place depends on the total change in length of all the sarcomeres in the muscle when they contract. Muscle contraction is a biochemical process and need not result in shortening of the muscle itself. It is more accurate to say that muscle contraction consists of the actin filaments tending to slide over the myosin filaments, and that this will result in shortening of the muscle under appropriate circumstances.

There are two main types of muscle fiber. Type I fibers have high endurance, can generate moderate forces and are utilized in many activities of daily living. They have large myoglobin stores and are sometimes referred to as "red" fibers. Type II fibers are larger and can quickly generate high forces for short periods. They are essential for forceful exertions in activities requiring strength or, for example, when regaining balance after a trip or slip.

Energy for Action

Energy for muscle contraction (and for many other bodily processes) comes from the breakdown of a substance known as adenosine triphosphate (ATP). By breaking one of the phosphate bonds, ATP is converted into adenosine diphosphate (ADP), and energy is made available inside the cell. ATP can be likened to a rechargeable battery pack—a short-term store of directly available energy. For the cell to continue functioning, the ADP must be reconverted to ATP so that energy can continue to be made available when required. A second phosphate compound, known as creatine phosphate, acts like a backup energy store to convert the ADP to ATP.

The actin and myosin filaments are always mutually attracted, and energy is required to break this attraction and allow the muscle to lengthen again. During the relaxation phase of muscle contraction, ATP is needed to weaken the attraction between the actin and myosin filaments. Endurance athletes sometimes experience muscle cramps when a muscle becomes fatigued during an event (Noakes, 1992) because the ATP needed to break the actin–myosin attraction is depleted.

The creatine phosphate and ATP stores are of very limited capacity and are reduced within seconds or minutes. The creatine phosphate system can provide energy for maximal muscle activity for about 8–10 s and therefore has to be continuously replenished. This takes place in structures inside the cells known as mitochondria. If the energy stores are depleted, the muscle will go into rigor (shortly after death, a state of rigor mortis sets in through the loss of all ATP).

In muscles, the mitochondria are placed close to the filaments. Carbohydrates and fatty acids are broken down—ultimately into carbon dioxide and water—and energy is released to form ATP. The basic reaction for the liberation of energy involves the oxidation of glucose as shown below:

$$C_6H_{12}O_6 + 6O_2 \rightarrow 6CO_2 + 6H_2O + \text{energy}$$

Carbon dioxide and water (so-called metabolic water) are the products of this chemical reaction. The carbohydrates and fatty acids used by the mitochondria are derived from the food we eat. They are obtained either from the bloodstream or from stores in the muscles or liver. Oxygen is obtained from the air ventilating the lungs and is transported to all parts of the body by the blood. The body has a very limited capacity to store oxygen, although muscles do contain a substance known as myoglobin, which is a short-term oxygen store. To use an electrical analogy, if the bloodstream is likened to an electrical reticulation network, the mitochondria can be considered as transformers plugged into the mains which keep the creatine phosphate and ATP battery packs fully charged.

Oxygen-Dependent and Oxygen-Independent Systems

If insufficient oxygen is available, the mitochondria are not able to metabolize the carbohydrates and fatty acids to produce energy, and an oxygen-independent system becomes the main source of ATP. Enzymes in the intracellular fluid surrounding the myofilaments and mitochondria use glycogen (the stored form of glucose) and blood sugar (but not fat) to produce energy. This backup system operates at all times, even at rest, but it is inefficient and produces much less energy per glucose molecule than the oxygen-dependent system. Furthermore, it produces waste products that cause the acidity of the muscle cells to increase, reducing the affinity between actin and myosin filaments and weakening the muscle. Oxygen is required to remove these waste products. If it is not available when the person is working, the waste products will accumulate. Under these circumstances, it is said that the person has built up an oxygen debt that must be paid back when work ceases. Fatty acids can be used as an energy source, particularly when muscles carry out low-intensity contractions for long periods.

As was described in Chapter 5, most tasks involve a mixture of static and dynamic work and the latter can usually be sustained longer than the former. Table 7.2 summarizes some of the physiological differences between static and dynamic work.

Implications

The oxygen-independent system can form ATP molecules about 2.5 times as fast as the oxygen-dependent system but can only provide energy for about 1.3 min of maximum muscle activity. It is a valuable system, which enables work to be carried out at a high rate for short periods interspersed with rest. Clearly, it is unrealistic to expect workers to carry out sudden, highly demanding tasks for more than a minute or so without rest, and in planning for high-workload emergency situations, backup must be provided to enable rapid alternation of work teams. The oxygen-dependent system can function as long as nutrients are available.

TABLE 7.2
Differences between Static and Dynamic Work

Static	Dynamic
Sustained muscular contraction	Repetitive muscle contraction/relaxation cycle
Reduced muscle blood flow	Increased muscle blood flow
No increase in muscle oxygen consumption	Increased muscle oxygen consumption
Oxygen-independent energy production	Oxygen-dependent energy production
Muscle glycogen ≥ lactate	Muscle glycogen > CO_2 + H_2O; muscle uptakes glucose + fatty acids from blood

Source: From Noakes (1992), with permission.

TABLE 7.3
Limiting Factors on Muscle Contraction

1. Demand for energy exceeds supply. This can occur when chemical fuel stored in the muscle is exhausted, and the rate of replenishment of oxygen or glucose is inadequate. The supply of fuel to working muscle depends on the capabilities of the circulatory system
2. Mechanical capabilities are exceeded. The force output of a muscle is finite and depends on the number of contractile elements (the maximum strength of a muscle is proportional to its cross-sectional area). A task may exceed the muscle's mechanical capacity. Strength training increases the cross-sectional area of a muscle and the ability to recruit muscle fibers to carry out muscular work
3. Accumulation of waste products impairs muscle function. This can occur during static muscle activity as described in previous chapters
4. The rate of heat production exceeds the body's thermoregulatory capacity. Excess heat can no longer be dissipated, and body temperature increases. Cardiovascular capacity is impaired

Efficiency of Muscle Contraction

The muscular system uses oxygen to convert chemical energy from foodstuffs (stored in the tissues or delivered by the bloodstream) into mechanical energy via the sliding filament mechanism. Muscles are, at best, only 20% efficient in converting chemical reactants to mechanical output due to friction and the viscosity of body parts. Heat is the main by-product. Metabolic heat production has important implications for the design of work in extreme climates, as will be seen in a later chapter. The system for converting chemical into mechanical energy can malfunction for a number of reasons, leading to fatigue (Table 7.3).

Muscle Function

Skeletal muscles exert tension between the bones to which they are attached when they contract, in proportion to their length and their maximum cross-sectional area. The term muscle contraction refers to the physiologically active state of the muscle, rather than its physical shortening. Muscles are able to contract eccentrically, isometrically, and concentrically:

1. *Eccentric Contractions*: The muscle lengthens while contracting (acts like a brake).
2. *Isometric Contractions*: The muscle length remains constant during contraction (maintains posture or stabilizes an object).
3. *Concentric Contractions*: The muscle shortens while contracting (moves the body or an object).

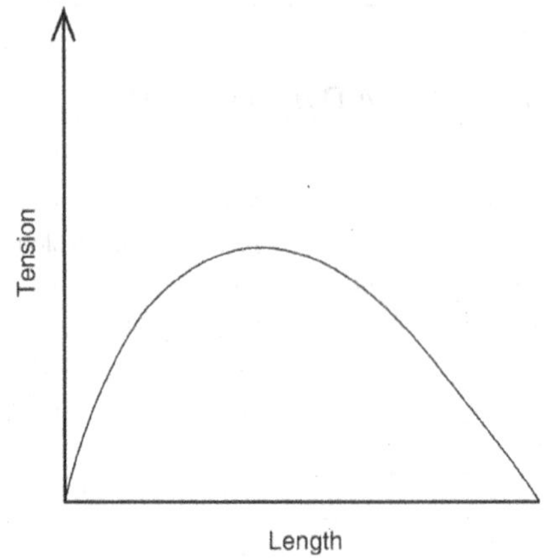

FIGURE 7.4 Active length–tension relationship. A longer muscle generates more tension when it contracts until its physiological length is exceeded.

The highest muscle forces are generated during eccentric contractions, so a high risk of injury is to be expected in occupations where slips and falls are prevalent or where personnel are exposed to destabilizing forces (e.g., users of small, fast boats exposed to shocks).

A close relationship exists between the length of a muscle and the tension it can exert. As the muscle is lengthened, increased tension is required to overcome the elastic resistance of the connective tissue, which surrounds the individual muscle cells and holds the muscle together. Muscles can exert tension both when in the active state and passively by their resistance to being lengthened. Muscles have an optimum length at which they are capable of exerting their maximum tension (Figure 7.4).

Muscles are not just simple mechanical actuators like motors or springs; they are reactive tissues. Their tension, length, and thermal, mechanical, and chemical states are integrated and controlled by means of complex neural circuitry at different levels of the central nervous system. In living organisms, muscles form part of a larger system of joints and ligaments, and the tension they exert is transmitted via their bony attachments. This results in a turning force, or torque, being exerted around the joint which the muscle spans. The magnitude of the torque depends not only on the length of the muscle but also on the geometry of the joint and the mechanical advantage of the muscles at a particular joint angle. Angle–torque curves can be plotted for different muscle–joint systems to identify those joint angles at which maximum torque can be exerted. Thus, manual tasks can be simulated and strong and weak body postures identified. For example, the flexion torque around the elbow joint is greatest in the middle portion of the joint range because the elbow flexors (such as the biceps muscles) have their greatest mechanical advantage in this position.

Control of Muscle Function

Skeletal muscle contains afferent and efferent fibers from the central nervous system. Some of these form complex feedback loops. Receptors in the muscle body provide information about the length of the muscle and its tension (Figure 7.5). This is conducted to the spinal cord and to higher levels of the central nervous system. The knee-jerk reflex is a well-known example of a feedback loop, which is made possible by a neural circuit between the stretch receptors in the muscle spindles and motor neurons in the spinal cord. When the muscle is lengthened, the stretch reflex comes into operation, causing the muscle to contract, that is, the reflex opposes the lengthening of the muscle. Muscle tension is detected by an organ known as the Golgi tendon organ, which is found at the point where small bundles of muscle fibers fuse with tendon fibers. The Golgi tendon organ forms a negative feedback system that prevents damaging levels of tension from building up in the muscle and possibly damaging the muscle fibers or the insertions of the tendons.

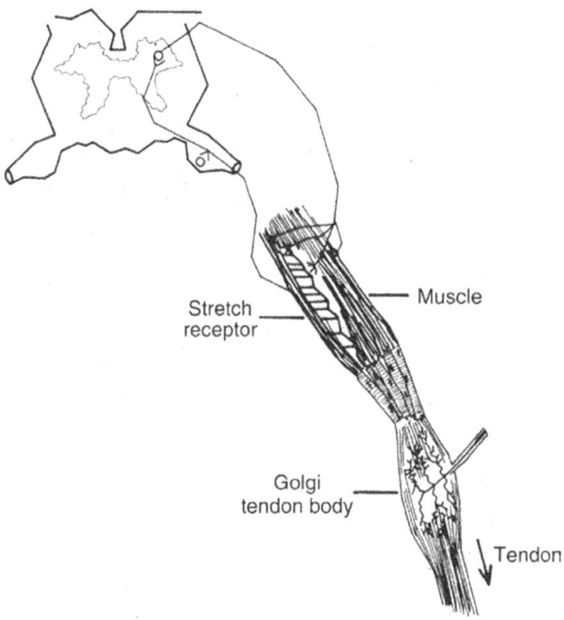

FIGURE 7.5 Positive and negative feedback comes from receptors in the muscles and a feedback loop at the spinal cord. A contraction/relaxation reflex exists between antagonistic muscles; when one muscle contracts, the other relaxes to allow movement at the joint. The contraction/relaxation reflex depends on the feedback from the muscle spindles and Golgi tendon bodies, respectively.

The neural circuits formed by these sensory organs in the muscles provide the central nervous system with information about the state of the muscle and the posture of joints. This is known as proprioceptive feedback and is essential for all but the most primitive of movements. Skeletal muscle also contains nerve endings that, when stimulated, lead to the sensation of pain. These pain receptors can be stimulated by chemical, thermal, or mechanical stressors. Nerve impulses induced by any of these causes can give rise to reflex muscle contraction via neural circuits in the spinal cord. This can further exacerbate the cause of the pain and lead to a positive feedback situation in which the muscle goes into spasm.

Cardiovascular System

The cardiovascular system collects oxygen from the lungs and food from the gut and delivers them to the cells of the body. It collects waste products and delivers them to the excretory organs. Hormones secreted by the endocrine glands in various parts of the body are transported to target cells in other organs by the blood. The cardiovascular system is also involved in thermoregulation—the control of body temperature.

An average adult has about 5 L of blood. Approximately 2.75 L of blood consists of plasma and 2.25 L of blood cells. There are three main types of blood cells. Red cells (erythrocytes) transport oxygen, white cells (leukocytes) fight infection, and platelets are required for blood clotting. The red cells are formed in bone marrow and contain a red pigment (heme) which is bound to a protein molecule (globin). Hemoglobin combines with oxygen to form oxyhemoglobin—the form in which oxygen is carried in the blood. Plasma consists of 90% water and 10% solutes—food, proteins, nutrients, and salts.

The heart is a four-chambered pump about the size of a fist (Figure 7.6). The right atrium receives blood returning from the tissues of the body and pumps it to the right ventricle, from where it is pumped to the lungs via the pulmonary artery. Gaseous exchange takes place in the lungs.

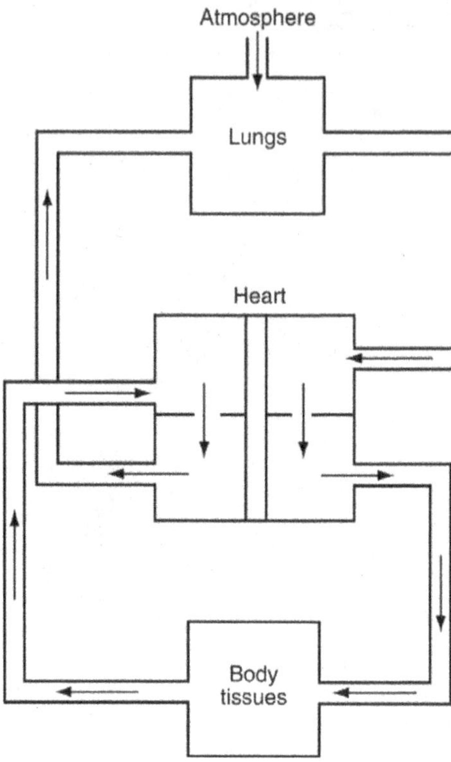

FIGURE 7.6 Model of the circulatory system.

Hemoglobin in the blood absorbs oxygen (a process known as oxygenation), and the oxygenated blood returns to the left atrium of the heart via the pulmonary vein. It passes to the left ventricle, where it is pumped via the aorta through a series of ever-branching blood vessels. Most tissues of the body are permeated by capillaries—tiny blood vessels with internal diameters not much larger than a red blood cell.

The system pumps oxygenated blood under pressure such that it permeates the body. Oxygen and food are delivered, and carbon dioxide and waste products are removed. After passing through the bed of capillaries in the tissues, the deoxygenated blood drains into the veins. The veins contain one-way valves that only permit the blood to flow in one direction—back to the heart. The output of the heart increases from around 5 L of blood per minute at rest to 25 L/min during heavy work. As workload increases, the volume of blood pumped per stroke (heartbeat) can be increased, as well as the heart rate itself. Furthermore, the flow of blood to the body can be increased on a selective basis by constricting and dilating different blood vessels. During physical work, more blood is directed to the muscles and to the skin under sympathetic control. After a meal, extra blood is directed to the intestines. Changes in acidity and the local concentration of metabolites provide feedback to the neural systems that control blood flow.

Blood Pressure

The heart pumps blood around the body in the manner depicted in Figure 7.6. The *systolic* blood pressure is the pressure on the arteries when the left ventricle pumps blood into the aorta. The diastolic blood pressure is the pressure when the heart rests between beats (the "background" blood pressure). Normally, these pressures fall below 120 and 80 mmHg, respectively. The *baroreflex* is the homeostatic mechanism that maintains blood pressure within a narrow range. Increases in blood pressure cause the heart rate to drop, leading to a drop in blood pressure. The reflex functions on a beat-to-beat basis, accounting for a natural *heart rate variability* due to constant fine-tuning of the system. Low heart rate variability can be a symptom of heart problems. There is evidence that constant exposure to mental stressors, or the failure to *appraise* these stressors appropriately, can suppress the baroreflex, causing chronic elevated blood pressure (Giarnos and Wager, 2015).

Respiratory System

The respiratory system is a gas exchanger, supplying the body with oxygen and removing carbon dioxide. It consists of the nose, pharynx, larynx, trachea, bronchi, and lungs. The nose filters incoming air, warms and humidifies it, and tests for substances that may be harmful if inhaled. The nasal epithelium is lined with mucus. Small particles in the incoming air are trapped in the mucus, which flows, under the action of cilia, to the pharynx, where it is swallowed. Before reaching the trachea, incoming air is warmed almost to body temperature and humidified to almost 100%.

The pharynx (throat) is a muscular tube that serves as the entrance to the respiratory and digestive tracts. It plays an important role in the formation of vowel sounds in speech (phonation). The larynx lies at the upper end of the trachea and just below the pharynx. It is part of the open passage through which air travels on its way to the lungs and protects the lungs against the ingress of solids. It is also the organ of voice production.

The trachea (windpipe) is a tube of cartilage and muscle about 2.5 cm in diameter extending from the larynx (in the neck) to the thoracic cavity (or chest), where it divides into two smaller tubes—the primary bronchi. The bronchi divide many times into increasingly smaller tubes, which finally terminate in microscopic sacs or alveoli. The alveoli are in contact with capillaries containing blood pumped from the heart via the pulmonary artery, and it is here that gaseous exchange takes place.

Physical Work Capacity

In principle, any increase in oxygen uptake over and above that required for basal metabolism can be used as an index of the physiological cost of work. When work begins, heart rate and oxygen consumption increase to meet the new demands. Since this response is not instantaneous, the immediate energy requirements are met by local (i.e., muscular) energy stores. When work stops, heart rate and oxygen consumption return to their initial levels slowly because extra oxygen is required to replenish the muscle stores and to oxidize the waste products of oxygen-independent processes (Figure 7.7), thereby paying back the oxygen debt.

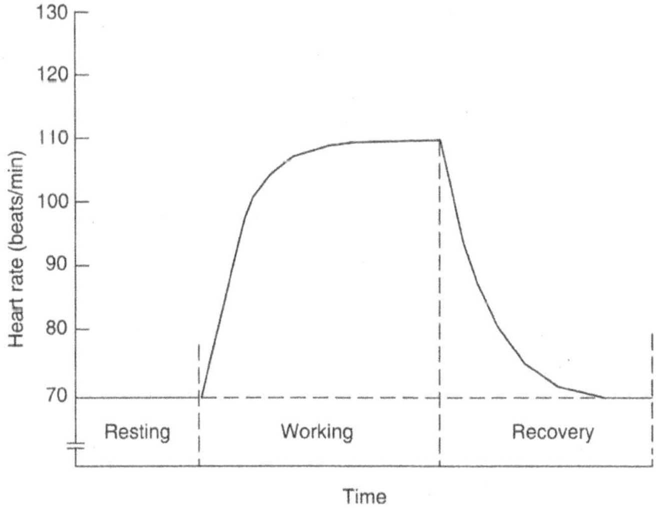

FIGURE 7.7 Physiological response to work. After the warm-up period, a steady state is reached where the task demands the match energy output. When work ceases, physiological processes return to resting levels during a warm-down period. Rest periods should be long enough to allow recovery.

In many industrial tasks, there is a warm-up period during which physiological processes adjust to meet the new demands, and the work rate increases steadily until a steady state is reached. The total physiological cost of work includes the energy expenditure during both the working period and the recovery period after work finishes. Assessment of the total physiological cost of work must begin with the subject at rest and continue until the physiological variables return to resting levels.

Physical work capacity refers to a worker's capacity for energy output. This depends primarily on the energy available to the worker in the form of food and oxygen and the sum of the energy provided by oxygen-dependent and oxygen-independent processes. The rate of energy consumption during physical work is the sum of the basal energy consumption and the metabolic cost of the work in terms of energy consumption. For continuous work at moderate intensities, oxygen-dependent processes usually make the major contribution to energy output. For each liter of oxygen consumed, about 20 kJ of energy is released. Work capacity depends on the ability to take up oxygen and deliver it to the cells for use in the oxidation of foodstuffs. The ability to work at a high rate is associated with a high oxygen uptake.

Maximum Oxygen Uptake

The maximum capacity to utilize oxygen (aerobic capacity) is described by the term VO_2 max. VO_2 max is estimated by having subjects run on a treadmill or pedal a bicycle ergometer while their oxygen uptake is measured. The speed is increased in increments, and oxygen uptake is measured every 3–5 min after the subject has adapted to each new work rate. Oxygen uptake increases as the work rate increases. The relationship is approximately linear (Figure 7.8).

Clearly, oxygen consumption and heart rate cannot continue increasing indefinitely—a point is reached where the person cannot increase the work rate further. When the aerobic capacity is reached, the work rate can only be increased if energy is provided by oxygen-independent processes, and only for a short period because these processes cause waste products to accumulate. Shortly after a person reaches a work rate that exceeds their VO_2 max, performance will decline dramatically.

There is great interindividual variability in VO_2 max, due largely to genetic factors. Training can improve an individual's VO_2 max, although only by about 10%–15%.

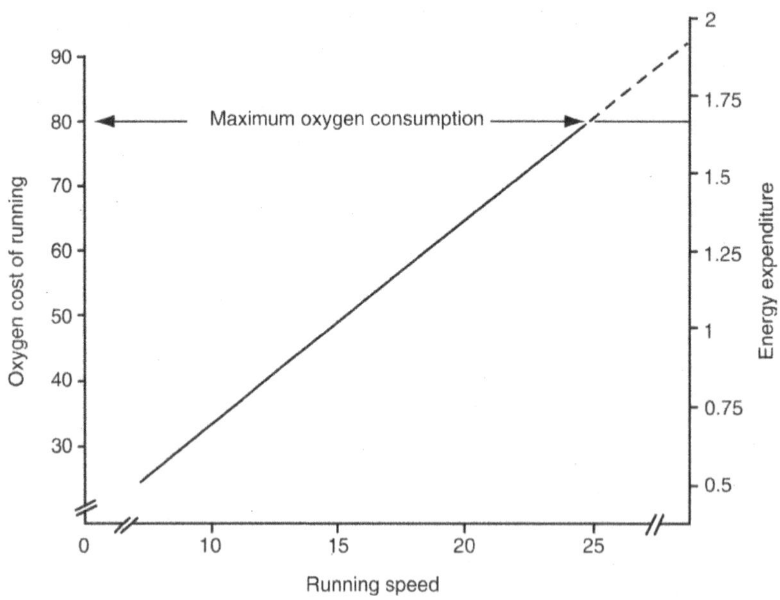

FIGURE 7.8 Relationship between workload and oxygen uptake.

Factors Affecting Work Capacity

Many factors can influence a person's capacity to carry out physical work. Table 7.4 summarizes the main occupational factors. Age is dealt with separately below.

TABLE 7.4
Factors Affecting Physical Work Capacity

Personal
Body Mass: Affects work capacity in all tasks where the operator has to use their own body. Excess body fat increases mass and lowers relative aerobic capacity
Sex: Women have a lower VO_2 max than men and usually have a higher percentage of body fat and less hemoglobin (Chapter 3). In general, women have lower upper body strength, controlling for body size. Male and female capacities should be considered as overlapping distributions, and selection tests should be based on the ability to carry out critical job functions
Alcohol: Alcohol increases cardiac output in submaximal work, thereby reducing cardiac efficiency. It affects liver function and can cause a predisposition to hypoglycemia (low blood sugar)
Tobacco: Tobacco smoke contains about 4% by volume carbon monoxide (CO). CO has an affinity for hemoglobin (combining to form carboxyhemoglobin) 200 times as powerful as oxygen. Smoking reduces the oxygen-carrying capacity of the blood and causes chronic damage to the respiratory system. Tobacco smoke also contains a very large number of toxic and carcinogenic chemicals, which have a depressing effect on physical capacity. Side-stream smoke, from the burning tip of the cigarette, affects nonsmokers who breathe it
Training: Physical training and job training in more efficient work methods. Specific training regimes can be developed to improve performance or prevent injury. Strength training requires that the body part in question be exercised to near maximal levels. Over a period of several months, the muscle fibers increase in size due to an increase in the number of myofibrils
Nutritional Status and General Health: Muscles with a high proportion of slow-twitch fibers use less ATP per unit of isometric tension than those with a high proportion of fast-twitch fibers. Undernutrition may bring about a reduction in the proportion of fast-twitch fibers in voluntary muscles, lowering the energy cost of daily activities but reducing the capacity for fast, dynamic work—an important consideration anywhere malnutrition is endemic
Motivation: Motivation is an extremely important determinant of work capacity. A worker's level of motivation may be affected by intrinsic factors, such as personality, personal and career goals, and the need for achievement at work, and by extrinsic factors such as work organization and the availability of alternative forms of employment. Piece- rate systems may motivate workers to work at increased intensity but have been associated with an increased risk of accidents
Cardiovascular Health: Many people eat a diet that causes raised plasma concentrations of cholesterol. Tiny crystals of cholesterol are deposited on the inside walls of the arteries, which eventually leads to a disease of the arteries known as atherosclerosis. The continued accumulation of cholesterol forms deposits called plaques, which eventually reduce the cross-sectional area of the artery and thus impede blood flow. In addition, the arteries lose flexibility (atherosclerosis or 'hardening of the arteries'). These changes in the structure of arteries can impede the flow of blood to the muscles and to the heart itself, resulting in decreased performance and an increased risk of heart attacks
Environmental
Air Pollution: Pollution may increase the resistance to airflow in the respiratory airways and cause damage to the lungs. Combustion of organic compounds, creating CO, can cause decrements in work capacity. If the concentration of CO exceeds 6.5 parts per million, it accumulates in the blood during submaximal exercise, reducing work capacity
Climatic Factors: Extreme environments can have significant effects on work capacity (Chapter 9)
Noise: Noise is a stressor that can elevate the heart rate and reduce cardiac efficiency
Altitude: The capacity for maximum muscular work remains unchanged up to an altitude of 1500 m above sea level. Above 1500 m, maximal work capacity decreases by about 10% per 1000 m. Submaximal work capacity is not affected below 3000 m above sea level but does require a larger percentage of the total work capacity compared to sea level. Short-term, high-intensity activities (such as lifting a heavy weight) are not influenced by altitude because their performance does not involve oxygen-dependent processes

(Continued)

TABLE 7.4 (*Continued*)
Factors Affecting Physical Work Capacity

Protective Clothing: Bulky clothing may limit the range of motion of certain joints and hamper work movements. It may decrease stride length and therefore reduce walking and running speed and efficiency. Heavy clothing assemblies or devices that have to be carried will reduce work output for the same level of effort in activities such as climbing or carrying. Protective clothing assemblies that trap heat may increase heart rate for the same level of effort, thus reducing work capacity or hastening fatigue

Breathing Equipment: Breathing equipment can affect performance in several ways. At rest, 10–15 L of air is needed per minute. During heavy exercise, 40–50 L/min may be needed. During speech, high acceleration and peak flows are observed, which the system must cater for (up to 200 L/min). Resistance to inspiration can be exacerbated by any factors that increase turbulence in the flow of incoming air—rough inner surfaces of pipes, sudden bends, joints, or changes in bore. Resistance to expiration depends on the valve characteristics. Under heavy resistance, the breathing pattern may be characterized by panting—shallow breaths repeated frequently. This brings with it the threat of hypercapnia (accumulation of carbon dioxide in the blood). The acidity of the blood increases and the heart rate increases. Certain designs of breathing equipment impose a physiological cost on the user due to the dead space through which air has to be breathed. An infinitely long pipe would eventually cause suffocation as the user continually rebreathed his own air. Shorter lengths (200 mm–2 m) can have significant effects on physiological variables both at rest and when working as part of the expired air is rebreathed. Dead space increases heart rate, tidal volume, and partial pressure of carbon dioxide and lowers work capacity

Pacing/Time Pressure: Paced work is often a feature of production line manufacturing systems, and the operator has little control over the work rate. It is thought to be more stressful than unpaced work, and it has long been known that older workers have difficulty tolerating this kind of work. Time pressure has mental and physical effects. Birch et al. (2000) found that time pressure in performing a low precision task with low mental demands caused increased EMG activity in the shoulder and forearm muscles

Removal of Waste Products

Sustained performance at a high level depends on the ability to remove metabolic waste products from the tissues. Endurance athletes do not perform at VO_2 max intensities but at a lower level (about 85% of maximum in marathon runners, according to Jones et al., 2021). Successful performance depends on the ability to remove lactate from the body tissues, as well as on aerobic capacity. The accumulation of lactate in the blood of an athlete signals the upper limit at which endurance exercise can be maintained. Unlike VO_2 max, the blood lactate response can be improved considerably by training. Successful performance of endurance tasks also depends on the ability to dissipate heat and the ability to utilize fat as a source of fuel.

VO_2 Max and Fatigue

The equation below describes the relationship between the amount of time (*T*) a task can be carried out continuously, until exhaustion, and the demands of the task (*P*) expressed as a percentage of VO_2 max (Louhevaara et al., 1986):

$$T = 5700 \times 10^{-0.031 \times P} \text{min} \tag{7.1}$$

VO_2 max can be estimated, very approximately, using the following equation:

$$VO_2 \max \approx \text{Hr}(\max)/\text{Hr}(\text{rest}) \times 14 \text{ mL}(\text{kg.min}) \tag{7.2}$$

Table 7.5 summarizes endurance times for continuous aerobic work as a function of an individual's VO_2 max.

TABLE 7.5
Time to Exhaustion When Working Continuously at a Percentage of VO_2 Max

Relative Work Demand (VO_2 Max)	Maximum Work Time
10%	>12 h
20%	>12 h
30%	11 h
40%	5:30 h
50%	2:40 h
60%	1:20 h
70%	40 min
80%	18 min
90%	9 min

STRESS AND FATIGUE

Stress and fatigue are sometimes used to refer to temporary states of parts of the body. At other times, they are used to refer to chronic states of the whole person. In both senses, stress arising from work demands causes fatigue when work capacity is exceeded.

Stress

In technical discussion, stress is merely the applied load or task demand. The classical physical model of stress is based on Hooke's law—if we place a load of 30 kg on a horizontal beam and the beam bends under the load, the stress is 30 kg and the strain is the deformation of the beam. Once the stress is removed, the beam will resume its former shape. Under high loads, the elastic limit of the beam will be exceeded, and the beam will not resume its original shape when the stress is removed; thus, the stress is excessive in relation to the beam's capacity.

The classical physiological model of stress is derived from the work of Selye (1956) in his book *The Stress of Life*. Selye was interested in the endocrinological responses to life events, and his key insight was that many very different stressors produce the same effects. This led Selye to coin the term general adaptation syndrome—a three-phase response to stress—consisting of an alarm reaction when the threat is perceived, then resistance, followed by adaptation, and finally exhaustion or death. Today, researchers use endocrine markers such as urinary catecholamine concentration and salivary cortisol levels to assess the overall level of work or life stress. The former indicates the level of stress and resulting physiological arousal, whereas the latter indicates the degree of emotional response to the stressful experience (Lundberg, 1995). Salivary cortisol levels taken early in the morning indicate recovery from stress—higher levels indicate delayed recovery—from which higher levels of work stress are inferred. Endocrine markers and other measures, such as blood pressure, are indices of strain from which the existence of stress is inferred.

Fatigue

The classical physical model sees fatigue in relation to cycles of loading ("number of repeated exposures" in epidemiological parlance).

$$\text{Cycles to failure} = \left(\text{Applied dynamic stress/static failure stress}\right)^{-S}$$

where S is an empirically derived constant for the material in question.

These models are too simplistic to be applied directly to biological systems because biological systems have the capacity to repair themselves (recover) when the load is removed. Conversely, if the rate of loading exceeds the rate of repair, the stress tolerance ("static failure stress") will be reduced.

The exponential nature of this function draws attention to the importance of *relative stress* in causing fatigue (applied stress as a function of maximum capacity to withstand stress) and how, around some threshold level, even small increases or reductions in load will have an enormous impact on endurance time. In biological systems, with the exception of traumatic injury, fatigue results from a resource demand imbalance, as the demands utilize resources more rapidly than they are renewed.

Fatigue in physical work is straightforward to measure when the nature of the deficit is well understood and when the work is of sufficient intensity and duration to create an imbalance between task demands and resources. However, for more complex but less physically demanding tasks, it is more difficult to measure fatigue directly because the nature of the accumulating deficit is not usually obvious.

There are at least three different meanings of the term "fatigue" as it is used in the modern literature. Sometimes it is used, incorrectly, to refer to sleepiness. It is also used as a synonym for tiredness (e.g., as in muscle fatigue after prolonged exertion or depletion of carbohydrate stores after running a marathon). Finally, it is used to refer to the kind of habituation to a mental task that occurs after prolonged performance and manifests itself as a desire to do something else. In essence, fatigue manifests itself as an increasing reluctance to continue with a task. As long as the reluctance can be overcome, performance continues, but with subjectively greater effort. It can be argued that the best early index of fatigue is not the performance of the task itself but a measure of the effort needed to maintain the level of performance.

Note that to impugn a state of fatigue implies the accumulation of some kind of deficit in the resources required to maintain performance. The nature of this deficit is better understood in physical than in mental work, and mental work, the evidence suggests that the fatigue may be quite specific. From this, a tighter definition of fatigue may be offered:

> Fatigue is a consequence of work that lowers the capacity for further work of the same kind (work that demands the same resources)

In the long term, there is evidence of a real accumulated deficit underpinning fatigue in the form of depletion of the neurotransmitter 5HT, which is linked to the regulation of tiredness. It seems that long-term exposure to demanding work can deplete neurotransmitters, and this is further evidence for the concept of resource demand imbalance that underpins the present discussion of stress, strain, and fatigue.

Biochemistry and Physiology of the Stress Response

In physical work, stress equates with the total work demands that place a physiological load on the body. The physiological response to these demands consists of a number of nervous system and endocrine changes designed to facilitate adaptation. These include activation of the sympathetic nervous system, causing the secretion of epinephrine (adrenaline) and cortisol. The sympathetic nervous system readies the body for action, while the parasympathetic nervous system seems to do the opposite (Table 7.6).

For example, on beginning heavy physical work, sympathetic nervous system activation increases the flow of blood to the skeletal muscles and decreases the flow to the skin and the gut. Epinephrine is released, and the blood sugar level rises. Glycogen stores are mobilized to produce glucose, and when these stores are depleted, fat is broken down into fatty acids, which are metabolized in the mitochondria of the muscle cells or converted into glucose to supply the brain.

TABLE 7.6
Effects of Activation of the Sympathetic and Parasympathetic Nervous Systems

	Effect of Sympathetic Stimulation	Effect of Parasympathetic Stimulation
Vision	Pupil dilated Ciliary muscle relaxed (far vision)	Pupil constricted Ciliary muscle constricted (near vision)
Sweating	Enhanced	Reduced except on the palms of hands
Heart rate	Increased	Decreased
Gut	Decreased peristalsis	Increased peristalsis
Liver	Glucose released	Slight synthesis
Bladder	Relaxed	Contracted
Basal metabolism	Increased	None
Metabolism	Release of glucose from the liver and muscles	None
Adrenal medullary hormone secretion	Increased	None
Mental activity	Increased	None
Skeletal muscle	Increased glycogenolysis Increased strength	None
Fat cells	Increased lipolysis	None
Circulation	Increased blood pressure	Little effect

Allostatic Load

Allostatic load is the result of repeated exposure to high demands over long periods of time, resulting in long-term health problems such as heart disease and reduced immunity to disease. Risk factors include multiple exposures to novel stressors or to stressors that are beyond the ability to cope.

SLEEP

Most people spend about 16 h/day being awake, and the remainder of the time either asleep or resting. Although people differ in the amount of sleep they need, sleep has a restorative function. The major restorative functions in the body, such as muscle growth, tissue repair, protein synthesis, and growth hormone release, occur mostly, or in some cases only, during sleep. When we are awake, a substance called "adenosine" accumulates in the brain as ATP is oxidized to ADP, and this is thought to cause sleepiness (the effect of adenosine is blocked by caffeine). ATP levels in the brain rise during sleep. Lack of sleep is harmful.

Sleep normally progresses through five stages (Stages 1–5 of increasing depth) and rapid eye movement (REM) sleep (Figure 7.9) in cycles lasting about 90 min. The stages are characterized by changes in electroencephalographic activity and ease of rousing. The earliest stages of sleep are light, and the sleeper wakes easily. Non-REM sleep (Stages 1–4) is characterized by reductions in heart rate, body temperature, and metabolic activity, and constitutes about 80% of all sleep. It has been suggested that one role of non-REM sleep is energy conservation as well as restoration of function. During the deepest stages of non-REM sleep, the sleeper is least responsive to external stimuli—this stage is thought to be the most restorative and the most important for preventing sleepiness the next day.

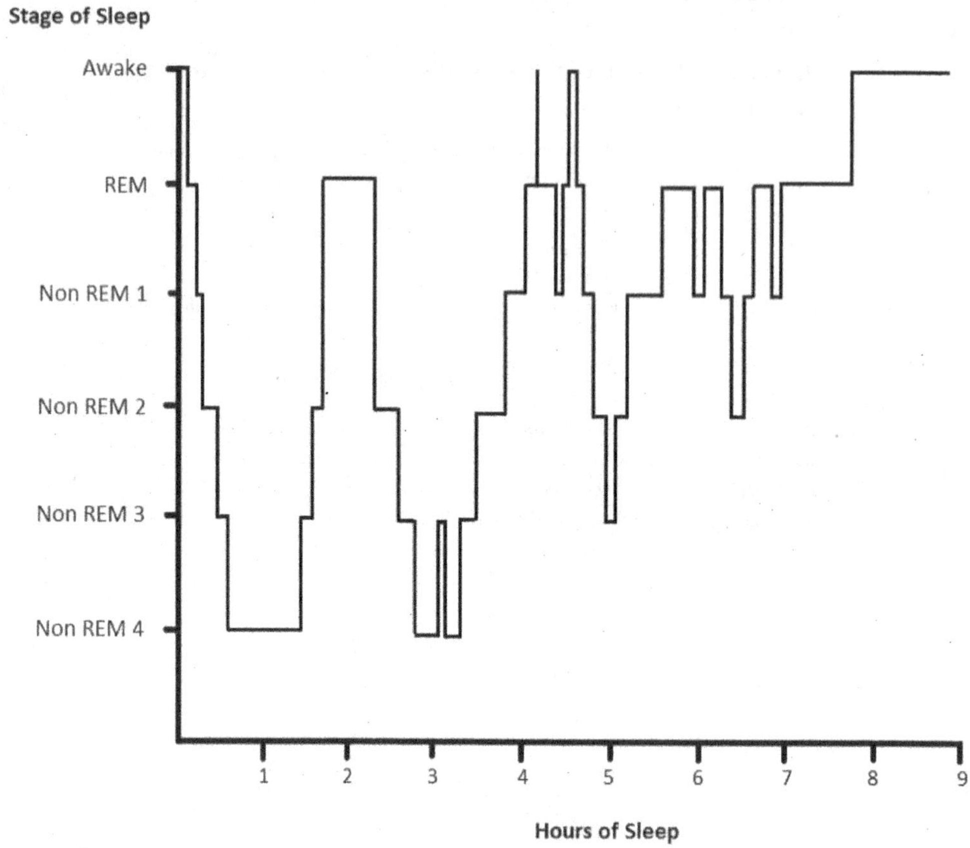

FIGURE 7.9 Main stages of the sleep cycle.

REM sleep occurs when the sleeper returns to Stage 1 sleep from Stage 4 and takes up a greater proportion of the sleep cycle during the latter part of the night. REM sleep is characterized by high-frequency ECG activity, muscle paralysis, and is the period of sleep when dreams occur. The sleeper is least easily woken during this phase. A lack of REM sleep appears to impair the ability to learn complex tasks.

Sleep Quality

A good night's sleep can be defined as a period of uninterrupted sleep through all stages of sleep a sufficient number of times. For most adults, 6–8 h seems to be required. Lack of sleep causes a sleep debt, analogous to hunger; the more deprived of sleep we are, the greater the debt and the longer we need to sleep to restore normal function. Disturbances during sleep may disrupt the normal cycling through the different sleep stages and result in insufficient REM or deep sleep. It is thought that chronic sleep deprivation is linked to the development of several conditions, including Type II diabetes. Psychological stress can disrupt sleep quality when problems at work or in life intrude into leisure time. Unresolved problems can lead to anxiety, resulting in disturbed sleep, as the sufferer is unable to stop thinking about the causes of the anxiety and is unable to consider solutions. Typically, the sleeper wakes at night and is unable to suppress intrusive thoughts, provoking further anxiety and preventing further sleep. This, in turn, can cause further anxiety and a vicious circle of disruption known as "rumination."

Sleep Inertia

Sleep inertia is a feeling of lassitude and detachment immediately after waking with diminished motor function. It is greater when waking after Stage 4 sleep and normally takes 30 min or more to dissipate. Sleep inertia is more common when sleep quality is diminished, such as after sleeping at times when one is normally awake or following disturbed sleep.

Circadian Rhythms

Many physiological processes exhibit a 24-h ("circadian") rhythm. Body temperature is low in the early morning and rises throughout the day before peaking in the late afternoon, then dropping again at night. Just after midday, many physiological variables that have been rising steadily since early morning suddenly dip to a lower level. This is known as the "post-lunch dip." Subjectively, we often feel overcome by lassitude in the early afternoon, which can be mistakenly attributed to lunchtime overindulgence. In fact, the post-lunch dip occurs even if lunch is missed.

Experimental subjects have been put into bunkers in which all time-of-day cues, as well as social contact, have been removed. Under these circumstances, body rhythms change to a slightly longer cycle. Physiological rhythms are thought to be genetically programmed but become entrained to the environment by external cues ("zeitgeber"). People living on the equator, where there is a 12-h day/night cycle throughout the year, have pronounced circadian rhythms. Social cues are also important in entraining circadian rhythms in humans, probably more so than light. People living in the Arctic, under total daylight in summer and total night in winter, also have identifiable rhythms due to the powerful effect of social cues.

Sleep and wakefulness follow a circadian rhythm entrained by daylight and social cues. The hormone melatonin plays a major role in setting the circadian rhythm and facilitating the onset of sleep, as well as having numerous other beneficial protective effects (Sieck, 2014).

Menopause/Menstruation

Menstruation is the part of the menstrual cycle in a woman when blood and other material from the lining of the uterus are discharged in the absence of pregnancy. It occurs at intervals of about one lunar month from puberty until menopause. Menopause is the time at which menstrual cycles cease, normally around 50 years of age. Both are natural processes and can have implications for work ability, absenteeism, well-being, and general health. Menstruation-related symptoms are common and include heavy bleeding, menstrual cramps, and mood disturbances, as well as reduced physical, mental, and social functioning (self-reported). Schoep et al. (2019) reported the findings of a social media survey of 32,748 women aged 15–45 years. Absenteeism during menstruation was common in 3.4% of women, with a mean of 1.3 lost workdays. Presenteeism (attending work when not fully functional, often resulting in lower productivity) was reported by 80.7% of women, on an average of 23.2 days/year. The daily productivity loss due to presenteeism was estimated at 33%. Almost two-thirds of women expressed the desire for more flexibility in their tasks and working hours.

Common symptoms of the menopause include hot flushes, night sweats, sleep problems, mood changes, brain fog (attentional deficits), headaches or migraines, and sore breasts. All these can impact productivity and the quality of working life. For further information, see Muir (2022).

BASIC APPLICATIONS

Recovery from Work Demands

Recovery can be defined as a state—as in a return to a prior state of readiness following cessation of an activity—or as a process—replenishment of resources both during and at the end of an activity. Different processes recover at different rates—ATP and creatine stores can be replenished in minutes; heart rate may take 30 min; lactate removal up to an hour; and it may take days or weeks to repair damaged muscle fibers.

Identification of recovery as a state (i.e., that the organism has recovered) requires that the nature of the deficit to be recovered from is known. Identification of recovery as a process requires that the process be described, particularly its dynamics.

Antecedents of Recovery

1. Demands have exceeded replenishment levels, and a resource deficit has occurred.
2. Capacity has dropped in relation to demands.
3. Performance has become less efficient.
4. A change in state is perceived.

In order to determine when recovery is complete, it is necessary to define the level of functioning when resources are not depleted and/or to observe the processes that give rise to recovery. In some cases, the functional level that is restored may exceed that which existed before, as when participating in a training program (*recovery plus*). Rehabilitation is defined as the restoration of normal function even though the functional level obtained may be less than that which existed before (*recovery minus*).

Even in the absence of injury, recovery is unlikely to be automatic and will require action by workers or those around them. The minimum requirement would be withdrawal from the task/setting and relocation to or engagement in a different task/setting with different demands, in an environment where the requirements for recovery (e.g., sleep) are met.

Can Recovery Be Facilitated?

Power athletes have long used creatine supplementation both to enhance their performance and to speed recovery in the short periods between high-intensity exertions involving anaerobic processes. The deficit to be recovered from is of creatine phosphate in the muscle fibers. Creatine phosphate is an intracellular compound used to convert ADP to ATP after an exertion. Dietary creatine supplementation increases the concentration of creatine in the muscle fibers. Recovery is facilitated by increasing the storage of creatine (Bean, 2000).

Conceptual Issues in the Identification of Recovery

Following Welford (1958):

1. Resource depletion may limit maximum potential, but in practice, task requirements do not require full use of available capacity. Acceptable performance does not always indicate that recovery has taken place or is not needed.
2. Alternative resources may be utilized to perform the task when capacity is limited, thereby masking the need for recovery.
3. Individuals may change the method they use in order to make additional capacity available (this depends on the constraints of the task), also masking the need for recovery.
4. To the extent that resource depletion may result in decrements in absolute capacity, several outcomes are possible when a task requiring the utilization of that capacity is performed:
   a. A task that could be done in the past can no longer be performed. Recovery is needed.
   b. A task cannot be done as easily as before because changes in method do not fully compensate for the loss of capacity. Former levels of achievement can only be maintained with increased effort, indicating the need for recovery.
   c. Performance of the task may not be affected because demands are still within capacity or because the change in method is fully effective.
   d. Overcompensation may occur so that performance is actually improved.

5. Measurement of decrements in capacity will not enable predictions to be made about complex task performance.
6. The means by which performance is maintained, rather than the level of performance on its own, may be a useful indicator of the need for recovery.
7. Since resource decrements develop at different rates within individuals, and the methods of compensating for deficits differ between people, we can expect to find a decrease in the range of activities that any one person can perform as he or she becomes fatigued, and an increase in the variability of tasks that can be performed in groups of people.
8. Many activities involve the utilization of multiple resources that recover at different rates and require different inputs.
9. Social factors will influence what people are prepared to do at work despite the need for recovery, in comparison with what they are capable of doing. Family responsibilities will motivate the behavior of men in their 30s and 40s. The desire to maintain status will motivate behavior in the decade or so after. We can expect to find disruptions in the performance of older people in maintaining their position in relation to failing performance, and in younger people whose responsibilities exceed their capacity.

Countering Sleep Disruption

Many bodily functions follow a circadian rhythm, including body temperature, heart rate, reaction time, blood pressure, the concentration of different hormones and, of course, sleep and wakefulness. The performance of many tasks also varies over the 24-h cycle. Lack of sleep and sleep disruption cause decrements in driving performance. Arnedt et al. (2001) compared the effects of sleep deprivation and alcohol intoxication on driving performance in a simulator. Performance after 22–24 h of wakefulness was similar to having a blood alcohol content of 0.08%–0.10%. This would be over the legal limit in most countries that have regulations about drinking and driving but no regulations about fitness to drive according to other criteria. The "Sustained Attention to Response Test" (SART) measures reaction time to repeated stimuli ("automatic response") as well as the ability to suppress inappropriate responses. Manley et al. (2002) found that subjects made approximately 10% more errors on the SART test at 01:00 and 07:00 h than at 13:00 and 19:00 h, indicating increasing failure to self-monitor (the "coffee in your cornflakes phenomenon"). Interestingly, subjects' reaction times when responding correctly were not influenced by time of day. This implies that for those working at night or early in the morning, increased use of failsafe, fail-soft, or forcing functions should be included in the design of interfaces and work procedures.

Task performance is better in the late morning and afternoon and worse at night and early morning—errors and accidents in industry are more common at these times than at others. Time-of-day effects have been found in the speed of reading gas meters or answering a switchboard, the frequency of car drivers falling asleep behind the wheel, and train drivers missing warning signals. These effects roughly track body rhythms—they are more common at night and during the post-lunch dip. Problems can occur when workers adapted to a particular cycle of light/dark and sleep/wakefulness have to adapt to a new cycle, as when flying to a different time zone—particularly when the difference is greater than 4 h.

Basic conditions for a good night's sleep:

- Room should be dark and quiet (<55 dB(A))
- Room temperature no greater than 18°C–20°C
- Avoid tea, coffee, and alcohol before sleeping
- Avoid large meals before bed
- Keep a regular sleep/wake schedule
- Exercise regularly but not immediately before retiring
- Develop rituals or activities to separate work from sleep

Jet Lag

Jet lag can cause physical and psychological disruption due to alterations in the timing of the sleep cycle, changes in mealtimes, and because long-haul flights can cause dehydration due to the low relative humidity in aircraft (~20%). Eastbound travel is often the most difficult because it is easier to lengthen the circadian rhythm (by staying awake longer after flying west) than to shorten it (sleeping earlier when flying east). Melatonin is available as a dietary supplement in the United States and Canada and is thought to be beneficial in helping to reset the body clock. General guidance for avoiding jet lag includes

- Make eastbound flights during daylight hours, leaving as early in the morning as possible
- Make westbound flights later in the day, arriving as close to bedtime as possible
- Drink plenty of water during the flight and eat light meals
- Stretch and walk regularly

Daylight Saving Time

Many countries in the Northern Hemisphere use daylight saving time to extend daylight hours in the winter. Typically, in the fall, the clocks are set one or more hours back so that it is lighter in the morning and dusk occurs earlier. Barnes and Drake (2015) recommend that daylight saving time be eliminated because it disrupts sleep and leads to negative health and work outcomes.

Design of Shiftwork Systems

Shiftwork has long been known to be associated with adverse health outcomes. About 20% of workers in the United States and other countries work shifts (rotating morning, afternoon, and night) compared to normal working hours (between 7:00 AM and 6:00 PM). Torquati et al. (2019) conducted a meta-analysis of seven high-quality studies and generated "effect sizes" (ES) to estimate the size of adverse outcomes of shiftwork on mental health (ES>1 = increased risk). The risk for all adverse mental health outcomes was greater in shift workers (ES = 1.28), and for depression (ES = 1.33). Female shift workers were at greater risk of depression (ES = 1.74). The increase in symptoms may be a direct result of the disruption of circadian rhythms due to changes in working hours, but might also be due to psychosocial stressors such as a lack of control over working life. Rivera et al. (2020) conducted a review of systematic reviews of research into the health effects of shift work, finding evidence for an increased risk of breast cancer in females and an increased risk of strokes in those working long hours. There was also weak evidence linking shift work with depression and cardiovascular problems. To minimize the adverse effects of shift work:

1. Adaptation to a new shift system takes place over at least a week and must be allowed for.
2. Where possible, minimize the change in working hours. If it is 4 h or less, workers will be able to accommodate the change with few problems.
3. If a voluntary, permanent night shift cannot be found, a rapidly rotating system of three 8-h shifts may be the best compromise: Day 1: early morning shift (e.g., 04:00–12:00), Day 2: day/evening shift (e.g., 12:00–20:00), and Day 3: night shift (20:00–04:00). It is pointless to try to adapt workers to night work if they revert to daytime wakefulness during their time off, and a rapidly rotating system may cause less disruption to circadian rhythms under these circumstances.
4. The permanent night shift with rapidly rotating day shifts may be the best compromise if enough people can be found to work the permanent night shift. Workers on the rotating day shift will be able to accommodate the change, since their hours of sleep and wakefulness are unlikely to be disrupted by more than 4 h.

5. Permanent night shift systems may not work, even with a willing workforce, if night workers live in noisy inner-city areas or in crowded shanty towns. Disturbances of daytime sleep may result in chronic sleep deprivation for night workers, which cancels out any health or performance benefits of their being adapted to the night shift.
6. Consider the social customs and circumstances of employees, including the time needed to travel to work. If night workers do not have personal means of transport, special arrangements may have to be made. Social, recreational, and entertainment facilities geared to the needs of night workers may assist them in maintaining their adaptation to night work during their free time.
7. It is sometimes suggested that the working week be compressed from 5, 8-h days out of every 7–2, 12-h days out of every 4. Rosa and Bonnet (1993) investigated performance and alertness on 8- and 12-h rotating shifts. Performance and alertness decrements were found on the 12-h system, as were reductions in sleep. Greater fatigue was apparent over the course of the 12-h shift than the 8-h shift, and the investigators suggested that critical activities be scheduled for the early part of 12-h shifts, particularly at night. However, not all investigations have found more problems with 12-h rather than 8-h shifts, and the former seem to be popular among workers because of the increased number of days off. Williamson et al. (1994) evaluated a 12-h shift system, which involved two 12-h day shifts followed by two 12-h night shifts followed by 4 days off work. The workers were computer operators, and the 12-h system replaced an irregular 8-h system. Data on mood state, sleeping and eating patterns, work quality, productivity, staff turnover, and sickness were collected when the 8-h system was in place and when the 12-h system had been worked for 7 months. The 12-h system produced positive changes in mental state and physical symptoms and improved sleep quality and quantity. There were no significant effects on absenteeism, staff turnover, or productivity.
8. Consider introducing "booster breaks" for nightshift workers. Booster breaks are 1-h breaks in the middle of a night shift in which night workers can take a nap for an hour. The exact timing of the break may be self-selected. Davy and Gobel (2013) found that a booster break during an 8-h night shift improved performance on a simple beading task, increased heart rate variability, and reduced self-reported sleepiness both when working and during the day. Recovery sleep was not impaired by the introduction of a booster break.
9. Night shifts in rotating shift systems:
    - Permit no more than four consecutive night shifts
    - Allow at least 2 days off after a night shift
    - Start after 7:00 AM, where possible
    - No more than four consecutive early shifts (starting before 7:00 AM)
    - Shorten the length of early shifts to avoid fatigue later n the day
    - Never combine early and late shifts on consecutive days (e.g., late nights followed by early starts)

Social factors may inhibit physiological adaptation to shift work—many individuals prefer to revert to the normal cycle on rest days and at weekends. Incomplete adaptation to shift work may be a cause of increased sickness absence. Degenfellner and Schernhammer (2021) reviewed 11 studies on "shift work tolerance." Some people can tolerate shift work better than others. It is well established that people differ in "morningness" and "eveningness," their "chronotypes"—feeling more alert and preferring to be active earlier or later in the day. There seem to be genetic underpinnings to many of the personal characteristics that characterize shift work tolerance. These include, but are not limited to: extroversion, rigidity, hardiness, sex, and low scores on self-esteem.

There is still considerable debate about how best to design shift systems to minimize unpleasant health and social effects.

Hours of Work

Hours of work are generally taken to include the time that an employee is at his or her employer's disposal (as opposed to "time on task"). Rosa and Bonnet cautioned that although workers may gladly tolerate increased fatigue in exchange for more free time on the 12-h system, there may be an increased risk of accidents, as illustrated below.

Hänecke et al. (1998) analyzed accident risk and hours of work for the German working population (Figure 7.10). As can be seen, accident risk increases exponentially after 8 h have been worked. Furthermore, there was an interaction between the hour of work and the time of day that work started. For shifts starting at 06:00, the accident risk rose rapidly after 13:00. For shifts starting at 14:00, the risk rose rapidly after 21:00, and for shifts starting at 22:00, the risk rose rapidly after 05:00. The biggest peaks at the ends of the shifts were for shifts starting at 14:00 and 22:00. This increase in the accident rate at night, after working for 7–8 h, is highly suggestive of an interaction between work hours and circadian rhythms and concurs with the predictions of a performance model developed by Spencer (1997), which is as follows:

- Performance is poorest when work starts in the late evening and continues through to mid-morning, but it only declines gradually over a work period of up to 16 h
- Performance is best when work starts from late morning to late afternoon but declines drastically after 8 h
- Do not work overtime on 12-h shifts
- No more than 12 h/day, including overtime

The explanation for this is that performance declines according to the hours worked, which is the main effect. However, for people who start work at night, the work duration effect *is countered by* the facilitating effect of the rising circadian rhythms during the morning and afternoon—people feel more awake, although they are not at their best. Those who start work around midday "start at their best," so to speak, and the effect of the circadian rhythms *is added to* the work duration effect, resulting in the largest drops in efficiency.

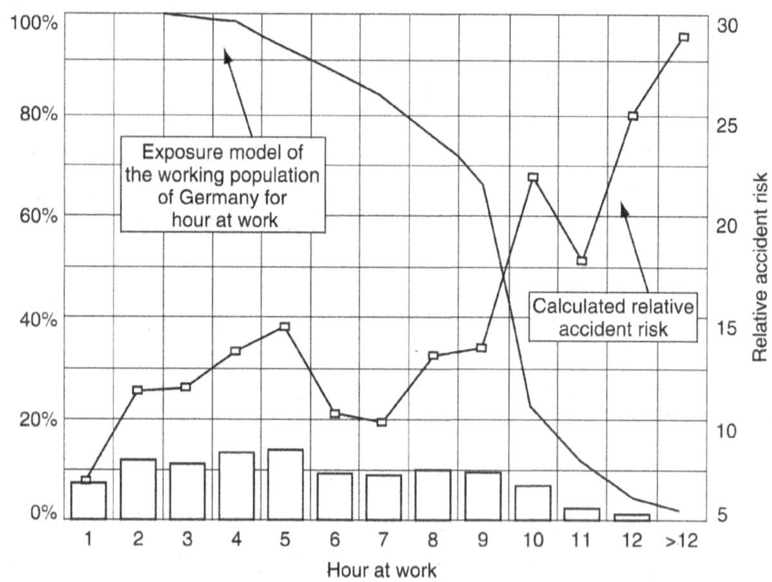

FIGURE 7.10 The relative risk of accidents increases rapidly after 8 h of work. (Redrawn from Hänecke (1998), with permission.)

Accidents and Fatigue

Acute or chronic undernutrition is often a cause of low blood sugar and is accompanied by feelings of tiredness and irritability. Studies have shown poorer performance in workers who skip breakfast than in those who don't. Kazar and Comu (2022) found an inverse relationship between blood sugar levels and accidents before lunch on construction sites. Midmorning and afternoon snacks seem to help workers sustain performance. On the other hand, blood sugar is often found to be low in victims of freeway automobile accidents. Conversely, other studies have shown that foundry workers have fewer accidents if given a high-energy glucose drink before work. For maximum work efficiency, calorie intake should be spread over the day to include a light breakfast, midmorning and afternoon snacks, lunch, and dinner.

> Generalizing from the above, it seems that to minimize accidents and for peak performance, 12-h shifts, if they are to be introduced at all, should start early in the morning.

Early research by the UK Industrial Fatigue Research Board found that long hours of work were counterproductive (Vernon, 1920). During World War I, workers in UK munitions factories worked more than 60 h/week. Reducing this to 50–55 h/week and introducing 10-min rest periods in the morning and afternoon did not reduce output and, in many cases, increased it (Vernon, 1920). Similar studies have been conducted in other countries and at different times, yielding similar findings, including lower absenteeism. The 8-h workday and 5-day week have been standard in many countries for over a hundred years. For those working long hours and likely to suffer sleep deprivation, expect decreased ability:

- To perform complex or monotonous tasks
- To perform tasks requiring sustained attention
- To perform tasks paced by a machine
- In short-term memory, attention, self-supervision, decision-making, and motivation
- In the capacity to carry out secondary tasks in addition to the main task (e.g., remembering to remove one's bank card when withdrawing cash from an ATM) and to design in
- Short, simple, well-learned tasks
- Standard procedures and written guidance
- Access to help
- Alerting displays (e.g., auditory displays)
- Error-tolerant interfaces and explicit feedback

After 3–4 days of sleep loss, few are capable of mental or physical work, and hallucinations are increasingly likely. In less extreme cases, sleepiness in the workplace can be countered by the introduction of a 15-min nap after lunch (Takahashi et al., 2004).

TOOLS AND PROCESSES

Heart Rate Recovery and Work Intensity

Brouha (1967) proposed a simple method for determining whether work intensity is excessive and whether sufficient rest is given. Based on the criterion that moderate physical work should cause the heart rate to increase by no more than 40 beats/min, the heart rate is measured between 30 s and 1 min after the cessation of work and again between 2.5 and 3 min. The first reading should not exceed 110 beats/min, and the second reading should be less by at least 20 beats/min. If the first reading is greater than 110 beats/min, the work is too arduous for the worker. If the first reading is less than 110 and the second reading is less than 20 beats/min lower, this suggests that the work–rest cycle contains too little rest (excessive waste products have accumulated, and hence the recovery in heart rate is delayed while blood is pumped to the tissues to remove them and to reoxygenate the tissues).

This is an approximate method. Many extraneous factors can also elevate heart rate and invalidate the method. These include thermal stress, anxiety, ingestion of coffee, tobacco smoking, and working after a heavy meal. Bearing this in mind, it can be used in everyday work situations.

Heart Rate Variability, Health and Fatigue

In healthy individuals, heart rate varies constantly (HRV). The intervals between beats change to maintain blood pressure, oxygen concentration, and so on under changing demands. Chronically low HRV may be a sign of ill health. Low HRV can also be a sign of fatigue.

Cardiac Stress Index

Heart rate differs between individuals due to many factors and within individuals depending on the time of day, food intake, etc. Heart rate data can be made more useful if they are normalized for each individual. The cardiac stress index (CSI) is given by

$$\text{CSI} = 100 \times \left(\frac{\text{HR during work} - \text{HR at rest}}{\text{HR max} - \text{HR at est}} \right) \tag{7.3}$$

The CSI takes into account differences in age and differences in resting heart rate. If the CSI is measured repeatedly over the workday on the same employees, it can be used to identify how work demands vary and highlight stressful periods or tasks.

Spade Design: Bridger et al. (1997) compared the physiological demands of digging with a conventional spade and the "levered" spade described in Chapter 6 (Figure 7.11). Subjects transferred 1.8 tons of sand in a laboratory task. Heart rate and oxygen consumption were measured throughout. It was hypothesized that the levered spade would be more energy efficient because it reduced the need to stoop and the need to flex and extend the trunk. The findings illustrate some important considerations in work physiology, principally the need to relate the physiological demands to the work outputs. Spade design had no influence on oxygen consumption but heart rate was higher when the levered spade was used, because of the more upright posture (the heart has to pump blood against gravity to the head). However, the digging task took longer and required more scoops with the levered spade, suggesting that the improvement in posture was associated with a reduction in digging efficiency. It is likely that the upright posture was physiologically more efficient, but that some of the benefits were offset by the addition of the second handle, which increased the weight of the spade.

The main conclusion was that the postural benefits of the levered spade were offset by the reduction in digging efficiency and the increased weight of the spade. Overall, there were no extra physiological costs, however.

Fatigue Monitoring and Detection

Under stressful conditions, breathing exceeds metabolic requirements. According to Schleifer and Ley (1994), this causes a drop in the carbon dioxide concentration in the end-tidal expired air (the last air to be expired when breathing out). Stress-induced hyperventilation decreases the concentration of carbon dioxide in the blood. This has a number of physiological effects. Vasoconstriction of blood vessels reduces the flow of blood to the heart and brain and the amount of oxygen that hemoglobin in the red blood cells can release to the tissues. This is accompanied by feelings of dizziness and heart palpitations. Theoretically, end-tidal carbon dioxide measurement should be a powerful way of measuring nonphysical stress. Unlike heart and ventilation rate measurements, which increase both under

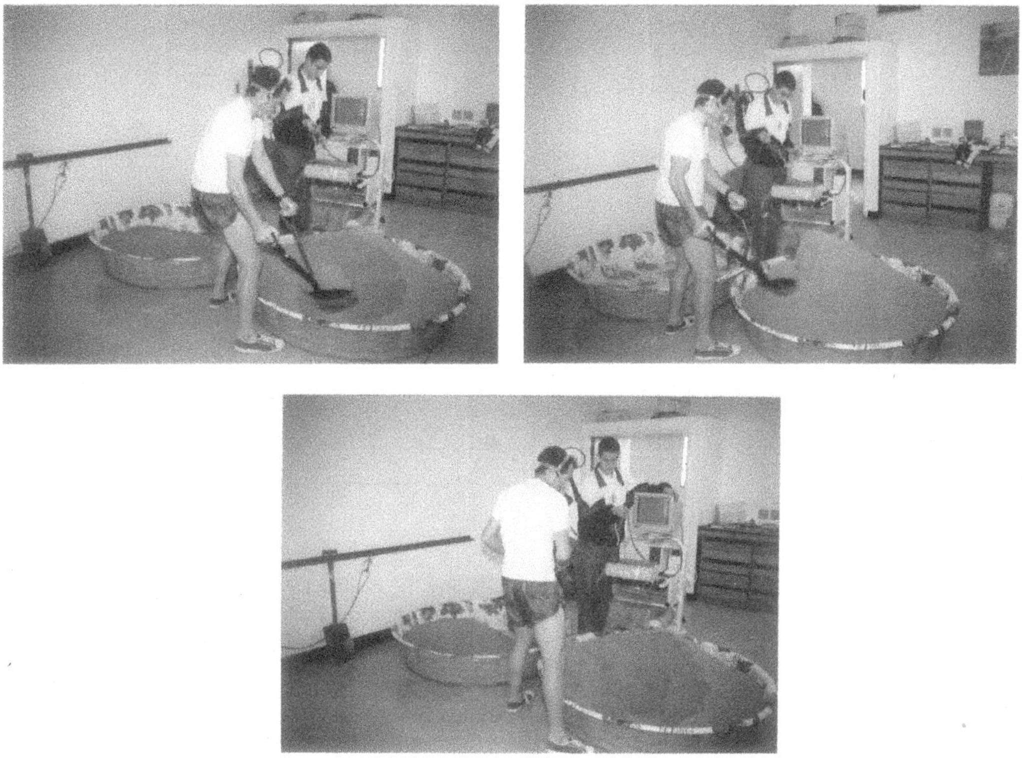

FIGURE 7.11 Measuring oxygen consumption and heart rate when using a levered spade. (From Bridger et al. (1997), with permission.)

stress and during exercise, end-tidal pCO_2 measurements decrease when the person is under stress but not when physical work is carried out (breathing does not exceed physical work requirements under these conditions). The authors compared these stress indices during self-relaxation, relaxation using a stress management technique known as progressive relaxation and during data entry work at a visual display terminal. As predicted, end-tidal pCO_2 more closely tracked self-reported mood states than the other variables. It seems to be a useful means of investigating the stress of interactive work and might be used as biofeedback in stress management programs.

Fatigue and Discomfort

Active muscles require a regular blood supply to replenish fuel and remove waste products. During rhythmic exercises involving eccentric and concentric contractions, blood flow is facilitated by the pumping action of the muscle. Fatigue occurs due to the depletion of metabolites. When muscles contract, they occlude the blood vessels within them and thus diminish their own blood supply. During sustained isometric contractions, the muscle is starved of oxygen, and waste products accumulate as oxygen-independent metabolic processes take place. Discomfort and fatigue occur rapidly during sustained isometric contractions for this reason.

Muscles fatigue rapidly under conditions of static loading, even at low workload (Figure 7.12). As can be seen, the endurance time for maximum contractions is on the order of a second; 50% of maximum can be held for a minute or less, and this is the rationale underlying ISO 1226. However, even though a static exertion such as standing with the arm fully extended at 90° to the torso

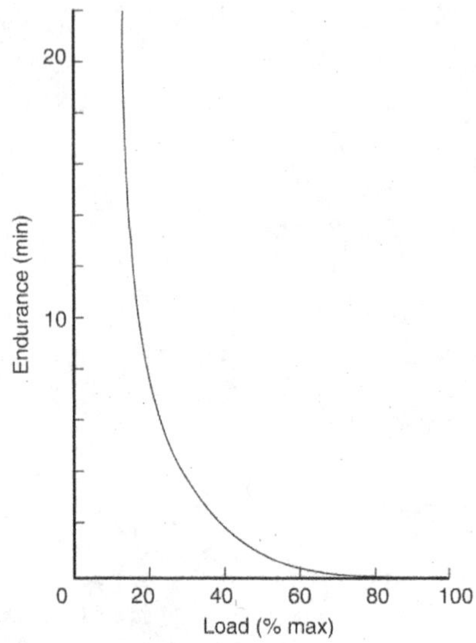

FIGURE 7.12 Relationship between the load on a muscle and the endurance time. Loads that require less than 20% of a muscle's maximum can be endured for minutes or hours. Loads of 80% or more can only be endured for a few seconds.

requires only 10% of maximum effort, musculoskeletal complaints will occur if the action has to be sustained all day at work. HFE Workshop 7.1 describes a method for estimating maximum endurance time for static exertions.

Fatigue and Pain

Repeated or sustained activities, rapid movements, and large forces can stimulate pain receptors in a muscle. Since skeletal muscle makes up 40% of the tissues of the body, it should come as no surprise that many of the aches and pains we experience in our daily lives are of muscular origin. The tension exerted by muscle can cause movement of a body part in carrying out work activities; it is also essential for resisting externally imposed forces, which might otherwise destabilize and damage a joint. The loss of muscle function that occurs with fatigue implies an increased risk of musculoskeletal injury. This indicates why a consideration of work–rest cycles, work durations, and forces is an essential part of the risk assessment process in job evaluation and injury prevention.

HFE WORKSHOP 7.1

ESTIMATING MAXIMUM ENDURANCE TIMES FOR STATIC EXERTIONS

According to Rohmert (1960), the value of an individual's maximum static exertion in a particular posture can be used to estimate their endurance time for a submaximal exertion in the same posture, using the following equation:

$$T = -90 + \frac{126}{P} - \frac{36}{P^2} + \frac{6}{P^3}$$

where P is the required exertion, expressed as a decimal fraction of the maximum exertion, and T is the predicted endurance time in seconds.

(Continued)

HFE WORKSHOP 7.1 (*Continued*)

For example, if an operator had to exert a grip force of 25 kg while carrying a milk crate and his maximum grip force was 50 kg, then the endurance time would be

$$\begin{aligned} T &= -90+\frac{126}{0.5}-\frac{36}{(0.5)^2}+\frac{6}{(0.5)^3} \\ &= -90+252-144+48=66\text{s} \end{aligned}$$

where $P=25/50=0.5$.

ESTIMATING ENDURANCE TIMES IN A POPULATION OF EMPLOYEES

Sometimes we want to know how long workers in the employee population can exert a force of a given magnitude, as in the following example.

The driver of an underground train has to hold a dead man's handle for safety reasons. If he lets go of the handle, as might happen if he were to die or have a heart attack, the train fails safe (i.e., it stops automatically). The handle requires a force of 30 N to be held in the open position, and it has to be held for up to 20 min at a time. A strength survey of 50 drivers reveals that the mean and standard deviation of the maximum pull forces on the handle are

$$\begin{aligned} \bar{x} &= 300\text{ N} \\ \text{sd} &= 50\text{ N} \end{aligned}$$

Clearly, we want to know if the weakest drivers will be able to hold the handle open for this period. We will select the first percentile pull force for the assessment.

From Appendix A, the value of z corresponding to the first percentile is –2.32; in other words, the first percentile strength is 2.32 standard deviations less than the mean strength.

$$1\text{st percentile strength} = 300-(2.32\times 50)=300-116=184\text{N}$$

The relative effort for a first percentile driver is given by

$$P = 30/184 = 0.16$$

Therefore, the maximum endurance time (T) for a driver with first percentile strength will be

$$\begin{aligned} T &= -90+\frac{126}{P}-\frac{36}{P^2}+\frac{6}{P^3} \\ &= -90+\frac{126}{0.16}-\frac{36}{(0.16)^2}+\frac{6}{(0.16)^3} \\ &= -90+787.5-1406.25+1464.84 \\ &= 756.1\text{ s} \\ &= 12\text{ min }36\text{ s} \end{aligned}$$

(*Continued*)

HFE WORKSHOP 7.1 (*Continued*)

COMMENTS

Clearly, the interpretation of the result depends on how long the journeys are and the length of rest periods between them. However, if the journeys were more than a few minutes, the arrangement would likely challenge many drivers over an 8-h day. In this case, we might investigate the design of the handle and the driver's posture. It might be possible to lengthen the handle to provide more mechanical advantage or to redesign it to lessen the required force.

To illustrate the scope for improvement, we might carry out some more calculations to find out how the endurance time would change if the force required to hold the lever was reduced to 15 N. Recalculating for a 1st percentile driver yields a maximum endurance time of 7578 s, which is over 2 h. A 50% reduction in force yields a tenfold increase in endurance time. More people will be able to do this job safely and with reduced risk of developing musculoskeletal pain as a result of localized muscle fatigue.

Cautionary Note

The method is a general one, whereas different muscle groups will fatigue differently. Van Dieen and Oude Vrielink (1993) argue that the method is unreliable. However, the method does have the advantage that the value of maximum voluntary contraction (MVC) can be based on tests carried out using the real task in the actual workplace, rather than on laboratory test data where the tools, equipment, and so on differ. It is unlikely that there will ever be a comprehensive database on human strength and endurance that is applicable to every situation. From a pragmatic point of view, we might argue that this method has some merits as part of a more comprehensive process, but that it should never be used on its own. Employees should be consulted, and a detailed investigation of the whole task should be carried out.

Maximum Exertions Over a Duty Cycle

Potvin (2012) provides an equation to estimate maximum acceptable exertions (MAE) over a duty cycle (DC), where DC is the duration of the exertion as a fraction of the cycle time. So, if a worker has to pull on a lever for 15 s every minute, DC=0.25.

$$\text{MAE} = 1 - \text{DC}^{0.24}$$

Potvin gives the following example:

Suppose a worker has to pull on a block with a pinch grip and DC=0.25. The maximum pinch grip is 70 N (maximum voluntary contraction, MVC).

$$\text{MAE} = 1 - 025^{0.24} = 1 - 0.72 = 0.28 \text{ of MVC} = 19.8\text{N}$$

Further examples of tools for estimating the strength of exertions can be found in Gibson and Potvin (2016) and Gibson et al. (2017).

Electromyography

Discomfort is a subjective experience that can result from a combination of physiological and psychological processes, including muscle fatigue. Muscle fatigue is a physiological phenomenon that can be observed using electromyography (EMG). Muscle cells are arranged in functional units known as motor units. Each motor unit is connected to a single nerve fiber, and all of the cells in that unit fire in response to an impulse from the fiber on an on–off basis. The number of muscle cells per motor unit varies considerably between muscles—calf (soleus) muscle motor units contain many hundreds of muscle cells, whereas motor units in the eye muscles contain as few as five. The latter muscles are capable of much finer, coordinated movement due to their superior innervation.

Muscle tension can be varied by increasing the number of motor units firing at a single time or by increasing the firing rate. Electrical activity in muscles can be detected using surface electrodes placed either on the skin overlying the muscle or by needle electrodes inserted into the muscle. EMG activity from fresh muscle resembles electrical noise—a smooth or buzzy trace usually appears. It has long been known that the frequency content of the myoelectric (ME) signal shifts toward the lower frequencies as a sustained contraction progresses to exhaustion (Tankisi et al., 2020). The mean and median frequencies of the EMG signal can be used as indices of fatigue. The initial median frequency (of the fresh muscle at the start of the contraction) can be taken as a cutoff, and the ratio between the root mean square (rms) voltages below and above the median is used to monitor the progress of fatigue. This parameter increases in magnitude as the muscle fatigues. In severely fatigued muscles, the availability of fresh motor units becomes limited, and visible tremor sets in (Figure 7.13).

EMG is used to identify tasks that cause rapid muscular fatigue, complementing subjective techniques in which workers indicate, on body diagrams or by questionnaire, the location and severity of musculoskeletal pain.

Rest Periods for Static Exertions

The rate of fatigue increases exponentially over time, and the rate of recovery decreases exponentially over time. Short, frequent breaks are usually better than infrequent, long breaks (because fatigue is minimized and recovery is maximized for a work–rest cycle of a given length). As a rule of thumb, for tasks requiring very high exertions, it is more effective to shorten the duration of work periods and provide more frequent rest to prevent exhaustion than to lower the work intensity (since lowering the work intensity only increases the length of the work period and provides less time for rest). For low-intensity tasks, it is more effective to lower the work intensity than to lower the duration of the work period

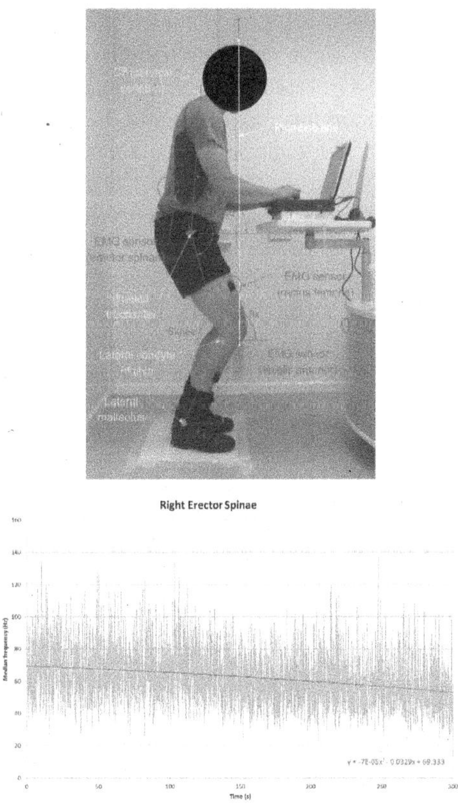

FIGURE 7.13 EMG activity during a sustained muscle contraction. The median frequency of the signal drops as fatigue sets in.

to prevent exhaustion (Rohmert, 1967). This is because endurance time increases greatly with small reductions in effort at the low-effort end of the scale. For any given task, it would be necessary to time the maximum intensities and durations to develop a practical recommendation.

Absolute VO_2

For some tasks, the energy demands are fixed and can be expressed in absolute terms, as a requirement for oxygen consumption. However, values of VO_2 max are usually normalized for body weight and can be misleading for the following reasons. Suppose we wish to select workers to carry a 40kg rucksack over rugged terrain. Worker A, weighing 100kg, has a VO_2 max of 40mL O_2/kg/min, and worker B, weighing 60kg, has a VO_2 max of 50mL O_2/kg/min. The first worker has a lower aerobic capacity relative to his own body weight but greater absolute capacity (4L of O_2 per minute as opposed to 3 L/min). If the carrying task (with no load) requires 50% of VO_2 max, worker A has 2L spare capacity compared to worker B with 1.5L spare capacity. Suppose that when the loaded carrying task is performed, an additional 1L of oxygen per minute is needed. Worker A will then be working at 75% of his VO_2 max (the oxygen requirement of the carrying task is 25% of his VO_2 max), whereas worker B will be working at 83% of his VO_2 max. From Table 7.5, we can estimate that worker A will be able to endure the task for around 25min before exhaustion sets in, whereas B will manage for less than 18min. This example illustrates that with fixed loads, the larger worker will have greater endurance. This explains why body size is often a good surrogate measure of an individual's capacity for physical work.

Indirect Measures of Energy Expenditure

Heart rate is easier to measure than oxygen consumption. It increases as a function of workload and oxygen uptake and is often used as an indirect measurement of energy expenditure. However, different individuals can have the same heart rate but completely different levels of oxygen consumption; on its own, heart rate cannot be used to estimate the energy requirements of a job unless the relationship between each worker's heart rate and oxygen uptake is known. Both variables have to be measured simultaneously in the laboratory at a number of different submaximal workloads. This is known as calibrating the heart rate/VO_2 relationship for a worker. Since the relationship between the two variables is linear, a worker's heart rate, when it is subsequently measured in the field, can be converted into an estimate of oxygen uptake by reference to the laboratory data. Estimates of energy expenditure during work can then be calculated from the oxygen consumption data.

Several researchers have validated this method (e.g., Ceesay et al., 1989) by comparing the heart rate estimates with estimates obtained using whole-body calorimetry (the gold standard for estimating energy expenditure). Close correlations between the two methods of estimation suggest that heart rate measurement of calibrated subjects can give valid estimates of energy expenditure. An introduction to the method is given in HFE Workshop 7.2.

HFE WORKSHOP 7.2

ESTIMATING OXYGEN CONSUMPTION FROM HEART RATE

The data yielded by the calibration procedure are in the form of paired heart rate VO_2 measurements made at several submaximal levels on each subject, that is:

HR	VO_2
X_1	Y_1
X_2	Y_2
X_n	Y_n

(Continued)

HFE WORKSHOP 7.2 (*Continued*)

Heart rate is taken to be the independent variable, and oxygen consumption the dependent variable (later on, we want to predict oxygen consumption from heart rate, not the other way around).

Linear regression analysis can be used to calculate the calibration curve for the data. We want the regression of y (oxygen consumption) on x (heart rate) of the form:

$$y = a + bx$$

where y is the oxygen consumption, x is the heart rate, and a and b are constants that give the intercept and slope of the line, respectively.

The formula for calculating the slope of the line is given by

$$byx = \frac{\Sigma xy - (\Sigma x \Sigma y/n)}{\left(\Sigma x^2 - \left[(\Sigma x)^2/n\right]\right)}$$

The formula for calculating the intercept is given by

$$ayx = \frac{\Sigma xy - byx\Sigma x}{n}$$

The analyses are done separately for each subject, and the equations derived are personal statements about each subject's cardiovascular/respiratory functioning. They are unique to each subject. The calibration procedure is completed by determining the upper and lower limits for heart rate and oxygen consumption.

The upper limit is the individual's predicted maximum volume of oxygen uptake in liters per minute. It is calculated from the calibration procedure by means of a predicted maximum heart rate given by

$$\text{HR}(\text{max}) = 200 - (0.65)\text{age}$$

Young persons from 20 to 30 years of age are assumed to have a maximum heart rate of 200 beats/min.

Maximum oxygen uptake is estimated from the calibration curve by extrapolation, under the assumption that the relationship remains linear at higher levels of heart rate.

Suppose a 25-year-old individual's heart rate/oxygen curve was as follows:

$$y = 0.028x - 1.49$$

y is the oxygen consumption in liters per minute, and x is the heart rate in beats per minute given that the maximum heart rate is taken to be 200 beats/min, then maximum oxygen uptake is given by

$$\begin{aligned} \text{VO}_2(\text{max}) &= 0.028(200) - 1.49 \\ &= 5.6 - 1.49 \\ &= 4.11\ \text{L/min} \end{aligned}$$

Resting heart rate (the lower limit) is defined according to predetermined criteria, in which a basal metabolic level is assumed to be reflected by an oxygen uptake of 0.25 L/min.

A second calibration curve must be calculated from the data. This is again of the form

$$y = a \pm bx$$

(*Continued*)

HFE WORKSHOP 7.2 (*Continued*)

However, y is now heart rate and x, oxygen consumption. That is, heart rate is the dependent variable and oxygen consumption the independent variable. This has to be done because we want to predict a theoretical resting heart rate from a criterion level of oxygen consumption.

Suppose the above subject has a calibration curve of

$$y = 28.2x + 66.9$$

For an oxygen consumption of 0.25 L/min, the resting heart rate will be

$$\begin{aligned} y &= 28.2(0.25) + 66.9 \\ &= 74 \text{ beats/min} \end{aligned}$$

Having obtained calibration curves for each of the subjects and established physiological limits for them, it is possible to evaluate the physiological cost of performing work using heart rate data taken in the work situation. When moving from the laboratory situation to the work situation, this estimate of resting heart rate can be compared with a measurement taken when the worker is at rest. If, in the work situation, the resting heart rate is higher, it may not be valid to use the procedure.

Aerobic Capacity of U.S. Workers: NIOSH Guideline

NIOSH (1981) has published data concerning the maximum aerobic capacity of U.S. workers. The fiftieth percentile male and female capacities, in terms of energy output, are approximately 63 and 44 kJ/min, respectively. The lower fifth percentile capacities are 52.3 and 33.5 kJ/min. For continuous work, NIOSH states that energy expenditure should not exceed 33% of an individual's maximum capacity (21 kJ/min for men and 14.6 kJ/min for women) over an 8 h shift. It can be seen that a fiftieth percentile male worker would be able to carry out heavy work for 8 h but not very heavy work. A fifth percentile worker would only be able to carry out moderate work for 8 h. By determining the energy requirements of different jobs, it is possible to identify those that can only be performed by the fitter members of the workforce. Where possible, such jobs can be redesigned to lower the required energy expenditure, thus making them available to a wider range of workers.

For occasional lifting tasks (1 h or less per day), NIOSH recommends that energy expenditure rates should not exceed 37.5 kJ/min for physically fit men and 27 kJ/min for physically fit women. The upper limit of energy expenditure is approximately 67 kJ/min for men and 46 kJ/min for women when sustained for up to 4 min.

VO_2 Max and Industrial Work

It appears that individuals can work for 8 h at a rate of 30%–50% of their maximum capacity, depending on the frequency and length of rest periods (see below). Industrial jobs must be designed so that the less fit members of the workforce can carry them out. For example, suppose two workers have VO_2 max values of 6 and 2 L of oxygen per minute, respectively, and the task requires the expenditure of 2 L of oxygen per minute. The first worker will have a greater work capacity because he will be able to work comfortably all day at only 33% of his VO_2 max. His energy will be met by oxygen-dependent processes, which means that he will be able to work in a physiologically steady state and obtain energy from the oxidation of fat as well as from carbohydrates (recalling that oxygen is required to enable fatty acids to be metabolized in the mitochondria). The second worker will be

TABLE 7.7
Severity of Work in Terms of VO_2, Heart Rate, and Energy Expenditure

Work Severity	VO_2 (L/min)	Heart Rate (beats/min)	Energy Expenditure (kJ/min)
Basal metabolism[a]	0.25		
Light work	<0.5	<90	<10.5
Moderate work	0.5–1.0	90–110	10.5–21
Heavy work	1.0–1.5	110–130	21–31.5
Very heavy work	1.5–2.0	130–150	31.5–42
Extremely heavy work	>2.0	150–170	>42

Source: Modified from Astrand and Rodahl (1977).
[a] Estimated value.

required to work at 100% of VO_2 max and will tend to utilize oxygen-independent processes. He will only be able to work for a few minutes and will require frequent rest breaks to metabolize the accumulated metabolic waste products. He will also be dependent on carbohydrates as a source of fuel.

It will be readily appreciated that in heavy manual work, which is remunerated on a piece-rate basis (i.e., in which the amount each worker earns depends on how much he produces), individuals with a high VO_2 max will be at an advantage, since they will require fewer rest breaks. Various attempts have been made to classify the severity of work in terms of oxygen uptake, heart rate, and energy expenditure. Table 7.7 is a representative example. HFE Workshop 7.3 describes a method for estimating the duration of hourly rest periods for tasks of different intensities that require a relatively steady level of effort.

HFE WORKSHOP 7.3

CALCULATION OF REST PERIODS IN MANUAL WORK

Murrell (1971) proposed that rest periods can be calculated according to the following empirical formula:

$$\text{Rest allowance} = \frac{w(b-s)}{b-0.3}$$

where w is the length of the working period, b is the oxygen uptake, and s is the standard uptake for continuous work.

The standard level for continuous work will depend on the population of workers for whom we are designing the task. Clearly, if the worker population is very fit, then a higher level will be needed than if the population is unfit. How might we decide what is an appropriate level for the standard oxygen uptakes? For U.S. workers, the maximum work capacities in terms of energy output are 63 and 44 kJ/min for males and females, respectively. If we take 33% of the maximum as our standard for an 8 h workday (see Table 7.5), then the work intensities are 21 and 14.6 kJ/min, respectively. Since 1 L of oxygen releases about 20 kJ of energy, the standard uptake for this group is given by

$$\text{Standard uptake for females} = 14.6/20 = 0.73\ \text{L}$$

$$\text{Standard uptake for males} = 21/20 = 1.05\ \text{L}$$

(*Continued*)

HFE WORKSHOP 7.3 (*Continued*)

The standard rates given above include normal rest breaks and meals for moderately paced work (Table 7.7).

Suppose the oxygen requirement for a manual task is 1.5 L/min, and personnel have to work for periods of 1/2 h. What is the required rest period?

$$\text{Rest allowance (males)} = \frac{0.5(1.5-1.05)}{1.5-0.3} = 0.19\ \text{h} = 11\ \text{min}\ 24\ \text{s}$$

$$\text{Rest allowance (females)} = \frac{0.5(1.5-0.73)}{1.5-0.3} = 0.32\ \text{h} = 19\ \text{min}\ 12\ \text{s}$$

ESTIMATING TIME TO EXHAUSTION

From Table 7.5, we can estimate time to exhaustion when carrying out the task. We can estimate the VO_2 max values for the U.S. males and females from the NIOSH maximum work capacities of 63 and 44 kJ/min, respectively. These convert to approximately 3.15 and 2.2 L/min, respectively. The required intakes of 1.5 L/min in relation to the percentage of maximum are

$$\text{Relative work demand (males)} = 1.5/3.15 = 48\%$$

$$\text{Relative work demand (females)} = 1.5/2.2 = 68\%$$

So, the task is 48% of the average male aerobic capacity and 68% of the average female capacity.

Time to exhaustion can be estimated using the equation below:

$$T = 5700 \times 10^{-0031 \times p}\ \text{min}$$

where p is the relative work demand.

$$\text{For males: } T = 5700 \times 10^{-0.031 \times 48} = 5700 \times 10^{-1.488} = 5700 \times \left(\frac{1}{30.76}\right) = 185.3\ \text{min}$$

$$\text{For females: } T = 5700 \times 10^{-0.031 \times 68} = 5700 \times 10^{-2.108} = 5700 \times \left(\frac{1}{128.23}\right) = 44.45\ \text{min}$$

WORKLOAD INDEX

Another way of expressing work demands is as a workload index (WI) with respect to a reasonable rate of work for 8 h, say 33% of VO_2 max. In our example above, we found 33% VO_2 max values of 1.05 L/min for U.S. males and 0.73 for U.S. females. The WI for the task is therefore

$$\frac{\text{Actual consumption}}{\text{Consumption at 33\%}} \times 100$$

$$\text{WI for males} = \frac{1.5}{1.05} = 1.43$$

$$\text{WI for females} = 1.5/0.73 = 2.05$$

(*Continued*)

HFE WORKSHOP 7.3 (*Continued*)

In both cases, the WI is greater than 1, and so the task should not be carried out for 8 h/day without additional rest periods or extra personnel.

INTERPRETATION

The equations enable us to estimate rest periods and maximum performance times for different tasks. By using the NIOSH data on the mean work capacity of U.S. workers, we can get an idea of what average people are capable of, how long they can work, and how much rest they will need. These data can then be used to design a system of work around a production schedule and are useful in estimating the number of workers needed to produce the required output. The general trend is to mechanize or automate tasks to increase production with fewer people. However, this is not always possible (in the case of the emergency and armed services, for example), and selection tests are used to ensure that all joiners have the required work capacity, which is normally higher than that of the general population. There is a difference in the work capacity of males and females, but the distribution of capacities overlaps. Therefore, gender-free selection tests are used to ensure that only males or females have the required physical capacity for the job.

Subjective Measures of Physical Effort

The Borg scale (Table 7.8) is the most widely used method in ergonomics (Borg, 1982). Subjects rate their perceived level of exertion (RPE) during or after performing the task on a scale from 6 to 20, corresponding to heart rates of 60–200 beats/min. High positive correlations between heart rate and RPE are usually found. The Borg scale is normally used with other measures, typically heart rate and oxygen consumption. It is not normally used as a substitute for these measures but is used in a variety of research areas, very often as part of a battery of subjective measures, such as ratings of regional body pain using body maps (Chapter 2).

TABLE 7.8
Borg RPE Scale

Rating	Interpretation of Rating
6	No exertion at all
7	
8	Extremely light
9	Very light
10	
11	Light
12	
13	Somewhat hard
14	
15	Hard
16	
17	Very hard
18	
19	Extremely hard
20	Maximal exertion

In general, subjective measures are best used with objective measures because, on their own, they are susceptible to experimental effects (Chapter 16). The Borg scale has the advantage of being a global measure—the ratings are influenced by all the demands in the work situation, whereas physiological and other subjective measures vary according to a subset of these demands.

Measuring Recovery from Work

A self-report scale measuring the "Need for Recovery After Work" has been developed (Van Veldhoven and Meijman, 1994, Sluiter, 2003). In cases where individuals do not feel they have sufficient opportunity to unwind and recuperate from work, this deficit may be carried over to the next working day and accumulate over time. A high need for recovery can be an indicator of increased susceptibility to health problems in a working population—it has been associated with the development of psychosomatic health complaints and an increased risk of cardiovascular disease in working populations (Table 7.9).

The rationale underpinning the scale is that the effect of work demands on personnel can be assessed by measuring the need to recover from work at the end of the working day. Indicators of fatigue assessed on the NFR scale include a lack of concentration and reduced motivation for activities at the end of the day. Participants are given dichotomous response options of "yes" or "no" to indicate whether or not they experience each indicator. The total number of indicators present for each individual is converted into a scale score ranging from 0 to 100, where higher scores show a greater need for recovery and thereby greater short-term work-related fatigue. The authors of the NFR scale have generated Dutch norms that can be used for the interpretation of surveys. No more than 50% of the sample should score 36 or more on the NFR scale (equivalent to reporting about four or more indicators of fatigue on the 11-item NFR scale). The NFR scale is a widely used,

TABLE 7.9
Need for Recovery from Work Scale

The following questions ask you to answer "yes" or "no" to a number of statements. It refers to how you generally feel *overall*—if you had to give a "yes" or "no" reply. Please circle the answer that best applies to you.

a.	Do you find it hard to relax at the end of a working day?	Yes	No
b.	At the end of a working day, do you feel worn out?	Yes	No
c.	Does your job cause you to feel rather exhausted at the end of a working day?	Yes	No
d.	Generally speaking, do you still feel fresh after your evening meal?	Yes	No
e.	Generally speaking, when the ship allows you time off to spend in your own way, do you find it hard to relax?	Yes	No
f.	Do you have trouble concentrating in the hours after your working day?	Yes	No
g.	Do you find it hard to show interest in other people when you have just finished work?	Yes	No
h.	In general, does it take you an hour or so to feel fully recovered after work?	Yes	No
i.	When you finish work, do you feel that people should leave you alone for some time?	Yes	No
j.	After a working day, do you find you are often too tired to start other activities?	Yes	No
k.	During the last part of the working day, do you find that on occasion you cannot perform your job properly because of fatigue?	Yes	No

reliable, and valid measure that detects early indicators of work-related fatigue and can be invaluable in the early identification of symptoms. The scale has previously been found to predict occupational health complaints in coach drivers, including sleeping problems and emotional exhaustion (Sluiter, 1999), and to predict sickness absence from work due to psychological complaints (Sluiter, 2003). It is sensitive enough to detect an increase in work-related fatigue when working hours are increased (de Croon et al., 2006) and correlates highly with other measures of fatigue (de Croon et al., 2006). The scale also has good test–retest reliability over 2 years if the work environment remains relatively stable (0.68–0.80; de Croon et al., 2006).

Karolinska Sleepiness Scale

Sleepiness is often measured using the Karolinska Sleepiness Scale (Akerstedt, 1990). Employees rate their level of sleepiness on a nine-point scale:

1 Extremely alert
2
3 Alert
4
5 Neither alert nor sleepy
6
7 Sleepy—but no difficulty staying awake
8
9 Extremely sleepy—fighting sleep

Measuring Fatigue

The Fatigue Assessment Scale (Table 7.10) provides an overall measure of global fatigue.

SYSTEM INTEGRATION

Physiological methods are well established, as are the techniques for ensuring that they give valid and reliable data.

TABLE 7.10
The Fatigue Assessment Scale[a]

Never Sometimes Regularly Often Always
I am bothered by fatigue
I get tired very quickly
I don't do much during the day
I have enough energy for everyday life
Physically, I feel exhausted
I have problems starting things
I have problems thinking clearly
I feel no desire to do anything
Mentally, I feel exhausted
When I am doing something, I can concentrate quite well.

[a] University of Pennsylvania School of Medicine

- The testing situation should be similar to the work situation in terms of climate and the clothing worn by the subject. Variables such as temperature can also increase heart rate independently of the energy expenditure due to the work itself. If a person is tested in a cool laboratory and then observed working in a hot environment, the observed heart rate would be higher in the heat than in the laboratory, and the oxygen consumption would be overestimated.
- The calibration method should resemble the work to be evaluated in terms of the type of muscles used. For some tasks, it may be more appropriate to use a bicycle ergometer than a treadmill or to use a method that requires the use of the upper as well as the lower body.

Heart rate variability (HRV) and the Galvanic skin response (GSR) are valid methods for detecting mental fatigue. Ouyang et al. (2023) measured GSR and HRV in construction workers carrying 15 kg loads on a treadmill at 4 km/h. Concurrently, a screen-based cognitive task involving the identification of hazards o construction sites was performed (with hazards in 20% of the stimuli). Under fatigued conditions, task reaction time increased, and both subjective concentration and task accuracy were lower. Algorithms were used to examine the specificity and sensitivity of GSR and HRV features in hazard detection. Using HRV features yielded a prediction accuracy of 88.33%; using GSR features alone achieved an accuracy of 76.67%; combining HRV and GSR features improved the accuracy to 96.67%. Figure 7.14 shows a monitor for HRV analysis.

Static strength endurance prediction is based on general models, whereas different muscle groups vary in size, type of muscle fiber, and size of motor unit, and so predictions are not very accurate. In practice, endurance depends greatly on motivation, which will not be the same in a testing situation as it is in the workplace. Physiological methods are very sensitive, which is both an advantage and a drawback, because it means that extraneous factors can affect the results if not controlled for.

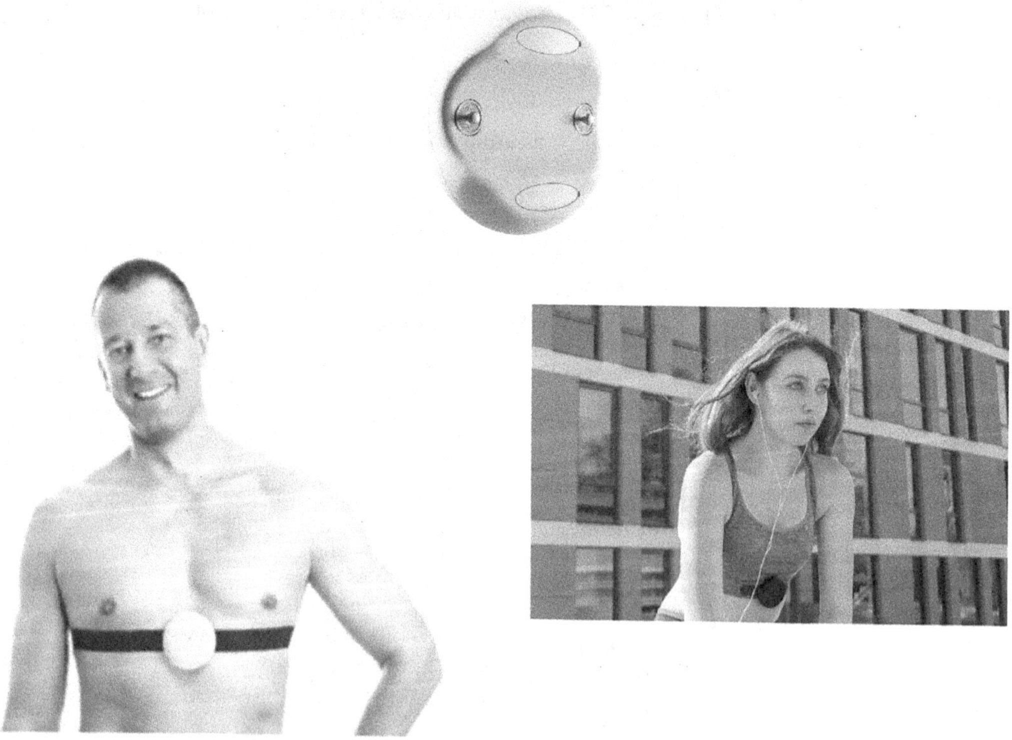

FIGURE 7.14 Compact sensor measuring 9D motion, ECG, HRV, HR, respiration, and temperature (https://web.fibion.com/fibion-vitals/).

Motivation is such a factor—highly motivated individuals may exhibit greater-than-expected work capacity, while unmotivated individuals may exhibit much less than expected. The methods are highly diagnostic in the sense that they disclose the reasons why a task should or should not be redesigned and the nature of any risks, but they are also highly intrusive and require that an ethically approved protocol be written before work can commence. The methods are relatively expensive—specialized equipment and trained personnel are needed.

The methods described in the workshops for static strength prediction and rest period calculation are old and based on the knowledge available at the time. For definitive recommendations to be made, modern methods using specialized equipment must be used. However, the old methods are relatively nonintrusive; they are inexpensive, easy to use, and suitable for rapid assessments of large numbers of tasks. Their validity can be improved by combining them with subjective assessment techniques (such as Borg scale measures), questionnaires (see Chapter 16), and, of course, consultation with real workers.

Exoskeletons to Reduce Static Postural Stress

Figure 7.15 shows a "chairless chair"—a lightweight exoskeleton attached to the waist and thighs that enables mobile workers to adopt sit-stand postures while working. When walking, the jointed exoskeleton follows the movement of the legs freely during normal gait. When the worker wishes to sit or perch to work at a machine or carry out a short task, the supports on the exoskeleton, jointed at the knee, can be locked into place for the duration of the task. This relieves the leg and back muscles of the effort that would be required to maintain a static posture appropriate to the task while working.

Static work postures have long been known to be associated with fatigue, discomfort, and injury in industrial work, and the International Standards Organization has published guidance in the form of time limits for the maintenance of a variety of static postures (ISO 11226, 2000). Bridger et al. (2018) evaluated the effects of wearing a passive exoskeleton to support a semi-standing posture while carrying out a computerized cognitive task for 5 min (the knees were flexed by 25°). The posture was based on Dul et al. (1993) and was expected to induce fatigue without exposing the subjects to the risk of injury. Subjects wore heart rate monitors for 30 s before, during, and after adopting the posture, and instantaneous heart rate was sampled every 5 s. Subjects rated task difficulty using the NASA TLX (see Chapter 12) and rated the self-control demands by answering three questions.

- It was difficult to resist the urge to move.
- Standing still with the knees bent took a lot of self-control.
- The semi-squatting task distracted me from the computer task.

Heart rate increased when semi-squatting both with and without the exoskeleton, from approximately 82 bpm (beats per minute) at the beginning, rising to 87 bpm with the exoskeleton and 96 bpm without the exoskeleton. Participants reported significantly less time pressure, less frustration, lower conflicting task demands, and lower mean self-control demands when the cognitive test was performed wearing the exoskeleton compared to not wearing it. Most of the ratings were in the middle of the scales, with the exception of physical demands, which were rated low. The maintenance of static postures has long been known to cause discomfort, fatigue, and injury, particularly if isometric muscular contractions are required. The cognitive demands of adopting these postures have received less attention. The findings of this study indicate that the maintenance of such postures may also consume mental resources and that there is potential for competition between the mental resources used to maintain the posture and those required for the cognitive demands of the task. The present findings show that the maintenance of these postures degrades performance on a cognitive task requiring sustained attention. It was found that the deficit persisted up to 15 min after cessation of the posture. This suggests that, in jobs where the work environment must be monitored

FIGURE 7.15 A "Chairless Chair"—passive exoskeleton for mobile workers. (Courtesy of AUDI AG and nonee Germany GMBH). "Chairless Chairs" fitted to the legs exert an extension moment around the knee joint or fully support the mass of the upper body when sitting or crouching.

or attended to for warnings or cues about the safety of the process, static postures may present a safety risk as well as a health hazard by diverting attention from the environment to the maintenance of internal bodily states and discomfort. Exoskeletons such as the "Chairless Chair" may improve work performance and safety by enabling static postures to be adopted with lower physical demands, thereby sparing the limited attentional capacity of employees.

Regulations Concerning Working Hours

In 1993, the European Union issued a "Working Time Directive," which limits the hours that citizens can work per week. The directive is controversial for many reasons; for example, that some people want to work longer hours, that for some organizations the limitations are impractical, and that it is difficult to enforce (e.g., with the self-employed). The key provisions are as follows:

- Maximum average working time per week: 48 h
- Daily rest: 11 consecutive hours (includes traveling time and unpaid work)
- Weekly rest: an uninterrupted period of 24 h once per week
- A compulsory break in any working day longer than 6 h
- At least 4 weeks paid leave per year

It is not known what effect the implementation of these regulations has had on the outcomes discussed above or to what extent long working hours were a problem before the directive was issued.

Reduce Long Work Hours and Badly Designed Shift Systems

The health effects of long working hours (>55 h/week) compared to standard working hours (35–40 h/week) were investigated in a cohort study by Ervasti et al. (2021). Only 4%–7% of over 5000 participants worked long hours. However, the risk of cardiovascular death was greater in those working long hours (hazard ratio 1.68), as were infections (1.37), diabetes (1.18), injuries (1.22), and musculoskeletal disorders (1.15). The risk of all-cause mortality (death for any reason) was not greater.

Landrigan et al. (2004) investigated medical errors made by interns in a large U.S. hospital. The interns worked a punishing shift system of 77–81 h/week. The system consisted of an 8-h shift on day 1, starting at 07:00, followed by a 28-h shift from 07:00 on day 2 to 12:00 on day 3. Day 4 was free, and day 5 consisted of a second shift of over 24 h starting at 07:00. This system was replaced by one in which interns worked a combination of 8 and 15:00-h shifts with two 15-h shifts rotating forward. Statistically significant reductions in medical errors were found when the new shift system was introduced. Interns made 36.5% more serious medical errors on the old system—20.8% more medication errors and 5.6 times as many diagnostic errors. Lockley et al. (2004), in a separate part of the same study, measured "attentional failure" in the interns working under the two systems. A technique known as electrooculography was used to detect the presence of slow, rolling eye movements in the interns—an indicator of attentional failure caused by severe fatigue. Interns working on the new system had half the rate of such failures as those working on the old system. Interns worked significantly fewer hours on the new system (65.4 h/week as opposed to 84.9 and slept 5.8 h/week more. Although this is an extreme example of long hours and a badly designed system, the findings illustrate that error rates are related to working hours and that by reducing working hours and providing more sleep, error rates can be significantly reduced.

Menopause

In many countries, menopause does not fall within the scope of company health and safety policies. Jack et al. (2016) reviewed the literature on employment aspects of menopause to identify the key physical and psychosocial factors that employers should consider. These factors can either exacerbate or ameliorate symptoms. Some physical factors that can exacerbate symptoms include: high temperature, crowding, and poor sanitary or rest facilities. Social factors include: interpersonal relations, and a lack of recognition or support by the organization. Jack et al. recommended that employers should review their health and safety and human resources policies to recognize that menopause is not to be regarded as a personal matter that does not require an employer's response. Educational support should be made available for those who require it to enable self-management of symptoms in the workplace. While being careful not to regard menopause as a disability, employers should undertake risk assessments of symptomatic employees and make reasonable adjustments to their conditions of work. These might include the location of workplaces or desks, provision of cooling fans and fresh water, easy access to sanitary facilities, or moving the workspace to a more open area. Employer health promotion programs should also consider the needs of women by providing information and advice to help them manage their symptoms, such as advice on diet, access to support groups, and leave to work from home as required.

The British Standards Institution has recently published a guide to menstruation, menstrual health, and menopause in the workplace (BS 30416:2023[1]). The guide takes a systems approach and considers:

- The workplace changes that can be made to support employees (such as dividing lunch breaks into shorter breaks); providing quiet spaces for rest and recuperation; ease of access to toilet facilities with washbasins inside cubicles, accessible from a seated position; allowing windows to be opened and fans to be used at the workplace; easy access to menstrual products; providing easy access to cooler spaces in buildings (e.g., stairways); ensuring uniforms, when worn, are made of breathable fabrics with elasticated waistbands; providing blinds, and dehumidifiers, and lighting with cooler color temperatures; and providing easy access to cool drinking water.
- Providing a supportive workplace culture
- Providing flexible working arrangements

The guide also contains a toolkit for line managers and HR personnel, including a workplace risk assessment.

RESEARCH ISSUES

Technology to monitor attention and alertness in professional drivers is currently under development. Huysamen et al. (2024) describe four main categories:

- *Behavioral Devices*: Monitor the driver's physical features or movements for symptoms of loss of attention or alertness. Eye movements, facial movements, head position, "chin drop" are examples.
- *Driving Performance Devices*: Monitor the driver's input to the vehicle or the vehicle's behavior. Loss of alertness may appear as sudden braking, oversteering, etc.
- *Physiological Devices*: Measure physiological signals that correlate with loss of attention or alertness. Heart rate variability is known to decrease with fatigue.
- *Galvanic Skin Response Devices*: Measure the electrical conductivity of the skin, which changes in response to autonomic nervous system activation.
- *Hybrid Devices*: Combine elements of the above categories.

Research on exoskeletons is continuing apace. At present, uptake is not widespread, although the evidence to date suggests that these devices will enable employees to perform more physical work with less physical effort. The benefits at the organizational level have yet to be demonstrated.

SUMMARY

Physiological mechanisms set limits to a worker's capacity for physical work. Quantitative techniques for predicting fatigue and endurance can be used to specify the design of tools, tasks, and work practices. Physical workload can be evaluated in terms of the physiological cost to the worker of carrying out the work. Oxygen consumption and heart rate are objective measures. Since people differ in their capacity for aerobic work, it is appropriate to consider the relative workload of a task—an individual's oxygen consumption while working can be expressed as a percentage of their VO_2 max. Workloads of 30%–40% of VO_2 max can be sustained for an 8 h shift. The requirement is to design the work so that it can be performed safely and without undue fatigue by the largest possible number of people in the workforce. This implies designing for individuals with lower aerobic capacity (such as older workers). If this is not possible, a physiological approach is needed to specify a minimum standard of fitness for the job.

Heart rate is a good indicator of the total stress of a job—the higher the heart rate when work is done, the more rest is needed over an 8-h shift

- Light work—1.5 h
- Moderate work—2 h
- Heavy work—2.75
- Very heavy work—3.25 h

Excessive work demands eventually lead to poor performance. The 40-h, 5-day week is the norm in most countries. Shiftwork has well-documented negative health effects, but the effects of working overtime (not involving shiftwork) are less clear. Fifty hours of work per week on a regular basis seems to be the upper limit—more than this, efficiency and output drop.

Wearable devices are now available to detect the onset of fatigue at work (using measures such as HRV and GSR).

TUTORIAL TOPICS

1. Is the work capacity of the population changing?
2. Should work be made less demanding through automation and the use of job aids?
3. What are the implications of demographic change for the design of physically demanding jobs?
4. At the link (https://www.youtube.com/watch?v=0hg_Nib6tZE), you will find a short video ('Tired all the time?')—a more medically orientated view of fatigue. Do HFE specialists need to understand the contents of this video when investigating fatigue in the workplace?

ESSAYS AND EXERCISES

1. Discuss what is meant by fatigue. How would an ergonomist go about investigating complaints of fatigue in manual work?
2. Mr. Smith's maximum oxygen uptake is 2.4 L/min. His body mass is 90 kg. What is his VO_2 max per kilogram of body mass?
3. Mr. Smith decides to fly to Spain for a holiday. His maximum grip strength is 250 N. His packed suitcase requires a grip force of 200 N when he is carrying it. How long will he be able to carry it before having to stop and rest or change hands? If he is walking at 3 km/h and the distance from the luggage reclaim carousel to the taxi rank is 300 m, how many times will he have to stop to change hands due to muscle fatigue?
4. Mr. Smith is 60 years old. His heart rate, while he is carrying the suitcase, is 160 beats/min. Is this excessive?

5. Mr. Smith's resting heart rate is normally 90 beats/min. After reaching the taxi rank, it drops from 160 to 140 beats/min after 30–60 s. After a few minutes, while he is waiting for the taxi, it has only dropped to 125 beats/min. Should he have paid a deposit of €1 and used a luggage trolley instead? Comment on the difference between the two heart rates.
6. Mr. Smith is a heavy smoker. To carry the suitcase at a walking speed of 3 km/h, he requires 2.25 L of oxygen per minute. How long will he be able to carry the suitcase without resting? Which has the longest endurance time for this task, his cardiovascular system or his finger flexors?
7. A hand-operated, high-pressure water jet for blasting rocks down mine stopes is operated by pushing together, with both hands, two handles, shoulder width apart and directly in front of the operator at chest height. The design engineer wants to know how long the operator will be able to operate the jet if the required force is 15 kg. You get data on the maximum force required by simulating the task using a bathroom scale sandwiched between two piles of books. Volunteers exert a maximum push force, which is read off the scale. The maximum push forces for 10 male volunteers are as follows:
   Push force (kg)
   25, 17, 30, 23, 23, 24, 19, 35, 34, 25
   Calculate the fifth and ninety-fifth percentile endurance times using Rohmert's equation.
8. Discuss whether an organization such as the army can introduce an entrance standard based solely on aerobic fitness. What other characteristics might be relevant?
9. A laborer is working at 45% of his VO_2 max. How long will he be able to work before exhaustion?
10. The use of physical selection tests implies the failure of ergonomics. Discuss.
11. The terms fatigue and stress should be banned from scientific discussion. Discuss.
12. A worker spends 15 min loading boxes onto a truck. His oxygen uptake is 2 L/min. If the standard is taken to be 1 L/min, what rest allowance should he be given?
    A worker weighing 90 kg has a VO_2 max of 50 mL O_2/kg/min. His work requires him to metabolize 2.5 L of O_2/min. Refer to Table 7.5 and specify approximately the maximum time he should be allowed to work without rest (assume there is no danger of heat stress). Hint: Use the data in Table 7.4 to draw a graph of maximum work time versus relative work demand. Calculate the relative work demand and interpolate using the graph.
13. A 25-year-old forestry worker has a VO_2/heart rate regression equation of

$$y = a + bx$$

    where y is the oxygen consumption (liters O_2/min), x is the heart rate (beats/min), $a = 1.42$, and $b = 0.025$.
    a. The worker's heart rate is monitored while felling trees and is found to be 150 beats/min. Use the regression equation to estimate the worker's oxygen consumption while performing this task.
    b. Use the equation in Workshop 7.3 to estimate the worker's maximum heart rate.
    c. Estimate the worker's VO_2 max from his estimated maximum heart rate using the regression equation above.
    d. Calculate the relative workload by expressing the oxygen consumption while working as a percentage of VO_2 max.
    e. Refer to Table 7.4 to estimate the time to exhaustion.
14. Worker A, weighing 60 kg, has a VO_2 max of 60 mL/kg/min, and worker B, weighing 100 kg, has a VO_2 max of 50 mL/kg/min. When walking steadily at 5 km/h with no extra load, A is working at 30% of his VO_2 max, and B at 40% of his VO_2 max. If each of them has to carry a 30 kg load that requires the consumption of one additional liter of oxygen per minute, which worker will tire more quickly? Provide a brief calculation to support your choice.

QUICK QUIZ

1. Carbon dioxide is: (a) a nutrient, (b) a product of the oxidation of glucose, (c) fuel for the body.
2. Muscles shorten when they contract eccentrically. True or false?
3. Oxygen is stored in the blood as: (a) myoglobin, (b) hemoglobin, (c) globulin.
4. Aerobic capacity decreases with age. True or false?
5. Fatigue is a linear process. True or false?
6. Electromyography is a measure of: (a) muscle activity, (b) muscle development, (c) muscle fatigue.
7. Actimeters can measure: (a) metabolic activity, (b) physical activity, (c) number of steps per day.
8. For heavy work you need rest during a shift of: (a) 1 h, (b) 2.75 h, (c) 3.25 hours.
9. Hear rate becomes more variable as fatigue sets in. True or false?
10. Melatonin is a hormone involved in: (a) blood sugar regulation, (b) stress reactions, (c) sleep-wakefulness.

NOTE

1 BS 30416 - Menstruation, Menstrual Health and Menopause in the Workplace can be downloaded from the BSI website (www.bsigroup.com).

8 Health and Fitness

GENERAL REQUIREMENTS FOR HUMANS IN SYSTEMS

1. Job demands must match the work capacity of employees.
2. Work should provide the optimal level of physical activity.
3. Facilities for healthy eating and exercise should be provided.

> A physically inactive lifestyle increases the risk of: obesity; heart disease; high blood pressure; high cholesterol: stroke: metabolic syndrome' Type 2 diabetes; certain cancers, including colon, breast, and uterine cancers.
>
> *World Health Organization*

CORE KNOWLEDGE: WORKLOAD, FITNESS FOR WORK, AND HEALTH

Health

Traditionally, health was defined as the absence of disease. More recently, wider definitions, such as that of the World Health Organization, have been proposed:

> Health is a state of complete physical, mental and social well-being and not merely the absence of disease or infirmity

A number of authors have criticized this definition (e.g., Saracci, 1997) on the grounds that it is really a definition of happiness rather than health. It also runs into contradictions when real human life is taken into account. For example, eating a diet of cheap, ultra-processed food might be mentally and socially pleasant for some, while leading to health problems associated with obesity. In other words, maintaining physical health may have social and psychological costs. Saracci proposed an alternative definition:

> Health is a condition of well-being free of disease or infirmity and a basic and universal human right.

From the perspective of HFE, health has employment implications, both in terms of what people can do and what they can be expected to do and be employed to do.

Fitness

There are several definitions of fitness. Most state that fitness refers to a state of the body in relation to activities of some kind, for example:

> The ability of your body systems to work together efficiently to allow you to be healthy and perform activities of daily living.

There are five components of physical fitness: body composition, flexibility, muscular strength, muscular endurance, and cardiorespiratory endurance.

Fitness and Health

A distinction can be drawn between fitness and health. Medical doctors tend to pronounce a patient healthy if they can find no evidence of disease. In fact, many athletes, while having a high level of fitness for their chosen sport, may actually be very unhealthy and suffer from frequent bouts of

DOI: 10.1201/9781032717210-8

colds and flu, as well as musculoskeletal and other ailments. Conversely, few healthy people are fit to run a marathon or to lift heavy weights without risk of injury unless they have undergone a prior program of training. As described below, a major problem in many countries is the decline in the fitness of the workforce due to lifestyle, poor diet, and demographic factors.

In this chapter, we will examine fitness and health in relation to work.

Fitness for Work

The definition above can be modified as follows:

> The ability of your body systems to work together efficiently to allow you to be healthy and perform activities to meet occupational requirements.

Thus, fitness for work can only be assessed if the demands of the job are understood. HFE has an important role to play in analyzing and quantifying job demands.

Energy Costs of Physical Activities

The Compendium of Physical Activities (Ainsworth et al., 2000) can be used in the interpretation of self-report and observational data about physical activity. Table 8.1 lists the physiological intensities of many work and leisure activities, expressed in units known as METs, where:

$$1 \text{ MET} = 4.184 \text{ kJ/kg/h} = \text{RMR} \quad (8.1)$$

where the RMR is the resting metabolic rate (the average rate when a person sits resting in a chair). RMR can be distinguished from the basal metabolic rate (BMR). The BMR is the rate of energy consumption necessary to maintain life. Individuals differ in their BMR. It is about twice as high in a child as in an adult. The decrease in BMR with age is one of the reasons for adult weight gain. The BMR can drop by about 20% in chronically malnourished individuals.

The MET calculations used to estimate energy expenditure in the Compendium are not intended to replace the kind of exact physiological measures described in Chapter 7. However, they can be used to identify activities that are *likely* to be of low, moderate, or high intensity in most people who engage in them. For example, light factory work has an intensity of 3 METs, so light factory work would consume up to three times as much energy over a working day as sedentary office work.

The Compendium enables different activities to be compared, but it takes no account of individual differences in capacity or skill. The calculations are useful in work involving large numbers of people for research, public health work, or making decisions about the design of a work system to optimize work demands, rest periods, and the energy content of any meals provided.

Glucose

In simple terms, the basic requirement for life is that oxygen and nutrients are metabolized by the cells of the body, where they undergo chemical reactions that result in the production of energy and waste products. During much of the day, muscle cells rely on the oxidation of fatty acids to produce energy, while the cells of the brain utilize glucose. Following the digestion of a meal rich in carbohydrates, glucose enters the bloodstream, and the blood glucose concentration rises, which triggers the release of the hormone insulin from the pancreas. Insulin causes glucose to enter the cells of the body, where it is metabolized in place of fatty acids. Insulin is therefore a fat-sparing hormone. Excess glucose that is not immediately required for either basal metabolism or the activity currently taking place is converted into glycogen for storage in the muscle cells or the liver. When these stores are full, and if the blood glucose concentration continues to rise, excess glucose is converted into fat.

When the blood glucose level falls below that needed to provide glucose to the brain cells, the pancreas releases a hormone known as glucagon, which causes glycogen to be converted into

TABLE 8.1
MET Values for Activities Sampled from the Compendium of Physical Activities

Activity	Work Intensity (METS)
Cycling	
Cycling (<15 km/h to work or for leisure)	4.0
Cycling (15–18 km/h slow, light effort)	6.0
Cycling (BMX or mountain bike)	8.0
Cycling (20–24 km/h vigorous effort)	10.0
Cycling (25–30 km/h)	12.0
Cycling (>30 km/h)	16.0
Exercise	
Light home exercises (moderate effort, stretching, back exercises)	3.5
Calisthenics (push-ups, pull-ups, sit-ups, heavy, vigorous effort)	8.0
Weight lifting (free weights or machines, heavy effort)	8.0
Stair treadmill ergometer (gymnasium)	9.0
Rowing machine (light effort)	3.5
Rowing machine (vigorous effort)	8.5
Water aerobics class	4.0
Aerobics class	6.0
Aerobics (dancing, low impact)	5.0
Dancing (Greek, flamenco, belly, swing, Middle Eastern)	4.5
Ballroom dancing (slow)	3.0
Home Activities	
Carpet sweeping, sweeping floors	3.3
Cleaning (e.g., washing a car)	3.0
Housework (multiple tasks, light effort)	2.5
Housework (multiple tasks, moderate effort)	3.5
Housework (multiple tasks, high effort)	4.0
Washing dishes (standing)	2.3
Vacuuming	3.5
Putting away groceries	2.5
Shopping for food (with or without a trolley)	2.3
Sweeping garage	4.0
Moving items upstairs/carrying boxes or furniture	9.0
Carrying small children	3.0
Care of the elderly	4.0
Washing or painting a fence	4.5
Spreading soil with a shovel	5.0
Wiring/plumbing	3.0
Garden	
Mowing the lawn (walk power mower)	5.5
Mowing the lawn (walk hand mower)	6.0
Tilling soil	6.0
Operating a snow blower (walking)	4.5
Shovelling snow by hand	6.0

(*Continued*)

TABLE 8.1 (*Continued*)
MET Values for Activities Sampled from the Compendium of Physical Activities

Activity	Work Intensity (METS)
Music Playing	
Conducting	2.5
Drums	2.0
Flute	2.0
Violin	2.4
Occupations	
Bakery (general, moderate effort)	4.0
Carpentry (general)	3.5
Carrying heavy loads (e.g., bricks)	8.0
Drilling coal/rock	6.5
Shoveling coal	7.0
Cleaning (polishing the floor with an electric polisher)	3.0
Digging ditches	8.5
General cleaning (moderate effort)	3.5
Factory work (standing: welding, auto repair, stocktaking)	3.0
Factory work (standing: heavy workpieces, rope twisting, etc.)	3.5
Factory work (standing: heavy lifting >20 kg, paper hanging, masonry work)	4.0
Farming (bailing hay)	8.0
Farming (driving a harvester)	2.5
Farming (feeding cattle and horses)	4.5
Farming (animal husbandry)	6.0
Farming (shoveling grain)	5.5
Forestry (axe chopping, fast)	17.0
Forestry (felling trees)	8.0
Forestry (carrying logs)	11.0
Forestry (hand planting)	6.0
Forestry (weeding)	4.0
Horse racing (galloping)	6.0
Locksmith	3.5
Machine tooling	2.5
Machine tooling (punch press)	5.0
Machine tooling (welding)	3.0
Masonry work	7.0
Masseuse (standing)	4.0
Office work (sitting, light work with computer, chemistry lab work, watch repair, etc.)	1.5
Office work (meetings)	1.5
Seated cab work (crane, LHD, etc.)	2.5
Orange grove work	4.5
Policeman (directing traffic)	2.5
Policeman (driving a squad car)	2.0
Shoe repair (general)	2.5
Standing (bartender, store clerk)	2.3

(*Continued*)

TABLE 8.1 (*Continued*)
MET Values for Activities Sampled from the Compendium of Physical Activities

Activity	Work Intensity (METS)
Occupational Activities	
Lifting 4–8 kg items continuously	4.0
Assembly line work (standing)	3.5
Steel mill work in general	8.0
Tailoring (cutting and sewing)	2.5
Tailoring (pressing)	4.0
Walking	
<3 km/h	2.0
3 km/h	3.3
3.5 km/h	3.8
<3 km/h carrying objects <11 kg	4.0
Pushing wheelchair	4.0
5 km/h carrying objects <11 kg	4.5
Walking (backpacking)	7.0
Walking (carrying infant or 7 kg bag)	3.5
Walking (carrying 1–7 kg upstairs)	5.0
Walking (carrying 7–11 kg upstairs)	6.0
Walking (carrying 11–23 kg upstairs)	8.0
Walking (carrying 23–33 kg upstairs)	10.0
Walking downstairs	3.0
Walking (marching rapidly)	6.5
Sports	
Canoeing, rowing for pleasure	3.5
Cross-country running	0
Badminton	7.0
Basketball	8.0
Billiards	2.5
Boxing	12.0
Cricket (batting and bowling)	5.0
Darts	2.5
Football	9.0
Golf (walking and carrying clubs)	4.5
Golf (walking and pulling clubs)	4.3
Golf (power cart)	3.0
Gymnastics (general)	4.0
Judo	10.0
Paddle boating	4.0
Skateboarding	5.0
Skating	7.0
Snorkeling	5.0
Soccer	10.0
Squash	12.0
Swimming (freestyle, light to moderate effort)	7.0
Swimming (general, light effort)	6.0
Swimming (treading water, moderate effort)	4.0
Tennis (doubles)	5.0
Tennis (singles)	8.0

Source: Extracted from Ainsworth et al. (2000), with permission.

glucose. Under conditions of food deprivation, or when engaging in activities with high metabolic demands, the glycogen stores may be depleted and fat stores have to be used to release fatty acids to provide fuel to the muscles and for conversion to glucose to supply the brain. Glucagon is involved in these processes. In extreme conditions, lack of food combined with work demands results in loss of fat and a net loss in body mass. This can happen in developing countries at times of the year when food is scarce and is described as negative energy balance—the energy required for life and work is greater than that obtained from food.

Factors Influencing Health and Fitness in the Population

There is a growing body of evidence that people who do physically demanding jobs are fitter and healthier than their less active colleagues. Conversely, the risk of a fatal heart attack is greater in those with less demanding jobs.

Physical inactivity is common in many parts of the world. In the United States, the Centers for Disease Control and Prevention publishes physical inactivity maps, where physical inactivity is defined as not participating in any physical activity outside of work (sports, gardening, walking, running, etc.). In 2022, the most inactive region was Puerto Rico (49.4% inactive), and the most active state was Colorado (17.7% inactive). In seven states in the South, 30% were inactive, followed by the Midwest, Northeast, and West (21%). At the same time, there has been a long-term change in employment in many countries, including the United States. At the turn of the 20th century, 30.9% of the U.S. workforce worked in agriculture and 38.9% in manufacturing, compared to 0.7% and 20.2% today. There has been an increase in service sector employment since 1980—inactive employees have been doing inactive office work.

Bedrest Studies

NASA conducted the famous "bedrest studies" to investigate the health effects of physical inactivity on astronauts during space flights. Initially, healthy subjects were inactive for long periods (Figure 8.1).

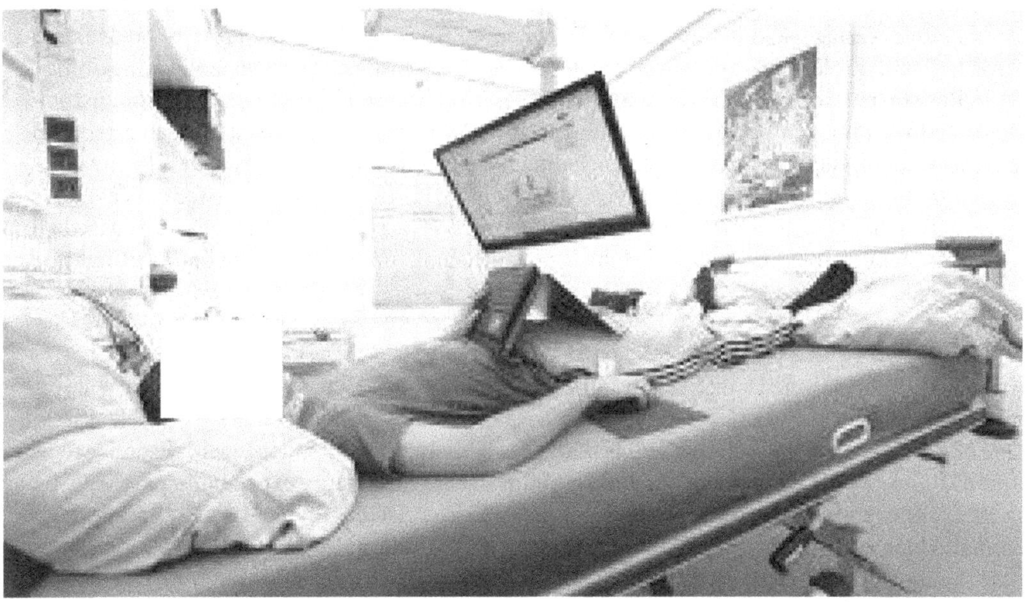

FIGURE 8.1 NASA bedrest studies to understand the effects of physical inactivity on otherwise healthy people.

One of NASA's main findings was that physical inactivity causes muscle atrophy and "metabolic inflexibility," which includes a reduced ability to metabolize fat. Note that reclining in an office chair for 8 h is a little more physically demanding than staying in bed all day.

Metabolic inflexibility has four components:

Insulin Resistance: There is normally around 1 g of glucose per liter of blood, and this concentration is controlled by the hormone insulin. When blood glucose levels rise after a meal, insulin enables target tissues such as the skeletal muscles and the liver to absorb glucose from the blood and either store it as glycogen for later use or use it for energy. The amount of glucose in the blood then falls. Insulin resistance is the reduced sensitivity to insulin, leading to high concentrations of glucose in the blood, which can damage nerves, blood vessels, and organs. In the long term, elevated blood glucose levels can cause a variety of health problems, including high blood pressure.

Changes in Blood Cholesterol: Blood triglyceride levels increase, LDL ("bad") cholesterol levels increase, and HDL ("good") cholesterol levels decrease.

Lowered Insulin Sensitivity in the Muscles: Physical inactivity promotes the storage of fat. It shunts the delivery of lipids in the blood plasma away from the muscles (where they would be used as an energy source) to adipose tissue (where they are stored as fat).

A Shift in Substrate Use toward Glucose: Human skeletal muscles normally use fatty acids as the "default" source of energy. In healthy people, "muscles burn fat." Metabolic inflexibility impairs the capacity to use fatty acids as an energy source when their availability increases. In all bedrest studies, glucose is used for energy to the detriment of lipid oxidation. Physical inactivity reduces the capacity of the body to use fat as an energy source, causes muscle loss, resistance to insulin, elevated blood sugar levels, and elevated levels of triglycerides and cholesterol.

Many of these features are also found in obese people.

Le Roux et al. (2021) report the findings of bedrest studies where a group of 'sedentary exercisers' combined moderate to vigorous activity (resistance exercises) with bedrest. These studies enable the protective effects of exercise to be disentangled from the adverse effects of inactivity. Resistance exercise helps maintain cardiorespiratory fitness, prevents muscle wastage, and maintains muscle function. However, it did not prevent the decrease in HDL cholesterol, the development of insulin resistance, or metabolic inflexibility. Resistance exercise was only partially protective because the energy expenditure associated with it is low. Combining aerobic exercise with resistance exercises further reduced aspects of metabolic inflexibility, including insulin resistance and carbohydrate utilization.

Implications for Work Design: These studies indicate that engaging in regular exercise will help prevent some of the negative effects of a sedentary lifestyle (discussed in Chapter 4). However, exercise alone is not sufficient, and reductions in daily sitting time are also indicated.

Sedentary Behavior and Exercise

Xu et al. (2019) conducted a meta-analysis of seven studies where sitting time and time spent watching television were analyzed in relation to all-cause mortality and moderate or light physical activity. One of the difficulties of such research is that these activities are not independent. Exercise reduces time available for sitting and vice versa. The findings were that energy expenditure of 7 MET-h/day in excess of energy expended on physical activity was associated with increased mortality risk. So, an individual who sits for 7 h with no physical activity would be at increased risk. Increasing the time spent on physical activity actually increases the threshold at which sitting time increases mortality risk. If the individual exercises for 2 MET-h, the sitting threshold becomes 9 MET-h. Chastin et al. (2021) reanalyzed the findings of seven accelerometer-based studies of physical activity, sedentary behavior, and sleep in relation to all-cause mortality. Time spent sleeping was not associated with mortality. Engaging in moderate to vigorous daily physical activity was associated with lower mortality in a curvilinear fashion (hazard ratio 0.63), with diminishing returns after 20 min/day, and this did not depend on whether the rest of the day was spent in light activity or sitting when activity was measured using wrist accelerometers. Data from studies using accelerometers on the hip also showed a protective effect of moderate to vigorous physical activity, but this was modified by how the rest of the day was spent (light activity also having a protective effect). The difference may be because hip-based accelerometers track movements of the whole body, which are likely to be higher in energy expenditure than movements of the wrist and therefore more beneficial.

These studies suggest that reducing daily sitting time and increasing light or vigorous physical activity are equally important. There is also evidence that breaking up long periods of sitting with short activity breaks is beneficial. Taken together, these studies support the use of activity-permissive workstations, as described in Chapter 4.

Recommended Activity Levels

The U.S. Centers for Disease Control and Prevention (CDC) and the American College of Sports Medicine (ACSM) recommend that adults engage in 150 min a week of moderate physical activity (above 3 METs) and two sessions of strength training. While both forms of exercise are beneficial, they are not beneficial in the same way, as indicated by the bedrest studies described above.

Table 8.2 summarizes the benefits of exercise.

Bridger et al. (2013a), in a longitudinal study, found that health risk (high BMI and waist circumference) measured in 2011 was predicted by lack of exercise in 2007. People with physically demanding jobs were less likely to be at risk in 2011. High BMI in 2007 was associated with the development of back pain, musculoskeletal pain, and high blood pressure in 2011.

Metabolic Demands and Food Consumption

Malnutrition in the modern world is a problem of affluence as much as poverty. In many parts of the world, people are malnourished because they lack sufficient food. Elsewhere, malnutrition occurs because of dietary and lifestyle imbalances. Most of the energy in food comes from fats, carbohydrates, and protein. Carbohydrates consist of starches, sugars, and cellulose (which humans cannot use as an energy source). Fat has the highest energy density at 37 kJ/g, followed by proteins and carbohydrates at 17 kJ/g. Starches are broken down into sugar in the gut, and all dietary carbohydrate is absorbed into the body in the form of glucose. Fat is broken down into fatty acids prior to absorption. Both fatty acids and glucose can be used as sources of energy. A healthy diet also contains essential vitamins and minerals, which are not used as sources of energy.

In the United States, the recommended energy intake of food is 11,300 and 8800 kJ for men and women between 31 and 50 years of age, at a physical activity level equivalent to walking about 2–5 km/day at 5–6 km/h (about 1 h at 4 METs). Assuming the man weighs 80 kg, his daily energy expenditure for basal metabolism is approximately $80 \times 4.184 \times 24$, or 8,033 kJ. Adding another 3 METs for 1 h ($4.184 \times 80 \times 3 = 1{,}004$ kJ) gives a total of 9,007 kJ, suggesting that the individual would be in positive energy balance if his energy intake was 11,300 kJ. However, it is unlikely that the individual would be completely inactive for the remaining 23 h, and a further 4 or 5 h might be spent in miscellaneous activities of 2 or 3 METs even in an inactive person, which would easily bring the total to 11,500 kJ (an extra 1.5 METs for 5 h, for example).

How much food do you need to eat to get 11,300 kJ? A widely available quarter-pound hamburger has an energy content of 2,300 kJ, and a medium portion of French fries adds another 1,540 kJ, for a total of 3,840 kJ, which is 34% and 44% of the daily requirement for men and women, respectively. An 80-kg man wanting to utilize the energy from the French fries would have to exercise at 8 METs for 46 minutes after eating them (by playing singles tennis, for example; see Table 8.1).

Several aspects of the modern diet appear to make food energy too readily available and too easy to consume. The *glycemic index* (GI) of food is a measure of its effect on raising blood sugar after it

TABLE 8.2
Benefits of Exercise for Health and Fitness

Benefits of Exercise	
Prevention of falls in older people	Improved cardiovascular fitness
Improved musculoskeletal function	Improved bone density
Prevention of Type II diabetes	Release endorphins–mood enhancer
Pain management (joint pain)	
Boosts energy	Better sleep
Better appearance	
Better cholesterol management	Better mobility
Regulates blood pressure	Prevents decline in cognitive ability
Lowers risk of certain cancers	
Improves longevity	Lower risk of infections
Help neurogenesis	Positive attitude
Better mood	Stress management
Better pain management	Better Weight management
Benefits Hypothalamus (better thermoregulation)	
Strengthens the intercostal muscles, making it easier to ventilate the lungs and increasing lung capacity	
Benefits of Cardiovascular Exercise	
Heart rate and stroke volume improved	
Longevity	
Avoidance of Type II diabetes	
Avoidance of overweight and obesity	
Greater work capacity	
Higher bone density	
Lower risk of osteoporosis	
Control of blood pressure	
Benefits of Musculoskeletal Exercise	
Maintain/increase bone density	
Improves balance and stability	
Facilitates the performance of everyday tasks	
Increases serotonin levels (improves mood)	
Increases dopamine	
Strengthens joints and protects against falls	
Balances hormones during menopause	
Weight control (replace fat with muscle)	
Release of myokines when muscle cells contract[a]	

[a] Myokines are proteins that regulate muscle metabolism and growth, improve glucose uptake, regulate the immune system, reduce inflammation. and protect against cancer.

has been eaten. Foods with a high GI contain carbohydrates that are digested easily, causing blood sugar levels to rise quickly. The fat and protein content of foods lowers the GI. Milk and dairy products have a low GI because they contain fat and protein, whereas potatoes, white rice, corn syrup, and fructose have a high GI. Many corn-based breakfast cereals have a high GI, whereas oatmeal in its natural form has a low GI. As blood sugar levels rise after a meal, insulin is secreted to enable the glucose to be either metabolized in the muscles or stored as glycogen or fat. People who eat a diet rich in foods with a high GI may develop *insulin resistance,* where the body no longer responds to insulin and blood sugar levels remain high, causing long-term damage to the arteries.

From the perspective of HFE, meals with a low GI would be expected to be more effective in sustaining employees in physically demanding jobs because the nutrients they contain are absorbed more slowly. However, some low-GI foods may cause a positive energy balance, resulting in weight gain if they contain high volumes of fat.

Much modern work is mentally demanding and physically undemanding, and there is now strong evidence that physical inactivity is a major threat to the well-being of the population. Wen et al. (2011) analyzed life expectancy in 416,175 healthy individuals in Taiwan in relation to self-reported lifestyle and other risk factors for mortality. They found a strong relationship between physical inactivity and increased mortality due to heart disease, diabetes, and cancer. Individuals in a low-volume physical activity group, who only exercised about 15 min/day, had a 14% reduction in mortality and a 3-year longer life expectancy than those in the inactive group. Every additional 15 min/day of exercise reduced all-cause mortality by 4%.

What is remarkable about this finding is that a small increase in metabolic activity appears to bring about a large reduction in mortality and an increase in life expectancy. For example, an inactive person weighing 80 kg will have an energy cost of approximately

$$4.184 \times 80 \times 24 \text{ kJ} = 8033 \text{ kJ}$$

in a 24-h period.

If the person engages in low-volume physical activity—say, walking at 3 km/h for 15 min/day, which is 3.3 METs according to Table 8.1—metabolic activity will have increased by 2.3 METs for 15 min. This is equal to

$$4.184 \times 80 \times 0.25 \times 2.3 \text{ kJ} = 192.4 \text{ kJ}$$

approximately, a 2.4% increase in metabolic activity over a 24-h period. How can such a small increase in metabolic activity have such a beneficial effect?

Physical Inactivity Is Harmful

Daily use of the musculoskeletal system, either at work or during leisure hours, is important to maintain fatty acid oxidation at a cellular level. The default substrate for muscle metabolism at rest is fatty acids—a low rate of fatty acid oxidation is associated with weight gain. Even short-term physical inactivity causes an increase in body mass and in the amount of visceral fat (fat deposited in the abdominal cavity). Total abdominal fat increases, as does the waist-to-hip ratio and LDL cholesterol. Psychological stress exacerbates this state of affairs—in physically undemanding but mentally stressful jobs—because cortisol levels are elevated. Cortisol is the primary stress hormone that increases blood sugar levels. It counteracts the effects of insulin. Cortisol levels are sometimes raised in highly stressed people, such that they develop *insulin resistance* and have chronically elevated blood glucose levels because the absorption of glucose into the tissues is impeded by the raised cortisol levels. In healthy people, the action of cortisol is to facilitate the conduct of physically demanding work.

Katzmarzyk et al. (2007) report the findings of a 15-year study of 1,543 Canadian adults who were free of diabetes at baseline. Over 15.5 years, 5% became diabetic. Adiposity and low physical fitness were predictors of diabetes. Those with a higher BMI and waist-to-hip ratio were 2.02 and 3.00 times more likely to develop diabetes, and physical fitness had a protective effect (odds ratio 0.28). Age, sex, parental history of diabetes, smoking, and alcohol consumption were controlled for.

The evidence is that stress increases cortisol secretion, and one effect of this is insulin resistance (IR). Chronic exposure to stress leads to chronic overproduction of insulin—eventually, the system can break down, leading to Type II diabetes.

Clearly, psychosocial stress at work combined with a lack of exercise, can have serious health implications for those exposed to it. The consequences of insulin resistance include increased levels of VLDL and LDL cholesterol particles in the circulation. Other effects include Type II diabetes and hypertension (high blood pressure). Over time, the amount of abdominal fat increases. IR syndrome is thought to be associated with a chain of biochemical reactions that produce plasmin—a substance that breaks down intravascular fibrin clots. Fibrin is a substance involved in the clotting of blood, and deficiencies in fibrinolysis are associated with elevated plasma fibrinogen levels (Vrijkotte et al., 1999). The presence of all of the above is sometimes referred to as *metabolic syndrome.*

Metabolic Syndrome

Greer et al. (2015) found that men who sat for 4–8 h or more than 8 h/day had a 65% and 76% greater risk of developing *metabolic syndrome* than those who were less sedentary. This is important because metabolic syndrome is a precursor to Type II diabetes, where the body becomes insensitive to insulin, and there is a risk of chronically elevated blood glucose levels causing damage to arteries and high blood pressure. The International Diabetes Federation defines metabolic syndrome as follows:

> The metabolic syndrome is a cluster of the most dangerous heart attack risk factors: diabetes and raised fasting plasma glucose, abdominal obesity, high cholesterol and high blood pressure. It is estimated that around 20%–25% of the world's adult population have the metabolic syndrome and they are twice as likely to die from and three times as likely to have a heart attack or stroke compared with people without the syndrome.
>
> In addition, people with metabolic syndrome have a five-fold greater risk of developing type 2 diabetes. They would add to the 230 million people worldwide who already have diabetes one of the most common chronic diseases worldwide and the fourth or fifth leading cause of death in the developed world. The clustering of cardiovascular disease (CVD) risk factors that typifies the metabolic syndrome is now considered to be the driving force for a new CVD (cardiovascular disease) epidemic.

The harmful effects of a sedentary lifestyle remained when physical activity and cardiorespiratory fitness were controlled for in the analysis (and vice versa). In other words, those who sat for long periods each day were more likely to develop metabolic syndrome than those who sat less, even if they were physically fit and active. At the same time, people who were more active and had better cardiovascular fitness were less likely to develop metabolic syndrome, even if they were sedentary. It is important to note that these findings indicate that the increased risk of metabolic syndrome in those who spent a lot of time sitting was not because these individuals were unfit or had cardiovascular problems that caused the sedentary lifestyle. *It was because they spent a lot of time sitting.* Or, to put it another way, people are more likely to be active when they are not sitting down! So, sedentary behavior and a lack of physical fitness are independent risk factors for metabolic syndrome.

Type II Diabetes

Type II diabetes usually occurs in adulthood due to a combination of factors, including genetic predisposition, high BMI, lack of exercise, and age. This is because the islets of Langerhans in the pancreas no longer produce sufficient insulin when blood sugar rises after a meal, or the tissues no longer respond to insulin. The main symptoms of Type II diabetes are:

- Urinating more often than usual, particularly at night
- Feeling very thirsty
- Feeling very tired

- Unexplained weight loss
- Itching around the penis or vagina, or frequent episodes of thrush
- Cuts or wounds that heal slowly
- Blurred vision (caused by the lens of the eye becoming dry)

Even low-level physical activity can have beneficial effects—not because it "burns calories," but because it helps maintain normal metabolic function, including the ability of cells to oxidize fatty acids as a source of energy while maintaining insulin sensitivity, providing proper control of blood glucose levels.

Changes in Work Capacity

Secular trends and the increased prevalence of obesity have rendered many working people less capable of performing normal work. Ciriello et al. (2008), for example, re-examined the maximum weights and forces acceptable to U.S. workers, finding a drop of almost one-third in the maximum acceptable weight for an 8-h day. Although not ill, an older workforce—with its high prevalence of metabolic disorders and sedentary lifestyle—is more at risk of many of the adverse physical and mental reactions to work demands that HFE seeks to prevent. The heaviest component of many manual handling tasks is the person performing the task—and people have gotten heavier over the last 30 years. All the more reason for avoiding manual handling as much as possible. Furthermore, a lack of physical fitness is a risk factor for all-cause mortality and ischemic heart disease when physical work demands are high (Holtermann et al., 2010), even when age, high blood pressure, BMI, diabetes, and smoking are controlled for. Thus, an unfit workforce is more at risk when the physical demands of jobs are high. Physical fitness protects against adverse cardiac outcomes. Employers of workers in physically demanding jobs should allow employees the opportunity for physical training during work hours and provide on-site exercise facilities.

Obesity and Social Trends

We saw in Chapter 3 the increase in BMI of the population over the last 40 years. The CDC in the USA reports that from 1999 to 2000 through 2017 to March 2020, U.S. obesity prevalence increased from 30.5% to 41.9%. During the same time, the prevalence of severe obesity (BMI > 35) increased from 4.7% to 9.2%. Prior to 1970, obesity was uncommon (the prevalence was about 10% in 1960, and extreme obesity was about 1%). There are many factors underlying the increased prevalence, particularly changes in diet and the availability of inexpensive, high-calorie food. Physical activity, partly due to changes in employment patterns, may be another factor. For example, the ratio of service sector employment to agriculture and manufacturing has almost reversed in the last 100 years (almost 80% work in the service sector nowadays) (Figure 8.2). An office worker with a physical activity level of 1.5 METs would have 60 MET.h of activity in a 40-h week. An assembly line worker would have 140 MET.h (3.5 METS × 40 h)—a difference of 80 MET.h, equivalent to 10 h of playing singles tennis per week (Table 8.1).

Sarcopenia and Osteopenia

Sarcopenia is age-related loss of muscle tissue, with lowered lean body mass, grip strength, and gait speed. Muscle shrinkage begins around 30 years of age in inactive people at about 1% per year, or 8% per decade. It accelerates after 65 years of age. In a systematic review of 263 studies, Petermann-Rocha et al. (2022) found the global prevalence to vary from 8% to 36% in people under 60 years of age and from 10% to 27% in people over 60. Inactive people will have lost about 25% of muscle tissue by the age of 65. This is accompanied by osteopenia—loss of bone mineral density. Physical inactivity results in the loss of Type II muscle fibers due to lack of use. Sarcopenia can be partially reversed and prevented by resistance training.

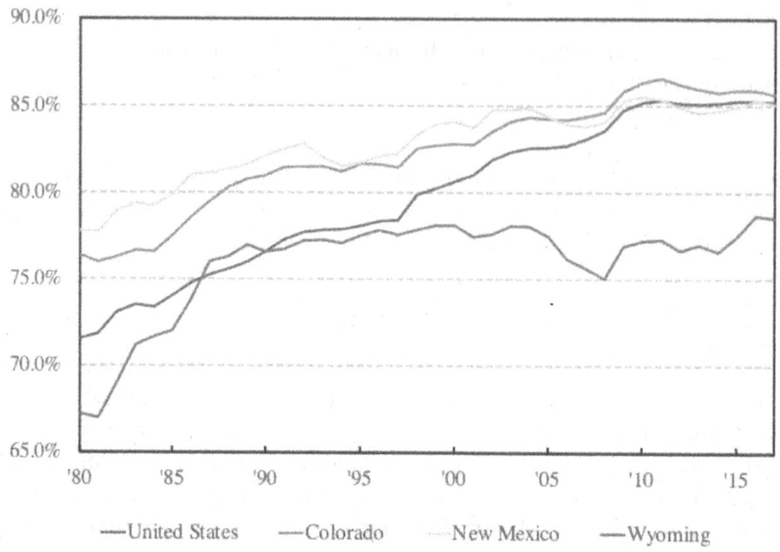

FIGURE 8.2 Service sector share of total employment in the United States (Bureau of Labor Statistics.).

Inflammation

Inflammation is how the body responds to harmful stimuli, be they pathogens or injuries. White blood cells go to the site of the stimulus, blood flow increases, the cause is eliminated, and the tissues are repaired. Low-level inflammatory responses can become chronic due to the following factors: age, obesity, poor diet, smoking, and sleep disorders. Apart from medication, improved diet and exercise are ways of reducing inflammation and avoiding the development of health complaints such as diabetes and cardiovascular disease.

The Aging Population

The mean age of the populations of many developed countries is increasing as people live longer and because birth rates have declined. Whether this has any implications for the capacity of the workforce is open to debate. It is clear that the physical capacity of the workforce as a whole will decline, but this is at a time when the contribution of manual labor to economic output is decreasing. Age may be a surrogate measure of physical strength and endurance, but for different adult age groups, there are distributions of ability, and these distributions overlap greatly (Figure 8.3). For most abilities, the differences within an age group are as great as or greater than those between age groups—excluding extreme old age, of course (Benjamin and Wilson, 2005).

Aging is a measure of the passage of time. The mean age of the populations of many developed countries is increasing as people live longer and because birth rates have declined. Whether this has any implications for the capacity of the workforce is open to debate. In terms of functional capacity and work performance, no simple conclusions can be drawn because:

1. Different processes age at different rates in the same person.
2. The same process ages at different rates in different people.
3. The residual capacity of older workers is partly a reflection of their capacity when younger.

Thus, an aging worker may have suffered considerable hearing loss but have good eyesight and be physically strong due to an active lifestyle when young and maintained over many years. A group of older workers may vary greatly in a particular ability, such as hearing. Thus, age in itself cannot be used to select people for work, and if selection is needed, it should be based on functional testing of the capacity to carry out realistic work tasks.

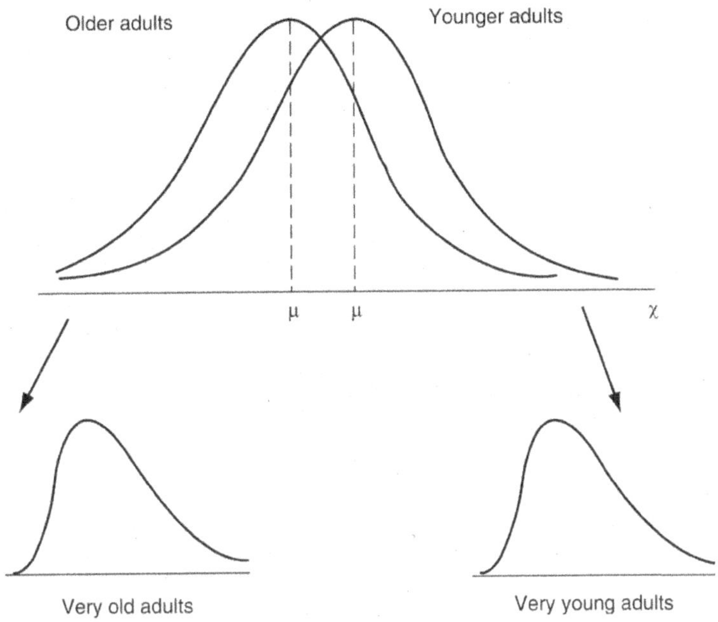

FIGURE 8.3 Relationship between age and work capacity. There is no simple cut-off age for work capacity declines. Rather, the mean of the distribution of capacity moves to the left of the horizontal axis. Many older workers remain fitter and stronger than their younger counterparts. In very young and very old populations, the distributions are skewed.

A few generalizations can be made. Weight gain in adult men is normal up to the age of 50 years due to the presence of a small positive energy balance. It tails off at about the age of 50 years. At the age of 45 years, the average body mass of men and women is 20% higher than it was 20 years before. Cardiorespiratory fitness declines with age, but the effect is not linear—it declines more rapidly after 45 years of age. The decline is accelerated by being overweight or obese. Jackson et al. (2009) found declines in metabolic capacity of 0.31 and 0.20 METs for men and women, respectively, for every unit increase in BMI. Smoking had a similar effect. Thus, the decline in cardiorespiratory fitness that occurs with age can be slowed by being physically active, not smoking, and maintaining a low BMI. In general, maximum heart rate decreases with age and can be estimated by:

$$HR(\text{max}) = 220 - \text{age bpm} \tag{8.2}$$

So, the maximum heart rate of a 60-year-old is 160 bpm.

In practice, when designing for an aging *workforce*, expect the following: decrements in aerobic capacity, general health, grip strength, lifting strength, balance, eyesight, hearing, reaction time, limb mobility, tissue and joint mobility and elasticity, tolerance for paced work, ability to recover from slips and trips, knee function, longer recovery time from physical work, short-term memory, tolerance for heat and cold, and joint function (pain/arthritis).

When selecting *individuals* for a job, age has little relevance (Figure 8.3)—employers should *not* recruit staff on the basis of age but should specify the qualifications and abilities required for the job and select the best applicant. Similarly, there is no scientific basis for the establishment of a compulsory retirement age for *individuals*. Compulsory retirement upon reaching a particular age can be justified for *populations* of employees because we know that so many physical and mental abilities decline with age in a *population*. However, because there is so much variability within age groups and so much overlap in the distributions of abilities (Figure 8.3), *individuals* should be allowed to continue as long as they can demonstrate the required ability or capacity.

Clearly, older workers as a group, are more heterogeneous than younger workers. Design for an older workforce must account for this by making reasonable allowances for those who need them, while maintaining sufficient challenge for all.

Psychological Stress and Health

Psychosocial hazards are those aspects of work design, the organization and management of work, and the social and environmental context that have the potential for causing social or physical harm (Cox and Griffiths, 1995). The social environment at work is thought to interact with the individual's psychology to create *psychosocial stress* that can trigger the conditions described below.

Table 8.3 summarizes the main work-related psychosocial hazards. In general, there is a social gradient in health and mortality—the lower down the social hierarchy, the higher the incidence of a wide range of diseases. Marmot and Brunner (2004) demonstrated that there is evidence for social gradients in insulin resistance and raised blood fibrinogen levels, and a clear relation with heart attack due to coronary thrombosis or stroke. Fibrinogen has a role in blood clotting, so the higher the levels, the more likely you are to have a heart attack or stroke due to the formation of blood clots. Exposure to stress raises cortisol levels in the blood. Cortisol is a stress hormone released from the adrenal glands. It increases blood glucose levels (it is an insulin antagonist) and the use of glucose in the brain. Cortisol levels can also be raised by a poor diet (an excess of high glycemic index foods and saturated fats). Chronically elevated cortisol levels can cause weight gain in the face, abdomen, and chest; high blood pressure; osteoporosis; and mood swings. Eventually, the psychological reactions become disabling (Table 8.4).

The Whitehall II studies (see below) have demonstrated that a lack of control at work is one of the causal factors (controlling for other risks such as smoking and diet).

Schaufeli and Bakker (2004) discuss the demand-resources model, which, although developed as a model of burnout, proposes a mechanism by which job strain may lead to adverse outcomes.

TABLE 8.3
Summary of Work-Related Psychosocial Stressors

Category	Risk Factors = Conditions
	Job Characteristics and the Nature of Work
Job content/demands	High physical/mental/emotional demands; lack of variety; short work cycles; fragmented or meaningless work; underutilization; high uncertainty; continuous exposure to people through work
Workload/workplace	Work overload/underload; machine pacing; time pressure; deadlines
Work schedule	Shiftwork, inflexible work schedules, unpredictable, long, or unsocial hours
Job control	Low participation in decision-making, lack of control over workload
Physical environment	Badly designed, inadequate or faulty equipment, poor environmental conditions
	Social and Organizational Context of Work
Organizational culture and function	Poor communication, low support for problem-solving, low scope for personal development, lack of clear objectives
Interpersonal relationships	Social/physical isolation, poor supervision, lack of social support at work
Role in organization	Role ambiguity, role conflict, level of responsibility
Career development	Stagnation, uncertainty, underpromotion or overpromotion, low pay, job insecurity, low social value of work
	Individual Risk Factors
Individual differences	Coping style, personality, hardiness
Home–work interface	Work–home conflict, low support at home

TABLE 8.4
Reactions to Psychosocial Stress

Healthy	Reacting	Injured	Ill
Normal mood, calm	Irritable/impatient/nervous Sad/overwhelmed	Anger, sadness, hopelessness	Angry outbursts/aggression
Good sense of humor Performing well	Procrastination Forgetfulness	Negative attitude, poor performance/ concentration	Cannot function at work or control behavior
Normal sleep patterns, few problems sleeping	Trouble sleeping, intrusive thoughts	Restless, disturbed sleep Nightmares	Can't sleep or stay asleep
Physically well. Good energy levels	Muscle tension/headaches Low energy	Fatigue Aches and pains	Constant fatigue Physically unwell
Physically and socially active	Decreased activity	Avoidance, withdrawal	Avoiding social contact
No/limited alcohol use or gambling	Regular, but controlled alcohol use	Uncontrolled alcohol use	Abuse of alcohol or gambling

High job demands deplete energy resources, resulting in withdrawal in order to conserve resources, and job performance drops as a result. Attempts to maintain performance (the "strain coping mode") are characterized by overactivation of the sympathetic nervous system, and this is the mechanism by which stressors in the work environment get "into the body." Long-term overactivation of the hypothalamic–pituitary–adrenocortical axis appears to be the mechanism, and it can have long-term health consequences (Miller et al., 2007). In the demand-resources model, autonomy and support from leaders and peers are resources that support active coping with work demands, thus lowering strain. Burnout is thought to be the endpoint of a process of energy depletion, so we would expect job strain to precede burnout (Demerouti et al., 2001).

BASIC APPLICATIONS

Build Movement into Sedentary Jobs

Fenery and Walker (2002) report that exercises for sedentary workers had short-term benefits. After doing brief exercises every 30 mins for 3–5-day shifts, employees reported less musculoskeletal discomfort and moved more while sitting. The main barrier to long-term adherence was reported to be the conflict between exercises and job performance.

After a meal, blood glucose levels increase as food is digested in the gut and nutrients are absorbed. In sedentary workers, this increase is exaggerated. Even a short walk of 15 mins after a meal can reduce postprandial glycemic excursion by up to 50% (Levine, 2014), suggesting that catering facilities should be sited some distance from office buildings to encourage walking after eating. Diaz et al. (2019) analyzed data from a cohort study of 30,239 people ≥45 years to investigate the effects on mortality of breaking up periods of sitting using breaks of moderate or light physical activity. Short bouts of either light or moderate physical activity (1–5 min every 30 min) were associated with lower mortality, although longer bouts (5–10 min every 30 min) were more strongly associated with risk reduction (hazard ratios of 0.76 as opposed to 0.86). Moderate activity had a greater effect than light activity (0.65 versus 0.83). The authors concluded that physical activity of any intensity at any time will mitigate the risks of sedentary behavior, although moderate-intensity activity will have a greater benefit than light-intensity activity. Loh et al. (2020) re-analyzed the findings of 37 studies that investigated the effects of breaking up long periods of sitting with physical activity breaks and the effects on blood

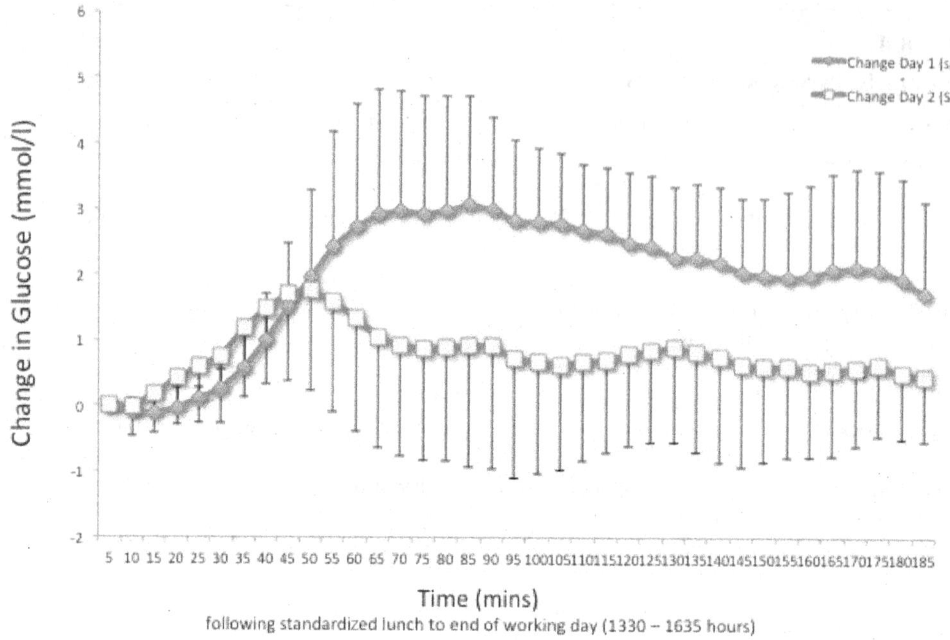

FIGURE 8.4 Change in blood glucose after working in a standing or seated position 185 min after eating a standard buffet lunch. (Adapted from Buckley et al. (2014), with permission.)

glucose, insulin levels, and triglyceride levels (a source of calories from fat) in the blood. Most of the studies were experimental over one or more days. Overall, physical activity breaks lowered glucose, insulin, and triglyceride levels after meals and had a larger effect in people with a high BMI. Intermittent breaks of physical activity had a greater effect on blood glucose levels than one continuous break, and the effects persisted overnight.

Buckley et al. (2014) monitored changes in blood glucose concentrations in subjects who either sat or stood for 3 h after lunch. The elevation in blood glucose concentration after lunch was attenuated when subjects stood and remained lower for the duration (Figure 8.4). Although energy expenditure was slightly higher when standing (0.83 kcal/min), the attenuation mechanism is likely to be better perfusion of tissues sensitive to insulin, leading to increased absorption of glucose. The circulatory demands of sitting are negligible, but they increase when standing. The large muscle groups of the lower limbs are better perfused when standing and are obvious absorption sites. These findings suggest that oxidative stress on the tissues would be reduced, with possible long-term health benefits.

Peddie et al. (2015), using actimeters to measure movement, found that even small movements (going to the bathroom, swapping sitting and standing, fidgeting, and so on) were sufficient to improve glucose metabolism in sedentary workers after they had eaten. Even these efforts were insufficient to cause an increase in heart rate but resulted in small reductions in glycemia and insulinemia, possibly because they increase the permeability of muscle cells to glucose.

Fitting the Job to the Obese Worker

In many countries, obese workers make up a significant proportion of the working population (up to 25%). Most employers are unable to discriminate against obese individuals seeking employment. They are required by law to make "reasonable arrangements" to accommodate obese employees. Some common problems of obese employees include:

- Difficulty Wearing Uniforms
- Difficulty wearing personal protective equipment
- Difficulty moving around at work
- Problems attending training courses
- Lack of fit with office furniture

Many of these problems are a reflection of the increased circumference at the hips, abdomen, and chest of obese individuals, coupled with a tendency for shortness of breath and abnormal gait. These changes in the anthropometry of the workforce are often not well reflected in the anthropometric standards used to design work equipment, and employers should take into account the prevalence of obesity in the target population when new equipment and facilities are procured.

Postural Stability: Obese people have a greater risk of falls because they have a higher center of gravity. Koep et al. (2015) analyzed six years of injury reports at a large organization in the USA. The BMIs of employees with slip, trip, and fall reports recorded at the medical center were compared with those not reporting such injuries. Sex was controlled for in the analyses. A total of 8,581 records were reviewed. The mean age of the workforce was 46.8±11.1 years, and the mean BMI was 26.6±5.3 kg/m^2: 34% of employees were obese, 42% were overweight, and 24% were normal weight. Over the study period, 189 falls were reported. The mean age was 47.9±10.8 years, with 79 falls occurring in women and 110 in men. There was a statistically significant difference in the percentage of injuries suffered in the obese group: 3.3% were injured in a slip, trip, or fall, whereas 1.6% of non-obese employees were so injured. It should be borne in mind that the data only indicate the *outcome*—an injury reported to the medical department as a result of a fall. It may be that non-obese people were equally likely to lose balance or slip but were more able to recover due to faster reactions and greater normalized strength. However, Miller et al., (2011) found no difference between obese and normal-weight subjects in recovery from small perturbations. Alternatively, falls suffered by the obese may have been more likely to result in an injury requiring medical treatment.

Strength: Bulbrook et al. (2021) investigated the effects of overweight and obesity on absolute strength and strength normalized for body mass. Strength tests included shoulder flexion and extension, internal/external rotation in the dominant arm, low back flexion and extension, one-handed upward and downward exertions, pushing and pulling one-handed, and bilateral lifting strength. Absolute strength did not differ between the groups, although males had 29%–62% greater strength for all tasks except pulling. However, when the strength data were normalized for body mass (N/kg or Nm/kg), significant and large differences were found between the obese and very obese (BMI>40) subjects and overweight/normalweight subjects. Body fat percentage correlated negatively with normalized strength but not with absolute strength.

Postural Stress: Obese workers typically have decreased muscular strength normalized for lean body mass, increased reach distance due to greater abdominal depth, and greater postural discomfort due to higher postural loads. Range of joint motion is often limited, and spinal compression is greater during many activities.

Park et al. (2009) compared the ratings of perceived exertion (Borg Scale) of obese (BMI>30) and non-obese subjects while holding a 5 kg box for 20 s in 84 different postures. The postures were defined using OWAS[1] (the Ovako Posture Assessment Scheme). OWAS categorizes postures using combinations of back, shoulder, and leg positions:

- Back: Straight, bent, twisted, bent and twisted
- Arms: Both below shoulder, one above shoulder, both above shoulder

- Legs: Sitting, standing, standing with weight on one leg, standing with both legs bent, standing with one leg bent, kneeling, and walking
- Forces: 10 kg or less, 10–20 kg, >20kg

There was a statistically significant difference in ratings of perceived exertion for the obese and non-obese subjects. Overall, obese subjects rated the tasks with a mean score of 5.6 ("Hard") compared to non-obese subjects with a mean of 2.5 ("Easy/Moderate"). All individual postures were rated as more demanding by the obese subjects. There were significant interactions between the postures themselves, as might be expected. Postures that combined deviations of the back, arms, and legs were rated as more demanding. The overall pattern of interactions was the same for both groups but with higher ratings of perceived exertion by the obese group. However, although combined postural deviations caused increases in perceived exertion in both groups, obesity amplified these effects. For example, postural deviations of the arms and legs combined might have increased perceived exertion by a certain amount in non-obese subjects (ΔRPE), while the same combination increased it in obese subjects by Δ1.6.

That the non-straight back, elevated arm(s), and bent leg(s) postures were identified as stress-increasing postural elements is not surprising, given that obese people have lower normalized strength and therefore reduced capacity to support the mass of their own bodies. In other words, few postures apart from standing and sitting can be considered of "low" or "moderate" stress for obese employees—*in effect, obesity places more constraints on the design of safe workstations.*

Task Performance: Baek et al. (2021) compared the performance of obese (BMI > 30) and non-obese (BMI < 25) subjects in an experimental foot-reaching task. Subjects sat upright on a chair in front of a computer screen with one foot on a pedal (the start position), which displayed the start position and 12 target positions arranged ahead and to either side of the start position. At the beginning of a trial, a target appeared, and the subject had to move the foot to that position as quickly as possible, with 24 foot reaches per trial. Reaction time (time to move the foot off the pedal after the target appeared), movement time (time to reach the target), and total time were recorded. Subjects also rated task difficulty on the Borg scale. As might be expected, target distance and direction affected movement time. For the obese subjects, movement time (time to move the foot to the target) was greater by 21% on average, and reaction time by 32%. No differences in Borg scale measures were found. The authors suggested that the difference might be due to the increased mass of the lower limbs, resulting in greater inertia when initiating and ending the movement. For obese workers, it was suggested that they might be better accommodated by having larger targets and reduced foot travel distances (see Chapter 12, Fitts' Law).

Don't Design an "Obesogenic" Workplace

HFE specialists are not medically trained and should not offer advice about weight management to obese individuals. However, they can offer advice to architects and managers about how to avoid building an "obesogenic" workplace (Table 8.5)—a workplace that makes it more likely that employees will become obese or remain obese.

Long shifts (12 h or more) may restrict the food choices available to employees to whatever is available, and this may also be obesogenic.

One approach is to design workplaces so that healthier options are either more salient or are the default options (Rothman et al., 2015). For example, the stairs in buildings (and ramps for wheelchair users) are more accessible than the elevators; in dining halls, salads are the default option or the "value meal," and diners have to ask for French fries, and so on; hard copy mail has to be collected from a post room and is not delivered to desks; smokers have to leave the building to smoke, and so on.

TABLE 8.5
Characteristics of the "Obesogenic" Workplace[a]

Easy access to unhealthy food options—that is, fast food, takeaways, vending machines
Lack of facilities for preparing food, that is, microwave oven, cooker, etc.
Lack of facilities for storing healthy food—that is, refrigerator
Lack of healthy food options in canteen
Lack of nutritional information in canteen or vending machine food
Lack of break—lunch areas causing employees to eat at their workstations, desks, cars, etc.
Occupational stress
Long working hours
Shift work
Continuous sedentary working behavior—spending a lot of time seated
Work schedule—not allowing enough time for lunch breaks
Lack of onsite physical activity facilities
Lack of stairs at the work
Stairs at the workplace not accessible and not properly illuminated
Continuous sedentary working behavior—spending a lot of time seated
Work schedule—not allowing enough time for lunch breaks
Lack of cycle storage facilities
Active access to the workplace discouraged by bad public transportation system
Limited availability of nearby (and safe) recreational areas, green spaces, parks, sports grounds, etc.
Lack of management commitment to promoting workplace wellness
Lack of a workplace health promotion policy
Lack of staff to implement workplace health promotion activities at the workplace
Lack of workplace health promotion programs and interventions promoting healthy nutrition and physical activity
Lack of awareness by employers concerning the adverse effects of obesity on job satisfaction, turn over, and productivity
Low employee participation in workplace health promotion programs
Lack of incentives for employees participating in obesity prevention programs
High cost to employers for providing physical activity opportunities
Not enough space—at the workplace—high levels of congestion at the workplace
"Pressure to look slim and well-trained" at work (preventative effect)

[a] From Guidelines for the Prevention of Obesity at the Workplace by Pania Karnaki, Dina Zota, and Professor Athena Linos from the Institute of Preventive Medicine, Environmental and Occupational Health, PROLEPSIS, Greece.

Evaluation of Organizational Interventions to Reduce Exposure to Psychosocial Hazards

The philosophy is to identify the causes of stress and modify them by redesigning jobs rather than treating the symptoms of stress. Effective interventions at the organizational level include:

- Increased participation in decision-making through higher levels of decision latitude and social support, and decreases in role conflict and role ambiguity, which lead to reduced emotional strain, job dissatisfaction, absenteeism, and intention to leave.
- Devolution of control over the pace of work, organization of rest pauses, allocation of work assignments and overtime, and increased control over work arrival and departure times (Wall and Clegg, 1981; Murphy et al., 1998).
- Changes in decision latitude and/or social support leading to improved cardiovascular risk patterns, decreased sleep disturbance and gastrointestinal complaints, and a decrease in the prevalence of pain in the upper spine.

Computer Stress

A study by Lazar et al. (2006) found that computer users wasted 42%–43% of their time on the computer, suffering frustrating experiences that prevented the achievement of work goals. Fifty computer users performed their typical work tasks. They completed a pencil-and-paper diary on which they recorded any frustrating experiences whenever an event occurred. The records included a description of the event itself, a description of the user's goal, the software or application being used when the event happened, a rating of the degree of frustration, and data on the time spent solving the problem and the time lost because of the problem, including the time lost trying to solve it.

The findings will not surprise anyone who uses a computer on a daily basis. Problems occurred most commonly with word processing, followed by email, web browsing, and other kinds of Internet use. Almost one-half of the problems experienced occurred at least once per week. Analysis of the solutions adopted by the users revealed that most of the time, they were able to solve the problem themselves. On only 10% of occasions were they unable to solve it, and on another 10% of occasions, help was sought from coworkers. Twenty-five percent of the solution strategies involved direct action by the user—mainly because the problem was recurrent and the user had already learned how to deal with it. On less than 1% of occasions were books or paper documentation consulted—proof, of a sort, of the old adage, "If all else fails, read the manual."

Users spent an average of 20% of the work session dealing with frustrating experiences and another 20% recovering lost work. Common causes of these frustrating experiences included lost Internet connections, Internet log-in failures, and browser failures. The Internet was not entirely to blame. System crashes caused by operating system failures were common, as were crashes caused by multitasking (trying to run several programs simultaneously). Interestingly, problems caused by the application software—including mature applications such as word processing—were also common and included unusable or difficult-to-use features and unpredictable responses by the program. The most time-consuming experiences involved the use of presentation software, spreadsheets, and databases.

There is an inverse relationship between the functionality of any device (what it can do for you) and its usability (how easy it is to use). Our capacity to do more with less increases continually in the IT domain, while our capability to deal with the additional demands, including complexity, lags behind.

Methods of Reducing Computer Stress

- *Training*: Training in the operation of generic application software, such as word processors and spreadsheets, should be mandatory but is not enough. Set standards for what is needed from these applications on the job. Avoid exploring the functionality of these systems during work time at all costs. There is always more than will ever be needed.
- *Review Internet Connectivity*: It has long been known that excessive system response time and response time variability cost money and frustrate users.
- *Use Templates*: Develop simple templates for corporate e-documents. Put clarity, ease of use, and simplicity ahead of style and sophistication.
- *Listen to Users*: If users complain of excessive downtime or unreliable systems, investigate the problem; don't blame it on "the system."
- *Improve the Design of Error Messages*: OK, something's gone wrong, so tell the user what to do next.
- *Use Nudging and Friction*: Design user interfaces and human–computer dialogs so that they are compatible with how people behave.
- Minimize changes to existing procedures when new capabilities are introduced.

TOOLS AND PROCESSES

Understanding the Risk: Hazard Ratios

Dunstan et al. (2010) investigated mortality in relation to the time people spent watching television in a sample of 8800 Australian adults, excluding the time spent doing other things while the television was switched on. They controlled for other health risks such as diabetes and tobacco smoking and divided the sample into three groups: those spending less than 2 h/day watching television; those watching 2–4 h/day; and those watching more than 4 h/day. *Hazard ratios were* calculated to determine whether television time predicted mortality in the following six years (a hazard ratio of "1" indicated no increased risk). For every additional hour spent watching television per day, the risk of mortality increased by 1.11 for all-cause mortality and 1.18 for cardiovascular mortality. People who had survived at a point in time after the beginning of the study and who watched television for more than 4 h/day were 1.80 times more likely to die subsequently than those spending less than 2 h/day watching television. Similar results were obtained by Foster et al. (2020), who analyzed data from 490,966 participants in a cohort study. A curvilinear relationship between television time and mortality was found, suggesting that television viewing time be limited to less than 2 h/day. Linear relationships were found between television time and the incidence of cardiovascular disease and stroke. For subjects walking half an hour per day, substituting 30 min of television time with 30 min of walking reduced mortality risk by 10%.

The authors point out that time spent watching television uses time that might otherwise be spent more actively, which provides further evidence of the link between physical inactivity and ill health. Designers should consider the possibility that work might be the only place where employees are active and that, by designing more activity into jobs, they may be able to counteract the health risks of sedentary leisure time. This provides further evidence for the provision of "activity-permissive workstations" for office workers.

Assessing Fitness for Work

Cardiovascular fitness can be estimated using simple "step tests" such as the Harvard Step Test. The subject steps up and down onto a 50 cm-high step every 2 s for 5 min (or until exhaustion if exhaustion occurs before 5 min). Stepping is paced by a metronome. After 5 min, or exhaustion if the subject cannot continue, the subject sits, and the heart rate is taken at 1-min intervals for 30 s at 1, 2, and 3 min. A "fitness index" is calculated as follows:

$$\text{Fitness} = T_e \times 100/h_b \times 2 \tag{8.3}$$

where

T_e = time to exhaustion in seconds
h_b = sum of the heart beats over the three intervals

The fitness index is interpreted as follows:

- Excellent > 96
- Good — 83–96
- Average — 68–82
- Low average — 54–67
- Poor < 54

Therefore, if a subject steps for 5 min and has 75 heartbeats for 30 s after 1 min, 55 heartbeats for 30 s after 2 min, and 45 heartbeats for 30 s after 3 min, his fitness index is

$$\text{FI} = 300 \times 100 / 175 \times 2 = 85.2 \tag{8.4}$$

The Harvard Step Test has good reliability over a 1-week test–retest period (Meyers, 1969).

FIGURE 8.5 (a) Heart rate monitor and (b) pedometer.

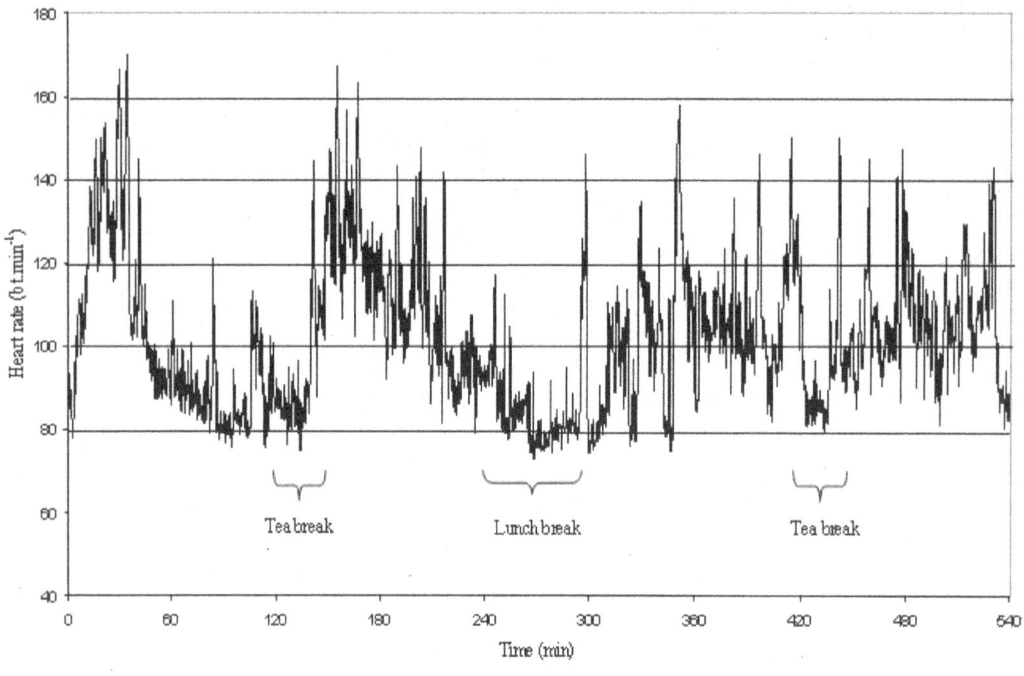

FIGURE 8.6 Heart rate over a workday on a self-paced task.

Assessment of Physical Work Demands: Ambulatory Monitoring

Ambulatory monitoring is used to assess physical activities and demands in everyday work situations. Heart rate monitors, actimeters, and pedometers (Figure 8.5) can be fitted to employees to capture data on daily activity and movement.

Figure 8.6 shows the heart rate recorded over an 8-h shift from a 45-year-old male. The shift ran from 08:00 (at time point 0 min) to 17:00 (at time point 540 min). The work was self-paced and involved physical and mental tasks in several different locations. Heart rate was variable (a range of over 100 bt/min) and peaks in heart rate (during physical activity) are interspersed with recovery periods. Heart rate quickly returned to baseline at the end of the shift, indicating good recovery

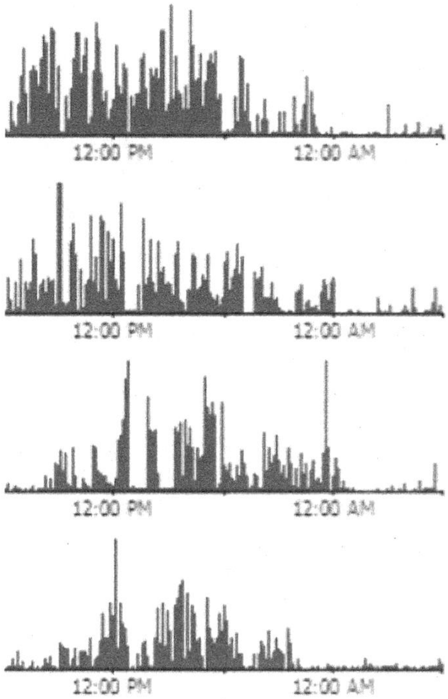

FIGURE 8.7 Actimeter data for two workdays and the weekend. (Courtesy of Professor Leon Straker, Curtin University.)

with no evidence of a deficit. Modern heart rate monitors are relatively nonintrusive; the sensor is affixed to the chest by means of a strap, and the data are transmitted to a receiver worn on the wrist.

Figure 8.7 shows actimeter data from Straker and Mathiassen (2009). Actimeters contain an accelerometer that simply measures the number of movements over a time period and summarizes them in a time series. The more one moves in a given time period, the higher the peak. Figure 8.8 shows oxygen consumption being measured while an employee carries out a maintenance task in a confined space.

Pedometers can be used to estimate how active people are both at work and at leisure. Research pedometers are normally worn on the waist to detect vertical motion of the hip joint. Many smartphones often have step counters for use in daily life. The U.S. Department of Health and Human Services recognizes the need for greater physical activity in the population, noting that for most people, more activity would be better, particularly vigorous activity. Tudor-Locke (2010) reported that U.S. adults are generally inactive and proposed targets for increased activity based on pedometer research (Table 8.6).

According to Tudor-Locke, adults in the United States are not active enough, taking 5,900–6,900 steps per day when they should be taking 8,900–9,900 steps per day on 5 days of the week, at least 3,000 of which should be of moderate intensity.

From the perspective of HFE, these findings have clear implications for how we plan and design work environments—even small changes, such as removing personal trash cans from desks in favor of a central disposal unit, can increase the amount of activity by making it essential for employees to walk to carry out routine tasks on a daily basis.

Activity Monitors and Smartphones

Activity monitors based on accelerometers are normally worn on the wrist, hip, or thigh. The former tends to be worn for longer and are less intrusive. Smartphones are almost ubiquitous in many

FIGURE 8.8 Measuring oxygen consumption in the workplace.

TABLE 8.6
Step Hierarchy for Classification of Physical Activity (Steps per Day)

<2500	Basal activity
2500–4999	Limited activity
5000–7499	Low active
7500–9999	Somewhat active
10,000–12,499	Active
≥12,500	Highly active

parts of the world and show promise, given that they record activity reliably and with validity. Smartphones can detect the number of steps taken at different walking speeds, but the accuracy depends on where the phone is carried or worn on the body. Höchsmann et al. (2020) compared the activity trackers of two smartphones (the iPhone SE and the Samsung Galaxy S6 Edge) with the StepWatch (a research-grade, ankle-worn device); the research-grade ActiGraph wGT3X+ accelerometer (worn on the hip and wrist); and the Garmin Vivofit 2, a consumer activity wristband. Thirty subjects wore all the devices for 3 days (the phones were kept in their pockets), and they kept diaries of their activities every 15 min. The StepWatch was used as the criterion measure, and the experimenters used the mean absolute percentage error of the other devices against the StepWatch as a measure of validity. The phones underestimated the number of steps by 27% per day (the Garmin and ActiGraph also underestimated step numbers but by smaller percentages). Underestimation was

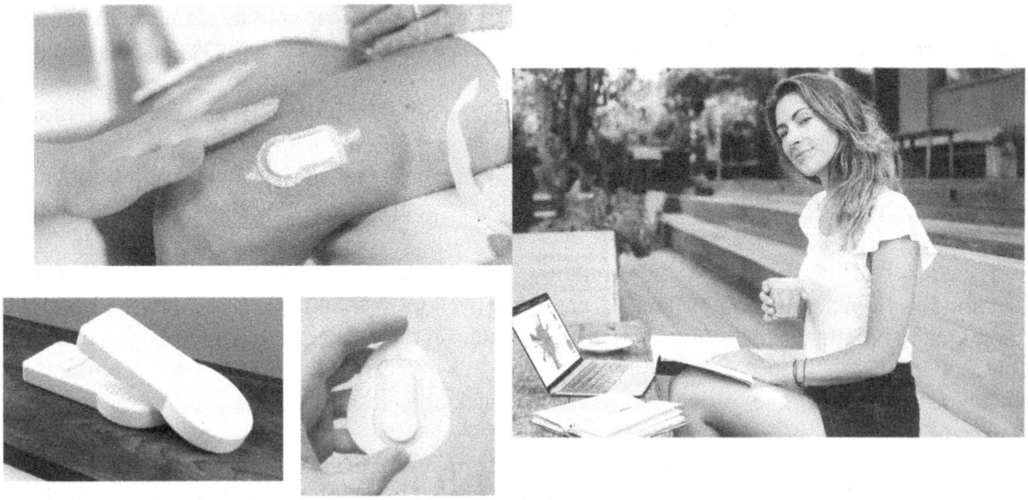

FIGURE 8.9 Fibion© wearable tri-axial activity tracker for monitoring daily activity including sedentary time. (Courtesy Fibion.)

greater on days when fewer steps were taken. During cycling, all devices underestimated activity by 55%. During other activities, such as sitting or using public transport, the phones gave comparable estimates to the StepWatch.

> The authors recommend that when using smartphones to record light, everyday activities, the number of steps is likely to be underestimated by about 20%.

Yang et al. (2018) tested the validity and reliability of the Fibion© activity tracker, a tri-axial accelerometer worn on the thigh or carried in a pocket, in comparison to an ActiGraph© monitor of known reliability and validity. Subjects followed a protocol of walking, standing, and sitting repeated over 90 min. In a validity trial, subjects carried out tasks for 12 h while their oxygen consumption was measured. Their activities were recorded by the experimenters. Both devices had low reliability for sitting, but the agreement for standing and walking was high when the Fibion was worn on the thigh. When worn on the thigh, the Fibion monitor accurately measured the types and intensity of different activities and energy expenditure (Figure 8.9).

When worn on the thigh, activity monitors are sensitive to posture and can differentiate between sitting and standing as well as movement. Alkalih et al. (2022) compared the Fibion activity monitor and the ActivPAL4 (the criterion measure) in a one-day trial. The two devices were worn on the same thigh while subjects carried out their everyday activities. Overall, there was good agreement between the devices, particularly for estimates of time spent walking.

One possible advantage of using smartphones in place of bespoke activity monitors is that the activity monitor is an extra "piece of kit" and may induce "participant reactivity" — participants' behavior changes because they are aware of the device and know their activity is being monitored by it (this is not a "Hawthorne Effect," by the way; see Chapter 1). This is less likely to happen if the activity is monitored by the participant's own phone.

Fitness Apps

There are many new tools for maintaining a healthy lifestyle that can be accessed via cell phones or tablets. Google Fit is an example that combines a smartphone activity monitor with GPS data and heart rate data if used with a device such as an Apple Watch. These applications or "apps," vary in the support they offer. Some are restricted to dietary advice and recording, while others give an indication of energy expenditure based on users' reports of their activities and relate these to user-reported food consumption. Others provide guidance via texting, either by counselors or online feedback, and some contain pedometers.

MyFitnessPal, for example, is a smartphone app that records activity and food intake. Users can enter the name of the food or search a food database to record their dietary intake. Users can also choose from a list of exercises to increase their daily energy requirements—essentially, the app helps them set and pursue fitness goals by optimizing their energy intake and expenditure.

Lyzwinski (2014) reviewed 17 evaluations of cell phone/tablet devices for weight loss and fitness. In general, mobile devices promoted weight loss and increased physical activity compared to baseline, to a greater extent than standard interventions or no-intervention controls. It is thought that the effectiveness of these devices lies in their ability to help users monitor their own behavior in real time, providing timely feedback about daily activity and its consequences for the user.

Muntaner-Mas et al. (2019) conducted a systematic review of the validity and reliability of fitness apps to assess their use in evaluating cardiovascular fitness. In addition, a Mobile App Rating Scale (MARS) was used to assess user engagement, functionality, aesthetics, information, and subjective and perceived impact. A total of 88 apps were reviewed. Five had been tested for validity and two for reliability, and most were judged to be of low to moderate quality.

Quick Activity Checklist

Table 8.7 lists items for a quick activity checklist that can be used to assess how physically active employees are and to decide whether there is a requirement to build more physical activity into the workplace (Holterman et al., 2010). One way to interpret responses to this checklist would be by reference to MET tables to give an approximate indication of the percentage of employees at risk due to inactivity.

Assessment of Physical Work Demands: MET Tables

HFE Workshop 8.1 provides some examples of how MET calculations might be used in practice. Table 8.8 lists representative activities listed in the compendium across the range of MET values.

TABLE 8.7
A Quick Assessment Checklist for Self-Reporting Physical Activity

	Yes	No
Physical Activity at Work		
1. I am mainly sedentary and do not walk around much at work		
2. I walk around quite a bit at work but do not carry loads or run		
3. Most of the time, I walk and often climb stairs or ladders and carry loads		
4. I have heavy physical work and carry heavy burdens often		
Physical Activity during Leisure Time		
1. I am mainly sedentary. I read, watch TV, or go to the movies		
2. I go for a walk, cycle a little, and am active for 4 h/week		
3. I am active—I run, play sports like tennis at least 3 h/week, or garden		
4. I am a competitive sports player (swimming, soccer, handball) or run long distances, several times per week		

HFE WORKSHOP 8.1

MET VALUES IN PRACTICE

Suppose we want to estimate the energy expenditure of a policeman directing traffic. It would be impractical to take direct measures of oxygen consumption because he or she would have to wear a mask to collect the expired gases. We could measure heart rate, but then the policeman would have to undergo exercise testing in the laboratory to establish his heart rate/VO_2 relationship.

We can estimate the energy expenditure by referring to the table of MET values, where the MET value for directing traffic is given as 2.5. We assume that 1 MET is the same as the policeman's basal metabolic rate, that is,

$$1 \text{ MET} = \text{RMR} = 4.184 \text{ kJ/kg/h}$$

If the policeman weighs 80 kg and directs traffic for 2 h/day, then the energy expended (EE) in this task is

$$\text{EE} = 4.184 \times 2.5 \times 80 \times 2 = 1673.6 \text{ kJ}$$

This corresponds to a rate of 9.3 J/min, which, according to Table 7.6, is classified as "light work."

We might then wish to investigate the other tasks that he carries out over the workday, and if we know the time spent on these tasks, we can estimate the total energy expenditure for the 8-h day by adding the different values. From this, we could then estimate the average energy expenditure.

There are several sources of error in making calculations such as those above—the RMR declines with age and is lower in malnourished people. If the real RMR is lower than the assumed RMR, we will underestimate the energy expenditure. However, if we have a good understanding of the daily work activities of people who do varied jobs and if we can obtain accurate estimates of the amount of time they spend doing different things, then we can make a fair initial assessment of the daily energy requirements using this approach. This method is likely to be useful in developing countries, where large numbers of people need to be screened, where there is a shortage of trained physiologists able to take accurate measurements of heart rate and oxygen consumption, and where malnutrition, in the form of a shortage of carbohydrates in food, might be a problem.

Work Ability Index

The WAI (Table 8.9) is a self-report measure of work ability that includes items on health complaints and on the respondent's perceptions of his or her ability to work compared to his or her lifetime best. Responses to the questionnaire are summed, giving a range of index values from 7 to 49:

- Work ability — poor
- Work ability — moderate
- Work ability — good
- Work ability — excellent

The WAI can be used to assess the current status of a workforce, to monitor trends over time, and to evaluate the effects of interventions designed to improve employee health and work ability.

TABLE 8.8
Energy Costs of Some Common Daily Activities[a]

Activity	Work (kJ/min)
Sitting	1.25
Kneeling	2.10
Squatting	0.5
Standing	0.6
Stooping	0.8
Walking unladen	2.1
Walking with 10 kg load	3.6
Cycling 16 km/h	5.2
Climbing stairs (30°)	13.7

[a] Older references often cite energy cost in "kilocalories or kcal." Convert to kJ as follows: 1 kcal – 4.184 kJ.

TABLE 8.9
Items of the WAI[a]

1. Current work ability compared with the lifetime best 0–10
2. Work ability in relation to the demands of the job 2–10
3. Number of current diseases diagnosed by a physician 1–7
4. Estimated work impairment due to diseases 1–6
5. Sick leave during the past year (12 months) 1–5
6. Own prognosis of work ability 2 years from now 1–7
7. Mental resources 1–4

[a] Finnish Institute of Occupational Health

The WAI is valid and reliable — it predicts ill health and occupational disability. In a 4-week reliability study, Zwart et al. (2002) found that 25% of respondents gave exactly the same score on both administrations of the WAI, and 66% of classifications into the categories above were replicated on the second administration. The overall WAI score for the group did not change over the test period.

In a study of aging seafarers (Bridger and Bennett, 2011), work ability was found to be good on average, and the mean BMI of the sample was 27.5 kg/m^2. The best predictor of WAI score was the interaction between BMI and age, in the expected direction. Older personnel with high BMI reported lower work ability. A statistical model of the effect of BMI on work ability found the effect to be large. Taking a WAI cut-off score of 43 to separate excellent work ability from good work ability, the model predicted that work ability would start to decline at 47 years of age in people with BMIs of 30 or more (the average age of the participants in this study was 47 years). For those with BMIs of <25, the predicted age of the same decline in work ability was 56 years—a difference of 9 years. Thus, although work ability does decline with age, it declines more rapidly in obese people.

The WAI is a powerful tool because it combines an individual's own perceptions of his or her ability to continue working with evidence about any existing health conditions that might degrade work ability. Conditions diagnosed by a doctor are weighted more heavily.

SYSTEM INTEGRATION

Successful integration requires a good understanding of the job demands, training requirements, and physical capacity of the workforce. In the developed world, the challenge is to integrate an aging, unhealthy workforce into increasingly interconnected and automated systems. The target audience description should be detailed enough to enable the health status of employees to be properly understood and the costs associated with health to be properly documented, so that a system of work can be designed to provide optimal and beneficial physical and mental stimulation.

Traditionally, HFE has been concerned with balancing employee characteristics and job demands. Demographic changes in work ability due to the aging population and the obesity epidemic have disturbed this balance. Many of the tools used to ensure a good "fit," and the evidence on which they were based, are over 30 years old and may now have been overtaken by these changes because they are generic and do not contain terms for quantifying employee characteristics. Rather than reducing job demands, a more modern focus might aim to optimize job demands. Many years ago, one of the founding fathers of HFE, Professor A. Chapanis, quoted a sign outside an elevator in an office building:

> *'Walk-up one floor'*
> *'Walk down two floors'*
> *'For better elevator service'*

A more appropriate version of this sign today might replace the last line with "To keep you fit and healthy."

In the last century, musculoskeletal ailments and conditions such as rheumatism were a major risk to the health of the workforce. The risk to today's workforce is metabolic. Jobs demand little physical activity but may be mentally demanding in an environment replete with inexpensive, calorie-dense food. The challenge for HFE is to find ways of designing work that is both physically and mentally stimulating.

Nudging and incentives are ways of increasing physical activity. In some parts of the world, special platforms have been installed in railway stations. In return for performing a set of squat exercises, travelers earn a free ticket:

https://www.youtube.com/watch?v=kHzQk59nfVc

COSTS OF ILL HEALTH

Ward et al. (2021) found that health care costs in the USA were lowest for females with a BMI of 20.5 and males with a BMI of 23.5. For every 1 BMI increase over a BMI of 30, excess health care costs increased by $235 per year. Overall, obesity was associated with excess costs of $1,861 per person, or $172.4 billion in annual expenditures. For severe obesity, the value was $3,097 per person/year.

Kowlessar et al. (2011) estimated the medical and productivity costs of 11 health risks for a large employer in the United States using data from over 60,000 employees. Table 8.10 summarizes the main findings. Most of the health risks were reflected in higher healthcare costs, greater absenteeism, and presenteeism (reduced productivity at work) in the high-risk groups. Lack of exercise, high blood pressure, and high blood glucose were the most costly.

WORK HARDENING PROGRAMS

Work hardening programs are a type of physical training program aimed at improving employees' physical capacity for a particular task or set of tasks, with the aim of reducing the risk of physical injury. To be effective, work hardening programs must

TABLE 8.10
Effects of Modifiable Health Risk on Medical Care Costs, Absenteeism, and Presenteeism in a Large Company in the United States[a]

	Medical Care Cost ($)	Absenteeism	Presenteeism
High body mass	992	0.4	0.7
High blood glucose	3362	0.6	1.2
High blood pressure	1702	0.2	0.4
High cholesterol	352	0.2	0.5
High triglycerides	1238	0.5	1.0
High stress	531	0.5	1.0
Low exercise	1359	0.6	1.6
Poor nutrition	−429	0.2	1.2
Poor safety	−631	0.0	0.9
Tobacco	97	0.5	1.8
Alcohol	−276	0.1	2.1

[a] The mean difference in costs, absenteeism, and presenteeism between "high-" and "low-"risk groups for each health risk.

- Be of greater intensity than that experienced on the job
- Be structured and graded, and take a progressive approach
- Use exercises that closely resemble the movements made on the job
- Be evaluated using tests that simulate the activities carried out during training

In many industries, work hardening takes place informally. Management frequently expects a "break-in period" of a few weeks before a new employee can work at the required pace. This is particularly true of jobs involving heavy or repetitive work.

Animal and human studies have shown that endurance training causes an increase in muscle stiffness (the ratio of tension developed in a muscle when it is stretched to the amount it lengthens) as a result of increased cross-linking of collagen fibers in the muscle. Animal studies have also shown an increase in the amount of stiffness and collagen in the muscles of elderly and sedentary rats. A lack of exercise, combined with age-related changes in muscle structure, may reduce flexibility, predisposing workers to injury. The three main components of a work hardening program are strength, flexibility, and endurance. Depending on the requirements of the task in question, the program must aim to achieve an appropriate balance between these components. This is particularly important when the training is part of a rehabilitation program for previously injured workers.

Effectiveness of Workplace Wellness Programs Based on Physical Activity

Marin-Farrona et al. (2023) carried out a systematic literature review of the effectiveness of work-site wellness programs. Seventeen publications from 16 studies (1997 onwards) were reviewed, covering a variety of occupational groups: healthcare workers, office workers, university employees, industrial workers, and one study that included cleaners and post office workers. All were randomized controlled trials (a study group and a control group for comparison). The studies varied in their contents, with some offering nutrition advice, pedometers or heart rate monitors to tailor the training to the participant. The effects of the programs were measured in a variety of ways. Productivity was measured using the Work Ability Index in six studies. Absenteeism was measured either by self-report or employee records in six studies. Nine studies measured productivity using self-report questionnaires. Health outcomes were measured in different ways, depending on the study: a self-report health questionnaire in seven studies; muscle strength was measured in four studies (maximum strength test, push-ups, or standing long jump); eight studies measured body mass before and after, as well as percentage muscle mass and body fat; blood pressure and blood profiles (e.g., cholesterol, triglycerides, and fasting blood glucose) were measured in three studies; musculoskeletal pain, measured using a variety of valid and reliable questionnaires (such as the Nordic questionnaire), was assessed in five studies; seven studies included daily physical activity, either using questionnaires or pedometers; eight studies measured cardiovascular fitness (see Chapter 7); four studies used a job stress questionnaire, and four included measures of anxiety or tension.

The effect of the programs was reported using a statistic known as Cohen's d, which measures the size of the difference between two groups. By convention, $d=0.2$ indicates that the difference is small; $d=0.5$ is medium; and $d \geq 0.8$ indicates that the difference is large. All studies showed improvements in many of the outcome variables measured, but the differences between the studies were large. For example, work ability was measured in six studies and improved in five of them (Cohen's d ranging from 0.08, 0.1, 0.2, 0.56 to 2.18). Only three of six studies reported improvements in absenteeism, possibly because absenteeism is a non-specific outcome (it can be caused by many other factors such as sickness of family members). Some studies did not find improvements in work performance, possibly because work performance was already good at the start of the trial (a "ceiling effect"). Combined exercise training programs (resistance training plus aerobics, resistance training alone, or walking carried out three times per week for 10–60 min during working hours) were all effective in improving productivity. All of the programs also improved cardiorespiratory fitness. As might be expected, large improvements were found with the more intense programs. Strength training improved musculoskeletal health in four of four studies. Body composition improved in six of eight studies, in those where it didn't improve, the programs may not have been long enough (8 weeks) for improvement to occur. Taken together, the studies provide evidence for the effectiveness of physical activity programs, particularly in the improvement of work ability and physical health.

Target High-Risk Groups

Although there is an economic case for worksite health promotion, simple statistical measures may overestimate the potential benefits of health promotion. This is because injury and illness claims are not evenly distributed in the workforce. In fact, the data are highly skewed, with a small percentage of the overall workforce accounting for a large percentage of the claims.

Viotti et al. (2020) conducted a survey of 1069 menopausal women working in public sector offices to identify work-related psychosocial factors impacting their work ability. In common with studies of many other occupational groups, work-family conflict (between the demands of work and family commitments) was negatively associated with work ability. Autonomy in the job and skill discretion were positively associated with work ability, possibly because they provide the *affordances* for an increased ability to manage job demands when faced with menopausal symptoms. Of particular importance was the finding that a health-oriented organizational climate was strongly associated with better work ability in menopausal employees — top management is seen to give a high priority to employee health, making it easier for menopausal employees to adjust to health challenges and express their needs.

Chronic Inflammation

A study by Miller et al. (2007) found that when people continually strive after goals that are not achievable, they experience chronic "systemic inflammation," part of the stress response described earlier that can lead to chronic ill health. In success-oriented cultures, such persistence is regarded as a strength and disengagement as a weakness, which may be one reason why seemingly successful individuals can suffer work-related ill health. No understanding of stress at work can be complete without understanding why people do the work they do in the first place—what they want from it, what they get from it, and what options they think they have.

Status of Physiological Methods in Risk Assessment and Task Design

In general, physical task stress is easier to measure objectively than psychosocial stress. This is because the instruments used to measure it are reliable and valid (see Chapter 7), although they are intrusive and relatively expensive. Questionnaires have been developed for measuring habitual activity, and these are useful for screening working populations. The Compendium of Physical Activities can be used to design interviews to find out about worker energy expenditure indirectly and to interpret the information received.

RESEARCH ISSUES

Escalating healthcare costs in developed countries have provided a stimulus for the investigation of factors influencing the health of the working population. Since work constitutes the major portion of an individual's waking life, it seems logical to evaluate the activities carried out at work and to suggest ways of incorporating health-promoting features into working life. Fitness programs and in-house gymnasiums are obvious examples, as are stress management programs. In high-technology jobs, the challenge is to avoid psychological stress by increasing employee control over work, introducing a more participatory style of management, and finding ways of increasing the physical activity of the predominantly sedentary workers. For HFE, a paradigm shift is needed—from lowering the physical demands of jobs to finding the optimum level of physical and mental demands.

SUMMARY

Dunstan et al. (2021) summarize the relationship between daily sitting time, physical activity, and risks to health (Table 8.11).

Increase in risk: Low — Baseline 1.0, Low/Medium — 1%–15% Medium >15% to <30%; Medium/High >30% to <45%; High >45%

Post-industrialized countries have an aging population, and many of them have an obesity epidemic. In the United States, the National Institutes of Health reported that in 2023, more than one in five children and adolescents ages 6–19 were considered to be obese. Obesity is a multifactorial problem that requires multidisciplinary interventions. Diet is a major factor whenever inexpensive, calorie-dense food is abundant ("you can't run off the effects of a bad diet"). The effects of age on work capacity should not be confused with the effects of ill health on work capacity, although age and ill health are correlated because many lifestyle diseases take time to develop. In the United States, there is both an aging workforce and a high prevalence of ill health—10% of the U.S. population had Type II diabetes in 2023.

The main issues center around designing work environments that promote health or at least enable health to be sustained.

How much daily physical activity do we need? Current guidelines recommend 150 min/week of moderate-intensity exercise and two sessions of vigorous exercise, such as resistance training.

TABLE 8.11
The "Risk-Act" Table (Dunstan et al., 2021)—All-Cause Mortality and Physical Activity

		Hours of Sitting per Day			
Daily Physical Activity (mins)		<4	4-6	6-8	>8
	>60	Low	Low	Low	Low
	30–60	Low	Low/Medium	Low/Medium	Low/Medium
	5–29	Low/Medium	Medium	Medium	Medium
	<5	Medium	Medium/High	Medium/High	High

Advice for Hybrid Working

1. If sitting for 8 h/day is a health hazard independently of leisure-time exercise, are we putting our office staff at risk if they can only work in a seated position?
2. Energy expenditure when standing is not much greater than when sitting, and substituting sitting for 2 h/day with standing is unlikely to result in weight loss in overweight or obese people. Standing more at work may bring other health benefits, though.
3. Has office design been "overtaken by events"? New evidence on health and sedentary work calls for a radical rethink in the way we work and the way offices are designed.
4. Time spent sitting per day and participation in exercise are independent risk factors for the development of health problems. To get the best outcome, most people should sit less and exercise more.
5. Limit sitting time at work to less than 4 h/day. Although this guidance is not yet definitive (the advice has not been tested in a prospective study), there is sufficient evidence that, together with sitting when commuting and at home, most of us would fall into the "moderate" sedentary group of 4–8 h/day where the health risks are lower (Katzmarzyk et al., 2009).
6. Build movement into sedentary jobs. Do:
   a. Encourage users of sit-stand desks to stand after lunch.
   b. Site catering facilities away from the office.
   c. Hold meetings at standing desks after lunch.
   d. Discourage employees from eating at their desks.

TUTORIAL TOPICS

1. How can we make better use of the "Internet of Things" to encourage and assist people to maintain a healthy lifestyle?
2. "There are too many labor-saving devices nowadays." Discuss.
3. How should we design cities in the future to encourage physical activity?
4. At the link (https://www.youtube.com/watch?v=cxpAVCan4CQ), you will find the video "Liverpool Wash House, 1959." Using Table 8.1, assume a housewife in 1959 spent 10 h/day doing housework at 2.5–4 METs. Estimate the weekly energy expenditure in MET.h. How does this compare with the energy expenditure of an office worker on a 40-h/week?

ESSAYS AND EXERCISES

1. Mr. Smith attempts the step test described above. After 1 min, his heart rate is 150 beats/min. After 2 min, it is 120 beats/min, and after 3 min, 100 beats/min. Is Mr. Smith healthy?
2. Try out the step test described in this chapter on yourself. Calculate your heart rate, measured immediately after the test, in beats per minute. Step for 9 min at a rate of 24 steps per minute. Which fitness category do you fall into?
3. Carry out the following activities for 5 min. Measure your pulse immediately afterward to estimate your heart rate. Lie down for 5 min prior to the activity and take your resting heart rate before starting.
   - Jogging
   - Walking up a flight of stairs briskly
   - Doing squats
   - Raising and lowering a 20 kg load from knee height to chest height at a rate of 1
   - Lifting cycle every 5 s
   - Standing still
   - Sitting

- Walking
- Walking with a 10 kg weight in each hand
- Walking with a 20 kg backpack
- Walking with a 20 kg load on your head
- Plot the heart rate data in graphical form to compare these different activities

4. Refer to Table 8.1. Find three activities that are low in energy expenditure, three that are moderate in energy expenditure, and three that are severe. Convert the MET values to kJ/kg/h and plot the results as a graph in order from lowest to highest energy expenditure. Comment on the reasons for the trends.
5. Assuming the basal metabolic rate is the same as the RMR, what is the MET rating of a task requiring an oxygen consumption of 1.70 L/min?
6. Mr. Smith has arrived in Spain and is walking up a flight of stairs at 3 km/h with a 20 kg suitcase in his right hand on his way to the taxi rank. His body mass is 90 kg. What is his total energy expenditure? How would you categorize the severity of this activity?
7. Mrs. Smith, on holiday in Spain with her husband, is determined not to put on too much weight while away. She decides to follow the advice of the U.S. Centers for Disease Control and Prevention (CDC) and take at least 30 min of moderate exercise per day. Refer to Table 8.1 to find three activities that would be suitable if followed for this length of time. Her body mass is 60 kg — about how many kJ will she expend?
8. Distinguish between health and fitness. What measures can organizations take to improve the health and fitness of their employees?
9. Mrs. Smith is on her last day of holiday and has too much ice cream for her dessert (amounting to 550 kJ/100 g). By how long will she have to extend her daily exercise to use up the extra energy?
10. "In developed countries, ill health is self-inflicted." Discuss.
11. Will the use of standing desks slow down or reverse the obesity epidemic?

QUICK QUIZ

1. The MET is a measure of: (a) work capacity, (b) body composition, (c) physical activity
2. Insulin resistance occurs when: (a) insulin levels in the blood are too low, (b) the pancreas stops secreting insulin, (c) the cells are insensitive to insulin
3. The prevalence of overweight and obesity has reached over 65% in some countries. True or false?
4. An obesogenic workplace has: (a) easy access to fast food, (b) limited access to stairs, only escalators, (c) exercise facilities or lunchtime classes
5. What is the maximum heart rate for a 70-year-old worker?
6. What range of heart rates (in bpm) would be found during heavy work?
7. How many METs is the work intensity when welding?
8. A hazard ratio is a measure of: (a) the incidence rate of an outcome, (b) a measure of risk, (c) the severity of an outcome
9. Fatigue is just a fancy name for sleepiness. True or false?
10. Inactive people should do two sessions of strength training per week. True or false?

NOTE

1 https://www.ergoibv.com/wp-content/uploads/2022/03/E21__12__Module-PosturesOWAS.pdf

9 Temperature and Climate

GENERAL REQUIREMENTS FOR HUMANS IN SYSTEMS

1. Operators should remain in thermal balance with the environment while carrying out critical and generic tasks.
2. Work must be designed so that skin temperatures remain within a range defined as comfortable.
3. Fluid should be readily available to prevent dehydration.
4. Ventilation should be sufficient to support human health and performance.
5. Workers should be protected from high radiant heat and ultraviolet radiation.
6. Internal environments should accommodate individual differences in the thermoneutral zone.

Dr. Michael Mosley died of "heat exhaustion" after "sitting down and losing consciousness" near a holiday resort on the Greek island of Symi, police have said.

Dimos Kotsidaras, police commander for Symi, told MailOnline: "It looks like cause of death was heat exhaustion after walking in high temperatures from St Nicholas to Agia Marina. Temperatures had almost reached 40°C at the time of Mr Mosley's death and it is believed that he may have sat down to gather himself after being overcome with exhaustion, according to the news website London, Evening Standard, 10 June 2024.

CORE KNOWLEDGE: FUNDAMENTALS OF HUMAN THERMOREGULATION

Thermal Balance

Thermoregulation is achieved by balancing the two main factors that determine body temperature: the metabolic heat produced and the rate of heat loss. The thermoregulatory goal is to maintain the core temperature at approximately 36°C–37°C. Core temperatures over 39.5°C are disabling, and those over 42°C are usually fatal; 35.5°C is the lower acceptable limit, while 33°C marks the onset of cardiac disturbances. Further drops in core temperature are extremely dangerous, and temperatures as low as 25°C are fatal. The temperature of the peripheral body tissues, particularly the skin, can safely vary over a much wider range. From a thermal point of view, the body can be considered to have a warm core where much of its heat is produced. This is surrounded by a shell of cooler, insulating tissues, particularly subcutaneous fat.

The principal sources of heat are the liver, the brain, the heart, and the working muscles. Muscular work is a source of heat because the mechanical efficiency of muscles is only about 20%. The body tissues are poor heat conductors. Heat is transferred to the skin from the deep body tissues by convection. Blood is an ideal medium for heat convection since its specific heat capacity and thermal conductivity are high. The fundamental thermodynamic processes involved in heat exchange are described by the equation given in Table 9.1.

Metabolic heat production can be measured directly using a method known as calorimetry. A resting person is placed in a sealed chamber, the temperature of which is maintained at a constant level by passing the air in the chamber through a cooling system (pipes in a water bath). The increase in the temperature of the water in the bath is directly related to the metabolic heat production of the

DOI: 10.1201/9781032717210-9

TABLE 9.1
Basic Equation of Human Thermal Balance

$$S = M - E \pm R \pm C - W$$

where

S is the heat gained or lost by the body
M is the metabolic energy production
E is the heat dissipated through evaporation (sweating)
R is the radiant heat to or from the environment
C is the convection to or from the environment
W is the work accomplished by the worker

S = 0 when the body is in thermal balance with the environment

person in the sealed chamber. Direct calorimetry is really only suitable for research purposes, and another method, known as indirect calorimetry, is often used in its place.

Since metabolic processes for energy production and basal metabolism require oxygen, the oxygen uptake of a person at rest can be measured and used to estimate M (metabolic energy production). The energy equivalent of metabolizing 1 L of oxygen is approximately 19–21 kJ (although it varies depending on whether glucose, fat, or protein is being used as fuel). However, if the person is working or exercising, not all of the oxygen consumed will give rise to heat; some will be used to carry out the work. If the amount of physical work (W) can also be measured, then the heat gained due to metabolic processes can be calculated (from the total metabolic energy production minus the energy used to carry out the work).

Heat may also be gained from the environment or lost to the environment by convection (C) or radiation (R). If the skin of a resting person is warmer than the surrounding air, the air at the skin surface is warmed by the conduction of heat from the skin. Since warm air rises, a flow of air around the person is established, and heat is convected away from the person's body. Cool air moves closer to the skin to replace the rising warm air. This airflow is known as a convection current. In practice, a resting person has a layer of slowly moving warm air at the surface of the skin. Movement of the body or wind can disrupt this warm layer and increase the rate of heat loss. Depending on the temperature of the air, this will be perceived as a chilly draught or a cooling breeze.

All objects at temperatures greater than absolute zero emit infrared radiation. This is sometimes referred to as "radiant heat." The human body is no exception to this; if its temperature is greater than that of its surroundings, radiant heat loss takes place. If the surroundings are hotter than the body, it experiences a net heat gain. Inside buildings, the walls and ceiling are common sources of radiant heat, whereas the sun is the major source outdoors. In steel and glass manufacture, furnaces are a major source of radiant heat.

Sweat production and evaporation (E) is a mechanism by which heat is lost to the environment. Since the body tissues are composed largely of water, it is unsurprising that water is lost to the environment by diffusion from the skin, the lungs, and the sweat glands. If the temperature of the surroundings is greater than that of the body, no heat can be lost by convection or radiation, and evaporative heat loss by sweating is essential to maintain thermal balance.

When the body is in a state of thermal balance, the terms on the right-hand side of the equation in Table 9.1 cancel each other out. This can occur in several ways. For example, in a cold environment, a state of thermal balance can be achieved by either increasing the rate of metabolic heat production (by shivering or by carrying out some fairly strenuous activity) or by reducing the heat lost by convection and radiation (by wearing more clothes). Metabolic heat production can be increased by a factor of about three when a person is working as opposed to resting. In cold conditions, a person would therefore require three times as much clothing insulation when resting as when working. Clothing reduces heat loss to the environment by trapping air in the clothing fibers between the garment and the skin. This disrupts the flow of convection currents and maintains a layer of warm air around the skin surface. Clothing also absorbs some of the radiant heat from the body. Both of these effects reduce the temperature gradient between the skin and its immediate surroundings and, therefore, the rate of heat loss. Water practically eliminates the insulating property of clothing because of its high heat conductivity and specific heat capacity.

In hot conditions (e.g., a manual worker in a blast furnace), there is a high rate of metabolic heat production as well as heat gain from the environment (by radiation). A state of thermal balance can be obtained either by reducing the rate of metabolic heat production (by reducing the work rate or introducing rest pauses) or by protecting the worker from the radiant heat (e.g., by means of a protective suit).

The design of clothing assemblies for work in extreme environments is more problematic than it might at first seem. If a clothing assembly for work in cold conditions traps air too well, then it may trap sweat as well. If the sweat accumulates, the insulation provided by the clothing will be degraded. Suits to protect workers from radiant heat may also trap air around the worker's body. Since the rate of evaporation of sweat depends on the humidity of the air, evaporative heat loss may steadily decline as the trapped air becomes more saturated with water vapor and less sweat evaporates.

Skin Temperature

The temperature of the skin can vary over a much wider range than that of the deep tissues of the body. Burning takes place at temperatures over 45°C. At lower temperatures, sweating may impair functional hand strength by reducing friction at the hand–handle interface. Conduction of heat to and from the skin by a solid object depends on the thermal properties of both the skin and the object—a metal surface at 100°C would feel much hotter than a cork surface. The threshold at which burning would take place can be estimated for different materials, assuming a momentary contact time of a quarter of a second (the reaction time to remove the part of the body touching the object). The threshold temperature for wood, for example, is estimated as 187°C, compared with 136°C for brick and 90°C for metal. On hot, sunny days, the surfaces of sidewalks and street furniture can exceed 50°C—hot enough to cause burns. Shoes have to be worn when going for a walk, and dogs can only be walked in the early morning or at night. The interiors of cars left in the sun can also be burn hazards on hot days.

Wind chill, the decline in effective temperature due to air movement, can have consequences for the extremities (hands, nose, cheeks, and ears), particularly if they are directly exposed to the wind. Skin at 20°C is one-sixth as sensitive as skin at 35°C. At 5°C, pressure and touch receptors do not respond to stimulation. This increases the risk of frostbite, as all sensation is lost at lower temperatures. In extremely cold environments, skin contact with bare metal is particularly hazardous—the skin may freeze immediately and stick to the metal, causing extensive tissue damage.

The reduction of skin sensitivity and loss of manual dexterity that take place at low skin temperatures are important from an industrial viewpoint. Allowance must be made for the fact that tasks may take longer to perform than would otherwise be the case.

> Controls and equipment should be designed out of wood or plastic rather than metal, and be usable with gross movements of the body.

Units of Clothing Insulation: Clo

The Clo is the unit of thermal resistance used to quantify the insulation provided by clothing:

$$\text{Clo} = 0.155\ \text{K} \cdot \text{m}^2/\text{W}$$

By definition, a naked person has a Clo value of zero. A man resting at 21°C with air movement of 6 m/min would be in thermal balance wearing 1 Clo, which would be obtained by wearing a suit, shoes, and light underwear. One Clo limits heat transfer to 6.45 W/m^2 of surface area per °C

TABLE 9.2
Insulation Required (Clo Units) to Maintain Thermal Balance

		Activity		
Temperature (°C)	Resting[a]	Slow Walking	Normal Walking	Fast Walking
21	1.5	0.7	0.4	0.3
10	3.1	1.5	0.9	0.7
–1	4.7	2.3	1.5	1.1
–18	7.2	3.5	2.3	1.

[a] When asleep, metabolic energy expenditure drops to 0.8 MET, so more insulation is required.

difference between skin and air temperatures, and 2 Clo allows 3.225 W/m^2 heat transfer, etc. The Clo value indicates the amount of insulation needed to balance metabolic heat production and heat loss. The greater the temperature difference between the body and the environment, the more rapidly heat is lost without adequate insulation. Table 9.2 shows the insulation needed with ambient temperature and activity. In practice, the Clo value of a clothing assembly will vary depending on how much the wearer is moving. Ski outfits are often fitted with vents under the arms and can deliver over 3 Clo when the skier is stationary and as little as 2 Clo when the vents are open and the skier is moving. The extra heat loss is accounted for by heat gain from muscle activity.

The Clo is a "two-way" unit of insulation—in hot conditions (e.g., a manual worker in a blast furnace), the worker must be insulated from environmental heat gain, so additional methods of preventing heat strain due to metabolic heat may be needed. A clothing assembly for cold work, if it traps air too well, may trap sweat as well. If the sweat accumulates, the insulation provided by the clothing will be degraded. Suits to protect workers from radiant heat may also trap air around the worker's body. Since the rate of evaporation of sweat depends on the humidity of the air, evaporative heat loss may steadily decline as the trapped air becomes more saturated with water vapor and less sweat evaporates.

The Tog is another unit of *thermal resistance* often used in carpet and bedding manufacture.

$$1\ \text{Tog} = 0.1\ \text{m}^2 \cdot \text{K/W}$$

Duvets are often sold with Tog ratings (minimal, 1.5–2.5 Tog; summer, 3.0–4.5 Tog; fall or spring, 7.5–10.0 Tog; winter, 13.0–12.5 Tog). One Clo is about 1.55 Tog.

Dry Bulb Temperature

Dry bulb temperature (DBT) is the temperature of the constituent gases of the air. Although DBT indicates the thermal state of the air, other factors have an equally important effect on eat gain or loss.

Relative Humidity and Wet Bulb Temperature

At all temperatures above freezing, water evaporates into the air. The rate of evaporation increases with temperature. At any given temperature, a point is reached where no more water vapor can evaporate. At this point, the air is saturated (with water vapor). When the water vapor pressure of the air reaches its maximum for a given temperature, evaporation from a wet object ceases. Evaporation resumes if the temperature increases. The saturated water vapor pressure of air increases with temperature. If warm air is cooled, a point will be reached where the water vapor pressure at the original temperature is equal to the saturated water vapor pressure at the new temperature. This is the dew point, when water condenses out.

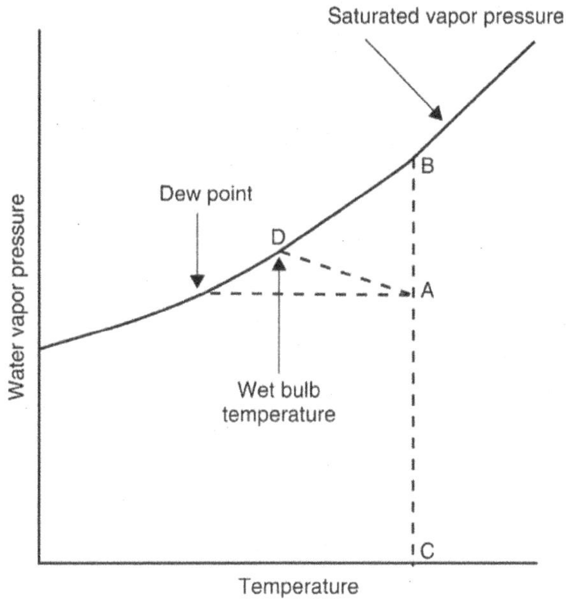

FIGURE 9.1 Relationship between temperature, vapor pressure, dew point, and WBT. The dew point is the temperature at which the vapor pressure at A equals the saturated vapor pressure. Relative humidity at A is given by AC/BC. The WBT is the temperature that would be reached if water were evaporated into the air. (Redrawn from Kerslake (1982), with permission.)

Relative humidity is the water vapor pressure at a given temperature, expressed as a percentage of the saturated water vapor pressure at that temperature. Figure 9.1 illustrates these terms.

Thermal assessment requires three measurements of the air temperature—DBT, wet bulb temperature (WBT), and globe temperature (GT). WBT is measured with a thermometer. A wetted cloth "sock" is placed over the bulb, and the measurement is made after allowing the thermometer and sock to stabilize at the ambient temperature. WBT depends on the DBT and the relative humidity of the air. Evaporation of water from the sock cools the thermometer bulb. The rate of cooling depends on the humidity of the surrounding air. WBT and DBT can be used to calculate the relative humidity using "psychrometric charts" (e.g., Kerslake, 1982).

When the air is completely saturated (i.e., the relative humidity is 100%), no water will evaporate from the WBT, no evaporative heat loss will take place at the bulb, and the WBT and DBT are the same. In all other circumstances, WBT will be lower than DBT.

Globe Temperature (GT)

GT accounts for the effects of radiant heat. Traditionally, the bulb of a mercury-in-glass thermometer is placed in a metal sphere painted matte black. Radiant heat (from the sun or from hot objects) is absorbed by the sphere and heats the thermometer. Radiant heat can prove to be a significant source of heat load on the worker (consider the beneficial effects of shade on a hot sunny day). GT measurement is often essential if the true nature of the thermal environment is to be evaluated.

Thermoregulatory Mechanisms

A number of physiological mechanisms exist for maintaining heat balance.

Peripheral Vasomotor Tone

Heat production and heat loss can be balanced within a fairly narrow range of skin temperatures by adjustment of the peripheral vasomotor tone. In hot environments, peripheral vasodilation occurs—the arterioles dilate and capillaries at the skin surface open, blood flow increases, and heat is conducted to the skin over a small distance from where it is dissipated to the environment. In the cold, vasoconstriction occurs, reducing blood flow in the cutaneous circulation. This increases the conductive distance and the insulation of deeper body tissues. Less heat is lost at the skin surface.

The skin can function both as an insulator and as a radiator of heat, depending on the peripheral vasomotor tone. An indication of the effectiveness of this system can be obtained by comparing its insulating capacity with that of clothes. The Clo is a common unit of clothing insulation. If a nude person has a thermal insulation of 0 Clo, a person wearing a business suit (plus shirt, underwear, etc.) has a thermal insulation of 1 Clo (Parsons, 1991). The insulation provided by the vasomotor system is effective over the range of approximately 0.2–0.8 Clo.

Counter-Current Heat Exchange

Counter-current heat exchange is a method of conserving heat, involving the exchange of heat between arteries and veins supplying the deep tissues. Arterial blood is precooled before it reaches the extremities, and venous blood is warmed before it returns to the vital organs. The efficiency of this system depends on the anatomical distribution of veins and arteries in the various bodily structures.

Sweating

There are about 1.6 million high-capacity sweat glands on the skin. These glands are only found on the hairless regions of other mammals that sweat principally through low-capacity (apocrine) glands, associated with hair follicles. Humans can lose about 500 g of sweat per square meter of skin, whereas horses and camels can only lose 100 and 240 g, respectively. Heat tolerance is well developed in all human populations.

Sweat is a dilute solution of various electrolytes, principally sodium, potassium, and chloride. Passive diffusion of water through the skin occurs most of the time as does "non-thermal" or "non-sensible" sweating because people are not aware that they are sweating. Sweating becomes sensible (can be sensed) when the rate of sweat production exceeds about 100 g/h. Sweat rates of 1–2 kg/h are possible. Sweat cools the body when it evaporates (due to the latent heat of vaporization of water), but in humid environments, evaporation diminishes and cooling efficiency is lost even though sweat continues to be produced. Another problem in humid environments is a kind of reverse sweating in which atmospheric water vapor condenses onto the skin, releasing its latent heat of condensation and warming the skin. Profuse sweating has two important disadvantages:

- Dehydration may occur if more water is lost than is replaced.
- Salt may be lost.

Symptoms of dehydration include fainting, dizziness, lack of energy, and confusion. Urine may have a deep yellow color. The adrenocortical hormone, aldosterone, is involved in the conservation of sodium, causing it to be actively reabsorbed in the sweat glands and in the kidneys. As the rate of sweat production increases, fewer sodium ions can be reabsorbed. Lower sodium concentration in the blood (hyponatremia) may be caused by excessive rehydration, resulting in increased heart rate, lower blood pressure, and a reduction in peripheral circulation.

People do not voluntarily replace all of the lost water when working in the heat, even if an ample supply of fluid is available. To prevent dehydration in the workplace:

- Ensure that workers "pre-hydrate" before starting work in the heat.
- Drink a cup of fluid every 20 min (OSHA, 1995).
- Avoid alcohol and caffeinated drinks while working.
- Provide fluid-rich foods such as soup, yogurt, and mousses.

In industrial situations, it can be taken as a rule of thumb that 1 L of fluid lost as sweat results in a 1 kg drop in body mass (i.e., the metabolic water gained due to metabolism and the mass loss due to the oxidation of substrates are negligible). Workers can be weighed before and after working in the heat to provide a rough estimate of the physiological challenge posted by the conditions of work. In general, dehydration increases the risk of heat exhaustion. Mass loss due to dehydration of

- 1% of body weight is the threshold for voluntary rehydration. The time it takes for this to occur can be taken as the maximum work duration without a break.
- 2% results in performance decrements and is unpleasant.
- 4% is the maximum allowable.
- 6% is a serious threat to health (Nevola et al., 2001).

Water is probably the best fluid to use for rehydration. Fruit juices and milk are absorbed more slowly than water, and although they contain carbohydrates, they may not be appropriate for the conditions. Sufficient salt should be provided by an ordinary diet, and salt tablets should not be given. It is probably simpler to deal with hydration and nutrition as separate issues, so sports drinks with carbohydrates for added energy should not be used. Such drinks may also contain excessive salt and may also contain caffeine.

Sweating is a response to thermal imbalance in hot conditions. It is also provoked by exercise to maintain thermal balance and dissipate metabolic heat. Sweating appears to be modulated by changes in non-thermal factors, including exercise intensity and current levels of hydration.

Shivering

Involuntary shivering is a thermoregulatory mechanism involving active heat production. Groups of motor units act out of phase with one another, and muscles act against their antagonists. Almost no movement occurs, resulting in an increase in heat production up to three times above resting levels. Fine motor control will be impaired.

Voluntary movement also increases heat production, up to ten times the resting levels, but is limited by the fitness of the individual. A disadvantage of movement is that insulating layers of air around the body are disrupted, increasing the rate of heat loss. Both shivering and voluntary movement increase oxygen uptake and cardiac output and can lower work capacity.

Heat Tolerance

Apart from acclimatization, several personal factors can influence a worker's ability to tolerate hot conditions (Table 9.3).

TABLE 9.3
Some Factors Influencing the Ability to Work in the Heat

1. Characteristics of the worker
   - Physiological heat tolerance
   - Age
   - Aerobic capacity
   - Degree of acclimatization
2. Thermal environment
   - Relative humidity
   - GT/radiant Heat/shade
   - Wind speed
3. Requirements of the task
   - Work rate
   - Provision of rest pauses
   - Provision of protective clothing

Age

Young children have less sweating capacity than adults. Older men are less able to tolerate heat stress, partly due to their higher skin temperature threshold for the onset of sweating. The increased mortality of the elderly during heat waves is thought to be due to the increased cardiovascular load rather than heat stress.

Sex

There are no qualitative differences between men and women in their response to, or acclimatization to, heat. There is some evidence that women begin sweating at a higher skin temperature and sweat less than men. Women have a higher proportion of body fat than men, which may also play a role.

Physical Fitness

Physical fitness improves heat tolerance because both rely on cardiovascular function and sweat production. Physically fit workers are less stressed by hot conditions, even if they are accustomed to a temperate climate.

The Thermo-Neutral Zone

The thermoneutral zone is the range of environmental temperatures where the body can maintain core temperature by regulating heat loss without sensible sweating or shivering. The zone varies between individuals. This range depends on the metabolic rate and the ability of the body to regulate the insulation of the tissues (by vasodilation or constriction, for example). Therefore, individual differences in metabolic rate will constrain the range of the zone. Older people typically have a lower metabolic rate than the young and, therefore, will find lower temperatures uncomfortable. Excess body fat in the obese degrades heat tolerance by increasing the mass-to-surface-area ratio of the body, by insulating the core tissues from the skin, and by reducing cardiovascular fitness. Adipose tissue contains less water than other tissue and has a lower specific heat capacity. Thus, the same heat load will cause a greater temperature increase in obese individuals compared to lean individuals. A study of 7564 office workers by Rupp et al. (2018) found that males, overweight people, and those working in air-conditioned offices were more likely to report feeling warm than females, non-overweight people, and those working in naturally ventilated buildings, who were more likely to report feeling cold.

Khovalyg and Ravussin (2022) in a laboratory study, found large individual differences in metabolic rate during everyday working/living activities. Standing to work at a desk as opposed to sitting resulted in no increase in metabolic rate in one subject and up to a 33% increase in others. Similarly, the thermal effect of eating a standardized meal varied greatly between subjects. These findings support the development of "smart" environmental control that caters to individual responses and may also save energy.

Variability in the thermoneutral zone challenges attempts to design environments that will be experienced as comfortable by all occupants.

Menopause—Hot Flashes

Menopausal women often experience "hot flashes"—sudden and unpleasant increases in skin temperature, usually in the face and chest, with peripheral vasodilation, increased (sensible) sweating, and a feeling of intense internal heat (Freedman, 2014). They can occur several times a day, often in the evening, and can disturb sleep. The immediate cause is a narrowing of the thermoneutral zone such that otherwise innocuous increases in core temperature cause an exaggerated thermoregulatory response (as if the increase in core temperature were too high). The cause is thought to be partly due to endocrinological changes from low estrogen levels and also an increase in noradrenaline (one of the stress hormones associated with sympathetic nervous system activation in the "fight or flight" response and with heat dissipation). Estrogen replacement therapy is effective.

Hot flashes are caused by small increases in core temperature; therefore, menopausal women should not be exposed to hot working conditions unless ventilation or other forms of cooling are provided. Similarly, working in crowded conditions, by windows in offices, or in direct sunlight should also be avoided. The provision of fans, cool drinks, and shade from direct sunlight is recommended.

Ultraviolet Light Exposure

Outdoor workers can be exposed to harmful levels of ultraviolet (UV) radiation from the sun, a major cause of skin cancer. Skin cancer is far more prevalent among outdoor workers than indoor workers. UV light lies between visible light and X-rays in the electromagnetic spectrum (Figure 10.1). About 10% of sunlight is UV. UV is categorized according to its wavelength: UV-A (or "soft UV" at 315–400 nm), UV-B (or "medium UV" at 280–315 nm), and UV-C (or "hard UV" at 100–280 nm). Much UV-C is absorbed by the ozone layer in the atmosphere. UV-B at 290–320 nm can cause skin reddening and sunburn, or in extreme cases, damage DNA, leading to skin cancer. UV-C can be produced by industrial processes such as arc welding and can damage the eye unless protective eyewear is provided. UV-B is necessary for the production of vitamin D.

Melanin is a skin pigment that protects against UV exposure. Light-skinned people are more at risk in strong sunlight during Northern summers and lower latitudes. Spring and early summer are high-risk times for many because melanin produced the previous year is lost over the winter.

Lansdown et al. (2020) carried out a randomized controlled trial to investigate the use of SMS (text) messaging to promote "sun-safe" behaviors in outdoor workers via a smartphone app. The trial took place in winter (to promote vitamin D supplementation using supplements made available at the time, together with dietary advice) and summer (to promote UV overexposure avoidance). During winter, the app contained links to sources of relevant information about vitamin D. During the summer, the messages contained location-specific information (e.g., "The UV index is high today. Cover up in the midday sun"). Workers received daily text messages during the trial periods in summer and winter. Vitamin D levels were measured using blood assays, and UV exposure using wearable dosimeters (230–320 nm). Overall, the SMS intervention did increase vitamin D levels, but there was less evidence for a reduction in UV exposure in summer. For further information on UV, see Cherrie and Cherrie (2022).

Carbon Dioxide—CO_2

CO_2 is one of the constituent gases in the air. Outdoors, the concentration is about 420 parts per million (ppm). Prior to the Industrial Revolution, the concentration was 280 parts per million. The increasing concentration, due to the use of fossil fuels, is a driver of climate change, the main effects of which are an increase in the number of extremely hot days per year and extreme weather events such as floods.

CO_2 monitors are available to enable indoor air quality to be maintained through ventilation.

BASIC APPLICATIONS

Work in Hot Climates

Heavy physical work in the heat imposes *conflicting demands* on the cardiovascular system. Peripheral vasodilation requires an increase in blood flow to the skin (up to 10 L/min). However, working muscle also demands increased blood supply (up to 25 L/min). Up to a point, blood flow to the skin and muscles can be increased by diverting blood from the viscera, but this reserve capacity is limited. Since cardiac output cannot exceed venous return, and the average maximum cardiac output is about 25 L/min, cardiac capacity is a limiting factor for muscular work in the heat. The cardiovascular system may be placed under considerable strain when a person is working in the heat, as output rises to meet the demands of both physical work and bodily cooling. A dangerous condition can arise if the heart is no longer able to meet both demands.

Relative Humidity

Whether workers can tolerate a hot environment depends on several classes of variables (Table 9.2). For example, DBTs of more than 38°C can be tolerated if the relative humidity is less than 20% because, at such low humidities, the cooling efficiency of sweating is high. However, a relative humidity of 90% at 32°C is only tolerable with air movement and if a low level of work activity is required. Work in hot environments can be made more tolerable by introducing job aids or rest pauses to reduce metabolic heat production, or by periodically removing the worker from the environment. Finally, workers differ in their ability to tolerate heat stress. At one extreme, hyper-heat-tolerant individuals can work in a hot environment without the need for prior acclimatization. Heat-intolerant individuals are never able to work safely in the heat. Identifying such individuals is an important part of worker selection in industries such as deep underground mining.

Heat Acclimatization

Heat acclimatization is a physiological process of adaptation rather than a psychological adjustment to life in a hot environment. It involves an increase in the capacity to produce sweat and a decrease in the core temperature threshold value for the initiation of sweating. The maximum rate of sweat production can double from 1 L/h in an unacclimatized person to 2 L/h in an acclimatized person. A state of acclimatization is best achieved by exercising in the heat and drinking plenty of fluid.

Increased sweat production enhances evaporative cooling of the skin and thus improves heat transfer from the deep body tissues to the periphery. The risk of dehydration and salt depletion is reduced in an acclimatized person due to expanded blood volume and a reduction in the salt concentration of the sweat. Furthermore, acclimatization reduces the skin's blood flow requirements, which reduces the cardiovascular load during work in the heat.

Heat acclimatization occurs naturally. Brown et al. (2022) carried out a systematic literature review of seasonal heat acclimatization studies carried out over the summer (29 studies with 561 participants overall). These studies varied in the summer temperatures their participants were exposed to—ranging from warm humid to hot arid conditions, with an overall mean summer WBGT of 25.2°C (range: 19.6°C–28.7°C) and in the methods used to assess acclimatization. However, the following changes to heat response tests over the summer period provide evidence for seasonal acclimatiztion: lower resting core temperature and heart rate; reduced sweat sodium concentration; and increased sweat rate.

Acclimatization may also be induced artificially. In deep mines, manual work has to be carried out in WBTs up to 33°C. In some mines, workers are artificially acclimatized to heat as a requirement of their employment. Surface acclimatization chambers are used, and workers exercise (by performing step-ups) in temperatures of 31.5°C WBT and 33.5°C DBT. Rectal temperatures are monitored, and rest periods are given in response to temperature increases over 38.3°C. Other methods of acclimatization include exercise while wearing a vapor barrier suit (which prevents sweat evaporation) and resting in baths at 41°C.

It has been suggested that an individual's ability to produce sweat depends on the climate they experienced in the early years of life (Diamond, 1991). It appears that the sweat reflex develops in the first 2.5 years of life. Although we are all born with a similar number of sweat glands in the skin, the number that develops or is "switched on" depends on the conditions experienced early in life. People who grew up in a hot environment develop a larger number of functional sweat glands than those who grew up in a cold environment and can tolerate hot conditions more easily. Due to climate change, summer temperatures are increasing in many parts of the world. However, outdoor workers who may have lost summer acclimatization over the cooler months of the year will be at increased risk of heat stress during unseasonal heat waves brought about by climate change.

Heat Illnesses

If a worker becomes dehydrated, sweat production decreases and core temperature increases. Rapid elevations in body temperature increase the metabolic rate. Since heat is a by-product of metabolism, the heat load is exacerbated, and a positive feedback situation may develop. In extreme situations, thermoregulation breaks down. If core temperature rises above 42°C, blood pressure drops and insufficient blood is pumped to the vital organs, including the heart, kidneys, and brain. Some of the main conditions that can arise when the body is unable to cope with the thermoregulatory challenge are as follows.

Heat Hyperpyrexia (Heat Stroke)

Thermoregulation fails, sometimes suddenly, and core temperature exceeds 41°C. It may be potentially fatal if untreated. The individual may collapse and be disoriented. The skin is hot, red, and dry. Active cooling on rescue from the heat is needed quickly, as there is only a 30-min window for preventing serious injury or death. Some key steps to rapidly reduce temperature include: cold water immersion cools the body most rapidly, followed by the TACO method (tarpaulin assisted cooling with oscillation,[1] which involves placing the person on a tarpaulin sheet, and covering them in ice and then lifting and rotating the sheet). Ice towels or ice applied to superficial arteries (in the groin, armpits, and wrists) can also be used. Rogerson and Brearley (2024) cite the following core temperature cooling rates for different methods:

1. Ice water immersion 0.350°C/min
2. TACO 0.170°C/min
3. Ice towels/sheets 0.160°C/min
4. Ice towels 0.110°C/min
5. Combined cooling* 0.036°C/min
6. Passive rest (underwear only) 0.027°C/min

*Cold packs on groin, neck, and axillae plus splashing body with water while evaporating with compressed air.

Heat Exhaustion

When thermoregulatory strain combines with cardiovascular strain, heat exhaustion can occur. The person feels weak and may be uncoordinated, and breathing may be shallow with a rapid, weak pulse. Dehydration contributes to heat exhaustion. Heat stroke can follow if the dehydration is not corrected. Removal from the heat, removal of clothing, and fanning may restore normal functioning.

Heat Syncope

Fainting occurs due to inadequate venous return. It is more common in unacclimatized individuals and can be fatal if the sufferer cannot lie down (e.g., when working in a confined space).

Heat Hyperventilation

This can occur when working in the heat while wearing protective clothing. Hyperventilation results in an excessive loss of carbon dioxide. It is treated by having the sufferer breathe into a small bag for a few minutes.

Prickly Heat

A fine, superficial skin rash associated with excessive sweating often occurs on areas of skin covered by clothing or protective equipment.

Kidney Disease Chapman et al. (2021) reviewed the literature on occupational heat exposure and the development of 'non-traditional kidney disease' (disease in the absence of other known risk factors). Chronic exposure to heat stress combined with physically demanding work appears to be a risk factor for disease development in occupations such as manual laborers and sugar cane workers. It is not known whether working in the heat is a causal factor or a catalyst for the development of the disease. The main risk factors appear to be higher core temperatures, dehydration, longer work durations, muscle-damaging exercise, and the consumption of beverages containing high levels of fructose.

Heat Stress Management

Some approaches to the elimination of heat stress are summarized in Table 9.4. Table 9.5 gives threshold WBGT values at which these precautions should be implemented (Figure 9.2).

TABLE 9.4
Some Basic Steps in Heat Stress Management

1. Reduce high relative humidity using dehumidifiers
2. Increase air movement using fans or air conditioners
3. Remove heavy clothing, issue loose-fitting overalls
4. Reduce the work rate
5. Include frequent rest pauses
6. Introduce job rotation
7. Carry out outdoor work at cooler times of the day (e.g., early morning)
8. Allow 2 weeks for acclimatization
9. Enforce rest breaks and provide drinking water or other fluids
10. Provide shade to reduce radiant heat load (plant trees, build awnings, and issue wide-brimmed hats)
11. In factories, build cool spots and refuges to lower worker exposure

TABLE 9.5
Recommended Threshold WBGT for Heat Stress Management

	Air Velocity	
Work Rate	<1.5 m/s	≥1.5 m/s
Light	30.0	32.0
Moderate	27.8	30.5
Heavy	26.0	28.9

Source: From OSHA 3123 (1993).

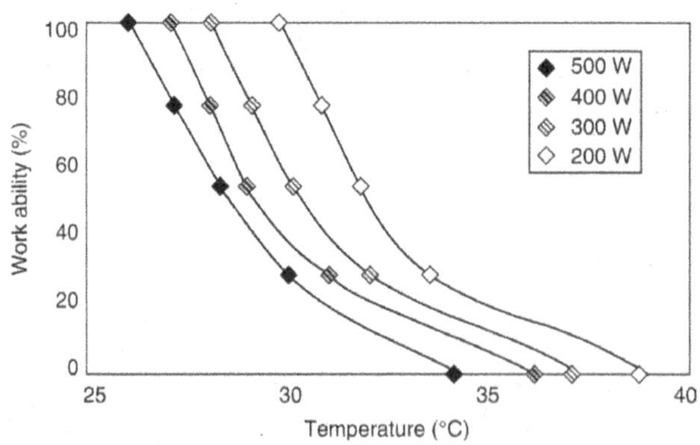

FIGURE 9.2 Describes the relationship between the ability to carry out work at different intensities for 20%–30% of the day and WBGT (Kjellstrom and Dirks, 2001). As can be seen, as temperatures exceed the 26°–30° range, work ability starts to diminish rapidly for all but the least physically demanding tasks.

According to OSHA, signs that thermoregulation is breaking down in hot environments are:

- Abnormal thinking or behavior
- Slurred speech
- Seizures
- Fainting
- Hot, dry skin
- Blurred vision

Work in Cold Climates

Core temperature can be maintained in the cold if the person is working and suitable clothing is provided. When manual work is not being performed, less metabolic heat is generated, and adequate insulation becomes increasingly important Metabolic heat production increases by a factor of three when a person is working compared to resting. When resting, therefore, three times the amount of clothing insulation is required. Adequate hydration is essential when working in a cold environment. Sweating may still take place, and moisture is lost as the lungs are ventilated by the cold, dry air. In extreme circumstances, a form of dehydration known as cold dehydration may occur. Cooling of the nasal epithelium, particularly during heavy breathing, may cause nosebleeds. Cold diuresis may also occur, increasing the loss of fluid.

Core Temperature in the Cold

If the core temperature of a person in a cold environment drops below 33°C, central nervous system temperature control becomes disrupted. At 29°C, hypothalamic core temperature control breaks down completely. Cooling of the core tissues lowers the metabolic rate and, therefore, the amount of metabolic heat production. If metabolic heat production cannot be elevated while heat is being lost, a vicious circle of further heat loss and still lower metabolic heat production develops.

Peripheral Temperatures and Repetitive Work

Cooling of the peripheral tissues, particularly in the hands and feet, causes reductions in strength and neuromuscular control, resulting in a loss of dexterity. Oksa et al. (2001) compared forearm EMG activity and maximum wrist flexion force in subjects doing light repetitive work in thermoneutral conditions and at 5°C. Cold increased muscle strain by 20%, and forearm flexor muscle fatigue was doubled. Rintamaki et al. (2001) surveyed 1490 workers in the meat processing and dairy industries; 51% of the workers worked at 0°C–5°C and 34% at 6°C–10°C. Complaints of cold in the hands and fingers were most prevalent (89%), followed by the wrist (58%), toe (59%), and shoulders (52%). Compared to warm conditions, the prevalence of musculoskeletal pain was greater in the hands, arms, and neck especially. Workers with musculoskeletal symptoms had lower skin temperatures in the neck/shoulder and forearm regions. Muscular strain, indexed by EMG, was negatively correlated with skin temperature. Low skin temperatures in the cold seem to be associated with increased muscular strain and pain in the tissues beneath the cooled skin. These findings indicate the desirability of preventing regional cooling of the skin at the extremities in those carrying out repetitive work in the cold.

Heus et al. (1995) present threshold temperatures for different tissues to ensure normal functioning (Table 9.6). These can be estimated from skin temperature. The minimum temperatures for work with the hands are skin 15°C and muscles 28°C.

Habituation to Cold

Local habituation to cold may occur in the extremities as a reduction in the peripheral vasoconstrictor response. Increased blood flow through the hands can occur after repeated exposure to cold conditions. Manual dexterity is preserved, but at the cost of increased heat loss from the body as a whole. It has been suggested that people accustomed to cold conditions have an increased basal metabolic

TABLE 9.6
Critical Temperatures for Peripheral Tissues

Tissue	Temperature (°C)
Nerves	20
Receptors	10
Joints	24
Muscles	
• Dynamic force	38
• Dynamic endurance	28
• Static force	28–38
• Static endurance	28
Skin	
• Local	15

rate and the ability to sleep while shivering. It may also be the case that those exposed to a cold climate increase their daily food intake, which would increase their metabolic rate and the thickness of the subcutaneous fat layer, resulting in a higher rate of heat production and better insulation of the core tissues. Yurkevicius et al. (2022) review habituation to cold, finding that habituation occurs after a few exposures to cold and reduces sensations of cold and shivering in experimental trials:

- Days 1–3: reduced thermal sensation and reduced cold pain
- Days 3–5: delayed onset of shivering and reduced cold shock response
- Days 5–10: Reduced levels of noradrenaline, blood pressure, and vasoconstriction

Apart from shivering, habituation also involves non-shivering thermogenesis by brown adipose and muscle tissue. The chemical processes involved in the digestion of food are also a source of heat. It is well known that BMR rises in people of all ages following a meal. Clothing provides a subtropical microclimate of warm, moist air between the skin and clothes. At rest, a skin temperature of about 33°C is perceived as comfortable—most people put on more clothes in cold conditions in order to maintain this microclimate and, for this reason, are rarely exposed to any stimulus for true acclimatization to the cold. Behavioral adaptation to the cold, through experience, is of great importance—wearing correct clothing and keeping "on the move" are examples. Peripheral vasoconstriction takes place in most areas of the body except the head, where up to 25% of all heat loss can take place when it is cold (Kroemer, 1991), so the head should be considered an extremity and protected in a similar way to the hands and feet.

Immersion in Cold Water

Sudden immersion in cold water produces a violent reaction characterized by initial hyperventilation, tachycardia, and peripheral vasoconstriction. This temporary reaction, occurring in the first few minutes of immersion, may incapacitate the victim and cause drowning, even in competent swimmers. The hyperventilation makes it almost impossible to hold one's breath, and submersion of the head for even short periods is extremely dangerous. This explains why the provision of life jackets is crucial. Even conscious victims who are competent warm-water swimmers may not survive cold water immersion. Should the victim survive the first few minutes, the next threat is from "swimming failure"—thought to be caused by direct cooling of the peripheral nerves, resulting in a loss of coordination and a failure to synchronize swimming strokes and breathing. In water at 10°C, stroke rate increases and stroke length decreases, indicative of a reduction in swimming efficiency (Tipton et al., 1999). The main threat to life following cold water immersion appears to be drowning. Hypothermia, where the core temperature drops below 35°C, becomes a serious threat only if the victim survives the first 20 min, and the chance of survival rapidly diminishes unless the person is rescued. Acclimatization to cold water immersion has been shown to occur experimentally after eight repeated exposures of 40 min in water at 15°C and is characterized by a reduction in hyperventilation and tachycardia (see Oakley, 2000 for a fuller discussion).

Perception of Cold

The perception of cold seems to depend on experience. People accustomed to a cold climate may feel comfortable when their deep body tissues are suitably insulated by clothing, despite local cooling of the skin on the cheeks, ears, nose, fingers, and feet. Those unaccustomed to wide disparities in skin temperature may confuse "being cold" (i.e., having a low core temperature) with "feeling cold" (i.e., experiencing low skin temperatures on the extremities). In fact, their core temperature remains perfectly normal, and in a physiological sense, they are not cold. Cold perception also depends on the set point of the hypothalamic temperature-regulating center. Bacterial infections elevate the set point for core temperature thermoregulation. The person, therefore, feels cold even though the core temperature is normal. Peripheral vasoconstriction takes place, and the person shivers. After several hours, the core temperature reaches the new set point, and the person again feels neither hot nor cold. When the factor that initially caused the increase in set point is removed, the person feels hot. Peripheral vasodilation and profuse sweating occur as the core temperature drops to the original set point.

Cold Injury

Cold exposure causes injuries to the fingers, toes, nose, cheeks, and ears. Clinicians distinguish between freezing and non-freezing cold injuries:

Freezing Cold Injury

Frostbite occurs when the tissues freeze, and the damage is caused by high concentrations of electrolytes left in tissue fluids when most of the water turns to ice (Oakley, 2000). The skin is white and hard when frozen and turns red on warming as blood flow returns. Frostnip is a mild form of frostbite. Only the superficial layer is frozen, and the tissues recover completely within 30 min of rewarming, with the return of sensation to the affected area. Rewarming is best achieved by placing the affected part against the warm skin of another person or under the armpit. Superficial frostbite is more serious; there is gross discoloration and sometimes blisters on rewarming. Gangrene may develop, and rewarming itself may be extremely painful. Rewarming should not be attempted until it is certain that further freezing will not occur. Deep frostbite is rarely seen in occupational settings and has serious medical consequences, including possible amputation of the affected limb.

Non-Freezing Cold Injury

These injuries occur as a result of longer-term (greater than 10 h) exposure to less severe temperatures in wet conditions (0°C–13°C) and are found in military populations (e.g., trench foot) and in survivors of shipwrecks. The cause of the injury is not known but the result is that the injured part becomes oversensitive to cold, in which exposure to innocuous cold stimuli for a short period causes peripheral vasoconstriction that persists for several hours afterward. Although the condition may resolve in the years following the exposure, it may worsen, as the cold sensitization response makes it more, rather than less, likely that the affected part will be cooled in the future. In the initial stages, gradual rather than rapid rewarming of the injured body part is recommended (Oakley, 2000).

That exposure to cold can result in such serious injuries leading to chronic disability highlights the importance of ergonomic interventions to prevent exposure to cold in the workplace.

The term *chilblain* is sometimes used to describe symptoms such as itching, redness, and burning of the skin on the upper parts of the fingers and toes following exposure to damp cold. These symptoms are reversible. At temperatures lower than –10°C, the head should be insulated using headgear designed for the purpose. This prevents heat loss from the head and makes more heat available to the other extremities. Mittens, in preference to gloves, should be worn in temperatures below 0°C, although for tasks requiring manual dexterity, they may need to be removed. Gloves should be worn when the temperature falls below 5°C.

Protection against Extreme Climates

When the work environment cannot be improved nor the task redesigned, methods of protecting the worker are needed.

Specify Safe Work–Rest Cycles

Welch et al. (1971) measured pulse rate, skin temperature, sweat loss, and rectal temperature of subjects working in hot, humid conditions. Rectal temperature, which is used as an index of core temperature, proved to be the only reliable indicator of the onset of heat exhaustion. Bell et al. (1971) investigated the time taken for subjects to reach a state of imminent heat collapse and specified safe exposure times for men working in hot environments. Figure 9.3 shows the limits of permissible exposure to hot conditions for workers engaged in different energy expenditures across a range of work–rest cycles (NIOSH, 1981).

Design Cool Spots

Thermal comfort can sometimes be improved by designing a "thermal refuge" for operators. Window coverings and screens can be installed to provide radiant heat "shade." In very hot conditions, a screen made of vertically hanging chains can be effective at blocking radiant heat while allowing access for those with portable protection. Sims et al. (1977) designed a "cool spot" for inspectors in a glassmaking factory. Discomfort caused by radiant heat and high air temperatures was reduced using glass and aluminum radiant heat shields and low-velocity cooling air.

Ohnaka et al. (1993) investigated heat stress among operators in the asbestos-removal industry. These workers wear protective clothing to prevent the inhalation of asbestos and the contamination of

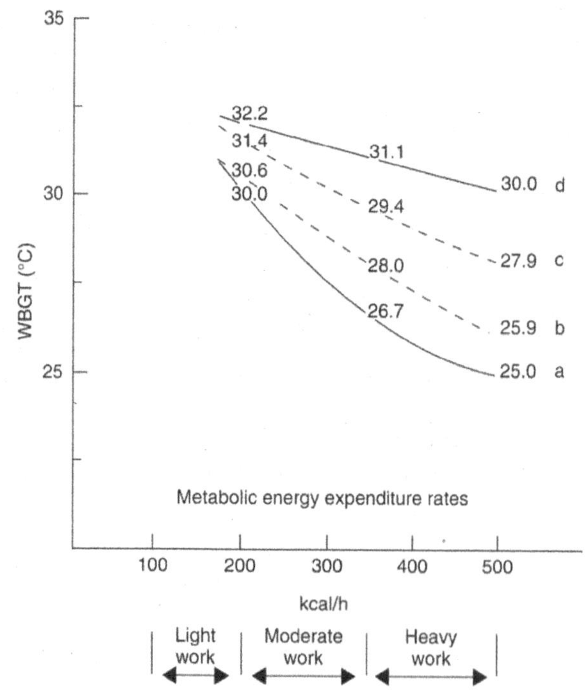

FIGURE 9.3 Limits for physical work in the heat at different work rates. (Redrawn from Kjellstrom and Dirks, 2001.)

personal clothing. The protective ensemble consists of impermeable garments and air masks, preventing heat loss and exacerbating the problems of working in hot weather. Several solutions to the problem were possible, including air-conditioning the work area, using air-cooled or liquid-cooled suits or ice vests, or constructing a cool room adjacent to the work area for rest breaks. The first option was too expensive and the second was impractical due to the restrictive nature of the clothing. The third option was trialed and evaluated using measures of heart rate, sweat rate, and rectal temperature of subjects working in hot conditions, cool conditions, and the hot–cool condition using the air-conditioned rest room. The hot–cool condition reduced physiological stress and discomfort significantly.

Issue Protective Clothing

Protective clothing can provide an acceptable microclimate. Liquid- and air-cooled suits have been designed for pilots. Air or a special cooling liquid is circulated around the suit to maintain thermal balance.

If the external temperature is greater than the core temperature (about 37°C), additional clothing is required to protect the skin from a net heat gain from the environment. In some hot countries, people traditionally wear long, flowing robes covering the whole body, presumably to maintain a layer of cool air between the skin and the clothing and to protect the skin from solar radiation. In more severe cases, protection from radiant heat can be achieved using clothing assemblies designed to insulate the worker from the heat. Helmets with a large air space between the shell and the operator can be used in conjunction with thick woolen suits and wooden clogs. Fireproof reflective materials are used in firefighting.

A major source of radiant heat is the sun. Wide-brimmed hats have been used by rural people for millennia and constitute portable shade—an effective first line of defense against the potentially debilitating effects of intense solar radiation. People who live in hot deserts cover their bodies with long, black, flowing robes. Black materials absorb more radiant heat than white materials but provide better protection against sunburn. In hot, humid environments, minimal clothing is worn to maximize evaporative heat loss. If there is also intense sunlight, this conflicts with the need for protection from solar radiation, and the only solution may be to restrict outdoor work to the early morning and late afternoon.

Cool the Extremities

A simple method of reducing heat stress in hyperthermic individuals is to immerse the extremities in cold water. According to DeGroot et al. (2013), arm immersion is more effective than hand or foot immersion in lowering core temperature and heart rate. Benefits of extremity cooling include:

- Subjects working in the heat while wearing firefighter protective clothing increased their total time in the heat and time spent working by 66% and 62%, respectively, when resting their hands and forearms in water at 17°C during a 50/30-min work–rest cycle.
- Reductions in core temperature and sweat loss have been found in some studies of extremity cooling.
- Thermal comfort is also improved, and immersing the hand in water at 10°C for 10 min is not a source of discomfort in these trials.
- Cooling water at 10°C and 20°C is more effective than at 30°C. During repeated cooling immersions, water starting at 10°C can be allowed to warm to 20°C and still be effective.

This simple method appears to be effective in treating mildly hyperthermic individuals in hot environments. However, the method is less likely to be effective in preventing hyperthermia in the first place. In individuals who are not suffering from heat stress, the natural response to immersion of the extremities in cold water is peripheral vasoconstriction (to prevent heat loss). When core temperature rises, peripheral vasodilation occurs under central nervous system control. Thus, when the extremities of a hyperthermic individual are placed in cold water, the peripheral blood vessels remain dilated, and convective heat loss takes place.

Cold Climate Protection

The main challenge is to provide a sufficient amount of insulation while allowing moisture from sweating to evaporate. Garments made of vapor-permeable materials such as Goretex can provide insulation and allow moisture to escape. This is very important where physical work is carried out in the cold. Sweat produced during work, if it collects in the interstices of the garment, will reduce the garment's thermal insulation properties and may even freeze within the garment when the work stops and metabolic heat production drops. The wearer will rapidly become cold. At temperatures below 0°C, gloves provide inadequate insulation, and mittens are needed, resulting in a major reduction in functional manual dexterity.

Comfort and the Indoor Climate

The climatic conditions inside a building depend primarily on the conditions prevailing outside and on the people and processes housed within. Humans feel comfortable when the skin temperature of the torso is around 33°C, with that of the extremities about 20% lower. Clothing, building design, work demands, and many other factors determine the effects of exterior conditions on those indoors. Of particular importance is the design of the building itself.

Building Design and the Indoor Climate

Building design determines the amount of solar penetration into the building and the radiant heat gain. The amount of insulation, particularly in the roof, has a large influence on heat exchange between the building and the environment. The construction materials influence a building's thermal performance via a mechanism known as the "flywheel effect" (see below).

Solar heat gain through uninsulated factory roofs and walls is an important cause of overheating. This is particularly important in developing countries, where cost-effective yet thermally efficient building concepts are needed. In hot countries, roof insulation is an important first step in combating overheating. Parts of the building that are exposed to direct sunlight can be painted bright colors to reflect solar energy. It has been estimated that reductions in external roof and wall temperatures of almost 20°C are obtained by whitewashing building exteriors. Awnings or covered walkways can be built around buildings to provide shade for both people and walls. If plentiful, water can be sprayed onto roofs, walls, adjoining courtyards, and sidewalks. In developing countries, people sometimes pour water onto nearby floors before sleeping to lower the air temperature in a room. Screens placed over doorways or windows may be wetted to cool incoming air.

Steel roofs in factories can be shaded with tiles, preferably with an air space between the roof and the tiles. Rectangular-shaped buildings can be constructed with their long walls facing north and south. In the summer, the sun shines on the shorter walls in the morning and evening and on the roof during the day. This minimizes the degree of heating of the building by the sun. Conversely, in the winter, when the sun is low in the sky, the long walls and roof are exposed to the sun during the day, which helps to warm the building.

Sunlight entering through windows is another major source of heat. In existing buildings, physically constrained workers should not have to work in direct sunlight for any length of time (this also applies in cold countries, since sunlight is a source of glare that degrades vision). The orientation of windows with respect to the sun is an important consideration in the design stage of a building. Special glazing, louvers, or shades can be retrofitted to reduce solar penetration. There is a modern trend to build buildings with large windows to save on construction materials. This, however, can lead to faster heat loss and gain.

The heat transfer properties of building materials vary considerably depending on their mass. Lightweight materials store very little heat themselves. Heat transfer through them depends on their thermal conductivities and the temperature differential across them. Heavyweight materials have a greater ability to store heat—heat gained during the day warms the material itself before being transferred to the air inside the building. Similarly, at night, heat is lost from the building materials themselves, and the interior of the building remains warmer for a longer period. The building material itself can be thought of as a buffer or heat store that lies between the internal and external thermal environments. Thus, heavy materials act like a "flywheel" to smooth out the effects on the indoor climate of daily oscillations in external temperature. To use an extreme example, the temperature inside a cave remains more or less constant throughout the year, whereas the temperature inside a tent change rapidly throughout the day in response to changes in the external temperature. The magnitude of the flywheel effect does not depend so much on the heat-conducting properties of the materials used in construction as on the mass of material enclosing the interior space. Thus, there is often a trade-off between the cost of constructing a new building and the later cost of controlling its internal temperature within acceptable limits. Money saved on construction materials and construction time in building a more lightweight structure may have to be spent later on additional air-conditioning (except in permanently hot, humid climates, where all buildings require artificial cooling to maintain a comfortable indoor temperature).

Thermal Comfort in Buildings

The thermal comfort of a factory or office worker depends on maintaining an average skin temperature of approximately 33°C, although colder temperatures at the extremities may be tolerated. Large disparities in skin temperature may lead to complaints of discomfort even if the average skin temperature is close to 33°C. Draughts, sunlight falling on an arm or the face, and sitting next to a cold wall are all causes of thermal discomfort due to uneven skin temperature distribution.

Modern approaches to the indoor climate have attempted to specify an acceptable range of conditions for the worker. ISO 9241 recommends winter temperatures of 20°C–24°C and summer temperatures of 23°C–26°C. Maximum values of relative humidity are 60%–80% at 20°C, 50%–70% at 22°C, 45%–65% at 24°C, and 40%–60% at 26°C. At temperatures of 24°C and above, workers may begin to feel lethargic. At temperatures of 18°C or lower, shivering may commence in sedentary or inactive workers unless extra clothing is worn.

Sex Differences in Thermal Comfort

According to Kingma and Lichtenbelt (2015), females typically prefer higher indoor temperatures than males (25°C as opposed to 22°C), although the skin temperatures perceived as 'comfortable' do not differ. This suggests that there are differences in resting metabolic rate between males and females and that building standards should reflect these differences. It may be that females have lower resting metabolic rates than males or that there are differences in insulation due to the distribution of subcutaneous fat. The authors argue that there is bias in thermal comfort predictions based on standard MET values because these values are based on male energy expenditures. For young adult females performing light office work, the energy expenditure is lower than the values used in predictive models, so energy expenditure is overestimated. Correcting for a lower female metabolic rate led the authors to recommend values of 23.2°C–26.1°C (for mean skin temperature ranging $32.4°C \leq T \leq 33.6°C$).

Sick Buildings Revisited

"Sick buildings" are typically sealed, mechanically ventilated buildings where some of the air is recirculated to save energy. Any contaminant in the air, from the furniture, the hardware, cleaning materials such as solvents, dust, or the occupants themselves, may degrade the air quality. Dissatisfaction with the indoor air quality, breathing difficulties, and absenteeism may all be symptoms of A "sick building." Good ventilation is vital for indoor air quality. Since people exhale CO_2, CO_2 concentration in indoor spaces such as offices is a good indicator of the amount of ventilation. At the time of writing, ambient CO_2 levels are approximately 420ppm (parts per million) but are normally higher indoors. For offices, the recommended levels are:

- 400–1000ppm: Indicates good ventilation
- 1000ppm: Maximum acceptable level by ASHRAE

- 1000–2000 ppm: Expect complaints that the air is 'stuffy' and the occupants feel drowsy
- 2500–5000 ppm.: Negative effect on health
- 5000 ppm: OSHA maximum recommended level for an 8-h day

Sick buildings can also arise when air conditioning systems are not properly maintained. Among the more famous cases is the spread of Legionella bacteria in a hotel in 1977, where 182 people contracted pneumonia and 34 died as a result. Niza et al. (2023) carried out a systematic literature review of 134 articles on the subject. Overall, the most common symptoms were respiratory symptoms, headaches, sore throats, eye pain, fatigue, and difficulty concentrating. Improved air quality and ventilation were the most effective countermeasures. There was some evidence that symptoms are more common among women than men, which may reflect increased sensitivity to airborne impurities such as those released from cleaning materials and synthetic products.

Kildeso et al. (1999) developed a visual analog scale (Table 9.7) that can be used to monitor occupants' perceptions of indoor air quality over time. They suggest using it once per week while interventions to improve air quality are carried out. Up to 4 weeks should be allowed for the interventions to have an effect and for symptoms to change.

TABLE 9.7
Scale for Detecting Changes in Indoor Air Quality

	How do You Feel Right Now?	
Too cold	______________________	Too hot
Too humid	______________________	Too dry
Draught	______________________	Too little ventilation
Bad air quality	______________________	Good air quality
Too dark	______________________	Too light
Too quiet	______________________	Too much noise
Nose blocked	______________________	Clear nose
Brittle nails	______________________	Normal nails
Dry eyes	______________________	Eyes not dry
Stinging, itching eyes	______________________	Normal eyes
Eyes ache	______________________	Eyes do not ache
Severe headache	______________________	No headache
Pressure in the head	______________________	Clear head
Dry nose	______________________	Runny nose
Dry throat	______________________	Normal throat
Dry mouth	______________________	Normal mouth
Dry lips	______________________	Normal lips
Dry skin	______________________	Normal skin
Dry, brittle hair	______________________	Normal hair
Dizzy	______________________	Not dizzy
Feeling bad	______________________	Feeling good
Tired, exhausted	______________________	In shape, fresh
Hard to concentrate	______________________	Easy to concentrate
Depressed	______________________	In a good mood
Today, I have been able to work 0%	______________________	100%

Source: From Kildeso et al. (1999), with permission.
Note: Lines are 100 mm long in actual scale. Respondents place a cross on the line to indicate how they feel about each item.

Ventilation

Ventilation is a determinant of thermal comfort and, more generally, of satisfaction with the indoor environment. The main purpose of ventilation is to provide fresh air and to remove accumulated noxious gases and contaminants. Ventilation helps remove heat generated in a working area by convection and cools the body. It is not always possible to lay down acceptable limits from the point of view of thermal comfort. However, air speeds of less than approximately 0.1 m/s will usually cause a sensation of staleness and stuffiness, even at relatively low temperatures. Air speeds greater than 0.2 m/s may be perceived as draughty. In hotter conditions (with a corrected effective temperature of more than 24°C), air speeds of 0.2–0.5 m/s will aid body cooling, particularly when the relative humidity is high. It is apparent that whether or not air movement is perceived as an irritating draught or a cool breeze depends on the ambient temperature and the relative humidity.

In hot countries, an important use of ventilation is to cool working areas and their contents by ventilating buildings at night, when the air outside has cooled. Apart from its effects on the cooling efficiency of sweating, the humidity of the air influences the comfort of people working indoors. Low relative humidity (less than 30% at office temperatures) causes bodily secretions to dry up. Under these conditions, the occupants may complain of dry, blocked noses and eye irritation. Contact lens wearers may experience eye discomfort, since proper adhesion of the lens to the eye depends on a continuous supply of lachrymal fluid to maintain a thin, moist film over the cornea. Static electrical charges are also more likely to build up in dry buildings, causing irritating shocks when an earthed conductor is touched. Somewhat counterintuitively, low relative humidity inside houses and offices can be a problem in cold, wet countries if artificially heated buildings are not ventilated adequately (which the occupants may be reluctant to do) and if humidifiers are not provided.

Thermal Comfort When Sleeping

Sleepiness during the day can occur in hot rooms because peripheral blood vessels dilate to dissipate heat, blood pressure drops, and blood flow to the brain decreases. Sleepiness occurs as a result of the reduction in oxygen supply to the brain. There is a circadian variation in body temperature, which is at its lowest at around 3:00 AM. Room temperatures greater than about 18°C can disturb sleep. Paradoxically, hot conditions can induce sleepiness during the day and keep us awake at night. One of the reasons for this is that the body's ability to thermoregulate is degraded during sleep, and we are therefore more sensitive to the ambient temperature in the room. Thermal comfort when sleeping depends on the ambient temperature in the room and on the amount of insulation provided by the bedding. The research cited below was conducted using light bedding (sheets and a blanket).

- The optimal room temperature for restorative sleep is in the range of 19°C–22°C. This is the thermoneutral range, where the temperature inside the bed does not vary (remaining at 29.6°C).
- In ambient temperatures of 16°C–25°C, the temperature inside the bed varies, and subjective discomfort increases with greater deviation from the thermoneutral range particularly when room temperatures rise above it.
- Body temperature is lower during sleep, and the ability of the body to thermoregulate is lessened during REM sleep, reducing the ability of the body to cool itself when temperatures are higher than the thermoneutral range.
- High ambient temperatures cause nighttime waking in the same way that fever causes the sufferer to wake.
- Ambient temperatures below 18°C are tolerable if extra insulation is provided, although there is a risk of mildew if the ventilation is poor.
- In general, temperatures below 13°C and above 25°C are to be avoided.

- Thermal comfort when sleeping is greatly affected by the ambient temperature and is essential for restorative sleep. Deviant temperatures cause insomnia-like symptoms. Such symptoms are known to degrade occupational performance and well-being in a wide range of occupations (Kucharczyk et al., 2012). They have been found to be consistently associated with reduced safety and productivity in the workplace, increased levels of sickness absence from the workplace, impeded career progression, and reduced job satisfaction among individual workers.

TOOLS AND PROCESSES

WBGT Thermometers

Modern heat stress monitors make use of thermistors (electrical transducers) instead of mercury-in-glass thermometers. DBT, WBT, and GT can be measured by the same instrument, and air movement can be taken into account. These devices often provide combined measurements of temperature known as heat stress indices. The "wet bulb GT" (WBGT) index is often used. Commercial heat stress monitors calculate the WBGT index as follows for measurements made outdoors and indoors, respectively:

$$\text{WBGT}(\text{out}) = 0.7\text{WB} + 0.2\text{GT} + 0.1\text{DB}$$

$$\text{WBGT}(\text{in}) = 0.7\text{WB} + 0.3\text{GT}$$

WBT is weighted most heavily in the WBGT index, demonstrating why measurements of DBT alone cannot provide an accurate measure of the stress involved in working in a hot environment. Figure 9.4 shows a modern heat stress monitor for measuring WBGT. HFE Workshop 9.1 provides basic advice on using heat stress monitors in workplace assessments.

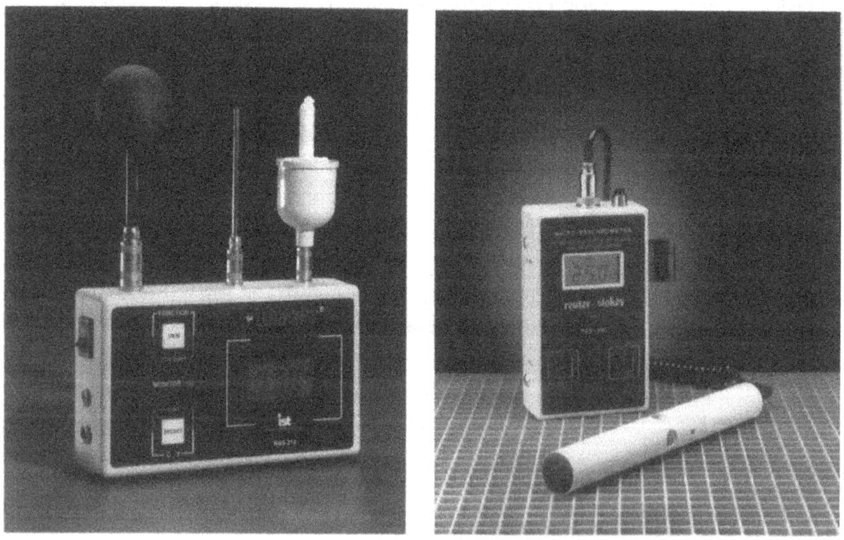

(a) (b)

FIGURE 9.4 Instruments for thermal environment evaluation. (a) Heat stress monitor (note three separate sensors for measuring DBT, WBT, and GT). (b) Micro-psychrometer, which gives direct readings of WB and DB temperatures, relative humidity, and dew point.

HFE WORKSHOP 9.1

BASIC ADVICE FOR HEAT STRESS SURVEYS[2]

1. Read the manufacturer's instructions for the heat stress monitor first.
2. Make sure the monitor has been calibrated recently.
3. Carry out a walk-around inspection first and familiarize yourself with the work system and any sources of heat, including radiant heat, that might be a problem.
4. Ask workers whether they have experienced any heat-related problems, whether any solutions have been attempted, and what the result was.
5. If using a battery-powered monitor, make sure the batteries are charged or new; if not rechargeable, take spare batteries with you.
6. Allow the monitor to reach room temperature in the area you are assessing.
7. Many heat stress monitors have a wet bulb thermometer in which the sensor is covered with a wick. The wick must be wetted with distilled water before use. The evaporation of water from the wick cools the thermometer to an extent dependent on the relative humidity of the air. The lower the relative humidity, the greater the evaporation of water from the wick.
8. Do not place the monitor in direct sunlight, on desks next to windows, for example, unless the aim is to measure the additional radiant heat load. In factory and office surveys, the aim is usually to assess the overall thermal environment. Sunlight usually only penetrates buildings at certain times of the day and can introduce error if not controlled for.
9. In buildings without air conditioning, take measurements throughout the day and throughout the year to get a representative sample of air temperatures.
10. Do not allow the wick on the wet bulb thermometer to dry out when conducting large surveys!
11. Take measurements as close as possible to operators' positions at the workplace.
12. For operators who move between different workplaces throughout the day, take measurements at each location where they work. A TASK ANALYSIS may be necessary to establish where these locations are and the duration of exposure at each workplace. Use the equation below to determine a weighted mean WBGT index for the overall exposure.
13. For work in hot conditions, include the temperature and time spent in cool spots.
14. OSHA recommends some simple worker monitoring techniques to include in risk assessments in cases where it is thought that the thermoregulatory challenge may be excessive and immediate action is needed:

    - Take the radial (wrist) pulse for 30 s at the beginning of the worker's normal rest period. If the pulse exceeds 110 beats/min, reduce the next work cycle by one-third keeping the rest periods constant.
    - Compare this pulse measurement with a second, one taken 2.5 min into the rest period. If the pulse is less than 90 after 2.5 min, recovery is satisfactory. If it is less than 10, recovery is not satisfactory.
    - If the oral temperature exceeds 37.6°C, shorten the next work period by one-third.
    - If the worker loses more than 1.5% of body mass over a work shift, fluid intake should be increased.

Measuring Thermal Comfort

Table 9.8 shows a thermal sensation scale that can be used when workers are expected to be exposed to a wide range of environmental conditions, both to assess how well they cope with the conditions and to evaluate the effectiveness of any interventions to mitigate thermal stress.

The Bedford Scale (Table 9.9) can be used to assess thermal comfort in buildings. For more detailed assessments, such as when evaluating the effects of protective clothing to be used in hot conditions, thermal comfort and sweating in different body regions can be investigated using the scale in conjunction with body maps.

Air Movement and Wind Chill

When evaluating the effects of temperature on the worker, it is important to take air movement into account. Air movement moderates the effects of high temperatures and exacerbates the problems of low temperatures (causing "wind chill"). Air movement can be measured using anemometers (in which the rate of rotation of a vane or the cooling of a hot wire is proportional to the velocity of the air flowing through it).

TABLE 9.8
Thermal Sensation Scale

Rating	Thermal Sensation	Shivering/Sweating	Clothing Wetness
1.	Very cold	Vigorous shivering	Dry
2.	Cold	Moderate shivering	Slightly damp
3.	Cool	Slight shivering	Damp
4.	Slightly cool	None	Wet
5.	Neutral	Slight sweating	
6.	Slightly warm	Moderate sweating	
7.	Warm	Heavy sweating	
8.	Hot		
9.	Very hot		

TABLE 9.9
The Bedford Scale

Value	Sensation
1	Much too cool
2	Too cool
3	Comfortably cool
4	Comfortable—neither cool nor warm
5	Comfortably warm
6	Too warm
7	Much too warm

Source: From Bedford (1950).

SYSTEM INTEGRATION

ISO Standards

For further information on environmental ergonomics, the reader is referred to the appropriate ISO standards. An overview of the available standards can be found in ISO 11399 (1995 – Ergonomics of the Thermal Environment). Exposure assessment for hot environments is dealt with in ISO 7243, and WBGT values are provided that can be used to make quick indoor and outdoor assessments. as long as there is no excessive solar load. More detailed assessments can be made using ISO 7933, which deals with thermal stress in conditions likely to lead to excessive sweat loss or core temperature. It also enables required modifications to be identified so that the threat of thermal strain is eliminated. Allowable exposure times can also be specified using this standard. ISO 7730 provides a method for predicting thermal sensations in different environments and for specifying thermal conditions that will be perceived as comfortable. Required insulation for cold environments is addressed in ISO 11079, and ISO 9920 outlines methods for determining the insulation value of clothing. ISO 9886 deals with the evaluation of thermal strain using physiological measurements (core temperature, skin temperature, heart rate, and body mass loss). ISO has drafted a standard on risk assessment (ISO FDIS 15265) for the prevention of stress and discomfort in the workplace under different thermal conditions. The standard was reviewed and confirmed in 2022. HFE Workshop 9.2 summarizes the main features.

ERGONOMICS WORKSHOP 9.2

FIRST STEPS IN RISK ASSESSMENT

The ISO has published a risk assessment strategy, based on observation. The first part of the strategy is presented below. It can be used in the initial screening of workplaces to identify any problem areas and assign priorities for further investigation. Each working area is visited in turn and is rated using a simple scoring system.

	Score
Air Temperature	
Generally freezing	–3
Generally between 0°C and 10°C	–2
Generally between 10°C and 18°C	–1
Generally between 18°C and 25°C	0
Generally between 25°C and 32°C	1
Generally between 32°C and 40°C	2
Generally greater than 40°C	3
Humidity	
Dry throat/eyes after 2–3 h	–1
Normal	0
Moist skin	1
Skin completely wet	2
Thermal Radiation	
Cold on the face after 2–3 min	–1
No radiation discernible	0
Warm on the face after 2–3 min	1
Unbearable on the face after more than 2 min	2
Immediate burning sensation	3

(Continued)

ERGONOMICS WORKSHOP 9.2 (*Continued*)

	Score
Air Movements	
Cold, strong air movements	–2
Cold, light air movements	–1
No air movements	0
Warm, light air movements	1
Warm, strong air movements	2
Physical Workload	
Office work, occasional movement at normal speed	0
Moderate work with arms or legs, steady walking	1
Intense work with arms and trunk	2
Very intense work at high speed, ladders	3
Clothing	
Light, flexible, and not interfering with work	0
Heavier clothing, interfering with work	1
Clumsy, heavy, especially for heat or cold	2
Special overalls, hoods, or gloves	3
Opinions of the Workers	
Shivering, great discomfort of the whole body	–3
Strong, local discomfort, general feeling of coolness	–2
Slight, local discomfort (cool)	–1
No discomfort	0
Slight discomfort, sweating, and thirst	1

Status of Methods Used in Risk Assessment and Task Design

Both the methods for measuring physiological function and those for assessing the physical environment are well developed. Physiological methods of risk assessment can be intrusive and require trained personnel, ethical approval, and specialized equipment. Heat stress and CO_2 monitors are commercially available and inexpensive. They are used routinely by occupational hygienists as part of health and safety assessments against national and international standards.

Limitations of WBGT

Budd (2008) discusses the history and limitations of WBGT as an index of heat stress. The main risk is that WBGT measurements may underestimate the risk of heat stress in conditions of high humidity and low air movement. At a WBGT of 31.7°C, heat stress indices of experimental subjects were higher when humidity was high and air movement low (restricting evaporation) than in conditions of low humidity and high air movement (and higher DBT). Where heat stress is suspected, air movement, activity, and clothing should be taken into account when WBGT temperatures are interpreted. Generally, the lower the opportunity for evaporative cooling, the more likely it is that the notional WBGT reading will underestimate the risk of heat stress.

Effectiveness and Cost-Effectiveness

There is no reason why extreme thermal environments should have any direct effects on performance unless they alter the thermal state of the individual. If thermal stress causes thermal strain, capacity may be reduced, and decrements in performance may follow. The goal of cost-effective interventions should be to eliminate thermal strain.

Physical Tasks

Figure 9.5 illustrates monthly variations in factory production levels as a function of the outdoor air temperature. Even in moderately hot environments, people lower their work rate to minimize thermal strain—work rate is a thermoregulatory mechanism.

Wyon (1974) re-analyzed data from the 1923 New York State Commission report on ventilation and was able to show that people consistently produced more at 20°C than at 24°C and 50% relative humidity. The task was a typing task, and the research was carried out over a series of days, with subjects spending whole days (up to 2 weeks) working in the laboratory. Furthermore, subjects were allowed to work at their own pace, in a relaxed manner, so that many aspects of the experimental design were similar to ordinary office work. Over a 3-day period, subjects working at 24°C produced 43%, 18%, and 49% less than subjects working at 20°C. When exposed to temperatures of 24°C, subjects who had worked at 20°C in the previous 3 days produced significantly less. Interestingly, subjects did not report feeling any difference in the different temperatures, suggesting that they had unconsciously lowered their work rate to maintain the same thermal state. These findings suggest that even light work, such as typing, may be sensitive to moderate thermal stress, with people lowering their work rates to reduce metabolic heat production and thereby compensate for the increased temperature. A temperature of 24°C is not particularly high and may be commonplace at certain times of the year in offices that are not air-conditioned. That it can result in reductions in typing performance of 40%–50% suggests that major improvements in productivity can be achieved through better building design and temperature control.

Mental Tasks

Figure 9.6 shows the upper limits for mental task performance as a function of exposure time.

A study in offices by McCartney and Humphreys (2002) found that there was no relationship between perceived productivity and the actual air temperature in offices. However, a "U-shaped" relationship was found between perceived productivity and whether the environment was perceived as being acceptable or not. This suggests that the performance of mental tasks is not affected by the indoor climate itself but by whether or not the climate is intrusive or noticeable in some way. According to Chen et al. (2020), temperatures of 39°C or more degrade the performance of cognitive tests in unacclimatized subjects, who completed the tests more quickly than at lower temperatures. In their study, subjects completed cognitive tests at 37°C, 70% RH, while walking at a moderate pace on a treadmill. Performance on cognitive tests (including the Stroop Test in Chapter 12) degraded at this temperature, but only after 45 min, when heart rate and core temperature increased. The authors recommended that 45 min should be the limit for continuous cognitively demanding work at this temperature.

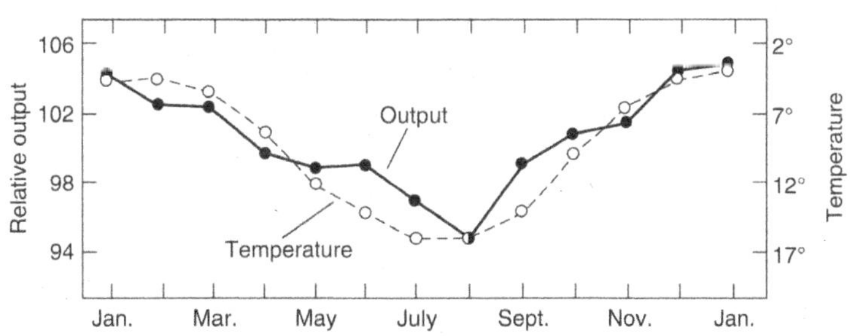

FIGURE 9.5 Yearly production figures from a factory circa 1918. (Adapted from Vernon, 1924.)

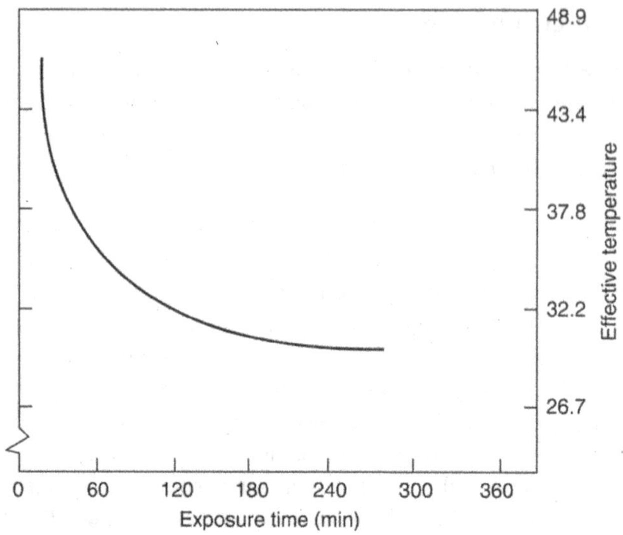

FIGURE 9.6 Upper limit for unimpaired mental performance as a function of exposure time to different effective temperatures. (Adapted from Sanders and McCormick (1992), with permission.)

Cost-Effective Interventions

Reducing workers' exposure to heat enables them to work longer or for the work to be done by fewer people. In foundries, a curtain of heavy chains can be installed to block radiant heat but allow the passage of workers or materials. Fans can be installed behind the curtain to further cool workers.

Feinstein and Crawley (1968) redesigned a shearer's cab in a steelworks. The job involved detecting defects at the ends of steel slabs prior to rolling. Defective ends were cut off. Slab surface temperatures resulted in a high radiant heat flux so the cab was positioned 10 m from the slab. The excessive distance, combined with an awkward field of view, meant that operators tended either to make several attempts at processing a slab (resulting in production bottlenecks) or cut off too much steel to ensure quality, resulting in unnecessary recycling of good steel. The cab was redesigned and fitted with gold laminate glass to block the radiant heat. It was then repositioned nearer to the shearer blades, and the angle of view was improved. The cost of these alterations was £10,000, and the savings due to reduced bottlenecks and a lower tonnage of recycled steel were £120,000 per annum (probably in excess of $1 million today). The redesign paid for itself in the first month.

Microclimate Cooling Garments

Microclimate cooling garments have been in use for many years to enable workers to work more safely and comfortably when climate control is not possible (such as in deep mines). Del Ferraro et al. (2024) list the main options for the design of such garments. Typically, jackets or vests have pipes fitted in them through which ambient air or cool air is circulated to create a convection current against the skin. Other options include the use of cooled liquids such as glycol, ice, or dry ice, which is carried in pockets within the garment; cooling is via the phase change of the dry ice. If there is a requirement for such a garment (because climate control is not possible and cool refuges cannot be used), then the choice of jacket will depend on factors such as the availability of compressors to pipe ambient or cooled air to jackets, electricity supplies to power fans or recharge "on-garment" batteries, etc. Choosing optimal garments will depend, in practice, on the logistical requirements needed to deploy them effectively.

Ventilation

Ventilation is important to stop the air from feeling stuffy. About one or two air changes per hour and an airspeed between 0.1 and 0.2 m/s are needed. If ventilation is inadequate, pollutants, including carbon dioxide, can accumulate. Wargocki (2005) carried out extensive tests on the effects of improved ventilation on productivity. In one experiment, ventilation was increased from 3 to 10, then to 30 L of fresh air per person per second. This corresponded to 0.6, 2, and 6 air changes per hour in the office under investigation. For every twofold increase in ventilation rate, performance on a text typing task improved by 1.7%. A 1% improvement in performance on office tasks was found for every 10% reduction in the number of workers dissatisfied with the air quality.

In a second study, which took place in a call center in a tropical climate, the temperature in the office was reduced from 24°C to 22°C, and the ventilation rate was increased from 10 to 23 L/s. The performance of the call center operators improved by 8%. The researchers carried out an analysis to assess the productivity benefits of improved air quality in relation to the increased costs. It was found that the payback period for the improvements would be less than 2 years, and the benefits 60 times greater over a 25-year period. The rate of return on the investment was seven times greater than the rate of return from the minimum acceptable interest rate at the time of the study. People differ in their sensitivity to indoor air quality, with smokers being particularly insensitive. As far as productivity is concerned, it is the level of satisfaction with the indoor air quality that seems to be important.

Mobile Workforces: Establishing Safe Systems of Work in the Heat

In many industries, employees work in different parts of the world for varying lengths of time and are exposed to wide variations in climate. Risk assessments should consider acclimatization, heat exposure, and protection from solar radiation to ensure that employees are safe on arrival. For example, employees from Northern Europe transferred to a hot equatorial or Southern Hemisphere work location in winter or spring will not be acclimatized to the heat and will have lost any suntan they may have acquired the previous summer. Risk assessment should consider exposure to high temperatures and radiant solar radiation, and mitigation through the provision of adequate rest periods, shade, and a ready supply of safe, cool drinking water (Figure 9.7).

Risk Management for Dealing with Extreme Events

Table 9.10 summarizes technological, organizational and personal measures for reducing the risks of exposure to extreme events.

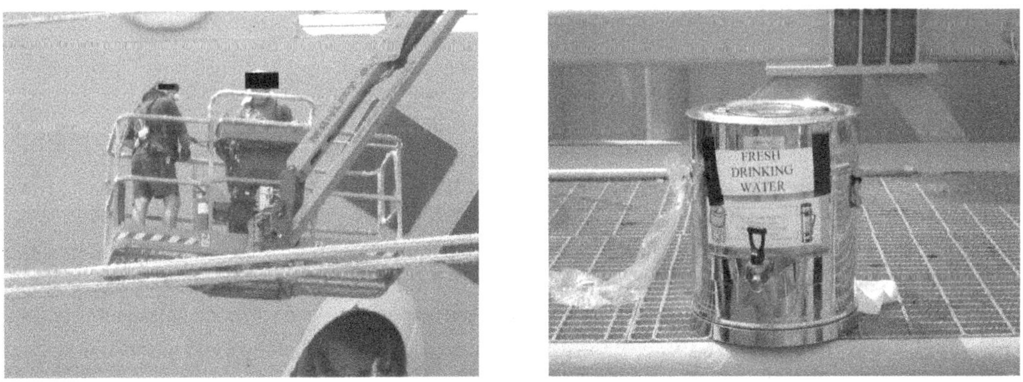

FIGURE 9.7 Fresh drinking water should be readily available, clearly marked, and regularly replenished. For high-intensity work at a WBGT of 34° or above, 1.25 L of water may be needed. For medium-intensity work at 20°–24.9°, 0.25 L/h may be sufficient.

TABLE 9.10
Measures to Reduce the Risks of Extreme Climatic Events

Technological Measures
Advanced monitoring systems to provide detailed and timely alerts for extreme events
Advanced monitoring systems to provide detailed and timely alerts for the allocation of resources (emergency response teams, infrastructure restoration, and emergency food, water, and shelter) for disaster situations
Infrastructure development to reduce the impacts of extreme events and to enable rapid evacuation where necessary
Organizational Measures
Flexibility in work-from-home arrangements to avoid unnecessary exposure to extreme events
Infrastructure planning to avoid locations likely to frequently experience extreme events
Contingency planning for unforeseen ("black swan") extreme events
Pre-disaster management planning
Post-disaster management planning
Personal Measures (This Will Vary Widely Depending on the Type and Intensity of Extreme Event)
PPE for extreme weather events
Choice of living and/or working locations to avoid extreme events
Design of personal environments and infrastructure to withstand extreme events
Design of equipment and tools to withstand extreme events

Source: Thatcher (2024).

RESEARCH DIRECTIONS

Further research is needed into the causes and treatment of thermal injuries and there is plenty of scope for developing and evaluating new methods of protecting people from extreme thermal environments. Although acclimatization to heat is well understood, there is still a research need to understand the specificity of acclimatization, how quickly it is lost, and the minimum requirements for acclimatizing people quickly so that they can work safely when suddenly exposed to hot climates.

Many people wear smartwatches and other devices that can measure their heart rate and skin temperature. This introduces new directions in research for "smart" environmental control that adapts to the user's thermal state. 'Smart desks' are being proposed to optimize local environmental conditions around the user (Aryal et al., 2019).

SUMMARY

The heat or cold stress on the worker depends on DBT, WBT, GT, and air speed, as well as the workload and the insulation value of clothing. Each of these factors can be manipulated to achieve thermal comfort.

People feel comfortable when surrounded by a subtropical microclimate of air and when the average skin temperature is 33°C. The ability to carry out manual work in extreme environments is limited. Most people are able to acclimatize to hot working conditions within 2 weeks, but the ability to adapt to old is much more limited. For these reasons, if work has to be carried out in extreme environments, the following steps should be considered:

1. Remove the operator by mechanizing the task.
2. Change the task or the environment.
3. Protect the operator.

High-risk groups for heat stress and thermal discomfort include pregnant women, menopausal women, the elderly, and the obese. As a rule of thumb, workers in hot conditions should drink one cup of water (or a sports drink) every 20 min to maintain hydration. Line managers and co-workers should be able to recognize the symptoms of heat exhaustion and heat stroke.

Heat Exhaustion	Heat Stroke
Faint or dizzy	Throbbing headache
Excessive sweating	No sweating
Cool/pale skin	Hot, dry skin
Rapid, weak pulse	Rapid, strong pulse
Muscle cramps	Loss of consciousness
Nausea/vomiting	Nausea/vomiting
	High body temperature

And take appropriate action. Heat exhaustion: Move to a cool place, drink water, and shower or use a cold compress to the skin. Heat stroke: Call emergency services; cool the person using any available means (e.g., shade, cold compress, etc.).

Temperatures ranging between 19°C and 23°C at 40%–70% relative humidity will accommodate most workers. Low relative humidity, still air, and draughts are some of the main causes of thermal discomfort in offices and factories. If nothing can be done to eliminate thermal stress, then changes in work organization and rest periods are needed. Approximate thresholds at which this should be done are 28°C–33°C for sedentary tasks, 23°C–28°C for moderately hard work, and 20°C–26°C for heavy work (Oakley, 2000).

Further guidance can be found in ISO/TC 159/SC 5: Ergonomics of the Physical Environment.

TUTORIAL TOPICS

1. Should thermal comfort considerations be traded off against energy efficiency?
2. Is the increased prevalence of obesity likely to affect the management of working conditions when climatic extremes are expected?
3. Will climate change affect life expectancy in the future?
4. How to survive a heatwave. After you have watched the video at the link, review this chapter, comment on the advice, and see if you can find any more helpful tips. https://www.youtube.com/watch?v=ter31IGeXzs

ESSAYS AND EXERCISES

1. Take DBT, WBT, and GT measurements in the following places (measure the external ambient temperature at the same time).
   Modern air-conditioned office
   Workshop on the factory floor
   Busy commercial kitchen
   Sauna room
   Cold room or cold store
   Park or beach at 8:00 AM, noon, and 4:00 PM on a sunny summer day
   Park or beach at midday in the shade
   Calculate the WBGT temperature and comment on your findings.
2. Measure the WBGT in a modern air-conditioned office containing 20 employees or more. Take measurements at various parts of the office and at different times of the day. Develop a small questionnaire to assess the occupants' satisfaction with the indoor climate. Suggest possible improvements and comment on any differences in the level of satisfaction among different people.

3. WBGT is only relevant to the assessment of thermal challenge in extreme temperatures. Discuss.
4. Why should workers in extremely cold environments take care to drink plenty of fluids?
5. Can heat strain occur in the cold? List four reasons why, if you think it might be a risk.
6. List five workplaces where radiant heat might be a problem.
7. Acclimatization to heat can take:
   1–2 days
   3–14 days
   6 weeks
   6 months
   You can never get used to the heat if you did not grow up in a hot country.
8. The body's core temperature should be maintained between:
   2°C and 40°C
   36.5°C and 37.5°C
   35°C and 40°C
   None of the above
9. Five workers are weighed in their undergarments (having eaten and visited the bathroom) at the start of a shift. They wear protective clothing and then work in a hot environment. After the work period, they remove the protective clothing and are weighed again before taking a shower. The exercise is repeated over several days, varying the work period each time. The conditions in the work environment are monitored and found not to vary over time. The table below shows the weight reduction that has occurred due to sweat loss (in kilograms).

   Refer to the discussion of sweat loss and heavy work, and comment on these data. Pay particular attention to the safety and ethics of carrying out this trial. Can you recommend a maximum safe work period based on these data?

	Length of Work Period in Heat (h)							
Worker Body Mass (kg)	**0.25**	**0.5**	**0.75**	**1.0**	**1.25**	**1.5**	**1.75**	**2.0**
70	0.30	0.74	1.00	1.30	1.50	1.75	1.70	1.70
65	0.10	0.30	0.55	0.75	1.00	1.35	1.70	2.50
60	0.20	0.40	0.60	0.80	1.20	1.50	2.00	2.80
80	0.10	0.30	0.60	1.00	1.50	2.50	3.00	3.70
85	0.10	0.20	0.40	0.60	0.80	1.20	1.40	1.60
90	0.20	0.40	0.50	0.70	1.0	1.30	1.60	2.00
95	0.43	1.05	1.43	1.85	2.14	2.50	2.42	2.42
100	0.15	0.46	0.85	1.15	1.54	2.08	2.62	3.85
105	0.33	0.66	1.00	1.33	2.00	2.50	3.33	4.17
110	0.13	0.38	0.75	1.25	1.88	2.56	3.75	4.63
115	0.12	0.24	0.47	0.71	0.94	1.41	1.65	1.88
120	0.22	0.44	0.56	0.78	1.11	1.44	1.56	1.78
Mean	0.23	0.54	0.84	1.18	1.60	2.08	2.56	3.12

10. A worker spends the following amounts of time in several workplaces while carrying out a task:

Time (min)	WBGT
20	28
10	21
15	33
15	30

a. Calculate the weighted mean WBGT.
b. If he spends 25% of the time resting and 75% carrying out heavy work, is his heat exposure above the threshold limit value?

11. Mr Smith is on holiday in Spain. After lunch on the beach, he decides to walk back to his hotel, which is 5 km away. The temperature is 35°C. What precautions would you advise Mr Smith to take?
12. Mr. and Mrs. Smith return from their holiday in Spain to England, where there is a cold spell. Mrs. Smith, much to Mr. Smith's discomfort, complains that her feet are cold in bed and that she cannot sleep, with the result that he cannot sleep either. Instead of buying her a pair of woolly bed socks, he buys her a woolly hat to wear in bed. Discuss the thermo-regulatory arguments for and against Mr. Smith's solution to this problem.

QUICK QUIZ

1. What is the normal core temperature of the body?
2. Above what skin temperature does burning take place?
3. Wet bulb temperature is normally higher than dry bulb temperature. True or false?
4. The conservation of sodium in the body is regulated by: (a) insulin, (b) aldosterone, (c) cortisol
5. Core temperatures below 29°C can be fatal. True or false?
6. People can never acclimatize to cold. True or false?
7. Chilblains are a form of cold injury. True or false?
8. There's a correct office temperature for all workers. True or false?
9. Hot flashes often occur during menopause. True or false?
10. It's more dangerous to touch a hot metal object than a hot wooden object of the same surface temperature. True or false?

NOTES

1 TARP assisted cooling with oscillation - demonstrations available online.
2 Further information can be found in the OSHA technical manual at http://www.osha.gov.

10 Light and Sight

GENERAL REQUIREMENTS FOR HUMANS IN SYSTEMS

1. Interior lighting should conform to ISO 8995:2002.
2. Reflectances of surfaces in rooms should comply with the recommendations of the American National Standards Institute (ANSI).
3. The level of illumination must be suitable for job demands.
4. Surface luminances in rooms should be balanced.
5. Glare must be minimized.
6. The color-rendering properties of light sources must support the color discrimination required by the task.

The eyes lead the body.

Dr. J. Sheedy
School of Optometry, University of California at Berkeley

CORE KNOWLEDGE: FUNDAMENTALS OF VISION AND LIGHTING

Light is electromagnetic radiation that is visible to the human eye. The electromagnetic spectrum is extremely wide, but the visible part is extremely narrow (Figure 10.1).

VISION AND THE EYE

The eye is a fluid-filled, membranous sphere that converts electromagnetic radiation into nerve impulses, which are transmitted to the brain along the optic nerve (Figure 10.2).

Light enters through a transparent outer covering called the cornea. The cornea plays a major role in refracting the light. Further refraction occurs as the light passes through the lens. The pupil works like the aperture of a camera to vary the amount of light entering the eye. In bright light, the iris contracts, the pupillary diameter decreases, and only the central part of the lens forms an image on the retina. In poor light, the iris expands, and a larger area of the lens is used. Since the

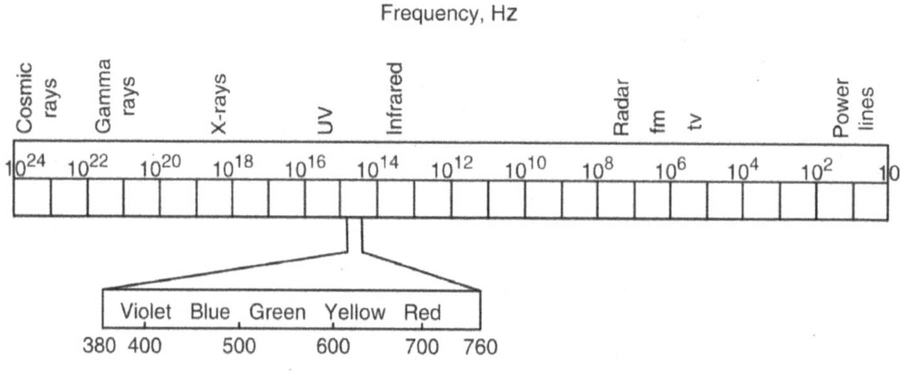

FIGURE 10.1 The electromagnetic spectrum. Visible light falls in the range 380–760 nm.

DOI: 10.1201/9781032717210-10

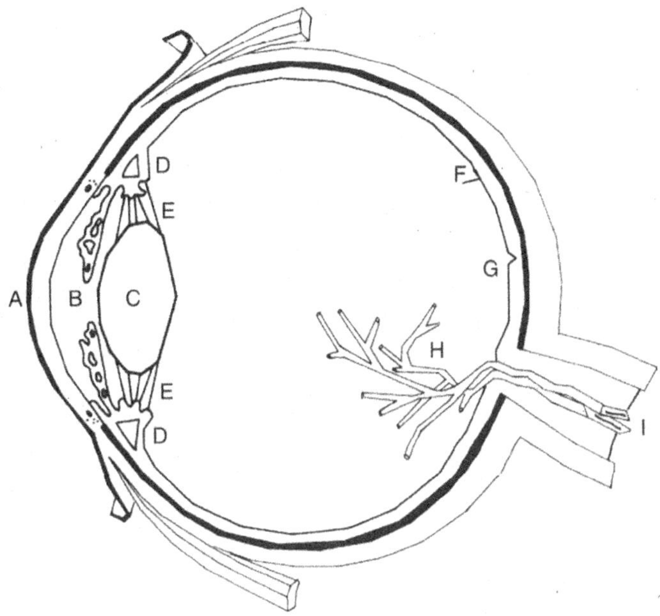

FIGURE 10.2 Basic structure of the eye. (a) Is the cornea, (b) Is the pupil, (c) Is the lens, (d) Is the ciliary muscle, (e) Is the suspensory ligaments, (f) Is the retina, (g) Is the fovea, (h) Is the blind spot, and (i) Is the optic nerve.

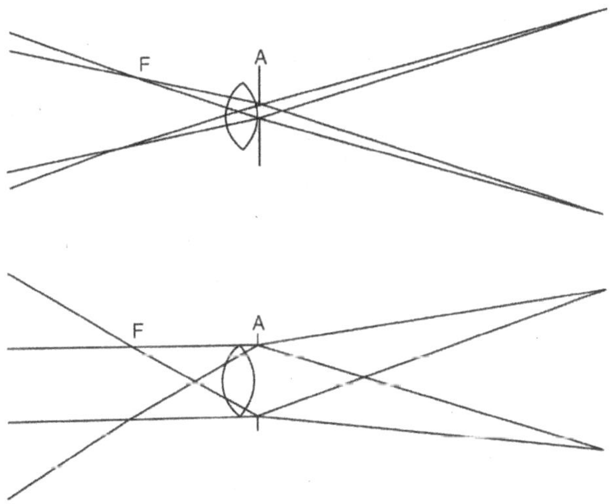

FIGURE 10.3 The depth of the field phenomenon. One of the reasons why we can see better in good rather than poor light. With a large aperture (A), only the image exactly at (F) is in sharp focus.

peripheral regions of the lens focus the light slightly in front of the image formed by the central part (a characteristic of all simple lenses, termed spherical aberration), slight blurring occurs when objects are viewed in poor light. This explains why the ability to discern detail (visual acuity) is reduced in poor light.

Using photographic terminology, it can be said that good lighting increases the depth of field of the eye. When the pupil is very small, the eye acts like a pinhole camera—all objects are focused on the retina irrespective of viewing distance. Increased depth of field reduces the need for optical adjustment by the refractive lens. It is in this sense that it can be said that good lighting reduces the load on the visual system. Figure 10.3 illustrates the optics of the depth-of-field phenomenon.

Refractive Apparatus of the Eye

The lens is held in place by a non-rigid, membranous sling. It divides the eye into two compartments. The smaller anterior compartment contains a watery fluid called the aqueous humor, which is secreted by the ciliary body. The posterior compartment contains the jelly-like vitreous humor. The humor helps maintain the structural integrity of the eyeball and lens and supply the lens with nutrients. The surface of the eye is covered by a transparent membrane, the conjunctiva, which supplies the cornea with nutrients. Dilation of blood vessels in the conjunctiva due to injury or infection causes the characteristic bloodshot or pink-eye appearance known as conjunctivitis.

The cornea, humor, and the lens are the refractive apparatus of the eye. The refractive power of the eye is measured in diopters (D). A diopter is the reciprocal of the focal length of a lens (1/focal length). A lens that can focus parallel light rays to a point 1 m from its axis has a refractive power of 1 D. A lens that can focus parallel light rays to a point 50 cm from its axis has a refractive power of 2 D; at 2 m, 0.5 D; and so on. Similarly, a lens that can focus parallel light rays to a point 10 cm from its axis has a power of 10 D. The eye is considered to have a single lens 17 mm in front of the retina. When focusing on a distant object, it has a refractive power of 59 D. About 48 D of the eye's total refractive power is due to the cornea, rather than the lens. This is because of the large difference between the refractive indices of the cornea and air, as opposed to the smaller difference between the refractive indices of the lens and the humors of the eye. If the lens is removed from the eye, its refractive power is about 150 D, but inside the eye it only contributes about 15 D when distant objects are viewed. This explains why removal of the lens of the eye (due to the presence of opacities or cataracts in the lens) does not lead to blindness—corrective lenses can be supplied to replace the lost refractive power and compensate for any optical abnormalities.

Blinking

Blinking is a reflex action that occurs every 2–10 s. It is also a voluntary, forced closure of the eye. The function of blinking is to stimulate tear production and flush out foreign objects (such as dust particles) from the surface of the eye. Tears, a dilute saline solution, lubricate eye movements and are mildly bactericidal. The eye is covered with a precorneal tear film (PTF) that protects it from the outer environment. The PTF consists of an outer lipid layer covering a water layer, below which is a layer of mucin. Instability in the PTF is one of the causes of discomfort in contact lens wearers (Graham and Lin, 2021). Eye blinking and tear production are essential for the maintenance of the PTF. Blinking is divided into three categories:

1. Voluntary blinking—a conscious decision to close the eyes
2. Involuntary blinking—at a frequency of approximately 12–20 blinks/min in subjects at rest
3. Reflex blinking—in response to mainly visual and ocular stimuli

The blink rate decreases under high mental workload and increases during conversation. The eyelids can be likened to windscreen wipers. At the inner corner of the lower eyelid is the nasolacrimal duct through which the tears are drained. Tasks requiring concentration can reduce the blink rate. This can cause particles of dust to accumulate and lead to drying and irritation of the surface of the eye (particularly if the relative humidity of the air is low, causing the PTF to break down). This may cause complaints of hot or rough eyes. It has been suggested that this abnormal drop in blink rate is common in work involving visual display terminals (VDTs) and is a contributory factor in the incidence of VDT-related visual problems. Tsubota and Nakamori (1995) measured the effects of ocular surface area on tear dynamics. Subjects blinked every 2 s when looking down, straight ahead, and when looking up. The area of exposed eye was 2.2 cm^2 when subjects were relaxed, 1.2 cm^2 when subjects read a book (with the angle of gaze cast downward), and 2.3 cm^2 when subjects looked straight ahead to operate a VDT. The rate of evaporation of tears increased with increasing ocular surface, leading the researchers to recommend that VDTs should be set at a lower height to prevent the eyes from drying out during VDT work. The reduced ocular surface when reading a book negated the drop in blink rate, which accounts for part of the increased prevalence of visual complaints when operating VDTs compared to reading.

Accommodation

Unlike the cornea, the lens has variable refractive power, enabling light from both distant and near objects to be focused sharply onto the retina—a process known as accommodation. Cameras usually have fixed focal-length lenses. Accommodation, or focusing, of a camera depends on adjusting the distance of the lens from the focal plane (the photographic film). The refractive power of the eye is varied by changing the shape of the lens. When distant objects are viewed, the lens assumes a flat disc shape. When near objects are viewed, the lens becomes fatter and rounder, with greater refractive power (Figure 10.4).

Most visual problems due to optical causes can be successfully corrected using appropriate lenses. When viewing a distant object, the incident light rays are approximately parallel. They are refracted by the cornea and lens and produce an image on the retina. In practice, any object more than about 6 m away can be considered to be at infinity. When viewing close objects, the incident light rays are divergent, and greater refractive power is required to produce a sharp image on the retina. In young people, the refractive power of the lens can increase from 15 to about 25 D to bring close objects into focus—the lens has about 14 D of accommodation in these individuals. Although this is quite a small proportion of the eye's total refractive power, it is necessary for seeing clearly both near and middle-distance objects (such as a VDT on a desk and a noticeboard on a wall across a room).

The mechanism of accommodation is as follows. The lens has a naturally convex shape. It is held in place by a capsule and a muscle known as the ciliary muscle. The ciliary muscle is situated around the equator of the lens to which it is attached by ligaments. When a near object is fixated, the ciliary muscle contracts and moves closer to the lens in a sphincter-like action. This reduces the tension in the ligaments and permits the lens to adopt its natural convex shape, increasing its refractive power. When fixating a distant object, the ciliary muscle relaxes and moves further away from the lens. Tension in the capsule is increased by the pull of the ligaments, and the lens is pulled into a flatter shape with less refractive power.

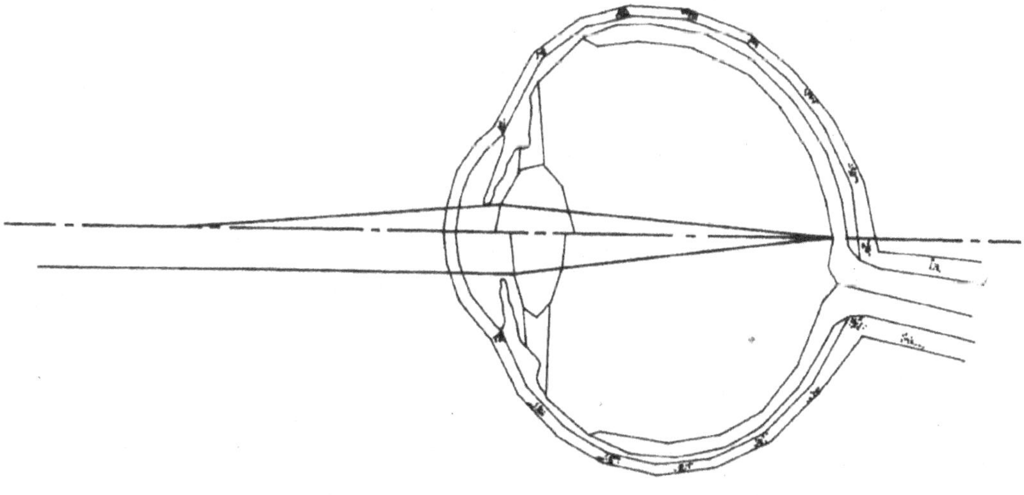

FIGURE 10.4 Accommodation. The lens assumes a spherical shape to focus divergent rays from near objects onto the retina. Parallel rays from distant objects can be brought into focus with a flatter lens.

Thus, the ciliary muscles have to contract to accommodate near objects, and it is in this sense that the visual workload can be considered to be greater when viewing near rather than distant objects. Visual workload in close tasks can be reduced by permitting micro-breaks every few minutes, in which the eyes are refocused on a distant object for a few seconds. This is known as visual relief. The near point of vision is the closest distance at which an object can be brought into sharp focus. A 16-year-old can focus on an object less than 10 cm in front of the eye. However, the lens loses elasticity with age, and, in practice, this results in a reduction in refractive power. Age-related reductions in the amplitude of accommodation are to be expected, with reductions of 0.3 D/year for people between 20 and 30 years, 0.25 D/year for 30–40-year-olds, and 0.2 D/year for people between 40 and 60 years being the norm.

By the age of 60 years, the near point may have receded to 100 cm. This is why older people often need reading glasses or, for example, have to hold the paper at arm's length when reading a newspaper. By the age of about 50 years, the lens has only about 2 D of accommodation left. After this, it can be regarded as completely non-accommodating—a condition known as presbyopia. The result of presbyopia is that the eye becomes focused at a fixed distance, which varies between different people depending on the characteristics and condition of their eyes. Frequently, the fixed viewing distance in the presbyopic eye is intermediate between the previous near and far points and the person has to wear bifocal lenses: the upper part is set for distant vision and the lower part for near vision (mainly for reading). Bifocal or trifocal lenses can restore a kind of stepwise accommodation to the presbyopic eye. In practice, if workplaces are adequately lit, the depth of the field of the eye is increased, and the net effect is to lower the requirements for accommodation. This explains why good lighting is important in all facilities used by older people. *The far point of accommodation of the lens is infinity. For practical purposes, the ciliary muscles are relaxed when viewing objects at a distance of 6 m or more.*

Optical Defects

Asthenopia is diminished visual acuity associated with eyestrain, pain in the eyes, and headache. It is common in people who carry out near visual work for long periods and naturally reverses shortly after the cessation of close visual work.

In a normal or emmetropic eye, there is a correct relationship between the axial (anteroposterior) dimensions of the eye and the power of its refractive system. Parallel light rays are focused sharply on the retina (Figure 10.5).

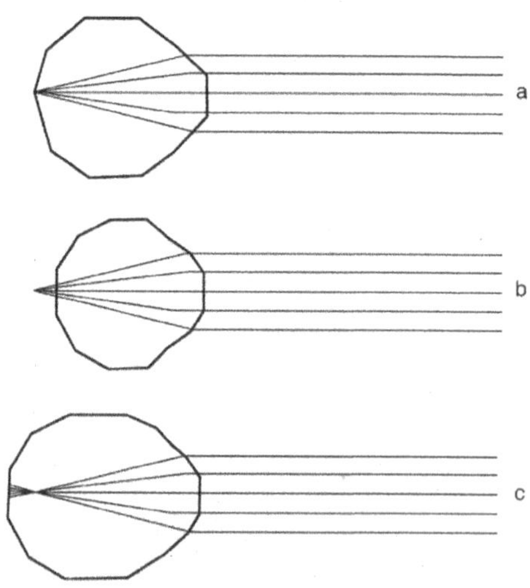

FIGURE 10.5 In the emmetropic eye (a), light rays are focused onto the retina. In hypermetropia (b), the eye lacks refractive power. In myopia (c), there is excessive refractive power.

In myopia, light rays entering parallel to the optic axis are brought into focus at a point some distance in front of the retina. This can be caused by the eye being too long anteroposteriorly or due to excessive power of the refractive system. Myopia is sometimes referred to as near-sightedness because the near point is closer to the eye in myopic people (for an equal amount of accommodation) than it is in a healthy eye. Myopic individuals cannot bring distant objects into focus. Temporary myopia often occurs after a near object has been viewed for a period of time—accommodation is not instantaneous because the lens requires time to change to a flatter shape when the ciliary muscle relaxes. Myopic individuals can often carry out close tasks such as VDT work or sewing with ease but experience difficulties with tasks such as driving, where target objects are more than 5–10 m away.

Myopia has a genetic component. In the developed world, the adult population prevalence is about 25%. In late childhood, it is about 12% if both parents are myopic and only 2.7% if neither parent is affected. Myopia is rare in hunter–gatherer societies and seems to be triggered by childhood exposure to close visual work. It was rare in Eskimo societies originally but reached 25% in the years following school attendance by Eskimo children. There is experimental evidence that blur is a stimulus to eye growth and can cause the system to overshoot, resulting in myopia (Nesse, 1995).

In hypermetropia, light rays entering parallel to the optic axis are brought into focus behind the retina. This can be caused by the eye being too short anteroposteriorly or by insufficient curvature of the refractive surfaces of the eye. Hypermetropia is sometimes referred to as farsightedness because the near point is further away from the eye (for an equal amount of accommodation) than it is in a healthy eye. Hypermetropic individuals can be said to lack refractive power and may tire quickly when carrying out work in which the viewing distance is short (such as using a VDT).

In astigmatism, there is an unequal curvature of the refractive surfaces of the eye such that the refractive power is not the same in one plane as in another. When an object of complex shape is viewed, the retinal image may be out of focus in one plane but not in others. Astigmatic individuals often perform quite well when given simple eye tests because the defect is corrected by the depth of focus of the eye. However, they may experience severe difficulties at night or when there is excessive glare.

These defects can be corrected using appropriate lenses. Myopic individuals can be given diverging lenses to reduce the total refractive power of their optical systems. Hypermetropic individuals can be given converging lenses to increase their refractive power. Several problems may occur if defective vision is not corrected. First, suboptimal vision may degrade performance. Second, excessive load on the muscles of the refractive system may cause visual fatigue. Third, the worker may adopt stressful body postures to orient the head in an attempt to see better. Neck and shoulder problems may result.

Chromatic Aberration

White light is a mixture of different wavelengths. A defect of all simple lenses is that the light of a given wavelength is focused slightly nearer or farther away than other wavelengths. This is known as chromatic aberration. Chromatic aberration can cause the outlines of objects to have purple or red fringes (the shortest and longest wavelengths). The refractive power of the eye is about 3 D greater for saturated blue than for saturated red, so blue objects are focused in front of red ones and both cannot be in focus at the same time. Purple letters or characters on VDT screens may appear to have fuzzy edges for this reason. In VDT, using less saturated colors, the difference may be only 1 D. However, this may still be enough to cause startling visual effects such as chromostereopsis (the illusion of depth).

Since blur is a stimulus for accommodation, unresolvable blur may destabilize the accommodation mechanism, making it "hunt" in vain to resolve the blur caused by the different focal points of the different colors. The onset of visual fatigue may be hastened. Reds, oranges, and greens can be viewed without refocusing, but cyan or blue cannot be viewed with red (because they are at opposite ends of the visible spectrum). Chromatic aberration is exacerbated when the pupillary diameter is large (as in poor lighting). Collins et al. (1991) measured steady-state accommodation of subjects viewing VDT screens with different colored backgrounds and text. They found a small but statistically significant increase in accommodation when subjects viewed high-contrast blue-on-red and red-on-blue screens compared to monochrome screens of either polarity. This suggests that it is more visually demanding to view such screens and that red/blue color combinations should be avoided.

For clarity and to avoid of strange visual effects, the conjoint use of saturated colors or colors from opposite ends of the spectrum should be avoided.

Convergence

Movements of the two eyes together (e.g., to track a moving object) are called "versions" or conjugate movements. Vergence occurs when the eyes move in opposite directions—to fixate on a target such that the images from the two eyes can be combined into a single image (binocular fixation). When looking at a distant object, such as a mountain several kilometers away, each eye receives a similar image because the lines of sight of the two eyes are parallel and the distance between the eyes is negligible compared to the distance of the viewed object.

When closer objects are viewed, the eyes converge on the object, that is to say, the lines of sight of each eye meet at the object. Convergence decreases with distance. When viewing close objects, the two eyes view the object from slightly different angles so that the images cast onto the two retinas differ slightly. The near point of convergence is the nearest point to the eyes where binocular vision can be maintained (about 6–10cm). It recedes with age and can make close work difficult for older workers at 25–30cm. The position of the eyes in their sockets is controlled by the vergence system. Lightly pressing one eye with the finger causes the perception of double images, which is known as diplopia. Long hours of close visual work may cause imbalances in the muscles controlling eye movement, a condition known as phoria, and increase the perceived effort required to carry out the task.

As with accommodation, the far point of convergence is infinity but for practical purposes it can be considered to be at a distance of 6m or more.

Resting Posture of the Eye

If people are asked to look ahead in the dark, their eyes assume their resting posture. Dark vergence and dark adaptation are the levels of binocular vergence and monocular accommodation that the eyes assume under such conditions. When looking straight ahead, the resting position of vergence is about 1m, and that of accommodation is at an intermediate distance. Prolonged viewing of near objects causes temporary, proximal shifts in vergence and accommodation, and these shifts are greater in people whose resting positions are farther away. The onset of these shifts is related to the onset of visual fatigue. Further exposure to high visual demands can create "vergence disparities," which express themselves as "double vision" caused by overconvergence to far objects and underconvergence to near objects (Heuer and Owens, 1989).

Jampel and Shi (1992) describe the primary position of the eyes as being 4°, on average, below the Frankfort plane (the Frankfort plane is described by lines passing through the upper margin of each external auditory canal and the lower margin of the orbit of the ipsilateral eye). In normal standing, the Frankfort plane is approximately horizontal, and the direction of gaze of the eyes in the resting position is slightly downward.

According to Jampel and Shi, the primary position of the eyes is the position to which the eyes reset automatically after we cease viewing something elsewhere in the visual field. This resetting occurs automatically and is remarkably constant. Subjects can reset their eyes to the resting position with an accuracy of a few degrees in lighted rooms, with or without fixating on objects, and in dark rooms. Furthermore, the resting position of the eyes in relation to the head (brain) is unaffected by body position (head supine, prone, tilted to one side, or when subjects are standing on their heads). The primary position is not affected by whether the eyes are open or closed. Subjectively, it is the natural position of the eyes, and no conscious effort is required to maintain it. According to Jampel and Shi, the primary position of the eyes is determined by neurological mechanisms in the reticular activating system of the brainstem rather than by the passive response of the eyes to mechanical forces and gravity.

Although the primary position of the eyes is fixed, the resting position of vergence is not. Heuer and Owens (1989) measured changes in the resting position of vergence in subjects viewing displays from 45° below the horizontal to 30° above. Subjects viewed the displays either by moving the eyes or by moving the head and keeping the position of the eyes constant with respect to the head. In both cases, the resting position of vergence increased with upward gaze (from about 80 cm at 45° below to 140 cm at 30° above). Furthermore, accommodative power increases with declination of the line of sight, by 20% when looking down at 40°.

Retina

The retina is the most complex part of the eye, consisting of a layer of light-sensitive cells connected to nerve fibers, and is sometimes considered to be an extension of the brain (a little piece of brain lying within the eyeball). Unlike photographic film, the retina actively processes incoming information before passing it on to the brain via the optic nerve.

Incident light causes chemical changes in the light-sensitive cells, which give rise to nerve impulses. The nerve fibers pass over the cells and converge to make up the optic nerve. The point at which the optic nerve leaves the retina is known as the blind spot. The retina can be likened to an array of electronic light detectors linked in complex ways that act in an on/off fashion when activated by incident photons.

The retina contains two types of light-sensitive cells, known as rods and cones. There are over 100 million rods and about 6 million cones. Rods are more sensitive to light than cones and are essential for scotopic (night) vision. Bright light bleaches the rods, which renders them ineffective, and the cone or photopic system then comes into operation.

Photopic and scotopic vision differ in several other ways. The retina has an uneven distribution of light-sensitive cells. Cones predominate in the central part of the retina (the fovea). Toward the periphery, the concentration of rods increases. Furthermore, each foveal cone cell has its own nerve fiber, whereas several rods may share one fiber. The structure of the retina is exceedingly complex, but it can be concluded that photopic vision is less sensitive to light but more acute than scotopic vision.

Melanopsin

The retinal ganglia contain a light-sensitive pigment called melanopsin (Lucas et al., 2020). Unlike rods and cones, there are very few melanopsin cells; they have low sensitivity to light, being most sensitive at about 480 nm (blue), low contrast sensitivity and temporal resolution (they don't respond quickly to changes in light intensity), and they are not involved in forming visual images. These cells appear to be important in determining pupil size and perceptions of brightness, which appears to be important in the entrainment of circadian rhythms. Exposure to blue light, in particular, suppresses melatonin secretion and increases alertness. Melatonin is a hormone that is secreted by the pineal gland in response to darkness.

Melanopic EDI: This is the melanopic equivalent of daylight illuminance. It reflects how much the light affects the melanopic system in humans. It indicates the amount of daylight that is needed to match the artificial light. To support good entrainment of circadian rhythms and alertness during the day, lighting systems should deliver a high melanopic EDI during the day. A standard daylight exposure has a melanopic EDI of 1, which depends on the illuminance and the spectral power distribution of the light. Daylight is taken as a standard because exposure to daylight entrains circadian rhythms. According to Trinh et al. (2023):

- During the day, the minimum melanopic EDI measured at 1.2 m above the floor (approximately sitting eye height) should be 250 lx.
- In the evening, for residential and other indoor environments, it should be 10 lx at 1.2 m (white light spectrum should be depleted of short wavelengths at the melanopic peak sensitivity).
- For sleep, both the light at the eye should have a melanopic EDI of 1 lx or lower.

  To calculate the melanopic EDI, both the spectral power distribution of the light source and the illuminance are required.[1]

Peripheral Vision and the Visual Field of the Stationary Eye

Subjectively, we are aware of an area of focal attention in bright light, which corresponds to the fovea or central, part of the retina, where there is a high concentration of cones. In dark conditions, the ability to resolve details is lost because the cones no longer function. The peripheral areas of the retina contain more rods. They are not sensitive to detail, but they are sensitive to changes in light and to movement, *so warnings can be placed in peripheral areas of a panel as long as flashing light or movement is used.* When looking straight ahead, we can detect targets up to 60° medially (toward the nose), 107° laterally, and 70° above and 80° below the horizontal. Interestingly, the sensitivity of peripheral vision depends on the location of the stimulus in relation to the midline of the body. Reaction time to visual stimuli is faster when these stimuli are "straight ahead" than when they are at the side, even when peripheral vision is controlled for (Durand et al., 2012). What this means is that when central vision is disengaged from frontal viewing (e.g., when glancing at something to the side of the body), objects straight ahead can only be detected using peripheral vision. Reaction time to these objects is faster than to objects in the periphery when central vision is engaged straight ahead (we can detect targets "out of the corner of our eye" more quickly when they are in front of us than to the side). Thus, in tasks where operators have to scan the environment using central vision, visual warnings should be placed in front of the operator and not to the side.

Retinal Adaptation

Retinal adaptation is the ability of the retina to change its sensitivity according to the ambient lighting. When walking from a darkened cinema into a sunny street, a temporary feeling of being dazzled is experienced. The diameter of the pupil decreases to reduce the amount of light entering the eye, the rods are bleached, and the photopic system quickly comes into operation. On going from a bright area to a dark one, the pupil increases in size, and chemical changes take place in the retina as the rods slowly come into operation. Full adaptation to dark conditions can take up to 20 min, so it is important not to expose the dark-adapted eye to sudden bright lights, since even brief exposures will degrade scotopic vision for some time afterward. This can happen to motorists driving along unlit roads at night when oncoming cars have badly adjusted or undimmed headlights. Extra light is usually provided at the entrances and exits of tunnels for similar reasons—to smooth the transition between light and dark and provide more time for retinal and behavioral adaptation.

Figure 10.6 depicts the sensitivity of the photopic and scotopic systems to light of different wavelengths. The systems are differentially sensitive to light of different wavelengths. Both are maximally sensitive to light in the middle of the spectrum (which is perceived as blue–green, green, and yellow). Violet and red are less readily sensed, which implies that for a red object to appear as subjectively bright as a green one, more illumination is required. It is unfortunate that red is normally used to signal danger, since the retina is less sensitive to red than to other colors at the same illumination level.

The scotopic system is several orders of magnitude more sensitive than the photopic system to shorter wavelengths. However, the two systems have similar sensitivity to the longer wavelengths. When the eyes are dark-adapted, long-wavelength (red) light should be used to illuminate objects

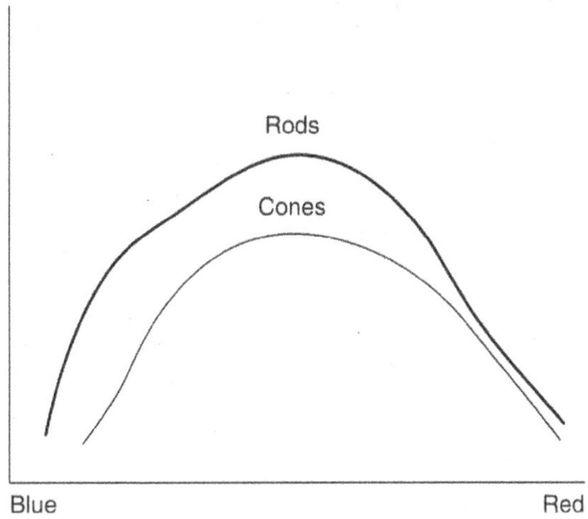

FIGURE 10.6 Sensitivity of rods and cones.

temporarily. The cones will be sensitive to these wavelengths at an intensity sufficiently low so as not to bleach the rods and degrade their state of dark adaptation. Night watchmen, sentries, etc., can be issued with torches containing long wavelengths for use, for example, when inspecting documents at night. Some guidelines for aiding vision while preserving dark adaptation are as follows:

1. Use wavelengths of 625–635 nm and exclude wavelengths less than 600 nm.
2. Use red light when demands on night vision are high (e.g., for sentries or watchmen).
3. Use a red light to facilitate the safe movement of people.
4. Achieve red illuminance of 1–20 lx.
5. Do not use red light if color vision is important—use low-intensity white light.
6. Personnel may prefer low-intensity white light to red.
7. More accommodation is required under red light than under low-level white light (0.5–10 lx).

In practice, a stage intermediate between purely photopic and scotopic vision can occur. This is known as mesopic vision and has characteristic features of both.

Color Vision

The output of the photopic system is experienced as color vision. Three types of cone cells with different spectral sensitivities have been identified. Since the absorption spectra of their light-sensitive pigments overlap, a given wavelength of visible light will cause the three types of cones to respond to varying degrees. The result is a pattern of cone outputs that signifies the particular wavelength. Different wavelengths give rise to different cone output patterns, which are interpreted as different colors by the brain. That the cone output patterns are transmitted to the brain, where they give rise to the color perception, seems straightforward and compatible with our everyday experience. That white light consists of light of many different colors (i.e., wavelengths) supports the component process explanation of color vision. For example, red and green light sources can be combined to give a yellow color, which matches monochromatic yellow even though the wavelength that gives rise to the perception of yellow is not present. These metameric matches are evidence for a component process view of color vision. The cone output pattern caused by the mixed-wavelength stimulus is identical to that caused by the monochromatic source, so the same color is perceived. According to this view, color vision is purely the result of the cone output pattern caused by the incident light—different wavelengths have characteristic cone output patterns, and if a combination of different wavelengths impinges on the retina, a cone output pattern intermediate between the two wavelengths, and in proportion to their respective intensities, will result. This type of reasoning is used to explain how people can distinguish between colors such as bluish-green, greeny-blue, etc.

Certain visual phenomena are more difficult to explain using simple color mixing theories. If a bright light is fixated for several seconds, a retinal afterimage remains when the viewer looks away. Blue lights produce yellow afterimages and vice versa, and red lights produce green ones. It has been suggested that at the retinal level, a component process exists, with functionally different cones responding maximally to different wavelengths. The outputs of this system are thought to fire into a so-called opponent process system. There are said to be three opponent process systems—a blue/yellow system, a red/green system, and a light/dark system. When long wavelengths are incident, the red/green system increases its firing rate above its spontaneous level, and red is perceived. As the incident wavelength is shortened, the system is inhibited and its firing rate decreases until it drops below the spontaneous level, and green is perceived. Thus, after viewing a bright red light for several seconds, a green afterimage of the object is seen due to a sudden drop in the output of the opponent process system. Certain cells in the brain appear to act like opponent process cells, increasing and decreasing their firing rates in this way.

Color vision is a complex phenomenon that can only be dealt with here at a superficial level. However, these theories of color vision can have implications for ergonomics. The notion that red and green, and blue and yellow, are related neurologically has interesting design implications. Generally, these opponent color combinations should be avoided in display design because of the afterimage problem—particularly in active displays such as VDT screens, and if the colors are highly saturated. A saturated color is pure in that it contains only the wavelengths that give rise to the perception of the particular color—"pillar box" red is an example. Unsaturated colors are those that have been "diluted" with white. Bright blue letters on a yellow background may make an attention-getting sign but would be unsuitable for use in a visual display screen. Sharp discontinuities between bright colors may produce startling visual effects—red letters on a blue background may vibrate, for example. Color should not be used for borders or to display messages on the periphery of a display, since cone density is very low at the periphery of the retina. The use of flashing lights or a moving cursor would be more effective. Some general guidelines for the use of color in visual displays are summarized in Table 10.1.

TABLE 10.1
Guidelines for Color Selection

In General

1. Choose compatible color combinations. Avoid red/green, blue/yellow, green/blue, and red/blue pairs
2. Use high color contrast for character/background pairs
3. Limit the number of colors to four for novice users and to seven for experts
4. Use light blue for background areas only
5. Use white to code peripheral information
6. Use redundant coding (shape or typeface as well as color). Around 6%–10% of males have defective color vision

For Screens

1. Luminosity diminishes in the order white, yellow, cyan, green, magenta, red, and blue
2. Use white, yellow, cyan, or green against a dark background
3. For reverse video, use nothing (i.e., black), red, blue, or magenta
4. Avoid highly saturated colors

Source: Adapted from Durret and Trezona (1982); Van Nes (1986).

The dark-adapted eye is more sensitive to shorter-wavelength (blue) than longer-wavelength (red) light, and most sensitive to green. These colors should be used to design displays or notices that have to be read in dark conditions. The light-adapted eye is most sensitive to middle-wavelength (yellow/green) light. In designing passive (reflective) displays for light conditions, maximum color contrast can be obtained by using yellow on black.

Color Blindness

Color blindness is normally an inherited condition that results in the inability to see different colors or to distinguish between certain colors. In monochromatism, there are no cones or only one type of cone, and only black, white, and shades of gray can be distinguished. In dichromatism, only two types of cone are available. In deuteranopia and protanopia, red or green cannot be distinguished, they both look brown. There are other forms of color blindness where the sensitivity of one type of cone is altered—colors can still be seen, but more subtle differences in hue cannot. Color blindness is more common in men than in women.

Color Perception

Color perception is a complex phenomenon. Not all colors can be duplicated by mixing three primary hues in the laboratory (brown, silver, and gold are examples). In color television sets, three-color mixing is used, yet viewers do perceive these otherwise unattainable colors. In laboratory conditions, it is possible to illuminate a dark object, such as a piece of coal, so that its luminance is greater than that of a piece of white paper. However, subjects will say that the paper looks brighter, as long as they can see enough to recognize the objects as paper and coal. This demonstrates that higher learning processes are involved in making judgments about color. Some colors are associated with familiar objects (as is indicated by sayings such as "as black as coal"), and judgments are influenced by other factors such as intensity and the reflectance of the object.

The Purkinje Shift

The Purkinje effect occurs when, under varying illumination, the relative contributions of the photopic and scotopic systems change. Under low light levels, humans are virtually color-blind, and the rods are most sensitive to light in the blue–green part of the visual spectrum. Under higher lighting levels, the cones are most sensitive to the green part of the spectrum. The color contrast that is perceived when objects are viewed changes depending on the lighting level. Ripe strawberries appear bright red against darker green leaves in strong sunlight, whereas at dusk the leaves will appear brighter than the strawberries.

The Pulfrich Pendulum

The Pulfrich pendulum is a visual illusion in which the lateral motion of the swinging pendulum appears to have a depth component. The effect occurs when a neutral density filter is placed in front of one of the eyes, which reduces the speed at which visual information from that eye is transmitted to the brain. The difference in reaction time creates a stereoscopic effect. Whether or not visual reaction time is reduced under poor light depends on a number of other factors—particularly the contrast between the target and the background.

Measurement of Light

Measurement of light is essential in the design and evaluation of workplaces. Since the eye adapts to light levels, automatically compensating for any changes in illumination, subjective estimates of the amount of light in a work area are likely to be misleading. Data concerning the visual response of the eye have been used to define lighting measures. The radiant flux arriving at the surface is weighted according to the eye's sensitivity to each of a number of wavelength intervals. The total incident radiant flux after weighting is known as the luminous flux.

TABLE 10.2
Derivation of Main Photometric Terminology

Name	Description	Units
Luminous intensity	Power of a source or illuminated surface to emit light	Candela (cd)
Luminous flux	Rate of flow of luminous energy	Lumen (lm)
Luminance	Light emitted by a surface	cd/m^2
Illuminance	Amount of light falling on a surface	Lx (lx)
Reflectance	Ratio of the luminance and illuminance at a surface	

The measurement of light is known as photometry. The main photometric units are luminous intensity, luminous flux, luminance, and illuminance. Their definitions are given in Table 10.2.

The SI unit of luminous intensity is the candela (cd). An imaginary point source of luminous intensity 1 cd will emit light in all directions. A source of greater intensity will emit more light. In both cases, we can imagine a sphere of light spreading out from the source. Clearly, the intensity of the source itself does not depend on the distance from which it is viewed. However, the strength of the light at the edges of the imaginary sphere will depend on the viewing distance.

Derivation of Terms

1. Luminous intensity (I) refers to the power of a source to emit light.
2. The light emitted is luminous flux (F), which is measured in lumens.
3. Flux (lumens) = Intensity × solid angle

$$F = I\omega$$

4. By definition, a light source of luminous intensity 1 cd emits 4π lm of flux.
5. If the flux is emitted in all directions, it forms a spherical wave front with total solid angle 4π st (the steradian (st) is the unit of solid angle).
6. Thus, 1 cd gives 1 lm/st.
7. The luminous flux per steradian does not vary with distance from the source.
8. If the light source of 1 cd is placed inside a sphere of radius (r) of 1 m, then: the illuminance (E) on the inside of the sphere will be 1 lm/unit area. The surface area of a sphere of radius 1 m is 4π m^2. Therefore, the illuminance will be

$$\begin{aligned} E &= \frac{4\pi \text{ lm}}{4\pi \text{ m}^2} \\ &= \frac{1 \text{ lm}}{1 \text{ m}^2} \\ &= 1 \text{ lx} \end{aligned}$$

9. Seen from below, a sphere of radius 1 m appears as a disc. The area of the disc is πr^2, and therefore, the luminance of the disc is

$$L = \frac{1 \text{ cd}}{\pi r^2} = \frac{1}{\pi}\frac{\text{cd}}{\text{m}^2} = \frac{1}{\pi}\text{cd} / \text{m}$$

10. A flat surface below the sphere is illuminated by luminous flux. The illuminance depends on the distance (d) of the surface below the sphere and is given by

$$E = \frac{I}{d^2}$$

11. The illuminance on a surface varies according to the cosine of the angle of incidence of the light and the normal to the surface. For horizontal surfaces,

$$E_h = \frac{I}{d^2}(\cos\theta)$$

For vertical surfaces,

$$E_v = \frac{I}{d^2}(\sin\theta)$$

The surface area of a sphere is 4Π st of solid angle. A steradian is defined as the solid angle that encloses a surface on a sphere equivalent to the square of the radius. By definition, a point source of luminous intensity 1 cd will emit a total luminous flux of 4π lm, or 1 lm/st.

Illuminance refers to the light falling on a surface. If we imagine a point source of 1 cd inside a sphere of 1 m radius, the illumination on the inside of the sphere is defined as 1 lx, or 1 lm/m^2. If we increase the radius of the sphere to 2 m but keep the intensity of the source the same, the illuminance on 1 m^2 of the inner surface will now be 1/4 lx. This is because when the radius of the sphere increases linearly, the surface area increases according to the square of the radius. Illuminance therefore follows an inverse square law, with intensity decreasing with the square of the distance from the source. So, if the radius of the sphere were 4 m, the area on the surface of the sphere enclosed by 1 st would be 16 m^2. The illuminance on this area would be 1/16 lx. If we increased the luminous intensity to 2 cd, we would have 8Π lm, or 2 lm/st. Since 1 lx is 1 lm/m^2, we would have an illuminance of 2 lm/16 m^2, or 1/8 lx.

The luminance of an object depends on the light it emits or reflects toward the eye. It is measured in candelas per square meter and corresponds roughly to brightness (although brightness perception depends on other factors such as contrast). In the example above, if the inside of the sphere is a perfect reflector of light, then the luminance will be the same as the illuminance. The percentage of the incident light reflected by a surface depends on the reflectance of the material. Reflectance is defined as the ratio of luminance to illuminance. White paper has a reflectance of about 95%, white cloth about 65%, newspaper about 55%, and plain wood about 45%. Matte black paper has a reflectance of about 5%. More formally, reflectance is given by

$$\text{Reflectance} = \frac{\text{Luminance} \times \pi}{\text{Illuminance}} \tag{10.1}$$

If the material is a perfectly diffuse reflector, then a surface illuminated by 1 lx has a luminance of 1/π cd/m^2.

If we know the illuminance of the surfaces in a room (e.g., by measuring it with a light meter), we can select materials of appropriate reflectances for each of the surfaces in order to achieve a balance of surface luminances in the room and ensure that the ratio of luminances of adjoining surfaces is not excessive. For example, suppose we place some pieces of white paper, wood, and matte paper on a desktop under an illuminance of 300 lx. What are the luminances of these materials?

1. White paper. Reflectance 95%. Luminance = 0.95 × 300/π = 90.7 cd/m^2
2. Wood. Reflectance 45%. Luminance = 0.45 × 300/π = 43.0 cd/m^2
3. Matte black paper. Reflectance 5%. Luminance = 0.05 × 300/π = 4.8 cd/m^2

A more detailed discussion of photometric terms, the relationships between them, and recommendations for the design of lighting systems can be found in ANSI/IES LS-1–22, Lighting Science: Nomenclature and Definitions for Illuminating Engineering.

BASIC APPLICATIONS

General Recommendations for Restful Viewing

The ability to view objects up to about 1.4 m depends on the interaction between the angle of gaze and the amount of vergence and accommodation, both of which increase when looking downward. For comfortable viewing:

- Most frequently viewed displays should be accessible with the head erect and the eyes in their primary position.
- Vergence effort should be minimized—the closer the display to the operator, the lower it should be (up to 30° below the horizontal), and vice versa.
- To facilitate accommodation, displays that require a high degree of visual acuity should be placed up to 40° below the line of sight and be directly in front of the viewer.
- To account for individual differences in the resting posture of the eyes, displays such as screens should be adjustable in both the horizontal and vertical planes, from eye height to over 1 m away from the eyes.
- No displays should be nearer to the operator than the resting point of vergence—110 cm for displays level with the eyes.
- Avoid accommodation traps, where objects placed close to the near point of accommodation (such as dirt on a car windscreen) cause involuntary focus and blurring of visual targets further away.

Hill and Kroemer (1986) investigated preferred sight lines by having subjects view displays while sitting erect (90°) and reclining at 105°, 130°, and 180°, respectively. Subjects sitting erect preferred a viewing angle of 29° below the Frankfort plane, which increased to 40° when reclining fully. This is almost certainly to correct the recession of the near point of vergence caused by the change in head position. Target distance also caused a change in the preferred line of sight, and this was independent of any changes due to head position. The preferred line of sight was greater for a task distance of 50 cm than for a distance of 100 cm, 38° as opposed to 30° below the Frankfort plane. The increased declination in the line of sight when viewing the 50 cm display was almost certainly to compensate for the increased accommodation required to view the closer display.

To manage the great variability in user preferences for VDT workstation layouts, Jaschinski-Kruza (1999) suggests a procedure for determining the individually comfortable position of visual displays relative to the eyes. On each of four successive days, users should work:

1. With the screen 90–100 cm from the eyes and as low as possible
2. With the screen 50 cm from the eyes at sitting eye height
3. With the screen 90–100 cm from the eyes and at sitting eye height
4. With the screen 50 cm from the eyes and as low as possible

During a test of this procedure, it was found that it did result in operators subsequently changing the viewing distance and gaze angle of their VDT screens. Some reduced the gaze angle and viewing distances, while others increased them, and the inter-operator variability was large. However, when asked to make these rearrangements on repeated occasions, individual operators made the same changes. Thus, viewing preference is variable across users but stable within users.

In organizations where users are issued laptop computers for travel, docking stations should be provided for office use. Adjustable workstations appear to be particularly important in organizations that practice hot-desking. For home workers, detachable keyboards should be provided for laptop users (Chapter 5).

Visual Acuity

Visual acuity is the ability to detect detail and is vital in any task requiring visual inspection of fine detail (e.g., the ability to detect cracks in pressure vessels or glass) or to visually separate distant objects, such as lights. Acuity depends on the luminance of the objects being viewed. Under scotopic conditions (10^{-3} cd/m^2), the minimum level of acuity is approximately 50′—almost 1° (360° make up a circle, each degree can be divided into 60 min, and each min into 60 s). Under mesopic conditions (about 0.1 cd/m^2), it is about 10′, and under photopic conditions (1000 cd/m^2), it is a little less than 1′.

Anyone who has ever had an eye test will be familiar with the Snellen charts and their successive rows of characters decreasing in size from top to bottom. Testing normally takes place with the patient 6 m (20 ft in the United States) from the chart. Normal vision is taken to be the ability to resolve a visual angle of 1′. In Snellen chart terminology, this equates to 20/20 vision (6/6 elsewhere). A person who could read at 20 ft what a normal person could only read at 10 ft would have 20/10 vision, and a person with poor acuity, who could read at 20 ft what a normal person could read at 25 ft, would have 20/25 vision.

Figure 10.7 depicts the visual angle when an object is viewed. This angle can be bisected to give two right-angled triangles. If we know the separation between the upper and lower parts of the object being viewed and if we also know the viewing distance, the visual angle (θ) can be estimated. The tangent of one of the right-angled triangles is given by half the separation between the object divided by the viewing distance. The total visual angle is the sum of the angles of the two right-angled triangles.

$$\tan\frac{\theta}{2} = \frac{1}{2}\frac{L}{D} \tag{10.2}$$

Ultimately, the visual acuity of the eye is limited by the density of cone cells in the retina. Even under ideal lighting conditions, 1′ is close to the lower limit. This is because an object that subtends 1′ is projected onto only 4–5 μm (thousandths of a millimeter) on the retinal surface. This area will contain only two to three cone cells, and at least two are needed to resolve detail at this level.

Taken together, these considerations enable us to carry out simple calculations to estimate maximum viewing distances for objects in the visual field or, if the viewing distance is fixed, the minimum size of the object to ensure that the necessary details will be resolved. HFE Workshop 10.1 provides a practical example.

Sanders and McCormick (1992) provide a simple (and quick) method for calculating visual angles of less than 10° when the height of the target object (H) and the viewing distance (D) are known.

$$\text{VA (min)} = \frac{3438 \times H}{D} \tag{10.3}$$

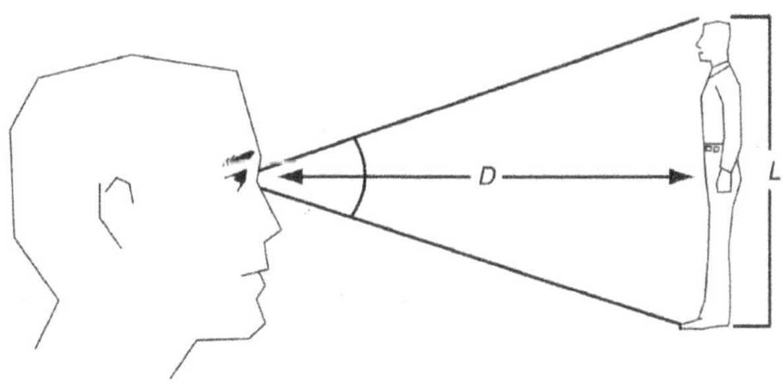

FIGURE 10.7 The visual angle (θ).

In the workshop example, $H=0.33$ and $D=6072$, therefore,

$$\text{VA (min)} = \frac{3438 \times 0.33}{6072} = 0.187$$

$$\text{VA (s)} = 0.187 \times 60 = 11.21$$

This is very close to the workshop estimate, which was obtained using trigonometry. It is accurate enough for design in ergonomics. The discussion of visual acuity has been limited to minimum separable acuity. The eye is able to resolve finer details; for example, a human hair held in the hand against a bright blue sky may subtend as little as 1″. However, several hairs held close together would fuse into a single image.

HFE WORKSHOP 10.1

VISUAL ANGLES

A naval architect is designing a new cruise liner and, to save money, wishes to place some of the navigation lights, used to enable other vessels to identify the ship, as close together as possible.

The legal requirement is for the two lights to be placed on a mast at a sufficient separation to be distinguishable from a distance of 3 nautical miles (1 nautical mile is 2024 yd). He wishes to know whether it will be possible to place the lights 1 ft apart (about 30 cm). One way to tackle this is to calculate the visual angle that would be subtended at the eye of an observer if the lights were so situated, and to assume ideal viewing conditions.

Let

Viewing distance $D=3\times2024=6072$ yd

Distance between lights $L=1$ ft

Visual angle subtended at the eye of an observer $=\theta$

Therefore,

$$\begin{aligned} \tan(\theta/2) &= \frac{0.5}{3\times6072} = \frac{0.5}{18216} = 2.75\times10^{-5} \\ \theta/2 &= 0.0016^\circ \\ \theta &= 0.0032^\circ \\ &= 0.0032\times3600 = 11.5\ \text{s} \end{aligned}$$

Reminder: There are 3 ft in a yard, so *L*/2 (0.5 ft) needs to be divided by 3 to ensure that all units are in yards.

INTERPRETATION

Assuming that these lights conform to international standards and are visible at a distance of 3 nautical miles, it is likely that they will not be discriminable and will merge to form a single image.

MINIMUM REQUIRED SEPARATION

There is no definitive limit on human visual acuity. However, most sources indicate that, under good viewing conditions, the minimum visual angle should be on the order of 1′. To achieve such an angle at a viewing distance of 3 nautical miles, the separation between the lamps is given by

(Continued)

HFE WORKSHOP 10.1 (*Continued*)

$$\tan\frac{\theta}{2}=\frac{L}{D}$$

where

$\theta=1'=1/60°=0.0167°$.
$D=3$ nautical miles $=18{,}228$ ft
$\tan(\theta/2)=1.454\times10^{-4}$
$L=10454\times10^{-4}\times2\times18.228=5.3$ ft

Required separation to achieve a viewing angle of 1′ at 3 nautical miles is therefore 5.3 ft.

COMMENTS

This is a minimum requirement, and no margin of safety has been allowed for viewers with visual defects or for weather conditions such as mist or fog that reduce transmission. To aid visualization of the degree of discrimination required, the proposed solution, yielding a viewing angle of 1 min, is equivalent to the task of discriminating two small lamps (similar, perhaps, to those found in Christmas decorations) situated on opposite sides of a U.S. quarter coin (United Kingdom 10p coin) at a distance of 90 yd. If possible, the separation should be increased beyond 5.3 ft.

It is likely that, under the present proposals, the signals from the two lamps will fuse because of inadequate physical separation. Alternating the lamps (separation in time) may be needed if physical separation is not possible.

Color and Visual Acuity

When stationary objects are being viewed, acuity is least if the object is blue. Long and Garvey (1988) investigated color and dynamic visual acuity. When moving targets are being viewed under photopic conditions, their color does not seem to influence visual acuity. However, dynamic visual acuity under scotopic/mesopic conditions is affected by target color, and blue targets are much more easily resolved than other colors, probably due to the differential sensitivity of the rod system. Long and Garvey (1988) suggest that we should be careful not to overgeneralize the usual ergonomic recommendations about color. Under some lighting conditions where there is a moving display or a moving observer, colors such as blue (which are commonly avoided for stationary targets) might be advantageous. There are possible implications here for the design of roadway warning signs, which will be used primarily at night.

Lighting Standards

Much effort has been applied over many years to the drafting of standards for the illumination of workplaces. Standards differ from country to country. Table 10.3 presents recommended illuminances for different work situations. Some other illuminance values found in more extreme situations have been included in the table to provide context.

It can readily be seen that the eye can operate under an extremely wide range of illuminance levels, due largely to the differential sensitivity of the photopic and scotopic systems.

Recommended illuminance levels vary both between countries and over time. Before the invention of electric lights, indoor workers depended on daylight, and building dimensions were limited

TABLE 10.3
Examples of Recommended and Naturally Occurring Illuminances

Area/Activity	Illuminance on a Horizontal Surface (lx)
Clear sky in summer	150,000
Overcast sky in summer	16,000
Performance of extremely low contrast tasks. Certain surgical operations	10,000
Textile inspection	1500
Office work (pencil handwriting, poorly reproduced documents)	1000
Precise assembly work	1000
Drawing offices, meat inspection, and chain stores	750
Office work (without VDUs)	500
Office work (data entry)	500
Office work (VDU conversational tasks)	300
Heavy engineering	300
Rough assembly work	200
Minimum illuminance for manual handling tasks (NIOSH)	150
Rarely visited places where little perception of detail is required (e.g., railway platforms)	50
Public spaces in otherwise unlit areas	20
Good street lighting	10
Emergency lighting	2
Moonlight	0.5

because daylight does not penetrate into a room beyond about 10m. Lighting standards have risen over the years and have been higher in the United States than elsewhere. In the early part of this century, investigations by various companies in the United States and by the Industrial Fatigue Research Board in Britain demonstrated that the performance of visually demanding tasks could be improved by increasing the level of illumination. Artificial lighting became accepted in factories, and standards rose under the assumption that "more is better." With the arrival of fluorescent lights, recommended levels could increase further because of the better luminous efficacy (lumens per watt of electricity), longer life, and reduced heat production of fluorescents.

A more recent trend has been to reduce the levels of illumination in workplaces, particularly offices. This has occurred partly because of the desire to conserve energy and also as a response to office automation. HFE specialists have recommended that illumination levels be lower in offices to avoid glare and reflectance problems (see below). In practice, the choice of an appropriate level of illumination depends not only on the task but also on the distribution of objects in the visual field and their luminances.

Contrast and Glare

The function of the eye is not so much to detect light as to detect luminance discontinuities between objects in the visual field. It is the difference in luminosity between an object and its surroundings that makes it visible, rather than the light the object reflects. A person wearing opaque spectacles can detect the presence of light but not the presence of objects.

The retina functions more like an edge detector than a light meter. Special cells (called horizontal cells) are part of a network of cells interposed between the rods and cones and the fibers of the optic nerve. If one part of the retina is stimulated by a bright light, the horizontal cells inhibit the adjacent photoreceptors. This has the effect of increasing the difference in the firing rates of the stimulated and unstimulated parts of the retina and serves to enhance the perceived contrast between the stimulus and its surroundings. This, in turn, sharpens the contours of objects, facilitating their detection (the importance of contrast and contour detection in design is discussed below and in Chapter 13).

The direction of gaze is involuntarily drawn to bright objects in the visual field. This is known as phototropism. Jewelry shops usually display their wares on black velvet cloth under bright lights to obtain maximum contrast—the intention being to stimulate a phototropic response in passersby.

One of the most important considerations in the design and evaluation of lighting systems is to ensure appropriate contrast between objects in the visual field. The luminance contrast between two surfaces is given by the difference between the luminance of the brighter and dimmer surface, expressed as a percentage of the brighter surface (note that contrast does not depend on the absolute brightness of the surfaces).

$$\text{Contrast} = \frac{100 \times (L_{\text{bright}} - L_{\text{dark}})}{L_{\text{bright}}}$$

In practical situations, for example, in offices where the lighting is fairly uniform throughout, contrast percentages can also be calculated using data on the reflectances of adjacent surfaces. In other words, it is not even necessary to measure the illuminance falling onto the surfaces; all that is required to get a fairly good estimate of contrast is the reflectance data of the materials used on adjacent surfaces. If the room has spotlights or the light is patchy, then this procedure is invalid, and surface luminance will have to be measured.

The luminance ratio is the ratio of the luminance of a work area to that of its surroundings. Recommended maximum luminance ratios have been proposed—3:1 between a task and its immediate surroundings, and up to 10:1 between the task and the walls, floors, etc. It is usually suggested that the task should be the brightest area in the visual field to take advantage of the phototropic response. This precludes the use of materials such as white Formica for desktop materials since the luminance ratio between white paper and desktop will be too low. Wood or pastel-colored finishes are preferable from this point of view.

The computer that I am using to write this chapter has a flat screen with a luminance of approximately 250 cd/m^2. The screen is housed in a matte black plastic casing, which includes the stand. The desktop has a light gray matte plastic surface that has a reflectance of approximately 60%. Assuming that the black plastic casing has a reflectance of 10%, are the luminance ratios in accord with the guidelines given above?

Assuming that the illumination in the office is fairly uniform at 300 lx, the luminance of the screen casing and desktop is as follows:

1. Casing $= 0.1 \times 300/\pi = 9.6$ cd/m^2
2. Desktop $= 0.6 \times 300/\pi = 57.3$ cd/m^2
3. Luminance ratio screen/casing $= 250/9.6 = 26$
4. Luminance ratio screen/desk $= 250/57.3 = 4.4$

The luminance ratio between the screen and the casing is far too large, and it would probably have been better to use a material of a lighter color for the casing. Interestingly, the luminance ratio between the desktop and the casing is also large (6.0). The result is that the casing is extremely intrusive visually and stands out both against the screen and against the desktop.

Interior decoration should be chosen to achieve a balance of surface luminances gradually diminishing from the task to the surroundings. Brightly colored walls, carpets, furniture, and fittings are contraindicated for this reason.

Glare occurs when there is an imbalance of surface or object luminances in the visual field, the brighter sources exceeding the level to which the eye is adapted. Sources of glare include the sun, bright or naked lamps, or reflections from shiny objects. Although the retina is able to adapt to different levels of luminance so as to operate over a wide range of conditions, it is not able to adapt selectively to large, simultaneous discontinuities in luminance in the visual field. For example, if the ambient luminance is high compared to the task luminance, the retina will adapt to the former rather than the latter, and the task will appear dim and will be more visually demanding. This can happen when VDT screens are placed against a window such that users face the window. If the illuminance in a room is very low, the retina will dark adapt and will be more vulnerable to the effects of glare from task elements or extraneous sources.

Glare is categorized in several ways. Disability glare increases task demands, whereas discomfort glare does not. Discomfort glare may occur in offices, for example, when one or more bright objects are seen peripherally. Disability glare occurs when objects brighter than the task interfere with the detection and transmission of visual task data—extraneous light sources may increase the adaptation level of the retina, making the task appear dimmer than it really is. Intense light may be inter-reflected by structures within the eye itself and reduce the contrast between the retinal image and the background. Older workers may be particularly prone to this type of glare due to age-related changes in the refractive media of the eye. Finally, bright lights may cause retinal afterimages, which have a veiling effect on the main task.

Glare may be direct or indirect, that is, it may be emitted by a source or reflected off an object. If reflected, the glare may be either diffuse or specular (mirror-like). White walls on the inside of a room may reflect sunlight into an operator's eyes and would be classed as a diffuse source. Chrome-plated controls or components such as bevels on dials and gauges may cause specular reflections. More detailed discussions of glare can be found in Cushman and Crist (1987) and Howarth (1990).

Dry Eye Disease (DED)

DED is a chronic condition that occurs when the production of tears provides inadequate lubrication of the eyes, leading to discomfort (an aqueous deficiency in the PTF or excessive evaporation from the surface of the eye such that the eyes feel "gritty," sore, or blurry). It can have many causes; it is more prevalent in older people and tobacco smokers and is sometimes associated with other chronic illnesses such as arthritis and with hormonal changes during menopause. Working in hot, dry environments with inadequate fluid intake can also cause DED. Fjærvoll et al. (2022) carried out a systematic literature review of 57 articles investigating DED and screen-based computer work. In 2019, 4 billion people used the Internet and were thus exposed to screen-based activities. This increased during COVID due to changes in work patterns.

DED is associated with computer work because of the reduced blink rate and increased evaporation, particularly if the screen is too high. The overall prevalence rate of DED in office workers ranged from 26% to 70%, compared to an overall global prevalence rate of 5% to 50%. Although DED prevalence rates increase with usage, there is no clear-cut "safe" exposure time. Estimates of 'safe' thresholds for screen-based work ranged from 2 h/day to 20–25 h/week. Clearly, DED prevalence increases with time spent on screen-based work.

In summary, eye complaints as a result of alterations to the PTF may be caused by:

1. Screen-based or other near work that is mentally demanding, reducing blink rate and causing dehydration of the PTF.
2. Prolonged upward gaze due to visual displays mounted too high (above sitting or standing eye height), widening the exposed ocular surface and increasing the rate of water loss or exposure to indoor air pollutants.
3. Low relative humidity increases the rate of water loss.
4. High air movement (>1 m/s) around the eye increases evaporation.
5. A slowing of upper eyelid movement and amplitude due to fatigue.
6. Contact lens use.
7. Interactions between indoor air pollutants, such as ozone and organic compounds, leading to the production of eye irritants.

Some suggestions for reducing discomfort in DED sufferers include:

- Avoid draughts directed at the eyes.
- Increase relative humidity in offices.
- Wraparound glasses may reduce evaporation from the eye.
- Take short breaks and look away from the screen.
- Position the top of the screen below eye level.

Digital Eye Strain

Otherwise known as computer vision syndrome (CVS), DES is defined as "a complex of eye and vision problems related to near work experienced when using digital electronic devices" (Lema and Anbesu, 2022). It is not restricted to VDT use but includes laptops, smartphones, and electronic games. Near work is known to cause visual fatigue and ocular symptoms, including temporary shifts in the near point of accommodation and the resting position of vergence. Recession of the near point suggests a loss of refractive power and is sometimes referred to as accommodation strain.

Piccoli et al. (1996) compared blink rate, refraction, phoria, near point of accommodation, convergence, and ocular symptoms in 14 female VDT operators when carrying out VDT and non-VDT office work. Ocular and visual symptoms occurred more frequently in the VDT work (71.5%) than in the non-VDT work (35.7%). Accommodation was more marked in the VDT work compared with the non-VDT work.

Lema and Anbesu (2022) carried out a systematic review of 49 articles on CVS. The pooled global prevalence of CVS was 66%, although the authors suggest that this prevalence may have been exaggerated du to an overreliance on self-reported symptoms. The main risk factors for CVS were: being female (odds ratio 1.74); incorrect posture (odds ratio 2.65); use of electronic devices outside working hours (odds ratio 1.66); long duration of use (odds ratio 2.02); short viewing distance (odds ratio 4.24); inadequate rest breaks (odds ratio 2.24); and generally poor workstation ergonomics (odds ratio 3.87). Knowledge of CVS and its causes was associated with a decreased risk of CVS (odds ratio 4.04). The lowest prevalence rate was found in Japan (12%).

Zalat et al. (2022) investigated CVS among 114 office workers. CVS prevalence was 82% and 53% among females who spent significant time outside work using smartphones. CVS was lower among those users who had correct visual ergonomics, regular cleaning of their screens, appropriate illumination, and the use of eye drops. The use of screen filters was not associated with the presence or absence of CVS.

The 20-20-20 Rule: Does it Work?

The 20-20-20 rule states that screen users should look away from their screens for a minimum of 20 s every 20 min and look at objects at least 20 ft (6 m) away. The eye muscles take 20 s to relax, and this breaks p static exertions of these muscles. Telens-Estarelles et al. (2023) carried out an experiment to test the 20-20-20 rule. Subjects were observed working at their laptops via software installed on the laptop camera. Breaks away from the screen were recorded over several weekly measuring periods. In the third week of testing, the software was activated to give prompts every 20 min to take a break. A CVS questionnaire was administered, as were tests of ocular function and measures of blink rate. Participants took more breaks when the prompts were activated in addition to spontaneous breaks, which did not change. Overall, breaks were shorter when the prompts were given, but bouts of continuous computer use were shorter. One aspect of visual function was improved—binocular accommodative facility (the speed of adaptation to a 2-diopter change), and scores on a CVS questionnaire were lower. No changes in the tear film were found. However, these benefits were not sustained post-trial. Although there is evidence that the rule is effective, there is no definitive evidence to suggest that the time periods (20 s every 20 min) chosen are optimal. Galinsky et al. (2000) evaluated the effects of supplementary rest breaks on eyestrain and upper body discomfort in VDT workers. Under the conventional schedule, workers received one 15-min break in the morning and in the afternoon, plus lunch. This was supplemented with an extra 20 min of breaks, and workers alternated between the schedules over 4-week periods. Eyestrain and upper limb discomfort increased over the day in both schedules, but the increase was lower when supplementary breaks were given.

As an 'aide-memoir' to advise employees how to avoid visual symptoms, there is evidence and a rationale to support the 20-20-20 rule.

Prevention of Visual and Upper Body Fatigue

Some general guidelines are summarized in Table 10.4. It seems that visual fatigue and upper body discomfort may be related—excessive loading of the accommodation and vergence systems during close visual work for long periods increases the tension in the muscles of the upper body (Lie and Watten, 1987, 1994). This connection between the ocular muscles and the muscles that control the position of the head is not surprising when it is remembered how the eyes and head move together to locate a stimulus in space. In fact, the two systems share common neural pathways.

Experiments have shown that the wearing of lenses to distort vision also increases upper body EMG activity. Alternatively, providing subjects with appropriate spectacles has been shown to reduce upper body EMG. These findings all serve to demonstrate the close connection between visual workload, visual fatigue, and pain in the upper body, including headaches, and support the practice of providing VDT users and others who engage in visually demanding work with regular eye tests and well-designed workstations. Users must be able to select their own preferred viewing distance when using a VDT. Since reading glasses are normally prescribed for a distance of 30 cm, users may need another set of glasses for VDT work.

TABLE 10.4
Guidelines for the Visual Design of Screen-Based Tasks

Workstation and Work Environment	Provide Visual Relief
	Never position screens against walls or screens
	Design workstations so that users can periodically change focal length and vergence by glancing out of a window or along the length of an open-plan office
	Provide pot plants or other visually complex objects as indoor visual relief. This facilitates automatic recovery from visual fatigue when operators are not looking at the screen
	Provide "activity-permissive" workstations
	Align screens in the correct relation to light sources
Terminal and screen	Use "reverse video" to minimize screen glare[a]
	Reduce remaining screen glare using filters
Job design	Include natural rest breaks
	Increase task variety
Operator selection and assessment	Eye-test new operators for existing visual defects
	Regular eye tests for frequent users[b]

[a] That is, make the screen brighter than the glare.
[b] See Stone (1986).

It is usually recommended that dedicated workers take periodic rest periods throughout the day (5 min/h).

TOOLS AND PROCESSES

Lighting Design Considerations

For visual comfort and to meet visual demands, the following should be considered:

1. Suitable level of illumination
2. Balance of surface luminances
3. Avoidance of glare
4. Temporal uniformity of lighting

The color-rendering properties of light sources might also be taken into account.

Illumination Levels

Although early research indicated that improvements in productivity were possible when illumination levels were increased, more does not necessarily equal better. In fact, high levels of illumination may increase glare and may wash out important visual details. If illumination levels are inadequate, increasing the levels may improve performance, but a region of diminishing returns will be reached as nonvisual limits to performance (such as motivation, fatigue, or manual dexterity) become increasingly important. Older workers generally require higher levels of illumination than younger workers due to the loss of refractive power and changes in the light-transmitting media of the internal structures of their eyes.

Lighting Surveys

HFE specialists are not illumination engineers, but they should be able to use a light meter (Figure 10.8). Light meters can be used to measure illuminance levels on the work surfaces in offices, shops, factories, etc., and the readings can be compared with the levels recommended in published standards or in Table 10.3. HFE Workshop 10.2 summarizes some basic steps in carrying out a lighting survey for ergonomic purposes.

Luminance meters are also commercially available and give the luminance of a small region of any surface they are directed at. Luminance meters are probably not necessary for basic ergonomic lighting surveys in well-lit facilities designed for photopic vision. However, retinal adaptation often depends on the luminances of objects in the visual field—as a rule of thumb, photopic vision requires objects to have luminances above 3 cd/m^2 and scotopic vision requires luminances of 10^{-3} cd/m^2. Mesopic vision occurs at intermediate illuminances.

Balance of Surface Luminances

In practice, a balance of surface luminances is achieved by specifying appropriate illuminances and corresponding reflectances of the surfaces in a room. Detailed specifications can be found in the publications of the IES and ANSI, and examples of this information are included here. Some additional considerations are described below.

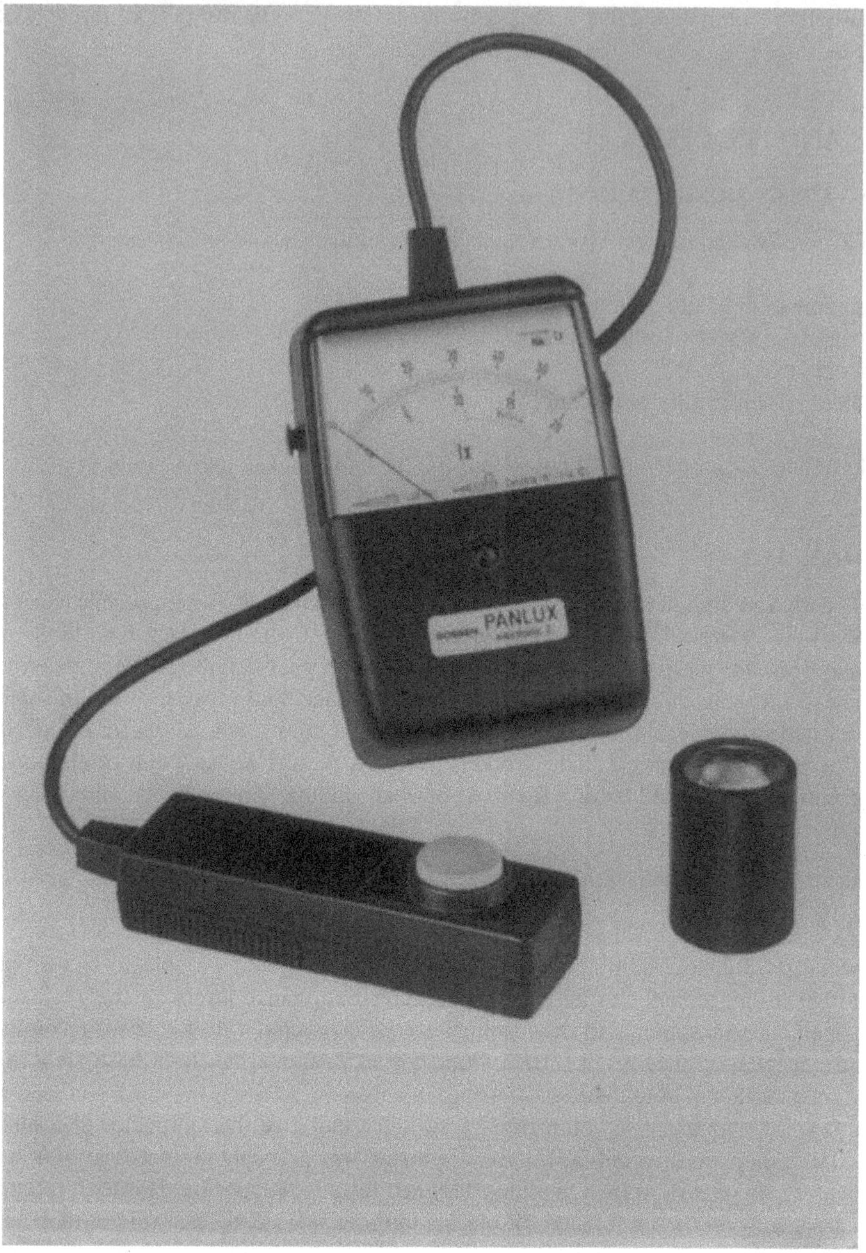

FIGURE 10.8 Light meters suitable for measuring illuminances. (Courtesy of Messrs Gossen-Metrawatt, Nuremburg, Germany.)

HFE WORKSHOP 10.2

BASIC STEPS IN LIGHTING SURVEYS

1. Read the manufacturer's instructions on the light meter first of all.
2. Make sure the light meter has been calibrated.

(Continued)

HFE WORKSHOP 10.2 (*Continued*)

3. Carry out a walk-around inspection and familiarize yourself with the work system and the main sources of light, including light from windows, which may vary throughout the day.
4. Familiarize yourself with the tasks being performed, the main tools, equipment, and materials. Estimate the approximate level of visual demand, including the need to distinguish between different colors. In some cases, a task analysis of particular jobs may be appropriate to gauge the amount of demanding visual work.
5. Make a note of the lighting system (e.g., uplighting fluorescent lamps with parabolic reflectors, luminaires with plastic glare shields, etc.) and its distribution, whether the ceiling is the main source, whether wall-mounted luminaires are provided, and whether there is any task lighting.
6. Note the type of lamps in use (e.g., cool white fluorescent) and the manufacturer's code numbers (this information may be useful later f you want to find out more about the color-rendering properties of the lamp).
7. Ask workers whether they have experienced any light- or vision-related problems, whether any solutions have been attempted, and what the results were.
8. If you are using a battery-powered light meter, make sure the batteries are charged or new, if not rechargeable, and take spare batteries with you.
9. Obtain a plan view map of the area or facility you wish to survey. If surveying an open-plan office, make sure the map is up to date and includes the location of all desks, windows, and all light sources. If a wide variety of visual tasks are conducted in different parts of the facility, note the locations of these tasks on the map.
10. Record the time of day the survey was conducted and the external conditions. Note whether all lamps were functioning properly at the time of the survey or whether any were broken. For surveys carried out in the morning, wait 20 min after the lights have been switched on (some fluorescent and some of the new energy-saving lamps take time to warm up).
11. The main goal of the survey is to check the illuminance levels at the worktops where personnel are working. Secondary goals are to investigate other areas, such as meeting rooms, beverage preparation or dispensing areas, storerooms, etc. Place the light meter on each work surface and record the illuminance in lx on the corresponding area of the plan view map. Repeat this for all work surfaces. Do not interpose yourself between the sensor on the light meter and any sources of light that illuminate the area being assessed.
12. As you visit each workstation, get as close as possible to the actual work area of each occupant and make a visual assessment of any sources of glare, including direct glare from unshielded lamps or indirect glare from the sun shining on a wall or from lamps shining onto reflective surfaces in the workplace. If possible, sit or stand in the actual work position and perform the task yourself for a short time. Note the location of any obstructions (office and industrial lighting systems are normally well designed and are installed before the desks and screens—poor desk/screen layout may be the real cause of any problems, not the lighting system itself).
13. Write the illuminance levels down on the plan view map of the facility.
14. On completion of the survey, relate the measured illumination levels to the position of windows, lamps, etc. Compare these levels with the levels given in Table 10.3. Note any areas where there is insufficient illumination.
15. Daylight usually has little effect on the illuminance inside buildings, except for people sitting next to windows. Repeated measurements may be needed in these cases,

(Continued)

HFE WORKSHOP 10.2 (*Continued*)

at different times of the day and on very sunny or overcast days. Try closing all the blinds and repeating measurements if in doubt about the contribution of daylight or turn off all the inside lights and repeat some of the measurements (the occupants of buildings normally overestimate the contribution of daylight from windows to the illumination inside the building).

16. In cases where the occupants have complained about the lighting, it may be worthwhile to consider issuing diaries where ocular and other symptoms can be recorded over a period and the effects of any improvements monitored. Close visual work and its interaction with the room lighting are often the cause of visual symptoms, so it may be worthwhile asking occupants to record the amount of time they spend each day on different activities.
17. If possible, find out whether the company has a policy on visual testing of the workforce (in Europe, eye testing of VDT operators is a legal requirement). Note whether the occupants wear glasses and when they last had their eyes tested (dissatisfaction with the lighting may be a sign that an eye test is needed).
18. Relate your findings to the standards and guidelines presented in this chapter. Recommendations may include improvements to the lighting system itself, relocation of workspaces/tasks, provision of additional/specialized lighting for particular tasks, improved glare shielding of screens and equipment, and improved curtains/blinds for the control of glare from sources outside the building.

Direct lighting is often used to illuminate sculpture because the shadows produced have a modeling effect. The areas of light and dark enhance the 3D appearance of the works by emphasizing differences in depth. Angled direct lighting is useful for enhancing the surface texture of materials such as cloth or wood and can be used in the production of these materials to aid inspection and quality control. However, direct light can reflect off desktops or surfaces and cause indirect glare and sharp luminance discontinuities between lit and unlit surfaces.

When indirect lighting is used, most of the light is directed onto the ceiling and walls, which reflect it back. Thus, objects are illuminated from many different directions at once, a smoother distribution of luminances is obtained, and shadows are reduced. Direct and indirect lighting can be combined to achieve a balance of surface luminances and minimum glare. In a well-illuminated office, for example, all large objects and major surfaces should have a similar luminance, whereas surfaces in the middle of the visual field should have a contrast ratio of no more than 3:1, and contrasts at the sides of the visual field should be avoided. A balance of surface luminances can best be achieved in practice by using materials with different reflectances in a room. The ANSI recommendations for the reflectances of surfaces in offices are given in Table 10.5.

TABLE 10.5
ANSI Recommended Reflectances for Offices

Surface	Reflectance (%)	Suitable Materials/Finishes
Ceiling	80–90	White paint (matte)
Furniture	25–50	Wood (matte, unpolished)
Upper walls (wall/ceiling border)	80–90	White paint (matte)
Walls	40–60	Neutral, unsaturated hues (e.g., pastel shades)
Business machines	25–50	Matte gray finish
Curtains/blinds	40–60	Cloth
Floor	20–40	Carpet (choose slightly darker colors, beige, brown, gray, etc.)

A balance of surface luminances is also an important design requirement for lighting systems in corridors, stairwells, and outdoor facilities such as railway platforms, which are used at night. Since shadow is a cue to depth, patterns of shadow on the floor can be misinterpreted as changes in ground level and increase the risk of people slipping, tripping, or falling.

Avoidance of Glare

Glare can be reduced by choosing a suitable combination of direct and indirect lighting. With direct lighting, most of the light is directed toward the target in the form of a cone (Figure 10.9). This produces hard shadows and sharp contrasts between illuminated and non-illuminated areas. Indirect lighting is reflected off other surfaces in a room, producing a smoother transition between surface luminances and reducing shadows.

No light sources should appear within the visual field during work activities. Glare control thus depends on both the design of the lighting system and the general workspace and task design. All lamps should have glare shields or shades, and the line of sight from the eye to the lamp should have an angle greater than 30° to the horizontal. Overhead lamps should not reflect off desktops or work surfaces into the operator's eyes. Conversely, reflective desktops and work surfaces should not be used. Wood or matte finishes are preferable. Fluorescent lamps arranged in rows should be arranged parallel to the line of sight to present the minimum luminous surface area possible (Figure 10.10). Generally, using a large number of low-powered lamps rather than a small number of high-powered lamps will result in less glare (Figure 10.11).

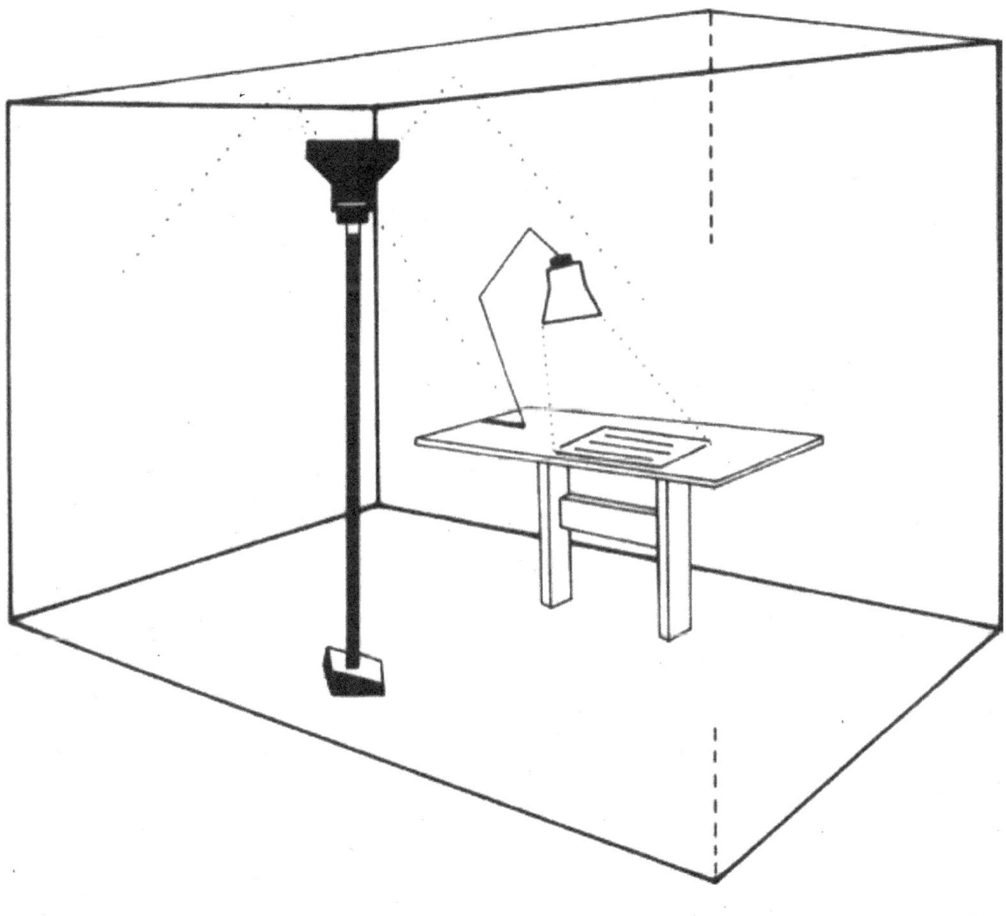

FIGURE 10.9 Direct and indirect lighting.

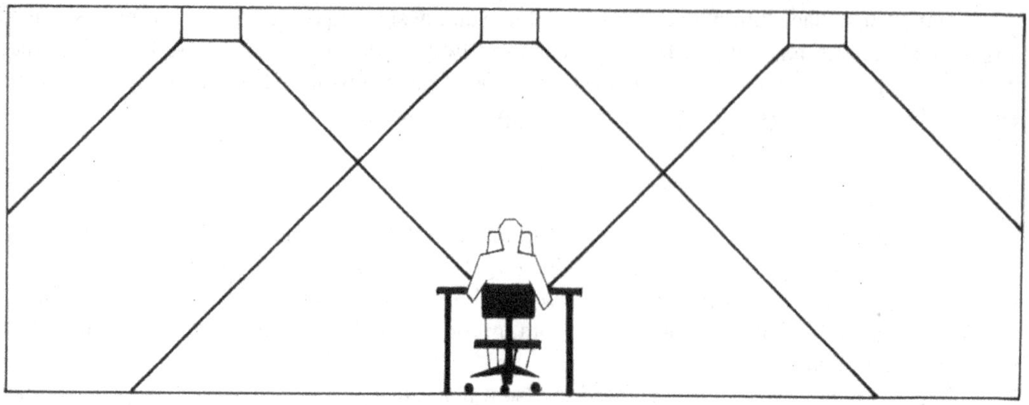

FIGURE 10.10 Place rows of luminaires parallel to the line of sight and the operator's side. Overlapping cones of light ensure uniform distribution of surface illuminances.

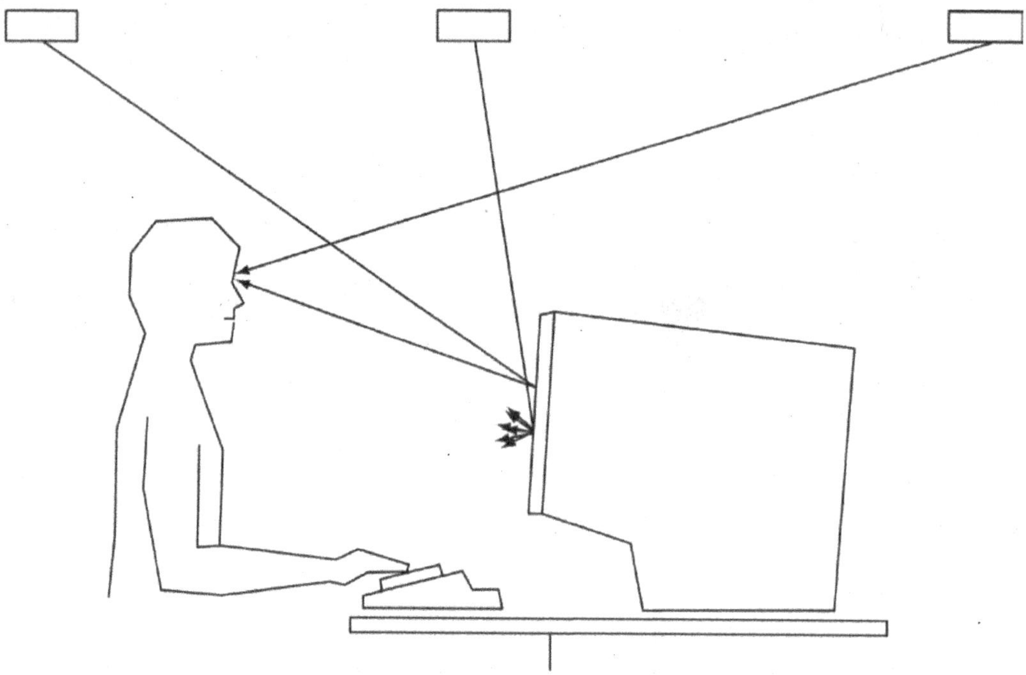

FIGURE 10.11 Avoid direct and reflected glare at VDT workstations.

Glare and Screen-Based Displays

Good lighting is particularly important in offices where VDTs are being used. Traditional offices designed for pencil-and-paperwork usually have rows of fluorescent lamps designed to provide an appropriate level of illuminance on the horizontal desktops below. Desks can usually be placed anywhere in the room since the spacing of lamps is such that the cones of light from each lamp overlap to provide a uniform distribution of illuminance. Since the direction of gaze is downward—toward the horizontal desktop—lamps on the ceiling are more than 30° above the line of sight of a seated worker and do not cause glare.

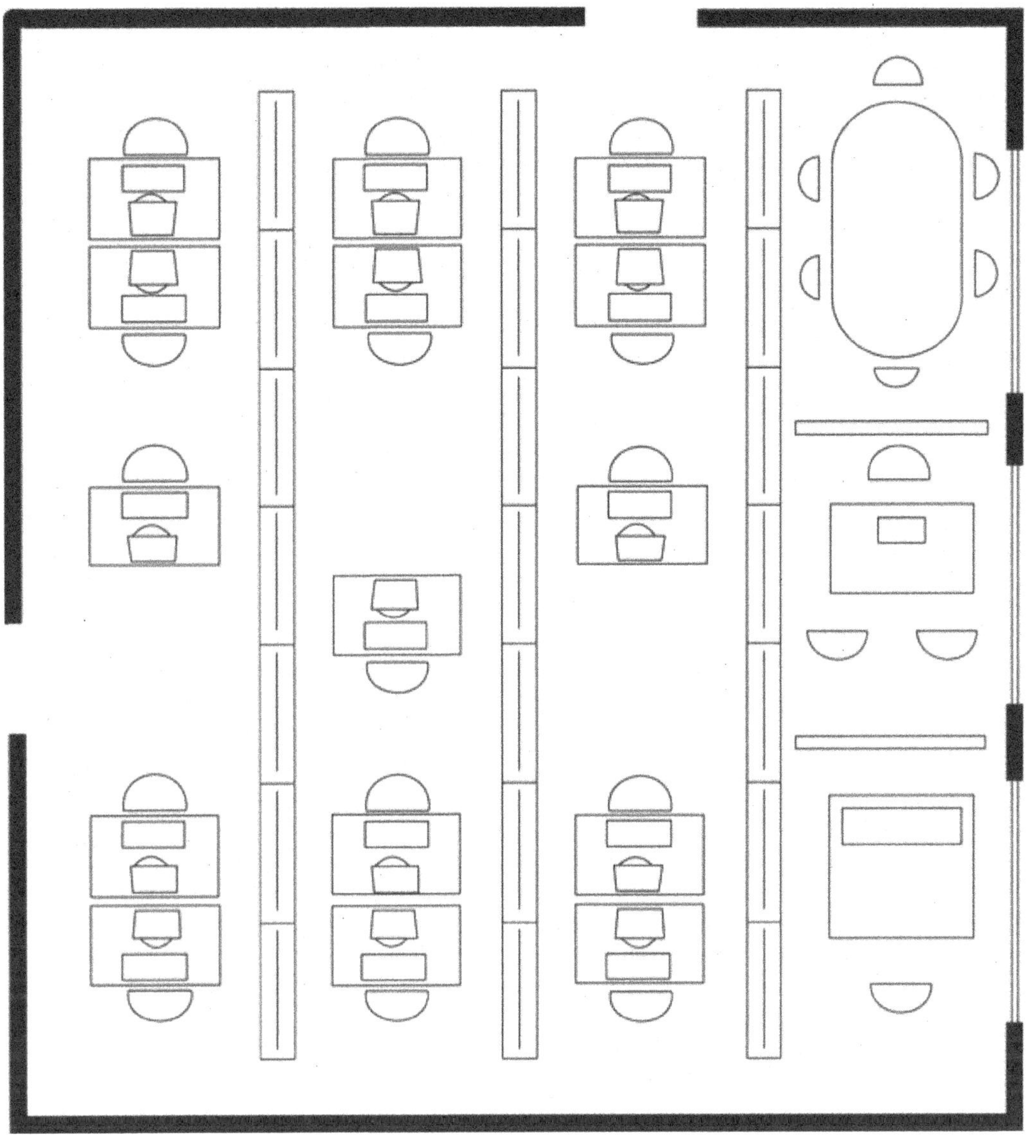

FIGURE 10.12 Layout for VDT and non-VDT work in an open office.

Since a user looks directly ahead at the screen, light from overhead lamps may cause direct glare or reflect off the screen, producing a "veiling effect"—reducing the contrast between the characters and the background. The correct positioning of workstations with respect to windows, lamps, and bright surfaces is therefore very important (Figure 10.12). Rows of fluorescent lights should be parallel to and on either side of the operator–screen axis to reduce direct and indirect glare. VDU workstations should be at right angles to windows, and more use should be made of indirect lighting.

At the workstation itself, sharp contrasts due to the dark screen (i.e., light characters against a dark background) and bright source documents can be problematic. Since the VDU screen emits its own light, general illumination can be lower in VDU workspaces than in traditional offices. Low-reflectance surfaces can also be used (particularly behind the screen), and task lighting (e.g., desk lamps) can be provided at each workstation to support the visually demanding aspects of the work.

Temporal Uniformity of Lighting

Fluctuating luminances can be more disturbing than static contrasts. The temporal uniformity of lighting can be influenced by both the characteristics of the light source and the requirements of the task. Incandescent bulbs radiate light fairly uniformly over time, whereas fluorescent lamps are known to flicker. Fluorescent lamps work by passing an electric current through gas in a glass tube. The gas emits visible (almost monochromatic) light and ultraviolet (UV) radiation periodically in accordance with the mains frequency. The UV radiation excites a phosphor on the inside of the glass tube that emits visible light—different phosphors are used to produce the various shades of commercially available lamps. The flicker frequency of correctly functioning fluorescent lights exceeds the threshold for the perception of flicker. However, malfunctioning old lamps can produce a visible flicker, which may cause visual discomfort. In factories, flicker is a hazard since it can have a stroboscopic effect—rotating or oscillating machine parts may appear stationary or move more slowly if their frequency is similar to that of the flickering source that illuminates them. Phase shifting of fluorescent tubes (using banks of two tubes and alternating the phase of each) and regular inspection/maintenance can prevent flicker problems.

Color Rendering and Artificial Light

The apparent color of an object depends on the spectral composition of the incident light and the wavelengths the object reflects. For example, in broad-spectrum (white) light, an object that reflects long wavelengths and absorbs the rest would appear red. The same object, when viewed under artificial light deficient in long wavelengths, would appear reddish gray or brown. An object that appears bright red under white light would appear black when illuminated by monochromatic blue light.

The light emitted by a source can be described in terms of the relative amounts of different wavelengths of which it is composed. Monochromatic sources emit a narrowband of wavelengths. Objects that reflect this wavelength will appear very bright when illuminated by the monochromatic source. Objects that do not reflect the wavelength at all will appear black. For example, sodium streetlights emit a narrow band of wavelengths and appear yellow. Most objects illuminated by them appear in various shades of yellow/gray and can only be distinguished by the contrast or brightness differences between them. A monochromatic light source can be said to have poor color-rendering ability because it does not reveal the way that objects differentially reflect incident light according to its wavelength composition.

To have good color-rendering properties, a light source must contain sufficient amounts of light across the visible spectrum. The term spectral energy distribution is used to describe the relative amounts of different wavelengths that a light source emits. Figure 10.13 depicts spectra for a variety of light sources.

Daylight, incandescent light, and some fluorescent tubes have broad emission spectra and therefore good color-rendering properties. The color-rendering properties of a light source cannot be determined subjectively by the appearance of the light itself. Although green and red lights can be mixed to give a yellow light of similar appearance to a monochromatic source, the color-rendering properties of the two sources are completely different. The color of the red and green mix will render the colors of ripening apples very well, whereas under the monochromatic yellow light, they will appear dull and grayish yellow.

Selection of light sources for good color rendering should be based on objective data about the emission spectra of the source. In practice, an index—the color-rendering index—has been devised to describe the color-rendering properties of light sources. Daylight is used as a standard and given a value of 100. The color-rendering properties of other sources are expressed with respect to this standard (Boyce, 1982; Howarth, 1990). The high value of the color-rendering index usually indicates that a light source has good color-rendering ability. The apparent color of an object depends very strongly on the properties of the incident light. Under red light, for example, white appears pink, red colors are greatly intensified, and blue appears purple.

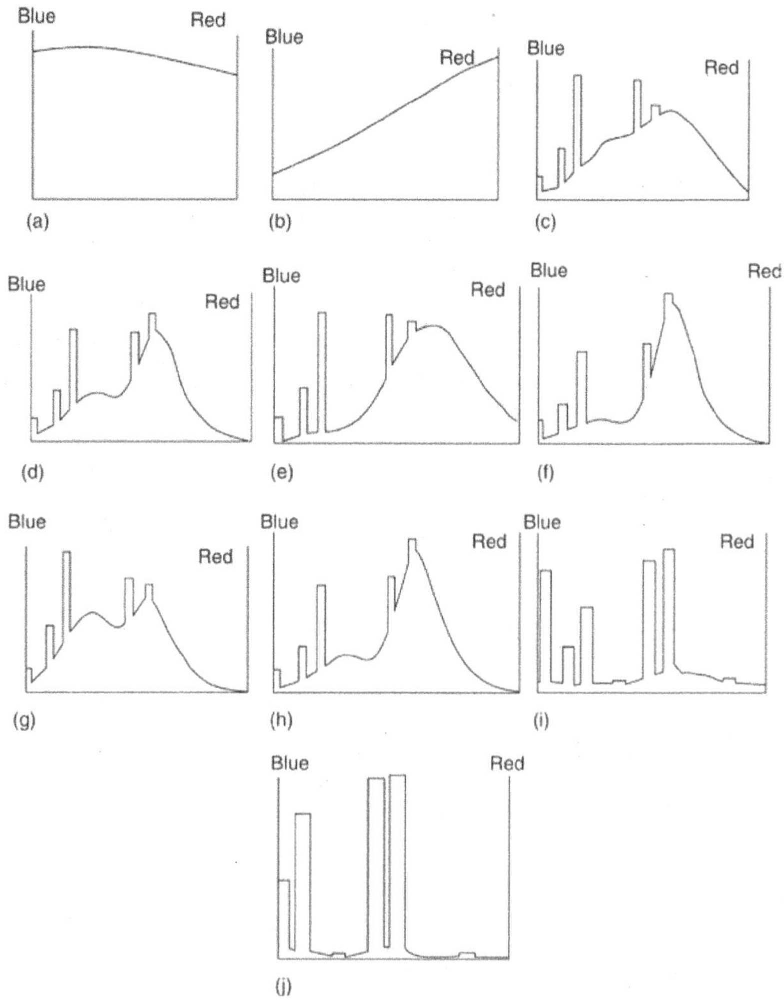

FIGURE 10.13 Spectra of a variety of light sources: (a) northern skylight, (b) incandescent lamp, (c) deluxe cool white, (d) cool white, (e) delxe warm white, (f) warm white, (g) daylight fluorescent, (h) white, (i) mercury lamp, and (j) mercury lamp. Note that mercury lamps (i and j) look white but have poor color rendering and are deficient in long (red) wavelengths. (Courtesy of Messers. Osram S.A., PTY Ltd.)

Incandescent lamps have good color-rendering properties and strengthen the appearance of red, pink, orange, and yellow objects, but bring out blue and green less well. Standard cool white, fluorescent lamps bring out the latter colors but not red, pink, and orange. A lamp's color-rendering property is an important consideration where color identification is a task requirement, as in the assembly and inspection of wiring looms, fruit inspection, and textile and garment manufacture. Dental laboratories, florists, and paint shops are examples of places where good color-rendering light is needed. In the retail industry, lighting is used to create displays that enhance the appearance of products such as clothing or food (meat is usually illuminated using lights that strengthen red).

The color-rendering properties of lights also influence the "atmosphere" in a room. Although there is still debate about whether environmental color has psychological effects, the ambient illumination can have social implications because of the way different light sources render the complexion of a person's face. Incandescent lights enhance skin tones and can facilitate social interaction through the creation of a warmer atmosphere. Owing to their lower luminous efficacy and higher heat production, incandescent lamps are not normally used for general lighting in offices

TABLE 10.6
Color Rendering and Luminous Efficacy of Lamps

Lamp	CRI	Luminous Efficacy	Applications/Comments
Incandescent	90–99	Low	Home, hotels, rooms dedicated to social interaction
		Fluorescent	
Color matching	85–100	Medium	Florists, paint shops, etc.
White	50–70	Medium	General purpose. Factories, shops, etc. Reds appear orange, violet/blue is dulled
Delxe warm white[a]	70–85	Low–medium	Violet/blue weakened. Creates a relaxed, informal atmosphere similar to incandescent
Tungsten–halogen	85–95	Low	Have many applications where good color rendering is needed
Mercury fluorescent	50–60	Medium	General illumination. Long life. Weakens red
High-pressure sodium	25–40	Medium–high	Background/outdoor illumination
Low-pressure sodium	Very low	High	Street lighting. Almost no color rendering
High-pressure metal halide	Low	High	Floodlighting, warehouse lighting, stadiums

[a] Fluorescent lamps last five times or more as long as incandescent lamps.

and factories. However, they may be used to advantage in interview and meeting rooms where social interaction occurs, and in areas such as lifts and toilets where mirrors are often fitted onto the walls. Table 10.6 summarizes some color-rendering properties and applications for different commercially available lamps.

Color Temperature and Color Rendering

Color Temperature: The color temperature of a light source, expressed in degrees Kelvin, is the temperature of an ideal black body radiator that would emit light of a comparable hue to the source. Lamps with low color temperatures appear "warm" (with more long-wave visible radiation), whereas lamps with a high color temperature appear "cool," as in the following examples: candles have a color temperature of about: 2000 K; daylight is typically 5500–6500 K. LEDs are sold in a variety of color temperatures: 2700, 2800, 3000, 4000 or 5000 K.

Color Temperature	Source of Light
1000 K	Candlelight
1700 K	Match flame, low-pressure sodium lamp
3000 K	Warm white fluorescent of LED lamps
5000 K	Cool white, fluorescent lamps
6500 K	Daylight (overcast)

In general, decisions about the color temperature to use in an application depend on the need for color contrast and the desired "atmosphere" in the room. Because the eye is less sensitive to light at the far ends of the visible spectrum, lamps with good color-rendering properties have had lower luminous efficacies than those with lower color-rendering properties (because, with the illuminance being equal, they emit more light at wavelengths to which the eye is less sensitive). Thus, in cases where high illuminance is required, poor color rendering will be more noticeable than in cases where low illuminance is required. Paradoxically, the apparent cost-effectiveness of using more efficient lamps for high-illuminance applications will be negated by the loss of color discrimination. In effect, people are far more sensitive to drops in color rendering than to changes in illuminance. As an example, Papamichael et al. (2016) demonstrate that a drop in color rendering index of a light source (from 100 to 80) is far more noticeable than a drop in illuminance from 1000 to 800 lx.

Surveying Satisfaction with Lighting

User satisfaction with indoor lighting can be investigated using questionnaire surveys or focus groups. Table 10.7 gives examples of items and issues to be investigated (Dianat et al., 2013).

Steenstra et al. (2009) devised a short questionnaire for assessing the prevalence of eye complaints in work with display screens (Table 10.8). The questionnaire is valid and reliable and can be used to assess the extent of any problems and the benefits of any interventions taken to improve the visual environment. The scores can be summed to give an overall measure of eye complaints (if one symptom is present in a group of employees, the others are likely to be present as well). If surveys are repeated to assess improvements, the questionnaire should be administered at the same time of day.

TABLE 10.7
Questions for Lighting Surveys and Focus Groups

How is the light in your workstation?
How satisfied are you with the lighting in your work environment?
How satisfied are you with the lighting in your workstation?
Does the lighting affect your work performance?
Does the light disturb your work performance?
Does the lighting level cause slips, trips, and falls?
Does the lighting cause tiredness in the eyes?
Do you need to change your posture for better viewing because of the lighting?
Additional artificial light would improve my work environment
More windows, natural light would improve my work environment
Better maintenance of lighting would improve my work environment
A better mix of natural and artificial light would improve my work environment
Better use of color would improve my work environment
Any other suggestions?

TABLE 10.8
Items from the Eye Complaint Questionnaire[a]

1. I have difficulty seeing
2. My eyelids feel heavy
3. My eyes hurt
4. My eyes water
5. My eyes burn
6. My eyes are tired
7. I have a strange feeling around my eyes
8. My eyes itch
9. My eyes are red
10. My eyes are dry

Note: Scored on a seven-point Likert scale, where 1 = "not at all" and 7 = "very much."

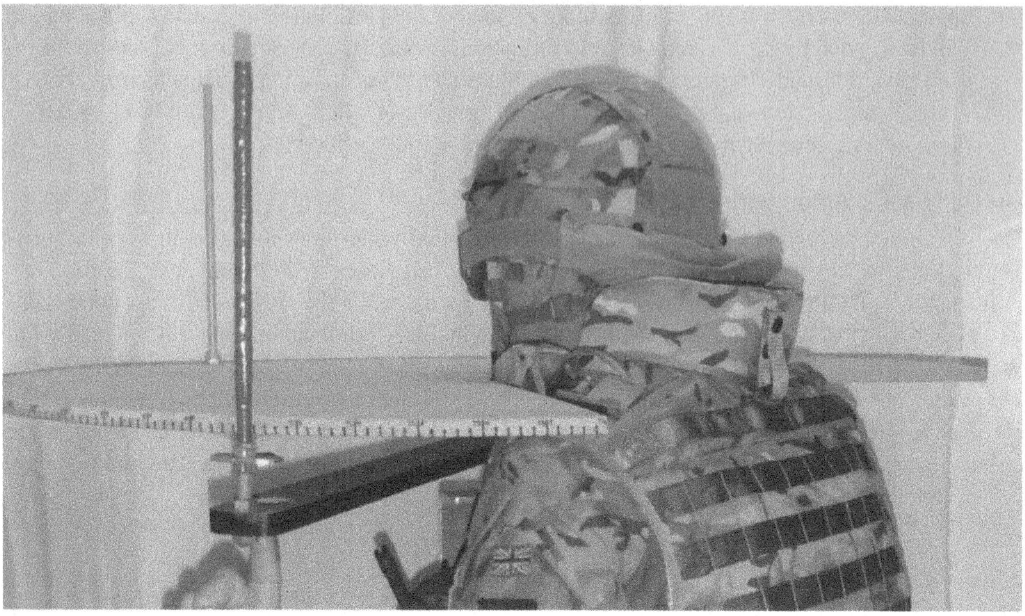

FIGURE 10.14 Testing the effect of headgear on peripheral vision. The subject fixates the rod directly ahead. The experimenter rotates the target toward the center, and the subject indicates when the target is first detected.

Perimetry: Peripheral Vision Testing

In most countries, there are minimum visual requirements for driving. Peripheral vision is one of them because drivers must not only be able to see the road ahead, but also be able to detect targets at the side while looking at the road ahead (e.g., vehicles joining the road, cyclists, and pedestrians). In the United Kingdom, the limits (minimum visual requirements) are 60° of peripheral vision on each side and 20° above and below. Perimetry is used to measure peripheral vision and can also be used to assess the effects of wearing different kinds of headgear, such as helmets and goggles (Figure 10.14). There is sometimes a trade-off between protecting the head and eyes of the wearer and preserving peripheral vision, the loss of which might incur new risks.

SYSTEM INTEGRATION

Illumination Levels

Studies were conducted by the Industrial Health (Fatigue) Research Board in the United Kingdom in the early 1900s. Weston and Adams (1927) studied the hourly output of linen weavers in relation to the intensity of daylight (the workplace had no artificial lighting). Not surprisingly, production decreased markedly with decreasing daylight. Similar results were found for typesetters. Note that both of these jobs are visually demanding.

The principle of diminishing returns was first demonstrated by Weston in 1945 in a laboratory task that involved the use of Landolt rings (subjects had to inspect a series of cards with circles drawn on them. Their task was to detect circles with small gaps in the circumference at the 12 o'clock position). Performance improved as the lighting levels were increased, then leveled off. Thus, for any task, an optimum level of illuminance can be specified that gives maximum performance.

Effects of Rest Breaks on Visual Symptoms and Productivity

Woods (2005) reports that intensive screen-based tasks such as data entry are associated with high yearly and weekly prevalence rates for musculoskeletal (87% and 56%, respectively) and visual symptoms (47% and 25%, respectively). Galinsky et al. (2000) investigated the use of supplementary rest breaks for data entry operators. Under the conventional system, operators worked 8.5 h/day with 30 min for lunch and a 15 min break each morning and afternoon. Every 4 weeks over the trial period, operators alternated between this and a system of supplementary breaks, consisting of four additional 5-min breaks (two in the morning and two in the afternoon on either side of the 15-min break). Significant reductions in musculoskeletal discomfort and eye soreness were found when the supplementary break system was used. Despite the 20-min reduction in the workday when the supplementary system was used, there was no reduction in the number of documents produced per day, nor in the number of errors. A possible reason is that the keystroke rate increased under the supplementary break system. It seems, then, that symptoms can be reduced by increasing the number of breaks without having a deleterious effect on productivity. It may be that the supplementary break system simply recognizes and formalizes what operators do anyway—take a break when they feel tired or uncomfortable.

Alerting Effects of Light

"Alertness" means more than just "wakefulness" (the opposite of sleepiness), it means the ability to sustain a level of performance on a task, normally a cognitive task. Mu et al. (2022) conducted a systematic literature review of studies on lighting levels and color temperature and their effects on alertness. Alertness was measured in some studies by self-report measures and in others by performance measures such as reaction time. Sixty-eight studies were selected in four categories: alerting effects of polychromatic light intensity; alerting effects of monochromatic light intensity; and alerting effects of polychromatic light spectrum. The findings of all these were very mixed, although some general conclusions can be drawn. In one study, increases in alertness at the end of a night shift occurred during a narrow range from 50 to 200 lx at the eye. Differences in subjective alertness under illuminance between 50 and 500 lx have been found. Studies that failed to find significant effects were *range-limited* in the sense that the lowest lighting levels were too high to reduce alertness. Increases in the intensity of polychromatic white light have been found to increase alertness in several studies, at a transition intensity of about 100 lx, irrespective of time of day or exposure time. Regarding monochromatic light, there is some evidence that reaction time is faster under blue monochromatic light (Souman et al., 2018). Based on their meta-analysis of 29 studies, Mu et al. (2022) concluded that higher color temperature light improves alertness better than lower color temperature light for both daytime and nighttime exposure.

Beyond Standards: Drop the Desktop

For many years, standards for office and industrial lighting focused on the delivery of a specified level of illuminance at the worktop in order to provide adequate task lighting cost-effectively. The reflectance and illuminance of the remainder of the workplace were specified to provide a balance of surface illuminances, devoid of glare, such that an optimal level of retinal adaptation could be maintained. Kelly et al. (2020) have recently questioned this approach for office lighting because most office workers work primarily with computers or laptops with their own illuminated displays. Therefore, the lighting level on the desktop is unlikely to affect the visibility of the screen-based task. They argue that a paradigm shift is needed and that the focus must shift to lighting the space rather than the task, partly because most office tasks nowadays don't require a high level of visual acuity. In addition, focusing on illuminating horizontal desktops that employees rarely need to look at is an inefficient use of energy and unlikely to be cost-effective. They argue that the focus must shift to ambient lighting rather than desktop lighting, with the emphasis on providing a pleasant, healthy visual environment.

Personalized Lighting

Juslen et al. (2007) investigated the effects of individually controllable task lighting on the productivity of assembly workers in a luminaire factory. The general lighting in the assembly hall delivered 100–380 lx to the workstations. Task lighting augmented the illuminance at the horizontal workstations to 700 lx. A controllable task lighting system was introduced in one assembly area so that workers in that area were able to increase the illuminance at their own workstations up to 3000 lx by pressing a button on an infrared controller. The daily output of workers with controllable task lighting was measured over a 16-month period and compared with that of workers with standard lighting. Workers' use of the infrared controller was also recorded: both the time spent pressing the control button and the illuminance achieved. There was a small positive correlation between worktop illuminance and productivity within the test group, and productivity was 4.5% greater, compared to workers with standard lighting. There was evidence that workers had preferred levels of illuminance and that they were able to select these levels consistently over the time period. The experimenters randomly alternated the dimming speed of the controllable system so that it took either 5 or 9 s to reach maximum illumination. Workers were found to reset the lighting to a slightly higher level (mean 1359 lx compared with 1181 lx) with the fast controller compared to the slow controller, but did press the slow controller for longer (about 5 s each time, compared to 3 s). This is evidence of intentional behavior aimed at reaching a target level of illumination, rather than just increasing the lighting for its own sake. Although the mean level of illuminance in the test group was greater than that in the comparison group (over 1000 lx compared with 700 lx), the standard deviations were large (over 600 lx), which suggests that there was great variation in preferred illuminance among the subjects in the test group. This, in itself, indirectly supports the concept of user-controllable task lighting (there is great variability across workers in their preferred levels of illumination, but individual preference varied little over time). Owing to the duration of the study, it is unlikely that the findings were due to the novelty effect of the controllable lighting or to the demand characteristics of the trial (see Chapter 16).

Unfortunately, the researchers did not carry out a cost–benefit study of the controllable lighting in relation to productivity.

There is no shortage of evidence that people differ in their preferences for lighting when carrying out tasks. Dick et al. (2024) experimented in a simulated office environment. Subjects worked at computers for two hours on two separate days with either fixed lighting (500 lx on the desk and 5300 K color temperature) or adjustable lighting in terms of illuminance or color temperature (60–6000 lx horizontally at the desk and between 2200 and 5300 K color temperature). In the personalized lighting condition, subjects could set the lighting levels on the desktop and the back wall behind the desk. Subjects carried out several tasks: a psychomotor vigilance task measuring reaction time, a memory task, and two tests of executive function. Heart rate and heart rate variability were also measured, as was sleepiness using a tool based on the Karolinska Sleepiness Scale and a measure of positive and negative effect (the PANAS scale). Symptoms of eye discomfort were also measured using Likert scales.

Most subjects selected warm white lighting for the backwall and the desk (means of 3259 and 3148 K, respectively). Mean self-selected illuminances for the back wall and desk were 1562 and 2603 lx, respectively. In general, subjects chose much higher illuminance levels compared to the control condition (mean preferred illuminance over 900 lx; mean melanopic EDI 381 lx). Under the control condition, the melanopic EDI was much lower (less than 100 lx). All subjects chose melanopic EDIs greater than in the control condition (at least double).

Personalized lighting had very little effect on the cognitive tests (although there was better performance on the memory task), and there were no statistically significant effects on heart rate, eye complaints or subjective responses. These findings may be due, in part, to the fact that subjects only worked for two hours at a time in each condition, which may not have been long enough for the differences in lighting to have any effects on alertness, mood, or the development of visual discomfort. Also, the subjects worked at a computer with a self-illuminated screen; therefore, the lighting was not "task lighting" as is generally understood (as in the study by Julsen, above), and any effects on task performance would only be indirect. Further, the lighting in the control condition *did* meet current standards and was therefore adequate for the task. However, a number of interesting conclusions can be drawn. First, the self-selected lighting levels were much higher than the 500 lx control condition, which may be a reflection of the fact that subjects chose warm white colors to achieve a mean melanopic EDI of 381 lx (greater than the recommended melanopic standard of 250 lx).

Chen et al. (2022) carried out a study of illuminance and color temperature in office workers in a smart building over two days. Subjects had a smartphone app that they could use to set the illuminance and color temperature of the lighting at their desks while carrying out cognitive tasks. Subjective and physiological measures were also taken. On the first and third of three half-day sessions, subjects worked under fixed lighting conditions (280 lx 3000 K; 210 lx 4000 K; 175 lx 5500 K in the first session, and 590 lx 3000 K; 420 lx 4000 K; 350 lx 5500 K in the third session). These conditions were converted into "Equivalent Melanopic Lx" or "EML" values of 120, 120, and 145 EML in the first session, and 250, 250 and 300 EML in the third session. In the first half of the second session, the subjects began with 100 lx 3000 K and could then set the illumination to their own preferences; in the second half, they set the illumination from starting conditions of 750 lx 5500 K.

Color temperature had no effect on task performance unlike illumination level which had a great effect (590 lx). For comfort, subjects preferred warm lighting (3000 K) and bright lights (590 lx). This has an EML level of 250 which exceeds the recommended value of 150. The mean values for self-selected lighting levels were 480 lx and 4200 K. Interestingly, preferred illuminance levels did not vary much depending on the starting position (high or low), suggesting that personal preferences for lighting in the office are fairly stable.

The benefits of personalized lighting for tasks where the display is illuminated (computer screens and mobile devices) are probably subtle. Most of the studies are short-term, and many of the existing lighting standards, which constrain the kinds of questions researchers can ask, are valid and satisfactory for the most part. Therefore, the benefits of personalized lighting are not easy to demonstrate. However, it seems clear that people do have preferences, which are likely to be stable over long periods, and in "smart" offices at least, there is increased scope for accommodating these preferences.

Light and Health

The findings of both laboratory and field studies indicate that people are satisfied with illumination levels of about 400 lx. Further increases in illuminance bring only modest increases in satisfaction. Glare is associated with dissatisfaction.

There is also evidence to suggest that people dislike uniform lighting. Light from windows varies throughout the day. Not only does the intensity change, but so does the spectral composition—sunlight differs from blue skylight, which differs from light from an overcast sky. Reflections from objects and buildings also add variety. It is often suggested that uniform lighting leads to boredom, whereas changes in light have a stimulating effect.

The contribution of daylight from windows to the illuminance on indoor surfaces is usually much less than it appears to be (although direct sunshine can cause severe glare on surfaces such as VDT screens, chromed surfaces, etc.). However, anecdotal evidence suggests that windowless workspaces are associated with dissatisfaction. Many indoor workers appear to believe that daylight is superior to artificial light or that working under artificial light is unhealthy. Windows are a valuable source of visual relief and seem to have psychological value. By conveying information about changes in the weather and day/night cycle, they provide continuity and contact with the outside world. Boubekri et al. (2014) compared office workers in a windowless office with those working in offices with more daylight. The former reported a poorer quality of life, less vitality, poorer sleep quality, and more sleep disturbances. Workers with windows at the workplace had more light exposure during the workweek, a trend toward more physical activity, and longer sleep duration as measured by actigraphy. The mechanism underlying the benefits of daylight exposure may be that daylight has a higher color temperature than artificial light and therefore has a greater effect on melanopsin causing better circadian entrainment. In a field-based experimental study, Viola et al. (2008) exposed office workers to blue-enriched artificial light (color temperature 17,000 K) and found improved self-reported alertness, performance, and sleep quality compared to exposure to light at 4000 K. Blue light (8000 K) can aid the entrainment of circadian rhythms (Vetter et al., 2010) compared to warmer white light (4000 K), and thus help shift workers remain alert during night shifts. To aid daytime sleeping, blue light should be avoided at all times (including on the journey home after sunrise, by wearing sunglasses that block the shorter wavelengths).

Seasonal Affective Disorder (SAD)

People living in temperate latitudes are subject to wide variation in daily exposure to natural light over the course of the year. In animals, the level of physical activity and behaviors such as reproduction and migration are thought to be triggered by the daily duration of sunlight. Changes in human mood and behavior may also exhibit seasonal variation. SAD is a mood disorder that occurs at specific times of the year. Depressive symptoms usually begin in late fall or winter and spontaneously disappear in the spring or early summer. It is more common in females, younger adults (18–30 years of age), and in identical rather than fraternal twins (suggesting a genetic component; Galima et al., 2020). Bright light therapy (BLT) is a standard treatment for SAD—typically 30–60 min/day of full-spectrum light at 2500–10,000 lx—and sufferers are encouraged to spend more time outdoors and to exercise. It is thought that a lack of bright daylight exposure (because people stay indoors, sunlight is weaker, and so on) disrupts the action of retinal basal ganglia and the production of melanopsin. It is thought that one effect is to disrupt the uptake in the brain of a hormone called serotonin, which controls mood.

Pjrek et al. (2020) conducted a meta-analysis of randomized-controlled trials of BLT therapy for SAD. Eighteen of 19 trials used lightboxes for BLT (light-emitting panels that the user sits in front of), with illuminances ranging from 1350 to 10,000 lx. Placebo treatments included lightboxes set at 400–500 lx and a sham negative ion generator was used in five studies. BLT was superior to placebo treatments, with a small to medium effect size (benefit).

View out of a Window

Ko et al. (2020) carried out an experiment to investigate the effects of a view out of a window on thermal comfort, cognitive performance, and emotion. It is believed that a view out of a window creates a positive response via a feeling of connectedness with the outside world. Thermal comfort, emotional state, eye symptoms, and perceived stress were measured while subjects worked in a room without a window and, in counterbalanced order, a room with a window. A battery of cognitive tests was also administered. The subjects were not allowed to bring personal electronic devices with them, since these might have been visually distracting. The visual and thermal environments were comparable in both offices. However, 12% more of those working in the office with windows reported thermal comfort than in the windowless office and 7% more reported that the thermal environment was pleasant, although participants' skin temperatures were the same in both conditions. Positive emotion scores were higher and negative scores lower when working in the office with windows. Scores on a test of working memory were 6% higher, and on a test of concentration, 5% higher in those working in the office with windows. Scores on a test of eye symptoms were 10% lower when working in the office with windows, although there was no difference in perceived stress scores. The authors offered a number of explanations for their findings. A view from a window may distract occupants from uncomfortable indoor conditions, or it may just compensate for these (i.e., people will accept less-than-optimal conditions indoors when they appreciate the view outside). The differences in cognitive performance may have been due to an *attention restoration* effect (Chapter 12). Although the authors caution that the effect sizes were small, it should also be borne in mind that the exposure durations were very short. To observe small effects with short exposures might be taken to imply that the effects are powerful, although slow-acting, and there may be long-term disbenefits of working in windowless offices.

Vitamin D and Calcium Metabolism

Light is known to have biochemical and behavioral effects on humans in addition to "driving" vision. Vitamin D is produced in the skin by the action of UV(B) radiation at 280–315 nm. Vitamin D is required for normal calcium metabolism in humans. Unlike other vitamins, it is synthesized in the human body and is absent from all natural foods except egg yolks, oily fish, cheese and mushrooms. It is formed by the action of sunlight on the skin. People who live beyond 37° latitude (north or south of the equator) are at risk of vitamin D deficiency at all times of the year except during the summer months. White-skinned people can obtain sufficient daily vitamin D following exposure of the hands and face to the sun for approximately 15–20 min/day. Vitamin D is stored in fatty tissue, and stores can last for months. However, people with dark skin, those who rarely go outdoors in the summer or at midday, and people who cover their hands and faces (with clothing or sunscreen) when they go outdoors are at risk of vitamin D deficiency. People living in temperate latitudes exhibit seasonal variation in intestinal calcium absorption with the lowest levels occurring in winter. This was one of the reasons for the fortification of dairy products with vitamin D, which was introduced to reduce the incidence of vitamin D deficiency bone disease ("rickets") in children.

A large proportion of the working population works indoors and is therefore at risk of vitamin D deficiency due to inadequate exposure to natural light. Most artificial lamps do not emit radiation of the required wavelength. However, "full spectrum" lamps and light boxes are commercially available.

SAFETY

Hom et al. (2022) report that approximately 20,000 workplace eye injuries occur annually in the United States, with an estimated yearly cost of 300 million dollars. Seventy-eight percent occurred in males and most of these injuries are preventable. Elsewhere, injury rates vary from 20% to 46%. The main causes were contact with objects and exposure to harmful substances.

Eye injuries can be very expensive, both in terms of the disablement of workers and loss of skills, and in terms of workers' compensation payments. Kyriakaki et al. (2021) reviewed 17 articles on occupational eye injuries. The high-risk occupational groups for injury were welders, farmers, metalworkers and grinders, construction workers, and manufacturing workers. Chemical burns required the longest recovery times. Most of the injuries could have been prevented by using suitable protective eye devices (PEDs). Several factors were identified that increased the risk of injury due to failure to use PEDs. These included situational factors, such as the availability and accessibility of the device at the required time. In some countries, affordability was also a factor. Climatic conditions that result in discomfort when PPE is worn were another factor, as was failure to comply with safety regulations in general. However, compliance seems to improve after an injury has taken place followed by recovery. Finally, the authors emphasize that it is important to educate workers about the risks and possible consequences of injury.

Target Detection on the Road

"Smombies" (smartphone zombies) are pedestrians who look at their phones while navigating urban spaces. New displays have been developed to alert them when crossing the road (Figure 10.15).

Direct Vision Standards for Heavy Goods Vehicles (HGVs): Following a number of accidents, sometimes fatal, involving HGVs and vulnerable road users (such as cyclists), a direct vision standard has been introduced. HGVs entering London (UK) require a permit, which is obtained if the vehicle meets the standard. Vehicles are rated according to a 5-star system depending on the driver's ability to detect other road users. Vehicles not meeting the standard (less than 3 stars) require retrofitting with safety features such as "slide under protection" so that vulnerable users cannot be swept underneath a passing HGV; cameras to monitor the nearside of the vehicle and additional mirrors; and auditory warnings to indicate turning. Figure 10.16 gives an example of one of the assessment criteria used by the standard to assign a star rating and determine the requirements for retrofitting (if any) in order to obtain a permit.

FIGURE 10.15 By mounting the display on the sidewalk, it is in the "straight-ahead line of sight" of a "Smombie" looking at his/her phone before crossing the road.

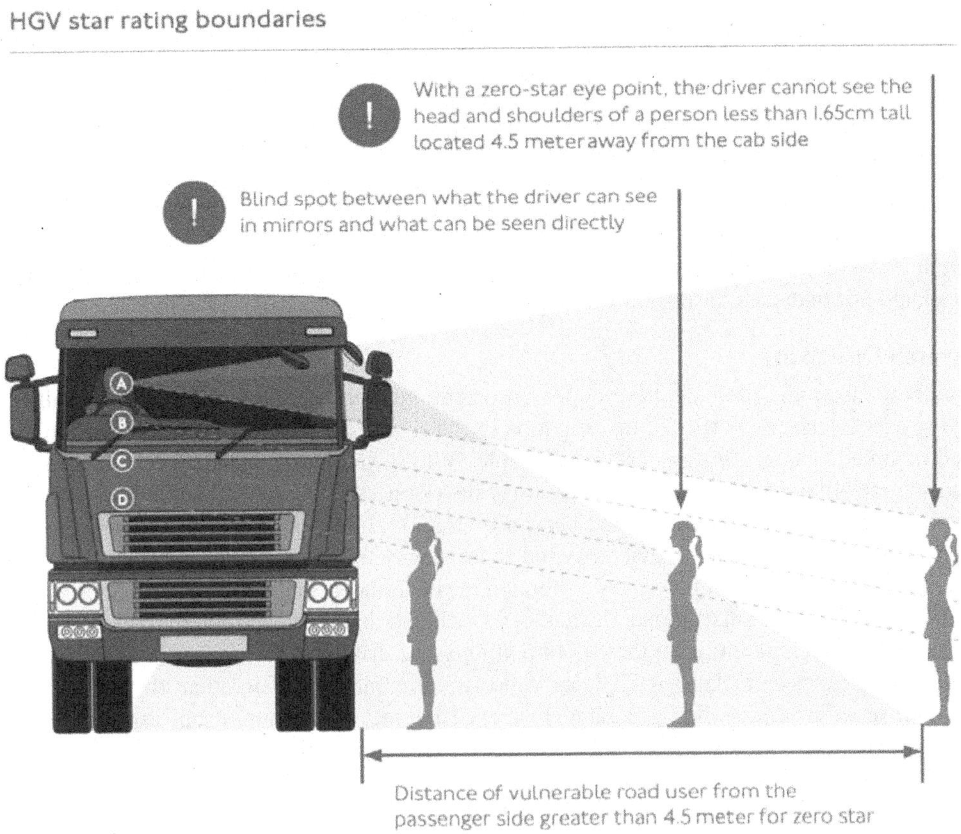

Sight line	Description
A	Zero-star eye point
B	One-star eye point
C	Three-star eye point
D	Five-star eye point

FIGURE 10.16 Lateral lines of sight to detect vulnerable road users. (From: *Direct Vision Standard: Guidance for Operators*. Property of Transport for London reproduced with kind permission of TfL.)

Effectiveness and Cost-Effectiveness

It amounts to common sense to say that light is essential for the completion of all tasks that depend on vision for their successful performance. The main economic considerations concern the cost of the lighting in relation to the benefits in terms of increased output and better-quality work. There is a U-shaped cost–benefit function that describes the relation between lighting and the performance of tasks that demand vision. With no light, performance is impossible. Performance increases rapidly as more light is provided but reaches a point of diminishing returns where additional light brings no further benefit. Further increases in lighting then cause glare, and performance deteriorates.

In developed countries, the cost of lighting is cheap in relation to the value added by the illuminated tasks, and so economic benefits can usually be expected when optimum lighting levels are provided. The illuminance levels recommended in this chapter have been designed to give optimum lighting for different situations—maximum performance with no waste. However, they do not account for individual differences in visual performance.

Status of Methods in Risk Assessment and Task Design

Valid and reliable instruments are available to assess visual function and the visual environment. Visual tests are intrusive and have to be carried out by qualified personnel. Instrumentation for lighting surveys must be checked for calibration. The surveys themselves are relatively unobtrusive and should be conducted during working hours.

Research Directions

Many new electronic products have appeared on the market, and they often have visually challenging user interfaces worthy of investigation in the search for new ergonomic design principles. Head-mounted optical displays are an example. While facilitating hands-free operation using "smartphone"-like interfaces, these devices may restrict peripheral vision and be hazardous to use when tasks such as driving are carried out. Sawyer et al. (2014) found that drivers texting using a smartphone or head-mounted device reacted more slowly to an unexpected obstacle and showed poorer vehicle control during recovery. Although the decrements in driving performance were even greater when using the smartphone, the authors concluded that the head-mounted display was not a panacea, although it did mitigate the risk of texting while driving.

Recent research on melanopic EDI, its importance in entraining circadian rhythms, alertness, and a sense of brightness has generated increased interest in the non-visual aspects of interior lighting that create a positive work experience. How best to illuminate interior spaces is receiving increased attention.

SUMMARY

The amount and type of light emitted by the light sources themselves determine the illuminance delivered to work and other surfaces. The furniture and materials, by virtue of their reflectances, determine the balance of luminances seen by workers and the amount of indirect lighting and glare. The visual demands of tasks should be analyzed when evaluating lighting, particularly with respect to visual acuity and the demands on accommodation and adaptation of the eye. Large visual targets should be positioned higher in the visual field and farther away, and small targets lower down and closer to the viewer.

The psychological and biological effects of light on people should be considered, and there is plenty of evidence that office workers prefer natural light, a view out of a window, and control over lighting (Galasiu and Veitch, 2006).

Digital Eye Strain is the term used to describe a collection of visual and somatic symptoms reported by those doing intensive screen-based work. On the basis of the available evidence, the provision of user-controllable task lighting, user-selected screen locations, and several short rest periods throughout the day would seem to be mandatory for operators carrying out intensive work with computers, laptops, and tablets.

TUTORIAL TOPICS

1. Are visual problems likely to become more prevalent as people interact socially with screen-based displays as well as at work?
2. How would you decide on the upper limit for a very high-resolution display (beyond which no further improvements in picture quality could be perceived)?

3. If you cannot read this, you need your eyes tested. Discuss.
4. Take a look at the video at the link: https://www.youtube.com/watch?v=0ZFtHzZqYDY
   Are there any myths about computers and vision you might come across as an HFE specialist?

ESSAYS AND EXERCISES

1. Measure illuminances on a horizontal plane in the following places:
   a. Outdoors, under a shady tree, on an overcast day
   b. Outdoors on a clear night (no moon)
   c. Outdoors on a clear night (full moon)
   d. Under urban streetlights at night
   e. In a fast-food outlet at night and during the day
   f. In a modern office at noon with the blinds open and the lights
      i. On
      ii. Off
   g. On workbenches in a workshop or factory, how do the measured values compare with your subjective impressions of light/dark?
2. Refer to the power spectra from four different, commercially available lamps (Figure 10.13). Which lamp has the best spectrum for illuminating a task involving the visual inspection of wiring looms?
3. Which of the following (a, b, or c) best describes the attenuation of light from a source? What range of desktop illuminances would be suitable for office work involving computers?
   a. Exponential decay
   b. Inverse square law
   c. Power function
4. Consider a light source in the middle of a sphere of radius 3 m. The source has a luminous intensity of 3 cd. What is the illuminance on the inside of the sphere? Seen from below, the sphere appears as a disc. What is its luminance?
5. Workers in a garment factory carry out a quality control inspection task under artificial lighting. The factory is illuminated by a lamp 4 m above the table, providing illumination of 700 lx. If the lamp is lowered to 2 m above the table, what will the new level of illuminance be?
6. A worker works at a table sorting rough diamonds into different categories according to color and size. The diamonds are delivered in small glass dishes and are manipulated with tweezers for inspection. The luminance of the task is 400 cd/m^2. The workbench has a reflectance of 40% and the illuminance is 500 lx. Is the luminance ratio appropriate for this task?
7. Refer to the example in Ergonomics Workshop 10.1. If the architect had misread the regulations and it then appeared that the maximum viewing distance for the navigation lights was 1 nautical mile (as opposed to 3), would he then be able to proceed with his proposal to place the lights 1 ft apart? Calculate the visual angle to support your argument.
8. Carry out a lighting survey in an office. Obtain or draw a plan map of the office showing the location of desks, luminaires, and windows. Measure desktop illuminances (at least two measurements per desk) and write them onto the map. Compare your measurements with published guidelines. At each workstation, take note of any glare sources. Ask employees whether they have any comments about the present lighting arrangements. Combine this information into a report and make suggestions for improvement.
9. Describe the structure and function of the eye. How can this information be used to analyze practical visual problems in the workplace?

10. "Lighting should be right, not bright." Discuss.
11. Mr. Smith is having trouble reading his newspaper at home in the evenings. His wife tells him that he must either have his eyes tested or his arms lengthened. The optometrist tests his eyes and describes Mr. Smith's vision as 20/30.
    a. What did Mrs. Smith mean by her remark?
    b. If he replied that the problem is the new energy-efficient lamp by his armchair, would you believe him?
    c. How good is Mr. Smith's vision?

QUICK QUIZ

1. Most of the refraction needed to focus light on the retina occurs in the cornea. True or false?
2. The ciliary muscles contract to make the lens: (a) fatter, (b) thinner, (c) there is no effect
3. Red and green are opponent colors. True or false?
4. Rods are needed to see in the dark. True or false?
5. "Blue light makes you sleepy." True or false?
6. What level of worktop illuminance is recommended for precise assembly work?
7. People work better when they can see out of a window. True or false?
8. What is the 20-20-20 rule?
9. Higher color temperatures look warmer. True or false?
10. What can cause rickets in children?

NOTE

1 https://www.lrc.rpi.edu/cscalculator/.

11 Hearing, Sound and Vibration

GENERAL REQUIREMENTS FOR HUMANS IN SYSTEMS

1. Operators should not be exposed to workplace noise likely to cause noise-induced hearing loss according to ISO 1999:2013 or a national equivalent standard.
2. Selection of PPE for those exposed to excessive noise must be made in consultation with operators and include an analysis of the context of use.
3. WBV exposure must be within safe limits.
4. Hand–arm vibration exposure should be minimized.
5. Noise must be attenuated as close to the source as possible: first, minimize the noise; if this is inadequate, protect the operator; if still insufficient, remove the operator.
6. Ambient noise should not interfere with task performance.

Are wind turbines making people sick? Or is it all psychological?

Plummer, 2012

CORE KNOWLEDGE: HEARING AND SOUND

Acoustic waves can be defined as pressure fluctuations in an elastic medium. Sound is the auditory sensation produced by these oscillations. In air, sound consists of oscillations about the ambient atmospheric pressure (Figure 11.1). Vibrating surfaces or turbulent fluid flow can act as the source, propagating alternately high- and low-pressure areas. The amplitude of the acoustic wave is expressed in Newtons per square meter or in pascals. The threshold of hearing (the lowest amplitude of pressure oscillations in air detectable by the ear) is 0.00002 N/m^2 at a frequency of 1000 Hz.

Vibration is the oscillating, reciprocating, or other periodic motion of a rigid or elastic body or medium forced from a position or state of equilibrium. Acoustic waves are an example of vibration.

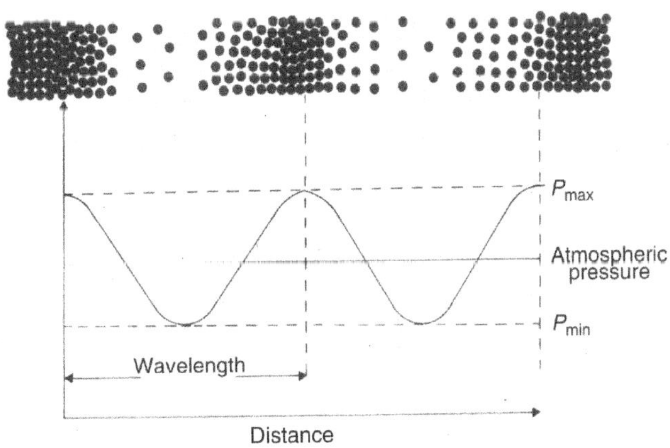

FIGURE 11.1 Sound is propagated in air as pressure oscillations above and below the ambient pressure.

DOI: 10.1201/9781032717210-11

Terminology

Frequency and amplitude define a pure tone. Their subjective counterparts are pitch and loudness. The perception of pitch also depends on the waveform, and the perception of loudness depends on frequency. A healthy, young person can hear sounds in the range of 16–20,000 Hz. Noise is usually defined as a sound, or sounds, at such amplitude as to cause annoyance or interfere with communication. Sound can be measured objectively, whereas noise is a subjective phenomenon.

The amplitude of sound is evaluated by measuring the sound pressure level (SPL). The range of SPLs to which the human ear is sensitive is so wide (0.00002–20 N/m^2) that linear scaling would present a problem. For this reason, a logarithmic scale—the decibel scale—is used. The decibel is a dimensionless unit related to the logarithm of the ratio of the measured SPL to a reference level (usually taken to be the threshold of hearing). The smallest noticeable difference in intensity between two sounds is about 1 dB. Commercial sound level (SL) meters measure and display a root-mean-square (rms) SPL, L_p, where

$$L_p = 20\log_{10}\left(p/p_r\right)\text{dB}$$

where

L_p is the SPL in decibels.
p is the sound pressure in Newtons per square meter.
p_r is the reference SPL (0.00002 N/m^2).

For example, if the rms SPL is 2 N/m^2, then

$$\begin{aligned}\log_{10}\left(p/p_r\right) &= \log_{10}2/0.00002\\ &= \log_{10}100{,}000\\ &= 5\\ \text{and } L_p &= 100\text{ dB}\end{aligned}$$

Table 11.1 depicts the range of human hearing in decibels.

Sound Transmission

Sound attenuates in a free field according to the inverse square law. The sound pressure drops according to the square of the distance from the source. So, if the distance from the source doubles, the reduction in SL is

$$\begin{aligned}10\log_{10}(2/1)^2 &= 10\log_{10}4\left(\text{since } 20\log_{10}\left(p/p_r\right) = 10\log_{10}\left(p/p_r\right)^2\right)\\ 10\log_{10}4 &= 6\text{ dB}\end{aligned}$$

In practice, many other factors influence the attenuation of sound, such as the layout of the work area and the presence and nature of objects between the source and the listener. The effect of these is described in a later section.

Frequency Analysis

Frequency analysis is sometimes carried out to obtain further information about the constituent frequencies of a sound source. Octave analysis involves dividing the frequency spectrum into bandwidths such that the lower boundary (f_1) is one half of the upper (f_2), where

$$f_2 = 2f_1$$

TABLE 11.1
Range of Human Hearing

Source	SPL (dB)	
Jet engine at 30 m	140	Ear damage
Pneumatic chipper	130	Onset of pain
Pneumatic rock drill	120	
Punch press	115	
Hydraulic rock drill	115	
Textile loom	105	
Powered lawn mower	100	
Newspaper press	95	
Milling machine	90	
Diesel lorry	85	
Very noisy street corner	80	
Crowded room	75	
Passenger car (at 15 m)	70	
Conversation	60	
Sound inside the car at 50 kph	55	
Private office	50	
Quiet room	40	
Library	35	
Whisper 1 m from the ear	20	

The center frequencies of the octave bandwidths are

$$31.5, 63, 125, 250, \ldots \text{Hz}$$

Further resolution can be obtained by using half or third octave bandwidths, where

$$f_2 = \sqrt[2]{f_1}\,(\text{half octave})$$

$$f_2 = \sqrt[3]{2f_1}\,(\text{third octave})$$

The center frequency of a bandwidth is given by the geometric mean of the upper and lower frequencies. Octave analysis and its variations are logarithmic scales and are used to take into account the ear's approximate logarithmic frequency selectivity. They have disadvantages when the goal is to analyze the sound source because resolution drops as frequency increases (i.e., the bandwidths get wider and all sounds within a bandwidth are added together). Narrowband constant-bandwidth analyses may then be used to determine intense sound at a particular frequency. Fortunately, the moving parts of many industrial machines, such as pistons, fans, or gears, are often the primary sources of noise. Since many of these rotate or cycle at known frequencies below 1000 Hz, their contribution to the total noise can be determined fairly easily.

The Ear

The ear converts sound waves in air into nerve impulses, which travel along the auditory nerve to the brain. Further processing in the brain results in the perception of sound and the recognition of auditory patterns. A simplified diagram of the ear is presented in Figure 11.2.

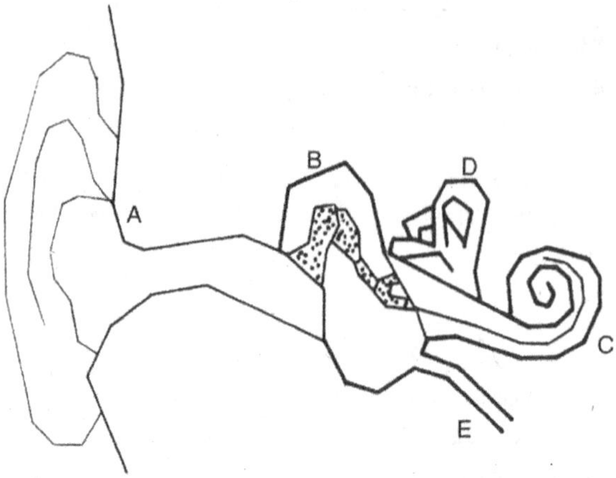

FIGURE 11.2 The ear consists of the outer ear (a), the middle ear (b), and the inner ear consisting of the cochlea (c) and the semicircular canals (d). The eustachian tube (e) is also shown.

Hearing requires three main steps. In each, energy is converted from one form to another:

1. Pressure variations due to longitudinal waves in the environment are guided into the external auditory meatus, where they cause a mechanical vibration of the tympanic membrane (eardrum). The tympanic membrane is connected to three small bones (ossicles) in the middle ear, which, in turn, are caused to vibrate.
2. Mechanical vibration of the auditory ossicles is converted to wave motion in the cochlear fluid at the oval window.
3. Wave motion in the cochlear fluid is converted to nerve impulses in the auditory nerve by hair cells in the cochlea.

Anatomically, the ear is divided into three parts where these energy transformations take place.

Outer Ear

The outer ear consists of the ear flap (pinna), the external auditory canal, and the eardrum. The pinna is like an ear trumpet that funnels sound into the canal. The pinna is relatively small and unimportant. When attempting to listen to a faint sound, it is best supplemented with a cupped hand. Sound waves travel to the eardrum (tympanic membrane). The sound causes the membrane to vibrate, and these vibrations are transmitted to the middle ear.

Middle Ear

The middle ear is an air-filled bony chamber connected to the outside atmosphere by the eustachian tube, which terminates at the back of the throat. Discrepancies in air pressure on either side of the eardrum are potentially harmful. Swallowing momentarily connects the middle ear to the atmosphere, enabling pressure differences to be equalized. The "popping" sensation of the ears, experienced when changing altitude, is caused by this sudden equalization of pressure.

Vibrations of the eardrum are transmitted to the inner ear via three connected bones, which comprise the ossicular system. The malleus is connected to the center of the tympanic membrane at one end and to the incus at the other end. Movements of the eardrum cause movements of the malleus that are directly transferred to the incus. The opposite end of the incus is connected to the third auditory ossicle—the stapes. The far end of the stapes is known as the "faceplate." It lies against the opening of the inner ear—the oval window.

The auditory ossicles are supported by ligaments. The tensor tympani muscle keeps the tympanic membrane under tension—like the diaphragm of a microphone. Movements of the tympanic membrane are transmitted from the auditory ossicles to the oval window of the inner ear. The ossicles play an important role in increasing the efficiency with which sound is transmitted to the sensitive hair cells of the inner ear. Since the inner ear is filled with fluid, sound energy has to be transmitted from air to fluid. The transmission loss from air to water is about 99% (which is why fishermen can listen to the radio while fishing without frightening the fish away).

The function of the auditory ossicles is to transmit sound energy from air to water. They act as a lever system in which displacements of the malleus at the tympanic membrane cause corresponding, but reduced, displacements of the faceplate of the stapes. This reduction in amplitude is accompanied by a corresponding increase in force of approximately 1.3. The surface area at the faceplate is approximately 17 times smaller than that at the tympanic membrane. The pressure of the air against the tympanic membrane is increased approximately 22-fold at the stapes against the oval window. Without the auditory ossicles, the threshold of hearing would rise by about 30 dB (Guyton, 1991).

Volume control is achieved by the tensor tympani and stapedius muscles. Intense sound (particularly at frequencies below 1000 Hz) causes these muscles to contract, pulling the malleus and stapes away from the interfaces of the outer and inner ear. This attenuates loud noise and protects the inner ear from damage. Differences in the effectiveness of this *attenuation reflex* may help to explain why some individuals are more susceptible to noise-induced hearing loss than others.

Inner Ear

The inner ear is known as the cochlea. It can be modeled as a fluid-filled tube separated into an upper and lower half by a membrane (Figure 11.3). At the end of the tube are two windows—the oval and round windows—on either side of the membrane. The structure of the cochlea is apparent if we imagine a tube folded back on itself and coiled like a shell. The central membrane is known as the basilar membrane, and it ascends along the length of the coiled tube from its base to its apex, where it is perforated by a hole (the helicotrema). The tube is tapered—it is widest at the helicotrema end.

Since the cochlea is sealed, inward movement of the stapes pushes the oval window inward and causes a column of fluid to move inward into the cochlea. At the other end of the cochlea, the round window moves outward. Fibers in the basilar membrane run along its length, projecting outward from the center of the cochlea. At the base of the cochlea, close to the oval and round windows, these fibers are short and thick. Toward the apex, they become longer and thinner. Sound vibrations at a particular frequency are transmitted by the stapes and cause motion in the fluid column in the cochlea. This sets up a traveling wave in the cochlea. At a certain point along the cochlea, the basilar membrane is just the right thickness to resonate at the particular frequency of the cochlear wave. At this point along the length of the cochlea, maximum vibratory amplitude will occur.

The basilar membrane is connected to an organ on its surface (the organ of Corti). This organ contains hair cells connected to afferent fibers of the auditory nerve. Movement of the hair cells causes depolarization of nerve endings and the transmission of a nerve impulse to the brain.

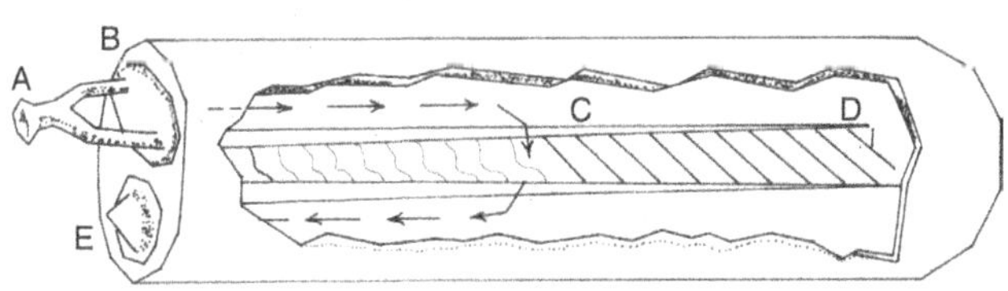

FIGURE 11.3 Simplified view of the cochlea "unwound" to reveal its tubular structure. (a) Is the stapes, (b) Is the oval window, (c) Is the basilar membrane, (d) Is the helicotrema, and (e) Is the round window. (From Guyton (1991), with permission.)

When the stapes transmits sound vibrations to the cochlea, they pass along its length until a point is reached where the frequency of the vibrations matches the resonant frequency of the basilar membrane. At this point, the movement of the basilar fibers "short-circuits" the incoming sound wave, which is converted to motion of the fibers. This causes the hair cells at that point to depolarize. Since the length and thickness of the basilar fibers vary along the length of the cochlea, different frequencies are detected at different parts of the cochlea. High frequencies cause resonance at the base of the cochlea, and lower frequencies are detected higher up. This theory, known as the place–pitch theory, explains how the cochlea separates a complex sound into its spectral components. In this sense, the basilar membrane and fibers are not unlike a harp lying along the length of the cochlea. Detection of sound energy at a particular frequency depends on there being a "string" available that will resonate at that frequency.

Discrimination between sounds of different frequencies takes place in the inner ear rather than in the brain. Supporting evidence for this comes from the observation that the cochlea is the only part of the human body that is the same size in an infant as in an adult. If the infant cochlea were a scaled-down version of the adult cochlea, infants would be deaf to low frequencies and sensitive to very high frequencies above the adult range and would probably experience difficulties in learning to speak (because they would not be able to hear their parents' voices properly). Although children are sensitive to higher frequencies than adults, the frequency range to which their ears are most sensitive is the same as that of adults because the cochlea is the same size.

Sensitivity of the Ear

Auditory sensitivity is greatest between 1000 and 4000 Hz—the frequency band in which speech is transmitted (note: the notes on a piano range from 26 to 4096 Hz). Figure 11.4 shows the sensitivity of the ear to different audible frequencies.

The loudness of a noise depends on its frequency as well as its SPL. Contours of equal loudness have been worked out using large numbers of subjects. A reference tone of 1000 Hz and, say, 60 dB can be presented to subjects. This is followed by a second tone at a higher or lower frequency. The subject then adjusts the intensity of the second tone until its loudness matches that of the reference. In this way, contours of equal loudness can be plotted for all frequencies with respect to a series of 1000 Hz reference tones from 10 to 120 dB. For example, a tone of 100 Hz at an intensity of approximately 70 dB would sound as loud as a 60 dB, 1000 Hz reference tone.

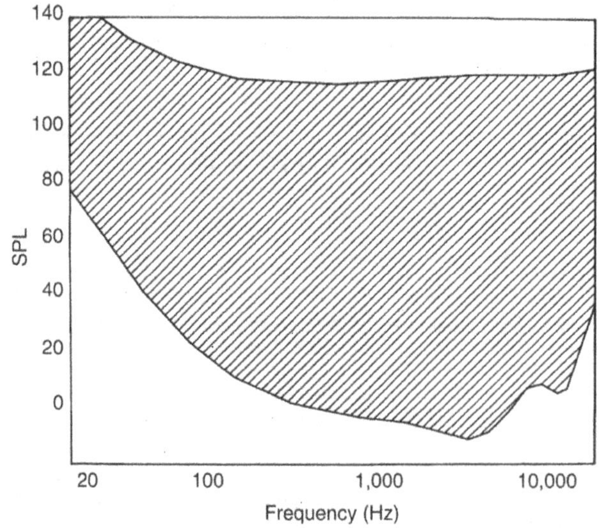

FIGURE 11.4 Sensitivity of the human ear. Sounds below the hatched area are inaudible. Sounds above cause damage to the ear.

TABLE 11.2
Median Threshold Shifts for Men at 4 kHz, Due to Age

Age (years)	Threshold Shift (dB)
35	5
45	12
50	22
65	35

Source: From Lutman (2000).

The unit of loudness is the phon and is derived from the equal loudness contour data. In the above example, the 70 dB, 100 Hz tone would have a loudness of 60 phons. A second loudness measure is the sone. One sone is defined as the loudness of a 1000 Hz tone at 40 dB. Since loudness perception is a logarithmic rather than a linear process, it can be said that a 10 dB increase in intensity represents a doubling of loudness. Thus, a 50 dB, 1000 Hz tone would have a loudness of 2 sones.

Old age is marked by compression of the audible frequency range from 16–20,000 to 50–8000 Hz, a condition known as presbycusis. The higher frequencies are usually lost first, and most aged people cannot hear sounds above 10,000 Hz, which is why it is better to lower the voice when attempting to communicate with older people. Raising the voice increases the pitch and is self-defeating.

The mechanism of hearing loss is at the neural/cochlear level rather than the ossicular level and is irreversible. Table 11.2 gives median age-related threshold shifts for men.

Localization of Sound

Humans have two ears, and if the listener is not facing the sound source, then the sound will arrive sooner and with slightly greater intensity at the ear closest to the source. These discrepancies in time and intensity are known as the interaural level and interaural time differences (ILD and ITD). In dry air at 20°C, the speed of sound is 343.2 m/s (1236 km/h), so the time difference is very small. Hearing loss degrades sound localization because it degrades the ability to detect small differences. The differences are greatest when the source is directly to the side of the listener rather than at 45° or straight ahead.

Echolocation

Echolocation is the ability to detect obstacles without vision and is often used by blind people to get around (although sighted people can also learn the technique; Thaler, 2015). Echolocators typically use mouth clicks of 3–15 ms around 6–8 kHz to interrogate the space around them. Nearby objects reflect (echo) these clicks, and the detection of the echoes enables the objects to be localized.

Noise-Induced Pathology of the Ear

Sudden noises of sufficient intensity may produce trauma resulting in a reduction in the efficiency of sound transmission to the cochlea or in the cochlear response itself. The tympanic membrane may be ruptured or the ossicular chain mechanically damaged. Surgical repair may be necessary. Sensorineural hearing loss occurs as a result of damage to the cochlea, auditory nerve, or central nervous system and is the most common type of hearing loss. Apart from presbycusis, it can also be caused by head injury, exposure to certain chemicals, or genetic factors. In the USA, there are about 66,000 new cases every year (Tanna et al., 2020). The prevalence of presbycusis doubles for every decade of life from 20 years of age. Exposure to loud noises is a major cause of occupational hearing loss.

Sensorineural hearing loss can also occur, particularly after chronic exposure to loud noises. The largest threshold shifts occur around 4kHz (Lutman, 2000), where the vibrational characteristics of the basilar membrane are greatest. The loss is irreversible. In audiometry, such loss is described as a "permanent threshold shift." ISO 1999 gives formulas that enable one to predict hearing loss in men and women for different intensities and durations of exposure to noise. Table 11.3 gives the median threshold shifts associated with different noise exposures. As can be seen, the loss at 80dB is negligible, but the risk rises rapidly around 85–90dB. Exposure to levels as high as 95dB poses a serious threat to hearing.

Audiometric testing determines the minimum intensity (the threshold) at which a person can detect sound at a particular frequency. Since sensitivity to particular frequencies is lost due to age or damage, the threshold increases. It is in this sense that hearing loss can be described as a threshold shift. Temporary threshold shifts can occur after exposure to loud noise (e.g., in clubs). The sensory cells in the cochlea remain intact, but the acute loss of afferent nerve function and delayed regeneration of the cochlear nerve can cause progressive loss of hearing, which is not initially detected by conventional hearing tests (Kujawa and Liberman, 2009). Repeated exposure leads to permanent threshold shifts (noise-induced deafness). Figure 11.5 shows example audiograms of noise-induced threshold shifts.

Although permanent threshold shifts occur with age, chronic exposure to noise (particularly over 90 dB[A]) hastens their onset and severity (some airborne chemicals, such as toluene, have a toxic effect on the auditory system and can cause hearing loss, but this is beyond the present scope).

Tinnitus

Tinnitus (ringing in the ears) is a common auditory effect of exposure to loud noise, both in the short and long term. The cause appears to be an increase in the gain of the central nervous system (Makar, 2021). Biswas et al. (2023) conducted a systematic literature review of 25 studies concerning the risk factors for tinnitus. The point prevalence in Europe was 16% and several otological and non-otological risk factors were identified. Tinnitus is associated with sensorineural hearing loss; occupational noise exposure (relative risk 1.70; 95% *CI*, 1.49–1.94); middle ear infections; and certain medical treatments. It was not associated with leisure-time noise exposure. According to a systematic review conducted by Makar (2021), noise exposure is the second biggest causal factor in the development of tinnitus, second only to noise-induced hearing loss.

That tinnitus is common in noise-exposed workers and that it interferes with activities of daily living suggests that information about tinnitus and its avoidance should be included in hearing conservation programs.

Psychosocial Aspects of Noise-Induced Hearing Loss

Noise-induced hearing loss can have serious implications for the quality of working life and employability. It causes social isolation, can prevent individuals from jobs where teamwork and

TABLE 11.3
Median Threshold Shifts at 4 kHz as a Function of Noise Level

Noise Level	Threshold Shift (dB)
80	2
85	7
90	15
95	27

Source: From Lutman (2000).

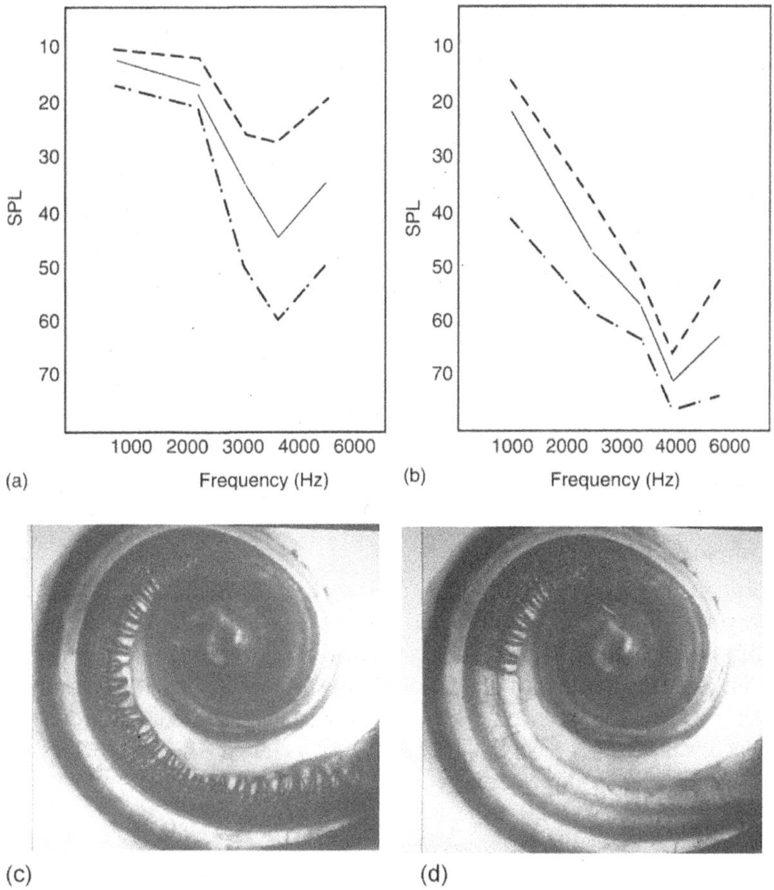

FIGURE 11.5 Noise-induced permanent threshold shifts. (a) After 5 years in a noisy job and (b) after 20 years in a noisy job. Sensitivity to the higher frequencies is lost first (centerline gives the patient's average sensitivity). The SPLs are the lowest that the person can hear at a given frequency. (c) Healthy cochlea with no loss of hair cells. (d) Loss of hair cells due to age, noise exposure, or both.

communication are required, and can damage promotion prospects by lessening the ability to manage others. Jobs such as sonar operation have precise hearing standards. Jobs such as air traffic control have precise hearing standards. Applicants must have no hearing loss in either ear of more than 25 dB at 500, 1000, and 2000 Hz, and no more than 20 dB loss in the better ear, using American National Standards Institute standards.

BASIC APPLICATIONS

Design of the Acoustic Environment

The alerting quality of noise makes it an ideal warning signal. However, noise can distract workers and is a major source of dissatisfaction with the environment. Noise can affect task performance directly, by masking or degrading auditory information needed to perform the task, or by diverting attention from the task. Individuals differ widely in their attitudes to and tolerance of noise. Many factors need to be considered if an optimal auditory environment is to be designed. Woodson (1981) has compiled the findings of many studies on the effects of noise on human performance. Although noise levels below 80–85 dB(A) are not a serious threat to hearing, they can degrade task performance and cause annoyance. Table 11.4 summarizes some of these effects.

TABLE 11.4
Effects of Noise below 85 dB(A)

Noise Level (dB[A])
80 Conversation is difficult. Upper limit for tasks where speech communication is essential
75 Telephone conversation is difficult. A raised voice is needed for face-to-face conversation
70 Upper level for normal conversation. Telephone conversation is difficult. Unsuitable for office work
65 Upper acceptance level when people expect a noisy environment. Upper level for low-demand supervisory work
60 Acceptable level for daytime living conditions
55 Upper acceptable level when people expect quiet. Upper limit for mental work and complex supervisory tasks
50 Acceptable by people who expect quiet. About a quarter of people will experience difficulty in falling asleep or be woken by the noise
45 Upper limit of creative work
40 Very acceptable for concentration. Few people will have sleep problems
<30 Low-level intermittent sounds become disturbing

Specification of Noise Levels

In practice, designers have to consider both the ambient and the transient noise levels in a facility. These levels vary according to the activities carried out and the usage patterns of different areas. It is not normally sufficient to specify a single ambient noise level because, in real work situations, unexpected or intermittent noise of short duration also occurs. Table 11.5 gives examples of the size of the differences between ambient and transient noise limits (the latter are the L_{eg} values for 5 s durations). Such values might be used in the design of a complex facility such as an offshore oil production platform or a large ship.

Reverberation

The noise level in a room depends on the intensity of any noise sources and on the reverberation characteristics of the room. The sound emanating from a source reaches the listener both directly and after being reflected off the walls and other objects (Figure 11.6).

Reverberation is the interreflection of sound waves inside a room. The more the reflection, the longer the reverberation time. If there is continuous sound, background noise caused by interreflection builds up to a steady level called the reverberant field. Reverberation is a very important acoustic characteristic that influences an occupant's perception of a room (Table 11.6).

TABLE 11.5
Upper Limits for Ambient and Transient (5 s) Noise Levels (dB[A])

Area/Section	Ambient Limit	Transient Limit
Unmanned machinery spaces	110	110
Unmanned storage facilities	90	100
Manned machinery spaces	80	90
Workshops	80	90
Food preparation/laundry	75	90
Offices/control rooms	65	75
Bathrooms/locker rooms	60	70
Medical complex	60	75
Sleeping quarters	55	70

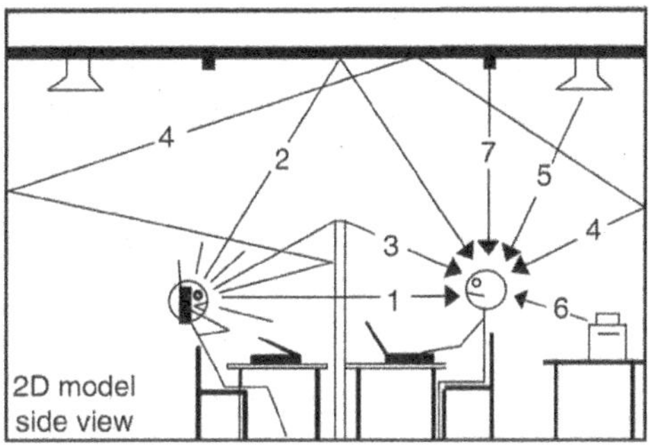

FIGURE 11.6 Reverberation. Sound in a room reaches the ear by direct and indirect routes. (Modified from Hongisto et al., 2004.)

TABLE 11.6
Some Characteristics of Reverberation in Rooms

1. Reverberation time in a room does not depend on the position of the source or the position of the listener
2. Excessive reverberation is a major cause of the blurring of acoustic signals such as speech
3. Room shape has little effect on reverberation time
4. Intensity of the reverberant field in a room depends on:
   SL of the source
   Volume of the enclosed space
   Amount of sound-absorbing material in a room

The reverberation time of a room is the time taken for the reverberant level to fall by 60dB. It is a function of room volume and total sound absorption. Excessive reverberation causes problems because, as the distance of the listener from the target sound source increases, the intensity of direct sound decreases (by about 6dB per doubling of distance). A point is reached where the intensity of the direct field is the same as the intensity of the reverberant field. This is known as the critical distance. The total SPL at this point will therefore be 3dB greater than the direct sound or the reverberant sound alone.

Reverberation can be minimized by increasing the amount of sound-absorbing material in a room. The occupants themselves (or rather their clothes) are sound absorbers, so the reverberation characteristics of meeting halls and conference rooms will differ when they are empty compared to when they are full, unless the seats are upholstered.

Speech Intelligibility and the Speech Transmission Index

The speech transmission index (STI) is a measure of the intelligibility of speech in environments where there are other sounds. It is calculated using specialized equipment that generates a test signal having many of the acoustic properties of real speech. The signal is transmitted in the environment of interest and compared with a received signal that consists of the test plus all the other ambient sound.

The STI can vary from 0 (speech is completely unintelligible) to 1 (speech is perfectly intelligible):

0.00–0.30 unacceptable
0.30–0.45 poor
0.45–0.60 fair
0.60–0.75 good
0.75–1.00 excellent

Reverberation can mask speech. Reverberation times of less than 1s are recommended for rooms where speaking is a main activity. Longer reverberation times are needed for listening to music because reverberation gives "depth." Rooms suitable for listening to music may not be suitable for conferences and vice versa. In open-plan offices, other people's speech may be distracting and so designers should aim for a low STI by minimizing direct and reflected speech sounds at each workstation. Screens around desks are rarely higher than 1.3 m, which is not enough to acoustically isolate desks from each other. Venetjoki et al. (2006) report that STIs less than 0.20 are needed in open-plan offices and that sound-absorbent ceiling tiles and wall panels can reduce speech from 55 to 40 dB(A), but that it is difficult to reduce speech by more than 10 dB(A) at a distance of less than 2.5 m. In many open-plan offices, 2.5 m will exceed the distance separating many of the desks, and distraction may be problematic.

Auditory Startle Response (ASR)

The ASR consists of freezing, increased heart rate, blinking and whole-body muscle contractions (Zheng and Schmid, 2023). It is thought to be an adaptive response that results in the rapid adoption of a defensive stance with maximum postural stability. The eyes close, the neck flexes, and the shoulders abduct with pronation of the forearms and clenching of the fists. The torso flexes forward, and the knees flex. Sudden auditory stimuli are powerful producers of the startle response, with a latency of 6–8 ms. The ergonomic relevance is that startle interferes with task performance and slows reaction time for other tasks. In the short term, operators will habituate to startling auditory stimuli after four or five exposures or over the course of a few minutes. Auditory stimuli with fast onset times have the advantage of being very attention-getting, but Sorkin (1987) cautions against onset rates of more than 10 dB/msec. For auditory stimuli to be alerting but not startling, onset rates of 1 dB/msec are recommended. Sudden loud noises can have serious health consequences for sensitive individuals, such as "startle epileptics," where they may trigger a seizure. Posttraumatic stress sufferers may exhibit an exaggerated startle response that is easily triggered.

Noise Control

Most developed countries have regulations requiring that noisy workplaces be assessed and that steps be taken to reduce the risk if necessary. If the noise level in the workplace exceeds the first action level (80 dB(A) in the EU), the noise must be reduced as far as possible, those exposed must be informed and provided with ear protectors, and noise hazard zones must be demarcated. If the noise exceeds the second action level (85 dB(A) in the EU and the USA), steps to protect the hearing of those exposed are mandatory and include the provision of ear protection and the reduction of exposure times. Employees who wish to undergo audiometric testing must be provided with it.

Several approaches to noise control can be identified:

1. Eliminate the threat to hearing by redesigning the machine or using a less noisy machine
2. Remove personnel from the noisy environment
3. Protect personnel by issuing earplugs or muffs or by building an acoustic refuge

Many machines have both primary noise sources (such as the power unit) and secondary noise sources (parts of the machine that resonate). The relative contribution of the primary and secondary sources to the total noise depends on the particular machine. For example, pneumatic machines are inherently noisy due to the exhaust gases. The energy loss due to noise in machines such as pneumatic rock drills is only about 0.08% (Holdo, 1958) because 87.5% of the noise is due to exhaust gases rather than the impact of the drill steel against the rock or the pistons and gears of the machine (Weber, 1970). Hydraulic pumps can be very quiet, yet the hydraulic system itself can be a source of considerable noise due to resonance if the pump is not securely mounted and mechanically isolated from fluid lines and other structures.

Noise from the primary sources is often periodic and below 1000 Hz, whereas secondary noise is usually of a higher frequency. Velocity gradients in compressed air create random vortices, which move and are dissipated in a chaotic manner, giving rise to wideband random noise (Beiers, 1966). Frequency analysis with the machine operating under representative loading is an important first step in machine redesign. The most intense frequencies can be identified and redesign can proceed in a systematic manner. Bartholomae and Kovac (1979) redesigned a pneumatic, rotary, percussive rock drill. These machines emit noise of 115–120 dB(A) at the operator's ear level. By muffling the exhaust, redesigning the drilling mechanism using nylon instead of metal components, and acoustically treating the drill steel, noise was reduced to 95 dB(A). Noise from pneumatic machines can also be reduced by piping the exhaust away from operators, although this, like muffling, reduces efficiency.

Some common problem areas and control methods are described below.

- *Fans*: Fans or blowers are much noisier when running at high rather than low speeds. The SL varies as the fifth power of the speed. Use a larger fan running at a lower speed.
- *Muffling*: Pneumatic tools such as paving breakers, screwdrivers, and dentists' drills produce noise due to the exhaustion of compressed air into the atmosphere. This can be reduced by piping the air away from the operator to an unoccupied area or by using mufflers, which reduce the escape velocity of the compressed air. Reductions of up to 6 dB are possible, but efficiency can drop by up to 15%. A third-octave analysis of a muffled and unmuffled pneumatic machine is presented in Figure 11.7.
- *Part Ejection*: Pneumatic ejectors are sometimes used to remove parts from presses. Mechanical ejectors are usually quieter.
- *Pneumatic Tools*: Hydraulic or electric equivalents are usually quieter.
- *Impact Tools*: Large electric solenoids for tripping clutches can be sources of impact noise. Resilient bumpers fitted to the point of impact can be beneficial.
- *Hydraulic Reticulation*: Cavitation (the release of dissolved air in a fluid due to sudden pressure drops) and turbulence can increase noise from pumps considerably. Cavitation can be reduced by decreasing fluid flow velocity, "smoothing" the system by removing sudden bends or junctions, and by avoiding sudden drops in pressure.
- *Vibration*: Machine vibration can be exacerbated by imbalance or eccentricity in rotating members, inadequate mountings, and wear. Resilient couplings can be used to isolate a source from surrounding objects. Regular maintenance and replacement of worn parts can prevent vibration. In offices, typewriters and daisy wheel printers can be placed on felt or rubber mats to block the transmission of vibration to the supporting table, which acts as a sound amplifier. Printer noise can interfere with speech.

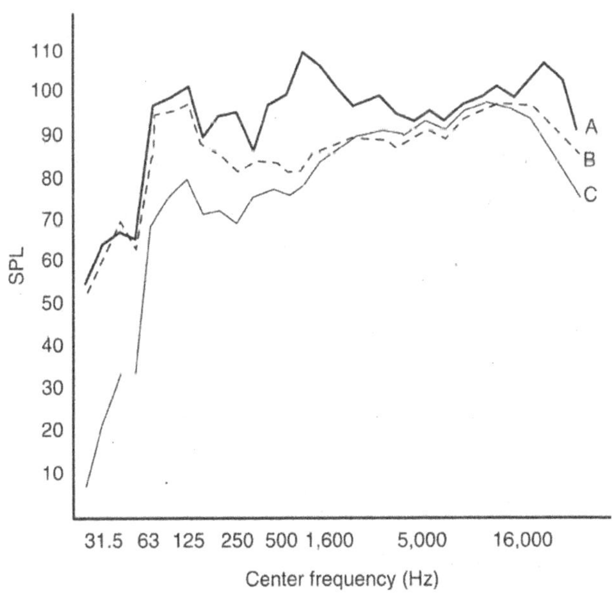

FIGURE 11.7 Effect of muffling a pneumatic machine. (a) Is a third octave analysis of an unmuffled machine, (b) Is attenuation when the exhaust is muffled, and (C) Is "A-weighting" of attenuated noise.

A change in process can also reduce noise. Riveting can be replaced with welding. "Dead" versus "live" materials are preferable in the construction of flooring, pipes, etc. Rubber, wood, and some types of plastic have high internal damping and do not transmit sound as readily as metals. Pneumatic drills can be replaced with electric drills.

Since sound attenuates with distance, attention should be paid to the location of machines or components such as fans or pumps. Locating these at floor level rather than at ear or waist height can significantly lower worker exposure to noise.

Noise Insulation

If machine redesign is not possible or produces inadequate reductions in noise, an acoustic enclosure can be built around the source or an acoustic refuge built for the operator. If the source is enclosed, noise is inter-reflected by the walls of the enclosure, losing energy as it does so. The inside of the enclosure may be lined with sound-absorbing materials to increase attenuation as the noise inter-reflects off the enclosure walls. The interreflection increases the noise level in the enclosure by up to 10dB without absorbent lining. ISO 23351-1:2020 deals with speech level reductions of furniture and enclosures. Measurements are made with and without the product to determine the speech reduction achievable. A-weighted sound power levels in octave bands from 125 to 8000Hz are measured to give an indication of the privacy obtained.

Sound-absorbing materials are usually porous and lightweight. As the sound waves travel back and forth within the tiny interstices of the material, their energy is converted to heat by friction. The sound-absorbing characteristics of a material are described by its absorption coefficient—the ratio of the energy absorbed to the energy striking the material. A perfect absorber has a coefficient of 1, whereas a perfect reflector has a coefficient of 0. The absorption coefficient of a material varies with frequency; therefore, frequency analysis of the problem noise is needed if an optimal choice of material is to be made.

Sound energy does not pass efficiently across the interface of two materials differing greatly in elasticity—a principle used in the construction of acoustic refuges. Noise striking the walls of the refuge is reflected away. The best barrier materials have a high density and are nonporous. Rubber seals must be used on windows and door frames to prevent leakage of sound into the refuge.

Concrete, which has absorption coefficients of 0.01–0.02 at frequencies from 125 to 4000Hz, is a good reflector and therefore a good noise insulator. Fiberglass has a coefficient of 0.48 at 125Hz rising to 0.99 at 1000Hz, and is a good noise absorber.

Partial enclosures can be effective provided that the operator's ear remains in the "shadow" provided by the enclosure. Partial enclosures are suitable for protection against higher-frequency noise—the dimensions of the enclosure have to be several times the wavelength of the sound.

Screens, Carpets, Curtains, and Tiles

In offices, upholstered screens, carpets, and acoustic ceiling tiles can be installed to absorb noise and block its transmission from one place to another. These materials are rated according to their noise reduction coefficient, or NRC. An NRC of 0 indicates that no ambient sound is absorbed by the material, and an NRC of 1 would mean that all the sound is absorbed. The absorption characteristics should be highest at the higher frequencies of noise found in offices (2000–4000Hz), since these are the most disturbing.

Thick pile carpets with heavy underlay can have an NRC of up to 0.7 over the frequency range 125–4000Hz. Carpets have the advantage of covering a large surface area and are also very effective at absorbing vibration. Without underlay, the NRC can be halved. Pleated curtains (drapes) can be placed over bare walls or over windows to reduce noise transmission and can be very effective because the pleating increases the surface area of material in contact with the air, and hence the scope for absorption of sound and the interreflection of sound from curtain to curtain rather than from curtain to air. NRCs of 0.55–0.60 have been reported for pleated curtains. It may also be less expensive to cover bare walls with curtains rather than acoustic tiles.

White Noise and Pink Noise in Offices

White and pink noise of low intensity are used to mask potentially distracting sounds such as speech and office machines. White noise is called "white" because it has a flat frequency spectrum across the audible range, in the same way that white light is a mixture of all wavelengths. It can be transmitted at around 40–50 dB in open-plan offices, depending on the characteristics of the office, such as the amount of noise insulation. In effect, the lower and higher frequencies are attenuated by the ear's frequency response, and distracting sounds around 2–4 kHz are masked. White noise has a "hissy" sound that may be distracting if too loud. Pink noise is similar in that it has a broad spectrum but is louder at lower frequencies (it is "pink" in the same way that light with a lower color temperature would look "redder" or 'warmer'). The intensity of pink noise reduces evenly as frequency increases. White noise has the same energy at each frequency, and pink noise has the same energy in each octave.

Active Noise Control

Active control is really only effective at lower frequencies but has many applications because much machine noise is of a periodic/cyclic nature. Noise cancellation works well for non-random noise sources. An antiphase version of the source, of equal amplitude, is produced. Fast processing of noise from one cycle can be used to predict the noise from the next cycle, which is propagated from a loudspeaker back toward the source. If the propagation of "anti-noise" is correctly synchronized, noise cancellation takes place. Noise cancellation systems can be combined with ear defenders to improve attenuation at low frequencies and can also be incorporated into intercom systems. Active vibration cancellation is used, for example, in some helicopters to eliminate the transmission of vibration from the rotors to the cockpit. Modern smartphones have more than one microphone—one for your voice and one placed further away (at the back of the casing) to capture ambient noise from traffic and so on. Sound from the two sources is compared, and the background noise is canceled.

Noise and Communication

Most work depends on verbal communication and concentration. Noise can interfere with communication in a manner analogous to the way glare interferes with vision. Registration of target sounds will be disrupted by background noise, and the disruption will be greater if the background noise shares similar frequencies with the target signal. Such disruption is known as *masking*, and there is evidence that masking is greater when the signal and the noise frequencies are similar. Effectively, the threshold for hearing a target sound increases in the presence of background noise. Table 11.7 gives guidance on the distance at which communication between a speaker and a listener is acceptable under different intensities of background noise.

The Cocktail Party Problem

Speech-on-speech masking occurs when the listener has to selectively attend to a speaker while others nearby are speaking (Wang and Xu, 2021). *Informational masking* occurs when the number of simultaneous speakers is small but the contents are similar (linguistically), so that the mask is misattributed to the target. *Modulation masking* can occur when there are many background speakers. The linguistic content of the masking speech is unclear, but it disrupts the ability to separate the speech signal into identifiable components. Spatial separation of target and masking speech helps to *unmask* the target speech. Similarly, differences in frequency between target and masking speech lower the threshold for the detection of target speech (see Wang and Xu, 2021, for further information).

TABLE 11.7
Maximum Speaker–Listener Distances for Intelligible Speech Communication

Speech Interference Level	Maximum Separation for Intelligible Speech Communication (Normal Conversation)	Maximum Separation for Intelligible Speech Communication (Raised Voice)
dB	Meters	Meters
35	7.50	15.00
40	4.20	8.40
45	2.30	4.60
50	1.30	2.60
55	0.75	1.50
60	0.42	0.85
65	0.25	0.50
70	0.13	0.26

Source: From ISO 3353 (1974), with permission.

Outdoor Noise

Several factors influence the transmission of sound outdoors. Climatic factors such as temperature gradients, wind, and relative humidity can attenuate and refract sound. Buildings can have a major effect by reflecting sound in a particular direction or by causing echoes (when the reflected sound reaches a listener more than 50 ms after the direct sound). Such factors should be considered when, for example, positioning loudspeaker systems outdoors.

Traffic noise can exceed 90 dB(A) in large cities or close to major roads. Two common approaches to the reduction of traffic noise are the use of barriers or embankments to shield residential areas from busy roadways, and the installation of windows with double glazing in residential and other buildings. Measuring the annoyance level of community noise is a complex area beyond the present scope. In addition to objective measurements of ambient and peak noise, survey and questionnaire data are needed to evaluate people's attitudes and reactions to noise.

Hearing Protection

Hearing protection devices (HPDs)—typically earmuffs or earplugs, singly or in combination—can protect the ear from excessive noise by providing an airtight (acoustic) seal between the atmosphere and the auditory canal. Reductions in SPL at the eardrum of up to 40 dB are possible. For very high noise levels (above 140 dB), simple ear protection is inadequate since excessive vibration can be conducted to the cochlea by bone or tissue. Figure 11.8 illustrates the effects of ear protection on the attenuation of different frequencies of sound from a pneumatic hammer.

Table 11.8 provides sample data on the attenuation likely to be achieved using well-fitting, commercially available earmuffs (Melamed and Bruhis, 1996).

Even if ear protection is worn, sound may still find its way into the auditory system. The protector may vibrate, or energy may pass through leaks—usually a problem with "homemade" earplugs. Some characteristics of effective ear protectors are given in Table 11.9.

Sized earplugs are reusable and are usually made of soft, flexible materials such as molded plastic. They come in different sizes and should be cleaned with soap and water after use. Formable earplugs are usually made of materials such as cotton or wax. The user molds the plug into a cone shape to conform to the dimensions of the ear canal. Protection depends on the material used—cotton

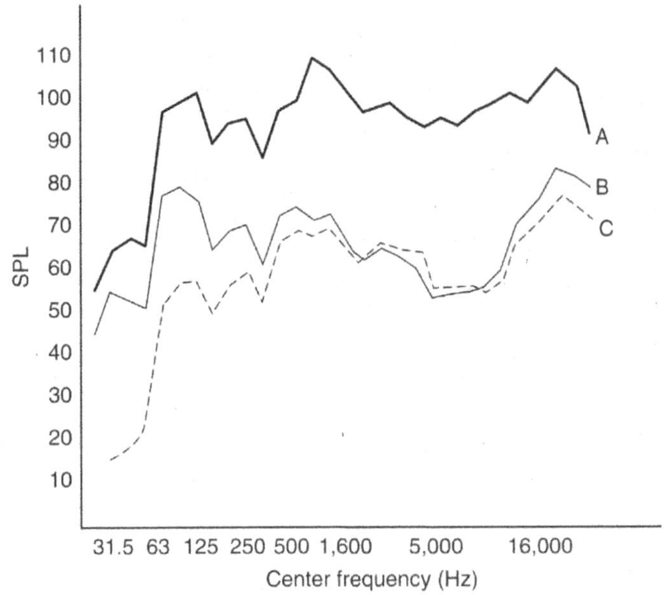

FIGURE 11.8 The effects of ear protection on the attenuation of different frequencies of sound from a pneumatic hammer. (a) Third-octave analysis. (b) Attenuation due to ear protectors. (c) A-weighting of attenuated noise.

TABLE 11.8
Attenuation Characteristics of Commercially Available Earmuffs

Frequency (Hz)	250	500	1000	2000	4000	6000	8000
Attenuation (dB)	16	21	26	30	34	34	33

Source: Adapted from Melamed and Bruhis (1996).

TABLE 11.9
Characteristics of Effective Ear Protectors

1. Impervious to air
2. Adaptable to the shape of the user's head or external auditory canal to form an airtight seal. Avoid pressure "hot spots" to ensure comfort
3. Remain firmly in place without causing pressure ischemia

provides less protection than wax and certain polymers. Earplugs have the advantages of being small and easy to carry, convenient to wear in confined spaces, and cheaper than earmuffs. They are also compatible with long hair or protective headgear. The main disadvantages are that they take longer to fit than earmuffs and are not always suitable for regular use because of hygiene problems. Unclean hands may transfer dirt to the ear canal via the plug. Earplugs cannot be worn by people with ear infections.

Nonlinear Earplugs

Perforated earplugs are used to protect against intermittent impulse noise. At low to moderate intensities, they offer little attenuation, particularly at low frequencies (up to 1 kHz), but offer attenuation of up to 26 dB for impulses over 160 dB and, at the same time, allow normal speech communication. Earmuffs should provide a good seal between the inside of the cup and the skin surrounding the pinna. They should be adjustable for a range of head sizes. They usually provide better and more consistent protection than earplugs and are perceived as more comfortable. Normally, a single-size, adjustable muff can fit most head sizes. Earmuffs can also be incorporated into other protective headgear, such as hard hats, and are less likely to be forgotten. However, they can cause skin rashes in hot environments because sweat accumulates at the muff–skin interface. In confined spaces such as mine shafts, they can be dislodged rather easily.

One possible disadvantage of hearing protection is loss of localization accuracy. Talcott et al. (2012) report reductions in the ability to localize gunshots when hearing protection is worn compared to no protection.

Training in the Use of HPDs

Green et al. (2021) report the findings of National Health Interview Surveys carried out in 2007 and 2014. Non-use of HPDs among almost 40,000 respondents exposed to loud noise at work was found to be prevalent at 53% overall. Non-use was defined as wearing HPDs "half the time," "seldom," or "never" when exposed to noise. Non-use of HPDs was highest in industries with a lower prevalence of noise exposure. HPD use was highest in mining, manufacturing and engineering and lowest in health care and education. Non-use of HPDs was more prevalent among younger workers. However, there were no trends in non-use of HPDs between 2007 and 2014.

Manufacturers of ear protection sold in the United States are required to provide data on the noise attenuation provided by their products (known as the NRR, or noise reduction rating). NRRs from 25% to 50% are needed depending on the intensity of the noise. NRR data are obtained under ideal testing conditions and less protection may be provided in practical settings.

Toivonen et al. (2002) divided 54 male subjects into two groups. One group was trained, by means of a lecture followed by practice, to fit earplugs correctly, and the other group was untrained. Both groups were then exposed to noise while wearing self-inserted earplugs. A-weighted noise attenuation was 21 dB(A) for the untrained group and 31 dB(A) for the trained group. At 1000 Hz, the attenuation was 24 dB for the untrained group and 30 dB for the trained group. These differences are both statistically and operationally significant. Exposed to noise of 110 dB, the trained group would be able to work safely for 8 h, whereas the untrained group would not (a net exposure of 79 dB(A) as opposed to 89 dB[A]). Workers must be trained in correct fitting techniques if proper protection is to be obtained. This applies to all kinds of ear protection, particularly compliant foam plugs. For tasks requiring a great deal of movement, foam plugs may offer the best protection over the workday if workers are trained to fit them properly. If wearers cannot be trained or use is occasional, earmuffs may be the best choice.

Stephenson (2009) summarized barriers to the use of HPDs expressed as concerns about comfort, convenience, cost, corporate climate, and communication.

Comfort: Concerns about comfort are common, but many are unaware of the large number of different designs available. Prior to training, only 53% of a focus group of apprentices agreed that earplugs could be comfortable. This increased to 83% after training on how to fit them properly.

Communication: It is commonly believed that HPDs hinder communication with colleagues. Many are unaware of "overprotection," wherein HPDs with excessive NRR are used, and that it is important to select an appropriate ear protector for the context of use. In extremely loud noise, where a high NRR is a requirement, voice communication without HPDs will also be hindered. Ear protectors do not necessarily interfere with speech communication because they lower the intensity of noise as well as of speech (the signal-to-noise ratio remains the same). The lowered SLs may reduce mechanical distortion of structures in the ear and actually improve speech recognition. Casali et al. (2004) report that, for people with normal hearing, HPDs have no effect on speech intelligibility in noise of 80 dB(A) and above. There may be detrimental effects on speech intelligibility in noise below 80 dB(A), but at these levels, HPDs are not required. Also, the ability to hear auditory alarms seems to be improved when HPDs are worn in background noise of 85 and 100 dB(A). Thresholds for detecting an alarm were between 3.2 and 4.4dB lower when earmuffs, earplugs, or active noise reduction headsets were worn compared to no hearing protection.

Training programs should focus on educating employees about the risks of hearing loss and then about the options available to protect their ears. Training in how to fit ear protectors correctly is also important to promote "self-efficacy"—the belief that one has the knowledge and skills to deal with the problem.

TOOLS AND PROCESSES

MEASUREMENT OF SOUND

SL meters provide several different measures of sound intensity. The dB (linear) scale is used to give the SPL. This is the rms sound pressure at the microphone. When sound is measured using the dB(A) scale, a weighting network selectively amplifies or attenuates different frequencies, similar to adjusting the tone control on a modern hi-fi amplifier. The sum of the pressure levels of the various frequencies is called the SL.

Many devices (e.g., electric fans or motors) produce intense sound energy at a particular frequency (centered around the 60Hz mains frequency or its harmonics). Although the SPLs might be high, the weighted SL is 20dB or more lower since the ear is not very sensitive to sound below 500Hz. Apart from the "A" scale, some SL meters have "C" and "D" scales. These specialized scales are used to evaluate noise containing intense pure tone components (e.g., noise from jet engines).

Speech Interference Level

The speech interference level (SIL) is the arithmetic mean of the SPLs of the octave bands that characterize the speech signal. The four-band SIL (500Hz, 1kHz, 2kHz, and 4kHz) indicates the extent to which ambient noise will mask speech and therefore degrade communication. The preferred speech interference level (PSIL) is a three-band measure of speech interference, where:

$$\mathrm{PSIL} = (L_{p500} + L_{p1000} + L_{p2000}) / 3\mathrm{dB}$$

and

L_{p500}—the octave-band pressure level at 500Hz
L_{p1000}—the octave-band pressure level at 1000Hz
L_{p2000}—the octave-band pressure level at 2000Hz

Table 11.10 gives the distances at which conversation can take place at different SILs.

Several Sound Sources

In practice, there are usually several sources of sound, and it is the combined effect of these on the worker that is of interest. The combined SPL from several sound sources can be calculated from first principles. For two sound sources of equal intensity, for example:

TABLE 11.10
Maximum Distances (m) for Satisfactory Communication in Noise

SIL (Four-Band)	Normal Conversation	Raised Voice
35	7.5	15.0
40	4.2	8.4
45	2.3	4.6
50	1.3	2.6
55	0.75	1.5
60	0.42	0.85
65	0.25	0.50
70	0.13	0.26

$$L_{p\,\text{combined}} = 10\log_{10}\left[(P_i/P_o)^2 + (P_i/P_o)^2\right]$$

$$= 10\log_{10}\left[(P_i/P_o)^2 \times 2\right]$$

$$= 10\log_{10}(P_i/P_o)^2 + 10\log_{10} 2$$

$$\text{Since } \log_{10} 2 = 0:301$$

$$L_{p\,\text{combined}} = 10\log_{10}\left[(P_i / P_o)^2 + 3.01\text{ dB}\right]$$

Put simply, to add two SPLs of equal intensity, just add 3 dB to the SPL. Combined SPLs for two sources that differ in intensity can be found by calculating the difference between the two loudest SPLs and adding the amount shown in Table 11.11 to the louder of the two.

If there are more than two sources of noise, the exercise can be continued by taking the combined SPL of the two loudest sounds and comparing this with the third loudest sound. The difference between these is then added to the estimate of the two loudest sounds to obtain a new estimate of the overall noise. This process can be repeated for other noise sources until the difference between the combined estimate and any remaining sources is 10 dB, at which point the addition of more noise sources will, in practice, have a negligible impact on the combined SPL.

That the addition of two equal sources increases the total SL by 3 dB reflects the logarithmic nature of the decibel scale. This also applies to the subtraction of noise levels. If three adjacent machines have noise levels of 90, 95, and 101 dB, the combined noise will be 102 dB. Removal of the 90 dB machine will still leave a noise level of 102 dB, whereas removal of the 101 dB source leaves a noise level of 96.5 dB. This has many practical implications for noise control, as will be seen below. In particular, it emphasizes that the best approach to noise control in a room with several noise sources is to begin with the loudest, even if it is not the most amenable to acoustic treatment. Another practical implication is that surveys of the noise produced by machines can be carried out in the presence of background noise if the latter is 10 dB or more below the machine noise.

Measuring Noise Exposure

A number of measures have been developed to enable employees' daily noise exposure to be measured. The equivalent A-weighted noise level, or "$LA_{eq,}$" has been developed for situations where noise levels fluctuate over the course of a day. It is an integrated value—the average level of sound energy over the measuring period.

TABLE 11.11
Summation of SPLs L_1 and L_2

Difference between L_1 and L_2	Amount to Increase Loudness of L_1 and L_2
0.0	3.0
0.5	2.8
1.0	2.5
1.5	2.3
2.0	2.1
2.5	1.9
3.0	1.8
3.5	1.6
4.0	1.5
4.5	1.3
5.0	1.2
5.5	1.1
6.0	1.0
6.5	0.9
7.0	0.8
7.5	0.7
8.0	0.6
8.5	0.5
9.0	0.4
9.5	0.4
10.0	0.3

The LA_{eq} is defined as the steady-state SL that would have the equivalent sound energy as the actual noise over the same period of time. The LA_{eq} can be thought of as "smoothing out" the peaks and troughs of daily noise exposure to give an average. If a worker's exposure times to different noise levels over the course of an 8h shift are known, the LA_{eq} can be calculated as follows:

$$LA_{eq} = 10\log_{10}\left[(T_1/8)\text{anti}\log(L_1/10)+(T_2/8)\text{anti}\log(L_2/10)+\cdots+(T_n/8)\text{anti}\log(L_n/10)\right]$$

where $T_1 \ldots T_n$,=exposure time (in hours) at SLs $L_1 \ldots L_n$.

Note that the time, T, spent at a given SL is expressed as a proportion of the total 8-h shift.

In practice, employees wear a small recorder or SL meter with the LA_{eq} facility during a shift. It continuously measures the noise to which they are exposed. Since the LA_{eq} takes into account infrequent but very loud noises, it can sometimes give a better indication of an employee's total noise exposure than a noise survey consisting of one or more short visits to measure the noise in an area, during which loud but infrequent noises may not occur.

Safe Exposure Levels

In the United States, the Occupational Safety and Health Administration (OSHA) has specified 90 dB(A) as the maximum permissible exposure to continuous noise for an 8-h shift. Mandatory hearing conservation programs, training, and audiometric testing are required if employees are exposed to 85 dB(A) or more per day. Continuous noise above 115 dB(A) is not permitted. The EU has a new physical agent (noise) directive that specifies a first action level of 80 dB(A), at which hearing protection becomes advisable. At 85 dB(A), the second action level, protection becomes mandatory. The maximum daily exposure (L_{ex}) is 87 dB(A). The peak action level, for all practical purposes, is now 135 dB(A). Note that the United States uses a halving rate of 5dB, whereas the EU uses 3dB, so the corresponding exposures for durations under 8h will differ even more between the two regions.

In general, *first action levels* indicate to employers that health surveillance is required and that employees must be informed of what they are being exposed to. *Second action levels* indicate, in addition, that action needs to be taken to reduce the exposure, and employees must be protected.

Noise Dosimeters

A noise dosimeter integrates the noise measured at the microphone over a period of time and expresses it as a percentage of the daily allowable noise dose (e.g., as a percentage of 90 dB(A) for 8 h). Dosimeters can be used to determine whether workers are being exposed to excessive noise and whether they require ear protection, as well as to specify how long per day they may be exposed to a particular noisy work situation (Figure 11.9).

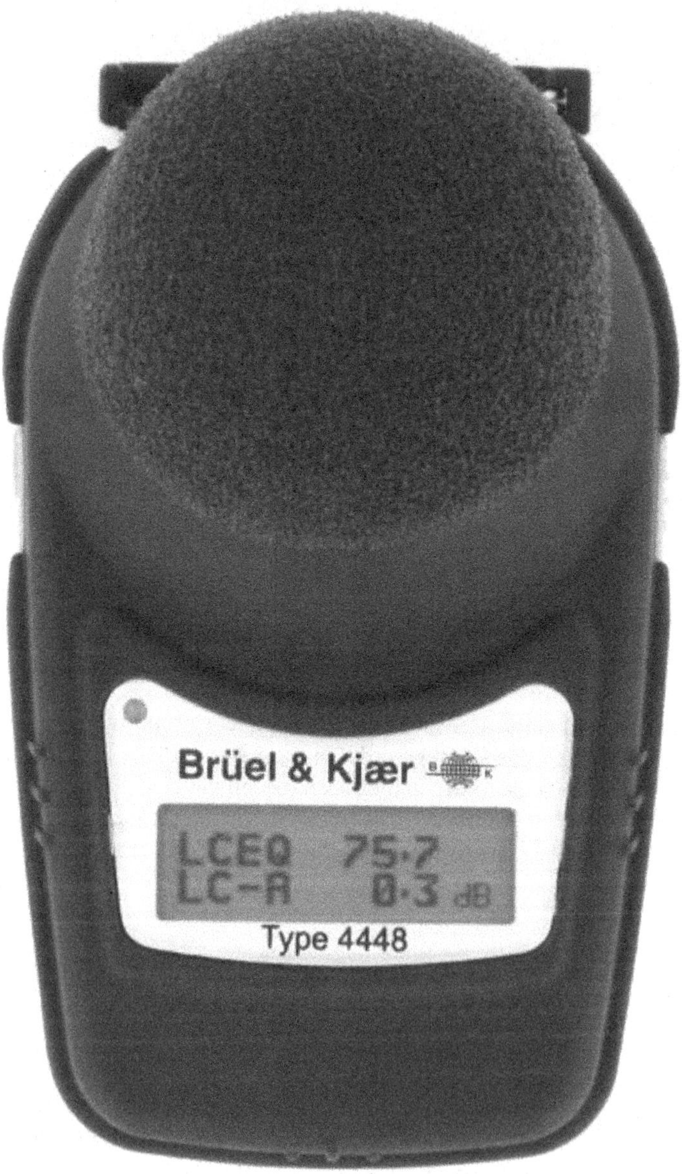

FIGURE 11.9 A noise dosimeter (Copyright © Hottinger Brüel & Kjær. All rights reserved).

LA_{eq} and noise dose measurements are particularly useful in hazard surveillance of occupations where the worker is mobile because they integrate the different intensities of noise, which can vary considerably over the course of the working day. For example, garbage collectors, factory supervisors, and agricultural workers are all exposed to varying noise levels during the course of their work, and an integrated value of their total noise exposure is often more meaningful than isolated measurements made during particular work activities.

Standards for noise exposure are a compromise between the desire to provide maximum protection for workers' hearing and the cost–benefits of noise reduction. Many industrial processes are intrinsically noisy, and although practical steps can usually be taken to reduce the noise, a point of diminishing returns is reached, where further reductions in noise become increasingly modest but cost much more. Also, individuals differ in their susceptibility to noise-induced hearing loss—it is often impossible to lower noise levels to ensure complete protection for all workers. A level of 90 dB(A) for 8h is OSHA's maximum permissible exposure level. An equal energy rule is used to specify maximum exposure times for louder noises. For example, using a halving rate of 5dB, a noise dose of 90 dB(A) for 8h is considered equivalent to 95 dB(A) for 4h or 100 dB(A) for 2h (a halving rate of 3dB is used in the EU). Standards usually specify a maximum steady noise level (overriding limit) that operators must not be exposed to at all. The OSHA limit is 115 dB(A). It can be seen that even very brief daily exposures to very high noise levels can render an otherwise innocuous work area a noise hazard. Once again, it must be emphasized that very loud noises (over 115 dB[A]) can make an inordinately large contribution to a worker's daily noise exposure even if they are of short duration. Eliminating exposure to noise at this level is usually one of the simplest and most straightforward ways of reducing daily noise exposure.

HFE Workshop 11.1 demonstrates the use of equations to calculate acceptable exposure times for noise exposure using U.S. and EU standards.

HFE WORKSHOP 11.1

CALCULATING DAILY EXPOSURE TIMES AND NOISE DOSES

Niebel and Freivalds (1999) provide some simple equations for calculating exposure times using U.S. standards. Given a maximum daily exposure of 90 dB(A) over 8h and a halving rate of 5 dB(A), the exposure time, *T*, for a particular noise exposure, *L*, is given by

$$T = 8/2^{(L-90)/5}$$

So, if the actual noise level were 95 dB(A), the maximum exposure time would be

$$T = 8/2^{(95-90)/5} = 8/2^1 = 4\text{ h}$$

For EU readers, estimates for an 85dB maximum noise dose and a 3 dB(A) doubling rate can be calculated in a similar manner, where

$$T = 8/2^{(L-85)/3}$$

Using these equations, maximum exposure times (hours) for the noise levels listed below are given using U.S. and EU standards:

(*Continued*)

HFE WORKSHOP 11.1 (*Continued*)

Clearly, the EU exposure limits are much stricter than their U.S. counterparts.

Noise Level (dB[A])	$T_{U.S.}$	T_{EU} (Upper Exposure Action Value)
85	16.0	8.0
86	13.9	6.4
87	12.1	5.1
88	10.6	4.0
89	10.1	3.2
90	8.0	2.5
91	7.0	2.0
92	6.1	1.6
93	5.1	1.3
94	4.6	1.0
95	4.0	0.8
96	3.5	0.6
97	3.0	0.5
98	2.6	0.4
99	2.3	0.3
100	2.0	0.25
105	1.0	0.08 (288 s)
110	0.5	0.025 (90 s)
115	0.25	0.008 (29 s)

Varying Exposures Throughout the Day

Again, Niebel and Freivalds provide a simple method from OSHA for calculating a daily noise dose, *D*.

$$D = 100 \times (C_1 / T_1 + C_2 / T_2 + \cdots + C_n / T_n)\%$$

where

C = Time spent at a specified noise level

T = Time permitted at a specified noise level

The values of T can be obtained from the table above.

Suppose a worker in the United States spends 2 h working at 85 dB(A), 3 h at 95 dB(A), and 1 h at 92 dB(A). What is his daily noise dose?

$$C_1 = 2 \quad C_2 = 3 \quad C_3 = 1$$
$$T_1 = 16.0 \quad T_2 = 4 \quad T_3 = 6.1$$

$$D = 100(2/16 + 3/4 + 1/6.1) = 100(0.125 + 0.75 + 0.16) = 100(1.035) = 103.5\%$$

The worker has just exceeded the daily noise for 8 h. Reducing the noise from 92 to 90 dB(A) would bring the daily noise dose to 100%.

(Continued)

HFE WORKSHOP 11.1 (*Continued*)

The same calculations can be carried out on these data using the EU standards. Only the permitted exposure times change, as follows:

$$C_1 = 2 \quad C_2 = 3 \quad C_3 = 1$$
$$T_1 = 8 \quad T_2 = 0.8 \quad T_3 = 1.6$$

$$\begin{aligned} D &= 100(2/8+3/0.8+1/1.6) = 100(0.25+3.75+0.625) \\ &= 100(4.625) = 462.5\% \end{aligned}$$

According to EU standards, the worker has received almost five times his daily noise dose.

Clearly, the standards chosen for daily noise doses can have large effects on the exposure levels permitted on a daily basis. In turn, this has major implications for the steps that need to be taken to comply with prevailing regulations. The difference between U.S. and EU standards lies not so much in the difference in the daily maximum allowable dose, but in the halving rate—the rate at which the permitted exposure time changes as the noise level deviates from the maximum daily noise dose. The rate is 5 dB(A) in the United States and 3 dB(A) in the EU.

Integrating SL Meters

To cope with the widely varying types of noise encountered in ndustry, commercial meters (Figure 11.10) usually have several response modes—normally "slow" and "fast," and sometimes an "impulse" response mode.

The slow mode is used to obtain an average reading when rapid fluctuations in intensity, measured in the fast mode, are greater than 3 or 4 dB. In fairly steady noise with fluctuations of only a few decibels, the fast response may be used, and a simple average can be taken by the person carrying out the measurements. For example, fluctuations from 85 to 91 dB could be interpreted as 88±3 dB. If the fluctuations are greater than 6 dB, the average SL can be estimated as 3 dB below the maximum value. Impulse noise cannot be measured accurately with basic meters because their response time is too slow. More sophisticated equipment is now available. As a rule of thumb, the impact noise level measured with a standard meter should be regarded as an underestimate of at least 5 dB.

Before measuring a sound source, the SL meter should be calibrated. Meters are commonly sold with a portable calibrator (which itself should be periodically checked). When measuring outdoors, it is advisable to attach a spherical windscreen (made of foam and about 5 cm in radius) over the microphone to prevent wind from distorting the measurements. When measuring operator exposure to noise, the microphone should be situated close to the operator's ear. Octave, half-octave, and third-octave sets are available, which can be used in conjunction with standard meters. Table 11.12 summarizes the information that should be obtained to complement the SL measurements.

Noise Surveys

Noise surveys are a useful method for evaluating the distribution of noise in a working area and the exposure of workers. Occupational hygienists use two methods—area sampling and personal sampling. In area sampling, the workplace is evaluated which works well if the noise is fairly constant and workplaces are fixed. If people move around at work, or if noise levels are closely tied to a worker's activity, personal sampling is necessary. In some industries, such as construction and mining, a combination of both approaches is best.

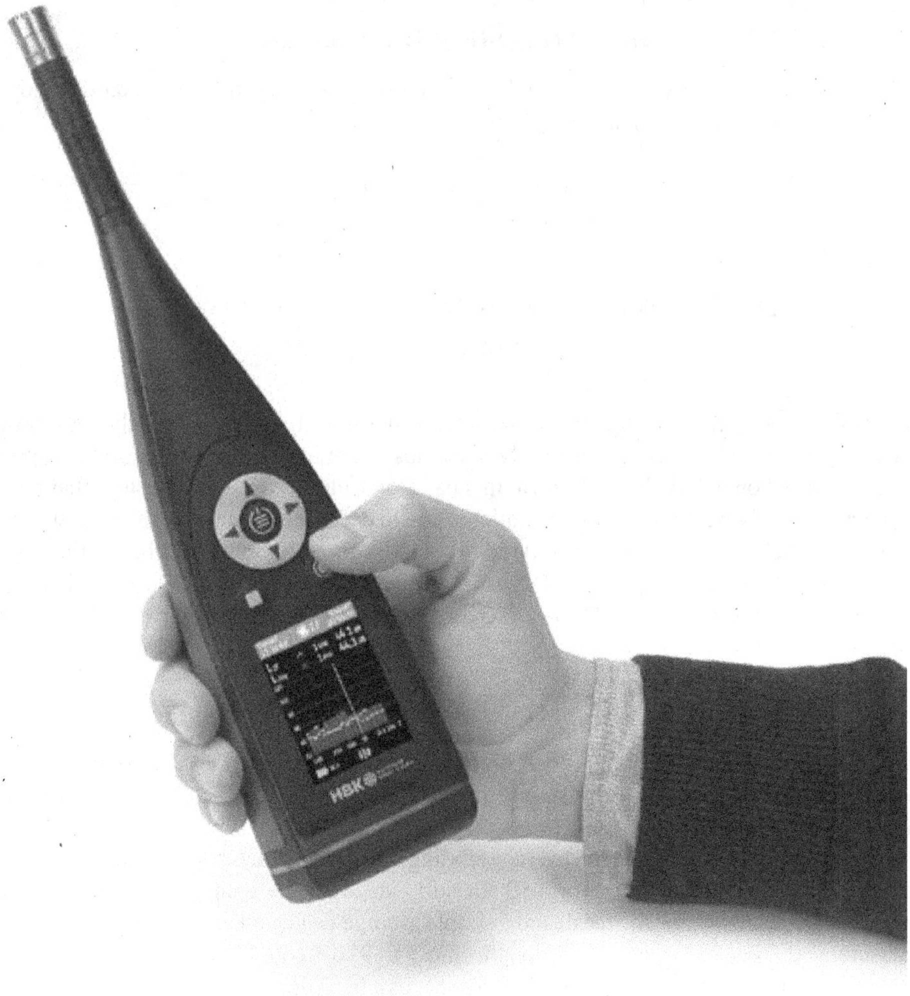

FIGURE 11.10 Modern SL meters. SL meter (Copyright © Hottinger Brüel & Kjær. All rights reserved).

TABLE 11.12
Information for Noise Surveys

1. Title of the assessment—purpose, location, date (if this is a reassessment, give details)
2. Assessor (name, post, company, signature)
3. Description of the space in which the measurements were made, its dimensions, background noise, and the presence of other noise sources (append sketch of the area, as required)
4. Description of the source itself. For example, data from the nameplate, such as the model number, operating speed, and power, the percentage of maximum load when the measurement was made, and, if different from this, the normal load
5. Summary of results:

Post/Officer	No. of Employees Exposed	First Action Level Exceeded?	Second Action Level Exceeded?	Peak Level	L_{eq}

6. Further specialist assessments required?
7. Make, model number, weighting network, and response mode of the SL meter

(Continued)

TABLE 11.12 (*Continued*)
Information for Noise Surveys

8. Date of last calibration
9. Background noise level
10. Position of the microphone with respect to the source
11. Extent of fluctuation of noise levels
12. "A" scale measurements at the operator's ear level and the positions of other personnel
13. Time spent at the machine by the operator each day
14. Results of any previous audiometric testing of workers
15. Previous attempts at noise control
16. HPD availability. Description of HPD and comments on compliance

TABLE 11.13
Some Basic Steps in the Management of Industrial Noise Exposure

1. Short-term measures	Issue earplugs/earmuffs
2. Medium-term measures	Reposition noisy machines
	Soundproof noisy machines
	Demarcate noisy areas with warning signs
	Rotate workers between "noisy" and "quiet" jobs
3. Long-term measures	Comprehensive noise reduction program
	Soundproof machines
	Replace with less noisy machines
	Change the process
	Build "acoustic refuges"
	Audiometric testing
	Implement rules and procedures for the wearing of ear protection
	Audiometric testing program

In area sampling, a plan map of the workplace is needed. SLs at various locations are recorded directly onto the map. Areas of high and low noise can be identified so that appropriate remedial action can be taken. Noise contours—regions of equal SL can be superimposed onto the floor plan. In conjunction with data about the positioning and movement of personnel, a number of steps can be taken to minimize excessive noise exposure (Table 11.13).

An example of a noise survey that was carried out in a colliery can be found in Bowler (1982). All areas where noise levels exceeded 90 dB(A) were identified, and the duration of exposure of persons working therein was determined. Special attention was paid to areas with noise levels greater than 105 dB(A). Within the previously determined 90 dB(A) contour, attenuation of noise was measured by taking repeated measurements up to 1 m from the source (intensity halved approximately as distance from the source doubled). Maximum and minimum noise levels were also measured at the operator's ear level at all fixed workstations. Together with data on exposure times, representative LA_{eq}'s were calculated. Operators suspected of being exposed to noise over 105 dB(A) were issued dosimeters so that more information on their exposure could be obtained. Noise exposure when traveling to and from working areas was also measured. In this way, both high-risk areas and high-risk jobs were identified.

Whether or not hearing protection is used can have large effects on daily noise exposures. HFE Workshop 11.2 describes a method known as activity sampling that can be used to estimate the amount of time employees really spend wearing ear protection when it is needed.

HFE WORKSHOP 11.2

ACTIVITY SAMPLING FOR HPD COMPLIANCE ASSESSMENT

Activity sampling is a method long used in work study to provide estimates of the amount of time in a day that a person is either carrying out an activity or not carrying it out, or the amount of time a machine is either working or not working. It can also be used to assess how long or what percentage of the day, a task takes. In this example, we are using this method to estimate the proportion of the day employees wear ear protection (HPD) when exposed to continuous noise. We can then estimate the daily noise dose and decide whether the protection is adequate.

The general approach is to observe, at random intervals over the workday, whether the ear protectors are being worn or not, and record the event. The proportion of occasions (p) spent not wearing ear protectors is then estimated as follows:

$$p = x / n$$

where

x is the number of occasions HPD was not observed being worn.
n is the total number of observations.

In this case, p is analogous to the mean time spent without HPD. Even though the employees were not observed continuously over their workday and no measurements of the time spent wearing ear protection were made, we will assume that, as long as the observations were made randomly, the proportion of occasions without HPD is a good surrogate measure of the time spent unprotected. So, if the workday is 8 h and the proportion of observations when HPD compliance was observed is 0.85, then the total time spent wearing them is estimated to be

$$T = 0.85(8 \times 60)\text{min} = 408 \text{ min or } 6 \text{ h and } 48\text{min}$$

and the time spent without HPD is 72 min.

As with any other kind of sampling procedure, the estimate of the proportion of time spent without HPD is only an estimate, and it has a degree of error associated with it. The standard deviation of p is given by

$$\text{SD}(p) = \surd\left(p(1-p)/n\right)$$

Since employees are either wearing or not wearing HPD, and if the probability of HPD noncompliance is p, then the probability of compliance is $1 - p$.

The 95% confidence interval for p—the range of values of p within which we can be 95% certain that the true value of p lies—is given by

$$p \pm 1.96 \text{ SD}(p) = p \pm 1.96(\surd\left(p(1-p)/n\right)$$

Accuracy and Power in the Estimation of p

The advantage of using activity sampling rather than continuous observation is that it should save time. This time saving should not be at the expense of accuracy. In this example, we are using the method to estimate HPD compliance, which will enable an accurate estimate of noise exposure to be made. Therefore, we need to make enough observations such that the 95% confidence interval associated with our estimate of p is small.

(Continued)

HFE WORKSHOP 11.2 (*Continued*)

Let:

S=the desired level of accuracy in the estimate of $p=5\%p=proportion$ of time without wearing HPDn=number of observations required to estimate p (±5%) $Sp = 1.96\sqrt{(p(1-p)n)}$

At this stage, there are too many unknowns. Both p and n are unknown. We can estimate p by conducting a pilot study over a day or two and perhaps by talking to supervisors. Suppose we obtain an initial estimate of

$$p = 0.25$$

This means that employees

This suggests that employees spend about a quarter of the working day in noise without HPD.

To obtain an estimate of the true proportion of time without HPD, with 5% accuracy:

$$Sp = 1.96\sqrt{(p(1-p)/n)}$$

$$0.05p = 1.96\sqrt{(p(1-p)/n)}$$

Therefore,

$$0.0025p^2 = 3.8(p(1-p)/n)$$

$$\begin{aligned} n &= 3.8p(1-p)/(0.0025p^2) \\ &= 3.8(1-p)/(0.0025p) \\ &= 520(1-p)/p \\ &= 4560 \end{aligned}$$

Thus, to gain a 5% accurate estimate of p if the true value of p is approximately 0.25, we need to make 4560 random observations.

If we only had time to make 3000 observations, we could estimate the accuracy of our value of p as follows:

$$\begin{aligned} Sp &= 1.96\sqrt{(p(1-p)/n)} \\ &= 1.96\sqrt{(0.25(1-0.025)/3000)} \\ &= 1.96\sqrt{0.0000625} \\ &= 0.015 \\ S &= 0.015/0.25 \\ &= \pm 0.062 \\ & \text{or} \pm 6\% \text{ accuracy} \end{aligned}$$

(*Continued*)

HFE WORKSHOP 11.2 (*Continued*)

INTERPRETATION

This example describes the use of the activity sampling method to estimate HPD compliance. The findings suggest that the 95% confidence interval for HPD compliance is between 23.5% and 26.5% of the workday. We can be 95% certain that employees spend at most 26.5% of the workday without ear protection. If we know the ambient noise level and have an idea of the NRR of the HPD, we can recalculate the actual noise dose using the method in Workshop 11.1.

Suppose the actual noise level is 95 dB(A) and the NRR of the HPD is 10 dB(A), then (using an 8 h dose of 90 and a 5 dB(A) halving rate) the noise dose with full HPD compliance would be 50% (actual dose of 85 dB(A) for 8 h).

Taking a worst-case scenario, employees fail to comply with HPD use 26.5% of the day. This means that they are exposed to 95 dB(A) for 26.5% of 8 h, that is, 2 h and 7 min, and for the rest of the time to an effective 85 dB(A). The actual noise dose is therefore:

$$D = 100(2/4 + 6/16) = 87.5\%$$

Refer to Workshop 11.1 and recalculate the noise dose based on an 8-h maximum of 85 dB(A) and a rate of 3 dB(A). Comment on the difference between the noise doses.

SAMPLING METHODS

It is important that observations are made randomly. Random number tables can be used but Barnes, with typical pragmatism, suggests that 480 small pieces of paper (representing each minute of a workday) be numbered from 1 to 480, placed into a hat, mixed, and drawn. Observations, say, 50 per day, might then be made at each minute of the workday at the times specified by the numbers on the slips of paper.

The advantage of the activity sampling method is that it enables us to make estimates from simple binary observations and to estimate the confidence of our estimates. It cannot be used where p is small (i.e., to estimate rare or unusual occurrences).

Vibration

A detailed treatment of vibration is beyond the present scope, although a few of the most important aspects in ergonomics can be briefly introduced. Vibration is defined as the oscillation of a body about a reference position and can be described, like noise, in terms of amplitude, frequency, and phase. Of most interest in ergonomics (HFE) are the effects of vibration on health, task performance, and communication.

The unit of vibration is the rms or peak acceleration of the oscillation. Vibration dose values are time-weighted averages of vibration magnitude. Vibration is measured using accelerometers placed in the workplace or on seats where exposure measurement is required. Accelerometers are placed such that vibration can be measured in three translational (backward and forward, up and down, and side to side) and three rotational axes (pitch, yaw, and roll). Vibration in the vertical plane is usually weighted most highly when the outputs of the different accelerometers are combined (see ISO 2631-1, 2008, for example).

TABLE 11.14
Human Response to Vibration

Axis	Frequency (Hz)	Effect
Vertical	0.5	Motion sickness, nausea, and sweating
	2	The whole body moves as one
		Difficulty positioning hands
	4	Vibration transmitted to the head. Lumbar vertebrae resonate
		Problems writing or drinking
	4–6	Resonance of the gastrointestinal system
	5	Maximum discomfort
	10–20	Voice warbles
	15–60	Vision blurred (resonance of the eyeballs)
Horizontal	<1	Increased postural sway
	1–3	Upper body destabilized
	>10	Backrest is a prime cause of vibration transmission to the body

Human Responses to Vibration

Table 11.14 summarizes some of the main effects. See Griffin (2012) for further information.

Vibration Exposure Limits

The limits vary in different countries, but they are similar to exposure limits for noise, being based on the concept of exposure to a "dose" of vibration. For WBV:

- Exposure to $0.5\,m/s^2$ (8 h) requires employers to take action to reduce exposure.
- Exposure to $1.15\,m/s^2$ (8 h) is the limit value which must not be exceeded.

Figure 11.11 is a nomogram that can be used to find permissible exposure times to vibration of different intensities.

For hand–arm vibration, the American Conference of Government Industrial Hygienists has the following threshold limit values (TLVs) (Table 11.15).

The values are frequency-weighted based on ISO 5349.

Vibration and Back Injury

Sandover (1988) proposed a fatigue model for predicting the effects of vibration on the spine. The model attempts to predict the number of loading cycles to failure of the spine using estimates of the applied dynamic load (or stress) and the static failure load of the particular spinal motion segment.

$$\text{Cycles to failure} = (\text{applied dynamic stress/static failure stress})^{-S}$$

where S is an empirically derived constant (for the spine, S is approximately 8).

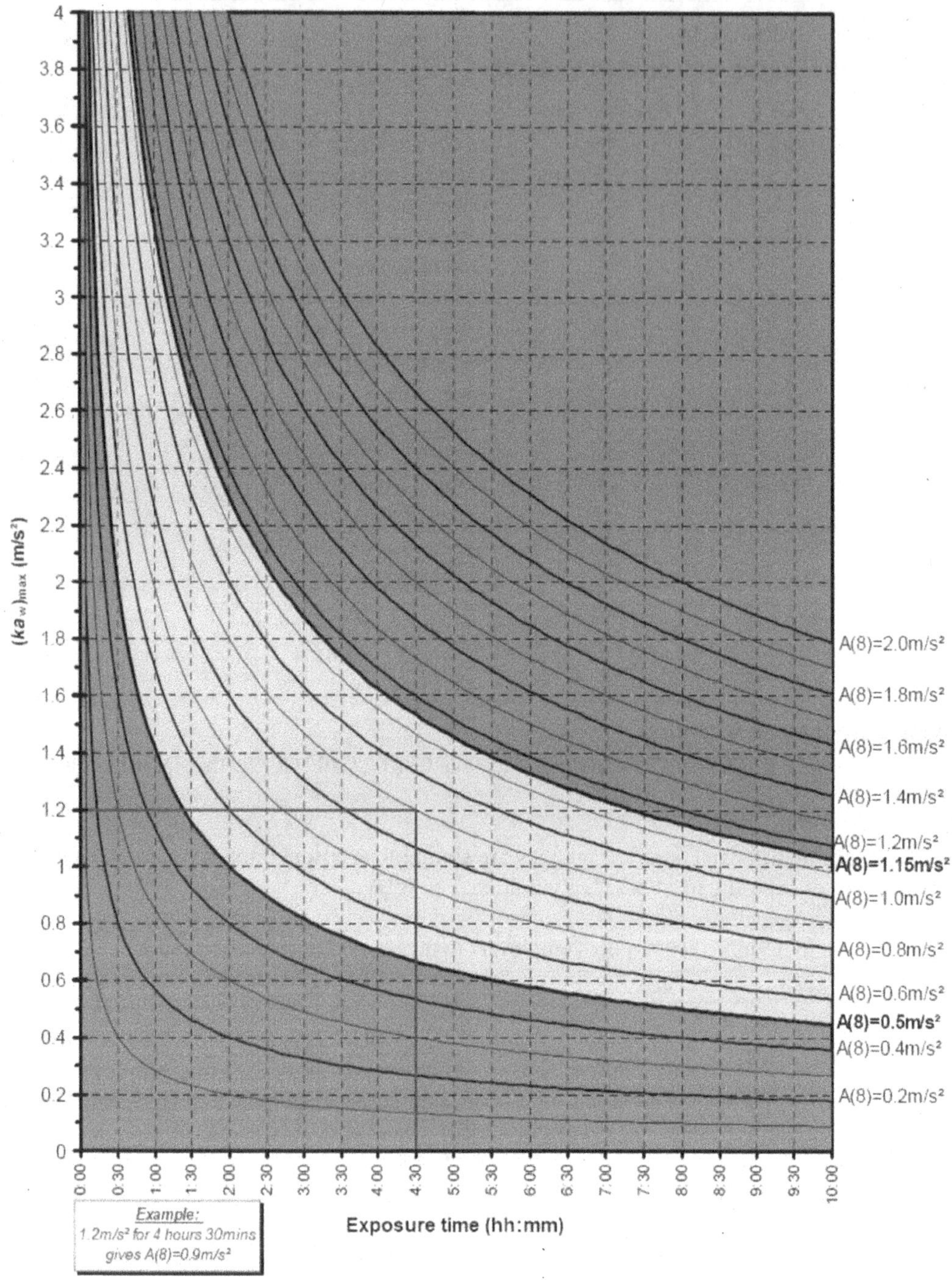

FIGURE 11.11 Exposure times for vibration.

The applied dynamic stress can be estimated from the acceleration data (vertical acceleration due to vibration in the z direction plus acceleration due to gravity acting on the upper body mass), and the static failure stress can be estimated from the equation of Genaidy in Chapter 2. For example, if the applied dynamic load is 2500 N and the static failure load of the L_3/L_4 lumbar motion segment is 7500 N, then the number of loading cycles before the lumbar spine fails is given by

TABLE 11.15
TLVs for Exposure of the Hand to Vibration in *X*, *Y*, or *Z* direction[a]

Total Daily Exposure Duration (h)	Maximum Value of Frequency Weighted Acceleration (m/s²) in Any Direction[a]
4 to less than 8 h	4
2 to less than 4 h	6
1 to less than 2 h	8
Less than 1 h	12

[a] Directions of axes in the 3D system.

$$\text{Cycles to failure} = (2500/7500)^{-8} = 6561$$

For a load of 250 N, the number of loading cycles before the lumbar spine fails is given by

$$\text{Cycles to failure} = (250/7500)^{-8} = 656{,}100{,}000{,}000$$

If the vibration frequency was 4 Hz, then the time to failure would be 6561/4 = 1640.25 s for the 2500 N dynamic load (about 27 min) compared with 164,025,000,000 s (over 5000 years) for the 250 N load. This illustrates that there is no simple dose–response relationship (i.e., the more the vibration, the worse it gets) and suggests that there is a threshold above which the risk of injury increases very rapidly, even with small increases in vibration. Despite the fact that the model is highly simplistic and does not take into account the effects of rest periods or the restorative power of the tissues, it has value in demonstrating the exponential nature of risk—that there may be a critical load above which failure becomes almost inevitable during occupational exposure, and that even small reductions in loading below this level may bring about very large increases in safety.

Exposure to Shock

"Shock" is the term used in some quarters when referring to "jerk," which is the third derivative of position (ms^{-3}) or the rate of change of acceleration. Jerk is dangerous for humans because the normal adaptive and balance responses do not react quickly enough to it, unless it is expected. Jerk can be experienced when a car gradually brakes under steady resistance and finally stops with the brake still applied. At that point, the driver may experience a slight jerk as the deceleration suddenly ceases. Compare this to gradually reducing the braking just before the car comes to a stop. In some situations, people are exposed to high levels of potentially injurious vibration and shock. High-speed craft (Figure 11.12) are an example of this. Ullman et al. (2022) sent questionnaires to 2460 ex-military high-speed boat operators to determine the rate of injury and its persistence post-career. The response rate was 32%. The self-reported injury rate was high—1.1 injuries per person/year—with 67% reporting at least one occasion of sick leave; 72% had applied for a veteran's disability rating; 39% reported having surgery during active duty; and 34% reported having surgery after leaving the service. Regarding the cause of injury, the most common impacts, in order, were: mainly vertical with some lateral impact as well; mainly lateral impact; purely vertical; head-on "frontal" impacts; seat "bottoming out"; and collisions. Injuries to the back and neck were the most common, and 33% reported loss of consciousness onboard.

FIGURE 11.12 High-speed craft users are often exposed to vibration and shock (a) a fast boat breaching a wave and (b) a crew workstation with a suspension seat. (Courtesy of Ullman Dynamics.)

Shock mitigation is normally achieved through the installation of better seats, such as the suspension seat shown in Figure 11.12, to protect the occupants from injuries such as the vertebral wedge fracture described in Chapter 2.

SYSTEM INTEGRATION

Effects of Noise on Task Performance

There are several main ways in which noise can affect task performance. For verbal tasks, noise can degrade or mask the auditory signal, or it may cause distraction. Loud noise may cause discomfort, which itself may be distracting. Venetjoki et al. (2006) carried out a laboratory study in which subjects carried out a variety of tasks under three conditions—continuous noise at 48 dB(A), masked speech at 48 dB(A), and speech (simulating offices with STIs of 0.00, 0.30, and 0.80). Exposure to speech while carrying out a proofreading task reduced error detection by about 6%. Performance on tasks such as reaction time, which had no cognitive component, was not affected. This suggests that rooms where cognitively demanding work is carried out (such as libraries) should be given priority if acoustical treatment of buildings is to be used cost-effectively.

Jahncke (2012) found that irrelevant speech interfered with short-term memory processes requiring the rehearsal of information, as well as tasks requiring memory for past events (causing reductions in performance of around 10%). Word fluency and counting were less affected by speech (with up to 5% reductions in the number of correct responses). In open-plan offices, then, whether or not background speech will interfere with task performance and be a source of dissatisfaction depends on the requirements of the task. Tasks involving short-term memory are likely to be the most susceptible.

A study on Norwegian naval vessels found that sustained exposure to noise above 85 dB(A) significantly impaired visual reaction time (by about 10–15 ms) compared to noise below 72 dB(A) (Irgens-Hansen et al., 2015). Unpredictable noise can also degrade the performance of the above tasks, as well as tasks involving mental calculation or short-term recall. Schlittmeier et al. (2015) found that traffic noise at 70 dB(A) degraded performance on a task requiring attention (the Stroop task) more than traffic noise at 50 dB(A), which did not degrade performance compared to silence. Tasks requiring short-term memory (e.g., mental arithmetic) were degraded by exposure to irrelevant speech, but not by traffic noise.

Dalton and Behm (2007) carried out a systematic literature review on the effects of noise on task performance, drawing the following conclusions:

- Noise can disturb sleep and degrade the performance of vigilance tasks the next day.
- Acute and continuous noise adversely affect vigilance and comprehension.
- Irrespective of sound type, loud volume intensities impair human performance during simple vigilance and simulated driving tasks.

Music While You Work

In developing countries, it is still common for people to sing while carrying out work. In developed countries, music is often played to workers in factories in the belief that it will improve task performance.

Dalton and Behm (2007),, following a review of research on music, came to the following conclusions:

- Overall, the findings are equivocal. Music may facilitate performance involving high levels of concentration and attention. Conversely, music has also been shown to be as distracting as noise during comprehension tasks.
- A fast music tempo increases the speed at which one completes a specific task. However, a fast music tempo also increases the number of mistakes during that task.
- Music is distracting with respect to peripheral stimuli during frustrating situations (i.e., heavy traffic congestion).
- Music can reduce stress and mild aggression.
- However, its distracting nature may impede simple vigilance performance during such situations (i.e., maneuvering to avoid another vehicle or pedestrian).
- A moderate level of music is optimal for activities requiring careful attention and concentration such as driving, because it closely resembles one's comfort level. However, the determination of a moderate level is subjective to the listener.
- Finally, irrespective of sound type, loud volume intensities impair human performance during simple vigilance and simulated driving tasks.

Industrial music may be contraindicated in already noisy factories if, in order to hear the music above the background noise, SLs exceeding 85 dB(A) are necessary.

Nonauditory Effects of Noise on Health

There is some evidence that exposure to loud noise affects health in other ways, apart from damaging the ear. The quality of much of the research is poor—many of the studies lack detailed information about how the noise exposures were measured, how the health outcomes were diagnosed (if they were diagnosed and not just self-reported), and how confounding factors were controlled for. Table 11.16 summarizes the main non-auditory health effects for which there is at least some evidence (Smith, 1991).

Virkkunen et al. (2005) studied the health of 6005 middle-aged male industrial workers in Finland over an 18-year exposure period. For men exposed to continuous noise over 80 dB (and impulse noise), the relative risk of coronary heart disease was 1.38 over 9 years (1.04–1.82 95% CI) and 1.54 after 18 years (1.28–1.86 95% CI). These findings remained when other relevant risk factors were controlled for in the analysis. The relative risk of coronary heart disease (CHD) for blue-collar workers not so exposed was lower. Chronic exposure to noise moderately increases the risk of CHD in otherwise healthy men. This increased risk was found to persist after retirement and, by implication, after chronic exposure to noise ceases.

TABLE 11.16
Nonauditory Effects of Noise on Health

Nonspecific Stress Effects	On RAS: Activation of Sympathetic NS, Adrenal Medulla, Cerebral Cortex
Cardiovascular effects	Increased blood pressure
Problems caused when communicating in noisy environments	Laryngopathies, laryngitis, vocal cord polyps
	Annoyance, social isolation, impaired teamwork
Factors increasing noise annoyance	Noise is perceived as unnecessary
	Those causing the noise are perceived to be unconcerned about the welfare of those exposed to it
	The noise is perceived to be harmful
	No control over the intensity of the noise
	The noise is believed to be harmful to health
	The noise is associated with fear
Effects of noise on sleep	Noise may prolong sleep onset time
	Noise may cause awakening once asleep
	Noise may interfere with returning to sleep
	Noise may cause sleep to be shallower
Effects on mental health	Acute symptoms may be worsened
	Noise annoyance appears to best predict adverse effects

Noise and Blood Pressure

Noise exposure is thought to increase blood pressure by causing the release of blood pressure-increasing hormones into the blood. These hormones increase heart rate and cause the narrowing of arteries, which increases blood pressure. One theory is that noise activates the sympathetic nervous system, and this causes an increase in blood pressure via catecholamine secretion by the adrenal glands. Bolm-Audorff et al. (2020) conducted a systematic literature review 24 high-quality studies on noise and its effects on blood pressure. Six were cohort studies, there was one case-control study, and the rest were cross-sectional studies. The authors concluded that, despite differences in the definition of "high" blood pressure across the studies, there was strong evidence for a causal relationship, both in terms of the intensity of the exposure and the duration (in years) of exposure. For noise exposure less than 80 dB(A), the effect of exposure was not statistically significant: for exposure >80–85 dB(A), the effect size was 1.77; and for exposure to noise >85–90 dB(A), the effect size was 3.55.

Noise and Stress

In noisy environments, both subjective and objective changes occur, indicative of stress. Melamed and Bruhis (1996) measured urinary cortisol, subjective fatigue, and post-work irritability in 35 textile mill workers under chronic exposure to noise over 85 dB(A). The workers did not normally wear ear protection but wore earmuffs for 7 days over the study period. Significant reductions in urinary cortisol at the end of the shift, post-work irritability, and subjective fatigue were found when the workers wore the earmuffs. Increased cortisol levels can occur when workers are exposed to stress from many different sources, including high workload, so the findings of this study suggest that noise is a general stressor and that the subjective reactions to it may be indicative of heightened physiological arousal. Although 51% of the workers in the study reported moderate to great discomfort when wearing the earmuffs, the inconvenience was not enough to offset the physiological and psychological benefits of a reduction in noise exposure.

Sander et al. (2021) conducted an experiment on noise and stress in simulated office work. Subjects carried out a proofreading task. Open-plan office noise varies on average between 52–58 dB(A), whereas in private offices it is as low as 39 dB(A). An "open-plan soundscape" was created (people speaking, walking, printing papers, ringing telephones, and keyboard typing noises). A low-noise condition of 36.3 dB(A) was created with sounds from air conditioning and computer fan noise. Subjects worked in both conditions in counterbalanced order. Open-plan office noise exposure had no effect on cognitive performance as measured by errors detected on the proofreading task. Phasic skin conductance, reflecting a change in the moisture level of the skin, was higher in the open-plan noise condition. Heart rate was slightly higher in the open plan noise than in the low noise condition, although the effect was not large. Mood, as measured by the PANAS scale, was 25% more negative in the open-plan condition (positive mood was lower and negative mood higher). However, these are short-term effects, as the task duration in both conditions was only 8 min. After longer periods of exposure, employees may habituate to the noise, or these reactions may become chronic. An increase in hybrid working following the COVID pandemic may prevent habituation to open plan office noise in hybrid workers.

These findings suggest that negative health effects may occur below the threshold of awareness, even under the relatively low exposure levels found in open-plan offices. An absence of complaints about noise, then, should not be taken to mean that all is well, and noise measurements should be regarded as mandatory in ergonomic surveys of offices, even when it is clear that there is no threat to hearing.

Noise and Satisfaction

Whether sound is perceived as noise depends as much on the listener and the context as it does on the sound itself. SLs of 100 dB(A) are commonly found in factories and are rightly termed noise. Noise is often associated with dissatisfaction with the environment. Office work usually involves conscious mental activities that fall prey to intrusion by noise. Several factors determine whether noise in offices will be a source of annoyance. First, if the noise level itself is too high, telephone or speech communication will be rendered more difficult and satisfaction with the environment will be reduced. If the noise level is too low (quiet enough to hear a pin drop), intermittent noises such as conversation, ringing telephones, loud traffic noise, etc., will be more likely to distract people. By a process known as habituation, people can block out regular noise (the experience of only becoming aware of the presence of a clock when it stops ticking indicates habituation). Inhibitory fibers exist in the neural pathway from the ear to the brain, which inhibit specific parts of the organ of Corti. Unexpected noise is more distracting than regular noise because it cannot be blocked by this mechanism. Attitudes toward a noise source play a major role in determining its disturbing effect. Approval of the source minimizes disturbance (e.g., noise from one's own or a subordinate's machine), whereas disapproval increases disturbance. The listener's control over the noise and the social distance between the listener and the noisemaker have a similar mediating effect on disturbance.

Wind Turbine Syndrome

Wind turbine syndrome (WTS) is the name given to a collection of symptoms including dizziness, disturbed sleep, and tinnitus associated with living close to industrial wind turbines. Wind turbines emit low-frequency noise and infrasound at intensities around 45 dB Lden. "Lden" is a metric of noise over a whole day (24-h L_{eq}) with a 10 dB(A) penalty for night-time noise between 11:00 PM and 07:00 AM. It is thought that the sound can cause cells in the cochlea to send signals to the brain, even though no sound can be heard. Nissenbaum et al. (2012) investigated self-reported sleep quality, daytime sleepiness, and general health in people living close to wind turbines. LAeqs of around 50 dB were measured at 150–250 m from wind turbines, dropping to around 35 dB at over 1 km. People living closer than 1.4 km from turbines reported worse sleep, more daytime sleepiness, and poorer general health than those living further away. These results held when attitudes to wind turbines were controlled for, suggesting that the current regulations for the location of wind turbines might need revision.

Van Kamp and van den Berg (2021) conducted a systematic literature review of 69 studies on the health effects of wind turbine sound. They found no evidence that exposure to WTS had direct effects on health, but that *annoyance* was associated with health effects. In other words, WTS has no effects on health that are any different from exposure to other noise sources, which are usually at much higher levels. Rather, contextual factors have a strong effect on annoyance, which, in turn, may cause sleep disturbance and so on. They suggest that those living near wind turbines should be able to participate in the planning process to bring about a sense of fairness and an appreciation of the benefits.

Vibration and Health

According to Griffin (2012), there is no disorder that is caused purely by exposure to vibration. Most people exposed to vibration are also exposed to other factors that may affect their health (e.g., bad posture or cold). The main disorders where vibration is a causal factor are summarized in Table 11.17. (See Johanning, 2015 for further information.)

Vibrations in the frequency range 4–8 Hz (the natural frequency of the trunk) are particularly hazardous. Seat design to minimize the transmission of vibration at these frequencies can reduce the risk of back injury. ISO (ISO 2631-1, 1997) specifies a limit of 0.32 m/s^2 beyond which exposure to vibration can have acute effects.

Raynaud's disease affects the blood vessels and nerves of the hands or feet. When associated with the use of vibrating tools, it is called vibration white finger (VWF). The modern term for the disorder is hand–arm vibration syndrome. It is characterized by local ischemia, pain, or numbness and is exacerbated by working in cold conditions. It is caused by operating hand tools with low-amplitude vibration below 500 Hz. Paving breakers, powered sanders, grinders, chipping hammers, and powered wrenches are examples of vibrating hand tools that can cause VWF.

Dong et al. (2021) carried out a review of studies on vibration VWF conducted by NIOSH since 2000. VWF is a consequence of the biodynamic response of the tissues to vibration exposure when the tissues can no longer restore normal function. Shear strains are particularly damaging to the superficial layers of the skin above 250 Hz (where many small blood vessels are located) but are not transmitted deeply. This is why the use of gloves can provide protection against VWF. Frequencies in the range of 63–250 Hz cause dynamic strain in deeper tissues, which can raise the

TABLE 11.17
Health Effects of Vibration Exposure

Disorder	Effect of Vibration
Low back problems	Microfracture of the vertebral end plate
	Schnorl's nodes (following sudden, intense shock)
	Increased disc pressure caused by higher back muscle activity
	Decreased disc height
	Increased radial disc bulge (after chronic exposure)
	Bad posture amplifies the effects of vibration
Gastrointestinal problems	Increased secretion of gastric juices causes acute stomach ache Possible link with gastric ulcers Nausea, vomiting, indigestion Hemorrhoids
Urogenital problems	Slight evidence for a link with WBV in women. Renal problems
Cardiovascular problems	No evidence for an association
Hearing problems Vertigo	Some evidence suggests that WBV combined with noise amplifies hearing loss by 6 dB.

threshold of mechanoreceptors in the hands and fingers. The resonant frequencies of the fingers increase from the proximal to the distal ends of the digits and with grip force. Vibration transmissibility also tends to increase with grip force. Increased grip force and non-neutral arm/hand postures pre-stress the tissues, further enhancing the transmissibility of vibration and the biodynamic response to vibration.

Prevention of VWF

If detected early, VWF is reversible. If not, it can cause permanent disability in the use of the hands. Control of employee exposure to vibration follows the same principles as noise control. Regular surveillance of exposed employees is important for early detection of the onset of problems. exposure times can be lowered by job design, vibration from tools can be measured to identify problem tools that cause high vibration doses, and these can be replaced. Changes in work design can be made to automate high-risk processes, and protection can be provided in the form of rubber/foam rubber handles to dampen vibration transmission at the hand–handle interface. The use of vibrating tools in the cold should be avoided. Vibration exposure can be reduced by issuing gloves, which also reduce cold.

Wearing gloves can prevent VWF, but Dong et al. (2021) concluded:

- Gloves can cause hand fatigue if they increase the required grip force on the handle.
- Vibration-reducing gloves do not attenuate well at frequencies below 25 Hz. Ordinary gloves are better when using tools such as rammers, tampers, and vibrating forks.
- Vibration-reducing gloves can reduce high-frequency components and sharp peaks.
- There are trade-offs when increasing the cushioning of gloves because higher grip forces may be needed, and operators lose dexterity.
- Gloves may increase hand dimensions and alter finger positions on the handle, thereby altering the vibration exposure.

Dupuytren's contracture occurs when the connective tissue in the palm thickens, causing one or more fingers to be permanently flexed. In an analysis of data from 4969 males, Palmer et al. (2014) found a doubling in the prevalence of the disease in men exposed to daily hand–arm vibration doses greater than 20.8 m/s^2.

Vibration and Public Transport

Mention must also be made of how people respond subjectively to vibration of different frequencies and intensities. This is particularly important in the design of public transport systems if passenger comfort is to be optimized in relation to engineering constraints. Individuals differ considerably in their susceptibility to motion sickness, but it seems to be caused by low--frequency (0.1–0.8 Hz) vibration in the vertical plane—not surprisingly, vibrations of this nature are commonly encountered at sea. Some general guidelines for the amelioration of motion effects are given in Table 11.17 (Bittner and Guignard, 1985). Impacts measured at the seats of occupants of high-speed craft during transits are shown in Figure 11.13. When a suspension seat was fitted, exposure to high-G impacts was reduced (Table 11.18).

Mitigating Exposure to Vibration and Shock

Myers et al. (2012) investigated the possible benefits of installing suspension seats to protect users of high-speed craft during sea transits of 3 h. Subjects completed a 12 m shuttle run 1 h before and immediately after the transits, which were conducted in both calm and rough sea states. Boat occupants sat either on rigid seats or on suspension seats designed to absorb shock and vibration. There was no difference in run time between the boat crews pre-transit. However, subjects sitting on fixed seats exhibited large reductions in run distance post-transit in both calm and rough sea conditions (~250 and 400 m, respectively). Heart rate during the transits was slightly lower when suspension

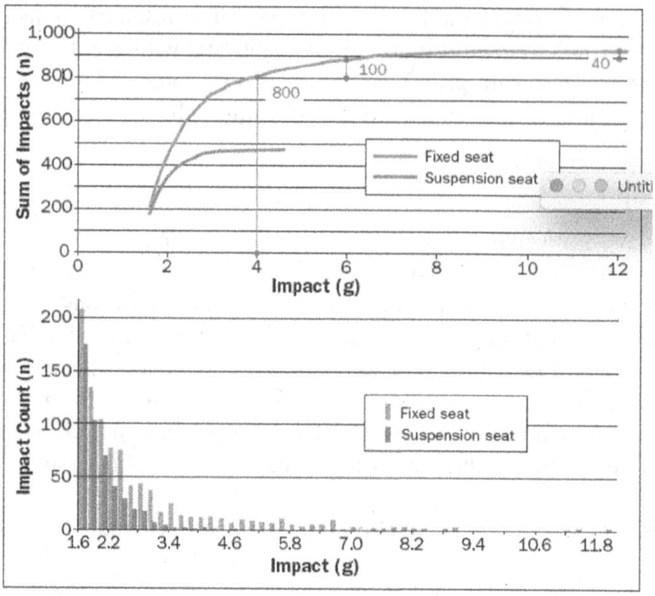

FIGURE 11.13 Distribution of impacts by G force during high-speed boat transits with and without a suspension seat. (Courtesy of Ullman Dynamics.)

TABLE 11.18
Principles for Minimizing Motion Sickness

1. Locate critical workstations near the effective center of rotation
   Minimize pitch and roll motions (<15°)
2. Minimize head movements
   Place tools within close reach (avoiding the need to stoop)
   Provide headrests on seats
   Place controls and displays in the area of focal attention (moving eye field)
3. Align operators with the principal axis of the vessel hull
   Reduce complexity of vestibular inputs
   Minimize passive head movements
4. Avoid combining provocative sources
   Nauseogenic stimuli tend to summate
   Avoid bad odors, medication, and disorientating displays
5. Provide an external visual frame of reference
   Provide a view of the outside world

seats were used and higher in rough compared to calm conditions. However, the rates were low and corresponded to "light" work. The vibration dose values at the seat pan were lower when the suspension seats were used in both calm (7.7 vs. 9.1 $m/s^{1.75}$) and rough conditions (33.8 vs. 36.8 $m/s^{1.75}$).

These findings demonstrate that effective vibration countermeasures can improve operational capability and performance, and improve workplace safety.

Office Pods

Office pods (Figure 11.14) are acoustic refuges that can be installed in offices or other communal spaces to enable occupants to work undisturbed by sounds such as conversations, telephones, office machines, and so on. Cell phone users can use the pod to make phone calls without disturbing

FIGURE 11.14 Office pods by Orangebox.

other occupants. Larger pods can be used for online meetings in hybrid workplaces or for discussing confidential or sensitive matters (Figure 11.15).

Pods enable designers to provide quiet spaces in open areas without having to acoustically treat the entire space. Pods are modular and mobile and can be moved as the office layout changes. A further advantage is that they offer employees more flexibility in deciding where they will work to best meet the requirements of different tasks, which may be particularly attractive for hybrid workers accustomed to privacy when working at home. According to Elena et al. (2023), speech level reductions of up to 30 dB guarantee privacy, whereas at 15–25 dB, speech privacy depends on the background noise level. Reductions of less than 15 dB do not guarantee privacy. Clearly, pods enable the requirement for privacy in open-plan offices to be met without having to acoustically treat the entire room and may be a cost-effective and flexible solution.

Status of the Methods

Valid and reliable instruments are available to assess noise, vibration, and the acoustic environment. Hearing tests are intrusive and have to be carried out by qualified personnel. Instrumentation for noise and vibration surveys has to be checked for calibration. The surveys themselves are relatively unobtrusive and should be conducted during working hours. There are real, practical, and theoretical difficulties in developing models to predict back injury following employee exposure to vibration. These difficulties include measuring vibration exposure in real work situations; determining the forces applied to the spine in these situations; developing indices of cumulative exposures over time; and controlling for confounding factors and effect modifiers such as posture, employee health, and stress, including psychosocial stress.

FIGURE 11.15 About here.

Effectiveness and Cost-Effectiveness

Several studies demonstrate the effectiveness and cost-effectiveness of implementing sound environmental ergonomic practices with respect to noise.

Advances in Hearing Conservation Programs

Moroe and Khoza-Shangase (2020) systematically reviewed 26 articles dealing with recent advances in hearing conservation. For present purposes, several of the themes they identified are relevant to HFE. One is the use of audiometric techniques to identify hearing loss in its early stages before it becomes measurable in traditional audiograms. This enables early detection of the risks of hearing loss so that conservation efforts can be made before irreversible damage occurs. Another theme is the development of wearable noise exposure meters, allowing employee exposure to be continuously monitored. Headsets with noise-embedded dosimeters and wireless communication transceivers can be used with threshold exposure levels set up in advance. The use of such devices eliminates the need for employees to report exposure to safety representatives.

Rabinowitz et al. (2018) assessed hearing conservation programs at 13 manufacturing plants in the United States. Site visits were conducted with structured interviews with management. A short questionnaire was administered to employees, covering perceptions of risk and management commitment to preventing hearing loss. Focus groups were also conducted with employees.

Access to medical records enabled NIHL to be assessed. The following factors were associated with higher rates of hearing loss:

- Age (older workers at greater risk)
- Sex (men at higher risk)
- Race (Black and Hispanic at higher risk)
- Tinnitus
- Hearing loss at baseline measurement (lower risk)

Several items relating to the management of hearing conservation programs also predicted hearing loss:

- Audiometric testing during or after a shift rather than before (because test results can be confounded by temporary threshold shifts)
- Hearing protection devices are required when noise levels exceed 85 dBA (vs >90 or other criteria). Lower risk of NIHL
- Noise monitoring conducted using both personal and area measurements (compared to personal measurements only). Lower risk of NIHL
- Work-relatedness of hearing loss determined by a professional supervisor (physician or audiologist) vs. other professional. Lower risk of NIHL
- Complaints about HPDs. Higher risk of NIHL

Finally, several employee responses also predicted hearing loss:

- When you had your most recent hearing test, did anyone talk to you about your hearing? (% yes ≥ 85%). Lower risk of NIHL
- Workplace level of commitment to hearing loss prevention (high or exceptional ≥ 70%). Lower risk of hearing loss

These findings indicate that there is a lot more to hearing conservation than just handing out HPDs and demarcating noisy areas. Good hearing protection is really nothing more than good management of occupational health and safety.

Costs and Benefits of Hearing Conservation Programs

Both the costs and benefits of such programs vary between countries and depend on how health care and workers' compensation are funded. However, there is an active debate about the cost-effectiveness of these programs, and several cost–benefit models have been put forward. Gibson and Norton (1981) in Australia put forward a simple model as follows:

$$\frac{\text{Benefits}}{\text{Costs}} = \frac{\text{Reduction in compensation} + \text{recurrent compensation}}{\text{Investment in noise control} + \text{hearing programs}}$$

The benefits of implementing a noise control program, according to the model are a reduction in the number of claims and a reduction in the size of payments to workers suffering permanent threshold shifts as a result of exposure to noise at work. This includes a reduction in the provision of hearing aids, for example. The costs are, first, the money spent to implement the noise control program, which consists of acoustic engineering of workplaces and machines, or the replacement of noisy machines with quieter ones. Second, there are costs associated with audiometric testing of employees and the provision of PPE.

According to Gibson and Norton's model, every A$1.9 spent on prevention would save only A$1 in compensation costs. According to this model, almost half of the money spent on noise reduction to a maximum of 90 dB(A) is a welfare payment rather than an investment. Hocking and Savage (1989) reviewed the limitations of the model, pointing out that it does not take into account the fact that investments in noise reduction will take time to produce benefits, since noise-induced hearing loss is an insidious process that can take place over many years. They propose a more comprehensive model that discounts costs over several years and takes into account the tax effects of paying for both the hearing program and compensation payments.

The costs associated with hearing conservation programs are due to:

- *Noise Assessment*: This consists of the equipment and labor needed to carry out a survey, either by in-company personnel or by external consultants.
- *PPE and Its Use*: This includes the cost of earmuffs (and their replacement every 2 years) and earplugs, as well as a hearing conservation education program (including instruction on the correct use of hearing protection).
- *Job Rotation*: There may be costs associated with changing the work organization so as to rotate workers from noisy to quiet areas (these may be offset by unexpected benefits such as reduced fatigue and fewer injuries).
- Retrofitting of noisy machines.
- *Audiometric Testing*: The costs include the audiometrist's fees and lost production (since workers are tested during working hours).
- *Administration of the Noise Control Program*: This includes the salary of the person managing the program plus sundry costs.

The benefits include:

- Reduction in compensation claims (including the cost of hearing aids)
- Reduction in lost time
- Reduction in the costs of administering the compensation scheme
- Possible improvements in productivity (e.g., due to better communication, less distraction, and the ability to concentrate for longer periods of time)

Cost-Effectiveness of Different Noise Control Strategies

Noise control strategies differ, as do their costs and benefits. Scannel (1998) points out that although earplugs and earmuffs are cheap, they can also be very ineffective at reducing noise exposure. For example, if earmuffs with an NRR of 25 dB are removed for only half an hour a day, they provide only about 12 dB of protection over the whole day. In practice, even good ear protectors may be of little use if the noise levels are over 100 dB. If employees only wear protectors in the presence of supervisors, the protection will be negligible. Rotating employees between noisy and quiet areas only lowers daily noise exposure if the noisier area is close to 90 dB(A) or if the time spent in it is extremely short (e.g., assuming a doubling rate of 3 dB, 1 h spent in 100 dB(A) is equivalent to 8 h in 91 dB(A), which is still too much even if the other 7 h are spent in total silence). Soundproof enclosures around machines can be effective, but even small gaps or holes lead to large reductions in sound insulation. A hole compromising 10% of the surface area of the enclosure reduces the maximum noise insulation from about 30 to 10 dB. Even a 1% hole reduces it to 20 dB (three times as much sound energy escapes from an enclosure when it has a hole that is 1% of its surface area).

Noisy Machines are Often Inefficient Machines

Noise control can reduce noise levels and extend machine life, since the same processes that give rise to noise also give rise to wear. Narrowband frequency analysis can be used to locate the part of the machine giving rise to the noise if the operating frequency of the components is known. Scannel gives some examples of cost-effective interventions:

- The flywheel of a 10 ton power press was found to generate bell-like tones, due to resonance. Dynamic vibration absorbers were attached to the flywheel to damp the resonant frequencies. The costs were $50 for materials and 1 day's pay for a fitter. The life of the machine was extended by 8 years, and the noise levels dropped from about 96 to 86 dB.
- Aluminum castings being worked on a capstan lathe emitted 94 dB(A) at the operator's ear. The cutting tool was causing the product to vibrate, which not only caused noise but also affected the smoothness of the cut. Vibration damping straps were made at $5 each and fitted onto the castings before they were machined. The noise dropped to 78 dB(A), the tool life was increased, and the machining cycle time was reduced.

RESEARCH ISSUES

The technology for evaluating the auditory environment and for dealing with troublesome noise sources is well established. There is an international trend to lower industrial noise exposures beyond their present levels to further safeguard the hearing of the workforce and improve working conditions. Applied research to investigate cost-effective ways of achieving these reductions is appropriate. Although vibration exposure is linked to many health effects, there is a lack of evidence for dose–response relationships. This hinders the drafting of exposure limits for the prevention of specific effects. The model of Sandover presented above is very much a simplification of what is really a very complex problem.

Noise has long been regarded as a hazard to hearing and is starting to be seen as a safety risk. Deshaies (2015) reviewed 788 reports of work-related fatal accidents. Noise was cited as one of a number of explanatory factors in 2.2% of the reports, particularly where moving vehicles and communication were involved. In other reports, the authors found that noise was not explicitly mentioned in 3.9% of reports when it should have been. Yoon et al. (2015) surveyed 1790 factories in Korea and found that occupational injuries were 1.68 times more likely when the noise was between 80 and 89 dB compared to below 80 dB, and 3.42 times more likely when the noise was greater than 90 dB (exposure to dust and shift work were controlled for, but several other risk factors were not). Given that noise lengthens reaction time to visual stimuli and degrades vigilance, communication, and short-term memory, the contribution of noise to accidents must be better understood.

Noise is known to affect cognitive performance, but the relationship between the intensity and nature of the noise and the kinds of cognitive functions that are susceptible to degradation requires further investigation.

SUMMARY

Noise can be evaluated by measuring SLs, noise doses, and the $LA_{eq,}$ and by carrying out frequency analyses. Unacceptably noisy machines should be quieted, replaced, or isolated from occupied areas. If this is not possible, operator exposure should be reduced or ear protection provided. Further information on the physical aspects of noise and noise control can be found in the publications of the International Standards Organization, the American Industrial Hygiene Association, and the *Handbook of Noise and Vibration Control.*

An ergonomic approach to the auditory environment should go beyond measurement of SLs and consider the satisfactoriness of the acoustic environment as a whole. Creative, intellectual work may suffer from noise more than routine work, in which operators have a high level of skill. A knowledge of people's attitudes toward noise is essential in the evaluation of physical measurements and in designing hearing conservation programs.

TUTORIAL TOPICS

1. Battery electric vehicles emit very little noise. Discuss the implications of the general adoption of battery electric vehicles (BEVs) for the design of the urban environment.
2. Whether a sound is perceived as noise depends on one's attitude toward the source. Discuss.
3. The residents of a small town in Texas report a variety of nonspecific health complaints following the installation of a wind farm nearby. How would you go about investigating these complaints?
4. Take a look at the videos at the links (https://www.youtube.com/watch?v=pPTwPZO1KgY and https://www.youtube.com/watch?v=YhrYQwnkQ6A). What implications does this have for traditional advice about the acoustic environment in offices?

ESSAYS AND EXERCISES

1. Noise is defined as
   - Loud sounds
   - Unnecessary sound
   - High-pitched sounds
   - Unwanted sound
   - Distracting sounds
   - Sound intensity greater than 85 dB(A)
2. Complete the sentence, "Frequency is to pitch as amplitude is to ..."
3. Age-related hearing loss is known as
   - Presbyopia
   - Phonophobia
   - Deafness
   - Presbycusis
4. Which of the following might not be related to the risk of noise-induced deafness?
   - Marital status
   - Family history
   - Number of children
   - Work history
   - Hobbies and sports
5. Carry out a noise survey in a small factory. Obtain or draw a plan map of the factory floor indicating the locations of machines and other noise sources. Indicate all fixed work positions.

   Write the measured SLs onto the map. Identify demarcated areas beyond which hearing protection should be worn. What other recommendations can you make?
6. Noise level in a bottling plant is 93 dB(A). Assuming an 8h exposure level of 85 dB(A), what is the maximum time an employee can work in this environment without ear protection, assuming a doubling rate of 5 dB(A)? If the doubling rate was 3 dB(A), how would it affect the exposure time?
7. One group of workers in a factory is exposed to a noise level of 100 dB(A) for 8h. In the adjoining work area, their colleagues are exposed to 70 dB(A) for 8h. Management proposes to introduce job rotation (everyone works for 4h in the noisy area and 4h in the quieter area) to bring the 8-h exposure down to 85 dB(A), on average, for everyone. Will management's proposal work? If not, suggest what the new noise dose will be.
8. A set of earmuffs has an NRR of 30 dB(A), and a set of plugs has an NRR of 20 dB(A). Workers in a car factory are exposed to 110 dB(A) for 8h. Before advising management on whether to purchase muffs or plugs, you do a small trial on the factory floor. Workers wearing the muffs take them off for about 1h a day while working due to discomfort, whereas plug wearers soon forget they are wearing them and keep them on all day. Which device will give the best ear protection in practice?

9. Refer to the section on ear protection and the paper by Toivonen et al. How long would the workers in the untrained group be allowed to work in a noise of 120 dB(A) according to
   - U.S. standards?
   - EU standards?
10. When people start to go deaf, due to noise exposure, which frequencies of sound do they lose sensitivity to first of all?
    - 40 kHz
    - 4 Hz
    - 40 Hz
    - 4000 Hz
11. A worker is exposed to 95 dB(A) for 30 min, 75 dB(A) for 3 h, and 63 dB(A) for 4 h 30 min. Calculate his noise dose for this 8-h period.
12. A worker is exposed to 91 dB(A). Assuming an 8-h exposure limit of 80 dB(A), what is the maximum permissible exposure time (assume a 3 dB doubling rate)?
13. Name three activities where the worker is exposed to WBV and three where the worker is exposed to hand–arm vibration. In no more than a few sentences, describe how such exposures can be effectively managed to minimize trauma.
14. Earmuffs and earplugs have advantages and disadvantages, depending on the context of use. List three situational factors that favor the use of earmuffs and three that contraindicate the use of earmuffs.
15. A factory has an ambient noise level of 100 dB(A). You are asked to advise on cost-effective ways of reducing the noise exposure of staff. The three noisiest machines have SPLs as follows:

Machine	SPL (dB[A])	Cost ($) of Retrofitting to Reduce Noise by 15 dB(A)	Cost of Sound Proof Enclosure Reducing Noise by 5 dB(A)
1	105	7000	5000
2	100	5000	3000
3	95	3000	1000

    Management is reluctant to spend more than $10,000 on the whole project, but could be persuaded to spend up to $3000 more if there were "significant advantages."

    What would you advise them to do?

    Give your reasons.
16. A small workshop has three machines running simultaneously. The SPLs are 83, 85, and 80 dB. What is the total SPL in the workshop? What assumptions did you make?
17. Mr. and Mrs. Smith are enjoying a lovely holiday in Spain. One day, they visit a bustling street market where the ambient noise level is about 60 dB. While inspecting the various stalls, Mrs. Smith accuses Mr. Smith of not listening to her. Given that she was speaking to him in a normal voice and standing about 1 m away, is this accusation justified?

QUICK QUIZ

1. Sound is transmitted in air by longitudinal or transverse waves. True or False?
2. How much does a sound level drop when you double the distance from the source? (a) 10 dB, (b) 50%, (c) 6 dB
3. The frequency of sound is detected by (a) the ear drum, (b) the stapes, (c) the cochlea?

4. At what range of frequencies is the ear most sensitive the sound?
5. What is age-related hearing loss called?
6. At what ambient sound level does conversation start to become difficult?
7. Reverberation adds “depth” to the auditory character of a room. True or false?
8. What is a safe noise exposure level for an 8-h shift?
9. Noise from wind turbines can damage the health of local inhabitants. True or false?
10. Exposure to sudden shocks (“jerk”) can damage the spine. True or false?

12 Thinking and Acting

GENERAL REQUIREMENTS FOR HUMANS IN SYSTEMS

1. The training burden for new systems must be minimized.
2. Interface design must be based on an analysis of legacy tasks to maximize the positive transfer of knowledge from legacy systems.
3. Mental workload for generic tasks must be sustainable and support multitasking.
4. Mental workload for critical tasks must not exceed the operator's mental capacity.

> If the organism carries a "small scale model" of external reality and of its possible actions within its head it is able to try out various alternatives, conclude which is the best of them, react to future situations before they arise, utilise the knowledge of past events in dealing with present and the future, and in every way to react in a much fuller, safer and more competent manner to the emergencies which face it.
>
> *Craik, 1943*

CORE KNOWLEDGE: PROCESSING INFORMATION IN EVERYDAY LIFE

THINKING

Human beings are said to have two separate modes of interacting with their surroundings. *System 1* is a mode of operation that is fast, effortless, and automatic. When we look at Figure 12.1, most of us will see a cube, even though we are really only looking at some lines on a flat piece of paper. One face of the cube appears to be in front of the other. With slight effort, we can bring *System 2* into operation by consciously reversing the figure, such that the face at the front is now at the back, and vice versa.

So how you look at something affects what you see. If we look at Figure 12.1 using System 1, we see a cube. If we use System 2, we can change the appearance of the cube. After a little practice, we can perform this reversal at will.

System 1 operates continuously in daily life when walking along a busy street, it scans the environment, causing us to step aside when we approach an obstacle; it draws our attention to salient features of the environment, such as pedestrian crossings or shops; it detects changes and generates basic impulses and impressions. Kahneman (2011) describes how System 1 enables us to navigate daily life with little conscious effort and how, under certain circumstances, it can lead us astray by creating false impressions and inaccurate assessments.

System 2 works more slowly and consists of all those cognitive processes that require attention—we know that our System 2 is in operation whenever we are aware of what we are doing. At the very least, System 2 can monitor the operation of System 1, and there is a view that so-called "mindfulness" (to "stay in the moment" and be aware of current activity rather than past events or future plans) is desirable. System 2 is active when, for example, we hold a password in memory while entering it into an application, or follow an unfamiliar recipe, do mental arithmetic, or respond to editorial comments when correcting a manuscript. System 2 is effortful, and mental effort requires the metabolism of glucose by the brain. The human brain uses a lot of energy; at 2% of body mass, it uses 25% of the available glucose. Tasks requiring mental effort (System 2) appear to be those most dependent on glucose. There is a transient *glucose facilitation effect* where supplementary

DOI: 10.1201/9781032717210-12

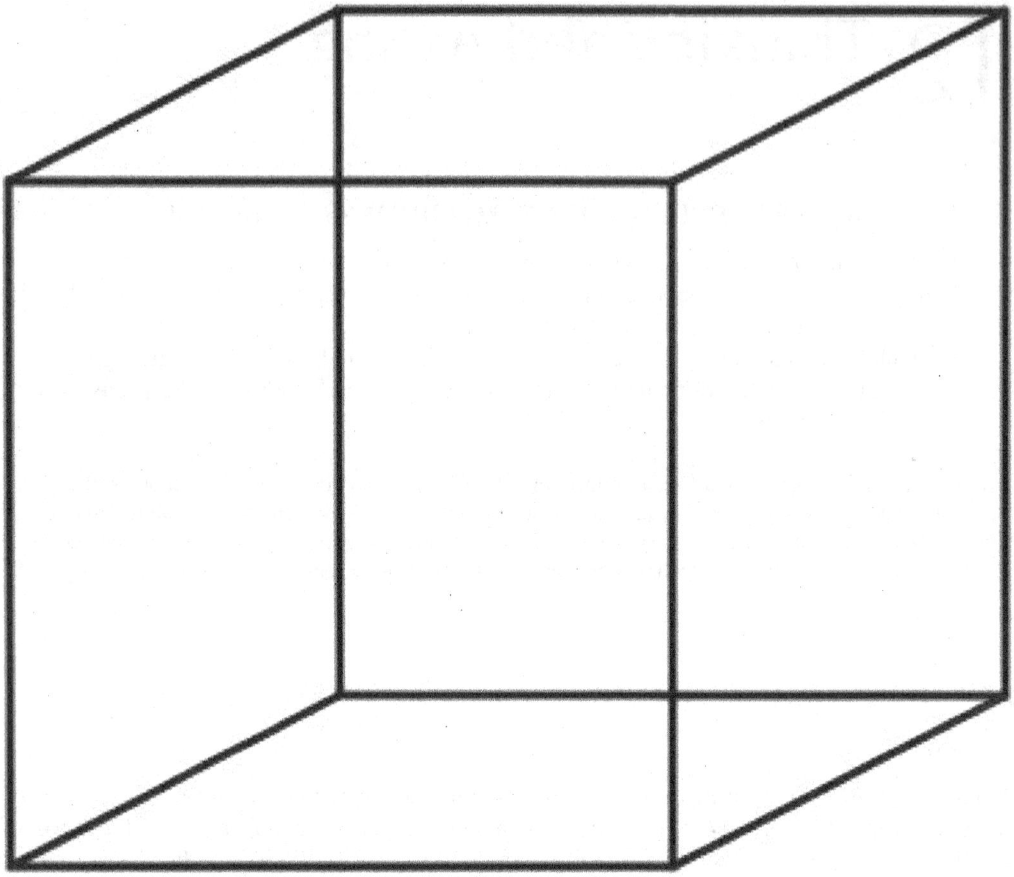

FIGURE 12.1 The Nekker cube.

consumption of glucose improves the performance of demanding cognitive tasks (Peters et al., 2020). Scholey et al. (2013), for example, found that the administration of glucose 10 min prior to a memory task improved performance when the task was effortful (made more difficult by the addition of a concurrent task) compared to a placebo. However, the beneficial effect of glucose was only apparent when the memory task was made effortful by the inclusion of the secondary task.

Some main features of the two systems are summarized in the following table:

System 1	System 2
Fast	Slow
Effortless	Effort required
High capacity	Limited capacity
Nonverbal	Involves language
Implicit	Explicit
Associative	Logical/procedural
Parallel processing	Serial processing

Processing of information in daily life involves the operation of both systems. Since much of our working life is regulated by external demands and contingencies, some key questions for HFE specialists are

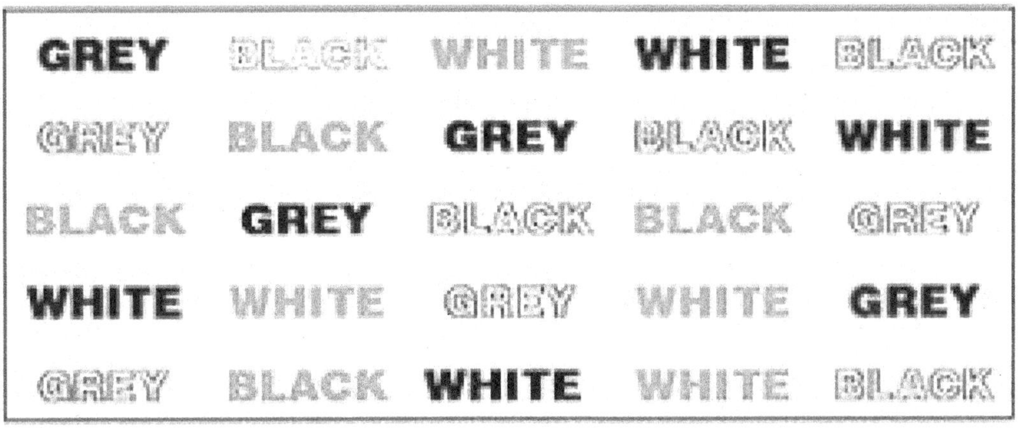

FIGURE 12.2 STROOP test stimuli.

1. How can we design the work environment to get the best out of both systems?
2. How can we design the work environment to avoid overloading these systems?
3. How can we design products and procedures to fit the way these systems operate?
4. How can we design interfaces to prevent conflict between these two systems?

It may come as a surprise to readers that the two systems can ever come into conflict. The STROOP test (Figure 12.2) demonstrates how conflict can arise.

The task is to state the color of the ink the word is written in, but System 1 automatically reads the word and prompts us to respond by saying the word. In the STROOP test, System 1 has to be overridden by System 2 when the color of the ink and the word are different. When presented as a sequential reaction time task, the STROOP task is effortful and tiring. There is a long history of research in HFE showing that human–machine interactions should be designed such that commands, control actions, and feedback have maximum consistency and compatibility.

Executive Control

Executive control relies on System 2 and is essentially the ability to monitor one's actions in relation to one's intentions—to pay attention to what one is doing. Any activity that requires memory, problem-solving, or reading complex language involves System 2.

Executive control is also essential for self-control, which consists of:

- Resisting distraction when carrying out a task
- Impulse control and self-monitoring
- Overcoming inner resistance
- Coping with emotional dissonance

The following scenario illustrates these demands:

Your job is to monitor a display in a control center to check that the system operating parameters are within limits; respond to queries from operators; and check that standard operating procedures are being followed. Your colleagues are talking about the tennis championship finals, which are of interest to you, but you must ignore their conversation. It is 30 min before your coffee break, and you must resist the desire to leave your workstation to get a drink. In 5 min, you must speak to one of your staff about a disciplinary matter that you find upsetting. During this time, you must remain dispassionate and unbiased, even though you know the individual is in the wrong and you dislike the individual personally.

The key point to note here is that both the process control task and the psychosocial and environmental demands require executive function—the demands are different, but they draw on the same (limited) inner resource.

Self-Regulation and Self-Control

Self-regulatory strength is analogous to will or persistence. People with high self-regulatory strength will continue to work toward a difficult goal longer than those with low self-regulatory strength. Self-control is needed to resolve conflicts between different goals (e.g., choosing a delayed reward in favor of an immediate one). Research on self-regulation (Baumeister et al., 2007; Inzlicht et al., 2021) argues that self-control is a limited psychological resource. Baumeister et al. make a straightforward analogy between self-regulatory strength and muscular strength. Acts of self-control cause short-term impairments in the ability to carry out further acts of self-control, in the same way that short, intense muscular exertions cause temporary reductions in muscular strength. In both cases, these reductions are not specific to the activity that caused the deficit—acts of self-control in a particular context cause temporary reductions in self-regulatory strength. For example, subjects in one experiment were asked to stifle an emotional response when watching an emotionally evocative film. Their subsequent performance on a test of handgrip endurance was lower than that of subjects who were not asked to stifle the response. In another experiment, subjects who were required to abstain from eating from a plate of freshly baked chocolate biscuits and were only allowed to eat from a plate of radishes were found to give up more quickly on a subsequent frustrating task than subjects who were allowed to eat the biscuits.

Self-Regulation and the Role of Glucose

Gailliot (2015) summarized research on self-regulation by stating that self-control appears to operate as if it relies on some kind of limited energy source. Acts of self-control deplete the energy stores by impairing subsequent attempts at self-control. The capacity for self-control is only regained after a period of engaging in activity that requires low self-control. Gailliot et al. (2007) summarize the evidence that blood glucose levels are the source of energy that fuels self-regulation. The evidence suggests that:

- Acts of self-control do cause blood glucose concentrations to drop.
- It is the self-regulatory effort that causes the reduced concentration, not the activity itself.
- It is the drop in glucose levels that lowers the capacity for further acts of self-control.
- A drop in blood glucose levels is a nonspecific response to any task requiring self-control. The drop lowers persistence rather than merely slowing task performance.
- Replenishment of glucose, in the form of a glucose drink, immediately after a task with high self-regulatory demands can restore self-regulatory strength in about 10 min.

It seems to be the case that the utilization of glucose is more effective at night and in the morning, and less effective in the early morning and evening. This may be a circadian effect and is likely linked to the circadian rhythm in cortisol. Taken together, these findings may help to explain why so many undesirable behaviors are not randomly distributed during the day, why people break their diets, make errors, forget safety procedures, and crash cars later in the day or when their circadian rhythms are disturbed.

Mental Effort

Embrey et al. (2023) define mental effort as a cognitive variable that mediates how well a task can be performed and how well it is actually performed. They give, as an example, that less effort is required to state roughly how many times 15 can be divided into 351, and more effort is needed to arrive at an exact answer. People tend to be "cognitive misers" in daily life and use tactics and strategies to minimize the mental effort needed to complete a task. Although System 2 can accomplish tasks with different mental demands, it appears to be the effort needed to complete the task and the incentives for completing it that determine how people act.

Fatigue

A general definition might describe fatigue as a consequence of work that lowers the capacity for further work of the same kind (i.e. work that has similar demands).

Gailliot (2015) summarizes evidence that cognitive fatigue occurs in System 2 as lowered attentional capacity following depletion of glycogen stores in the brain. Recent research by Wiehler et al. (2022) investigated cognitive fatigue in daily office work. They hypothesized that high-demand tasks would lead to the accumulation of metabolic waste products in the brain—markers of cognitive fatigue. They measured levels of glutamate (a substance important for the function of brain cells that is normally tightly controlled within the cells) in the lateral prefrontal cortex (a part of the brain active during cognitive control tasks). Subjects performed easy or difficult versions of a cognitive control task for several hours, interspersed with an economic decision-making task. Subjects performing the demanding task made faster, less costly (to themselves) decisions suggesting that their cognitive control capacity was depleted by the demands of the main task. Subjects in the high-demand group had higher glutamate levels in the lateral prefrontal cortex but not in the primary visual cortex (the visual demands were the same in both groups). Normally, the waste products that accumulate in the brain are removed during sleep. Thus, cognitive fatigue seems to occur in the prefrontal cortex due to metabolic changes resulting from increased neural activity—an increase in extracellular glutamate concentrations. Subjects in the high-demand group were less likely to subsequently choose more demanding options and had lower patience. Thus, cognitive fatigue is a "state of the brain" rather than a "state of mind." In demanding occupations such as air traffic control and long-distance lorry driving, regular breaks are compulsory. When cognitive fatigue occurs, continued task performance becomes more reliant on System 1, which results in predictable and specific changes in the way tasks are performed, as described in detail below:

- Forgetting to carry out critical subtasks on time
- Mistiming the execution of one task in relation to others
- Attentional "stickiness" (inability to shift attention from one subtask to another)
- Lack of persistence
- Breakdown of peripheral attention (e.g., failing to monitor displays in the periphery of the visual field)
- Breakdown of incidental learning (failure to get the "hang of things")
- Reluctance to carry out nonmandatory, unprogrammed, or novel tasks
- Errors of commission (incorrect operations)

Sustaining Task Performance

Cognitive fatigue is a deficit in executive resources. If performance must be maintained, this deficit can be overcome in one of two ways, according to Hockey (2005):

- Increase the effort expended to perform the task
- Reduce the goal such that it matches what can be achieved with the current level of effort

This is known as "compensatory control." If performance ("effectiveness") has to be maintained at the same level, despite the deficit, then the "efficiency" of performance will be reduced at a cost to the operator of increased effort. This may manifest itself as

- Decrements in the performance of secondary tasks (e.g., the pilot forgets to check the fuel gauge)
- Strategy changes (e.g., a long-distance lorry driver switches to a passive "lane tracking" strategy, rather than an active strategy of reading the road ahead)

- Regulatory costs appear—increased heart rate, reduced heart rate variability, and increased anxiety
- After effects—disinclination for nonmandatory tasks or less inclined to help others

Essentially, when routine tasks become difficult due to a high workload or external stressors, increased *effort* is applied to the main task, and task performance is maintained. The effect of exposure to the stressor is that performance on the main task remains the same, but the *efficiency* of task performance is reduced. The reduced efficiency can manifest itself as an increase in the perceived difficulty of the task; as physiological changes such as reduced heart rate variability; or as a reduction in the capacity to carry out other activities at the same time as the main task. Hockey has described these manifestations as *latent deficits.*

What these studies show is that primary task performance is relatively robust in conscious individuals, no matter how tired they say they are. Although subjects may complain of fatigue, if the primary task performance is unchanged, it is not immediately obvious what the nature of the deficit, assumed to be caused by the fatigue, really is.

Acting

Figure 12.3 shows an action sequence for use in the design of tasks (Norman, 1981). In daily life and at work, action can be understood as a series of stages:

1. *Goal Formation*: Deciding what to do or being told what to do. Environmental cues may automatically activate goals (e.g., slowing down when you see a red traffic light).
2. *Action Formation*: Deciding on an action or series of actions to reach the goal. Well-learned actions may be executed automatically under System 1. Novel actions may require System 2—conscious planning.
3. *Action Sequence*: Ordering the actions in the correct sequence.
4. *Execution*: Carrying out the actions in the correct order.
5. *Evaluation*: Comparing the outcome to the goal. The resources needed to evaluate the outcome may depend on the complexity of the task and the clarity of the goal. Good feedback can make goal attainment more explicit (e.g., proximity sensors to aid reversing).
6. *Deciding on the Next Goal*: This often depends on the more general consequences of goal attainment.

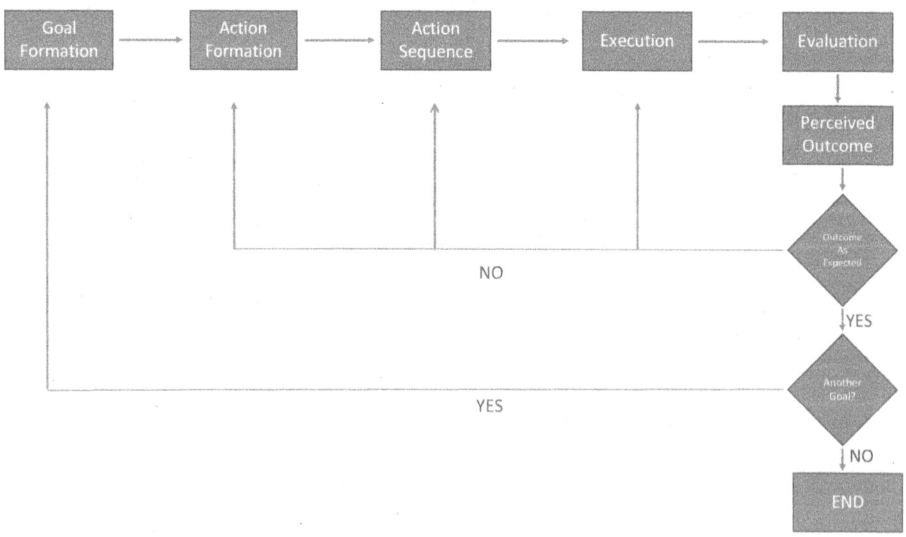

FIGURE 12.3 A task action sequence. (From Bridger (2021), with permission of the author.)

As can be seen, action sequences can fail if errors are made at any stage. Recovery from failure can also fail if remedial actions are re-applied at the wrong stage:

- *Slips* are actions not carried out as intended or planned, for example, pushing the wrong button when dialing or making "Freudian slips" when saying something.
- *Lapses* are missed actions and omissions, that is, when somebody has failed to do something due to lapses of memory and/or attention. When the lapse involves losing the goal from memory, the action sequence stops completely (as in looking in a cupboard without knowing what one was looking for).
- *Mistakes* result from faulty plans/intentions, that is, indicating a left turn when the correct action is to continue going straight. Slips often occur at the task execution stage, lapses at the memory stage, and mistakes at the training or planning stages.
- *Violations* are deliberate illegal actions, that is, somebody did something knowing it to be against the rules.

MAKING LIFE EASIER FOR SYSTEM 2: HEURISTICS AND BIASES

Tversky and Kahneman (1974, 1981) and Todd and Gigerenzer (2007) describe common heuristics (rules of thumb) that people use to simplify and speed the mental operations required to make a decision. McCammon (2004) analyzed cases where skiers had been victims of avalanches. He noted that in almost all cases, warning signs of increased risk of avalanches were ignored (e.g., above-freezing air temperatures, rain, cracks in the snow). Moreover, accidents were not random. People were more likely to be in an avalanche in familiar rather than unfamiliar terrain, when they were committed to a route or exercise, when they were in groups, and when they were led by a charismatic or experienced skier. These rules of thumb enable people to make correct decisions very rapidly but can lead to systematic and predictable errors, and are known as *heuristic traps.*

Recognition

To choose or discriminate between two options, people will select the one they recognize. For example, to decide which of two tennis players will win a match, people will choose the player they know (assuming that one of the players is unknown to them). This can work well because good players tend to be better known. However, it is insensitive to contradictory information (e.g., he lost his last five matches).

Take the Best

The "take the best" heuristic is used when people have to make a decision using multiple cues. Instead of searching for all relevant cues, people will search for cues in order of their validity. They then base their decision on the one best cue they can find and the other cues are ignored. This works well and fast if the cue really is valid and if there is high redundancy between different cues. It works less well if the cues are not correlated. For example, when selecting players for a football team, we might select based on body mass if physical features are the only cues we have. However, selecting this cue leads us to ignore other potentially useful cues, such as playing experience.

Choose the Default

Todd and Gigerenzer note that 28% of Americans become organ donors, whereas 99.9% of French people become donors. This says little about the attitudes and values of the two nationalities and a lot about the differing policies adopted by their governments. In the United States, the default choice is not to donate organs, whereas in France, people are automatically donors and have to actively opt out if they do not wish to donate their organs. Thus, in designing safe systems, it seems sensible to choose default options that are intrinsically safe, rather than relying on operators to choose safe courses of action (e.g., doors that automatically lock behind you when you leave a secure area).

Tallying

When deciding which of several houses to buy, for example, buyers might make a list of the positive features of each one and buy the one with the highest "score." Tallying is a simple way of conducting an informal variate analysis that can lead to valid choices if the cues are not highly correlated (e.g., number of bedrooms in the house, proximity to schools and shops, size of plot, secure garage, etc.).

Representativeness

People often have to make judgments that really depend on the probability that an event or object belongs to a particular class or is the result of a certain process. The representativeness heuristic replaces judgments about probability with judgments about whether the event is representative of the class or process. For example, in trying to decide whether an individual is a farmer or a librarian, the similarity of the individual to the stereotype of a farmer or a librarian is used. Although judgments about similarity and probability are often the same, the use of the rule leads to systematic and predictable errors. In order to come to a decision about someone's occupation, as in the above example, it would be useful to know about the relative proportions of librarians and farmers in the population (the prior probability of a person being a librarian or a farmer), but judgments about similarity do not take data of this type into account. In the countryside, farmers probably outnumber librarians, so it is more likely that someone living there will be a farmer irrespective of any other considerations, such as whether he looks like a farmer. In the middle of a large city, the opposite is likely to be the case.

To use an example more relevant to ergonomics: if a manual worker complains of arm pain at work, it is often concluded that the pain is caused by the work, based on its similarity to occupational arm pain. However, to make a correct decision, it is important to know the probability of someone experiencing the pain both at work and outside of work. Failure to do this causes the role of occupational factors to be overestimated. Representativeness also leads to misconceptions of chance. People expect sequences of random events to be similar to the process that gave rise to them, even if the sequence is short. In tossing a coin, for example, people will say that a sequence such as

H T H T T H H T

is more likely than

H H H H T T T T

or

H H H H H H H T

Although it is true that when a fair coin is tossed, there will be an equal number of heads and tails over the long term (because the probability of either is 0.5), short sequences of tosses need not resemble long sequences. This misconception underpins the well-known Gambler's Fallacy, or Law of Averages, in which random events are seen as self-correcting processes. The misconception states, for example, that after a long run of heads, a tail is now "due" to restore the appearance of the process. Unfortunately, a fair coin does not "know" that the sequence of prior tosses "looks wrong," and the probability of a head or tail remains the same each time the coin is tossed. Similarly, people often expect sequences generated by a random process to "look" random—few lottery players will deliberately gamble the numbers 1, 2, 3, 4, 5, 6 when allowed to bet on any 6 numbers between 1 and

50—because they expect the winning sequence to "look random." Perhaps fewer people would bet on lotteries if they realized the probability of the 1–6 sequence is equally unlikely as any other!

The representativeness heuristic can lead to systematic biases when people accept that a small sample is representative of the process that gave rise to it. Even engineers and scientists are prone to this over-reliance on small samples and make unwarranted extrapolations from them. Many complex processes, in economics or meteorology, are affected by overlapping trends with varying periodicities. For example, a person may decide, after experiencing five increasingly long and gloomy winters, that the weather is getting colder. However, this conclusion is an artifact of extrapolation from a short time series, which may not be representative of long-term trends (in the case of Britain, current weather may be slightly cooler than 15 years ago but is definitely warmer than 200 years ago, definitely cooler than 1000 years ago, and definitely much warmer than 10,000 years ago at the end of the last ice age). In process industries (and other complex situations) where historical plant data can be presented using computer graphics, designers must use a knowledge of system dynamics to ensure that variables are sampled, presented, and displayed so that real trends, rather than random fluctuations or noise, are presented to operators.

Availability

When people use the availability heuristic, they judge the likelihood of events according to the ease with which instances or occurrences of the event can be brought to mind. Things that are difficult to imagine are deemed to be unlikely. Sometimes this is a useful cue for making probabilistic judgments because instances of large classes of objects can indeed be imagined faster and more frequently than instances of small classes. However, availability, the ease with which events can be imagined or recalled, is influenced by variables that do not influence probability. For example, people may judge one activity to be more dangerous than another because it conjures up more vivid images.

Judgments based on availability may be biased by the effectiveness of a search strategy. If people are asked whether more English words have the letter "r" as their first or third letter, many will choose the former, even though it is incorrect. It is easier to retrieve from memory words beginning with "r" than words with "r" as the third letter.

Burgess (2007) examined the reasons for the belief that using a cell phone while refueling a vehicle at a gas station might cause a fire or an explosion. There are many reasons why it is advisable not to use a cell phone while refueling—cell phones are distracting, full attention should be given to the refueling task, and the user might take longer and inconvenience others. However, it seems unlikely that the cell phone itself could emit static electricity and directly ignite gasoline vapors. Burgess suggests that this is a "phantom risk" that many people believe in. Belief in the hazard seems to relate to its imaginability, and this is reinforced by safety propaganda such as warning notices instructing customers not to use their cell phones at the gas station. These notices make the phantom risk more available and, therefore, more probable in the minds of gas station users than it really is. The availability of the risk is further inflated by imaginary fears resulting from reports in the mass media about other possible health hazards of cell phone use (emission of non-ionizing radiation, brain cancer, and so on). In the United States, up to 2000, there were 115 automobile refueling fires (or 100 fires out of 120 billion annual refuelings in the United States). Thus, the real risk of a fire, for any reason, is very small. Burgess suggests that the most likely causal mechanism for user-induced fires is the practice of re-entering the vehicle while refueling (leaving the pump on latch before re-entering the car). As the user slides in and out of the car, friction of clothing against the car seat induces a static charge, which may then cause an electrical spark when the user touches an earthed object after leaving the vehicle (such as the metal trigger of the pump). Burgess points out that the practice of issuing warnings that only reinforce belief in phantom risks may, in itself, be hazardous by distracting users from real dangers and by inducing inappropriate avoidance behaviors.

Adjustment and Anchoring

When people are required to make quantitative estimates or evaluations, they often begin at an initial value and adjust it to arrive at a final value. As Tversky and Kahneman point out, the amount of adjustment and the final value can be influenced by the size of the initial value (the "anchor"). Salesmen and advertisers often use this as a technique to influence people's perception of the fair price of goods (e.g., "How much would you expect to pay for this packet of washing powder, $5.00, $4.50? Why not take advantage of our special offer of $2.50?").

In one experiment, subjects were first asked whether the percentage of African countries in the UN was greater or less than a particular value (the anchor). They were then asked to estimate the actual percentage. With an anchor of 10%, an estimate of 25% was obtained. With an anchor of 65%, an estimate of 45% was obtained. In another demonstration, subjects were asked to estimate 8! written in two ways:

$$8\times7\times6\times5\times4\times3\times2\times1 \text{ or } 1\times2\times3\times4\times5\times6\times7\times8$$

and gave estimates of 2250 in the first case and 520 in the second (the correct answer is 40,320).

The section on anchoring and adjustment highlights the need for careful design of training programs for operators of systems, since it is at these early stages of learning that anchoring and adjustment biases can be inadvertently set.

Heuristics and Biases: Pros and Cons

For much of the time, though, heuristic principles can produce essentially correct decisions and have considerable power in enabling complex problems to be dealt with quickly.

BASIC APPLICATIONS

Practically everything we do at work involves manipulating objects and processing information to attain a goal. First, we formulate the goal (decide what we want to do), then we execute the actions needed to achieve the goal, and as we do this, we evaluate the effect of our actions on the system state. In simple tasks, most of this takes place under System 1. In more complex tasks, System 2 is increasingly involved, and we subjectively experience these tasks as more mentally demanding.

Perception: How the Design of the World Affects the Ease with Which We Interact with It

Sensation occurs when environmental stimuli give rise to neural events. System 1 constructs a model of the event that gave rise to the initial sensation, which depends both on the neural "hardware" and individual factors such as personality and mood, previous experience and expectations, and the context in which the sensation takes place. For example, a pilot landing an aircraft in difficult conditions might see a series of lights below and ahead of the aircraft. Whether these are perceived as lights on the runway or streetlights will depend on individual factors such as the pilot's expectations, level of training, and external factors such as visibility and the visual cues that distinguish between the two types of lights.

The distinction between sensation and perception is by no means an academic one—the lower the level of processing required for a task, the quicker the operator's reaction time. The more intense the stimulus, the more likely the operator is to register it at a sensory level. In some U.S. states and various other countries, legislation has been introduced compelling motorcyclists to use their headlights during the day as well as at night. Langham et al. (2002) found that drivers took longer to identify a parked police car as a hazard if it was parked in the direction of the flow of traffic than if it was parked at an angle. The effect was more pronounced when drivers performed a secondary reasoning task. Clearly, during tasks such as driving, sensations do not always become perceptions. The factors that affected this in Langham's experiment were whether a "false hypothesis" was possible ("that car is pointing in the direction of traffic flow, therefore it is moving"). Langham suggested that emergency vehicles be parked at an angle to the flow of traffic and with their emergency lights switched off to better convey information that the vehicle is stationary.

Keeping Stuff in Your Head: Paying Attention to Short-Term Memory

Short-term memory (STM) is like a buffer where incoming information that has been attended to is held and is either acted on or lost through decay. There are three main modes of short-term storage—an articulatory loop, based on inner speech; an episodic buffer, for storing recent events; and a visuospatial sketchpad, based on visual imagery (when we look at a display and then close our eyes, we can retain a mental image of the display). The capacity of STM is about seven symbolic items—an item can be a word, a number, a letter, or an object (Miller, 1956). Items can only be maintained in STM if they are rehearsed (normally repeated either out loud or using inner speech). If not rehearsed, the items are either decayed or replaced by new items. When remembering a series of items, STM exhibits *primacy* and *recency* effects (Murdock, 1962). The recency effect seems to work on a "last-in/first-out" basis where people recall the last items in the list first. The primacy effect depends on the fact that the items heard first have been rehearsed more than subsequent items and that they are more distinctive, having no predecessors. People typically forget or confuse items in the middle of the list (Figure 12.4).

It is effortful to retain items in STM. All items decay in seconds if not rehearsed. The probability of correct recall diminishes as the retention interval increases—from about *80% after a delay of 3 s to 10% after a delay of 18 s* if items are not rehearsed.

The visuospatial sketchpad is used for storing visual material. Shopping malls often have floor plans laid out to indicate the relative locations of different shops. If the floor plan is not aligned with the building, shoppers may have to use the visuospatial sketchpad to mentally rotate an image of the floor plan to determine the correct direction to walk. The more the image has to be rotated, the greater the mental effort. Visual working memory appears to have a fixed number of slots—we can retain images of the same number of items regardless of the complexity of the items (Awh et al., 2007). However, the greater the complexity of the items, the lower the resolution of their representation in working memory. When people have to detect changes in multiple complex displays, they are more likely to detect them when the difference is large (because the resolution of the mental images

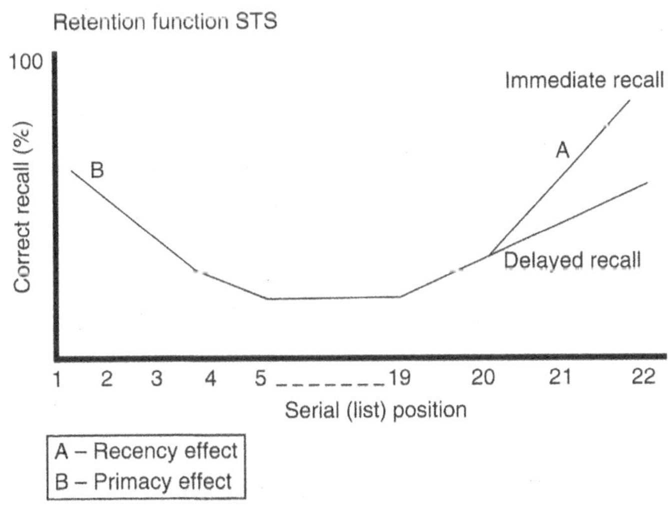

FIGURE 12.4 The serial position effect in short-term storage.

of the complex displays is low). This suggests that changes in system state should be indicated by changing the whole display and not just part of it (e.g., change the entire background color of the display, or change the color of all the text or outlines).

There is a tendency to focus on the fragility of STM. Although it may be a handicap in some situations, it is probably advantageous in daily life. We often notice things going on around us and act upon them with little thought. That STM is "fragile" and of "limited capacity" means that redundant or extraneous items which find themselves in STM are soon gone when we stop thinking about them, leaving us free to concentrate on the business at hand. In this sense, we might say that, instead of being "fragile," STM has an automatic delete facility.

Chunking

Chunking increases the amount of information that can be held in STM by assembling separate items of information into larger units. A word is a way of chunking letters. For example, codes such as telephone numbers can be printed to explicitly encourage people to store them in a particular way. An 11-digit number such as 27214712503 appears less daunting when it is revealed to be the telephone number 27-21-471-2503, with 27 as the country code, 21 the city code, and 471 the area code. By making the structure explicit, an 11-digit string is reduced to seven items (the three chunked codes and the unique four-digit user number).

Many tasks involve STM in one way or another—mental arithmetic, for example. Since STM requires attention, it can be disrupted by distractions. For example, rehearsing a password while entering it into an application might be disrupted by the conversations of other people or by a colleague asking whether you would like a cup of coffee. *STM limitations have important implications.* People who have to use STM in their jobs are easily distracted by interruptions of even a few seconds—particularly by conversation—or by having to carry out verbal secondary tasks (such as answering the phone while doing mental arithmetic). Tasks that require STM storage from multiple sources can be said to involve a high *mental workload.*

Attention: A Usable Resource That Gets Used Up

Attention is the activity directed to facilitate the processing of an expected stimulus. It is essential for the operation of System 2 and is regarded as a limited resource that is channeled to "drive" processes such as working memory, response execution, etc. It can be thought of as the spotlight that illuminates the information world or, to use a computer analogy, as the CPU (central processing unit) responsible for scheduling tasks and allocating them to subroutines. Attention can be directed externally (to the world around us) or internally (to our own thoughts).

The time taken to switch attention from one target to the next is known as the "attentional blink" and takes about half a second (Olivers et al., 2007). Experimental research on the attentional blink often requires subjects to monitor a rapidly streaming list of characters. Most of the characters are not relevant, but within the stream are target characters that subjects have to detect. If two targets appear close together in time (from about 100 to 500 msec), detection of the second is usually worse than detection of the first. The proposal is that detection of the first target has "drained" the available attentional resources, which require about 500 msec to recover. Other research has demonstrated the existence of the opposite effect—attentional enhancement. If two data streams are presented to an observer, one of letters and the other of digits, and a target letter is used to cue subjects to switch from scanning the stream of letters to the stream of data, then attention to the digit stream is enhanced. The enhancement is greatest for the digits appearing closest in time to the appearance of the target letter and fades over a period of about 500 msec.

This research does *not* imply that it will take automobile drivers 500 msec to press the brake pedal after first seeing a red traffic light—a driver's reaction time is faster than this. What it does imply is that, in complex automobile environments where multiple data streams are presented to drivers (e.g., the view of the road, traffic data from the radio, in-vehicle warning systems and

TABLE 12.1
Characteristics of the Multichannel View of Attention

1. Processor and subroutines are fed by limited capacity, modality-specific channels
2. Processor handles data in a way that roughly equates with consciousness
3. Subroutines are built up through practice

The system can break down under a number of conditions

1. No subroutines available. Multiple inputs then compete for executive processes, which may be insufficient
2. Competition for subroutines. More than one task is being carried out simultaneously, and the tasks are similar in nature
3. The existence of subroutines implies skill. If the processor tries to take over the operation, the subroutine is disrupted

SATNAV signals, and searching for songs on MP3 players), the driver's attention to the road can be seriously disabled by inappropriate targets (if attention switches to "relevant" SATNAV data, "irrelevant" incoming visual data may be suppressed for up to 500 msec). Conversely, roadside information appearing within 500 msec of the target information in the road may be effective in ensuring that this information is attended to (e.g., "warning signs directly ahead").

Humans are multimodal in the sense of being able to process and represent information in many modalities (e.g., visual, auditory, and semantic), and many purposeful activities, such as walking and standing, can be carried out without being consciously attended to. Attention is seen as a problem of allocating processing resources to tasks (Table 12.1).

How Many Tasks Can We Do Simultaneously?

Multitasking can succeed when

1. The demands of the tasks do not exceed capacity.
2. Time-sharing is possible. Although two tasks may both be continuous, attention switching between them may be possible if competing demands can be dealt with consecutively. If the demands of one or both require concurrent processing, performance breaks down if the modalities are the same.
3. The tasks do not interfere with each other (compete for the same resources).
   a. They are in different modalities.
   b. One or both are skill-based.

A skilled operator has subroutines to handle the low-level aspects of task performance, which spare the executive processes. The subroutines that handle motor actions are of particular importance in integrating different muscular actions into a unified whole. The skilled golfer or tennis player conceives of and executes a swing or a serve as one smooth action, whereas the novice has to assemble the action out of a set of more primitive movements. For the skilled operator, detailed actions are derived in a "top-down" fashion from higher-level representations. For the novice, purposeful actions are assembled from smaller components in a "bottom-up" fashion.

The model is intuitively attractive in that it can account for how we can walk and talk at the same time but might have difficulty in walking and juggling or in talking and reading simultaneously. Activities that occupy different modalities are fed by different channels to dedicated subroutines and can be carried out simultaneously. Activities that occupy a single modality can only be carried out simultaneously if the channel capacity is not exceeded.

Wickens (1980) has described how human attentional resources are subdivided along spatial/verbal and other lines. Voice control is of benefit in tasks requiring a large amount of visual/spatial processing (such as computer-aided drafting and graphic design) but can interfere when the primary task is verbal (such as text creation). Haptic displays can be used to alert operators when vision and hearing are occupied (Figure 12.5).

FIGURE 12.5 "Rumble strips" on highways are effective at alerting drivers when vision and hearing are occupied.

Using Cell Phones in Public Places

Evidence has emerged of the dangers of using cell phones while walking (Naser and Troyer, 2013). Phone users (both hands-free and hands-on) are less aware of their surroundings and have divided attention that reduces their capacity to deal with multiple inputs. This applies to both pedestrians and drivers. People tend to cross the road more slowly when using their phones and are less attentive to traffic. In a naturalistic study, Hyman et al. (2010) interviewed people walking across a square after they had crossed. Some were using their phones, others had MP3 devices, and the rest had no devices at all. In the middle of the square, there was a clown on a unicycle cycling around a sculpture. Phone users walked more slowly, changed directions more frequently, and were less likely to acknowledge other people. When asked if they had seen anything unusual while crossing the square, only 8% of phone users mentioned the clown compared to 32% of MP3 users and people with no devices. When asked directly if they had seen the clown, 25% of phone users said they had, compared to 50% and 60% in the other two groups. It seems that cell phone use drains attentional resources, leading to attentional blindness.

Cell Phone Use While Driving?

In 2005, the Human Factors and Ergonomics Society summarized the findings of research into new technologies in cars as follows:

- Cell phone conversation makes drivers more likely to miss traffic signs and other important information.
- Hands-free devices are not always easy to use and may be even more distracting than handheld devices.
- Information presented by voice competes for the driver's attention to a far greater extent than when the information is presented on a display.
- The magnitude of the effect of distractions depends on when they occur. For example, answering a cell phone is likely to be more dangerous during maneuvers like merging to exit a freeway.

Zhu et al. (2021) analyzed the effectiveness of bans on cell phone use while driving in the USA. Comprehensive bans on the use of handheld cell phones while driving were found to be associated with lower driver fatality rates, while non-driver and total fatality rates were unaffected. Partial bans (e.g. on calling only or texting only) did not reduce fatality rates.

TABLE 12.2
Factors Influencing Vigilance

- *Signal Conspicuity*: Auditory signals—increased signal amplitude (signal-to-noise ratio)—improve detection. The number of missed signals drops approximately by a factor of 10 when the signal is 5.1 dB greater than the noise, as compared to 2.1 dB (double the intensity of the signal and get a 10-fold improvement). Vigilance decrement goes away.
- *Signal and Noise Intensity*: Boosting the intensity of both signal and noise in an auditory task can abolish the vigilance decrement (probably due to the alerting aspect of noise).
- *Signal Duration*: Increasing signal duration to 2–3 s leads to almost 100% detection in visual tasks.
- *Number of Events*: Quality of sustained attention is inversely related to the total number of events. Mackworth suggests that when there are lots of events and only some of them are signals, we habituate to them more rapidly, which explains the poorer performance.
- *Knowledge of Results*: Feedback can abolish the vigilance decrement and enhance the speed and frequency of detections. On simple vigilance tasks, the beneficial effects of KOR are probably motivational; people know how well they are doing and do not get bored. In more complex tasks, the feedback may reinforce detection of signal cues.
- *Ways of Sustaining Attention*: Synthetic targets with feedback, secondary tasks.
- Provide amphetamines, caffeine, and methylphenidate, but there are problems with side effects.
- *Other Factors*: Effects of elevated body temperature are equivocal. Effects of cold, noise, and vibration are difficult to predict in practical situations because the observer is an active participant in the environment and can take steps to modify exposure or enhance performance.
- *Techniques for Improving Performance*: Reduce uncertainty (about what constitutes a target and where it will occur). If possible, provide specifications of what to look for (reference standards, if possible).
- *Provide Redundant Information*: For example, visual and auditory cues.
- *Signal Enhancement Techniques*: Signal processing techniques exist for modifying displayed data. For example, instead of displaying raw EEG data, you can display changes in voltage over time in different sites, electrical power rather than voltage, or smooth out noise using running means or other statistical techniques that operate in real time.
- *Reduce Spatial Uncertainty*: Redesign display to fit into a smaller area (e.g., eye field as opposed to head and field).
- *Introduce Job Enlargement*: For example, inspection of coins in batches can be interspersed periods of fetching and loading/unloading a conveyor. Combine radar monitoring with navigation (useful in teams where tasks can be rotated dynamically).
- Provide regular short breaks of around 5 min.
- *Motivation*: Subjects perform better when told the task is a selection test for a well-paid job than when they believe it is a psychological experiment (Nachreiner, 1977).

Source: From Warm (1984).

Sustained Attention: Vigilance as a Self-Control Problem

The ability to sustain attention over long periods is known as vigilance. Research on vigilance dates from the Second World War, when there was great interest in the ability of operators to monitor radar and sonar displays. For a variety of reasons (e.g., boredom, distraction, and sleepiness), operators are not very good at performing these tasks, particularly when they are alone. Some of the factors that influence vigilance task performance are summarized in Table 12.2.

Aiding Vigilance Task Performance

In aviation, in response to crashes caused by "human error," attempts have been made to use automatic systems such as the "ground proximity warning system." Aircraft have crashed when, under a heavy workload, the pilot stops looking out of the cockpit and, when the autopilot disengages, crashes in a perfectly controlled manner into the ground. The problem with automated warnings is that they may malfunction. The result is an exchange of one type of vigilance task for another. Type 1 errors (warning you when there is no danger) and Type 2 errors (not warning you when there is) can occur, and there is a trade-off between the two. E.L. Wiener has argued that alerting devices are nothing more than substitute vigilance tasks.

LTM: Our Model of the World

Cognitive processes—those which have to do with knowing—are of great importance for the design of usable products. For example:

1. Learning what facilities a new word processor offers and how to use them to create documents
2. Deciding which of a number of software packages to buy based on an analysis of the usefulness of their attributes and cost
3. Solving comprehension problems that arise when learning to use a system

Long Term Memory (LTM) is organized on the basis of meaning. If people are given lists of random words to remember, they tend to recall them in semantically similar clusters, which suggests that they have actively organized the information in this way when storing it. Bartlett (1932) showed people pictures or stories which they repeatedly recalled up to 6 years later. He found systematic differences between the original stimuli and the recalled material and noted that people abstracted out the details, integrated them with other information, and omitted "irrelevant" information, producing a condensed form of the original material. People also made additions to the original material in an *effort after meaning,* in which *we fill in the lowlands of our memory with the highlands of our imagination.* This has important implications for the assessment of eyewitness testimony in court cases and the need to avoid the use of leading questions when interrogating witnesses.

A fundamental distinction can be made between episodic and semantic memory (Tulving, 1972). Episodic memory is memory about personal experiences and events, whereas semantic memory contains world (or general) knowledge. The former is more easily forgotten than the latter.

How to Make Things Easier to Remember and to Recall

The way an item is encoded in memory plays an important role in determining how easily it can be recalled. In fact, some theorists regard memory as a by-product of processing rather than a store in which items are placed. Paradoxically, in order to retrieve something from LTM, we first have to know what it is we are trying to remember. A "retrieval cue" acts like a description of the to-be-remembered item. If the cue is sufficiently detailed, the recall probability is increased. In addition, the desired item has to be discriminated from other similar items in memory in order to be recalled, and this depends on how it was encoded when it was stored. Similar items that were encoded in sensory or contextually different ways can be discriminated more easily than those that were encoded in a similar way. Elaborative rehearsal (Craik and Lockhart, 1972) is a method that increases the number of retrieval cues for an item. Items can be encoded multidimensionally (e.g., visually, verbally, and semantically) to increase information redundancy and meaningfulness. Most *mnemonic systems* rely on some form of elaborative rehearsal to code new information.

Mnemonics, Verbal Elaborative Processing, and Visual Imagery

The use of mnemonics has a long history. Ancient Greek orators used a system known as the "method of loci" to help them remember the details of a long speech. In rehearsing the speech, a familiar street would be imagined. At the start of the speech, the orator would imagine himself at the beginning of the street. Each important section of the speech would be associated with a particular part of the street such as a shop. By "walking down" the imaginary street during the speech, each section of the speech would be recalled in the order it was memorized. The method works by linking new information (the parts of the speech) to an intrinsically sequenced set of previously memorized objects, which act as retrieval cues.

Verbal elaborative encoding can be used to assist in the initial stages of learning. Reduction usually involves taking the first letters of each word to be remembered and using them to create a new word. The original words can be derived from their first letters, which are stored together as one item. For example, many students of trigonometry use the acronym "SOHCAHTOA" to remember the relationship between the trigonometric functions and the sides of a right-angled triangle (Sine = Opposite/Hypotenuse, Cosine = Adjacent/Hypotenuse, and Tangent = Opposite/Adjacent). Expanded mnemonics usually involve the addition of meaning to the to-be-remembered material, as when learning the lines of the treble clef in music. E, G, B, D, F can be expanded to "every good boy deserves food." At first, it may seem paradoxical that one of the best ways to improve the ability to remember something is to increase the amount of information to be remembered. The paradox disappears when it is remembered from memory theory that the process of storing an item in memory involves linking the new item to a previously established associative network of meaningful information. Expansion works by increasing the number of links between the new item and the network.

Information in LTM is stored at a semantic level but is linked to words, rules, and other structures. Many concepts can be represented in terms of other symbols or sensory modalities at will. In this way, it is possible to talk about the imagery value of words. Words such as justice and philosophy are linked to abstract concepts, which are difficult to link with information received from the senses. In contrast, words such as balloons, birthday cake, and candy floss are linked to more concrete concepts—we obtain knowledge about them directly via our senses, and they evoke strong visual and other images when we hear them. There is evidence that abstract words are more difficult to learn than concrete words (Baddeley, 1983), which may explain some of the difficulties experienced by novice computer users in learning to interact with a system in which objects are represented in an abstract way.

Network Theories of Memory

Modern theories of semantic memory emphasize its associative nature—all items stored in memory are ultimately connected to all other items on the *basis of their meaning*. LTM is modeled as a *semantic network* (Figure 12.6). The problem facing these theories of memory is twofold—to describe the organization of information stored in memory and account for the ways it can be accessed (Simon, 1979).

Since all items in semantic memory are ultimately linked (memory is associative), they can be retrieved by searching along the links that connect them. Thus, in trying to remember something, we may search for a cue—some other item known to us that is closely connected to the required item. A good retrieval cue can amount to a full description of the desired item of information, which distinguishes it from all

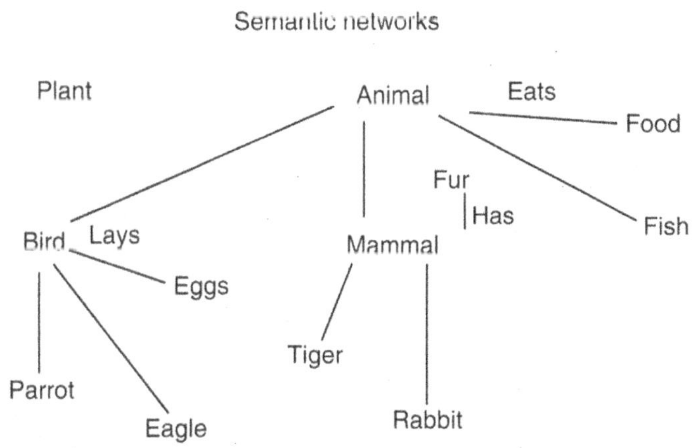

FIGURE 12.6 Sample representation of LTM as a semantic network.

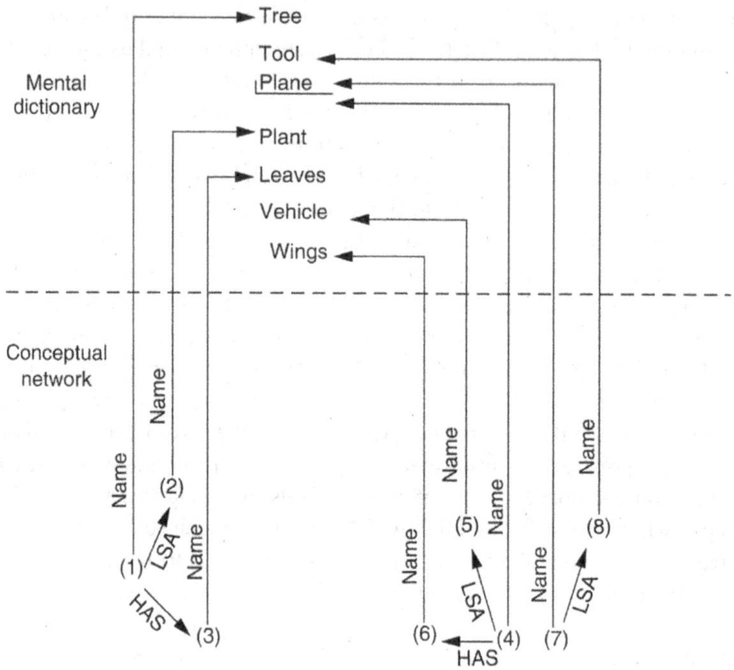

FIGURE 12.7 Concepts are stored in an associative manner in the semantic network. They are linked to words that are stored separately in a mental dictionary. (Redrawn from Collins and Quillian, 1972.)

other items. Discriminability is a major determinant of the ability to remember something. Memory can be modeled as a series of nodes (concepts) interconnected by relations. The nodes themselves are not words but are connected to words in a mental dictionary. Several words may be linked to the same concept (e.g., car and automobile), or the same word may be linked to different concepts (e.g., plane).

Figure 12.7 shows an example of a network theory that was developed by Quillian (1969). In order to understand a sentence, each word is decoded, and the appropriate concept is flagged. A parallel search then takes place in all directions. Thus, understanding the sentence "Canaries can fly" would involve searching the network for the existence of interconnections between the different nodes. If the nodes are found to be connected, then the sentence is understood and can be verified. Despite some evidence supporting the theory (Collins and Quillian, 1972), it could not account for certain features of sentence comprehension.

Anderson (1983) distinguished between declarative knowledge ("knowing that…") and "procedural" knowledge ("knowing how…"). The former is stored as a network of linked nodes, a small part of which can be active at any one time (which corresponds roughly to working memory). Activation can spread out from a node to other nodes—thinking about a particular person may remind us of other related people, or when trying to recall the details of a report, we may think of a similar report whose details we do remember. The more nodes through which activation spreads, the more slowly it spreads. Since declarative knowledge is said to be stored in the form of abstract propositions, the theory explains why people only remember the "gist" of things. Procedural knowledge is stored as a set of production rules specifying what should be done if a given set of conditions becomes active (e.g., if the light is green, then cross the street).

Relationship between STM and LTM

To use a computer analogy, STM can be likened to the assignment of a particular value to a variable in a computer program for the duration of an operation. After the operation is completed,

TABLE 12.3
Differences between LTS and STS

	STS	LTS
Storage capacity	7 items ±2	Extremely large
Retention interval	5–30 s	Many years
Mechanism of information loss	Trace decay Displacement by new items	Inability to retrieve item
The way information is coded	Phonetic/articulatory	Semantic

Note: LTS, long-term storage; STS, short-term storage.

the value is no longer needed and is lost. The variable itself, however, remains in the program (in LTM) and new values can be assigned to it the next time the operation is invoked. Unlike in a computer, however, storing information in STM requires considerable mental resources, particularly conscious attention. Storage is fragile, and the information is easily lost if the operator is distracted by unexpected events or overloaded by new information. For example, when driving home in busy traffic, a driver may forget details of traffic conditions just heard on the car radio when distracted by a call on his car phone. The STM trace of these items is displaced by new information, but the route information, which is stored in LTM, is not forgotten.

Response Selection and Execution

Response selection and execution can occur as the output of a specific subroutine—an automatic or "reflex" response—or may be initiated by the executive processor as a measured or calculated decision to act.

Feedback

The lag between executing an action and receiving feedback is one of the most important variables affecting our ability to carry out continuous tasks. The classic experiment in which a subject is given delayed feedback about his own voice illustrates the importance of feedback in the production of speech. Even slight delays severely disrupt the ability to speak. Modern approaches to user-interface design stress the importance of appropriate feedback (of the system communicating with the user). For example, when executing a command, some immediate feedback must be given to indicate that the command has been received and is now being executed.

Bank statements, bills, and receipts are examples of the complex array of feedback individuals receive in their everyday lives. An interesting and potentially important area, in the context of increasingly integrated modern communication and data processing systems, is the design of feedback about everyday behavior. Feedback, in this context, is only useful if users recognize it as such, are motivated to use it, and if it is in a form that they can easily map onto their own behavior. McClelland and Cook (1979), in an early study, investigated the effects on domestic energy consumption of providing continuous feedback (in cents per hour) in 101 all-electric homes. They found an average energy saving of 12%, with the greatest improvement in the more temperate months, suggesting that the savings were due to changes in energy use other than energy for heating and cooling.

The design of feedback requires careful consideration. If the grain size of feedback is too small, major trends may be obscured by minor variations. Alternatively, the patterning of feedback over time may be just too complex for the user to extract meaning from it. If the grain size is too large, people may not be able to map the information in the feedback onto specific behaviors amenable to self-modification. Thus, there is a trade-off between two extremes. At one extreme, detailed feedback that can be related to particular behaviors must not cause information overload. At the other extreme, general feedback may be easily grasped but if too general, cannot be related to specific behaviors.

Returning to the domestic energy consumption example, most people do not regularly monitor how much electricity they are using. Often, monthly or even quarterly bills are the only feedback they receive. Alternative energy purchasing and payment schemes have been developed, including the "pay as you use" system, in which users buy electricity "credits" that are entered into their domestic energy meters using a plastic card similar to a credit card. The meter displays the amount of electricity remaining before more credits have to be purchased. It seems likely that users of this type of system would be much more able to modify their energy usage because of the much closer mapping between consumption behavior and feedback.

Cognitive Processing and Physical Stress

Physical stressors can degrade cognitive performance if they have cognitive effects. Alomari et al. (2015) had subjects hold their hand in cold water for three minutes while carrying out a sustained attention to response test (SART). The SART requires attention to be focused and sustained on a rapidly changing sequence of stimuli for four minutes (Robertson et al., 1997). The impulse to respond to infrequent (one in every nine) targets has to be continuously suppressed. Self-control is required to keep the hand in the cold water because it is distracting and uncomfortable. The urge to remove the hand has to be suppressed. Compared to a control condition, the dual-task combination resulted in poorer performance on the SART, suggesting that both tasks competed for similar processes (attention).

Dietrich and Sparling (2004) argued that because demanding physical activities have large motor and sensory costs that require energy, these activities will interfere with energy-intensive cognitive processes in the prefrontal cortex (System 2), but not with processes that require little resource. They conducted an experiment in which subjects ran or cycled on a treadmill at 70%–80% of their maximum heart rate for 45 minutes, completing cognitive tests after the first 20 min. Performance on a test of cognitive function known to be associated with the prefrontal cortex (the Wisconsin Card Sorting Task) was lowered when running or cycling compared to a control condition, whereas performance on a test of general intelligence was not. They concluded that "endurance exercise selectively compromised prefrontal-dependent executive functions such as working memory, sustained attention, and the ability to inhibit habitual responses." Indeed, these findings rule out alternative explanations (such as "fatigue," because only one type of cognitive test was affected by endurance exercise).

TOOLS AND PROCESSES

Mental Workload

ISO 10075 (1991) Parts 1 and 2 deal with ergonomics and mental workload. The mental workload of a task is nothing more than the sum of the mental demands placed upon the operator, in a manner analogous to physical workload or stress. However, the effort required to meet the demands and carry out the task effectively will depend on many factors, including operator training and skill, the duration of task performance, and the strategies used by the operator to cope with the demands. If the effort is too great or cannot be sustained, mental strain may result. The role of ergonomics is twofold: to modify the demands in accordance with the characteristics and limitations of human information processing and to schedule the task such that human performance can be sustained in accordance with system requirements.

According to ISO 10075, mental stress is defined as "any external influence impinging on human beings and affecting them mentally." Mental strain is defined as the immediate reaction to stress, and its consequences may be facilitatory and induce learning, or they may result in fatigue and human error. Table 12.4 summarizes the consequences of operator experience of mental strain.

Factors Affecting Mental Workload

Many of the factors that affect mental workload are under the control of designers and ergonomists. Table 12.5 gives examples (ISO 10075).

TABLE 12.4
Consequences of Mental Strain

Warming-Up Effect: A reduction in effort follows initiation of the activity.
Activation: The level of mental functional efficiency with respect to the task.
Mental Fatigue: Temporary impairment of mental functioning in proportion to the duration, intensity, and temporal pattern of exposure to mental stress. A temporary reduction in the capacity to perform the tasks or tasks of a similar kind.
Fatigue-Like States: Monotony, lack of vigilance, and mental satiation occur as a result of overexposure to situations offering little variety. These are activity-specific and disappear on a change of activity.
Monotony: A slowly developing state of reduced activation that may occur during long, repetitive tasks lacking variety. Associated with drowsiness, tiredness, and a decrease in performance.
Reduced Vigilance: A drop in signal detection during monitoring of tasks.
Mental Satiation: Strong emotional rejection of further participation in a repetitive mental task.
Practice Effect: Facilitation of performance as a result of successful coping with mental strain.

Source: From ISO 10075 (1991).

TABLE 12.5
Factors Affecting Mental Workload

- *Ambiguity of the Goal*: Can be divided into ambiguity of the goal itself or ambiguity in the nature of the sufficiency criteria used to decide how well the goal has been achieved (e.g., was the purpose of an ergonomic intervention to improve productivity or health and safety?). Workload can be reduced by reducing goal ambiguity, for example, by making explicit the criteria used to assess performance.
- *Competing Goals*: Particularly when there are problems assigning priorities when goals are in conflict.
- *Serving Strategy*: Mental workload increases when there is no clear strategy or rule for dealing with incoming tasks. Workload can be reduced by training operators in strategies for dealing with different tasks. "First-in, first-out" is probably the simplest strategy. Prioritization can be made in advance according to deadlines or importance.
- *Task Complexity*: A lack of complexity can lead to underload and problems with vigilance and engagement with the task. Excess complexity may overload the operator with decisions.
- *Adequacy of Information*: Insufficient or excessive information can both increase mental workload. Information adequacy can be assessed in relation to both routine task performance and when dealing with unexpected/unprogrammed events.
- *Signal Discriminability*: Reduce workload by using ergonomic principles to improve display design (Chapter 13). Improve signal to noise ratios and provide appropriate context (other information relevant to the signal).
- *Redundancy*: Provide more than the minimum information to detect the signal (e.g., for warnings, change the color of a displayed signal to red, flash the display, and activate an auditory alarm).
- *Compatibility*: See Chapter 13.
- *Parallel versus Serial Processing*: Serial displays are recommended for sequential tasks that are not strongly interrelated. Parallel displays can reduce mental workload when comparisons have to be made.
- *Time-Sharing*: Time-sharing (switching attention between two concurrent tasks) can increase mental workload if the tasks compete for the same processing resources or if one or both require a high level of attention. Workload can be reduced by relaxing the time standard for the work.
- *Delays*: System response delays can increase mental workload if information has to be held in memory until the system can receive it. Lag can cause uncertainty if feedback, preferably with an indication of when the system will be ready again, is provided.
- *Mental Models*: Provision of an explicit mental model of the process/system appropriate to the operator's level of involvement can reduce mental workload by improving the operator's ability to understand current and predict future system behavior.
- *Absolute versus Relative Judgment*: Absolute judgment can be supported by displaying reference values or standards for comparison with currently displayed values. Gauges can be marked with zones of different colors, for example, green (safe operation), amber (caution), and red (danger or overload).

(*Continued*)

TABLE 12.5 (*Continued*)
Factors Affecting Mental Workload

- *Working Memory Load*: Mental workload may become excessive if the flow of to-be-remembered items is too rapid, and reduced if items are held in the system's memory as a time series; for example, LTM load can be reduced if information retrieval facilities are provided with an explicit system for coding and retrieving items.
- *Recognition versus Recall*: Presenting operators with a list of options is preferable to requiring that the options be recalled from memory.
- *Controllability*: Ease of operation can reduce mental workload. Ergonomic considerations include: control–display ratio; compatibility; design in accordance with population stereotypes (Chapter 13). Error tolerance: See Chapters 14 and 15. Common ergonomic recommendations for system design include: make actions reversible; design the system to reject obviously illegal entries (e.g., reversed codes); accept upper and lower case entries; confirmation of concurrent entries and their effects; use of forcing functions to prevent illegal actions.
- *Environmental Design*: Optimize noise, lighting, and climate for the work being performed (Chapters 9–11). Minimize intrusive stressors (such as task-irrelevant speech).
- *Social Interaction*: Provide opportunities for communication and collaboration, task sharing, etc.
- *Time Pressure*: Workload can be reduced by relaxing the time standard for the job.

Source: From ISO 10075 (1991).

Boredom and monotony can be caused by social isolation, absence of rest pauses, lack of physical activity, lack of task variety, unvarying climate, and unvarying acoustical stimulation. The kinds of coping strategies adopted by operators can also affect the mental workload. Many of these, such as work scheduling and error recovery, have to be taught. In general terms, *avoidance coping* consists of doing something else when the workload becomes high (e.g., carrying out a secondary or less demanding task) or ignoring the problem. Avoidance coping strategies may work as a last resort (e.g., fleeing the scene before the reactor melts down) but are not generally *problem focused* enough to maintain system performance. In the short term, they may reduce the mental workload but at the cost of the system's operation and increased workload in the future. Approach coping strategies consist of finding out more about the problem, speaking to others about it, trying to "see the bigger picture," etc., and are more likely to lead to solutions, even though the mental workload may increase in the short term.

Broadbent (1979) argued that fatigue, both after sustained performance and after exposure to abnormal environments (such as physical stressors), caused a breakdown in the organization of the performance of complex tasks, characterized by:

- Forgetting to carry out critical subtasks on time
- Mistiming the execution of one task in relation to others
- Attentional "stickiness" (inability to shift attention from one subtask to the next)
- Lack of persistence
- Breakdown of peripheral attention (e.g., failing to monitor displays in the periphery of the visual field)
- Breakdown of incidental learning (failure to "get the hang of things" when a system changes)
- Failures of peripheral attention (narrowing of the attentional field)
- Reluctance to carry out nonmandatory, unprogrammed, or novel tasks
- Errors of omission (forgetting to do something when it needs to be done)
- Errors of commission (an incorrect action or operation)

Table 12.6 summarizes the characteristics of poor performance associated with exposure to excessive work demands.

TABLE 12.6
Some Characteristics of Poor Performance Resulting from Excessive Work Demands

1. Slower reaction time
   Increased time to complete the task
2. Increased time needed to learn new information and procedures
   Increased forgetfulness
3. Failure to take all information into account when making decisions
4. Increased need to repeat steps in a chain of reasoning
   Inability to concentrate
5. Lapses in attention
   Missing or misinterpreting signals
   Omitting steps in a procedure
6. Slowness in noticing changes in task requirements
   Slowness in initiating new tasks as required
   Increased need for instructions
7. Narrowing of attention
   Increased concentration on individual subtasks
   Neglect of other tasks
8. Erratic operation of controls
   Mistiming of actions
   Inappropriate gauging of responses
9. Making unnecessary responses
   Carrying out irrelevant tasks
10. Increasing tolerance of own errors
    Problems recovering from errors
    Acceptance of lower performance standards
11. "Chin drop"

Modeling the Breakdown of Task Performance

Hockey (2005) has discussed the breakdown of human performance using an "Operator Functional State" model, the capacity for sustained performance on a task, which is a function of operator background factors such as sleep loss and fatigue, stable operator characteristics such as level of expertise and coping style, and the mode of interaction with the task (e.g., the priorities, deadlines, etc.). The model enables a distinction to be made explicitly between the *effectiveness* of task performance and the *efficiency* of performance. Efficiency here refers to the costs incurred by the operator in maintaining a target level of performance. When task demands are high, when the environment interferes with the maintenance of performance, when operators are fatigued, or when there is exposure to physiological stressors, performance on the main task may be maintained at high cost to the operator. The efficiency of performance can be assessed best not by measuring the main task, but by searching for *latent decrements* in performance. Such decrements occur as a result of the increased effort to protect performance on the main task and include:

- Decrements in the performance of secondary tasks (e.g., a pilot might forget to check the amount of fuel remaining in the fuel tank while flying in difficult conditions; a sleep-deprived doctor might forget to check the dosage before prescribing a little-known medicine)
- Strategy changes (a fatigued long-distance driver drives to his destination using a passive "lane-tracking" strategy rather than an active strategy of reading the road signs and looking at the road ahead)

- Regulatory costs (mental strain may cause increased heart rate and reduced heart rate variability, or increased anxiety)
- After-effects (e.g., disinclination to carry out further complex or challenging tasks)

The practical application of the model in safety-critical systems is to monitor operator performance with the aim of detecting the emergence of these latent decrements before an accident happens. When decrements are detected, automatic safety systems are activated that either reduce the operator's workload or lower the goal. One way in which decreased efficiency in performance manifests itself is that the task becomes more demanding subjectively. A wide variety of methods have been developed to measure the mental workload subjectively. HFE Workshop 12.1 describes some common methods.

HFE WORKSHOP 12.1

THREE METHODS FOR SELF-REPORTING OF MENTAL WORKLOAD

Mental workload assessment tools are essential tools for ergonomists. They can be used for a wide variety of purposes: to evaluate the effects of improvements to the design of user interfaces aimed at optimizing the workload, to identify the peaks and troughs of the workload in the design of complex tasks, and to assess and improve the usability and learnability of products, and ultimately, their acceptance by users.

All three methods described here are self-report methods designed for use in practical settings.

NASA TLX (TASK LOAD INDEX)[1]

The TLX (Hart and Staveland, 1988) is completed following the performance of a task.

Step 1

Six different scales are used to rate the different aspects of workload. The scales are 10-cm wide and are divided into 20 intervals. The ratings are converted to scores that vary from 0 to 100.

Mental demand Workload score

|_|_|_|_|_|_|_|_|_|_|_|_|X_|_|_|_|_|_|_|_| 60

Low High

Physical demand

|_|_|_|_|X_|_|_|_|_|_|_|_|_|_|_|_|_|_|_|_| 20

Low High

Temporal demand

|_|_|_|_|_|_|_|_|_|_|_|_|_|_|_|_|X_|_|_|_| 80

Low High

Effort

|_|_|_|_|_|_|_|_|_|X_|_|_|_|_|_|_|_|_|_|_| 40

Low High

Performance

|_|_|_|_|_|_|_|_|X_|_|_|_|_|_|_|_|_|_|_|_| 35

Low High

Frustration

|_|_|X_|_|_|_|_|_|_|_|_|_|_|_|_|_|_|_|_|_| 10

Low High

(Continued)

HFE WORKSHOP 12.1 (*Continued*)

The definitions of the scale items are as follows:

- *Mental Demand*: How much mental and perceptual activity was required (e.g., thinking, deciding, calculating, remembering, looking, searching, etc.)? Was the task easy or demanding, simple or complex, and exacting or forgiving?
- *Physical Demand*: How much physical activity was required (e.g., pushing, pulling, turning, controlling, activating, etc.)? Was the task easy or demanding, slow or brisk, slack or strenuous, and restful or laborious?
- *Temporal Demand*: How much time pressure did you feel due to the rate or pace at which the task or task elements occurred? Was the pace slow and leisurely or rapid and frantic?
- *Performance*: How successful do you think you were in accomplishing the goals of the task set? How satisfied were you with your performance in accomplishing these goals?
- *Effort*: How hard did you have to work (mentally and physically) to accomplish your level of performance?
- *Frustration Level*: How insecure, discouraged, irritated, stressed, and annoyed versus secure, gratified, content, relaxed, and complacent did you feel during the task?

Step 2

Next, the different aspects of workload for the particular task are rated by means of pairwise comparison. This means that each aspect of workload is compared to every other aspect of workload. A value of "1" is assigned to each aspect of workload every time it is rated as more important than the aspect to which it is being compared.

For example:

	Mental	Physical	Temporal	Performance	Effort	Frustration	Total
Mental		1				1	2
Physical						1	1
Temporal	1	1		1	1	1	5
Performance	1	1			1	1	4
Effort	1	1				1	3
Frustration							0

In this example, the temporal demands of the task are the greatest, and the demands caused by frustration arc thc lcast.

Step 3

The ratings are multiplied by the weightings to give a set of adjusted workload ratings. These are then summed and divided by 15 to give an overall TLX value for the task.

Aspect of Workload	Rating	Weighting	Rating × Weighting
Mental	60	2	120
Physical	20	1	20
Temporal	80	5	400
Performance	40	4	160
Effort	35	3	105
Frustration	10	0	0
Total	245		805

(Continued)

HFE WORKSHOP 12.1 (*Continued*)

Unweighted TLX score = 245/6 = 40.8
Weighted TLX score = 805/15 = 53.7

The weighted score is higher and reflects the high temporal demands and low physical and frustration demands.

TASK DIFFICULTY INDEX

The task difficulty index is a simple technique for comparing the mental demands of different tasks. Task difficulty index scores have been shown to correlate positively and significantly with subjective workload measures of different tasks. The advantage of the index, apart from its simplicity, is its diagnosticity. In principle, it can tell us something about why one task might be more difficult than another task. Furthermore, the sources of difficulty are described in such a way as to be compatible both with human information processing models and with the kinds of design interventions used in ergonomics. The main drawbacks include the need to train users in what the dimensions really mean and how to describe tasks in terms of the dimensions, the fact that the dimensions of difficulty may vary during a single task, the lack of a dimension to account for task duration, and the assumption that there is a simple additive relationship between the different dimensions. This latter point can be illustrated by considering two tasks, one with a difficult single task and the second with two easy dual tasks. The second task is rated more difficult, even though it may be easier. Owing to its simplicity, the index can be used together with other techniques without undue administrative burden.

Task Difficulty Index

Sum of Scores on Four Dimensions	
Familiarity of stimuli	0 = Familiar (e.g., letters)
	1 = Unfamiliar (e.g., spatial patterns and unfamiliar map)
Number of concurrent tasks	0 = Single task
	1 = Dual/multitask
Task difficulty	0 = Memory set size 2 or less
	1 = Memory set size 4 or more, delayed recall
Resource competition	0 = No competition
	1 = Competition for either modality of stimulus (e.g., visual and auditory) or central capacity

Source: From Gawron (2000), with permission.

MEASURING MENTAL WORKLOAD

Measurement of spare mental capacity is based on the rationale that the operator has a limited capacity to process information. As task difficulty increases, this capacity is used up, and less is available to be allocated to other tasks if they have to be carried out at the same time. If the operator is given a secondary task to do concurrently with the main task, the level of performance on the secondary task will reflect the difficulty of the main (primary) task. If the main task is not too difficult, processing capacity will be available for performing the secondary task as well. If the difficulty of the main task increases, less capacity will be left for the secondary task, and performance on the secondary task will suffer. For this reason, secondary tasks are designed so that performance can be measured objectively by means of performance time and/or number of errors.

In the secondary task paradigm, a change in primary task workload is indicated by a reduction in the level of performance of the secondary task (secondary task performance is sacrificed to divert mental resources to the maintenance of the primary task). If there is a reduction in primary task load, the performance of both tasks may improve.

Brown carried out a number of classic experiments on driving using the secondary task technique (Brown and Poulton, 1961). Brown (1966) found that a secondary task could predict success and failure in trainee bus drivers. Trainees were given a subsidiary task to do while driving (checking eight-digit strings for changes). The trainees' scores on the subsidiary task while driving were expressed as a percentage of their scores while only doing the subsidiary task. The percentage score was taken as an index of the amount of "spare mental capacity" while driving. Previous driving experience was found to be a good predictor of whether the trainees were successful in passing a final driving test. Secondary task performance while driving was also found to be a good predictor of future success, but only if performance on the secondary task was investigated at the beginning of the training program. These findings were interpreted as follows:

1. Those trainees who initially experienced the smallest decrements in secondary task performance while driving already had driving skill (either naturally or through previous experience). This meant they had spare processing capacity to carry out the secondary task even during the early stages of learning to drive a bus.
2. Those whose secondary task performance suffered while driving had little skill (i.e., few subroutines) and had to use conscious mental capacity to drive during the early stages of training. This left less mental capacity for the subsidiary task.

The Bedford Scale (Figure 12.8) is a self-report measure of mental workload based on the assessment of spare mental capacity (Roscoe and Ellis, 1990).

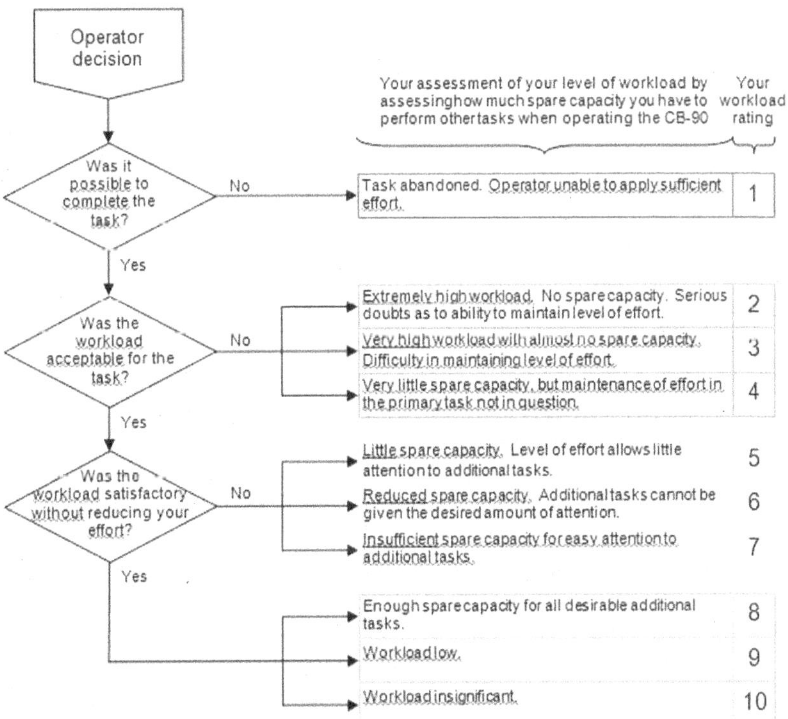

FIGURE 12.8 The Bedford scale.

It has been widely used in aviation to measure pilot workload. HFE Workshop 12.2 gives an example of a secondary task measure based on the estimation of time intervals.

Some problems with secondary tasks are that they may not be as sensitive as required; they may disrupt the performance of the primary task or even be perceived as the main task by the operator. This, of course, violates the logic of the procedure and renders it invalid. However, the secondary task procedure is intuitively appealing in the sense that at certain times, when the primary task load is high, operators will be too busy to perform the secondary task well. Some commonly used secondary tasks are:

- Mental arithmetic
- Time estimation
- Shadowing (visually presented digits have to be named—used to estimate visual free time)
- Digit classification (odd or even)
- Reaction time to memorized target digits

Performance of the main task is a source of information about the rate of work from which the mental workload of the operator may be inferred. In tasks such as driving, specially instrumented vehicles may be used in which control actions (e.g., steering, braking, and accelerating) can be recorded. Although these variables objectively quantify task performance, they do not necessarily have a one-to-one relationship with mental workload. A skilled operator may be able to cope with a wide range of task loadings with minimal demands on conscious processing capacity. Subjectively, these may all be perceived as being equally easy to perform. In conjunction with users, the investigator may carry out an analysis of critical points in task performance and assess the workload. Primary task performance can be assessed by developing performance metrics that specify the standards required or, in safety research, for example, by generating lists of critical and generic behaviors or actions. Quantification of the metrics and specification of the behaviors are best done in consultation with experts in the task domain.

Executive Control in Skilled Performance and the "Dysexecutive Syndrome"

Complex tasks are often carried out under executive control, in which the limited capacity executive processor monitors overall task performance and allocates different operations to dedicated subroutines. Executive control during skilled performance is vital for

- Forward planning
- Resistance to distraction
- Attention to relevant features
- Response inhibition

Loss of executive control is essentially the loss of the ability to monitor one's own behavior in relation to one's intentions (to pay attention to what one is doing). Studies of patients with brain injury, as well as EEG investigations of normal subjects carrying out sustained work, suggest that it is the right prefrontal dorsolateral cortex of the brain that is involved in this aspect of attention (Datta et al., 2003). Thus, the breakdown of skilled behavior in complex tasks is a real neurological phenomenon and not just a sign of sloppiness, weakness, or lack of motivation.

"Dysexecutive syndrome" is a term used in the clinical literature to describe adults with impaired executive function due to brain damage or disease, and also children with Attention Deficit Hyperactivity Disorder (ADHD). In healthy adults, mild, reversible loss of executive function occurs predictably. There is a 24-h rhythm in executive control in which the propensity to commit errors of commission (as measured by a sustained attention to response test, the SART; see Chapter 8) is greater at 01:00 and 07:00 h than at 13:00 and 19:00 h (Manley et al., 2002). This change in executive function may explain why accidents and errors are more likely to happen in the early hours than during normal daytime working hours. People who occupy very stressful jobs for long periods under conditions of social isolation, with conflicting demands or high emotional demands, suffer "burnout," which is characterized by a drop in work efficiency, emotional exhaustion, and depersonalization (difficulties relating to others). Burnout sufferers also have difficulties concentrating, loss of executive control, and report more "cognitive failures" in daily life (van der Linden et al., 2005). It is noteworthy that burnout sufferers made more errors of commission on the SART test than controls, but their reaction time on the test (time to make a correct response) did not differ. This suggests that the deficit was an attentional one and not due to a global reduction in ability. Van der Linden et al. proposed that chronic exposure to high stress might lead to chronic reductions in stress hormone levels, resulting in the mechanisms of executive control going "off-line" to be replaced by automatic response patterns.

HFE WORKSHOP 12.2

ASSESSMENT OF WORKLOAD USING INTERVAL PRODUCTION AS A SECONDARY TASK

Interval production is a secondary task that requires people to report when they think a target interval of time has elapsed. They do this continuously while performing a primary task or when doing nothing if baseline data are required. They continue to estimate the passage of time until sufficient data have been gathered to assess the task at hand. The experimenter records the actual passage of time for comparison.

Interval production requires that attention be paid to the passage of time while the main task is carried out. It utilizes a number of System 2 resources—inner speech if the passage of time is counted, or STM of the previous report of the passage of a time interval. At the same time, interval production requires no special equipment; that is, it does not require the use of the hands or the senses and, in this respect, is a secondary task of a general cognitive nature but responses can be made either verbally or by using foot pedals or buttons, enabling automatic data capture of interval estimates. If the primary task is verbal or requires use of the voice, response by hand or foot is recommended, and by voice if the task is manual. For driving, for example, voice seems best.

The rationale for the interval production task is the same as that for all secondary tasks—under conditions of low workload, spare mental resources will be available for the secondary tasks, and the interval production will be more accurate and less variable. As workload increases, interval production performance will decline.

Grant et al. (2013) provide advice for choosing an optimal interval. The interval length should not correspond to the frequency of repetitive parts of the primary task, because then the primary task itself acts as a cue to the passage of the time intervals to be estimated. It should be sufficiently short so that at least 5–10 estimates are made for each assessment to enable performance metrics to be generated, typically less than 30 s. Subjects are normally given three practice trials of about 90 s on the secondary task and a "warm-up" of one trial with both tasks before data are collected.

Performance Metrics for Interval Production

Calculate both the percentage absolute error (PAE) and coefficient of variation (CV) as performance metrics.

$$\text{PAE} = (\text{MD/Ti}) \times 100$$
$$\text{CV} = \text{SDPi/MPi}$$

(Continued)

HFE WORKSHOP 12.2 (*Continued*)

where Ti is the target interval, MD is the mean difference between the actual produced intervals and the Ti(|Ti–Pl), MPi is the mean of the produced intervals, and SDPi is the standard deviation of the produced intervals.

The main drawback of the method is that it is not suitable for tasks of long duration since it adds to the workload of the main task. Interval production will be degraded by boredom or fatigue, and not by the primary task workload.

EXAMPLE

A driver exits a freeway and enters a busy intersection, at which point her SATNAV gives her directions as she approaches and turns left at a junction.

Location	Ti	Pi	|Ti–Pi|	PAE
Freeway	30	35	5	16.7
Ramp	30	30	0	0.0
Street	30	28	2	6.7
Intersection	30	26	4	13.3
Turn left	30	23	7	23.3
Suburban avenue	30	32	2	6.7
Σ	180	174	20	

$$\begin{aligned} \text{MD} &= 20/6 = 3.33 \\ \text{PAE} &= (3.33/30)\times 100 = 11.1\% \\ \text{CV} &= 4.29/29 = 0.15 \end{aligned}$$

So, the PAE in the estimation of 30 s time intervals was 11.1% and the coefficient of variation was 0.15.

We could repeat the experiment with the driver cruising at a constant speed along the freeway. Assuming that the workload was lower, we would expect the PAE in the estimation of time intervals and the coefficient of variation to be lower under these conditions.

The tendency for people to go onto "autopilot" in familiar surroundings is familiar to anyone who has arrived home after the daily commute and is surprised to be unable to remember anything of the journey. People experiencing these states of high "automaticity" can deal with unprogrammed events, such as obstructions, automatically, but are less able to deal with events that require attention. The author remembers buying a 3.08 kg tin of green olives by mistake, thinking it contained black olives (the label on the tin had the words "Green Olives" printed on it, but both the olives pictured on the label and the words "Green" and "Olives" were printed in black ink).

Table 12.7 summarizes some characteristics of poor performance that can be caused by attentional failures. Probably the biggest and most easily measured feature of attentional failure is the inability to carry out more than one task at a time. This has important safety implications—if a task

TABLE 12.7
Some Characteristics of Poor Performance Caused by Attentional Failure

Goal Setting: Personnel act as if "future blind"	*Organizing*: Losing important papers or possessions, unrealistic scheduling
Planning: Forgetting vital equipment Sequencing: Steps missed, "jumping the gun," losing track of time	*Initiation*: Cannot get started *Inhibition*: Easily distracted, going for short-term gain at the expense of long-term goals
Prioritizing: Waste time on small details Miss important signals	*Task Shifting*: Cannot be switched between subtasks in relation to changing work demands
Self-Monitoring: Does not check adequacy of own performance or monitor work pace in relation to time pressure *Time-Sharing*: Inability to carry out several tasks at once	*Emotional Control*: Overreacting, failure to complete tasks

analysis reveals that dual-ask performance is required at critical times, it would be wise to consider the effects of attentional failure on system functioning and consider ergonomic aids that might be used to prevent either attentional failure or its effects.

The Cognitive Failures Questionnaire

Many mistakes at the skill-based level are the result of "slips" (failures of attention) and "lapses" (failures of memory). Individuals with good executive control demonstrate the ability to monitor their actions in relation to their intentions. Those with poor executive control tend to lapse into a mode of behavior characterized by increased "automaticity" (carrying out simple tasks without thinking, allowing the mind to wander, etc.), are less able to resist sudden impulses (to take short-cuts, to deviate from a planned course of action in return for temporary gain, etc.), and are more easily distracted. There is empirical evidence, for example, that computer users with high scores on a questionnaire designed to measure cognitive failure in daily life (the "Cognitive Failures Questionnaire," CFQ, Broadbent et al., 1982) are more likely to lose computer work because of a failure to save it (Jones and Martin, 2003). The error occurs when the user's attention is captured by inappropriate stimuli, suggesting that people with high CFQ scores are less able to resist distraction. Wallace and Chen (2005), for example, found that several indicators of organizational safety were associated with CFQ scores—including supervisor ratings of individuals' safety behavior. They also suggested that it would be worthwhile to examine the influence of anxiety and stress on cognitive failure.

The CFQ has recently been updated, the CFQ 2.0, by Goodhew and Edwards (2024). It is free to use if the authors are acknowledged (Table 12.8). People are more likely to miss targets when they are rare, for example, detecting weapons in airport baggage scanning or detecting cancers in diagnostic imaging. Goodhew and Edwards (2023), in an experiment on target detection, found that people with high cognitive failure scores were more likely to miss rare targets.

Table 12.9 offers tentative guidance for designing "failsoft" environments—environments designed to minimize the consequences of human error caused by acute or chronic problems with executive control. The guidance has not been validated but is based on the author's interpretation of a variety of sources.

TABLE 12.8
The CFQ 2.0 (Goodhew and Edwards, 2024)

1. Do you find you forget why you went from one part of the house to the other?
2. Do you find you forget whether you've turned off a light or a fire, or locked the door?
3. Do you fail to hear people speaking to you when you are doing something else?
4. Do you find you forget appointments?
5. Do you find you accidentally throw away the thing you want and keep what you meant to throw away?
6. Do you find you forget what you came to the shops to buy?
7. Do you forget to take your keys with you?
8. Do you forget passwords you need to remember?
9. Do you receive a delivery and forget what you ordered?
10. Do you call people by the wrong name?
11. In conversations, do you forget people's names even when you know them well?
12. Do you get the dates of things mixed up?
13. Do you find you go to put objects in the wrong place, like putting something from the cupboard in the fridge?
14. Do you leave home and later realize you have forgotten something you intended to bring with you?
15. Do you think something is missing, only to later see it where you had already looked?

Items are scored as follows:

Never (0), Very Rarely (1), Occasionally (3), Quite Often (3), Very Often (4).

TABLE 12.9
Tentative Guidelines for Work Environment Design When Attentional Failure Is Likely[a]

- *Eliminate Clutter*: Stow away items that "might be useful someday." "A place for everything and everything in its place" is a good rule.
- *Simplify*: Focus on what is needed and ignore nonessential tasks. Avoid trying to carry out each task "perfectly" when performance only needs to be adequate.
- *Develop Standard Routines*: Encourage the automation of task performance to preserve spare mental capacity.
- *Reduce Complexity*: Break large, complex tasks into small, highly specific tasks and do one at a time (e.g., change "clean kitchen" to "wash dishes") to reduce the chances of interruption or attentional "stickiness."
- *Compatibility*: Do compatible tasks together—for example, where the cues from one task are always available while the other task is being performed (e.g., wash dishes while dinner is cooking, combine fishing with watchkeeping).
- *Be Pragmatic*: If a daily work routine has not become familiar and largely automated after 1 week, change it.
- *Seek Help*: If outside help is not available, use internally generated cues—set reminders, post-it notes, e-mail, or scheduler cues to remind yourself what you should be doing.
- *Checklists*: Develop checklists and procedures for critical tasks to reduce the load on memory.
- *Associate Tasks with Locations*: The location acts as a cue for the task.
- *Develop Rituals*: Lone operators/shift workers/operators in extreme environments should develop wake-up, daytime, and night-time rituals (e.g., for remembering to take medication, food supplements, etc.).
- *Record Keeping*: Keep a log of what has been achieved today and carry over unfinished items to the next day.
- *Reduce memory Load*: Choose displays that indicate safe levels.
- *Contextualize Warnings*: For example, put "Diesel Only" signs on the inside of fuel filler flaps or on the cap itself.

[a] Guidelines are tentative because they are based on scientific knowledge but have not been validated.

SYSTEM INTEGRATION

Behavioral Design: Nudging and Friction

Behavioral design operates by manipulating the activation of Systems 1 and 2 in goal-directed behavior. Nudging makes it easier for people to do the right thing by facilitating particular actions, and friction makes it difficult to do the wrong thing by introducing barriers into the interaction. Many undesirable behaviors, such as smoking or eating unhealthy food, are difficult to change because we engage in them without thinking. Friction works by forcing us to make a conscious effort to engage in the activity—diverting control of the action sequence from System 1 to System 2. Nudging tends to do the opposite.

We have seen that much of our daily work and leisure activity occurs under the control of System 1. We spend much of our lives "on autopilot," reacting instinctively to well-learned cues and environmental stimuli (attention is, after all, of limited capacity and necessarily selective, whereas human behavior is rich and complex). Many of our habitual activities of daily living are influenced by the salience and ease of processing of the cues available to us. For example, Sullivan et al. (2015) found that people processed information about the tastiness of food more rapidly than they processed information about its healthiness (about 195 ms faster in a choice task). Since cues about tastiness were processed first, they were considered for longer before the choice was made and weighted more strongly. The time difference was greatest in people with poor dietary self-control. Self-control might therefore be improved by increasing the salience of information about the healthiness of food by designing food packaging differently or by forcing longer decision times to give people more time to think about the healthiness of the options. In a world where we have to make decisions about products and processes, often trading off information from a variety of cues and information sources, these findings demonstrate a clear role for HFE to identify the most salient cues in any complex task and ensure that they are designed for ease of processing.

Organ donation tends to be higher in countries with an "opt-out" system of consent (citizens are automatically considered to be donors and have to apply to be removed from the list) rather than an "opt-in" system (citizens are automatically regarded as non-donors and have to volunteer to donate their organs). Since opting in or out requires a conscious decision, it is more effortful than doing nothing. Organ donation is greater when citizens do nothing to be donors. Some examples of nudging and friction are given in Table 12.10.

Learning to Use New Systems

The general idea of knowledge being represented as a semantic network is useful in characterizing and analyzing many human–machine interaction issues, such as how to represent complex systems to users and how users learn to operate new systems. Existing concepts can be used to assist in the learning of new ones by drawing analogies between the new system and systems users already understand.

How well we remember something depends on the processes we used to learn it—because memory is a by-product of processing information. Storage of items in LTM is slow because new items have to be "linked" in with the existing network. Linking a new item, such as a computer command word, could be achieved by rehearsing it over and over again or by associating its sound with another known word (finding a rhyme for it). Alternatively, the user could concentrate on the meaning of the word in relation to the functions it invokes and to other command words and their functions. Semantic processing such as this would be predicted to lead to better

TABLE 12.10
Nudging and Friction in Behavioral Design

		Nudge	Friction
Desirable goal/action	Organ donation Investing for retirement Increased physical activity Safer driving	Automatic "opt-in" Automatic "opt-in" Locate stairs at the entrance and elevators at the rear of the building. Remove personal trash bins so employees have to walk to a central bin to dispose of trash Replace speed cameras with signs that light up with a friendly warning	Limit the speed of elevators and escalators—make it quicker to walk Speed humps in roads
Undesirable goal/action	Tobacco smoking Unhealthy eating Not donating organs Driving too fast Suicide by overdose Suicide by gassing	Ban tobacco cigarettes, but not e-cigarettes or vaping Make the healthy option the "value meal" Reduce the size of plates and cups Roadside display indicates speed to driver (strengthens external cues)	Ban smoking in pubs to weaken the link between alcohol and tobacco. Ban the display of cigarettes so smokers have to ask for them Place French fries at the end of the self-service buffet so that the plate is already almost full. Introduce "sugar tax" "Opt-out" to become a non-donor "Haptic" road surfaces to amplify feedback ("rumble strips" ridge sequences in road) Sell paracetamol in 8 g packets at retail outlets and 16 g at pharmacies Design vehicle exhaust pipes so that hoses are difficult to fit

Source: Rothman et al. (2015).

retention. There is evidence that this is the case: deeper or more elaborative rehearsal of new material does lead to better retention (Craik and Lockhart, 1972). This implies that when trying to learn something new, such as commands, the act of trying to understand them in relation to the system will itself lead to them being committed to memory.

There is evidence that well-learned information really is stored in this kind of multidimensional manner. For example, when searching for a particular word, we may experience the tip-of-the-tongue phenomenon. Unable to retrieve the word at a lexical level, we access the code that describes how it is articulated or spoken. When unable to remember how to spell a word, we may write down differently spelled versions of it to see which one "looks right."

Massed versus Spaced Practice

In general, people learn new skills more effectively when training takes place over a larger number of short practice sessions rather than a smaller number of long sessions. It may take longer to train people using spaced practice than massed practice (because the training sessions are spaced out), but the total time dedicated to training may be less with spaced practice, or the level of performance attained will be higher. Learning new skills requires System 2—sustained attention—which, as we have seen, is a limited resource. Spaced practice allows recovery, and it also allows for consolidation of what has been learned (as new subroutines are built up between training sessions). Spaced practice sessions, if the material to be learned is added progressively, provide the opportunity for more rehearsal of newly learned material and a stronger memory trace.

Transfer of Learning from One System to Another

Many mechanical or electromechanical products have been replaced by electronic products. Physical similarity between them encourages users to apply familiar ideas about the old product to the operation of the new one. Where this happens successfully and is appropriate, we can say that *positive transfer* of learning has taken place. Where it is inappropriate, *negative transfer* has occurred.

The primary functions of new electronic devices are often automated, and many new secondary functions have been introduced. For example, in smartphone cameras, many of the photographic and optical tasks that were a key part of the operation of the legacy devices are now automated.

Cell phones have positive and negative transfer features when compared to old-fashioned telephones. For example, the "ON-OFF" function on old phones was integrated with the lifting and replacing of the handset. With cell phones, a button has to be pressed or a PIN number may have to be entered before the device can be used. Cell phones offer users additional features not found on older models of telephones, such as text messaging, and there is little positive transfer between older models of telephones and a cell phone. With cell phones, you can phone a person directly, whereas with old phones you could only phone the place where you believed them to be. However, many new electronic devices share similar features both within and between products of their own generation. Having learned to use one make of a cell phone, users will have little difficulty in learning to use a different make. Many of the problems of learning to use these new electronic devices are transitory and are diminishing, or may have already disappeared, as a new generation of "Digital Natives" emerges, who have only had experience with the electronic versions.

Demands on Attention of Mobile Electronic Devices

Many of the mobile personal electronic devices available today enable us to accomplish multiple goals in tandem often by switching attention between different tasks on a time-sharing basis. MP3 players are a good example of mobile devices that we can interact with in daily life while engaged in other activities. When driving automobiles, however, these devices may distract users from the goal of driving safely by capturing the driver's attention. Lee et al. (2012) compared the effects of searching for songs on an MP3 player with a radio tuning task and a control condition on driving performance in a simulated driving task. Drivers did not hesitate to initiate these secondary tasks in all conditions of driving difficulty and, in general, the radio tuning and MP3 search tasks had little effect on overall driving performance. However, the length of the MP3 playlist to be searched and the number of songs to be scrolled through degraded driving performance and caused drivers to spend longer glancing at the device than at the road. Long playlists (580 songs) with long lists (100–130 songs) degraded driving performance. An "aftermarket controller" designed to improve the usability of MP3 players when driving was not found to be beneficial.

Understanding how different components, tasks, and goals can interact during concurrent operation is critical to successful integration. Lee et al. explained their findings using a "goal activation" hypothesis. When we choose a goal to work toward, the goal becomes more active the longer we work on it, and competing goals weaken, which accounts for the effect of long but not short playlists on driving performance. For successful integration, then, long secondary tasks should be minimized and interfere as little as possible with the main task. Adaptive feedback to alert drivers should make use of sensory channels that are independent of the secondary task. For example, "rumble strips" at the sides of roads can be used to alert drivers when their line tracking deviates from safety while engaged in distracting secondary tasks.

Preventing Driver Distraction

Current trends in automobile design point to increased integration of the car control systems with other devices with functions separate from the main task. Victor (2011) discussed integration as a means of preventing driver distraction. In general, the more complex and cognitively demanding the secondary task, the more likely it is to distract drivers from the main task. Even when looking at the road ahead, cognitively demanding tasks reduce the range of visual scanning. Embedding eye-tracking technology in the automobile control system is an approach to preventing driver distraction as the vehicle monitors the driver. In addition, *interaction managers* are being developed to prevent driver distraction from happening in the first place. Incoming calls may be blocked if the driver is braking or negotiating a roundabout, for example. Text messages may default to voice messages if the driver's hands or eyes are busy. If the driver's head is rotated to the side or if the driver is talking, distraction alerts activate with flashing LEDs on either side of the windshield to remind the driver to scan ahead. These innovations are in addition to the proximity warning indicators already installed in automobiles and represent an exciting shift toward a more symbiotic relationship between the automobile and the driver. In the past, the driver actively monitored the road ahead and the state of the automobile. In the future, in addition, the automobile will monitor the driver, providing warnings and feedback to reduce distraction and improve safety.

Interacting with Websites Using Passwords

Many users experience problems interacting with websites because of a lack of consistency across websites in the criteria for an acceptable password and the lack of guidance for password construction at the first encounter. Moshfeghian and Ryu (2012) found that many websites failed to provide sufficiently detailed instructions about the criteria for a good password and failed to provide clear instructions about what to do next when the user failed on the first attempt. Many users are unwilling to create the kind of elaborate passwords that websites require and find it mentally demanding to generate a combination of letters, numbers, and symbols in mixed cases that they believe they could ever remember. It is a matter for debate whether website registration should be standardized across applications—perhaps it should be mentally demanding to prevent users from developing insecure habits. Nonetheless, explicit and well-designed instructions for new users would seem to be a sensible requirement.

Attention Restoration Theory: Designing for Recovery of Mental Resources

Berman et al. (2008) distinguish between environments that impose "top-down" and "bottom-up" demands on attention. The former are normally goal or task-driven (e.g., checking that a budget is reasonable and balanced), while the latter are not (e.g., strolling in a forest). In the former, attention is directed by the demands of the task, whereas in the latter, it is undirected but can be captured by objects of interest (to System 1). In general, the former kinds of environments are demanding, and after excessive exposure to them, we suffer fatigue. The latter are generally relaxing and restorative, but after excessive exposure to them, we may become bored.

Loss of self-regulatory strength is a true fatigue effect, inasmuch as it is an effect of task performance that results in reduced capacity to perform further work of the same kind. Irrespective of the nature of the task, self-regulation is the component of task performance that is diminished and that has to recover before efficiency can be restored. The evidence to date is that recovery depends on rest, which, in the present context, means that tasks requiring self-monitoring, directed attention, and willpower are avoided. Berman et al. (2008) gave subjects a demanding memory task in the laboratory and then compared the restorative effects of a 50-min walk in an arboretum or a 50-min walk in the city, after which they performed further memory tasks in the laboratory. The restoration of performance was found to be greater after a walk in the arboretum than after a walk in the city.

The authors concluded that the improved recovery was due to differences in attentional demands—walking in the arboretum placed low demands on attention, allowed "the mind to wander," and exposed subjects to a wide variety of nonintrusive stimuli. This essentially passive activity can be contrasted with walking in the city, where attention has to be directed toward hazards such as cars or other pedestrians, and subjects are exposed to a large number of highly intrusive stimuli.

Attention restoration theory explains why even simple design features, such as a view out of a window or displays of plants or outdoor scenes are beneficial. They capture attention without demanding it and enable recovery in mentally demanding jobs. In a recent systematic literature review, Barragan-Jason et al. (2023) concluded that exposure to nature was beneficial for physical and mental health, and the psychological benefits include both attention restoration and stress reduction. Connectedness with nature includes being outside in natural spaces and engaging in activities such as gardening.

Status of Mental Workload Methods Used in HFE

The TLX has high *convergent validity*, in that TLX scores correlate very highly with scores on other tests of mental workload. The TLX also has high *concurrent* validity, in that it correlates highly with actual performance on real tasks. Psychophysiological methods are minimally intrusive and have the advantage of enabling concurrent measurements to be made during actual task performance. Their main disadvantage is one of reliability—because variance in the measures is not *specific* to the mental demands of the task. The methods are usually very sensitive to factors unrelated to the task being studied (e.g., whether the operator drank alcohol the night before, is anxious for other reasons, or has consumed coffee, tea, and cigarettes, etc.).

Subjective methods are intrusive if used during continuous tasks because the operator has to stop doing the task to rate its difficulty. Where natural rest breaks occur, or where the method can be used at the end of a task, intrusiveness may be less of a problem with self-report workload assessment techniques. Young and Stanton (2005) report that the pairwise comparison procedure and calculation of a TLX score based on weighted workload ratings may be superfluous. Use of a raw TLX score would render the method even less intrusive. Bridger and Brasher (2011) identified a further problem with the TLX—it lacks internal reliability. Many cognitively demanding jobs have little physical workload or variability in workload. They found that a modified version of the TLX, the *cognitive TLX*, had better psychometric properties when the physical demands scale was replaced by a scale measuring conflicting work demands.

The advantages of the Bedford Scale are its simplicity and focus on a single, easily understood conception of mental workload. It has construct validity, but data on its criterion validity and reliability are less well understood. As with all unitary measures of complex tasks, it lacks diagnosticity—the ability to pinpoint the source of high workload.

RESEARCH ISSUES

HFE research is focused on how to optimize the interaction between operators and users and increasingly complex and interconnected systems. Users and operators now have to deal with a multiplicity of stimuli arriving simultaneously from different systems. At the same time, users and operators now carry out many tasks simultaneously—it is no longer just a requirement to stay "on task," but to be able to switch between tasks as required. Recent memory research shows that limitations in working memory capacity are strongly associated with mind wandering in tasks requiring concentration—off-task thoughts may intrude as primary task items decay from memory (Kane et al., 2007). This poses new challenges for HFE. With increasingly adaptive and intelligent automation, the challenge is to ensure that, of the information that can be presented to users, the information that is presented is most salient and distinctive in relation to the users' current goals. That much modern mental work requires concentration and attention, and that both require self-control, is well recognized. The notion that self-control is a limited resource, and that when it is used it is used up due to a lowering of glucose levels in the brain, is more controversial and has been challenged (e.g., see Vohs and Baumeister, 2016), while still providing a practical framework for understanding mental workload in everyday life.

New developments in psychometric testing include the use of smartphone apps to deliver tests. This makes it much easier to conduct testing in everyday situations without the need for special equipment. Perzl et al. (2023) developed smartphone versions of the SART, a psychomotor vigilance test, and a digital substitution test, with promising results for validity and reliability over two weeks of testing.

SUMMARY

Tasks that require mental effort are those that cause users to fatigue most quickly. The mechanism appears to be metabolic—the utilization of glucose—which explains why the administration of sugary drinks or other stimulants that cause blood sugar levels to rise can facilitate performance in the short term. Glucose enhances complex cognitive functions that require divided attention. Furthermore, glucose administration provides the brain with sufficient metabolic resources to function in situations of high cognitive demand. Reliable and valid tools exist to measure mental workload, such that the most demanding tasks can be recognized. The breakdown in human information processing during the performance of a task begins with an increase in the mental effort required and is followed by increased error propensity and deterioration in the organization of task execution. The performance of human operators can be sustained by careful design aimed at minimizing the effects of the decrements known to occur.

TUTORIAL TOPICS

1. Is "mental workload" a unitary concept?
2. "Time passes more quickly when you are busy." Discuss the implications of this statement from the perspective of HFE.
3. If memorization is a skill, are we losing it because of easy access to the Internet?
4. At the link, you will find a short interview with "Missy Cummings" From Piloting Fighter Jets to Developing AI' - https://www.youtube.com/watch?v=KPfA_YlOO_c. Will cars ever be 'driverless'?

ESSAYS AND EXERCISES

1. Describe the elements of human information presented in this chapter. Illustrate each part with examples from everyday life.
2. What is the capacity of STM?
3. Give three reasons why somebody might be able to walk, listen to an MP3 player, and follow written instructions simultaneously, but be unable to spell-check a document and carry out a telephone conversation at the same time.
4. The "THOG" problem.

   The THOG problem is famous in psychology. The problem states that something is a THOG if it is either the correct shape (triangle versus circle) or the correct color (red vs. green), but not both. If the red triangle is a THOG, your task is to say which, if any, of the other shapes are also THOGS.

   Give the THOG problem to 20 students to solve and time how long it takes them to solve it. Ask them to complete the NASA TLX when they arrive at their solution. Plot the TLX scores against time to solution. Next, plot the TLX against time to solution for those students who got the right answer and separately for those who answered wrong (for the correct answer, see the instructor's manual). Comment on your findings and whether you think that the NASA TLX is a *valid* measure of mental workload for this kind of task.

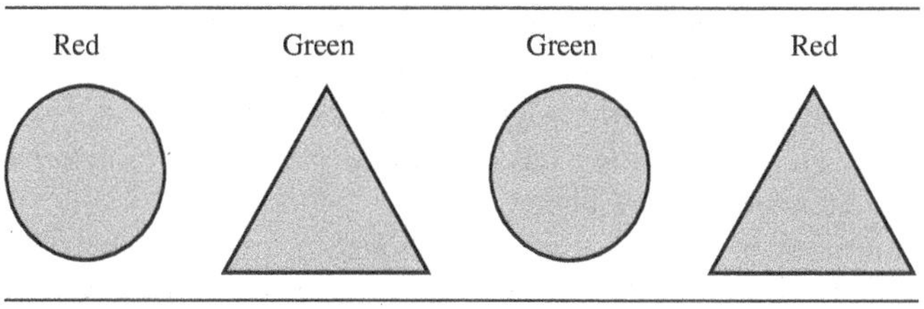

5. Use the NASA TLX scale to assess the mental workload of a variety of everyday tasks (e.g., washing dishes, looking after young children, and buying the weekly groceries in a supermarket). Describe and discuss your findings.
6. Ask colleagues or friends to memorize the recipe below without taking notes. Allow them 10 min study time. Ask them to complete the NASA TLX and WP Technique at the end of 10 min
    a. Ask them to recall the recipe the next day, after 1 week, and after 1 month. Describe the differences and similarities between the recalled and original versions. In particular, look for errors of omission and sequence and for additions and simplifications. Develop a system for categorizing these errors. Does previous culinary experience influence recall?
    b. Draw a series of scatter plots of each subject's mental workload score against the task performance metrics at the different time periods. Does self-reported task difficulty affect recall?
    c. Compare and contrast the difficulty of the task as reflected by the different workload assessment scores.

PEDRO RUIZ'S PAELLA RECIPE

Obtain 12 large prawns and remove their heads. Simmer the heads in 3 L of chicken stock for 2 h together with a large whole peeled onion. Add two teaspoons of salt to the stock. If you have any fresh fish heads, add them to the stock as well.

Debone, skin, and cut into small, lean pieces one large fresh chicken. Remove the fat from 1 kg of tender pork, either chops, leg, or steak, and cut into small pieces. Using a large flat-bottomed frying pan (or a "paellera" if you have one), add two tablespoons of olive oil and heat. Finely chop three large cloves of garlic and add to the olive oil. Fry the garlic for several minutes.

Add the diced chicken and pork to the pan and stir, making sure that it cooks evenly. When the meat is nearly ready (i.e., until almost cooked on the inside), add two large red and two green peppers that have been previously cut into thin strips. Fry them with the rest of the ingredients until soft on the outside, but still crisp.

Next, add the contents of 2 tins of peeled "plum tomatoes" and stir. Heat until the mixture simmers and reduces. Remove the stock pan from the burner and strain it to remove the solids. Put 3.5 heaped cups of rice into the frying pan and cover it with 1.5 L of stock. Simmer gently for 10 min.

Add more stock as needed to prevent the mixture from sticking to the bottom of the pan. Season to taste and add 0.5 g of saffron. Add 500 g of mussels (or clams), 500 g of diced, deboned hake, and 300 g of calamari rings. Make sure there is sufficient stock to cover the ingredients. Finally, arrange the prawns nicely around the circumference of the pan. Simmer gently for 5 min or until the prawns are cooked.

7. What is the difference between skilled and unskilled performance? Discuss the ergonomic implications of the difference.
8. Administer the CFQ to three groups of students: those who have just arrived at University/College for the first time, those in the middle of their courses, and those studying for their final exams. Can you detect any differences in the scores?

9. Write an essay on design for human reliability using the tentative guidelines given in Table 12.9 as a framework and information in this book. Describe the psychological mechanisms by which each guideline *might* contribute to a safer work environment—for example, guideline: *eliminate clutter*, mechanism: *less distraction, better figure-ground differentiation of task objects* (see Chapter 13).
10. Ask a friend to accompany you as you drive around a familiar town. Plan a route, in advance, that will allow you to travel through areas differing in traffic density/route complexity. Have your friend time the duration of the journey. Your secondary task is to estimate the passage of time in 60-s intervals, while your friend records the actual time on each occasion that you estimate 60 s has passed. Does time seem to pass more quickly or more slowly when you are stuck in slow-moving traffic, compared to other traffic situations? Can you identify any situations where time seems to pass quickly? Discuss possible reasons for your findings.
11. "Mental workload can only be measured by its effects." Discuss.
12. "In terms of the human information processing requirements, in-car satellite navigation systems are the same as cell phones, so people shouldn't be allowed to use them while driving." Do you agree with this statement? Give your reasons.
13. Mr. Smith is driving home from the airport after returning from his holiday in Spain. As he negotiates a complicated traffic intersection (during the rush hour), his wife asks him if he likes the dress she bought in the street market they had visited. When Mr. Smith does not answer immediately, his wife accuses him of never listening to her. Which of the following explanations seems most likely:
    a. Mr. Smith never listens to his wife
    b. He has an STM deficit caused by jet lag
    c. Auditory masking of her voice by traffic noise
    d. Task incompatibility
    e. He is too busy to answer
    f. He has an acute lack of spare mental capacity to process the question, due to the immediate traffic conditions
    g. He has a response bottleneck due to high demand from both tasks
    h. He does not like the dress

QUICK QUIZ

1. Fatigue can be defined as: "a consequence of work that lowers the capacity for further work of the same kind (i.e., work that has similar demands)." True or false?
2. What is the capacity of short-term memory?
3. People are more likely to donate their organs when they die if they have to "opt out" rather than "opt in." True or false?
4. Slips are actions not carried out as intended. True or false?
5. Lapses are failures of attention. True or false?
6. Should drivers be allowed to use cell phones when driving?
7. If you don't rehearse items in short-term memory, how long are they retained?
8. How is information stored in STM?
9. What does the Bedford Scale measure?
10. How can we encourage people to use the stairs more, by nudging or friction?

NOTE

1 The TLX is available on a APP at https://humansystems.arc.nasa.gov/groups/TLX/tlxapp.php.

13 Displays and Controls

GENERAL REQUIREMENTS FOR HUMANS IN SYSTEMS

1. The conspicuity and salience of displays should be maximized.
2. Use grouping principles to configure consoles and screens.
3. Conduct early task and human factors analyses to determine display/control layouts.
4. Input and control devices shall be appropriate to the characteristics of user tasks, including the frequency, importance, sequence, and urgency of use; required standards of task performance; and the environmental conditions in which tasks are expected to be performed.
5. Data entry and control devices shall be self-descriptive in the context of the task and environment.
6. Control–display relationships should conform to population stereotypes and user expectations.
7. Early prototyping and user consultation are mandatory.

> This disregard for human factors in the control rooms was appalling. In some cases, the distribution of displays and controls seemed almost haphazard. It was as if someone had taken a box of dials and switches, turned his back, thrown the whole thing at the board, and attached things wherever they landed. For instance, sometimes 10–15 feet separated controls from the displays that had to be monitored while the controls were being operated. Also, sometimes no displays were provided to present critical information to the operators. There were many instances in which information was displayed in a manner that was not usable by the operators, or else was misleading to them. A textbook example of what can go wrong in a man–machine system when people have not been taken into account.[1]
>
> *O'ÄôHara et al., 2006*

CORE KNOWLEDGE: INTERACTION AT THE INTERFACE

Figure 13.1 depicts a basic model of human–machine interaction. The user carries out a task by interacting with the device via an interface consisting of displays and controls. Progress is determined by the design of the displays and controls, by the internal processing in the device, and by the user's skills and knowledge of the task at hand (Systems 1 and 2, respectively).

At the most basic level, the object being worked on is its own display, and the effectors we use to work on it are our controls (imagine the "look and feel" of pastry when we are kneading it by hand). The smartphone screen is an artificial display that uses icons to represent applications and tools abstractly. A speedometer in a car is an analog display, while the external environment contains road signs and other displays. Similarly, controls can be part of the device itself, such as keyboards or steering wheels, or they may be the device (e.g., a rolling pin). In any action-based activity of this kind, user behavior is modified by feedback about the activity as it progresses. We spend much of our lives scanning the environment, or the devices we are interacting with, to detect information (feedback, cues, warnings, etc.) to guide our actions as follows:

DOI: 10.1201/9781032717210-13

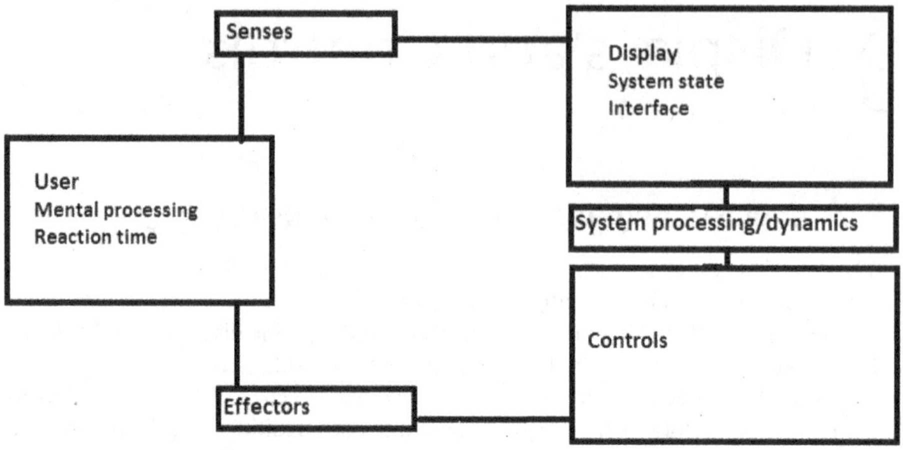

FIGURE 13.1 Display control model: the senses detect the target, it takes time to react while processing, control actions, machine processing of control actions, and display update.

- *Scan Environment:* Search for targets
- *Detect Target (Sensation):* The target and its location are registered
- *Perceptual Processing:* Target is processed using top-down and bottom-up processes
- *Response Selection:* Readiness to respond
- *Execution of Response:* Action
- *Scan for Feedback:* Detection and registration of the outcome

Sensation

How can we tell whether two things are the same or different? It depends on the size of the difference, of course. Weber and Fechner (see Wixted, 2020) found that sensation is not linearly related to stimulus intensity; rather, the intensity of the stimulus must be increased geometrically for the sensation to increase arithmetically. For example, if you hold 20g of chocolate in one hand and 5g are added, you'll probably notice the difference. However, if you are holding a 20kg barbell in two hands and 5g are added, you probably won't notice the difference. Weber's Law states:

$$I/\Delta I = k \tag{13.1}$$

where:

I=The intensity of the stimulus
ΔI=Change in intensity needed for the change to be noticeable
k=a constant

In other words, the more intense the stimulus, the larger the change has to be for it to be noticed.

If the values of k and I are known, ΔI—the *just noticeable difference* (JND)—can be calculated. Fechner related sensation to detection as follows:

$$S = \lambda \ln(I/I_0) \tag{13.2}$$

where:

S=sensation
I=stimulus intensity
I_0=threshold for sensation

λ = a constant
ln = natural logarithm

Accordingly, when $I = I_0$ ("1") $S = O$, and there is no sensation. For a stimulus to be detected, it must exceed the threshold.

Is There Anything There? Signal Detection Theory

These equations have important design implications. To design effective displays, we need to know about the size of the thresholds for sensing changes and the size of the differences that must be noticed. However, the concept of an absolute threshold and a fixed JND was called into question when it was realized that the distribution of sensations created by neural noise in the brain overlaps with the threshold for conscious awareness (Wixted, 2020). When detecting weak stimuli or small changes in displays, neural noise can give rise to sensations of its own—the "threshold" is not fixed.

Signal detection theory (SDT) is based on the idea that the detection of target stimuli takes place in a "noisy" environment, both internally and externally (random fluctuations may give the appearance of a stimulus when there is no stimulus). An internal response is generated whenever a target stimulus is presented to the user. When a doctor examines an X-ray in the search for a tumor, SDT defines four possible outcomes:

1. *Hit*: The doctor detects the tumor.
2. *Miss*: The doctor fails to detect a tumor when the tumor is present.
3. *False Positive*: The doctor detects a tumor when no tumor is present.
4. *Correct Rejection*: No tumor is present, and none is detected.

The first and last outcomes are correct, and the second and third are incorrect—SDT in its basic form is limited to these four outcomes.

Figure 13.2 presents this graphically. Note that the *x*-axis represents the value of a hypothetical "decision" variable, dependent on the intensity of the sensation and used to decide whether a target is present. Note that the distributions overlap because sensations created by noise can occur above the threshold for conscious awareness.

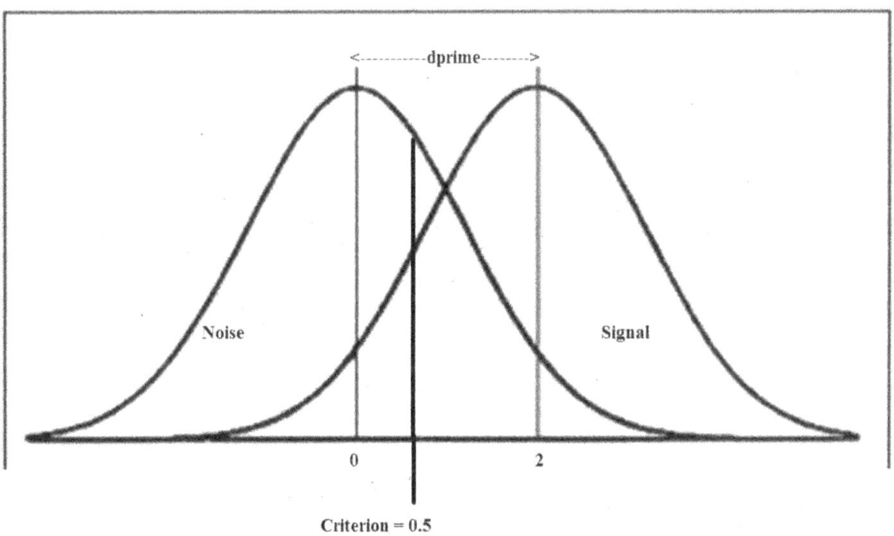

FIGURE 13.2 Signal detection in noise.

Let us suppose the decision variable has a mean of 0 and a standard deviation of 1 when noise is present, and a mean of 2 and a standard deviation of 1 when a target is present. What will happen if the criterion for reporting a target is 0.5?

Since the x-axis in Figure 13.2 is numbered in Z scores, we can see that the criterion lies 1.5 standard deviations below the mean for the target distribution and 0.5 standard deviations above the mean for noise. In many repeated trials, then, and referring to the table of Z scores in Appendix A, we will correctly identify 93.32% of targets ("hits") with 30.85% false positives. We will miss 6.68% of targets with 69.15% correct rejections.

Trade-Offs

We can see immediately that there is a trade-off between the hit rate and the false-positive rate. The further to the left our criterion variable is, the greater our hit rate, but with more false positives as well. The more sensitive the observers, the more likely they are to detect real targets and to report false positives.

Suppose we change our criterion value from 0.5 to 1.0 (assuming means for the noise and target distributions of 0 and 2, respectively, with SD=1). Our hit rate will drop to 84.13%, and our false-positive rate will drop to 15.87%. We will miss 15.87% of our targets. This is regarded as the "neutral" or *unbiased* criterion value.

Sensitivity

The sensitivity of the observer can be understood from Figure 13.2 as the separation between the two distributions. This is known as d′ where

$$d' = Z(H) - Z(F) \tag{13.3}$$

where H=hit rate and F=false alarm rate.

In Figure 13.2, for example,

$$d' = 1.5 - (-0.5) = 0.2$$

As long as there is overlap between the distributions of targets and of noise, there will always be incorrect decisions (misses and false positives). If misses have more serious consequences than false positives, then it makes sense to be biased in favor of hits (relax the criterion value by moving it to the left in Figure 13.2), and vice versa. For example, it may be more important to detect a highly infectious and serious disease, even if it means treating healthy people (false positives), if the cost of treatment is low and the treatment is relatively harmless. This may explain the use of lockdowns and compulsory vaccination programs in some countries during the recent COVID pandemic.

Sensitivity does not depend on the criterion value of the observer. The criterion value chosen depends on the consequences of making incorrect decisions, whereas d' depends on the design of the displays as well as working conditions, operator training, and experience. The last point is where HFE comes in, of course.

In the next section, applications of HFE in display design will be described. From the SDT perspective, HFE offers the means to increase d′ by improving the conspicuity of target stimuli—optimizing the size of the *just noticeable difference.* The effect of doing this is to increase the hit rate without increasing the false-positive rate.

Reaction Time When a Target is Detected

Having detected a target, the user has to react in some way. The reaction time to a stimulus depends on the nature of the stimulus, the processing demands, and the selection and execution of a response. In general, *simple* reaction time to single auditory stimuli is about 150 and 190 ms for visual stimuli. This is the total time to, for example, detect when a green light comes on and press a button. Reaction time is faster when the stimulus is expected and is slowed by other factors such as age.

Suppose a display has four possible lights, each with its own button, and the task is to press the button as quickly as possible when the corresponding light comes on. *Choice reaction time*, under these conditions, is higher than simple reaction time and depends on the number of lights. *Hick's Law* defines choice reaction time as follows:

$$T = b.\log_2(n+1) \tag{13.4}$$

where n=number of choices, b=a constant derived empirically from the experiment, and (NB: a logarithm to the base 2 is the power to which the number 2 must be raised to give the value of n. So, if $n=2$, then $\log_2 n=1$; if $n=4$, then $\log_2 n=2$; and if $n=16$, then $\log_2 n=4$.)

So, if we did a choice reaction time experiment and found that b=200 ms, then with 3 targets, T, the choice reaction time would be 400 ms; with 7 targets, T would be 600 ms, and with 31 targets, T would be 1000 ms—plus movement time (MT), of course. In practice, MT depends on the design of the workstation or console, the size and location of the buttons; in other words, it depends on human factors and ergonomics!

Hick's Law has many real-world applications and explains why restaurant owners do not give diners excessively long menus (they want them to choose quickly) and why menus on software applications tend to be short. Hick's Law describes the relationship between the number of choices and the time to select a response and is therefore applicable to any real-world application where people have to respond quickly. It explains why the uncontrolled growth of roadway safety signs can become a hazard by slowing driver reaction time (Figure 13.3).

Hick's Law neither applies to highly complex tasks demanding high levels of cognitive processing (System 2), such as multiple-choice examination questions, nor to highly practiced tasks. For multiple stimuli arriving sequentially, there is a psychological refractory period, where response to a second stimulus arriving immediately after is slower than the response to the first.

Reaction time is strongly influenced by many design features of the device and the environment in which it is used, particularly *Stimulus Response Compatibility*. Reaction time is faster and

FIGURE 13.3 Hick's Law explains why too many safety signs can be dangerous!

responses are more often correct under compatible arrangements. When these arrangements are learned over time following exposure to technology, they are referred to as "population stereotypes."

Population Stereotypes

People tend to attribute meaning to the movements of objects and have a tacit understanding of mechanical cause–effect relationships. Displays or controls that behave in a manner consistent with the prevailing stereotype are said to be *compatible.* One of the most common stereotypes is the "clockwise-to-increase" stereotype. The clockwise movement of a pointer on a dial signifies the passage of time, an increase in the speed of a motor vehicle, or an increase in pressure or flow. An upward movement of a pointer on a vertical scale indicates an increase, as does a rightward movement of a pointer on a linear scale. The development of a repertoire of stereotypes seems to take place over a period of years and is accompanied by *an increasing lack of flexibility and ability to adapt to new situations.* By adulthood, certain stereotypes are well learned. Although new relationships can be acquired, they merely overlay the originals rather than replacing them. In stressful situations, previously learned stereotypes replace even well-learned skills. If the operator lacks the processing capacity to monitor their own performance, old subroutines containing the original learning will emerge, leading to reversal errors.

It is for this reason that, when it is clear that a stereotype exists, designers should ensure that their design is compatible.

A number of exceptions to the clockwise-for-anything rule may be noted. Faucets are a good example of an exception to this rule, and people appear to be able to learn this exception as a special case. In some circumstances, the ability to correctly interpret the movement of a dial or a pointer will depend on the operator's knowledge of the system or of the displayed variable. For example, the dial on a vacuum gauge moves anticlockwise to indicate an "increase" in vacuum (i.e., a drop in atmospheric pressure). Whether operators will be able to correctly interpret instructions such as "increase the vacuum" will depend on their understanding of what a vacuum is.

People frequently have strong expectations about how the movement of a control will influence the direction in which a display moves. These expectations are referred to as motion stereotypes. For example, clockwise movements of a rotary knob are associated with upward or rightward movements of display pointers. Turning the steering wheel to the right should result in the vehicle moving to the right. When tuning a radio, a clockwise movement of the knob should cause the dial to move to the right, or up, and so on.

Operators may not have strong stereotypes about the controls and controlled parts of novel machines, for example, where the controlled part moves in a different plane from the controls or in a complex way. Alternatively, stereotypes may not exist for novel control devices such as palm wheels, voice, or the movement of remotely controlled vehicles. Under these circumstances, the designer may have to undertake empirical research to learn more about people's expectations and preferences, ensuring that if there are any strong expectations, they are not violated. Whatever control logic is then decided upon should be used in a consistent manner throughout the design to facilitate the learning of any new stereotypes and avoid interference between mutually incompatible ones.

Many activities of daily living involve the chaining together of learned behaviors and associations. When these are violated or the environment is altered, increased errors and accidents may occur.

BRITONS ABROAD ARE THE WORST DRIVERS IN EUROPE[2]

A recent study by a team at the University of Granada in Spain found that British drivers are more likely to cause an accident when driving on Spanish roads than drivers of any other nationality, despite the fact that the UK road accident rate is the lowest in Europe.

The relative risk of causing an accident was 2.16 for British drivers, compared to 1.0 for Spanish drivers. Other European countries had relative risks intermediate between these values, but most were closer to the Spanish than the British levels.

> Of all the groups compared, only the British drivers habitually drove right-hand drive vehicles on the left-hand side of the road. It seems that the increased risk is not due to a difference in driving skill, but rather to British drivers being unfamiliar with left-hand drive vehicles and driving on the right-hand side of the road. During emergencies, they may become confused and revert to visual scanning patterns appropriate to right-hand drive vehicles.

People also have stereotypes about color (Table 13.2).

Control Order and System Dynamics

So far, we have looked at target detection and reaction time, which for simple tasks, involve internal processing time prior to the execution of a response. Much of the complexity in human–machine interaction can come from the way the system responds to operator inputs. There are four main control orders of systems:

- *Zero Order/Position Control:* Movement of the control produces a corresponding movement of the controlled object as when pointing a bow and arrow at a target or moving a cursor across a screen using a mouse. The movement may be optimized by adjusting the gain (amplification, sometimes referred to as "sensitivity") of the control. With zero gain, moving the mouse 5 cm horizontally would result in the cursor moving 5 cm horizontally across the screen.
- *First-Order/Rate (Velocity) Control:* Movement of the control affects the rate of change in output of the controlled object, as when the accelerator pedal of a car is depressed.
- *Second-Order/Acceleration Control:* Movement of the control affects the rate of change in velocity of the output, as when the rudder and flaps of an aircraft are moved.
- *Third-Order/Jerk Control:* Movement of the control affects the rate of change of acceleration of the output. Submarine depth controls work this way.

In all cases, the operator has to perform what is known as a "tracking task." In "pursuit" tracking tasks, the control actions are made to ensure that the controlled object follows the target as closely as possible. The movement of the target is under outside influence. Aiming a rifle at a moving target is an example of a pursuit tracking task. In "compensatory" tracking tasks, the operator has to maintain the output of the system at a given reference by compensating for fluctuations caused by outside influences. Maintaining the velocity of a vehicle at 50 kph is an example of a compensatory tracking task. Operator performance is normally assessed using measures of tracking error aggregated in some way over time, where the error is simply the difference between the actual tracking performance at any time and the ideal (where the movement of the controlled object matched that of the target perfectly).

In general, the higher the control order, the more demanding the task becomes because it is difficult to predict how a particular control action will affect the controlled object subsequently.

Cursor Control Devices

The efficiency of user control actions can sometimes be improved by combining control orders in a single device. For example, a trackball may act like a zero-order control when moved slowly, but like a higher-order control when moved quickly—smooth movements cause the cursor to move across the screen in proportion, whereas a flick of the trackball causes the cursor to accelerate rapidly across the entire screen.

Position Effects When Viewing Simultaneous Displays

In many areas of life, we are presented with many potential targets, and our task is to locate and choose between them. Choosing between items on a restaurant menu, numbers to bet on a roulette wheel, or the correct answer items in a multiple-choice examination is not random (Bar-Hillel,

2015). Two common (and seemingly contradictory) biases occur, and these are known as "centrality" or "edge" effects. People tend to choose items at the beginning or end of restaurant menus, rather than items in the middle. The primacy and recency effects discussed in Chapter 12 make these items more available and more salient. Although all the items in the menu are presented to the diner simultaneously, they are read sequentially; therefore, items at the beginning and end of the menu are easier to consider. The centrality effect occurs when there is uncertainty and when cognitive processing is required. For example, people tend to bet on numbers that seem representative of random sequences (e.g., choosing 12, 33, 42, 54, 17, 5 as the winning numbers in a lottery, as opposed to 1, 2, 3, 48, 49, 50), or the middle items in a multiple-choice examination.

BASIC APPLICATIONS: DESIGNING DISPLAYS AND CONTROLS TO SUPPORT SYSTEM 1

Key Principles for Display Design

Table 13.1 summarizes some key ideas.

TABLE 13.1
Perceptual Principles for Use in Display Design

Common Region: Elements contained within the same region or area are grouped together	XxY ZZZ PPP C 12478 99 "£$66^^^
Proximity: Items close together are grouped together	1 1 1 1 1 1 1 1 1
Horizontal proximity groups elements as rows, not as columns	11111111111111 11111111111111 11111111111111
Similarity: Items similar in color, size, or shape are grouped together	0 0 0 1 1 1 0 0 0
Symmetry: Elements in an array are grouped to form symmetrical figures	1 1 1 1 1 1 versus 1 1 1 1 1 1 11 1 1 1 1 1 1 1 1
Closure: The tendency to see complete figures even when presented with partial information	
Continuity: All parts of an array having a common feature are seen as separate from other parts, even if a common point is shared	1 101010101010 1 0 0 0 0 1 1 1 1 0 0 0 0 1 1 1010101010101001010101010
Visual Apprehension Limits: We can "see by counting" up to 5 or 6 items	11 versus 000000000 1 1 1 1 versus 0000000 0000000000000000

Design to Promote Figure–Ground Differentiation

Figure 13.4 contrasts designs with strong versus weak figure–ground differentiation.

- The figure has form while the background is formless.
- The figure appears to stand out against the background.

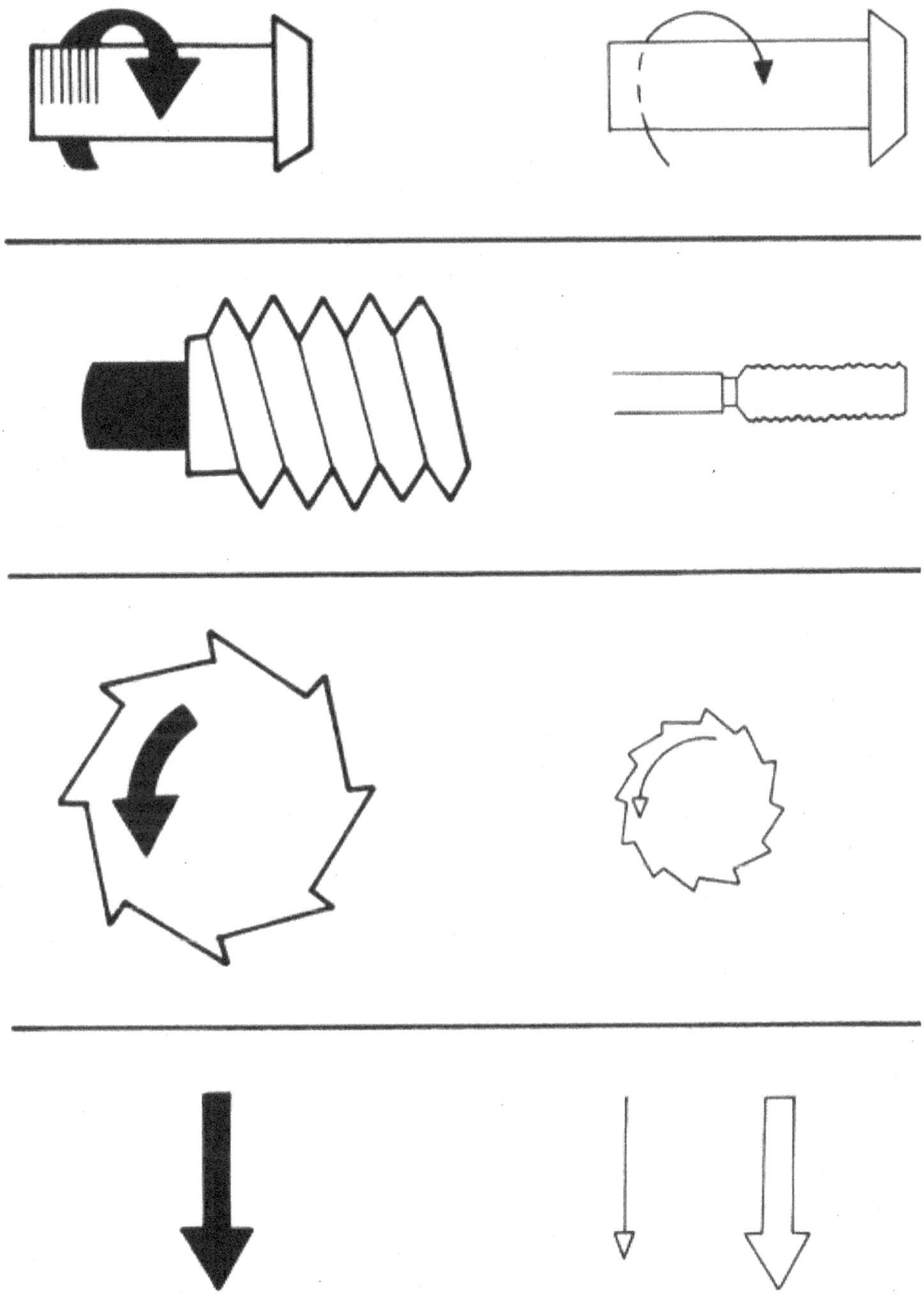

FIGURE 13.4 Strong contour enhances figure–ground differentiation and the visual impact of a figure. (Adapted from Easterby (1970) with permission.)

Figure–ground differentiation reduces complexity—information in the figure receives preferential processing at the expense of ground information, although the latter is not entirely lost. Contour, closure, and previous experience influence the differentiation of a stimulus array into figure and ground. Contour is a characteristic of a stimulus array that provides cues for figure–ground differentiation—sharp contrasts enhance contour detection and produce strong figures.

Enhance Contours

Sudden changes in brightness or color provide natural contours between objects. Contour enhancement is a useful method for helping operators differentiate parts of a display panel or recognize a symbol. It is known that well-designed road signs with strong figures are more easily recognized than written messages, giving drivers of all ages almost twice as much time to respond to them (Kline et al., 1990).

Promote Closure

Closure can be described as the *rendering of form*—a tendency to complete or "close" a figure. Closure is the tendency to produce a meaningful percept from incomplete cues and is the mechanism that prevents us from being aware of the effects of the retinal blind spot on vision. The ability to create meaningful percepts from incomplete cues depends on past experience and is a characteristic that distinguishes the skilled from the unskilled operator. When first viewing a map, the initial impression may be of a bewildering complexity of unrelated parts. Following study and use, closure may be achieved, and the map seen as a whole.

Designing Icons for Ease of Recognition

Icons are ubiquitous nowadays. Icons should be made to resemble their referent objects as much as possible. Stotts (1998) compared search and select times for graphically concrete or graphically abstract icons (Figure 13.5a) that were either functionally representative of their referents or were not. Concrete icons were found almost twice as quickly as abstract icons (in about 4 s), and the effect was strongest for experienced users. Concreteness was most effective in speeding search and select times when the icons were functionally representative of their referents. It seems, then, that search time depends on the detection of features that can be mapped onto the user's internal representation of the referent, and that the more concrete these features are, the quicker the icon is recognized.

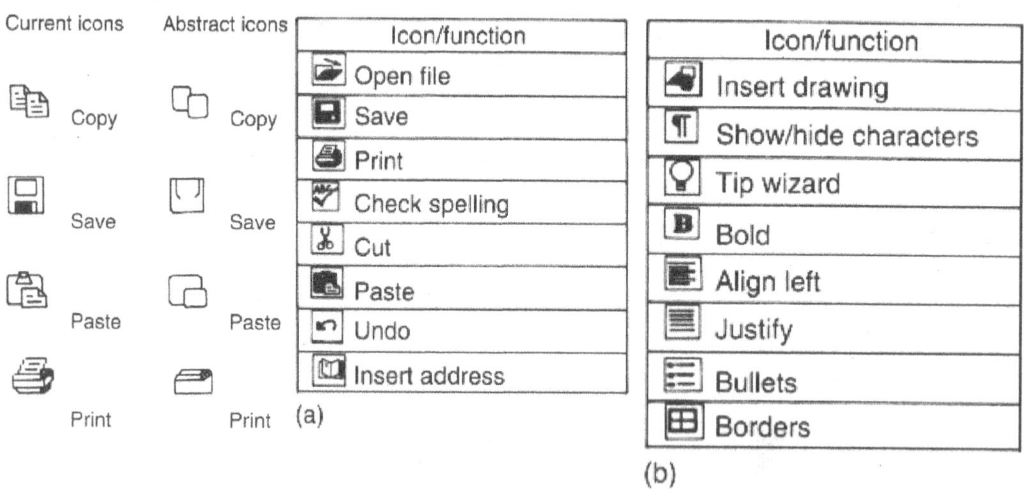

FIGURE 13.5 (a) Concrete and abstract icons compared by Stotts (1998). (b) Icons of the kind investigated by Sears et al. (1998) for effectiveness and identifiability. (Reprinted from Stotts (1998) with permission.)

Sears et al. (1998) investigated the quality of the icons used in Microsoft® Word 7.0. Three hundred and eighty-four volunteers rated the identifiability and perceived effectiveness of the icons found on Word's toolbar (Figure 13.5b). Effectiveness was assessed by presenting subjects with an icon, a verbal description of its function, and a seven-point rating scale, then asking them to assess how effectively the icon represented its referent function. Identifiability was assessed by presenting each icon and asking subjects to identify its function.

Subjects also indicated how frequently they used each icon. Icons that were perceived as effective and were identified correctly tended to be used to perform frequently performed functions. These included Print, Bold, Italic, and Underline. Icons that were rated as ineffective and were not identified tended not to be used. These included Auto Format and Insert Address. A number of icons were identified that were used less frequently than the functions they represented (because users employed the pull-down menus or used other techniques to access the functions). Since the purpose of using icons instead of commands is to increase usability, this suggests that either these icons are redundant or that they should be redesigned. Examples of such icons include File, Save, Print Preview, Check Spelling, Cut, Copy, Paste, Undo, and Insert.

Skeuomorphs

A skeuomorph is an ornamental feature of an object that was originally a functional necessity. The shutter sound of a mechanical camera is often incorporated into digital cameras, which would otherwise be silent. Icons representing cameras used in traffic control are often designed with a "bellows" feature that was once an essential part of old cameras (Figure 13.6). Pages on e-diaries are often designed to look identical to ring-bound pages in paper diaries. Originally thought to aid recognition by resembling the objects they represent, many younger users may never have seen the original objects. In modern digital applications, the use of skeuomorphs is probably still beneficial for creating strong images to represent increasingly abstract functions.

Dynamic Iconography

Cian et al. (2015) found that warning signs (e.g., children crossing the road, unstable terrain, dangerous chemicals, etc.) were responded to more quickly if the graphics presented dynamic situations (e.g., children running, drops of chemical falling onto a hand) than if the images were static. Similarly, in the UK, radar-activated traffic *calming* signs are often placed at the entrances to towns and villages. These signs display the driver's speed (in green if below the speed limit or red if above), together with icons (e.g., a smiley face) or a message such as "Thank you."

FIGURE 13.6 Skeuomorphs—an image with strong figure-ground differentiation. Many smartphone users have never seen a camera that the skeuomorph represents—but how else would you represent the camera function of a smartphone?

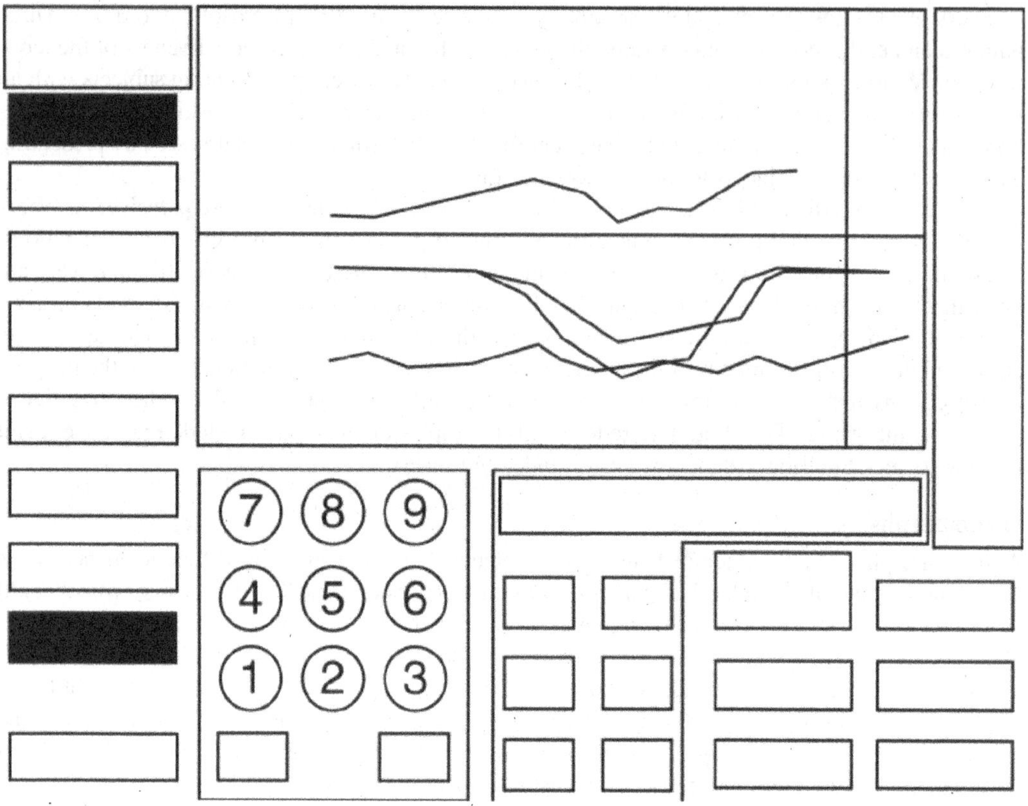

FIGURE 13.7 Good application of grouping principles. Proximity, similarity, symmetry, and continuity have been used to create a coherent display, which naturally decomposes into subsystems and different functional areas.

Grouping

Table 13.1 describes basic perceptual principles that determine how an array of separate elements is "grouped" to form a complete percept.

According to Rock and Palmer (1990), these laws of grouping are fundamental and have general applicability. Operators first seem to group based on common region and then proximity. Further grouping occurs if elements in the proximal cluster differ. Similarity based on shape or color can override proximity and provide a common boundary where one is lacking. This is why color is useful for retro-designing badly laid-out panels. Figure 13.7 shows a display with a variety of elements that has been designed with a grouping mind. The whole display seems to decompose quite naturally into different regions, suggestive of different functions. If the different regions do support different functions, then we can say that the perceptual layout is compatible with the systems structure.

Good mappings between display layouts and system functions can increase the efficiency of visual scanning. Kotval and Goldberg (1998) compared eye movements of subjects scanning the icons, as shown in Figure 13.8. The same icons were grouped in four different ways. In functional grouping, the icons were arranged based on common function to give three sets of icons dealing with editing, drawing, or text. In majority grouping, most of the icons in each group had the same general function, but one differed. In physical grouping, the icons were clustered into groups irrespective of their function. Finally, there was a no-grouping condition in which all the items were

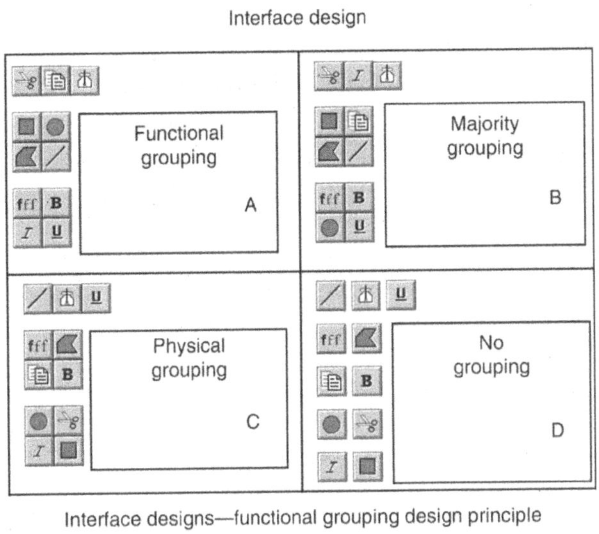

FIGURE 13.8 Functional groupings support effective visual scanning of screens. (From Dr. X. Kotval, Nokia Bell Labs, Murray Hill, New Jersey, USA. With permission.)

laid out equally spaced. Subjects were prompted to find and select an icon, and their eye movements were recorded. Not surprisingly, subjects' visual scanning was most efficient when the icons were functionally grouped, as reflected in shorter scan paths and fewer saccades (ballistic eye movements used in visual search). Interestingly, the no-grouping condition was the next most efficient, and the majority and physical groupings were the least efficient. *Incorrect grouping seems to be worse than no grouping at all,* and so it should only be used when designers are certain about functional relationships between different components. Ha and Hong (2024) confirmed that search times are reduced when icons are grouped according to their familiarity.

Color

People have a very strong tendency to perceive similarly colored objects as belonging together (van Nes, 1986), as long as *no more than three or 4four colors* are used together. Color can be used to provide conceptual grouping of text in itineraries, timetables, and so on. For example, station names and cities can be presented in one color, destinations in another, and departure and arrival times and on-board facilities in a third.

Some of the advantages and disadvantages of the use of color and its traditional interpretation in Western cultures are given in Table 13.2.

Generally, red is used to mean "Stop" or "Danger," green to mean "Go" or that the system is running normally, and orange to signify caution—unfortunately because approximately 5%–8% of people have color vision defects (so-called "color blindness"). The most common deficit is an inability to distinguish red and green! (They both look brown.) Approximately 7 million drivers in North America have this problem (University of California, 1993), and some studies suggest that color-deficient drivers have problems seeing traffic lights. According to Clement (1987), because 4% of males cannot distinguish green and red, the optimal design for traffic lights is to use red/orange stop indicators and blue/green go indicators. Atchison et al. (2003) compared reaction times to simulated red, yellow, and green traffic signals. For green lights, response times were similar. For red and yellow lights, reaction times increased between 35% and 85%, depending on the type of color blindness.

TABLE 13.2
Guidelines for the Use of Color in Display Design

Advantages	Disadvantages
Draws attention to specific data	8% males are colorblind
Faster uptake of information	May cause confusion
Can reduce error	May cause fatigue
Can separate closely spaced items	May cause unwanted groupings
May speed reaction	May cause errors
Adds another dimension	Can cause afterimages
Seems more natural	Appears more frivolous

Color	Interpretation	Associations
Red	Danger	Stop
	Emergency	Do not proceed
	Failure	Hot/fire/live
Green	Safe	Proceed
	On	Healthy
		Normal
Yellow/orange	Caution	Delay
		Pause
		Warm
Blue	Mandatory	Cold
		Off
		Water
		Drinking water
White	Operational	Pure
		Cold
		New

Some possible improvements to increase traffic signal reaction time include:

- Use of unsaturated colors to improve the discriminability of colored signals for color-blind users. Unsaturated colors emit light at a dominant wavelength (from which the perception of color arises) but contain light from other parts of the spectrum as well. Differences in the composition of these latter wavelengths may improve discriminability.
- Consider redundant coding in other dimensions—use shape as well as color.
- Use blue/green instead of green, as disorders with the perception of blue are rare.

Abd Ellfattah (2014) discussed web design for color-blind users. For a viewer with normal vision, "Merry Christmas" written in green on a red background appears as such. For a person with no color vision at all, it is almost illegible—there is very limited contrast between the gray letters and the background because these two colors have similar "lightness." For red-green color blindness, it appears as dark olive letters on a light olive background, and for blue-yellow color blindness, as teal letters on a magenta background. By manipulating the hue, saturation, and choice of colors, displayed objects can still be distinguished by color-blind viewers. Blue-green, green, yellow, and orange are "light colors," whereas red, purple, violet, and blue are "dark." By making "light" colors "lighter" and "dark" colors "darker," displayed objects can be better distinguished by color-blind viewers. Adding texture and other cues will further aid detection.

Resolution of Detail: Object Size and Viewing Distance

The ability to resolve detail depends on the accommodation of the lens of the eye, the ambient lighting, and the visual angle—the angle subtended at the eye by the viewed object. Visual angle is a more useful concept than absolute object size since it takes into account both the size of the object and its distance from the viewer (see Chapter 10). Under good lighting, a minimum visual angle of 15 min of arc is recommended for displays, although 20 is preferable. (This is considerably greater than the values quoted in Chapter 10—the minimum angle for *detection* of a stimulus is much smaller than the minimum angle for *perception of detail* and the extraction of useful information.) If the viewing distance is known, then minimum target sizes can be specified. For alphanumeric characters, for example—

Viewing Distance Minimum (m)	Character Height (mm)
<0.5	2.3
0.5–1.1	4.7
1.1–2.1	9.4
2–4	19
4–8	38

Many instruments, such as dials and gauges, are designed to be read at a reference distance of about 70 cm. If the reference distance is known, the required change in object size to equalize the visual angle when the object has to be viewed at a different distance can be calculated by

$$\text{Size at new distance} = \frac{\text{Size at reference} \times \text{new distance}}{\text{Reference distance}} \tag{13.3}$$

For example, if the reference distance is 70 cm and the new distance is 140 cm, the size of the object would have to be doubled to maintain the visual angle. Murrell (1971) carried out detailed investigations of the variables influencing dial design. Figure 13.9 summarizes some of the design principles that were derived from this work.

Color Coding of Dials

Color coding makes target values more conspicuous and saves operators from having to recall critical values from memory.

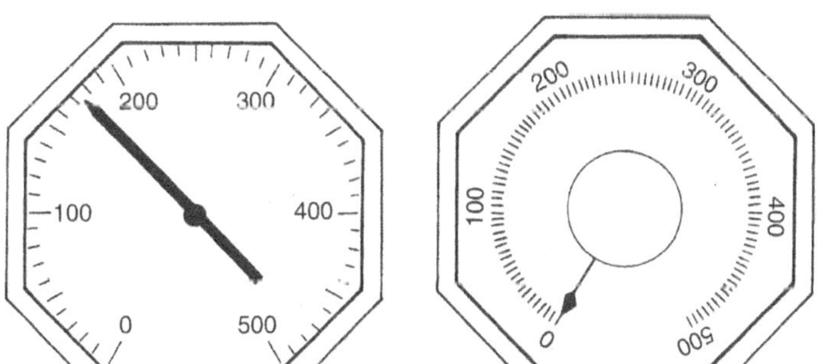

(a) (b)

FIGURE 13.9 Design of scales. Dial at (a) is easier to read than dial at (b). Dial (a) has been made easier to read by having fewer, bolder graduation marks. The scale length has been increased by putting the numerals on the inside. The numerals themselves are larger and level. The whole of the pointer is visible. (Redrawn from Galer, 1989.)

Digital versus Analog Displays

Digital displays should be used in situations where highly accurate reading of displayed quantities is required. Digital displays are inferior to analog displays for conveying information about rates of change; therefore, in those cases where both types of information are required, a compromise design can be used in which a digital counter is superimposed on the dial face.

"Dark Mode"—Negative Polarity Displays

Many devices, such as laptops and cell -phones, offer negative polarity ("dark mode") display options wherein light characters are displayed against a dark background (Figure 13.10). According to Sethi and Ziat (2023), some of the advantages claimed for dark mode viewing are reduced visual fatigue and preserved battery life. Dark screens for desk-based work may be more susceptible to reflected glare from luminaires or windows. Positive polarity displays do seem to offer advantages such as better comprehension, greater visual acuity, and improved legibility. In low ambient light, dark mode displays can offer shorter response times and less visual fatigue (possibly due to lower demands on retinal adaptation due to the absence of sharp contrasts). Another of the claimed advantages of dark mode displays is reduced exposure to blue light (which reduces melatonin secretion, as described in Chapter 10).

Clearly, the advantages of dark mode displays will depend on the viewing conditions and the requirements of the task. The visual demands of casually browsing web pages will be lower than in tasks such as proofreading text or bookkeeping. Further, any task requiring the discrimination of color will require a suitable level of illumination for photopic vision. Sethi and Ziat (2023) investigated the effects of positive polarity and negative polarity ('Dark Mode') displays in bright and dark (15 lux) conditions. Their subjects completed a writing task and a word search task and varied in age. Overall, use of the positive polarity displays seemed the most advantageous. For older adults, search time was longer in bright light when in dark mode, whereas for younger adults, search time was greater in dark mode in dim light. Mental effort was lower for both older and younger adults in both lighting conditions with positive polarity displays. Overall, there seem to be few advantages of dark mode displays under the conditions studied. However, it was observed that pupil size was greater in dark conditions (more so in younger subjects) and when the dark mode display was used. Thus, dark mode displays may offer the greatest benefits in tasks where the eye must be dark-adapted and the user's attention is divided between the screen and the environment (as when driving at night).

Multiple Displays and Control Rooms

The ability to gather information quickly by shifting focal attention around a complex display depends on several factors. Sanders (1970) demonstrated that as the visual angle between two targets increased, the reaction time and number of errors needed to make a "same/different" comparison between them also increased. Sanders identified three functional attentional fields. Objects placed within 30° of each other fall into the stationary field and can be detected with the stationary eye. The eye field encompasses a region from 30° to 80°, and eye movements are required to compare objects. Beyond 80°, head and eye movements are required to detect targets. Focal attention can be

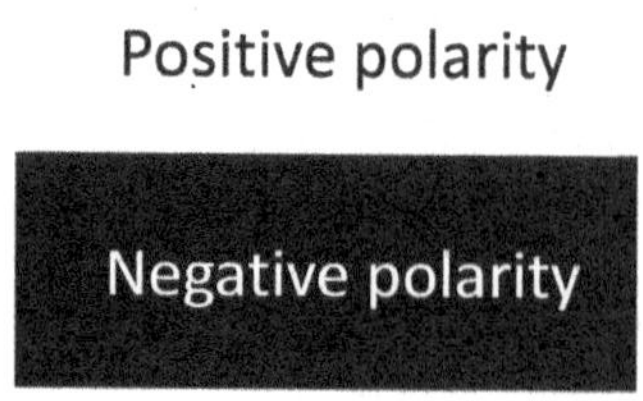

FIGURE 13.10 Positive and negative polarity displays.

TABLE 13.3
Common Criteria for Evaluating the Layout of Complex Panels

1. Eliminate unnecessary or complex movements
2. Locate controls and displays to reduce postural load
3. Reduce movement complexity, encourage muscle memory by consistent choice of movements
4. Control actions should not obscure important display areas
5. Avoid spatial transformations
6. Ensure spatial proximity of displays and controls used frequently in a sequence of tasks

likened to a "spotlight," which illuminates parts of a complex display making them accessible to higher cognitive processes. Parts of the display within the spotlight can be processed simultaneously, whereas those outside the spotlight require shifting of attention and search.

Data on the extent of the various functional attentional fields are useful in designing complex control panels. If an operator has to monitor several displays simultaneously and make judgments using the data from all of them, they should be placed in the stationary or eye fields with respect to a predefined point of observation (such as the operator's desk). Careful analysis of the operator's task and operational priorities, including emergency recovery procedures can be used to assist in optimizing a panel layout. Some examples of criteria for evaluating prototype layouts are given in Table 13.3.

Guiding Visual Search in Environments

Multimodal cueing is often used to improve the detection of targets or changes in system state. Cell phones, for example, can vibrate as well as chime when a call arrives. Some cell phone cameras emit a sound when a photo is taken (this is, in fact, an auditory skeuomorph mimicking the noise made by the shutter of a camera). Baldwin et al. (2012) reviewed the use of multimodal cueing, which is most effective under conditions of high perceptual load. Some key points for the design of multimodal cueing systems are as follows:

- The cues should come from the same direction or preferably the same location (e.g., an auditory alarm should come from the same location as a flashing light).
- In uncertain environments or when sensitivity is low, multimodal cues can confirm signal validity and facilitate rapid response.
- Auditory or haptic cues can lower the demands on visual attention in large display configurations where different displays are far apart.
- Vibro-tactile cues are highly intuitive and require little cognitive effort to detect.
- In some cases, smell can be used as a multimodal cue (gas used for domestic cooking has a chemical odorant added to give it a distinctive and unpleasant smell).
- Auditory warnings are omnidirectional, whereas visual warnings are unidirectional and demand visual attention.
- Multimodal warnings are perceived to be more important than unimodal warnings.
- The tactile sense is often underused or not used at all in many tasks. Therefore, the tactile "channel" is underutilized and has plenty of spare capacity to process tactile warnings. However, tactile cues have limited perceptual resolution compared to visual and auditory cues.

Bolia et al. (1999) evaluated the effects of a spatial audio display on the performance of pilots searching a complex visual scene with distractors. The auditory cue was presented spatially via a "grid" of four loudspeakers. Modulating the intensity of sound from each produces an auditory "image" of the location of the target in the region of space where the pilot is visually searching for the target. The aiding was found to reduce search time by a factor of six or more in complex searches, with no reduction in the accuracy of the search.

Multimodal cueing systems should be designed on the basis of a task analysis and for a target population. Auditory warnings are fleeting, whereas visual information can be presented on screens or paper, lowering the demands on working memory. Auditory cues may be less effective in noisy environments or for older populations with presbycusis.

Maps and Navigation Aids

Butler et al. (1993) compared the effectiveness of a number of navigation aids (referred to as "wayfinding aids") in helping newcomers find their way around a complex building. Wayfinding signs (and signposts) were found to be more effective than "You are here" maps. Signs and signposts seem to result in superior navigation because they place the directional and spatial information required for navigation into the environment itself. The information conveyed by the sign is used when needed and can then be dispensed with. On the other hand, "You are here" maps impose a considerable mental load on the user. The cues needed to navigate must first be identified on the map, then held in memory, and finally recognized in the environment. Additional improvements can be obtained by numbering the objects in the space to be navigated and incorporating the numbers into the sign system. Butler et al. use the example of hotels, where bedrooms are often numbered and are easy to find, whereas rooms containing other facilities are only named and are often harder to locate. The navigability of many shopping centers could no doubt be improved in a similar way by placing more emphasis on the use of numbers, rather than names, as cues to the location of particular shops.

Dong et al. (2021) compared the use of smartphone-based navigation aids by pedestrians navigating an unfamiliar route. The first navigation aid was a 2D map on the phone with the route highlighted in blue. The second used augmented reality (AR) to superimpose a virtual blue track on the video stream provided by the smartphone camera. All users of the AR aid had to do was follow the blue route. Although there were few differences in performance between the two navigation aids, there were differences in behavior. Users of the 2D map spent more time looking at the surrounding buildings than AR users, who spent more time looking at people. Overall, they spent less time looking at their phones. After the experiment, all subjects sketched maps of their journey, and it was found that AR users had poorer spatial memory of the route than 2D users. The authors concluded that the AR navigation aid lowered the spatial processing load of navigation but at the cost of lower spatial memory. Overall, it would seem that the AR navigation aid lowered the cognitive demands of wayfinding while enabling users to arrive at their destination without getting lost. However, AR users were less able to remember how they got there!

Three-Dimensional Displays

Our experience of a three-dimensional (3-D) visual world is spontaneous and is dependent on a number of cues, both internal and external to our senses. A knowledge of these cues is essential in the design and analysis of 3-D displays.

Monocular Depth Cues: These cues can be detected using one eye only but are also operative when both eyes are used:

- *Accommodation:* Proprioceptive feedback from the ciliary muscles provides valuable information about the distance of an object. However, this feedback is most valuable in judging the distance of objects up to approximately 6 m.
- *Movement Parallax:* Depth perception is strongly influenced by head movements, which give rise to the relative movement of close objects against an unmoving background. When traveling by automobile, distant objects appear to travel in the same direction as the automobile and close objects in the opposite direction.

- *Interposition:* A near object interposed between a more distant one and the viewer provides depth cues when it obscures part of the far object.
- *Relative Size of Familiar Objects:* The size of the retinal image can be used as a depth cue if the size of the object is known. Familiar objects exhibit *size constancy.* A vehicle or a house is perceived as being distant if its image on the retina is small. Young children soon learn that people do not shrink when they walk away.
- *Texture Gradient:* Seen in perspective, the surfaces of many objects present a texture gradient with texture changing from rough to smooth as distance increases.
- *Linear Perspective:* Perhaps, the most obvious example of perspective is the way that parallel railway lines appear to converge as distance increases or the way an object's shape appears to change as it is rotated about an axis.
- *Casting of Shadows:* A light source in the direction of a near object will cause it to cast a shadow on objects farther away. This is often used as a cue to depth on computer systems equipped with graphic interfaces.

Binocular depth cues depend on the use of both eyes:

- *Convergence:* Kinaesthetic feedback from the vergence system is a cue to depth, which functions up to about 15 m.
- *Retinal Disparity:* Since the eyes are approximately 6 cm apart, the retinal images differ slightly when near objects are viewed. The images are "fused" to create a single percept characterized by depth. This is known as stereopsis. Retinal disparity gives the perception of depth in the absence of any other cues and is used in stereophotography and "3-D" films. The most common method is to use two cameras, a small distance apart, to create a pair of images. Special spectacles (differing in color or plane of polarization) can be used to view the differently colored or polarized members of the stereo pair such that only one of the images is presented to each eye, in much the same way as when the real object is viewed. At distances over about 20 m, the retinal images are very similar because the sight lines of the eyes are almost parallel.

Burks et al. (2014) compared active, passive, and autostereoscopic 3-D displays for use with laptops. Passive displays require the user to wear polarizing spectacles, whereas active displays use a "shutter" principle to present separate images to each eye and are thought to provide superior 3-D quality. Autostereoscopic displays eliminate the need to wear spectacles. Users rated the quality of the displays on a 9-point Likert scale based on smoothness, clarity, crispness, sharpness, ghosting, level of quality, and consistency of 3-D images from different angles after 15-min viewing sessions with each display. They also rated their level of discomfort when viewing.

Overall, the autostereoscopic display was rated lowest for quality and viewing comfort. The active display was rated higher for quality, and the passive display was highest for comfort (the active spectacles weighed 56 g, whereas the passive spectacles weighed 21 g). Although the prospect of 3-D viewing without spectacles is attractive, autostereoscopic displays appear to require further development.

Head-Mounted Displays

Head-mounted displays (HMDs) seem to have a number of advantages over CRTs or fixed panels if the update of the display can be closely synchronized temporally and spatially with head movements. Disparate views of the virtual environment (VE) can be presented to each eye to give the perception of depth. Individual differences in interpupillary distance do not seem to pose much of a problem. However, the screens that display these images may be only 50–70 mm from the eye. Light from binocular HMD displays passes through a lens before reaching the eye, such that the light rays are collimated (parallel) when they reach the eye, as they would be if the displays were at infinity. Under normal viewing conditions, the accommodation and vergence systems are linked. Accommodation is a cue for vergence and vice versa. When looking at a far object, the eyes diverge and, simultaneously, the ciliary muscles relax to flatten the lens. As a distant object approaches, the eyes begin to converge and the ciliary muscles contract. So, accommodation, together with disparity (double images), is a cue for vergence, and vergence, together with blur is a cue for accommodation. As a distant object approaches the viewer in the real world, the demand for accommodation and vergence increases. In VE, however, the viewing distance may be infinity, but vergence may be required to avoid seeing a double image and to get a stereoscopic view. Alhassan et al. (2021) call this the "accommodation-vergence conflict." Other problems can occur if there is "crosstalk" or "spillover" between the separate images. Excessive retinal disparity may cause "vergence strain," as can image distortion. Common symptoms provoked by HMD use include general discomfort, headache, vertigo, disorientation, pallor, fatigue, vomiting, nausea, stomach awareness, eye strain, difficulty focusing, blurred vision, increased salivation, and sweating.

Alhassan et al. (2021) conducted an experiment where subjects played video games using HMDs for 45 min. Before-and-after tests of accommodation, vergence, strabismus (correct alignment of the eyes), and scores on a simulator sickness questionnaire[3] were recorded. Significant changes in accommodation and vergence were found together with increased symptoms of discomfort. Changes in near relative accommodation were found (the ability to stimulate accommodation while maintaining vergence), as well as a tendency to *overaccommodate* to visual stimuli. The authors concluded that VR game players have a reduced capacity to make rapid vergence changes because HMD use requires continuous changes to the accommodative and vergence systems. Although the visual system exhibits a certain amount of plasticity, the time scale and the scope for adaptation are not yet known.

Auditory Displays

The auditory sense seems to be intrinsically directional and largely pre-attentive, which makes it compatible with demands for focal attention in visual search. Some advantages of auditory displays are:

- More alerting than visual displays; they are "eyes-free" and "hands-free," capturing attention during the performance of other tasks
- Suitable when the message is short, not need be referred to later, and the user's job requires movement (e.g., among banks of complex control panels). They are also useful for communicating with people working in dark places (such as mines) or at night.

Auditory Warnings and Cues

An effective auditory warning should be of sufficient intensity to stand out above background noise and differ in pitch, waveform, etc., so as to be discriminable. The ear is most sensitive to the frequency range 500–3000 Hz; hence, warnings should ideally be in this range, with high intensities at the lower frequencies because these are not deflected by objects and travel further than high frequencies. Warnings should not be excessively loud so as to increase workers' daily noise dose above acceptable levels (louder is not "better" beyond a certain point).

ISO 7731 (2003) recommends that the warning be clearly audible if the effective masked threshold of the warning is exceeded. For this to happen, the ambient noise and acoustic properties of the space must be understood, as well as the use of hearing protection devices by workers. The warning should be 15 dB above ambient noise in the whole signal reception area, and the main frequency spectrum should be between 500 and 1500 Hz. Sound attenuates according to an inverse square law, so the further away the listener is from the source, the greater the attenuation. Less attenuation takes place in confined spaces than in open spaces because of the interreflection of sound waves. In practical situations, then, the following considerations are essential in the design of auditory cues:

1. Intensity of background noise at the point(s) where the cue is to be heard. This requires a noise survey of the premises at times when the background noise is at its loudest to determine background sound levels.
2. Frequency of background noise. A frequency analysis of the background noise can be carried out to assist in the selection of an auditory cue of a different frequency. Much machinery produces complex sounds with many different frequencies, so the warning selected must be more intense than the higher-intensity frequencies, particularly if these are of similar frequency to the warning. In the case of very high-intensity background noise (e.g., the pneumatic rock drills described in Chapter 11, in particular, refer to Figure 11.8), auditory cues may not be used at all.
3. Attenuation of cue intensity from the source. If the cue is produced at a different place from where it is intended to be heard, it will be necessary to determine by how much the source intensity of the cue is attenuated at the points where it is to be heard. It is at the listening points that the cue must be distinguished from the noise.

Apart from intensity and frequency, the cue must have high discriminability—it must be different from any other sounds the operator is likely to hear at work. Sirens, for example, can be made to warble or fluctuate in intensity or frequency. Auditory cues and warnings should persist for at least 100 ms.

Some of these design considerations are illustrated by Miller and Beaton's (1994) analysis of the problems in detecting emergency vehicle sirens by motorists. In the United States, emergency vehicle sirens have an intensity of 118 dB(A) at a distance of 10 m from the source. This is a maximum limit that was set to prevent ear damage and limit noise pollution. Although pedestrians will usually hear an oncoming emergency vehicle at a great distance, drivers of other vehicles will not. This is because the interiors of many modern automobiles attenuate sound by 20–30 dB(A) if the windows are closed. Thus, at 10 m from an emergency vehicle, the intensity of the sound at the driver's ear will be about 98 dB(A). At 50 m, it will be about 74 dB(A). Since the intensity of sound in a car with the radio turned off is about 70 dB(A), the driver is unlikely to hear the siren of an oncoming emergency vehicle until it is about 25 m away. If the emergency vehicle is traveling at a speed 50 km/h faster than the vehicle, the driver will have 1.8 s to take evasive action. Clearly, this is not long enough and explains why sirens alone are inadequate to prevent emergency vehicles from being impeded by traffic.

Voice Warnings

Wogalter and Young (1991) investigated behavioral compliance to voice and print warnings. They found that for simple warnings, compliance was greater to voice than to print, and greater still when the two were combined. They argue that voice warnings have been relatively unexploited and offer many advantages over other types of warning. Voice is attention-getting and omnidirectional and can orient people to hazards such as "slippery floors" in complex and visually distracting environments, such as malls. Compared to other auditory warnings, voice conveys information directly. However, voice is less useful for conveying complex warnings because of its serial nature, transmission load, and fragility. Wogalter and Young suggest that it may be a useful supplement to complex visual warnings, capturing attention and providing effective contextual orientation.

Baldwin and Struckman-Johnson (2002) point out that auditory warnings are likely to be used in situations in which operators are under high mental workload. Demand is placed on the attentional resources required to carry out the task and also to detect and interpret the warning. In conditions of low mental workload and low background noise, warning intensities ranging from 30 to 80 dB give almost 100% correct identification of the spoken word. However, in dual-task situations (e.g., detecting a warning while driving a vehicle), warning intensity becomes critical. Subjects in their experiment had to judge the veracity of sentences derived from Collins and Quillian (1972), such as "birds can fly" and "dogs have five legs" (refer to Chapter 12, Figure 12.7). Response time dropped from 1.05 s when the warning intensity was 45 dB to 0.9 s at 65 dB, while errors dropped by two-thirds. It seems that the detection of "quiet" warnings does indeed demand additional attentional resources compared to "noisier" warnings, and in dual-task situations, *a minimum intensity of 65 dB is needed.*

Representational Auditory Warnings

Representational auditory warnings are attention-getting if of sufficient intensity. The relationship with the referent can be *symbolic* (arbitrary, having to be learned like the sound of an ambulance siren), *metaphoric*, where the sound shares common features with the referent (a fall in pitch to represent a reduction in quantity), and *nomic*, where the sound mapping is based on physical causation (the sound of garbage falling into a can when a file is deleted). Nomic icons are the most direct and might be expected to give the quickest reaction with the least learning. The requirements for the design of effective icons are as follows:

- The sound must be identifiable (experienced in the work context and sufficiently different from other sounds).
- The mapping between the icon and its referent must be intuitive and close.
- The physical parameters of the sound must be appropriate to the context (e.g., the sound must not be too long).
- The sound quality must be adequate.
- The sound must lead to an appropriate emotional response from the user.

Nees and Liebman (2023) conducted a systematic review and meta-analysis of representational warnings:

- *Auditory Icons:* e.g., sound of a vehicle crashing
- *Earcon*: e.g., three ascending tones
- *Speech*: "headway closing!"
- *Spearcon*: e.g., accelerated phrase "headway closing"
- *Hybrid*: e.g., earcon plus speech

Overall, speech spearcons were superior in terms of reaction time and accuracy, with the caveat that they may be less effective in noisy environments and susceptible to interference by concurrent speech.

Auditory Alarms: Compatibility with Other Auditory Displays

What happens if the work environment becomes too noisy due to all the electronic systems and products in our lives? Wiese and Lee (2004) investigated the interaction between auditory collision warnings in automobiles and auditory in-vehicle e-mail alerts. Warning onset and offset rates, duration, and urgency were used to manipulate the perceived urgency of warnings. Urgent warnings produced faster accelerator pedal release than less urgent warnings but also produced higher mental workload. E-mail alerts occurring 1000 ms before the collision avoidance warning enhanced the response to the collision warning. E-mail alerts occurring 300 ms before the collision avoidance warning had the opposite effect. The authors suggested that the effectiveness of the collision avoidance warning system would be improved if e-mail alerts were suppressed when there was any evidence that a collision might occur.

Haptic ("Tactile") Displays

Tactile displays can have the same parameters as sounds—frequency, intensity, waveform, and so on. Potentially, they are a powerful means of communicating information to users and may be embedded in the device being used, in the working environment, or attached to the user. From an SDT perspective, they would be expected to be most effective in circumstances where the haptic background noise was minimal. For example, rumble strips on the side of freeways alert drivers to poor lane tracking by creating low-frequency sound and vibration as the tyres roll over the corrugated strip at the edge of the lane. These stimuli are highly conspicuous on a modern freeway, where low-frequency sound and vibration are rarely encountered. Their sudden onset is alarming. Rumble strips would be less effective on dirt roads where the uneven surface is itself a source of low-frequency sound and vibration. In SDT terms, d′ for sound and vibration targets is smaller on dirt roads than on freeways.

Hancock et al. (2015) have reviewed some of the main considerations in the design of tactile displays:

- Tactile displays work better when people are standing still than when engaged in strenuous activity.
- Tactile cues are more alerting when they are salient (the detection threshold is low and their meaning is clearly defined).
- Tactile belts have been developed to aid the navigation of soldiers in complex environments and to provide "earth-reference" information for pilots engaged in challenging maneuvers.
- Tactile cues can augment other displays in degraded environments (e.g., glare on screens or excessive background noise).

Myles and Kalb (2013) investigated the use of head-mounted tactile displays for displaying navigational information to soldiers. The forehead, sides, and back of the head were most sensitive to tactile stimuli. A further advantage of tactile displays is that they neither emit light nor require it to work and are eminently suitable for use in the dark and when dark adaptation of the eye must be maintained. The main considerations for the optimal use of tactile displays are:

- Selecting an appropriate body region where sensitivity to the tactile stimulus is sufficient.
- Selecting a body region where tactile stimulation would not degrade the performance of the main task.
- Designing a tactile display that can be donned and doffed easily while wearing clothing required for the task.

Vehicle Controls

Steering wheels, joysticks, and pedals are commonly used to control vehicles. Control usability can be affected by the resistance of the control, which should be operable using forces that are a fraction of the operator's maximum voluntary contraction. However, the control should offer some resistance to movement so that "bumping" errors and muscle tremor do not cause control errors.

With position controls, the displacement of the control is in direct relation to the change in the controlled part. Pressure (sometimes called "isometric") controls are almost immovable, and the force on the control stick provides the control signal. Pressure controls are useful in controlling higher-order systems, which exhibit lag (see below). Some options for control design are control stick friction, preload, control/display ratio, and control system hysteresis. Static friction can aid performance in environments in which operators are exposed to vibration. Inertia and viscous damping can be built into controls to assist with the execution of smooth control actions.

Control Distinctiveness

In many instances, numerous controls are grouped on a panel, and the designer's task is to ensure that operators can easily distinguish between different controls. Some of the display design guidelines mentioned above are relevant here. In addition, manual controls provide the operator with tactile feedback, which can be used to give a distinctive identity to a control or a related set of controls.

Avoid Spatial Transformations

The domestic stove is a good example of a device in which operational problems can arise due to a lack of spatial correspondence between the layout of the switches and the layout of the burners. This lack of correspondence renders the true control/display relationships ambiguous and leads to error. Two typical arrangements for a four-burner stove are given below (Figure 13.11).

Problems arise because the burners are arranged in a plane (which has two dimensions), whereas the switches are arranged linearly. Since this 2-D array cannot be projected onto one dimension without loss of information, spatial incompatibility is almost inevitable. The loss of the front/rear information makes switch selection for the front and rear burners arbitrary. Two possible solutions are to either stagger the layout of the burners to create a 2-D array that can be projected unambiguously onto one dimension, or to arrange the switches themselves in a plane arranged to be spatially compatible with the burner plane (Figure 13.12).

Chapanis and Lindenbaum (1959) found the staggered burner arrangement to be freer of operating error than other arrangements. Many different burner/switch arrangements have been developed by manufacturers over the years, and none appears to be optimal. Fisher and Levin (1989) investigated this problem using an unconstrained approach in which subjects were asked which of the switches of a stove burner simulator would turn the burners on. The CABD arrangement was found to be the most consistently chosen and preferred. However, other arrangements were also found to be acceptable, although less so. The fundamental incompatibility between the traditional burner/switch layouts probably explains why so many different designs are commercially available and why

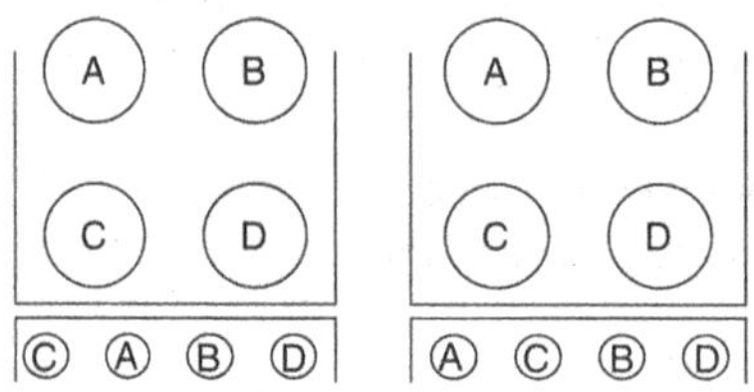

FIGURE 13.11 Two common arrangements for the layout of controls and burners on a domestic stove.

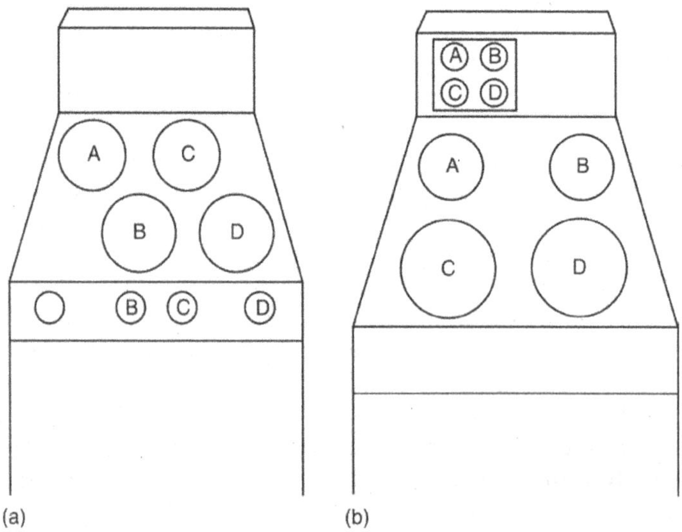

FIGURE 13.12 Possible solutions to the stove control problem.

designers have resorted to techniques such as the use of logos, overlays, and mimic displays in an effort to make explicit the functional relationships between the controls and the hobs (Figure 13.12).

Keyboards

Most keyboards have the letters of the alphabet arranged in the QWERTY layout. This layout was designed in the nineteenth century to physically separate the type bars of common letters on mechanical typewriters and so prevent them from jamming. The result was that many common letters were placed at a distance from the home row of keys, and this is now thought to be inefficient. The QWERTY layout has costs—it overworks the left hand and some of the fingers and underutilizes the home row of keys. Since the arguments in favor of the QWERTY layout no longer apply, several investigators have evaluated alternatives. The most well-known are the Alphabetic layout, in which the keys are arranged in alphabetical order over three rows, and the Dvorak layout (Figure 13.13 with the QWERTY layout for comparison), which was developed in the 1930s using data from time and motion studies.

The Dvorak layout places the most common letters in the middle of three rows. The left hand types the vowels, a, o, e, u, and i, and the right hand types the most common consonants, d, h, t, n, and s. Thus, the two hands tend to work in sequence much of the time. Tanebaum (1996) reports that, using the Dvorak layout, typists make 37% fewer finger movements, with the right hand making 56% of the keystrokes (the right hand being stronger in most users). The forefinger and middle finger of both hands type the more common letters e, u, h, and t, whereas with QWERTY, these fingers type d, f, j, and k. Norman (1982) suggested that the Dvorak layout only brings about a 5% improvement in typing speed, and this may not be large enough to justify the expense of retraining millions of users. Tanebaum (1996) reported that the Dvorak keyboard was evaluated by the U.S. Navy in 1944. Despite the positive feedback from typists, training on Dvorak required over 50 h. Although productivity improvements might be expected to justify the expenditure, initial training costs likely act as a barrier to widespread adoption of the layout.

The alphabetic layout is often assumed to be better than the QWERTY layout for naïve users ("hunt and peck" typists). Norman (1982) compared several alphabetic keyboards with a random keyboard and with the QWERTY keyboard. Non-typists typed for 10-min sessions on each of the keyboards. Typing speed was 7.1 words per minute on the random layout, 7.8 words per minute on the alphabetic layout, and 13.0 words per minute on the QWERTY layout. Norman suggests that part of the reason for the superiority of QWERTY was that subjects had, at least, some familiarity with it and that its layout is more efficient than that of the alphabetic and random layouts. The alphabetic layout was superior to the random layout by about 1 word per minute. However, Norman suggested that the alphabetic layout did not yield large benefits because naïve subjects do not use their knowledge of the alphabet to type with it. Rather, they visually scan the keyboard, looking for

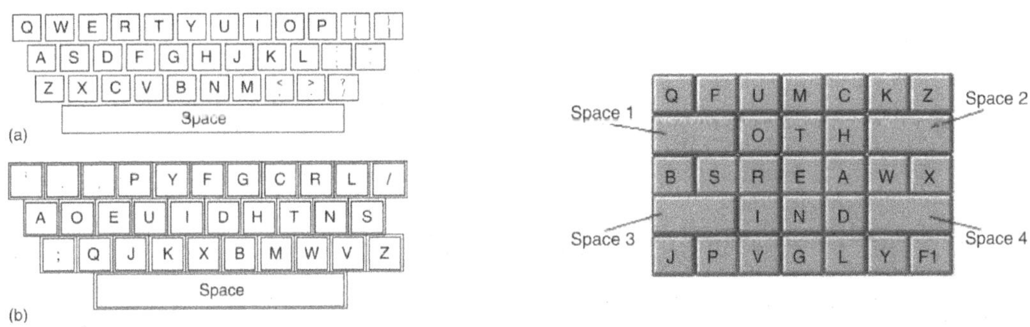

FIGURE 13.13 Rival keyboard layouts: (a) QWERTY and (b) Dvorak. QWERTY is used because it arrived first. The OPTI soft keyboard layout.

the next key in the same way as with the random keyboard. For naïve typists, the *mental workload* of reading and typing text leaves little spare capacity for recall and rehearsal of the alphabet.

Both the QWERTY and DVORAK layouts were designed for use by "touch typists" (trained operators who type with all fingers of both hands without looking at the keyboard). Few users are so trained nowadays, and modern devices have predictive text capability, which reduces the need to press keys and also reduces errors.

In general, the QWERTY layout seems best. However, hardware and touchscreen QWERTY layouts present a problem when used on small handheld devices because the keyboard is so small. The gap between the keys should be about the same as the width of the tapping device (fingertip or pen). Wright et al. (2000) compared the keying performance of older users on handheld computers with touchscreens or keyboards. Older users, in particular, were slower when using a touchscreen and pen than a keyboard, and 88% of them preferred to use the keyboard.

Tablets and phones use soft keyboards, and text entry rates are much lower than on laptops and desktop computers. Predictive text and word selection are ways of increasing text entry speed on these devices. Roy et al. (2021) investigated the effect of these on texting efficiency. Word prediction systems either include word completion suggestions or next-word prediction. Texting efficiency was investigated on a desktop computer, a landscape-oriented tablet, or a phone used with one hand, in relation to the number of word suggestions suggested and the number of keystrokes saved when they were used. Subjects transcribed phrases from a set designed to be representative of the English language. The main findings were that the usage of word suggestions increased with their accuracy and decreased relative to the amount of typing saved. Highly accurate suggestions were used 50% of the time across devices, but up to 70% of the time on the phone. Word suggestions did not improve text entry speed on desktop computers, even when they were very accurate. They did improve text entry speed on phones and tablets when accuracy was high. The authors concluded that the more usable the interface, the less likely users are to rely on automated support such as word suggestions.

Pointing Devices

The mouse is the most common pointing device in everyday use. Most of the ergonomic issues are at the physical level. The mouse is designed to be placed at the side of the keyboard and requires abduction of the shoulder, increasing the muscular load.

Fitts' law applies to most computer input devices. MacKenzie and Buxton (1993) describe test software that enables the coefficients in Equation 13.4 to be estimated based on keyboard trials of 4–5 min. Input devices with the lowest values of a and b will give the fastest MTs. In general, the mouse tends to give better performance than the trackball for dragging and pointing tasks, whereas the interactive tablet plus stylus can give better dragging performance than the mouse. Mice tend to be rated higher on overall satisfaction, comfort, and operation than either trackballs or joysticks (Woods et al., 2003).

Table 13.4 presents a checklist for the assessment of, possibly novel, non-keyboard input devices.

Touchscreens

Touchscreens are used in an increasingly wide range of devices—from smartphones to supermarket checkouts. The icons displaying the options are often the buttons controlling the interaction. Many of the display design principles discussed above also apply to touchscreen technology. Claypoole et al. (2016) describe interface design principles for older users, which appear to be widely applicable. For ease of interaction

- Place important icons in the middle of the screen
- Consider secondary modes of selection (e.g., voice or hard keys)
- Basic functions should be analogous to their counterparts in the real world (direct mappings)
- Tolerate imprecise gestures (older or disabled users may have slower reaction times or lack dexterity)

TABLE 13.4
Checklist for the Assessment of Non-Keyboard Input Devices

Name:	**Occupation:**	**Date of Assessment:**
Terminal/PC/Laptop: Name		**Wrist Support (Y/N)**
Device: Name		**Time Used in Assessment:**
Daily Time Working on Computer	**Intensity of Use**	**Breaks from Computer**
<2 h	Occasional	Regular (>2/h)
2–4 h	Regular	Occasional (<2/h)
4+ h	Continual (hand constantly on device while working)	Very rare (>1/h)
Have you had training to use the device?	Yes	No
Is the device within the ZCR?	Yes	No
Is the device suitable for the desk layout?	Yes	No
Device Performance		
Does the device respond as you would expect (i.e., same for similar tasks)?	Yes	No
Does the device give you adequate feedback when you press a button?	Yes	No
Is the device responsive or sensitive enough?	Yes	No
Do you have adequate control of the device (easy to place/move the pointer to select information)?	Yes	No
Does the device move smoothly and easily?	Yes	No
Device Operation		
Is the device easy and intuitive to use?	Yes	No
Do you know the function of all the controls/buttons on the device?	Yes	No
Can you work at the speed you want to using the device?	Yes	No
Does the device require an acceptable amount of effort to operate?	Yes	No
Do the cables interfere with the operation of the device?	Yes	No
Do you find the device works well with the software you use?	Yes	No
Is the size of the mouse mat appropriate?	Yes	No
Is the material of the mouse mat appropriate?	Yes	No
Device Design		
Is the size of the device suitable for your hand?	Yes	No
Is the shape of the device satisfactory?	Yes	No
Is the grip comfortable (is it easy to grasp)?	Yes	No
Can the device be positioned easily and quickly?	Yes	No
Is inadvertent button operation a problem with this device?	Yes	No
Can you easily reach all the device buttons or other controls?	Yes	No
Is the device stable (does it slip or rock?)?	Yes	No
Daily time working on computer	Intensity of use	Breaks from computer
Device Comfort		
Are your hands/fingers in a comfortable position when you use the device?	Yes	No
Does the device cause pressure points that result in discomfort?	Yes	No
Do you experience finger fatigue when using or after using the device?	Yes	No
Do you experience wrist fatigue when using or after using the device?	Yes	No
Do you experience arm fatigue when using or after using the device?	Yes	No
Do you experience shoulder fatigue when using or after using the device?	Yes	No
Do you experience neck fatigue when using or after using the device?	Yes	No

(Continued)

TABLE 13.4 (*Continued*)
Checklist for the Assessment of Non-Keyboard Input Devices

Name:	Occupation:	Date of Assessment:
Terminal/PC/Laptop: Name		Wrist Support (Y/N)
Device: Name		Time Used in Assessment:
Daily Time Working on Computer	Intensity of Use	Breaks from Computer
Is the device suitable for your dominant hand?	Yes	No
Can the device be used without undue deviations of the wrist from neutral?	Yes	No
Can the device be used without undue deviations of the fingers from neutral?	Yes	No
Can the device be used without undue deviations of the shoulder from neutral?	Yes	No
Can the device be used without undue deviations of the arm from neutral?	Yes	No
Can the device be used without undue deviations of the neck from neutral?	Yes	No

Source: Adapted from Woods et al. (2003), with permission.

- Use detailed, concrete icons to maximize transparency and offer a textual description when hovering over the icon
- Provide visual, auditory, or tactile feedback when the command is accepted
- Limit the number of options per screen (following Hick's law)
- Use legible fonts for text messages (e.g., sans serif)

There is currently discussion about the use of touchscreens in cars. Touchscreens reduce the need for physical buttons and provide flexible display/control options. However, there are concerns that they can distract the attention of drivers and lack the haptic feedback of buttons for tasks such as scrolling.

Voice Control

Control of machines by voice has benefits for disabled users and for those in "eyes-busy, hands-busy" occupations, such as surgery. Voice can

- Provide an extra communication channel, which may take some of the load from more conventional channels
- Free the hands to conduct other activities
- Lower attentional demands if the command set is well known

The processing requirements for issuing voice commands would not be expected to conflict with those for manual control and, therefore, voice is often thought to have the potential to speed up task performance or increase an operator's information-handling capacity. Wogalter et al. (2021) give the following recommendations for voice control:

- Avoid using long codes that exceed short-term memory limits, or design long codes that are easily "chunkable."
- Use easily pronounced words and codes (one-syllable versus two- or more-syllable items).
- When using voice menus, allow users to respond immediately rather than having to wait until the final menu item.

Other advantages of voice control include: more responses can be given in a short period of time (most people can speak faster than they can type); categorical or symbolic information can be transmitted efficiently by voice (e.g., "up," "down") whereas graded commands (e.g. 'left a little') are better made using manual controls.

Problems the Voice Recognizer Faces

Speech is a complex acoustic signal containing bands of energy centered around 500, 1500, 2000, and 3000 Hz. These are known as *formants*, consisting of an initial transient segment followed by a steady-state segment. The lower two formants are sufficient for the perception of speech, which enables bandwidth compression to be implemented to reduce the information load of speech transmission systems and speech synthesizers.

Speech perception depends on the interplay of both "top-down" and "bottom-up" processes. Bottom-up (analytic) processes detect particular features in the speech signal as it is received. Combinations of features detected in speech form the characteristic signatures of basic speech sounds such as phonemes, syllables, or words. When matched with a stored representation of known features, the speech can be said to have been recognized. Top-down (synthetic) processes use higher-order information to synthesize possible items that can be matched against the incoming signal (trying to "guess what's coming next"). Knowledge of grammar, syntax, and context is required to drive these higher-order interpretive processes.

Real speech is a "messy" signal. It is mixed in with other noises that humans make (e.g., coughs, grunts, sneezes, etc.). The human speech perception system seems to be able to disentangle connected verbal and nonverbal utterances with ease, but it is not clear how this is accomplished or how a machine might best be programmed to do this. People experience *voice fatigue* if they have to speak over the course of a day. This adds to the variability in speech that the voice recognition system has to be able to cope with. Stress, alcohol, colds and flu, and other factors also increase the physical variability of a speaker's utterances.

Connected speech is characterized by a lack of segmentation—the building blocks of speech blur into one another. It is not possible to identify a unique set of building blocks at one level. Many elements of speech are co-articulated—that is, the acoustic form of an utterance differs depending on what precedes and follows it. For example, the spoken form of the word "bag" is different from that of the word "beg." It is not possible to record "bag" on audiotape and remove the middle "a" and replace it with "e"—the whole utterance has to be changed. Similarly, compare the shape of the lips when pronouncing the "st" in "stew" and "stay." The human mouth, lips, and tongue move slowly, have inertia, and exhibit hysteresis when speech is produced. Although continuous speech is perceived consistently, it is not produced consistently. Parts of an utterance may sound the same to a human listener but differ acoustically depending on the words that surround them.

Modern voice recognition systems, such as Siri™ and Alexa™ have advanced recognition systems. Such systems are less proficient in understanding people with voice disorders (such as vocal dysphonia, a rough, strained voice). Rohlfing et al. (2021) found that several commercially available systems had high recognition scores for passages read by subjects without dysphonia (over 90% accuracy), dropping to 70% or less in dysphonic voices. In safety-critical systems, voice control may need to be supplemented with alternative control modalities.

Hearing Lips and Seeing Voices

Voice recognizers are at a disadvantage when it comes to decoding human speech. When people converse with each other, they adjust their perceptions of speech to take account of the way other people produce speech—speech perception can be conditioned by contextual factors. In a classic study, McGurk and MacDonald (1976) videotaped people repeating the syllables "ba-ba," "ga-ga," "pa-pa," and "ka-ka." They then spliced the soundtrack onto the video so that "ba-ba" was heard with the lip movements for "ga-ga," and so on. When listeners were only presented with the auditory track, they correctly recognized the syllables. When they watched the discordant video, they *heard* different syllables. "Ba-ba" combined with "ga-ga" fused into "da-da." "Ga-ga" combined with "ba-ba" gave "bagba" or "gaba." "Pa-pa" (lips) combined with "ka-ka" (voice) gave "ta-ta." Thus, the sounds that people hear when listening to other people speak depend on nonauditory factors. This may help explain why lip reading is possible and why speech communication in noisy environments is facilitated by sight of the speaker's face, particularly the mouth.

These examples demonstrate some of the difficulties of constructing machines to recognize fully connected speech in a natural way.

TOOLS AND PROCESSES

MacKenzie and Zang (1999) used Fitts' law to design an optimum layout for such a keyboard. Motor performance on many interactive tasks can be modeled using Fitts' law (Fitts, 1954), which states that

$$\text{Movement Time} = a + b[\log_2(2.A/W)] \tag{13.4}$$

where a and b are device-dependent constants, A is the distance between the starting position and the target (the amplitude of movement), and W is the width of the target. The MT increases predictably as the distance increases or the target gets smaller. When using a soft keyboard with a single stylus, typing can be modeled as pointing. The key just pressed is taken as the starting position, and the next key as the target. In this way, the optimum keyboard layout can be determined by minimizing the distance traveled between successive key presses (including pressing the space bar and pressing the same key twice for repeated letters, and letter space bar presses for word transitions). The width of keys ("W" in the equation) can be taken as fixed by the space constraints of the screen and the need to display 26 letters in English, plus the space bar, on separate keys.

MacKenzie and Zang used data on the frequency of letters and letter pairs ("digraphs") in written English to produce an optimized keyboard layout known as "OPTI" (Figure 13.13). They did this by running user trials to produce a Fitts' law model of soft keyboard operation to find the values of the device-dependent constants. Next, they calculated the MTs for all possible digraphs and included empirically determined time estimates for double letter tapping (the mean time was about 127 ms) and for the use of the space bar and a single letter. These MTs were then weighted using the data on digraph frequency in English to produce a mean MT for the optimized keyboard. Note that the OPTI layout is designed to maximize word entry rate by minimizing the distance the stylus has to travel between each successive digraph. The most common keypress in typed English is the space bar, and the most common digraph is "e-space." Not surprisingly, the OPTI keyboard has the letter "E" in the middle, equidistant from four space keys. By having four space keys arranged around the keyboard, the distance from he space key to any other letter is minimized. Note also that common digraphs such as "TH" and "IN," "CH" and "RE" are implemented using adjacent keys. Letters with a high frequency of occurrence are placed in the center of OPTI, and uncommon letters, such as "Q" and "Z," are placed on the periphery.

Subjects practiced with both the QWERTY and OPTI keyboards over 20 sessions of 45 min, split equally between QWERTY and OPTI practice. Initially, QWERTY was superior in terms of entry speed in words per minute (wpm). After about 10 sessions, it was equal to QWERTY, and after 20 sessions, OPTI was over 10% faster in wpm and had a consistently lower error rate. HFE Workshop 13.1 presents a brief introduction to Fitts' Law and its use in the design of workstations.

HFE WORKSHOP 13.1

FURTHER EXPLORATIONS OF FITTS' LAW

According to Fitts (1954), MT is given by

$$\mathrm{MT} = a + b[\log_2(2A/W)] \tag{13.5}$$

where

a and b are constants.
A is the amplitude of movement (distance to the target from a starting position).
W is the width of the target.

Although it seems intuitively obvious that MT increases with distance to the target, the increase is not linear—doubling the distance does not double the MT, as is demonstrated below. Access to calculators with logarithms to the base 2 will assist in following the calculations.

Some readers will not have access to calculators with logarithms to the base 2. This is easily remedied. Logarithmic conversion is given by

$$\log_b(x) = \frac{\log_k(x)}{\log_k(b)}$$

Most calculators have logarithms to the base 10, so to find the logarithm to the base 2of any number, x, first find its logarithm to the base 10 and then divide by the logarithm of 2 to the base 10. For example,

$$\begin{aligned} \log_2(32) &= \frac{\log_{10} 32}{\log_{10} 2} \\ &= \frac{1.5051}{0.3010} \\ &= 5 \end{aligned}$$

(or $2^5 = 32$).

$\log_d>(2A/W)$ is called the "index of difficulty" (I_d; Fitts, 1954). For a constant target area, doubling the amplitude of movement only increases I_d by 1. In other words, MT increases linearly with I_d and not with distance itself. This has very important implications for designers.

- Associated controls do not necessarily have to be close together to be operated quickly in succession, as long as the width of the control itself is large enough (in relation to the direction the hands are moving to operate the control).
- MT can be held constant when controls are separated by increasing the size of the controls (to keep the A/W ratio constant).

Readers familiar with statistics will recognize Equation 13.6 as the equation of a straight line of the general form (see Ergonomics Workshop 7.2 for more on this equation and how to estimate its parameters).

$$y = a + bx \tag{13.6}$$

(Continued)

HFE WORKSHOP 13.1 (*Continued*)

where y is the dependent variable, x is the independent variable, and a and b are empirical constants.

The table below shows data from Fitts (1954) from a pin transfer task. Subjects transferred pins from one pin board to an adjacent board. The variables manipulated were the distance between the boards, the diameter of the pins, and the diameter of the holes in the pin board where

A=distance between the two pin boards
W=difference in diameter between the pins and the holes into which they were inserted

Tolerance and Amplitude Conditions	Movement Time		
W[a]	A[a]	I_d	t(s)
0.03125	1	6	0.673
0.03125	2	7	0.705
0.03125	4	8	0.766
0.03125	8	9	0.825
0.03125	16	10	0.959
0.0625	1	5	0.484
0.0625	2	6	0.518
0.0625	4	7	0.580
0.0625	8	8	0.644
0.0625	16	9	0.768
0.125	1	4	0.388
0.125	2	5	0.431
0.125	4	6	0.496
0.125	8	7	0.557
0.125	16	8	0.668
0.25	1	3	0.326
0.25	2	4	0.357
0.25	4	5	0.411
0.25	8	6	0.486
0.25	16	7	0.592

[a] Cited using inches as in the original paper.

For example, in the top row, I_d has a value of 6 because

$$\begin{aligned} I_d &= \left[\log_2(2A/W)\right] \text{ and } A = 1, W = 0.03125 \\ &= \log_2(2/0.03125) \\ &= \log_2(64) \\ &= 6 \end{aligned}$$

MODEL TO PREDICT MT

Most statistics packages contain programs for regression analysis. I_d and MT can be entered into such a program to obtain the values of a and b in Equation 13.2. With I_d as the independent variable and MT as the dependent variable, the following equation is obtained from the data above:

(*Continued*)

HFE WORKSHOP 13.1 (*Continued*)

$$\text{MT} = 0.0025 + 0.086(I_d)$$

The correlation between I_d and MT is close to +0.94. So, for any combination of movement amplitude and target size, we can calculate I_d and use the equation to estimate average MT. If W=0.2 and A=10, then

$$I_d = \log_2(20/0.2) = \log_2(100)$$

$$log_{10}(100) = 2.0$$

$$\log_2(100) = 2.0/0.3010 = 6.598$$

Substituting these values into the equation above

$$\text{MT} = 0.0025 + 0.086(6.598) = 0.57\text{s}$$

LIMITATIONS AND ASSUMPTIONS

Application of Fitts' Law is bounded by various assumptions including

- As with all regression modeling, predictions and interpolations can only be made within the range of independent variable values in which the model was developed.
- Assumptions of linearity beyond the measured range may break down because larger target distances may require qualitatively different movement patterns or different joint recruitment.
- It really applies to actions that are controlled by feedback from the environment (closed-loop motor actions) and not ballistic movements, such as kicking a ball, which may follow a simple linear model.

Visual inspection of the data from Fitts' pin transfer task shows that the model fits well when the tolerance is fairly large (W is large). For small W (0.03125 of an inch or 0.08 mm), the relationship is not so good, and MT is longer (see the three outlying points in the upper, centermost part of the graph corresponding to data from the first three rows in the table above).

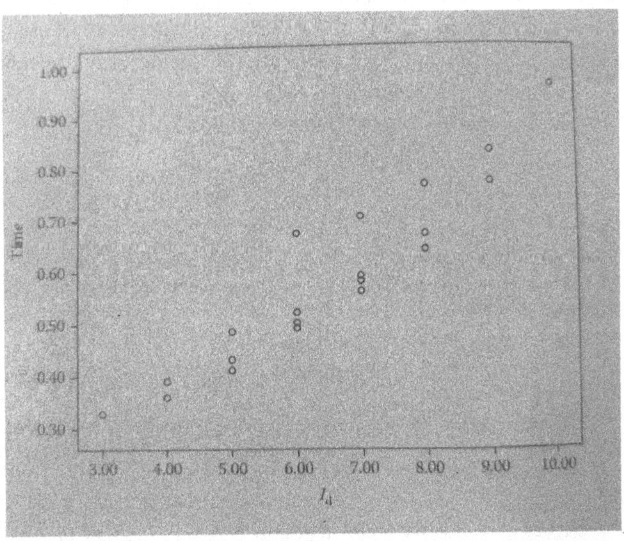

RECEIVER OPERATOR CHARACTERISTIC CURVES—ROC CURVES

Earlier, we saw in signal detection theory that signals and noise often overlap. We have to decide on a decision point or "cut-off" value to determine whether a signal (be it a diagnostic test result meaning that the patient has the disease, a value on a display panel indicating a system fault, or whether a blip on a radar screen is really a UFO or just a flock of birds) is real or just noise. ROC curves are plots of the false positive rate (where noise is classified as a signal) against the true positive rate (a 'hit' or correct identification). Using historical data, we can explore the effects of varying the cut-off to decide how accurate our test is and determine the value we should use in practice to identify targets. The diagonal red line in the middle of Figure 13.14 indicates that the test cut-off is useless—no better than tossing a coin. The outer blue line is better because, for a high true positive rate, there are very few false positives.

Eye Tracking Systems

Eye tracking systems are used to study operator interactions with display panels in aviation, health care, driving, and control room operations. By mapping eye movements onto task performance, a better understanding of the attentional requirements of the task can be obtained and related to the design of the displays and controls. Blundell et al. (2023) used eye tracking to evaluate head-up displays in the aircraft cockpit. Eleven pilots taxied in a flight simulator using a head-up display and deviations from the centerline and speed were taken as performance indicators. It was found that performance was related to gaze behavior—during turns, gaze directed at an undercarriage lateral deviation indicator was associated with better lane tracking.

SYSTEM INTEGRATION

Control Room Design

Control room design requires the application of HFE from almost all chapters of this book. It is not just a question of integrating displays and controls. Control rooms are normally separate from the process under control and all sources of information are secondary—operators have no direct contact with the plant itself. The layout of the control room is critical. Normally, several operators will have their own workstations to control a particular aspect of the process, so each workstation will have its own controls and displays. Control rooms often have a large display on one wall, giving an overall representation of the state of the system at any one time and visible to all operators. Individual workstations need to be arranged to enable all operators to see this display as well as their own displays and to communicate with each other. Consideration must be given to both routine

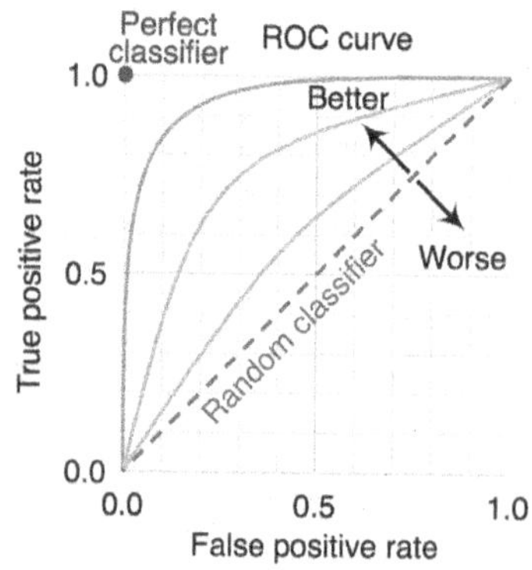

FIGURE 13.14 ROC curves to decide on decision points. (via Wikimedia Commons.)

operation of the plant and emergencies. This goes beyond the safety of the operators themselves and extends to the ability to maintain control in less-than-optimal conditions. The flexibility and programmability of electronic displays and keyboards offer great advantages in terms of display design and are space-efficient. However, old-fashioned knobs and dials may offer advantages in emergencies—if the control room is full of smoke and visibility is poor, operators may still use tactile memory and feedback to operate these controls.

The control room should be seen as a system, consisting of several operators and several workstations all sharing information to achieve a common understanding of the state of the system being controlled. Link and task analysis can be used to determine the requirements for each workstation and the verbal and nonverbal communication needs of the operators when they work as a team. At the same time, the minimum space between operators should be at least 700 mm or more, and the layout should consider the requirements for supervision and differences in status and responsibility as well as space for storage of personal items. Further information can be found in ISO/DIS 11064 "Ergonomic Design of Control Centers," parts 1–5.

Status of HFE Principles Used in Control and Display Design

Fitts' Law seems to have remarkably good validity and reliability. It has high predictive power and wide application, and even applies to imagined as well as real movements. Furthermore, it applies to the perception of the movements of other people or of machines such as robots (Grosjean et al., 2007). In other words, viewers' predictions of whether a person can move at a perceived speed without missing a target are as predicted by Fitts' Law. Possible applications of this finding include using Fitts' Law to ensure that virtual reality systems have high fidelity. Although Fitts' Law models do exist, the coefficients are often application-dependent, and some applications may require that trials be conducted to determine their values.

Fitts' Law was used recently by Wu et al. (2023) to investigate icon size and position on smartphones. The icons were accessed with one hand from a starting position using the right or left thumb. Movement time for women increased with smaller icon sizes (less than 110 rpx, where rpx was defined as responsive pixels to compensate for any differences in the size of the smartphone screens) and increased task distance. For men, this effect was found for right-hand use only. The screen was divided into a grid (four cells wide and seven deep), and fast, medium, and slow regions were identified for each hand (Figure 13.15).

Many of the design guidelines described above have high *face validity.* This is particularly true of population stereotypes that govern our expectations of what displays mean and the effects that control operations have on controlled objects. Ergonomic assessments of controls and displays are not intrusive because much of the work can be done offline. However, general guidelines should not be applied without an understanding of the task, so some kind of task analysis or link analysis is normally a prerequisite.

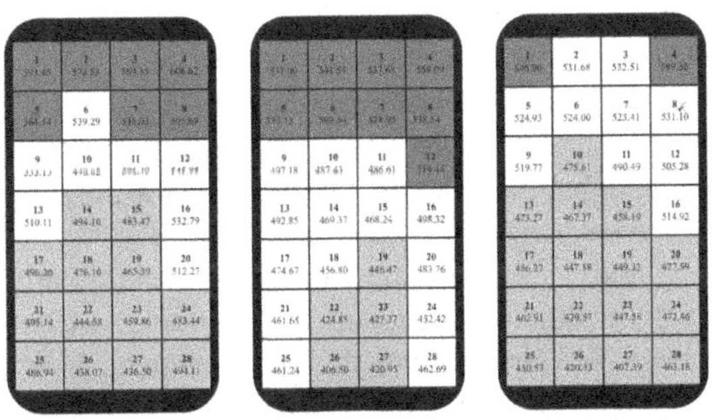

(a) All (b) Left hand (c) Right hand

FIGURE 13.15 Smartphone regions for one-handed operation: white, fast; gray, medium; and dark gray, slow. (From Wu et al. (2023) with permission.)

Effectiveness and Cost-Effectiveness

That ergonomic design of controls and displays to optimize human–machine interaction has sound financial benefits is demonstrated by a research literature going back nearly 70 years. Singleton (1960) redesigned the speed control of an industrial sewing machine, and in a 1-month comparative trial on the shop floor, estimated that the potential production improvement would be 10%–15%. Given that some of the increased production would result in an increase in earnings of 5%, it was estimated that the new controls would pay for themselves in 7 years.

Murrell and Edwards (1963) compared the operation of lathes fitted with traditional Vernier-style gauges and digital gauges. The digital indicators resulted in a marked reduction in setting time—from 11.5% to 1.2% of the total time spent processing orders, an increase in machine utilization time by 11%.

Warnings

Clement (1987) illustrates the potential benefits and likely costs of warning signs and instructions. Using an industrial example (warnings in forklift trucks about the dangers of hydraulic failure), the main costs and benefits are as follows:

Benefits

- Avoidance of major injuries or death
- Avoidance of minor injuries
- Enhanced perception that the manufacturer is safety-conscious
- Increased attention to maintenance by the user company
- Reduced liability insurance costs

Costs

- Cost of producing instructions
- Cost of fitting warning plates on all trucks
- Perception that the product is dangerous
- Perception that the product may need increased outside maintenance

Assuming that there had been one serious back injury in the company in the previous 10 years, the estimated cost of a further injury would have been $300,000 (the 1987 purchase price of a 30-year policy paying the injured party a total of $1 million). If the probability of the company being found guilty of negligence was 0.3, the cost that would be avoided by proper warnings and instructions would be $90,000 over the next 10 years (this is an underestimate, as it does not adjust for inflation). For minor injuries, a cost of $2400 was estimated. The better safety material might result in one more sale per year (with a profit of $500, yielding $5000 over 10 years). Savings in the cost of liability insurance were estimated at $5000 per year for 10 years. The total benefit was estimated by Clement to be $90,000+$2400+$5000+$2500+$50,000=$149,000.

The costs of developing the warnings, including extra printing costs, might be $1000, and for the warning plates $10,000. Assuming a cost of $25,000 over 10 years due to a loss of sales as the product is perceived as being dangerous, the total cost of the warnings would be $25,000+$10,000+$1000=$36,000.

No additional benefits would accrue to the company by deciding not to provide warnings but the costs of not doing so can be seen to be substantial. According to Clement's estimates, the ratio of benefits to costs is almost 5–1. This ratio can, of course, change, dropping if lower-cost liability insurance becomes available and rising if new legislation is introduced that increases employer liability for health and safety (i.e., if the regulatory authorities increase their propensity to prosecute or if new precedents increase the likelihood of private litigation).

Analyses of this nature provide companies with a rational choice for deciding whether to proceed with the incorporation of safety features and thereby incur extra costs. As soon as the benefit/cost ratio drops below 1, a rational decision can be made to do nothing.

RESEARCH DIRECTIONS

Human–machine interaction and display design are areas subject to considerable "technology push." E-readers are becoming increasingly popular, although in their current form, they have lower legibility than traditional paper books (Kim et al., 2014), and traditional paper books are more comfortable to read and are preferred by readers. Hancock et al. (2016) report that e-readers are preferred for their portability, web access, perceived environmental friendliness, and search function, whereas traditional books are preferred for their relative inexpensiveness, unrestricted use, and freedom from power supplies. Clearly, the pros and cons of e-readers and traditional books will change with technology.

New interactive devices are increasingly permeating everyday work and private life. There is often a lack of integration of these devices, and there are nontrivial problems of compatibility between the use of these devices themselves and other concurrent activities, such as driving automobiles. Miniaturization of products and tools brings with it miniaturization of their user interfaces and new challenges for implementing usability into an ever-decreasing space. Additional interaction modes, such as voice, may help to overcome these challenges, while, at the same time, in developed countries, the aging user population exhibits decreased functionality with respect to the use of such devices.

SUMMARY

A systems view, centered on control–display user-task integration, is needed. Many ergonomic guidelines are available, going back to the 1950s, to assist with the design of individual components, and these should not be ignored. However, people operate machines to achieve work goals, and these goals and their task context need to be understood before such guidelines can be effectively utilized. Basic task and link analysis can support human–machine integration at the control–display level.

Tables 13.5 and 13.6 summarize guidance for the design of displays and controls.

Table 13.7 summarizes general principles for control–display integration.

Further guidance can be found in ISO/TC 159/SC 4 Ergonomics of Human–System Interaction.

TABLE 13.5
Display Design Guidelines

Use dials to display rates of change of the displayed variables
Use alphanumeric displays when the detection of discrete states or values is required
Use color coding of dials when absolute judgment is required
Use shape coding to differentiate functions
Use flashing or sounds to alert operators to changes in system state
Use non-reflective materials to prevent glare on displays and screens
Built-in illumination when ambient lighting is low
Use trend displays for historical information and slow changes
Use pictorial displays to integrate data from different subsystems

TABLE 13.6
Control Design Guidelines

Controls should be resilient to inadvertent operation or misuse
Push Buttons: Use in high-frequency applications when binary control actions are frequent
Foot Buttons: Use only when the hands are occupied or for noncritical tasks (toes or ball of feet only)
Toggle Switches: Use when more than two or three settings are needed, with clear labeling of toggle values
Push–Pull Controls: Use for "on–off" functions, such as handbrakes
Levers and Pedals: Use when large forces are required
Rotary Switches: Use when 3–24 detented positions are to be selected
Thumbwheels: Compact cursor control devices for occasional use only
Knobs: For precise adjustment of continuous variables (e.g., dimmer switches)
Handwheels: Two-handed operation for continuous variables (e.g., valves)

TABLE 13.7
General Principles for Control–Display Integration

Principle	Description
Visibility	Controls should not obscure displays or sight lines
Importance	Most important items must be in the most advantageous position
Frequency of use	Most frequently used items must be in the most advantageous position
Function	Use grouping principles to group items by function
Location compatibility	Locate controls near their corresponding displays
Movement compatibility	Indicator movements should be compatible with control movements
Conceptual compatibility	Layout and use of controls should be consistent with user/population stereotypes
Sequence	Use link analysis to optimize layout in relation to the sequence of movements in real tasks
Effort balance	Share workload between dominant and nondominant hands

TUTORIAL TOPICS

1. Discuss the use of skeuomorphs to represent functions in interactive devices. Does it matter if the object represented by the skeuomorph bears no physical resemblance to the referent?
2. If BEVs are almost silent when running, could you use auditory icons to represent the state of the system and its main components? Can you think of some original examples?
3. Does the QWERTY keyboard layout have a future?
4. Take a look at the video at the link https://www.youtube.com/watch?v=KLOcj5qvOio

Do you think the metaverse is the future of education in HFE?

ESSAYS AND EXERCISES

1. Analyze the design of examples of the following domestic appliances. Identify and comment on the design of the controls, the displays, and the relationship between them.
   A four-burner stove
   A videocassette recorder
   A food mixer
   A microwave oven
   What feedback do the devices provide users to indicate response adequacy?

2. Describe the design options available for human–machine interfaces. Discuss the advantages and disadvantages using examples from everyday life.
3. Use the material in this chapter to write a checklist for evaluating the design of complex displays.
4. Find the logarithms to the base 2 of the following numbers:
   2
   8
   16
   1024
   36
5. The equation below was developed using Fitts' Law to describe the operation of the foot pedals (accelerator, brake, and clutch) of a large forklift truck. The model was developed with particular reference to the time taken for the right foot to move from the accelerator pedal to the brake (and back again) when driving the truck on uneven ground. The model was built by observing real drivers in a simulator in which the values of A (movement amplitude) and W (pedal width) were systematically manipulated. The truck speed rarely exceeded 20 km/h. In order to achieve the required stopping distance, the driver has to be able to react in 0.25 s. Find the value of I_d that will deliver an MT to the brake pedal of 0.25 s.

$$\text{MT} = 0.1 + 0.154\log_2(A/W)$$

6. Using your answer to question 5, find the ratio of pedal width to interpedal distance that will deliver an MT of 0.25 s.
7. The company decides to issue all forklift truck drivers new safety boots. Some tests are carried out in a simulator to determine whether the new boots affect driving performance, particularly emergency braking. Below are the mean MTs to brake for five drivers wearing the new boots. Use the Fitts' Law model in question 5 to determine whether the boots have affected performance. What, if anything, needs to be done?

I_d	MT (New Boots)
2.0	0.4
2.5	0.5
3.0	0.8
3.5	1.0
4.0	1.4

8. Carry out a link analysis at a facility of your choice and produce one or more link diagrams, distinguishing between communication links, manipulation links, and attentional links.
9. Design a control box for remote controlled driving of a motorized drilling rig for underground mining. The drills are mounted on columns on a four-wheeled, rectangular carriage. The hydraulic-powered carriage can move in either direction, and both axles can be steered independently. The vehicle stands between the operator and the rockface and can move left to right or right to left. It can curve in either direction or "crab" toward or away from the face. Aim for compatibility and consistency using the minimum number of controls.
10. Mr. Smith wants to go on a motoring holiday in the south of France. He lives in London and has only ever driven a right-hand drive vehicle in England, where they drive on the

left-hand side of the road. He cannot make up his mind whether to put his own vehicle on the ferry to France (where they drive on the right-hand side of the road in left-hand drive vehicles) or to fly to France and hire a left-hand drive vehicle when he gets there.

a. What are the ergonomic principles involved in choosing between these two options?
b. What would you advise Mr. Smith to do?

11. Mr. Smith's cousin, Mr. Jones, lives in the United States, where he drives a left-hand drive vehicle with an automatic gearbox. He wants to visit Mr. Smith and hire a vehicle while he is in England, but has doubts about his ability to drive a right-hand drive vehicle with a manual ("stick-shift") gearbox and clutch.
    a. What do you think the main psychological challenges would be if he did hire the vehicle?
    b. What additional hazards, if any, would occur as a result of these challenges?
    c. What background information would you need to conduct a risk analysis of the proposal?
    d. What would you advise Mr. Jones to do?

QUICK QUIZ

1. Is the relationship between sensation and stimulus intensity? (a) linear, (b), geometric, (c) logarithmic?
2. Which is better for displaying rates of change? (a) rotary dials, (b) digital displays, (c) neither
3. Displays that are close together seem to belong together. True or false?
4. "Clockwise for increase" always applies. True or false?
5. Is this a skeuomorph[4]?

6. What percentage of the population is colorblind?
7. What aspects of a panel layout most affect movement time between controls?
8. How much louder should an auditory warning be above background noise?
9. An ideal detection system has a high true positive rate and a low false positive rate. True or false?
10. You can use as many colors as you like when designing displays, as long as they look nice. True or false?

NOTES

1 Charles O. Hopkins commenting on the design of the control room at the Three Mile Island Nuclear power station.
2 Matthews (2002).
3 Kennedy et al. (1993).
4 Icon by Freepik.

14 Usability and User Experience

GENERAL REQUIREMENTS FOR HUMANS IN SYSTEMS

1. Interaction must be based on a clear and consistent model of the task.
2. Language must be designed for ease of comprehension.
3. System states must be visible to users.
4. The operation of automation must be understandable to users.
5. Modes must be clearly defined and explicit, with clear indications of mode changes as they occur.
6. Users must be able to influence the level of automation in any task.
7. Cybersecurity must be usable.

Where once we used to *access the Internet* or be *on the Internet*, we are now encircled by the Internet. We *live inside the Internet.* We *inhabit* the Internet. And as with all inhabitations, we are shaped by, and in turn shape, the Internet.

Elizabeth F. Churchill, personal communication

CORE KNOWLEDGE: USABILITY AND USER EXPERIENCE

As technology develops, the interfaces between users and task objects become increasingly sophisticated. Human-machine interaction often takes place via interfaces with embedded intelligence or adaptive capability. Successful interaction requires usability and implies *complementarity* of function.

Usability in a World of Intelligent Devices

ISO 9241-11 (Ergonomics of human–system interaction—Part 11: Usability: Definitions and concepts) defines usability as:

The extent to which a system, product, or service can be used by specified users to achieve speci fied goals with effectiveness, efficiency, and satisfaction in a specified context of use.

Usability is not a characteristic of a device as much as a characteristic of the use of a device to achieve a goal. The physical and mental effort needed to use the device is traded off against the benefits of achieving the goal. When the physical costs—operating the controls, reading the displays, and the mental costs—the load on System II, are less than the benefits of attaining the goal, the device is usable.

To master any device, users have to understand:

- The characteristics of the device—what it is
- The tasks to be performed with the device—what it does
- How to perform the tasks using the device—how to use it to achieve task goals
- Whether and how it links with the other devices and with the Internet

In order to speed learning, designers aim for positive transfer from everyday life using quasi-analog models of the real world:

DOI: 10.1201/9781032717210-14

- Making a new task similar to the one that is already known
- Striving for conceptual and operational consistency with known tasks

In practice, designers need a good understanding of the task requirements and of the knowledge, expectations, and needs of users and *what they already know:*

- *Syntactic Knowledge:* Users' understanding of device-dependent details (e.g., understanding the operation of function keys).
- *Semantic Knowledge:* Knowledge of task objects (the things you need to carry out a task) and how they are represented on the computer (e.g., caches, files, and folders).
- *Procedural Knowledge:* What you have to do with the objects to carry out a task (e.g., create a file, enter data, edit, and save to an appropriate location).

According to Weichbroth (2019), the *attributes* of usability (which designers should strive for in design) are:

- Efficiency (time taken to complete a task or achieve a goal)
- Satisfaction (user's level of satisfaction and fulfilment of needs and expectations)
- Ability to complete a task in the context of use of a device
- Learnability (usability at first use and effort needed to gain proficiency)
- Memorability (retention of proficiency regardless of use once gained)
- Mental effort (demands on working memory throughout the task action sequence)
- Errors (number and type of errors and ease of error recovery)
- Simplicity (number of steps in the action sequence, number of button presses needed, and so on)
- Ease of use (overall level of effort in relation to the utility of the outcome)
- Navigation (accessing buttons or navigating modes to complete the task—"finding one's way around the interface")

The *key performance indicators* used to assess the usability of a system are:

- Success rate when the task is attempted
- The time taken to perform a task
- The frequency of errors
- User satisfaction

Usability and Cybersecurity

Is there a trade-off between the level of cybersecurity in a system and its usability for critical and generic tasks? Generating, remembering, and entering passwords can be seen as a secondary task that can interfere with the performance of the main task, either concurrently or on a time-sharing basis (e.g., security issues may interrupt performance of the main task). When the requirements for usability conflict with the requirements for security, cybersecurity issues may arise. Cybersecurity and HFE are discussed at the end of this chapter.

Mental Models

A mental model is the representation of a system in which the essential features of the system and their interrelationships are stored in memory. Users build up models through experience and refine them with practice. When you imagine yourself reversing your car into a parking space, you are "running" your mental model of the task. Conversely, system designers begin with a complete model of the system at the earliest stages. If a new system is built around an existing model that users already understand, positive transfer of learning will occur; whereas, if a task has to be learned by rote, without any possibility of developing a mental model, the user will be unable to cope with novel or unprogrammed events. Users with fully developed models of the system and the operational context will be able to "run" the model to predict the outcome of alternative actions.

When the mapping between the mental model and the real system is poor or incomplete, user difficulties and errors may result.

Mental Models of the World Wide Web

Whereas the Internet is the global system of computer networks connected by a common protocol (a "network of networks"), the World Wide Web is an information space consisting of documents and other resources connected by hyperlinks. The World Wide Web can be accessed via the Internet. Web pages are documents formatted and annotated using hypertext mark-up language (HTML). Like documents on a word processor, they can contain images, video, and other media. The hyperlinks enable users to navigate between web pages. Each web page has a Uniform Resource Locator (URL), informally termed a web address. The URL can be entered to locate the desired page, if the URL is known. More often, users employ search engines to locate web pages.

A search engine is basically an information retrieval system designed to locate web pages on the World Wide Web.

One way of organizing a search for information based on a query is by means of Boolean operators (Figure 14.1).

The words "AND," "OR," and "NOT" can be used to focus a search and retrieve more targeted information. In modern web search engines, "AND" (or "+") is usually unnecessary because it is the default operator when more than one term is used—for example, if you enter "mental model.pdf," the search engine only returns PDF files about mental models.

Utility of Mental Models

A knowledge of users' models of the WWW should help us explain their actions and understand why users behave as they do. A simplistic model of web searching might be nothing more than the rule, "you enter some words into the box and the computer finds all the pages that have those words." This model sees the web as an online library and the search engine as an online librarian. This model lacks any understanding that search engines are databases, differ in the information they contain, and that the search engine delivers a ranked list with the most salient pages at the top of the list. A more complex model might conceive of the web as a multidimensional database that can be interrogated using a search engine. The closer the mapping between the keywords used to search and the most salient features of the targets, the more likely the search will be successful (Table 14.1).

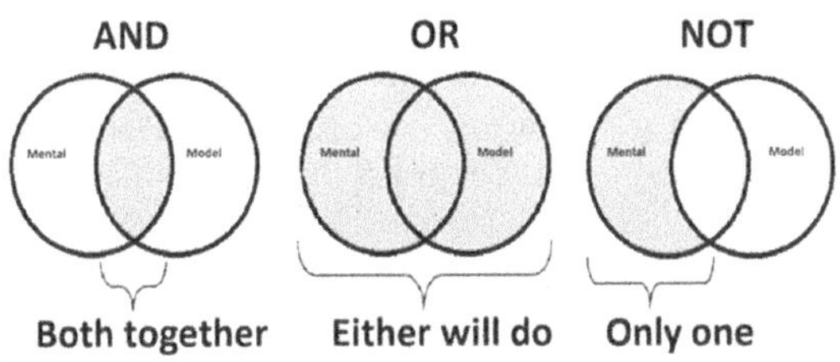

FIGURE 14.1 Boolean operators.

TABLE 14.1
Common Web Search Tactics

Beginning moves	Enter Term(s) for the Next Search
New concept	
Reduce the size of the search set	
Add concept	Add a new term to the query
Combine AND	Use the AND operator to combine terms
Narrow search space	Replace a term with a more specific term (e.g., replace "animals" with "mammals")
Increase the size of the search set	
Delete concept	Delete a term from the previous query
Broaden term	Replace a specific term with a more general term (e.g., replace "Bank" with "Savings Bank")
Combine with OR	Combine two concepts using the "OR" operator
Other	
Replace term	Replace with a synonym

Modes

Modes are nothing more than particular states of systems, but states in which the system behaves differently in response to user actions. Modes can boost the efficiency of a device. Cars have a "disengage" mode when the clutch is depressed that automatically reverts to "engaged" mode when the clutch is released. Driving a car with a manual transmission requires operation through a fairly strict sequence of modes—the "disengage" (clutch depressed) mode alternating between each drive mode (gear). Traditional telephones also have modes. You pick up the receiver, and the dial tone automatically prompts you to dial a number. Note that the act of picking up the phone initiates the transition from "off" mode to "dial" mode—the transition is embedded in the task action sequence. Cell phones are different, and the transition to "dial" may have to be made explicitly. These examples may seem trivial until it is remembered that the usability of many everyday items depends on users' knowledge and expectations about them.

Modes can enhance interaction if, at the semantic level, users understand the modes; at the syntactic level, it is clear how to alternate between modes; and at the physical device level, the mode states are explicit.

Automated Systems and Mental Models

Bainbridge (1982) describes several *ironies of automation*:

1. The automated system is supposed to perform the task better than the operator. Ironically, the operator is supposed to determine whether the machine is performing optimally and to take over when the machine fails.
2. If the operator only assumes control when the machine breaks down or in emergencies, they will be less skilled than if they always controlled the system. In automated systems, then, operators are required to assume control at the most difficult times and with suboptimal control skills.
3. Second-generation operators (who have never controlled the system) will have no skills (subroutines), only knowledge.
4. Operators of poorly designed automated systems will have better control and recovery skills than those of well-designed systems.

TABLE 14.2
Levels of Knowledge in Language Comprehension

1. The sound/spelling of words
2. The normal meanings of words
3. The grammatical rules for acceptable utterances
4. Metaphorical/idiomatic usages
5. Contextual knowledge
6. General (world) knowledge

One of the drawbacks of automation is that the manual control skills used before automation (System 1) are no longer utilized, or the loss of skill adds to the load on System 2 when the operator has to take over. Another drawback is that with increasingly complex automation, there is a lack of transparency—the operator may not understand the behavior of the automated system, particularly when the automated system has several modes. These considerations are increasingly important in the design of new systems such as self-driving automobiles and in human-robot interaction, where the robot may have several modes of operation differing in the degree of autonomy. Quenehen et al. (2023) define four modes of human-robot operation for industrial assembly tasks depending on the degree of human-robot separation.

Designing Language

Language comprehension is a cognitive process involving both top-down and bottom-up processes, requiring several layers of knowledge (Table 14.2).

This applies to all communication between speakers and between users and computers. Human–computer dialogue is usually limited because most systems can communicate only at the lowest levels.

Sound and Spelling of Words

Speech is thought to be recognized by an analysis of the constituent features and by synthetic processes that use contextual and other information. *Prosodic* cues (emphasis and rhythm) aid speech comprehension but are usually lacking in synthetic speech. In an experiment by Hardee et al. (2007), older adults were found to have more difficulty understanding and recalling words produced by synthetic speech than younger adults. The intelligibility rate for synthetic speech was 71% for younger adults and 51% for older adults. Synthetic speech can be improved through better design of its prosodic features. One of the main differences between synthetic speech and natural speech is that speech synthesizers don't breathe. Székely et al. (2020) found that the intelligibility of synthetic speech was improved by the incorporation of synthetic breathing patterns to modulate the flow of synthetic speech.

Normal Meanings of Words

The meaning of a word is understood when the stimulus is matched with its stored representation in the mental dictionary and the stored version is linked to a node in the semantic network.

Difficulties can arise when users lack knowledge of system objects or the vocabulary used to describe the objects. In the first case, the user's knowledge of the word is "empty"—not linked to a node in the semantic network. The user may simply learn to respond in an appropriate way when the word is presented without ever developing anything more than a superficial understanding of the system. Sometimes, this may be sufficient—it is not usually necessary to understand how an electric motor works or what the field strength of the magnet is in order to use a vacuum cleaner or a food mixer. However, if the task at hand *does* require knowledge of the concepts, a user who only has a superficial understanding of the words will experience problems; for example, in knowing what to do when presented with a message such as "mouse driver not installed."

> Designers must understand how users conceptualise a system and be familiar with the vocabulary they use to describe it.

Alternatively, the user may understand the appropriate concepts, but the system may use unfamiliar words to describe them, and the user may have to infer the meaning of the words from their knowledge of first principles and the current system behavior. Novices are unlikely to possess either the appropriate concepts or to know the words that refer to them. It is for this reason that they need to be shielded from technical feedback.

Grammatical Rules for Acceptable Utterances

Grammar is defined as a set of rules for generating all grammatical sentences in a language and no ungrammatical ones (Chomsky, 1957). Speakers of English, for example, can tell which of these sentences follow the rules of English grammar:

1. Press the red button in an emergency.
2. In an emergency, the red button press.

Syntax refers to the rules for combining words in an order that will be recognized as "grammatical" by a native speaker of a language. In the example above, it is the syntax of the two sentences that is different. Note that we are not necessarily talking about the meaning of these sentences. In an emergency, the second version would be understood, despite its incorrect syntax.

In order to understand a sentence, the basic syntactic relationships between the words have to be determined. The actual sequence of words (called the surface structure of a sentence) has to be analyzed in order to determine these basic syntactic relationships. Any grammatical sentence can be said to have a deep structure in which the syntactic relationships are represented, and it is from the deep structure that the semantic analysis of a sentence takes place.

Syntactic analysis to describe the sequence of nouns and verbs is known as "parsing." Further processing is needed to identify the basic syntactic relationships between the words (e.g., subject, verb, and object). Semantic processing of the contents of the sentence then takes place. When designing verbal material or analyzing comprehension problems, it is important to distinguish between the syntactic and semantic aspects of language comprehension. In HFE, language comprehension can be regarded as a mental task, which, like any other mental task, has a workload associated with it, and we should attempt to design verbal material such as warnings, recovery, and emergency procedures that can be comprehended with a minimal amount of mental effort.

Metaphorical/Idiomatic Usages

One of the remarkable aspects of the use of language by humans is the tendency to find new ways of using words to convey ideas. Everyday speech is also characterized by this type of word usage, as in "John climbed rapidly up the ladder, grasping every opportunity with both hands." In order to comprehend utterances in a language, the listener often has to decide between a literal or metaphorical approach to interpretation.

Contextual Knowledge

Information about context can be an important determinant of comprehension. Specific knowledge about people and events is often essential to disambiguate the usage of particular words. For example, Bransford and Johnson's (1973) sentence, "The haystack was important because the cloth ripped," could only be understood when accompanied by the heading "The Parachutist." Similarly, the use of "top-down" processes in word comprehension is exemplified in the modern practice of text messaging ("txtng") using cell phones. Redundancy in spelling is minimized to shorten word length, and knowledge of the context of the message is used by the reader to decode received text messages. For example, "I 1ly hv i's 4 u" is better understood if sent on Valentine's Day. Hardee and Mayhorn (2007) found that synthetic speech was better understood in stories than in isolated words, particularly by older adults.

> Document designers should use headings and subheadings to provide local context for readers of extended documents.

General (World) Knowledge

The everyday use of a language by native speakers is underpinned by assumptions about shared beliefs and experiences. Speakers expect listeners to share similar beliefs and to keep them in mind during a conversation (we all expect others to use their "common sense" when listening to us). Of particular importance to the construction of user-friendly language are the assumptions made by the writer about what constitutes common sense on the part of readers. If these assumptions are invalid, comprehension problems and misunderstandings may result.

BASIC APPLICATIONS: SUPPORTING SYSTEM 2 IN EVERYDAY LIFE

Design of Visible Language

Easily understood English sentences have four characteristics:

1. They contain few clauses.
2. They utilize the active rather than the passive voice.
3. They are affirmative rather than negative.
4. They are concrete, rather than abstract.

Complex phrases are often found in official and technical documents; for example,

> Pull out the left-hand ignition element retaining screw pin.

It is not readily apparent what is being referred to, the left-hand ignition element or the retaining screw on the left of the element. A more easily understood version is, "pull out the left-hand pin on the screw that retains the ignition element" (Wright, 1978).

Table 14.3 summarizes checkpoints for the design of user-friendly language.

Few Clauses

A complex sentence such as "The man who met the clerk that made the loan that saved the old firm is in prison" (Johnson-Laird, 1988) may place excessive STM load on the reader. The danger is that listeners will forget what the sentence is about before it finishes.

Problems arise when clauses are embedded within each other in a sentence, as in, "The man that the girl saw stood up." The phrase "that the girl saw" merely specifies something about the man who stood up, but because of its position in the sentence, nothing about the man can be understood until the end of the sentence. The alternative version, "The girl saw the man. The man stood up," involves less memory load and communicates information about its subject more directly. People do seem to be able to cope when two ideas are brought together in a single sentence, despite this, although performance breaks down rapidly with increasing complexity. Wright gives an even more complicated version: "The man that the girl that the dog chased saw stood up."

Active versus Passive Voice

Passive sentences rarely occur in everyday spoken or written English and lack idiomatic force, as in "The notes should be read" as opposed to "Read the notes."

TABLE 14.3
Some Checkpoints for Evaluating the Design of User-Friendly Language

Level of Analysis	Checkpoints
Sound/spelling/appearance of words	Signal/noise ratio of the spoken message and background noise
	Digitized speech has better quality than synthesized
	Typeface of letters (e.g., uppercase or lowercase)
	Print size (minimum 8 point, preferably 10 point)
	Color contrast of letters and background
	Luminance of active displays
	Illuminance in the work environment
	Visual angle
Normal meanings of words	Frequency of usage in everyday English
	Frequency in a particular industry
	Avoid technical jargon in a nontechnical context (e.g., substitute "paid work" for "gainful employment")
Grammatical rules for clauses, long sentences	Complexity of syntax (embedded utterances) Number of transformations to convert the sentence into subject/verb object (SVO) format
	Avoid overly officious style (e.g., unnecessary use of complex grammar and words such as hereby, heretofore)
Metaphorical/idiomatic usages, Contextual knowledge	Replace "purge files" with "delete files," for example
	Is the context made clear in opening remarks, instructions, or screen design (or the change in context with each new screen in a menu-driven system)
	Does the text have sufficient headings
General knowledge	Has the language designer made unwarranted assumptions about the general knowledge of the reader? (e.g., use of words such as "debentures" assumes some knowledge of investments, business, and the workings of the financial services sector)
	Can general knowledge intrude in an inappropriate way?

Passive constructions have a special place in language—when the speaker wishes to avoid mentioning, or does not know the agent of a verb, and are commonly used by politicians and scientists. For example, "taxes were increased" as opposed to "the government increased the taxes." The passive voice usually places a greater comprehension load on the reader but can sometimes be useful in maintaining continuity in a body of text. Edworthy et al. (2004) found that compliance with instructions on warning labels was improved if the personal pronoun was added, as in

"*You* should read the notes."

Negative versus Affirmative Sentences

Sentences with negative elements are more difficult to understand than affirmative sentences, as in "Is the book unavailable?" To answer this question, many people have to convert it into a positive question ("Is the book available?"), answer it, and then reconvert the answer. The negative may be missed altogether through hurried reading or noisy speech (Wickens, 1992).

Double negatives can be even more problematic, for example, "Do not attempt to adjust this equipment if a technician is not present." The alternative, "Only adjust this equipment in the presence of a technician," is more easily understood. The value of negative statements is to correct incorrect assumptions held by readers or to issue commands, such as

Do NOT prescribe to children under 5 years of age.

To guard against the reader mistaking the negative sentence for a positive one, the word "not" can be highlighted, printed in bold or uppercase, or in a different color. Fillenbaum's "perverse" sentence illustrates the challenge to comprehension posed by double negatives combined in a conditional sentence:

> Don't print that or I won't sue you.

which the majority of people do not understand.

Concrete versus Abstract Words

People seem to cope better with complex syntax when the topic is concrete or when concrete words are used. Unnecessarily abstract words should be avoided in the design of forms or instructions, particularly if complex syntax has to be used. A more concrete, "imaginable," and perhaps less perverse version of the previous example serves to illustrate this point:

> Don't wear that, or I won't take you to dinner.

Instructions and Warnings

Wogalter et al. (2021) offer the following general guidance for when to warn:

1. A significant hazard exists.
2. The hazard, its consequences, and appropriate safe modes of behavior are not known by the people exposed to the hazard.
3. The hazards are not open and obvious; that is, the appearance and function of the environment or product do not convey them.
4. A reminder is needed to assure awareness of the hazard at the proper time. This concern is especially important in situations of high task loading or potential distractions.

An effective warning must be delivered via an appropriate communication channel (e.g., text on reflective displays may not be effective in poor lighting). Warnings should provoke an "attention switch" due to their salience, conspicuity, or intensity. Multimodal warnings can be very effective (e.g., haptic feedback and low-frequency noise from a rumble strip with white ridges will alert a sleepy driver more effectively than visual cues alone). Warnings should be designed with an understanding of the competence of the receiver (education, literacy level, and language proficiency, for example) and should maintain attention once captured. Brightly colored visual warnings or loud, clear verbal messages are examples, as are dynamic images in pictographs. Warnings should also motivate the receiver to respond appropriately (e.g., by washing their hands or donning a face mask). In general, warnings are more likely to be followed when the cost of compliance is low in relation to the perceived hazard.

Research has shown that it is easier to search memory for the attributes of a particular object (e.g., to decide whether a particular animal has stripes) than to talk about an attribute and instantiate objects that possess that attribute (name stripy animals). As a general rule, then, objects should be followed by attributes, but this does depend on the context for which the instructions or documents are being prepared. When writing a tutorial in a user manual, the object–attribute approach might be best—telling users what the various controls or commands enable them to do. However, when preparing documentation to be used in fault-finding or recovery from abnormal operating conditions, it may be better to emphasize system functions rather than physical characteristics. Documents suitable for an operator (who is concerned with the system functions and attributes rather than technical details about its components) may not be suitable for the maintenance or design engineer.

People are inclined to carry out tasks in the order in which they read about them. For example, instructions such as "Before completing this section, read note 3" would be more compatible if rephrased as "Read note 3 before completing this section." Edworthy et al. (2004) found that compliance with warnings was improved when the warnings were placed in the "Directions for Use" section of a label rather than in a separate precautions section (presumably because users are more likely to read the former than the latter).

According to ANSI (2006) Z535, "DANGER" (white print on a red background) should be used when death or serious injury will occur if the receiver does not comply. "WARNING" (black print on a yellow background) should be used when serious injury might occur in the event of non-compliance. "CAUTION" (black print on an orange background) should be used when less serious injury or damage to property might occur.

Composite Signs Work Better

Wogalter et al. (1997) compared elevator users' compliance with four signs, three of which were from the classic article by Chapanis (1965). The signs were intended to improve the efficiency of elevator usage by exhorting people to use the stairs for short trips (one floor up or two floors down). The signs are shown in Figure 14.2.

Subjects rated the signs according to how understandable they were and their willingness to comply. Sign A was rated lowest on both criteria, and Sign D was rated highest. A field study confirmed that Sign D was the most effective of all the signs—the proportion of people complying was 0.91 when Sgn D was placed outside all the elevators in a building, compared to 0.67–0.77 for the other signs. Therefore, when badly designed signs are used, up to one-third of users may not comply, compared to one in ten when a better design is used.

Design of Alphanumeric Codes

Codes are strings of letters, numbers, or both that are used to represent system elements. The main considerations in code design are (Bailey, 1982):

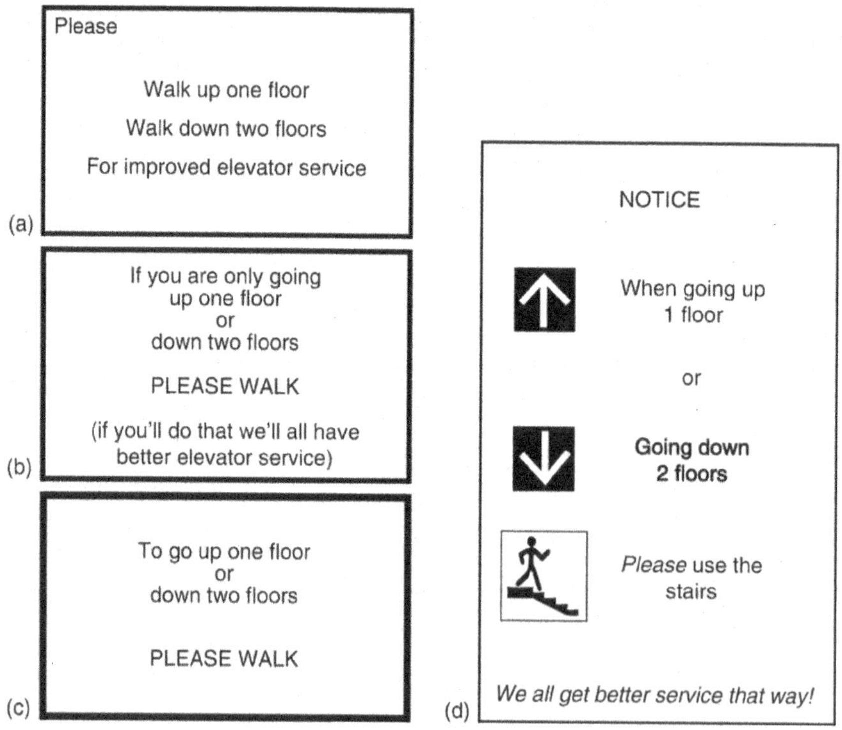

FIGURE 14.2 Composite sign evaluated by Wogalter et al. (1997).

1. Length
2. Sequence
3. Content
4. Chunking
5. Errors

Codes should not exceed the capacity of STM (about seven items). Bailey (1982) pointed out that every person in the world could be given a unique identifier using only a 10-character code, but the then New Jersey driver's license number was 15 characters long—even though there were only four million drivers. Codes such as these are handwritten and read in many situations and invite unnecessary errors.

The middle items of a code have the highest probability of incorrect recall (due to the primacy and recency effects). In shorter codes, this can lead to the penultimate items being recalled worst. It may be advantageous to sequence the letters and numbers together to increase the "chunkability" of the code by increasing the discriminability of its component parts (e.g., GA2 as opposed to G2A).

Abbreviations

Some rules for the construction of abbreviations are given in Table 14.4.

Reversal Errors

Reversal errors are common in data entry tasks and can be minimized by avoiding the use of character combinations that utilize adjacent keys. For small domains, reversed forms of legal codes should not be recognized.

Code Content

People can remember mixed-category lists better than single-category lists, favoring mixtures of letters and numbers. Special symbols (!, $, #) should be avoided in most keyboard work because they require the use of the "shift" key, increasing the likelihood of psychomotor errors.

Codes may be designed to increase meaningfulness by capitalizing on already learned associations (e.g., QU vs. QV) and by designing "rehearsable" codes ("BORT" vs. "TBRO"). My student number when I was a student was "BRDROB003"—a meaningful alphanumeric mix that naturally decays into three easily retrievable chunks.

Coding in Errors by Design

Visual errors are more likely if physically similar characters are included in the character set used to generate codes (e.g., F,E, C,G, B,3, O,0, and 1,I). The quality of the various media used to display codes can be evaluated, with particular attention paid to the appearance of easily confusable characters. If handwriting as well as print is involved, the following characters are likely to be confused:

TABLE 14.4
Guidelines for the Design of Abbreviations

1. Determine the required number of letters in the code
2. Remove suffixes (e.g., ed, er, ing)
3. Select the first letter and the last consonant (e.g., Boulevard B…D)
4. Fill the remaining spaces with consonants (e.g., BLVRD, BLVD)
5. Avoid double consonants
6. Use vowels only if there are insufficient consonants

Source: Bailey (1982).

Z—2
U—V,W
Y—7,4
s—5,8
G—9,6
T—J,7
Q—O,2
D—0

Confusions can be minimized by removing 0, I, Z, Y, S, and G from alphanumeric character sets and U, Y, Q, T, and J from alphabetic character sets.

Text written in uppercase letters is usually less legible than lowercase text (for most commercial typefaces, at least) because the ascenders and descenders in lowercase provide additional visual cues for letter discrimination, which may be particularly beneficial when viewing conditions are poor.

Auditory confusions usually occur when codes have to be held in STM. Items that look very different but sound the same can be confused (e.g., e,D, c,B). STM storage is usually in the acoustic or articulatory modalities. If codes have to be held in STM, while people speak over the phone, for example, similar-sounding letters and numbers should be excluded from the character set.

Communicating with Codes

The phonetic alphabet is often used to communicate codes verbally in order to minimize confusion. It is particularly useful for communicating codes over the telephone or when the ambient noise is high (Table 14.5).

Synthetic Speech Messages

Hardee et al. (2007) offer the following guidance for the use of synthetic speech:

- Messages should be presented within context for all listeners and should not be presented in isolation.
- When ideas must be presented in isolation (in the absence of contextual cues), designers should consider the limitations of working memory and hearing in older adults. The rate of speech should be carefully monitored, together with the number of words and their ease of comprehension, so as not to overload the cognitive abilities of the listener.

TABLE 14.5
Standard Phonetic Alphabet for Verbal Communication of Codes

A	Alpha	J	Juliet	S	Sierra
B	Bravo	K	Kilo	T	Tango
C	Charlie	L	Lima	U	Uniform
D	Delta	M	Mike	V	Victor
E	Echo	N	November	W	Whiskey
F	Foxtrot	O	Oscar	X	X-ray
G	Golf	P	Papa	Y	Yankee
H	Hotel	Q	Quebec	Z	Zulu
I	India	R	Romeo		

- Messages should be preceded by warnings or cues to redirect the listener's attention and reduce mental demands.
- Consider using prosodic cues—the subtle intonations and cues present in natural speech but not in synthetic speech.

Waiting in the Web: System Response Time

Clearly, what is tolerable and what is acceptable may well depend on the context of use and the alternatives available to the user for obtaining the desired information. However, some general recommendations for web use can be made. Fui-Hoon-Na (2004) found that most users are willing to wait 2 s for simple information retrieval. This is consistent with the literature—delays of longer than 2 s are intrusive and can divert attention from a task. For delays between 2 and 4 s, users will modify their intentions, perhaps abandoning the task to achieve alternative goals. The upper bound appears to be about 15 s for systems that do not provide feedback. Feedback increases tolerable waiting time by increasing the user's sense of control. Delays of 0.1 s or less are hardly perceived at all, and there is little to be gained by reducing response time further.

Implementation Modes for Human–Computer Interaction

Shneiderman (1988, 1991) identified five main interaction styles for dialog design (Table 14.6).

Menu Selection

Makes the options explicit and reduces the load on STM. The main ergonomic issues in menu design are:

TABLE 14.6
Advantages and Disadvantages of Different Interaction Styles: HCI Styles

Advantages	Disadvantages
Menu Selection	
Little training required	Rapid display rate needed
Reduces keystrokes	Danger of too many menus
Structures decision making	Skilled users may be slowed down
System can "drive" naïve users	
Can be used with minimal keyboard skills	
Form Fill-In	
Simplifies data entry	Consumes screen space
Little training needed	Typing skills needed
Shows context for the activity	
Command Language	
Flexible. User "drives" the system	Substantial training needed
Appeals to sophisticated users	Language has to be learned, retained, and recalled
Macro commands increase power	Invites errors due to forgetting
Natural Language	
No new computer syntax needs to be learned	Natural language can be ambiguous
Clarification may be needed	Typing skills needed
Direct Manipulation	
What you see is what you get	Additional pointing devices needed
Encourages the use of an explicit "model" of the task	Visually impaired users may have problems
Good for disabled users who can point but cannot type	

Source: Adapted from Shneiderman (1988) with permission.

- Mapping the task domain onto the menu structure
- Compatibility of the menu structure with the user's conception of the task
- Compatibility with the way the task is performed
- Navigation through the menu without "getting lost"
- Backtracking
- User's memory load

Highly structured tasks lend themselves to the construction of unambiguous menu trees with simple choices at each level. Care should be taken to ensure that the structure of the menu is compatible with the users' conception of the domain (e.g., should the *Financial Times* newspaper be accessed via "Business" or "Entertainment"?). It should also be compatible with the task itself (e.g., users should not need to navigate four layers of a hierarchical menu just to turn off the computer, although some systems do come close to this with continual reminders and warnings before shutting down). Thus, menu designers need:

- An understanding of the task domain and the task semantics
- An understanding of the users' conception of the domain (hence the need to consult users)
- To carry out a task analysis in order to arrive at a description of the task

According to Shneiderman, shallow menu structures are to be preferred because few levels have to be traversed to reach the goal. One disadvantage is that there may be too many options per level, and users may have forgotten some of the choices by the time they finish reading the list. Deep menu structures have the advantage that the user has fewer choices each time, but the user may get lost in the system, and backtracking to previous states may be time-consuming. For complex domains, the compromise position seems to be to have more depth at the root (beginning) of the menu and fewer intermediate choices. Ideally, users should quickly develop a mental model of the menu, particularly if it is compatible with their conception of the domain. Problems occur with domains in which categories are not mutually exclusive, and it may then be necessary to provide "jump ahead" facilities.

The final choice depends on the ease with which the task domain can be represented hierarchically.

STM limitations are also potentially problematic in auditory menus used for telephone access to databases because

1. The user cannot control the rate of information presentation.
2. The user does not know the number of items that will be presented.
3. The presentation mode is sequential, and the user cannot scan the list of items.
4. There are no visual cues to indicate the menu structure.

Auditory menu design for telephones is discussed in ISO 13714:1995. Sample guidelines include:

- System prompts should be short enough not to frustrate expert users, yet clear enough for novices.
- No more than four choices should be given in a single prompt.
- The command sequence must be designed to minimize the steps needed to achieve the goal (menus three levels deep or less).
- Functions should be assigned the same key in different parts of the menu (e.g., "for account enquiries, press 1").
- Choices should be offered in their natural order or in order of frequency of occurrence (most popular options first).
- Choices should be made by single key presses wherever possible.
- "Time-outs" should be indicated using an auditory prompt and should give the user time to respond after presentation of the prompt.

- Phones with the keypad on the handset may not be appropriate, particularly for older users.
- Telephones should be fitted with volume control.
- System speech rates in excess of 170 words/min should be avoided.
- "Barge-in" facilities should be allowed during option presentation so that the user does not need to process the entire message.
- The system should be tested with a wide range of users, including older users.

If these guidelines are adhered to rigidly, then many applications will not be suitable for implementation in auditory form using telephones because the dialogue will be too complex and too time-consuming. Older users and people with certain disabilities, such as limited mobility, may be most challenged by such complexity, yet they are very likely more in need of telephone menu systems in order to access services (see Sharit et al., 2003).

Form Fill-In

Form completion requires keyboard skills and, therefore, may not be appropriate for certain types of users. One advantage is the increased density of items that can be displayed on a single screen compared with menus. Many of the design issues center around the basic requirement for screen legibility (similar to optimizing the layout of any visual display or "pencil-and-paper" form).

Command Languages

Command languages are less used nowadays due to the emergence of GUIs. A discussion of some of the design principles can be found in the first edition of this textbook.

Natural Language

It is easy to assume that natural language will ultimately be the best mode of interaction between a user and a computer, but this is debatable.

Le Bigot et al. (2004) compared spoken versus written natural language dialogs when information retrieval tasks were performed. They found that mental workload, assessed using the NASA TLX, was greater with spoken rather than written communication. Furthermore, the written interactions were more efficient, suggesting that there may be few benefits to being able to "talk to your computer," except where the hands are not free or the user has a manual disability. Le Bigot et al. (2007) compared different modes of interaction during a search task—phone (speaking and listening), Web (typing and reading), and mixed (speaking and reading or typing and listening). Speaking and listening had higher mental demands than typing commands and reading displayed text. Satisfaction was lower for speaking and listening, and there were more errors. Speaking to issue commands and reading textual prompts and feedback seems to be the most effective use of natural language interaction for such tasks.

Direct Manipulation

Much of the software now developed for personal computers and cell phones is based on direct manipulation platforms, with menu selection or form fill-in used to support specific tasks.

Rauterberg (1992) found that direct manipulation may be a more efficient mode of interaction than menu selection. Novice and expert subjects were given 10 database interrogation tasks to carry out on both types of interfaces. It was found that both groups needed less time to carry out the tasks in desktop-interface mode than in menu mode (regardless of differences in the number of keystrokes required to operate the two interfaces).

Communicating with Gestures Carfì and Mastrogiovanni (2021), following a review of 83 studies of gesture control, offer the following definition:

> Gestures are body actions that humans intentionally perform to affect the behaviour of an intelligent system.

In everyday life, we use gestures to interact with devices such as smartphones directly via the interface (e.g., swiping the screen) and to communicate with others (e.g., holding up one hand and mimicking writing on it with the other to attract the attention of a waiter when asking for the bill). A "vocabulary" of gestures can consist of hand movements, or upper limb movements, and these have to be detected by the system (e.g., using cameras, accelerometers or goniometers). Carfì and Mastrogiovanni represent the human body as a series of linkages (Chapters 2 and 3). The system detects angles between the links at the joints and maps the derived posture onto the gesture vocabulary. *Poses* are static postures, whereas gestures are dynamic. In either case, they are detected from measurements of one or more joint angles and the changes in these angles over a specified range and time period.

A pose, q, can be defined at a time point t as the sum of the joint angles in three dimensions between adjacent body links:

$$q(t) = (q_1 \ldots q_2 \ldots q_n) \tag{14.1}$$

Gestures are defined as trajectories between two poses (at the start and end) of the gesture, taking into account the angular velocity and acceleration around the joint. Thus, the system has to identify the body linkages, the angles between them, and the rates of change in those angles. The trajectory of these angles defines the gesture—Equation 14.1 with added terms for angular acceleration and velocity. Note that gestures can also be detected in virtual reality applications using an avatar.

Gestures can be subdivided as follows:

- *Referential Gestures:* In common with icons, the gesture is linked to a referent (as in the restaurant example above or placing one's palm against one's forehead to indicate exasperation).
- *Modalizing Gestures:* Used to complement a message (e.g., holding one's hand above one's head to indicate the stature of a tall colleague).
- *Mimetic Gestures:* Involve imitation (e.g., running one's finger and thumb across one's lips as when closing a zip to indicate silence).
- *Deictic Gestures:* Pointing at an object or area to direct attention to it.

Gestures can be conversational or used to control objects (e.g., pointing at a target object). There are several more fine-grained categorizations of gestures depending on the context of use. As with all other communication modalities, the usability of gesture control depends on the system reaction time—100 ms is probably the maximum.

According to Carfì and Mastrogiovanni, the main HFE issues are system responsiveness, user adaptability, feedback, learnability, accuracy, intuitiveness, and low-mental workload. Clearly, there is a trade-off between learnability, mental workload, and the size of the gesture dictionary.

Communicating with gestures shows great potential because it provides an extra channel with different attentional demands than speech, or in tasks where speech is not the ideal form of communication (e.g., when communicating with cobots or in noisy environments).

Immersive Technologies

Immersive technologies create a virtual environment or digitally extend the physical environment such that the user is "immersed" in the digital world created.

Virtual Reality (VR): According to Lanier (2017), virtual environments are "the substitution of the interface between a person and their physical environment with an interface to a simulated environment." Virtual environments isolate the user from the real world.

Augmented Reality (AR): Our experience of the real world is augmented by the addition of digitally created augmentations. Real-world data can be projected into our experience of the world using smartphones or other devices. Applications such as Google Maps have AR features to support route finding in real time. "Snapchat features" are another example of augmented reality.

Mixed Reality: This is a combination of the first two, involving the projection of AR material onto the real world—such as digital holograms of real objects.

Virtual environments are interactive models of systems implemented on computers and can be displayed in three dimensions via a head-mounted display (HMD). There need be no similarity to the real world. Applications include computer games, training skills in virtual simulators, and education. The intention is to create the experience of being "in" the environment, of being able to move around and explore it, and of being able to manipulate objects encountered in it. The main uses of VEs are as design tools, training simulators, and visualization aids. The degree of fidelity between the model and the real system may vary. The display may be on a computer screen or may be viewed through a stereoscopic head-mounted display (HMD) or wall-mounted screens. In the case of a "cave," the user is surrounded by screens. In many modern systems, the user can explore a three-dimensional representation of the real world or of a fictional reality. When wearing a head-mounted display, there is a one-to-one mapping between the user's head movements and the change in scene. Users' subjective reactions to systems can be positive, with the experience of "presence" or of being "immersed" in the system, or negative in the case of nausea, disorientation, visual discomfort, and postural problems.

The feeling of presence seems to depend on a number of variables, including system response time to user actions, reflected in the time taken to update the display, the optic flow as the user moves through the system, and the internal geometry of virtual objects and how well it changes with viewing distance and angle compared to the real world. According to Kalawsky (1999), the key usability factors for VEs are

- *Functionality:* The interface provides the level of control the user expects to complete a task.
- *Natural Inputs:* The user can interact with the system in a natural way.
- *Displays:* Information is understood, relevant to the task, and unambiguous.
- *Help:* Is available online.
- *Consistency:* System operation is consistent with the user's understanding and with convention.
- *Flexibility:* Users are not artificially constrained by system limitations.
- *Model Fidelity:* Interaction takes place in the context of a model of the system or task.
- *Error Correction:* Errors can be corrected/undone before a permanent change is made.
- *Sense of Immersion:* The user feels part of the system.
- *Usability:* Interaction should be intuitive (consistent with the user's model).

According to Hou et al. (2012), presence in screen-based VEs was greater when playing 'Tomb Raider' viewed via an 8-inch screen compared to a 12.7-inch screen. People with a greater ability to generate spatial imagery ("visualizers") tend to have greater sensations of presence in VEs.

As VE technology becomes more sophisticated, more powerful, and faster, new concepts have been developed to define constructs such as "presence" and "immersion" more precisely. VEs must offer what are known as "sensorimotor contingencies" appropriate to being present in the VE. The possibilities for action should be appropriate to the VE, the task, and the context of use (e.g., when simulating a surgical operation) and the outcome of the user's actions. Outcomes should be contingent on actions in a convincing way. Slater (2009) proposed the concept of "Place Illusion"—"the strong illusion of being in a place in spite of the sure knowledge that you are not there." Place illusion depends on valid sensorimotor contingencies. The "Plausibility Illusion" is determined by the extent to which the system can produce events that directly relate to the participant and the overall credibility of the scenario being depicted in comparison with user expectations. In other words, the place illusion depends on the fidelity of the simulation throughout the task action sequence, and the plausibility illusion depends on the affordances created by the simulation throughout the task action sequence. The feeling of "being there" therefore depends on the physical mapping between the task as carried out in the real world and the task as carried out in the VE. In support of these suggestions, Gibbs et al. (2022) found that presence in an immersive task (ball bouncing) was improved when visual feedback was supplemented by haptic feedback.

Presence, then, is the sensation of "being there" of being in the VE rather than the immediate physical environment. Weber et al. (2021) distinguish two components of presence. The first is a sense of "being there" and the second is perceived realism—which depends on fidelity, both temporal and spatial (for example, fidelity of movement; whether objects in the VE cast shadows in the way they would in the real world; whether perspective is realistic; and whether there are cues to the perception of a three-dimensional world, such as motion parallax). "Being there" implies the direction of attentional resources to the VE, away from the immediate physical surroundings. The important point to note is that these two components of presence are independent—it is quite possible to experience being in a virtual environment that is plausible even though it is not perceived as "real." However, according to Weber et al., perceived realism facilitates higher cognitive processing of the VE, such as "metathoughts" about the immersive experience.

Virtual Environments and "Gamification" Virtual environments are used for playing computer games; providing "online" or remote training, for meetings or to provide a "safe space" for training in high-hazard industries (where exposure to the real environment is dangerous). Gamification is the application of game elements and terminology to real tasks and is an effective means of education. Virtual simulations are the gamification of an operational environment to develop transferable skills. It is a form of VE training where "serious games" in VEs simulate dangerous situations (such as in military combat operations, flight training, and surgery). Trainees can work alone or in teams.

Cyber Sickness

Interaction with VEs may produce three kinds of unpleasant effects, collectively known as cyber sickness:

- Nausea and vomiting
- Oculomotor disturbances and headache
- Disorientation ataxia (postural stability problems)

Cyber sickness is similar to motion sickness and is associated with movement in the virtual world. It is thought to be due to discrepancies between the optic flow experienced in the VE system and the dynamics of the manual controls, particularly the haptic feedback. When entering the VE for the first time, users may have to learn to "recalibrate" their control actions in relation to the visual feedback they receive. When leaving the VE, there may be a delay in readaptation to the real world. Conflict can occur when the scene changes to indicate movement of the user in the VE, but the user's head remains still (in the real world). Thus, the expected vestibular feedback does not match the visual feedback. For users wearing HMDs, the VE can be explored by moving the head. Sensors in the HMD detect this movement, and the scene is changed accordingly. However, update lags mean that the scene remains still as the head begins to move, and continues to change when the head stops moving. *Vection* is another cyber sickness symptom that seems to be due to exposure to visual cues in the VE, in the absence of any real movement. Vection is the illusion of movement in the opposite direction. It seems to be similar to the experience one sometimes gets after looking out of the window of a moving train—when the train comes to a halt, there is the illusion of moving in the opposite direction. Vection itself can be nauseogenic. So et al. (2001) investigated nausea and vection in subjects navigating a VE (a city road network). Ratings of nausea and vection increased with the root mean square navigation speed. The range 3–10 m/s showed the greatest increase in the ratings. Above 10 m/s, the ratings leveled off up to 59 m/s. To avoid upsetting virtual tourists, it seems that either very slow or fairly fast virtual tour buses are needed! In general, the incidence of cyber sickness can be reduced by minimizing the sensory conflict.

Caserman et al. (2021) conducted a systematic review of 49 studies of cybersickness. Symptoms appear to be less frequent and less severe with more modern systems and usually abate within 15 min after the user leaves the virtual environment. Sensory conflict caused by mismatched stimuli (e.g., using a joystick to walk in the VE while remaining stationary) remains an important cause of cybersickness. On average, 7.11% of users left the VE when exposed to mismatched stimuli compared to 3.21% with matched stimuli (actually walking or "teleporting" from one location to another in the VE). In one study, participants had negligible symptoms of cybersickness when they could physically walk in the real world with a one-to-one mapping between their movements and the visual feedback on the HMD. When using a joystick to walk and remain stationary, symptoms increased dramatically. Latency, when there is a lag between head movements and visual feedback in the HMD, is another cause of cybersickness. Latency of more than 75 ms causes cybersickness, and a maximum delay of 20 ms has been recommended. Cybersickness can also occur when using driving simulators and appears to be more common when using HMDs to view the road than when viewing flat-screen displays. Overall, cybersickness can be reduced by reducing sensory mismatches and ensuring good mappings between head and body movements and the VE response. For example, a 120° movement of the head should be accompanied by a 120° shift in the scene displayed on the HMD.

VE Technological Limitations are HFE Issues

The usability and health aspects of VE have much to do with technological limitations. HMDs have a *refresh rate* of 60 Hz (the frequency at which the image on the screen is repeated). Higher refresh rates can improve image quality. The screen update rate is the frequency at which the screen image can change and depends on the capacity and processing speed of the system. The faster the update rate, the smoother and more realistic the representation of user interaction with the virtual world is. System lag is the delay between the user performing an action and the perception of it in the VE. The shorter the lag, the more likely the user is to experience "presence." System lag may induce sensory conflict and cyber sickness. For an enhanced sense of presence and greater comfort, VE systems require

- High resolution
- Fast refresh rates
- 60 or more fps (some systems now offer 90 or 120 fps)
- Minimum system lag

Autonomous Vehicles

Most automobiles have had some degree of automation for a long time. Cruise control is an obvious example and is selected by the driver depending on the driving conditions. Further development of the automation of specific functions and integration of these functions is continuing. For example, cruise control can be adapted to restrict the speed of vehicles to the legal limit for the road. Sensors of different kinds can be used to assist, or even carry out, maneuvers such as reverse parking.

One of the drivers of in-car automation is the desire to reduce accidents caused by driver error. However, as Leveson (2011) has pointed out, when a system is automated, the easiest of the task elements tend to be automated first, which can have the unexpected effect of making the difficult parts more difficult. By taking the driver "out of the control loop" and into a more supervisory role, the demands are then to monitor and understand an automated system while still being responsible for the safety of the vehicle. Recalling the discussion of attention in Chapter 12, several tasks can be carried out simultaneously if they do not compete for the same resources. Therefore, steering the vehicle and braking can be carried out by a skilled driver without competing with the requirement to monitor the road ahead or read traffic signs. The challenge, therefore, is to find an appropriate level of automation that maintains driver awareness of the vehicle and the road ahead. One way of doing this is to enable the driver to select the desired level of automation, which is clearly displayed.

Banks and Stanton (2016) assessed a prototype vehicle with driver-selected automatic longitudinal, lateral, and auto-overtake functions. Overall, drivers were less trusting of the automated system, possibly due to unfamiliarity with its capabilities. Improved automation should:

- Remove the automatic overtake capability. Drivers should make this decision, with the system overriding it if the overtake cannot be completed safely (e.g., a fast-moving vehicle approaches from behind in the intended lane).
- Remove the facility to change more than one lane at a time (to prevent unexpected lane changes).
- Mode confusion was more likely than in manual driving. Greater transparency is required in the interface.
- Better design of warnings to alert the driver when manual takeover was required.
- System initiation control to be placed within easy sight and reach (the control was placed under the indicator stalk on the steering wheel).

Richards and Stedmon (2016) argue that although autonomous vehicles—vehicles that can perceive and act on salient aspects of the environment to execute a goal—can be built, the challenge is to identify the appropriate level of automation, given that the driver is always responsible for safety. The Society of Automated Engineering defines six levels of automation (Table 14.7).

- Virtual boundary around an area where automated driving is supported

At present, developers are faced with a number of challenges. These include the development of fast and accurate environmental monitoring systems such that the automation has a clear model of the conditions in and around the vehicle and can respond accordingly. The use of artificial intelligence is one option to interpret information from the vehicle's light detecting and ranging technology (LIDAR) sensors. At present, these systems cannot detect and interpret many of the hazards vehicles are likely to encounter on the roads (such as pedestrian behaviors or the behavior of other road users, such as cyclists).

According to Cummings (2017), several challenges remain, including the ability of these systems to detect and respond appropriately to stationary vehicles. All of the sensors used by these vehicles have shortcomings and require processing after their feedback is received. Machine

TABLE 14.7
Levels of Automation for Vehicles

The driver monitors the environment	0	No automation. The driver is in control of all tasks
	1	Driver assist – e.g., cruise control or lane assist
	2	Partial automation of steering and speed. The driver can take control
The system monitors the environment	3	Conditional automation. The system monitors the environment. Driver override is required
	4	High automation. The system takes control in geofenced zones. The driver can override the system
	5	Fully automated, driver attention not required

learning systems may be able to interpret these signals correctly, but under some circumstances (e.g., bad weather), they may not because they rely on pattern recognition and probabilistic reasoning. At present, these systems are unable to understand aspects of pedestrian behavior such as gestures, the understanding of which is important in busy urban spaces. A major challenge is how to ensure that drivers understand the behavior of the automated systems in control of their vehicles—that there is a good mapping between the driver's model of the automation and the designer's model (i.e., how it works and why it behaves as it does).

Once an appropriate level of automation has been implemented, HFE is presented with several new and specific challenges:

- How to avoid mode errors by designing the human–machine interface to ensure that the operating mode is transparent with respect to the level of automation in operation.
- Implementing the automation such that its workings are transparent at all times (i.e., the driver understands what the vehicle is doing).
- Deciding on the criteria and method of enabling and alerting the driver to take over when needed (volitional takeover or takeover due to a fault or hazard).

Essentially, the challenge is to design systems such that the driver maintains a mental model of the driving task in the context of use, and that this model has close correspondence with what the automated system is doing and the external driving conditions. Attaining the appropriate level of trust in the automated vehicle will be essential; if drivers are too trustful, they may expect the automation in their own vehicle to deal with events beyond its programming. If drivers distrust the system, the automation may be underutilized, and the proposed safety benefits of automation in reducing "human error" will not be achieved. If drivers trust the automation in their own vehicle but not the automation in vehicles around them, how will this affect their behavior and expectations? How will they know whether vehicles around them are under automatic or manual control?

TOOLS AND PROCESSES

ISO 13407 defines usability as "the extent to which a product can be used by specified users to achieve specified goals with effectiveness, efficiency, and satisfaction in a specified context of use."

User Consultation

Early involvement of users is recommended. HFE Workshop 14.1 provides guidance for running focus groups to gain a user-centered understanding of the design issues. User-centered design is based on the following principles (according to ISO 13407):

Active Involvement of Users and a Clear Understanding of the User and Task Requirements

Shneiderman (1991) lists the following as advantages and disadvantages of user involvement:

- More accurate information about tasks
- Opportunity to discuss design decisions early on
- A sense of participation fosters involvement and acceptance

Possible risks include

- Longer implementation period
- Antagonism of users who are not involved (not everyone can be)

- Antagonism of people whose ideas are rejected
- Designers forced to compromise design to satisfy incompetent participants
- Build opposition to implementation
- Possible rejection of novel ideas because users cannot appreciate the potential of the technology

HFE WORKSHOP 14.1

GUIDANCE FOR RUNNING FOCUS GROUPS

Focus groups are used as exploratory tools to obtain information about users' attitudes and opinions on a topic. Many of the requirements for a good focus group survey are also the requirements for a good questionnaire survey; therefore, readers should also refer to the appropriate sections of Chapter 16 on questionnaire and survey design.

COMPOSITION OF FOCUS GROUPS

A single group should consist of a representative sample of users, with the proviso that large discrepancies in rank or status be avoided. Lower-ranked personnel may be inhibited by the presence of their line managers. More senior personnel may feel the need to direct the interactions and dominate the group. If the users span a large organization, several focus groups may be needed, differentiated by rank. Some individuals may feel intimidated in a large group (6–10 individuals per group seems to work best).

DEFINE THE SCOPE AND CONTEXT OF THE FOCUS GROUP SESSION

It may be useful to user model or theory to determine the scope of the session (e.g., the job stress theory of Karasek in Chapter 8 or the computer hassles or usability items in this chapter). Focus group leaders should have a detailed list of items and prompts to present to the group at the beginning.

GETTING STARTED

As with questionnaire design for attitude surveys, start with open-ended questions of a general nature (e.g., What do you think about the Internet?). Start with neutral questions and avoid using cues or prompts in the initial stages, as these may artificially restrict the range of responses.

OPEN-ENDED PROMPTING AND PROBING

Use open-ended prompts to explore issues that are raised early on (e.g., and how does this apply to the kinds of software programs you work with on a daily basis?). Try to ask positive or neutral questions in the early stages to prevent the session from becoming too focused on negative points early on (negative issues may be more difficult to develop). As the discussion develops, probe questions can be inserted to obtain further information, as in "You mentioned the problem with spam e-mail, how often do you get unsolicited e-mails?"

CLOSURE

Focus group leaders can use verbal techniques to bring a topic of discussion to a conclusion. For example, "So, we have talked about the problems with the intranet system that the company has at the moment. Can we now put them in order of priority for change … what needs

(*Continued*)

HFE WORKSHOP 14.1 (*Continued*)

to be done first of all?" Or, "So what are the main advantages and disadvantages of having a computerized system for conducting the yearly staff appraisals as opposed to the existing system using paper forms?"

LEADING QUESTIONS SHOULD BE AVOIDED

A leading question implies its own answer, as in "We have had a lot of spam recently, have we not?"

VALUE-LADEN QUESTIONS SHOULD BE AVOIDED

"Do you agree that the company should really spend money upgrading the currently unreliable Internet access, which some people might find stressful?"

AVOID COMPLEX QUESTIONS

Double-barreled questions contain two related parts, for example, "Are you trained in using spreadsheets, and which is your favorite?" and are better asked separately.

SUMMING UP

Focus group leaders should bring the session to a close by summarizing the main points and adding comments such as, "Is there anything else that we have missed here?" Or "Do you think we have covered the ground today?"

RECRUITMENT AND DATA CAPTURE

Potential focus group members must be invited well in advance of the session, preferably by someone of the same or lower rank in the organization or by an outside consultant. Written instructions must be provided, explaining the purpose of the meeting and the membership of the group. Potential members must be told why they have been invited and assured that participation is voluntary and that, should they agree, they are free to withdraw at any time without penalty. They should also be assured that any records of the meeting will be anonymized. If the meeting is to be recorded (videotaped), the informed consent of the participants is mandatory and, in some countries, the tapes must be transcribed afterward, anonymized, and then destroyed or stored under lock and key.

FOCUS GROUPS

There are several variations on the general format for focus groups. Online focus groups can be cost-effective and bring people together even though they work in different offices. Online groups offer the possibility of users sharing information in electronic format. Dual-moderator focus groups share the load such that one person ensures the discussion goes smoothly and another ensures that all the topics are discussed.

ANALYZING DATA FROM FOCUS GROUPS

Focus group data can be analyzed using a variety of qualitative techniques. One of the simplest ways is by using *a priori* categories. These categories are derived from the model or checklist used to structure the group sessions in the first place. A simple method is simply to count the number of times comments appear in each category to identify the most common issues. More complex exploratory techniques are also available, such as cluster analysis, in which the data are analyzed to identify groups of issues that tend occur together.

Measuring the Mind: Psychometrics

Questionnaires and rating scales can be used if they have good *psychometric properties*, as described below.

Questionnaires are used to gather information from users across a wide range of topics. Unlike physical measurement in HFE (where we can see what we are measuring), questionnaires are often used to measure states of mind that are not directly observable. This introduces the problem of circularity (if it is not observable, how do you know it is there? If you hadn't asked users, would they have even thought about it in the first place?). Some of the questions we have to ask ourselves are:

- Do these constructs exist (independently of the questionnaire)?
- Can the questionnaire measure them?
- Can the questionnaire measure them reliably?
- Are they worth measuring?

Psychometric methods, originally developed by psychologists to measure personality and intelligence, can help. The classic approach to questionnaire design in psychology consists of several stages:

- Generate items for the questionnaire using focus groups, theory, or the opinions of subject matter experts.
- Develop a pilot questionnaire and test it using a sample from the target population.
- Use data reduction techniques to identify the underlying *constructs* in the domain.
- Group the items into sections and identify those that are most representative of different constructs.
- Score the questionnaire using *summative scales* for each construct (see below).

How Do People Think About Cell Phones?

Suppose we want to investigate cell phone design by means of a questionnaire survey. Modern cell phones are complex and are used for a variety of purposes; there are many variables—how do we identify the important ones? Where do we start? Answer: we find a valid and reliable questionnaire designed by somebody else and use that (see, e.g., Ryu and Smith-Jackson, 2006).

Suppose we decide we need to update the questionnaire with the advent of smartphones and cloud computing. Where would we start?

We could start with a literature review, then conduct some focus groups or structured interviews, talk to retailers about the popularity of features and/or feedback they receive from customers, and develop a pilot questionnaire of, say, 50 items with 5-point Likert scales to capture responses (e.g., 1 = "strongly disagree" and 5 = "strongly agree"). The pilot questionnaire could be sent to, say, 300 people. It is very unlikely that there would be little covariance in the data (i.e., that all of the correlation coefficients between all the item scores from all the respondents would be close to zero), because this would imply that users' perceptions of every aspect of cell phone design dealt with in the questionnaire were unrelated to every other, which is unlikely, given that phones, although they have many different features, are designed to offer users a smaller number of highly specific functions.

Data Reduction

In Chapter 3, we saw how a large number of anthropometric measures could be condensed into three or four main features using a technique called *principal components analysis (PCA)*.

Suppose we carried out a PCA on the data from our pilot study and obtained a component matrix with five components. We might summarize the PCA findings in a table (Table 14.8) by grouping the questionnaire items according to the component "loadings"—the strength of the association

TABLE 14.8
Components (and Percentage of Total Variance) of Cell Phone Design (from a 50-Item Pilot Questionnaire with Items Numbered 1–50). Items are Grouped according to their Highest Loading onto Each Component[a]

Component 1 (32%)	Component 2 (23%)	Component 3 (10%)	Component 4 (7%)	Component 5 (6%)
1. Communication	7. Style	2. Predictive text	26. Battery life	17. Vibrating alerts
14. Games	8. Appearance	33. Button feedback	27. Recharge time	18. Alarms
22. Internet	9. Weight	42. Ease of texting	28. Charger use	3. Screen brightness
23. Apps and office	10. Screen quality	29. Ease of correcting mistakes	29. Charge duration	25. Screen color
39. Camera operation	11. Cover/carrier	30. One-handed operation?	12. Battery removal	5. Screen layout
50. Sense of control	24. Durability	45. Ease of use	31. Charge level indicator	37. Volume control
4. Torch	44. Pride in ownership	49. Overall satisfaction		43. Picture quality
47. Pleasure in use	48 Personalization	20. Frustration		41. Legibility
13. Call logging	46. Quality of interaction	19. Range of options		32. Keypad design
	21. Color and finish			6. Button size

[a] nb When the pilot questionnaire was sent out, the 50 items were randomly ordered to prevent response bias due to the researchers' preconceptions of the item groupings. Items 15, 16, 28, 34, 35, 36, 38, and 40 do not appear in the table because they do not load onto the components. They might be included as single items in a final questionnaire, if required.

between an item and a component (in this example, items are listed under a component if the loading on that component is greater than 0.5).

Not all items load onto the five components, though. This might be because these items have little variability. For example, perhaps everyone answered "agree" or "strongly agree" to the following items:

"My phone is an essential part of my social life."
"My phone is an essential part of my business life."
"I'd be lost without my phone."

After considering the groupings in the table, we can invent some names for our components. Let us call component 1: "utility"; component 2: "appearance and style"; component 3: "usability"; component 4: "battery"; component 5: "interface ergonomics." The PCA implies that users have higher-level conceptions of cell phones beyond the individual cell phone features. In the same way that we can consider stature and sitting height as indices of a more general characteristic of "linear body size," so can we consider access to apps and office tools, Internet access, and so on as indices of a more general conception of cell phone "utility" (what you can do with it). In the same way that "linear body" size is independent of "heart risk" and "complexion," so, in our example above, "utility" is independent of "appearance and style," "usability," and "ergonomics."

The components account for a total of 78% of the total variance, so together, they tell us a lot about how users conceive of cell phones. Although component 1 accounts for 32% of the total variance and component 3, 10%, this does not mean that "utility" is three times more important to users than "usability." The answers to the utility items might just be more variable.

The analysis offers several interesting insights into the user experience of cell phones. For example, "pleasure in use" is a part of utility, suggesting that the pleasure is derived from, for example, interacting with others, social networking, using apps, etc., whereas pride in ownership depends on stylistic factors. Frustration links with usability (or the lack of it).

Reducing the Items and Constructing Composite Scales

One could argue that the "utility" of a cell phone depends on the goals it enables a user to pursue, whereas "usability" reflects some kind of "cost-benefit" or "effort-reward" trade-off in achieving the goals. Rather than focus on specific features of cell phone design, we might decide to focus on the more general dimensions implied by the PCA. How would we do this? Which of the 44 items in Table 14.8 would we choose to include in a questionnaire, and how would we choose them?

First, we have to decide whether to choose items that represent all of the different dimensions implied by the components in the table. Is the questionnaire going to measure utility, appearance and style, usability, battery issues, and phone ergonomics? Maybe battery issues are not relevant to our study, and we can leave this out.

Second, we can examine the interrelationships between the different items and the construct using a statistic known as "Cronbach's alpha" (HFE Workshop 14.2).

HFE WORKSHOP 14.2

DEVELOPING PSYCHOMETRIC SCALES FROM QUESTIONNAIRE ITEMS—CRONBACH'S ALPHA[1]

In our cell phone example (Table 14.8), we summarize the PCA for the cell phone questionnaire, finding five main components. Let us look at component 3, which we have called *usability*. Nine items from the pilot questionnaire load onto this component, but our goal is to produce a short cell phone questionnaire, with around five items on usability. How do we decide which items to keep and which to remove?

Remember, the PCA implies that all nine items, as well as being interesting in themselves, reflect the same underlying idea of usability. A statistic known as "Cronbach's alpha" can be used to find out which of the items are most representative of usability. The alpha statistic is a measure of the internal consistency of the scale items and enables us to identify the most representative items and combine these into a *summative scale.*

Below are the Likert scale responses for 10 respondents (in the example, we used 300, but to save space and enable readers to test the data themselves, only 10 sets of responses are shown below).

	Item No.								
Subject	2	33	42	29	30	45	49	20	19
1	4	3	4	5	4	4	1	1	2
2	2	3	4	5	3	4	5	4	3
3	3	3	3	3	3	3	2	4	4
4	2	2	2	3	3	3	4	2	5
5	4	4	5	3	4	3	5	4	3
6	3	3	5	4	4	3	4	4	3
7	1	2	2	1	3	3	2	5	5
8	1	2	2	1	3	3	5	3	2
9	2	3	4	5	4	4	4	5	4
10	2	3	4	5	4	4	4	4	4

- See Table 14.8 for item descriptors.

Alpha is given by

$$\text{Alpha} = N\bar{r}/(1+(N-1)\bar{r}$$

(Continued)

HFE WORKSHOP 14.2 (*Continued*)

where

N is the number of test items = 9

r is the Pearson product-moment correlation coefficient between each item

$\bar{r}$ = the mean of all the Pearson product-moment correlation coefficients between the items

ANALYSIS

Most statistical packages provide measures of scale reliability, such as alpha. Running the data above gives an alpha value for 9 usability items of 0.60. This is low—in general, values over 0.7 indicate adequate internal reliability. The table below can be used to identify which of the nine items are the best indicators of usability.

Item No.	Cronbach's Alpha if Item Deleted
2	0.574
33	0.498
42	0.399
29	0.466
30	0.535
45	0.568
49	0.652
20	0.622
19	0.697

If we remove items 49, 20, and 19, the value of alpha increases to 0.854. The remainder of the items are therefore better indicators of usability.

Item No.	Cronbach's Alphaif Item Deleted
2	0.832
33	0.820
42	0.781
29	0.843
30	0.834
45	0.858

When alpha is low (<0.7), reliability analysis can be used to identify items that are most representative of the construct we wish to assess. In this example, we have reduced a 9-item scale to a 6-item scale that gives a more reliable measure of usability. This has the advantage of shortening the questionnaire in which the scale is used or making space for items to measure other aspects of cell phone use.

Internal Reliability

Suppose we use alpha to analyze the data from our pilot study of 300 people, and we find that the value of alpha for the nine items grouped under component 1 (utility) is 0.69. What does this mean?

It means that the item scores correlate quite well and can all be regarded as measures of utility (the "gold standard" value of alpha is 0.70). Most statistical packages that offer this kind of analysis enable us to identify the "best" and "worst" items in a set (the items most and least representative of all the others together). If we wanted to reduce our nine "utility" items to five, we could use alpha to identify the five "best" items and remove the others (achieving a higher value of alpha in the process).

Having identified the best five items, we can treat these as a "scale measure" of utility and sum the scores from each item when we analyze real survey data. We can use alpha to construct scales for the other components and reduce the original questionnaire from 50 items to, say, 25 items, with 5 scale scores of 5 items each.

Benefits of Using Summative Scales

The main benefits are that we have shortened our questionnaire to 25 items, and the data can be summarized using five scores by summing or averaging the scores on each item. We can then compare different cell phones on 5 dimensions, rather than 25 features. If the scales all have 5 items, we have 5 scores, ranging from 5 to 25 (from the 5-point Likert scales), which simplifies the data analysis and increases the statistical power by reducing measurement error (error is assumed to be random, so an error in rating a given item is likely to be negated by the opposite error in rating another item in the same scale). Scores from composite scales are often normally distributed, so the data are amenable to a wider range of statistical analyses.

Psychometric considerations are only part of the story, though. Whether or not we exclude *any* of the original items depends on the purpose of the study and the need for a short rather than a long questionnaire. The value of using statistics such as Cronbach's alpha to construct scale measures is that these methods enable us to identify and remove questionnaire items that are *redundant*.

External Reliability

Questionnaires are measuring instruments. So how do we decide whether a questionnaire is a reliable instrument? If we get the same answers every time we use it under the same conditions, then the questionnaire is reliable. The reliability of a questionnaire is not the same as the stability of the construct it measures, so questionnaire designers have to decide on the time period over which they expect the underlying construct to be stable in order to find out whether the questionnaire is reliable. In the same way that we would expect body size to be stable over a week, but not over several years (some people gain weight or diet, but it takes time), so might we expect users' satisfaction with their phones to be stable over a week, but not over several years, because the expectations that drive satisfaction may change in response to the emergence of new products.

There are two types of external reliability for any measuring instrument, be it a questionnaire, rating scale, or an actual device such as an inclinometer. Test–retest reliability is a measure of how repeatable a scale is—if we administer it to the same people under the same conditions do we get the same answers? How do we analyze the data?

One drawback of using correlation to examine the reliability of the summated scale scores is that correlation coefficients only take account of shared variance—they ignore scale agreement. This means that they are insensitive to sensitization or other confounding effects, such as mood. Sensitization occurs when the first administration of the questionnaire causes raters to reassess the issue or change the way they think about the topic and they respond differently to the second administration. Correlations should be supplemented with other tests including visual inspection of scatterplots and checks to see whether the mean or median scores have changed over the testing period.

We could do a paired *t*-test on the data to see whether users' ratings were stable over the week. HFE Workshop 14.3 gives an introduction to the assessment of test–retest reliability.

HFE WORKSHOP 14.3

TEST–RETEST RELIABILITY

Test–retest reliability is a measure of the stability of a scale or questionnaire when used on separate occasions. There are two main factors that can affect the reliability of a scale. The first is that the constructs being assessed are ephemeral—they change quickly over time. For example, opinions about the appearance of cell phones might be influenced by mood swings. Attitudes about one's salary might be influenced by a sudden, unexpected cost (e.g., sudden loss of one's wallet). Returning to the trust scale in the previous workshop, we might be intervested to know whether trust in self-driving cars (or the lack of it) is a temporary phenomenon caused by most drivers having already learned to drive on the previous generation of cars and, therefore, lacking experience, or whether further research is needed on the automation of transport. Will new drivers, who have only ever used self-driving cars, have the same level of trust?

Irrespective of the answer, it would be helpful to have a reliable questionnaire to assess trust in vehicle automation. Suppose we carry out an experiment on a test track and administer the trust scale to 20 drivers on two occasions. The data below set out three different scenarios (items were scored on a five-point Likert scale where higher scores indicate greater trust—data are means for 20 drivers).

Item No.	T1 (day)	T2 (day)	T2 (1 yr)	T2 (night)
1	2.45	2.55	3.60	1.50
2	3.01	3.11	4.00	4.50
3	3.76	3.56	4.50	3.60
4	2.86	2.76	4.20	2.10
5	2.45	2.33	3.50	2.00
6	3.00	2.87	3.55	2.10
7	2.87	2.98	3.50	3.70
8	2.67	2.84	2.68	2.80
9	3.21	3.33	3.13	3.30
Item No.	T1 (day)	T2 (day)	T2 (1 yr)	T2 (night)
10	3.10	3.23	4.75	2.80
11	3.20	3.22	4.67	3.80
12	2.55	2.45	3.38	2.10
13	3.30	2.90	3.68	3.80
14	2.90	2.83	2.80	2.20
15	2.70	2.89	2.89	3.80
16	2.55	2.66	3.45	3.70
17	2.76	2.77	2.89	1.80
18	2.87	2.85	2.99	1.30
19	2.77	2.80	3.33	4.00
20	2.67	2.70	3.50	2.60

T1 is the first administration of the scale—on the test track during the day. In T2 (day), the same drivers repeated the test during the day 1 week later and completed the scale a second time. In T2 (1 yr), the interval was 1 year. In T2 (night), they did the test 1 week later, but in the dark.

If the scale is reliable because it measures stable constructs, then it should be reliable under all three scenarios. How can we determine whether this is the case?

(Continued)

HFE WORKSHOP 14.3 (*Continued*)

LIMITS OF AGREEMENT

A simple, graphical method for examining the repeatability of a measure is illustrated in the plots below (sometimes referred to as "Bland–Altman" plots). First, we calculate the difference in scores (T1–T2) for each item. Second, we calculate the corresponding mean score, (T1+T2)/2. We then plot the data with the mean on the *X*-axis and the difference on the *Y*-axis. If the scale were completely reliable, T1–T2 would always be 0, and we would see only a horizontal line crossing the *Y*-axis at 0.

In the graphs below, we can compare the reliability of the trust scale in the three different scenarios.

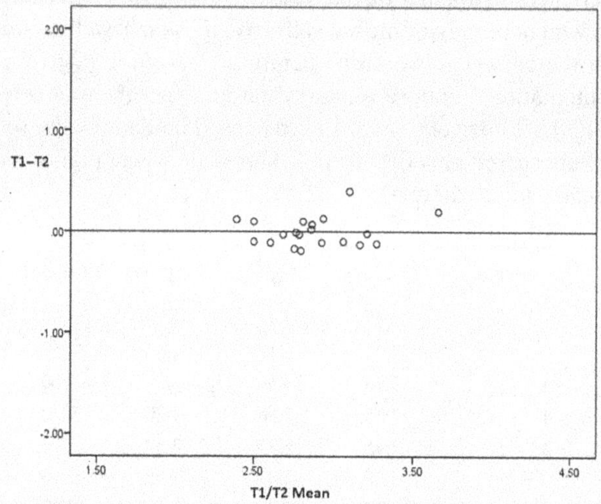

Therefore, for the two test drives carried out during the day, 1 week apart, the drivers gave similar ratings of their trust in the vehicle automation. All of the T1–T2 differences are small (less than one point on the Likert scale), and there is no significant trend—the difference score does not vary systematically with the mean score. This suggests that most of the deviations from the horizontal line are due to questionnaire (or random) variation.

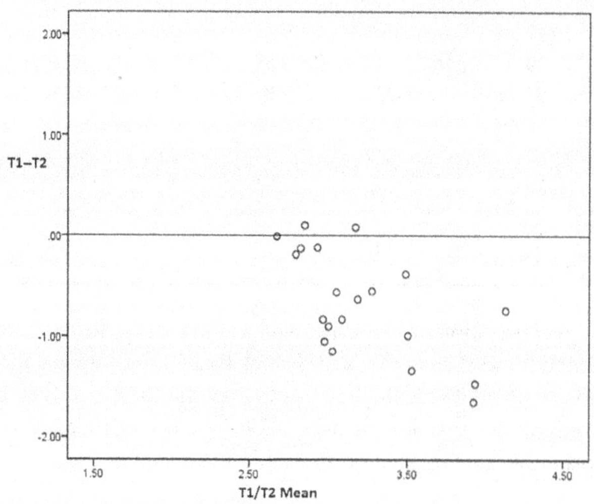

(*Continued*)

HFE WORKSHOP 14.3 (*Continued*)

For the test drives conducted one year apart, we can see that there are large differences between the ratings—more than 1-point Likert on most items. Also, most of the deviations are below the centerline, suggesting that the degree of trust has increased over the year. From this, we need not conclude that our scale is unreliable over the year. It is more plausible that the drivers had gained experience in using automated vehicles and came to trust them more—in other words, it is the underlying construct that has changed.

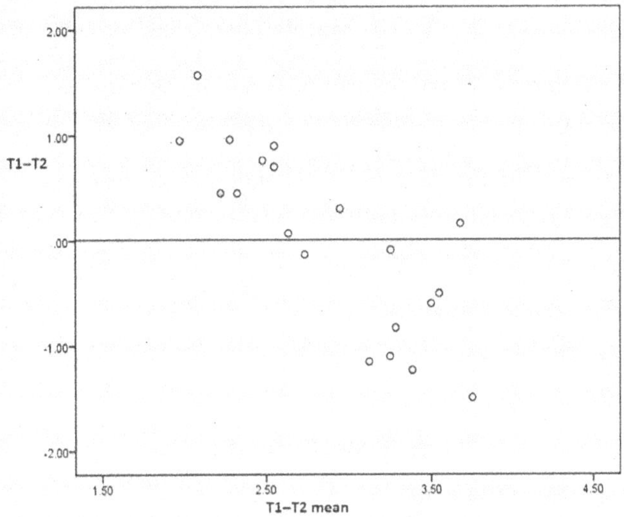

The data for driving at night presents a very different picture. Here, we can see a large difference in trust scores on either side of the centerline. The scale may be unreliable when used under changing environmental conditions, possibly because trust in vehicle automation is not absolute—it depends on the context in which the automation operates. Thus, the scale cannot be used to assess whether drivers always trust the automation, only whether it is trusted in a particular context.

COMPARISON WITH CORRELATION COEFFICIENTS

The method is more precise and more informative than bivariate correlation. Pearson product-moment correlation coefficients were calculated as follows:

Test Scenarios	Correlation
T1 (day): T2 (day)	0.89
T1 (day): T2 (1 yr)	0.51
T1 (day): T2 (night)	0.42

As can be seen, a very high correlation was obtained for the 2-day test scores 1 week apart. If a questionnaire has good test–retest reliability, the scores will correlate highly. For the tests separated 1 year apart, a high correlation coefficient of 0.51 was obtained, but the corresponding graph shows clear evidence of bias. This is because the size of a correlation coefficient depends on covariance, not agreement. Therefore, correlation is insensitive

(*Continued*)

HFE WORKSHOP 14.3 (*Continued*)

to systematic bias effects on retest and should never be used in isolation. Trust scores for the day–night test did not correlate highly, indicating that if a questionnaire is unreliable under different conditions of use, its scores will not correlate highly. However, the corresponding graph above gives a much better indication of the nature of the lack of reliability.

Another measure of external reliability is *inter-rater* reliability. Do two or more assessors agree when they use the scale to assess a device? We could administer the questionnaire to different raters and use a statistic called Cohen's kappa (HFE Workshop 14.4). There are several other methods of assessing reliability. In functional anthropometric research, the intra-class correlation coefficient is preferable to bivariate correlation coefficients such as Pearson's r, but further discussion is beyond the scope of this book.

In summary, *test–retest reliability* is used to determine whether items in a scale or questionnaire are reliable over time, and *inter-rater reliability whether* the items are stable between people. If a scale has poor test–retest reliability over a short period of time, then it is unlikely to predict future behavior (whether users will actually buy a particular cell phone, for example). If a scale has poor inter-rater reliability when used to assess a device, it may mean that judgments about the device depend very much on the experiences of the raters rather than the device itself.

HFE WORKSHOP 14.4

INTER-RATER RELIABILITY

Cohen's Kappa: Inter-Rater Agreement

Inter-rater reliability is a measure of the agreement between two or more assessors when they rate the same product under the same conditions and with the same instructions. Before we can say that a scale is reliable, we need to demonstrate that it produces the same kind of assessments when used by different people to rate the same product. There are many reasons why a scale might not be reliable—there may be comprehension difficulties, poor layout, and so on.

Cohen's Kappa is a statistic that measures the agreement between two assessors. Suppose we are developing a new design of a self-driving vehicle and, before we test it on real motorists, we want to know whether the automation is trustworthy. Two test drivers are hired to test the vehicle on a track designed to simulate the intended context of use. The test drivers work independently and then assess the vehicle using the trust scale below:[2]

	Driver 1	Driver 2
The automation is deceptive	Y	Y
The vehicle behaves in an underhanded manner	Y	Y
I am suspicious of the automation's intent, action, or outputs	Y	Y
I am wary of the vehicle	N	N
The vehicle's actions will have a harmful or injurious outcome	N	N
I am confident in the automation	Y	Y
The automation provides security	0	0
The automation has integrity	N	N
The vehicle is dependable	N	0

(*Continued*)

HFE WORKSHOP 14.4 (*Continued*)

The vehicle is reliable	0	0
I can trust the vehicle	Y	N
I am familiar with the system	N	N
The automation's behavior is always understandable	Y	Y
The vehicle state is clearly displayed	0	N
I always know what the vehicle is going to do next	0	Y
I was never surprised by the automatic system	0	0
I know I can take over whenever I want to	0	0
The vehicle is full of surprises	0	0
I trust the vehicle not to do anything dangerous	Y	Y
I feel safe and secure when the automation has control	0	0

We can recast the data into the table below:

		Driver 1			
Driver 2		**Y**	**0**	**N**	**Row Totals**
	Y	6	1	0	7
	0	0	6	1	7
	N	1	1	4	6
Column Totals		7	8	5	20

The diagonal cells show the number of items the drivers agree on—16/20 (80% agreement). This looks impressive, but it fails to take chance agreement into account. The advantage of Cohen's Kappa is that it corrects for agreement due to chance. It does this by calculating the *expected frequencies* given the row and column totals. For example:

Expected frequency (Cell YY) $= 7 \times 7/20 = 2.45$
Expected frequency (Cell 00) $= 7 \times 8/20 = 2.80$
Expected frequency (Cell NN) $= 6 \times 5/20 = 1.5$

We can recast the data into the table below:

		Driver 1			
Driver 2		Y	0	N	Row Totals
	Y	6 (2.45)	1	0	7
	0	0	6 (2.80)	1	7
	N	1	1	4 (1.5)	6
Column Totals		7	8	5	20

The diagonal cells show the number of items the drivers agree on—16/20 (80% agreement). This looks impressive, but it fails to take chance agreement into account. The advantage of Cohen's Kappa is that it corrects for agreement due to chance. It does this by calculating the expected frequencies given the row and column totals. For example:

(*Continued*)

HFE WORKSHOP 14.4 (*Continued*)

Expected frequency (Cell YY)=7×7/20=2.45
Expected frequency (Cell 00)=7×8/20=2.80
Expected frequency (Cell NN)=6×5/20=1.5

Kappa, K, is given by:

$$K = \sum a - \sum ef / N - \sum ef$$

$$\sum a = 6 + 6 + 4 = 16$$

$$\sum ef = 2.45 + 2.80 + 1.5 = 6.75$$

$$K = 16 - 6.75/20 - 6.75 = 9.25/13.25 = 0.70$$

Kappa can be interpreted as follows:

Interpretation of Kappa

Kappa	Agreement
<0	Less than chance agreement
0.01–0.20	Slight agreement
0.21–0.40	Fair agreement
0.41–0.60	Moderate agreement
0.61–0.80	Substantial agreement
0.81–0.99	Almost perfect agreement

In this example, the agreement is good—both drivers agree that some aspects of the automation may be untrustworthy and may warrant further investigation.

Kappa has an advantage over other methods in that it takes chance agreement into account. Correlation coefficients measure covariance, which is not the same as agreement. For example, suppose we used a 5-point Likert scale (agree/neutral/disagree) to rate the items in the questionnaire above and one driver consistently rated one point higher than the other (e.g., neutral rather than "agree")—the correlation coefficient would still be high, but the value of Kappa would be much lower.

User Experience Questionnaire (UEQ)[3]

The UEQ is a good example of a questionnaire that can be used to give a measure of the overall experience of using a device (Laugwitz et al., 2008).

After using a product, participants are given the questionnaire with the following instructions:

> For the assessment of the product, please fill out the following questionnaire. The questionnaire consists of pairs of contrasting attributes that may apply to the product. The circles between the attributes represent gradations between the opposites. You can express your agreement with the attributes by ticking the circle that most closely reflects your impression. Please assess the product now by ticking one circle per line:

	1	2	3	4	5	6	7		
annoying	○	○	○	○	○	○	○	enjoyable	1
not understandable	○	○	○	○	○	○	○	understandable	2
creative	○	○	○	○	○	○	○	dull	3
easy to learn	○	○	○	○	○	○	○	difficult to learn	4
valuable	○	○	○	○	○	○	○	inferior	5
boring	○	○	○	○	○	○	○	exciting	6
not interesting	○	○	○	○	○	○	○	interesting	7
unpredictable	○	○	○	○	○	○	○	predictable	8
fast	○	○	○	○	○	○	○	slow	9
inventive	○	○	○	○	○	○	○	conventional	10
obstructive	○	○	○	○	○	○	○	supportive	11
good	○	○	○	○	○	○	○	bad	12
complicated	○	○	○	○	○	○	○	easy	13
unlikable	○	○	○	○	○	○	○	pleasing	14
usual	○	○	○	○	○	○	○	leading edge	15
unpleasant	○	○	○	○	○	○	○	pleasant	16
secure	○	○	○	○	○	○	○	not secure	17
motivating	○	○	○	○	○	○	○	demotivating	18
meets expectations	○	○	○	○	○	○	○	does not meet expectations	19
inefficient	○	○	○	○	○	○	○	efficient	20
clear	○	○	○	○	○	○	○	confusing	21
impractical	○	○	○	○	○	○	○	practical	22
organized	○	○	○	○	○	○	○	cluttered	23
attractive	○	○	○	○	○	○	○	unattractive	24
friendly	○	○	○	○	○	○	○	unfriendly	25
conservative	○	○	○	○	○	○	○	innovative	26

User experience is measured in two main parts: the *pragmatic* quality of using the device (its efficiency, perspicuity, and dependability) and the *hedonic* quality of using it (stimulation and novelty). The raw scores are combined to give scale scores as follows:

Efficiency: slow–fast, inefficient–efficient, impractical–practical, cluttered–organized
Perspicuity: not understandable–understandable, difficult to learn–easy to learn, complicated–easy, confusing–clear
Dependability: unpredictable–predictable, obstructive–supportive, complicated–easy, confusing–clear
Stimulation: inferior–valuable, boring–exciting, not interesting–interesting, demotivating–motivating
Novelty: dull–creative, conventional–inventive, usual–leading edge, conservative–innovative

Note, when scoring items, that some of the items are reversed to control for response sets.

Six additional questions are also available:

English version

Scale	Question
Attractiveness	The product should look attractive, enjoyable, friendly and pleasant.
Efficiency	I should perform my tasks with the product fast, efficient and in a pragmatic way.
Perspicuity	The product should be easy to understand, clear, simple, and easy to learn.
Dependability	The interaction with the product should be predictable, secure and meets my expectations.
Stimulation	Using the product should be interesting, exiting and motivating.
Novelty	The product should be innovative, inventive and creatively designed.

These are rated on a 7-point scale, Not important at all (1) and Very important (7). A calculator is available at ueq-online to give an overall score for the product.

Mixed Methods Research

Mixed methods research involves quantitative and qualitative methods. The former might be performance data from user trials or questionnaire scores, and the latter might be an analysis of data from interviews or focus groups. *Thematic analysis* is a method of analyzing data to uncover common *themes* or patterns in the data. Normally, themes are ideas or comments that appear repeatedly in recordings or transcripts of participants' responses. Thematic analysis can be carried out by one researcher or by several independently who then meet to discuss their findings. It has six main stages but can be recursive if an earlier stage is returned to following later findings:

1. *Familiarization:* The researcher reads the transcripts and tries to understand the main points and form an initial idea of what the main themes might be
2. *Coding:* The researcher then assigns a code to segments of text or recordings that appear to deal with a theme.
3. *Generating Themes:* Once the data has been coded, the coded material is collated into the body of the theme. Different themes are generated in this way, and the relationships between them can be explored.
4. *Review:* The themes are then reviewed against the body of coded extracts.
5. *Defining Themes:* The themes are then given names or defined more carefully. They are not just the names of the individual codes but should encompass the coded material within them.
6. *Reporting:* The qualitative report is similar to any scientific report in content but the themes form the basis for the "Results" section. The themes may be represented graphically, showing the interrelationships between them. Common themes may be presented as larger than less common themes, and themes may be represented hierarchically.

In mixed methods research, qualitative analysis may be done at the beginning of a research project to generate hypotheses for quantitative testing or to assist with experimental design. Qualitative methods are also useful to inform the design of questionnaires used in surveys. They may also be used *post hoc* following user trials, when participants are debriefed in order to gain further insights into the users' *experience* of the device or product.

Tools such as NVIVO™ are available for thematic analysis. In HFE, mixed methods research is widely used. Carayon et al. (2015) conducted a systematic literature review of 58 studies of HFE in healthcare finding that mixed methods research was widely used. The most common combination of methods involved interviews first (qualitative), followed by surveys (quantitative). Thus, HFE practitioners require skills in both qualitative and quantitative methods. Further information on mixed methods research can be found in Dawadi et al. (2021).

Readability Metrics

Some word processors can generate readability statistics for written material. Two common metrics are the *Flesch–Kincaid readability tests*, which consist of a test of reading ease and a test of grade level. The latter is particularly useful when designing text for schoolchildren or for occupational groups in which the reading age is low. The reading ease test uses a scale of 0–100; the higher the score, the easier it is to read the text. The formula is

$$\text{FRES} = 206.835 - 1.015(\text{total words/total sentences}) - 84.6(\text{total syllables/total words})$$

Text scored 90–100 is considered readable by fifth-grade schoolchildren, whereas eighth- or nineth-graders could understand text scored 65. Graduate-level texts might score 30 or lower. Note that the test is really a test of sentence and word length, and the assumption is that text with long sentences and long words is more difficult to read than text with short words and sentences. While this may be true for long passages, it may not be true for small passages, since not all short words are easy to read or understand (as in the *perverse* sentences discussed above).

The grade level formula is

$$\text{FGL} = 0.39(\text{total words/total sentences}) + 11.8(\text{total syllables/total words}) - 15.59$$

A score of 7.1 would indicate that the material could be understood by an average seventh-grade student.

Wizard of Oz Experiments

A Wizard of Oz experiment enables usability to be trialled before the system is ready. Users believe that they are interacting with a real computer system and carry out generic or critical tasks. Unknown to the user, the system is actually operated, or partially operated, by a member of the team (the Wizard) acting in the role of the completed system. In this way, early testing of prototypes can be carried out cost-effectively, and the findings fed back into the design process.

SYSTEM INTEGRATION

User interface design is highly iterative and participatory, as designers work from a general design concept to a working prototype:

- Collect information
- Define system requirements and semantics (system objects, their properties, etc.)
- Design syntax and support facilities (e.g., online help)
- Specify physical devices (screens, keyboards, pointing devices, etc.)
- Develop software
- Integrate the system and disseminate to users
- Nurture the user community
- Prepare an evolutionary plan for further development

The design ideas are evaluated along the way with focus groups of users or experts. The process can begin before any software or hardware has been built. Designers may present drawings or sketches of their ideas to user groups in the form of "storyboards" (a series of screens from a sample transaction) or other mock-ups, and gain feedback, which is used to develop the ideas further. The goal is to test out design proposals against real-world scenarios in order to arrive at a robust and effective system.

Summary of Design Guidelines for Usability in HCI

Table 14.9 lists usability design guidelines in the form of questions that might be asked when evaluating prototypes or for use in surveys (from Lin et al., 1997).

Status of HFE Principles Used in Human–Computer Interaction

Consulting users is almost conventional wisdom among HCI practitioners. However, the validity and reliability of the techniques used to consult users are open to question. Focus groups, for example, are thought to be sensitive to a number of confounding factors. For example, the information from

TABLE 14.9
General Guidelines for Usability

Feedback
Provides useful feedback during use
Prompts attention during critical actions
Warns about potential errors
Warns about potential hazards
The device helps users understand how to use it

Flexibility and Adaptability
Can be used in ways most efficient for the user
Can be used in whatever ways are effective for the user
Provides choices about how it is used
Provides choices about the speed of use
Works as expected
Is straightforward to use
Menus can be bypassed by commands
Direct manipulation capability
User customization possible
Flexible arrangements for data entry
Menu options dependent on task context

Intuitiveness
Does not require instructions
Uses language that is always understood

Simplicity
Simple to use
Has easily identified features
The most important features are most obvious
The most used features are most easily accessed

Error Tolerance
Undo errors facility
Correct errors facility

Compatibility
Effects of control actions compatible with expectations
Controls matched to user skill
Coding compatible with conventions
Familiar wording

Consistency
Words and codes consistent across displays, menus, and modes
Consistent display formats
Consistent feedback
Consistent labeling

Learnability
Menu options grouped logically
Menu ordered logically
Command names meaningful

Minimal Memory Load
Mnemonics for abbreviations and codes
Highlight selected data
Indicates current position in menu structure
Icons combined with names

User Guidance
Helpful error messages
CANCEL and UNDO options

the group can depend, to an extent, on the person leading the group and on the questions or topics chosen for discussion. Because the researcher is also a participant in the group, the researcher's own interests, lapses of attention, sensitivity to body language, and other nonverbal forms of communication may influence the outcome. Focus groups are intrusive in that they involve questioning people and disrupting normal work patterns. However, they offer high diagnostic capabilities in identifying user difficulties and problems that would be more difficult and much more expensive to investigate through observation. Several users can, at once, summarize their experiences during long periods of computer use. Furthermore, some problems may occur quite rarely and might easily be missed during short periods of observation.

If It's Hard to Read, It's Hard to Understand and Hard to Do

There is evidence that people make inferences from the ease of reading instructions about the ease of actually following them. Song and Schwarz (2008) gave subjects instructions for an exercise routine

.... written in Arial 12-point font or
... *Brush 12-point font*

They found that Arial was reported to be easier to read than Brush. Although there was no effect of font on the recall of the instructions, subjects estimated that the exercises would take less time when they were written in Arial and were more willing to include them in their exercise routines.

In a second study, subjects read instructions for making a Japanese roll

...printed in Arial 12-point font or
...*Mistral 12 point font*

Arial was again reported to be easier to read, but memory for detail did not differ. Subjects reported that the recipe would take less time when printed in Arial and that it would require more skill from the chef when printed in Mistral.

VEs for Training

Simulators have been used for many years in aviation, in particular, and also in shipping and other high-hazard industries. However, these simulators are expensive and lack flexibility due to the physical limitations of their hardware. VEs, or any type of extended reality training system, have largely generic hardware—only the application software needs to be application-specific. Evidence is mounting that training time and costs can be reduced using VEs to simulate critical tasks or tasks that are unpleasant or dangerous to carry out. Flight simulators have been in existence for many years—without them, it would be extremely expensive to train pilots on real aircraft—not least because more aircraft would be needed just to train the pilots. It seems to be the case that as long as the level of realism is high enough for the particular task, learning takes place interactively and transfers positively to the real world. At the same time, VE costs are dropping, and simple environments can be implemented on commercially available Windows-based systems.

One of the main arguments in favor of using extended reality applications such as VE, AR, and MR for training is that even a low-fidelity interactive simulation of a task environment (such as a battlefield) is likely to be better than a lecture hall. Kaplan et al. (2021), in their meta-analysis of 25 high-quality papers, list some of the benefits of using extended reality in training:

- People can be trained in situations that are too dangerous in the real world.
- People can be trained to deal with situations that don't yet exist (e.g., landing on the planet Mars).
- If there is a good mapping between the extended reality environment and the task environment, then extended reality training may reduce concerns about "encoding specificity." Encoding specificity occurs when something memorized in one environment is not recalled as well in a different environment.

- The cost of failure is lower than in the real environment (e.g., crashing a driving simulator).
- There are also considerations about cost, particularly for high-volume training.

Kaplan et al. reviewed the evidence for the transfer of training from extended reality to the real task, finding evidence for transfer in cognitive tasks (where something had to be recalled or used in the real task), physical tasks (the acquisition of psychomotor skills), and mixed tasks (where physical skills needed to be learned and applied according to standard procedures). Somewhat disappointingly, the authors found no difference in the transfer of training between extended reality and traditional training, although this may only indicate that the traditional training and the extended reality training were equally effective. From this point of view the benefits of extended reality training can be achieved with no adverse effect on training outcomes. As such, it is likely to be the case that bulk training requirements can be met more easily, more cost-effectively, and more safely using extended reality training.

Cybersecurity: The Dark Side of Usability?

According to the Chartered Institute of Ergonomics and Human Factors (CIEHF, 2022), the main ways cybersecurity can be breached are as follows:

1. Access to a system or data is unauthorized
2. Unauthorized use of a system for processing or storing data
3. Changes to hardware, software, or firmware without the owner's consent
4. Malicious disruption or denial of service

Cyber incidents are often attributed to human error and may be unintentional and not malicious or intentional, whereas cyber-attacks are both intentional and malicious. In one example, cybercriminals entered the branch of a high street bank and persuaded the staff that they were from the IT department. They then accessed the computer system and withdrew £1.25 billion. Although they were caught and the staff might have been blamed for not checking their credentials properly, this incident really discloses weaknesses in the bank's policy for managing visitors.

Following our discussion of memory in Chapter 12, it can be seen that passwords that are easy to remember are also weak—they are easy to crack. Systems with multiple access points requiring different passwords may overburden user memory such that the same or similar (weak) password is chosen by the user for multiple applications. Conversely, strong passwords are often difficult to remember because they lack redundancy. In some applications, users are able to choose their own passwords, but the system has in-built requirements for password strength. Many systems have "blocklists"—a list of items in a set that are not accepted. In computer security, a blocklist may be used to exclude which set to detect, quarantine, block, or perform security scans on. Acceptable passwords must meet requirements such as item length, character type, and so on. According to Tan et al. (2020), password length and character requirements do not necessarily lead to strong passwords (e.g., "Password!" or "passwordpassword" are easily cracked). According to the National Institute of Science and Technology:

- Password length matters the most
- Allow 64-character passwords
- Check passwords against a blacklist
- Make special character rules optional
- Provide feedback to users explaining password rejections
- Remove hints and security questions
- Use password managers safely

"Phishing" and It's Variations

Phishing is the fraudulent practice of sending emails or other messages purporting to be from reputable companies to induce individuals to reveal personal information, such as passwords and credit card numbers.

- *Spear-Phishing:* Emails that appear to be from a known or trusted sender are targeted at individuals to reveal confidential information.
- *Vishing:* Phone calls or voice messages appear to be from reputable companies to obtain personal information.
- *SMiShing:* SMS or instant messages on the victim's smartphone appear to be sent by a trusted source (e.g., your bank) and invite the victim to click a link, as in the case of more traditional phishing, or even to download an attachment. In the latter case, the attachment installs malware, giving the attacker access.
- *Pharming:* Malicious code is installed on a personal computer or server, which misdirects users to fraudulent sites without their knowledge or consent.

User Behavior and Cyber Security

Coventry et al. (2014) reviewed behavioral aspects of cybersecurity. More recently, Desolda et al. (2021) conducted a systematic literature review, finding 52 articles on HFE and phishing attacks. The main areas of vulnerability were as follows:

Lack of Knowledge: Users may be unaware of the threat and unable to recognize that they are under attack. This can be countered in two ways: first, by training and providing online support, and second, by providing information at the time it is needed (e.g., that a URL is suspicious).

Lack of Resources: Small businesses, home users, and hybrid workers may lack access to the required resources or support to maintain security. No one person may be responsible for cybersecurity. Even larger organizations may lack a coherent cybersecurity policy that is transparent and regularly promulgated to employees.

Lack of Awareness: Employees may be unaware of the risks and what they look like. Even cyber experts can fall prey to attacks if overconfidence reduces their awareness of the risks. A phishing e-mail may be processed automatically by System 2; the user responds quickly without inspecting the message more carefully.

Habituation: Users are accustomed to having access to the Internet in their daily lives, both at work and outside work. Being connected seems natural, and risks are not expected. People are habituated to click the 'agree' button without fully understanding what they are agreeing to.

Cost-Benefit Trade-Offs: Secure behavior is a secondary task that may incur costs both in attentional demands and in time. Easy "work-arounds" may be chosen to reduce these costs just to "get the job done." More abstract notions of risk may be ignored in favor of immediate gains.

Distraction and Pressure: Users may be distracted by unexpected or inviting communications that capture their attention. Attackers often place pressure for a quick response (e.g., "his offer expires in one hour").

Constant Change: In some organizations, passwords and procedures are changed regularly, placing a demand on users to learn new information. This may add to the perceived costs of compliance and encourage the use of workarounds.

Norms: In some organizations, employees may be more willing to share passwords and less likely to lock their screens when they leave their desks. Organizational-level interventions should seek to establish norms that increase cybersecurity.

Lack of Assertiveness: Related to norms, some employees may be unwilling to challenge insecure behavior or deny requests for passwords or other sensitive information. Supportive channels to increase assertiveness and assertive communication may counter this.

Cybersecurity Pitfalls

Haney (2023) lists six pitfalls in cybersecurity relating to HFE and how to overcome them:

Users Are Not Stupid: The tendency among system designers to consider users as the "weak link" in the cybersecurity chain is misplaced, if users are really suffering from "security fatigue," being overburdened. Security measures can be likened to secondary tasks, placing an additional burden on users' attentional resources. Users may also experience cognitive biases—such as the availability bias ("I can't recall anything bad happening in the past, so I'm not worried now"). This pitfall can be countered by empowering users and understanding their difficulties to design better ways of embedding security into their main tasks.

Communication Problems: Information should be tailored to the audience. For example, research scientists typically have a culture of open communication and knowledge sharing. They may be less amenable to communications about security than, for example, human resources personnel, who are very familiar with data protection and privacy issues and the legislation and practices surrounding these matters.

Poor Usability may Create Insider Threats: Security measures should be designed with usability in mind because poor usability can lead users to devise "work-arounds" and less secure ways of working. Complex password policies and the requirement to use multiple passwords may cause users to write passwords down on sticky notes or in unprotected Word files. Usability testing on new security measures can help identify user difficulties that may threaten security. Even tests with small groups of representative users may help improve security. For example, one way to increase security is to replace passwords with biometric identification to reduce the burden on users. Some systems have automatic screen locks when the user stops using the keyboard or mouse after a specified time interval. The user then has to use a password to unlock the screen. This is done to prevent unauthorized access to the system when the user is away from the desk. By understanding a little about users' work schedules and habits, an appropriate time interval for automatic lockout can be determined to balance usability and security. For example, a lockout after one minute might be acceptable for a deskbound user with one main task who rarely leaves the desk, but not for a user in a customer service role who multi-tasks customer interaction and database interrogation continuously (i.e., short periods of computer interaction are frequently interspersed with customer interaction, meaning that quick lockouts would be intrusive and inefficient).

Too Much Security: Excessive or fragmented systems may lead to complexity and slow response when security is breached, leading to non-malicious negligence. A risk assessment prior to the introduction of new measures should be conducted to determine an appropriate level of security.

Depending on Punitive or Negative Measures to Ensure User Compliance: This pitfall is based on the assumption that users will always "do the right thing," and then punishing them when they do not. In the short term, disabling user accounts or publicly shaming them may have benefits, but in the longer term, deeper problems will remain. Alternatively, designers should attempt to understand why a user fell prey to a malicious attack. For example, phishing attacks are a common cyber threat. Phishing e-mails often resemble the company they purport to be from but may contain small spelling errors or cues that reveal their true nature. Users are sometimes more likely to be deceived by such attacks if the context is familiar to them (e.g., they often deal with that particular company in their work). In one case, an attack on Facebook and Google resulted in a loss (subsequently recovered) of \$100 million when the phishing e-mail was sent to employees who regularly dealt with the company. Training users to increase awareness of phishing e-mails and how to recognize them is recommended, as is a shift from punishing phishing "clickers" to rewarding those who successfully counter phishing attacks.

Collecting "Lagging Indicators" of Effectiveness: Typical KPIs for security assessment include risk assessment scores, number of malware infections, and time taken to remediate risks.

Haney recommends more use of "leading indicators," such as regular staff security training with feedback from employees about the effectiveness of the training and their experience of it in relation to their daily work. This, together with open feedback channels for employees, will create a more "user-centered" approach to the design and management of system security.

One reason why people fall for cyber-attacks is simply because they lack the resources to properly evaluate the risks. System 1 responds to cues automatically, and by then it is too late. Designing systems so that the most secure way of working is also the easiest way is recommended.

Cybersecurity Recommendations

Coventry et al. (2014) offer the following guidance for cyber best practice:

- Use strong passwords and manage them securely (no sharing).
- Use and update anti-virus software and firewalls.
- Use the latest version of software.
- Log out of sites when the task is completed.
- Use trusted and secure connections, sites, and services.
- Provide the minimum personal information and protect your identity.
- To prevent habituation to warnings about security threats, use polymorphic warnings (see Figure 14.2 for an example).
- Be aware of your surroundings (be aware of threats such as "shoulder surfing").

Integrating Cybersecurity in Organizations: Cyber Maturity

CIEHF (2022) presents a maturity framework to enable the integration of HF issues in organizational cybersecurity to be assessed. Extracts are given below:

Level 1: Reactive	Many cybersecurity incidents (including poor behaviors) are seen as unavoidable Minimal cybersecurity incident sharing Inconsistent execution of cybersecurity processes
Level 2: Repeatable	Management understands the overall process Repeatable execution of processes Performance measured in terms of lagging (retrospective) indicators (instead of the number of control measures)
Level 3: Defined and managed	Majority of staff believe cybersecurity is important Regular training exercises (role play) Formal cyber security incident sharing
Level 4: Sustained	Employees are confident in cybersecurity processes Constantly striving to do better in cybersecurity and improve controls. All employees believe cybersecurity is critical to their job, and accept that prevention of cybersecurity incidents is important
Level 5: Optimized	Innovative ideas and techniques are piloted on an ongoing basis Optimal service levels are achieved Prevention of cybersecurity incidents (at work and home) is a core company value, and the company invests significant effort to promote it

For further information, see Widdowson (2025).

RESEARCH DIRECTIONS

There is currently research on the design of VEs using novel display modalities such as real-time or near-real-time 3D computer-generated graphics and novel control devices. VEs can be used for remote collaboration, where a worker on-site wearing an AR headset is supported by a remote expert using VE while carrying out a maintenance task, for example. The remote expert and the worker view the same workspace via their headsets. In one experiment, Bai et al. (2020) investigated the effect of visual and gesture cues on the performance of subjects in an experimental task (finding and sorting colored bricks in an office while receiving guidance from an "expert" in an adjacent office). Both the "expert" and the "worker" were presented with the same 360° view of the office. Gaze cues were presented to the worker via an avatar head showing the field of view and line of gaze of the other. The expert's gesture cues (e.g., a hand pointing at a location) were displayed on the worker's headset. In effect, combined gaze and gesture cues gave a better experience of "co-presence," improved spatial awareness, and reduced workload for the worker.

Automation is often driven by the availability of technology in an attempt to improve efficiency and reduce "human error." However, when systems are automated, the user is often removed from the loop (direct participation in the task–action sequence). By removing the parts of the interaction that are easier to do (because they are accomplished using System 1), there is always a risk that the remaining tasks will be more difficult (because they require System 2). The challenge is to automate while retaining the optimum degree of user engagement in the task–action sequence.

According to Elizabeth Churchill, Head of User Experience at Google, whereas a few years ago we talked about "accessing" the Internet, most of us live inside it nowadays. There are over three billion Internet users in the world today, and 90% of U.S. adolescents have access to cell phones. There are many interesting research questions, as well as concerns, about the emerging world of interconnected people and interconnected things. According to George and Odgers (2015), many parents commonly voice fears about the health and safety of their adolescent children. The most common fears include: interacting with and sharing information with strangers; cyberbullying and "stranger danger" online; constant connectivity and disengagement with "real life"; a new digital "generation gap"; alternate online identities that may damage the sense of self; constant multitasking impairing cognitive performance; and loss of sleep due to nighttime use of the device.

How well-founded are these fears? George and Odgers point out that parents have always been worried about the effects of new technologies and trends on their children, and this is documented as far back as the 1940s with the emergence of radio, comic books, and later, television and pop music. Much adolescent behavior in the digital world merely mirrors behavior in the real world. Children with more online relationships report higher offline friendship quality. Most online interaction is positive or neutral, and there is a high correlation between online and offline bullying. The quality of online interaction between parents and their children mirrors offline interaction. However, there is some evidence that constant multitasking impairs cognitive performance by increasing distractibility. Using smartphones, laptops, and so on at night or in bed (e.g., reading an e-book as opposed to a paper book) impairs sleep quality because the emitted light interferes with melatonin secretion. In general, offline factors tend to predict online outcomes, and these fears can be balanced by new initiatives to deliver positive information and messages to users, such as fitness apps.

SUMMARY

In general, offline and online usability depend on the right amount of conceptual knowledge about the semantic, syntactic, and physical features of the application, the relations between them, and their mappings onto the system-level implementation of the application (Figure 14.3).

European regulations about work with display screen equipment present general principles of software ergonomics, as does EN ISO 9241-10:1996. See also ISO/DTS 16982:2000, Ergonomics of Human System Interaction—Usability Methods Supporting Human-Centered Design. The Human Factors and Ergonomics Society has a draft standard with ANSI, *Human Factors Engineering of Software User Interfaces*, in five parts (2005): (1) introduction, (2) accessibility, (3) interaction techniques, (4) interactive voice response, and (5) visual presentation and use of color.

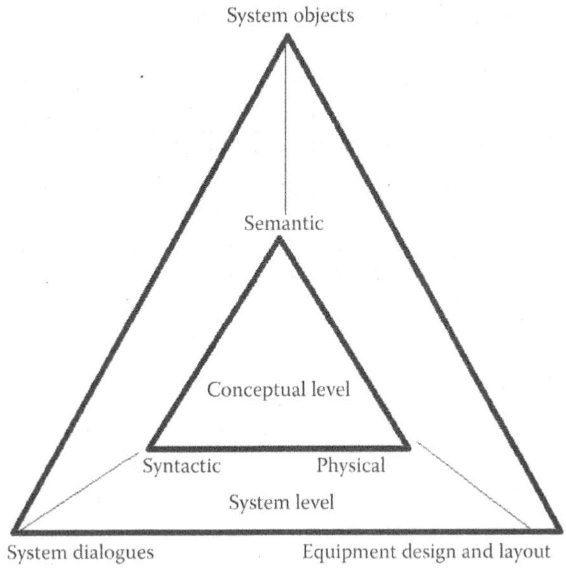

FIGURE 14.3 The "virtual triangle." Conceptual and system-level mappings for usable applications.

TUTORIAL TOPICS

1. Much of the debate in HFE to date has been focused on a simple human–machine model of a single driver in a car. What new properties and hazards will emerge at the aggregate level with millions of automated cars interacting with the roads and with each other remains to be seen?
2. Do new technologies generate new user requirements over and above the usability and utility of the technologies themselves?
3. Search engines use complicated algorithms that go beyond merely aiding users in their searches. Is it true that, as Elizabeth Churchill describes it, we are shaped by the Internet in terms of the reality it presents us with?
4. What do you think is the future of self-driving vehicles? Watch: The two opposite futures of self-driving cars, youtube.com. Self-driving private vehicles, self-driving buses (public transport on fixed routes), or self-driving long-distance trucks? All three or something else?

ESSAYS AND EXERCISES

1. The school outing to Hampton Court Palace ought to have been a fairly simple journey … But the driver chose to rely exclusively on his satellite navigation system, and 60 children spent the whole day being driven around in circles after it directed him to a narrow street in the north of the Capital … The problem began when the driver punched "Hampton Court" into his vehicle's satnav, which directed him to a road of that name in Islington.

   "Every driver should have a little bit of knowledge about where they are going…" said a spokesman.

   The Daily Mail, March 23, 2007

   Apart from consulting a map before starting the journey, what else might the driver have done before setting off on his journey to Hampton Court Palace?

2. Evaluate the operating instructions of some common consumer products. Use the language comprehension framework described in Table 14.2 as a guide. Describe the knowledge required to understand the instructions.
3. Evaluate a telephone menu-based user interface of your choice. Comment on the number of levels, the hierarchical structure and representation of the domain, the provision of backtracking, and any other facilities for navigation.
4. Survey product warning labels on commonly used consumer products.
5. Calculate Cronbach's alpha for the ratings of 20 raters on 5 aspects of computer stress in an office (a score of 1 = strongly agree, 5 = strongly disagree, 3 = neutral):

	Subject No.									
Item	**1**	**2**	**3**	**4**	**5**	**6**	**7**	**8**	**9**	**10**
Too much spam email	1	2	3	5	4	1	3	2	3	2
Excessive monitoring	1	3	3	4	4	2	3	2	2	3
Intrusive reminders	1	3	3	2	4	1	1	2	2	4
Response time too slow	1	4	4	1	4	1	2	1	1	2
Poor Internet access	1	4	2	3	4	1	1	5	5	5

Hint: You will find it easier if you have access to a statistical software package to calculate alpha, or, at least, to calculate the Pearson product-moment correlation coefficients.

6. Below are Pearson product-moment correlation coefficients for all pairs of correlations between the items shown in question 6 (assume data are from a separate study from that above). Calculate the value of alpha.

   Pearson product-moment correlation coefficients
   For item pairs (no. of raters = 30)

Items	1 vs. 2	1 vs. 3	1 vs. 4	1 vs. 5	2 vs. 3	2 vs. 4	2 vs. 5	3 vs. 4	3 vs. 5	4 vs. 5
R	0.71	0.42	0.75	0.70	0.33	0.79	0.77	0.45	0.50	0.78

7. "The desktop metaphor has failed on modern user interfaces because of the slow response time and response time variability of the graphic user interfaces of modern PCs." Discuss.
8. Survey warning labels on a variety of products used in the home. Suggest ways that the labels might be redesigned.
9. Type the paragraphs of text (below) into a computer using a capable word processor and generate readability statistics for each. Next, give colleagues each section to read.

Caesar once, seeing some wealthy strangers in Rome, carrying up and down with them in their arms and bosoms, young puppy dogs and monkeys, embracing and making much of them, took occasion, not unnaturally, to enquire whether the women in that country were not used to bear children; by that Prince-like reprimand, gravely commenting upon those who spend and lavish upon brute beasts, that affection and kindness which nature has implanted within us to be bestowed on those of our own kind.

Plutarch, The Life of Pericles, in Dryden, 1990

Back in September, we asked how you really felt about yourselves in our Love the Skin You're in survey with Olay. As you know, our whole Mantra is "being the best you can be" in all areas of our lives—but sometimes this can be really hard. There are times when we all look in the mirror and hate everything we see, and would quite happily spend the day under the duvet. Other days, we feel like we can take on the world … and do. I'm not sure how big a part hormones, sunshine or other people being nice to you play a part in all this.

Cosmopolitan Magazine, 2007

> A theory was proposed to reconcile paradoxical findings on the invariance of personality and the variability of behavior across situations. For this purpose, individuals were assumed to differ in (a) the accessibility of cognitive-affective mediating units (such as encodings, expectancies and beliefs, affects, and goals) and (b) the organization of relationships through which these units interact with each other and with psychological features of situations. The theory accounts for individual differences in predictable patterns of variability across situations (e.g., if A. then she X, but If E then she Y), as well as for overall average levels of behavior, as essential expressions or behavioral signatures of the same underlying personality system.
>
> *From Mischel and Shoda, 1995*

Next, ask them to rate the readability on a five-point scale where a score of 1="very easy to read," 2="easy to read," 3="neutral," 4="difficult to read," and 5="very difficult to read." Plot the readability scores against the readability statistics and comment on your findings. What other factors would need to be controlled for to arrive at some definitive conclusions about the usefulness of the online readability statistics?

10. Mr. and Mrs. Smith are trying to decide where to go for their next holiday. After looking at different destinations on the Internet, they independently rank each destination within their budget as either positive (a place worth visiting), negative (a place to avoid), or "maybe." Analyze their responses and state the level of agreement between them.

Destination	Mr. Smith's Vote	Mrs. Smith's Vote
Florida	+	+
California	+	+
Spain	+	+
Italy	+	+
Alaska	+	–
Malaysia	–	+
Singapore	–	+
Germany	–	–
Greece	+	+
England	+	M
Scotland	–	–
Ireland	–	–
France	+	+
New England	+	+
Canada	+	–
South Africa	M	+
Australia	M	M
New Zealand	M	M
Turkey	+	+

Examine the pattern of destination preferences and briefly comment on the findings.

QUICK QUIZ

1. All products have modes. True or false?
2. How many levels of automation are there?
3. Automated vehicles do not always understand pedestrian behavior. True or false?
4. Car drivers should always remain "in the loop" (a) true. (b) false. (c) depends on the context
5. How many KPIs of usability are there?
6. If it's easier to read, it's easier to understand? True or false?

7. Correlation coefficients are not the best way of assessing the reliability of a measuring instrument. True or false?
8. Bland Altman plots measure test-retest reliability (repeatability). True or false?
9. Secure passwords are always easier to remember. True or false?
10. Using smartphones, laptops, etc. at night can impair sleep quality because… (a) you stay awake too late and sleep less. (b) playing games and using apps is exciting. (c) light from these devices suppresses melatonin secretion

NOTES

1 https://datatab.net/statistics-calculator/reliability-analysis/cronbachs-alpha-calculator.
2 Y = Yes, 0 = Neutral, and N = No.
3 https://www.ueq-online.org.

15 Safety and Accidents

GENERAL HFE REQUIREMENTS FOR SAFETY

1. Risks and hazards are understood and controlled.
2. Design safety constraints to block illegal actions.
3. Provide feedback on the operation of the system and the effectiveness of safety constraints.
4. Designers should consider the kinds of errors operators are likely to make.
5. Safety constraints must be designed to prevent operator error from causing accidents.
6. Use legacy data from previous safety occurrences to improve system safety.

It is the nature of every human to err, but to persevere in error is only the act of a fool.

Marcus Tullius Cicero. Roman Senator, 106-43 BC

CORE KNOWLEDGE: SAFETY AND ACCIDENTS

Safety can be considered both as a *state* and as a *process*. Safety as a state implies that there are no hazards or, if there are, the risk of being exposed to them is zero. Safety as a process implies that all hazards have been identified and all necessary controls to prevent exposure are in place and working properly.

Safety as a permanent state is impossible because there are always hazards that can't be controlled or foreseen. For example, in 2011, a 15 m tsunami disabled the power supply and cooling of three Fukushima Daiichi reactors, causing a nuclear accident that began on March 11, 2011. All three cores largely melted in the first 3 days.

Safety as a process is never perfect because we can never be certain that all necessary controls are in place and working properly. There may even be too many controls or controls that are incompatible with each other. As a process, designers aim to reduce risk to levels that are as low as reasonably practicable (ALARP), although there is always room to debate what is reasonable and what is practicable. On November 19, 2016, at 06:17 AM in heavy rain, a tram overturned at a sharp bend approaching a junction. The inquiry found that the most likely cause was that the driver experienced a "microsleep" on the approach, noting, among other things, the lack of signage and the design of the shift system.

A *hazard* is anything that can cause harm to people, equipment, or the environment. Noxious chemicals are hazards; working at height, and long working hours are hazardous. Heavy loads are hazardous if they are difficult to lift. Work objects, equipment, procedures, and processes can all be hazardous. A *risk* is the likelihood that a hazard will have an adverse effect on the system (either as a numerical probability, such as a "one in a thousand chance of failure," or a verbal guestimate, "the risk of exposure to the hazard is negligible"). Some authorities combine the two, so that risk reflects both the likelihood of exposure to the hazard and the severity of the adverse effects. This is not really a "risk"—it's the *negative expected utility* of the outcome when things go wrong.

So, it is not correct to believe that a system is safe only because nothing has gone wrong yet. Rather, safety is a control problem that depends on what, when, and how hazards are controlled and whether all the hazards have been identified in the first place. From this point of view, safety is all about controlling a process, and this means putting constraints in place at the design stage. One way to control

DOI: 10.1201/9781032717210-15

"what happens next" in a sequence of actions is to limit "what can happen next," which depends on the way state transitions are designed (how the state of the system can change over time and/or in response to changes in its operation). Barriers can block unsafe transitions, and controls can be put in place to limit what can happen (e.g., cut-outs on domestic electric kettles operate when the water boils, pressure relief valves prevent pressure inside vessels from reaching dangerous levels, and so on). It is quite common, though, to find that hazards are only properly understood after they have made themselves known in *safety occurrences,* of which there are several kinds:

Accidents: Unplanned events (they are "accidental") that cause damage to people or property
Incidents: Unexpected events that may result in property damage but don't cause serious (i.e., notifiable) injury
Near Misses: Incidents that did not cause damage but *might have* under similar circumstances, and *might do* so next time.

Safety occurrences need to be reported and investigated as part of safety management because they are evidence that the risk is not ALARP.

Accidents

H.W. Heinrich (1886–1962) conducted some of the first systematic studies of accidents, concluding that 88% of all accidents and injuries were caused by "man-failure." The "Heinrich Triangle" was developed to illustrate that every major accident is preceded by many more minor accidents and unsafe acts (Figure 15.1). While this is often true, the implication that reducing minor incidents will also reduce the number of serious accidents assumes that serious accidents share common causes with minor incidents. In reality, organizations have to comply with regulations and standards in different domains to manage hazards of different kinds, using a variety of different tools under different constraints. For example, a chemical industry may comply completely with regulations regarding the control of hazardous substances but not with legislation concerning other aspects of safety (such as manual handling safety or noise exposure). In other words, minor incidents may be a symptom of a minor safety management problem, but they are not necessarily a symptom of a more serious problem. However, Yorio and Moore (2018), in a retrospective study of reportable accidents in the U.S. mining industry, *did* find some evidence in support of the triangle. Their study was restricted to injuries in the mine only and included fatalities, occurrences such as explosions of dust or gas, roof falls and rock bursts, and fires. Outcomes included fatalities, permanently disabling injuries, days-lost injuries, reportable injuries, and reportable non-injuries. There were

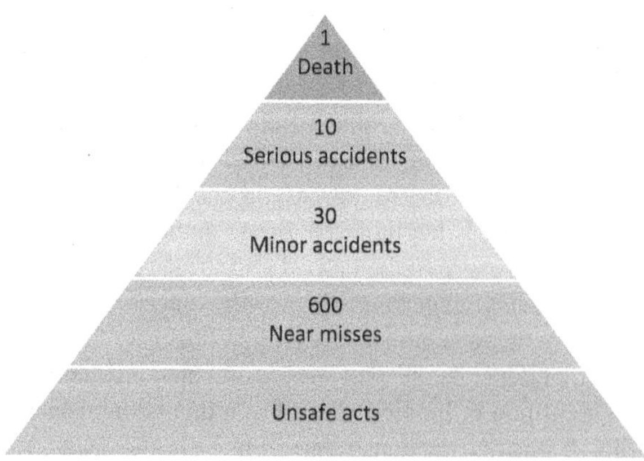

FIGURE 15.1 The Heinrich triangle.

statistically significant associations between these outcomes. Fatal injuries in a given year were predicted by permanently disabling injuries in the preceding year (OR=6.63), days lost injuries (OR 1.10), reportable injuries (OR=1.19), and reportable non-injuries (OR=1.08). Although these data provide some evidence for the safety triangle, it seems that non-reportable injuries were not strongly associated with fatalities, whereas permanently disabling injuries were.

HFE in Safety Management and Safety Investigations

HFE supports safety management and safety investigations by contributing knowledge about working conditions and how they affect human behavior, as summarized by the model in the Preface to this book prEN 14386: 2002(E), pinpointing where HFE can help investigators identify hazards and assess risks.

This book provides an introduction to the competencies needed to investigate the areas in the diagram. One way to start applying HFE in an investigation is to refer to the requirements listed at the beginning of each chapter and assess how well they were met before the accident happened.

What Is Human Error?

In Chapter 12, we looked at Norman's task action cycle and at slips and lapses. Figure 15.2 shows a simple flowchart for deciding whether an action was an error or a violation (a deliberate act not in accordance with correct procedures).

Human–machine interaction can be conceived of as a stream of behavior that is governed by the user's intentions, expressed as goals and subgoals, and executed using "pockets" of well-learned behaviors, known as plans. Norman (1981, 1984) calls these sequences of actions "action schema" (a kind of "memory-motor program" that can be triggered by particular events to enable a person to carry out a sequence of actions—such as tying one's shoelaces). Errors can occur when the wrong schema is activated, when the correct schema is activated but executed incorrectly, or when the schema is not activated when it should have been. Action schema is usually well learned and carried out using System 1, once initiated. Many of the errors that people make in everyday life can be explained once it is understood *that intentions are usually disassociated from the behaviors that are executed to satisfy them*. Errors can occur when the intention to act is correct but the wrong action schema is activated due to memory failure or task interference. Sometimes people lose the goal after activating the schema (e.g., finding themselves searching through a drawer but not knowing what they are looking for).

Error Categorization

A distinction is often made between errors of omission, commission, and psychomotor or "bumping" errors. Errors of omission involve an operator not doing something he was supposed to have done (e.g., forgetting to lock the door when leaving the house). Errors of commission involve executing a correct action at the wrong time or performing an incorrect action. Psychomotor errors involve accidentally operating a control or executing a sequence of actions in the wrong order.

Mistakes may be due to a lack of knowledge or skill (using the wrong rule or deciding on a course of action that is incorrect). Violations might be routine, to meet targets or for personal benefit, or due to recklessness or sabotage. They may also occur in exceptional circumstances when the operator has no safe way out but to break the rules. Error may also be viewed as a process that develops over time as follows:

- Error production
- Error detection, particularly self-detection
- Error identification
- Error recovery
- Error prevention

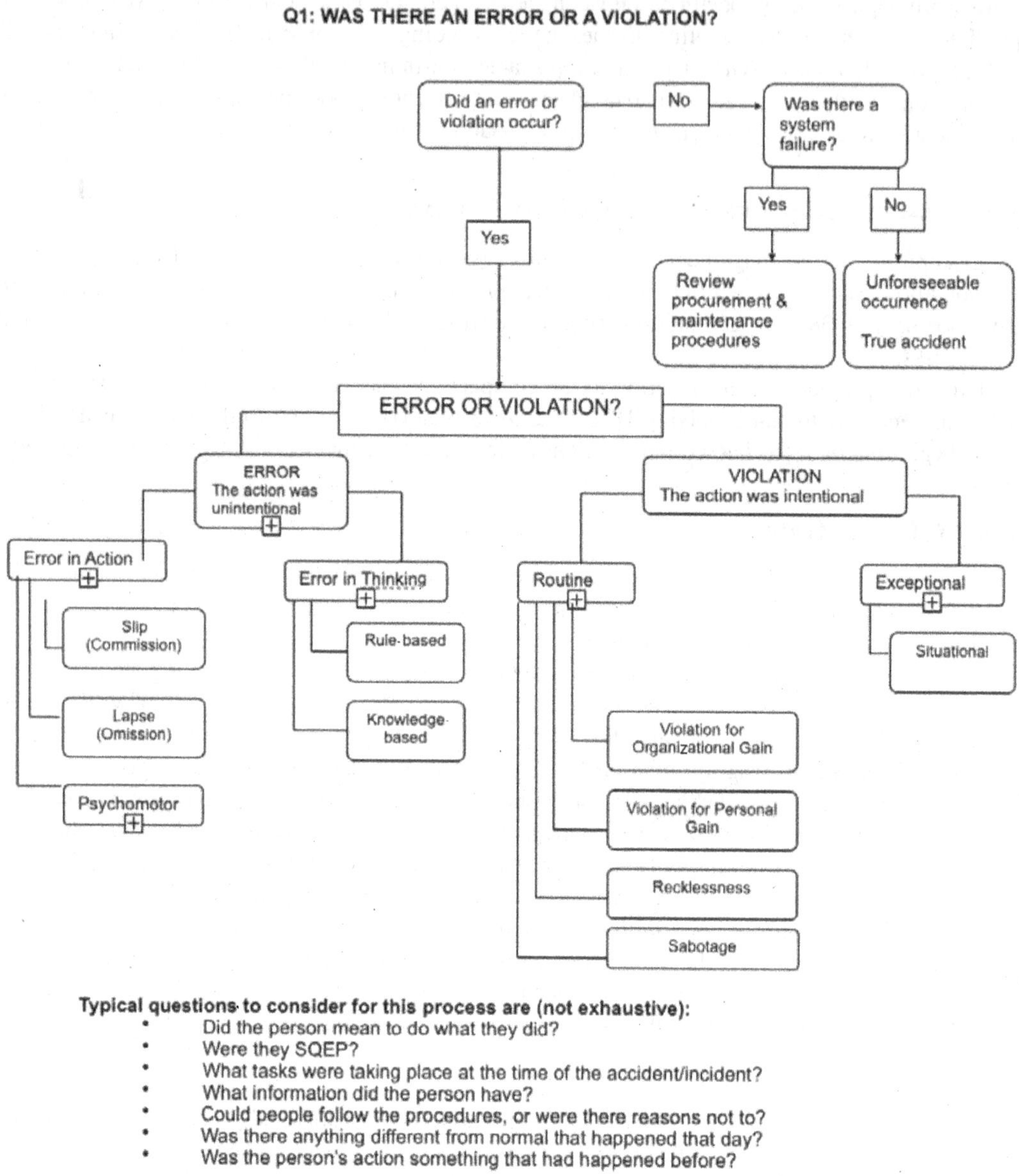

FIGURE 15.2 Flowchart for identifying errors and violations in accident investigations. Questions to consider when using the flowchart include whether the action was intentional, whether the person had the right training, what information operators had or what instructions they had been given, whether the person had done the same thing before, and whether anything unusual was happening at the time. (Adapted from Bridger et al., 2012.)

Error Production

1. Errors in the formation of the intention to act (e.g., choosing the wrong goal).

   Mode Errors: Involve misclassifying the situation. When mode errors are made, the actions are correct but are applied in the wrong situation. Mode errors may be difficult to detect until it is too late (as in cleaning the trigger of a gun when the safety catch is off).

   Description Errors: Occur when the intention is insufficiently specified, perhaps due to memory limitations or high workload. The late Douglas Adams described how he bought a packet of biscuits and a newspaper at a railway station and sat down at a table to eat the biscuits. A man sitting opposite him started to eat one of the biscuits. Adams was horrified but said nothing and took a biscuit from the packet. The man did the same. When the packet was empty, the man went away. Adams folded his newspaper only to find his unopened packet of biscuits on the table in front of him.

2. Faulty activation of the action schema (e.g., looking in the wrong place):
    - *Unintentional activation* occurs, for example, when a familiar habit is invoked at an inappropriate time. The psychologist William James described going to the bedroom to change for dinner and ending up in bed! Well-learned behaviors will often be triggered by environmental cues when attention is diverted elsewhere—when daydreaming or when distracted by sudden events, we are particularly susceptible to these errors.
    - *External activation* of action schema occurs when stimuli in the environment trigger unintended actions. Norman gives an example of someone who, after one failed attempt to buy cigarettes, tried again after getting more coins but put the coins into the adjacent coffee vending machine.
    - *Associative activation* of action schemas is similar to external activation and occurs when the stimulus is strongly associated with the desired response. Associative activation often occurs in speech when people misname a person or thing, but the name used is strongly associated with the one intended.
    - *Loss of Activation*: The classic example of loss of activation is when one finds oneself standing in a room or looking in a cupboard without knowing why one is there, or what one is looking for. The cause is often due to interference with memory, either by external events or by internally generated thoughts. Loss of activation, when it happens in the middle of a stream of closely linked behaviors, may result in one or more steps being missed; for example, pouring boiling water into an empty teapot. Loss of activation may result in steps being repeated if the achievement of the goal has not been registered (e.g., brushing one's teeth twice in the morning).
3. Faulty triggering of the schema (e.g., looking in the right place at the wrong time)

The traditional focus on human error and system safety is in cell A in Table 15.1—an error is made, the system crashes, so the blame is attributed to the error. If no error can be detected, the blame is directed elsewhere (cell C). Less attention is paid to the fact that people make errors all the time, and nothing happens (cell B). The failure to understand the kinds of errors that are routinely made and why the system is resilient most of the time is a major knowledge gap, and it is why many high-risk organizations encourage "near miss" reporting. Cell D is perhaps the most important of all and the most easily taken for granted—when all is going well—understanding the key ingredients that make a system safe.

TABLE 15.1
Human Error and System Safety

	System Crashes?	
Human Error?	**Yes**	**No**
Yes	A	B
No	C	D

A—Blame the user
B— Unknown Unknown
C—Blame something else
D—Taken for granted

Operator actions that are regarded as "error" when an accident has happened, and with the benefit of hindsight, may be nothing more than normal coping responses to the pressures in place at the time, seen as errors only because they are noticed. As we saw in the opening chapter, there are many interactions within systems, with the result that a small change to one part of a system (a work role or a device) can sometimes have unanticipated consequences. Change itself can be a hazard, depending on how well it is managed. Paradoxically, the more barriers put in place to prevent accidents, the greater the likelihood that one will fail.

Auto-Detection of Error

Sellen (1994) found three mechanisms by which people detected their own errors and also identified situations in which the errors were detected by others. The main mechanisms of self-detection are *action-*, *outcome-*, and *process-based* (in which the constraints in the external world prevent further action).

Action-based detection involves "catching oneself in the act." The incorrect actions are detected as they are being performed. In manual activities, the person may detect the error by means of tactile feedback. In speech, the error may be detected immediately when the word uttered does not match the word in mind. Action-based detection depends on the person paying attention to the task or to feedback from the task. If, for example, the wrong action schema has been initiated, this will only be detected if the person is monitoring the task and the intention (or goal state) is in working memory. Action-based detection only works when you are paying attention to what you are doing.

Outcome-based detection comes into play if action-based detection fails. Sellen gives an example where a person placed a teabag into a cup, intending to pour boiling water over it and poured coffee over it instead. The error was only detected when the teabag floated on the surface of the coffee. If the intention itself is wrong, and if the action is appropriate to the intention, then detection of the error based on the action is not possible. Sellen gives the example of a person walking down some stairs who thought there was only one more stair to go, when there were really two more. The feeling of the descending leg being in the air for too long and the painful recovery of balance was the only cue that an error had occurred. Outcome-based detection may be inadequate to prevent an accident or injury from occurring. There are several requirements for successful detection of error on the basis of outcome. First, the person must have expectations about what the outcome of the behavior is likely to be. Second, the effects of the actions must be perceptible. Third, the behavior and the system must be monitored throughout the action sequence. Fourth, the person must attribute any mismatches between desired and actual outcomes to their own actions and not to the hardware or to extraneous factors. Outcome-based detection works in two ways. The person detects a mismatch between the desired outcome and the expected outcome. This depends on the mismatch being visible, or indeed being alerting, in the case of a person who is not expecting anything to go wrong. If the person *is conscious of the possibility of error*, they may actively evaluate the outcome when it occurs. In January of each year, many people are conscious of the fact that they may write or enter the date incorrectly, using the number of the previous year instead of the new year, and actively check to see whether this has happened.

Process-based detection occurs when constraints in the world "short-circuit" the action schema, preventing further activity. Norman calls these constraints "forcing functions" and gives, as an example, the practice of putting a barrier on the ground floor landing of stairwells in tall buildings. In emergencies, when occupants evacuate these buildings using the stairs, they have been known to become disoriented and continue descending into the basement. Unable to escape, they may suffocate if the building is on fire. The erection of a barrier on the ground floor landing breaks the action sequence and acts as a cue for exit on the ground floor. Another example, this time in HCI, is the practice of interrupting the user when they wish to close a file that has not been saved. The system interrupts the action and switches to a warning mode in which the only legal key presses are those that will answer a prompt such as, "Do you want to save the changes to Chapter 15?" The key difference between forcing functions and other methods is that the error detection is cued by the design of the environment, not by user expectation or evaluation of feedback.

Latent Errors

Errors that don't always have immediate consequences are known as "latent errors." In areas such as aircraft maintenance and health care, there is often a delay between the error occurring and its consequences. Thus, timely identification and correction of the error can prevent serious outcomes. Saward and Stanton (2017) report a study of aircraft maintenance errors. Latent error detection was more likely:

- Within two hours of the error being made
- When the person is in the same location as where the error was made (appropriate cues may "jog" the person's memory)
- When the person is alone and is not distracted by others
- When "signing off" the task as close to where the error was made as possible (in health care, carrying out patient handovers at the bedside).

Human Factors Accident Classification System—HFACS

HFACS was developed by Dr. Scott Shappell of the Civil Aviation Medical Institute and Dr. Doug Wiegmann of the University of Illinois at Urbana-Champaign. HFACS classifies errors according to unsafe acts, preconditions for unsafe acts, unsafe supervision, and organizational influences (Figure 15.3).

Essentially, HFACS prompts us to look for *performance shaping factors* (PSFs) in accident investigation. PSFs are not causes; they are characteristics of the system, its operation at the time, and the environmental conditions that may have influenced the likelihood of error. Essentially, failure to meet any of the requirements is a possible PSF.

What Causes Human Error? A Closer Look at Performance Shaping Factors

Figure 15.4 shows how PSFs can influence the probability of error and the probability of an error leading to a safety occurrence.

PSFs can influence the severity of exposure to the hazard, the risk of exposure to the hazard, the operator's response to the exposure, whether the response was correct or not, whether an incorrect response was detected and appropriate action taken, and the severity of the outcome. There are also several individual and interpersonal factors that shape performance and need to the considered, as is illustrated below.

The Dunning–Kruger Effect

The Dunning–Kruger effect (Dunning, 2011) is a metacognitive bias in which low-ability individuals overestimate their competence. The very lack of competence in conducting the task is the source of individuals' inability to accurately rate their competence. Dunning put it this way:

> If you're incompetent, you can't know you're incompetent. … the skills you need to produce a right answer are exactly the skills you need to recognize what a right answer is.

High-ability individuals can fall foul of a similar bias—they overestimate the ability of others to conduct tasks that they themselves find easy. The existence of these biases underlines the need to formulate and communicate explicit standards of competence, provide training that delivers the required standard of performance, and conduct objective testing by an outside authority or third party. Because the ability of individuals can change over time, regular retesting may be indicated in high-risk industries. What appears to an outside observer to be a violation may really be an error caused by the illusion of competence on the part of the operator.

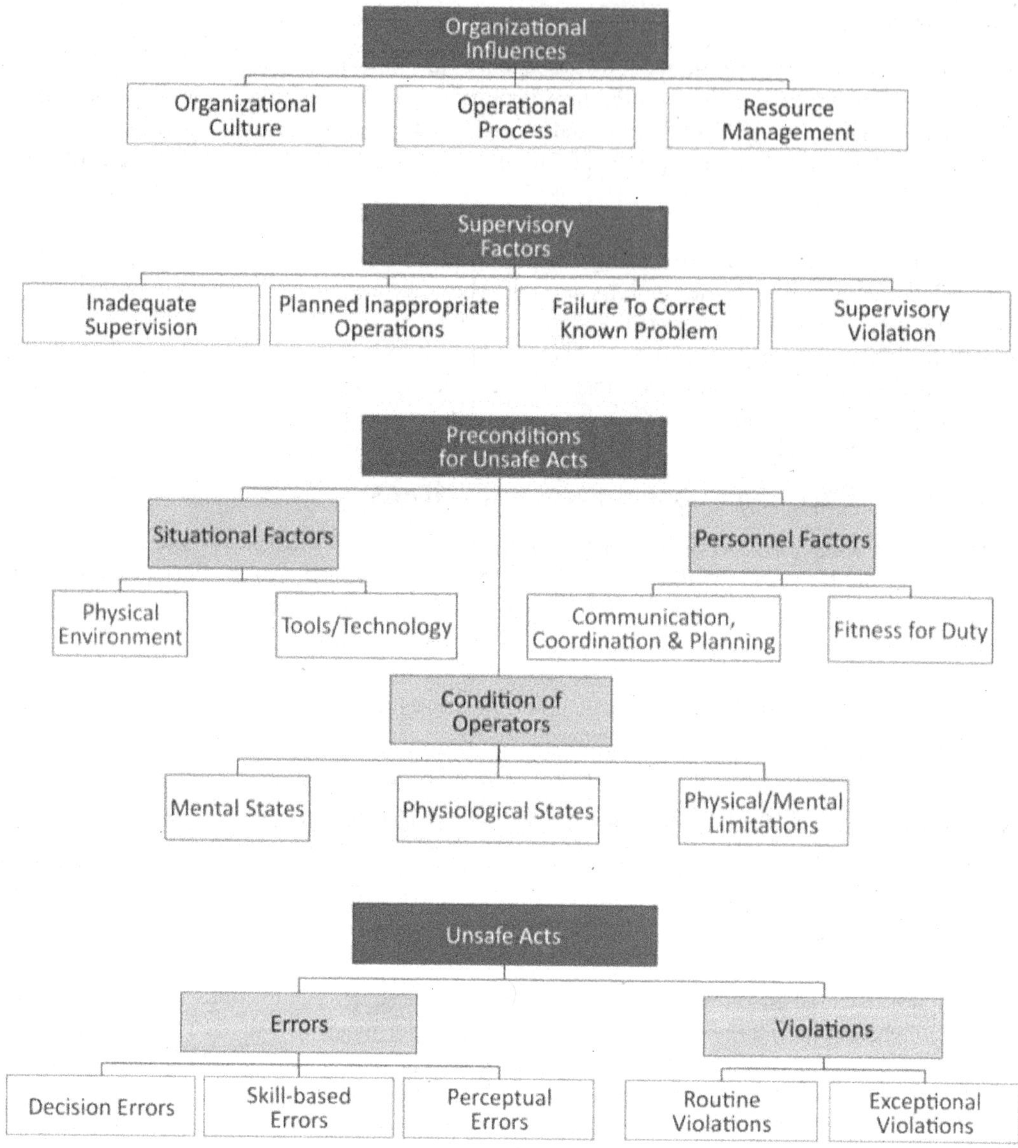

FIGURE 15.3 Human factors accident analysis and classification system.

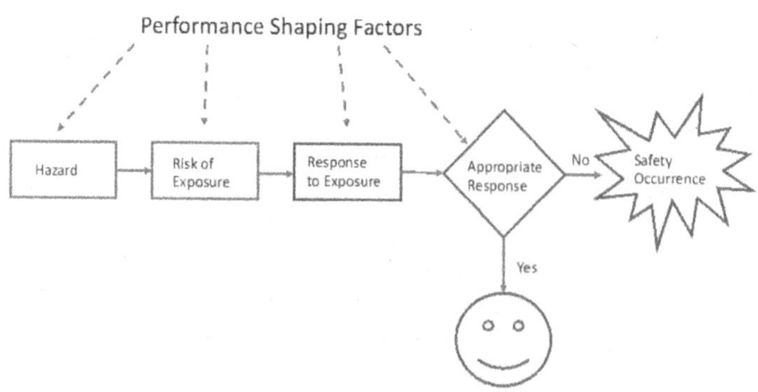

FIGURE 15.4 PSFs are everywhere. (From Bridger (2021), with permission.)

Situational Awareness

Accidents are often said to have occurred because operators lost "situational awareness." Endsley (1995) defines three levels of situation awareness:

Level 1, the perception of task-relevant elements in the environment
Level 2, the comprehension of their meaning in relation to task goals
Level 3, the prediction of future events

To be "situationally aware" means having a mental model of the task, "running" the mental model, and updating it as the task progresses. When you imagine yourself carrying out a task, you are "running" your mental model of that task (imagine yourself reversing your car into a parking space, for example). "Running" the model requires monitoring one's actions in relation to one's intentions and the state of relevant features of the task environment. Loss of "situation awareness" is nothing more than a mismatch between one's model of the task status at a particular point and the actual status of the task.

Accidents are often said to have occurred because individuals lacked "situation awareness." Really, this is just a verbose way of saying that they did not know what was going on and didn't do what, with the benefit of hindsight, should have been done to keep things safe.

Terms like "situation awareness" are only descriptive, and they do not describe much. They should not be used as a final explanation for what happened, as they do not explain *why* individuals did what they did or why the accident occurred. What is important is understanding the actual awareness that operators had at the time, and why this may have differed from what they should have been aware of, with the benefit of hindsight.

Professor Herbert Simon (quoted in Todd and Gigerenzer, 2007) used a metaphor that illustrates situation awareness: the mind and the world fit together like the blades of a pair of scissors. The fit between the blades determines whether the scissors cut properly. When we are examining human behavior and the reasons for it, merely focusing on human operators and attributing accidents to "human error" or "loss of situation awareness" explains nothing in the same way that just looking at one blade of a pair of scissors will not allow us to determine how well the scissors can cut. Understanding the world in which operators find themselves is vital if we are to understand their actions (Figure 15.5).

In Figure 15.6, we can see a telegraph pole in a parking bay in a parking lot. Drivers do not expect to find obstacles in the middle of unoccupied parking bays, and it is no surprise that so many people have reversed into this particular pole that the owners of the parking lot have installed protective padding, hazard tape, and a traffic cone to make the pole more conspicuous and lessen damage to cars when they do reverse into it. You could say that people reversed into the pole because they lacked situation awareness, *or* you could say that the accident happened because there was a hazard in the parking bay. Which explanation is most likely to lead to recommendations to prevent recurrence?

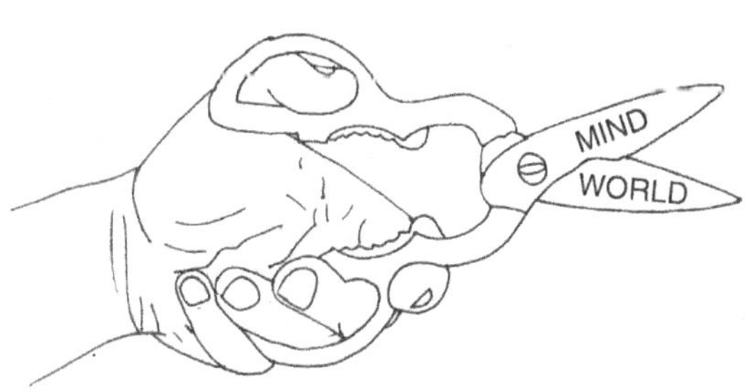

FIGURE 15.5 The scissors analogy. The mind and the world fit together like the blades of a pair of scissors. To understand why people behave as they do, we need to understand the environment that they are in as well as what is in their minds.

FIGURE 15.6 When drivers reverse into the telegraph pole, to whom should we attribute a lack of "situation awareness"? The driver, the owner of the parking lot, or the utility company? For a critique of the concept of Situation Awareness, see Dekker (2015).

Individuals are often blamed for unfortunate results, even though their incorrect actions may have been reasonable given the information they had at the time. If the investigator can find gaps in the operator's model, this may explain why the accident occurred. The next step is to explain *why* there were gaps in the model. In the telegraph pole example, two leads of investigation may help explain *why:*

Driver Expectation: Drivers don't expect to find telegraph poles in empty parking bays
Fatigue: The car park is on a route where many drivers would have driven for a long time prior to stopping, and they may be tired when they stop.

Driver expectation and fatigue are examples of PSFs which, although not root causes, help investigators to reconstruct and understand what happened—from the driver's point of view.

Accident Proneness

The concept of accident proneness emerged around the turn of the twentieth century, when researchers found that statistical models of industrial accident frequency did not follow a Poisson distribution, as had been anticipated (see Day et al., 2012). Rather, it seemed that individuals differed in accident propensity, leading to the notion of accident proneness. The idea that some people have a tendency to have more accidents than others lost favor in the latter part of the twentieth century, and researchers focused more on the design of the work environment and of the safety systems in place.

More recently, however, several studies have been published, which show that a measure of cognitive lapses and slips (Broadbent et al., 1982; Goodhew and Edwards, 2024) predicts safety behavior, slips, and lapses in daily working life. The cognitive failures questionnaire (CFQ, Chapter 12) is of interest because it measures people's discernment of their perception, action, and memory. Indeed, the everyday failures measured by the CFQ mirror the typologies used by many safety researchers to classify human error (e.g., omission, commission, and psychomotor errors). Accidents can result from distractibility, mental error, and poor selective attention (Wallace and Vodanovich, 2003). People with high scores on the cognitive failures questionnaire are more likely to have accidents due to lower attentiveness, which implies that they are always at greater risk than those with low CFQ scores. In other words, accident proneness equates with a tendency toward cognitive failure in daily life.

However, an addition to this line of thought is that accidents occur because people under stress are more susceptible to cognitive failures, and these lapses then cause accidents. According to this view, stress coupled with cognitive dysfunction is the major cause of accidents rather than the accident proneness of the employee per se. Öhman et al. (2007) highlighted the issue that patients with perceived chronic stress also report cognitive dysfunction. Specifically, they found that this dysfunction was a selective deficit mainly affecting episodic memory and divided attention.

These findings introduce the idea that accident proneness is both a trait (a stable feature of some individuals) and a state (a condition that can affect all, if under sufficient stress or time pressure). Day et al. (2012) found evidence that this was the case. Individuals who were stressed were more likely to have an accident in the workplace if they had a high propensity for cognitive failures.

It has been argued that the concept of accident proneness has little practical use because employers in many industries have to make the best of the workforce they have. A selection test for accident proneness would be difficult to validate and might exclude people who were suitable for the job in many other respects. However, the concept is useful when the role of stress is considered, indicating that a closer link is needed between health monitoring and safety management, and that workplaces can be made safer if they are designed to suit accident-prone employees. Additionally, systems should be designed for safe operation when attentional failure is likely (see Table 12.9 for guidance).

What Causes Violations? Risk Homeostasis Theory

Accidents sometimes happen because people deliberately break the rules or take risks. Whereas people don't go to work to make errors, violations are defined an *intentional* actions or decisions.

Violations are conscious actions not to conform to procedures, rules, or instructions and involve a deliberate deviation from the rules. Violations tend to be classified as routine or exceptional:

- *Routine Violations*: Noncompliance can become the "norm." These are habitual by nature and are often tolerated by the governing authority for a variety of reasons.
- *Violation for Organizational Gain:* Deliberately ignoring the rules so that the organization benefits in some way, for example, cutting corners in order to sail on time.
- *Violation for Personal Gain:* Deliberately ignoring the rules so that the individual benefits in some way, for example, cutting corners to finish early or "show off."
- *Recklessness:* Ignoring risks and the potential consequences for themselves and others.
- *Sabotage:* Occurs when there is intent for both action and consequence, for example, malicious damage.
- *Exceptional Violations:* Isolated departures from authority; not following the rules in unforeseen or highly unusual circumstances. This often occurs when something has gone wrong.
- *Situational Violation:* Rules could not be followed due to situation-specific factors, for example, excessive time pressure or unsuitable tools to complete the task.

Why do people do this? RHT was developed during research into road safety by Wilde (1994) and states that, when allowed to do so, individuals will modify their behavior so that the level of risk they are exposed to is compatible with their target (desired) level of risk. If there is a discrepancy between the target level of risk and the actual level, people modify their behavior to bring the actual level in line with the target level. For example, a driver on a long journey may increase their cruising speed when arriving at a motorway, only to slow down later when it starts to rain. *Risk compensation is simply behavioral adaptation to changing circumstances to match the perceived level of risk to one's target level (or comfort zone).*

RHT is a behavioral theory and only applies to situations in which people have sufficient control over their behavior to influence the level of risk. It only applies to situations in which people get sufficient feedback to enable them to estimate the real risk. RHT can be applied to car driving but not to paced industrial jobs where the worker has no control over the work rate or the work environment.

There is empirical support for the concept of risk compensation, if not for RHT itself. Wilde cites many examples of risk compensation having unexpected and unwanted effects. For example, in California in the 1970s, the introduction of compulsory child-proof containers for storing medicines led to an increase in the number of child poisonings in the following year (presumably because parents were less vigilant in storing medicines in "child-proof" bottles, and the degree of safety was relative rather than absolute).

What Determines the Target Level of Risk?

The target level of risk is determined by the aggregate value of the costs and benefits associated with risky and safe behavior.

According to Wilde, the best way to reduce accidents is to lower the target level of risk in a population of users:

- Where risk perception is faulty, for example, due to ignorance, by training people to improve their perception of risk
- By providing incentives to increase the benefits associated with safe behavior (e.g., free driving licenses for accident-free drivers)
- By increasing the costs associated with the unsafe acts (e.g., having dangerous drivers interviewed by a psychiatrist and recording this fact on their medical records)

RHT is controversial but has heuristic value—instead of blaming operators for following standard operating procedures (SOPs), we should inquire whether there is something in the procedure or the work organization that provides a perverse incentive to break the rule. Operators may "consume" any additional safety as a performance benefit, so the accident rate may remain unchanged. According to RHT, then, better roads and safer cars will result in more miles traveled, more quickly, for the same number of road deaths.

Seen from the perspective of RHT, violations are just another form of error, but at the organizational rather than the individual level.

Problem-Solving in System Operation and Incident Investigation.

Problem-solving can break down for a number of reasons (Table 15.2).

Wason has shown that one of the main weaknesses of human problem solvers is the preference for confirmatory evidence over evidence that will refute their ideas (Wason and Johnson-Laird, 1972). Preference for *confirmatory evidence* can have serious consequences for fault diagnosis and correction in process control tasks. For accident investigators, a one-solution fixation on the part of the investigation team can result in a biased report and an unjust outcome. The tendency to blame operators may lead investigators to focus too strongly on immediate events and not on the situational, organizational, and managerial processes that created the situation in which the operators found themselves at the time of the accident. One solution is to train investigators to recognize confirmatory behavior and provide them with better, specific problem-solving skills, including how to test their hypotheses by searching for evidence that will refute rather than confirm them.

TABLE 15.2
Breakdown of Problem-Solving Behavior

1. *The One Solution Fixation:* The problem solver may develop a "one solution fixation." After failing to solve the problem, they return to the same starting point, and use the same premises or approach again. The problem solver is unable to think of alternative approaches. This is sometimes accompanied by the irrational belief that it is the problem, rather than the solution, which is incorrect.
2. *Getting "Stuck in a Loop":* The problem solver may get "stuck in a loop"—repeating a set of moves that lead nowhere except back to the starting point. When this occurs, it can be particularly difficult to break out of.
3. *Inability to Think Ahead:* This is often due to memory limitations. In the missionaries and cannibals problem, and in games such as chess and draughts, the number of alternative configurations soon becomes unmanageable when the problem solver tries to consider all the possibilities even two or three steps ahead.
4. *Unwillingness to Consider Counterintuitive Actions:* Unwillingness to consider counterintuitive moves. In many problems, there are certain essential subgoals that must be executed in order for the main goal to be attained. Sometimes, these may appear to be in direct conflict with the main goal.

Safety investigators should be particularly aware of these weaknesses when they investigate safety occurrences. Excessive focus on those involved in a specific location at the time of the occurrence may lead to a "one-solution fixation," such as a search for the "root cause" and the tendency to blame the operators. Excessive focus on "finding the root cause," in the belief that if it can be found and fixed, the system will be safe, will narrow the scope of the investigation, and other, latent hazards may remain unexamined. A related problem is known as *bounded rationality*, which is essentially a failure to appreciate the "big picture" – a failure to take a systems approach using a range of competencies to understand what went wrong. Bounded rationality restricts the problem solver's efforts to a limited part of the problem space and reduces the chances of finding evidence to refute incorrect hypotheses. Avoidance of confirmation bias is particularly important in investigations. Instead of seeking evidence to confirm what is believed to be the cause, investigators should seek evidence to refute it. Hindsight bias (the belief that the incident was foreseeable and preventable) may lead investigators down a path where they seek confirmatory evidence rather than evidence to open up new perspectives on what happened.

MacLean (2022) discusses eight sources of bias that can affect the accuracy of safety investigations (Table 15.3).

According to MacLean, "bias cascade" occurs when an initial bias influences the capture and interpretation of new information. Thus, new information is not independent of what was discovered before. To manage bias, MacLean suggests the following:

- *Blinding*: Irrelevant information is withheld from the investigators. This includes material that is not relevant to the task. In scientific peer review, for example, the names and affiliations of authors are often withheld from reviewers.
- *Sequential Unmasking:* Information that might bias the investigation is withheld. Data from the occurrence should be examined prior to the examination of reference materials.
- *Independent Assessment and Verification:* Local investigators may be biased by their knowledge of the organization and the task. Independent assessors with no personal connection to the organization or employees can be blinded to the findings of the internal investigation.
- *Information Saliency to Combat Base Rate Neglect:* Investigators may be biased by inaccurate beliefs about the likelihood of events. The provision of accurate information can correct this bias.
- *Supportive Organizational Culture:* Organizational support for thorough investigations can reduce bias by ensuring that investigators have access to all relevant materials and the time to scrutinize them thoroughly.
- *Documentation:* Investigators should document their activities in real time, keeping proper records at all stages. This prevents the kinds of "mission creep" that can occur during a lengthy process, as past thoughts and insights are re-interpreted in the light of new information.

TABLE 15.3
Sources of Bias in Safety Investigations

	Source of Bias
Data	Data is always incomplete since investigations are conducted "post hoc." As such, readily available data may be perceived as more salient than it really is.
Reference materials	Manuals, checklists, and "models" may bias investigations because of what they encourage investigators to focus on. Once focused, investigators interpret subsequent information in accordance with their prior understanding
Context	Contextual information may be irrelevant. Leading questions may bias witness statements (e.g., How fast was the forklift travelling when it crashed into the office?)
Base rate	Experience with similar occurrences may provide useful contextual information to aid in the assessment of operator actions (e.g., false alarms may be common and real alarms more likely to be ignored). Informal working practices may be better understood by local teams of investigators.
Organizational factors	Tight deadlines, long working hours, and cognitive fatigue may promote System 1 thinking over System 2 and reduce the accuracy of the analysis. The working relationship between investigators, operators, and injured persons may affect judgment.
Education and training	Training in accident investigation and human factors supports a systems approach to the investigation.
Personal factors	Personality factors such as "conscientiousness"and being open-minded support the search for understanding. Low tolerance for ambiguity may lead to premature/closed thinking.
Cognition	Heuristics and biases (Chapter 12). "Jumping to conclusions" (System 1 takes over). Confirmation bias.

Source: After MacLean (2022).

Safety occurrences present an opportunity to direct human and financial resources to investigate the safety of a system. The purpose of safety investigations is to improve *system safety*, not to find out who or what was to blame. Where operators or suppliers *are* thought to be blameworthy, a parallel investigation is sometimes conducted independently by a different team with a different skill set. Even if someone is to blame, an incident is always an opportunity to investigate *system* safety.

Rationalization

Rationalization is the post-hoc construction of beliefs or desires to make past behavior seem rational. Human behavior is influenced by instincts, norms, and habits—often automatically or at least without careful consideration of what to do or how to respond (Cushman, 2020). If someone reacts automatically in daily life (under System 1) and then seeks to understand why, an explanation may be constructed to make the behavior seem rational. For example, a man, while enjoying a carefree evening out drinking with his friends, buys another round of drinks and arrives home late. When he arrives home, his wife asks him why he is home so late. He replies, "I had to buy another round of drinks because it was my turn." This, of course, is a post-hoc rationalization. Rationalization is a way of explaining non-rational or instinctive acts after the event. Rationalization underpins cognitive dissonance, which is a state of anxiety and discomfort when one's behavior does not align with one's core beliefs. Dissonance is reduced by rationalizing either the behavior or changing the beliefs to achieve a state of consonance.

When investigating safety occurrences, both investigators and operators (during an interview) may fall prey to rationalizing their actions. Confirmation bias may occur because investigators wish to rationalize the next steps taken in an investigation (information that is easy to find is seen as more important than information that is difficult to find). Operators, when asked to explain their actions at the time of the accident, may construct rational explanations when, in fact, they acted instinctively or out of habit (e.g., "I didn't check the pressure gauge because the pressure relief valve indicator was working OK"). It is for this reason that investigators should try to understand how employees actually behave at work, and this is one reason why it is often useful to have someone on the investigation team who has direct experience of how the work is done in practice.

Work as Imagined (by Whom?) Versus Work as Done

The author once worked on an investigation where an engineering technician was electrocuted while carrying out tests on a piece of high-voltage equipment. A highly choreographed standard operating procedure (SOP) was supposed to be followed in a repetitive, stepwise fashion to ensure that the door to a cabinet containing the equipment was not opened until the power had been cut off and the interior earthed, to ensure that the operator was not exposed to 50,000 volts. If the procedure was adhered to, the task was perfectly safe, but it was repetitive, taking about one hour to complete. After each of several tests inside the cabinet, the cabinet door had to be closed and the system reactivated before shutting it down and earthing it before opening the door again for the next test. By taking shortcuts through the procedure, the technician was able to complete the task in about 20 min. However, he lost track of where he was in the procedure and opened the door before the system was earthed, and he was electrocuted (surviving after receiving emergency medical attention and a stay in hospital). Several performance-shaping factors were identified, including not only the fact that the interlock on the cabinet door (which automatically earthed the system if the door was opened) had been disabled, but that anyone could disable it, making the system open to sabotage. A wider investigation then took place, and a team of psychologists trained in interviewing techniques conducted interviews with similarly trained technicians. Although all of them denied ever conducting work on high-voltage electrical equipment without strict adherence to safety protocols, several stated that "they knew people who did."

So, the accident occurred primarily due to a violation for personal gain (to simplify the task and save time), but the investigation found that the barriers (safety interlocks on the cabinet door) were inadequate, making the violation easier to do. Furthermore, the equipment had not been designed for ease of use when tests were done, probably due to missing requirements in the system's requirement document.

Work as imagined versus work as done is a term used to illustrate the fact that employees do not always follow all the procedures and rules in their job descriptions for various reasons, including:

- SOPs are cumbersome (see above example)
- New rules may not be compatible with old rules
- Too many procedures degrade performance
- Work pressure and deadlines make shortcuts necessary
- Equipment and/or staff shortages lead to "workarounds"

A good question to ask is, "Imagined by whom?" In most industries, it is expected that employees use their own *common sense and initiative* to get the job done safely. Management understands, employees understand it, and it sometimes becomes clear when people *actually follow all the rules.*

Working to Rule: It is a form of industrial action by trade unions, short of going on strike, where employees follow all the rules literally and strictly, often leading to drops in efficiency. Working to rule may be done with malicious intent to cause productivity problems but can also be seen in a positive light. In many organizations, the "job gets done" only because employees go beyond their contractual obligations and work more and/or harder than they need to. In one example, nurses were found to work unpaid overtime and during their rest breaks out of dedication to their vocation. In a survey, 59% of respondents said that, if industrial action were required, they would refuse to work unpaid overtime, and 53% would insist on leaving the ward or clinic for breaks. In another example, the trade union of a large number of UK teachers advised members not to send or respond to emails outside of working hours, not to provide cover for absent colleagues, and not to complete any admin duties such as photocopying or run extracurricular activities. In both cases, working to rule was chosen in place of striking. What is really interesting in these examples is that they demonstrate that the nurses and teachers did work unpaid overtime and through their breaks—an excellent example of 'work as done' and not in strict accordance with the rules

The above examples are relevant to safety and to accident investigation because they illustrate that many organizations function effectively due to the goodwill and dedication of their employees and that informal ways of working are, if not the norm, then very common in many industries. Therefore, to claim in an investigation report that the accident happened because the correct procedure was not followed really begs the question – it explains nothing. A better question to ask might be "What's wrong with the procedure?"

Danger! A Plethora of Safety Models and Methods

Wienen et al. (2017) conducted a systematic literature review of 216 articles, making a useful distinction between the models and methods described in the articles.

Method: A sequence of steps to lead to a conclusion (like a recipe)

Model: A conceptual structure by which an incident is understood (like a "cover story")

So, a method is like a recipe that says nothing about what a safety occurrence is. A model is a representation of the safety occurrence—an explanation or "cover story" as it were, of what the safety occurrence really was. Wienen et al. categorized the models and methods as follows:

Linear: Linear models see the safety occurrence as the end point of a series of events (one thing leads to another, and then the accident happens).

Epidemiological: These models see accidents in a similar way to how epidemiologists see diseases—as a result of the interplay of different risk factors that can interact in complex ways.

Systemic: Systemic models see the interactions between system components and the surrounding environment, focusing on how these components interact and how these interactions are affected by feedback. For HFE, prEN 14386: 2002(E), as detailed in the Preface, is a good model to use to consider how different system components interact.

Other: Models and methods not in any of the above categories.

A total of 63 different models and methods were found. Some examples are given below:

Linear: Linear models see the accident as the result of a "a series of unfortunate events"—one thing leads to another as the risk increases over time (Figure 15.7). Factors contributing to the event are analyzed according to their role in accident progression (e.g., Ambulance crew fatigued AND ambulance delayed LEADING TO heavy traffic OR diversion LEADING TO Patient losing blood AND rare blood group LEADING TO Patient dies). Fault tree analysis (FTA) is often based on probability theories or Boolean logic. FTAs can have barriers but don't have feedback loops and the method lends itself to safety occurrences where the facts are well understood and the probability of failure is known (see Workshop 15.2). In process control industries, where control is event-based and dynamic, FTAs can give a detailed summary of what happened and what kinds of interventions or modifications could improve safety.

Epidemiological: These methods take *latent conditions* into account (similar to risk factors in epidemiology). These latent conditions can be environmental, organizational or have to do with resources and economic drivers. As such, the analyses take longer than FTAs but take socio-technical factors into account. According to the review by Weiner et al., ACCIMAP (Figure 15.8) is one of the most well-known of such methods (Svedung and Rasmussen, 2002).

Accimaps show the flow of events at different levels in a hierarchy of PSFs, directing investigators to search for high-level factors as well as immediate events around the accident.

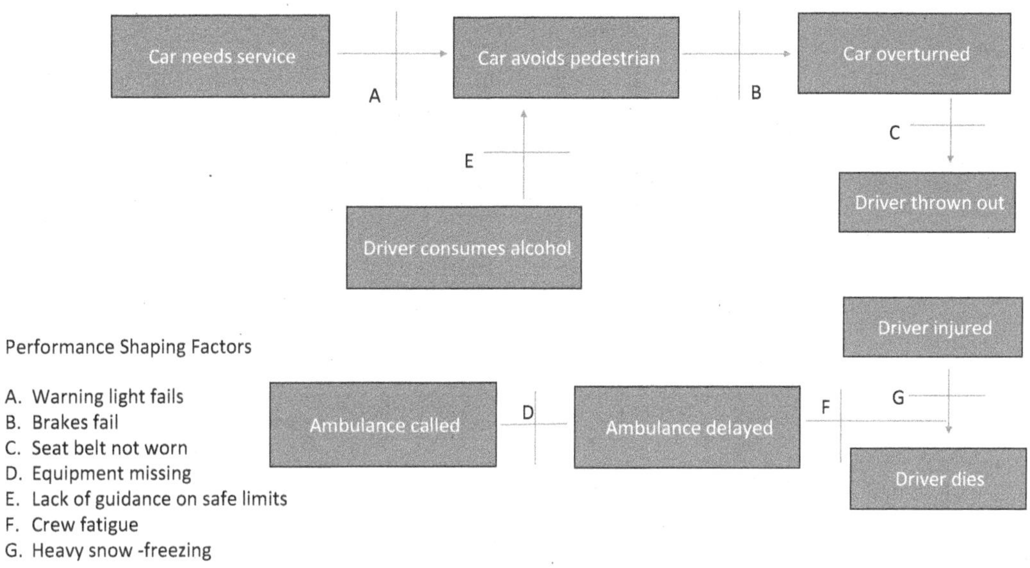

FIGURE 15.7 A linear model.

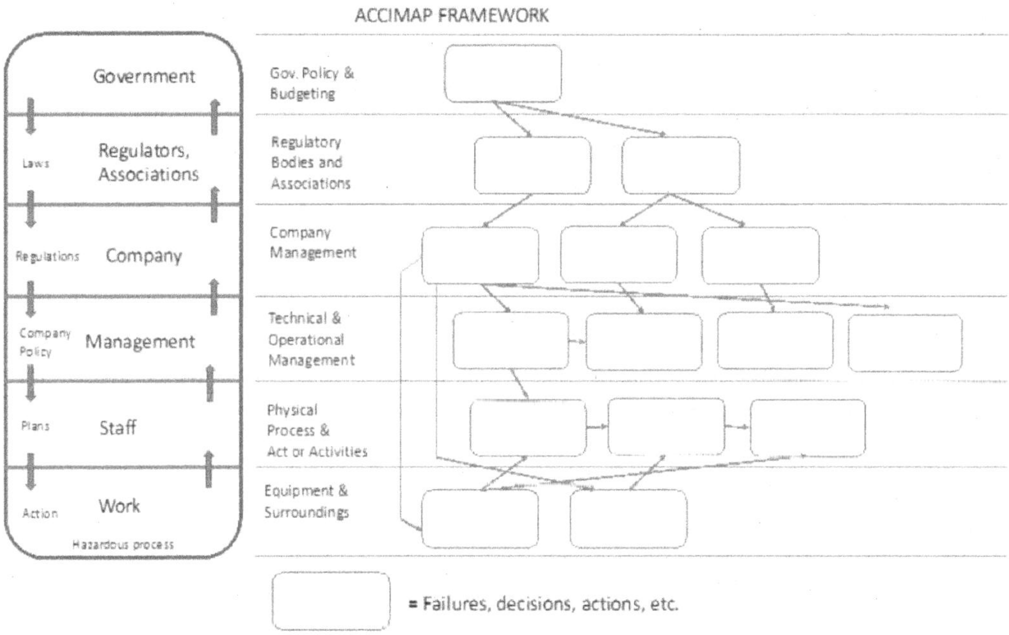

FIGURE 15.8 Accimap (after Svedung and Rasmussen, 2002).

The analysis also requires an analysis of how different factors interact with each other. Safety recommendations may therefore extend to managerial, supervisory, or educational issues.

Systemic: STAMP (Systems Theoretic Accident Model and Process) was found to be the most frequently cited systemic model by Weiner et al. Leveson (2011) offers a model that views safety as a control problem (Figure 15.9). Safety is about designing what *can* happen so as to control what happens next.

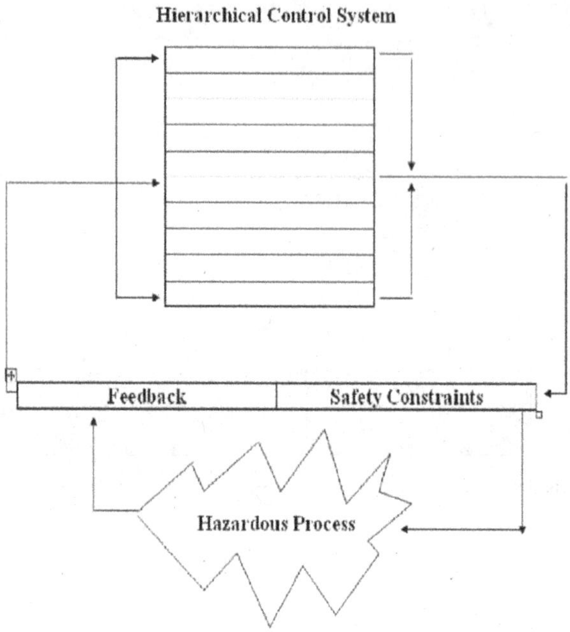

FIGURE 15.9 Safety as a problem of control (a visual summary of some key ideas from Leveson, 2011).

We start with the assumption that all systems are hazardous and will only be safe if they are properly controlled:

- Low-level controls include, for example, fuses, relief valves, and other engineering solutions.
- Intermediate-level controls include, for example, thermostats and devices that monitor aspects of the process.
- Higher levels of control include, for example, standard operating procedures, maintenance programs, and training programs.
- The highest levels of control include, for example, safety committees, wider policy and legal frameworks, national and international standards, and legislation.

In a safe system, a hazardous process is controlled by

- A hierarchical control system
- Safety constraints
- Feedback about the operation of the process and the operation of the constraints

Therefore, when investigating an accident, we can ask ourselves the following questions:

1. Were all the appropriate constraints in place?
2. Was the process controlled correctly?
3. Was the correct feedback received?

As can be seen, STAMP links very tightly with the basic HFE model presented in the Preface to this book. It focuses on the immediate events during a very thin slice of time at a critical moment, seeking to find *risk factors for human error* in the proximal workspace of the operator (the inner part of prEN 14386: 2002I in the Preface). Many of these risk factors have been described in previous chapters. It also looks more widely at the organization in which the operators found themselves and at the events leading up to the accident, seeking to discover *why* the risk factors were present in the environment when the accident took place—the outer part of prEN 14386: 2002I. In the preceding chapters, we have already reviewed many of the ideas needed to conduct an analysis to understand why critical signals might not have been detected, why information might have been misinterpreted, or instructions misunderstood.

Weiner et al. point out that all three methods have commonalities as well as differences:

- Finding all events that have a causal relationship with the accident.
- Describing the history of the accident by linking these events.
- Finding all the conditions that provoked these events, including events that led to those conditions (only in Epidemiological and Systemic methods).
- Identifying components, feedback mechanisms, and control mechanisms that played a role during the development of the accident (only in Systemic methods).
- Identifying points at which the accident could have been prevented, and analyzing if this can be generalized.
- Drawing conclusions and proposing improvement actions.

Wienen et al. concluded that, although there are a great many models and methods, most of them were cited only a few times in the literature they reviewed. They found no evidence for a general incident or accident investigation model, although they acknowledged that there would be many advantages for an organization that chose one method or model as a standard format for investigating and reporting safety occurrences. It almost certainly boils down to fitting the model to the accident and the resources available to investigate it. Despite their shortcomings, linear models, such as FTA, provide a structured and targeted approach that enables incidents to be analyzed and remedial action to be taken quickly. Epidemiological and systemic approaches are more time-consuming and are probably better suited to major catastrophes where an official inquiry is underway.

BASIC APPLICATIONS

STAMP sees a safety constraint as a mandatory safeguard that places limits on what can happen. Designers must consider which constraints to include in a new system, and accident investigators should seek to understand whether the constraints in place were sufficient to prevent the accident from occurring:

A Safety Constraint was not Provided: Mosenkis (1994) reported that some medical devices can be misconnected due to the design of their leads and plugs. An apnea monitor had an electrode at one end (placed on an infant's chest) and a plug at the other that could be attached to the AC mains electricity supply. Several infants were electrocuted because people saw unconnected leads close to power connectors and assumed they should be connected. Similar problems have occurred with enteral feeding tubes. Liquid feed is slowly pumped into the patient's stomach via a gastric tube. These pumps have inadvertently been connected to the cuff inflators on endotracheal tubes (providing an airway to the lungs). When the cuff inflator bursts, fluid is pumped into the lungs. Unconscious patients have been known to drown in their beds when this happens. Such errors happen because of the *affordances* offered by unrelated components. In both these examples, different parts of different systems shared the affordance that they could be connected and so sometimes they were connected. *Protective incompatibility* is a constraint that can prevent disastrous results!

A Safety Constraint(s) was not Applied: In 1987, The Ferry, Herald of Free Enterprise, was *at* Zeebrugge in Belgium. It was designed for the Dover–Calais route with a double linkspan at both ports. Vehicles could be loaded simultaneously onto G and E/F vehicle decks through vertically hinged watertight doors. She had not been designed for this port, and only one ramp could be used for loading vehicles. To load the higher decks, the bow ballast tanks were filled because the ramp was not high enough to reach the E deck. After all the vehicles had been loaded, the tanks were NOT emptied, meaning the bow was lower in the sea than normal. The assistant Boatswain was supposed to close the G deck bow doors BEFORE the ship slipped her moorings. He had, however, gone for a nap after cleaning the deck. The first officer was supposed to remain on G deck until after the doors had closed but is believed to have been under pressure to get to his station on the bridge, believing that the assistant boatswain was on his way. *The bow doors could not be seen from the bridge.* At 18:05 h, believing that the bow doors had been closed, and unable to see them from the bridge, the Captain gave the order to depart. The ship had 80 crew, 459 passengers, 80 cars, 3 buses, and 47 trucks. 18:24: The ferry entered the open sea. When the ferry reached 18.9 knots (21.7 mph), water began to enter through the G deck doors. The ferry capsized in 90s. The electrical and power systems failed immediately, leaving the interior in darkness. The ferry floundered in shallow water onto a sandbar that prevented her from sinking completely. Despite immediate attempts at rescue, 189 people died. The water temperature was 3°C—most drowned after being disabled by immersion in freezing water.

A Safety Constraint(s) was Misapplied: On October 31, 2014, SpaceShipTwo (SS2) operated by Scaled Composites LLC, broke up into multiple pieces during a rocket-powered test flight and impacted terrain over a 5-mile area near Koehn Dry Lake, California. Scaled Composites was operating under an experimental permit issued by the Federal Aviation Administration (FAA) Office of Commercial Space Transportation (AST). The pilot received serious injuries, and the copilot received fatal injuries. SS2 was destroyed. It had been released from its launch vehicle, WhiteKnightTwo (WK2), about 13 s before the structural breakup. SS2 had been designed with a "feathering" mode on its tail fins to slow reentry. On "feathering," the tail fins rotated to slow the descent. The system was operated in two stages—the lever to deploy feathering mode had first to be manually unlocked. After release from WK2 at an altitude of about 46,400 ft, SS2 entered the boost phase of flight. During this phase, SS2's rocket motor propelled the vehicle from a gliding flight attitude to an almost vertical attitude, and the vehicle accelerated from subsonic speeds to supersonic speeds. At this time, the copilot unlocked the feather just after SS2 passed through a speed of 0.8 Mach, and the feather extended uncommanded, causing structural breakup. SS2's feather control system prevented inadvertent operation of the feathering system by the copilot in the "locked" mode but was unsafe when in the "unlocked" mode at certain speeds and altitudes. There was no interlock or other method of preventing deployment of the tail feathers or preventing unlocking them at certain speeds and altitudes.

A Safety Constraint(s) could not be Activated: Users may assume that if they know how to use one type of device, they know how to use all types. Problems occur when different makes have different control logic. Good practice standardizes the procedures for operating different types of products to enhance positive transfer of learning. Dixon (1991) demonstrated that even if there is a large conceptual difference between two systems, procedural similarity enhances the learning of the other (e.g., landing an aeroplane and shutting down a nuclear power station). There have been several instances where patients have died in hospitals because the clinical personnel did not know how to operate a piece of safety equipment. In one case, the mayor of a small town in South Africa had a heart attack while undergoing routine surgery. A new defibrillator was brought into the operating theater, but the staff did not know how to turn it on, and the patient died.

A Safety Constraint(s) was Applied too Late: Gould et al. (2006) report that a Norwegian offshore patrol vessel hit a submerged rock after an inexperienced navigator lost control over his exact position. Failing to observe a waypoint, the vessel was late in turning and hit a submerged rock. This was a knowledge-based error caused by a lack of training in the identification of waypoints.

A Safety Constraint(s) was Removed too Soon or Applied for too Long: On October 31, 2014, at 1007:32 Pacific daylight time, Virgin Galactic's SpaceShipTwo (SS2) operated by Scaled Composites LLC, broke up into multiple pieces during a rocket-powered test flight. SS2 had been released from its launch vehicle, WhiteKnightTwo (WK2), N348MS, about 13 s before the structural breakup. Scaled was operating SS2 under an experimental permit issued by the FAA's Office of Commercial Space Transportation. After release from WK2 at an altitude of about 46,400 ft, SS2 entered the boost phase of flight. Although the copilot made the required 0.8 Mach callout at the correct point in the flight, he incorrectly unlocked the feather immediately afterward instead of waiting until SpaceShipTwo reached the required speed of 1.4 Mach.

A Safety Constraint(s) was Unsafe: A Norwegian coastguard vessel grounded when the retractable sonar dome was damaged as the vessel entered shallow waters. The sonar indicator was only visible from one side of the bridge, leaving the navigator unaware of the vessel's depth. The navigator could not see the indicator from his position because of the design of the bridge and failed to retract the dome when the vessel entered shallow water.

A Safety Constraint was Ignored: Prior to leaving the Port of Zeebrugge, the boatswain—the last person on G Deck of the Herald of Free Enterprise—said that he did not close the bow doors of the ferry because it was not his duty.

A Safety Constraint was Inadequate: Mosenkis gives the example of a contact lens package that contained two almost identical bottles, one containing a caustic lens cleaning fluid and the other containing saline to wash the fluid off when the lens had been cleaned. Small, difficult-to-read labels were the only clues as to the contents of the bottles. Eye damage occurred when users cleaned the lenses with saline and then washed them with cleaning fluid. The example illustrates the failure to take into account the users' capacity throughout the task—at the critical point, the user's vision is at its poorest. Shape and color could be used to distinguish the two bottles (e.g., a red, hexagonal bottle for the cleaner and a blue, smooth one for the saline).

Feedback about Safety Constraints was Inadequate, Misleading, or Missing: During the Three Mile Island nuclear accident, the operators commanded a valve to open and received feedback that the valve had opened as a result. However, the valves were wired to provide feedback that power had been applied to the valve, but not that it had actually opened. This feedback was not only inadequate but also misleading, since the information it provided to the operators was false.

Unbounded Rationality: Why Were Risk Factors Present?

The inquiry into the loss of SS2 determined that the probable cause of the accident was Scaled Composites' failure to consider and protect against the possibility that a single human error could result in a catastrophic hazard to SS2. This failure set the stage for the copilot's premature unlocking of the feather system as a result of time pressure, vibration, and loads that he had not recently experienced, which led to uncommanded feather extension and the subsequent aerodynamic overload and in-flight breakup of the vehicle.

Human factors were identified as contributing to the error:

- *Design Flaws*: The feather control system prevented inadvertent operation by the copilot in the "locked" mode but was unsafe when in the "unlocked" mode at certain speeds and altitudes.
- There was no interlock or other method of preventing deployment of the tail feathers or unlocking them at certain speeds and altitudes.

- *Inadequate Training*: Test pilots did not properly understand the risks of unlocking feathering early.
- *High Workload*: The copilot was under high workload and had to recall tasks from memory under time pressure.
- The pilots were not accustomed to the levels of vibration and acceleration to which they were exposed.

Why was the Feather Control System Unsafe in Certain Flight Modes? Because

- SCC engineers had assumed that human error would be a response to system failure, not a cause of it. The systems safety analysis correctly identified the hazards of uncommanded feathering due to multiple system failures. It did not, however, consider that a single human error could cause uncommanded feathering.
- The FAA had issued experimental permits after failing to recognize that Scaled Composites' hazard analysis did not meet regulatory requirements to identify hazards caused by human error.
- FAA failed to consider whether Scaled Composites' mitigations would adequately address human error and its consequences.
- FAA managers exerted pressure to approve permits quickly.

Why was the Systems Safety Analysis Inadequate? Because

- There was no formal process for human error identification and analysis
- The development process with a "Test and Fix" philosophy was in place

Why were the pilot and copilot exposed to levels of vibration and acceleration to which they were unaccustomed? Because

- There had only been one previous 16-s rocket-powered flight to 55,00 ft in the previous April.
- SCC's flight simulator was static and could not simulate the accelerations to which pilots were exposed during flight.
- Aircrews were not issued G-suits.

Overall, there was a lack of regulatory oversight from the Federal Aviation Authority. FAA inspectors were assigned to operators, not to missions, and consultation did not begin early enough in the design phase to support proper oversight over the CADMID cycle.

NTSB Chairman Christopher Hart stated, *Many of the safety issues that we will hear about today arose not from the novelty of a space launch test flight, but from human factors that were already known elsewhere in transportation.*

In both of these examples, decisions and actions made long before created the conditions which made it possible for the accident to happen.

Work as Done—a Patient Safety Example

The author recently provided HFE support to a healthcare investigation. A 78-year-old woman who had suffered a stroke (a blood clot blocking the flow of blood to the brain) was admitted to the emergency department of the hospital. After diagnosis, she was admitted to an acute stroke unit for initial treatment, as set out in the SOP governing the treatment of stroke patients.

A locum doctor then conducted an initial risk assessment, in accordance with the policy, which led to the *decision* to fit an intermittent pneumatic compression (IPC) device to her legs (Figure 15.10). IPC devices prevent the formation of blood clots in the legs of immobile patients by squeezing the legs every 30 s. Recall that the calf muscles are often called the "second heart" because their contractions are needed for the return of blood. The IPC squeezes the legs of bedridden patients, pushing blood back toward the heart. This prevents blood from pooling in the legs and the subsequent

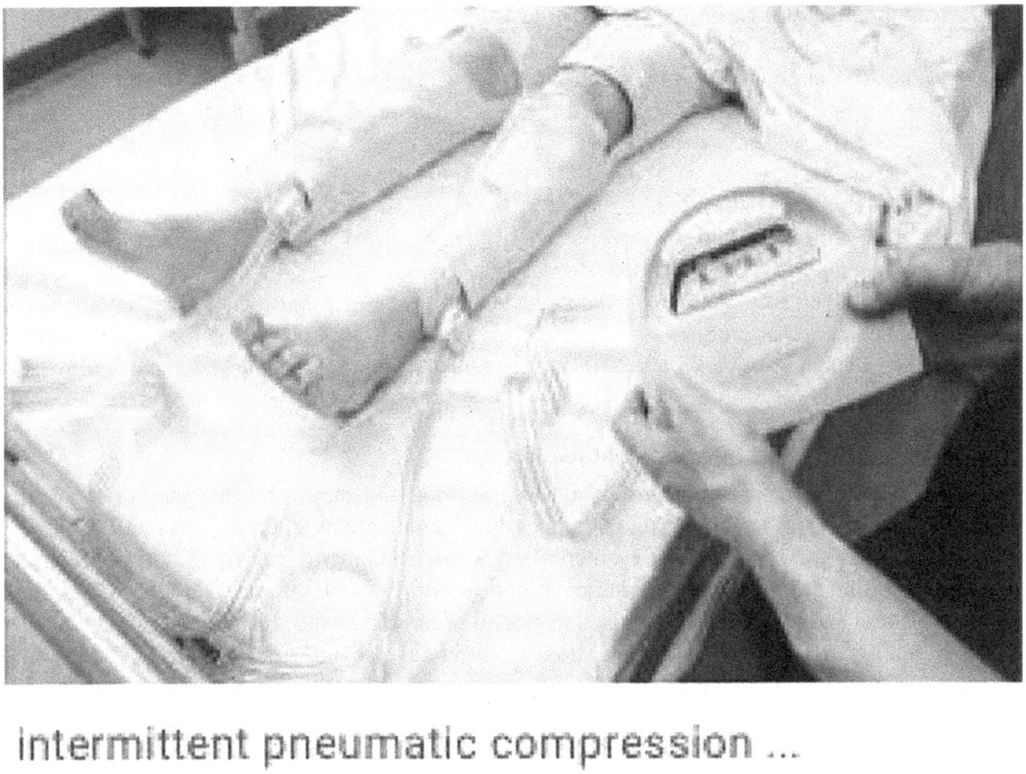

FIGURE 15.10 An intermittent pneumatic compression device.

formation of blood clots. These blood clots can migrate to other parts of the body, posing a serious health risk. When they lodge in the lungs, this is called a "venous thromboembolism."

The doctor completed a risk assessment in accordance with the SOP. This was done using a standard flowchart, which led to the decision to fit the IPC device to the patient. The next step in the SOP was to write a prescription for the IPC device on the patient's treatment board (a treatment board is a board where all the medicines prescribed for the patient are written down for action by nurses or other responsible specialists). However, the doctor did not write the prescription to the board, and the requirement to fit the device was not recorded. Because of the division of responsibilities in this case (doctors treat patients; nurses care for patients), it was not the doctor's responsibility to fit the device. It was the nurses' responsibility to administer the treatments listed on the treatment board. However, because the prescription for the device was not written on the board, the device was not fitted. The risk assessment was supposed to be repeated after 12 h according to the SOP. This was not done, and the IPC device was not fitted.

A few weeks later, the patient started to have trouble breathing after a blood clot appeared in her lungs.

What was Supposed to Happen?: The use of IPC devices for stroke patients is conducted in accordance with a policy. The application of the IPC device to the patient's legs, if the device is required according to the risk assessment, is the last action in a sequence of actions following admission and diagnosis. After diagnosis, thrombolytic drugs are first administered to dissolve the clot in the brain. The patient is then seen by a doctor, who completes the risk assessment form (in fact, this form was really a decision flowchart that led to a decision, such as whether or not to apply the IPC device). If the IPC device is not contraindicated, the IPC is to be prescribed by the doctor. If prescribed, the doctor is to write the prescription on the treatment board. When the nurse sees the prescription, the device is then fitted to the patient's legs.

The "risk assessment" is then to be completed again within 12 h.

What Happened?: Everything went well initially. The doctor completed the risk assessment form. The IPC was prescribed, then:

- The doctor forgot to write the IPC device on the treatment board.
- The device was not fitted.
- The risk assessment was not repeated 12 h later.
- A second opportunity to apply the device was missed.
- About 2 weeks later, the patient developed trouble breathing—a blood clot had appeared in her lungs.
- The patient survived after further treatment but required a *much longer stay* in the hospital.

Hazards and Risks: Absolute and Relative Risks

Being immobile in bed for any length of time is hazardous for many reasons, one of which is the formation of blood clots in the legs and other possible complications. A clinical trial of IPC devices showed that IPCs can reduce adverse outcomes in stroke patients overall by 6% during recovery. This is the absolute reduction in the risk of complications. Over 50,000 people in the UK suffer a stroke each year and are admitted to hospital. Not all of these patients are immobile for long; many do not develop blood clots in the legs, nor do they suffer adverse outcomes.

However, the *relative risk* of an adverse outcome (for example, blood clots forming in the lungs) in patients at risk is lower when IPCs are fitted (RR=0.59–0.75). Since the health care team doesn't know whether a patient is at risk of an adverse outcome, it makes sense to apply the device to all patients unless use of the device is *contraindicated* (the devices are inexpensive compared to the cost of treating a patient in hospital for an extra day).

The contraindications to the use of the device are fairly obvious to caregivers (for example, sores on the legs), and they are also infrequent. The application of the device is indicated for most patients.

Human Factors Analysis

The process outlined in the organization's policies meant that the application of Intermittent Compression Devices was distributed among team members with different roles and responsibilities. If key steps were omitted, the *process defaulted to no treatment*, rather than to fitting the device anyway, despite the fact that contraindications were clearly defined, easy to identify, and assessed a priori.

When multiple goals are combined in the same process, goal-sequencing issues may cause *goal substitution*. In this case, the doctor's task is completed when the IPC prescription is written on the treatment board. However, the completion of an administrative goal (filling in the risk assessment form and handing it in at the desk) might have triggered the end of the action sequence prematurely, thereby short-circuiting the delivery of treatment.

The risk assessment was really a decision flowchart, which gave no direction about fitting the IPC. The policy dictated that the IPC could only be applied if written on the treatment board (the nurses were not qualified to apply the device unless it was prescribed by a doctor).

Therefore, if the nurse did not see the prescription for the IPC written on the board, they might be expected to conclude that the IPC was *not needed* because it had not been prescribed. In the present case, the doctor completed the risk assessment but didn't complete the task by writing the IPC prescription on the treatment board, and so the device was not fitted.

Default Failure Mode under Human Error

As we saw in the section on human error, there are several kinds of errors that people often make:

Errors of Omission: In this case, not fitting the device when it should be fitted is an error of omission. In the context of the investigation:

- The consequences of not fitting the device are potentially *very serious* (note the relative risk statistic quoted above from the clinical trial).
- Most patients requiring the device are put at risk by this error of omission.
- *Process-based* detection of the error by the doctor or nurse is unlikely *until* the patient experiences an *adverse outcome* (outcome-based detection), which is too late from the perspective of patient safety and efficient delivery of health care services.
- Possible outcomes are that the patient suffers a pulmonary embolism (blood clot in the lungs) survives after further treatment, or dies.
- Outcome-based detection of error is detection *too late*.

Errors of Commission: An alternative scenario is that the nurse fits the device even though it is contraindicated (this was not supposed to happen under the policy in place at the time). The error of commission, in this case, is not serious—no real harm is done, although the IPC may be uncomfortable if the patient has other conditions, such as sores on the legs. Data from the investigation revealed that:

- Few patients are contraindicated anyway (<5%).
- Process-based detection is more likely to occur very early on (e.g., the nurse will see the contraindications when fitting the IPC, such as sores visible on the patient's legs).

Figure 15.11 illustrates these two kinds of errors and their implications for the safety of the population of stroke patients.

Root Cause

This case has a root cause. Only one thing needed to change to protect over 50,000 stroke patients admitted to hospital in the UK each year. The policy should dictate that the device is fitted automatically unless contraindicated. Remove the step in the sequence that requires writing to the treatment board and remove the IPC decision from the risk assessment (i.e., the "Risk Assessment" form should deliver an assessment of risk, not a decision about IPCs).

Work as Done

As often happens in safety investigations, the investigation team visited stroke units in several other hospitals to learn more about the perceptions of nursing staff and doctors. Several consultants endorsed the investigation team's view of the default failure modes of the current policy. The ward sister at one of the hospitals put it this way: "It's not rocket science, a nurse prescriber could do it. No harm in fitting it to everyone." In another unit, the ward sister explained that they, the nurses, applied the device when the patient was admitted. When the doctor arrived, they told the doctor to write the IPC on the treatment board and complete the risk assessment. Exactly the opposite of the policy!

Safety Recommendations: In terms of HFE, this case shifts the focus from "How can we stop doctors forgetting to write things to the treatment board," to "How can we design a better procedure where patient safety doesn't depend on the memory of staff?"

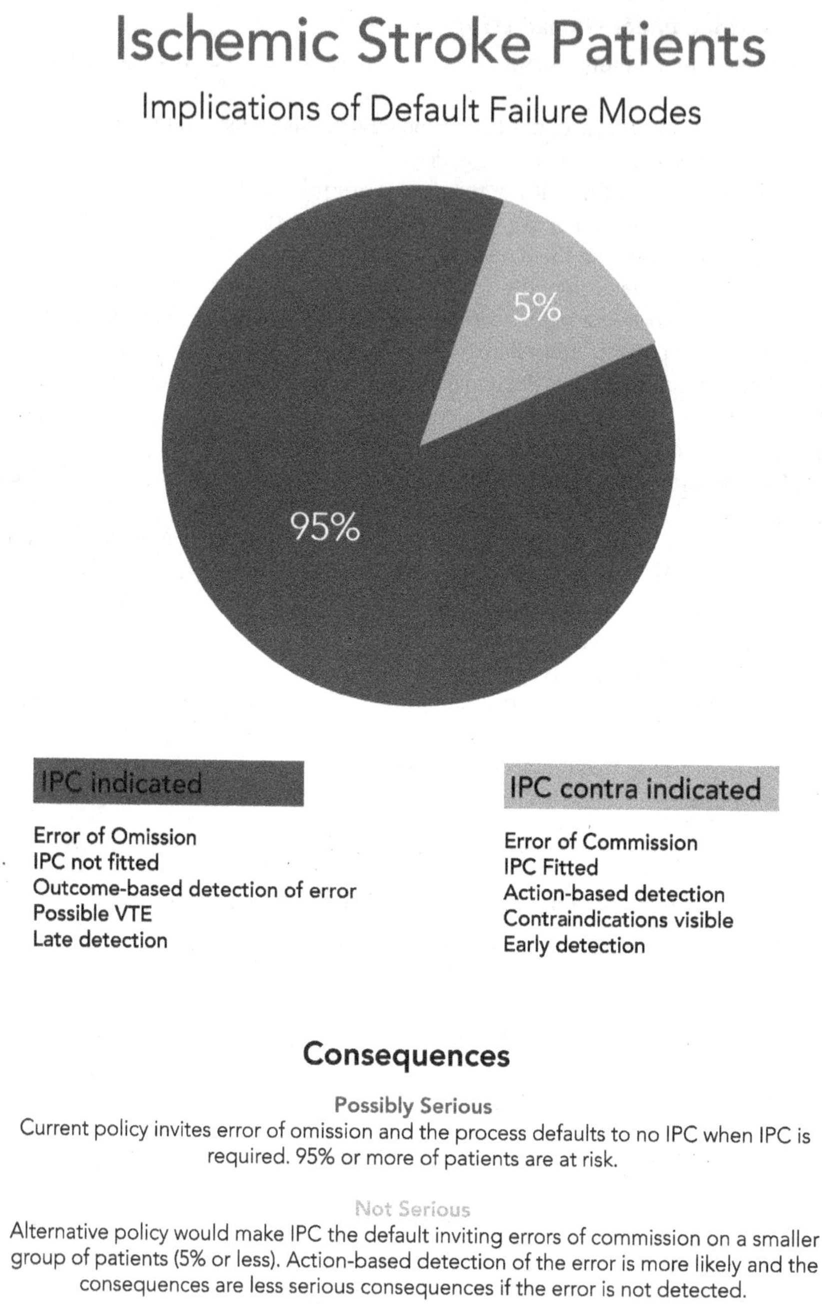

FIGURE 15.11 Contrasting the consequences of errors of omission and errors of commission in the treatment of stroke patients.

Postscript: In this example, "Work as imagined" refers to work as mandated by policy. "Work as done" refers to the actions taken by nursing staff to ensure that patients get the care they need. Further investigation revealed that the policy was out of date, having been originally implemented during a clinical trial of IPC devices, where it was necessary to record their use on the treatment board. Given that the trial was successful, this was no longer necessary. It seems that SOPs for safety can themselves be hazards when they are no longer up to date!

Deepwater Horizon—Bad Risk Management or Confirmation Bias?

On April 20, 2010, an incident occurred on the Deepwater Horizon drilling rig. The Deepwater Horizon was an ultra-deepwater, semi-submersible drilling rig. The Macondo oil field, which was located 4km below the seabed, lay beneath 1.5 kkm of sea. The field was 50 miles off the coast of Louisiana in the Gulf of Mexico. A heavy fluid known as "drilling mud" had been injected into the shaft to prevent oil and gas from rising to the surface. The drilling had been completed, and the oil well was in the process of being abandoned so that the oil could be recovered at a later date. Rigs such as Deepwater Horizon are mobile and can be moved around the world to complete tasks for major oil companies. On this occasion, Deepwater Horizon was 6 weeks late due to delays in the drilling process.

Part of the process of abandonment involved the installation of a cement plug to prevent hydrocarbons from escaping from the oil field. However, the plug had not been correctly installed meaning that the well had not been safely plugged, and the crew had misinterpreted test results. Part of the process of sealing the well was to monitor the pressure in the well to ensure that it was not increasing and that the cement barrier was working effectively as drilling mud was removed and replaced with seawater. However, oil had been gathering at the bottom of the well, and the drilling mud was not replaced with an equal volume of seawater. The crew were not monitoring the amount of seawater and drilling mud in the well, believing that the well had been sealed. As the drilling mud was removed, hydrocarbons passed through the failed cement barrier, rising toward the drilling rig for almost one hour without being detected by either the crew or by automatic controls. The pressure continued to rise, with methane gas entering the well from below and pushing the drilling mud onto the rig. This was followed by the methane gas itself, which then exploded, killing 11 men and injuring many others. The only barrier to prevent a blowout at the critical time was a device known as a "blowout preventer." This is a large mechanical device that can seal the pipe or shear it and clamp it to prevent the escape of hydrocarbons. However, operation of the blowout preventer depended on the ability of the crew to detect a "kick" – a difference in pressure between the wellbore and the rock. Only when oil and gas entered the rig did the crew activate the blowout preventer. Oil and gas that had already passed the preventer entered the rig, and it seems likely that the attempt to shear the drill pipe was also unsuccessful due to faults in the blowout preventer. The rig burned for 36h before sinking, and it is estimated that 4 million barrels of oil flowed into the Gulf of Mexico over the next few months, polluting 800km of coastline in Louisiana, Florida, and Alabama.

According to the report issued by British Petroleum in 2010:

- The cement and shoe track barriers—and in particular the cement slurry that was used—at the bottom of the Macondo well failed to contain hydrocarbons within the reservoir, as they were designed to do, and allowed gas and liquids to flow up the production casing.
- The results of the negative pressure test were incorrectly accepted by BP and Transocean, although well integrity had not been established.
- Over a 40-min period, the Transocean rig crew failed to recognize and act on the influx of hydrocarbons into the well until the hydrocarbons were in the riser and rapidly flowing to the surface.
- After the well flow reached the rig, it was routed to a mud-gas separator, causing gas to be vented directly onto the rig rather than being diverted overboard.
- The flow of gas into the engine rooms through the ventilation system created a potential for ignition, which the rig's fire and gas system did not prevent.
- Even after the explosion and fire had disabled its crew-operated controls, the rig's blowout preventer on the seabed should have activated automatically to seal the well. But it failed to operate, probably because critical components were not working.

In the case of a major disaster such as this, the official report is very extensive and takes into account human, technical, and organizational failures at all stages. It is not just a failure of "barrier management" or a lack of situational awareness on the part of the crew. Some of the main points in relation to the contents of this book are:

- *Confirmation Bias*: The well integrity test to check that the cement barrier was working correctly was unlikely to indicate that the barrier had failed. The crew looked for evidence that it was working, not that it was defective.
- This then influenced the way subsequent events were interpreted.
- *Fatigue*: Several key crew members were on the last day or two of 21, 12-h shifts.
- Inappropriate expectations about human behavior on the part of management (CIEHF, 2021):
    - That the crew would detect gas influx into the "riser" soon after it occurred.
    - That they would quickly realize that the gas was of sufficient volume to need diversion overboard.
    - That they would quickly complete the correct sequence of actions to send the gas overboard.
- *Risk Management*: A blowout on another rig earlier had not been effectively communicated so that lessons could be learned

TOOLS AND PROCESSES

Methods of Data Capture for the Analysis of Cognitive Tasks

There are several verbal methods of capturing data. Some basic "online" and "offline" methods are noted below. System state/action state diagrams can be used offline. The operator is presented with a matrix of possibilities concerning a controlled variable (such as temperature). At each cell in the matrix, the operator is asked what action (if any) they would take if the system behaved in the way described in the matrix. Although the method provides a structured approach to data capture, it has a number of limitations, such as the validity and reliability of the answers and whether the operator really comprehends or can imagine the occurrence of all the states presented.

Structured interviews can be used to obtain data on the rules and general principles that operators utilize and can be accompanied by the use of static simulation, in which the operator is given a number of hypothetical situations that they must talk the interviewer through.

These methods have the advantage that the most interesting aspects can be isolated. The disadvantages include an absence of realism and a lack of time pressure. A wide sample of situations must be investigated to characterize the operator's knowledge.

Verbal protocol collection is an online method of capturing data on the operator control strategy. The operator is asked to "think out loud" while carrying out the job, and what they say and do is recorded. The method attempts to "get inside the operator's mind" and is intuitively appealing. For example, it may be difficult to determine the difference between a good car driver and a highly skilled car driver using objective measurements alone. Unless exposed to extremely demanding situations, most of what they do will seem the same. However, the running commentary of the two drivers might differ considerably, with the highly skilled driver showing more evidence of higher cognitive activity, such as thinking ahead and detecting potentially dangerous situations well in advance, etc.

The verbal protocol technique has a number of advantages over other verbal methods. Objective data about system functioning can be recorded simultaneously with the protocol. The protocol itself can be used to design follow-up questions for use in an offline interview. However, the technique has been questioned, particularly in relation to its validity:

1. It makes assumptions about the operator's ability to "put into words" their control strategy.
2. Operators may neglect to mention things that seem obvious to them but are not obvious to anyone else.
3. Most people can think more quickly than they can talk.

4. The process of thinking aloud may distract the operator—it can be seen as a secondary task which, depending on the modality and processing requirements of the primary task, may compete for common processing resources (car drivers often break off in the middle of a conversation when carrying out a difficult maneuver).

Support Decision-Making—Provide Decision Support Systems

Current thinking in HFE can be applied equally well to understand why an accident happened ("what went wrong") and how to prevent accidents from happening in the future ("what could go wrong?"). Good HFE is all about asking the right questions—some basic tools are described below.

- Signal detection support consists of a computer-generated probability that a signal appearing on a complex array is a target rather than noise.
- ascading" of alarms and warnings can overload the operator. Decision support systems use a model of the process to process large amounts of fault data rapidly.
- Diagnostic problem-solving aids (see Shortliffe et al., 1979; Kleinmuntz, 1984; Shortliffe, 1984 for further information). Multiple faults can have a single cause (e.g., poor roadholding and fuel consumption due to too-low tyre pressures), and single faults can have multiple causes (e.g., uneven engine performance due to worn-out spark plugs, incorrect ignition timing, a dirty air filter, and leaky gaskets on the inlet and outlet manifolds). Until all faults are fixed, the problem remains—each remedial action, although correct in itself, is not followed by the expected improvement in performance. Diagnostic aids aggregate the signs and symptoms to determine the most likely disorder.
- Decision trees consist of a hierarchy of questions with a limited number of responses (sometimes only "yes" or "no"). The user is prompted by a series of questions and gleans data from the system to arrive at an answer, which then leads to a question at the next level down in the hierarchy.
- Statistical pattern recognition techniques are based on survey data about the prevalence of a problem and the variables associated with its manifestation (the signs and symptoms in medicine). Survey data are subjected to statistical analysis using regression and other multivariate techniques. This produces a mathematical model of the disease pattern. The model gives the probability that a patient has a particular disease, given that certain signs or symptoms are present.
- Bayesian decision aids help decision makers handle conditional probabilities, which are not easy for humans to deal with intuitively. A judgment about the likelihood of an event depends on some other set of conditions, which also have an associated likelihood, and there is an absence of other types of information to help the decision maker come to a conclusion (HFE Workshop 15.1).

HFE WORKSHOP 15.1

BAYES' THEOREM

Bayes' theorem deals with conditional probabilities—it tells us the probability that an event has taken place, given some other event that also has a particular likelihood. Human decision makers often fail to account for conditional probabilities and treat the event as if it were independent of anything else. Their judgments of probability are often incorrect for this reason.

(Continued)

HFE WORKSHOP 15.1 (*Continued*)

EXAMPLE

Mr. Smith is on holiday in Spain. In the dark, he goes to his hotel room to get his contact lenses, which are resting in their container. He has soaked them in contact lens cleaner after a hard day at the beach. Before putting his contacts in, Mr. Smith remembers to soak them in saline. However, because his eyesight is poor and the lighting in the bathroom is low, he takes the first bottle he can find on the counter.

He rinses the contacts with the contents of the bottle and puts his lenses onto his eyes.

There are three bottles on the counter. One is saline, the other is contact lens cleaner, and the third contains soda (i.e., effervescent water, not caustic soda).

After a short while, Mr. Smith experiences a stinging sensation in both eyes. What is the probability that he rinsed his contact lenses in contact lens cleaner, as opposed to saline or soda water?

Let B = symptoms, A = cleaner, and A' = saline or water.

The probability of having the eye discomfort immediately after rinsing the lenses in cleaner and replacing them is

$$P(B/A) = 0.8$$

The probability of having discomfort after using saline or water is

$$P(B/A') = 0.05$$

If the three bottles had an equal chance of being selected, the probability that Mr. Smith selected the cleaner, instead of one of the safer alternatives, is

$$P(A) = 0.33$$

and the probability of selecting the saline or water is

$$P(A') = 0.67$$

Bayes' theorem states that the probability that Mr. Smith has the symptoms because he used the cleaner instead of the water is given by

$$\begin{aligned} P(A/B) &= P(B/A)P(A)/(P(B/A)P(A) + P(B/A')P(A')) \\ &= [0.8 \times 0.33]/([0.8 \times 0.33] + [0.05 \times 0.67]) \\ &= 0.264/(0.264 + 0.0335) \\ &= 0.89 \end{aligned}$$

Therefore, Mr. Smith would be well advised to remove his contact lenses and go rinse his eyes immediately! Note that although the probability that he selected the cleaner is less than 0.5, it is highly likely that this is what happened given his symptoms.

Expert systems attempt to mimic the decision-making behavior of the expert. In diagnostic tasks, judgment is made on the basis of gross chunks of knowledge rather than detailed facts. Heuristic rules are applied to these knowledge chunks to arrive at a decision. The rule-based approach directs attention to pertinent areas and obviates the need for detailed analysis of the problem. Knowledge is in the form of production rules—simple conditional statements that relate observations to inferences (such as "if a bacterium is a gram-positive coccus growing in chains, then it is apt to be streptococcus"). The inference engine decides which rules to use and how they should be chained together to make decisions. The system's knowledge structure can be modified by changing the rules or adding new ones, and the rules themselves can be used to explain how the system came to a particular decision.

These aids help to prevent poor decision-making because they are free of the biases that affect human decision makers.

Building Reliability into Systems

There are many techniques for estimating the reliability of systems. Most of these techniques require some knowledge of mathematical probability theory and are beyond the scope of the present study. HFE Workshop 15.2 provides an introduction to some of the basics of probability theory and a simple example that illustrates how human reliability analysis can be incorporated into the assessment of system reliability.

HFE WORKSHOP 15.2

QUANTIFYING HUMAN RELIABILITY: INTRODUCTION

If we know the number of actions open to an operator in any situation, we can estimate the likelihood that each action will be performed. We can check our estimates using *activity sampling* (Chapter 11). In the simplest case, suppose that when a red warning light comes on, the operator can either shut down the machine or leave it running. Furthermore, suppose that either action is equally likely. The probability of shutting down the machine when the red light comes on is ½. This is exactly analogous to tossing a coin—if it is a fair coin, the chances of getting heads or tails when the coin is tossed are 0.5 (or 50% or "50/50" in lay language).

Suppose the red light comes on twice per day at random interval. What is the probability that the operator will leave the machine running *at least once?*

There are four possible pairs of outcomes where ShD is "Shut down" and LvE is "Leave running."

ShD, ShD
ShD, LvE
LvE, ShD
LvE, LvE

On three out of four possible outcomes, the machine is left running at least once. Therefore, the probability of leaving it running at least once is 0.75. Clearly, the only way to improve matters is to reduce the probability that the operator leaves the machine running when the red light appears.

Let us now suppose that there are six possible courses of action, such as

Ignoring the light, leaving the control room, shutting down the wrong machine, shutting down the correct machine, phoning the supervisor, and activating the alarm system.

If all of these are equally likely and independent events, then the probability of any of them happening on any one occasion is 1/6.

If the first three actions are hazardous and the second three are safe, what is the probability of a safe action occurring?

(Continued)

HFE WORKSHOP 15.2 (*Continued*)

The general law states that if any one of a number (N) of different outcomes is equally likely and a subset (n) of these outcomes corresponds to outcome A, then

$$P(A) = \frac{n}{N}$$

Therefore, in the example above, since three of six actions are safe

$$P(\text{safe}) = \frac{3}{6} = \frac{1}{2}$$

A second law states that the probability of one or the other of two independent events occurring is the sum of the probabilities of either

$$P(\text{ignore light OR shutdown wrong machine})$$
$$= P(\text{ignore light}) + P(\text{shutdown wrong machine})$$

$$= \frac{1}{6} + \frac{1}{6} = \frac{2}{6} = \frac{1}{3}$$

A third law states that the probability of two independent events is the product of both. For example, suppose that the probability that Mr. Smith will forget to bring his passport when packing to go to the airport is 0.1. The probability that Mrs. Smith will forget to remind him to bring his passport is 0.00001.

Therefore, the probability that he arrives at the airport without passport is

$$\begin{aligned} P(\text{no passport}) &= P(\text{forgets}) \times P(\text{not reminded}) \\ &= 0.1 \times 0.00001 \\ &= 0.000001 \end{aligned}$$

or one in a million. This is mainly thanks to Mrs. Smith, who is clearly a great deal more reliable than her husband.

EXAMPLE

An operator has to monitor the pressure in a pressure vessel in a distillery. The level and pressure of the contents of the vessel have to be kept within a narrow range of safety. Pressure is indicated by dials on a control panel. If the pressure increases above the safe limit, a relief valve opens automatically and lowers the pressure. The valve stays open until the pressure is safe. If the relief valve fails to open when it should, an auditory alarm sounds to alert the operator. This automatic system might fail for several reasons:

- The valve is stuck in the closed position: probability 0.005
- There is a power failure to the valve mechanism: probability 0.0005

(*Continued*)

HFE WORKSHOP 15.2 (*Continued*)

"It is the operator's job to monitor the displays and respond to any alarms." The operator might fail for a number of reasons:

- The auditory alarm fails: probability 0.001
- Operator responds incorrectly: probability 0.01
- Operator is absent from the control room at the time: probability 0.005

What is the probability that the safety system will fail on those occasions when the pressure exceeds the operating range?

First, we can draw a fault tree to summarize the data we have been given

$$P(\text{automatic system fails}) = 0.005 + 0.0005 = 0.0055$$

$$P(\text{manual system fails}) = 0.01 + 0.005 + 0.001 = 0.016$$

$$P(\text{pressure exceeds critical, system fails}) = 0.0055 \times 0.016 = 0.000088$$

$$= 8.8 \times 10^{-5}$$

INTERPRETATION

Clearly, the risk of the pressure exceeding the critical level is quite high. The safety system seems to rely heavily on the operator, who can be expected to fail approximately 1.5 times out of every hundred that the pressure exceeds the safe range. To understand the implications of our probability calculations, we need to know how often these events occur. If the pressure exceeds safe limits twice per week, then the weekly probability of system failure is $8.8 \times 10^{-5} \times 2$ or 1.67×10^{-4}.

One of the main limitations of mathematical approaches, such as that described in the workshop, is that the values of the probabilities associated with component and human failure are not stable over time or context. Therefore, the estimates are likely to be indicative rather than definitive of reliability. However, this approach can be very useful in estimating the likely benefits of improvements. If we introduce compound actions to ensure safety, the likelihood of a mishap is much reduced (e.g., the probability of both Mr. and Mrs. Smith forgetting to take their passports to the airport is lower than the probability of only one of them forgetting to do so). However, if either forgets their passport, the holiday is ruined, so it makes sense for Mrs. Smith, who is least likely to forget her passport, to take both passports. More generally, systems can be made more reliable by building in multiple checks and oversight. This only works, however, when the probabilities of error are truly independent. Having a copilot or pilot check the actions of the other only increases safety if they are likely to challenge the actions of the other when necessary. In a later section, we will briefly review safety culture—for cultural reasons, having several operators check the actions of the other does not always improve safety.

Problems in Accident Investigation

Dekker (2006) argues that it is of fundamental importance to understand *why* an error was made. To suggest that an accident occurred because the operator made an error, lost "situational awareness," or suffered stresses such as "information overload" explains nothing, and neither does generating lists of PSFs that were present at the time. Accident investigators should attempt to reconstruct what happened as accurately as possible and then try to understand the sequence of events from the operator's point of view. *Hindsight bias* can obstruct our ability to analyze accidents because:

- Accident investigators know the outcome of the sequence of events that led to the accident, whereas the operators did not know what was about to happen. This causes investigators to lay out what people *should* have done, distracting attention from asking the really interesting question, such as why the operators did what they did.
- Investigators know much more about the events during and surrounding the accident than the operators did at the time. This can make it difficult to understand operator actions. According to Dekker, "loss of situation awareness" is nothing more than the difference between what investigators know having written their report and what operators knew at the time.
- Investigators focus on judging the operators and attributing blame rather than on understanding why the operators thought that their actions at the time were reasonable. Dekker points out that most people do not go to work to make mistakes or to cause accidents, and that simply attributing blame to individuals is unlikely to yield the kind of information that will lead to a better-designed system in the future. Issuing more instructions, rules, and regulations is equivalent to stating, "don't make mistakes."
- Hindsight bias tends to cause investigators to focus on those people who were closest in time and space to the accident when it occurred and to "cherry-pick" data that proves an outcome that could only be known after the accident happened. Analysis of the wider system will yield more information for improving safety in the future.

Dekker recommends that accident data be put in context so as to understand:

- How the world looked to the operators at the time of the accident
- How the situation unfolded and what cues the operators received and when
- The operators' goals at the time (not knowing what was about to happen)

Organizations normally have formal safety management systems in place to conduct proactive assessments of hazardous operations. There is no one way of doing this, but a minimal requirement is the formation of a safety committee that meets to conduct systematic analysis of hazards. These analyses are often based on *failure mode analysis*, and the acronyms HAZID and HAZOPS are sometimes used for *hazard identification* and *hazardous operations*. HFE Workshop 15.3 exemplifies the approach.

HFE WORKSHOP 15.3

SOME BASIC TOOLS IN SAFETY MANAGEMENT

In this example, a safety committee is considering the hazards involved in conducting oceanographic surveys on a new vessel. Personnel will spend 2 months at sea in the southern oceans conducting oceanographic survey work. Hazardous operations on deck will have to be carried out at all times of the day and in all weather conditions. These include the manual handling of buoys, both for deployment overboard and for retrieval from the sea. An ad hoc committee is established, consisting of a health and safety officer, an ergonomist, an occupational medical officer, an oceanographic expert, a navigator, and an experienced ship's captain. The committee's goal is to establish a safety case for the trial. The case is made when all possible hazards have been identified and all means of mitigation have been determined. The safety case is then checked (audited) by an independent safety advisor to verify whether all the correct procedures have been followed.

(*Continued*)

HFE WORKSHOP 15.3 (*Continued*)

ACTIVITIES TO BE INCLUDED IN THE SAFETY CASE

The first step is to determine the scope of the analysis. The activities to be considered are

- The daily survey work itself
- Loading and offloading trial equipment such as buoys
- Daily activities involving additional (trial) personnel over and above the ship's complement
- Harbor pilotage, if part of the survey work
- Operation of helicopters as part of the trial (transits involving trial personnel)
- Any non-trials time spent in abnormal sea conditions
- Hazards associated with the trial instrumentation and equipment

ACTIVITIES EXCLUDED FROM THE SAFETY CASE

Already Covered in the Vessel Safety Case

- Life-saving
- Anchoring, berthing, and mooring arrangements
- Boat handling
- Docking arrangements
- All activities and operations already covered in the vessel safety case

Beyond the Scope of Either Safety Case

- Hazards from terrorist attacks
- Activities of other projects carried out at the same time

Activities to Be Excluded

- Activities associated with repair/maintenance
- Carriage of hazardous goods
- Operations in areas with sea ice

AREAS SELECTED FOR HAZID

- Hydrodynamics
- Vessel structure
- Survey operations
- Ergonomics
- Helicopter support

A classification scheme is developed to standardize the committee's risk assessments.

Severity	Effect on Personnel	Effect on Ship	Effect on Environment
Catastrophic	Multiple deaths	Loss of ship	Severe environmental damage
Critical	Single death/multiple injuries	Loss of major system Long-term disablement	Long-term damage (e.g., oil slick)
Marginal	Single injury or multiple minor injuries	Significant damage/short-term disablement of ship	Minor damage easily repaired (e.g., small oil slick)
Negligible	Minor injury or occupational illness	Minor damage to systems Mission unaffected	Minor breach of regulations (e.g., release of gray water)

(*Continued*)

HFE WORKSHOP 15.3 (*Continued*)

Each risk that is identified is first categorized according to its severity using the classification scheme above. Next, it is categorized according to its likelihood using the categorization scheme below.

Probability	Definition	*P* value
Frequent	Likely to occur continuously	100,000 per 10^6 trialing hours
Probable	Likely to occur often	10,000 per 10^6 trialing hours
Occasional	Likely to occur several times	100 per 10^6 trialing hours
Remote	Likely to occur sometime	1 per 10^6 trialing hours
Improbable	Unlikely to occur. Exceptional	0.01 per 10^6 trialing hours
Incredible	Extremely unlikely to occur at all	<0.01 per 10^6 trialing hours

Finally, each hazard that is identified is rated according to its severity and likelihood of occurrence, where A is intolerable, B is undesirable/must be mitigated before trial starts, C is tolerable following institutional review, and D is tolerable following peer review.

Classification	Catastrophic	Critical	Severe	Marginal	Minor/Negligible
Frequent	A	A	A	A	B
Probable	A	A	A	A	C
Occasional	A	A	A	B	C
Remote	A	B	B	C	C
Improbable	B	C	C	D	D
Incredible	C	D	D	D	D

ACTIVITY HAZARD ANALYSIS: ERGONOMIC ANALYSIS

Below are examples from an ergonomic analysis of some of the key activities required during the trial:

Activity	Possible Failure Mode	Outcome	Classification	Mitigation
Buoys (stowage)	Restraints fail Buoys fall onto deck	Personnel struck by buoy(s)	A	Stow below decks Use extra restraint Stow away from walkways Demarcate with hazard warnings
Man overboard	Loss of balance	Drowning/hypothermia	A	Don lifelines on deck Work in teams Extra illumination for night workers
Manual handling of buoys	Back injury	Personnel temporarily unable to work	B	Lift with teams NIOSH assessment Provide handles
Manual deployment and recovery of buoys	Man overboard	Drowning/hypothermia	A	As above Use lifting/ lowering aids
Transit on deck	Slips, trips, and falls	Orthopedic trauma	B	Footwear Lighting Deck covering
Helicopter landing operations	Noise exposure	Acute hearing damage	C	Noise assessment

(Continued)

HFE WORKSHOP 15.3 (*Continued*)

Activity	Possible Failure Mode	Outcome	Classification	Mitigation
		Impaired communication		Active noise reduction
Helicopter landing operations	Crashing into deck	Multiple injuries/death damage to ship Loss of helicopter	A	Night vision goggles for landing officer Douse lights on bridge

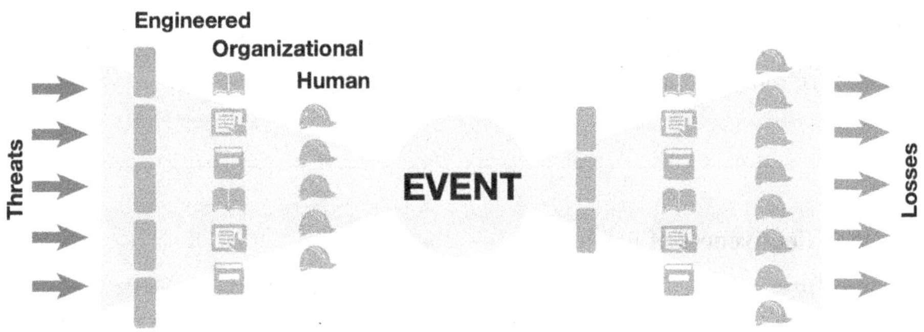

FIGURE 15.12 Conceptual model of a Bow Tie analysis. (Chartered Institute of Ergonomics and Human Factors, with permission.)

Bow Tie Analysis

Bow Tie analysis is a method for analyzing the links between threats, adverse events, and losses in relation to the controls that are in place to control threats and mitigate losses (Chartered Institute of Ergonomics, 2021). The emphasis is on barrier design and management. Basic analyses are essentially linear (Figure 15.12) and link the adverse event to the threats and the subsequent losses. On either side of the event are controls to prevent the event due to exposure to threats and, if the event occurs, to mitigate the losses. Thus, the approach also requires us to look at recovery after an event when it does occur.

The three main categories of barrier, in order of importance, are engineering barriers, organizational barriers, and human barriers. In each category, there may be several different barriers, but they can all be defeated. Engineering barriers can fail due to wear or lack of maintenance, organizational barriers can fail because of non-compliance, and human barriers can fail due to sleep deprivation, error, and so on (Figure 15.13). These barriers are put in place both before the event (to prevent it from happening) and after the event, should it occur, to minimize losses. Bow Tie analysis distinguishes between barriers (such as lock-outs in electrical circuits) and safeguards (such as alarms or warning signs), which make adverse events less likely. Bow Tie diagrams can be layered; for example, if human error defeats a barrier, we can conduct a Bow Tie analysis of the error (as the event) and examine the barriers in place to prevent error and mitigate its effects. Thus, Bow Tie analysis provides a hierarchical, top-down approach to hazard identification and control and can be used to summarize the findings of a hazard identification exercise. Bow Tie analyses can deal with multiple threats simultaneously.

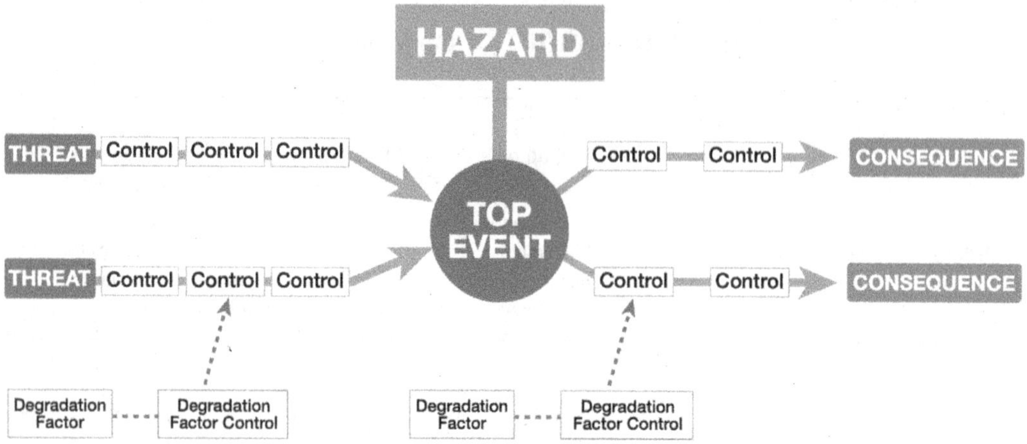

FIGURE 15.13 Layered Bow Tie analysis with degradation of controls. (Chartered Institute of Ergonomics and Human Factors, with permission.)

TABLE 15.4
Assessing the Effectiveness of Barriers

Effective Human Barriers	Ineffective Human Barriers
The task is clear and specific	The task is vague. Not clear how to initiate the task or whether the goal was achieved
It is clear who is responsible	Responsibility not assigned to a specific role
Operator knows what to do and under what circumstances	Conflict between safety and productivity, divided responsibilities
Expectations are clear and realistic:	Context of operation was not considered
Knowing what to do and when	
Knowing what needs to be done	
Having the resources to do it	
Knowing that it had the desired effect	
Safety/performance trade offs not needed	Unrealistic expectations of operator ability under the circumstances
Can be monitored	Depends on operator situation awareness—of hazards and risks
Easy to tell if the barrier is in place and working	Requires coordination between individuals who may have different personal or organizational responsibilities and objectives
Criteria for work systems needed to support the function are defined and have been implemented	No allowance for the unexpected or ambiguity
Clear feedback on success	Potential for conflict between what is expected for effective barrier performance and personal or organizational incentives
Initiated by strong signals	Relies on people identifying and correctly interpreting early signs of trouble that may be perceptually weak, ambiguous, or unclear

Bow Tie analysis provides analysts with a clear indication of the threats, hazards, and controls and how they are linked together. The next step is to assess the effectiveness of the barriers (Table 15.4).

Bow Tie analysis can also be used for incident investigation and in safety management to assess current arrangements for safety.

SYSTEM INTEGRATION

In most organizations, safety is a managed process. Safety committees are responsible for:

- Policy making
- Setting targets for safety
- Planning the implementation of policies
- Monitoring the operation of policies in place
- Measuring the performance of the policies
- Auditing safety performance
- Feeding back the results to inform the first three functions

Safety committees often rely on "lagging indicators" to monitor safety performance. These are retrospective measures of safety performance, such as quarterly records of safety occurrences. Lagging indicators confirm long-term trends (and whether targets set earlier were met) but they don't predict future outcomes. "Leading indicators" look forward and include the implementation of training courses (in HF, for example), employee participation initiatives, and equipment inspection findings. HFE specialists can support all of these activities.

HFE can be integrated into all stages of an investigation, from initial visits to the scene and data gathering, to planning interviews, follow-up visits, reviewing literature, and report writing.

- Promptly responding to the incident.
- Planning the investigation.
- Gathering data and accounts from eyewitnesses.
- Visiting the incident scene for further work.
- Analyzing collected data.
- Creating an accident report and disseminating the findings.
- Making safety recommendations.

Different sectors often have their own or preferred approaches to system safety, and some examples are given below.

Systems Engineering Initiative for Patient Safety (SEIPS)

SEIPS is a systems-engineering approach to patient safety that seeks to analyze the "journey" of a patient along a health care pathway, while taking into account the needs of health care providers (Carayon et al., 2020). SEIPS has three main areas of analysis: work systems, work processes, and work outcomes (Figure 15.14).

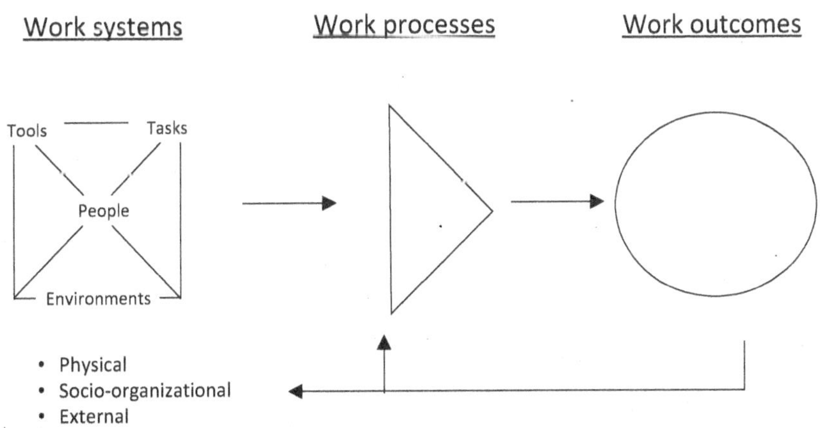

FIGURE 15.14 SEIPS model for patient safety.

Work systems consist of people who carry out tasks using tools in an environment. All of these must be described and understood as part of the analysis, together with the interactions between them.

Work processes consist of how the work is done and how it flows as treatment progresses, taking into account the physical, cognitive, and socio-behavioral characteristics of the process.

Work outcomes (positive or negative) are not only the outcome of treatment but also the outcome for the patient and staff.

The model has feedback loops between outcomes, processes, and systems.

Holden and Carayon (2021) present a simplified version of SEIPS in seven steps, which are summarized below:

1. *PETT Scan (People, Environments, Tools, and Tasks):* First, we need to understand the work system, the people who carry out the work, the tasks they perform, and the tools that they use. This includes an analysis of the barriers and facilitators to task performance. A PETT scan facilitates the application of HFE to analyze the work system by providing a detailed frame of reference. HFE can play a major role here in assessing equipment and workplaces in relation to the system requirements described in this book (e.g., noise interferes with communication; equipment is difficult to move; handwritten instructions are illegible).
2. *People Map:* People maps are diagrams used to represent the main players and how they relate to each other. Different shapes of different sizes can be used to code individuals (e.g., patients, nurses, doctors). Arrows can be used to illustrate how different individuals interact with each other and the relationships between various stakeholders.
3. *Task and Tools Matrices:* This is a kind of task analysis in which tasks, tools, and task-tool interactions are described. The matrix summarizes which tools are used for which tasks and can be useful when deciding on equipment upgrades or changes (for example, by increasing the use of smartphones for some tasks).
4. *Outcomes Matrix:* An "outcome" in SEIPS is not a single event. There may be several outcomes for the treatment of one patient or condition, and there may be different outcomes for different stakeholders. Outcomes can be ranked in order of priority while, at the same time, balancing the outcomes against the needs of different stakeholders. For example, while there is a need to reduce hospital-acquired infections, this should not present obstacles for other staff in the pursuit of their own goals.
5. *Journey Map:* This is a process matrix that describes the process illustrated in the SEIPS figure above in a linear fashion. For example:
   a. Admission (who is involved; tasks of the family; tasks of healthcare team; physical environment)
   b. Pre-rounds (who is involved; tasks of the family; tasks of the healthcare team; physical environment)
   c. Before round (who is involved; tasks of the family; tasks of the healthcare team; physical environment)
   d. During round (who is involved; tasks of the family; tasks of the healthcare team; physical environment)
   e. After round (who is involved; tasks of the family; tasks of the healthcare team; physical environment)
6. *Interactions Diagram:* Having described the work system, the interactions diagram shows how interactions take place between the tools, tasks, and stakeholders. These diagrams have a variety of uses ND can be used to identify hazards or to evaluate new proposals for equipment, environments, or ways of working.
7. *System Story:* This is a way of summing up how things happen in the system and how work is managed. We can contrast, for example, a top-down, hierarchical way of working with a more decentralized style of management (see the section on soCio-technical systems theory in the Prologue).

Accident Route Matrix (ARM)

The accident route matrix (Harris, 2016) was developed in the military and is particularly useful for reporting safety occurrences in military operations and in aviation. This is because these operations tend to be highly sequential, with one problem leading to another as the incident develops. One of the features of the ARM is that it embeds the investigation of events and PSFs into a timeline—the point in the sequence of operations at which each of the PSFs came into play. The ARM is not really a method or a model, but it can certainly influence the way an incident is investigated. Some questions that it might promptinvestigators to ask are, for example:

1. *Number of Personnel:* Were there enough personnel in specific roles or with specific qualifications? Was the complement correct?
2. *Training and Currency:* Had up-to-date training been provided to personnel to enable them to perform effectively in their roles?
3. *Documents and Procedures:* A common issue is the high volume of rules, procedures and regulations. Was this issue identified in the analysis? Did it increase the risk of information being forgotten or ignored and so contributing to the occurrence? Were issues identified with the content of documentation or displays, such as errors or information spread across multiple documents/displays?
4. *Overall Workload:* Was the workload excessive?
5. *Situation Awareness (SA):* Were there difficulties in developing and maintaining SA? Were system states and state transitions visible?
6. *Distraction:* Were personnel distracted at the point of the occurrence?
7. *Fatigue and Pressure:* Was there a risk of fatigue or time pressure?
8. *Experience:* Did personnel have sufficient experience to operate the system properly? Is there a training standard?
9. *Recovert:* Were recovery procedures timely and successful?
10. *Human Factors Integration:* Were the requirements for fault detection and diagnosis specified at the design stage?

Below is an example of an ARM analysis in reporting a safety occurrence.

	Entry Conditions	Readiness	During Occurrence	Post Occurrence
Organization	Sufficient personnel?[1] Excessive job demands? Badly designed jobs? Procedures and rules?	Task demands on personnel[7]	Teamwork effective. Staff shortages[8]	Emergency procedures followed?
Supervision and Risk Management	Training and competence?[2] Communication? Procedures and rules?[4]	Supervision on the day	Checking conducted?	Situation assessed correctly and responded to appropriately?
Task	Workload? Time on task?	Actual task as planned?	Violation of safety constraints? Task performed correctly?	Recovery and completion of mission?
Equipment	Human-machine interface Workplace design	Equipment working as required?	System state displayed correctly? System state explicit?[5]	Faults identified and corrected?
Environment	Bridge layout Weather/sea states	Environmental conditions as expected?	External indications and warnings visible?	Environmental conditions enable recovery?

(*Continued*)

(*Continued*)

	Entry Conditions	Readiness	During Occurrence	Post Occurrence
Behaviors and actions	Rest, nutrition, mental models, errors, understanding?	Correct procedures followed?[3] Performance to required standard?	Communications understood? Teamwork effective?	Assistance sought was appropriate and timely?
Operator Conditions	Fatigue, time pressure, motivation?	Required competencies available at the time?	Actual actions?[6] Correct/ appropriate	Injuries dealt with correctly?
HFI	HFI in procurement of systems and system implementation	Fault and error management facilities embedded in system?	System state displayed and visible? Errors and malfunctions recognized and mitigated?	HFI investigation commissioned?

One of the advantages of the ARM as a reporting tool is that it provides a one-page summary of the safety occurrence. If adopted as a reporting standard in an organization, this would facilitate annual reviews of multiple incidents and make it easier to identify common PSFs.

Mining Industry: Control-Focused Incident Investigation Process

Dodshon (2023) has developed a comprehensive incident investigation process for use in the mining industry. The process is based on the identification and assessment of controls, as a response to the research-practice gap and the lack of usable methods that take a human-centered approach:

> In particular, the aim is to assist practitioners to identify, analyse, assess and generate operator-based risk controls........ The overall objective of this research is to enhance incident investigation processes that are currently being used in high-risk industries.........”the scarcity of academic research that focuses on incident investigation processes and the scarcer research focusing on incident investigation effectiveness from the practitioners' perspective.

The Process is Operationalized via Six Charts

Chart 1: Control-focus chart. The process begins with this flowchart, which directs the investigator to identify the controls in place and assess their effectiveness. If ineffective due to human actions, the investigator is directed to Chart 2.

Chart 2: This flowchart prompts an assessment of operator competencies and fitness for work: the operator's assessment of the situation and the appropriateness of subsequent actions. If the assessment is found to be incorrect, the investigator is directed to Chart 3; and, if the selection and execution of remedial action were lacking, to Chart 4.

Chart 3: Prompts an analysis of the situation to be assessed, the cues available, and the facilitators and distractors.

Chart 4: Prompts an analysis of the decision to be made and the workplace factors that facilitate or prevent appropriate action.

Chart 5: Is an organizational analysis tool that prompts an analysis of control support activities at all levels of the organization.

Chart 6: Prompts an in-depth analysis of operator decision strategies and the recommendations that might improve decision-making.

The tool embeds human factors in incident investigations by prompting investigators to focus their attention on working conditions rather than operator behavior. For example, in Chart 3, situation assessment, the following are considered:

- *Element 1:* What are the triggers that alert the worker, or that the worker detects, to indicate that there is a potential need to act?
- *Element 2:* What cues or information need to be observed, noticed, and comprehended?
- *Element 3:* What potential interpretations could be made of the situation?

These are good questions that will lead investigators to consider workplace ergonomic factors and identify design deficiencies that create unsafe conditions and areas where control is lacking.

In summary, the process is based on a coherent set of charts, prompting investigators to ask the right questions during an investigation, the answers to which generate useful information to improve workplace safety. The charts comprise a control-focused, self-contained tool that can be integrated into existing investigation processes, uncontaminated by the plethora of under-specified and untested models found in the safety literature.

Safety Culture

Traditional approaches to safety rely on "lagging indicators" to evaluate the safety of an organization (previous accidents, injuries, and so on). Safety audits and measures of safety climate and "safety culture" are examples of leading indicators.

There is no generally accepted definition of safety culture. However, if we define culture in general terms as

> ....a shared understanding within a group of the social significance of events and behaviours at a symbolic level.

Then, the safety culture in an organization is the shared understanding that people have about the meaning of safety and of activities related to safety.

Safety climate is a related term that refers to perceptions of the safety culture under a given set of circumstances.

Safety culture is made up of many elements (Table 15.5). Good management is key to creating a positive safety culture. All organizations will display a variety of cultural beliefs related to hazards and their management—these can be formally laid down (rules and procedures) or more informally embedded in everyday working practices. It is easy to see the rules and procedures, but everyday practices are often taken for granted, so looking at how things are *actually* done, rather than how they are expected to be done, may play a role in understanding why people do what they do.

Safety culture is not independent of other organizational systems and processes. So, if, for example, people are thought to have taken unnecessary risks prior to an event, investigators should enquire whether such behavior was seen as acceptable by those around them. Was it rewarded in some way—was risk-taking and "getting away with it" a way of gaining prestige and respect among one's peers?

Measurement of Safety Culture

Safety culture can be measured in several different ways: using questionnaire surveys, focus groups, and interviews, and, rarely, using auditors. There are many different safety culture questionnaires developed in different industries, and the contents vary (some focus more on perceived compliance with rules, while others focus more on the perceived importance of safety). Table 15.5 summarizes items from the main psychometric scales found in such questionnaires.

Safety Culture Maturity

Safety culture maturity refers to the extent to which safety is embedded in the day-to-day management of an organization (Parker et al., 2006).

TABLE 15.5
Findings from a Preliminary Factor Analysis of Safety Culture Questionnaire Items with Four Main Components

- *Factor 1:* Adherence to safety rules/regulations/procedures
  - I understand how to report an accident
  - Safety warnings and notices are easy to understand
  - I am clear about what my responsibilities are for safety
  - I fully understand the health and safety risks of the work for which I am responsible
  - Sometimes, I am unsure what to do to ensure safety in the work for which I am responsible
  - I know who to report safety issues to
- *Factor 2:* Noncompliance with safety rules/regulations procedures
  - If safety is ignored, it is none of my business
  - Workers are not encouraged to report safety concerns
  - There are too many rules and restrictions around the work for which I am responsible
  - Safety rules and procedures do not reflect how the job is done
  - Safety procedures and rules are only there to protect against legal action
  - My immediate superior turns a blind eye when rules are bent
- *Factor 3:* Safety under pressure
  - I cannot always get the equipment I need to work safely
  - Safety rules are adhered to, even under pressure
  - Staff shortages sometimes prevent the job from being done safely
  - The boss would stop us from working if it wasn't safe, even if it meant the job didn't get done
  - I am always informed of the outcomes of meetings where safety is discussed
- *Factor 4:* Positive views of immediate superior and safety
  - My boss is genuinely concerned about health and safety at work
  - Safety information is always brought to my attention by my superior
  - The boss talks to us about safety while we are working
  - I don't think the boss does enough to ensure work is safe
  - I am satisfied with the way I am kept informed about safety issues at work

A *pathological* organization has the lowest level of safety commitment, where safety is regarded as an inconvenience. Safety initiatives tend to be described as "job-stopping," "bureaucratic," and "tick-box."

A reactive organization will only react when things go wrong and often only to safeguard a brand name or reputation. Much of the response will be to present an image of a healthy safety culture. Typically, such an organization will be a "repeat offender."

A *calculative* organization realizes that compliance is the minimum acceptable state and will invest in demonstrating compliance at this level. However, calculative compliance does require concerted effort. The appearance on the surface can be impressive, but underlying this, there is no real understanding of, or commitment to, the benefits.

A *proactive* organization recognizes that there are real benefits to proactive safety management in saving lost time, money, and reputation. It is typified by "above the line behaviors," where excelling in safety management is praised and rewarded.

In a *generative* organization, a healthy paranoia exists throughout that continuously checks and provides feedback at all levels. This is only achieved when training and education have embedded a self-sustaining, healthy attitude toward safety that actually requires only occasional direct engagement by senior management. The individual approach is intuitive and proportionate to the risk.

The following extracts from the report into the capsize of the Herald of Free Enterprise illustrate a lack of safety culture maturity. Captain Blowers, one of the company's ferry masters, wrote a memorandum to a senior manager with the following suggestion (extract from the official report No. 8074):

> 4. Mimic Panel—There is no indication on the: bridge as to whether the most important watertight doors are closed or not. That is the bow or stern doors. With the very short distance between the berth and the open sea on both sides of the channel this can be a problem if the operator is delayed or having problems in closing the doors. Indicator lights on the very excellent mimic panel could enable the bridge team to monitor the situation in such circumstances.

Some of the responses from the senior managers are reported below:

> The answers which Mr. Develin received will be set out verbatim. From Mr. J.F. Alcindor, a deputy chief superintendent: "Do they need an indicator to tell them whether the deck storekeeper is awake and sober? My goodness!!" From Mr. A.C. Reynolds: "Nice but don't we already pay someone!" From Mr. R. Ellison: "Assume the guy who shuts the doors tells the bridge if there is a problem." From Mr. D.R. Hamilton: "Nice!"

By the time these comments were made, the company was already aware that ferries had left port with the bow doors inadvertently left open—in one case, a ferry traversed the English Channel with the doors open and only remained afloat due to the benign conditions at the time. Modern approaches to safety would regard this as a "near miss," and a formal investigation would have been carried out to ensure that it never happened again.

Reliability of Safety Culture Maturity Models

Ashford et al. (2016) investigated the model to see if it was internally reliable. It isn't. Basically, the levels in the hierarchy aren't mutually exclusive—you can be increasingly informed and trustworthy about some things, but not others, so the framework falls to pieces.

Learning from Adverse Events

Accident and incident reports should enable organizations to learn from the event. Safety recommendations should be drafted in such a way that real, directed changes can be made. This means focusing on the system as it actually operates and how work is actually done. In addition to analyzing what happened in a thin slice of time in the immediate environment, the working conditions and management's role in creating them have to be understood.

The Chartered Institute of Ergonomics and Human Factors has recently published a white paper on learning from adverse events. The white paper sets out, and discusses in detail, nine key principles:

- Be prepared to accept a broad range of types and standards of evidence.
- Seek opportunities for learning beyond actual loss events.
- Avoid searching for blame.
- Adopt a systems approach.
- Identify and understand both situational and contextual factors associated with the event.
- Recognize the potential difference between the way work is imagined and the way work is actually done.
- Accept that learning means changing.
- Understand that learning will only be enduring if change is embedded in a culture of learning and continuous improvement.
- Do not confuse recommendations with solutions.

The organization has the willingness and competence to draw the right conclusions from its safety information system, and the will to implement changes when their need is indicated.

Safety I, Safety II, and Safety III: Approaches to Safety Management

Safety I, II, and III are contrasting approaches to safety management that, although not mutually exclusive, focus on safety in different ways. There is much debate about these approaches—often of a philosophical rather than a practical nature. Table 15.6 summarizes these approaches, together with the approach of modern safety engineering (Leveson, 2020).

TABLE 15.6
Main Features of Safety I, II, and III

	Safety I	Safety II	Safety Engineering	Safety III
Definition of Safety	As few things as possible go wrong	As many things as possible go right	Safety is usually defined as freedom from unacceptable losses as identified by the stakeholders, but may be defined in terms of acceptable risk or ALARP in some fields. The goal is to eliminate, mitigate, or control hazards, which are the states that can lead to these losses.	Safety is defined as freedom from unacceptable losses as identified by the system stakeholders. The goal is to eliminate, mitigate, or control hazards, which are the states that can lead to these losses.
Management Principle	Reactive, respond when something happens, or is categorized as an unacceptable risk	Proactive, continuously trying to anticipate developments and events	Concentrates on preventing hazards and accidents but does learn from accidents, incidents, and audits of how system is performing.	Concentrates on preventing hazards and losses, but does learn from accidents, incidents, and audits of how system is performing
Explanations of accidents	Accidents are caused by failures and malfunctions. The purpose of an investigation is to identify causes and contributory factors.	Things basically happen in the same way, regardless of the outcome. The purpose of an investigation is to understand how things usually go right as a basis for explaining how things go wrong.	Accidents are caused by linear chains of failure events. The purpose of investigation is to identify the chain of events and the root cause.	Accidents are caused by inadequate control over hazards. Linear causality is not assumed. There is no such thing as a root cause. The entire socio-technical system must be designed to prevent hazards; the goal of investigation is to identify why the safety control structure did not prevent the loss.
Attitude to the human factor	Humans are predominantly seen as a liability or a hazard.	Humans are seen as a resource necessary for system flexibility and resilience.	Humans are expected to prevent or respond to hazards and to be flexible and resourceful when they occur.	The system must be designed to allow humans to be flexible and resilient and to handle unexpected events.
Role of performance variability	Harmful, should be prevented as far as possible	Inevitable but also useful. Should be monitored and managed.	The primary reason for performance variability is to enhance productivity and system requirements. Procedures are provided when response time, and information is limited. Effort is put into providing appropriate controls and interfaces to allow operators to prevent or respond to hazards. Design so that when performance (of operators, hardware, software, managers, etc.) varies outside safe boundaries, safety is still maintained.	Design the system so that performance variability is safe and conflicts between productivity, achieving system goals, and safety are eliminated or minimized. Design so that when performance (of operators, hardware, software, managers, etc.) varies outside safe boundaries, safety is still maintained.

Source: From Leveson (2020), with permission.

Safety I is presented as an approach to safety management where the focus is on risk assessment and the investigation of safety occurrences. Attention is focused on negatives and employees are seen as a source of risk, either through error, violations, or other factors such as fatigue. The approach focuses on prevention. Safety II, on the other hand, focuses on how employees cope successfully with challenges and adapt to changing circumstances. In investigations, a Safety I approach seeks to find the causes of the incident and the likelihood of it happening again (the risk). Safety II seeks to understand why incidents are prevented from happening (work as done). The two approaches are seen as compatible rather than mutually exclusive inasmuch as a reactive approach (what went wrong that caused the accident) is more useful when combined with a proactive approach (what are the behaviors that lead to success). In this way, safety recommendations can include both preventative measures and ways to reinforce adaptive behaviors.

Leveson (2020) has criticized the above formulation in several ways. First, that Safety I does not really exist—it is a caricature of safety engineering that does not see employees as a source of risk but as a system component. Further, Safety II, rather than taking a sociotechnical approach, has an undue focus on human operators at the expense of technology. An undue focus on "what goes right" also runs the risk of confirmation bias if it encourages investigators to focus on successful performance:

> To summarize this section so far, Safety-I does not exist, i.e., it is a misrepresentation of engineering in general and safety engineering specifically. Safety-II, on the other hand, violates what we know about engineering and is more likely to decrease safety or, at best to have no impact, than to increase safety
>
> *Leveson, 2021*

In Safety III, Leveson describes STAMP and the systems approach to safety, which is the approach that aligns most closely with HFE.

RESEARCH ISSUES

As technology becomes increasingly reliable and autonomous, the focus is shifting to the human element of system safety. Some of the emerging themes in safety research are concerned with the arrangements for safety in organizations and how to integrate groups of people so that the opportunities for safety are maximized and those for failure are minimized. Hence, the focus on constructs such as "safety culture" and "safety culture maturity." There is no single best way of researching or implementing these ideas, which offers great scope for applied research in real organizations.

Todd and Gigerenzer (2007) suggest that real advances will be achieved when we have a better understanding of *ecological rationality*—the kinds of environments the human mind evolved to work in and the kinds of problems it evolved to solve. For HFE, the real challenge, to paraphrase Todd and Gigerenzer (2007), is to design *environments that make us smart.*

SUMMARY

The focus of HFE is the system, particularly the boundaries between human operators and components; the interfaces and the equipment being used, the organization of work, and the arrangements for safety. Cook (2000) discusses the ways that complex systems can fail and cautions against approaches that see system safety as nothing more than the aggregate level of the safety of the components. Complex systems are intrinsically hazardous and depend on multiple, hierarchical controls. When they do fail, there is normally no single point of failure (or "root cause"); rather, failure occurs at multiple points that can interact in unexpected ways. In reality, component failures occur all the time, but due to redundancies built into the system, the results are not catastrophic. Resources are always limited, and there is always competition for resources in organizations; hence, safety management aims for ALARP status, where "reasonably practicably" really means as low as possible given the limitations of time and resources.

Accident reports are neither "data" nor an objective account of why an accident happened—they are narratives constructed after the event by people who were not there at the time. The information included in a report may often be as much a reflection of the background and motives of those investigating as a true reflection of the accident itself. That is why it is important to have reports reviewed independently.

Three basic questions to begin your HFE investigation of an accident are:

- What happened?
- What's supposed to happen?
- How's it supposed to happen?

A complete list of questions in all competency areas of HFE can be found in Bridger (2021).[1]

Well, we began this chapter with a quote from Cicero, so perhaps we should leave the last word to him:

> Is anyone unaware that Fortune plays a major role in both success and failure?

TUTORIAL TOPICS

1. When accidents happen, it is often said that those involved were "complacent." Is this really just another way of saying that they were too trustworthy of the system (the technology and coworkers)?
2. Is it possible to compare different organizations in terms of their safety culture?
3. Is "human error" to blame for accidents? Can it ever be?
4. Take a look at the video at the link (https://www.youtube.com/watch?v=9NQ8LehUWSE). Is this evidence of a hazard making itself known?

ESSAYS AND EXERCISES

1. Observe users operating a vending machine. Time how long it takes them to perform the task and note any errors and difficulties. Try to relate the errors to the design of the machine and its dialog.
2. Carry out a task analysis of the task of paying for a parking ticket using a machine in a multi-storey car park. Use such a machine yourself and identify all its possible states. List all the syntactic and semantic errors that *could* occur on the basis of the task analysis. Observe people performing the task. Time how long the operations take and note any errors. Compare the actual distribution of errors with all the errors that could occur and relate any differences to the ergonomics of the machine.
3. Is the concept of fatigue of any use in accident investigation?
4. A utilities company employs inspectors to read the water meters of houses in a city. The probability of misreading a meter is 1 in every thousand. When inspectors have been drinking alcohol, the probability of misreading typically increases to 1 in 500. An inspector misreads 50 of the last 10,000 m he reads. Is he likely to have been drinking? How would you find out?
5. Refer to Workshop 15.2. If the pressure exceeds the safe limits approximately three times per week, what is the annual probability of safety system failure? If the consequences of failure are life-threatening, do you find this acceptable?
6. Refer to Workshop 15.2. The company decides to improve the safety of its pressure vessel monitoring and control system. It does this by introducing an automatic level control system that drains the tank if the pressure exceeds critical (which will happen if the relief valve fails). So, if the relief valve is stuck closed or if there is a power failure to the valve, the tank will be drained. If the probability of failure of the level control system is 0.003, how will this improve the safety of the overall system?

7. Use the data below to estimate the probability that Mr. and Mrs. Smith will fail to board their flight for their holiday.

Event	Probability
Mr. Smith forgets his passport	0.1
Mrs. Smith forgets her passport	0.000001
Their taxi is delayed in a traffic jam	0.01
The flight is canceled	0.05

Draw a fault tree diagram of this scenario.

8. Lottery contestants compete in a lottery by selecting 5 numbers between 1 and 50. If the chances of winning the lottery are 1 in 17 million, what is the probability that the winning sequence on any one draw will consist of the numbers 1, 2, 3, 4, and 5?
9. Mr. and Mrs. Smith are waiting for a taxi to take them to the airport. The probability that Mr. Smith will forget to bring his passport with him is 0.1. The probability that he will forget to bring his driving license is 0.05, and the probability that he will leave the plane tickets at home is 0.07. What is the probability that he will leave home without one of these items? What assumptions did you make?
10. Mr. Smith has just returned from a holiday in Spain. His wife is not talking to him and does not reply to anything he says. The probability that she is not talking to him because he has said something to upset her is 0.8. The probability that she is not talking to him for some other reason is 0.05. Given that the probability of him having said something that would upset her is 0.33, what is the probability that she is not talking to him because of something he said?

QUICK QUIZ

1. Safety is all about people. True or False?
2. Ninety percent of accidents are caused by human error.? True or False?
3. Models lie at the heart of human factors. True or False?
4. You can't stop people making mistakes. True or False?
5. Safety doesn't always depend on following the procedures. True or False?
6. If a person experiences an injury at work, the injury was caused by the work. True or False?
7. Fatigue is not the same as sleepiness. True or False?
8. Human factors investigations of accidents should not seek to find out who is to blame. True or False?
9. All accidents have root causes. The aim of human factors is to find that cause. True or False?
10. If you improve system safety, people will take more risks. True or false?

NOTE

1 https://www.amazon.com/Introduction-Human-Factors-Accident-Investigation-ebook/dp/B09F3BZN59/ref=sr_1_1?crid=5L7B051PTVQ0&dchild=1&keywords=introduction+to+human+factors+in+accident+investigation&qid=1632036809&sprefix=introduction+to+human+factors+%2Caps%2C231&sr=8-1.

16 System Stability and Sustainability

GENERAL REQUIREMENTS FOR SUSTAINABLE SYSTEMS

1. Use of resources, including human resources, must be sustainable.
2. Secondary costs of production must be internalized.
3. Personnel management should be human-centered.
4. Design should take a "life-cycle" approach so that the system can be commissioned and decommissioned in a sustainable way.

All things are bound together
All things connect
What happens to the Earth
Happens to the children of the Earth
Man has now woven the web of life
He is but one thread
Whatever he does to the web
He does to himself
How can one sell the air?
Attributed to the Native American Chief Seattle

CORE KNOWLEDGE: GENERAL CHARACTERISTICS OF STABLE AND SUSTAINABLE SYSTEMS

Sustainable systems are in equilibrium with their environments. Their inputs are the outputs of other systems and vice versa.

Humans in Equilibrium with the Rest of the Ecosystem

Human beings evolved in equilibrium with the natural environment, hunting and gathering what they needed to survive and reproduce. They were part of the natural environment. When the European settlers first encountered Native Americans in the Pacific Northwest, they asked if they could buy some land. The chief replied that the land did not belong to them—they belonged to the land.

Hunter-gatherer societies are the earliest form of human social organization that utilize key intellectual tools such as language, the manufacture of artifacts, and the transmission of a shared culture across generations. Kates (1994) reported that the population of the globe increased over the last 1 million years in three stages, in which rapid growth was followed by a plateau. The first stage occurred following the advent of toolmaking, and the population increased from about 150,000 to about 5 million. The second occurred about 10,000 years ago with the development of agriculture, resulting in an increase from 5 to 500 million. The rise of industrial civilization brought about the third stage, with an increase from 500 million to over 6 billion people.

According to Clark and Haswell (1964), hunter-gatherers require 15–30 km^2 of land per person in temperate climates with good rainfall in order to live sustainably. Primitive agriculture required

DOI: 10.1201/9781032717210-16

about 1.5 km^2 of land per person. Not surprisingly, the growth of the population from these times has resulted in the extinction of many other species, and mankind now lives outside of the natural environment, which has become less diverse over time.

In the short term, the sustainability of a system depends on its internal stability. In the long term, it depends on the stability of the ecosystem in which it is embedded.

System Stability

System stability can be considered in relation to:

- *Inertia:* the ability of the system to withstand sudden changes in the external environment.
- *Resilience:* the ability to operate over a range of conditions.
  - *Amplitude:* the width of the operational range.
  - *Elasticity:* the rate at which the system can adapt to changing demands.
- *Succession:* planned and unplanned obsolescence and replacement of elements.
- Sensitivity to feedback.
- Self-renewal and evolution.

These characteristics apply to both the human subsystem and the technical subsystem, but the subsystems have different parameter values for stability, which must be understood at the design stage. When designing any system, it is a requirement that the critical parameter values of the human and technical subsystems are identified and met. This requires that the context of use, the surrounding sociotechnical system, and the environment are understood so that the correct parameter values are identified and met in the final design.

Inertia

Systems with high inertia are able to withstand external shocks. For example, a cruise ship at sea must be structurally strong enough and seaworthy enough to sail safely in high sea states and strong winds. There have been several notable accidents where even large vessels, such as container ships, have floundered after being hit by "freak" waves—waves larger than those anticipated by the designers. Long before the stability of the ship is at risk, those on board the ship will begin to experience physical instability, difficulty keeping balance, and difficulty completing tasks.

- The stability of this system depends on the stability of the people operating it—the human subsystem.
- The effect is not linear; there is a "tipping point"—a motion threshold above which instability increases rapidly. Knowledge of this threshold can be used to inform ship design decisions at the very earliest stages (Figure 16.1).

If the threshold is known, designers can build a more stable system by taking due consideration of the parameter values for human physical stability in the design of the vessel.

Resilience

A resilient system is one that can operate under a wide range of external conditions and demands. Resilience has two properties: amplitude and elasticity. Amplitude refers to the range over which the system can operate, and elasticity refers to the speed with which the system can react to external demands. For example, the electricity supply network is designed to meet demands that are typically of a periodic nature. Early in the morning during the working week, people get up, make breakfast, take a shower, and so on, resulting in a peak demand for energy. During the winter, in temperate regions, the overall demand increases, with very high peaks during short bouts of very cold weather. A resilient system is able to respond to sudden changes in demand, both expected and unexpected, and to return to normal when demands revert to normal.

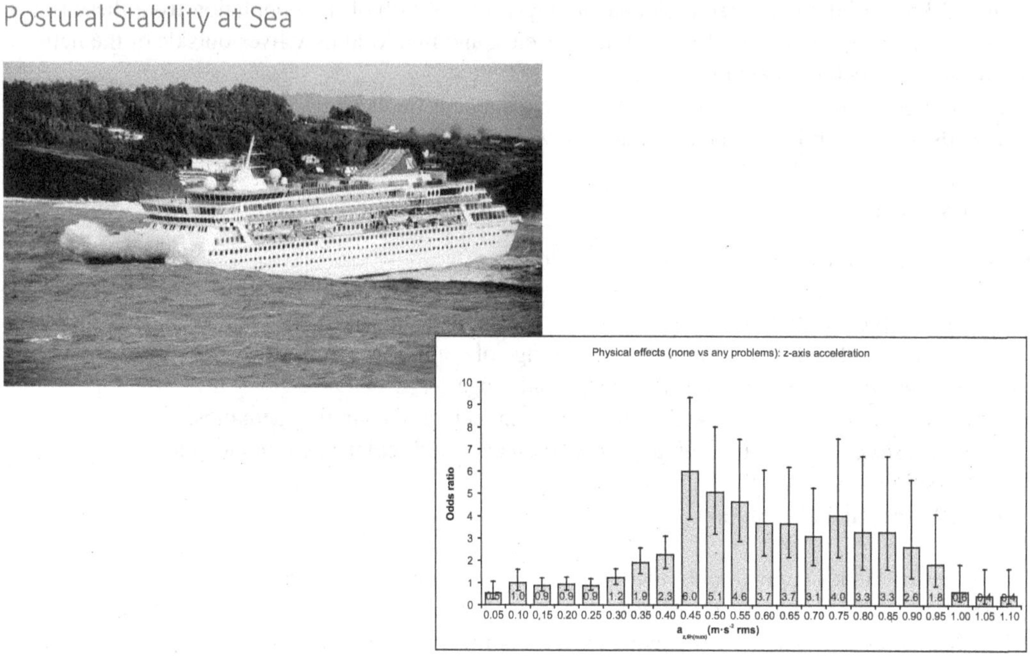

FIGURE 16.1 The inertia of a ship as a system depends on the ability of the crew to maintain stability. The effect of deck motion on stability is not linear—there is a "tipping point" (6-h rms acceleration of about 0.45 m/s^2) beyond which the incidence of problems carrying out physical tasks increases sixfold. (Adapted from Pisula et al. (2012). Crown Copyright, contains public sector information licensed under the Open Government Licence v2.0.)

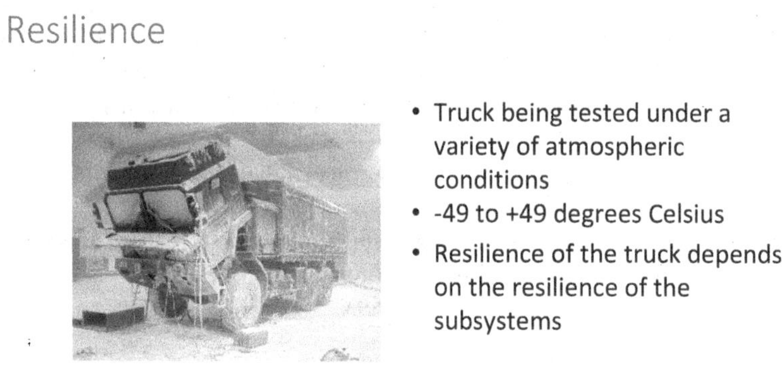

FIGURE 16.2 Truck being tested for resilience at extreme temperatures. (Courtesy of Mr. Bob Davies, QinetiQ, UK.)

Figure 16.2 shows a truck being tested in an environmental testing facility at temperatures ranging from −49°C to +49°C. The testing protocol might involve deactivating all subsystems on the truck, waiting for the truck to meet the external temperature, and then activating each of the subsystems in turn to check that they all work. This can be done at a range of temperatures, with operators in the cab itself to determine whether the cab can be kept habitable under all conditions. In this example, given that the cab is habitable, the resilience of the system depends on the resilience of the truck—the ability of its subsystems to operate across the temperature range (amplitude 98°C) while maintaining a habitable environment for the occupants.

Succession

In natural ecosystems, succession refers to the replacement of one species by another. Invariably, the new species do not simply occupy the niche of the species it replaces—it introduces new properties that have an effect on other species or introduces subtle changes to the stability or resilience of the system itself.

Succession also occurs in artificial systems. Figure 16.3 shows some simple devices that help us to perform arithmetical calculations. The abacus has been succeeded by the mechanical calculator, the electronic calculator, and now calculator functions on other devices such as cell phones. At the time of writing, apps with speech recognition and natural language processing are beginning to supersede calculators for simple arithmetic.

Succession also occurs in more complex systems. Railways predate the invention of the steam locomotive by several thousand years and had long been used in a variety of industries such as mines, dockyards, and even in theaters (to move props around). These railways were not extensive and never formed networks of the kind we are familiar with today.

In the eighteenth century in England, steam engines were developed to pump water out of tin mines (the "Newcomen" engine was the first example). In the coal mining industry at that time, railways were used both inside the mines themselves to pull cars loaded with coal (pulled by donkeys or, in confined spaces, by children) and outside the mines, where teams of horses were used to pull wagons filled with coal over tracks called "wagon ways."

The Napoleonic Wars during the early nineteenth century caused an increase in the price of fodder for horses, and coal mine owners started using steam engines to pull coal along railways instead of using horses. There were a number of consequences. First, the replacement of horses with steam engines enabled mine owners to use their own product (coal) to transport coal, thus lowering the price of transporting coal and increasing their independence from outside suppliers of fodder. Second, steam engines, although inefficient, were more powerful than horses and were not subject

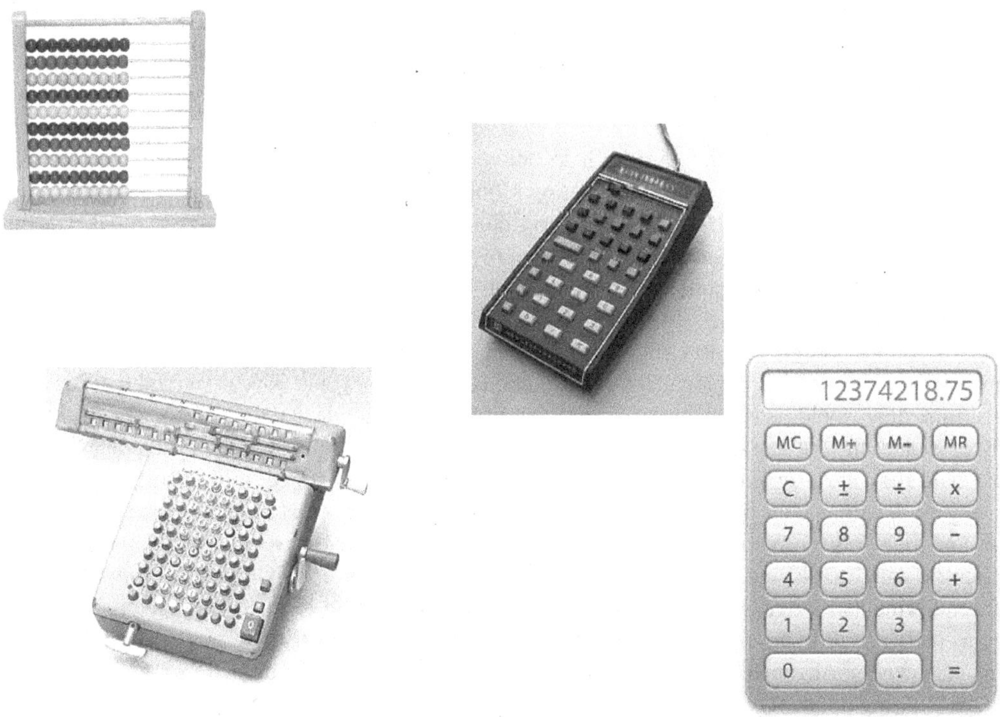

FIGURE 16.3 New products can succeed old ones rapidly to meet requirements more efficiently.

to fatigue. The increased capacity to deliver coal over longer distances stimulated the development of railways and improved access to the market for coal. Third, with the development of a market for steam engines, engineers began to experiment and build more powerful and more efficient steam engines. Fourth, as the rail network grew in size, it enabled other products and people to be transported. Horses, as a means of pulling wagons along railways, were succeeded by steam engines, and with the emergence of a new transportation network, steam engines were themselves succeeded by diesel and diesel-electric locomotives.

Thus, an invention that was originally developed to meet a very specific requirement (control of water in tin mines) grew as it began to meet other requirements for which it was not originally developed. Succession occurs when a new system element meets a requirement better than the element currently in place.

Sensitivity to Feedback

The expansion of the rail network occurred because of the positive feedback loops created by the use of coal to transport coal. As steam engines succeeded horses and the network grew, it began to satisfy other requirements, and further growth occurred.

Microprocessors, originally developed for data processing, preceded the Internet by many years. The first home computers had limited capacity and were used by enthusiasts for writing programs, running applications such as word processors and spreadsheets, playing computer games, and so on. They were not typically connected to any networks.

One of the features of today's laptops and cell phones is the affordances created by the Internet and the large number of requirements met by a single device. Today, we live in two parallel worlds—a world of post offices, retail outlets, and local communities, and an electronic world of e-mail, online commerce, and social media.

Explosive growth and evolution can occur after succession when a new element starts to meet requirements for which it was not originally developed.

System Evolution

Systems thinking gives us a common framework and a vocabulary for analysis across disciplines and technologies. It draws our attention to the fact that systems are not static; they interact all the time with the external environment, changing and adapting in response to external demands and conditions.

Systems are purposeful—they are built to deliver capability. Evolution occurs when requirements change, when new elements meet existing requirements better, and when new elements start to meet requirements for which they were not originally developed. System evolution can also occur when new properties emerge with the introduction of new elements, as in the following examples.

With rail, small networks came first and grew into bigger networks when horses were succeeded by steam engines, then by diesel-electric/electric trains. With the arrival of the Internet, increasingly powerful PCs were connected to each other and then used to satisfy other requirements. Few users, writing software on a 65 K personal computer in the early 1980s, would have imagined that in a few years' time they would be using a very similar, but more powerful, device to go shopping, read their mail, or look at the photographs of the restaurant dinner their friends in New Zealand ate last night. Nor would they have imagined how little of their time they would spend writing software instead of using applications built by others. In systems terms, the Internet created many positive feedback loops and the opportunity to use microprocessors to meet many existing requirements in new ways.

Electric cars are closing the gap with internal combustion engines (ICE) by meeting all of the user requirements. Battery technology is improving rapidly both in terms of storage capacity and time to recharge. Purchase price may still constrain the use of electrically powered vehicles for many car owners. Advances in cost-effective battery technology and improved infrastructure may create a "tipping point" where positive feedback leads to rapid replacement of existing vehicles with electric vehicles. Legislative changes in some countries are setting deadlines to end the manufacture and sale of ICE vehicles.

Electric vehicles can store up to 100 kWH of electrical energy at the time of writing. With millions of such vehicles in operation, all connected via the "Internet of Things," an enormous amount of capacity for storing surplus electricity at the national level becomes available, and owners might be able to "rent" their idle battery capacity to electrical utilities given appropriate infrastructure via the "Internet of Things." This, in turn, might have implications for the use of renewable but intermittent sources of energy, such as solar power, resulting in a new kind of linkage between the transportation system and the electricity-generating system. In some ways, this is analogous to the use of coal-powered steam engines to deliver coal at the dawn of the Industrial Revolution—coal producers used their own product to deliver itself. Owners of electric vehicles might get paid to store their own fuel.

One of the main conclusions that can be drawn from this brief review is that requirements are far more stable than the technologies that meet them. Good system design and analysis should be based on an analysis of requirements rather than technological solutions for this reason.

Systems of Systems

Networks give rise to the interconnection of different systems to form "systems of systems," as in the example below.

One of the main barriers to international trade in previous centuries was the need to move goods from one transport modality to another. This required a great deal of manual handling, human effort, and expense as goods were moved from trucks to ships, to trains, and so on. In 1955, Malcolm McLean, owner of a transport business in the United States, came up with the idea of loading the entire back section of his trucks, including the wheels, onto the ship instead of unloading the cargo. The cab would then drive back to the depot, and another empty cab would meet the load at the destination. This idea eliminated the need for dockworkers to handle the cargo itself. Nowadays, container ships carry thousands of containers that are handled at either end by special cranes and forklifts. The use of containers kills many birds with one stone—it is no longer necessary to build wooden crates to protect fragile items; the containers can be locked and provide secure storage both on deck and at the quayside; fewer workers are required to move them; and a truck delivering a container at the dockyard can collect another to transport back—irrespective of what the load in the container is. Similarly, transporting goods using a combination of transport modalities along the way (train, ship, truck, etc.) is greatly simplified because of the speed and ease of transferring a container and tracking it as it moves from one modality to another (Figure 16.4). The container revolution caused a massive reduction in transport costs. Although manual handling is still required to load and offload the containers, the total amount of handling needed to transport goods from a factory in one country to a retail outlet in another has been greatly reduced.

A User-Centered Approach Is Not Enough

Sustainable systems use resources not just to satisfy user needs but to satisfy them in a sustainable way. How will users know that their needs are sustainable and that their interactions with technology are compatible with environmental constraints?

Levels of Sustainability

Sustainability depends on the use of resources by the system and, from the HFE perspective, on the constraints on user behavior when the system was designed. Different degrees of constraints are possible:

- The user has no control—the system is designed to be sustainable both in the way it meets user needs and in the way users can interact with it.
- The user has control but is nudged toward sustainability.

FIGURE 16.4 "Multi-modal transportation"—a global system of previously independent localized transport systems is now the norm for transporting most manufactured products. International trade now takes place via long and sometimes complex supply chains. A new discipline "Supply Chain Management" has emerged.

- The user can choose among options but must be aware of what the options are and their resource implications. The user must have a good "model" of system functioning with respect to sustainability.
- The user is given incentives: tariffs are set to reward sustainable use and penalize overconsumption.
- The user has complete control: sustainability is a free choice (whether to buy a gas guzzler or an electric bicycle).

BASIC APPLICATIONS

Green Ergonomics

Thatcher (2023) defined "Green Ergonomics" as follows:

> Green ergonomics is defined as HFE with a pro-nature focus aimed at realising the reciprocal benefit between humans and the natural world.

Norton et al. (2021) cite four principles of green ergonomics:

1. Promote eco-efficiency, eco-effectiveness, and eco-productivity to obtain a sustainable balance between system inputs and outputs. This includes optimal demands on employee mental and physical resources combined with opportunities for rest and recovery.
2. Promote workplace ecological resilience to respond to change while maintaining structure and function. Employees have the scope to adapt working arrangements to meet changing demands.

3. Promote the use of local solutions to problems using a participatory approach.
4. Minimize the use of natural resources in achieving organizational outputs. A restorative approach aims for a net positive effect on the natural environment by restoring, renewing, and/or revitalizing the natural resources consumed.

Although HFE is always "human-centered," green ergonomics also focuses on the ecological consequences of the application of HFE, which includes consideration of sustainability. In practice, HFE interventions must take a wider view and consider ecological aspects such as restorative design; resource efficiency in design; recycling; and end-of-life cycle considerations.

Green ergonomics is more than the application of HFE to the design of equipment and jobs in green industries such as recycling, waste treatment plants, and wind farms. It should be applied to all applications of HFE and include considerations of the efficient use of resources and the scope for changing user behavior in a more sustainable direction. Essentially, green ergonomics sees humans in context—as part of a wider ecosystem.

Connection with Nature

Pocock et al. (2023) conducted an experiment to investigate whether activities requiring interaction with the natural world led to improvements in well-being and nature connectedness—the subjective sense of a relationship with nature. Increased nature connectedness is thought to lead to improved mental health, particularly in those initially less connected. Those experiencing higher levels of nature connectedness are more likely to engage in pro-environmental (more sustainable) behavior. Five hundred volunteers were allocated to one of 5 experimental groups or to a waiting list control group. In groups one and two, participants engaged in Citizen Science activities (group 1 spending 10 min counting insects in a 50 × 50 cm flowerbed, and group 2, spending 10 min looking for and recording butterflies in a specified location). These activities were repeated 5 times/day over 8 days. In group three, participants were asked to spend 10 min in a natural green space and record three good things they had noticed (e.g., natural diversity, beautiful plants, etc.). Groups 4 and 5, in addition to performing the Citizen Science activities, were also asked to record three good things they had noticed.

The outcome variables were: measures of nature connectedness including a six-item Nature Relatedness Scale; a measure of physical health (one item); a Pro-Nature Conservation Behavior Scale (containing items relating to behaviors such as picking up litter or doing wildlife-friendly gardening); measures of the experience of engaging in the activities (such as "I felt close to nature through my senses when taking part"); additional items on their engagement in the tasks; and open-ended questions about their likes and dislikes of their participation. All groups engaging in the nature-based activities showed improvements in nature connectedness, well-being, and pro-nature conservation behavior compared to the control group. There were no differences in physical health after the study, possibly because of its short duration (1 week). Measures of participation in the nature-based activities showed that those taking part more frequently had a stronger change in pro-nature conservation behaviors. Females showed a greater increase in happiness and satisfaction with life than males.

The authors concluded that contact with nature can have reciprocal benefits—for the mental health of those engaging and for the environment by promoting eco-friendly behaviors. It would appear that it is contact with nature itself that had the beneficial effects, rather than the tasks themselves. The Citizen Science activities are "top-down" attentionally demanding tasks, whereas the second task (noting "three good things") was a "bottom-up" task, possibly with lower attentional demands. So, irrespective of the task demands, contact with nature had positive effects.

Domestic Energy Consumption and Home Heating

In heating the home, users have control over the amount of energy they use but may not be aware of the resource implications of their energy use. According to Revell and Stanton (2016), users will only be able to optimize their use of energy for domestic consumption if they understand key components of domestic central heating systems and how these components interact with each other (i.e., they have a good model of the system). These components include the programmer and how to control the pattern of energy use over time; booster functions (if fitted) that override a heating cycle for a predetermined period; the thermostat that estimates the temperature in the house from the temperature of the air surrounding it and controls the operation of the boiler; the boiler itself; and the thermostatic control valves on radiators that determine whether hot water flows through them depending on the temperature of the room the radiator is in. The boiler provides the heat, the thermostat controls the temperature in the rooms, and the radiators distribute the heat in each room.

Such systems can be controlled in a number of ways. The output of the boiler can be increased or decreased, the setting on the thermostat can be varied, and the individual radiators can be turned on and off, with their set points altered. In a simple model of this system, these control options are seen as independent. The user might decide to lower the overall temperature in the house and turn off one of the radiators in one of the rooms. The expected result, according to the model, is that the overall temperature in the house would be lower than normal, and the temperature in one room would be cold. To further complicate matters, there are lags in the operation of home heating systems. Raising the temperature setting of the thermostat might cause the temperature in an occupied room to increase beyond a comfortable level if the air temperature at the thermostat has not yet reached the target level. However, in real central heating systems, the control devices are not independent. Turning one radiator off can increase the flow of hot water to the remaining radiators. Since the thermostat only detects the temperature in its immediate vicinity, the rest of the house might heat up. A "higher-fidelity" mental model of the system would represent the interdependencies between the control devices and might lead to a better control strategy—for example, lowering the setting on the boiler when turning off one of the radiators.

Peffer et al. (2016) report that programmable thermostats are theoretically capable of saving 5%–15% of the energy needed to heat a house but save little in practice due to usability problems resulting from non-intuitive interfaces with poorly designed menus (too shallow or too deep), difficulties setting the timer and with mode switching, and largely digital interfaces with small and difficult-to-use buttons.

At present, there appears to be a gap between the design model of domestic heating systems and the user's model, such that users may have difficulty understanding the behavior of the system and, therefore, whether or not their use of energy is efficient.

Using Locally Produced Energy

Kobus et al. (2016) investigated homeowners' attitudes toward a smart energy management system (EMS) designed to optimize the use of electricity produced by the homeowners' solar panels. Much renewable energy production is intermittent and, therefore, traditional approaches to energy efficiency are not optimal—merely reducing consumption is not enough; it is more important to match consumption to production (from intermittent sources such as solar panels). Much domestic consumption is time-critical (such as heating the home or watching TV), whereas the operation of appliances such as dishwashers and washing machines is not. The EMS collected data on electricity production from solar panels along with data on domestic consumption and forecasts for solar radiation to identify time periods when solar electricity was abundant and available to run appliances. Users were able to set up the appliance in advance, specifying only when the process should be completed.

A number of factors were identified that appear to be important if such systems are to be accepted more widely. Feedback about consumption and production should be given frequently, providing appliance-specific breakdowns to give users a feeling of control over their own behavior so that changes in behavior could be mapped onto changes in consumption. *Feedforward* information about solar radiation was important to reduce uncertainty and enable users to plan alternative courses of action in appliance use, helping foster a sense of self-reliance at the level of their own household. Barriers to effective utilization of the system included the usability (or lack thereof) of the EMS and a lack of trust in the information provided. Several other factors were important determinants of the acceptability of the system. These included the amount of time people spent at home—those who were away more during the day were more likely to use the autonomous system; those with higher self-efficacy regarding information technology were more likely to use the EMS; and the system was less likely to be used in houses with more occupants (e.g., children), where the ability to overrule the system was important.

From an understanding of these factors, the requirements for more efficient localized energy systems can be derived. These requirements will specify combinations of feedback linked to incentives, transparency of operation, and usability for the more efficient use of locally produced renewable electricity.

Electric Vehicles: Range Anxiety

Franke et al. (2016) report an experimental study in which new users of electric vehicles completed a 94-km journey designed to place them in a critical range situation. The aim was to understand the factors associated with range anxiety in order to better understand the situations in which range anxiety is likely and how to minimize it in potential new users. The following factors were investigated:

- *Route Familiarity:* user familiarity with the route prior to the trip
- *Trust in the Range Estimation System:* trust in the kilometer-based system displaying the range remaining at any time
- *System Knowledge:* knowledge of relevant aspects of BEV technology
- *Subjective Range Competence:* confidence in one's skills to control range-influencing factors
- *Emotional Stability:* low neuroticism thought to increase resistance to range stress
- *Control Beliefs:* belief in one's ability to control the technology and events associated with it
- *Range Stress:* adverse emotional state in relation to the range regarding the remaining trip distance

Range stress was associated with lower route familiarity and lower levels of trust in the range estimation system. Weaker associations were found with system knowledge, range competence, emotional stability, and control beliefs.

Price, range, and charging infrastructure are the biggest barriers to the uptake of EVs, and the possible loss of autonomy can be a source of anxiety. Rainieri et al. (2023) conducted a systematic literature review of 17 studies of range anxiety, which occurs when transport demands (range required) do not match the driver's perceived control (mediated by feedback on the state of charge and remaining range of the BEV). Other factors that mediate range anxiety include route uncertainty.

In order to encourage the ownership and wider usage of BEVs, the authors suggest that every effort be made to reduce range stress on users' first encounters with BEVs, bearing in mind that the available range for a fully charged BEV is more than adequate for most daily usage. This can be done by encouraging use on familiar routes, increasing users' knowledge of BEV technology and system characteristics, and improving their subjective competence and trust in the interface (particularly the state of charge of the battery). Clearly, range stress will be higher under conditions of uncertainty, such as unexpected detours that lengthen routes, and stress could be mitigated by advance warnings and more advice about the impact of detours on travel time to the destination and remaining range. "Smart Freeways" that control aggregate levels of traffic flow to prevent delays are a "top-down" method of reducing range anxiety.

Green Buildings

Many countries now encourage new buildings to have sustainability features such as energy efficiency, rainwater harvesting, rooftop gardens, and greater use of daylight to illuminate workspaces. The United States has a Green Building Rating system (LEED) to certify buildings with key sustainability features. In a survey of working conditions in two green office buildings in the United States, Hedge and Dorsey (2016) found large differences between the buildings in occupant satisfaction with working conditions, particularly air temperature, air quality, and the amount of fresh air. Satisfaction with noise levels was low in both buildings. However, occupants did rate working conditions as superior to those in their previous buildings (83.4% and 70% in favor in the two green buildings, respectively). To gain maximum benefit from green buildings in the future, the authors concluded that the certification system should include mandatory measures of the quality of ergonomics in the design and occupant experiences of working in the building. Nurick and Thatcher (2023) carried out a qualitative investigation of productivity in green buildings in relation to indoor environmental quality. The main drivers of productivity were access to dedicated workspace; acceptable ambient noise levels in an open-plan office environment; quality lighting (natural and artificial); temperature control (or lack thereof); the physical quality of the office environment, including the furniture; building management support services (e.g., repairing a broken light); access to breakaway zones and refreshment stations (location and amenities) that can be used as informal collaborative spaces access to activity-permissive workstations such as standing desks; knowing that there are external amenities within walking distance (e.g., groceries); and access to privacy when required. Access to nature was also cited as a driver of productivity (e.g., views of natural scenes outside).

TOOLS AND PROCESSES

There are many ways to investigate the stability and sustainability of systems. Key performance indicators can be defined and measured and include output, wage costs, employee turnover and absenteeism, energy usage overall or for specific functions, compliance with standards and regulations, and so on.

Behavioral research is increasingly important in order to understand human behavior in relation to energy use and attitudes toward more sustainable alternatives, so that barriers can be removed in the design of new systems. Nickerson (2014) argued that people fail to engage in environmentally sustainable ways of working and living because of a lack of knowledge or a lack of motivation. Providing users with feedback in the form of meter readings or product labels can potentially remove the knowledge barrier, as long as the feedback is correctly interpreted and of an appropriate "grain size" and regularity so that users can map the feedback onto their own behavior.

Transportation accounts for about one-quarter of the energy consumed in the United States. and, therefore, the increase in home and hybrid working following COVID-19 lockdowns will reduce resources used in commuting while raising many new challenges, including job redesign, supervision, social isolation of homeworkers, and the impact on family life.

In HFE, questionnaire surveys and interviews are commonly used to understand people's attitudes and opinions about a range of issues, including working conditions, new ways of working, and new technologies.

Questionnaire Design in HFE

We have already been introduced to some of the first steps in questionnaire design in Chapter 14, where we reviewed the construction of psychometric scales. Questionnaires normally consist of several different scales, plus information on respondent demographics and characteristics.

Questionnaire design is a science (see Fife-Shaw, 2000; Brewerton and Millward, 2001, for example). A questionnaire is a measuring instrument and must conform to the same standards as any other instrument. This means that the questionnaire must be valid, reliable, and sufficiently sensitive to measure the construct under investigation. The word "construct" is used here to refer, for example, to a process, idea, personality trait, outcome, or response.

A questionnaire is reliable if it elicits the same responses when completed by the same person under the same conditions. It is valid if it measures what it is supposed to measure. A questionnaire can be reliable but not valid if it reliably measures the wrong construct. If it is not reliable, it cannot be valid. Reliability is, therefore, a prerequisite for validity. The reliability of a questionnaire should be distinguished from the *stability* of the construct it is designed to measure.

For example, suppose we want to design a questionnaire to measure attitudes toward telecommuting or BEVs. We might discover that our questionnaire is reliable—it elicits the same responses when users complete it on a Monday morning as it does on a Friday afternoon. However, the questionnaire will not be valid if users make their decisions on the basis of other considerations not measured by the questionnaire. In this case, use of the questionnaire will consistently lead to incorrect conclusions. Attitudes toward telecommuting might be more positive if the questionnaire is completed on a Monday morning, prior to a formal staff meeting, than if it is completed on a Friday afternoon, at an informal meeting in the pub after work. This would be a reflection of the instability of the constructs under investigation rather than a reflection of the reliability of the questionnaire.

Questionnaire validity depends very much on the assumptions questionnaire designers have about what motivates people to make the choices that they do, or what factors influence people's perceptions and attitudes. Questionnaire design is best preceded by conducting structured interviews or focus groups (see Chapter 14) in order that questionnaire designers can see things from the users' point of view.

More about Reliability and Validity

If an HFE expert were to claim that he had designed a device that could read people's minds, nobody would believe him. Yet countless decision-makers rely on the findings of questionnaire surveys of all kinds without ever questioning the measurement properties of the instruments used. Reliability and validity are the two key *psychometric* properties of a questionnaire that we need to know about before we can use a questionnaire to assess psychological constructs.

- *Test–Retest Reliability:* The extent to which the questionnaire elicits the same responses when used in identical situations by the same persons.
- *Internal Reliability:* The extent to which items within sections of the questionnaire are consistent (see Chapter 14 for a discussion of the construction of psychometric scales).
- *Face Validity:* Appropriateness of use with the population; whether the questionnaire appears relevant to the issues it purports to address.
- *Construct Validity:* Whether the questionnaire content reflects current theoretical thinking in an area. For example, a questionnaire designed to measure exposure to occupational stressors should demonstrate congruence with models such as Karasek's. Construct validity depends on the control of confounding factors; for example, if we want to find out if beliefs about global warming influence the sustainable use of electricity in the home, the questionnaire items must be designed to be *specific* to the outcome under investigation and insensitive to other causes, such as a sudden increase in price. Construct validity can be subdivided into:
    - *Discriminant Validity:* Does the questionnaire have sufficient resolution to distinguish between key groups? For example, if we are interested to know whether people who have installed solar panels on their houses have different views about global warming than those who did not, we might first test a set of questions on the two groups of homeowners to identify questions with high discriminant validity. If we had chosen questions with low discriminant validity, we might erroneously conclude that there is no difference in beliefs between the two groups.
    - *Convergent Validity:* A questionnaire with high convergent validity will demonstrate that constructs that should be related to each other are related. For example, a usability questionnaire should be able to demonstrate that two products of equal usability, but with different appearance, price, and manufacturer, have the same usability.

- *Content Validity:* Whether the questionnaire covers the area of investigation adequately, including the appropriateness of the phrasing of statements/questions/wording for use in a particular environment. See Chapter 14 for a discussion of written language design and readability indices.
- *Concurrent Validity:* Whether the responses elicited by the questionnaire agree with other measures taken at the same time. For example, people who believe strongly in anthropogenic climate change, according to the questionnaire, might be expected to be more frugal in their use of energy.
- *Predictive Validity:* Whether the questionnaire correctly predicts the outcome of interest. For example, those expressing an interest in having solar panels installed on the roofs of their houses might be expected subsequently to make inquiries to vendors, seek information on solar panels, or even have them installed.
- *Criterion Validity:* Whether the questionnaire findings agree with some external "gold standard" of performance. For example, people scoring highly on a questionnaire about "green beliefs and behaviors" might be expected to engage in a wide variety of behaviors, and not just those measured in the questionnaire (e.g., owning solar panels, BEVs, or bicycles; engaging in home composting or gardening, etc.).

Related issues in questionnaire design include the following:

- Information on the norm group that the questionnaire was developed on. The findings of questionnaire surveys are often difficult to interpret unless there are published norms or standards for the issues under investigation. For example, the fact that 20% of employees in a business complain that the work is stressful and that they intend to change jobs in the next year tells us little unless we know the background prevalence of self-reported stress and intentions to leave for employees of a similar age and socioeconomic status. The population prevalence of acute anxiety and depression is about 18% in the United Kingdom; hence, surveys of occupational stress in the United Kingdom should use standard questionnaires so that their findings can be related to this benchmark. There are many standardized questionnaires on a variety of topics, ranging from job satisfaction and computer stress to self-reported musculoskeletal pain, that can be used in ergonomic studies.
- Inter-rater reliability for open-ended measures. Some questionnaires have an open section at the end for "any other comments." The information provided in these sections often has to be interpreted, categorized, and summarized. It is a wise precaution to have two independent investigators summarize a subset of the survey returns so that the level of agreement between the researchers can be determined. For open-ended questions, categorization can proceed on an ad hoc basis or using "a priori" categories selected by the researchers. In either case, some measure of the level of agreement between independent raters should be given (see Chapter 14).

There is a great deal of discussion of the different facets of validity in the psychological literature. For HFE, the two most important aspects of validity are probably "discriminant validity" and "criterion validity."

A questionnaire with high discriminant validity would be good at differentiating user responses to different products or at differentiating the responses of different user groups to the same product. We could use such a questionnaire to compare different phones against one another or against an existing phone acknowledged to be of good design (and chosen as a "gold standard" for comparison). We might, for example, require that any new design must be at least as good, if not better, than the standard on one or more of the dimensions. A cell phone questionnaire with high "criterion validity" would be valid in the sense that users' responses relate to some criterion measure of use.

The criterion might be an affective state, such as satisfaction, an intention (such as the intention to change phones in the next 6 months), or a real outcome (such as buying a new phone).

Questionnaire designers are often tempted to analyze their data using some of the items as if they were outcomes suitable for criterion validation; for example, by looking for correlations between ratings of product features and ratings of overall satisfaction. The main drawback of doing this is that the correlation coefficients from one administration of the questionnaire are inflated by *common method variance.*

In practice, psychometric techniques can be very useful in making complex datasets easier to analyze and interpret and, by exposing the underlying structure, they can give us new insights into what the questionnaire is really measuring. As with physical anthropometrics, sometimes we need data on single variables such as stature or legibility of screens, and psychometric reduction is not needed. However, when it comes to asking more general, or even vague, questions such as "Is the population changing in some way?" or "How do people think about their phones these days?" the psychometric approach to handling complex databases can be very useful.

Assessing Reliability and Validity: Pilot Studies Are Essential

Proper testing of questionnaires is an essential part of the questionnaire design process. Draft questionnaires should be tested with real users who are representative of those who will complete the final version. An initial goal of user-centered questionnaire design is to optimize the set of items in the questionnaire and to ensure that people understand them. Following this, the questionnaire should be evaluated for reliability. A small group of volunteers should be given the questionnaire to complete on at least two separate but identical occasions. If the questionnaire is reliable, the respondents should provide approximately the same answers. For example, if we want to know whether a questionnaire can measure the usability of a smart electricity meter, we might first give it to a small group on two separate occasions under as identical conditions as possible and then compare the answers. If each individual gives similar answers on the second occasion compared with the first, the agreement will be high, and the questionnaire will be deemed reliable. Next, we might give the questionnaire to a small group consisting of people known to be computer-literate with high self-efficacy in the use of interactive products, and another group of people known to be less computer-literate. Comparing their responses should enable the validity of the questionnaire to be assessed. Barriers to Reliability Include:

- Respondents have not formed stable opinions about the items in the questionnaire (e.g., they haven't had time, don't know, or don't care). The act of completing the questionnaire causes them to think about the issues, and they give different answers the next time. This is called *post-test sensitization.*
- Respondents have difficulty recalling the information from memory or making the required assessments of the product.
- The information is not available to respondents because it was never stored in memory (e.g., "How many times per day do you scratch your nose?").
- The investigator's assumptions about the stability of the phenomena in question are incorrect.
- The answers depend on factors that change over time (e.g., the weather). If the investigator does not control for this, the answers will vary.
- The person changes over time (e.g., the mood of workers may be lower on Monday morning than on Friday afternoon). If the questionnaire items deal with issues sensitive to mood, responses may vary depending on the day the questionnaire is completed.

Barriers to Validity Include:

- Respondents cannot answer the questions because they do not understand them.
- Respondents cannot answer the questions because they do not know the answer.
- Respondents do not want to answer the question.
- Respondents do not give an honest answer because they do not care about the topic covered by the questionnaire, resent the process, or find it intrusive.
- People do not give a full answer because they worry about how the information will be used.
- The questions have superficial validity but do not truly reflect what people think or believe.
- The questions have superficial (face) validity (they appear to capture the right attitudes) but do not accurately reflect what people actually do (i.e., how they behave rather than how they want others to think they behave).
- Response bias: For example, respondents give socially desirable answers rather than stating what they really think (e.g., "Are you a racist?" Yes/No).
- Response sets: People give the same answer irrespective of the question (e.g., always ticking the right-hand response box).

If the questionnaire contains scales (where multiple items are summed), Cronbach's alpha should be calculated for each scale (see Chapter 14). If barriers to validity and reliability are identified at the piloting stage, it strongly suggests that the investigators need to conduct further investigations and organize focus groups to gain a better understanding of the psychological issues in question and the constructs that can be meaningfully measured using a questionnaire.

A Structured Approach to Questionnaire Design

A structured approach consists of the following:

- Generation of an item set for the questionnaire
- Testing the item set on a subset of the target population
- Identification of constructs
- Construction of scales
- Item selection for the scales
- Testing scale reliability
- Pilot studies for comprehension and wording
- Test–retest reliability
- Validation

Generation of Items for Questionnaires

There are three main ways of generating items for questionnaires. They are not mutually exclusive:

- Review existing questionnaires and checklists in the general area to be researched, and select a subset of items in the domain of interest.
- Generate an item set using focus groups or structured interviews.
- Review the literature and current theories.

Wording of Items

Many of the design issues relating to focus groups also apply to questionnaire design (Workshop 14.1).

Avoid leading questions such as, "To what extent does global warming cause you to reduce your electricity consumption at home?"

- This is a leading question on two counts:
  1. It implies that the respondent is aware of or believes in global warming.
  2. It implies that any such beliefs affect domestic energy consumption.

A better approach would be

"Are you aware of claims that modern domestic energy usage is causing global warming?"
Yes No
"If yes, does this affect your use of energy at home?" Yes No
To what extent does it cause you to use less energy at home?
No effect Minor extent Moderate extent To a great extent

Response sets are best detected at the piloting stage. Given that there are a number of items dealing with the same issue, word them so that a positive response is given by providing a positive answer to some questions and a negative one to others. For example:

	Strongly Agree		Neutral		Strongly Disagree
I am worried when I travel more than 50 miles in my BEV	1	2	3	4	5
My BEV has adequate range when I travel long distances	1	2	3	4	5

The existence of response sets, when a mix of positive and negative items is used, is indicated by low agreement between all the items in a scale (a low value of Cronbach's alpha), indicating a contradictory pattern of responding:

- Provide context. Questions such as, "Are you satisfied or dissatisfied with your smart energy meter?" are ambiguous because they are nonspecific and lack context. If the question is exploratory, a better alternative is the open-ended, "What are your likes and dislikes about your smart energy meter?" A more specific alternative would be, "Are you satisfied or dissatisfied with the frequency of the information updates?"
- Use basic vocabulary. Replace "consider" with "think," "initiate" with "start," "reside" with "live," etc.
- Avoid hypothetical questions, such as "Would you like central heating controllers that are easier to use?" Most people are unlikely to say "no."
- Separate opinions from facts. Response rates to opinion-based questions can be very different from response rates to factual questions and may cast doubt on the predictive value of the former. For example, "Do you think it is wise to have your boiler serviced at regular intervals?" is likely to be answered very differently from, "Have you had your boiler serviced in the last 5 years?" Or "Do you think it is sensible to pay more for good quality clothes?" compared to, "How many items of clothing do you own that cost less than $20.00?" Questionnaire designers should carefully consider the extent to which they really want to know about respondents' opinions, or the extent to which the survey should enable predictions to be made about future behavior.

HFE Workshop 16.1 provides a basic introduction to statistical techniques for questionnaire validation.

HFE WORKSHOP 16.1

DETERMINING SAMPLE SIZES FOR SURVEYS

Sample size determines the accuracy of a survey and the error associated with the findings. Krejcie and Morgan (1970) offer the following equation for calculating sample sizes (returned questionnaires from surveys) when categorical (e.g., "Yes/No") items are used.

$$n = \chi^2 * N * P(1-P)/\left(ME^2 * (N-1)\right) + (\chi^2 * P(1-P))$$

Where:

n = sample size
χ^2 = Value of the Chi-square statistic for 1 degree of freedom at a specified confidence level (3.841 at 0.05 in this example)
N = population size
P = Population proportion
ME= desired margin of error

Suppose we ask all 300 members of a small ergonomics society whether they receive value for money from their membership fees. We assume that the proportion agreeing will be 0.5 (50% say "yes").

$$n = 3.841(300)(0.5)(0.5)/(0.5)^2(299) + 3.841(0.5)(0.5)$$

$$= 169$$

If we do not know how many will agree or disagree, then set P at 0.5 because this gives the largest value of n (if we assumed that more would be satisfied with their fees, we could set P at 0.7), and n is 158. In either case, we need a response rate of over 50%.

The table below gives sample sizes for continuous and categorical variables at $P = 0.05$.

N	Continuous	Discrete
100	55	80
200	75	132
300	85	169
400	92	196
500	96	218
600	100	235
700	102	249
800	104	260
900	105	270
1000	106	278
1500	110	306
2000	112	323
4000	119	354
6000	119	362
8000	119	367
10,000	119	370

A sample size calculator for surveys can be found at the link.[1]

Sources of Bias in the Design of Experiments, Surveys, and Field Trials

Participant Reactivity: Participant reactivity is the term used to describe changes in behavior brought about by participation in experiments or field trials. Social psychologists have done a great deal of research on the behavior of subjects during experiments and the factors that can bias the findings. In general, participants in experiments or trials are conscious of the fact that they are being observed and often feel that they themselves are being evaluated (even if they are not). The result is that they are usually compliant and make an effort to perform well. When new equipment or methods are being tested, the benefits may be overestimated because of this. Reactivity can take several forms:

A Response to the "Demand Characteristics" of the Experiment: Demand characteristics are all the cues from the experimental situation that influence subjects' behavior. Orne (1962) coined the term after conducting a classic experiment in which subjects were asked to carry out a tiresome and meaningless task. Orne was impressed by the subjects' diligence and how readily they accepted the task. Subjects may try to guess the experimental hypothesis and confirm it by behaving in what they perceive to be the "correct" way. The demand characteristics of experiments can have serious implications for the assessment of the effects of mild stressors on task performance. This is probably one of the reasons why research concerning the effects of noise on performance is inconclusive (experimental subjects work harder to overcome the distracting effects of the noise). In real work situations, mild stressors can have a large impact on the performance of work tasks because people are not so readily motivated to continue working in unpleasant conditions. They may even have the right to stop work or go home if work conditions exceed certain limits.

Novelty and Disruption Effects: Novel interventions may have short-term effects on performance because they are novel, not because they work better. The novelty creates interest, diverts attention from problems, and may even heighten arousal and therefore boost performance. New office furniture, for example, may lead to reductions in musculoskeletal complaints in the short term as everyone experiments with it and gets used to the change. In situations where change is common, novelty effects may be less likely than in situations where change is rare.

Pretest Sensitization: The administration of questionnaires or interviews before a trial may "prime" people to respond in ways that differ from how they would respond had the pretest not been administered.

Experimenter Effects

The experimenter may unintentionally bias the trial. Subjects may infer, from the experimenter's behavior, what is expected of them or what the hypothesis might be. The experimenter may unconsciously reinforce certain behaviors by means of body language or attention. Subjects may then behave in such a way as to confirm the hypothesis. It is also possible that subjects may try to sabotage the experiment if the experimenter effect is negative.

Placebo Effect

A placebo is something that is not a medicine but is made to appear like one and is given to patients who are told that it is a medicine. Patients receiving placebos who believe they are being given an effective treatment often improve. Up to 40% of the benefits of some medicines are thought to be due to the placebo effect. Placebo effects are good for consultants but not for researchers, or as Blau (1992) put it:

> The doctor who fails to have a placebo effect on his patients should either become an anaesthetist or a pathologist.

Placebos are nearly always used in clinical trials of new treatments; the treatment is only judged to be efficacious if it brings about a larger improvement than that due to the placebo. Hrobjartsson and Gotzsche (2001) evaluated the evidence for the placebo effect and found evidence for beneficial effects on subjective outcomes and pain levels. However, placebos were ineffective in having beneficial effects on objective outcomes or on subjective outcomes when assessed using binary (e.g., "yes/no") measures.

Reduction of Bias

There are many ways to reduce bias in field research. The double-blind, randomized controlled trial is probably the "gold standard." A real treatment and a control or placebo treatment are randomly allocated to subjects. Neither experimenters nor subjects are aware of which one is being administered. An alternative to the placebo is to compare a new treatment with the best conventional method. In the evaluation of seats and other artifacts, "blinding" of subjects and experimenters is not possible. Experimenter bias can sometimes be avoided by using naïve experimenters who are not aware of the hypothesis.

Much modern ergonomics research, particularly in the psychosocial, cognitive, and epidemiological areas, makes use of questionnaires. Proper design of questionnaires using formal psychometric methods is one of the best ways to reduce bias in ergonomics surveys.

Survey Design in HFE

The reader is referred to specialized statistics texts for detailed information. However, broad guidance for survey design can be summarized as follows. Surveys should be based on random samples, stratified according to the variables of interest—for example, sex and age. Stratification simply means that the population is divided into groups, and respondents are selected from each group in proportion to the size of the group in the population. Random sampling techniques are used so that the results of the survey, conducted on a small sample of the population of interest, can be *generalized* to the population as a whole.

Sample Size and Participant Response Rates

Surveys should be designed to achieve high response rates (e.g., 75%). If the rate is less than 50%, the findings may be of limited value due to *response bias.* Response bias can invalidate surveys because the respondents are not representative of the target population (i.e., even though a random sample was selected and questionnaires were sent, a nonrandom subset has responded by returning completed questionnaires). For example, if we want to estimate the percentage of commuters interested in buying a BEV, a survey with a low response rate (say, 30%) might lead us to overestimate the true percentage.

If response rates are too low, the data will yield inaccurate estimates of the outcomes of interest. This is because the standard errors of the estimates will be too large (see Chapter 3 for further discussion of standard errors).

In general, for 5% accuracy and assuming no response bias (response rate of over 50%), the recommended sample sizes are as follows:

- 500 or fewer—All workers to be included in the survey
- 501–1000—A minimum of 500 selected randomly
- 1001–2000—A minimum of 650
- 2001–3000—A minimum of 700
- 3000 or more—A minimum of 800

It is unwise to restrict the size of a sample on cost-benefit grounds because sample size determines the discriminant validity of the survey. It is often the case that the purpose is to detect changes over time (e.g., suppose a piece of software is redesigned and we want to know whether usability has increased). The larger the sample, the smaller the error associated with the initial estimate and, therefore, the greater the likelihood that a change can be detected subsequently. For example:

- Suppose 50% of 100 commuters responded to a survey expressing an interest in BEVs at time point 1. The 95% confidence interval was 37%–63%. Therefore, at time point 2, after an increase in the price of gasoline, with the same sample size, the rate has to be less than 37% or greater than 63% to be significantly different.
- For 250 responding, the confidence interval would narrow to 42%–58%.
- For 1000 responding, the confidence interval would narrow to 45%–55%.

Note that the improvement in statistical accuracy is far greater when the sample is increased from 100 to 250 compared to an increase from 250 to 1000. Selection of an appropriate sample size should therefore be based on statistical, and not just cost, considerations.

Reasons for Nonresponse and How to Improve Response Rates

- *Questionnaire Fatigue:* The population of interest already receives questionnaires on different topics and lacks motivation to respond to another.
- *Lack of Motivation to Complete the Questionnaire:* An opening paragraph should be constructed to explain the purpose of the survey, why it is important, and how the information will be used.
- *Not Expecting the Questionnaire:* Respondents can be "primed" by sending a letter a week or two in advance explaining the purpose of the study and that they will soon receive a questionnaire.
- *No Return Facilities:* A stamped, addressed envelope should accompany the questionnaire. Electronic surveys can be conducted via a website.
- *Distrust of the Process:* Where anonymity is ensured, this must be spelled out clearly and early on. For all surveys, an explanation of the measures taken to ensure data security must be given. It must describe how the data will be stored, for how long, and who will have access to it. In the United Kingdom, the Data Protection Act covers these requirements. Respondents must be informed that they can withdraw at any time, even after having returned their questionnaires.
- *Inadequate Instructions:* Details of when and how to return the questionnaire, and the name(s) and position/job title of the investigators, are not included.

Tactics to Maximize Response Rates

The higher the response rate, the better, assuming that no coercive measures are taken to encourage recipients to respond. Table 16.1 summarizes tactics for increasing the response rates of surveys.

Distribution of questionnaires by Internet is becoming increasingly common due to convenience and cost-effectiveness considerations. The statistical requirements for Internet surveys are exactly the same as for postal surveys. Depending on the requirements of the particular survey, investigators might consider respondents' need for privacy and the practicalities of completing screen-based survey questionnaires in an open-plan office. However, the Internet does seem to offer several advantages over postal methods (e.g., not having to mail the returns), and there are few general disadvantages. For further information, see the International Test Commission (2005).

Reduction of Bias in Surveys

There are several sources of bias in ergonomic surveys. *Sampling bias* can occur if the sample is not random and key groups are under- or overrepresented. This can be avoided by using stratified sampling techniques and large samples. *Response bias* can occur if key groups are less likely to respond to the survey. The risk can be minimized by using the tactics provided in Table 16.1. It is sometimes possible to reduce response bias post hoc by "re-balancing" the sample of respondents. This involves randomly removing respondents from other demographic or user groups so that the sample of respondents again mirrors the population of interest. Unfortunately, this wastes valuable data.

Common method variance or *common method bias* occurs when the same instrument is used to measure different outcomes. For example, suppose we ask 100 people to rate on a 5-point scale whether they experience role conflict at work and how dissatisfied they are with their jobs. We find that the scores on the two measures correlate highly—those with high role conflict are also highly dissatisfied. According to measurement theory, the variance in the scores consists of true variance and random error variance; for example,

$$\mathrm{VAR}_{\text{job dissat}} = \mathrm{VAR}_{\text{true}} + \mathrm{VAR}_{\text{random}} \tag{16.1}$$

However, many instruments used in ergonomics also have *systematic* variance, which is due to the way the instruments are designed and how users interact with them. This is called *method variance:*

$$\mathrm{VAR}_{\text{job dissat}} = \mathrm{VAR}_{\text{true}} + \mathrm{VAR}_{\text{random}} + \mathrm{VAR}_{\text{method}} \tag{16.2}$$

TABLE 16.1
Tactics for Maximizing Response Rates in Questionnaire Surveys

Tactic	For Example
Make multiple contacts	Mail an opening letter to inform people in advance. Send the questionnaire 2 weeks later with the second letter requesting assistance. Re-send questionnaire, plus a reminder letter 3 weeks later.
Increase the salience of topic	The initial contact letter and the instructions on the front page of the questionnaire must capture the interest of the recipients. The findings of previous surveys and any initiatives resulting from them might be mentioned.
Incentives	Encourage recipients to see themselves as participants in a two-way process in which the results of the survey are fed back to them over time by means of a newsletter.
	Lotteries: Completed questionnaires entered into a lottery when returned. The evidence for lottery effectiveness is equivocal. Direct incentives work better (cash or a gift enclosed with the questionnaire).
Questionnaire design lengthy	Use brightly colored paper. Avoid, poor layout, small print, questionnaires, difficult/intrusive questions, inadequate instructions. Do not increase length by including items in which the information is already known or easily obtainable (e.g., recipient's current address).
Increase trust	Mark questionnaires "Private and Confidential".
Engage and maintain interest	Send questionnaires in colored envelopes (except brown). Begin with sections dealing with most interesting aspects (rather than demographics) as these items are more likely to capture the interest of recipients.
Target low response groups	Send envelopes addressed using handwriting, rather than printed labels, as this has been shown to increase response rates in postal surveys.
Return envelopes	Include a prepaid return envelope. Stamped, rather than business reply envelopes, work better.
Inter/Intranet surveys	Distribute questionnaires via the Internet to save time and cost.

Source: Church (1993); Cui (2003); Porter and Whitcomb (2003).

If we correlate ratings of job satisfaction with ratings of role conflict in our hypothetical example, any covariance in VAR_{true} might be inflated by covariance in VAR_{method}. The effect of common method variance is to create spurious correlations between variables and to inflate the strength of true interrelationships.

Causes of common method variance include:

- Social desirability bias—where respondents give responses that enhance their self-image or are "politically correct." In health surveys, short males typically overreport their stature and overweight people underreport their body mass, leading to underestimation of the prevalence of overweight and obesity. A survey on the island of Svalbard (Norway) found that self-reported alcohol consumption underestimated actual bar and liquor store sales by 40%!
- Acquiescence bias—agreeing with the presumed intentions of the survey designers and overreporting.
- Mood state—response patterns biased by mood.
- Stoicism—denial of any problems or difficulties.
- Common scale formats for different variables.
- Response sets and item grouping.

Burton-Jones and Straub (2004) report that, despite the fact that common method variance can inflate correlation coefficients by significant amounts, the practice of using only one instrument in information systems research was common—over 50% of 116 published studies did so. These authors provide advice that is generally applicable to much research in ergonomics—use questionnaires and rating scales to measure *subjective* states and use observation or automatic computer logging to measure *behavior*. For example, use self-reports to measure beliefs in climate change, but use hard data on energy consumption or direct assessment to evaluate home insulation and automobile usage. Use rating scales to obtain data on user preferences or pleasure in use and task analysis or performance measures to assess user–product interactions such as errors and difficulties. Use rating scales to measure strain, dissatisfaction, and role conflict, and hard data to measure absenteeism, turnover, and sickness.

How to Deal with Low Response Rates

Suppose we send 100 questionnaires on global warming to a random sample of people in a town. The questionnaire has only one question:

"Do you like completing questionnaires on global warming?" Yes/No

Forty percent respond and, of these, 80% (i.e., 32 out of 40) report that they like completing questionnaires on global warming.

If there were no response bias, we could use our 40% respondent sample to describe the population and conclude that 80% of people in the town like completing questionnaires on global warming. Alternatively, maybe everyone who likes completing these questionnaires did so. The nonrespondents do not like completing questionnaires on global warming, which is why they did not respond. Therefore, the true proportion would not be 80%. Unless we can understand *why* some people did not respond, our survey has told us very little—that the true percentage is anything between 32% and 80% (i.e., 80% of 40 respondents = 32 respondents).

Nonresponse can be due to "non-coverage" (some people did not receive the questionnaire) or it can be due to bias. Bias is a serious problem when the reason for nonresponse is related to the topic under investigation. There are two main models of response bias (Groves, 2006):

The Deterministic Model

$$\text{Bias}(\bar{Y}_r) = [M / N][\bar{Y}_r - \bar{Y}_m] \tag{16.3}$$

where:

$\bar{Y}_r$ = The sample mean achieved by the survey
$\bar{Y}_r$ = The mean of the respondents in the target population
$\bar{Y}_m$= The mean of the nonrespondents in the target population
M = The number of nonrespondents in the target population
N = The total number in the target population

Bias increases as nonresponse increases and in proportion to the difference between respondents and nonrespondents on the measured variable. According to the deterministic model, low response rates are only a problem if nonrespondents think differently from respondents. Low response rates (less than 60%) can be dealt with by following up nonrespondents, possibly with a shortened version of the original questionnaire using selected items from different scales, or by using telephone interviews to determine the size and statistical significance of $Y_r - Y_m$. The nonresponse bias can then be estimated using Equation 16.3 to assess the implications for the interpretation of the data.

The Stochastic Model

$$\text{Bias}(Y_r) = \frac{\sigma_{yp}}{p} \tag{16.4}$$

where σ_{yp} is the covariance between the propensity to respond and the survey variable, and p is the mean propensity to respond.

This model proposes that bias is independent of response rate but increases according to the correlation between the propensity to respond and the response to the attribute measured. This implies that response bias in single surveys distorts the estimates of some parameters but not others.

The variable p, the propensity to respond, is likely to have *state-like* and *trait-like* components ("I'm too busy to respond to this survey at the moment" as opposed to "These surveys are just a way of spying on us"). Arguably, the latter is more likely to covary with stable attitudinal item scores, and the former with items dealing with transient factors such as workload.

According to the stochastic model, we can deal with low response proactively by including survey items to measure response propensity. There are several ways of doing this. In the questionnaire itself, include items such as:

> I would complete this questionnaire again, if I received it in a future survey
> or
> I usually complete surveys when I receive them

We can also investigate response propensity *post hoc* by comparing early and late responders (early responders are assumed to have greater response propensity).

Where p is found to correlate with substantive scores, there is evidence of response bias. The magnitude of the bias in substantive scores can be estimated in these cases. Even with low response rates, if p does not correlate with an item or scale score (or with a demographic variable), then this suggests that there is no bias. Strictly speaking, it suggests that the null hypothesis—that the survey is unbiased—cannot be rejected.

Demographic Checks and Balances: A final method for dealing with low response rates is to check for demographic bias. Suppose we carry out a survey of attitudes toward domestic solar panels in a small town because we want to estimate the size of the market for solar panels. We find that younger people in our respondent sample have more positive attitudes toward solar panels than older people, believe global warming is more of a threat to the planet, and say they are more likely to fit solar panels to the roof of their house. If younger people were more likely to respond to the survey than older people, then the aggregated survey data are biased—they reflect more positive attitudes than they should, and we might overestimate the size of the market. We can then correct for this bias *post hoc* by re-balancing the respondent sample (by excluding some of the younger people so that the proportions of different age groups in the sample represent those in the town).

Computer-Administered Questionnaires and Internet Surveys

Distribution of questionnaires via the Internet is becoming increasingly common due to convenience and cost-effectiveness considerations. The statistical requirements for Internet surveys are exactly the same as for postal surveys. Depending on the requirements of the particular survey, investigators might consider respondents' need for privacy and the practicalities of completing screen-based survey questionnaires in an open-plan office. However, the Internet does seem to offer several advantages over postal methods (not having to mail the returns, for example), and there are few general disadvantages.

Richman et al. (1999) reviewed 61 studies to determine whether computerized administration affected social desirability distortion. Overall, no effects of computer administration were found. The suitability of computerized administration is moderated by a number of factors. For minimum social desirability distortion in computerized administration of questionnaires:

- Respondents should be alone or have physical privacy.
- Respondents can backtrack and change previous responses, if desired.
- Respondents are not identified.

It seems that there is less distortion in computerized interviews than in face-to-face interviews. For further information, see the International Test Commission (2005), *International Guidelines on Computer-Based and Internet delivered testing, version 1.0*, January 2005.

Big Data

"Big data" refers to extremely large data sets acquired from cell phones, Internet commerce, and social network sites. It can be used to track trends, disclose patterns, and is sometimes available in real time. It is a potentially very powerful tool for monitoring system performance, predicting future behavior, and so on. However, large samples do not guarantee statistical accuracy, as the following example demonstrates.

In the run-up to the 1936 U.S. Presidential Election, The Literary Digest magazine polled 10 million potential voters and predicted that Alfred Landon would win with 57% of the vote against Franklin D, Roosevelt's 43%. In fact, Roosevelt won with 62%. The Gallup organization polled 50,000 voters using a stratified sampling method and predicted that Roosevelt would win with 56%.

The Digest survey fell prey to two common problems. First, the sample was biased from the outset, being selected from telephone directories, club memberships, magazine subscribers, and so on. It was not a random sample. Second, the response rate was only 24%, so it is likely that Roosevelt supporters were less likely to respond than Landon supporters.

Large samples do not guarantee correct predictions—big data *can* cause big mistakes.

Sustainability Questionnaires

A variety of questionnaires have been developed to assess sustainability at the individual and organizational levels. Gericke et al. (2019) developed a sustainability consciousness questionnaire of 49 items loading onto 9 factors, expressed as 9 valid and reliable subscales (Table 16.2).

TABLE 16.2
Sample Items from the Sustainability Consciousness Questionnaire

		Sample Item
Sustainability knowingness	Environmental	Sustainable development requires a shift to renewable natural resources.
	Social	Reinforcing girls' and women's rights and increasing equality around the world is necessary for sustainable development.
	Economic	Sustainable development requires that companies act responsibly toward their employees, customers, and suppliers.
Sustainability attitudes	Environmental	I think that it is important to take measures against problems that have to do with climate change.
	Social	Think that the government should provide financial aid to encourage more people to make the shift to green cars.
	Economic	I think that companies in rich countries should give employees in poor nations the same conditions as in rich countries.
Sustainability behaviors	Environmental	I recycle as much as I can.
	Social	I support an aid organization or environmental group.
	Economic	I often purchase second-hand goods over the Internet or in a shop.

Questionnaires such as this can be used to assess the level of awareness of sustainability issues in the population or in particular groups, and subsequently, the effects of sustainability initiatives or the response to sustainability propaganda.

SYSTEM INTEGRATION

Traditional approaches to industrial organization have viewed the economy as a closed system, with matter and energy circulating between producers and consumers. In this model, nothing leaves the system, and the environment is regarded as an open system with an inexhaustible capacity to supply raw materials and absorb pollution. The modern view regards the economy as an open system, which draws from and leaks into a finite ecosystem. The concept of efficiency means different things depending on the view of the economy that is taken. In the former case, sustainable resource utilization is seen as inefficient and unnecessary because resources are assumed to be infinite. In the latter case, sustainability is seen as efficient because the long-term costs of overconsumption and waste are avoided. A key consideration is to attach appropriate economic "costs" to individual and corporate behavior in order to differentiate, in economic terms, sustainable versus unsustainable practices.

As the global economy increases in size, its structure and behavior must come to resemble more closely the ecosystem of which it is part. At the same time, the global economy is undergoing rapid change, which has the capacity to both facilitate and derail attempts to utilize resources more sustainably.

There has been a long-term trend for the cost of raw materials to decrease in comparison to the price of the goods manufactured. Innovation, dynamism, and the capacity for organizations to restructure and upgrade are replacing cheap labor and abundant natural resources as the driving forces for prosperity. The shift from trading commodities to adding value has put pressure on countries and companies to restructure. This explains why so many seemingly simple products now have excess functionality (adding value by offering consumers more "choice") and are now being absorbed into the "Internet of Things." This is part of a global trend, which began in the 1990s and is driven by the same motive that led many developing countries to switch from exporting grapes to exporting wine—the profit is in the value added to the commodity, not the commodity itself.

Many of the more profitable industries in present times require very little start-up capital compared with primary and secondary manufacturing—Internet companies are a prime example. Paradoxically, Internet and software-based companies are labor-intensive compared to, for example, automobile assembly, where a $30,000 vehicle can be assembled with a few hours of manual labor, whereas much software development is now carried out in developing countries with lower wage costs.

Daly (1993) pointed out that, in the absence of controlling legislation, companies have an economic incentive to pass on as many of the costs of doing business as possible. Costs can be *externalized* by polluting the environment rather than treating waste products or paying for somebody else to dispose of them safely. In addition to pollution from waste products, employee injury and ill health can also be externalized in some countries. In countries with established social welfare systems, companies have an incentive to ignore work environment issues since the state, rather than the company, will pay for the cost of work-related accidents and ill health rather than the company itself. In essence, national health services and related forms of social welfare can subsidize unsafe or unhealthy work practices and are a disincentive to the uptake of HFE, in the same way that subsidies of raw materials cause overconsumption and discourage energy and resource use efficiency.

Companies may all be taxed in the same way in a country, but those whose operations have high costs due to secondary effects of doing business have a disproportionately large incentive to externalize these costs, with the result that the rest of society is burdened with them—unless, of course, legislation is in place to prevent them from doing so. The resources used in caring for the sick are always a cost to society, irrespective of whether the state pays for them or whether the employer pays, because if the person had not been sick, the resources could have been used for something else. Compensation paid to individuals, either by the state or by the employer, is not a cost to society since it is a transfer of income between people.

Herein lies the rationale for legislation to encourage and/or enforce, as required, proper standards of occupational health and safety and the sustainable use of resources. Seen in this light, regulations about HFE are part of a general trend to force companies to internalize secondary costs by improving the overall efficiency with which they conduct their business. When companies comply with such legislation, they are put under pressure to become more efficient because all of their costs of doing business are included in the price of their products (including the cost of the facilities, the workstations, the equipment, and health insurance).

An alternative, from the point of view of the company, is to relocate to a country with less restrictive legislation and/or lower wage costs. This is sometimes seen as a retrograde step because it is the socially and environmentally inefficient companies that have the most to gain by relocating—affluent countries exporting the "dirtier" parts of their economies to the developing world, which receives all the pollution and ill health of these businesses in return for low-pay, low-skill jobs. According to this view, the developing countries to which these businesses relocate are effectively subsidizing overconsumption in primarily affluent countries, who avoid paying the "real costs" for products imported from the developing world. This is essentially an argument against free trade.

Not all economists agree with this view. Bhagwati (1993), for example, countered that the needs of the developing world are very different from those of developed countries, and any investment that helps to finance development goals and the delivery of basic services such as clean water and basic education is to be welcomed. The disbenefits may be worth incurring in the short to medium term. Over the longer term, increasing affluence in developing countries will create its own impetus for higher standards.

This is a view that sees workplace controls (and HFE practices that support these goals) as relative rather than absolute and dependent on the level of industrial development of the country concerned.

Economic Growth and Environmental Pressure: Putting HFE "Back in the Box"

In the prologue, we saw how HFE works best when it is placed in a "bigger box," such as systems engineering. This applies to sustainability, and therefore, HFE should be seen as an enabling discipline that makes sustainability initiatives work better.

Economic activity in a country has both proximal and global components. To compete in the global economy, businesses engage in a relentless drive for efficiency. The proximal economy consists of locally produced products and services that are consumed locally. Hairdressers, dentists, and schoolteachers work in the proximal economy, whereas commodity brokers, automobile assembly workers, and arms manufacturers work in the global economy. Inefficiency can be tolerated in the proximal economy and may even be socially desirable—many countries provide "free" healthcare for those who cannot afford to pay the real cost of the service. However, proximal resources are limited, and those that are used for one purpose cannot be used for another. This is why countries with highly developed social welfare systems introduce legislation to stop companies from externalizing their production costs, and it is a strong economic argument for HFE in the proximal economy. The "bigger box" for HFE in the proximal economy is, therefore, legislation and governance including legislation for occupational health and safety. Occupational ill health and accidents are seen as unacceptable costs to society rather than an unavoidable part of working life.

Supply chain management presents an opportunity for HFE to influence global economic activity via global trade rules. Although the World Trade Organization has a stated aim to protect the environment and health, at the time of writing, there appears to be little or no engagement with the HFE community. Supply chains are often complex and span different countries. The cost of transporting goods is low in relation to their value and gets lower as value is added at source, as is illustrated in the example below.

Containerization—A Worked Example

Freight costs for containerized transport are low even in relation to fairly unsophisticated products such as table wine. Consider a vineyard owner in Stellenbosch, South Africa, wanting to ship a container of wine to a wine merchant in the United Kingdom. The cost of delivering the container to the vineyard, loading it, and returning it to the dock is approximately $250, handling charges included (1998 prices). The sea freight charge is $600. If we assume that the cost of off-loading the container at the other end and transporting it to the customer's warehouse is double the local charge, that is, $500, then the total cost is $1350. A 6 m container holds 1200 cases of wine (14,400 bottles). The transport cost is, therefore, 9.3 cents per bottle. If the retail price of the wine is $6.00 per bottle, the transport cost per bottle is approximately 1.5% of the price of the bottle. In relation to other factors that influence the cost of the wine to the consumer (and therefore the competitiveness of the product), such as import duties, value-added tax, etc., the transport cost is trivial. Even if the above estimate leaves out significant incidental costs, it is clear that the transport costs are small in relation to the likely profit margins of the producers, wholesalers, and retailers. It is easy to see why high value-added products such as compact discs or computers can be made in one part of the world and shipped to other countries, and why containerization is one of the key drivers of the globalization of the economy.

Influencing What Happens at the End of the Chain

One way to improve conditions at the end of the supply chain (the country where the raw materials and basic commodities or products are produced) is from the top—the end user places constraints on the conditions under which the goods are produced and transported throughout the chain. This would include, for example, refusing to source goods or services from overseas companies that employ child labor, use toxic or banned herbicides or insecticides, or lack basic occupational hygiene. The International Labour Organisation has begun to address some of the issues that fall within its terms of reference—in particular, the use of child labor.

Barriers to Sustainable Behavior: Temporal Discounting

Temporal discounting is the tendency for individuals and organizations to discount the value of rewards the more distant they are in the future. For example, most people would prefer to receive

\$10.00 today than \$10.00 in a month's time. However, some people might be happy to wait a month if the reward were increased to \$20.00. There are some interesting research questions aimed at understanding the point at which people's preference changes from accepting a "smaller sooner reward" (an SSR—\$10.00 now) to accepting a "larger later reward" (an LLR—\$20.00 in the future). The two variables of most interest here are the time interval and the difference between the SSR and LLR. Kunreuther and Heal (2012) summarize temporal discounting as follows:

Let

C = cost of investing in domestic energy efficiency and home generation

T = any given year in the future

p = probability of an energy crisis and increase in costs

B = benefit of increased energy self-sufficiency

β^t = homeowner's discount rate (the discount rate is the rate at which the value of future benefits is discounted—the value of the benefit reduces as the time horizon recedes)

If the cost of the initial investment is less than the discounted stream of expected benefits, the homeowner is likely to invest in domestic self-sufficiency, or in more formal terms, if

$$C < \sum_{t=1}^{t} (pB)\beta^t \tag{16.5}$$

The implications for HFE are that if we want to design systems where people will make sustainable choices, we need to understand how they conceive of costs and the likely benefits of making these choices. Despite the mathematics in Equation 16.5, these are exactly the kinds of questions that can be answered using questionnaire surveys. One of the problems in changing individual and organizational behavior in the direction of sustainability is that the expected benefits to the decision maker (or, in this case, the person or organization paying the bills) are not always apparent.

One way to approach this matter is to consider what a "perfect" system would look like. What would be required to cause people and organizations to make sustainable choices in their daily lives? Clearly, the costs of doing so would need to be affordable. Therefore, in the example above, the householder would need to be able to afford to have, for example, solar panels installed. Thus, there might be scope for subsidies or special loans to minimize the costs or link repayments to energy savings. Second, householders must have a good understanding of the value of B—the expected benefits in financial terms—and, therefore, claims about the likely savings should be realistic. Third, information about expected fuel price increases would also be useful, particularly if prices were expected to rise. If fuel price rises were believed to continue in the long term, the lower the discount rate would be. Additionally, feedback about the likely benefits should be tangible and immediate.

There are many other reasons why people might invest in energy efficiency and behave more sustainably—deeply held beliefs about anthropogenic climate change and the need to modify behavior, coupled with emotional reactions to sustainability propaganda about melting polar ice caps and starving polar bears, etc. However, Equation 16.5 provides a sociotechnical framework for designing and evaluating sustainability initiatives. As we saw in the prologue, sociotechnical systems theory emphasizes the need to control variance in a system (in this case, patterns of energy usage) as close as possible to the source of variance (i.e., end users), providing information to where it is needed (in the end user's home) with a congruent energy supply system that reinforces the desired behavior (perhaps by setting differential tariffs according to usage).

Underweighting the Future

There is evidence that discounting is hyperbolic—that discounting is disproportional as the time horizon recedes. Attempts to change behavior by appealing to the need to save the planet for future generations may be emotionally satisfying but are less likely to change consumer behavior because the immediate costs of changes in consumption loom disproportionately large.

Myopic Behavior: Decision makers may defer decisions because the expected benefits will not be gained over the life of the measure. For example, householders may decide not to install solar panels because they intend to sell the house in a few years' time, and the increase in the sale price following the installation of the panels is expected to be less than the cost of the panels.

Procrastination: Automobile owners may defer the decision to replace their vehicles with a BEV because batteries are constantly improving in their capacity to store energy, and the price continues to decline. They therefore postpone the decision in anticipation of a better deal in the future. Householders may defer the decision to replace incandescent bulbs with energy-efficient bulbs because they see the replacement of a working bulb as a loss (preferring to wait until the incandescent bulb fails before replacing it). Procrastination is best dealt with by providing accurate information about the savings (e.g., "If you buy a BEV now, the fuel savings will pay for half the cost of the next one you buy in 10 years").

Population Growth and Immigration

The world population has increased from approximately 5.7 billion in the mid-1990s to about 8.5 billion at the time of writing, and the trend is expected to continue with an annual growth rate of 1.3%–1.5%. This is partly due to increased longevity in many countries, bringing attendant problems of designing for an aging workforce.

The aging of the population in some countries is also due to lowered birth rates. In the United Kingdom, the birth rate peaked in 1961 and has fallen ever since. In many developed countries, the birth rate is lower than the population replacement rate. The effects of aging populations have been mitigated over the last 50 years by greater participation in the workforce by women, particularly in jobs that were previously male-centric (such as the armed forces), and the age at which people marry and start families has increased. Immigration to developed countries has also slowed the aging of the workforces of these countries. Immigration can prevent population shrinkage, but it cannot prevent population aging in the long term because, within a generation or two, the birth rates of high-fertility immigrant groups approach those of the host population (Coleman, 1995).

Coupled with economic growth and improved living standards, these trends are bringing about large increases in energy consumption and agricultural output, all of which have an environmental impact. As countries develop beyond a basic subsistence level, it is how people and resources are used that determines their long-term competitiveness and sustainability. These environmental and demographic changes will create further demand for better products and services and work that can be performed equally well by young and old, male or female employees—strong motivating factors for HFE as defined and described in the preceding pages of this book.

RESEARCH DIRECTIONS

From the perspective of sustainability, there are a number of research directions for HFE. First, to engage with all initiatives aimed at improving resource utilization and consumption because resources are used by people as they interact with technology in their daily lives. Essentially, this means making sustainable behaviors easy to engage in, easy to understand, and effective. Second, to support initiatives to change work patterns in a more sustainable direction (homeworking and teleconferencing to reduce commuting increased greatly following the COVID-19 pandemic). Third, to engage more fully with the proximal and global economic systems and enable constraints to be designed and placed on the way that supply chains are managed so as to establish more sustainable practices at source. HFE must take a broader view of systems based on the concept of the system life cycle (the CADMID cycle described in the opening chapter), so that sustainability issues are considered at the design stage and include sustainability in manufacture and use, and sustainability at disposal.

HFE in the future will be faced with many new challenges, and at this stage, it is not clear what research questions we will need to ask to achieve optimal human–machine interaction in the future. Trust in automated systems and human interaction with robots and AI are likely to be growth areas that will pose interesting questions about how "human-like" these intelligent systems should appear to be. Sustainable economic behavior consumes the interest of the natural and social capital available to it and not the capital itself (Zink and Fischer, 2013).

SUMMARY

HFE is focused on human–machine interaction but, in the future, needs to focus on system interaction—how humans in systems contribute to system sustainability. HFE is needed at all stages of the design process—from requirements specification and concept generation to post-implementation evaluation and decommissioning.

Questionnaire surveys can be of use in all areas of HFE. Regarding sustainability research, surveys can be used to gauge user understanding of sustainability issues, propensity to engage in resource-efficient behavior, and attitudes toward new technologies. In all cases, problems such as sampling bias, reflected by low response rates to questionnaire surveys, must be taken very seriously. There is a worrying trend for researchers to assume that sampling strategies are valid and that nonrespondents are "missing completely at random"—that the reasons for nonresponse are completely unrelated to the topic under investigation. This is a dangerous assumption that may lead to incorrect conclusions. A better approach is to assume that there will always be some degree of response bias (i.e., nonresponse and some or all of the survey items share common variance) and to carry out post hoc checks on nonrespondents or include measures of response propensity in questionnaires.

TUTORIAL TOPICS

1. Should "sustainability" be considered a key performance indicator (KPI) for a new system?
2. If you carried out a questionnaire survey and got a 40% response rate, what evidence would encourage you to believe that the findings could or could not be generalized?
3. Under what circumstances would it be better to interview people about their attitudes toward BEVs rather than use a questionnaire?
4. Strong sustainability will decrease our standard of living. What is Sustainability at: https://www.youtube.com/watch?v=iI779TrHT1M

ESSAYS AND EXERCISES

1. What factors determine the implementation of ergonomics in practice?
2. Discuss possible future technological developments of your choice and the ergonomic problems that will have to be solved before they can be successfully implemented as systems or products.
3. Discuss the macro-ergonomic implications of the following global issues. Explain how these issues might influence the design of work practices, the management of organizations, and the training of future generations of ergonomists:
    - Global warming
    - Massive enlargement of the hole in the ozone layer
    - Continued demographic aging of populations on a global scale
    - The replacement of an oil-based energy supply with a solar/hydrogen-powered energy base

4. The data below are from a study of cell phone use by older users. The left-hand column gives ratings of ease of use on a 7-point scale (where a score of "1" indicates "very difficult" and a score of "7" indicates "very easy"), and the right-hand column gives the time taken to enter a new name and phone number in the cell phone's memory. Does the rating correlate with the performance time?

Ease of Use Rating	New Name Entry Time (s)
5	57
5	45
3	91
7	23
1	82
2	73
4	50

5. Which of the following tasks would you allocate to a person and which to a machine when designing a system? Briefly state your reasons.

   Checking sheet glass for flaws as it emerges from a furnace.

   Landing an aircraft in stormy weather at night.

   Packing 1,000 apples per hour into a cardboard box.

   Grading apples into three quality bands—export, sale at local supermarkets, and bulk sale to juice producers.

   Medical screening of applicants for life insurance premium calculations. Screening of applicants for bank loans.
6. Refer to the opening quote by Captain Smith of the Titanic:

   I cannot imagine any condition which would cause a ship to founder. I cannot conceive of any vital disaster harming this vessel. Modern shipbuilding has gone beyond that.

   Attributed to Captain Smith, Commander of the Titanic

   To what extent might it be true to say that Captain Smith did not understand "the scope of the context of use" of the Titanic?
7. Below is a list of abilities, originally generated as two separate lists by Paul M. Fitts in 1953. The two lists contrasted the abilities where "humans appear to surpass present-day machines." Consider twenty-first-century "machines" and generate your own lists: List 1: Humans Superior, List 2: Machines Superior. Briefly state your reasons and give an example from working life.
   - Ability to perform repetitive routine tasks
   - Ability to detect small amounts of visual or acoustic energy
   - Ability to store very large amounts of information for long periods and recall relevant facts at the appropriate time
   - Ability to reason inductively
   - Ability to exercise judgment
   - Ability to respond quickly to control signals
   - Ability to store information briefly and then erase it completely
   - Ability to reason inductively, including computational ability
   - Ability to improvise and use flexible procedures
   - Ability to perceive patterns of light or sound
   - Ability to handle highly complex operations, that is, to do many different things at once
8. Mr. and Mrs. Smith are trying to decide where to go for their next vacation. They decide to rate five destinations in terms of preference, where "100" indicates "excellent" and a score

of "0" indicates "unacceptable." Calculate Spearman's rho for these rankings. Do Mr. and Mrs. Smith really have the same preferences?

	Mr. Smith	Mrs. Smith
Florida	50	80
Mexico	30	65
Spain	60	90
France	30	60
Italy	55	95
Greece	70	100

9. Repeat the analysis in Question 8 using the data below. This time, use the partial correlation technique to control for the effect of price on the preference data.

	Mr. Smith	Mrs. Smith	Price per person
Spain	50	50	US$1000
Italy	80	90	US$1100
California	70	85	US$2500
Florida	45	80	US$1750
Hawaii	10	80	US$3750
French Polynesia	15	85	US$7500
Ireland	50	50	US$500
Greece	60	60	US$800
Scotland	45	40	US$900

10. Two raters use the general ergonomics checklist in Chapter 1 to assess 20 jobs in a food processing factory. They then use the 5-point prioritization scheme for change in Chapter 1 to rate the urgency of redesigning each job. The ratings are given below, where a score of 1 = "Change immediately" and a score of 5 = "Change when new equipment is built or purchased." The raters' scores for each job are shown below:

    a. Calculate Cohen's Kappa statistic to determine whether the raters agree.
    b. Calculate the median rating for each rater. Comment on any differences in view of the fact that Rater 1 is a health and safety officer and Rater 2 is a production manager.

Job	Rater 1	Rater 2
1	2	3
2	3	4
3	1	5
4	1	4
5	2	3
6	2	2
7	2	1
8	4	4
9	4	4
10	5	5
11	5	5

(Continued)

(Continued)

Job	Rater 1	Rater 2
12	4	5
13	1	3
14	2	3
15	3	4
16	2	3
17	1	1
18	2	2
19	1	1
20	1	1

QUICK QUIZ

1. What are the two main types of questionnaire reliability?
2. In how many ways can a questionnaire be shown to be reliable?
3. What is a leading question?
4. What is the minimum sample size when sending out questionnaires to a population of $N = 500$?
5. The larger the sample size, the better. True or False?
6. It is unethical to offer incentives to participate in surveys. True or False?
7. It's a waste of time to follow up with non-respondents to surveys to see if they will respond a second time. True or False?
8. Cash rewards for survey participation will bias the sample. True or False?
9. People will always prefer an immediate reward rather than having to wait. (a) True, (b) False, (c) Depends on the size of the reward and the waiting time.
10. As long as an incandescent light bulb is working, it's best to keep it rather than replace it with an energy-efficient one. True or False?

NOTE

1 https://www.smartsurvey.co.uk/resources/calculators/sample-size-calculator

Epilogue

The Future of Work: From Requirements to Affordances

GENERAL REQUIREMENTS FOR FUTURE SYSTEMS

1. For control tasks, the operator must remain in the loop.
2. AI must be explainable to users.
3. AI limitations must be understood by users.
4. Remote workers have the right to disconnect.

Author: Are you intelligent?

ChatGPT: I'd say I'm more knowledgeable than intelligent! I can process information and provide answers based on what I've learned, but I don't have feelings or consciousness like a human does. How can I help you today?

Author: So, are you just a talkative version of Google?

ChatGPT: I can see how it might seem that way! While I can provide information like a search engine, I also aim to have more natural, conversational interactions. I'm here to help with questions, brainstorm ideas, or just chat about whatever's on your mind. What are you curious about?

Author: How do you do this?

ChatGPT: I work by processing text and using patterns from a wide range of information I've been trained on. When you ask a question or make a comment, I analyze it and generate a response based on context and relevance. It's all about understanding language and providing helpful, coherent replies. If you have any specific questions about how I function, feel free to ask!

Author: What is a large language model?

ChatGPT: A large language model (LLM) is a type of artificial intelligence that's trained on vast amounts of text data to understand and generate human-like language. These models learn patterns, grammar, facts, and even some reasoning skills from the data. They can perform a variety of tasks, such as answering questions, writing stories, translating languages, and more, all by predicting the next word in a sentence based on the context. The "large" part refers to the extensive size of the model, which allows it to capture a broad range of language nuances.

Author: Do you always tell the truth?

ChatGPT: I aim to provide accurate and reliable information based on my training, but I can make mistakes or be outdated, especially if the information has changed since my last update. It's always a good idea to double-check important facts or consult multiple sources. If you ever spot something that seems off, let me know!

Author: So, you don't know whether your responses are correct or not?

ChatGPT: Exactly! I don't have real-time awareness or the ability to verify facts after my last training cutoff. I generate responses based on patterns in the data I was trained on, but I don't have a way to confirm their accuracy. If you have specific concerns about something I say, I encourage you to check it against trusted sources!

Author: So, you don't understand the difference between right and wrong?

ChatGPT: I don't have personal beliefs or a moral compass like a human does. My understanding of "right" and "wrong" is based on patterns in the data I was trained on, including cultural norms and ethical considerations present in that data. I can discuss these concepts and provide information about them, but I don't have personal judgments or feelings about them.

Author: So, you don't know what you don't know?

ChatGPT: That's a good way to put it! I can provide information based on what I've been trained on, but I don't have self-awareness or the ability to recognize gaps in my knowledge. If there's something I can't answer or if you ask about a very specific topic, I might not have the information you're looking for. It's always okay to ask more questions or clarify!

Author: Well, at least you're honest about it!

ChatGPT: I appreciate that! Transparency is important, and I'm here to help as best as I can. If you have more questions or topics you want to explore, just let me know!

CORE KNOWLEDGE

There's no time like the present to talk about the future. At the individual level, the future of work is the future of human augmentation. At the systems level, the future of work is the systems that augmented humans will work in, including the emergence and design of new jobs.

There is nothing new about this.

THE AUGMENTED HUMAN

When the ancestors of modern humans learned to control fire, wear clothes, and build shelter, the ability to live in a wider range of environments was augmented. Language, first spoken then written, augmented the ability to communicate over distance and time. Agriculture augmented the ability to control the food supply and sustain larger communities. Systems of transportation then connected these new communities, further improving communication and the exchange of resources, and tradeable goods. All these changes led to new forms of social organization, specialization, and development. The Industrial Revolution enabled machines, powered by fossil fuels, to manufacture products, making slavery outdated and rendering many traditional craft skills obsolete or restricted to specialized niches. Today, we live in an ecosystem that is largely artificial, embedded, unsustainably, within a natural ecosystem (Chapter 16). Many of us now live in parallel worlds–the physical world and a virtual world via the Internet. This has augmented our ability to access information, entertainment, education, to communicate, and enabled us to join virtual communities. Augmentation continues apace and presents new opportunities for HFE, since it is the design of the augmentation that HFE will support, as well as the design of the systems that the augmented humans will work and live in. Specification of the requirements for the optimal design and operation of these new systems will require major input from HFE.

SYSTEMS OF AUGMENTED HUMANS

New technologies both destroy jobs and create them. New jobs are needed to build new systems, and more still to operate them. Between 1750 and 1850 in the United Kingdom, thousands of miles of canals were dug by hundreds of thousands of workers ("navvies"). Manufactured goods were transported on narrow boats pulled by horses via locks, manned by lock-keepers. Canals were superseded by railways, then by roads. In the nineteenth century, many industrial and mine workers did not own clocks or watches. It was the job of "knocker-uppers" to walk the streets in the early morning, tapping on the bedroom windows of these workers to wake them for the morning shift. Affordable alarm clocks put an end to this job. At the turn of the twentieth century, 31% of the workforce in the United States worked in agriculture, and about 39% worked in physically demanding jobs such as manufacturing. In 2024, less than 3% work in farming, and about 10% in agricultural activities, and about 8.5% in manufacturing. In 1950, the United Kingdom employed 700,000 people in the mining industry. Unemployment was about 2%. At the time of writing, about 4% of the workforce in the United Kingdom is unemployed, and only about 3500 are employed as miners. The percentage of the workforce employed in the service sector has increased by about 60% in the last 100 years.

Over half of today's jobs did not exist 100 years ago.

Most of the jobs we see around us are a reflection of the technology in operation today. Many jobs (such as Gas Meter Reader, VHS Recorder Repairman, Copy Typist, Punch Card Operator, and Elevator Attendant) no longer exist, either the technology has disappeared or the requirement for the job has disappeared. Jobs likely to disappear (or already starting to disappear) in the next 20 years include Data Entry Clerk, Cashier, Customer Service Advisor). It is possible that some jobs, or much of the routine work currently performed in occupations such as accountancy and programming, will be automated.

Unemployment

The introduction of new technologies does not cause unemployment (this is known as the "Luddite Fallacy"). This is not to say that new technologies are not disruptive. Technological change is often extremely disruptive. During the Industrial Revolution in the United Kingdom, the introduction of new machines for the textile industry was met with fierce resistance by skilled craftsmen (the "Luddites") because the new machines were faster and operable by low-skilled, lower-paid workers. Acts of Parliament were introduced to prevent the destruction of machines by protestors. In the twentieth century, much criticism was directed at the work of F.W. Taylor (see Chapter 1). Production systems based on work simplification and standardization (de-skilling) lowered employment costs and made workers more easily replaceable.

Discussion of the causes of unemployment is outside the present scope, although one factor is a lack of demand for goods and services. While some jobs become obsolete, the introduction of new technologies creates new goods and services and therefore new job opportunities, such as AI Annotators.[1] These are known as "substitution effects." New goods create new consumer wants, and economic growth combined with more leisure time provides increased scope for consumption. Almost everyone has their own cell phone nowadays, which they can use while visiting a "nail bar," "vape store," or "tanning booth" on the high street (services that didn't exist in the era of landline phones). Annoyingly, some people even use their phones in many of the gymnasia that can be found in almost any town or city nowadays.

Mental Well-Being in the Workplace

Mental well-being is more than the absence of an adverse stress response (see Chapter 8). It is a state of emotional, psychological, and social well-being. At work, it manifests as a feeling of being able to cope with work demands and being actively engaged. There are many components of working life that can contribute to or diminish mental well-being. These include: commitment to the organization and the role; autonomy and control over work demands; a clear role without conflict; work-life balance; support from peers and colleagues; support from line managers and/or supervisors; sufficient resources to work; a satisfactory work environment; and appropriate work hours for the role (see Bridger et al., 2012). Mental well-being at work depends on a balance between work demands and employee-perceived control, and on whether these other work arrangements are, at least, satisfactory. It also depends on a balance between employee effort and the rewards received in the form of compensation, other benefits and recognition. A good balance of work and life demands results in an effective and engaged workforce—employees with high work engagement experience high energy, absorption, and dedication to their work. HFE will be essential for understanding how these components operate in the jobs of the future to create effective engagement in employees.

Future-Proofing HFE?

The relationship between work, working conditions, and the work environment has undergone a dramatic shift, making many of our scientific and legislative responses to work design inadequate, inapplicable, or obsolete, while introducing many new challenges.

Digital platforms now control many existing jobs and make new work roles possible; the gig economy being an example, with food delivery and taxis, but extending to other areas such as logistics and distribution. New technologies like smart glasses can augment human performance in applications such as stock picking, but the long-term effects of using these and many other new devices are poorly understood. A renewed emphasis on mental well-being at work is underway, as home and hybrid working change the psychosocial environment, and its constraints and opportunities, taking into account what makes work fun and rewarding, while maintaining the traditional focus on the identification and amelioration of psychosocial risks, including newer risks such as "technostress." Workers may no longer be "tied to a machine" on the factory floor, but they may be tied to the Internet by mobile devices, with little control over work demands or scheduling.

Much of our knowledge about human factors is based on the assumption that most people work full time in one job or career, possibly for many years. Will the rise of a new generation of "slashers" (one person/multiple roles), where individuals combine different jobs and alternate between them constantly, render many of our ideas obsolete?

Technology Readiness Levels (TRLs)

The level of maturity of a new technology is rated by its TRL:

- *TRL 1:* basic principles observed and reported
- *TRL 2:* technology concept or application formulated
- *TRL 3:* analytical and experimental critical function or characteristic proof-of-concept
- *TRL 4:* technology basic validation in a laboratory environment
- *TRL 5:* technology basic validation in a relevant environment
- *TRL 6:* technology model or prototype demonstration in a relevant environment
- *TRL 7:* technology prototype demonstration in an operational environment
- *TRL 8:* actual technology completed and qualified through test and demonstration
- *TRL 9:* actual technology qualified through successful mission operations

The contribution of HFE to any new technology will depend on its TRL—from user-consultation, mixed methods research, lab studies, or field trials.

BASIC APPLICATIONS

Future of Industrial Work

The Fourth Industrial Revolution is well underway. Some recent developments are described below:

- *Exoskeletons:* Exoskeletons have reached TRLs of 5, 6, and 7, according to the systematic literature review by Ahmad et al. (2024)
- *TRL-5:* Validation of the exoskeleton during a simulated work task in a laboratory with novice subjects.
- *TRL-6:* Simulation of the task in a relevant environment with the exoskeleton by expert workers.
- *TRL-7:* Demonstration of the effectiveness of the exoskeleton in an operational environment with experienced workers.

Manufacturing industries have the highest adoption of exoskeletons, possibly due to the high incidence of musculoskeletal injuries. Other industries where exoskeletons have been trialed include waste treatment, recycling, construction, health and social care, agriculture, forestry and fishing, and transport and storage. In general, the findings are favorable, however, at present, widespread uptake is limited. Whether the augmentation of human physical capacity by exoskeletons becomes widespread, bringing a reduction in workplace injuries, remains to be seen.

Cobots (Collaborative Robots): In contrast to traditional industrial robots, which are separated from operators both physically and functionally, cobots work with operators. According to Knudsen and Kaivo-Oja (2020), cobots work alongside human operators to complete tasks. Independent collaboration takes place in a shared workspace where the robot and the operator complete separate tasks. In simultaneous collaboration, both complete separate operations on the same workpiece at the same time to speed up the production process. In sequential collaboration, the operator and the robot work on the same workpiece in sequence. There are time dependencies between them, and the robot may undertake the more arduous or tedious part of the operation. In supportive collaboration, there are dependencies between the operator and the robot, and both are needed to complete the task. The TRLs of the first two modes of collaboration are high, while the latter are under development. Ultimately, the goal is to achieve the optimum combination of both human and robot capabilities to maximize safety and productivity.

Social Robotics: This field deals with the social aspects of human-robot interaction. Anthropomorphism is a common human tendency—to ascribe human-like intentions, emotions, or characteristics to non-human things, including animals. Understanding the social characteristics of human-robot interaction is a growing area of research.

Smart Glasses: According to Chondromatidis et al. (2022), manual stock picking in warehouses constitutes 55%–65% of the total warehouse cost. Stock pickers walk the aisles of the warehouse with a list of items needed to complete the customer's order. Items are loaded onto a trolley in sequence and delivered for dispatch. Pick by vision systems (a form of augmented reality) guide the stock picker to the location of items and confirm by scanning the barcode before leading the stock picker to the next item on the list, leaving the hands free to handle items. Many of the design parameters for these systems fall squarely in the realm of HFE. These include display design, battery position, scanner position, focal position, field of view, workload, and mass of the equipment. The TRL of these systems is high, and they appear to increase efficiency without incurring significant workload costs.

Stock picking is but one area where wearable devices can be used to enhance performance.

Computer-generated imagery can also be used to identify risks and support employees in the practice of difficult tasks. The long-term effects on users remain to be investigated.

Industrial workers are becoming part of the Internet of Things, completing their tasks in a cyber-supportive environment rich in easily accessible information and lower in physical demands.

Future of Office Work

With a shift in focus from office workers sitting at desks to teams communicating via networks, designers and facilities managers are starting to question whether offices with desks and chairs are needed at all and will consider the requirements for designing the offices of the future. Perhaps in 40 years' time, nobody will work in offices and the QWERTY keyboard, like the mechanical typewriters it was originally designed for, will be a distant memory in the minds of none but the oldest HFE experts.

What Are Offices For? An office is defined as a space where the employees of an organization perform administrative work in order to support and realize the various goals of the organization. Like factories, people went to the office because that is where the work was. Twentieth-century offices were designed to support work with technologies completely different from those of today. Much office work can be done remotely nowadays, and home and hybrid working are now commonplace. This changes the essential function of the office, which now has to support collaborative working and meetings. Instead of putting employees into private boxes in the "bullpen" designs of 1980s "automated offices," the layout has to change to a more collaborative space—with "office pods" (Chapter 11) for those needing privacy. Fiorentino et al. (2022) cite a number of factors driving change. The Internet has decentralized much office work, so entire offices can be moved to regions with cheaper property prices, while subsidiary functions can be outsourced. Employees may have shorter, less expensive commutes and buy property in less expensive areas, with implications for salary and benefit adjustments. Home and hybrid working is another driver of change. Studies of organizational productivity during and post-COVID lockdowns have shown no adverse impact of working from home, leading to a re-evaluation of the purpose of offices. Future offices may well be (20%–30%) smaller, with a hybrid workforce (office vacancy rates in Central London more than doubled between 2019 and 2022). Post-COVID, Fiorentino et al. summarize the purpose of offices as follows:

> The main purpose of the office is expected to shift in a way that reinforces centralisation. As the domestic sphere is expected to be the location for more individual and 'focused work', the central office is expected to continue to provide a hub for employees to create, connect, collaborate, learn, build corporate culture and meet clients.

Third Places: Third places are neither the home nor the office. The coffee houses of eighteenth century London (where merchants and businessmen went to drink coffee and do business) are making a comeback. A town or a city can, itself, support office functions, while providing a healthy work environment and social affordances. In many parts of the world, libraries, coffee shops, and other spaces are being used to create local "hubs" where people from different organizations can work close to home, but away from it. The main requirements for a successful third place are accessibility, access, a digital platform, privacy, and opportunities for relaxation and social contact. Third places are part of the trend toward "multi-local" working for "Shoulder Bag Workers" and "Digital Nomads" (see Di Marino and Lapintie, 2020, for an interesting case study). Researchers from Swinburne University of Technology found that about one-half of remote workers used third places. Cafes and coffee shops were the preferred location, but remote workers also used libraries, pubs, parks, and co-working spaces, typically visiting them 2–3 times/week for periods ranging from 15 min to 4 h. Third place users fell into four categories:

- *Device Disconnectors:* visiting third places to disconnect
- *Caffeine Creatives:* making a change of scene to boost creativity
- *Suburban Socialites:* home workers visiting as a respite from social isolation at home
- *Lunchtime Liaisons:* meeting for lunch with colleagues or clients

Home and Hybrid Working: For most of history, most people worked from home or very close to home because that was where the work was—in adjoining fields, workshops, and buildings. Several studies of home working during COVID-19 were conducted. Ipsen et al. (2021) found that more than half in their study were positive, overall, about working from home. The main positives were savings on commuting time and flexibility regarding mealtimes and breaks. The main disadvantages were missing colleagues, not getting out of the home enough, and dissatisfaction with home office arrangements. Overall, responses were highly variable around the middle values of the scales used to rate the items, suggesting that the experience varies greatly depending on individual circumstances. The six main factors that need individual attention to optimize home working were found to be: work–life balance; work efficiency; work control (all positive): home office constraints (ergonomics); work uncertainties; and inadequate tools (all negative).

Regarding hybrid working, which is becoming the "new normal," Eng et al. (2024) found that, when optimally combined, home and office working could counteract the shortcomings of each other, supporting work effectiveness. They reiterated the need for suitable ICT equipment and connectivity for home workers while pointing out that not all workers feel the need to work with others all the time in the same office. Home working was also reported to be less stressful to improve work-life balance. Time saved commuting enabled work to be done more flexibly and improved work-life balance by enabling work to be intertwined with other activities of daily living. As with home working, great individual variability was found, suggesting that there is no "one size fits all" solution for optimum hybrid arrangements.

Most recent studies have found that employee experiences of home working are highly variable, depending on personal circumstances, workspace ergonomics in the home, and work-life balance. Personality characteristics may also play a role. Introverted people need less stimulation than extroverts, particularly social stimulation, and may find working alone more agreeable and less stressful than working in a team in an open-plan office.

Many workers in telephone and Internet-based occupations are able to work from home.

Home workplaces may still need formal assessment (including HFE assessment) because in some countries, national health and safety legislation states that it is the employer's responsibility to ensure that working conditions at home meet national standards.

Homeworkers may face more or less distractions depending on their personal circumstances, but all may be at risk of social isolation. Social isolation can be reduced by requiring homeworkers to attend regular office meetings and by ensuring that they can access the office intranet. For "hot-desking," where employees share workstations on the days they attend the office, furniture should be easy to adjust, as each employee may have to adjust it in order to use it. The TRL for home working is at level 9 for many organizations; others, at levels 7 or 8, may require further trials in order to achieve productive and safe working.

Home Working and Health: Working at home all day, while saving time spent commuting, may result in less daily physical activity (see Chapter 8 for guidance). Suka et al. (2022) investigated the effect of remote working on the health and lifestyle of 130,000 workers in Japan during the COVID-19 pandemic. The one-year incidence of overweight, hypertension, hyperglycemia, and liver damage increased by 15% during this period. Health problems were related to weight gain, which, in turn, was related to an increase in unhealthy habits (snacking, heavy drinking, physical inactivity, and sleep deprivation).

Remote Work Performance: Evidence is emerging that remote working can be as productive as working in a traditional office. Two key determinants appear to be "social presence" and workspace ergonomics control. Conrad et al. (2023) carried out a questionnaire survey of 400 remote workers to investigate factors influencing their perceived work performance—whether they considered that they had completed their duties adequately, whether they had done what was expected of them according to their job description and so on. Social presence was assessed using the following items:

- Online or web-based communication is an excellent medium for social interaction.
- I felt comfortable conversing through the online medium.
- I felt comfortable participating in workplace discussions.
- I felt comfortable interacting with coworkers.
- I felt comfortable disagreeing with coworkers while still maintaining a sense of trust.

- I felt that my point of view was acknowledged by my coworkers.
- Online discussions help me develop a sense of collaboration.

Social presence did not depend on the frequency of meetings, but on the *efficacy* of the tools for collaboration:

- I was able to use electronic collaborative platforms to help in remote working.
- I used electronic collaborative platforms to communicate with my coworkers.
- I used electronic collaborative platforms to manage teams.

Workplace ergonomic control was measured as follows:

- I was able to personalize my workspace.
- I was able to control the temperature or airflow in my home workspace.
- I was able to determine the organization/appearance of my work area.
- I was able to control the lighting level in my work area.

These two factors (social presence and ergonomic control) accounted for 30% of the variance in perceived work performance. In order to get the most out of remote workers, organizations should implement effective collaboration tools and provide training in their use, as required. Assisting employees with home workspace ergonomics is also recommended.

Zoom Fatigue: Zoom fatigue is a term used to describe fatigue due to the use of all types of videoconferencing tools. Attendance at virtual meetings places higher demands on the attention of participants, in part because of the degradation of visual social cues (such as eye contact) and time-sharing between the screen and the chat function. Participants may not be looking directly at the camera, only the head and upper body are visible, and the picture quality is limited by the camera and screen size. Videoconferencing is therefore a low-fidelity version of natural face-to-face communication, particularly in relation to facial expressions and body language. In addition, participants can view their own image on the screen (like looking in a mirror), which may increase self-awareness, disrupting the automatic processes in normal face-to-face communication. Further demands on attention may result from multitasking (e.g., reading e-mails while videoconferencing). Riedl (2022), following a review of 44 articles, defined Zoom fatigue as follows:

> Zoom fatigue (synonym: videoconference fatigue) is defined as somatic and cognitive exhaustion that is caused by the intensive and/or inappropriate use of videoconferencing tools, frequently accompanied by related symptoms such as tiredness, worry, anxiety, burnout, discomfort, and stress, as well as other bodily symptoms such as headaches.

Riedl suggested a variety of coping strategies to mimimize Zoom fatigue:

- *Organizational*: Avoid overstimulation – with voice-only interaction or use the phone
- Schedule meetings and circulate agendas 24 h in advance
- Have 10 min breaks every hour and breaks between different meetings
- Limit meeting time
- Use mute when not speaking to avoid interruptions
- Avoid multitasking
- Have good Internet connections
- Illuminate faces front (make facial expressions visible)
- Log-in early
- Stand and stretch

Affordance Spaces: Offices of the future will use technologies not yet invented. With a shift of focus from a requirement for office workers to sit at desks interacting with their keyboards and screens, to teams communicating via networks, designers and facilities managers may start to question whether offices with desks and chairs are needed at all and will consider the requirements for designing the offices of the future and the affordances these offices should offer. If the affordances of desks and chairs are no longer needed, will future offices look like this (Figure E.1)?

FIGURE E.1 Concept design for an affordance workspace supporting the spontaneous adoption of natural postural behavior, as described in Chapter 4. (Courtesy of RAAAF architects (Rietveld Architecture Art Affordances). Concept design by RAAAF and the artist Barbara Visser.)

Neurodiversity and New Ways of Working

Neurodiversity is the term used to describe natural variability in cognitive function. *Neurominorities* have atypical function and include the following groups:

- *Attention Deficit Hyperactivity Disorder (ADHD):* People with ADHD are easily distracted, lack attention to detail, and have difficulty staying focused. They tend to have difficulty following instructions and sustaining concentration.
- *Autism Spectrum Disorder (ASD):* People with ASD have difficulty communicating with others, understanding social cues, and a narrow range of interests. They like routine and order in life and have tight boundaries around themselves socially. They make little eye contact with others, dislike small talk (e.g. gossip), are not adept at detecting social cues, and communicate very directly.
- Aspberger's syndrome is part of the autism spectrum disorder and is characterized by many autistic traits. People with Asperger's syndrome often excel in specific occupations.
- *Dyslexia:* Dyslexics have difficulty reading, writing, and spelling.
- *Dyscalculia:* Difficulties with numbers and number-based learning.
- *Dyspraxia:* Difficulties with movement and coordination. May be seen as clumsy and have poor hand-eye coordination.

None of the above characteristics of neurodiverse employees (known collectively as "neurominorities") imply low intelligence or low employability. However, all of these characteristics imply that the neurodiverse may respond differently to working conditions than neurotypicals. Loud noises, complex social interactions, and environmental stress may be particularly troublesome for neurodiverse employees.

Szulc et al. (2023) conducted interviews with neurominority employees to learn about their experiences of working from home during COVID-19 lockdowns. Overall, benefits and disbenefits were reported. Benefits included the ability to structure their own workspace, control their surroundings and follow their own rituals, and reduction of the sensory and social distractions of working in an open-plan office. Disbenefits included working from home blurred the lines between home and work, difficulty following discussions in virtual meetings because of the impoverished social interaction, and degraded feedback on the performance from managers during online meetings, creating ambiguity. It was recommended that managers receive training in how best to support neurominorities and, during remote work in particular, agree on how best to communicate and deliver feedback.

Hours of Work

Working hours and how to optimize them have come into focus recently due to the emergence of disruptive technologies such as Artificial Intelligence (AI) and their implications for employment.

A Shorter Working Week?

Recently, proposals have been made to shorten the traditional working week from 5 days to 4. This is not a new idea. Sir Thomas More (1478–1535) was a philosopher who served as Lord High Chancellor of England. In his famous work, *Utopia,* a conception of an ideal society, nobody would work more than 6 h/day (three before and three after lunch), spending the rest of the day in educational activities. In 1930, the economist John Maynard Keynes predicted that the working week in 2030 would be only 15 h long, which may have been correct if consumption had remained at 1930s levels (Crafts, 2022). In fact, average hours worked per year have reduced from 2500 in 1930 to 1840 in 2011. Keynes was correct, though, in predicting over fivefold growth in real GDP per capita (gross domestic product), with the result that people had a greater capacity to consume newly emerging goods and services.

For much of its history, HHE has focused on reducing work demands and the emphasis has been on redesign to avoid the negative consequences of demanding work. However, employment itself has many benefits for employees over and above a salary. These include routine, social contact, physical and mental variety, activity, shared goals, and identity. Unemployment has many negative consequences for individuals and for society. Kamerāde et al. (2019) surveyed 156,734 people to determine the minimum working hours and the optimal working hours for mental and physical well-being (19,555 were unemployed or otherwise economically inactive). The outcome variables measured were anxiety and depression symptoms (measured by the General Health Questionnaire, GHQ-12) and ratings of overall life satisfaction. The survey was repeated longitudinally in 8 waves from 2009 to 2018. Moving from unemployment to a working week of as little as 1–8 h was associated with lower GHQ scores. For previously unemployed men, life satisfaction improved even when working hours were small (1–16 h/week) and again at >24 h/week. Compared to working standard hours (36–40 h/week), shorter working hours were not associated with poorer life satisfaction or mental health. Working 40–44 h/week was associated with lower life satisfaction and mental health in women. The study controlled for confounding factors such as pay. Not all work durations were associated with better outcomes. The authors caution that since the study was conducted in the United Kingdom, low levels of employment can affect access to state benefits.

Controlling for pay, occupational groups, and employment contract, there was no evidence that the standard working week of 36–40 h was optimal in regard to mental health and life satisfaction. However, the findings do support the view that, if the requirements for human labor change, due to the introduction of disruptive technologies, there is scope for redistributing work with a shorter working week.

Fan et al. (2023) report data from 2,134 employees in 123 organizations working a 4-day week with no reductions in pay (and agreement to maintain outputs). Productivity is maintained by changes in work organization, such as reducing the number of meetings. Employees reported improvements in work ability, better sleep, and less fatigue. Issues remain about how a 4-day week should be organized. For example, working longer hours on the 4 days may reduce flexibility for those accustomed to working a 5-day week with flexible working hours. Implementation of shorter working hours may require a participatory approach.

O'Kelly et al. (2022) present the findings of 4-day week trials in 12 small- to medium-sized companies in Ireland (predominantly in the service sector). In general, the findings were positive. Nine of the 12 companies wished to continue with the 4-day week; company performance was rated highly, and the experience of the trial was positive. All employees wanted to continue the work schedule; self-rated work performance was high; stress, burnout, fatigue, and work-family conflict were reduced; average sleep time increased; and the schedule was particularly successful for women.

Gig Economy

The gig economy makes work available on a part-time or ad-hoc basis via digital platforms. Tasks are performed by individuals working as contractors rather than employees. Work is often carried out on a task-by-task basis and can provide an efficient and flexible way of working for both companies and workers. However, by treating gig workers as independent contractors, companies forego the provision of many traditional employee rights (such as paid leave and sick leave). Related to this are "zero-hours contracts" where employees have a contract with an employer but no fixed hours of work. Staff are only paid for the hours they work and work only when they are needed. In both cases, these new forms of working may offer flexibility to so-called "slashers"—people who work in multiple roles or use gig work to supplement other income streams. The gig economy has grown rapidly with the introduction of digital platforms, particularly in sectors such as transport, delivery of goods and products, and cleaning.

Taylor et al. (2023) conducted a systematic literature review of physical and psychosocial hazards in the gig economy. Out of 1963 articles, 34 were selected for review. A systems approach was used to classify hazards identified in the articles:

Government/Regulatory Hazards: Hazards identified (22% of the total) related to factors such as low pay, breaches of the law (particularly labor protection law), unfair employment practices, privacy, and data protection issues.

Company Level Hazards: These were most common (53%) and included worker safety, insufficient platform support, asymmetric control of platforms in favor of the company.

Gig Worker Level: Hazards relating to the disconnect between the company and gig workers manifested as a lack of gig worker control and companies seeing themselves as "platform providers" rather than employers, resulting in poor engagement and satisfaction and, in some cases, a lack of training. Excessive monitoring and supervision of gig workers were also found to be hazards.

Gig workers are sometimes referred to as the "Precariat" (rather than the proletariat) due to the hazards outlined above. The companies they work for may regard themselves as "platform providers" rather than employers.

Artificial Intelligence (AI)

Artificial intelligence refers to the ability of machines to complete tasks that are usually thought to require (normally human) intelligence. Tasks such as problem-solving, making judgments, describing, and classifying, for example. This does not necessarily mean that the tasks are completed by the machine in a "human-like" way. AI is a large and growing field. Two examples are given below because they both have high TRLs and have contrasting implications for HFE.

Expert Systems: Expert systems have been in use since the 1980s in specific domains such as computer configuration, medicine, and banking. They are used to configure computer systems for customers, to aid medical diagnosis, support clinical decision-making in medicine, and assess loan applications and provide investment advice in banking. They consist of a user-interface, a knowledge base, and an inference engine (and sometimes, an explanation module). They are suitable for use in tightly configured domains where there is an extant knowledge base and a clear set of rules for making a decision. Information, for example, about a loan applicant, is entered into the system, which then applies the bank's rules to process the application and make the decision. MYCIN is an expert system for diagnosing bacterial infections, and DENDRAL is used for chemical analysis. Often, these systems use the same rules as human experts and can therefore explain themselves in terms appropriate to the user. The rules are often in the "if→then" form and can be chained together in decision trees.

Knowledge Acquisition Is One Of The Difficulties In Designing Expert Systems

Neural Networks: A neural network is a form of artificial intelligence consisting of layers of nodes (or artificial neurons). There is an input layer, one or more inner layers, and an output layer. Neural networks are trained by human operators or experts by presenting data in the domain of interest to the network and providing the system with the result or decision in each case. For example, suppose we want to train a neural network in a fruit packaging plant to classify apples into four categories (export, local market, food processing, animal feed) depending on their quality. The system has to be able to detect defects that determine how an apple is classified (data from cameras is coded and is fed into the network). The system is trained by exposing it to many apples in each category and telling it how they are classified. Variations in the input data affect nodes in the input layer, which may or may not send data to the next layer. Patterns of activation pass through one or more layers to the output layer, which, in this example has four output nodes (the classification of the apples). Once trained, for any apple, nodes in the input layer either respond to the inputs and feed this forward to the next layer, or they do not. These patterns of response among many nodes in the layers are weighted and differ depending on the quality of the apple, which is then categorized at the output layer.

Large Language Models: As ChatGPT (ChatGPT is an AI chatbot) explained at the beginning of this Epilogue, large language models (LLMs) are very deep networks trained on a vast amount of data. Neural networks are sometimes referred to as "black boxes" in the sense that it is not apparent how they arrive at decisions or conclusions, what the role of individual neurons was, or the internal workings of the layers. LLMs may be "knowledgeable" depending on how they were trained, but are not necessarily "experts." ChatGPT is an example of *generative AI*—it can create original content from the material it has been trained on.

According to Hicks et al. (2024), LLMs are designed to produce human-like language. They do not have *extra-linguistic* goals in the way that humans do and are incapable of understanding whether their outputs are accurate. LLMs are trained using enormous amounts of text, which is used to construct a large number of nodes with associated probability functions, essentially a large statistical model that predicts the next word from the context provided by those preceding it. The model can choose randomly between likely words to give the appearance of natural language. In other words, the goal of an LLM is to produce responses that "look right" rather than to transmit information accurately. Essentially, they are akin to linguistic pattern recognizers with no capacity to reason or calculate. This is why, at the beginning of this chapter, ChatGPT denied that it was "intelligent."

Explainable AI: It is intuitively appealing to suggest that AI-supported decision-making by users will be improved if the AI can explain its decisions. Feature-based explanations give the features of the problem that led the AI to a prediction or decision. Rule-based explanations give the explanation in the form of if→then rules. However, current research findings are mixed: some studies show improved decision- making when the AI decision is accompanied by an explanation, and others do not. Alufaisan et al. (2021) compared participant decision-making when it was unaided, aided by AI, and by AI with an explanation. In one study, participants were presented with data on offenders in criminal cases (e.g. age, offence, previous convictions, risk score, sex, race) and had to decide whether the person would re-offend. Unaided, participants' decisions were better than chance (50:50) so they weren't just guessing and the decisions were further improved by AI support. AI with explanation did not lead to better decisions however. One reason might be the nature of the "explanations" given by the AI. The "explanations" around the decision to re-offend or not were restricted to the features that the AI had used to make its prediction (e.g., race, age, sex). It may be that from the decision-maker's point of view, simply stating that the person would not re-offend because she was a 50-year-old female with no previous convictions and a low-risk score was not enough of an "explanation" (even though, from a statistical point of view, these were predictors in the AI's database). Merely listing features of a problem that led to the AI's decision may not be sufficient as explanations for users who conceptualize the matter differently. From the perspective of HFE then, there may be interesting research avenues to understand what makes "explainable AI" explainable to users—what is the nature of an "explanation" that really "explains" the answer from the user's point of view?

AI Hallucinations: The investigation of human error has a long history in HFE. AI also makes mistakes, and it is worthwhile to illustrate the kinds of errors that these systems can make. They are often labeled "Hallucinations." While in humans, hallucinations manifest themselves as distortions of sensory inputs or sensations in the absence of external stimuli, AI systems do not have senses and therefore the term is potentially misleading. AI doesn't "hallucinate" in the way that humans do, and there is no analog of mental illness in AI. Maleki et al. (2024) conducted a systematic literature review of papers on the topic, finding the following "hallucinatory" aspects of AI behavior (particularly LLMs):

1. *Conflabulation:* AI-generated responses that appear plausible but are incorrect
2. *Delusion:* False responses
3. *Stochastic Parroting:* Training data are repeated back to the user with no factual understanding or reasoning ("word play" or "waffling" around a topic)
4. *Factual Errors:* Unintentional information or statements that are false
5. *Fact Fabrication:* Inaccurate information is invented and presented in a lucid fashion
6. *Fabrication:* Outputs that sound plausible but are false, the result of predictive text selection in the absence of reasoning
7. *Mistakes and Blunders:* Answers that are fabricated when query or training data are insufficient
8. *Hasty Generalizations/False Analogy*: Inferences that do not follow from the premises

According to Maslej et al. (2024) in one study, ChatGPT fabricated unverifiable information in approximately 19.5% of 35,000 samples across a range of fields such as language, climate, and technology.

Hallucinations or "Bullshit"?: Hicks et al. (2024) argue that the outputs of LLMs, although often correct, are nothing more than "bullshit." Bullshitting is defined, at the lowest level, as speech or text produced without any concern for the truth but with no intent to mislead the audience about the utterer's attitude toward the truth. To state that LLMs "hallucinate" is incorrect because it anthropomorphizes them and creates inaccurate perceptions of their abilities and how they function. When humans hallucinate, they perceive things that aren't really there—hallucinations are not products of normal human functioning. When LLMs produce incorrect outputs ("hallucinations"), they are using the same processes that produce correct responses. LLMs do not hallucinate because they do not misperceive the source of information – they are functioning in the way they are intended to function: to produce responses that "seem right" because they look like natural language. They do not have a model of the world or of the subject matter that they can "misperceive" and are therefore indifferent to the truth of the matter.

Human-AI Interaction Guidelines: Amershi et al. (2019) offered the following guidelines for human-AI interaction. These guidelines include many types of AI-infused services, such as web-based algorithms and smart apps on cell phones:

1. Make it clear what the system can do.
2. Make clear how well the system can do what it can do.
3. Provide timely interactions/guidance.
4. Show contextually relevant information (time/location nearby of a product or service).
5. Match relevant social norms in interaction (e.g., tone of voice, formality).
6. Mitigate social biases (do not reinforce undesirable stereotypes).
7. Support efficient invocation of AI support (easy to access AI when needed).
8. Support efficient dismissal.
9. Support efficient correction (edit or refine AI outputs).
10. Scope services when in doubt (provide options in place of autocomplete).
11. Make clear system behavior (easy access to explanations).
12. Remember recent interactions (AI has short-term memory).
13. Learn from user behavior (update search strategy in a timely manner).
14. Update and adapt cautiously.
15. Encourage granular feedback (user can dictate preferences going forward).
16. Convey consequences of user actions (effect on future AI behavior).
17. Provide global controls (user can customize AI behavior).
18. Notify users about changes (new features highlighted as "pop-ups").

As we will see in the section on system integration, AI is a large and growing area with rapid development in many domains. This section serves mainly to introduce some of the emerging HFE issues as AI is "infused," into many of our online tools and applications. Cummings et al. (2024) have criticized the way AI is designed and introduced, stating that AI algorithms released to the public are often knowingly error-laden, trained with partial and incomplete data, and have undergone very little testing. Further, members of the public may be exposed to them unknowingly and without their consent, introducing, among other matters, issues of privacy and data protection.

Autonomous Vehicles (AVs)

We have already looked at the automation of vehicles in Chapter 14. As the technology matures, questions have been raised about the safety of the sensors used to guide fully or highly automated vehicles, as well as the control systems that govern the behavior of the vehicle on the road. Currently, fully autonomous vehicles are at the TRLs of 6–7 approaching 8.

How do Self-Driving Cars "Read the Road?": A number of sensors and other technologies are used. LIDAR (laser detection and pulsing) uses laser pulses to create a 3D model of the car's surroundings (e.g., distance from road edges, although not all such vehicles use LIDAR). RADAR is also used to monitor the position of nearby vehicles. Ultrasound is used to detect obstacles (similar to the way bats use sound to fly at night or the use of active SONAR in submarines, sound pulses are sent out, and if an "echo" is detected, it gives information on the position and distance of an object). Cameras are also used to detect objects such as traffic lights and road signs. Electronic control systems use feedback data on speed, velocity, and centrifugal force to monitor driving performance. Neural network-based AI is used to identify and classify objects around the vehicle, and the control system drives the car.

Concerns about Fully Autonomous Vehicles: Reliability of the sensors is only one of the concerns expressed about fully autonomous vehicles (for example: how well does LIDAR work in heavy rain or snow? How do we ensure that the sensors do not interfere with each other when the roads are full of autonomous vehicles?). There are also concerns about the control systems and the AI itself. Cummings et al. (2024) cite data showing that self-driving cars are six times more likely to be involved in accidents resulting in injury than cars driven by people. In one example cited by Cummings et al., a lady was pushing her bicycle across a four-lane road when she was hit (and killed) by an Uber taxi undergoing testing. The safety driver in the taxi was not paying attention and failed to slow the vehicle. The vehicle's sensors detected the lady 5.6 s before impact but failed to classify her correctly, misclassifying her first as a car and then as a bicycle. In another example, a self-driving vehicle collided with the rear end of an articulated bus in San Francisco. Although the sensors detected the bus, the software was unable to interpret its movements correctly (the rear end of an articulated bus does not move along the same path as the front end due to the articulation), and the length of the bus was miscalculated.

How Intelligent is AI?: Cummings et al. point out that the AI in autonomous vehicles is limited to rule-based reasoning. If a STOP sign is detected, it will stop as required. However, if the STOP sign is partially obscured, the AI is unable to process the uncertainty. Whereas a human driver would use knowledge-based reasoning to infer that it was a STOP sign, the AI will default to a minimum risk maneuver (i.e., stop). AI systems also struggle when there is a plethora of signs in one location (Figure 13.3) and no clear rule about how to proceed. AI systems are limited to "bottom-up" reasoning by aggregating data from different sensors. Unlike human drivers, they do not learn while they are operating. If the data are lacking (e.g., lane markers are faded), they may fail to respond appropriately. According to Cummings et al., humans can use "top-down reasoning" to deal with complex situations and uncertainty. This may include goal-oriented problem-solving, deciding on an action on the basis of the overall intention or the context, whereas AI systems exhibit "bottom-up reasoning," building a model of the world by aggregating low-level data and lacking the ability to consider causality or context.

Theory of Mind and Metaknowledge: "Theory of Mind" is a cognitive ability of humans. We all understand that other people have different knowledge, skills, abilities, and intentions to our own. This enables the mental state of others, including their intentions, to be inferred from their behavior, facilitating predictions about future behavior. Human drivers can therefore cope with busy areas with many pedestrians in a way that self-driving cars cannot (although AI with theory of mind capability is under testing). This explains why concerns have been expressed about the efficiency and safety of these cars in busy city centers. As ChatGPT explained at the beginning of this Epilogue, LLMs do not know what they don't know.

TOOLS AND PROCESSES

Many of the tools referred to in previous chapters can be used to assess the safety and efficacy of human augmentation in the workplaces of the future. New tools are emerging to meet new requirements. Some examples are given below.

Taxonomy for AI Hazard Analysis (TAIHA)

Cummings et al. (2024) adapted the Swiss Cheese Model of industrial accidents for the analysis of accidents involving AI-infused systems. In the classic Swiss Cheese model, the model has four main "layers" of protection, all of which have weaknesses or omissions (holes in a slice of cheese). If the holes in the layers are lined up, an accident trajectory is created, and a hazard can cause an accident. The classic Swiss Cheese layers are: organizational influences such as bad policy or board decisions; line management problems such as inadequate supervision; local failures or stressors; unsafe acts (human error); and inadequate defenses. As with STAMP and Accimaps (Chapter 15), human error is only part of a bigger picture. In TAIHA, Cummings et al. propose the following:

Inadequate AI Oversight: Lack of regulation and pressure to introduce new products. At the organizational level, companies, under pressure to market their products early, may have a reactive safety culture (only acting after an adverse event). Lack of regulation may allow companies to trial new products before reaching an appropriate TRL. Cummings et al. point out that humans are not allowed to drive cars unless they have passed a driving test and a basic vision test, whereas no such requirements apply to self-driving cars. How "good" is "good enough"?

Inadequate AI Design: Lack of AI sensor fusion or neural network training. For example, the AI is unable to distinguish between an ordinary bus and an articulated bus.

Inadequate AI Maintenance: Rushed software updates to meet unrealistic deadlines. Failure to update the training of the AI (e.g., to distinguish between a bicycle and a person pushing a bicycle).

Inadequate AI Testing: Lack of thorough "real world" testing or overreliance on simulation for testing.

Evaluating Human-Robot Interaction

Several instruments have been developed over the years. Spatola et al. (2021) developed the Human-Robot Interaction Evaluation Scale (HRIES) to measure how robots were perceived and whether human-like characteristics were attributed to them. The scale was developed in a series of studies: participants rated pictures of 20 robots on a set of test items with factor analysis (Chapter 3) yielding main components of anthropomorphism. This was repeated using videos of robots and, finally, in interaction with a robot in both a passive and an active state. The factor analysis was confirmed, and the scale items were reliable (Chapter 14). The final scale with items rated from 1 = not at all to 7 = totally, is given in Table E.1.

TABLE E.1
HRIES Scale (Spatola et al., 2021)

Items	Factor
Warm	Sociability
Likeable	Sociability
Trustworthy	Sociability
Friendly S	Sociability
Alive	Animacy
Natural	Animacy
Real	Animacy
Human-like	Animacy
Self-reliant	Agency
Rational	Agency
Intentional	Agency
Intelligent	Agency
Creepy	Disturbance
Scary	Disturbance
Uncanny	Disturbance
Weird	Disturbance

Agency refers to the tendency to view the robot as an intentional agent.

Stress in Human-Robot Interaction: Taking a systems approach, Babamiri et al. (2024) developed a valid and reliable questionnaire of 46 items to evaluate stressors in human-robot interaction. The questionnaire has scales for: Human-Related Factors (situation awareness, resilience, self-efficacy, trust competency, affinity, and fear of substitution); Task-Related Factors (remaining up to date, job demands environmental conditions, skill utilization, technical support, and decision complexity); Robot-Related Factors (usability, level of automation, safety, and software/hardware problems).

Chatbot Usability

Chatbots are widely used nowadays, often in customer-relationship management roles where the chatbot is linked to a specific service. While chatbots offer flexibility, 24-hour real-time access to a service provider, their usability is key to a satisfactory user experience. Borsci and Schmettow (2024) examined the psychometric properties of the chatBot Usability Scale (BUS-11) in Table E.2.

Factor analyses of the responses of 678 participants using 44 chatbots yielded a four-factor solution with standardized scores for each summated (5-point) scale (Table E.3).

TABLE E.2
The chatBot Usability Scale (BUS-11) for Short-Term Interactions (5-Point Likert Ratings)

1. *Ease to Start a Conversation:* Chatbot seems able to find ways to respond appropriately even when it encounters situations or arguments it is not equipped to handle
2. *Access to Chatbot:* Functions and location of the chatbot on the screen are visible and accessible
3. *Expectation Setting:* Ability of chatbot to make clear its capabilities and to not create false expectations in the end –users
4. *Flexibility and Communication Effort:* Chatbot seems able to manage and adapt to different conversational styles of the end-users, minimizing conversational efforts for the end user
5. *Ability to Maintain a Themed Discussion:* Chatbot maintains a conversational theme once introduced and keeps track of the context to understand the user's utterances
6. *Reference to the Service:* Chatbot seems designed to use the environment (information, options, buttons on screen, etc.) to guide the user toward its goal
7. *Users' Privacy and Security:* Chatbot appears to be able to protect user's privacy and make appropriate decisions on behalf of the user
8. *Recognition and Facilitation of Users' Goal and Intent:* Chatbot seems able to recognize the user's intent and guide the user to its goals
9. *Relevance of Information:* The chatbot provides relevant and appropriate information/answer to people at each stage to make them closer to their goal
10. *Maxim of Quantity:* The chatbot responds in an informative way without adding too much information
11. *Resilience to Failure:* Chatbot seems able to find ways to respond appropriately even when it encounters situations or arguments it is not equipped to handle
12. *Understandability and Politeness:* The chatbot seems able to understand input and convey correct statements and answers without ambiguity and with acceptable manners
13. *Perceived Conversational Credibility:* The chatbot responds in a credible and informative way without adding too much information
14. *Speed of Answer:* The chatbot is perceived as able to respond to requests and solve issues in a timely manner
15. *Meet the Neurodiverse Needs:* Chatbot seems able to meet needs and be used by users independently from their health conditions, well-being, age, etc.

TABLE E.3
Components of Chatbot Usability

Scale	Very Poor	Poor	Average	Good	Very Good
Accessibility	0–3.5	3.5–3.7	3.7–3.92	3.92–4.19	4.19–5
Functional Interactive. Conversation	0–3.32	3.32–3.53	3.53–3.68	3.68–4.05	4.05–5
Privacy	0–2.52	2.52–2.63	263–2.77	2.77–3.18	3.18–5
Responsiveness	0–4.03	4.03–4.23	4.23–4.38	4.38–4.58	4.58–5

AI Content Detection ("Fraudbusters?")

LLMs (large language models) offer many affordances that can be put to use in different ways. Content generation, in itself, may improve productivity and provide employees with support. However, less ethical uses are fraudulent misrepresentation of one's abilities (e.g., in education) and plagiarism. AI content detectors are programs based on natural language processing that can detect whether a text has been produced by a human or by AI. Teachers and examiners are obvious users of content detectors, but anyone (such as a magazine editor) who has commissioned an article or report might want to ensure that the work has been produced by a person and not by an LLM.

Content detectors are trained on human- and AI-generated texts. Some key differences between AI-generated text and text written by people are: lack of variability in AI-generated text; simplistic syntax; repetition of words and phrases; and cliched phrases. Text generated by people often has greater variability and "burstiness" (variation in sentence length) and may be more "perplexing" for the content generator (text that differs from the text that the AI content detector was trained on).

Chaka (2024) reviewed 17 articles investigating the ability of AI detection tools to detect AI-generated texts. The articles tested between 2 and 6 content detectors, using text produced by different versions of ChatGPT. Overall, all of the content detectors were found to be unreliable yielding false positive results or failing to classify one way or the other. Chaka points out that AI detectors are vulnerable to adversarial attacks. The output of the LLM can be edited (by a person) to appear more "human-like" (more use of synonyms, removal of commas, paraphrasing, and rewording, for example). Chaka cautioned that AI content detection tools cannot detect plagiarism and recommended that AI content detectors be used with plagiarism detection tools and overseen by human reviewers.

Measuring Workplace Stressors and Mental Well-Being

There are many self-report questionnaires that can be used to measure employee perceptions of work stress. One commonly used questionnaire is based on the demand/control model of Karasek and Theorell (1990). Karasek's model predicts adverse mental outcomes when work demands are high and when employees have little control over their work. Social support from management and/or coworkers acts as a buffer to mitigate the effects. Siegrist (1996) proposed the "effort-reward imbalance" model, where adverse mental health outcomes result from an imbalance between rewards (monetary and social rewards) and work efforts. Sample items from Siegrist's questionnaire (of 16[2] or 22 items) are given below:

- I have constant time pressure due to a heavy workload.
- Over the past few years, my job has become more and more demanding.
- Considering all my efforts and achievements, I receive the respect and prestige I deserve at work.
- My current occupational position adequately reflects my education and training.
- Considering all my efforts and achievements, my salary/income is adequate.
- My job security is poor.

The questionnaire also measures a construct known as "overcommitment." Overcommitted employees have problems disengaging from work, lowering their capacity to recover from work demands (Chapter 7). Example items are given below:

- As soon as I get up in the morning, I start thinking about work problems.
- Work rarely lets me go, it is still on my mind when I go to bed.
- If I postpone something that I was supposed to do today, I'll have trouble sleeping at night.

Questionnaires such as these are valid and reliable tools that can be used to assess the demands of new ways of working. Overcommitment, for example, might be particularly relevant to home and hybrid workers, where, in the absence of a daily commute, there is no longer a physical buffer between work life and domestic life. There is also a "demands/control/ resources" model (Demerouti et al., 2001), which takes into account the resources provided at work that support employee efforts. These include the physical, psychosocial, and organizational resources that support the achievement of work goals and reduce job demands. This focus on support might be particularly relevant to the assessment of the working lives of home workers at risk of social isolation.

Anxiety and Depression: Anxiety and depression (or, chronically, burnout) can be a response to stressors in working life beyond the ability to cope. There are several survey instruments for measuring symptoms of anxiety and depression. One of the most common is the General Health Questionnaire (12-item version, Goldberg and Williams, 1991). It measures symptoms of social dysfunction, anxiety, and depression. For example:

- Have you been able to face up to problems?
- Have you been able to enjoy your usual day-to-day activities?
- Do you feel you have been playing a useful part in things?
- Have you been thinking of yourself as a worthless person?

The questionnaire items are scored on rating scales and coded 0/1. Those scoring more than 4 symptoms out of 12 are regarded as "strain cases" suffering significant anxiety and depression and in need of action. Combined with information on workplace stressors, changes to work organization can be made.

Person-Focused Interventions for Employee Support

Employee well-being programs are common in many organizations. They have a long history, going back to the Hawthorne counseling program of the 1930s and 1940s (Chapter 1). Workplace counseling and assistance for individuals with specific needs are beyond the scope of this book. Preventative approaches directed or offered at the group level are relevant, though. Essentially, these approaches focus on "fitting the person to the job" (FPJ, Chapter 1) and not on changing working conditions. Fleming (2024) reviewed studies from surveys of 46,336 employees in 233 organizations to determine whether these well-being interventions were effective. Among the support programs evaluated were: mindfulness training; resilience training; stress management training; healthy lifestyle interventions; relaxation classes; volunteering or charity work; and well-being apps. Unlike the support described in the job-demands-resources model, these approaches attempt to improve individual capacity to deal with work demands rather than changing the demands themselves.

Overall, Flemming found no differences between non-participants and participants in the following well-being programs: relaxation practices; time management; coaching; financial well-being programs; well-being apps; online coaching; and sleep apps. Volunteering was the only type of intervention to suggest benefits for workers' well-being. There was some weak evidence, overall, for increased engagement in participants. Similarly, overall, participation in the interventions did not improve perceptions of the work environment, except for volunteering. These findings were similar across different groups according to gender, ethnicity, and income.

There are several possible explanations for these findings. The most obvious is that the participants did not lack resilience, time, or stress management skills in the first place, and therefore the training had no benefit. Another is that the studies were cross-sectional and therefore selection bias may have occurred (e.g., happier, more engaged people may have chosen volunteer programs, and those with poor time or stress management skills or more stressful jobs may have chosen to participate in these programs, which raised them to the level of non-participants). In general, FPJ approaches such as this are lacking because they offer generic solutions to (often) situation-specific problems; they make inappropriate assumptions about the factors that degrade well-being at work; and they assume that the imbalance between demands and outcomes can be corrected by changing employees rather than changing work conditions.

Autism/Asberger's Syndrome

In the same way that HFE focuses on assessing worker conditions in order to fit the job to the person (FJP, Chapter 1) there is a new focus on making reasonable adjustments to working conditions in order to accommodate neurodiverse employees. This involves taking into account workplace exposures that neurodiverse people find stressful, including the need to behave like neurotypical colleagues. This is why many neurodiverse employees prefer working alone rather than in groups—the self-control demands (see Chapter 12) are lower when they are alone because they no longer have to concentrate on staying "in-role." There is an "Autism Spectrum Quotient"[3] that can be completed by self-report, enabling the reader to gain insight into the nature of these stressful exposures and also the kinds of activities and environments that neurodiverse people find preferable—insights that can be used to inform the design of their jobs and working conditions.

Support for Neurominorities: Hall et al. (2024) investigated employment support for neurominorities. A traditional approach to increasing their employment opportunities has been the provision of job coaches to help them learn new skills and deal with situations that challenge them, and by adjusting work tasks according to their abilities. However, this is very labor-intensive and can require 100–140 h of workplace support per employee. An alternative approach sees neurodiversity as a deficit and seeks to reduce deficiencies by correcting behavior or improving skills. Hall et al. criticize this approach because it views neurodiversity as a problem that needs to be corrected, and because practising the new skills (e.g., making eye contact) may itself be stressful. Furthermore, the approach does not make the best use of existing skills and may disempower the neurodiverse employee. An alternative approach has used assistive technology such as PDAs ("personal digital assistants," iPADS, iPhones, and video-based task sequencing prompts. Supporting a structured workday in this way can increase job retention and independence. This approach can also reduce coaching time considerably. Hall et al. conducted a preliminary study to explore the use of a virtual job coach based on ChatGPT to provide personalized coaching support. This support included increasing the employee's sense of agency (being able to complete a task independently); motivating the employee by re-iterating past successes; providing decision support by providing lists of options; and improving self-management at work by assisting with work scheduling, timekeeping, and so on.

SYSTEM INTEGRATION

The future of work will depend on the emerging technological landscape, as new technologies develop rapidly and enter increasingly interconnected workplaces. These changes are having "supply side" effects in economic terms—easier access to information, goods, and services at lower prices. This increases the scope for the emergence of new goods and services and new occupations, as consumers have greater purchasing power for other goods and services.

In the industrial age, work took place in dedicated locations using tools designed for a specific purpose. Work requirements were straightforward to meet because the affordances offered by the technology were limited, and job design options were constrained by technological limitations. The emerging technological landscape now offers many more affordances. How do we choose between these to design the organizations of the future?

The Future of Jobs

Clearly, new and disruptive technologies are going to make some jobs redundant, create new jobs and change the tools that many of us use to do our existing jobs. This trend has been happening for hundreds of years and will continue. A report by the World Economic Forum (2023) detailed findings from 803 companies employing over 11 million workers worldwide, finding that over 85% of companies planned to adopt new technologies to transform their organizations while applying social and environmental standards more broadly. The green transition was also found to be a major driver of changes, with the introduction of green technologies and plans for shortening supply chains (Chapter 16). These plans are all drivers of the creation of new jobs in environmental management and cybersecurity (Chapter 14). Humanoid and non-humanoid robots were expected to displace many existing jobs. Overall, *churn* was expected in 23% of jobs (jobs either disrupted by new technologies or transformed by them). At present, 66% of business tasks were reported to be performed by humans and 34% by technologies—an increase in the use of technology over humans of only 1% in 3 years. To date, new and disruptive technologies have not increased overall unemployment. While clerical and administrative jobs are expected to disappear, jobs in education, agriculture, and digitally-enabled roles are expected to grow. In particular, almost half of the companies expected to be investing in education and training of their employees and cited creativity and problem-solving skills as key areas for future personnel development.

Growth and Spread of AI

At the time of writing, investment in AI is growing rapidly. Between 2023 and 2024, the number of LLMs doubled worldwide, with LLMs trained on increasingly large datasets. According to a recent report from Stanford University (Maslej et al., 2024), AI can equal or exceed human performance on a variety of tasks, including image classification, basic reading comprehension, visual reasoning, and natural language inference. However, it falls short in visual commonsense reasoning, abstract reasoning, and advanced-level mathematical problem-solving.

AI is being introduced into many sectors, including healthcare, sales and marketing, and manufacturing. It is also being infused into many applications, such as office software. Many organizations are investing in AI products and in training ("upskilling") to enable employees to work in AI-augmented roles. In 2022, 2.0% of AI job advertisements in the United States required skills in AI. These trends are occurring widely across the economies of many countries in sectors such as information services, finance and insurance, manufacturing, public administration, utilities, forestry and agriculture, transportation and warehousing, and retail. Utilization of AI produces both cost savings and revenue gains in manufacturing and services—faster completion of higher-quality work—and the main uses are as follows (Maslej et al., 2024):

- Text creation (e.g., advertising/sales, contracts)
- Summarizing documents
- Creating images and videos
- Identifying customer needs
- Customer service chatbots
- Creating first drafts of documents
- Specialized AI for tasks such as program development

These developments have led many countries to introduce legislation to govern the use of AI. For example, the European Union regulation requires "right to explanation" for any algorithms that make decisions significantly impacting users with user-level predictors (EU Artificial Intelligence Act, Article 86).

An emerging theme is that of "Responsible AI," which has the following characteristics:

- *Data Governance*: Policies are in place to ensure data protection, controlled access, and the ethical use of data from, for example, social media platforms and healthcare databases. Informed consent for the use of personal information is a key issue.
- *Explainability*: The AI can explain its outputs in ways that are understandable by users.
- *Fairness*: The platform is free of bias in its outputs. This will depend to an extent on how it was trained and the material used to train it.
- *Privacy*: Personal data is stored in a confidential way, anonymized, as required.
- *Security and Safety*: This includes both cybersecurity (Chapter 14) and protection against system errors (such as AI hallucination).
- *Transparency*: This relates to how the system works, how it is being introduced, and for what purpose.

AI is developing and spreading rapidly, with billions of dollars being invested in the development of new systems and their implementation worldwide.

Language Modelers- Effects on Jobs: Felten et al. (2023) analyzed the effects of language modelers such as ChatGPT on occupations using an AI Occupational Exposure measure (AIOE). Organizations or jobs with high AI occupational exposure are those where the job could be modified or substituted with AI. The measure essentially takes an index of the abilities needed in an occupation to determine its future exposure to AI. AIOE scores are listed for 774 occupations. Table E.4 lists the AIOE scores for the top and bottom 10 occupations. Jobs with higher AIOE scores have higher mean and median salaries.

This does not mean that jobs with high scores will be replaced by AI; it indicates that they are more likely to be infused with AI in the future. A study by KPMG (2024) reported that 2.7% of jobs requiring a high school diploma will be affected by AI, compared to 21% of jobs requiring a degree. The long-term effects of generative AI and robotics on

TABLE E.4

AI Exposure of Different Occupations (High-Middle-Low, from Felten et al., 2023)

Hight AIOE	Middle AIOE	Low AIOE
Telemarketers	Engineering Technicians	Fence Erectors
English Language and Literature Teachers[a]	Veterinarians	Cement Masons and Concrete Finishers
Foreign Language and Literature Teachers[a]	Respiratory Therapy Technicians	Terrazo Workers and Finishers
History Teachers[a]	Nuclear Technicians	Iron and Steel Workers
Law Teachers[a]	Chemical Technicians	Plasterers
Philosophy and Religion Teachers[a]	Athletic Trainers	Iron and Rebar Workers
Sociology Teachers[a]	Booth Cashiers	Fallers
Political Science Teachers[a]	Museum Technicians	Dancers
Criminal Justice Teachers		Brickmasons, Stonemasons, Tile Setters
Sociologists		Pressers, Textile and Garment Workers

[a] Postsecondary

jobs are not well understood, although the introduction of generative AI is expected to increase GDP in the United Kingdom by 1.2% over the next decade and account for 30% of hours worked in the United States by 2030 (KPMG, 2024). A study of 4,197 employees in Australia, Canada, Germany, the United Kingdom, and the United States found that 66% of employees expected new technologies to increase their productivity in 3 years, although there were concerns about being overloaded with technology and job insecurity. Negative impacts on mental health were reported, with 31% citing a degraded work-life balance. There were also reports of a lack of uninterrupted time to focus due to intrusions from e-mail and virtual meetings. The introduction of new technologies should be based on an understanding of how work is actually conducted on a day-to-day basis ("work as done," Chapter 15), rather than as mandated by policy or job descriptions. For many organizations, generative AI should be considered a tool with a TRL of 5 to 6—trials are needed in-house to identify the best way to use the affordances of AI to meet organizational requirements.

AI Support Functions: Many new jobs have been created to support the development of AI algorithms. "Click workers" are sometimes employed like gig workers and recruited via applications such as Amazon's Mechanical Turk. Thousands of people work in roles such as "AI Annotators." Annotators receive training on the company platform and spend their time categorizing data to train the system—for example, highlighting an object in a complex image and labeling it using a target descriptor. Sometimes, the work is outsourced to developing countries where labor and regulatory costs are lower (see Chapter 16). "Content Moderators" work in social media to identify and remove content that breaches the platform's guidelines. This is potentially a stressful job because moderators are exposed to harmful content, which may include, but is not limited to: hate speech, incitement, pornography, material from terrorist groups, and content promoting suicide and self-harm. Part of their work is used to train the platform AI to recognize such material.

Home and Hybrid Working—The Right to Disconnect

Working hours are the time that an employee is at an employer's disposal (Chapter 7). Outside of working hours, the physical separation between employees and employers has been replaced, and for many remote workers, communication is conducted via technology. The International Labour Organisation (Goshesh, 2022) defines "telework" as:

> the use of information and communications technology – such as smartphones, tablets, laptops and computers – for the purposes of work outside the employer's premises (either from the worker's home or from another location)

The right to disconnect refers to a worker's right to be able to disengage from work and refrain from engaging in work-related electronic communications, such as e-mails, texting, or other communications, during non-work hours. The main components of the right to disconnect are:

- The right to switch off communication devices
- The right to not be disturbed during free time
- The right to not be penalized for not responding to work outside of normal hours
- The right to not be routinely expected to work outside of normal hours
- The duty to respect other people's right to disconnect

In some countries, the right to disconnect is already embedded in regulations about working hours (the EU Working Time Directive, for example). Nevertheless, remote working blurs the distinction between work and home life and arrangements are necessary to ensure that "core hours" are agreed upon (when the employee is always at the employer's disposal), together with a method for accounting for working time outside these hours and a stronger focus on outputs, by prior agreement.

Greater clarity is also needed about rest periods and time away from the screen during core hours, compared to working in an office.

This has led to legislation in many countries concerning the right to disconnect. Most regulations focus on the right to disconnect (not the right to refrain from working) or how to implement the right in practice. Technological interventions are one option; for example, blocking or delaying e-mails at the server outside of core hours. The right to disconnect is not a new right in any sense; it is partly an extension of the right to rest and leisure and also the right to privacy—the right to be left alone.

Popularity of Hybrid Working: Combining home and office work is increasingly accepted, and employees are starting to see it as an essential part of a job offer rather than an entitlement, influencing their decision to accept a job offer or stay in a job (KPMG, 2024). Whereas 34% reported that they valued fair pay among other components such as organizational culture, 33% valued flexible working. Two main reasons for wanting to leave an organization were poor work-life balance (reported by 35% of respondents) and inflexible working hours (32%). Clearly, hybrid working is going to be a key component of future job design, and HFE will have a role to play in formulating and evaluating policies and practices for hybrid working.

The Future of Workplaces: Healthy Buildings, Third Places, and Smart Cities

Smart cities are cities with high levels of interconnectivity to maximize the sustainable use of services. Some characteristics of smart cities relevant to HFE include energy efficiency, mobility, the design of urban spaces, and an interconnected transport network. As discussed earlier in this chapter, third places are emerging as workplace options for hybrid workers. Many of our current cities are obesogenic environments. Commuters often have few options and have to use cars or trains to get to work. This increases the time spent sitting or standing still and reduces the time available for other activities. According to Verde et al. (2024), the design of the layout and infrastructure of cities has a major influence on the health of the people who live and work in them. Some *obesogenic design features* of cities include:

Artificial Light at Night (ALAN): Urbanization and the development of artificial light sources have disrupted the natural circadian rhythms of people dwelling in urban environments. ALAN suppresses melatonin secretion and sleep activity. One consequence of this is to change the way lipids are normally metabolized. ALAN also facilitates non-stop work activities that are also disruptive of normal work-rest cycles. Light pollution is strongly associated with obesity rates in different parts of the world.

Lack of Green Areas and Walkable Neighborhoods: The layout of many cities limits the ability of occupants to walk or cycle to their destinations. Coupled with sedentary lifestyles and workstations that do not permit activity (Chapter 4), this places constraints on daily physical activity levels. Engaging in physical activity to maintain health has long been seen as the responsibility of the individual. In smart cities, the infrastructure provides the affordance for walking or cycling. Verde et al. cite evidence that obesity rates are lower in areas with green spaces that offer affordances for recreation. Further, there is evidence that air pollution is linked to obesity rates, possibly because it causes chronic inflammation (Chapter 7). Green areas for recreation may also reduce mental stress, facilitate attention restoration (Chapter 12), help normalize the circadian rhythm in cortisol, and reduce the risk of developing insulin resistance (Chapter 8).

Few Transport Options: Smart cities control access to and use of the infrastructure by cars by integrating roads with walkways and cycle lanes. A total of 150 min/week of brisk walking meets the target for moderate activity levels to maintain health and is achievable if part or all of the daily commute is completed on foot. Verde et al. cite evidence from China that the purchase of a motor vehicle doubles the odds of developing obesity.

"Digital Nomads" and "Shoulder Bag Workers" can work almost anywhere, and the concept of a fixed workplace no longer applies. However, one goal of the development of smart cities is to provide a healthy and safe workplace for those who work in them—wherever they are.

Healthy Buildings: The design of future buildings is likely to be influenced by new methods of certification and new standards. Buildings that meet the criteria can be considered healthy buildings. Fitwell Certification and the Well Building Standard[4] are examples. The Well Building Standard assesses buildings on:

- Air
- Water
- Nourishment
- Light
- Fitness
- Comfort
- Mind

Within each of these categories, there are detailed assessment criteria numbering 100 overall; under "Air," for example, a smoking ban, pesticide control, and cleaning of surfaces; under fitness, for example, active furnishings (Chapter 4) and structured fitness opportunities. The reader is referred to source documents for further details, and some examples of healthy building characteristics are given below:

- Is there a public transport stop within 800 m of the building entrance?
- Are showers and lockers available for regular occupants (some employees may like to cycle or run to work)?
- Is there a secure and covered bicycle parking area within 30 m of a building entrance for a minimum of 5% of regular occupants?
- Is there an accessible stairwell as visible or more visible than elevators or escalators from the main entrance?
- Is access to active workstations provided?
- There are views of nature from a majority of workspaces.
- Are there common break areas accessible to all regular occupants so that people can eat away from their desks?
- Is there a dedicated exercise room accessible, free of charge, to all regular occupants?
- Are healthy food options encouraged using price incentives?

Future workplaces will be healthier and accessible using a variety of transportation modes, enabling the working day to be more physically active and to combine work with exercise and recreation.

The Future of Transport

The Technology readiness level of automated vehicles (AVs) is high (7–8) in some applications. There are five levels of automation (Chapter 14). Level 1 automation (e.g., lane assist) is already in use and is a legal requirement for new vehicles in some countries. Level 2 and 3 automation is available, but the driver is legally responsible for safety. Automated vehicles (Level 4) are in use in "geofenced" regions where the road infrastructure has been modified for their use (e.g., traffic lights that communicate directly with the AV). AV taxis are being trialed in San Francisco, USA, at the time of writing. Minibuses traveling defined routes are another example where AV TRLs are high. Integration into the transport network requires further development, however.

Despite concerns about the safety of AVs, it should be remembered that manually controlled vehicles are not safe, given the high rate of accidents and injuries on the roads. AVs are likely to be very much safer. One of the main challenges is to integrate them into the transport network, not least by installing sufficient charging points. A recent report (KPMG, 2020) outlines some of the braking factors on AV integration:

Safety: Despite the increased safety, the public has a very low tolerance of accidents involving AVs.

Privacy: One of the key advantages of a transport system consisting almost entirely of AVs is that they can be connected to the local authorities' traffic management systems. With real-time picture of all traffic, the speed and routes of vehicles can be altered to avoid congestion and optimize overall road usage. However, in some cultures, privacy is highly valued, and this information would need to be anonymized and, in some countries, comply with data protection legislation (such as the General Data Protection Regulations in the EU and UK).

Digital Infrastructure: Although Level 5 AVs would not require additional infrastructure (such as external sensors or smart traffic lights), geofencing is required for lower levels of AV automation to work autonomously. There are other concerns about efficiency. Because AVs default to safety, their widespread use might cause congestion or increase travel times.

Cross-Border Travel: Before AVs can fully replace existing vehicles, agreements between countries will be needed about many of the issues described above. Minimum standards for AV design and use will have to be accepted in different jurisdictions. At the time of writing, countries still vary in their readiness for AV integration.

As can be seen, AVs may be the future of transport in the "Smart Cities" of the future, but safety is only one of the concerns facing the widespread integration of AVs.

Old Wine in New Bottles: Green Industries and Cobots

Despite the general trend for a reduction in physically demanding jobs, new, physically demanding jobs are appearing in sectors such as waste recycling and offshore wind generation (where employees work at height exposed to the elements). Engkvist et al. (2011) reported that workers in Swedish waste recycling centers had 3–5 times more accidents than the general worker population. These accidents occurred most often during the manual handling of waste. The main risk factor for musculoskeletal disorders was heavy lifting. The hand, shoulder/arm, and leg were the most frequently injured body parts. Handling of objects, over-exertion, or falling/flying objects were the most frequent causes of reported accidents. Heavy and/or awkward objects such as fridges, freezers, and televisions were often being handled when the injury or accident occurred. Recommendations to improve safety were made, including provision of manual handling aids and lifting equipment, better gloves, and better housekeeping (to remove trip hazards from floors). Clearly, the demand for traditional ergonomic interventions is likely to continue in some of the newly emerging industries (see Chapters 2, 4–6).

Collaborative Robots and Occupational Injuries: In 2017, collaborative robots represented a mere 2.8% of new industrial robot installations, a figure that climbed to 9.9% by 2022 (Maslej et al., 2024).

Cobots work alongside human operators to complete tasks collaboratively. They have the potential to perform the more hazardous parts of a task or to perform tasks requiring high forces, endurance, or at faster pace. Higher productivity is one of the reasons companies invest in robotics. Burtch et al. (2024) found mixed results for the reduction in injuries when cobots were introduced with some studies showing a reduction in injuries and others an increase. From the perspective of HFE, the benefits or disbenefits of cobots will depend on how they are integrated into the job, how the pace of work and exposure to stressors is changed and the effects on the role of the human collaborator. Job satisfaction may decrease if the repetitiveness of tasks increases or if the employee feels that the cobot is a threat to job security. Burtch et al. analyzed data on injuries from 26 Amazon Fulfillment Centers (FCs or warehouses) to test the hypothesis that FCs with cobots would have a lower rate of severe injury than legacy centers (without cobots). They also tested the hypothesis that centers with cobots would have a higher rate of non-severe injury than legacy centers. The analysis consisted of data from OSHA (2016–2020) on work-related injuries. Cobotic centers used autonomous robots to complete large amounts of daily tasks (e.g., stock picking) alongside human operators, who performed the parts of the job that were difficult to automate. Sixty percent of the FCs were robotic. The most common injuries were strains or sprains (about 75%), followed by bruises (about 10%), with cuts, scratches, concussion, fractures, and crushing making up the remainder. Employee perceptions of job demands were also obtained, with 63% reporting that stock picking was safer at robotic FCs, although perceptions of the work itself as being more interesting were mixed. Employees reported walking less at robotic FCs, but also found the job more monotonous. Severe injuries were categorized as those that resulted in lost workdays. Less severe injuries were those that resulted in employees being temporarily reassigned to other duties or placed on restricted duties without lost days due to injury.

Several analyses of the data were carried out to enable injury rates at the robotic and legacy FCs to be compared, controlling for headcounts and hours worked. Overall, there was a 40% reduction in lost time injuries in robotic FCs and a 70% increase in injuries involving restriction of duties or reassignment. There was evidence that injuries in robotic FCs increased during periods of high operational load, probably associated with increased pace of work, repetitive motion, and stress. These findings held when the total number of employee hours worked in the FCs was controlled for. The findings also held in a separate analysis in which subsequent ("lagged") injury rates were analyzed with respect to initial injury rates before the introduction of cobots.

It seems that the severe (lost days) injuries are the result of accidents, and therefore, the introduction of cobots makes warehouse work safer. The less severe injuries are the result of normal work exposures, and are therefore amenable to task or job redesign. Burthch et al. suggest that the risk of injury due to the increase in work pace brought about by cobot working can be mitigated by HFE interventions such as task rotation and job enlargement (such as training to operate machinery such as forklift trucks). In the author's view, cobot working is highly amenable to improvement using traditional HFE methods (as described in Chapters 3, 5, and 6).

HFE and the Fourth Industrial Revolution

Reiman et al. (2021) reviewed 37 articles dealing with HFE and the Fourth Industrial Revolution. Key challenges identified included:

- *Technological:* New technological systems bring new safety challenges in the technology itself and the work environments created to house it, including the usability of interfaces and the interpretation of more complex data (including "big data" Chapter 16). Demand for upskilling will increase.
- *Organizational:* New organizational skills will be needed to manage highly automated systems linked to the Internet and big data.
- *Personal:* Tasks will become more complex, increasing the demands on personal, cognitive, and organizational skills. Demands on personal autonomy and self-management skills will increase in the face of skills shortages and the ageing workforce.

Human-Centered Organizations

We saw in the opening chapter that one of the requirements of any new system is to optimize whole life costs through the appropriate use of people to use human resources in a sustainable way.

ISO 27500 is a new standard on the Human-Centered Organization. Human-Centered Organizations have the following characteristics:

- Capitalize on individual differences as an organizational strength
- Make usability and accessibility strategic business objectives
- Adopt a total system approach
- Ensure health, safety, and well-being are business priorities
- Value employees and create a meaningful work environment
- Be open and trustworthy
- Act in socially responsible ways

Good Work Design

Much of the traditional focus of HFE has been on the control of hazards and the reduction of risks. However, this does not make work either interesting or fun. "Good Work Design" is an approach to job design that aims to achieve the best out of employees by actively engaging them in the process to maximize motivation and innovation (Crawford et al., 2023). We have already seen ("Work as done" in Chapter 15) that employee initiative is often the solution to faulty procedures or organizational inefficiency and variance is best controlled as close to the source as possible (Table P.5, Prologue).

Good work design has three phases (Figure E.2).

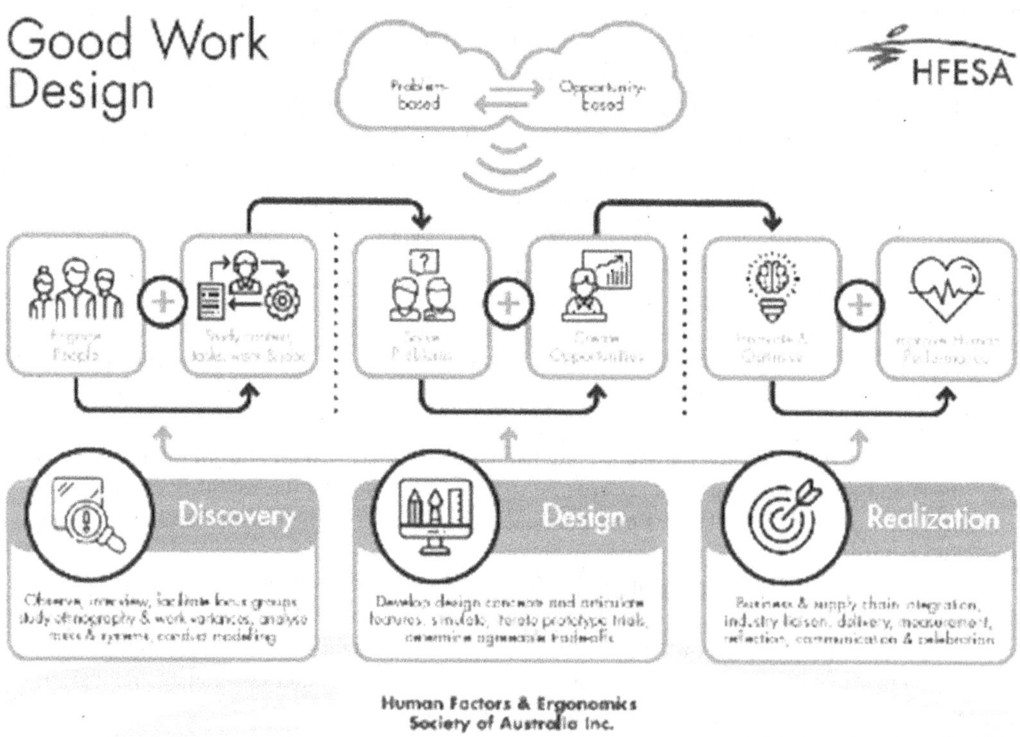

FIGURE E.2 Phases of Good Work Design (HFESA (2020) with permission of the Human Factors and Ergonomics Society of Australia.)

Essentially, a participatory approach (Chapter 1) is needed to understand the work as done, the hazards, constraints, and the context of operation. When the capability need and the system requirements are understood, the design process can proceed, including the design of jobs. The final phase requires integration and post-implementation evaluation. Good work design is not a method. Its features are outlined in Table E.5.

New technologies don't design jobs—jobs have to be designed.

TABLE E.5
Key Features of Good Work Design (Crawford et al., 2023)

	Good Work Design
Is	• A framework for undertaking workplace (re)design
	• Human-centered
	• Is propped by human factors and ergonomics approaches
	• Regenerative because it builds design capability (skills and resources) and capacity (ability to host and support design projects) throughout the organization and the supply chain
Appreciates	• The evolutionary and ecological aspects of variable human performance
	• The "just right" balance of "joy work" and "work-work" design-in-use or the applications and spontaneous adaptations of work design in business
	• The role of humans in highly automated systems
	• That the design process is as important as the outcomes
	• The positive emotional experiences associated with creation through design versus the fear associated with needs to contain all that can go wrong
Needs	• Facilitation by a work design strategist
	• Support by subject matter experts
	• Collaboration with "conventional" designers
	• Resilience engineering strategies to inform and test work designs
	• Access to successful "workarounds" or "near rights" (versus "near misses")
	• To leverage on design-ready changes
	• Systems of transparent and defensible decision-making in work governance
Does	• Reconciles the varieties of human work
	• Designs for diversity
	• More than "consider business needs"; it realizes business objectives in a competitive, pioneering, and sustainable manner, while maintaining and promoting worker health and performance
	• Involves discovery, (iterative) design, and realization of good work
	• Implements effective change management practice to test ideas, manage iterative trials, and launch progressively larger and more ambitious design campaigns throughout an organization or cross-industry
	• Addresses safety-critical, material, unwanted, high-consequence, and other types of unfavorable events (what is not wanted) but also focuses on design for what is wanted for human performance across a spectrum of needs
	• Prospects new design opportunities on an ongoing basis
	• Tolerates a degree of fallibility to promote innovations
	• Focuses on storytelling to promote shared learning and tacit knowledge in a business
	• Enables cross-industry learnings and continual insights about a changing world of work
	• Celebrates design successes in a resounding way
Creates	• A visibly human-centered organization
	• Unified business strategies among departments
	• Positive experiences of work, the effects of which extend beyond work
	• A sense of coherence, meaning, and manageability to work a culture of innovation
	• Design that either works well or stands out because it is magnificent
	• Tacit knowledge about design to enable self-efficacy while building confidence to ideate, experiment, and innovate design-related change in supported or structured ways

RESEARCH DIRECTIONS

We began this book with a discussion of human factors and its place in systems engineering—Human Systems Integration. We can end by summarizing some of the main challenges for HSI and for SE itself. Karwowski et al. (2025) have identified the following challenges.

AI: Integrating HFE principles into the design and operation of these systems in view of their limitations and characteristics outlined in this chapter (e.g., bias and explainability).

Cyber-Security: Future challenges extend beyond personal security and data protection, as outlined in Chapter 14, to the security of cyber-physical systems such as automated vehicles, specifying the role of the user.

Sustainable Energy: Energy supply networks have become dependent on a variety of traditional and novel technologies while, at the same time, have become disaggregated. How humans interact with this infrastructure and with the natural environment is a challenge for HFE.

Health: Wearable devices are becoming increasingly used to monitor a variety of markers in both healthy users and patients. Data privacy issues aside, there are challenges for integrating these data into the delivery of interventions and providing feedback to users that prompts appropriate behaviors or behavior modifications.

This a just a summary of the many challenges identified by Karwowski et al.

Digital Natives: Rapid and widespread technological change introduces many new challenges and research questions, not all of which are foreseeable. One of the many future research questions for HFE is the intergenerational effects of technological change and how best to design the jobs of the future for future generations. Generation Z (born after 1999) are known as "digital natives." They differ from preceding generations because they grew up "online," augmented by the Internet, and have the following characteristics (Hernandez-de-Menendez et al., 2020):

	Before Gen *Z*	Gen *Z*
Internet access	A privilege	A human right
Influence	Purchased	Earned
IP[a]/patents	Valued creators	Barriers
Failure	Avoided	Embraced
Gaming	Non-value-added play	Foundation for engagement
Uncertainty	Prepared	Predicted
Retirement	A destination	A journey
Connectivity	A luxury	A necessity

[a] Intellectual property.

There is debate about the effects of exposure to the Internet and to AI-infused technology, particularly on the thinking skills of digital natives. Gerlich (2025) investigated the use of AI tools on critical thinking skills (self-reported). Critical thinking was defined as the ability to analyze, evaluate, and synthesize information to make reasoned decisions, thought to be affected by "cognitive offloading" when tools are used (using tools to reduce the load on working memory). A negative correlation between AI tool use and critical thinking ability was found, mediated by cognitive offloading. The effect was greater in younger subjects because they made more use of such tools. No doubt, future studies will explore these matters in more detail.

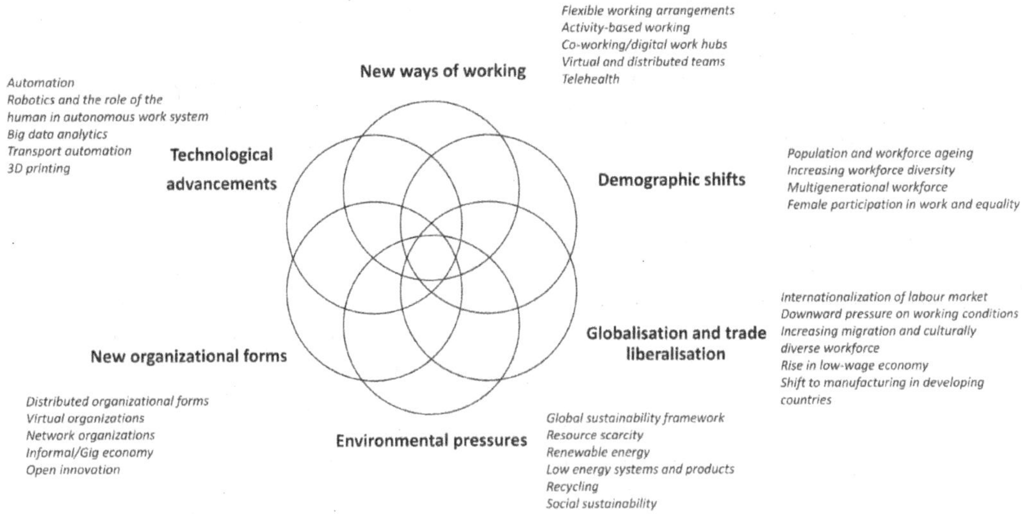

FIGURE E.3 Mega trends and the future of work. (From Bentley et al. (2020) with permission, www.tandfonline.com.)

SUMMARY

Well, yesterday's future just arrived! Figure E.3 summarizes some of the developments determining the future of work.

Let's leave the last words to ChatGPT.

Author: What's the future of work?

ChatGPT: The future of work is evolving rapidly, influenced by technology, societal changes, and shifting global trends. Some key aspects that are shaping this future include:

1. *Remote and Hybrid Work:* The COVID-19 pandemic accelerated the shift toward remote work, and many organizations are adopting hybrid models, where employees split their time between working from home and in the office. This trend is likely to continue, driven by advancements in collaboration tools and a growing preference for work-life balance.
2. *Automation and AI:* Automation, AI, and machine learning are transforming many industries. Tasks that were previously manual or repetitive are increasingly being automated, allowing workers to focus on higher-level creative, strategic, and interpersonal tasks. While automation will reduce the need for certain types of jobs, it will also create opportunities for new roles in AI management, data analysis, and tech development.
3. *Gig Economy and Freelancing:* More people are moving away from traditional 9-to-5 jobs in favor of freelance or gig work. Platforms like Uber, Upwork, and Fiverr enable people to find short-term, flexible work arrangements. The gig economy is expected to continue growing, with more people seeking autonomy over their schedules and career paths.
4. *Skill Development and Lifelong Learning:* As industries evolve, there will be a greater emphasis on continuous skill development. The rapid pace of technological advancements means workers will need to constantly adapt and learn new skills. This might involve reskilling or upskilling programs to stay competitive in the job market.
5. *Diversity and Inclusion:* The future of work will likely see a stronger emphasis on diversity, equity, and inclusion in the workplace. Companies are increasingly recognizing the

value of diverse perspectives and are striving to create more inclusive environments. This includes promoting equal pay, eliminating bias in hiring practices, and fostering workplaces where all employees feel valued.

6. *Collaboration and Digital Tools:* With remote and hybrid work models becoming more common, collaboration tools like Slack, Zoom, and Microsoft Teams will continue to play a crucial role in connecting teams. Additionally, virtual and augmented reality could become more integrated into work processes, allowing for more immersive and interactive meetings and collaborations.
7. *Workplace Flexibility and Mental Health:* There is growing awareness of the importance of mental health and well-being in the workplace. Employers are increasingly offering flexible schedules, mental health resources, and initiatives that support employees' work-life balance. This will likely become a bigger priority for organizations in the future.
8. *Global Talent Pool:* Remote work allows companies to tap into a global talent pool. Organizations can hire skilled professionals from different countries, which can help them access new ideas, perspectives, and markets. However, this also raises issues like managing cross-cultural teams and ensuring fair compensation practices across different regions.
9. *Sustainability:* The push for sustainable practices is becoming a major factor in shaping the future of work. Many businesses are focusing on sustainability in their operations, reducing carbon footprints, and finding innovative solutions to environmental challenges. This focus on sustainability will likely shape industries like energy, transportation, and manufacturing.
10. *Employee Empowerment:* There's an increasing trend toward empowering employees to take ownership of their roles and decisions. The hierarchical structure in many companies is being replaced with more collaborative and decentralized models. This shift is motivated by the desire for more innovation, creativity, and employee satisfaction.

In summary, the future of work is likely to be more flexible, tech-driven, and dynamic, with a focus on remote work, continuous learning, automation, and mental health. How businesses adapt to these changes will depend on their ability to embrace new technologies, create inclusive cultures, and empower their workforce.

NOTES

1 Annotators help train AI algorithms to recognize images, data, or images captured from cameras in automated vehicles.
2 https://www.uniklinik-duesseldorf.de/fileadmin/Fuer-Patienten-und-Besucher/Kliniken-Zentren-Institute/Institute/Institut_fuer_Medizinische_Soziologie/Forschung/PsychometricProperties.pdf.
3 https://embrace-autism.com/autism-spectrum-quotient/.
4 International Well Building Institute.

Appendix A
Probabilities Associated with Values of z in Normal Distribution

z	0.00	0.01	0.02	0.03	0.04	0.05	0.06	0.07	0.08	0.09
−0.0	0.5000	0.4960	0.4920	0.4880	0.4840	0.4801	0.4761	0.4721	0.4681	0.4641
−0.1	0.4602	0.4562	0.4522	0.4483	0.4443	0.4404	0.4364	0.4325	0.4286	0.4247
−0.2	0.4207	0.4168	0.4129	0.4090	0.4052	0.4013	0.3974	0.3936	0.3897	0.3859
−0.3	0.3821	0.3783	0.3745	0.3707	0.3669	0.3632	0.3594	0.3557	0.3520	0.3483
−0.4	0.3446	0.3409	0.3372	0.3336	0.3300	0.3264	0.3228	0.3192	0.3156	0.3121
−0.5	0.3085	0.3050	0.3015	0.2981	0.2946	0.2912	0.2877	0.2843	0.2810	0.2776
−0.6	0.2743	0.2709	0.2676	0.2643	0.2611	0.2578	0.2546	0.2514	0.2483	0.2451
−0.7	0.2420	0.2398	0.2358	0.2327	0.2296	0.2266	0.2236	0.2206	0.2177	0.2148
−0.8	0.2119	0.2090	0.2061	0.2033	0.2005	0.1977	0.1945	0.1922	0.1849	0.1867
−0.9	0.1841	0.1814	0.1788	0.1762	0.1736	0.1711	0.1658	0.1660	0.1635	0.1611
−1.0	0.1587	0.1562	0.1539	0.1515	0.1492	0.1469	0.1446	0.1423	0.1401	0.1379
−1.1	0.1357	0.1335	0.1314	0.1292	0.1271	0.1251	0.1230	0.1210	0.1190	0.1170
−1.2	0.1151	0.1131	0.1112	0.1093	0.1075	0.1056	0.1038	0.1020	0.1003	0.0985
−1.3	0.0968	0.0951	0.0934	0.0918	0.0901	0.0885	0.0869	0.0853	0.0838	0.0823
−1.4	0.0808	0.0793	0.0778	0.0764	0.0749	0.0735	0.0721	0.0708	0.0694	0.0681
−1.5	0.0668	0.0655	0.0643	0.0630	0.0618	0.0606	0.0594	0.0582	0.0571	0.0559
−1.6	0.0540	0.0537	0.0526	0.0516	0.0505	0.0495	0.0485	0.0475	0.0465	0.0455
−1.7	0.0446	0.0436	0.0427	0.0418	0.0409	0.0401	0.0392	0.0384	0.0374	0.0367
−1.8	0.0359	0.0251	0.0344	0.0336	0.0329	0.0322	0.0314	0.0307	0.0301	0.0294
−1.9	0.0287	0.0281	0.0274	0.0268	0.0262	0.0256	0.0250	0.0244	0.0239	0.0233
−2.0	0.0228	0.0222	0.0217	0.0212	0.0207	0.0202	0.0197	0.0192	0.0188	0.0183
−2.1	0.0179	0.0174	0.0170	0.0166	0.0162	0.0158	0.0154	0.0150	0.0146	0.0143
−2.2	0.0139	0.0136	0.0132	0.0129	0.0125	0.0122	0.0119	0.0116	0.0113	0.0110
−2.3	0.0107	0.0104	0.0102	0.0099	0.0096	0.0094	0.0091	0.0089	0.0087	0.0084
−2.4	0.0082	0.0080	0.0078	0.0075	0.0073	0.0071	0.0069	0.0068	0.0066	0.064
−2.5	0.0062	0.0060	0.0059	0.0057	0.0055	0.0054	0.0052	0.0051	0.0049	0.0048
−2.6	0.0047	0.0045	0.0044	0.0043	0.0041	0.0040	0.0039	0.0038	0.0037	0.0036
−2.7	0.0035	0.0034	0.0033	0.0032	0.0031	0.0030	0.0029	0.0028	0.0027	0.0026
−2.8	0.0026	0.0025	0.0024	0.0023	0.0023	0.0022	0.0021	0.0021	0.0020	0.0019
−2.9	0.0019	0.0018	0.0018	0.0017	0.0016	0.0016	0.0015	0.0015	0.0014	0.0014
−3.0	0.0013	0.0013	0.0013	0.0012	0.0012	0.0011	0.0011	0.0011	0.0010	0.0010
−3.1	0.0010	0.0009	0.0009	0.0009	0.0008	0.0008	0.0008	0.0008	0.0007	0.0007
−3.2	0.0007									
−3.3	0.0005									
−3.4	0.0003									

Note: The table shows the area under the normal curve to the left of z. For example, for $z = -1.0$, the area to the left of z is 0.1587. This corresponds roughly to the sixteenth percentile. The data for the left-hand portion of the normal distribution (i.e., values of z where x is less than or equal to the mean) are shown. For values greater than the mean, subtract the area to the left of z from 1. For example, for $z = +1$ (one standard deviation above the mean), the area to the left of $z = 1 - 0.1587 = 0.8413$. This corresponds to the eighty-fourth percentile. For $z = \pm 1.64$, the areas are 0.0505 and $1 - 0.0505 = 0.9495$. These correspond to the fifth and ninety-fifth percentiles.

Glossary

"A" scale: System for weighting sound energy to reduce high and low frequencies to approximate the frequency response of the human ear.

Abdomen: Anterior part of the torso from the diaphragm to the pelvis, including the intestines and the overlying muscles.

Abdominal circumference: Surface measure at the point of maximum protrusion of the abdomen, used as a measure of girth.

Abdominal depth: Sagittal plane measure from the back to the front at the level of the waist.

Abduction: Movement of a body part away from the midline of the body.

Abductor: Muscle that causes abduction of a body part when it contracts concentrically.

Ability: Capacity to perform a task to a required standard.

Acceleration: Rate of change of velocity with time. SI unit: m/s^2.

Accelerometer: Device for measuring acceleration, often used in studies of vibration exposure.

Acclimatization: Physiological adaptation to new environmental conditions.

Accommodation: Change in the refractive property of the eye to bring objects into focus on the retina.

Accuracy: Proximity of a measurement to a true or known value.

Acoustic: Referring to sound.

Acoustics: The study or measurement of sound.

Acromial height: Vertical distance from the floor or seat surface to the acromion.

Actin: Protein found in strands in muscle tissue, involved in muscle contraction.

Activity sampling: Statistical method used to estimate the amount of time a worker spends performing an activity based on the binomial distribution.

Acute: Condition of short onset and duration (less than 3 months in medical conditions such as back pain).

Adduction: Movement of a body part toward the midline of the body.

Adductor: Muscle that causes adduction of a body part when it contracts concentrically.

Adipose tissue: Tissue consisting mostly of fat.

Administrative control: Procedures and methods set up by the employer that significantly reduce exposure to risk factors by altering the way in which work is performed. Examples include employee rotation, job task enlargement, and adjustment of work pace.

Adrenalin: Hormone secreted to facilitate the "fight or flight" response.

Aerobic capacity: Capability of the body to take up and utilize oxygen to provide energy for the performance of an activity.

Aerobic metabolism: Metabolic process that requires the utilization of oxygen as one of the reactants.

ALARP: Acronym used in systems design—"As Low As Reasonably Practicable." Normally relates to risk.

Allocation of function: The decision-making process whereby the responsibility of system functions is partitioned between operators and machines.

Anaerobic metabolism: Metabolic process that takes place when there is insufficient oxygen for aerobic metabolism.

Angular motion: Rotational motion measured as a change of angle around an axis over time (deg/s).

Annulus: Cartilaginous structure that forms part of the intervertebral disc.

ANSI: American National Standards Institute. A private, nonprofit membership organization that coordinates voluntary standards activities. ANSI assists standard developers and standard users from the private sector and government to reach agreement on the need to standards and establish priorities.

Anthropometry: The study of the dimensions and shape of the human body.

Anti-gravity muscles: The postural muscles that maintain upright posture against gravity — soleus, quadriceps, iliopsoas, gluteals, and back and neck extensors.

Anxiety: A negative emotional state triggered by the perception of dangerous or challenging events.

Apoptosis: Programmed cell death to eliminate unwanted cells.

Aptitude: Ability to acquire skill or knowledge.

Arithmetic mean: Sum of a set of observations divided by the number of observations.

Arousal: Readiness to respond to internally or externally generated stimuli.

Arthritis: General term for inflammation or degeneration of the joints. Sometimes used as a euphemism for musculoskeletal pain of unknown origin.

Astigmatism: Unequal curvature of the optical apparatus of the eye, resulting in variations in focal length in different planes.

Attitude: Projection of core beliefs onto a simplified view of the world.

Audiogram: Graphical plot of an individual's ability to detect sounds of different frequencies (detection thresholds).

BMI: *See* **body mass index.**

Bandwidth: Range of frequencies that can be processed by a system.

Basal metabolic rate: Rate of energy expenditure when an individual is resting.

Basilar membrane: Part of the inner ear that resonates when exposed to sound energy transmitted from the middle ear via the oval window.

Biceps brachii: Large muscle on the anterior part of the upper arm.

Biomechanics: Field of science in which principles of mechanics are applied to the study of living things.

Body mass index: Mass (kg) divided by stature (height in meters) squared (<25 normal, 25–30 overweight, 30+ obese).

COG: *See* **center of gravity.**

Candela: SI unit of luminous intensity of a light source.

Carbohydrate: Chemical substance composed of carbon, hydrogen, and oxygen atoms, consisting of nutrients (starches and sugars) and non-nutrients (cellulose).

Carpal tunnel: Passage in the wrist between the carpal bones through which pass the tendons, nerves, and blood vessels that supply the hand.

Carpal tunnel syndrome: Set of symptoms including numbness and tingling in the fingers and weakness of the hand, caused by compression of the median nerve.

Catecholamine: A class of substances that serve as neurotransmitters and/or hormones in the body.

Center of gravity: The point at which the weight of a body can be considered to act.

Cervical spine: The seven uppermost vertebrae in the human spine.

Chromatic aberration: Error in the resolution of an optical system caused by unequal refraction of light of different wavelengths, resulting in blur.

Ciliary muscle: Muscle surrounding the equator of the lens of the eye that contributes to accommodation.

Clo Unit of the thermal insulation provided by clothing: One clo is the amount of insulation needed to ensure thermal comfort for a person resting in a room at 21°C with 50% relative humidity.

Cinderella effect: The hypothesis that chronic muscle pain is due to overuse of Type 1 muscle fibers at a low level.

Coefficient of friction: Ratio of the force acting parallel to a contact surface (boundary) that resists motion of one body on another, to the normal (perpendicular) force acting on the contact surface. If the parallel force is large when the normal force is small, the coefficient of friction is high, and thus the resistance to motion is also high.

Cognitive processes: Higher mental functions, including problem-solving and reasoning.

Cohort study: A study in which a group is followed over a period of time to determine how many experience an outcome (e.g., a disease, death) and the variables associated with that outcome.

Concentric muscle contraction: Muscle contraction in which the muscle shortens as it contracts (usually to perform work).

Cone: Light-sensitive cell found in the retina of the eye, responsible for high visual acuity and color vision.

Confidence interval: Range of values within which a value will fall with a specified probability.

Construct: An aggregation of behaviors or psychological processes.

Cooper–Harper scale: A scale used for the subjective assessment of task difficulty.

Core temperature: Temperature of the deep body tissues and the brain.

Coronal plane: Vertical plane that divides the body into anterior and posterior sections.

Cortisol: The main stress hormone. It increases blood glucose. Chronic excess cortisol leads to weight gain, diabetes, and muscle weakness.

Cumulative trauma disorders (CTDs): Term used for injuries that occur over a period due to repeated trauma or exposure to a specific body part, such as the back, hand, wrist, and forearm. Muscles and joints are stressed, tendons become inflamed, nerves are pinched, or blood flow is restricted. Common occupational disorders in this class include carpal tunnel syndrome, epicondylitis (tennis elbow), tendinitis, tenosynovitis, synovitis, stenosing tenosynovitis of the finger, de Quervain's syndrome, and low back pain.

Cycle time: Time required to complete all the elements of a specified task in the correct order.

Dark adaptation: Change in the retina of the eye that increases sensitivity to light.

Density: The mass of a specified volume of matter. The SI unit is kilograms per cubic meter.

Dew point: Temperature at which condensation begins to appear when air at relative humidity above zero is cooled.

Diurnal: Pertaining to daytime.

Dorsal: Relating to the back or upper part of something (e.g., dorsal spine — the part of the spine forming the back of the body).

Dose: The amount of something to which a person is exposed.

Double-blind: an experimental design involving a control condition and an experimental condition in which neither the experimenter nor the participant knows the nature of the condition at the time of exposure.

Dry bulb temperature: The temperature of the constituent gases of the air.

Dystonia: A disorder involving abnormal muscle tone.

Earcon: an acoustic symbol used to convey a specific meaning.

Eccentric muscle contraction: Muscle contraction in which the muscle lengthens as it contracts (usually as an agonist or to perform a "braking" function).

Ectomorph: Lean and long, with difficulty building muscle.

Effective temperature: Temperature in a 50% relative humidity environment that results in the same heat loss as the actual environmental temperature.

Electromyography: Measurement and display of the electrical activity of muscles.

Emmetropia: Normal vision.

End plate: Cartilaginous layer separating a vertebral body from an intervertebral disc.

Endomorph: Big, with high body fat — often pear-shaped — with a high tendency to store body fat.

Endurance: Ability to maintain performance over time.

Energy: The capacity for work.

Engineering control: Physical changes to jobs that control exposure to risk. Engineering controls act on the source of the hazard and reduce employee exposure without relying on the employee to take self-protective action. Examples include changing the handle angle of a tool, using a lighter-weight part, or providing a chair with adjustability.

Epinephrine: *See* **adrenaline.**
Equilibrium: State of balance.
Equivalent form: Two or more forms of a test that are designed to differ in terms of their surface features or appearance but not in terms of their content.
Equivalent sound level: Constant sound level that, over a specified period, has the same acoustic energy as a varying source. Used for assessing the effects of exposure to time-varying noise sources.
Ergonomic program: A systematic process for anticipating, identifying, analyzing, and controlling ergonomic risk factors.
Ethnicity: Group identity based on cultural and linguistic ties.
Etiology: The study of the causes of diseases.
Evaporative heat loss: Heat loss due to the evaporation of fluid from a surface. In humans, evaporative heat loss occurs due to the evaporation of sweat from the skin.
Expectation: Mental state in which an individual anticipates a particular outcome under specified circumstances.
Experimenter effect: Effect that confounds the results of an experiment due to unexpected or uncontrolled influence by the experimenter(s).
Extension: Straightening of a limb around a joint such that the angle between the body parts increases.
Facet joint: *See* **zygapophyseal joints.**
Fascia: A sheath of stringy connective tissue that surrounds every part of the body. It provides support to muscles, tendons, ligaments, tissues, organs, nerves, joints, and bones
Fatigue: (i) Degradation in readiness to continue an activity after prolonged exposure to it, often accompanied by subjective symptoms of discomfort and tiredness. (ii). Failure of a structure or sample under load.
Fault tolerance: Extent to which a system continues to function if a fault develops.
Femur: Thigh bone.
Fibula: Small bone in the lower leg.
Field of view: The visual area accessible from a defined position.
First class lever: A lever in which the opposing forces are located on either side of the fulcrum.
Flexion: Movement of body parts around one or more joints such that the angle between the parts decreases.
Focus group: A group of users assembled to discuss a topic in relation to a set of predefined issues.
Force: The action of one body on another, either directly (as when pulling a lever) or indirectly (as in the gravitational attraction between the Moon and the Earth).
Fovea: The central part of the retina of the eye with the greatest visual acuity.
Frankfort Plane: A plane used in craniometry that is determined by the highest point on the upper margin of the opening of each external auditory canal and the lowest point on the lower margin of the left orbit. It is used to orient a human skull or head, usually so that the plane is horizontal. Also called **eye–ear plane, Frankfort horizontal, or Frankfort plane.**
Friction: A tangential force between two bodies in contact that acts to resist motion between them. If the two bodies are stationary, the force is termed static friction. If the bodies are moving in relation to each other, the force is termed kinetic friction. Friction is independent of contact area but proportional to the normal force (the force acting perpendicular to the two surfaces). The units of force are newtons. The coefficient of static friction is the ratio of the frictional force to the normal force.
Function: an activity that must be performed for a system to attain a goal.
G: The acceleration of a body under the influence of the Earth's gravity. 1 g = 9.81 m/s2.
Gait: Style of locomotion.
Galvanic skin response (GSR): A change in the electrical resistance of the skin due to a change in ion concentrations at the cellular level.

Gender: The social manifestation of maleness and femaleness rather than biological sex

Gluteal muscles: A large group of muscles over the posterior portion of the pelvis whose main function is to extend the hip joint and prevent the trunk from "jackknifing" forward over the legs.

Glycogen: The form in which glucose is stored in the liver and muscles.

Golgi tendon organ: A stretch receptor found in the musculotendinous unit that provides feedback to the central nervous system on muscle tension.

Goniometer: A device for measuring joint angles.

Graphical user interface (GUI): A user–computer interface based on visual and auditory symbols rather than words, supporting interaction via gestures such as pointing.

Hamstring muscles: A group of muscles in the posterior thigh that cross the hip and knee joints. They are involved in extension of the hip joint and flexion of the knee joint.

Haptic sense: The sense of touch or pressure.

Hawthorne effect: No precise definition exists or is universally accepted, but it roughly equates to unexpected subject or participant reactivity.

Hazard ratio: An estimate of the relative risk of an event. Ratios less than one indicate reduced risk; ratios greater than one indicate increased risk.

Hazop analysis: A safety procedure for systematically identifying hazardous operations in a process.

Head-up display: A display projected onto a transparent surface through which the operator looks while carrying out a task, typically within the straight-ahead line of sight.

Healthy worker effect: Occurs when workers in a particular occupation are healthier or more robust than the general population due to the operation of nonrandom selection processes.

Hearing conservation programs: A set of activities aimed at preventing hearing loss among those working in noisy environments. Typically includes monitoring, assessment, treatment, and prevention.

Heat acclimatization: Physiological adaptation to living and working in hot conditions that takes approximately two weeks and involves increased capacity to sweat, greater sweat volume, and a lowering of the skin temperature threshold for the onset of sweating.

Heat exhaustion: A syndrome including muscle weakness, nausea, and headache following dehydration when exposed to heat stress.

Heat stroke: A sudden rise in body temperature following failure of the body's thermoregulatory system.

Homeostasis: The process by which physiological variables—such as heart rate and body temperature—are allowed to fluctuate within narrow limits despite widely varying external or internal demands. Note that the limits themselves may vary depending on the demands.

Hue: The color perceived as a result of the interaction between the wavelength of light reaching the retina from the object and the wavelengths from the rest of the visual field.

Human factors: A term sometimes considered synonymous with ergonomics. Often used more broadly in industry to include a wider range of psychological and personnel management issues.

Hypothesis: A proposition testable by experiment. (From the ancient Greek meaning "foundation").

Icon: A nonverbal representation of a system object.

Iliac crest: The lateral rim of the ilium of the pelvis.

Illuminance: The light falling on a surface. (SI unit: lux).

Incandescent lamp: A light source in which light is generated by passing an electric current through a wire within a sealed glass bulb.

Incidence: The number of new cases over time. (Incidence rate refers to the number of new cases over a specified time period).

Inguinal hernia: A condition in which a portion of the bowel or fatty tissue protrudes through a weak spot in the abdominal wall.

Insertion: The distal attachment of a muscle.

Integration plan: A plan outlining the HFE activities needed at each point in the development process and how they are to be integrated into the overall engineering, assurance, and project management process.

Integrate: To find the area under a curve.

Intensity: A measure of the strength of a stimulus.

Interface: A common boundary between two parts of a system.

Internal consistency: The positive interrelationship between a set of stimuli, exposures, or items in an experiment, measuring device, or questionnaire.

Intervertebral disk: A fibrocartilaginous structure that separates the vertebral bodies. it consists of a central nucleus pulposus, which is largely fluid, and is surrounded by an annulus fibrosus made of layered cartilage.

Intraclass correlation: A statistic that describes the consistency of repeated measures of the same quantity values between 0.5 and 0.75 indicate moderate reliability, values between 0.75 and 0.9 indicate good reliability, and values greater than 0.90 indicate excellent reliability.

Inverse square law: A principle stating that the intensity of a stimulus varies inversely as the square of the distance from its source.

Iris: The colored area around the pupil of the eye.

Ischemia: Lack of blood supply to a tissue.

ISO: International Organization for Standardization.

Isometric muscle contraction: A physiological change in the muscle that causes it to exert tension without changing its length, such as when holding a weight in the hand to keep it at a fixed height above a surface without moving one's body.

Isotonic muscle contraction: The lengthening or shortening of the muscle while exerting constant tension.

Iterate: Repetition of an operation to bring the current system state closer to a desired state.

Jerk: The rate of change of acceleration.

Job enlargement: Increasing the complexity of a job by introducing more varied tasks with the aim of increasing satisfaction.

Job enrichment: Increasing the complexity of a job by providing more responsibility, autonomy, and control.

Job rotation: The assignment and periodic reassignment of one or more workers to different tasks throughout the work shift.

Joint: The region of articulation between two bones.

Joint capsule: The connective tissue that surrounds a joint.

Joint range of motion: The angle in a given anatomical plane through which a joint can be moved without encountering resistance due to the stretching of tendons or ligaments.

Joystick: A control lever with at least two degrees of freedom.

KPI: Key performance indicator of system function.

Knowledge of results: Feedback given to an employee about some aspect of his or her work performance.

Kyphosis: A convex curvature of any part of the spine, typically refering to the natural curvature of the thoracic spine.

Labor turnover: A record of the number of employees leaving and joining a workplace within a given time period.

Lamp: A source of light.

Lateral: Pertaining to the side of the body.

Length–tension relationship: The relationship between the tension a muscle can exert and its length when the tension is exerted. Tension is normally maximal at a muscle's optional physiological length and decreases as the muscle shortens or is stretched beyond its normal physiological limits.

Lever arm: The distance between the point of application of a force and the fulcrum in a lever system.

Light meter: A device used to measure the amount of light falling onto a surface—the illuminance.

Likert scale: A 5-point scale used in questionnaire and rating scale design to measure the level of agreement or disagreement with a proposition.

Link analysis: The identification of the spatial sequence of interactions between a worker, his or her workplace, and any tools or equipment when carrying out a specific task.

Long-term memory: A memory store in which remembered items remain indefinitely below the level of consciousness. Items are accessible only through a conscious retrieval strategy or an external stimulus.

Longitudinal study: A research paradigm in which data are recorded sequentially from the same individuals over time, allowing analysis of changes or trends within subjects and across time periods.

Lordosis: A concave curvature of any part of the spine. it normally refers to the natural curvature of the lumbar and thoracic spines.

Loudness: The subjective perception of the intensity of a sound source.

Lumbar lordosis: The concave curve of the lumbar spine—the lowest portion of the spine above the pelvis, arising from the angle of the sacrum and consisting of five vertebrae.

Lumbar spine: The region of the spine between the sacrum and the thoracic spine, composed of five lumbar vertebrae.

Lumen: The luminous flux passing through one steradian of solid angle from a light source with an intensity of one candela.

Luminaire: A lighting fixture consisting of the lamp, the cover, glare shield or lampshade, and the fixtures for installation and use.

Luminance: Luminous flux per unit solid angle.

Luminous efficacy: The amount of light produced by a source per unit of electrical power, in watts, consumed.

Luminous intensity: The power of a light source to emit light.

Lux: The SI unit of illumination.

Manual material handling: Lifting, carrying, and moving materials without mechanical aide. Mass the internal property of a body (SI unit kilogram), where inertia refers to resistance to change of state of motion.

Maximum oxygen uptake: Limit of the capacity of the body to take up and utilize oxygen during the performance of an activity. Point at which no further increases in oxygen uptake are observed despite increase in work rate (abbreviation, VO2 max).

Maximum voluntary contraction: Greatest voluntary instantaneous effort of a muscle or group of muscles when applied to a task. Sometimes used to estimate endurance times for sub-maximal exertions.

Mean: Average value calculated by summing every individual value and dividing the result by the number of values. Used to estimate the central tendency of values in a population.

Mechanical advantage: Ratio of force applied to a mechanical device to the force produced. Normally used when describing the mechanical properties of levers and pulleys.

Median: The value in a set of scores that has the same number of greater and lesser scores around it.

Median nerve: Nerve innervating the forearm and the thumb, arising from the brachial plexus.

Melatonin: A hormone involved in the sleep/wake cycle. Levels increase at night.

Mental workload: Mental effort required to perform a task.

Mesomorph: Muscular and well-built, with a high metabolism and responsive muscle cells.

Mesopic vision: The process of seeing in intermediate levels of illumination that is achieved by the activation of both rods and cones.

Methodology: The scientific study of methods of investigation or a set, or family, of related methods used in a given branch of science.

Methods engineering: The systematic study of ways of working when carrying out a specified task. Normally used to enable improved ways of working to be arrived at.

Moment: The turning effect produced by a force. It is the product of the force acting at a point and the distance of the point from the axis of rotation. Units are newton per square meter.

Moment of inertia: In rotational mechanics, the equivalent of mass—resistance to a change in the state of rotation. The moment of inertia I is the product of the mass and its distance from the axis of rotation (units are kilogram per square meter).

Monochrome: Use of one color only in the design of a display.

Monotonic increase: Increase or decrease in a dependent variable through all the measured range of an independent variable to which it is related.

Monotony: Boredom.

Muscle tone: Baseline level of muscle "tightness."

Muscular fatigue: Reduction in tension generated by a muscle during the performance of a task, usually accompanied by an accumulation of lactic acid in the muscle and the sensation of pain.

Musculoskeletal disorders (MSDs): Injuries and disorders of the muscles, nerves, tendons, ligaments, joints, cartilage, and spinal discs; examples include carpal tunnel syndrome, rotator cuff tendonitis, and tension neck syndrome.

Myoglobin: Substance found in muscles that stores oxygen and supplies it to the muscle tissue to support aerobic metabolism.

Near vision: The process of seeing objects close to the eye, dependent on accommodation of the lens.

Neuropathy: Pathology of the nervous system.

Newton-meter: SI unit of torque. A force of 1 N acting at a distance of 1 m from a point of rotation produces a torque of 1 N m.

NIOSH: National Institute for Occupational Safety and Health in the United States. The institution that provides scientific data upon which OSHA makes recommendations.

Noise: 1. Unwanted sound. 2. Measurement error. Noise may be random, as when measuring a process that has random fluctuations caused by extraneous factors or it may be systematic, due, for example, to bias in the measuring system.

Normative: Manner of carrying out an activity according to a predefined method.

Obesity: Excess body fat often defined as a body mass index greater than 30 kg/m2.

Occipital cortex: Portion of the cerebral cortex of the brain responsible for processing nerve impulses from the visual system.

Occupational illness: Any abnormal condition or disorder, other than one resulting from an occupational injury caused by exposure to factors associated with employment. It includes acute and chronic illnesses or diseases, which may be caused by inhalation, absorption, ingestion, or direct contact. Tcategories of occupational illnesses are skin diseases and disorders, dust diseases of the lungs, respiratory conditions due to toxic agents, poisoning (systemic effects of toxic materials), disorders due to physical agents other than toxic materials, and disorders from repeated trauma.

Occupational injury: Any injury such as a cut, fracture, sprain, amputation, etc., that results from a work-related event or from a single instantaneous exposure in the work environment.

Octave: A range of sound frequencies in which the upper value of the range is double the lower value.

Odema: Accumulation of fluid in the body.

Optic nerve: Nerve from the retina of the eyes to the occipital cortex.

Orthogonal: Independent, not related.

Orthostatic hypotension: Orthostatic hypotension—also called postural hypotension—is a form of low blood pressure that happens when standing after sitting or lying down. Orthostatic hypotension can cause dizziness or light headedness and possibly fainting.

OSHA: Occupational Safety and Health Administration of the United States having an executive/enforcement function. The mission of the OSHA is to save lives, prevent injuries, and protect the health of America's workers.

OSHA 200 Log: An OSHA-required form for employers to record and classify occupational injuries and illnesses, and note the extent of each case.

Osteoporosis: Loss of bone strength due to loss of calcium; normally age-related.

Oval window: Part of the ear where the middle ear meets the inner ear and where vibration of the stapes is transmitted to the cochlea.

Ozone: Toxic gas in which the molecules are composed of three oxygen atoms.

PTF: Pre-corneal tear film. Water and lipid layer covering the eye.

Paced work: Job or task in which the rate of work is beyond the worker's control.

Parameter: Statistical feature of a population.

Patella: Kneecap.

Pathology: The study of disease.

Percentile: A measure used in statistics indicating the value of a variable below which a given percentage of observations in a group of observations falls. For example, 1550 mm might be the 20th percentile stature in a user group—20% of users would be shorter than 1550 mm in this population.

Perception: The means of awareness of external objects and events.

Permanent threshold shift: Partial or complete irreversible deafness at one or more frequencies or across a range of frequencies.

Photopic vision: Process of seeing under moderate to high levels of illumination that depends on the operation of cones and makes possible the perception of color.

Placebo: A noneffective treatment given to patients in a trial to control for extraneous effects related to the act of participation.

Plantar Fasciitis: Inflammation of the ligament that supports the arch of the foot (also known as "Policeman's Heel").

Popliteal fossa: Posterior part of the leg at the level of the knee joint.

Population: All members of a group or instances of an event having a specified feature.

Postural muscles: Muscles responsible for maintaining normal standing posture, chiefly the plantar flexors, iliopsoas, back and neck extensors, and the muscles around the temporomandibular joint.

Posture: The average orientation of the body over time.

Power: The rate at which work is done. An increase in power implies that a given amount of work can be done in a shorter time.

Presbycusis: Loss of hearing with age.

Prevalence: The number of cases of a condition in a given population at a specified time interval.

Primary task: The main task to which an operator must apply himself or herself.

Process control task: Task in which the operator is required to oversee a process with the intention of keeping specified process parameters within predetermined operating ranges.

Productivity: A measure of the output of a process in relation to the cost of the resources consumed.

Pronation: Rotation of a limb or part of a limb such that the ventral part of the limb is uppermost.

Proprioception: Process by which data on the mechanical position and state of the body are transmitted by sensory nerves to the central nervous system.

Psychometrics: The science of quantifying and measuring the psychological attributes of individuals.

Psychosocial factor: A construct reflecting the interplay of individual psychological factors in a social environment.

Pupil: Opening at the front of the eye through which light passes.

Quadriceps muscles: Muscles at the front of the thigh that extend the knee.

Quality control: Management method carried out to ensure that the output of a process meets a required standard.

Questionnaire design: The process of producing and choosing a set of questions for use in an instrument to obtain information about subjective phenomena that are not readily measurable by other means.

Questionnaire survey: A means of obtaining information about the attitudes, beliefs, or opinions of the members of a population.

Radian: Unit of planar angle such that there are 4π radians in a circle.

Radiculopathy: Pinching of the nerves at their roots in the spine

Radius: Lateral forearm bone.

Rating of perceived exertion: Subjective measure of physical workload used as a secondary measure in physiological research.

Recency effect: Bias in recall of information such that items presented recently are better recalled than items presented earlier.

Recovery time: Recovery time is the length of rest between exertions. Short work pauses can reduce discomfort. Inadequate rest periods between exertions can decrease performance. As the duration of the uninterrupted work increases, so does the amount of recovery time needed.

Refraction: Change in direction of a light wave as it passes from one medium to another.

Regression equation: Equation in which the value of a dependent variable can be estimated from the value(s) of one or more independent variables.

Rehearsal: Maintenance of one or more items in conscious awareness for immediate use or with the intention of improving their storage in memory for subsequent recall.

Relative humidity: Ratio of the water vapor pressure of the air at a given temperature to the saturated water vapor pressure of the air at that temperature.

Reliability: A characteristic of an instrument that responds in the same way when exposed to the same level of to-be-measured variable.

Repetition: Number of similar exertions performed during a task. A warehouse worker may lift three boxes per minute from the floor to a countertop; an assembly worker may make 20 units/h. Repetitive motion has been associated with injury and worker discomfort. Generally, the greater the number of repetitions, the greater the degree of risk. However, there is no specific repetition limit or threshold value (cycles/unit of time, movements/unit of time) associated with injury.

Response set: Form of bias in which an individual's responses to stimuli are influenced by extraneous factors (such as the design or layout of the stimulus array rather than the stimuli themselves).

Response time: Time interval between a stimulus or control action and the operator or system reaction.

Resultant force: The force obtained when each of the several forces acting upon a body is combined. The resultant force has both a magnitude and a direction.

Retina: Lining at the back of the eye containing light-sensitive cells.

Risk factor: Actions in the workplace, workplace conditions, or a combination thereof, that may cause or aggravate work-related musculoskeletal disorders; examples include forceful exertion, awkward postures, repetitive exertion, and environmental factors such as temperature.

Rotator cuff: Muscles surrounding and supporting the shoulder joint.

Sacrum: Bone at the base of the spine connected to the ilia of the pelvis.

Sagittal plane: Side-on view of a body or workstation.

Sample: Number of individuals selected from a population to participate in a survey or experiment.

Scalar value: A quantity that has only magnitude (e.g., mass).

Scale: System for weighting sound energy so as to reduce high and low frequencies to approximate the frequency response of the human ear.

Sciatic pain: Pain in the lower limb or foot caused by entrapment of the sciatic nerve in the lumbar spine.

Scotopic vision: Process of seeing in low illuminance dependent only on the operation of rods and with no perception of color.

Seat reference point: Intersection of the lines formed by the seat and backrest of a chair.

Secondary task: An additional task, often of lower priority than the primary, or main, task.

Secular trend: Change in a parameter of the general population of a country over time.

Segmental vibration (hand–arm vibration): Vibration applied to the hands/arms through a tool or piece of equipment. This can cause a reduction in blood flow to the hands/fingers (Raynaud's disease or vibration white finger). Also, it can interfere with sensory receptor feedback leading to increased handgrip force to hold the tool. Further, a strong association has been reported between carpal tunnel syndrome and segmental vibration.

Selective attention: Conscious effort to filter aspects of the task environment that are not germane to the primary task or activity.

Semicircular canals: Organs of the vestibular system of the inner ear.

Short-term memory: Process of retaining information in the mind for short periods.

Short-term memory store: Hypothetical store for retaining items for short periods and characterized by limited capacity of approximately seven items, rapid decay of nonrehearsed items, and susceptibility to interference by verbal stimuli.

Sign: Evidence of a disease observed by the examiner.

SMART: Acronym used in management science to signify that an objective is "Specific/Measurable/Appropriate/Realistic/Timely."

SMOMBIES: Smartphone zombies (walking while texting).

Spectral sensitivity: Sensitivity of the eye to light of different wavelengths.

Spondylosis: Abnormal wear on the cartilage and bones of the neck

Spondylolisthesis: Medical condition involving anterior movement of a lumbar vertebra due to weakness of the pars interarticularis.

Standard deviation: Square root of the mean of the squared differences between individual scores and the mean of a sample or population. A measure of dispersion. The larger the standard deviation, the more variable the scores in the population.

Stapedius muscle: Muscle in the posterior wall of the middle ear connected to the head of the stape. The muscle contracts in response to loud noises pulling the stapes away from the oval window.

SOP: Standard operating procedure (mandated way of performing a task).

Statistic: Estimate of the value of a population parameter by measurement of a subset, or sample, of the population.

Strain: General term for the response of a body or person to an applied load.

Stress: General term for applied load.

Symptom: Any aspect of abnormal human function inferred from the patient's report or his/her subjective experience of it.

Syndrome: A set of symptoms that occur together.

Synovial fluid: Fluid found within joints and some tendons that reduces the friction between adjoining surfaces.

Synthesis: Collection and combination of individual data or items of information to form an overall score or solution.

Systems engineering: The process of conceptualizing, assessing, designing, manufacturing, using, and decommissioning a facility or device.

System response time: Time taken for a computer system to respond following a command.

Task: A discrete, bounded activity in which a set of operations is carried out to achieve a goal.

Task analysis: A formal and systematic method of representing a task as a series of subtasks while making explicit the interactions and dependencies between them.

Tendon: Cartilaginous tissue that connects muscle to bone.

Tendonitis: Swelling of a tendon.

Tennis elbow: Swelling at the musculotendinous junction of the wrist extensor muscle at the humerus, and pain on the lateral part of the elbow, which is exacerbated when the fingers are flexed to grasp an object.

Tenosynovitis: Inflammation of a tendon and its sheath.

Thermoregulation: Homeostatic mechanism used to maintain the body in thermal equilibrium with its environment.

Third class lever: Lever in which the lever arm and the moment arm are on the same side of the fulcrum, and the lever arm is shorter than the moment arm.

Thoracic kyphosis: Convex curve of the 12 thoracic vertebrae.

Threshold: Minimum level at which a stimulus can be detected.

Tibialis anterior: Lower leg muscle lateral to the shin that causes dorsiflexion of the ankle. This muscle can fatigue, for example, when maintaining foot position during pedal operation.

Time and motion study: Branch of methods engineering.

Tinnitus: Ringing or buzzing in the ears in the absence of corresponding stimuli.

Torsion: Twisting of an object about an internal axis when a torque is applied.

Torso: Upper body above the lower limbs but excluding the arms and the head.

Trachea: Upper part of the respiratory tract between the throat and the first division of the tract.

Trackball: Computer interface device consisting of a freely rotating sphere in a housing. The upper part of the sphere rests in the palm of the hand, and rotational movements of the sphere are converted into linear movements of a cursor on the computer screen.

Transverse plane: View of the body from the front or back.

Tympanic membrane (eardrum): Thin layer of tissue that transmits sound energy from the external auditory canal to the middle ear.

Ulna: Medial bone of the forearm.

Valgus angle: Angle where the distal end of a bone deviates laterally. In the case of the femur, the deviation is inward.

Validation: Proof that a test or questionnaire measures what it is designed to measure.

Validity: The degree to which an instrument measures what it claims to measure.

Variance: Square of the standard deviation.

Varicose veins: Swollen, uncomfortable veins in the legs and feet, often blue in color

Varus angle: Opposite of valgus angle.

Vasoconstriction: Decrease in the cross-sectional area of a capillary.

Vasodilation: Increase in the cross-sectional area of a capillary.

Vector: A quantity that has magnitude and direction (e.g., a force).

Ventral surface: The upper part of an object.

Vergence: Movement of the eyes in opposite directions to fixate a target.

Vestibular system: System that provides the sense of orientation and balance in relation to gravity.

Vibration white finger: Vascular disease of the fingers characterized by peripheral vasoconstriction following chronic exposure to vibration.

Virtual environment: A three-dimensional, computer-generated, and fully interactive representation of a real or imagined environment.

Viscoelastic property: Property of a material that exhibits both elastic deformation (in which complete reversal of strain occurs when the load is removed) and viscous deformation (in which the material progressively deforms under sustained loading).

Visual acuity: The ability of the eye to resolve detail.

Weight: Action of a mass under the influence of gravity. A mass of 1 kg on the surface of the Earth has a weight of 9.81 N. The term is often used as a synonym for "static force"—the force applied to a supporting surface by the action of gravity on a mass at rest on the surface.

White noise: Broadband noise containing all parts of the sound spectrum audible to humans.

Whole body vibration: Exposure of the whole body to vibration (usually through the feet/buttocks when riding in a vehicle). Whole body vibration may increase the risk for injury, including low back pain and internal organ disruption.

Work: Work takes place when a force moves an object over a distance. The unit of work is the joule (J).

Work–rest cycle: Total amount of time needed to complete an operation, task, or assignment that is repeated over time. Rest, as distinct from recreation, is calculated to allow recovery and is part of the working period.

Work study: General term for time and motion study and any related techniques.

Zygapophyseal joints: Joints posterior to the vertebral bodies that limit the coarticulation of adjacent bodies. Known colloquially as "facet joints."

References

Aaras, A. 1994. The impact of ergonomic intervention on individual health and corporate prosperity in a telecoms environment. *Ergonomics, 37*: 1679–1696.

Abd Ellfattah, M.T. 2014. Web design for color blind persons. *International Design Journal, 4*(4): 37–46.

Adams, M.A. and Hutton, W.C. 1983. The effect of posture on the fluid content of the intervertebral discs. *Spine, 8*: 665–671.

Adams, M.A. and Dolan, P. 1995. Recent advances in lumbar spinal mechanics and their clinical significance. *Clinical Biomechanics, 10*(1): 3–19.

Adams, M.A. and Hutton, W.C. 1983. The effect of posture on the fluid content of the intervertebral discs. *Spine, 8*: 665–671.

Adams, M.A., Bogduk, N., Burton, K., and Dolan, P. 2000. *The Biomechanics of Back Pain.* Churchill Livingstone, Edinburgh.

Agyekum, E.K. and Ma, K. 2015. Heel pain: A systematic review. *Chinese Journal of Traumatology, 18*(03): 164–169.

Ahmad, J., Fanti, V., Caldwell, D.G., and Di Natali, C. 2024. Framework for the adoption, evaluation and impact of occupational exoskeletons at different technology readiness levels: A systematic review. *Robotics and Autonomous Systems, 179*: 104743.

Ainsworth, B.E. 2000. Compendium of physical activities: An update of activity codes and MET intensities. *Medicine and Science in Sports and Exercise, 32*: S418–S516; *Journal of Neuroscience*, 52: 29–37.

Akerstedt, T. 1990. Subjective and objective sleepiness in the active individual. *International Journal of Neuroscience, 52*: 265–273.

Alhassan, M., Alhamad, F., Bokhary, K., and Almustanyir, A. 2021. Effects of virtual reality head-mounted displays on oculomotor functions. *International Journal of Ophthalmology & Visual Science, 6*(10): 10–16.

Alkalih, H.Y., Pesola, A.J., and Arumugam, A. 2022. A new accelerometer (Fibion) device provides valid sedentary and upright time measurements compared to the ActivPAL4 in healthy individuals. *Heliyon, 8*(10): e11103.

Allread, W.G., Marras, W.S., and Burr, D.L. 2000. Measuring trunk motions in industry: Variability due to task factors, individual differences and the amount of data collected. *Ergonomics, 43*: 691–701.

Alomari, R.A., Fernandez, M., Banks, J.B., Acosta, J., and Tartar, J.L. 2015. Acute stress dysregulates the LPP ERP response to emotional pictures and impairs sustained attention: Time-sensitive effects. *Brain Sciences, 5*(2): 201–219.

Alufaisan, Y., Marusich, L.R., Bakdash, J.Z., Zhou, Y., and Kantarcioglu, M. 2021. Does explainable artificial intelligence improve human decision-making? *Proceedings of the AAAI Conference on Artificial Intelligence, 35*(8): 6618–6626.

Alzahrani, H., Mackey, M., Stamatakis, E., Zadro, J.R., and Shirley, D. 2019. The association between physical activity and low back pain: A systematic review and meta-analysis of observational studies. *Scientific Reports, 9*(1): 8244.

Amershi, S., Weld, D., Vorvoreanu, M., Fourney, A., Nushi, B., Collisson, P., Suh, J., Iqbal, S., Bennett, P.N., Inkpen, K., and Teevan, J. 2019. Guidelines for human-AI interaction. In: *Proceedings of the 2019 CHI Conference on Human Factors in Computing Systems.* ACM, New York, pp. 1–13.

Anderson, J.R. 1983. *The Architecture of Cognition.* Harvard University Press, Cambridge, MA.

Anderson, S., Thygesen, L.C., Davidsen, M., and Helweg-Larsen, K. 2012. Cumulative years in occupation and the risk of osteoarthritis in men and women: A register-based follow-up study. *Occupational and Environmental Medicine, 69*: 325–330.

Anema, J.R., Cuelenaere, B., van der Beek, A.J., Knol, D.L., De Vet, H.C.W., and Van Mechelen, W. 2004. The effectiveness of ergonomic interventions on return-to-work after low back pain; a prospective two year cohort study in six countries on low back pain patients sicklisted for 3–4 months. *British Medical Journal, Occupational and Environmental Medicine, 61*(4): 289–294.

Anonymous prEN 14386. 2002. *Safety of Machinery. Ergonomic Design Principles for the Operability of Mobile Machinery.* Swedish Institute for Standards, Sweden.

Anonymous prEN 614. 2003. *Safety of Machinery. Ergonomic Design Principles—Part 2: Interactions between the Design of Machinery and Work Tasks.* Swedish Institute for Standards, Sweden.

Anonymous prEN 614-1. 2003. *Safety of Machinery. Ergonomic Design Principles—Part 1: Terminology and General Principles; German version.* Swedish Institute for Standards, Sweden.

Anonymous. 2007. *Cosmopolitan Magazine.* https://www.cosmopolitan.com/.

ANSI/HFS 100. 1988. *American National Standard for Human Factors Engineering of Visual Display Terminal Workstations.* Human Factors Society Inc., Santa Monica, CA.

Arnedt, J.T., Wilde, G.J.S., Munt, P.W., and MacLean, A.W. 2001. How do prolonged wakefulness and alcohol compare in the decrements they produce on a simulated driving task? *Accident Analysis and Prevention, 33*: 337–344.

Arshad, R., Pan, F., Reitmaier, S., and Schmidt, H. 2019. Effect of age and sex on lumbar lordosis and the range of motion. A systematic review and meta-analysis. *Journal of Biomechanics, 82*: 1–19.

Aryal, A., Becerik-Gerber, B., Anselmo, F., Roll, S.C., and Lucas, G.M. 2019. Smart desks to promote comfort, health, and productivity in offices: A vision for future workplaces. *Frontiers in Built Environment, 5*: 76.

Ashford, A., Fisher, I., and Bridger, R.S. 2016. *Reliability of a Safety Culture Assessment Tool during Safety Culture Auditing.* INM Report 2016.028, p. 42.

Aspden, R.M. 1989. The spine as an arch: A new mathematical model. *Spine, 14*: 266–274.

Astrand, P.O. and Rodahl, K. 1977. *Textbook of Work Physiology.* McGraw-Hill, New York.

Atchison, D.A., Pedersen, C.A., Dain, S.J., and Wood, J.M. 2003. Traffic signal color recognition is a problem for both protan and deutan color-vision deficients. *Human Factors, 45*: 495–503.

Awh, E., Barton, B., and Vogel, E.K. 2007. Visual working memory represents a fixed number of items regardless of complexity. *Psychological Science, 18*: 622–628.

Ayoub, M.A. and Mital, A. 1997. *Manual Materials Handling.* Taylor & Francis, London.

Ayoub, M.M. 1973. Work place design and posture. *Human Factors, 15*: 265–268.

Babamiri, M., Heidarimoghadam, R., Ghasemi, F., Tapak, L., and Mortezapour, A. 2024. Going beyond general stress scales: Developing a new questionnaire to measure stress in human-robot interaction. *International Journal of Social Robotics, 16*: 1–17.

Baber, C. and Young, M.S. 2022. Making ergonomics accountable: Reliability, validity and utility in ergonomics methods. *Applied Ergonomics*, 98: 103583.

Baddeley, A.D. 1983. Working memory. *Philosophical Transactions of the Royal Society of London B: Biological Sciences, 302*(1110): 311–324.

Baek, S., Jung, J., Moon, P., and Park, W. 2021. Obesity impacts on task performance and perceived discomfort during seated foot target reaches. *Ergonomics, 64*(12): 1569–1578.

Bai, H., Sasikumar, P., Yang, J., and Billinghurst, M. 2020. A user study on mixed reality remote collaboration with eye gaze and hand gesture sharing. In: *Proceedings of the 2020 CHI Conference on Human Factors in Computing Systems.* ACM, New York, pp. 1–13.

Bailey, R.W. 1982. *Human Performance Engineering: A Guide for System Designers.* Prentice-Hall Inc., Englewood Cliffs, NJ.

Bainbridge, L. 1982. Analysis, design and evaluation of man–machine systems. In: *Proceedings of IFAC/IFIP/IFORS/IEA Conference, Ironies of Automation.* Elsevier, Baden-Baden, Germany, pp. 151–157.

Baldwin, C.L. and Struckman-Johnson, D. 2002. Impact of speech presentation level on cognitive task performance. *Ergonomics, 45*: 61–74.

Baldwin, C.L., Spence, C., Bliss, J.P., Brill, J.C., Wogalter, M.S., Mayhorn, C.B., and Ferris, T.K. 2012. Multimodal cueing: The relative benefits of the auditory, visual, and tactile channels in complex environments. In: *Proceedings of the Human Factors and Ergonomics Society Annual Meeting.* SAGE Publications, Los Angeles, CA. Vol. *56*, No. 1, pp. 1431–1435.

Banks, V.A. and Stanton, N.A. 2016. Keep the driver in control: Automating automobiles of the future. *Applied Ergonomics, 53*: 389–395.

Bar-Hillel, M. 2015. Position effects in choice from simultaneous displays: A conundrum solved. *Perspectives on Psychological Science, 10*: 419–433.

Barnes, C.M. and Drake, C.L. 2015. Prioritizing sleep health: Public health policy recommendations. *Psychological Science, 10*: 733–737.

Barragan-Jason, G., Loreau, M., de Mazancourt, C., Singer, M.C., and Parmesan, C. 2023. Psychological and physical connections with nature improve both human well-being and nature conservation: A systematic review of meta-analyses. *Biological Conservation, 277*: 109842.

Bartholomae, R.C. and Kovac, J.G. 1979. USBM develops a low-noise percussion drill. In: *Coal Age Conference and Expo V.* Heritage Hall, Louisville, KY.

Bartlett, F.C. 1932. *Remembering.* Cambridge University Press, New York.

Baumeister, R.F., Vohs, K.D. and Tice, D.M. 2007. The strength model of self-control. *Current Directions in Psychological Science, 16*(6): 351–355.

Bayona, J.A., Caballer, A., and Peiró, J.M. 2020. The relationship between knowledge characteristics' fit and job satisfaction and job performance: The mediating role of work engagement. *Sustainability*, *12*(6): 2336.

Beach, T.A.C., Parkinson, R.J., Stothart, J., and Callaghan, J.P. 2005. Effects of prolonged sitting on the passive flexion stiffness of the in-vivo lumbar spine. *The Spine Journal*, *5*: 145–154.

Bean, A. 2000. *The Complete Guide to Sports Nutrition*. 3rd Edn., A&C Black, Ltd, London.

Beaudette, S.M., Briar, K.J., Mavor, M.P., and Graham, R.B. 2020. The effect of head and gaze orientation on spine kinematics during forward flexion. *Human Movement Science*, *70*: 102590.

Bedford, T. 1950. Environmental warmth and human comfort. *British Journal of Applied Physics*, *1*: 2.

Beiers, J.L. 1966. A study of noise sources in pneumatic rockdrills. *Journal of Sound and Vibration*, *3*: 166–194.

Bell, C.R., Crowder, M.J., and Walters, J.D. 1971. Durations of safe exposure for men at work in high temperature environments. *Ergonomics*, *14*: 733–757.

Benites-Zapata, V.A., Jiménez-Torres, V.E., and Ayala-Roldán, M.P. 2021. Problematic smartphone use is associated with de Quervain's tenosynovitis symptomatology among young adults. *Musculoskeletal Science and Practice*, *53*: 102356.

Benjamin, K. and Wilson, S. 2005. *Facts and Misconceptions about Age, Health Status, and Employment.* Health and Safety Laboratory, Buxton, UK.

Bentley, T., Green, N., Tappin, D., and Haslam, R. 2020. State of science: the future of work – ergonomics and human factors contributions to the field. *Ergonomics*, *64*(4): 427–439. https://doi.org/10.1080/00140139.2020.1841308

Berman, M.G., Jonides, J., and Kaplan, S. 2008. The cognitive benefits of interacting with nature. *Psychological Science*, *19*(12): 1207–1212.

Bernoux, P. 1994. Participation: A review of the literature. In: *P+, European Participation Monitor, The Economics of Participation*, M. Gold, ed. European Foundation for the Improvement of Living and Working Conditions, Dublin, Ireland.

Bhagwati, J. 1993. The case for free trade. *Scientific American*, *269*(5): 18–23.

Bilzon, E., Chilcott, P., and Bridger, R.S. 2000. *Investigation of Anthropometric Criteria for Escape through a Kidney Hatch, an Escape Hatch and for Passage through a Bulkhead Door.* INM Report No. 2000.037, July.

Birch, L., Juul-Kristensen, B., Jensen, C., Finsen, L., and Christensen, H. 2000. Acute response to precision, time pressure and mental demands during simulated computer work. *Scandinavian Journal of Work Environment An Health*, *26*: 299–305.

Bisdorff, A.R., Bronstein, A.M., Wolsley, C., Gresty, M.A., and Young, D.A. 1999. EMG responses to free fall in elderly subjects and akinetic rigid patients. *Journal of Neurological and Neurosurgical Psychiatry*, *66*: 447–455.

Biswas, R., Genitsaridi, E., Trpchevska, N., Lugo, A., Schlee, W., Cederroth, C.R., Gallus, S., and Hall, D.A. 2023. Low evidence for tinnitus risk factors: A systematic review and meta-analysis. *Journal of the Association for Research in Otolaryngology*, *24*(1): 81–94.

Bittner, A.C. and Guignard, J.C. 1985. Human factors engineering principles for minimising adverse motion effects: Theory and practice. *Naval Engineers Journal*, *97*: 205–213.

Black, N.L., Tremblay, M., and Ranaivosoa, F. 2022. Different sit: Stand time ratios within a 30 minute cycle change perceptions related to musculoskeletal disorders. *Applied Ergonomics*, *99*: 103605.

Blanchonette, P. 2009. *JACK Human Modelling Tool: A Review*. Air Operations Division, Defence Science and Technology Organisation, DSO-TR-2364, Victoria, Australia.

Blau, J.N. 1992. Migraine: Theories of pathogenesis. *The Lancet*, *339*: 1202–1297.

Blundell, J., Collins, C., Sears, R., Plioutsias, T., Huddlestone, J., Harris, D., Harrison, J., Kershaw, A., Harrison, P., and Lamb, P. 2023. Multivariate analysis of gaze behavior and task performance within interface design evaluation. *IEEE Transactions on Human-Machine Systems*, *53*: 875–884.

Boas, F. 1910. *Changes in Body Form of Descendents of Immigrants. Senate Document 208, 61st Congress, Second Session* (Final Report). Washington, DC.

Bobjer, O., Johansson, S.E., and Piguet, S. 1993. Friction between the hand and handle: Effects of oil and lard on textured and non-textured surfaces; perception of discomfort. *Applied Ergonomics*, *24*: 190–202.

Boix Rodríguez, N., Formentini, G., Favi, C., and Marconi, M. 2021. Engineering design process of face masks based on circularity and life cycle assessment in the constraint of the COVID-19 pandemic. *Sustainability*, *13*(9): 4948.

Bolia, R.S., D'Angelo, W.R., and McKinley, L. 1999. Aurally-aided visual search in three-dimensional space. *Human Factors*, *41*: 664–669.

Bolm-Audorff, U., Hegewald, J., Pretzsch, A., Freiberg, A., Nienhaus, A., and Seidler, A. 2020. Occupational noise and hypertension risk: A systematic review and meta-analysis. *International Journal of Environmental Research and Public Health*, *17*(17): 6281.

Bonin, D., Ackermann, A., Radke, D., Peters, M., and Wischniewski, S. 2023. Anthropometric dataset for the German working-age population using 3D body scans from a regional epidemiological health study and a weighting algorithm. *Ergonomics*, *66*(8): 1057–1071.

Borg, G.A.V. 1982. Psychophysical bases of perceived exertion. *Medicine and Science in Sports and Exercise*, *14*: 377–381.

Borsci, S. and Schmettow, M. 2024. Re-examining the chatbot usability scale to assess user experience with customer relationship management chatbots. *Personal and Ubiquitous Computing*, *28*: 1033–1044. https://doi.org/10.1007/s00779-024-01834-4

Boubekri, M., Cheung, I.N., Redi, K.J., Kuo, N.-W., Wang, C.-H., and Zee, P.C. 2014. Impact of windows and daylight exposure on overall health and sleep quality of office workers—A case–control pilot study. *Discolure*, *1*(*9*): 20.

Bowler, J.N. 1982. Welbeck colliery noise level survey. *The Mining Engineer*, *7*: 159–162.

Bransford, J.D. and Johnson, M.K. 1973. Consideration of some problems of comprehension. In: *Visual Information Processing*, W. Chase, ed. Academic Press, Cambridge.

Brewerton, P. and Millward, L. 2001. *Organisational Research Methods*. Sage, London.

Bridger, R.S. 1988. Postural adaptations to a sloping chair and worksurface. *Human Factors*, *30*: 237–247.

Bridger, R.S. 2021. *Introduction to Human Factors in Accident Investigation*. https://www.amazon.com/Introduction-Human-Factors-Accident-Investigation-ebook/dp/B09F3BZN59/ref=sr_1_1?crid=5L7B051PTVQ0&dchild=1&keywords=introduction+to+human+factors+in+accident+investigation&qid=1632036809&sprefix=introduction+to+human+factors+%2Caps%2C231&sr=8-1

Bridger, R.S., Ashford, A.I., Wattie, S., Dobson, K., Fisher, I., and Pisula, P.J. 2018. Sustained attention when squatting with and without an exoskeleton for the lower limbs. *International Journal of Industrial Ergonomics*, *66*: 230–239.

Bridger, R.S. and Bennett, A.I. 2011. Age and BMI interact to determine work ability in seafarers. *Occupational Medicine*, *61*(*3*): 157–162.

Bridger, R.S. and Brasher, K. 2011. Cognitive task demands, self-control demands and the mental well-being of office workers. *Ergonomics*, *54*: 830–839.

Bridger, R.S. and Freidberg, S. 1999. Managers' estimates of safe loads for manual handling: Evidence for risk compensation? *Safety Science*, *32*: 103–111.

Bridger, R.S. and Orkin, D. 1992. Effect of a footrest on standing posture. *Ergonomics SA*, *4*: 42–48.

Bridger, R.S., Brasher, K., and Bennett, A. 2013a. Sustaining person-environment fit with a changing workforce. *Ergonomics*, *56*(3): 565–577.

Bridger, R.S., Caborn, N., Goedecke, J., Rickard, S., Schabort, E., Westgarth-Taylor, C., and Lambert, M.I. 1997. Physiological and subjective measures of workload when shovelling with conventional and two-handled ("levered") shovels. *Ergonomics*, *40*: 1212–1219.

Bridger, R.S., Day, A.J., and Morton, K. 2013b. Occupational stress and employee turnover. *Ergonomics*, 56: 1629–1639.

Bridger, R.S., Green, A., Oakley, H., and Bilzon, E. 2002. *Sick Bay Design for Autonomous Rescue and Recovery Craft*. Unpublished MOD Report. Ministry of Defence, London, UK.

Bridger, R.S., Orkin, D., and Henneberg, M. 1992. A quantitative investigation of lumbar and pelvic postures in standing and sitting: Interrelationships with body position and hip muscle length. *International Journal of Industrial Ergonomics*, *9*: 235–244.

Bridger, R.S., Ossey, S., and Fourie, G. 1990. The effect of lumbar traction on stature. *Spine*, *15*: 522–525.

Bridger, R.S., Pisula, P., and Bennett, A. 2012. *A Guide to Understanding Human Factors in Safety Management and Accident Investigation*. British Crown, BC, Canada, p. 40.

Bridger, R.S., Sparto, P., and Marras, W.S. 1998. Spade design, lumbar motions, risk of low back injury and digging posture. *Occupational Ergonomics*, *1*: 157–172.

Bridger, R.S., Verweckken, B., Whistance, R.S., and Adams, L.P. 1994. A prototype standing workspace. In: *Contemporary Ergonomics*, S.A. Robertson, ed. Taylor & Francis, London.

British Standards Institution. 2023. Menstruation, menstrual health and menopause in the workplace: A Guide. BS 30416:2023.

Broadbent, D.E. 1979. Is a fatigue test now possible? *Ergonomics*, *22*: 1277–1290.

Broadbent, D.E., Cooper, P.F., FitzGerald, P., and Perkes, K.R. 1982. The cognitive failures questionnaire and its correlates. *British Journal of Clinical Psychology*, *21*: 1–16.

Brouha, L. 1967. *Physiology in Industry*. Pergamon Press, New York.

Brown, C.B., Barrett, T.S., Long, C., Corbridge, S., Braeger, A., Zollinger, B.J., Harrison, K., Poulsen, S.L., Boman, T., Dakin, C.J., and Harper, S.A. 2023. Step edge highlighters and illuminance changes influence stair descent in a real-world setting. *Ergonomics*, *66*(9): 1219–1228.

Brown, H.A., Topham, T.H., Clark, B., Smallcombe, J.W., Flouris, A.D., Ioannou, L.G., Telford, R.D., Jay, O., and Périard, J.D. 2022. Seasonal heat acclimatisation in healthy adults: A systematic review. *Sports Medicine*, *52*(9): 2111–2128.

Brown, I.D. 1966. Subjective and objective comparisons of successful and unsuccessful trainee drivers. *Ergonomics*, *9*: 50–56.

Brown, I.D. and Poulton, E.C. 1961. Measuring the spare mental capacity of car drivers by a subsidiary task. *Ergonomics*, *4*: 35–40.

Buchman-Pearle, J.M., Gruevski, K.M., Gallagher, K.M., Barrett, J.M., and Callaghan, J.P. 2023. Defining the lumbar and trunk-thigh neutral zone from the passive stiffness curve: Application to hybrid sit-stand postures and chair design. *Ergonomics*, *66*(3): 338–349.

Buckley, J.P., Mellor, D.D., Morris, M., and Jospeh, F. 2014. Standing-based office work shows encouraging signs of attenuating post-prandial glycaemic excursion. *Occupational and Environmental Medicine*, *71*(2): 109–111.

Budd, G.M. 2008. Wet-bulb globe temperature—Its history and limitations. *Journal of Science and Medicine in Sport*, *11*: 20–32.

Bulbrook, B.D., La Delfa, N.J., McDonald, A.C., Liang, C., Callaghan, J.P., and Dickerson, C.R. 2021. Higher body mass index and body fat percentage correlate to lower joint and functional strength in working age adults. *Applied Ergonomics*, *95*: 103453

Bunzli, S., Smith, A., Schütze, R., Lin, I., and O'Sullivan, P. 2017. Making sense of low back pain and pain-related fear. *Journal of Orthopaedic & Sports Physical Therapy*, *47*(9): 628–636.

Burgess, A. 2007. Real and phantom risks at the petrol station: The curious case of mobile phones, fires and body static. *Health, Risk and Society*, *9*: 53–66.

Burgess, R. and Neal, R.J. 1989. Document holder usage when reading and writing. *Clinical Biomechanics*, *4*: 151–154.

Burgess-Limerick, R. 2020. Human-systems integration for the safe implementation of automation. *Mining, Metallurgy & Exploration*, *37*(6): 1799–1806.

Burks, R., Harper, C., and Bartha, M.C. 2014. Examining 3-D technologies in laptop displays. *Ergonomics in Design*, 22: 17–22.

Burt, S., Crombie, K., Jin, Y., Wurzelbacher, S., Ramsey, J., and Deddens, J. 2011. Workplace and individual risk factors for carpal tunnel syndrome. *Occupational and Environmental Medicine*, *68*(12): 928–933.

Burtch, G., Greenwood, B.N., and Ravindran, K. 2024. *Lucy and the Chocolate Factory: Warehouse Robotics and Worker Safety*. George Mason University School of Business Research Paper. https://ssrn.com/abstract=4389032 or http://doi.org/10.2139/ssrn.4389032

Burton-Jones, A. and Straub, D.W. 2004. Minimizing method variance in measures of system usage. In: *Proceedings of the Southern Association for Information Systems (SAIS)*, R. Papp, ed. Savannah, GA, pp. 336–342.

Buruck, G., Tomaschek, A., Wendsche, J., Ochsmann, E., and Dörfel, D. 2019. Psychosocial areas of worklife and chronic low back pain: A systematic review and meta-analysis. *BMC Musculoskeletal Disorders*, *20*: 1–16.

Butler, D.L., Acquino, A.L., Hissong, A.A., and Scott, P.A. 1993. Wayfinding by newcomers in a complex building. *Human Factors*, *35*: 159–173.

Butler, D.S. 1991. *Mobilisation of the Nervous System*. Churchill Livingstone, Edinburgh.

Canetti, E.F., Schram, B., Orr, R.M., Knapik, J., and Pope, R. 2020. Risk factors for development of lower limb osteoarthritis in physically demanding occupations: A systematic review and meta-analysis. *Applied Ergonomics*, *86*: 103097

Carayon, P., Kianfar, S., Li, Y., Xie, A., Alyousef, B., and Wooldridge, A. 2015. A systematic review of mixed methods research on human factors and ergonomics in health care. *Applied Ergonomics*, *51*: 291–321.

Carayon, P., Wooldridge, A., Hoonakker, P., Hundt, A.S., and Kelly, M.M. 2020. SEIPS 3.0: Human-centered design of the patient journey for patient safety. *Applied Ergonomics*, *84*: 103033.

Carfì, A. and Mastrogiovanni, F. 2021. Gesture-based human–machine interaction: Taxonomy, problem definition, and analysis. *IEEE Transactions on Cybernetics*, *53*(1): 497–513.

Carter, J.T. and Birrell, L.N. 2000. *Occupational Health Guidelines for the Management of Low Back Pain at Work—Principal Recommendations*. Faculty of Occupational Medicine, London.

Casali, J.G., Robinson, G.S., Dabney, E.C., and Gauger, D. 2004. Effect of electronic ANR and conventional hearing protectors on vehicle backup alarm detection in noise. *Human Factors*, *46*: 1–10.

Caserman, P., Garcia-Agundez, A., Gámez Zerban, A., and Göbel, S. 2021. Cybersickness in current-generation virtual reality head-mounted displays: Systematic review and outlook. *Virtual Reality*, *25*(4): 1153–1170.

Ceesay, S.M., Prentice, A.M., Day, K.C., Murgatroyd, P.R., Goldberg, G.R., and Spurr, G.B. 1989. The use of heart rate in the estimation of energy expenditure: A validation study using indirect whole-body calorimetry. *British Journal of Nutrition*, *61*: 175–186.

Chaffin, D.B. and Andersson, G.B.J. 1984. *Occupational Biomechanics*. John Wiley & Sons, New York.

Chaka, C. 2024. Reviewing the performance of AI detection tools in differentiating between AI-generated and human-written texts: A literature and integrative hybrid review. *Journal of Applied Learning and Teaching*, *7*(1): 1–12.

Chany, A.M., Marras, W.S., and Burr, D.L. 2007. The effect of phone design on upper extremity discomfort and muscle fatigue. *Human Factors*, *49*(4): 602–618.

Chapanis, A. 1965. Words, words, words. *Human Factors*, *7*: 1–17.

Chapanis, A. and Lindenbaum, L.E. 1959. A reaction time study of four control display linkages. *Human Factors*, *1*: 1–7.

Chapman, C.L., Hess, H.W., AI Lucas, R., Glaser, J., Saran, R., Bragg-Gresham, J., Wegman, D.H., Hansson, E., Minson, C.T., and Schlader, Z.J. 2021. Occupational heat exposure and the risk of chronic kidney disease of non-traditional origin in the United States. *American Journal of Physiology-Regulatory, Integrative and Comparative Physiology*, *321*(2): R141–R151.

Chartered Institute of Ergonomics and Human Factors. 2021. *Barrier Management*. https://ergonomics.org.uk/resource/human-factors-in-barrier-management.html.

Chartered Institute of Ergonomics and Human Factors. 2022. *The Role of Human Factors in Delivering Cyber Security*. CIEHF, p. 28. https://ergonomics.org.uk/resource/the-role-of-human-factors-in-delivering-cyber-security.html.

Chastin, S., McGregor, D., Palarea-Albaladejo, J., Diaz, K.M., Hagströmer, M., Hallal, P.C., and Van Hees, V.T. 2021. Joint association between accelerometry-measured daily combination of time spent in physical activity, sedentary behaviour and sleep and all-cause mortality: A pooled analysis of six prospective cohorts using compositional analysis. *British Journal of Sports Medicine*, *55*(22): 1277–1285.

Chen, J.C., Dennerlein, J., Chang, C.C., Chang, W.R., and Christiani, D.C. 2005. Seat inclination, use of lumbar support and low back pain of taxi drivers. *Scandinavian Journal of Work Environment and Health*, *31*: 258–265.

Chen, R., Tsai, M.C., and Tsay, Y.S. 2022. Effect of color temperature and illuminance on psychology, physiology, and productivity: An experimental study. *Energies*, *15*(12): 4477

Chen, Y., Tao, M., and Liu, W. 2020. High temperature impairs cognitive performance during a moderate intensity activity. *Building and Environment*, *186*: 107372.

Cherns, A. 1976. The principles of sociotechnical design. *Human Relations*, *29*: 783–792.

Cherrie, J.W. and Cherrie, M.P. 2022. Workplace exposure to UV radiation and strategies to minimize cancer risk. *British Medical Bulletin*, *144*(1): 45–56.

Cholewicki, J. and Vanvliet Iv, J.J. 2002. Relative contribution of trunk muscles to the stability of the lumbar spine during isometric exertions. *Clinical Biomechanics*, *17*(2): 99–105.

Chomsky, N. 1957. *Syntactic Structures*. Mouton and Co., The Hague.

Chondromatidis, N., Gialos, A., and Zeimpekis, V. 2022. Investigating the performance of the order-picking process by using smart glasses: A laboratory experimental approach. *Logistics*, *6*(4): 84.

Church, A.H. 1993. Estimating the effect of incentives on mail survey response rates. *Public Opinion Quarterly*, *57*: 62–79.

Cian, L., Krishna, A., and Elder, R.S. 2015. A sign of things to come: Behavioral change through dynamic iconography. *Journal of Consumer Research*, *41*(6): 1426–1446.

Ciriello, V.M. and Snook, S.H. 1995. The effect of back belts on lumbar muscle fatigue. *Spine*, *20*(11): 1271–1277.

Ciriello, V.M., Dempsey, P., Maikala, R.V., and O'Brien, N.V. 2008. Secular changes in psychophysically determined acceptable weights and forces over 20 years for male industrial workers. *Ergonomics*, *51*: 593–601.

Ciriello, V.M., Snook, S.H., and Hughes, G.H. 1993. Further studies of psychophysically determined maximum acceptable weights and forces. *Human Factors*, *35*: 175–186.

Ciriello, V.M., Webster, B.S., and Dempsey, P.G. 2002. Maximal acceptable torques of highly repetitive screw driving, ulnar deviation, and handgrip tasks for 7-hour workdays. *AIHA Journal*, *63*(5): 594–604.

Clark, C. and Haswe II, M.R. 1964. *The Economics of Subsistence Agriculture*. McMillan Publishing Co., London.

Claypoole, V.L., Schroeder, B.L., and Mishler, A.D. 2016. Keeping in touch: Tactile interface design for older users. *Ergonomics in Design*, *24*: 18–24.

Clement, D.E. 1987. Human factors, instructions, warnings and product liability. *IEEE Transactions on Professional Communications*, *30*(3): 149–156.

Clemes, S.A., Haslam, C.O., and Haslam, R.A. 2010. What constitutes effective manual handling training? A systematic review. *Occupational Medicine*, *60*(2): 101–107.

Coch, L. and French, J.R.P., Jr. 1948. Overcoming resistance to change, *Human Relations*, *1*: 512–532

Cohen, S.P., Chen, Y., and Neufeld, N.J. 2013. Sacroiliac joint pain: A comprehensive review of epidemiology, diagnosis and treatment. *Expert Review of Neurotherapeutics*, *13*(1): 99–116.

Coleman, D. 1995. Human migration: Effects on people, effects on populations. In: *Human Variability and Plasticity*, C.C.N. Mascie-Taylor and B. Bogin, eds. Cambridge University Press, Cambridge.

Collins, A.M. and Quillian, M.R. 1972. Experiments on semantic memory and language comprehension. In: *Cognition in Learning and Memory*, L.W. Gregg, ed. John Wiley & Sons, New York, pp. 309–351.

Collins, M.J., Brown, B., Bowman, K.J., and Caird, D. 1991. Task variables and visual discomfort associated with the use of VDT's. *Optometry and Vision Science*, *68*: 27–33.

Conrad, C., Klesel, M., Oschinsky, F.M., Mayhew, K., O'Neil, K., and Usai, F. 2023. Quality is more important than quantity: Social presence and workplace ergonomics control predict perceived remote work performance. In *Proceedings of the 56th Hawaii International Conference on System Sciences*. University of Hawai'i at Mānoa, Honolulu, HI.

Cook, R.I. 2000. *How Complex Systems Fail*. Cognitive Systems Laboratory, University of Chicago, Chicago, IL, Vol. *5*.

Cook, T.M. and Neumann, D.A. 1987. The effects of load placement on the activity of the low back muscles during load carrying by men and women. *Ergonomics*, *30*: 1413–1423.

Coventry, L., Briggs, P., Blythe, P., and Tran, M. 2014. *Using Behavioural Insights to Improve the Public's Use of Cyber Security Best Practices*. Government Office for Science, URN GS/14/835, p. 20.

Cox, T. and Griffiths, A.J. 1995. Psychosocial hazards: Assessment issues. In: *Proceedings of the International Symposium from Research to Prevention: Managing Occupational and Environmental Health Hazards*, J. Rantanen, ed. Helsinki.

Crafts, N. 2022. The 15-hour week: Keynes's prediction revisited. *Economica*, *89*(356): 815–829.

Craik, F.I.M. and Lockhart, R.S. 1972. Levels of processing: A framework for memory research. *Journal of Verbal Learning and Verbal Behaviour*, *11*: 71–84.

Craik, M. 1943. *The Nature of Explanation*. Cambridge University Press. Cambridge.

Crawford, E., Pazell, S., and Karanikas, N. 2023. Envisaging regenerative futures through good work design. In: *Contemporary Ergonomics & Human Factors 2023: Proceedings for the Chartered Institute of Ergonomics and Human Factors Annual Conference*, D. Golightly, N. Balfe, and R. Charles, eds. Kenilworth, UK.

Crea, S., Beckerle, P., De Looze, M., De Pauw, K., Grazi, L., Kermavnar, T., Masood, J., O'Sullivan, L.W., Pacifico, I., Rodriguez-Guerrero, C., and Vitiello, N. 2021. Occupational exoskeletons: A roadmap toward large-scale adoption. Methodology and challenges of bringing exoskeletons to workplaces. *Wearable Technologies*, *2*: e11.

Criado-Perez, C. 2019. *Invisible Women: Data Bias in a World Designed for Men*. Abrams, New York.

Cui, W.W. 2003. Reducing error in mail surveys. *Practical Assessment, Research and Evaluation*, *8*: 442–453.

Culver, C.C. and Viano, D.C. 1990. Anthropometry of seated women during pregnancy: Defining a fetal region for crash protection research. *Human Factors*, *32*: 625–636.

Cummings, M. 2017. The brave new world of driverless cars. *TR News*, *308*: 34–37.

Cummings, M., Noorman, M.E., and Verdiccio, M. 2024. Identifying AI hazards and responsibility gaps. In: *Computer Ethics Across Disciplines: Applying Deborah Johnson's Philosophy To Algorithmic Accountability And AI*, M. Noorman and M. Verdicchio, eds. Springer Nature, New York.

Curry, D.G., Zebrich, S., and Kirk-Braland, K. 2004. I fell, ergo, I slipped: Fact or fantasy? *Occupational Health and Safety*, *73*(*7*): 80–94.

Cushman, F. 2020. Rationalization is rational. *Behavioral and Brain Sciences*, *43*: e28.

Cushman, W.H. and Crist, B. 1987. Illumination. In: *Handbook of Human Factors*, G. Salvendy, ed. John Wiley & Sons, New York, pp. 670–695.

Cutlip, R.G. and Marras, W.S. 2000. Soft tissue pathomechanics and its application to ergonomics. In: *Ergonomics for the New Millennium, International Ergonomics Association XIVth Triennial Congress*. The Society, San Diego, CA.

Daams, B.J. 1993. Static force exertion in postures with different degrees of freedom. *Ergonomics*, *36*: 397–406.

Dalbøge, A., Frost, P., Anderson, J.H., and Svenden, S.W. 2014. Cumulative occupational shoulder exposures and surgery for subacromial impingement syndrome: A nationwide cohort study. *Occupational and Environmental Medicine*, *71*: 750–756.

Dalton, B.H. and Behm, D.G. 2007. Effects of noise and music on human and task performance: A systematic review. *Occupational Ergonomics*, *7*(3): 143–152.

Daly, H. 1993. The perils of free trade. *Scientific American*, *269*: 24–29.

Dannen, H.A.M. and Psituka, A. 2018. 3D body scanning. In: *Automation in Garment Manufacturing*, R. Nayak and R. Padhye, eds. Elsevier, Amsterdam, Netherlands, pp. 1–27

Datta, A.K., Hawkins, K., Heutnik, J., Manly, T., Cusack, R., Rorden, C., Davison, B., and Robertson, I. 2003. An electrophysiological predictor of human error. *Journal of Physiology, PC107*: 555.

Datta, S.R. and Ramanathan, N.L. 1971. Ergonomics comparison of seven modes of carrying loads on the horizontal plane. *Ergonomics, 14*: 269–278.

Davy, J. and Gobel, M. 2013. The effects of self-selected nap opportunity on the psychophysiological, performance and subjective measures during a simulated industrial night shift regimen. *Ergonomics, 56*: 220–234.

Dawadi, S., Shrestha, S., and Giri, R.A. 2021. Mixed-methods research: A discussion on its types, challenges, and criticisms. *Journal of Practical Studies in Education, 2*(2): 25–36.

Day, A.J., Brasher, K., and Bridger, R.S. 2012. Accident proneness revisited: The role of psychological stress and cognitive failure. *Accident Analysis & Prevention, 49*: 532–535.

De Carvalho, D., Greene, R., Swab, M., and Godwin, M. 2020a. Does objectively measured prolonged standing for desk work result in lower ratings of perceived low back pain than sitting? A systematic review and meta-analysis. *Work, 67*(2): 431–440.

De Carvalho, D.E. and Callaghan, J.P. 2023. Effect of office chair design features on lumbar spine posture, muscle activity and perceived pain during prolonged sitting. *Ergonomics, 66*(10): 1465–1476.

De Carvalho, D.E., de Luca, K., Funabashi, M., Breen, A., Wong, A.Y., Johansson, M.S., Ferreira, M.L., Swab, M., Kawchuk, G.N., Adams, J., and Hartvigsen, J. 2020b. Association of exposures to seated postures with immediate increases in back pain: A systematic review of studies with objectively measured sitting time. *Journal of Manipulative and Physiological Therapeutics, 43*(1): 1–12.

de Croon, E.M., Sluiter, J.K., and Frings-Dresen, M.H.W. 2006. Psychometric properties of the need for recovery after work scale: Test-retest reliability and sensitivity to detect change. *Occupational and Environmental Medicine, 63*: 202–206.

De Groot, D.W., Gallimore, R.P., Thompson, S.M., and Kenefick, R.W. 2013. Extremity cooling for heat stress mitigation in military and occupational settings. *Journal of Thermal Biology, 38*(6): 305–310.

De Vries, A. and De Looze, M. 2019. The effect of arm support exoskeletons in realistic work activities: A review study. *Journal of Ergonomics, 9*(4): 1–9.

de Wall, M., van Riel, M.P.J.M., and Snijders, C.J. 1991. The effect on sitting posture of a desk with a 10 degree inclination for reading and writing. *Ergonomics, 34*: 575–584.

De Winter, J.C. and Dodou, D. 2014. Why the Fitts list has persisted throughout the history of function allocation. *Cognition, Technology & Work, 16*: 1–11.

Degenfellner, J. and Schernhammer, E. 2021. Shift work tolerance. *Occupational Medicine, 71*(9): 404–413.

Dekker, S. 2006. *The Field Guide to Understanding Human Error*. Ashgate Publishing, Aldershot.

Dekker, S.W. 2015. The danger of losing situation awareness. *Cognition, Technology & Work, 17*(2): 159–161.

Del Ferraro, S., Falcone, T., Morabito, M., Bonafede, M., Marinaccio, A., Gao, C., and Molinaro, V. 2024. Mitigating heat effects in the workplace with a ventilation jacket: Simulations of the whole-body and local human thermos-physiological response with a sweating thermal manikin in a warm-dry environment. *Journal of Thermal Biology, 119*: 103772.

Delleman, N. and Berndsen, M.B. 2002. Touch typing VDT operation. *Ergonomics, 45*: 514–535.

Demerouti, E., Bakker, A.B., Nachreiner, F., and Schaufeli, W. 2001. The job demands–resources model of burnout. *Journal of Applied Psychology, 86*: 499–512.

Dempsey, P.G., Burdorf, A., Fathallah, F.A., Snook, G.S., and Hashemis, L. 2001a. Influence of measurement accuracy on the application of the 1991 NIOSH equation. *Applied Ergonomics, 32*: 91–99.

Descatha, A., Roquelaure, E., Chastang, J.F., Evanoff, B., Melchior, M., Mariot, C., Ha, C., Imbernon, E., Goldberg, M., and Leclerc, A. 2007. Validity of Nordic-style questionnaires in the surveillance of upper-limb work-related musculoskeletal disorders. *Scandinavian Journal of Work Environment and Health, 33*: 58–65.

Deshaies, P. 2015. Noise as an explanatory factor in work-related fatality reports. *Noise & Health, 17*(78): 294.

Desolda, G., Ferro, L.S., Marrella, A., Catarci, T., and Costabile, M.F. 2021. Human factors in phishing attacks: A systematic literature review. *ACM Computing Surveys (CSUR), 54*(8): 1–35.

Di Marino, M. and Lapintie, K. 2020. Exploring multi-local working: Challenges and opportunities for contemporary cities. *International Planning Studies, 25*: 129–149.

Diamond, J. 1991. Pearl harbour and the Emperor's physiologists. *Natural History, 12*: 5–7.

Dianat, I., Bagherzade, J., Jarafabadi, M.A., and Stedmon, A. 2013. Objective and subjective assessment of lighting in a hospital setting: Implications for health and performance. *Ergonomics, 56*: 1535–1545.

Diaz, K.M., Duran, A.T., Colabianchi, N., Judd, S.E., Howard, V.J., and Hooker, S.P. 2019. Potential effects on mortality of replacing sedentary time with short sedentary bouts or physical activity: A national cohort study. *American Journal of Epidemiology, 188*(3): 537–544.

Diaz, K.M., Howard, V.J., Hutto, B., Colabianchi, N., Vena, J.E., Safford, M.M., Blair, S.N., and Hooker, S.P. 2017. Patterns of sedentary behavior and mortality in US middle-aged and older adults: a national cohort study. *Annals of Internal Medicine*, *167*(7): 465–475.

Dick, F.D., Gravelling, R.A., Munro, W., and Walker-Bone, K. 2011. Workplace management of upper limb disorders: A systematic review. *Occupational Medicine*, *61*: 19–25.

Dick, M., Pohl, W., Lackner, H.K., Weiss, E.M., and Canazei, M. 2024. Effects of personalized lighting on subjective ratings, cognitive performance, and physiological stress response in a simulated office environment. *LEUKOS*, *20*: 1–20.

Dietrich, A. and Sparling, P.B. 2004. Endurance exercise selectively impairs prefrontal-dependent cognition. *Brain and Cognition*, *55*(3): 516–524.

Dodshon, P. 2023. *Development of a Control-Based Incident Investigaition Process to Reduce Workplace Injuries*. Ph.D Thesis, University of Queensland, Australia.

Dong, R.G., Wu, J.Z., Xu, X.S., Welcome, D.E., and Krajnak, K. 2021. A review of hand–arm vibration studies conducted by US NIOSH since 2000. *Vibration*, *4*(2): 482–528.

Donisi, L., Cesarelli, G., Coccia, A., Panigazzi, M., Capodaglio, E.M., and D'Addio, G. 2021. Work-related risk assessment according to the revised NIOSH lifting equation: A preliminary study using a wearable inertial sensor and machine learning. *Sensors*, *21*(8): 2593.

Doraisamy, R., Ramaswami, K., Shanmugam, J., Subramanian, R., and Sivashankaran, B. 2021. Genetic risk factors for lumbar disc disease. *Clinical Anatomy*, *34*(1): 51–56.

Dryden, J. 1990. *The Lives of the Noble Grecians and Romans*. Vol. *13*. Encyclopaedia Britannica Inc., NewYork.

Ducharme, R.E. 1977. Women workers rate "male tools" inadequate. *Human Factors Society Bulletin*, *20*: 1–2.

Duke, K., Mirka, G.A., and Sommerich, C.M. 2004. Productivity and ergonomic investigation of bent-handle pliers. *Human Factors*, *46*: 234–243.

Dul, J., Douwes, M., and Miedema, M. 1993. A Guideline for the prevention of discomfort in static postures. In: *Advances in Industrial Ergonomics and Safety*, R. Nielsen and K. Jorgensen, eds. Taylor & Francis, London, pp. 419–429.

Dunning, D. 2011. The Dunning–Kruger effect: On being ignorant of one's own ignorance. *Advances in Experimental Social Psychology*, *44*: 247.

Dunstan, D.W., Barr, E.L.M., Healy, G.N., Salmon, J., Shaw, J.E., Balkeau, B., Magliano, D.J., Cameron, A.J., Zimmet, P.Z., and Owen, N. 2010. Television viewing time and mortality: The Australian diabetes, obesity and lifestyle study. *Circulation*, *121*: 384–391.

Dunstan, D.W., Dogra, S., Carter, S.E., and Owen, N. 2021. Sit less and move more for cardiovascular health: Emerging insights and opportunities. *Nature Reviews Cardiology*, *18*(9): 637–648.

Dunstan, D.W., Kingwell, B.A., Larsen, R., Healy, G.N., Cerin, E., Hamilton, M.T., Shaw, J.E., Bertovic, D.A., Zimmet, P.Z., Salmon, J., and Owen, N. 2012. Breaking up prolonged sitting reduces postprandial glucose and insulin responses. *Diabetes Care*, *35*(5): 976–983.

Durand, J.-B., Camors, D., Trotter, Y., and Celebrini, S. 2012. Privileged visual processing of the straight-ahead direction in humans. *Journal of Vision*, *12*(6): 34–34.

Durrett, J. and Trezona, J. 1982. How to use colour displays effectively. *Byte*, *7*: 50–53.

Easterby, R.S. 1970. The perception of symbols for machine displays. *Ergonomics*, *13*: 149–158.

Edworthy, J., Hellier, E., Aldrich, K., and Lee, A. 2004. Linguistic and location effects in compliance with warning labels for amateur and professional users. *Human Factors*, *46*: 11–31.

Elena, N., Ivica, S., Haritina, S., Simona, D.M., and Anačkova, M. 2023. Methodology development and performance testing of an acoustically isolated booth using the ISO 23351-1:2020 standards. *Mechanical Engineering Scientific Journal*, *41*(2): 99–107.

Embrey, J.R., Donkin, C., and Newell, B.R. 2023. Is all mental effort equal? The role of cognitive demand-type on effort avoidance. *Cognition*, *236*: 105440.

Endsley, M.R. 1995, Towards a theory of situation awareness in dynamic systems. *Human Factors*, *37*: 32–64.

Eng, I., Tjernberg, M., and Champoux-Larsson, M.F. 2024. Hybrid workers describe aspects that promote effectiveness, work engagement, work-life balance, and health. *Cogent Psychology*, *11*(1): 2362535.

Engkvist, I.L., Svensson, R., and Eklund, J. 2011. Reported occupational injuries at Swedish recycling centres–based on official statistics. *Ergonomics*, *54*(4): 357–366.

Ervasti, J., Pentti, J., Nyberg, S.T., Shipley, M.J., Leineweber, C., Sørensen, J.K., Alfredsson, L., Bjorner, J.B., Borritz, M., Burr, H., and Knutsson, A. 2021. Long working hours and risk of 50 health conditions and mortality outcomes: A multicohort study in four European countries. *The Lancet Regional Health–Europe*, *11*: 100212.

EU Artificial Intelligence Act, Article 86: Right to Explanation of Individual Decision-Making. 2 August 2026.

Faber, G.S., Kingma, I., and van dieen, H. 2007. The effects of ergonomic interventions on low back moments are attenuated by changes in lifting behaviour. *Ergonomics*, *50*: 1377–1391.

Falowski, S., Sayed, D., Pope, J., Patterson, D., Fishman, M., Gupta, M., and Mehta, P. 2020. A review and algorithm in the diagnosis and treatment of sacroiliac joint pain. *Journal of Pain Research*, *8*: 3337–3348.

Fan, W., Schor, J., Kelly, O., and Gu, G. 2023. *Does Work Time Reduction Improve Workers' Well-Being? Evidence from Global Four-Day Workweek Trials*. University College Dublin, Dublin, Ireland.

Farris, D.J., Harris, D.J., Rice, H.M., Campbell, J., Weare, A., Risius, D., Armstrong, N., and Rayson, M.P. 2023. A systematic literature review of evidence for the use of assistive exoskeletons in defence and security use cases. *Ergonomics*, *66*(1): 61–87.

Fatoye, F., Gebrye, T., Ryan, C.G., Useh, U., and Mbada, C. 2023. Global and regional estimates of clinical and economic burden of low back pain in high-income countries: A systematic review and meta-analysis. *Frontiers in Public Health*, *11*: 1098100.

Feinstein, J. and Crawley, J.E. 1968. *The Ergonomic Design of a Slab Shear Pulpit at British Steel*. BISRA Report OR/HF/8/68.

Felten, E., Raj, M., and Seamans, R. 2023. How will language modelers like chatgpt affect occupations and industries?. *arXiv preprint arXiv:2303.01157*.

Fenery, A. and Walker, J.M. 2002. Short-term effects of workstation exercises on musculoskeletal discomfort and postural changes in seated video display unit workers. *Physical Therapy*, *82*: 578–589.

Feng, Y., Grooten, W., Wretenberg, P., and Arborelius, U.P. 1999. Effects of arm suspension in simulated assemble line work. *Applied Ergonomics*, *30*: 247–253.

Ferguson, S.A., Marras, W.S., and Burr, D.L. 2004. The influence of individual low back health status on workplace trunk kinematics and risk of low back disorder. *Ergonomics*, *24*: 1226–1237.

Fife-Shaw, C. 2000. Questionnaire design. In: *Research Methods in Psychology*, G. Breakwell, S. Hammond, and C. Fife-Shaw, eds. Sage, London.

Fiorentino, S., Livingstone, N., McAllister, P., and Cooke, H. 2022. The future of the corporate office? Emerging trends in the post-Covid city. *Cambridge Journal of Regions, Economy and Society*, *15*(3): 597–614.

Fisher, J. and Levin, N. 1989. Display control compatibility in the design of domestic cookers for the South African population. *Ergonomics SA*, *1*: 29–41.

Fitts, P.M. (ed). 1951. *Human Engineering for an Effective air Navigation and Traffic control System*. National Research Council, Washington, DC.

Fitts, P.M. 1954. The information capacity of the human motor system in controlling the amplitude of movement. *Journal of Experimental Psychology*, *47*: 381–391.

Fjærvoll, H., Fjærvoll, K., Magno, M., Moschowits, E., Vehof, J., Dartt, D.A., and Utheim, T.P. 2022. The association between visual display terminal use and dry eye: A review. *Acta Ophthalmologica*, *100*(4): 357–375.

Fleming, W.J. 2024. Employee well-being outcomes from individual-level mental health interventions: Cross-sectional evidence from the United Kingdom. *Industrial Relations Journal*, *55*(2): 162–182.

Fontana, D., Ardito, C., Costa, G., Boschetto, B., and d'Errico, A. 2023. Predictive validity of an indicator of exposure to unfavorable ergonomic working conditions on work-related musculoskeletal disorders. *Quality & Quantity*, *57*(5): 4545–4561.

Forouzandeh, B. and Wright, R.A. 1998. Do back support belts cause gastroesophageal reflux? *Journal of Clinical Gastroenterology*, *27*: 47–49.

Foster, H.M., Ho, F.K., Sattar, N., Welsh, P., Pell, J.P., Gill, J.M., Gray, S.R., and Celis-Morales, C.A. 2020. Understanding how much TV is too much: A nonlinear analysis of the association between television viewing time and adverse health outcomes. *Mayo Clinic Proceedings*, *95*(11): 2429–2441.

Foster, R.S., van Geems, B., Adams, L.P., and Bridger, R.S. 1998. Postural adaptations in standing computer work. *Occupational Ergonomics*, *1*(2): 145–156.

Fournier, D.E., Yung, M., Somasundram, K.G., Du, B.B., Rezvani, S., and Yazdani, A. 2023. Quality, productivity, and economic implications of exoskeletons for occupational use: A systematic review. *PLoS One*, *18*(6): e0287742.

Franke, T., Rauh, N., and Krems, J. 2016. Individual differences in BEV drivers' range stress during first encounter of a critical range situation. *Applied Ergonomics*, *57*: 28–35.

Freedman, R.R. 2014. Menopausal hot flashes: Mechanisms, endocrinology, treatment. *The Journal of Steroid Biochemistry and Molecular Biology*, *142*: 115–120.

French, F.J. 1953. Experiments in field settings. In: *Research Methods in Behavioural Sciences*, L. Festinger and D. Katz, eds. Holt, Rinehart and Winston; Dryden Press, New York, pp. 98–135.

Fryar, C.D., Gu, Q., and Ogden, C.L. 2012. *Anthropometric Reference Data for Children and Adults; United States, 2007–2010*. US Department of Health and Human Services, Centers for Disease Control and Prevention. National Center for Health Statistics: DHHS Publication No. (PHS) 2013-1602, Hyattsville, MD.

Fui-Hoon-Na, F. 2004. A study on tolerable waiting time: How long are web users willing to wait? *Behaviour and Information Technology, 23*: 153–163.

Gagnon, M., Plamondon, A., and Gravel, D. 1993. Pivoting with the load. *Spine, 18(11)*: 1515–1524.

Gailliot, M.T. 2015. Energy and psychology. *Global Journal for Research Analysis, 4*: 253–255.

Gailliot, M.T., Baumeister, R.F., DeWall, C.N., Maner, J.K., Plant, E.A., Tice, D.M., Brewer, L.E., and Schmeichel, B.J. 2007. Self-control relies on glucose as a limited energy source: Willpower is more than a metaphor. *Journal of Personality and Social Psychology, 92*(2): 325.

Galasiu, A.D. and Veitch, J.A. 2006. Occupant preferences and satisfaction with the luminous environment and control systems in daylit offices: A literature review. *Energy and Buildings, 38*: 728–742.

Galer, I. (ed). 1989. Applied Ergonomics Handbook. Butterworth, London.

Galima, S.V., Vogel, S.R., and Kowalski, A.W. 2020. Seasonal affective disorder: Common questions and answers. *American Family Physician, 102*(11): 668–672.

Galinsky, T.L., Swanson, N.G., Sauter, S.L., Hurrell, J.J., and Schliefer, L.M. 2000. A field study of supplementary rest breaks for data-entry operators. *Ergonomics, 43*: 622–638.

Gallagher, S. and Barbe, M.F. 2022. *Musculoskeletal Disorders: The Fatigue Failure Mechanism.* John Wiley & Sons, New York, p. 464.

Gallagher, S. and Schall, M.C. 2020. Musculoskeletal disorders as a fatigue failure process: Evidence, implications and research needs. *New Paradigms in Ergonomics, 60*: 105–119.

Gallagher, S., Sesek, R.F., Schall Jr, M.C., and Huangfu, R. 2017. Development and validation of an easy-to-use risk assessment tool for cumulative low back loading: The Lifting Fatigue Failure Tool (LiFFT). *Applied Ergonomics, 63*: 142–150.

Garg, A., Moore, J.S., and Kapellusch, J.M. 2017. The Revised Strain Index: An improved upper extremity exposure assessment model. *Ergonomics, 60*(7): 912–922.

Garrett, G., Benden, M., Mehta, R., Pickens, A., Peres, S.C., and Zhao, H. 2016. Call center productivity over 6 months following a standing desk intervention. *IIE Transactions on Occupational Ergonomics and Human Factors, 4*(2–3), 188–195.

Garrone, B. 2001. Retail falls: The whole story. *Occupational Health and Safety, 70*(9): 196–200.

Gawron, V. 2000. Guides to measuring workload and situational awareness. In: *XIVth Triennial Congress of the International Ergonomics Association and 44th Annual Meeting of the HFES.* The Society, San Diego, CA.

Genaidy, A.M., Waly, S.M., Khali, T.M., and Hidalgo, J. 1993. Spinal compression tolerance limits for the design of manual material handling operations in the workplace. *Ergonomics, 36*: 415–434.

George, M.J. and Odgers, C.L. 2015. Seven fears and the science of mobile technologies may be influencing adolescents in the digital age. *Perspectives on Psychological Science, 10*: 832–851.

Gerard, M.J., Armstrong, T.J., Rempel, D.A., and Wooley, C. 2002. Short term and long term effects of enhanced auditory feedback on typing force, EMG and comfort while typing. *Applied Ergonomics, 33*: 129–138.

Gericke, N., Boeve-de Pauw, J., Berglund, T., and Olsson, D. 2019. The sustainability consciousness questionnaire: The theoretical development and empirical validation of an evaluation instrument for stakeholders working with sustainable development. *Sustainable Development, 27*(1): 35–49.

Gerlich, M. 2025. AI tools in society: Impacts on cognitive offloading and the future of critical thinking. *Societies, 15*(1): 6.

Gerr, F., Marcus, M., Ensor, C., et al. 2002. A prospective study of computer users: I. Study design and incidence of musculoskeletal symptoms and disorders. *American Journal of Industrial Medicine, 41*: 221–235.

Gharajedaghi, J. 1985. *Towards a Systems Theory of Organisation.* Intersystems Publications, Seaside, CA.

Giarnos, P.J. and Wager, T.D. 2015. Brain body pathways linking psychological stress and physical health. *Current Directions in Psychological Science, 24*: 313–321.

Gibbs, J.K., Gillies, M., and Pan, X. 2022. A comparison of the effects of haptic and visual feedback on presence in virtual reality. *International Journal of Human-Computer Studies, 157*: 102717.

Gibson, J.J. 2014. The theory of affordances:(1979). In *The People, Place, and Space Reader.* Routledge, pp. 56–60.

Gibson, D. and Norton, M. 1981. The economics of industrial noise control. *Noise Control Engineering, 16*: 126–133.

Gibson, M. and Potvin, J.R. 2016. An equation to calculate the recommended cumulative rest allowance across multiple subtasks. In: *Association of Canadian Ergonomists (ACE) Conference*, Niagara Falls, ON.

Gibson, M., Potvin, J.R., and Seattle, W.A. 2017. Calculating the maximum acceptable effort for an isolated subtask while still accounting for the demands of all other subtasks. In: *The XXIXth Annual Occupational Ergonomics and Safety Conference.* ISOES, Orlando, FL, pp. 88–91.

Gilchrist, A. and Pokorná, A. 2021. Prevalence of musculoskeletal low back pain among registered nurses: Results of an online survey. *Journal of Clinical Nursing, 30*(11–12): 1675–1683.

Gilson, M. 1946. Review of management and the worker. *The American Journal of Sociology, 46*: 98–101.

Glaister, D.H. 1978. Human tolerance to impact acceleration. *Injury, 9*: 191–198.

Goggins, R.W., Spielholz, P., and Nothstein, G.L. 2008. Estimating the effectiveness of ergonomics interventions through case studies: Implications for predictive cost-benefit analysis. *Journal of Safety Research, 39*(3): 339–344.

Goldberg, D. and Williams, P. 1991. *A User's Guide to the General Health Questionnaire*. NFER-Nelson, Windsor.

Goodhew, S.C. and Edwards, M. 2023. Subjective assessments of cognition and affect and their relationship with objective performance: Individuals with high levels of cognitive failures or negative affect miss more rare visual targets. *Journal of Experimental Psychology: General, 152*(3): 749.

Goodhew, S.C. and Edwards, M. 2024. The cognitive failures questionnaire 2.0. *Personality and Individual Differences, 218*: 112472.

Gorsche, R., Wiley, J.P., Renger, R., Brant, R., Gemer, T.Y., and Sasyniuk, T.M. 1998. Prevalence and incidence of stenosing flexor tenosynovitis (trigger finger) in a meatpacking plant. *Journal of Occupational and Environmental Medicine, 40*: 556–560.

Goshesh, N. 2022. *Telework and the Right to Disconnect: International Experiences*. ILO.org.

Gould, K.S., Knappen-Roed, B., Koefoed, W., Bridger, R.S., and Moen, B. 2006. Performance shaping factors associated with navigation accidents in the Royal Norwegian Navy. *Military Psychology, 18*: S111–S129.

Govaerts, R., De Bock, S., Provyn, S., Vanderborght, B., Roelands, B., Meeusen, R., and De Pauw, K. 2023. The impact of an active and passive industrial back exoskeleton on functional performance. *Ergonomics, 67*: 1–22.

Graham, A.D. and Lin, M.C. 2021. The relationship of pre-corneal to pre-contact lens non-invasive tear breakup time. *PLoS One, 16*(6): e0247877.

Grandjean, E. 1973. *Ergonomics of the Home*. Taylor & Francis, London.

Grant, R.C., Carswell, C.M., Lio, C.H., and Seales, W.B. 2013. Measuring surgeons' mental workload with a time-based secondary task. *Ergonomics in Design: The Quarterly of Human Factors Applications, 21*(1): 7–11.

Grasser, T., Borges Dario, A., Parreira, P.C.S., Correia, I.M.T., and Meziat-Filho, N. 2023. Defining text neck: A scoping review. *European Spine Journal, 32*(10): 3463–3484.

Graveling, R. (ed). 2018. *Ergonomics and Musculoskeletal Disorders (MSDs) in the Workplace: A Forensic and Epidemiological Analysis*. CRC Press, Boca Raton, FL.

Green, D.R., Masterson, E.A., and Themann, C.L. 2021. Prevalence of hearing protection device non-use among noise-exposed US workers in 2007 and 2014. *American Journal of Industrial Medicine, 64*(12): 1002–1017.

Greendale, G.A., Han, W., Finkelstein, J.S., Burnett-Bowie, S.A.M., Huang, M., Martin, D., and Karlamangla, A.S. 2021. Changes in regional fat distribution and anthropometric measures across the menopause transition. *The Journal of Clinical Endocrinology & Metabolism, 106*(9): 2520–2534.

Greer, A.E., Sui, X., Maslow, A.L., Greer, B.K., and Blair, S.N. 2015. The effects of sedentary behaviour on metabolic syndrome independent of physical activity and cardiorespiratory fitness. *Journal of Physical Activity and Health, 12*: 68–73.

Gregg, A.C., Foley, R.C., Livingston, L.A., and La Delfa, N.J. 2021. A biomechanical evaluation of different footrest heights during standing computer work. *Ergonomics, 64*(3): 342–353.

Greig, M.A., Village, J., Dixon, S.M., Salustri, F.A., and Neumann, W.P. 2019. Assessing human factors and ergonomics capability in organisations–the Human Factors Integration Toolset. *Ergonomics, 62*(10): 1254–1272.

Griffin, M.J. 2012. Vibration and motion. In: *Handbook of Human Factors and Ergonomics*, G. Salvendy, ed. John Wiley & Sons, New York, pp. 616–637.

Grooten, W.J.A., Mulder, M., Josephson, M., Alfredsson, L., and Wiktorin, C. 2007. The influence of work-related exposures on the prognosis of neck/shoulder pain. *European Spine Journal, 16*(12): 2083–2091.

Groothuis, J.T., Van Dijk, N., Ter Woerds, W., Wieling, W., and Hopman, M.T. 2007. Leg crossing with muscle tensing, a physical counter-manoeuvre to prevent syncope, enhances leg blood flow. *Clinical Science, 112*(3): 193–201.

Grosjean, M., Shiffrar, M., and Knoblich, G. 2007. Fitt's law holds for action perception. *Psychological Science, 18*: 95–99.

Groves, R.M. 2006. Nonresponse rates and non-response bias in household surveys. *Public Opinion Quarterly, 70*: 646–675.

Guest, D., Knox, A., and Warhurst, C. 2022. Humanizing work in the digital age: Lessons from socio-technical systems and quality of working life initiatives. *Human Relations, 75*(8): 1461–1482.

Gull, M.A., Bai, S., and Bak, T. 2020. A review on design of upper limb exoskeletons. *Robotics*, *9*(1): 16.

Gustafsson, E., Johnson, P.W., Lindegård, A., and Hagberg, M. 2011. Technique, muscle activity and kinematic differences in young adults texting on mobile phones. *Ergonomics*, *54*(5): 477–487.

Guyton, A.C. 1991. *Textbook of Medical Physiology*. 8th Edn., W.B. Saunders, Philadelphia, PA.

Gyi, D., Masson, A., and Hignett, S. 2019. Plus size and obese workers: Anthropometry estimates to promote inclusive design. *Ergonomics*, *62*(9): 1234–1242.

Ha, Y. and Hong, S.K. 2024. Effects of Icon Grouping based on Familiarity on Search Times and Eye Movements (Not Peer-Reviewed). https://doi.org/10.20944/preprints202408.0152.v1.

Haapakangas, A., Hallman, D.M., and Bergsten, E.L. 2023. Office design and occupational health–has research been left behind? *Scandinavian Journal of Work, Environment & Health*, *49*(1): 1.

Hagberg, M. 1987. *Occupational Shoulder and Neck Disorders*. The Swedish Work Environment Fund, Stockholm.

Hall, K., Arora, P., Lowy, R., and Kim, J.G. 2024. Designing for strengths: Opportunities to support neurodiversity in the workplace. In: *Proceedings of the CHI Conference on Human Factors in Computing Systems*. ACM, New York, pp. 1–14.

Hamilton, A. 1949. Brains that click. *Popular Science Magazine*, p. 258.

Hancock, G.M., Schmidt-Daly, T., Fanfarelli, J., Wolfe, J.L., and Szalma, J.L. 2016. Is E-reader technology killing or kindling the reading experience. *Ergonomics in Design*, *24*(1): 25–30.

Hancock, P.A., Lawson, B., Cholewiak, R., Elliot, L.R., vanErp, J.B., Mortimer, B.J., Rupert, A., and Redden, E.S. 2015. Tactile cueing to augment multisensory human-machine interaction. *Ergonomics in Design*, *23*(2): 4–9.

Hänecke, K., Tiedemann, S., Nachreiner, F., and Grzech-Sukalo, H. 1998. Accident risk as a function of hour or work and time of day as determined from accident data and exposure models for the German working population. *Scandinavian Journal of Work Environment and Health*, *24*(3): 43–48.

Haney, J. 2023. Users are not stupid: Six cyber security pitfalls overturned. *Cyber Security: A Peer-Reviewed Journal*, *6*(3): 230–241.

Hardee, J.B. and Mayhorn, C.B. 2007. Re-examining synthetic speech: Intelligibility and the effects of age, task, and speech type on recall. In: *Proceedings of the Human Factors and Ergonomics Society Annual Meeting*. SAGE Publications, Los Angeles, CA. Vol. *51*, No. 18, pp. 1143–1147.

Harris, S. 2016. Errors and accidents. In: *Ernsting's Aviation and Space Medicine*. Taylor and Francis Group, Boca Raton, CA, pp. 707–722.

Hart, S.G. and Staveland, L.E. 1988. Development of NASA TLX (Task Load Index). In: *Human Mental Workload*, P.A. Hancock and N. Meshkati, eds. Elsevier, Amsterdam, Netherlands, pp. 139–183.

Hassan, A., Beumer, A., Kuijer, P.P.F., and van der Molen, H.F. 2022. Work-relatedness of carpal tunnel syndrome: Systematic review including meta-analysis and GRADE. *Health Science Reports*, *5*(6): e888.

Healey, E.L., Burden, A.M., McEwan, I.M., and Fowler, N.E. 2011. Diurnal variation in stature: Do those with chronic low-back pain differ from asymptomatic controls? *Clinical Biomechanics*, *26*(4): 331–336.

Health and Safety Commission. 1991. *Manual Handling of Loads: Proposals for Regulation and Guidance*. Health and Safety Commission, London.

Hedén, K., Anderson, V., Kemmerlt, K., Samdahl-Høden, L., Seppännen, H., and Wickström, G. 1993. Model for the assessment of repetitive monotonous work. In: *The Ergonomics of Manual Work*, W.S. Marras, W. Karwowski, J.L. Smith, and L. Pacholski, eds. Taylor and Francis, London.

Hedge, A. (ed). 2016. *Ergonomic Workplace Design for Health, Wellness, and Productivity*. CRC Press, Boca Raton, FL.

Hedge, A. and Dorsey, J.A. 2016. Green buildings and ergonomics. *Ergonomics*, *56*: 492–506.

Hellebrandt, F.A. 1938. *Standing as a geotropic reflex. The mechanism of the asynchronous rotation of motor units. American Journal of Physiology*, *121*: 471–474.

Hermans, V., Hautekiet, M., Haex, B., Spaepen, A.J., and Van der Perre, G. 1999. Lipoatrophia semicircularis and the relation with office work. *Applied Ergonomics*, *30*: 319–324.

Hernandez-de-Menendez, M., Escobar Díaz, C.A., and Morales-Menendez, R. 2020. Educational experiences with Generation Z. *International Journal on Interactive Design and Manufacturing (IJIDeM)*, *14*(3): 847–859.

Heuer, H. and Owens, D.A. 1989. Vertical gaze direction and the resting posture of the eyes. *Perception*, *18*: 363–377.

Heus, R., Daanen, H.A.M., and Havenith, G. 1995. Physiological criteria for functioning of hands in the cold. *Applied Ergonomics*, *26*: 5–13.

Hewes, G.W. 1957. The anthropology of posture. *Scientific American*, *196*: 122–132.

Hicks, M.T., Humphries, J., and Slater, J. 2024. ChatGPT is bullshit. *Ethics and Information Technology*, *26*(2): 1–10.

Hill, S.G. and Kroemer, K.H.E. 1986. Preferred declination of the line of sight. *Human Factors*, *28*: 127–134.

Höchsmann, C., Knaier, R., Infanger, D., and Schmidt-Trucksäss, A. 2020. Validity of smartphones and activity trackers to measure steps in a free-living setting over three consecutive days. *Physiological Measurement*, *41*(1): 015001.

Hockey, G.R.J. 2005. Operator functional state: The prediction of breakdown in human performance. In: *Measuring the Mind, Speed, Control and Age*, J. Duncan, L. Phillips, and P. McLeod, eds. Oxford University Press, Oxford, pp. 373–393.

Hocking, B. and Savage, C. 1989. Cost–benefit models of hearing conservation. *Journal of Occupational Health and Safety (Australia and New Zealand)*, *5*: 518–525.

Hogan, D.A.M., Greiner, B.A., and O'Sullivan, L. 2014. The effect of manual handling training on achieving transfer, employee's behaviour change and subsequent reduction of work-related musculoskeletal disorders: A systematic review. *Ergonomics*, *57*: 93–107.

Holden, R.J. and Carayon, P. 2021. SEIPS 101 and seven simple SEIPS tools. *BMJ Quality & Safety*, *30*(11): 901–910.

Holdo, J. 1958. Energy consumed by rockdrill noise. *The Mining Magazine*, *99*(2): 73–76.

Holterman, A., Mortensen, O.S., Burr, H., Sogaard, K., Gyntelberg, F., and Suadicani, P. 2010. Physical demands at work, physical fitness, ischaemic heart disease and all-cause mortality in the Copenhagen Male Study. *Scanidnavian Journal of Work Environment and Health*, *36*: 357–365.

Hom, G.L., Kalur, A., Iyer, A., and Singh, R.P. 2022. Ocular occupational injuries in the United States between 2011–2018. *Occupational Medicine*, *72*(4): 255–259.

Hongisto, V., Keränen, J., and Larm, P. 2004. *Joint Baltic-Nordic Acoustics Meeting*, Mariehamn.

Hoogendoorn, W., Bongers, P.M., De Vet, H.C.W., Douwes, M., Koes, B.W., Miedema, M.C., Ariens, G.A.M., and Bouter, L.M. 2000. Flexion and rotation of the trunk and lifting at work are risk factors for low back pain. *Spine*, *25*: 3087–3092.

Hou, J., Nam, Y., Peng, W., and Lee, K.M. 2012. Effects of screen size, viewing angle, and players' immersion tendencies on game experience. *Computers in Human Behavior*, *28*(2): 617–623.

Howarth, P.A. 1990. Assessment of the visual environment. In: *Evaluation of Human Work: A Practical Ergonomics Methodology*, J.R. Wilson and E.N. Corlett, eds. Taylor & Francis, London, pp. 446–478.

Hrobjartsson, A. and Gotzsche, P. 2001. Is the placebo powerless? *New England Journal of Medicine*, *344*: 1594–1602.

Hug, F. and Tucker, K. 2017. Muscle coordination and the development of musculoskeletal disorders. *Exercise and Sport Sciences Reviews*, *45*(4): 201–208.

Hunter, G.R., McGuirk, J., Mitrano, N., Pearman, P., Thomas, B., and Arrington, R. 1989. The effects of a weight training belt on blood pressure during exercise. *Journal of Applied Sports Science Research*, *3*: 13–18.

Huysamen, K., Leach, P., Vereker, A., Coombes, C., Hyatt, T., Bayliss, J., and Tailor, A. 2024. Exploring effectiveness of driver attention and alertness monitoring devices for GB railway. In: *Proceedings of the Annual Conference of the Chartered Institute of Ergonomics and Human Factors*. CIEHF Publications, New York, pp. 300–307.

Huysmans, M., Blatter, B.M., and van der Beek, A.J. 2012. Perceived muscular tension predicts future neck-shoulder and arm-wrist symptoms. *Occupational and Environmental Medicine*, *69*: 261–267.

Hwang, J., Kim, I.S., Madasu, C., and Lee, C. 2023. Effects of chair-based fidgeting and age on prefrontal brain activity and cognitive performance. *Journal of Speech, Hearing and Language Disorders, 32*(1): 91–98.

Hyman Jr, I.E., Boss, S.M., Wise, B.M., McKenzie, K.E., and Caggiano, J.M. 2010. Did you see the unicycling clown? In attentional blindness while walking and talking on a cell phone. *Applied Cognitive Psychology*, *24*(5): 597–607.

Ilic, I., Ristic, B., Stojadinovic, I., and Ilic, M. 2023. Epidemiology of hip fractures due to falls. *Medicina*, *59*(9): 1528.

INCOSE ed., 2023. *INCOSE systems engineering handbook*. John Wiley & Sons.

International Organization for Standardization (ISO). 2008. *Evaluation of Human Exposure to Whole Body Vibration—Part 1: General Requirements*. Geneva (ISO 2631-1).

International Test Commission. 2005. ITC Guidelines on Computer-Based and Internet Delivered Testing1st July, 2005, Version 1.0: ITC-G-CB-20140617.

Inzlicht, M., Werner, K.M., Briskin, J.L., and Roberts, B.W. 2021. Integrating models of self-regulation. *Annual Review of Psychology*, *72*(1): 319–345.

Ipsen, C., van Veldhoven, M., Kirchner, K., and Hansen, J.P. 2021. Six key advantages and disadvantages of working from home in Europe during COVID-19. *International Journal of Environmental Research and Public Health, 18*(4): 182

Irgens-Hansen, K., Gundersen, G., Sund, E., Baste, V., Harris, A., Bratveit, M., and Moen, B. 2015. Noise exposure and cognitive performance: A study of personnel on board Royal Norwegian Navy vessels. *Noise and Health, 17*: 320–327.

Irvine, C.H., Snook, S.H., and Sparshatt, J.H. 1990. Stair way risers and treads: Acceptable and preferred dimensions. *Applied Ergonomics, 21*: 215–225.

ISO 10075. 1991. *Ergonomic Principles Related to Mental Work-Load—General Terms and Definitions*. International Standards Organization, Geneva, Switzerland.

ISO 11226. 2000. *Ergonomics – Evaluation of Static Working Postures*. International Standards Organization, Geneva.

ISO 11399. 1995. *Ergonomics of the Thermal Environment—Principles and Application of Relevant International Standards*. International Standards Organization, Geneva, Switzerland.

ISO 2631-1. 1997. Mechanical vibration and shock. https://www.iso.org/standard/7612.html

ISO 3353. 1974. *Aerospace — Lead and Runout Threads — Part 1: Rolled External Threads*. International Standards Organization, Geneva.

ISO 7731. 2003. Danger signals for public and work areas. https://www.iso.org/standard/33590.html

Jack, G., Riach, K., Bariola, E., Pitts, M., Schapper, J., and Sarrel, P. 2016. Menopause in the workplace: What employers should be doing. *Maturitas, 85*: 88–95.

Jackson, A.S., Sui, X., Hebert, J.R., Church, T.S., and Blair, S.N. 2009. Role of lifestyle and physical activity on the longitudinal change in cardiorespiratory fitness. *Archives of Internal Medicine, 169*: 1781–1787.

Jäger, M. 2018. Extended compilation of autopsy-material measurements on lumbar ultimate compressive strength for deriving reference values in ergonomic work design: The Revised Dortmund Recommendations. *EXCLI Journal, 17*: 362.

Jahncke, H. 2012. Open plan office noise: The susceptibility and suitability of different cognitive tasks for work in the presence of irrelevant speech. *Noise and Health, 14*: 315–320.

Jampel, R.S. and Shi, D.X. 1992. The primary position of the eyes, the resetting saccade and the transverse visual head plane. *Investigative Opthalmology and Visual Science, 33*: 2501–2510.

Jaschinski-Kruza, W. 1999. A procedure to determine the individually comfortable position of visual displays relative to the eyes. *Ergonomics, 42*: 535–549.

Jastrzebowski, W. 1857. *An Outline of Ergonomics or the Science of Work*. Central Instititute for Labour Protection, Warsaw, 2000.

Johanning, E. 2015. Whole-body vibration-related health disorders in occupational medicine–an international comparison. *Ergonomics, 58*(7): 1239–1252.

Johnson-Laird, P.N. 1988. *The Computer and the Mind—An Introduction to Cognitive Science*. Harvard University Press, Cambridge, MA.

Jones, A.M., Kirby, B.S., Clark, I.E., Rice, H.M., Fulkerson, E., Wylie, L.J., Wilkerson, D.P., Vanhatalo, A., and Wilkins, B.W. 2021. Physiological demands of running at 2-hour marathon race pace. *Journal of Applied Physiology, 130*(2): 369–379.

Jones, G.V. and Martin, M. 2003. Distribution of attention and failure to save computer work. In: *Human Computer Interaction INTERACT 03*, M. Rauterberg ed. IOS Press IFIP, Amsterdam, The Netherlands, pp. 789–792.

Jones, J.A. 1933. Effect of posture work on the health of children. *American Journal of Diseases of Children, 46*(1): 148–154.

Juslen, H., Wouters, M., and Tenner, A. 2007. The influence of controllable task lighting on productivity: A field study in a factory. *Applied Ergonomics, 38*: 39–44.

Kahneman, D. 2011. *Thinking Fast and Slow*. Penguin Random House, London.

Kalawsky, R.S. 1999. VRUSE–A computerised diagnostic tool. *Applied Ergonomics, 30*: 11–25.

Kamerāde, D., Wang, S., Burchell, B., Balderson, S.U., and Coutts, A. 2019. A shorter working week for everyone: How much paid work is needed for mental health and well-being?. *Social Science & Medicine, 241*: 112353.

Kane, M.J., Bronwn, L.H., McVay, J.C., Silvia, P.J., Myin-Germeys, I., and Kwapil, T. 2007. For whom the mind wanders and when: An experience sampling study of working memory and executive control in daily life. *Psychological Science, 18*: 614–621.

Kapandji, I.A. 1982. *The Physiology of the Joints*. Vols. *1–3*. Churchill Livingstone, Longman Group, Edinburgh.

Kaplan, A.D., Cruit, J., Endsley, M., Beers, S.M., Sawyer, B.D., and Hancock, P.A. 2021. The effects of virtual reality, augmented reality, and mixed reality as training enhancement methods: A meta-analysis. *Human Factors*, *63*(4): 706–726.

Karasek, R.A. and Theorell, T. 1990. *Healthy Work: Stress, Productivity and the Reconstruction of Working Life*. Basic Books, New York.

Karwowski, W. 2005, Ergonomics and human factors: The paradigms for science, engineering, design, technology, and management of human-compatible systems. *Ergonomics*, *48*(5): 436–463.

Karwowski, W., Salvendy, G., Albert, L., Kim, W.C., Denton, B., Dessouky, M., Dolgui, A., Duffy, V., Kumara, S., Li, J., and Madni, A.M. 2025. Grand challenges in industrial and systems engineering. *International Journal of Production Research*, *63*: 1–46.

Kates, R.W. 1994. Sustaining life on earth. *Scientific American*, *271*: 114–117.

Katzmarzyk, P.T., Cgurch, T.S., Craig, C.L., and Bouchard, C. 2009. Sitting time and mortality from all causes, cardiovascular disease and cancer. *Medicine and Science in Sports and Exercise*, *41*(5): 998–1005.

Katzmarzyk, P.T., Craig, C.L., and Gauvin, L. 2007. Adiposity, physical fitness and incident diabetes: The physical activity longitudinal study. *Diabetologia*, *50*: 538–544.

Kazar, G. and Comu, S. 2022. Exploring the relations between the physiological factors and the likelihood of accidents on construction sites. *Engineering, Construction and Architectural Management*, *29*(1): 456–475.

Keegan, J.J. 1953. Alterations of the lumbar curve related to posture and seating. *Journal of Bone and Joint Surgery*, *35A*: 589–603.

Keir, P.J., Farias Zuniga, A., Mulla, D.M., and Somasundram, K.G. 2021. Relationships and mechanisms between occupational risk factors and distal upper extremity disorders. *Human Factors*, *63*(1): 5–31.

Kelly, K., Boyce, P., Cuttle, K., and Raynham, P. 2020. The ambient lighting manifesto. *Level 3*, *15*(1): 2.

Kennedy, R.S., Lane, N.E., Berbaum, K.S., and Lilienthal, M.G., 1993. Simulator sickness questionnaire: An enhanced method for quantifying simulator sickness. *The International Journal of Aviation Psychology*, *3*(3): 203–220.

Kermavnar, T., de Vries, A.W., de Looze, M.P., and O'Sullivan, L.W. 2021. Effects of industrial back-support exoskeletons on body loading and user experience: An updated systematic review. *Ergonomics*, *64*(6): 685–711.

Kerslake, D.M. 1982. Effects of climate. In: *The Body at Work*, W.T. Singleton, ed. Cambridge University Press, Cambridge, pp. 235–297.

Keyserling, W.M. et al. 1988. Trunk posture and back pain: Identification and control of occupational risk factors. *Applied Industrial Hygiene*, *3*: 87.

Khovalyg, D. and Ravussin, Y. 2022. Inter individual variability of human thermoregulation: Toward personalized ergonomics of the indoor thermal environment. *Obesity*, *30*(7): 1345–1350.

Kildeso, J., Wyon, D., Skov, T., and Schneider, T. 1999. Visual analogue scales for detecting changes in symptoms of sick building syndrome in an intervention study. *Scandinavian Journal of Work Environment and Health*, *25*: 361–367.

Kim, D.S., Won, Y.H., and Ko, M.H. 2022. Comparison of intermittent pneumatic compression device and compression stockings for workers with leg edema and pain after prolonged standing: A prospective crossover clinical trial. *BMC Musculoskeletal Disorders*, *23*(1): 1–14.

Kim, J.-Y., Min, S.-N., Subramaniyam, M., and Cho, Y.-J. 2014. Legibility differences between e-books and paper books by using an eye tracker. *Ergonomics*, *57*: 1102–1108.

Kim, Y.H., Choi, J.H., Chung, Y.K., Kim, S.W., and Kim, J. 2019. Epidemiologic study of hand and upper extremity injuries by power tools. *Archives of Plastic Surgery*, *46*(01): 63–68.

Kingma, B. and van Marken Lichtenbelt, W. 2015. Energy consumption in buildings and female thermal demand. *Nature Climate Change*, *5*(12): 1054–1056.

Kjellstrom, T. and Dirks, K.N. 2001. Heat stress and work ability in the context of climate change. *Proceedings of the Australian Physiological and Pharmacological Society*, *32*(2, Supplement 1): 23.

Klausen, K. 1965. The form and function of the loaded human spine. *Acta Physiologica Scandinavica*, *65*: 176–190.

Klausen, K. and Rasmussen, B. 1968. On the location of the line of gravity in relation to L5 in standing. *Acta Physiologica Scandinavica*, *75*: 45–52.

Kleinmuntz, B. 1984. Diagnostic problem solving by computer: A historical review and the current state of the art. *Computers in Biology and Medicine*, *14*: 255–270.

Kline, T.J., Ghali, L.M., Kline, D.W., and Brown, S. 1990. Visibility distance of highway signs among young, middle-aged and older observers: Icons are better than text. *Human Factors*, *30*: 609–619.

Knight, A.P. and Baer, M. 2014. Get up, stand up: The effects of a non-sedentary workspace on information elaboration and group performance. *Social Psychological and Personality Science*, *5*(8): 910–917.

Knudsen, M. and Kaivo-Oja, J. 2020. Collaborative robots: Frontiers of current literature. *Journal of Intelligent Systems: Theory and Applications*, *3*(2): 13–20.

Ko, W.H., Schiavon, S., Zhang, H., Graham, L.T., Brager, G., Mauss, I., and Lin, Y.W. 2020. The impact of a view from a window on thermal comfort, emotion, and cognitive performance. *Building and Environment*, *175*: 106779.

Kobus, C.A., Mugge, R., and Schoormans, J.P.L. 2016. Washing when the sun is shining! How users interact with a household energy management system. *Ergonomics*, *56*: 451–464.

Koepp, G.A., Moore, G., and Levine, J.A. 2017. An under-the-table leg-movement apparatus and changes in energy expenditure. *Frontiers in Physiology*, *8*: 318.

Koepp, G.A., Snedden, B.J., and Levine, J.A. 2015. Workplace slip, trip and fall injuries and obesity. *Ergonomics*, *58*(5): 674–679.

Kolose, S., Stewart, T., Hume, P., and Tomkinson, G.R. 2021. Cluster size prediction for military clothing using 3D body scan data. *Applied Ergonomics*, *96*: 103487.

Kompier, M. 2006. The Hawthorne effect is a myth, but what keeps the story going? *Scandinavian Journal of Work Environment and Health*, *32*: 402–412.

Kong, Y.K., Kim, J.H., Shim, H.H., Shim, J.W., Park, S.S., and Choi, K.H. 2023. Efficacy of passive upper-limb exoskeletons in reducing musculoskeletal load associated with overhead tasks. *Applied Ergonomics*, *109*: 103965.

Konz, S. 1986. Bent hammer handles. *Human Factors*, *28*: 317–323.

Konz, S., Bandla, V., Rys, M., and Sambasivan, J. 1990. Standing on concrete versus floor mats. In: *Advances in Industrial Ergonomics and Safety II*, B. Das, ed. Taylor & Francis, London, pp. 991–998.

Korpinen, L., Pääkkönen, R., and Gobba, F. 2013. Self-reported neck symptoms and use of personal computers, laptops and cell phones among Finns aged 18–65. *Ergonomics*, *56*: 1134–1146.

Kotowski, S.E. and Gibson, S.L. 2023. *Ergonomic Assessment Toolkit*. American Industrial Hygiene Association, Falls Church, VA.

Kotval, X. and Goldberg, J.H. 1998. Eye movements and interface component grouping. In: *Proceedings of the Human Factors and Ergonomics Society Annual Meeting*. SAGE Publications, Los Angeles, CA. Vol. *42*, No. 5, pp. 486–490.

Koukli, M., Siegmund, F., and Papageorgopoulou, C. 2023. Stature estimation in Ancient Greece: Population-specific equations and secular trends from 9000 BC to 900 AD. *Archaeological and Anthropological Sciences*, *15*(5): 53.

Kowlessar, N.M., Goetzel, R.Z., Smith Carls, G., Tabrizi, M., and Guindon, A. 2011. The relationship between 11 health risks and medical and productivity costs for a large employer. *Journal of Occupational and Environmental Medicine*, *53*: 468–477.

Kozycki, R. and Gordon, C.C. 2002. Applying human figure modeling tools to the RAH-66 Comanche crewstation design. *VDI Berichte*, 1675: 191–202.

KPMG. 2020. *Autonomous Vehicles Readiness Index*. KPMG International. https://assets.kpmg.com/content/dam/kpmg/xx/pdf/2020/07/2020-autonomous-vehicles-readiness-index.pdf.

KPMG. 2024. *The Future of Work*. https://kpmg.com/uk/en/home/insights/2024/02/future-of-work.html.

Krause, N., Rugulies, R., Ragland, D.R., and Syme, S.L. 2004. Physical workload, ergonomic problems and incidence of low back injury: A 7.5 year prospective study of San Francisco transit operators. *American Journal of Industrial Medicine*, *46*: 570–585.

Krejcie, R.V. and Morgan, D.W. 1970. Determining sample size for research activities. *Educational and Psychological Measurement*, *30*: 607–610.

Kucharczyk, E.R., Morgan, K., and Hall, A.P. 2012. The occupational impact of sleep quality and insomnia symptoms. *Sleep Medicine Reviews*, *16*(6): 547–559.

Kuijer, P.P.F.M., Hondebrink, D., Hulshof, C.T.J., and Van der Molen, H.F. 2020. Work-relatedness of inguinal hernia: A systematic review including meta-analysis and GRADE. *Hernia*, *24*: 943–950.

Kujawa, S.G. and Liberman, M.C. 2009. Adding insult to injury: Cochlear nerve degeneration after "temporary" noise-induced hearing loss. *Journal of Neuroscience*, *29*(45): 14077–14085.

Kunreuther, H. and Heal, G. 2012. *Managing Catastrophic Risk (No. w18136)*. National Bureau of Economic Research, New York.

Kyriakaki, E.D., Symvoulakis, E.K., Chlouverakis, G., and Detorakis, E.T. 2021. Causes, occupational risk and socio-economic determinants of eye injuries: A literature review. *Medicine and Pharmacy Reports*, *94*(2): 131.

Lafave, M., LaPorta, A.J., Hutton, J., and Mallory, P.L. II. 1995. History of high-velocity impact water trauma at Letterman Army Medical Center: A 54-year experience with the Golden Gate Bridge. *Military Medicine*, *160*(4): 197–199.

Lahiri, S., Markkanen, P., and Levenstein, C. 2005. The cost effectiveness of occupational health interventions: Preventing occupational back pain. *American Journal of Industrial Medicine*, *48*(6): 515–529.

Lai, X., Lee, Y.C., Hong, X., and Rau, P.L.P. 2024. Watch your step: A pilot study of smartphone use effect on young females' gait performance while walking up and down stairs and escalators. *Applied Ergonomics*, *114*: 104130.

Landers, M., Barker, G., Wallentine, S., McWhorter, J.W., and Peel, C. 2003. A comparison of tidal volume, breathing frequency and minute ventilation between two sitting postures. *Physiotherapy Theory and Practice*, *19*: 109–119.

Landrigan, C.P. 2004. Effect of reducing interns' work hours on serious medical errors in intensive care units. *The New England Journal of Medicine*, *351*: 1838–1848.

Langham, M., Hole, G., Edwards, J., and O'Neil, C. 2002. An analysis of 'looked but failed to see' accidents involving parked police vehicles. *Ergonomics*, *45*: 167–185.

Lanier, J. 2017. *Dawn of the New Everything: A Journey Through Virtual Reality*. Random House, Ireland.

Lansdown, T.C., Cowan, S., Nioi, A., Cowie, H., Wendelboe-Nelson, C., Rashid, S., Ritchie, P., and Cherrie, J.W. 2020. Vitamin D and UV exposure in construction workers—a randomized control trial using text messaging to promote positive behaviours. *Journal of Public Health*, *42*(3): 594–601.

Lassek, W.D. and Gaulin, S.J. 2022. Substantial but misunderstood human sexual dimorphism results mainly from sexual selection on males and natural selection on females. *Frontiers in Psychology*, *13*: 859931.

Laugwitz, B., Held, T., and Schrepp, M. 2008. Construction and evaluation of a user experience questionnaire. In: *HCI and Usability for Education and Work: 4th Symposium of the Workgroup Human-Computer Interaction and Usability Engineering of the Austrian Computer Society, USAB 2008, Graz, Austria*. Springer, Berlin, pp. 63–76.

Lazar, J., Jones, A., and Schneiderman, B. 2006. Workplace user frustration with computers: An exploratory investigation. *Behaviour and Information Technology*, *25*: 234–239.

Le Bigot, L., Jamet, E., and Rouet, J. 2004. Searching for information with natural language dialogue systems: A comparison of spoken versus written modalities. *Applied Ergonomics*, *35*: 557–664.

Le Bigot, L., Rouet, J.F., and Jamet, E. 2007. Effects of speech-and text-based interaction modes in natural language human-computer dialogue. *Human Factors*, *49*(6): 1045–1053.

Le Roux, E., De Jong, N.P., Blanc, S., Simon, C., Bessesen, D.H., and Bergouignan, A. 2021. Physiology of physical inactivity, sedentary behaviours and non-exercise activity: Insights from the space bedrest model. *The Journal of Physiology*, *600*(5): 1037–1051.

Le Roy, B., Martin-Krumm, C., and Trousselard, M. 2023. Mindfulness for adaptation to analog and new technologies emergence for long-term space missions. *Frontiers in Space Technologies*, *4*: 1109556.

Lee, D.L., Fleischer, J., and Mcloone, H.E. 2007. Alternative computer mouse design and testing to reduce finger extensor muscle activity during mouse use. *Human Factors*, *49*: 573–582.

Lee, J.D. 2012. Scrolling and driving how an MP3 player and its aftermarket controller affect driving performance and visual behavior. *Human Factors: The Journal of the Human Factors and Ergonomics Society*, *54*(2): 250–263.

Lee, K.S., Chaffin, D.B., Herrin, G.D., and Walker, A.M. 1991. Effect of handle height and lower-back loading in cart pushing and pulling. *Applied Ergonomics*, *22*: 117–123.

Lee, Y.H., Her, L.L., and Tsuang, Y.H. 1999. A comparison of sitting posture adaptations of pregnant and non-pregnant females. *International Journal of Industrial Ergonomics*, *23*: 391–396.

Lema, A.K. and Anbesu, E.W. 2022. Computer vision syndrome and its determinants: A systematic review and meta-analysis. *SAGE Open Medicine*, *10*: 20503121221142402.

Leveson, N. 2011. *Engineering a Safer World: Systems Thinking Applied to Safety*. MIT Press: Cambridge.

Leveson, N. 2020. Safety III: A systems approach to safety and resilience. *MIT Engineering Systems Laboratory*, *16*: 2021.

Levine, J.A. 2014. Lethal sitting: Homo Sedentarius seeks answers. *Physiology*, *29*: 300–301.

Li, J.Q., Kwong, W.H., Chan, Y.L., and Kawabata, M. 2022. Comparison of in vivo intradiscal pressure between sitting and standing in human lumbar spine: A systematic review and meta-analysis. *Life*, *12*(3): 457.

Liberty Mutual Research Center for Safety and Health. 2004. *From Research to Reality*. Annual Report, Hopkinton, MA.

Lie, I. and Watten, R. 1987. Oculomotor factors in the aetiology of occupational cervicobrachial diseases (OCD). *European Journal of Applied Physiology and Occupational Physiology*, *56*: 151–156.

Lie, I. and Watten, R. 1994. VDT work, oculomotor strain and subjective complaints: An experimental study. *Ergonomics*, *37*(8): 1149–1160.

Lim, J. and Hoffman, E. 1997. Appreciation of the zone of convenient reach by naïve operators performing an assembly task. *International Journal of Industrial Ergonomics*, *19*: 187–199.

Lin, H.X., Choong, Y.Y., and Salvendy, G. 1997. A proposed index of usability: A method for comparing the relative usability of different software systems. *Behaviour and Information Technology*, *16*: 267–278.

Lockhart, T.E., Smith, J.L., and Waldstad, J.C. 2005. Effects of age on the biomechanics of slips and falls. *Human Factors*, *47*: 708–729.

Lockley, S.W. 2004. Effect of reducing interns' weekly work hours on sleep and attentional failures. *The New England Journal of Medicine*, *351*: 1829–1837.

Loh, R., Stamatakis, E., Folkerts, D., Allgrove, J.E., and Moir, H.J. 2020. Effects of interrupting prolonged sitting with physical activity breaks on blood glucose, insulin and triacylglycerol measures: A systematic review and meta-analysis. *Sports Medicine*, *50*: 295–330.

Long, G.M. and Garvey, P.M. 1988. The effects of target wavelength on dynamic visual acuity under photopic and scotopic viewing. *Human Factors*, *30*: 3–13.

Lord, M.J., Small, J.M., Dinsay, J.M., and Watkins, R.G. 1997. Lumbar lordosis: Effects of standing and sitting. *Spine*, *22*: 2571–2574.

Lotz, J.C. and Chin, J.R. 2000. Intervertebral disc cell death is dependent on the magnitude and duration of static loading. *Spine*, *25*(12): 1477–1483.

Lovejoy, C.O. 1988. Evolution of human walking. *Scientific American*, *259*: 118–125.

Lucas, R.J., Allen, A.E., Milosavljevic, N., Storchi, R., and Woelders, T. 2020. Can we see with melanopsin? *Annual Review of Vision Science*, *6*: 453–468.

Ludwig, O., Dindorf, C., Kelm, S., Kelm, J., and Fröhlich, M. 2024. Muscular strategies for correcting the pelvic position to improve posture—an exploratory study. *Journal of Functional Morphology and Kinesiology*, *9*(1): 25.

Lundberg, U. 1995. Methods and applications of stress research. *Technology and Health Care*, *3*: 3–9.

Lutman, M.E. 2000. What is the risk of noise-induced hearing loss at 80, 85, 90 dB(A) and above? *Occupational Medicine*, *50*: 274–275.

Lyzwinski, L.N. 2014. A systematic review and meta-analysis of mobile devices and weight loss with an intervention content analysis. *Journal of Personalised Medicine*, *4*: 311–385.

MacDermid, J.C., Walton, D.M., Avery, S., Blanchard, A., Etruw, E., Mcalpine, C., and Goldsmith, C.H. 2009. Measurement properties of the neck disability index: A systematic review. *Journal of Orthopaedic & Sports Physical Therapy*, *39*(5): 400–417.

MacFarlane, G.J., Hunt, I., and Silman, A.J. 2000. Role of mechanical and psychosocial factors in the onset of forearm pain: Prospective population-based study. *British Medical Journal*, *16*: 676–678.

McKenzie, I.S. and Buxton, W. 1993. A tool for the rapid evaluation of input devices using Fitts' law models. *SIGCHI Bulletin*, *25*(3): 58–63.

MacKenzie, I.S. and Zang, S.X. 1999. The design and evaluation of a high performance soft keyboard. In: *CHI-99 Proceedings, ACM SIGCHI Conference on Human Factors in Computing Systems*. David Lawrence Convention Center, Pittsburgh, PA, pp. 25–31.

MacLean, C.L. 2022. Cognitive bias in workplace investigation: Problems, perspectives and proposed solutions. *Applied Ergonomics*, *105*: 103860.

Magnusson, M., Pope, M.H., and Hansson, T. 1996b. Does a back support have a positive biomechanical effect? *Applied Ergonomics*, *27*: 201–205.

Magnusson, M.L., Aleksiev, A.R., Spratt, K.F, Lakes, R.S., and Pope, M.H. 1996a. Hyperextension and spine height changes. *Spine*, *21*: 2670–2675.

Magnusson, S.P., Simonsen, E.B., Aagaard, P., Sørensen, H., and Kjaer, M. 1996c. A mechanism for altered flexibility in human skeletal muscle. *The Journal of Physiology*, *497*(1): 291–298.

Makar, S.K. 2021. Etiology and pathophysiology of tinnitus: A systematic review. *The International Tinnitus Journal*, *25*(1): 76–86.

Maleki, N., Padmanabhan, B., and Dutta, K. 2024. AI hallucinations: a misnomer worth clarifying. In: *2024 IEEE Conference on Artificial Intelligence (CAI)*. IEEE, New York, pp. 133–138.

Mandal, A.C. 1991. Investigation of the lumbar flexion of the seated man. *International Journal of Industrial Ergonomics*, *8*: 75–87.

Manley, T., Lewis, G.H., Robertson, I.H., Watson, P.C., and Datta, A.K. 2002. Coffee in the cornflakes: Time of day as a modulator of executive response. *Neuropsychologia*, *40*: 1–6.

Margaritis, S. and Marmaras, N. 2007. Supporting the design of office layout meeting ergonomics requirements. *Applied Ergonomics*, *38*: 781–790.

Marin-Farrona, M., Wipfli, B., Thosar, S.S., Colino, E., Garcia-Unanue, J., Gallardo, L., Felipe, J.L., and López-Fernández, J. 2023. Effectiveness of worksite wellness programs based on physical activity to improve workers' health and productivity: A systematic review. *Systematic Reviews*, *12*(1): 1–13.

Marmot, M. and Brunner, E. 2004. Cohort profile: The Whitehall II study. *International Journal of Epidemiology*, *34*: 251–256.

Marras, W.S., Allread, W.G., Burr, D.L., and Fathallah, F. 2000. Prospective validation of a low-back disorder risk model and assessment of ergonomic interventions associated with manual materials handling. *Ergonomics*, *43*: 1866–1886.

Marras, W.S., Lavender, S.A., Leurgans, S.E., Rajulu, S.L., Allread, W.G., Fathalla, F.A., and Ferguson, S.A. 1993. The effects of dynamic three dimensional trunk position and trunk motion characteristics on risk of injury. *Spine*, *18*: 617–628.

Marras, W.S., Rangarajulu, S.L., and Lavender, S.A. 1987. Trunk loading and expectation. *Ergonomics*, *30*: 551–562.

Maslej, N., Fattorini, L., Perrault, R., Parli, V., Reuel, A., Brynjolfsson, E., Etchemendy, J., Ligett, K., Lyons, T., Manyika, J., Niebles, J.C., Shoham, Y.,Wald, R., and Clark, J. 2024. *The AI Index 2024 Annual Report*. AI Index Steering Committee, Institute for Human-Centered AI, Stanford University, Stanford, CA.

Mathiowetz, V., Kashman, N., Vollnad, G., Weber, K., Dowe, M., and Rogers, S. 1985. Grip and pinch strength: Normative data for adults. *Archives of Physical Medicine and Rehabilitation*, *66*: 69–74.

Matthews, R. 2002. *The Sunday Telegraph*, April 28, p. 4. https://www.telegraph.co.uk/news/uknews/4182184/Britonsabroad-are-the-worst-drivers-in-Europe/html.

Maughan, R.J. and Shirreffs, S.M. 2019. Muscle cramping during exercise: Causes, solutions, and questions remaining. *Sports Medicine*, *49*(2): 115–124.

Mawston, G., Holder, L., O'Sullivan, P., and Boocock, M. 2021. Flexed lumbar spine postures are associated with greater strength and efficiency than lordotic postures during a maximal lift in pain-free individuals. *Gait & Posture*, *86*: 245–250.

McCambridge, J., Witton, J., and Elbourne, D.R. 2014. Systematic review of the Hawthorne effect: New concepts are needed to study research participation effects. *Journal of clinical epidemiology*, *67*(3): 267–277.

McCammon, I. 2004. Heuristic traps in recreational avalanche accidents: Evidence and implications. *Avalanche News*, *68*: 1–10.

McCartney, K.J. and Humphreys, M.A. 2002. Thermal comfort and productivity. *Proceedings: Indoor Air*, *2002*: 822–827.

McClelland, L. and Cook, S.W. 1979. Energy conservation effects of continuous in-home feedback in all-electric houses. *Journal of Environmental Systems*, *9*: 169–173.

McCook, A. 2001. Napolean's revenge. *Scientific American*, *16*: 27.

McCullogh, C.E., Paal, B., and Ashdown, S.P. 1998. An optimisation approach to apparel sizing. *Journal of the Operational Research Society*, *49*: 492–499.

McEwan, B.T., McDonald, D.J., and Burr, J.F. 2015. A systematic review of standing and treadmill desks in the workplace. *Preventative Medicine*, 70: 50–58.

McGill, S. 2002. *Low Back Disorders: Evidence-Based Prevention and Rehabilitation*. Human Kinetics Publishers, Champaign, IL.

McGill, S., Norman, R.W., and Sharratt, M. 1990. The effect of an abdominal belt on trunk muscle activity and intra-abdominal pressure during squat lifts. *Ergonomics*, *33*: 146–160.

McGill, S.M. and Norman, R.W. 1992. Low back biomechanics in industry. In: *Current Advances in Biomechanics*, M. Grabiner, ed. Human Kinetics Publishers, Longman, IL.

McGurk, H. and MacDonald, J. 1976. Hearing lips and seeing voices. *Nature*, *264*(5588): 746–748.

McMorris, T., Myers, S., Dobbins, T., Hall, B., and Dyson, R. 2009. Seating type and cognitive performance after 3 hours travel by high-speed boat in sea states 2–3. *Aviation, space, and Environmental Medicine*, *80*(1): 24–28.

Melamed, S. and Bruhis, S. 1996. The effects of chronic industrial noise exposure on urinary cortisol, fatigue and irritability. *Journal of Occupational and Environmental Medicine*, *38*: 252–256.

Melhorn, J.M., Wilkinson, L., and Riggs, J.D. 2001. Management of musculoskeletal pain in the workplace. *Journal of Occupational Medicine*, *43*: 83–93.

Messing, K., Tissot, F., and Stock, S. 2008. Distal lower extremity pain and work postures in the Quebec population. *American Journal of Public Health*, *98*: 705–713.

Meyers, C.R. 1969. A study of the reliability of the Harvard step test. *Research Quarterly. American Association for Health, Physical Education and Recreation*, *40*: 423.

Middlemist, R.D., Knowles, E.S., and Matter, C.F. 1976. Personal space invasions in the lavatory: Suggestive evidence for arousal. *Journal of Personality and Social Psychology, 33*: 541–546.

Miller, E.M., Matrangola, S.L., and Madigan, M.L. 2011. Effects of obesity on balance recovery from small postural perturbations. *Ergonomics, 54*(6): 547–554.

Miller, G.A. 1956. The magical number seven, plus or minus two: Some limits on our capacity for processing information. *Psychological Review, 63*: 81–97.

Miller, G.E., Chen, E., and Zhou, E.S. 2007. Of it goes up, must it come down? Chronic stress and the hypothalamic–pituitary–adrenal cortex in humans. *Psychological Bulletin, 133*: 25–45.

Mischel, W. and Shoda, Y. 1995. A cognitive-affective system theory of personality: Reconceptualizing situations, dispositions, dynamics and invariance in personality structure. *Psychological Review, 102*(2): 246–268.

Miyamoto, K., Linuma, N., Maeda, M., Wada, E., and Shimizu, K. 1999. Effects of abdominal belts on intra-abdominal pressure, intramuscular pressure in the erector spinae muscles and myoelectric activity of the trunk muscles. *Clinical Biomechanics, 14*: 79–87.

Moore, J.S. 1997. De Quervain's tenosynovitis. *Journal of Occupational and Environmental Medicine, 39*: 990–1002.

Moore, S.J. and Vos, G.A. 2005. The strain index. In: *Handbook of Human Factors and Ergonomics* Methods, N. Stanton, A. Hedge, K. Brookhuis, E. Salas, and H. Hendrick, eds. CRC Press, Boca Raton, FL, pp. 1–10.

Moorhouse, K.M. and Granata, K.P. 2007. Role of reflex dynamics in spinal stability: Intrinsic muscle stiffness alone is insufficient for stability. *Journal of biomechanics, 40*(5): 1058–1065.

Moroe, N.F. and Khoza-Shangase, K. 2020. Recent advances in hearing conservation programmes: A systematic review. *South African Journal of Communication Disorders, 67*(2): 1–11.

Morris, J.S., Link, J., Martin, J.C., and Carrier, D.R. 2020. Sexual dimorphism in human arm power and force: Implications for sexual selection on fighting ability. *Journal of Experimental Biology, 223*(2): jeb212365.

Mosenkis, R. 1994. Human factors in design. In: *Medical Devices: International Perspectives in Health and Safety*, C.W.D. van Gruting, ed. Elsevier, New York, pp. 41–51.

Moshfeghian, S. and Ryu, Y.S. 2012. A passport to password best practices. *Ergonomics in Design, 20*(2): 23–29.

Mrugalska, B. and Karwowski, W. 2024. *Anthropometry: Human Body Measurements and How to Use Them.* Routledge, London.

Mu, Y.M., Huang, X.D., Zhu, S., Hu, Z.F., So, K.F., Ren, C.R., and Tao, Q. 2022. Alerting effects of light in healthy individuals: A systematic review and meta-analysis. *Neural Regeneration Research, 17*(9): 1929.

Muir, K. 2022. *Everything You Need to Know About the Menopause (but were too Afraid to Ask).* Simon and Schuster, New York.

Muntaner-Mas, A., Martinez-Nicolas, A., Lavie, C.J., Blair, S.N., Ross, R., Arena, R., and Ortega, F.B. 2019. A systematic review of fitness apps and their potential clinical and sports utility for objective and remote assessment of cardiorespiratory fitness. *Sports Medicine, 49*: 587–600.

Murdock, B.B. 1962. The serial position effect in free recall. *Journal of Experimental Psychology, 64*: 482–488.

Murphy, E., Dingwall, R., Greatbatch, D., Parker, S., and Watson, P. 1998. Qualitative research methods in health technology assessment: A review of the literature. *Health Technology Assessment, 2*: 3–9.

Murrell, K.F.H. 1971. *Ergonomics: Man and His Working Environment.* Chapman and Hall, London.

Murrell, K.F.H. and Edwards, E. 1963. Field studies of an indicator machine tool travel with special relevance to the ageing worker. *Occupational Psychology*, 37: 267–275.

Muscio, B. 1921. Is a fatigue test possible? *British Journal of Psychology, 12*: 31–46.

Myers, S.D., Dobbins, T.D., King, S., Hall, B., Holmes, R., Gunston, T., and Dyson, R. 2012. Effectiveness of suspension seats in maintaining performance following military high speed boat transits. *Human Factors, 54*: 264–276.

Myles, K. and Kalb, J.T. 2013. Head tactile communication: Promising technology with the design of a head-mounted tactile display. *Ergonomics in Design, 21*: 4–8.

Nachemson, A. 1966. The load on the lumbar discs in different positions of the body. *Clinical Orthopaedics, 45*: 107–122.

Nachreiner, F. 1977. Experiments on the validity of vigilance experiments. In: *Vigilance: Theory, Operational Performance and Physiological Correlates*, R.R. Mackie, ed. Plenum Press, New York, pp. 665–678.

Nasar, J.L. and Troyer, D. 2013. Pedestrian injuries due to mobile phone use in public places. *Accident Analysis & Prevention, 57*: 91–95.

NCD Risk Factor Collaboration. 2016. Trends in adult body-mass index in 200 countries from 1975 to 2014: A pooled analysis of 1698 population-based measurement studies with 19.2 million participants. *The Lancet, 387*(10026): 1377–1396.

Nees, M.A. and Liebman, E. 2023. Auditory icons, earcons, spearcons, and speech: A systematic review and meta-analysis of brief audio alerts in human-machine interfaces. *Auditory Perception & Cognition*, *6*(3–4): 300–329.

Nesse, R.M. 1995. *Technology Review*, May–June: 34–37.

Neuhaus, M., Eakin, E., Straker, L., Orwen, N., Dunstan, D., Redi, N., and Healy, G.N. 2014. Reducing occupational sedentary time: A systematic review and meta-analysis of evidence on activity-permissive workstations. *Obesity Reviews*, *15*(10): 822–838.

Nevola, V.R., Staerck, J., and Harrison, M.H. 2001. *Fluid Intake for Optimal Performance During Military Operations in the Heat: A Reference Guide for Commanders*. DERA Report CG/Nutrition/v2.0/Mar2001.

Nguyen, P., Le, L.K.D., Nguyen, D., Gao, L., Dunstan, D.W., and Moodie, M. 2020. The effectiveness of sedentary behaviour interventions on sitting time and screen time in children and adults: An umbrella review of systematic reviews. *International Journal of Behavioral Nutrition and Physical Activity*, *17*(1): 1–11.

Nickerson, R. 2014. Is global warming a challenge to human factors? *Ergonomics in Design*, *22*: 4–7.

Niebel, N.W. and Freivalds, A. 1999. *Methods, Standards and Work Design*. 11th Edn., McGraw-Hill, New York.

Niebel, N.W. and Frievalds, A. 1993. *Methods, Standards and Work Design*. 10th Edn, WCB, McGraw-Hill, New York.

NIOSH. 1981. *Work Practices Guide for Manual Handling. NIOSH Technical Report No. 81–122*. U.S. Department of Health and Human Services, National Institute for Occupational Safety and Health, Cincinnati, OH.

NIOSH. 1994. Applications manual for the revised lifting equation. By Waters TR, Ph.D., Putz–Anderson V, Ph.D., Garg A, Ph.D. US Department of Health and Human Services, Centers for Disease Control and Prevention, National Institute for Occupational Health. DHHS Publication No. 94-110_Revised 90 2021, Cincinatti, OH. https://doi.org/10.26616/NIOSHPUB94110revised092021

Nissenbaum, M.A., Aramini, J.J., and Hanning, C.D. 2012. Effects of industrial wind turbine noise on sleep and health. *Noise and Health*, *1*: 237–243.

Niza, I.L., de Souza, M.P., da Luz, I.M., and Broday, E.E. 2023. Sick building syndrome and its impacts on health, well-being and productivity: A systematic literature review. *Indoor and Built Environment*, *33*: 218–236.

Noakes, T. 1992. *The Lore of Running*. Oxford University Press, Cape Town.

Nodooshan, H.S, Daneshmandi, H., Choobineh, A.,Yazdani, F., Razeghi, M., Shahnazar, N., and Khales, N. 2023. Effect of tilt-adjustable mouse pad on forearm muscle activity and upper limb discomfort during a computer mouse task. Journal of Health Sciences and Surveillance System, *11*(3): 500–507.

Noguchi, M., Zehr, J.D., Tennant, L.M., Fok, D.J., and Callaghan, J.P. 2023. Increasing movement during office work at sit-stand workstations: A novel seating device to facilitate transitions. *Applied Ergonomics*, *111*: 10404.

Norman, D.A. 1981. Categorisation of action slips. *Psychological Review*, *88*: 1–15.

Norman, D.A. 1982. Why alphabetic keyboards are not easy to use: Keyboard layout doesn't matter much. *Human Factors*, *24*: 509–519.

Norman, D.A. 1984. Stages and levels in human–machine interaction. *International Journal of Man–Machine Studies*, *21*: 365–375.

Norris, C.M. 1995. Spinal stabilisation. *Physiotherapy*, *81*(2): 64–72.

Norton, T.A., Ayoko, O.B., and Ashkanasy, N.M. 2021. A socio-technical perspective on the application of green ergonomics to open-plan offices: A review of the literature and recommendations for future research. *Sustainability*, *13*(15): 8236.

Nurick, S. and Thatcher, A. 2023. Examining the impact of indoor environmental quality on individual productivity of knowledge workers in green certified buildings. *Journal of Corporate Real Estate*, *25*(4): 307–324.

O'ÄôHara, J.M., Parsons, S.O., and Seminara, J.L. 2006. History of human factors/ergonomics in power systems. In *International Encyclopedia of Ergonomics and Human Factors*, Informa Healthcare, ed., CRC Press, Boca Raton, FL.

O'Kelly, O., Schor, J., Fan, W., Bezdenezhnykh, T., Gu, G., and Bridson Hubbard, N. 2022. *The Four Day Week: Assessing Global Trials of Reduced Work Time with no Reduction in Pay: Evidence from Ireland*. University College Dublin, Dublin, Ireland, p. 32.

Oakley, E.H.N. 2000. Heat and cold. In: *Hunter's Diseases of Occupations*, P.J. Baxter, ed. Oxford University Press, Inc., New York.

Oakman, J., Lambert, K.A., Rogerson, S., and Bell, A. 2024. We know it doesn't work: Why do we still use how to lift training for the prevention of musculoskeletal disorders?. *International Journal of Industrial Ergonomics*, *100*: 103542.

Occhipinti, E. and Colombini, D. 2004. The OCRA method: Updating of reference values and prediction models of occurrence of work-related musculo-skeletal diseases of the upper limbs (UL-WMSDs) in working populations exposed to repetitive movements and exertions of the upper limbs. *Medical Laboratory*, *95*(4): 305–319.

Öhman, L., Nordin, S., Bergdahl, J., Birgander, L.S., and Neely, A.S. 2007. Cognitive function in outpatients with perceived chronic stress. *Scandinavian Journal of Work, Environment & Health*, *33*: 223–232.

Ohnaka, T., Tochihara, Y., and Murumatsu, T. 1993. Physiological strains in hot-humid conditions while wearing disposable protective clothing commonly used by the asbestos removal industry. *Ergonomics*, *36*: 1241–1250.

Okholm Kryger, K., Thomson, A., Tang, A., Brown, N., Bruinvels, G., Rosenbloom, C., Carmody, S., Williamson, L., Datson, N., Jobson, E., and Mehta, R. 2022. Ten questions in sports engineering: Technology in elite women's football. *Sports Engineering*, *25*(1): 25.

Oksa, J., Ducharme, M.B., and Rintamaki, H. 2001. The combined effect of repetitive work and cold on muscle strength and fatigue. *Proceedings of the Australian Physiological and Pharmacological Society*, *32*(Supplement 2): 118.

O'Leary, T.J., Wardle, S.L., Rawcliffe, A.J., Chapman, S., Mole, J., and Greeves, J.P. 2020. Understanding the musculoskeletal injury risk of women in combat: the effect of infantry training and sex on musculoskeletal injury incidence during British Army basic training. *BMJ Military Health*, *169*(4): 57–61.

Olivers, C.N.L., van der Stigchel, S., and Hulleman, J. 2007. Spreading the sparing: Against a limited-capacity account of the attentional blink. *Psychological Research*, *71*(2): 126–139.

Orne, M.T. 1962. On the social psychology of the psychological experiment with particular reference to demand characteristics and their implications. *American Psychologist*, *17*: 776–783.

OSHA 3123. 1993. *Ergonomics Program Management Guidelines for Meatpacking Plants*. U.S. Department of Labor, OSHA, Washington, DC.

OSHA 9516. 1995. *Protecting Workers in Hot Environments*. U.S. Department of Labor, OSHA, Cincinatti, OH.

Ostberg, O., Warell, B., and Nordell, L. 1984. ComforTable—A generic desk for the automated office. *Behaviour and Information Technology*, *3*: 411–416.

Ouyang, Y., Liu, M., Cheng, C., Yang, Y., He, S., and Zheng, L. 2023. Monitoring inattention in construction workers caused by physical fatigue using electrocardiograph (ECG) and Galvanic Skin Response (GSR) sensors. *Sensors*, *23*(17): 7405.

Oxenburgh, M. 2004. *Improving Productivity and Profit through Occupational Health and Safety*. 2nd Edn., CIH, Sydney.

Oye-Somefun, A., Azizi, Z., Ardern, C.I., and Rotondi, M.A. 2021. A systematic review and meta-analysis of the effect of treadmill desks on energy expenditure, sitting time and cardiometabolic health in adults. *BMC Public Health*, *21*(1): 1–8.

Page, E.B. 1963. Ordered hypotheses for multiple treatments: A significance test for linear ranks. *Journal of the American Statistical Association*, *58*: 216–230.

Palmer, K.T. and Smedley, J. 2007. Work relatedness of chronic neck pain with physical findings—a systematic review. *Scandinavian Journal of Work, Environment & Health*, *71*: 165–191.

Palmer, K.Y., D'Angelo, S., Sydall, H., Griffin, M.J., Cooper, C., and Coggon, D. 2014. Dupuytren's contracture and occupational exposure to hand-transmitted vibration. *Occupational and Environmental Medicine*, *71*: 241–245.

Papamichael, K., Siminovitch, M., Veitch, J.A., and Whitehead, L. 2016. High color rendering can enable better vision without requiring more power. *Leukos*, *12*(1–2): 27–38.

Park, W., Singh, D.P., Levy, M.S., and Jung, E.S. 2009. Obesity effect on perceived postural stress during static posture maintenance tasks. *Ergonomics*, *52*(9): 1169–1182.

Parker, D., Lawrie, M., and Hudson, P. 2006. A framework for understanding the development of organisational safety culture. *Safety Science*, *44*(6): 551–562.

Parnell, K.J. and Plant, K.L. 2024. How can human factors close the gender data gap?. *Human Factors and Ergonomics in Manufacturing & Service Industries*, *34*(1): 63–75.

Parsons, K. 1991. Human response to thermal environments: Principles and methods. In: *Evaluation of Human Work: A Practical Ergonomics Methodology*, J.R. Wilson and E.N. Corlett, eds. Taylor & Francis, London.

Pearcy, M.J. 1993. Twisting mobility of the human back in flexed postures. *Spine*, *18*: 114–119.

Pearse, S., Léger, M., Albert, W.J., and Cardoso, M. 2024. Active workstations: A literature review on workplace sitting. *Journal of Bodywork and Movement Therapies*, *38*: 406–416.

Peddie, M.C., Cameron, C., Rehrer, N., and Perry, T. 2015. Postprandial metabolic effects of accelerometer measured spontaneous low-level activity. *Journal of Physical Activity and Health*, *12*: 1133–1138.

Peebles, L. and Norris, B. 1998. *Adult Data. The Handbook of Anthropometry and Strength Measurements.* Department of Trade and Industry, UK. DTI/Pub 2917/3k/6/98/NP. URN 98/376.

Peffer, T., Perry, D., Pritoni, M., Aragon, C., and Meier, A. 2016. Facilitating energy savings through programmable thermostats: Evaluation and guidelines for the thermostat user-interface. *Ergonomics, 56*: 463–479.

Perzl, J., Riedl, E.M., and Thomas, J. 2023. Measuring situational cognitive performance in the wild: A psychometric evaluation of three brief smartphone-based test procedures. *Assessment, 31*: 1270–1291.

Petermann-Rocha, F., Balntzi, V., Gray, S.R., Lara, J., Ho, F.K., Pell, J.P., and Celis-Morales, C. 2022. Global prevalence of sarcopenia and severe sarcopenia: A systematic review and meta-analysis. *Journal of Cachexia, Sarcopenia and Muscle, 13*(1): 86–99.

Peters, R., White, D., Cleeland, C., and Scholey, A. 2020. Fuel for thought? A systematic review of neuroimaging studies into glucose enhancement of cognitive performance. *Neuropsychology Review, 30*(2): 234–250.

Pheasant, S.T. 1986. *Bodyspace*. Taylor & Francis, London.

Pheasant, S.T. and O'Neill, D. 1975. Performance in gripping and turning—A study in hand/handle effectiveness. *Applied Ergonomics, 6*: 205–208.

Pheasant, S.T. and Scriven, J.G. 1983. Sex differences in strength—Some implications for the design of hand tools. In: *Proceedings of the Ergonomics Society's Conference*, K. Coombes, ed. Taylor & Francis, London, pp. 9–13.

Piccoli, B., Braga, M., Zambelli, P.L., and Bergmaschi, A. 1996. Viewing distance variation and related opthalmological changes in office activities. *Ergonomics, 39*: 719–728.

Pisula, P.J. 2015. *Load Carriage Using a Manual Pallet Truck at Sea: Maximum Push Forces and Allowable Loades*. Unpublished MoD Report, pp. 56, London.

Pisula, P.J., Lewis, C.H., and Bridger, R.S. 2012. Vessel motion thresholds for maintaining physical and cognitive performance: A study of naval personnel at sea. *Ergonomics, 55*: 636–649.

Pjrek, E., Friedrich, M.E., Cambioli, L., Dold, M., Jäger, F., Komorowski, A., Lanzenberger, R., Kasper, S., and Winkler, D. 2020. The efficacy of light therapy in the treatment of seasonal affective disorder: A meta-analysis of randomized controlled trials. *Psychotherapy and Psychosomatics, 89*(1): 17–24.

Pocock, M.J., Hamlin, I., Christelow, J., Passmore, H.A., and Richardson, M. 2023. The benefits of citizen science and nature-noticing activities for well-being, nature connectedness and pro-nature conservation behaviours. *People and Nature, 5*(2): 591–606.

Podrekar, N., Kozinc, Ž., and Šarabon, N. 2020. The effects of cycle and treadmill desks on work performance and cognitive function in sedentary workers: A review and meta-analysis. *Work, 65*(3): 537–545.

Porter, J.M., Gyi, D.E., and Robertson, J. 1992. An evaluation of a tilting computer desk. *Contemporary Ergonomics, 1992*: 54–59.

Porter, S. and Whitcomb, M.E. 2003. The impact of lottery incentives on student response rates. *Research in Higher Education, 44*: 389–407.

Potvin, J.R. 2012. Predicting maximum acceptable efforts for repetitive tasks: an equation based on duty cycle. *Human factors, 54*(2): 175–188.

Pritchard, M.H., Pugh, N., Wright, I., and Brownless, M. 1999. A vascular basis for repetitive strain injury. *Rheumatology, 38*: 636–639.

Punnett, L., Fine, L.J., Keyserling, W.M., Herrin, G.D., and Chaffin, D.B. 2000. Shoulder disorders and postural stress in automobile assembly work. *Scandinavian Journal of Work Environment and Health, 26*: 283–291.

Putz-Anderson, V. 1988. *Cumulative Trauma Disorders: A Manual for Musculoskeletal Diseases of the Upper Limbs*. Taylor & Francis, London.

Quenehen, A., Klement, N., Abdeljaouad, A.M., Roucoules, L., and Gibaru, O. 2023. Economic and ergonomic performance enhancement in assembly process through multiple collaboration modes between human and robot. *International Journal of Production Research, 61*(5): 1517–1531.

Quillian, M.R. 1969. The teachable language comprehender: A simulation program and theory of language. *Communications of the Association for Computing Machinery, 12*: 459–476.

Quinter, J.L. and Elvey, R.L. 1993. Understanding RSI: A review of the role of peripheral neural pain and hyperalgesia. *The Journal of Manual and Manipulative Therapy, 1*: 99–105.

Rabinowitz, D., Bridger, R.S., and Lambert, M. 1998. Lifting technique and abdominal belt usage: A biomechanical, physiological and psychological investigation. *Safety Science, 28*: 155–164.

Rabinowitz, P., Cantley, L.F., Galusha, D., Trufan, S., Swersey, A., Dixon-Ernst, C., Ramirez, V., and Neitzel, R. 2018. Assessing hearing conservation program effectiveness: Results of a multi-site assessment. *Journal of occupational and environmental medicine, 60*(1): 29.

Raghuraman, R.N., Upasani, S., Gonzales, A., Aviles, J., Cha, J., and Srinivasan, D. 2023. Manufacturing industry stakeholder perspectives on occupational exoskeletons: Changes after a brief exposure to exoskeletons. *IISE Transactions on Occupational Ergonomics and Human Factors*, *11*: 1–10.

Rainieri, G., Buizza, C., and Ghilardi, A. 2023. The psychological, human factors and socio-technical contribution: A systematic review towards range anxiety of battery electric vehicles' drivers. *Transportation research part F: traffic psychology and behaviour*, *99*: 52–70.

Ramazini, B. 1717. De Morbis Artificium. https://www.cdc.gov/niosh/hhe/reports/pdfs/1990–0385–2173.pdf

Rasmussen, J. 1983. Skills, rules and knowledge. Signals, signs and symbols and other distinctions in human performance models. *IEEE Transactions on Systems, Man and Cybernetics. SMC*, *13*: 257–266.

Rauterberg, M. 1992. An empirical comparison of menu-selection (CUI) and desktop (GUI) computer programmes carried out by beginners and experts. *Behaviour and Information Technology*, *11*: 227–236.

Reilly, T. and Davies, S. 2013. Effects of a weightlifting belt on spinal loading during performance of the dead lift. In: Sport, Leisure and Ergonomics, G. Atkinson and T. Reilly, eds. E. & F.N. Spon, London, p. 136.

Reiman, A., Kaivo-oja, J., Parviainen, E., Takala, E.P., and Lauraeus, T. 2021. Human factors and ergonomics in manufacturing in the industry 4.0 context–A scoping review. *Technology in Society*, *65*: 101572.

Remaud, A., Boyas, S., Caron, G.A.R., and Bilodeau, M. 2012. Attentional demands associated with postural control depend on task difficulty and visual condition. *Journal of Motor Behaviou*, *44*: 329–340.

Revell, K.M.A. and Stanton, N. 2016. Mind the gap—Deriving a compatible user mental model of the home heating system to encourage sustainable behaviour. *Applied Ergonomics*, *57*: 48–61.

Reynolds, K.L., White, J.S., Knapik, J.J., Witt, C.E., and Amoroso, P. 1999. Injuries and risk factors in a 100-mile march. *Preventative Medicine*, *28*: 163–173.

Richards, D. and Stedmon, A. 2016. To delegater or not to delegate: A review of control frameworks for autonomous cars. *Applied Ergonomics*, *53*: 383–388.

Richards, D.C. 2010. Relationship between speed and risk of fatal injury: Pedestrians and car occupants. Road Safety er Publication No. 16. Transport Research Laboratory. ISBN 9781 906581924.

Richman, W.L., Weisband, S., Kiesler, S., and Drasgow, F. 1999. A meta-analytic study of social desirability distortion in computer-administered questionnaires, traditional questionnaires and face to face interviewing. *Journal of Applied Psychology*, *84*: 754–775.

Riedl, R. 2022. On the stress potential of videoconferencing: definition and root causes of Zoom fatigue. *Electronic Markets*, *32*(1): 153–177.

Riel, P. and Imbeau, D. 1995. Economic justification of investments for health and safety interventions: Part I: Cost typology. *International Journal of Industrial Engineering*, *2*: 45–54.

Rintamaki, H., Korhonen, E., Rissanen, J., Oksa, H., and Peinimaki, T. 2001. Cold problems and upper limb muscular strain in food processing industry. *Proceedings of the Australian Physiological and Pharmacological Society*, *32*(2): 127.

Rivera, A.S., Akanbi, M., O'Dwyer, L.C., and McHugh, M. 2020. Shift work and long work hours and their association with chronic health conditions: A systematic review of systematic reviews with meta-analyses. *PLoS One*, *15*(4): e023103

Roberts, D.A., Bridger, R.S., Kilminster, S., and Etherington, J. 2015. Predicting low back pain outcome following rehabilitation for low back pain. *Journal of Back and Musculoskeletal Rehabilitation*, *28*: 119–128.

Roberts, D.F. 1995. The pervasiveness of plasticity. In: Human Variability and Plasticity, C.G.N. Mascie-Taylor and B. Bogin, eds. Cambridge University Press, Cambridge, pp. 1–17.

Robertson, I.H., Manly, T., Andrade, J., Baddeley, B.T., and Yiend, J., 1997. Oops!': Performance correlates of everyday attentional failures in traumatic brain injured and normal subjects. *Neuropsychologia*, *35*(6): 747–758.

Rock, I. and Palmer, S. 1990. The legacy of Gestalt psychology. *Scientific American*, *6*: 48–61.

Rogerson, S. and Brearley, M. 2024. Suspected exertional heat stroke; A case study of worker cooling in a hot and humid field environment. *Work, (Preprint)*, *79*: 1–6.

Rohlfing, M.L., Buckley, D.P., Piraquive, J., Stepp, C.E., and Tracy, L.F. 2021. Hey Siri: How effective are common voice recognition systems at recognizing dysphonic voices?. *The Laryngoscope*, *131*(7): 1599–1607.

Rohmert, W. 1960. Ermittlung von Ehrolungspausen fuer statische arbeit des Menchen. *Internationale Zeitschrift Angewandte Physiologie*, *18*: 123–164.

Rohmert, W. 1967. Physiological and psychological workload measurement and analysis. In: *Handbook of Human Factors and Ergonomics*. G. Salvendy, ed., John Wiley & Sons, New York.

Roman-Liu, D., Grabarek, I., Bartuzi, P., and Choromański, W. 2013. The influence of mental load on muscle tension. *Ergonomics*, *56*(7): 1125–1133.

Roman-Liu, D., Kamińska, J., and Tokarski, T. 2023. Differences in lumbar spine intradiscal pressure between standing and sitting postures: a comprehensive literature review. *PeerJ*, *11*, e16176.

Rosa, R.R. and Bonnet, M.H. 1993. Performance and alertness on 8 h and 12 h rotating shifts at a natural gas utility. *Ergonomics*, *36*: 1177–1193.

Roscoe, A.H. and Ellis, G.A. 1990. A subjective rating scale of assessing pilot workload in flight: A decade of practical use, A Technical Report, Procurement Executive, Ministry of Defence Farnborough, Hampshire.

Rothlisberger, F.J. and Dickson, W.J. 1939. *Management and the Worker*. Harvard University Press, Cambridge, 615.

Rothman, A.J., Gollwitzer, P.M., Grant, A.M., Neal, D.T., Sheeran, P., and Wood, W. 2015. Hale and hearty policies: How psychological science can create and maintain healthy habits. *Psychological Science*, *10*: 701–705.

Rovniak, L.S., Cho, J., Freivalds, A., Kong, L., Bopp, M., Sciamanna, C.N., and Rothrock, L. 2022. Effects of desk pedalling work rate on concurrent work performance among physically inactive adults: a randomized experiment. *Medicine and Science in Sports and Exercise*, *55*(1): 20–31.

Roy, Q., Berlioux, S., Casiez, G., and Vogel, D. 2021. Typing efficiency and suggestion accuracy influence the benefits and adoption of word suggestions. In: *Proceedings of the 2021 CHI Conference on Human Factors in Computing Systems*. ACM, New York, pp. 1–13).

Rucker, N. and Moore, J.S. 2002. Predictive validity of the strain index in manufacturing facilities. *Applied Occupational and Environmental Hygiene*, *17*: 63–73.

Rumbo-Rodríguez, L., Sánchez-SanSegundo, M., Ferrer-Cascales, R., García-D'Urso, N., Hurtado-Sánchez, J.A., and Zaragoza-Martí, A. 2021. Comparison of body scanner and manual anthropometric measurements of body shape: a systematic review. *International Journal of Environmental Research and Public Health*, *18*(12): 6213.

Rupp, R.F., Kim, J., de Dear, R., and Ghisi, E. 2018. Associations of occupant demographics, thermal history and obesity variables with their thermal comfort in air-conditioned and mixed-mode ventilation office buildings. *Building and Environment*, *135*: 1–9.

Rydevik, B., Lundborg, G., and Bagge, U. 1981. Effects of graded compression on intraneural blood flow. *Journal of Hand Surgery*, *6*: 3–12.

Rys, M. and Konz, S. 1994. Standing. *Ergonomics*, *37*: 677–687.

Ryu, S.Y. and Smith-Jackson, T.L. 2006. Reliability and validity of the mobile phone usability questionnaire (MPUQ). *Journal of Usability Studies*, *2*: 39–53.

Sander, E.L.J., Marques, C., Birt, J., Stead, M., and Baumann, O. 2021. Open-plan office noise is stressful: Multimodal stress detection in a simulated work environment. *Journal of Management & Organization*, *27*(6): 1021–1037.

Sanders, A.F. 1970. Some aspects of the selective process in the functional visual field. *Ergonomics*, *13*: 101–117.

Sanders, M.S. and McCormick, E.J. 1992. *Human Factors in Engineering and Design*, McGraw-Hill Inc, New York.

Sandmark, H., Hogstedt, C., and Vingard, E. 2000. Primary osteoarthrosis of the knee in men and women as a result of lifelong physical load from work. *Scandinavian Journal of Work Environment and Health*, *26*: 20–25.

Sandover, J. 1988. The fatigue approach to vibration and shock: Is it a viable and practical way of predicting the effects on people? *Journal of Sound and Vibration*, *215*: 699–721.

Santra, D., Goswami, S., Mandal, J.K., and Basu, S.K. 2021. Low back pain expert systems: Clinical resolution through probabilistic considerations and poset. *Artificial Intelligence in Medicine*, *120*: 102163.

Saracci, R. 1997. The world health organisation needs to reconsider its definition of health. *BMJ*, *314*(7091): 1409.

Saraceni, N., Campbell, A., Kent, P., Ng, L., Straker, L., and O'Sullivan, P. 2022. Does intra-lumbar flexion during lifting differ in manual workers with and without a history of low back pain? A cross-sectional laboratory study. *Ergonomics*, *65*(10): 1380–1396.

Saraceni, N., Kent, P., Ng, L., Campbell, A., Straker, L., and O'Sullivan, P. 2020. To flex or not to flex? Is there a relationship between lumbar spine flexion during lifting and low back pain? A systematic review with meta-analysis. *Journal of Orthopaedic & Sports Physical Therapy*, *50*(3): 121–130.

Sathya, P. and Tamboli, S.A. 2020. Prevalence of text neck syndrome in young-adult population. *International Journal of Medical and Exercise Science*, *6*: 749–759.

Saward, J.R. and Stanton, N.A. 2017. Latent error detection: A golden two hours for detection. *Applied Ergonomics*, *59*:104–113.

Sawyer, B.D., Finomore, V.S., Calvo, A.A., and Hancock, P.A. 2014. Google glass: A driver distraction cause or cure? *Human Factors*, *56*(7): 1307–1321.

Scannel, K. 1998. Low-cost methods of noise-control at source increase production and reduce the risk of hearing damage. *Journal of Occupational Health and Safety (Australia and New Zealand)*, *14*: 493–503.

Schaufelli, W. and Baker, A.B. 2004. Job demands, job resources and their relationship with burnout and engagement: A multi-sample study. *Journal of Organisational Behaviour*, *25*: 293–315.

Schilling, R. and Anderson, N. 1986. Occupational epidemiology in developing countries. *Occupational Health and Safety (Australia and New Zealand)*, *2*: 468–471.

Schleifer, L.M. and Ley, R. 1994. End-tidal pCO_2 as an index of psychophysiological activity during VDT data entry work and relaxation. *Ergonomics*, *37*: 245–254.

Schlittmeier, S., Feil, A., Liebl, A., and Hellbruck, J. 2015. The impact of road traffic noise on cognitive performance in attention-based tasks depends on noise level even within moderate ranges. *Noise and Health*, *17*: 148–157.

Schneider, S. and Wahl, G. 1998. Ergonomic intervention has a return on investment of 17 to 1. *Applied Occupational and Environmental Hygiene*, *13*(4): 212–213.

Schoep, M.E., Adang, E.M., Maas, J.W., De Bie, B., Aarts, J.W., and Nieboer, T.E. 2019. Productivity loss due to menstruation-related symptoms: A nationwide cross-sectional survey among 32 748 women. *BMJ Open*, *9*(6): e026186.

Scholey, A., Macpherson, H., Sünram-Lea, S., Elliott, J., Stough, C., and Kennedy, D. 2013. Glucose enhancement of recognition memory: Differential effects on effortful processing but not aspects of 'remember-know' responses. *Neuropharmacology*, *64*: 544–549.

Schröder, C. and Nienhaus, A. 2020. Intervertebral disc disease of the lumbar spine in health personnel with occupational exposure to patient handling—a systematic literature review and meta-analysis. *International Journal of Environmental Research and Public Health*, *17*(13): 4832.

Schultz, T.Z. 1969. *The Life of the Primates*. Weidenfield and Nicholson, London.

Schuna, J.M., Hsia, D.S., Tudor-Locke, C., and Johannsen, N.M. 2019. Energy expenditure while using workstation alternatives at self-selected intensities. *Journal of Physical Activity and Health*, *16*(2): 141–148.

Sears, A., Jacko, J.A., Brewer, B., and Robelo, L.D. 1998. An empirical perspective on icon design. In: *Proceedings of the Human Factors and Ergonomics Society 42nd Annual Meeting*. SAGE Publications, Los Angeles, CA, pp. 448–452.

Sebo, P., Beer-Borst, S., Haller, D.M., and Bovier, P.A. 2008. Reliability of doctors' anthropometric measurements to detect obesity. *Preventive Medicine*, *47*(4): 389–393.

Sellen, A.J. 1994. Detection of everyday errors. *Applied Psychology: An International Review*, *43*: 475–498.

Selye, H. 1956. *The Stress of Life*. McGraw-Hill, New York.

Sethi, T. and Ziat, M. 2023. Dark mode vogue: Do light-on-dark displays have measurable benefits to users? *Ergonomics*, *66*(12): 1814–1828. https://doi.org/10.1080/00140139.2022.2160879

Shafizadeh, M., Parvinpour, S., Balali, M., and Shabani, M. 2021. Effects of age and task difficulty on postural sway, variability and complexity. *Adaptive Behavior*, *29*(6): 617–625.

Sharan, D. 2023. *How to Perform an Ergonomic Workplace Analysis and Select the Best Tools?* PREMUS, Bengalaru, India.

Sharit, J., Czaja, S.J., Nair, S., and Lee, C.C. 2003. Effects of age, speech rate and environmental support in using telephone menu systems. *Human Factors*, *45*: 234–251.

Shneiderman, B. 1988. Designing a better world. In: *Proceedings of the 10th congress of the International Ergonomics Association*. ACM, New York.

Shneiderman, B. 1991. A taxonomy and rule base for the selection of interaction styles. In: *Human Factors for Informatics Usability*, B. Shackel and S. Richardson, eds. Cambridge University Press, Cambridge.

Shortliffe, E.H. 1984. Reasoning methods in medical consultation systems: Artificial intelligence approaches. *Computer Programs in Biomedicine*, *18*: 5–14.

Shortliffe, E.H., Buchanan, B.G., and Feigenbaum, E.A. 1979. Knowledge engineering for medical decision making: A review of computer-based decision aids. *Proceedings of the IEEE*, *67*: 1209–1224.

Siegrist, J. 1996. Adverse health effects of high effort/low reward conditions. *Journal of Occupational Health Psychology*, *1*: 27–41.

Siemens, G., Marmolejo-Ramos, F., Gabriel, F., Medeiros, K., Marrone, R., Joksimovic, S., and de Laat, M. 2022. Human and artificial cognition. *Computers and Education: Artificial Intelligence*, *3*: 100107.

Simon, H.A. 1979. Information processing models of cognition. *Annual Reviews of Psychology*, *30*: 363–396.

Sims, M., Gillies, G., and Drury, R. 1977. Using a cool spot to improve the thermal comfort of glassmakers. *Applied Ergonomics*, *8*: 2–6.

Singleton, W.T. 1960. An experimental investigation of speed controls for sewing machines. *Ergonomics, 3*: 365–375.

Singleton, W.T. 1974. *Man–Machine Systems*. Penguin Books, London

Slater, M. 2009. Place illusion and plausibility can lead to realistic behaviour in immersive virtual environments. *Philosophical Transactions of the Royal Society B: Biological Sciences*, 364(1535): 3549–3557.

Slaughter, P.R., Rodzak, K.M., Fine, S.J., Ice, C.C., Wolf, D.N., and Zelik, K.E. 2023. Evaluation of US Army Soldiers wearing a back exosuit during a field training exercise. *Wearable Technologies, 4*: e20.

Sluiter, J.K. 1999. The influence of work characteristics on the need for recovery and experienced health: A study on coach drivers. *Ergonomics, 42*(4): 573–583.

Sluiter, J.K. 2003. Need for recovery from work related fatigue and its role in the development and prediction of subjective health complaints. *Occupational and Environmental Medicine, 60*(Suppl. I): i62–i70.

Smith, A. 1991. A review of the non-auditory effects of noise on health. *Work and Stress, 5*: 49–62.

Snijders, C.J., Slagter, A.H.E., van Strik, R., Vleeming, A., Stoeckart, R., and Stam, H.J. 1995. Why leg crossing? The influence of common postures on abdominal muscle activity. *Spine, 20*: 1989–1993.

Snook, S.H. and Ciriello, V.M. 1998. The design of manual handling tasks: Revised tables of maximum acceptable weights and forces. *Ergonomics, 34*: 1197–1213.

Snook, S.H., Vaillancourt, D.R., Ciriello, V.M., and Webster, B. 1995. Psychophysical studies of repetitive wrist flexion and extension. *Ergonomics, 38*: 1488–1507.

Snook, S.H., Webster, B.S., McGorry, R.W., Fogleman, M.T., and McCann, K.B. 1998. The reduction of chronic nonspecific low back pain through the control of early morning lumbar flexion: a randomized controlled trial. *Spine, 23*(23): 2601–2607.

Snyder, K.M. 1991. *A Guide to Software Usability*. IBM Internal Publication, Armonk, NY.

Snyder, R.G. 1965. Human tolerance limits in water impact. *Aerospace Medicine, 36*: 940–947.

So, R.H.Y., Lo, W.T., and Ho, T.K. 2001. Effects of navigation speed on motion sickness caused by an immersive virtual environment. *Human Factors, 43*: 452–461.

Solomonow, M. and D'Ambrosia, R. 1987. Biomechanics of muscle overuse injuries: A theoretical approach. *Clinics in Sports Medicine, 6*: 241–257.

Sommer, I., Teufer, B., Szelag, M., Nussbaumer-Streit, B., Titscher, V., Klerings, I., and Gartlehner, G. 2020. The performance of anthropometric tools to determine obesity: a systematic review and meta-analysis. *Scientific Reports, 10*(1): 12699.

Sommerich, C.M., McGlothlin, J.D., and Marras, W.S. 1993. Occupational risk factors associated with soft tissue disorders of the shoulder: A review of recent literature. *Ergonomics, 36*: 697–717.

Song, H. and Schwarz, N. 2008. If it's hard to read, it's hard to do processing fluency affects effort prediction and motivation. *Psychological Science, 19*(10): 986–988.

Sonne, M.W. and Potvin, J.R. 2015. A psychophysical study to determine maximum acceptable efforts for a thumb abduction task with high duty cycles. *Ergonomics, 58(1)*: 118–127.

Sorkin, R.D. 1987. Design of auditory and tactile displays. In: *Handbook of Human Factors*, G. Salvendy, ed. John Wiley & Sons, New York, pp. 549–576.

Soule, R.G. and Goldman, R.F. 1969. Energy cost of loads carried on the head, hands and feet. *Journal of Applied Physiology, 27*: 687–690.

Souman, J.L., Tinga, A.M., Te Pas, S.F., Van Ee, R., and Vlaskamp, B.N. 2018. Acute alerting effects of light: A systematic literature review. *Behavioural Brain Research, 337*: 228–239.

Spatola, N., Kühnlenz, B., and Cheng, G. 2021. Perception and evaluation in human–robot interaction: The Human–Robot Interaction Evaluation Scale (HRIES)—A multicomponent approach of anthropomorphism. *International Journal of Social Robotics, 13*(7): 1517–1539.

Spencer, M.B. 1997. The influence of irregularity of rest and activity on performance: A model based on time since sleep and time of day. *Ergonomics, 30*(9): 1275–1286.

Spencer, E.A., Appleby, P.N., Davey, G.K., and Key, T.J. 2002. Validity of self-reported height and weight in 4808 EPIC–Oxford participants. *Public Health Nutrition, 5*(4): 561–565.

Srinivasan, D. and Mathiassen, S.E. 2012. Motor variability in occupational health and performance. *Clinical Biomechanics, 27*(10): 979–993.

Stagner, R. 1982. The importance of a historical context. *American Psychologist, 37*: 856.

Stanton, N., Salmon, P.M., and Rafferty, L.A. 2013. *Human Factors Methods: A Practical Guide for Engineering and Design*. Ashgate Publishing, Ltd, Farnham.

Steenstra, I.A., Sluiter, J.K., and Frings-dresen, H.W. 2009. The eye complaint questionnaire in a visual display unit work environment: Internal consistency and test retest reliability. *Ergonomics, 52*: 334–344.

Steindler, A. 1955. *Kinesiology of the Human Body under Normal and Pathological Conditions*. Charles C. Thomas, Springfield, IL.

Stephenson, M.R. 2009. Hearing protection in the 21st century: They're not your father's earplugs anymore. *Seminars in Hearing*, *30*(01): 056–064.

Stone, P.T. 1986. Issues in vision and lighting for users of VDU's. In: *Health Hazards of VDU's?* B. Pearce, ed. John Wiley & Sons, New York.

Stotts, D.B. 1998. The usefulness of icons on the computer interface. In: *Proceedings of the Human Factors and Ergonomics Society 42nd Annual Meeting*. ACM, New York, pp. 453–457.

Straker, L. and Mathiassen, M. 2009. Increased physical work loads in modern work—A necessity for better health and performance. *Ergonomics*, *52*: 1215–1225.

Straker, L., Abbott, R., Collins, R., and Campbell, A. 2014. Evidence-based guidelines for wise use of electronic games by children. *Ergonomics*, *57*(4): 471–489.

Straker, L.M., Coleman, J., Skoss, R., Maslen, B.A., Burgess-Limersick, R., and Pollock, C.M. 2008. A comparison of posture and muscle activity during tablet computer, desktop computer and paper use by young children. *Ergonomics*, *51*: 540–555.

Stranden, E. 2000. Dynamic leg volume changes when sitting in a locked and free floating tilt office chair. *Ergonomics*, *43*: 421–433.

Stroop, J.R. 1935. Studies of interference in serial verbal reactions. *Journal of Experimental Psychology*, *18*: 643–662.

Suka, M., Shimazaki, T., Yamauchi, T., and Yanagisawa, H. 2022. Increased health risk in office workers in the COVID-19 era: Comparison of one-year incidence of health problems before and during the COVID-19 pandemic. *Journal of Occupational and Environmental Medicine*, *64*(4), pp.271–277

Sullivan, N., Hutcherson, C., Harris, A., and Rangel, A. 2015. Dietary self-control is related to the speed with which attributes of healthfulness and tastiness are processed. *Psychological Science*, *26*: 122–134.

Svedung, I. and Rasmussen, J. 2002. Graphic representation of accident scenarios: mapping system structure and the causation of accidents. *Safety science*, *40*(5): 397–417.

Swain, A.D., Shelton, G.C., and Rigby, L.V. 1970. Maximum torque for small knobs operated with and without gloves. *Ergonomics*, *13*: 201–208.

Swain, C.T., Pan, F., Owen, P.J., Schmidt, H., and Belavy, D.L. 2020. No consensus on causality of spine postures or physical exposure and low back pain: A systematic review of systematic reviews. *Journal of Biomechanics*, *102*: 109312.

Swenson, E.E., Purswell, J.L., Schlegel, R.E., and Stanevich, R.L. 1992. Coefficient of friction and subjective assessment of work surfaces. *Human Factors*, *34*: 67–77.

Snyder, R.G. 1965. Human tolerance limits in water impact. *Aerospace Medicine*, *36*: 940–947.

Székely, É., Henter, G.E., Beskow, J., and Gustafson, J. 2020. Breathing and speech planning in spontaneous speech synthesis. In: *ICASSP 2020–2020 IEEE International Conference on Acoustics, Speech and Signal Processing (ICASSP)*. IEEE, New York, pp. 7649–7653.

Szulc, J.M., McGregor, F.L., and Cakir, E. 2023. Neurodiversity and remote work in times of crisis: Lessons for HR. *Personnel Review*, *52*(6): 1677–1692.

Takahashi, M., Nakata, A., Haratani, T., and Ogawa, Y. 2004. Post lunch nap as a worksite intervention to promote alertness on the job. *Ergonomics*, *47*: 1003–1013.

Talcott, K.A., Casali, J.G., Keady, J.P., and Killion, M.C. 2012. Azimuthal auditory localisation of gunshots in a realistic field environment: Effects of open-ear versus hearing protection-enhancement devices. *International Journal of Audiology*, *51*: S20–S30.

Talens-Estarelles, C., Cerviño, A., García-Lázaro, S., Fogelton, A., Sheppard, A., and Wolffsohn, J.S. 2023. The effects of breaks on digital eye strain, dry eye and binocular vision: Testing the 20-20-20 rule. *Contact Lens and Anterior Eye*, *46*(2): 101744.

Tan, J., Bauer, L., Christin, N., and Cranor, L.F. 2020. Practical recommendations for stronger, more usable passwords combining minimum-strength, minimum-length, and blocklist requirements. In: *Proceedings of the 2020 ACM SIGSAC Conference on Computer and Communications Security*. ACM, New York, pp. 1407–1426.

Tanaka, H. and Uetake, T. 2005. Characteristics of postural sway in older adults standing on a soft surface. *Journal of Human Ergology*, *34*(1–2): 35–40.

Tanebaum, D. 1996. Dvorak keyboards: The typists long-lost friend. *Technology Review*, *99*(5): 21–23.

Tankisi, H., Burke, D., Cui, L., de Carvalho, M., Kuwabara, S., Nandedkar, S.D., Rutkove, S., Stålberg, E., van Putten, M.J., and Fuglsang-Frederiksen, A. 2020. Standards of instrumentation of EMG. *Clinical Neurophysiology*, *131*(1): 243–258.

Tanna, R.J., Lin, J.W., and De Jesus, O. 2020. *Sensorineural Hearing Loss*. StatPearls Publishing, Treasure Island. FL.

Taylor, F.W. 1911. *The Principles of Scientific Management.* Harper and Brothers Publishers, New York and London.

Taylor, K., Van Dijk, P., Newnam, S., and Sheppard, D. 2023. Physical and psychological hazards in the gig economy system: A systematic review. *Safety Science, 166*: 106234.

Thaler, L. 2015. Using sound to get around. *Observer, 28*: 24–27.

Thatcher, A. 2013. Green ergonomics: definition and scope. *Ergonomics*, 56(3): 389–398.

Thatcher, A. 2023. Education, sustainability, and human factors and ergonomics. *Rivista Italiana Di Ergonomia, 25*: 1–11.

Thatcher, A., 2024. Climate Ergonomics: Preparing for the Next 50 Years. In: *International Ergonomics Conference.* Cham: Springer Nature Switzerland, pp. 25–32.

Thatcher, A., Alexander, T., and Kirby, B. in press. Human factors and ergonomics' responses to climate crises (Chapter 38). In: *Helander's Compendium of Human Factors and Ergonomics: Contemporary Issues in Product and Systems Design for Effective Human Use*, H.M. Khalid, ed. CRC Press, Boca Raton, FL.

Thomsen, J.F., Gerr, F., and Atroshi, I. 2008. Carpal tunnel syndrome and the use of computer mouse and keyboard: A systematic review. *BMC Musculoskeletal Disorcers, 9*: 134.

Tile, M. 1984. *Fractures of the Pelvis and Acetabulum.* Williams & Wilkins, Baltimore, MD, London.

Tipton, M., Eglin, C., Genser, M., and Golden, F. 1999. Immersion deaths and deterioration in swimming performance in cold water. *The Lancet, 354*: 626–628.

Tittiranonda, P., Rempel, D., Armstrong, T., and Burastero, S. 1999. Effect of four computer keyboards in computer users with upper extremity musculoskeletal disorders. *American Journal of Industrial Medicine, 35*: 647–661.

Todd, P.M. and Gigerenzer, G. 2007. Environments that make us smart. *Current Directions in Psychological Science, 16*: 167–171.

Toivonen, M., Paakkonen, R., Savolainen, S., and Lehtomaki, K. 2002. Noise attenuation and proper insertion of earplugs into ear canals. *Annals of Occupational Hygiene, 46*: 527–530.

Tomei, F., Baccolo, T.P., Palmi, S., and Rosati, M.V. 1999. Chronic venous disorders and occupation. *American Journal of Industrial Medicine, 36*: 653–665.

Torén, A. 2001. Muscle activity and range of motion during active trunk rotation in a sitting posture. *Applied Ergonomics, 32*(6): 583–591.

Torquati, L., Mielke, G.I., Brown, W.J., Burton, N.W., and Kolbe-Alexander, T.L. 2019. Shift work and poor mental health: A meta-analysis of longitudinal studies. *American Journal of Public Health, 109*(11): e13–e20.

Trinh, V.Q., Bodrogi, P., and Khanh, T.Q. 2023. Determination and measurement of melanopic equivalent daylight (D65) illuminance (mEDI) in the context of smart and integrative lighting. *Sensors, 23*(11): 5000.

Trist, E.L. and Bamforth, K.W. 1951. Some social and psychological consequences of the longwall method of coal getting. *Human Relations, 4*: 3–38.

Tsubota, K. and Nakamori, K. 1995. Effects of ocular surface area and blink rate on tear dynamics. *Archives of Opthalmology, 113*: 155–158.

Tuchsen, F., Krause, N., Hannerz, H., and Burr, T.S. 2000. Standing at work and varicose veins. *Scandinavian Journal of Work, Environment & Health, 26*(5): 414–420.

Tudor-locke, C. 2010. Steps to better cardiovascular health: How many steps does it take to achieve good health and how confident are we in this number? *Current Cardiovascular Risk Reports, 4*: 271–276.

Tulving, E. 1972. Episodic and semantic memory. *Organization of Memory, 1*: 381–403.

Tversky, A. and Kahneman, D. 1974. Judgement under uncertainty: Heuristics and biases. *Science, 185*: 1124–1131.

Tversky, A. and Kahneman, D. 1981. The framing of decisions and the psychology of choice. *Science, 211*: 453–458.

Tynan, O. 1980, Improving the quality of working life in the 1980's. WRU Occasional Paper 16, November, Work Research Unit, London.

U.S. Manual Handling Regulations. 1992. *The authoritative version is the Queen's Printer copy published by The Stationery Office Limited as the Manual Handling Operations Regulations* 1992.

U.S. Mil Std-1472. 2015. MIL-STD 1472 is considered one of the most comprehensive standards document for Human Factors.

U.S. National Safety Council. 2008. National safety council applauds house passage of safety of seniors act. https://www.reuters.com/article/pressRelease/idUS230178+08-Apr-2008+PRN20080408

U.S. Occupational Safety and Health Act. 1970. Law passed by the United States Congress that created the Occupational Safety and Health Administration to Prevent employees from being Injured or Contracting Diseases in the course of their employment.

Ullman, J., Hengst, D., Carpenter, R. and Robinson, Y. 2022. Does military high-speed boat slamming cause severe injuries and disability?. *Clinical Orthopaedics and Related Research, 480*: 10–1097.

University of California. 1993. Not Seeing Red. *University of California at Berkeley Wellness Letter*, 9 August: 2.

Van der Linden, D., Keijsers, G.P.J., Eling, P., and van Schaijk, R. 2005. Work stress and attentional difficulties: An initial study on burnout and cognitive failures. *Work and Stress, 19*: 23–36.

Van der Ploeg, H.P., Chey, T., Korda, R.J., Banks, E., and Bauman, A. 2012. Sitting time and all-cause mortality risk in 222,497 Australian adults. *Archives of Internal Medicine, 26*: 494–500.

Van Deursen, D.L., Goosens, R.H.M., Evers, J.J.M., van der Helm, F.C.T., and van Deursen, L.L.J.M. 2000. Length of the spine while sitting on a new concept of office chair. *Applied Ergonomics, 31*: 95–98.

Van Dieen, J.H. and Oude Vrielink, H.H.E. 1993. An investigation into the relevance of the pattern of temporal activation with respect to erector spinae muscle endurance. *European Journal of Applied Physiology, 66(1)*: 70–75.

Van Duijvenbode, I.C. 2009. Lumbar supports for prevention and treatment of low back pain. *Journal of Evidence-Based Medicine*, *2*(1): 63–64.

van Kamp, I. and van den Berg, F. 2021. Health effects related to wind turbine sound: An update. *International Journal of Environmental Research and Public Health, 18*(17): 9133.

Van Nes, F.L. 1986. Space, colour and typography on visual display terminals. *Behaviour and Information Technology, 5*: 99–118.

Van Veldhoven, M.J.P.M. and Meijman, T.F. 1994. *The Measurement of Psychosocial Job Demands with a Questionnaire (VBBA)*. NIA: Amsterdam.

Vasiliadis, A.V. and Itsiopoulos, I. 2017. Trigger finger: An atraumatic medical phenomenon. *The Journal of Hand Surgery (Asian-Pacific Volume)*, *22*(02): 188–193.

Venetjoki, N., Kaarlela-Tuomaala, A., Keskinen, E., and Hongisto, V. 2006. The effect of speech and speech intelligibility on task performance. *Ergonomics, 49*: 1068–1091.

Verde, L., Barrea, L., Bowman-Busato, J., Yumuk, V.D., Colao, A., and Muscogiuri, G. 2024. Obesogenic environments as major determinants of a disease: It is time to re-shape our cities. *Diabetes/Metabolism Research and Reviews, 40*(1): e3748.

Vernon, H.M. 1920. Industrial efficiency and fatigue. In: *The Industrial Clinic*, E.L. Collis, ed. John Bale, London, pp. 51–74.

Vernon, H.M. 1924. The influence of rest pauses and changes of posture on the capacity for muscular work. *The Medical Research Council Report, 29*: 28–55.

Vernon-Roberts, B. 1989. Pathology of the intervertebral discs and apophyseal joints. In: *The Lumbar Spine and Back Pain, III*, M.I.V. Jayson, ed. Churchill Livingstone, Oxford, Edinburgh.

Vetter, C., Juda, M., Lang, D., Wojtysaiak, A., and Roenneberg, T. 2010. Blue-enriched office light competes with natural light as a zeitgeber. *Scandinavian Journal of Work Environment and Health, 37*: 437–455.

Victor, T. 2011. Distraction and inattention countermeasure technologies. *Ergonomics in Design, 19*: 20–22.

Vieira, J.B., Pierzchajlo, S.R., and Mitchell, D.G. 2020. Neural correlates of social and non-social personal space intrusions: Role of defensive and peripersonal space systems in interpersonal distance regulation. *Social Neuroscience, 15*(1): 36–51.

Villarroya López, A., Rivas Moar, J.M., and Louzán Mariño, R. 2023. Main challenges in the implantation of exoskeletons according to applicability criteria. Selection, Evaluation and Validation Process. International Symposium on Occupational Safety and Hygiene, 20–21 July 2023.

Viola, A.U. 2008. Blue-enriched white light in the workplace improves self-reported alertness, performance and sleep quality. *Scandinavian Journal of Work, Environment & Health, 34*: 297–306.

Viotti, S., Guidetti, G., Converso, D., and Sottimano, I. 2020. Fostering work ability among menopausal women. Does any work-related psychosocial factor help?. *International Journal of Women's Health, 12*: 399–407.

Virkkunen, H., Kaupinnen, T., and Tenkanen, L. 2005. Long term effect of occupational noise on coronary heart disease. *Scandinavian Journal of Work Environment and Health, 31*: 291–299.

Vohs, K.D. and Baumeister, R.F. (eds.). 2016. *Handbook of Self-Regulation: Research, Theory, and Applications*. Guilford Press, New York.

Vrijkotte, T.G.M., van Doornen, L.J.P., and de Geus, E.J.C. 1999. Work stress and metabolic and hemostatic risk factors. *Psychosomatic Medicine, 61*: 796–805.

Wærsted, M., Koch, M., and Veiersted, K.B. 2020. Work above shoulder level and shoulder complaints: a systematic review. *International Archives of Occupational and Environmental Health, 93*: 925–954.

Walker-Bone, K., Palmer, K., Reading, I., Coggon, C., and Cooper, D. 2012. Occupation and epicondylitis: A population-based study. *Rheumatology, 51*: 305–310.

Walker-Bone, K., Reading, I., Coggon, D., Cooper, C., and Palmer, K.T. 2006. Risk factors for specific upper limb disorders as compared with non-specific upper limb pain: Assessing the utility of a structured examination schedule. *Occupational Medicine, 56*: 243–250.

Wall, T.D. and Clegg, C.W. 1981. A longitudinal study of group work redesign. *Journal of Occupational Behaviour, 2*: 31–49.

Wallace, J.C. and Chen, G. 2005. Development and validation of a work specific measure of cognitive failure: Implications for organisational safety. *Journal of Occupational and Organisational Psychology, 78*: 615–632.

Wallace, J.C. and Vodanovich, S.J. 2003. Workplace safety performance: Conscientiousness, cognitive failure, and their interaction. *Journal of Occupational Health Psychology, 8(4)*: 316.

Walsh, L.K., Restaino, R.M., Martinez-Lemus, L.A., and Padilla, J. 2017. Prolonged leg bending impairs endothelial function in the popliteal artery. *Physiological Reports, 5*(20): e13478.

Wang, M., Jiang, C., Fekete, G., Teo, E.C., and Gu, Y. 2021. Health view to decrease negative effect of high heels wearing: A systemic review. *Applied Bionics and Biomechanics, 2021*: 6618581.

Wang, X. and Xu, L. 2021. Speech perception in noise: Masking and unmasking. *Journal of Otology, 16*(2): 109–119.

Wang, Y.X.J. and Griffith, J.F. 2010. Effect of menopause on lumbar disk degeneration: potential Etiology. *Radiology, 257*(2): 318–320.

Ward, Z.J., Bleich, S.N., Long, M.W., and Gortmaker, S.L. 2021. Association of body mass index with health care expenditures in the United States by age and sex. *PLoS One, 16*(3): e0247307.

Wargocki, P. 2005. High indoor air quality improves office work and provides economic benefits. In: *Proceedings of the 11th International Conference on Environmental Ergonomics*. ACM, New York, pp. 182–185.

Warm, J.S. 1984. *Sustained Attention in Human Performance*. John Wiley & Sons, Chichester.

Wason, P.C. and Johnson-Laird, P.N. 1972. *Psychology of Reasoning: Structure and Content*, Vol. *86*. Harvard University Press, Harvard.

Waters, T.R., Putz-Anderson, V., Garg, A., and Fine, L.J. 1993. Revised NIOSH equation for the design and evaluation of manual lifting tasks. *Ergonomics, 36*: 749–776.

Weber, B.H. 1970. Silencing of hand-held percussive rock drills for underground operations. *Canadian Mining and Metallurgical Bulletin, 63*(694): 163.

Weber, S., Weibel, D., and Mast, F.W. 2021. How to get there when you are there already? Defining presence in virtual reality and the importance of perceived realism. *Frontiers in Psychology, 12*: 628298.

Weichbroth, P. 2019. A mixed-methods measurement and evaluation methodology for mobile application usability studies. *Architecture, 43*(44): 45.

Welch, R.B., Langley, E.O., and Lamaev, O. 1971. The measurement of fatigue in hot working conditions. *Ergonomics, 14*: 85–90.

Welford, A.T. 1958. *Ageing and Human Skill*. The Nuffield Foundation, Oxford University Press, Oxford.

Wen, C.P., Wai, J.P.M., Tsai, M.K., Yang, Y.C., Cheng, T.Y.D., Lee, M.C., Chan, H.T., Tsao, C.K., Tsai, S.P. and Wu, X. 2011. Minimum amount of physical activity for reduced mortality and extended life expectancy: A prospective cohort study. *The Lancet, 378*: 1244–1253.

Wentzel, E. and McKune, A. 2013. Effect of a foot pump device on lower leg swelling in physically inactive office workers. *Ergonomics SA, 25*: 22–31.

Wergeland, E.L., Veiersted, B., Ingre, M., Olsson, B., Åkerstedt, T., Bjørnskau, T., and Varg, N. 2003. A shorter workday as a means of reducing the occurrence of musculoskeletal disorders. *Scandinavian Journal of Work, Environment & Health, 29*: 27–34.

Weston, E.B., Aurand, A., Dufour, J.S., Knapik, G.G., and Marras, W.S. 2018. Biomechanically determined hand force limits protecting the low back during occupational pushing and pulling tasks. *Ergonomics, 61*(6): 853–865.

Weston, H.C. and Adams, S. 1927. *Industrial Health Research Board Report No. 40*. HMSO, London.

Whistance, R.S. 1996. An investigation into the ergonomics of standing. PhD Thesis, University of Cape Town.

Whistance, R.S. and Bridger, R.S. 1995. Postural adaptations to workbench modifications in standing workers. *Ergonomics*, *38*: 2485–2503.

Wickens, C.D. 1980. The structure of attentional resources. *Attention and Performance VIII*, *8*: 239–257.

Wickens, C.D. 1992. *Engineering Psychology and Human Performance*. 2nd Edn., Pearson Education, New York.

Widdowson, A. 2025. *Humans and Cyber Security: How Organisations can Enhance Resilience through Human Factors*. Taylor & Francis, New York.

Wiehler, A., Branzoli, F., Adanyeguh, I., Mochel, F., and Pessiglione, M. 2022. A neuro-metabolic account of why daylong cognitive work alters the control of economic decisions. *Current Biology*, *32*(16): 3564–3575.

Wienen, H.C., Bukhsh, F.A., Vriezekolk, E., and Wieringa, R.J. 2017. *Accident Analysis Methods and Models-A Systematic Literature Review*. Centre for Telematics and Information Technology, The Netherlands.

Wiese, E.E. and Lee, J.D. 2004. Auditory alerts for in-vehicle information systems: The effects of temporal conflict and sound parameters on driver attitudes and performance. *Ergonomics*, *47*: 965–986.

Wilde, G.J.S. 1994. *Target Risk*. PDE Publications, Toronto, Canada.

Wilensky, J.L. and Wilensky, H.L. 1951. Personnel counselling: The Hawthorne case. *The American Journal of Sociology*, *57*: 265–280.

Williamson, A.M., Gower, C.G.I., and Clarke, B.C. 1994. Changing the hours of shiftwork: A comparison of 8 and 12 hour shift rosters in a group of computer operators. *Ergonomics*, *37*: 287–298.

Wilmot, E.G., Edwardson, C.L., Achana, F.A., Davies, M.J., Gorely, T., Gray, L.J., Khunti, K., Yates, T., and Biddle, S.J. 2012. Sedentary time in adults and the association with diabetes, cardiovascular disease and death: Systematic review and meta-analysis. *Diabetologia*, *55*: 1–11.

Wilson, J. and Morris, W. 2004. Participatory design of work equipment: Lessons learned and suggestions for future actions. *TUTB Newsletter*, *24–25*: 22–25.

Wixted, J.T. 2020. The forgotten history of signal detection theory. *Journal of Experimental Psychology: Learning, Memory, and Cognition*, *46*(2): 201.

Wogalter, M.S. and Young, S.L. 1991. Behavioural compliance to voice and print warnings. *Ergonomics*, *34*: 79–89.

Wogalter, M.S., Begley, P.B., Scancorelli, L., and Brelsford, J.W. 1997. Effectiveness of elevator service signs: Measurement of perceived understand ability, willingness to comply and behaviour. *Applied Ergonomics*, *28*: 181–187.

Wogalter, M.S., Mayhorn, C.B., and Laughery Sr, K.R. 2021. Warnings and hazard communications. In: *Handbook of Human Factors and Ergonomics*, G. Salvendy, ed. John Wiley & Sons, New York. pp. 644–667.

Woods, V. 2005. Musculoskeletal disorders and visual strain in intensive data processing workers. *Occupational Medicine*, *55*: 121–127.

Woods, V., Hastings, S., Buckle, P., and Haslam, R. 2003. Development of non-keyboard device checklists through assessments. *Applied Ergonomics*, *34*: 511–519.

Woodson, W.E. 1981. *Human Factors Design Handbook*. McGraw-Hill Inc., New York.

World Economic Forum. 2023. *Future of Jobs Report*, May 2023. https://www.weforum.org/reports/the-future-of- jobs-report-2023.

World Health Organization. 1995. *Physical status: The Use and Interpretation of Anthropometry*. WHO Technical Report Series 854. World Health Organization, Geneva.

World Health Organization. 2021. *Step safely: Strategies for Preventing and Managing Falls Across the Life-Course*. World Health Organization, Geneva.

Wright, P. 1978. Feeding the information eaters: Suggestions for integrating pure and applied research on language comprehension. *Instructional Science*, *7*: 249–312.

Wright, P., Bartram, C., Rogers, N., Emslie, H., Evans, J., Wilson, B., and Belt, S. 2000. Text entry on handheld computers by older users. *Ergonomics*, *43*: 702–716.

Wu, J., Wang, X., Li, J., and Tang, R. 2023. Effect of icon size, icon position and sex on clicking motion when operating smartphones with single hand. *Ergonomics*, *66*(7): 1031–1041.

Wyon, D.P. 1974. The effects of moderate heat stress on typewriting performance. *Ergonomics*, *17*: 309–318.

Xiong, J. and Muraki, S. 2014. An ergonomics study of thumb movements on smartphone touchscreen. *Ergonomics*, *6*: 943–955.

Xu, C., Furuya-Kanamori, L., Liu, Y., Færch, K., Aadahl, M., Seguin, R.A., LaCroix, A., Basterra-Gortari, F.J., Dunstan, D.W., Owen, N., and Doi, S.A. 2019. Sedentary behavior, physical activity, and all-cause mortality: Dose–response and intensity weighted time-use meta-analysis. *Journal of the American Medical Directors Association*, *20*(10): 1206–1212.

Yadegaripour, M., Hadadnezhad, M., Abbasi, A., Eftekhari, F., and Samani, A. 2021. The effect of adjusting screen height and keyboard placement on neck and back discomfort, posture, and muscle activities during laptop work. *International Journal of Human– Computer Interaction*, *37*(5): 459–469.

Yang, B. and O'Connell, G.D., 2019. Intervertebral disc swelling maintains strain homeostasis throughout the annulus fibrosus: a finite element analysis of healthy and degenerated discs. *Acta biomaterialia*, 100, pp.61–74.

Yang, Y., Schumann, M., Le, S., and Cheng, S. 2018. Reliability and validity of a new accelerometer-based device for detecting physical activities and energy expenditure. *PeerJ*, *6*: e5775.

Yazuli, Z.A., Mohamad Jamil, P.A.S., Mohammad Yusof, N.A.D., Karuppiah, K., Perimal, E.K., Naeini, H.S., Sambasivami, S., and Perumal, P.A. 2022. Effect of anti-fatigue mat on leg muscle discomfort and muscle activity due to prolonged work in upright position among production workers. *Journal of Health & Safety at Work*, *12*(3): 483–498.

Yoon, J.-H., Hong, J.-S., Roh, J., Kim, C.-N., and Won, J.-U. 2015. Dose–response relationship between noise exposure and the risk of occupational injury. *Noise and Health*, *17*: 43–47.

Yorio, P.L. and Moore, S.M. 2018. Examining factors that influence the existence of Heinrich's safety triangle using site-specific H&S data from more than 25,000 establishments. *Risk Analysis*, *38*(4): 839–852.

You, H., Simons, Z., Freivalds, A., Kothari, M., Naidu, S., and Young, R. 2004. The development of risk assessment models for carpal tunnel syndrome: A case referent study. *Ergonomics*, *47*(6): 688–709.

Young, M.S. and Stanton, N.A. 2005. Mental workload. In: *Handbook of Human Factors and Ergonomics Methods*, N. Stanton, A. Hedge, K. Brookhuis, E. Salas, and H. Hendrick, eds. CRC Press, Boca Raton, FL, pp. 39–41.

Yurkevicius, B.R., Alba, B.K., Seeley, A.D., and Castellani, J.W. 2022. Human cold habituation: Physiology, timeline, and modifiers. *Temperature*, *9*(2): 122–157.

Zacharkow, D. 1988. *Posture: Sitting, Standing, Chair Design and Exercise*. Thomas, Springfield, IL.

Zalat, M.M., Amer, S.M., Wassif, G.A., El Tarhouny, S.A., and Mansour, T.M. 2022. Computer vision syndrome, visual ergonomics and amelioration among staff members in a Saudi medical college. *International Journal of Occupational Safety and Ergonomics*, *28*(2): 1033–1041.

Zhang, F., Zhao, X., Shen, H., and Zhang, C. 2016. Molecular mechanisms of cell death in intervertebral disc degeneration. *International Journal of Molecular Medicine*, *37*(6): 1439–1448.

Zheng, A. and Schmid, S. 2023. A review of the neural basis underlying the acoustic startle response with a focus on recent developments in mammals. *Neuroscience & Biobehavioral Reviews*, *148*: 105129.

Zhu, M., Shen, S., Redelmeier, D.A., Li, L., Wei, L., and Foss, R. 2021. Bans on cellphone use while driving and traffic fatalities in the United States. *Epidemiology*, *32*(5): 731–739.

Zink, K.J. and Fischer, K. 2013. Do we need sustainability as a new approach in human factors and ergonomics? *Ergonomics*, *56*(3): 348–356.

Zipp, P., Haider, E., Halpern, N., and Rhomert, W. 1983. Keyboard design through physiological strain measurements. *Applied Ergonomics*, *14*: 117–122.

Zwart, B.C.H., Frings-Dresen, M.H.W., and van Duivenboden, J.C. 2002. Test–retest reliability of the work ability index questionnaire. *Occupational Medicine*, *52*: 177–181.

Index

B

C

D

E

F

G

H

I

J

K

O

P

Q

R

T

W

Y

Z